What Sustains Life?

What Sustains Life?

Consilient Mechanisms for Protein-Based Machines and Materials

Dan W. Urry

Springer

Dan W. Urry
Professor of Biophysics
University of Minnesota
Twin Cities Campus
BioTechnology Institute
St. Paul, MN 55108-6106
USA
urryx001@tc.umn.edu

Library of Congress Control Number: 2005926819

ISBN-10: 0-8176-4346-X
ISBN-13: 978-08176-4346-1

Printed on acid-free paper.

© 2006 Springer Science+Business Media, LLC.
All rights reserved. This work may not be translated or copied in whole or in part without the written permission of the publisher (Springer Science+Business Media, LLC, 233 Spring Street, New York, NY 10013, USA), except for brief excerpts in connection with reviews or scholarly analysis. Use in connection with any form of information storage and retrieval, electronic adaptation, computer software, or by similar or dissimilar methodology now known or hereafter developed is forbidden.
The use in this publication of trade names, trademarks, service marks, and similar terms, even if they are not identified as such, is not to be taken as an expression of opinion as to whether or not they are subject to proprietary rights.

Printed in Singapore.   (BS/KYO)

9 8 7 6 5 4 3 2 1

springer.com

To my mentor, Henry Eyring, whom I yet strive to represent well,

To my parents for their striving to instill values and a strong work ethic,

To my wife, Kathleen, for her support enabling me to practice the work ethic required by this vocation,

To my children—Kelley, David, Douglas and Weston—for all they have taught,

To the Office of Naval Research and Program Officers, Michael Marron and Keith Ward, for the financial and other support that provided the foundation for the research on which this volume is based.

Preface

In the sixth century BC, Thales of Miletus, father of the Ionian Enlightenment, setting aside the mythic views of Homer and Hesiod asked, "What is the world made of?" and thereby became the first physicist.[1] He answered that water is the basis of all matter and thereupon became an often cited example of early Greek reasoning gone astray. However, not only did Thales initiate scientific inquiry, but also, with reference to living things, he was substantially correct. Living organisms are composed mostly of water, but the unique role of water in living organisms has been wanting for adequate description. From the perspective of D.H. Lawrence, "Water is H_2O, hydrogen two parts, oxygen one part, but there is a third thing that makes it water and nobody knows what that is." From our perspective and as advanced in this volume, the interaction of water with dissimilar groups comprising each protein molecule, the competition for water between these disparate substituents along chain molecules of living organisms, and the freedom of motion that water gives protein chains combine to provide the physical basis of Life.

This book, *What Sustains Life? Consilient Mechanisms[2] for Protein-Based Machines and Materials*, at its foundation is a monograph. It arises from the design of elastic-contractile model proteins to convert energy from one form to another and thereby constitutes a limited element of natural history. Such a seemingly specialized focus, however, unfolds to present insight into how living organisms access energy in the environment to become thriving entities of progressively increasing complexity and diversity. Thus, it becomes an account of the means whereby Life can flourish as a physical–chemical entity. In addition, this book unfolds a journey of personal Ionian Enchantment,[3,4] for it contains elements of a memoir. The content draws principally upon three decades of collaborative research and its analysis: it recounts the original design and demonstration of model proteins interconverting the set of six energies interconverted by living organisms, and it arrives at interlinked pervasive consilient mechanisms.

Quite apart from one's theological perspective, a general fascination exists with the physical basis of Life. On the one hand, a living organism seems so improbable and so fragile, and yet, on the other hand, Life can seem so hardy and resilient. Living organisms are complex blends of interdependent and cooperative molecular machines with an innate facility to evolve toward greater complexity of structure and diversity of function. What are the physical forces and mechanisms whereby Life can exist? In our view, at its foundation is a competition for hydration between disparate chemical groups distributed along a protein chain and a resultant phase separation of oil-like components of proteins from water.

Our view of "What Sustains Life?" allows for a relatively nontechnical level of description. At the start, the perspective presented here does not require complex mathematical analyses and specialization. Rather, the concepts grow from commonplace experience such as noting the immiscibility of oil and water or, perhaps even more appropriately, the separability of oil from vinegar. Accordingly, an effort has been made, within the author's near icarian[5] tendencies and other limitations, to reach a broader audience. Should interest develop, original publications and more in-depth analyses within Chapter 5 can be sought for greater detail. Also a more in-depth volume is under development through presentations of a graduate course of 24 lecture hours on contractile protein-based machines and materials.[6]

Chapter 1 introduces the message of the book in terms of four assertions. The first assertion, the phenomenological assertion, derives from the design of a series of model proteins capable of converting the family of energies interconverted by living organisms by controlling association of oil-like domains. A set of five axioms result from the phenomenological assertion. The second assertion, the mechanistic assertion, presents a single molecular process of association of oil-like domains whereby model proteins can function as machines capable of accessing available energies and converting them to those energies utilized in sustaining Life. An elemental understanding of elasticity couples with the association of oil-like domains to complete the second assertion. The third assertion, treated in Chapters 7 and 8, refers to the wider biological literature and presents argument for the dominance of a common unifying mechanism in the functioning of proteins, the molecular machines of Life. Chapter 7 contains examples such as oxygen transport, blood clotting, formation of the amyloid deposits of Alzheimer's disease, and the protein-based machine, chaperonin, that unfolds and refolds proteins having inappropriate association of oil-like domains. Chapter 8 provides an example for each of the three fundamental energy conversions of living organisms: (1) electron transfer that concentrates acid (protons) on one side of the inner mitochondrial membrane with Complex III of the electron transport chain as the enlightening example, (2) the use of the protons by the vital rotary motor called ATP synthase to produce ATP (adenosine triphosphate), the energy currency of biology, and (3) protein motors that use ATP as their energy source to perform the work of the cell. The prominent example of the latter features the linear motor of muscle contraction that requires ATP as its energy source. The common underlying pair of physical processes that provide for the disparate energy conversions have been named consilient mechanisms in that they "create a common groundwork of explanation," which E.O. Wilson gives as the definition of consilience.[3] Accordingly, the new insight into energy conversion provides interlinked consilient mechanisms[7] for the diverse energy conversions that sustain and define Life.

The thesis of the third assertion begins, *"Biology thrives near a movable cusp of insolubility."* and, as one of its hallmarks, describes ATP production by ATP synthase as the result of a three step rotary motor having as one of the three sides of its rotor an insoluble face that rotates from the insolubility of oil-like association with sleeve to solubility due to repulsion with the sleeve containing vinegar-like precursors, ADP (adenosine diphosphate) and P_i (inorganic phosphate), to drive formation of ATP, the universal energy coin of biology. Analogous effects occur in ATPases, the molecular machines that use ATP as their energy source. In muscle contraction the binding of ATP and its breakdown to ADP and P_i converts insoluble domains of the contracted state to soluble domains, and their release returns oil-like domains to the insoluble contracted state to drive linear motors.

The fourth assertion applies the understanding of the new mechanism in the development of protein-based materials to improve health care, to decrease healthcare costs, and to assist in alleviating additional major problems of society.

Chapter 2 presents an overview of the energy conversions that sustain Life and steps further toward the molecular processes involved. Chapter 3 notes highlights of the grand, yet at its outset proscribed, pilgrimage that has given us an understanding of components and products of living things. The enigma of Life erodes further as the premier, yet erstwhile improbable, molecular machines of Life, the proteins, emerge as described in Chapter 4, to be almost energetically wasteful in construction. With knowledge of the biochemical details of protein synthesis,[8] the protein molecule transforms from a highly improbable entity to an example of extravagant expenditure of energy in the process of construction. Significantly, however, protein machines exhibit remarkable efficiency in function, and it is an energy-fed march toward more diverse and efficient molecular machines that constitutes the basis for evolution through natural selection.

Chapter 5 is the scientific core of the book. It begins with familiar phase transitions of ice-to-water and water-to-vapor, both of which demonstrate increased disorder with increased temperature. It then brings in the unique phase transitions of the two-component system, model proteins in water, in which the protein component progresses to increased order and the water component to decreased order with increased temperature. Next, it experimentally derives five axioms for protein function as molecular machines by means of designed model proteins. This chapter describes the development of diverse molecular machines, utilizing a single, consilient mechanism to achieve diverse function. The approach begins with a particularly compliant molecular building block. Rules are demonstrated whereby this building block of elastic model protein can be made to perform mechanical work simply by heating, and then mechanical and a number of other forms of work performed by living organisms can be made to occur at physiological temperature. Herein lies your author's personal Ionian Enchantment.

Following the historical development, initially, contraction to lift a weight is demonstrated by heating cross-linked elastic model protein. Shortly thereafter this performance of mechanical work is demonstrated without the use of thermal energy, but instead with the chemical energy of salt on the parent model protein and then with acid acting on a designed modification of the initial model protein. Additional model proteins are designed, each utilizing a common underlying mechanism, and found capable of the many different energy conversions extant in biology. Each designed model protein, once constructed and tested, functioned as intended in diverse energy conversions.

Reversing the trend to examine smaller protein components, a particularly compliant model protein system had been chosen to which new functional capacities could be added. It has been a simple exercise of the scientific method with specific elements of an axiomatic approach of Bacon with awareness of the illusion of knowledge,[9] of Descartes with development of graphical methods[10,11] of Galileo with experimental exploration and verification,[12,13] and of so many others. In particular, temperature-induced oil-like aggregation of model proteins gave rise to a hypothesis. The hypothesis was tested by sequential construction of modified model proteins and followed by experimental verification. There was simple extension of a basic concept and its translation into new model protein constructs. The result was development of protein-based machines for diverse energy conversions not previously demonstrated in model proteins, and the family of energy conversions obtained were those found in living organisms. Experimental testing directly followed, with the result that the new model protein constructs worked as expected, each capable of a different type of energy conversion and, within a type, capable of different efficiencies for energy conversion. The common underlying mechanism for all of the energy conversions, the consilient mechanism, was found to be competition for hydration between oil-like and vinegar-like groups constrained to coexist along the protein chain molecule that exhibits as a repulsion between oil-like and vinegar-like groups.

Preface

The molecular players in the drama of sustaining Life are introduced in Chapter 3, the biosynthetic process of protein construction is explained in Chapter 4, and the diverse energy conversions and their mechanism are demonstrated in Chapter 5. With this background, we can now address the question of the evolution of protein machines. In Chapter 6, we respond to the challenge of Behe: "if you search the scientific literature on evolution, and if you focus your search on the question of how molecular machines—the basis of Life—developed, you find an eerie and complete silence."[14] The silence that Behe reports is borne of a lack of awareness of the broader scientific literature, which is not silent on the issue of how molecular machines can evolve. As part of our thesis here, the literature reviewed in Chapter 5 contains experimental demonstration of *de novo* designed protein-based machines of increasing diversity, complexity, and efficiency. Given the consilient mechanism of energy conversion detailed in Chapter 5, the evolution of protein-based molecular machines unfolds as a remarkably simple process. Chapter 6 demonstrates how an elementary mutation, a single base change, allows a thermally driven protein machine to become a chemically driven machine, that is, to access a new energy source. Another single base change can allow access to yet another source of energy. In fact, each new energy source in the set of energy conversions of living organisms demonstrable by these designed model proteins is accessible by a single base change. Furthermore, in each case a subsequent single base change results in a more efficient molecular machine. Because of this simple access to new energy sources with greater efficiencies, each step toward more diverse and efficient protein-based machines occurs without an additional cost to the organism to produce. Evolution toward more efficient and more complex protein machines becomes a natural step-by-step process when viewed in terms of the consilient mechanism of energy conversion and in terms of the genetic code and protein biosynthesis. In the simplest case, one might consider two organisms each requiring the same amount of food, one to produce a less efficient and less effective machine and the other to produce a more useful and more efficient machine. Because both less-efficient and more efficient protein-based machines cost the same amount of energy to produce, in the competition for survival the improved organism wins. This provides elementary insight into natural selection.

Chapter 7 begins with the assertion that biology thrives near a movable cusp of insolubility and that excursions too far either direction into the realm of insolubility or of solubility spells disease and death. For example sickle cell disease represents an excursion too far into the realm of insolubility, as do the more ominous prion diseases of Alzheimer's disease, and mad cow disease. Blood clotting is a special example of thriving at the cusp of insolubility. This chapter also discusses molecular chaperones whose role is to bring proteins back from the realm of incorrect insolubility and ends with a brief example of shifts toward excess solubility due to oxidative processes with the result of degradation and disease such as pulmonary emphysema.

Chapter 8 continues with the thesis that biology thrives near a movable cusp of insolubility whereby the forces that, in a positively cooperative manner, power the molecular machines of biology drive spatially localized oil-like regions of protein back and forth across movable water solubility–insolubility divides, that is, back and forth between being associated (water insoluble) and dissociated (water soluble). Life's commonplace energy source is ATP. Life's workhorse protein-based machines are ATPases, where ATP binding and its breakdown to ADP and P_i acts like cocking of a gun and P_i release functions like the discharging of the gun. Binding ATP, and especially its breakdown to ADP and P_i, moves the cusp of insolubility, the water solubility–insolubility divide, to solubility, and the release of P_i functional components (hydrophobic domains) of the machine to insolubility. Remarkably, even one part of the two-part rotary protein machine that produces ATP from ADP and P_i functions in reverse as an ATPase that breaks ATP down to ADP and P_i to perform work with high efficiency.

The primary examples in Chapter 8 represent three key aspects of energy conversion in animals. The first explicitly described protein-based machine pumps protons across a membrane to create acid on one side of the membrane; it is Complex III of the electron transport chain of mitochondria, the energy factory of the cell. The second protein-based machine uses the return of the protons back across the membrane to make ATP, the energy coin of living organisms; it is ATP synthase, as noted immediately above. The third protein-based machine uses the ATP to produce motion; it is the myosin II motor of muscle contraction, categorized as an ATPase. The latter presents a transparent demonstration of the consilient mechanism whereby binding ATP (adenosine tri*phosphate*) disrupts association of oil-like domains and loss of *phosphate* re-establishes association of the oil-like domains to provide the power stroke of a contraction. In addition, further scrutiny of motion-producing protein-based machines couples the cusp of insolubility aspect of the consilient mechanism to a "common groundwork of explanation" of an increase in elastic force resulting from a decrease in motion along the backbone of a single protein chain. In doing so, this sets aside a popular concept of polymer elasticity that required random chain networks and a Gaussian distribution of end-to-end chain lengths (See Appendix 1).

In Chapter 9, the capacity to design protein-based molecular machines and materials turns toward the development of useful applications for improving individual health and environmental health. With insight as to how the protein-based machines of biology work, and, because we know how they work, we can design materials for the future as never before! A principal medical application considered here employs a consilient approach to tissue engineering; it utilizes materials and mechanisms natural to the tissue to be restored or augmented and couples these properties to the natural capacity of cells to function, for example, as mechano-chemical transducers. This area involves such specific health concerns as prevention of urinary incontinence, prevention of postsurgical adhesions, intervertebral disk restoration, and temporary functional vascular and urological scaffoldings with the potential to remodel into natural tissues.

Another general medical application with great promise is the area of drug delivery also referred to as the controlled release of pharmaceuticals, including natural proteins and genes. Specific areas under development include transdermal delivery for the prevention of pressure ulcers, drug addiction intervention, programmed adjuvantcy and release of vaccines, and analgesics and anesthetics. Although there are numerous nonmedical applications, the development of programmably biodegradable plastics, of specific transducers and biosensors, and of new sound-absorbing materials, among others, are addressed.

After the more technical aspects of the book are integrated, the Epilogue places the message of the book into broader context. It considers Schrödinger's *What Is Life?*[15] and the Prigogine argument for biology functioning as creating order out of chaos under far-from-equilibrium conditions. It speaks to the oft-noted paradox of time's arrow for the universe pointing toward less order and uniformity, whereas time's arrow for Life and Society points toward increasing structure and diversity. In short, explicit examples of protein machines are given whereby living organisms utilize products of photosynthesis to create structure by means of a pair of efficient consilient mechanisms.

Thus, life is a complex integration of mutually dependent protein machines with a unique energy-driven capacity to reverse time's arrow for the wider universe. Life originated and continually evolves by employment of machines, composed of large protein polymers, to capture available energy and by use of the energy in many small, not far-from-equilibrium steps to create the structures of Life. Life is not well represented by transient dissipative structures like tornadoes. Rather, Life provides examples of seeds and spores that can lie dormant for years and even centuries only to spring to life on receiving the proper modest energy inputs. We conclude this molecular machine perspective of Life with an apparent message. Time's arrow for Life and for Society points toward greater complexity of organization and diversity of function only as long as there remain adequate energy sources.

Dan W. Urry
Stillwater, MN
2006

References

1. D.J. Boorstin, *The Seekers: The Story of Man's Continuing Quest to Understand His World*. Random House, New York, 1998, pp. 22, 111, 112.
2. The word consilient is used as the adjective form of the noun consilience. As listed in *Webster's Third New International Dictionary*, consilience contains the sense that there exist pervasive, fundamental laws of nature underlying related disciplines, which provide a common groundwork of understanding. Here we refer to the consilient mechanism as a water dependent, pervasive, fundamental process by which the macromolecules of living organisms function and evolve. The term originates from William Whewell in *The Philosophy of the Inductive Sciences*, London, 1840.
3. E.O. Wilson, *Consilience, The Unity of Knowledge*, Alfred E. Knopf, New York, 1998, pp. 4–5. On these pages Wilson describes his youth in southern Alabama as being immersed in the rich fauna and flora, principally insects, of the region. On introduction at the University of Alabama in Tuscaloosa, Alabama to the concept of Darwinian evolution as advanced by Ernst Mayr's 1942 *Systematics and the Origin of the Species*, Wilson describes the excitement of his own unfolding Ionian Enchantment as he became aware of the relationships among the many species of his personal experience.
4. G. Holton, in *Einstein, History, and Other Passions* American Institute of Physics Press, Woodbury, NY, 1995.
5. E.O. Wilson, *Consilience, The Unity of Knowledge*. Alfred E. Knopf, New York, 1998, p. 7. (Reference to the Greek myth of Deadalus and Icarus.)
6. This course has been taught at the Ludwig-Maximilians-Universtät, München, Winter Term 2000–2001 as "Biomolecular Machines and Materials: Phenomenology, Physical Basis, and Applications," at the Universidade do Minho, Portugal, May–June 2002, as an Advanced Course of the Departamenti de Biologia & Engenharia de Polímeros with the same title with a repeat in 2005, and is currently being presented at the Kyushu Institute of Technology, Wakamatsu, Kitakyushu, Japan, under the title of "Biological Functions and Engineering."
7. E.O. Wilson, "Evolution and Other Disciplines." In *Darwin*, Third Edition, P. Appleman, Ed., W.W. Norton. New York, 2001, pp. 450–459.
8. See, for example, the general text, D. Voet and J.G. Voet, *Biochemistry*, Second Edition, John Wiley & Sons, New York, 1995.
9. P. Zagorin, *Francis Bacon*. Princeton University Press, 1999.
10. J. Cottingham, Editor, *The Cambridge Companion to Descartes*. Cambridge University Press, 1992.
11. J.F. Scott, *The Scientific Work of Rene Descartes*. Taylor and Francis, London, 1952.
12. P. Machamer, Editor, *The Cambridge Companion to Galileo*. Cambridge University Press, 1988.
13. J.H. Randall, Jr. *The School of Padua and the Emergence of Modern Science*, Editrice Antenore-Padova, 1961.
14. M.J. Behe, *Darwin's Black Box: The Biochemical Challenge to Evolution*. Simon and Schuster, New York, 1996, p. 5.
15. E. Schrödinger, *What Is Life?: The Physical Aspect of the Living Cell with Mind and Matter and Autobiographical Sketches*. Cambridge University Press, 1944.

Acknowledgments

Your author gratefully acknowledges the contributions of:

T.M. Parker and L.C. Hayes of Bioelastics Research, Ltd., were instrumental in preparing the figures and obtaining references and additional experimental data utilized in the book as unpublished.

The ever-challenging, original editor for this volume, Alla Margolina-Litvin stimulated discussions that lead to the broader scope of the present volume rather than to a more staid volume on the engineering of protein-based machines and materials. Alla's challenge was to write a follow up to the Schrödinger book, *What Is Life?* of six decades ago. Drawing from the recognition that our deigned elastic-protein-based polymers could interconvert the energies interconverted by living organisms and that accessing energy in the environment was the key to sustaining Life, I believed that the present title was warranted. This set a path of developing an analytical approach of the data on our elastic-contractile model proteins in order to considering more explicitly the protein-based machines of biology and a path to the broader implications of the discovery of energy conversion by the inverse temperature transition. Attempting to deliver a manuscript worthy of the title has taken much longer than originally planned and includes new data and analyses.

Many coworkers contributed over the past several decades during which the foundation for the consilient mechanisms developed. The many inadvertent shortcomings within the book itself should be recognized as mine rather than theirs.

Dan W. Urry
2006

Contents

Preface vii

Acknowledgments xv

Chapter 1 Introduction 1

Chapter 2 What Sustains Life? An
           Overview 28

Chapter 3 The Pilgrimage: Highlights
           in Our Understanding of
           Products and Components of Living
           Organisms 70

Chapter 4 Likelihood of Life's Protein Machines:
           Extravagant in Construction Yet *Efficient
           in Function* 94

Chapter 5 Consilient Mechanisms for Diverse
           Protein-based Machines: The Efficient
           Comprehensive Hydrophobic Effect 102

Chapter 6 On the Evolution of Protein-based
           Machines: Toward Complexity of Structure
           and Diversity of Function 218

Chapter 7 Biology Thrives Near a Movable
           Cusp of Insolubility 239

Chapter 8 Consilient Mechanisms for the
           Protein-based Machines of Biology 329

Chapter 9 Advanced Materials for the Future:
           Protein-based Materials with Potential to
           Sustain Individual Health and Societal
           Development 455

Epilogue 541

Appendix 1 Mechanics of Elastin: Molecular
            Mechanism of Biological Elasticity and
            Its Relationship to Contraction
            Dan W. Urry and Timothy M. Parker 574

Appendix 2 Development of Elastic Protein-based
            Polymers as Materials for Acoustic
            Absorption
            *Dan W. Urry, J. Xu, Weijun Wang,
            Larry Hayes, Frederic Prochazka, and
            Timothy M. Parker* 598

Index 611

1
Introduction

1.1 About the Title

1.1.1 What Sustains Life?

An account of *what sustains Life* is an effort more modest in scope than either an attempt to explain *what is Life*[1] or a challenge to recount *where or how Life began*.[2] The latter focus more on heredity and address the transition from a prebiotic to a biotic earth, which concern the more profound transition from the absence to the presence of self-sustaining, self-replicating Life. With present knowledge it seems possible that the origins of Life could have involved far-from-equilibrium conditions of the Prigogine focus.[3]

From the work of many gifted biochemists, however, we do know that the creation of the chain molecules of the living organism, the nucleic acids and proteins required to duplicate and sustain Life, does not, as has been proposed,[3] require far-from-equilibrium conditions and is not the result of dramatic dissipative processes. Building of Life's great molecules of heredity, the nucleic acids, occurs by means of one reversible small energy step after another, but with a special trick. The translation of these nucleic acid sequences into protein sequence also occurs, one reversible small energy step after another, but again with the special twist. As presented in Chapter 4, the relentless drive toward these otherwise improbable macromolecules of Life derives simply from the enzymatic removal of a reaction side product. Removal of a reaction product blocks reversal of chain growth. Each individual reaction required in the process of adding each nucleic acid residue to form the polynucleotides of DNA and RNA, or of adding each amino acid residue to form protein of hundreds, and even thousands, of residues, represents a small energy step of discarding some 8 kcal/mol reaction. Nonetheless, in sum, with the repeated discarding of the energy within a reaction product, the production of a protein represents an enormous expenditure of energy that, once recognized, removes any sense of protein improbability.

Except for the singular events of the absorption or emission of a photon of light, the energy conversions that sustain Life occur by relatively small energy steps, one-tenth the magnitude of the energy of a single initiating photon from the sun. The present effort, however, begins with the living organism and seeks to describe how these relatively small steps in energy sustain Life and allow for continual evolution into more diverse, complex, and efficient entities.

1.1.2 Access to Energies That Sustain Life Derives from a Special Phase Separation Exhibited by Proteins

Having temporarily addressed the question of how Life's great molecules come into existence, we address the question posed in the title to this book. The answer necessarily resides in the underlying function of these remarkable macromolecules. The following two statements, each derived from its subsequent bulleted train of thought provide a phenomenological glimpse into our perspective of biomolecular functions that sustains Life.

Accessing energy from the environment sustains Life!
- What makes it possible for Life to exist?
- No biological problem is more profound!
- Yet, at one level, the answer is quite simple and even well known.
- It is the capacity to utilize energy sources, for example, food, in the environment!

A phase separation process accesses available energy!
- The above answer, however, defers the currently challenging question.
- What converts available energy in food and in oxygen to energies essential for Life?
- As presented here, the answer again becomes quite simple!
- Protein-based machines undergo a special kind of *phase separation* that converts energy from one form to another!

1.1.3 A Consilient Mechanism: A Unifying Thesis for *What Sustains Life*?

1.1.3.1 Thesis

Biology thrives near a movable cusp of insolubility, and the forces that, in a positively cooperative manner, power the molecular machines of biology drive paired oil-like domains of proteins back and forth between association (insolubility) and dissociation (solubility), and excursions too far in either direction into the realms of insolubility or solubility spell disease and death.

1.1.3.2 Recognition of a Consilient Mechanism

Whether the function of a molecular machine is to produce motion or convert one form of chemical energy into another or to perform any of several other energy conversion functions of biology and whether dysfunction occurs that results in disease and/or death, in our view, there exists a "common groundwork of explanation,"[4] that is, there occurs a consilient mech-

anism.[5] From our perspective, the process of phase separation from water by association of oil-like domains, that is, of insolubilization, provides a consilient mechanism for diverse energy conversions that sustain Life.

As demonstrated by elastic-contractile model proteins, the consilient mechanism achieves essentially all of the energy conversions that sustain Life. Knowledge of the mechanism arose *not* from analysis and extension of a particular biological energy-converting system, such as that of muscle contraction. Instead, the mechanism originated by *de novo* design, by designing entirely new energy-converting functions into a component of the mammalian elastic fiber never known or previously considered for such functions. On the contrary, the role of the mammalian elastic fiber is to store the energy of deformation and to use the stored energy to restore the nondeformed state as the deforming force recedes.

1.1.3.3 Oil-Like Groups Are Primarily Hydrocarbons

The most representative oil-like groups are the elemental hydrocarbon units, $-CH_2-CH_2-$, and $=CH-$; these are the most common chemical groupings of lubricating oils, of gasoline and heating oils and gases, and, of course, of biological fats and lipids. When these hydrocarbons combine to form the side chains, the R-groups, of a protein chain molecule, the protein with a balanced occurrence of these side chains is soluble in water at low temperature, but on raising the temperature these oil-like groups associate, that is, they become insoluble. There is a particular *temperature interval* over which insolubility develops. For reasons considered in detail in Chapter 5, this transition from solubility to insolubility is called an *inverse temperature transition*.

1.1.3.4 Definition of the Cusp of Insolubility

For our model elastic-contractile proteins in water, a plot of heat absorbed on increasing temperature exhibits an abrupt rise and then a more gradual decline as the temperature reaches the start of and passes through the tran-

1.1 About the Title

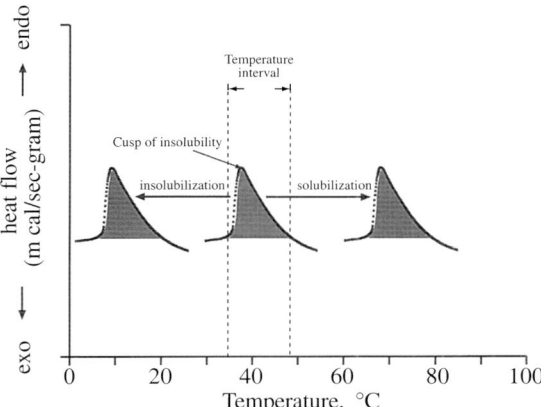

FIGURE 1.1. Schematic representation of the *movable cusp of insolubility* based on actual experimental data of the heat absorbed as the temperature is raised through the range of the phase separation for association of oil-like domains within an elastic model protein. Starting at the tip of the cusp and tracing to lower temperatures gives solubility of oil-like groups, seen as swelling within a cross-linked matrix. On the other hand, tracing to higher temperatures gives insolubility of oil-like groups, seen as contraction for a cross-linked matrix and the capacity to *pump iron*. Additionally, instead of changing the temperature, many different energy inputs can move the cusp to lower temperatures to drive contraction or can move the cusp to higher temperatures to cause relaxation (swelling).

sition to insolubility. As shown in Figure 1.1, the trace of the curve looks like the cross section of an incisor, of a cuspid. The trace presents a *cusp of insolubility*.

As defined in *Webster's Third New International Dictionary*, a cusp is "a fixed point on a mathematical curve (graph) at which a point tracing the curve would exactly reverse its direction, that is, a point at which traversing in opposite directions has opposite consequences."[6] Accordingly, starting from the peak of the curve and going to lower temperatures gives solubility, whereas starting from the peak of the curve and going to higher temperatures gives the opposite consequence of insolubility. As is also discussed in Chapters 5, 7, and 8, every energy input of relevance to biology, other than heating itself to surmount the cusp, moves the location of the *cusp of insolubility* along the temperature axis either to lower or higher temperatures. Hence, the phase separation of oil-like domains from water behaves as a *movable cusp of insolubility*. As shown in Figure 1.1, energy inputs that lower the temperature range of the *cusp*, that is, that lower the onset temperature for the phase separation, cause insolubility, and energy inputs that raise the onset temperature for the phase separation result in solubility. In other words, the *cusp of insolubility* simply marks the movable boundary between solubility and insolubility of an oil-like domain, a cluster of oil-like side chains, of a protein.

1.1.3.5 Excursions Too Far from the Solubility–Insolubility Boundary

The plaques of Alzheimer's disease and the fibrous state of the prions of mad cow disease (both with resulting brain destruction), the thrombi of stroke (cerebral thrombosis) and of heart attack (myocardial infarction), and the familiar manifestation of death (rigor mortis) represent excursions too far in the direction of protein insolubility. The favorable actions of antioxidants keep proteins from becoming so soluble (unfolded) that protein function disappears and proteolytic degradation ensues. Of course, the lack of blood clotting, hemophilia (the lack of clotting proteins to become insoluble by association of oil-like domains), results in death. Such devastations result from loss of proper balance between solubility and insolubility. They represent excursions too far from the *cusp of insolubility*, that is, too far from the boundary between insolubility and solubility.

1.1.3.6 Excursions Back and Forth Across the Solubility–Insolubility Boundary

1.1.3.6.1 Linear Motors of Biology

As is argued in Chapter 8, small, often reversible, energy excursions back and forth across the boundary between associated (water-insoluble) and dissociated (water-soluble) oil-like domains (clusters of oil-like groups) drive the protein-based machines of biology. Biology often achieves mobility by many linear motors comprised of protein, such as represented by muscle. In our view, oil-like regions, covered by organized water within motion-producing components of a protein motor, associate on loss of charge with release of water. The association of oil-like domains stretches interconnecting dynamic chain segments. This results in an increase in entropic elastic force due to a decrease in internal chain motions.[7] The entropic elastic force, thus developed, converts to motion as the extended chain retracts and the internal chain motions of interconnecting chain segments regain the increased amplitudes of the relaxed state.

1.1.3.6.2 Rotary Motors of Biology

For rotary motors within biological membranes, the oil-like domains of the rotor function much like special cleats on the rim of a tractor wheel passing into the oil-like domain of the membrane sleeve. In short, oil-like domains associate on loss of charge (on loss of vinegar-like species). Association of oil-like groups represents a *pull* component of force. There also, however, occurs a *push* component of force in the repulsion between oil-like and vinegar-like groups in the coupled extramembrane component of the rotary motor that produces almost 90% of biology's energy currency in the form of ATP (adenosine triphosphate). In particular, the oil-like domains either become repulsed by the presence of charge or, when externally forced to confront charge, repulse charge and force the charged entity to become less charged. As is introduced below and developed in Chapters 5 and 8, the latter is proposed to cause the most charged state, bound ADP (adenosine diphosphate) plus phosphate, of biology's ATP synthase rotary motor to combine to become a less charged ATP, the energy coin of biology. This production of ATP, in our view, occurs through the combination of *push* and *pull*.

1.1.4 Protein-based Machines Push and Pull!

The consilient mechanisms in relation to protein-based machines of biology are introduced. Before the four major assertions that form the basis of this book are elaborated, the next paragraphs build a mental framework for further elements associated with the rejoinder to the question posed in the title.

1.1.4.1 Elemental Contractile Event as One Aspect of Protein-based Machines

A change in the three-dimensional shape of a chain-like molecule necessarily changes the distance between two of the repeating units (or links) forming the chain molecule. If one of the two units is fixed in space and the other is attached to a weight, then the change in shape of the chain-like molecule moves the weight. Proteins are chain molecules where each link is decorated by any one of twenty different chemical side groups with each link derived from any one of twenty different amino acids. Accordingly, the change in shape of a protein represents an elemental contractile event whereby an object could be moved, and there can be *pull* and *push* elements to the mechanical event. Because of the many different repeating units available and because of the absolute control of their sequence in the protein chain, there are many ways with which to achieve the change in shape. Immediately below, four lists state the arguments for the roles of contractile proteins in biology.

1.1.4.2 Contractile Machines That Perform Mechanical Work by Pulling

- Machines perform useful work by converting energy from one form or location to another.
- Energy inputs cause shape changes in protein-based machines because of changes in association of oil-like domains.
- Such shape changes cause contractile protein-based machines to perform mechanical work.
- By this means, contractile protein-based machines can lift or pull weights; they can *pump iron*.

1.1.4.3 Contractile Linear Motors of Protein-based Machines

- The protein chain of a linear motor may be attached at both ends.

- Association of oil-like groups of certain chain segments between attachments stretch other interconnecting chain segments.
- On stretching (pulling), the interconnecting chain segments develop elastic force.
- The stretched interconnecting chain segments either retract and pull the attachment sites closer together or increase the force sustained at immovable attachment sites.

1.1.4.4 Contractile (Push–Pull) Rotary Motors Using Acid to Produce ATP

- Changing oil-like associations between fixed sleeve and rotary components drives rotation, whether intramembrane or in an attached extramembrane location.
- Reduction of charge in a transmembrane cleat of an otherwise oil-like *rotor* rotates a newly formed oil-like cleat into the cell membrane's sea of oil.
- By rotation, the most oil-like side of the extramembrane rotor faces off through a cleft of water with the most charged state of sleeve to create maximal (oil-like–vinegar-like) repulsion.
- The most charged state, ADP+P, relaxes repulsion (the push component) coming from the most oil-like side of the rotor by forming less-charged ATP.

1.1.4.5 Contractile Machines That Perform Work Other Than Mechanical

- The contractile protein machines of Life, however, accomplish more than the mechanical work of *pumping iron* or of rotation.
- By proper choice of sequence, they can pump protons; they can perform chemical work.
- By further design variation, they can pump electrons; they can perform electrical work.
- In the process of contracting, they can perform even additional kinds of work essential for Life.

Thus, the perspective is that contractile parts of protein-based machines sustain Life.

1.1.5 Protein-based Materials!

Contractile protein-based machines become useful materials:

- Knowledge of how contractile protein machines can sustain Life provides the capacity to design protein-based materials.
- Designed and synthesized protein-based materials have the potential to restore and sustain individual health and societal health.
- Designed contractile protein materials can perform functions beyond that which evolution has called upon them to do.
- They can be designed to deliver drugs, to prevent postsurgical adhesions, to restore diseased tissue, and to be environmentally friendly biodegradable thermoplastics.

Accordingly, protein-based materials hold promise of restoring and sustaining individual and societal health.

1.2 Four Principal Assertions of "What Sustains Life?"

This book stands on four principal assertions:

1. The phenomenological assertion
2. The mechanistic assertion
3. The assertion of biological relevance
4. The applications assertion

Introductions to these assertions follow.

1.2.1 Assertion 1: The Phenomenological Assertion

Designed elastic model proteins exhibit diverse functions that mimic biological functions by diverse means of controlling association of oil-like domains. As a result, five experimentally derived axioms phenomenologically categorize means by which energy conversions occur through control of association of oil-like domains.

A diverse set of energy conversions that sustain life can be experimentally demonstrated by *de novo* design of elastic-contractile model proteins under the precept of a single, pervasive, mechanism, that is, by a consilient mechanism that "creates a common groundwork of explanation."[4,5] It is a mechanism that achieves function by controlling association of oil-like domains in an environment of water. Five experimentally based axioms characterize the phase separation process of an inverse temperature transition for association of oil-like domains (see Chapter 5, Section 5.6.3). The axioms provide a phenomenological foundation with which to begin an understanding of protein function and with which to design protein-based machines and materials.

Phase separation occurs as oil-like groups separate from water!

- What is the nature of the *phase separation*?
- The answer lies within the repulsive energies responsible for the adage that "oil and vinegar don't mix"!
- Oil-like and vinegar-like adornments, constrained to coexist along the protein chain, cannot similarly separate.
- Instead, separation occurs by chain folding and assembly whereby the oil-like groups associate, separate from vinegar-like groups, and close off from water!

1.2.2 Assertion 2: The Mechanistic Assertion

1.2.2.1 With the Proper Balance of Oil-like Groups and Charged (Vinegar-like) Groups, There Exists a Competition for Limited Water Between Oil-like and Vinegar-like Groups Constrained to Coexist Along a Protein Chain

Development of too much structured water around emerging oil-like groups causes oil-like domains to reassociate. Decrease of organized water surrounding newly emerged oil-like groups, as charged groups successfully compete to add water to their own hydration shells, causes oil-like domains to dissociate. The energy change represented by association, or dissociation, of oil-like domains can be quantified using the information of the heat change represented by the cusp of insolubility in Figure 1.1.

These concepts of mechanism, developed in Chapter 5, are introduced immediately below by two statements, each followed by a list that develops the statement.

1.2.2.1.1 The Development of Too Much Organized Water Surrounding Oil-like Groups Causes Phase Separation!

- Vinegar-like groups, often as ions, prefer being surrounded by organized water.
- Oil-like groups, however, become surrounded by differently organized water, unsuited for ions.
- Development of too much water organized around oil-like domains renders them insoluble, and they associate, that is, separate from water.
- To control the association of oil-like domains is to control protein function and to allow for design of remarkably efficient and diverse energy conversions.

1.2.2.1.2 The Competition for Water Between Oil-like and Vinegar-like Groups Controls Separation!

- Charged vinegar-like groups steal organized water from around emergent neighboring oil-like groups and allow solubility of otherwise insoluble oil-like domains.
- Abundant structured water surrounding oil-like groups can force neutralization of charged groups and allow protein folding and assembly by association of oil-like domains.
- In addition to contraction by association of oil-like domains, such forced neutralization results in the performance of chemical and electrical work that sustains Life.
- More intense competition for water between oil-like and charged groups yields more positive cooperativity and correspondingly more efficient energy conversion.

1.2.2.1.3 Properties Once Considered Unique and Essential to Life Are Unexpectedly Found in Designed Model Proteins

The particular designed elastic model proteins, through which the mechanistic assertion developed, use a repeating five-amino-acid residue sequence propagated by *translational symmetry*. Of the five axioms noted above under the phenomenological assertion, the fifth axiom describes the conditions for increased positive cooperativity and the result of increased effi-

1.2 Four Principal Assertions of "What Sustains Life?"

ciency. Remarkably, positive cooperativities exhibited by these designed elastic model proteins, at their best, markedly exceed generally discussed biological examples of positive cooperativity. This was unforeseen.

As Monod states in his treatise, *On Symmetry and Function in Biological Systems*, "One may set aside the simple problem of fibrous proteins. Being used as scaffolding, shrouds or halyards, they fulfill these requirements by adopting relatively simple types of *translational symmetries*."[8] Therefore, it was not anticipated that positive cooperativity, the effect Monod thought to be "the second secret of life," second only to "the structure of DNA,"[9] would be most beautifully demonstrated by designed variations of a repeating sequence of the mammalian elastic fiber based on translational symmetry.

Based on a series of designed elastic-contractile model proteins, Figure 1.2 exhibits a family of curves whereby stepwise linear increases in oil-like character give rise to supra-linear increases in curve steepness, that is, in positive cooperativity. More oil-like phenylalanine (Phe, F) residues with the side chain $-CH_2-C_6H_5$ replace less oil-like valine (Val, V) residues with the side chain $-CH-(CH_3)_2$. Here the structural symmetry is translational with as many as 42 repeats (Model protein v) of the basic 30-residue sequence, and the structure is designed beginning with a repeating five-residue sequence of a fibrous protein, the mammalian elastic fiber.

Figure 1.2 demonstrates a series of increasingly shifted and increasingly steep acid–base titration curves that correspond with linear

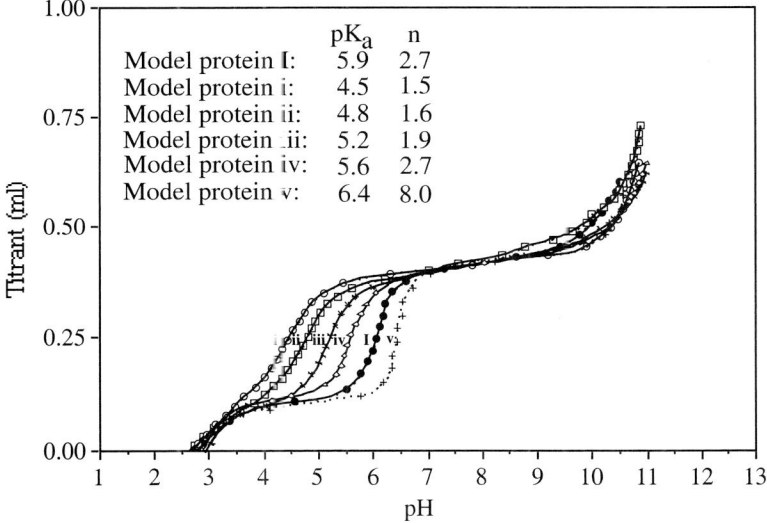

FIGURE 1.2. Plots of the acid–base titration curves for the following series of model proteins:

Model Protein I:	(GVGIP GFGEP GEGFP GVGVP GFGFP GFGIP)$_{26}$(GVGVP)	2E/5F/2I
Model Protein i:	(GVGVP GVGVP GEGVP GVGVP GVGVP GVGVP)$_{36}$(GVGVP)	E/0F
Model Protein ii:	(GVGVP GVGFP GEGFP GVGVP GVGVP GVGVP)$_{40}$(GVGVP)	E/2F
Model Protein iii:	(GVGVP GVGVP GEGVP GVGVP GVGFP GFGFP)$_{39}$(GVGVP)	E/3F
Model Protein iv:	(GVGVP GVGFP GEGFP GVGVP GVGFP GVGFP)$_{15}$(GVGVP)	E/4F
Model Protein v:	(GVGVP GVGFP GEGFP GVGVP GVGFP GFGFP)$_{42}$(GVGVP)	E/5F

Stepwise increases in oil-like character, as when the mildly oil-like Val (V) residue is replaced by the very oil-like Phe (F) residue, cause the acid–base titration curves to be shifted to higher pH values and to be steeper. The energy required to drive the model protein from the phase separated, contracted state to the swollen, relaxed state is proportional to the width of the curve, that is, inversely proportional to the steepness of the curve. Accordingly, the model protein with the steepest curve exhibits the most efficient function for performing the work of lifting a weight.

stepwise increases in oil-like character of the model proteins. Here the low pH side (the more acidic, uncharged side) of the curve represents the contracted (the oil-like, phase-separated, insoluble) state, and the high pH side is the unfolded soluble state. The energy required to drive contraction is inversely proportional to the steepness of the curve. In particular, a change in pH ($\Delta pH = \Delta G_{CE}/2.3RT$) is proportional to a change in chemical energy, ΔG_{CE}, and, therefore, a smaller change in pH required to go from the relaxed to the contracted state means a more efficient energy conversion. Accordingly, increased steepness signifies more efficient function.

A means of quantifying steepness uses Hill plots, as given in Figure 1.3, where the slope of the Hill plot is the Hill coefficient, n. In the absence of cooperativity the Hill coefficient is 1, whereas the highest positive cooperativity exhibited by the set of elastic-contractile model proteins in Figure 1.3A is 8. As further noted below and shown in Figure 1.3B, the Hill coefficient is 1.0 for myoglobin that binds oxygen in the tissues and 2.8 for hemoglobin, the protein that transports oxygen from the lungs to the

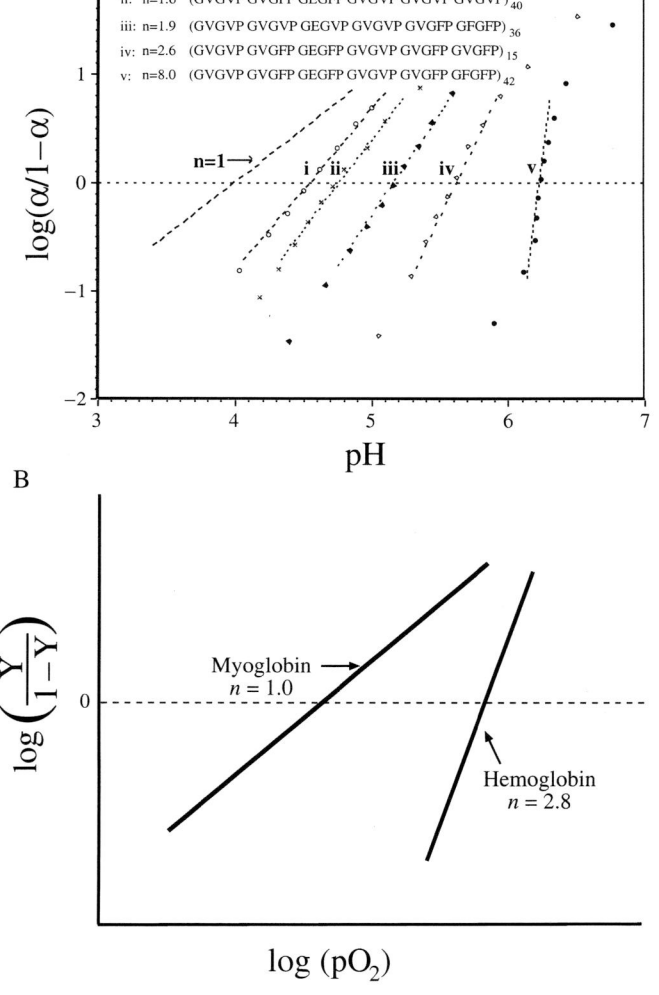

FIGURE 1.3. **A** Hill plot of the set of designed elastic-contractile model proteins shown in Figure 1.2 with Hill coefficients, n, ranging from 1.5 to 8.0. **B** Hill plot of myoglobin (n = 1) and hemoglobin (n = 2.8). It is shown that the vaunted hemoglobin positive cooperativity is relatively small compared with that of designed elastic protein-based polymers and, in particular, of designed Model protein v.

tissues. Myoglobin exemplifies the simple binding of oxygen, whereas hemoglobin is perhaps the most commonly considered example of positive cooperativity in biology, which is essential to its oxygen transport role in biology. The consilient explanation for this difference between myoglobin and hemoglobin is given in Chapter 7, Section 7.2.4.

Most significantly, however, the molecular basis for positive cooperativity and the result of increased functional efficiency in designed elastic-contractile model proteins has been experimentally determined to be the competition for water that occurs between oil-like domains and charged groups constrained to coexist within a protein structure (see immediately below and Chapter 5, section 5.1.7.4). This represents the principal statement of the Mechanistic Assertion.

1.2.2.2 An Explicit Demonstration of the Mechanistic Assertion in Terms of Monod's "Second Secret of Life"

Central to the mechanistic assertion is competition for hydration between oil-like and vinegar-like groups constrained to coexist along a chain molecule. In the process, the disparate groups each reach for water unaffected by the other. This results in an effective repulsion between oil-like and vinegar-like groups, that is, the groups physically get as far away from each other as possible in their thirst for water unaltered by the other. When the sequence does not allow the oil-like and vinegar-like groups to get far enough apart, repulsion exists even in the totally unfolded state. Relaxation of the repulsion occurs when the vinegar-like group becomes less vinegar-like, as could occur on neutralization (e.g., by protonation) or partial neutralization (e.g., by ion pairing) of the charge. As a result, structurally related oil-like groups lose solubility; they develop too much special structured water around them, which becomes the driving force for folding by association of oil-like groups.

The classic vinegar-like group is the carboxylate $-COO^-$, and it becomes less vinegar-like on adding acid, H^+, to become the carboxyl $-COOH$. The measure of the acid, the pH (= $-\log[H^+]$), required for neutralization of any particular carboxylate is the pKa. This is the pH at which the concentration of carboxylates and carboxyls undergoing change are equal; in other words, $[-COO^-]/[-COOH] = 1$. The pKa is 4.0 or less for glutamic acid (Glu, E) with a side chain of $-CH_2-CH_2-COOH$ or aspartic acid (Asp, D) with a side chain of $-CH_2-COOH$ when not involved in repulsive interactions.

1.2.2.2.1 More Detailed Consideration of the pKa Shifted State Arising from Competition Between Oil-like and Vinegar-like Groups

With the preceding background, Figure 1.4 considers the pKa values relevant to Model protein v using the special way of plotting the data of Figure 1.3. Remarkably, the pKa of the first carboxyl to form carboxylate on raising the pH, on decreasing acid, is 7.0 due to the water-mediated repulsion between oil-like groups and charged carboxylate. As the ionization proceeds, the pKa decreases until the last (the 42nd) carboxyl to form carboxylate of the 42 carboxylates in the chain of 42 repeats of 30 residues does so with a pKa of 5.7. Accordingly, ionization of the last carboxyl becomes more than 20 times more likely than the first one because of the progressive and cooperative destruction of the structured water around exposed oil-like groups.

The repulsion between oil-like groups and charged carboxylates is relaxed by 1.8 kcal/mol-carboxylate as more and more of the special hydration around oil-like groups is disrupted. Formation of the first carboxylate occurs only when it can destructure sufficient structured water around oil-like groups as they become exposed. With the first carboxylate having destructured sufficient structured water around exposed oil-like groups, a subsequent carboxylate with an overlapping sphere of influence can form more readily as it has less structured water around exposed oil-like groups to destructure in order to form its own hydration shell. This constitutes positive cooperativity.

Of perhaps even greater significance in demonstration of the mechanistic assertion is that when the oil-like association is fully disrupted and the carboxylates and the oil-like groups are fully exposed to water in Model protein v (see Fig. 1.2), there still exists a repulsion between the oil-like and vinegar-like groups that raises the pKa from 4.0 to 5.7. This amounts to a repulsion of 2.4 kcal/mol-carboxylate. *The carboxylates are still unable to achieve full hydration, because they cannot get sufficiently removed from the oil-like groups.* Even with the less oil-like Val (V) residues of Model protein i, a small residual pKa shift remains. When carefully analyzed, the data in Figures 1.2 through 1.4 clearly identify the basis for positive cooperativity and provide an understanding of what Monod has called the "second secret of life."[9] In short, positive cooperativity results from the competition for hydration between oil-like and vinegar-like groups constrained to coexist along a chain of defined sequence and/or within the folded and assembled state of the chain or chains.[10]

1.2.2.2.2 Two Aspects of the Fully Charged State

The fully charged state lies bare the tense, taut, or cocked state with two aspects. The fundamental aspect is the repulsion between vinegar-like and oil-like constituents along the chain molecule, as reflected in pKa shifts and positive cooperativity. The other aspect is the resulting elastic deformation of chain segments due to repulsion between vinegar-like and oil-like components within the model protein structure. The vinegar-like and oil-like components of the model protein each reach out for water unused by the other. The model protein becomes constrained, even in the more disordered state, as the result of being limited from arrangements where oil-like and vinegar-like constituents would be in closer proximity.

1.2.2.2.3 Analogy Between Fully Charged Carboxylate (–COO$^-$) State of Model Proteins and the ATP^{-4}-bound State of Protein-based Machines

In Figure 1.4, one carboxylate in each 30 residue repeat, (GVGVP GVGFP G$\underline{E}$GFP GVGVP GVGFP GFGFP)$_{42}$(GVGVP), sustains a repulsion of 2.4 kcal/mol-carboxylate

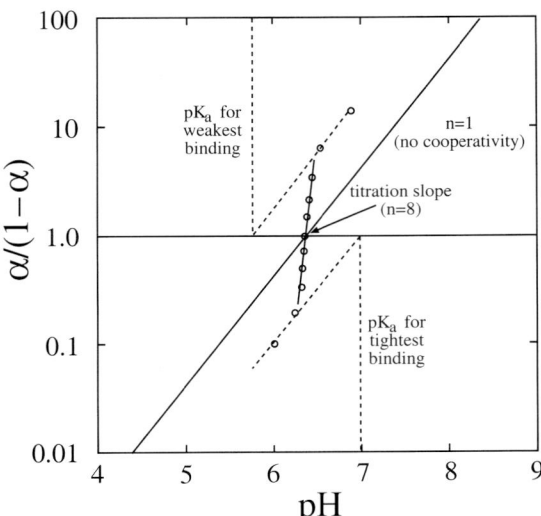

FIGURE 1.4. Hill plot, $\log[\alpha(1 - \alpha)]$ vs. pH, for Model protein v: (GVGVP GVGFP G$\underline{E}$GFP GVGVP GVGFP GFGFP)$_{42}$. The normal pK for a glutamic acid residue (Glu, E) with E having the ionizable carboxyl group –COOH → –COO$^-$ + H$^+$ occurs at a pH of about 4. This plot shows for Model protein v that the first COOH pK occurs at 7.0 and the 42nd COOH pK occurs at about 5.7. Even when completely unfolded and largely disordered, the pK is shifted from 4.0 to 5.7. With the glutamic acid residues separated by 29 uncharged residues in an unfolded, largely disordered model protein, charge–charge repulsion cannot be responsible for the increase in pK. Nonetheless, a repulsion of 2.4 [= (5.7 − 4.0) × 2.3RT)] kcal/mol remains. Noting in Figure 1.2 that the pK shift increases with increases in the oil-like nature of the model protein, the conclusion (along with other data in Ch. 5) is that the pK shift is due to competition for hydration between oil-like and vinegar-like groups constrained to coexist along the amino acid sequence of Model protein v. Of further interest is the change in pK from 7.0 to 5.7 that occurs during the unfolding and dissociation of the oil-like groups within the assembled model protein. During an unfolding fluctuation, oil-like groups become surrounded by special structured water, and, for the first –COO$^-$ to form, the emergent carboxylate must rearrange water around the oil-like groups and organize it for its own hydration. The first carboxylate, having destructured the most water around oil-like groups, makes it easier for the second carboxylate to form, and so on. This is the physical basis for positive cooperativity, which was of particular interest to Monod.

even when the model protein is completely unfolded. This repulsion arises almost entirely due to the presence of the phenylalanine (Phe, F) residues. Half of the repulsion, that is, 1.2 kcal/mol-carboxylate, remains, even when the two most sequence-proximal F residues are replaced by V. ATP, on the other hand, exhibits four negative charges instead of the single negative charge of a carboxylate. Because of this, ATP produces a much greater repulsion between oil-like and charged groups, even when ion paired with magnesium ion (Mg^{+2}). ATP, therefore, reaches out substantial distances to destroy the structured water around exposed oil-like groups that allows separation of existing ion pairs and to destroy structured water forming around emerging oil-like groups that promotes dissociation of oil-like groups and domains. The central arguments that the development of too much hydration around oil-like groups causes insolubility and that the presence of charge disrupts hydration around oil-like groups to give solubility are developed in Chapter 5 (sections 5.1.3.3, 5.1.7.4, 5.3.3.3, and 5.7.9.2), and illustrated in Chapters 7 (sections 7.2 and 7.5.1.4) and 8.

1.2.3 Assertion 3: The Assertion of Biological Relevance

Biology thrives near a movable transition for insolubilization of oil-like domains, and the forces that, in a positively cooperative manner, power the molecular machines of biology drive-paired oil-like domains of proteins back and forth between association (water insolubility) and dissociation (water solubility); excursions too far in either direction into the realms of insolubility or solubility spell disease and death.

Phenomena exhibited during protein function and dysfunction (disease) parallel those phenomena exhibited during function of designed model proteins. Examples of this coherence of phenomena follow in a cursory introductory form. Detailed considerations are given in Chapters 7 and 8 based on the physical mechanisms developed in Chapter 5.

1.2.3.1 By the Consilient Mechanism Protein-based Machines Require Water to Function

A protein-based machine without water as an integral part of its structure could not function by the consilient mechanism. In other words, water is required in at least one of the two states in order to have a "movable cusp of insolubility," and in order for competition for hydration to be relevant there must be adequate water present. The first prerequisite, therefore, in addressing the biological relevance of the consilient mechanism is to assess whether or not water exists within or between the changing structural elements of a protein motor during function.

1.2.3.1.1 The Myosin Motor Domain of *Dictostelium discoidium*

The myosin motor is an ATPase, because it is driven by cyclic ATP binding to cause detachment, splitting to form ADP and P_i, and the sequential release initially of P_i with contraction and then of ADP in readiness for a new ATP to bind. The organism *Dictostelium discoidium* provides a convenient motor domain for studying muscle contraction because of its near identity to the myosin motor of skeletal muscle.

Protein motors are three-dimensional. Therefore, being able to see in three dimensions is very helpful in order to understand structure and the changes in structure that drive function. A convenient means of visualizing in three dimensions can be obtained by a "stereo view." Two views of the structure of interest are given with one view rotated a few degrees on its vertical axis from the other. Rotation one direction allows for a three-dimensional view when looking with crossed eyes (i.e., as though the image were midway between the printed paper and your eyes), whereas rotation in the opposite direction allows a three-dimensional view when looking wall-eyed (i.e., as if the pair of images were at a distance). Figure 1.5 provides a stereo view of the motor domain arranged for cross-eye viewing.

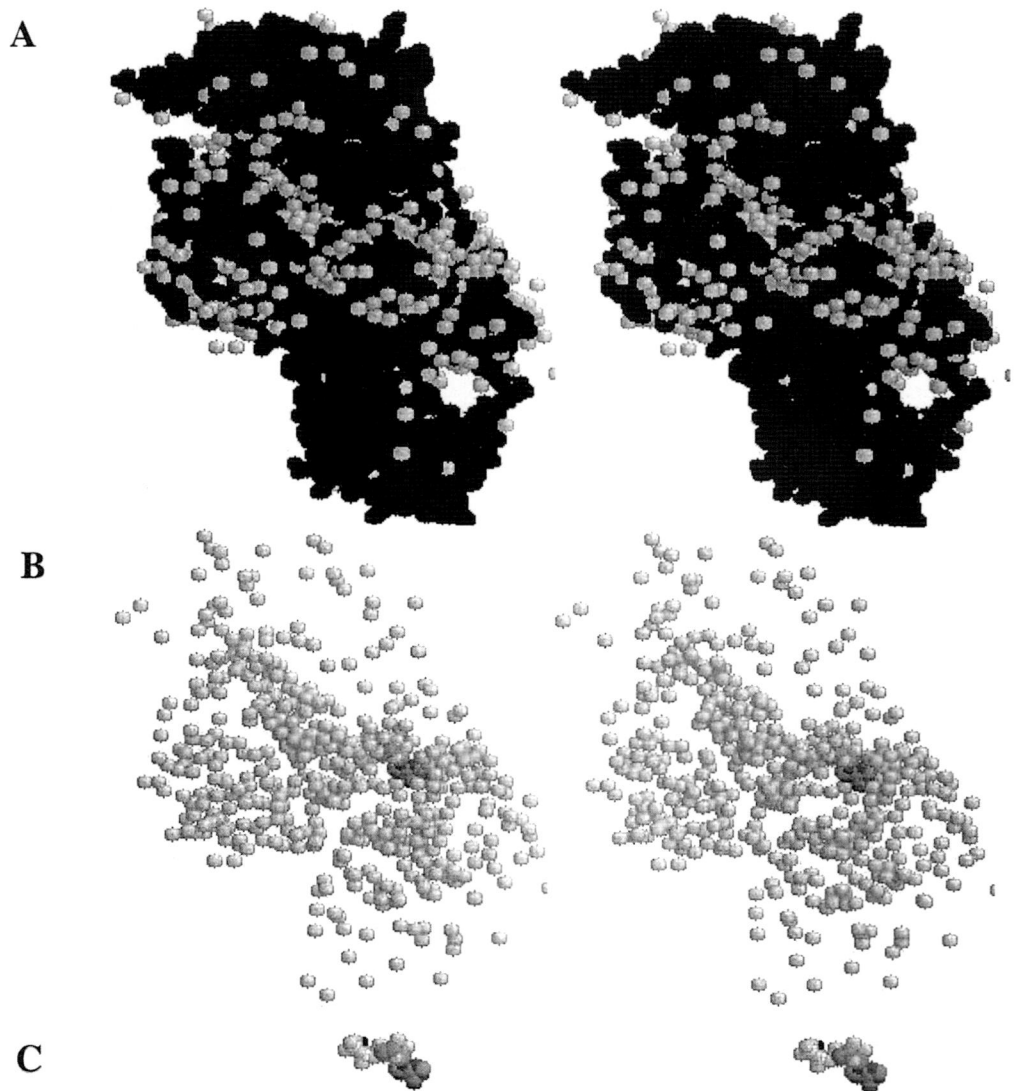

FIGURE 1.5. The "waters of Thales" of the myosin II motor. Stereo views of the crystal structure of the myosin motor domain of *Dictostelium discoidium* in presence of ATP are shown. **A** Space-filling display showing water molecules on a sculptured surface of the myosin II motor. **B** Those water molecules throughout the entire structure that are sufficiently fixed in space to be seen by X-ray diffraction are shown by removal of the protein component. These waters and additional water in the surface crevices are the "waters of Thales" required for function by the consilient mechanism. **C** The ATP is shown in the same orientation with all other atoms hidden. (Figure preparation was based on the crystallographic results of Bauer et al.[11] as obtained from the Protein Data Bank, Structure File 1FMW.)

Our special interest at this point is to gain insight into the presence and distribution of water within and around a motor domain, as would be required for the consilient mechanism to be relevant to motor function. Because it shows a recently determined structure,[11] Figure 1.5 represents an up-to-date capacity for locating water molecules. This structure of the myosin motor contains ATP, as shown in Figure 1.5C, with all other atoms hidden.

1.2 Four Principal Assertions of "What Sustains Life?"

1.2.3.1.2 The Sculpted Appearance of the Myosin Motor Domain

Especially when seen in three dimensions, as in Figure 1.5A, the stereo view of the myosin motor domain has the appearance of a sculpted surface. The surface contains crevices and depressions, as though formed from sandstone that had been weathered by wind and rain. Only a relatively few water molecules are seen in these surface recesses, because the majority of water molecules are too mobile to be observed by X-ray diffraction. Yet these surface crevices and depressions can be filled with water molecules that, by the consilient mechanism, contribute to the energy considerations of motor function. In this regard, it should be appreciated that only 10% to 20% of the existing water molecules are sufficiently fixed in space to be located by X-ray diffraction.[12]

1.2.3.1.3 The "Waters of Thales"

When the space-filling protein component is removed, it becomes possible to view the located water molecules within the myosin motor. As shown in Figure 1.5B, an impressive number and distribution of the detected water molecules appear. It can also be expected that there are many more water molecules relevant to function of the myosin motor that are too mobile to be seen by X-ray diffraction, just as is apparent in the crevices and recesses of the surface. By the consilient mechanism these water molecules (seen in Fig. 1.5B and the additional unseen water molecules) are essential to motor function. These water molecules, which in our view are essential for Life, we choose to call the "waters of Thales."[13,14] Thus, as required for this protein motor to function by the consilient mechanism, internal water molecules do exist. Accordingly, in our view, this fundamental protein motor that produces motion contains ample water as part of the structure in order to function in the competition for water between oil-like and vinegar-like groups, which competition expresses as a repulsion between these groups.

1.2.3.2 ATPase, Biology's Workhorse Protein-based Machine

In general, ATP (adenosine triphosphate or an equivalent nucleoside triphosphate, NTP) powers Life's protein-based machines. Specifically, the breakdown of ATP to form ADP (adenosine diphosphate) and P_i (inorganic phosphate, PO_4^{-3}) provides the energy that powers protein-based machines. As will be argued in Chapter 8, section 8.1.11.2, the large amount of energy released on ATP breakdown results from the limited availability of water to ATP. Also, by the consilient mechanism, if a pair of oil-like surfaces forms too much special oil-like hydration during a transient separation, they reassociate. Should ATP bind with its multiply charged and thirsty triphosphate tail directed toward the pair of dissociable oil-like surfaces during transient separation, the triphosphate tail recruits the water adjacent to the oil-like surfaces for its own hydration; too much oil-like hydration no longer forms, and the pair of oil-like surfaces remain dissociated. This is the essence of the consilient mechanism; charged (vinegar-like) groups compete with oil-like groups for hydration. Therefore, we ask, might the consilient mechanism dictate a common structural motif for ATP binding to ATPases?

In fact, the ATPase of skeletal muscle exhibits a commonly recognizable structural feature for the ATP bound state.[15] So in an initial illustration (Fig. 1.6), we approach ATPases with the most simplistic cycle, that of an "idling" motor. Then, in Chapter 2 we take a first step toward useful function by attachment to and detachment from a surface and progress from there to more complete depictions. Once the basic science is laid out in Chapter 5, a detailed molecular description is given in Chapter 8.

Like a car in neutral with its motor running, an idling motor runs and consumes energy without producing useful motion. Figure 1.6 depicts an "idling" ATPase motor by means of a cross section of a globular protein that contains the ATP binding site. ATP binding opens a cleft, a cleft that had been closed, or partially so, by association of paired oil-like domains.

"Idling" ATPase Motor

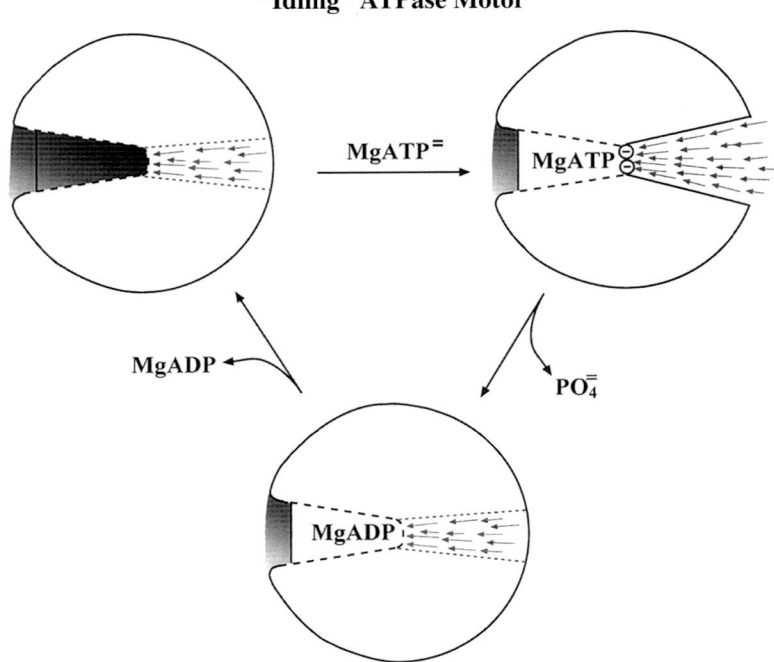

FIGURE 1.6. The "idling" ATPase motor consumes the energy released on conversion of MgATP to MgADP and P_i without the performance of useful work. On binding MgATP to the globular apoprotein, a cleft, partially or entirely closed by association of oil-like groups, opens near the base of the nucleotide triphosphate (NTP) binding site, with the super vinegar-like triphosphate tail at the interface. Release of phosphate allows some recovery of significant oil-like association, and, on release of MgADP, the apoprotein is recovered to complete the cycle.

The ATP molecule orients with its charged triphosphate tail at the base of the cleft such that it can access the water of the cleft. Because of this, water, which would normally form around the oil-like groups and effect reclosure, becomes oriented toward the phosphates and maintains the cleft in the open state. On hydrolysis of ATP and phosphate removal, the cleft partially or completely closes. Release of the ADP molecule recovers the original state to complete the cycle. Accordingly, energy has been consumed, but in this "idling" ATPase motor no work is accomplished.

1.2.3.3 Muscle Contraction

Again, consider statements grouped as bulleted lists with each list defined by a heading.

1.2.3.3.1 Structural Description of the Sliding Filament Model of Muscle Contraction

- Overall in muscle contraction, calcium-ion-binding triggers release of phosphate (the paramount vinegar-like group) with the consequence of contraction.
- The shortening of contraction results from thick filaments sliding past thin filaments.
- A cross-bridge from the thick filament causes the movement by cyclic attachment to and detachment from the thin filament.
- Specifically, the cross-bridge detaches from the thin filament, reaches forward, reattaches to the thin filament and contracts sliding the thin filament past the thick filament.

1.2 Four Principal Assertions of "What Sustains Life?"

1.2.3.3.2 Muscle Contraction as Inferred from Model Protein Studies

- Stepwise, ATP binding to the cross-bridge causes detachment from the thin filament by bringing about dissociation of oil-like domains.
- Binding of ATP also causes separation of oil-like domains within the cross-bridge in a way that allows forward extension of the cross-bridge.
- Positive calcium ions pair with negative vinegar-like groups associated with the thin filament to uncover an oil-like domain for a new cross-bridge attachment site.
- Phosphate release reestablishes oil-like association between cross-bridge and thin filament and the power stroke occurs on reestablishment of oil-like associations within the cross-bridge.

This perspective, developed on applying the elastic model protein studies to crystal structures of scallop muscle, brings us to the stage longingly sought by Perutz in 1990 with the statement "that one day we shall learn how chemical energy is generated and turned into motion."[16] Interestingly, at the time Perutz penned these words, the demonstration of chemical energy turned into motion by the unexpected source of designed elastic-contractile model proteins of primary focus here had been in the literature for little more than one year.[17] Figure 1.7 gives the key results of that publication in part A for acid-driven contraction at constant force (isotonic contraction) and in part B for acid-driven force development at constant length (isometric contraction).[17] The fundamental process, regardless of the experimental conditions, is acid-driven association of oil-like domains.[18]

The movable cusp of insolubility represented in Figure 1.1 provides a means of visualizing the molecular process. Addition of acid (proton, H^+) converts the vinegar-like carboxylate group

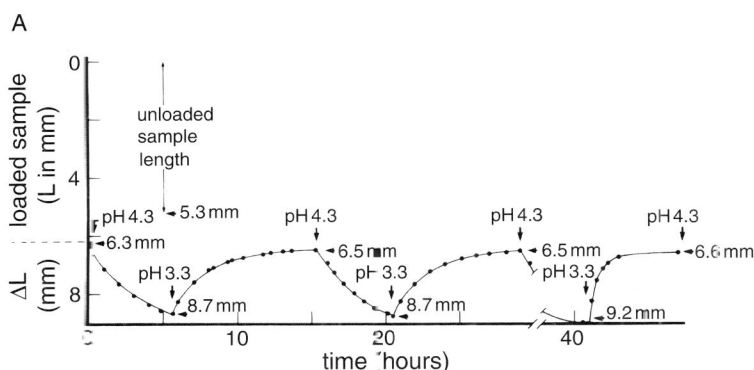
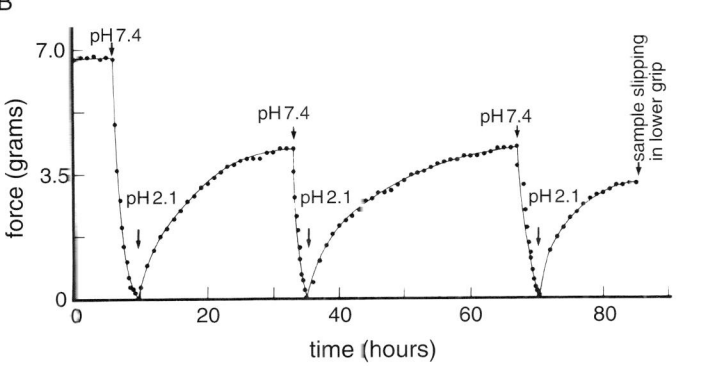

FIGURE 1.7. Shown are the first reported data[17] of the conversion by an elastic-contractile model protein of chemical energy due to an increase in concentration of acid into the mechanical work of contraction. **A** Length changes at constant force (isotonic contraction) in phosphate-buffered saline. **B** Force changes at constant length (isometric contraction) in phosphate-buffered saline. (Reproduced from Urry et al.[17])

-COO⁻ to the more oil-like –COOH group. This lowers the cusp of insolubility along the temperature scale and results in association of oil-like groups, which is a key element of the contraction. When at fixed load, the association (insolubilization) of oil-like domains lifts the weight; when at fixed length, the association (insolubilization) of oil-like domains stretches interconnecting chain segments, resulting in an increase in elastic force.[7]

1.2.3.4 Energy Conversion in Mitochondria

The mitochondria are the energy factories that produce ATP (adenosine triphosphate), the energy currency of biology. In particular, a single proton-driven protein-based machine in the mitochondria produces 32 of the 36 ATPs resulting from the oxidation to CO_2 and H_2O of a single glucose molecule (the molecule that best exemplifies a food source). This energy conversion in mitochondria occurs in two steps: (1) the oxidation of glucose-reduced carriers of the electron transport chain that pumps protons from one side to the other of the inner mitochondrial membrane and (2) the flow of protons back across the inner mitochondrial membrane through a membrane-bound rotary motor that drives the rotor extending into the extramembrane rotary motor to produce the ATP from ADP and P_i, inorganic phosphate.[19,20] The doubled rotary motor is descriptively named *ATP synthase*. The following four lists of bullets provide succinct introductory statements of this fundamental energy conversion of biology.

1.2.3.4.1 The General Perspective

- ATP represents the energy coin of the biological realm.
- Mitochondria, biology's intracellular energy factories, produce almost all of the ATP.
- Mitochondria produce ATP by a two-step energy-converting process.
- The energies interconverted by both of these steps are demonstrable by model proteins designed to function by the consilient mechanism of controlling phase separation.

1.2.3.4.2 Step One for Energy Conversion in Mitochondria as Interpreted from Model Protein Studies

- In the first step at the cytosolic side of the inner mitochondrial membrane, electron removal forms a positively charged chemical group with releasable protons.
- The positive charge disrupts hydrophobic association, allowing proton release to that side.
- In a second step at the matrix side of the inner mitochondrial membrane, electron addition forms a negatively charged group capable of proton uptake.
- The negatively charged group disrupts hydrophobic association, allowing proton uptake from the matrix side to result in net transport of protons across the inner mitochondrial membrane.

1.2.3.4.3 Step Two for Energy Conversion in Mitochondria in Two Parts

1.2.3.4.3.1 Part 1: The F_0-motor: Ten Hairpin Loops of Protein Helices, with One Side of the Loop Arranged as a Cleat of a Tractor Tread, Constitute a Transmembrane Rotary Motor

- Excess proton on one side of the membrane enters a channel; in transit to the other side the proton binds to vinegar-like group of a transmembrane helix and makes it more oil-like.
- The more oil-like helix, one of the ten protein elements of a transmembrane rotary motor, rotates into the oily layer of the membrane.
- The next helix rotates into position for a half-length channel site and releases its proton to the opposite side of the membrane.
- This process rotates an oil-like double helical rod that extends like an axle out of the center of the wheel-like cluster of transmembrane helices of the membrane-bound rotary motor.

1.2.3.4.3.2 Part 2: The F_1-motor: Transmembrane Rotary Motor Drives Rotation of Oil-like Protein Rotor Within an Orange-Shaped, Six-Subunit Protein Structure, Causing ADP and P_i to Form ATP.

- The rotating axle extends to form the stem and core of an orange-shaped protein struc-

1.2 Four Principal Assertions of "What Sustains Life?"

ture of six sections, with three symmetrically arranged sections permanently bound to ATP.
- The most oil-like side of the asymmetric protein rotor, rotated by the F_0-motor, begins to oppose one of three symmetrically arranged catalytic sections that contains ADP and P_i.
- Repulsion between the approaching most oil-like side of the rotor and the most super vinegar state of ADP and P_i produces ATP, and complete exposure expels the ATP.
- Then ADP and P_i add to the emptied catalytic site to form the most vinegar-like state as that site faces the least oil-like side of the rotor, and so on.

1.2.3.5 Molecular Aspects of the F_1-motor

The F_1-motor component of ATP synthase is composed of a γ-rotor and three α-subunits and three β-subunits arranged as three αβ pairs, that is, $(αβ)_3$. Figure 1.8 schematically represents three states of the F_1-motor in cross section. Near their junctions with a β-subunit, each of the three α-subunits contains one molecule of ATP that is constant and does not change during the catalytic process. Nonetheless, the α-subunit ATPs fill an essential role in the consilient mechanism. From the crystal structure data of Abrahams et al.,[21] the three β-subunits are found with one catalytic site empty, a second site contains ADP, and the third site contains ATP. For purposes of our discussion, ADP plus P_i (inorganic phosphate, which at physiological pH would have as a dominant species HPO_4^{-2}) replaces ATP. The nucleotides reside near the interfaces between the αβ and βα pairs with the adenine component most clearly defining its subunit, α or β. The polar phosphate components reside at the interfaces between the α- and β-subunits. In this case the sides of the α- and β-subunits form the cleft that emanates from phosphate to the γ-rotor, as sketched in Figure 1.8. Thus, the structural feature of the charged phosphates of the nucleotides residing at the base of a water-filled cleft occurs with the F_1-motor just as it does with the generalized ATPase considered above and represented in Figure 1.6.

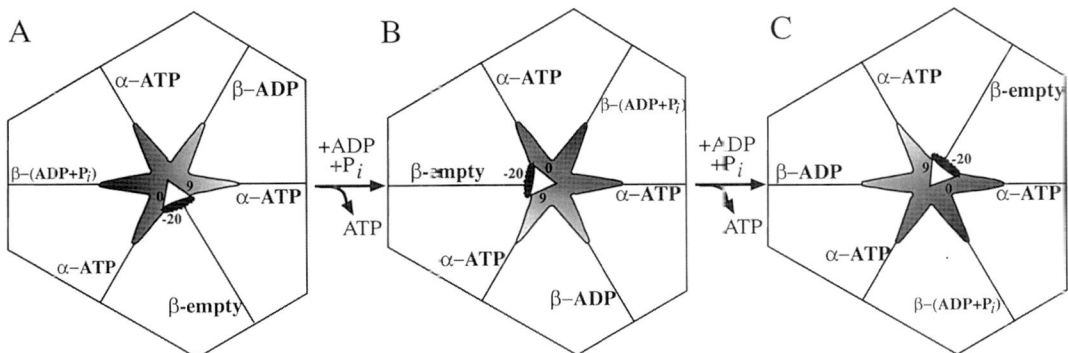

FIGURE 1.8. Illustration of a movable (rotating) cusp of insolubility representing the cross section of the F_1-motor of ATP synthase and depicting an asymmetrical rotor with one of its three sides being very oil-like. An oil-like association is shown to occur each time the most oil-like side of the rotor, rotated by the F_1-motor, resides at an empty site. This provides a literal, spatial demonstration of the "movable cusp of insolubility." On rotation of the oil-like side of the rotor from the empty site to an ATP-filled or an (ADP plus F_i)-filled site, the oil-like association would be lost. The metaphorical representation becomes the "movable cusp of insolubility" with the graphical equivalence of the leftmost (low-temperature) cusp of Figure 1.1 being raised to the rightmost (high-temperature) cusp. Thus, the "movable cusp of insolubility" at which biology thrives can, by the F_1-motor of ATP synthase, be given both literal and metaphorical representation.

1.2.3.6 The F_1-Motor's Cusp of Insolubility in Both Assembly and Function

1.2.3.6.1 When Assembled, the F_1-motor Functions as a Rotating Cusp of Insolubility

The "movable cusp of insolubility" of the F_1-motor becomes both literal and metaphorical. The metaphorical representation occurs as the oil-like side of the rotor passes from association with the oil-like empty β-subunit where the cusp of insolubility, represented in Figure 1.1, resides at low temperature (below physiological temperature) to high temperature, that is, above physiological temperature to result in oil-like dissociation of rotor from sleeve. Thus, the metaphorical representation occurs as the oil-like side of the γ-rotor is rotated by the F_0-motor from association with an oil-like empty β-subunit to dissociation from a vinegar-like α-subunit that contains ATP.

For the appropriate concentrations of ADP and P_i, Figure 1.8 demonstrates a literal visual representation of going from association to dissociation and back to association for each 120° rotation. This literal "movable cusp of insolubility" results from the competition for hydration that exists between oil-like and vinegar-like groups as in "*oil and vinegar don't mix.*" Expressed as a repulsion, the same competition for hydration occurs involving interaction of the super vinegar-like group of ATP and the most super vinegar-like state of ADP plus P_i with the oil-like side of the γ-rotor in Figure 1.8. An analogous situation occurred between the very oil-like phenylalanine groups of Model protein v in Figure 1.4 and the classic vinegar-like carboxylate group with its single negative charge. In the latter case the repulsion was 2.4 kcal/mol. In Figure 1.8, the repulsion would be many times greater as the nucleotide phosphates would have more negative charges and the change in the oil-like character of the three faces that define the central γ-rotor are so striking, namely, −20, 0, and +9 kcal/mol (see Chapter 8, section 8.4.4.3, and Table 8.2).

1.2.3.6.2 The Cusp of Insolubility for the F_1-motor, Alias the F_1-ATPase When Running in Reverse

When the rotor of ATP synthase is driven clockwise by the F_0-motor, as shown in the progression from states A to B to C in Figure 1.8, synthesis of nearly 90% of biology's ATP occurs. When the F_1-motor is separated from the F_0-motor, it can function in reverse as an "idling" rotary ATPase called the F_1-ATPase, or it can have a filament attached to it and perform the work of rotating the filament.[22] As occurs with many enzymes, F_1-ATPase cold denatures,[23] that is, its subunits—the γ-rotor and an accompanying ε-subunit together with the three αβ components within which the rotor rotates—dissociate on lowering the temperature. In other words, the oil-like associations between the subunits of F_1-ATPase separate on lowering the temperature and recombine on heating to physiological temperature. Therefore, the left-hand cusp of Figure 1.1 represents the cusp of insolubility for the oil-like assembly/disassembly of the functional F_1-motor.

1.2.3.7 Considered Steps in the F_1-motor Synthesis of ATP as Shown by the Consilient Mechanism

1.2.3.7.1 Consilient Steps in the Synthesis of ATP by the F_1-motor

The F_1-motor structure due to the Walker group[21] functions ideally to demonstrate the steps whereby the rotating cusp of insolubility would result in the synthesis of ATP. The γ-rotor may be represented by the relative oil-like character of its three faces that are calculated in Chapter 8, section 8.4.4.3, shown in Figures 8.31 to 8.33, and listed in Table 8.2. As represented in Figure 1.8, there is a very oil-like face (marked by −20 and shown directed toward the empty catalytic site), a neutral face (indicated by 0 and directed toward the β-ADP+P_i site), and a relatively vinegar-like face (indicated by +9 and directed toward the site indicated in Figure 1.8 as occupied by β-ADP).

1.2 Four Principal Assertions of "What Sustains Life?"

Starting with the structure and occupancy states of Figure 1.8 and given the insight of the consilient mechanism, P_i would add to the β-ADP site while facing the +9 side of the γ-rotor and would do so as the −20 face of the γ-rotor began clockwise rotation. Then the γ-rotor, driven by the F_0-motor, would continue rotation in a clockwise direction until the very oil-like, −20 face would begin opposing the β-(ADP plus P_i) site. As the repulsion between the oil-like side of the rotor and the most super vinegar-like (ADP plus P_i) state would build, the repulsion would be relieved by formation of the lesser but yet super vinegar-like ATP product. On continuing the 120° rotation to complete the face-off against the most oil-like side of the rotor, this most oil-like face of the γ-rotor would force the ATP product out of the catalytic site. As the oil-like face of the rotor begins rotation from the recently emptied site, ADP and P_i could enter with the least oil-like face (+9) in apposition, and the beginning of the next 120° rotation would bring in the intermediate oil-like face (0) to close the site before facing off with the most oil-like face (−20).

1.2.3.7.2 The Functional Role of the Three Symmetrically Arranged, Noncatalytic, ATP-containing α-subunits

The ATP-containing α-subunits are arranged with threefold symmetry around the γ-rotor. By the consilient mechanism, the α-ATP molecules become a trigonal force array focused on the central γ-rotor that would serve to keep the asymmetrical rotor more central and would serve to prevent insolubilization involving a more oil-like side of the rotor and the sleeve. As required for efficient function, the three α-ATP molecules would assist in preventing a viscous drag or a freezing-up by hydrophobic association between rotor and the central parts of the α- and β-subunits. Accordingly, by the consilient mechanism of water-mediated repulsion between oil-like elements of the γ-rotor and the super vinegar-like phosphates of the α-ATP molecules, the ATP-containing noncatalytic sites form a crucial function by providing a structure capable of a more efficient energy conversion.

1.2.3.8 The "Waters of Thales" Within the F_1-motor of ATP Synthase

To be workable, the consilient mechanism of repulsion between oil-like and vinegar-like (e.g., charged) groups, technically called an apolar (oil-like)–polar (e.g., charged) repulsive free energy of hydration, requires the presence of adequate water to form between oil-like and vinegar-like groups. Therefore, the first test of relevance of the consilient mechanism is to demonstrate sufficient water within the protein-based machine. As considered above for the myosin motor, X-ray diffraction methods for protein structure determination have reached the stage where at least some of the water molecules in the structure can be seen. As the observed water molecules must be fixed in position during the long period of time required for collecting the X-ray diffraction pattern, more mobile water molecules are not seen. Thus any observed water molecules represent a minimal number of the actual water molecules present.

A top view of the F_1-motor is shown in cross-eye stereo view in Figure 1.9A, where the three αβ pairs are given in space-filling representation and the γ-rotor and its attached ε-subunit are given in a backbone-only representation. The light dots are the water molecules observable on the surface of the F_1-motor. When seen in stereo, the observed water molecules almost entirely reside in clefts and recesses of the surface that in reality are totally filled with water molecules. This provides an opportunity to obtain a sense of how few water molecules are observed out of those actually present.

Showing exactly the same view, but with the three pairs of αβ-subunits removed, the stereo view in Figure 1.9B, allows the number of observed water molecules within the F_1-motor to be seen. Therefore, these observed water molecules become the "waters of Thales"[13,14] required to give this most important protein-based machine of biology the capacity to function.[24] Living organisms are "kept alive" by the "waters of Thales"[13]; they are the waters through which the repulsive force between oil-like and vinegar-like groups becomes expressed.

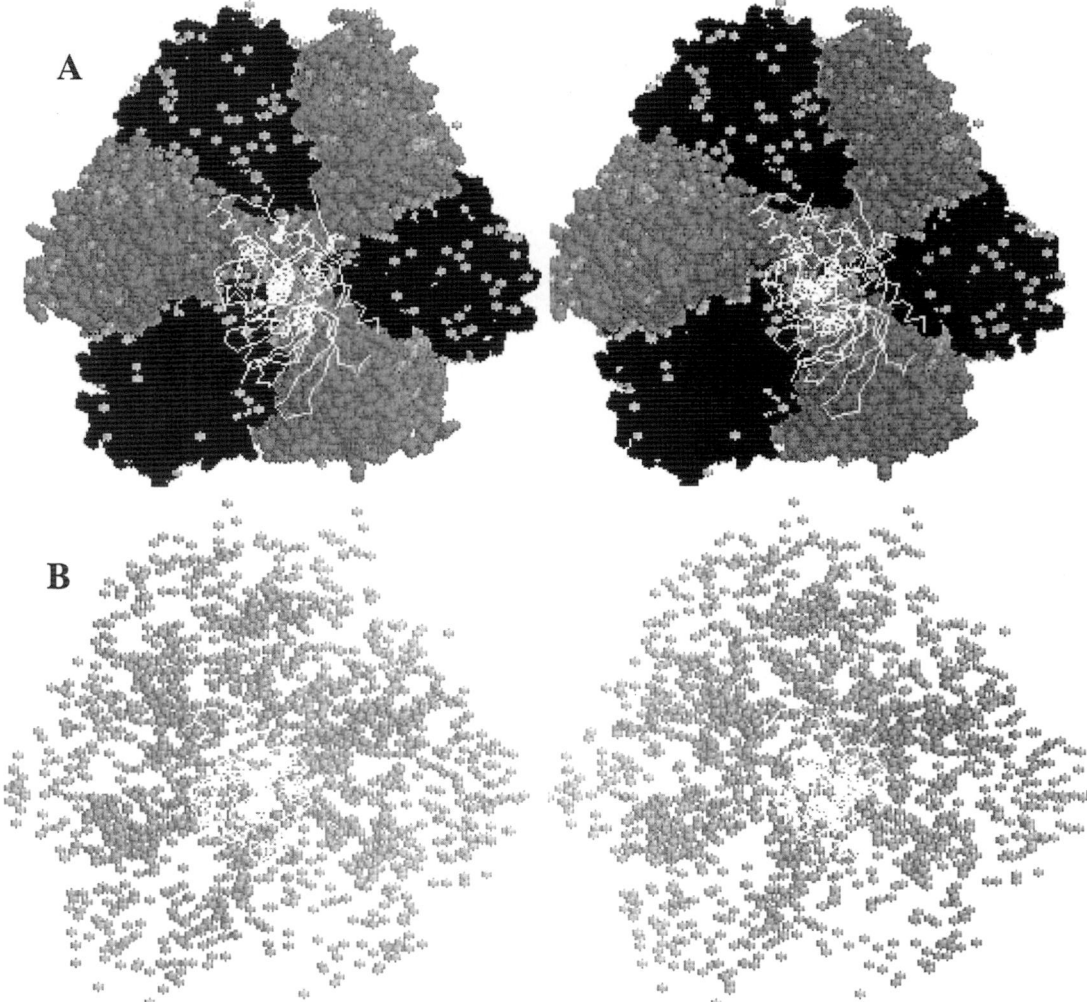

FIGURE 1.9. Observation of the "waters of Thales" of the F_1-motor of ATP synthase. The crystal structure of the F_1-motor of ATP synthase is shown in cross-eye stereo view from the top axis having all sites filled with ATP and with the rotor given by the fine lines of a backbone representation. **A** A space-filling model of $(\alpha\beta)_3$ shows detected surface waters. **B** A space-filling view of all water molecules detected by X-ray diffraction demonstrates the internal "waters of Thales" available to function by the consilient mechanism for protein-based machines. (Figure preparation was based on the crystallographic data of Abrahams et al.[21] obtained from the Protein Data Bank, Structure File 1BMF.)

Immediately above, we have noted key stages of the consilient basis for the energizing and de-energizing processes. In Chapter 2, with all sites containing nucleotide, a slightly more detailed illustration of the ATP-forming part of the rotary motor than in Figure 1.8 adds further structural perspective of a cusp of insolubility that is externally driven to rotate. In Chapter 8, and in a separate publication,[25] arguments at molecular detail address this remarkable rotary motor that synthesizes most of the energy currency required by living systems.

1.2.3.9 Hemoglobin Transport of Oxygen from the Lungs to the Tissues, in Two Parts

Immediately below, we note the historical case of oxygen transport by hemoglobin from lungs to the tissues and note how this would be discussed in terms of the consilient mechanism. This was the first biological system appreciated for its positive cooperativity, the cooperative binding of oxygen molecules that is key to successful transport.

1.2.3.9.1 Part 1: Happenings in the Arterioles of the Lungs

- In the lungs, the first of four vinegar-like oxygens binds poorly to one of four nearly identical units comprising hemoglobin because oxygen binding disrupts transient hydration that maintains oil-like domain association.
- The second oxygen binds more easily as there is now less oil-like hydration to disrupt.
- The third and fourth oxygens bind more easily yet, to complete loading of the hemoglobin transport system with its oxygen cargo.
- Thus nearly empty hemoglobin entered the lungs where there is enough oxygen to load fully the hemoglobin molecule, which leaves the lungs in transit to the tissues.

1.2.3.9.2 Part 2: Happenings in the Arteriovenous Junction of the Tissues

- In the tissues, CO_2 and biphosphoglycerate, a super vinegar-like molecule with many negative charges, bind at positive sites, the latter with many positive charges where four subunits of hemoglobin meet.
- Association of many negative with positive charges neutralizes their charge effect and results in a shift toward more oil hydration and its consequence of dominance of oil-like domains.
- The increase in oil hydration shifts toward release of vinegar-like oxygen, and spatially localized oil-like domains between protein subunits associate.
- In this way, hemoglobin provides an efficient oxygen transport vehicle by being able to release more of its load in the tissues.

1.2.3.10 Hemoglobin Mutation of a Single Amino Acid from Vinegar-like to Oil-like Causes Dysfunction by Shifting the Cusp of Insolubility Too Far Toward Insolubility

- The disease sickle cell anemia provides another example of coherent phenomena.
- Sickle cell anemia is due to a mutation of a vinegar-like to an oil-like group on the surface of hemoglobin.
- The resulting increase in oil hydration poises the hemoglobin molecules for aggregation.
- Limited oxygen supply allows a shift toward sufficient oil hydration to trigger hemoglobin phase separation and the onset of the symptoms of the disease.

1.2.3.11 Coherence Between Assertions 2 and 3 and Monod's "Second Secret of Life"

As noted by Perutz,[9,16] Monod believed that his discovery of allosteric effects, the existence of two or more states in equilibrium that give rise to the efficient function of positive cooperativity, constituted the "second secret of life," second only to "the structure of DNA." Monod's discovery was exciting; it represented a phenomenological observation, an axiom, describable by a mathematical formalism. The phenomenological observation and the associated mathematical representation were, however, devoid of a physical basis for the conversion between structural states. The Mechanistic Assertion of this book fills that void. Furthermore, Monod impugned the simple problem of translational symmetry of fibrous structures to be "set aside" as not having relevance to "the second secret of life." Thus, to find the lauded allostery in model proteins designed from a repeating sequence of the fibrous protein elastin was unanticipated in the community at large. Furthermore, these elastic-contractile model proteins have been designed to perform the family of energy conversions extant in biology. Therefore, the remarkable finding of coherence of phenomena of positive cooperativity between designed elastic-contractile model proteins and biology's proteins requires that the mechanism whereby these model proteins perform their diverse functions be given serious consideration for function of biology's proteins.

The Hill plots in Figure 1.3B compare the Hill coefficients of hemoglobin and myoglobin. The celebrated positive cooperativity of hemoglobin exhibits a Hill coefficient, n, of 2.8. This positive cooperativity evolved for the efficient transport of oxygen from lungs to the tissues. On the other hand, myoglobin, with the role of storage and diffusion of an innocuous state of oxygen in the tissues, exhibits a Hill coefficient of 1, indicating the presence of neither positive nor negative cooperativity. As will be shown in Chapter 7, the replacement in myoglobin of many of protein's most vinegar-like residues, the carboxylate-containing glutamic acid (Glu, E) residues, by more oil-like residues results in the oil-like association of four myoglobin-like chains, two α-chains and two β-chains, to form hemoglobin, designated as $\alpha_2\beta_2$. The allosteric structural change on oxygenation occurs at the oil-like interface between pairs of αβ dimers. In our view, this change in oil-like association at the $\alpha^1\beta^1-\alpha^2\beta^2$ interface is responsible for the positive cooperativity of hemoglobin.

Larger Hill coefficients indicate more efficient function. Interestingly, the set of designed elastic-contractile model proteins in Figure 1.3A exhibit Hill coefficients with an extraordinary range from 1.5 to 8.0. Thus, these *de novo*–designed elastic-contractile model proteins exhibit a comprehensible range of efficiencies that depend on the degree of oil-like character of the model protein and that result from the competition for hydration between oil-like and vinegar-like functional groups, as developed in Chapter 5. It is not inconsequential, therefore, that the understanding developed as the mechanistic assertion of this book gives rise to parameters for measuring efficiency that can exceed those widely acclaimed in the function of biology's multisubunit proteins.

Thereby, the objective of this book is to demonstrate the phase separation mechanism of oil-like domains separating from water and related controlling phenomena at the molecular level as relevant to specific biological protein-based machines.

1.2.4 Assertion 4: The Applications Assertion

1.2.4.1 The Comprehensive Capacity to Control Association of Oil-like Domains Provides the Necessary Understanding for Meaningful Engineering of Protein-based Materials for Diverse Medical and Nonmedical Applications

The families of model proteins by which the fundamental mechanism is illustrated become designable materials with potential for sustaining both individual health and an increasingly complex and populous society. This includes all polymers operating under the consilient mechanism where adequate control over composition and sequence exists.

The first assertion, as presented in Chapter 5, results in five rules (axioms) for the design of model proteins for energy conversion by the phase separation mechanism. We began these developments with a simple hypothesis visualized as cusps of insolubility, as shown in Figure 1.1: Any change that makes the elastic model protein more oil-like lowers the onset temperature for association of oil-like domains, that is, lowers the temperature for the cusp of insolubility. Similarly, any change that makes the elastic model protein more vinegar-like raises the onset temperature for association of oil-like domains, that is, increases the temperature range for the cusp of insolubility. On this basis we systematically designed model proteins with varying composition for the purpose of achieving different energy conversions. Testing of these designed model proteins indeed demonstrated the intended energy conversions. Five phenomenological axioms for energy conversion resulted. In the fourth assertion these rules, buttressed by development of an understanding of the underlying mechanism (the second assertion) and by the demonstration that biology's protein-based machines function using this mechanism (the third assertion), substantiate the blueprint for designing new model proteins to fill needs in health care and to support an increasingly complex and populous society.

In the design focus of the conventional architect, gravity is the force of primary concern. For architectural design at the molecular level in the aqueous milieu of biology, however, the forces associated with the separation of oil from vinegar and of oil from water dominate. The medium for design at the molecular level does not involve the steel, cement, brick, mortar, wood, glass, or even petroleum-based plastics of the traditional architect, and it is not the more closely related petroleum-based material of the typical bioengineer and biomaterials scientist. Rather, our medium for design is the same as that of biology itself, protein-based materials in water. Skillful use of the forces controlling protein folding, assembly, and function allows for the design of medical and other useful devices and materials.

The approach of the fourth assertion is one of utilizing biology's own natural processes to solve health-care and broader societal problems. This inherent need rings in the concluding paragraph of E.O. Wilson's book *Consilience: The Unity of Knowledge*, where he states "To the extent that we depend on prosthetic devices to keep ourselves and the biosphere alive, we will render everything fragile. To the extent that we banish the rest of life, we will impoverish our own species for all time."[26]

There are as yet no commercial sales of elastic and plastic protein-based polymers, but recent progress in correcting relevant intellectual property rights by the U. S. Patent and Trademark Office will hopefully allow for commercial uses in the near future. *Thus, the applications noted below must yet be seen in terms of potential rather than as realized utility.*

1.2.4.2 The Remarkable Biocompatibility of These Elastic Model Proteins

- These elastic model proteins exhibit remarkable biocompatibility due to an elasticity different from that of disordered petroleum-based elastomers and natural latex rubbers.
- These elastic model proteins form regular dynamic structures that exhibit mechanical resonances at low frequencies.
- To approach and identify these elastic materials as foreign requires direct interaction, for example, with antibody.
- The mechanical resonances present a barrier to interaction, as their coherent motions must be stopped in order for antibodies to interact with the biomaterial and to identify it as foreign.

1.2.4.3 The Design of Medical Care Devices for Soft Tissue Restoration

1.2.4.3.1 Prevention of Adhesions and Correction of Urinary Incontinence

- The soft tissue of mammals, from which our model proteins were developed, is a compliant tissue.
- Consequently, soft tissue restoration becomes a manifest area of application of elastic model proteins.
- Preventing scar tissue formation after spinal surgery holds promise to correct the problem of failed back surgery syndrome and its consequence of painful disability.
- Inducing tissue generation at the base of the bladder can restore urinary continence from a state that constrains activity and limits quality of life.

1.2.4.3.2 Function/Adaptive-Restructuring Cycles in Tissue Restoration

- At the cellular level, *function/adaptive-restructuring cycles* can be quite explicit.
- Cells in an arterial wall have multiple attachments to surrounding connective tissue.
- Through these attachments, arterial cells become stretched and relaxed as blood pulses through the artery.
- Stretching/relaxing provides the energy stimulus that transforms into a chemical energy output for remodeling an arterial wall in order to better sustain the pulsing forces.

1.2.4.3.3 Restoration of Arteries

- Tissue restoration can utilize this principle of *function/adaptive-restructuring cycles*.
- Synthetic elastic tubes formed from elastic model protein can match the elastic properties of the natural artery.
- The synthetic elastic tubes can also contain cell chemoattractant and attachment sequ-

ences to provide attachment sites for the natural arterial cells.
- In such temporary functional scaffoldings, natural cells can be induced into the scaffold, attached, and stretched due to the pulsing blood; as a consequence, cells remodel the synthetic scaffold into a natural tissue sufficient to sustain essential mechanical function.

1.2.4.4 The Design of Medical Care Devices for Controlled Drug Release

1.2.4.4.1 Drug Delivery by the Phase Separation Mechanism

- From loading to release, drug delivery can be achieved by the phase separation mechanism.
- Drugs, containing charge, pair with oppositely charged vinegar-like groups of the elastic model protein.
- Increase in oil hydration due to the association of opposite charges causes phase separation.
- The separated phase of designed protein plus drug constitutes the drug delivery vehicle.

1.2.4.4.2 Increasing Oil-like Character of Charged Model Protein Enhances Ion Pairing

- Above by protein design, the drug provides the chemical energy for its own packaging.
- The protein can be designed with a range of oil-like groups while maintaining charge.
- The more oil-like association on pairing of oppositely charged drug and model protein, the more tightly the drug is held.
- The more tightly held the drug, the slower the release.

1.2.4.4.3 Strength of Ion Pairing Controls Rate of Drug Release

- A large number of drugs have a charged, vinegar-like part and an oil-like part.
- Examples are the many analgesics and anesthetics (painkillers), including narcotics.
- The loaded drug delivery vehicle can be implanted in the body.
- By design of model protein, release from the implant can be sustained at a desired level.

1.2.4.4.4 Controlled Release of Narcotic Antagonists from Subcutaneous Implants

- There are also narcotic antagonists that block the action of drugs like heroin and cocaine.
- The drug delivery vehicles containing narcotic antagonists can be implanted in the body.
- Release from the implant would block, for example, the action of a large dose of heroin.
- The potential is for sustained delivery of narcotic antagonist, no high, no redependence, and the delivery vehicle can be designed to gracefully disappear as its task is completed.

1.2.4.5 The Design of Medical Care Devices for Vaccine Delivery

Nanoparticles, with different affinities for vaccine and with potential to be adjuvants, by means of a single inoculation could replace spaced multiple immunizations.

- The drug delivery vehicle can be formed as charged nanoparticles of different oil-like composition with the potential for functioning as an adjuvant.
- The different nanoparticles would take up vaccine with different binding affinities and different release profiles.
- The vaccine-loaded nanoparticles could be introduced into the body through epithelial lining without the need for injection.
- The requirement for multiple vaccinations, spaced over many months, could be replaced with a single treatment using a family of nanoparticles each with a different release level and period to function as multiple immunizations.

1.2.4.6 Design of Nonmedical Applications: Release of Agricultural Enhancement Factors

Controlled release can enhance factor effectiveness and limit harmful environmental side effects.

- Controlled release of agricultural enhancement factors is analogous to drug delivery.

1.2 Four Principal Assertions of "What Sustains Life?"

- Herbicides, pesticides, fertilizers, and growth factors can be released by the phase separation mechanism.
- Time, location, and amount of release can be controlled by the phase separation mechanism.
- Such effective and efficient use can limit extent of harmful side effects to the environment.

1.2.4.7 Design of Nonmedical Applications: Biodegradable Plastics

1.2.4.7.1 Problems of Present Plastics

- Presently, plastics are made from petroleum, an exhaustible resource.
- Preparation of petroleum-based plastics utilizes toxic and noxious chemicals.
- Both preparation and the plastic products themselves pollute and deface the environment.
- Because they are not biodegradable, plastics disposal at sea means death to marine life.

1.2.4.7.2 Protein-based Materials Can Be Thermoplastics

- With proper design, proteins can be plastics and even thermoplastics.
- Protein-based plastics can be designed to melt 100°C below their decomposition temperatures.
- Such protein-based thermoplastics can be processed as melts.
- Protein-based thermoplastics can be molded as desired or extruded and pulled into fibers (see Fig. 1.10).

1.2.4.7.3 Protein-based Plastics Can Be Programmed for Different Rates of Biodegradation

- Protein-based plastics can be prepared by living cells, renewable resources, without the usual necessity of resorting to toxic and noxious chemicals.
- Protein-based plastics can be designed for a given environment to biodegrade in time periods ranging from days to decades.

A
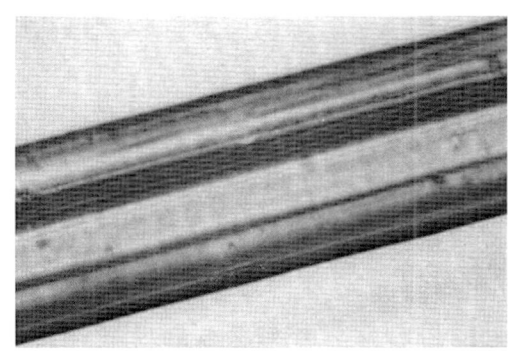

B
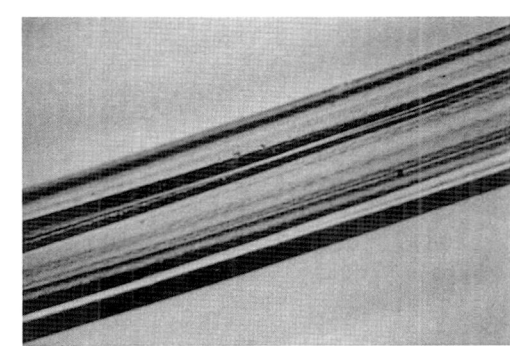

C
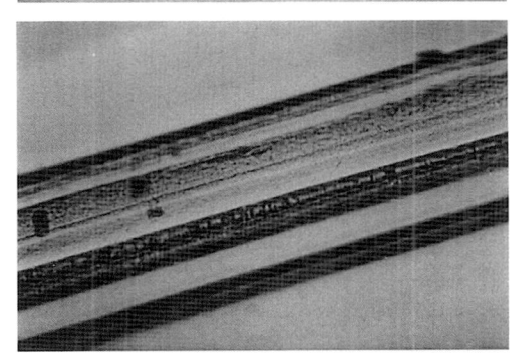

FIGURE 1.10. Thermoplastics are polymers that melt at a temperature well below thermal decomposition such that they can be melted and formed into a desired shape as a melt. Shown are three thermoplastics, two are plastics of our daily use and a third is a designed protein-based thermoplastic that melts at 160°C and does not decompose until the temperature is raised above 250°C. The protein thermoplastic can be programmed to biodegrade with half-lives ranging from days to decades when in an aqueous environment.

- Global environmental concerns result in ever-mounting costs for disposal of solid waste.
- Solid waste of non-degradable petroleum-based plastics, and the toxic and hazardous

chemicals of their production buttress concerns.
- The result is demand for polymers, like protein materials, that are biodegradable, of benign production, and from renewable resources.
- The Martime Pollution (MARPOL) Treaty of 1995 and the Plastic Pollution Research and Control Act of 1987 prohibit, disposal of plastics at sea due to damage to the marine environment.
- Protein-based plastics can be environmentally friendly and could even represent healthful products from tobacco plants.
- Rather than death to marine life, programmably biodegradable plastics can mean food for the fishes.

1.2.4.8 Design of Nanosensors: The Ultimate in Sensors, Detection of a Single Molecule

- Using atomic force microscopy, it is possible to obtain the smooth force-extension profile of a single elastic model protein chain.
- Insertion of a globular protein sensing element within an elastic model protein chain results in a force-extension profile that contains the unfolding peak for the single globular protein.
- Binding of a single molecule to a highly selective site on the globular protein component can change the unfolding peak in the force-extension profile.
- When the selective site binds a single molecule of nerve gas or of an explosive, the changed profile would provide the ultimate in detection, that of detection of a single molecule, with the renowned selectivity of protein to limit false positives.

In summary, there is promise that one can harness the consilient, phase separation mechanism in the design of materials to extend and improve quality of life at both individual and societal levels.

References

1. E. Schrödinger, *What is Life?* Cambridge University Press, Cambridge, England. First published in 1944, Canto edition with "Mind and Matter" and Autobiographical Sketches, forward by R. Penrose, 1992.
2. Origin of Life reference: M. Eigen, *Steps towards Life: A Perspective on Evolution.* Oxford University Press, Oxford, England, 1992.
3. I. Prigogine and I. Stengers, *Order Out of Chaos: Man's New Dialogue with Nature.* Bantum Books, New York, 1984, p. 13.
4. E.O. Wilson, *Consilience: The Unity of Knowledge.* Alfred E. Knopf, New York, 1998, p. 8.
5. *Consilient* is the adjective form of the noun *consilience*. As defined in *Webster's Third New International Dictionary*, the term *consilience* contains the sense that there exist pervasive, fundamental laws of nature underlying related disciplines that provide a common groundwork of understanding. Here we refer to the consilient mechanism as a water-dependent, pervasive, fundamental process by which the macromolecules of living organisms function and evolve. The term originates from William Whewell in *The Philosophy of the Inductive Sciences*, London, 1840. In Chapter 5 a second consilient mechanism is introduced that relates to the mechanism for a near ideal elasticity that is important to efficient protein-based machines wherein mechanical energy becomes important.
6. *Webster's Third New International Dictionary*, Merriam-Webster, Inc., Springfield, MA, 1993, p. 557.
7. D.W. Urry and T.M. Parker, "Mechanics of Elastin: Molecular Mechanism of Biological Elasticity and its Relevance to Contraction." *J. Muscle Res. Cell Motil.*, **23**, 541–557, 2002. Special Issue, *Mechanics of Elastic Biomolecules*. H. Granzier, M. Kellermayer, W. Linke, Editors.
8. J. Monod, "On Symmetry and Function in Biological Systems." In *Nobel Symposium 11: Symmetry and Function of Biological Systems at the Macromolecular Level*, A. Engstrom and B. Strandberg, Eds. Almqvist & Wiksell Forlag AB, Stockholm, 1968, p. 1527. Also reprinted in *Selected Papers in Molecular Biology* by Jacques Monod, A. Lwoff, and A. Ullmann, Eds., Academic Press, New York, 1978, p. 708.
9. M.F. Perutz, "Mechanisms of Cooperativity and Allosteric Regulation in Proteins." *Q. Rev. Biophys.*, **22**, 139–236, 1989. Reprinted in book form as referenced below (in ref. 16).
10. These data stand as a challenge to those many computational programs presently in use for calculation of protein structure and function that treat water as a uniform dielectric without more specific consideration of the energetics of

hydration. From our studies, treatments of detailed water structure and interactions are required to properly describe protein structure and function, including the competition for hydration, which is expressible as a repulsion between oil-like and vinegar-like (e.g., charged) groups of protein.

11. C.B. Bauer, H.M. Holden, J.B. Thoden, R. Smith, and I. Rayment, "X-ray Structures of the Apo and MgATP-bound States of *Dictostelium discoidium* Myosin Motor Domain." *J. Biol. Chem.*, **275**, 38494–38499, 2000.

12. E. Martz, *Front Door to Protein Explorer 1.982 Beta*. Copyright 2002. Website: proteinexplorer.org.

13. As noted in the opening paragraph of the Prologue and in the initial paragraphs of Chapter 3, of this book, *Thales of Miletus* launched scientific investigation in the sixth century BC with the inquiry "What is the world made of?" His simple answer of "water" has been suggested as deriving "perhaps from seeing that the nutriment of all things is moist and kept alive by it...."[14] In recognition of Thales' having initiated our quest and because of the central role that water fills in function of the protein-based machines that sustain Life, we refer to this key water as the "waters of Thales."

14. D.J. Boorstin, *The Seekers: The Story of Man's Continuing Quest to Understand His World.* Random House, New York, 1998, p. 22.

15. A.J. Fisher, C.A. Smith, J. Thoden, R. Smith, K. Sutoh, H.M. Rayment, and I. Rayment, "Structural Studies of Myosin: Nucleotide Complexes: A Revised Model for the Molecular Basis of Muscle Contraction." *Biophys. J.*, **68**, 19s–28s, 1995.

16. M.F. Perutz, *Mechanisms of Cooperativity and Allosteric Regulation in Proteins*. Cambridge University Press, Cambridge, England, 1990, Preface, p. x.

17. D.W. Urry, B. Haynes, H. Zhang, R.D. Harris, and K.U. Prasad, "Mechanochemical Coupling in Synthetic Polypeptides by Modulation of an Inverse Temperature Transition." *Proc. Natl. Acad. Sci.* U.S.A., **85**, 3407–3411, 1988.

18. These results of 18 years ago, demonstrating the capacity of *de novo*–designed model protein-based machines for the conversion of chemical energy into mechanical work, remain unexplained by the computational methodologies currently in use to describe the function of protein-based machines of biology.

19. P.D. Boyer, "The Binding Change Mechanism for ATP Synthase—Some Probabilities and Possibilities." *Biochim. Biophys. Acta*, **1140**, 215–250, 1993.

20. P.D. Boyer, "Catalytic Site Occupancy During ATP Synthase Catalysis." *FEBS Lett.*, **512**, 29–32, 2002.

21. J.P. Abrahams, A.G.W. Leslie, R. Lutter, and J.E. Walker, "Structure at 2.8 Å Resolution of F_1-ATPase from Bovine Heart Mitochondria." *Nature*, **370**, 621–628, 1994.

22. K. Kinosita, Jr., R. Yasuda, and H. Noji, "F_1-ATPase: A Highly Efficient Rotary ATP Machine." In *Essays in Biochemistry: Molecular Motors*, G. Banting and S.J. Higgins, Eds., Portland Press, London, England, 2000, pp. 3–18.

23. P.L. Privalov, "Cold Inactivation of Enzymes." *C. R. Biochem. Mol. Biol.*, **25**, 281–305, 1990.

24. It may be noted that present calculations of protein structure and function neglect the presence of so much internal water. In calculations, a quantity called the *dielectric constant*, required in electrostatic calculations of the energy of interaction between charges, is utilized. The value of the dielectric constant of bulk water is about 80, and commonly the assumption is made that the value shifts at the surface of the protein from 80 to 5 or less within the protein. The presence of the "waters of Thales" raises concern about such assumptions. Similarly, in Chapter 5, particularly in Figure 5.30, the experimental values could only be approximated by electrostatic calculations when a value of 5 or less was used, whereas direct measurement of the dielectric constant for the model protein system required that the value within the model protein motor be no less than 65. These points provide substantial support for the consilient mechanism.

25. D.W. Urry, "Function of the F_1-motor (F_1-ATPase) of ATP synthase by Apolar-polar Repulsion through Internal Interfacial Water." *Cell Biol. Int.*, **33**(1), 44–55, 2006.

26. E.O. Wilson, *Consilience: The Unity of Knowledge*. Alfred E. Knopf, New York, 1998, p. 298.

2
What Sustains Life? An Overview

2.1 Introductory Comments

2.1.1 Dilemma of a Disordering Universe Yet Life Systems of Increasing Order

The physicist looks at the expanding universe, disassembling toward disorder, and views increasing disorder as the inevitable flow in nature. The chemist burns the oils that have accumulated over the millennia in the earth's crust and energizes molecules to react. On a smaller scale, the chemist sees the march toward disorder; the oils become dispersed and disordered gases, and the excited new molecules become dormant. Both physicists and chemists look at living organisms—propagating, assembling, growing—and wonder, how can living matter act in such an inverse way to the nonliving matter of their experiences? On the other hand, biochemists and biophysicists look at molecular systems of a dissected organism and successfully describe a still functional component in terms of the equilibrium laws of physics and chemistry. So, how can living matter seem so different from nonliving matter?

The timely access to forms of energy—light, carbon dioxide, and water for plants and food, oxygen, and water for animals (and plants in the dark)—has been known for more than a century to provide the nourishment for living organisms. The energies accessible to the living organism, transformed by components of the living organism itself, result in the characteristics that define Life—metabolism, growth, reproduction, responsiveness, adaptability, and motion. How can the energy sources of light and food, along with the additional chemical energies represented, for example, by supplies of oxygen, carbon dioxide, and water, result in these attributes of living organisms?

2.1.2 Familiar Energy Conversions

The sun's energy unevenly heats the surface of the earth and evaporates water from the seas. Solar energy drives the energy consuming water-to-vapor phase transition and creates moisture-laden warm air masses. Clashes between cold and warm moist air masses cause energy-releasing condensation of water vapor to rain drops. Resulting winds as extreme as hurricanes and tornadoes disfigure the earth's surface—uproot trees, demolish man-made structures, and cause waves that transform shorelines. The rains become rivers that carve mountain ranges into grains of sand and remodel flood plains. These processes continue the march toward less order.

Looking at man-made machines, ignited fuels create heat and expanding, disordering gases that propel rockets and jet planes and drive pistons of the internal combustion engines of our motor vehicles. These familiar, natural and man-made energy conversions, which produce motion and perform mechanical work, are all associated with changes in temperature and increases in disorder. The primary energy conversions of living organisms, however, occur spontaneously within, without significant changes in temperature and often with a component of increasing order.

2.1.3 The More Enigmatic Energy Conversions of Living Organisms

The seemingly mysterious motions of Life "come spontaneously from within, in the blink of an eye, in the twitch of a muscle, in the sudden kick of a runner, in the jerk and press of a weight lifter,"[1] and in the abrupt burst of motion of the discus thrower.

The discus thrower stands motionless, discus in hand. In response to an entirely internal signal, he bursts into circular motion of ever increasing rotational speed; he releases the discus that soars along an arching trajectory and falls to the ground. The discus thrower again becomes motionless except for the heaving of his chest, taking in molecular oxygen (O_2) and giving off carbon dioxide (CO_2) and water (H_2O). Conversion of glycogen from chemical energy storage depots in the muscles to glucose and oxidation of this carbohydrate molecule give rise to CO_2 and H_2O and replenish ATP (adenosine triphosphate), the immediate universal biological energy currency, utilized for the muscle contractions that propelled the discus. All of this happens without a significant change in temperature.

2.1.4 Proteins Catalyze Energy Conversions

Proteins perform the energy conversions that sustain Life; they consist of carbon (C), hydrogen (H), oxygen (O), nitrogen (N), and a more limited number of sulfur (S) atoms. Proteins are polymers, long-chain molecules. The individual links of the chain are amino acids joined in formation of the peptide bond, $-CONH-$. Accordingly, this chain structure may also be called a polypeptide. The resulting protein polymers are sequences of 20 different monomer units, the amino acid residues, to form poly($-HNCHRCO-$), as depicted in Figure 2.1A.

Different side chains, designated as **R**-groups, distinguish the 20 naturally occurring amino acid residues. An **R**-group can be *oil-like*, containing only carbon and hydrogen, for example, $-CH_2CH(CH_3)_2$; or the **R**-group can be *vinegar-like*,[2] with a *functional* group, for example, $-CH_2COOH$, in which the carboxyl, $-COOH$, part can dissociate to give acid, H^+ (a proton with its positive charge), and the negatively charged carboxylate $-COO^-$. The **R**-group can even be neutral like the simple hydrogen, $-H$, of glycine. (See Table 5.1 for the molecular structures of the 20 naturally occurring **R**-groups.)

The constrained alignment of *oil-like* and *vinegar-like* groups along a protein chain, as shown in Figure 2.1B, confounds the familiar adage "oil and vinegar don't mix." Nonetheless, the same forces that cause oil and vinegar to separate affect the chain's behavior. The forces that exist between oil-like and vinegar-like groups distributed along the chain compel the chain to fold in such a way as to separate oil from vinegar. This brings oil-like side chains together, as demonstrated in Figure 2.1C, in a manner that separates oil-like groups from vinegar-like groups. In the process this association of oil-like groups shortens the chain length. The complete separation of oil-like side chains from water occurs by association of chains or chain segments, as indicated in Figure 2.1D. These structural changes are referred to as *hydrophobic* (water fearing) *folding* (Fig. 2.1C) and *hydrophobic assembly* (Fig. 2.1D). In short, the sequence of amino acid residues controls the way a protein folds[3] and functions as the **R**-groups variously attempt to segregate, as oil from vinegar and as oil from water.

The tendency to fold and assemble hydrophobically in this way varies remarkably with the presence and state (charged or uncharged) of certain **R**-groups. The vinegar-like side chains can transform to become more oil-like. Changes in chemical groups of proteins, including vinegar-like **R**-groups and other bound chemical groups, control folding and assembly through shifting the balance between being more *oil-like* and being more *vinegar-like*. In fact, these reversible chemical changes in vinegar-like side chains and additional bound chemical groups provide the means whereby proteins catalyze energy conversions.

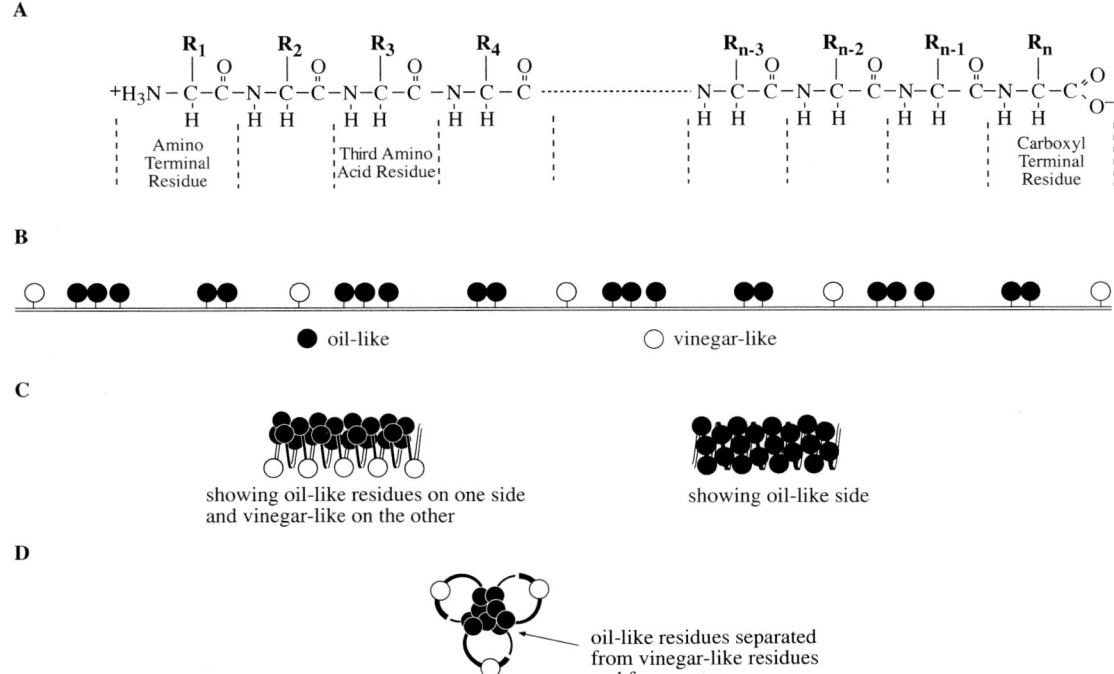

FIGURE 2.1. The phase transition of oil-like groups separating from water produces motion. **A** A protein is described as a polypeptide chain molecule with different side chains, **R**-groups. **B** The side chains delineated simply as oil-like and vinegar-like (e.g., charged), are distributed along the chain. **C** The separation of oil-like groups from vinegar-like groups and the association of oil-like groups to form oil-like domains result in folding that shortens the chain. **D** The folded chains associate to complete the separation of oil-like domains from water while maintaining the vinegar-like groups in water at a distance from the oil-like domains. In reality, the steps shown in C and D occur simultaneously and in a cooperative manner.

2.1.5 Boundless Potential of Biosynthesis to Produce Proteins of Varied Sequence

Proteins perform the work that sustains Life. The genius of biological synthesis of proteins, however, is that it matters relatively little how complex the linear alignment of the available 20 **R**-groups, and it matters only in a limited way how energetically unfavorable the resulting linear alignment, that is, the sequence, of **R**-groups that form the protein. The information for the protein sequence resides independently in the nucleic acid sequence, and the energy required to produce a nucleic acid sequence is independent of the protein sequence it specifies and, therefore, independent of the interaction energies within the resulting protein sequence.

Separating the complexities and probabilities of preparation from the complexities of performance strikes advantage readily appreciated by peptide chemists and thermodynamicists.

In our experience with the model proteins of the type considered below, a wide range of sequences can be designed, constructed, and expressed to a useful degree in the organism *Escherichia coli*, a generally symbiotic bacterium of our intestinal tract. Therefore, setting aside the perennial issue of sufficient research funds, the design of model proteins to develop and demonstrate concepts of energy conversion becomes limited primarily by one's imagination. The fidelity and diversity of sequences possible with genetic engineering is without parallel among all known chain molecules. Accordingly, an extraordinary potential now

2.1 Introductory Comments

exists to develop an understanding of the forces involved in energy conversion (i.e., in protein function). Such an understanding also has the potential to provide advanced materials for the future.

The example of the basic sequence from which the designed set of diverse protein-based machines evolved follows. By means of well-established techniques of recombinant DNA technology, many genes, that is, many different DNA sequences, have been constructed for the many designed model protein-based machines. For example, one gene has been prepared for the basic sequence that encodes for 251 repeats of a five-residue sequence. It is the elastic-contractile model protein, designated as (glycyl-valyl-glycyl-valyl-prolyl)$_{251}$ and abbreviated as $(GVGVP)_{251}$. The G stands for glycine (Gly) with the hydrogen atom, –H, for the **R**-group; the V for valine (Val) with the $-CH_2(CH_3)_2$ side chain; and the P for proline (Pro) with the $-CH_2-CH_2-CH_2-$ **R**-group that bridges from the attachment backbone C to the N of the same residue, as shown in Figure 2.2. The structure of this pentamer with a 10-atom hydrogen bonded ring, called a β-turn, and the single chain folding pattern for $(GVGVP)_{251}$, a helical structure called a β-spiral, are given in Figure 2.2.[4]

Production of $(GVGVP)_{251}$ gives an example of successful genetic engineering to produce the designed model proteins that provide the experimental basis for the perspectives reported in this volume. Figure 2.3 shows an example of the results from a single fermentation run of *E. coli* that has been genetically transformed to produce $(GVGVP)_{251}$. At the end of the fermentation period the cells are broken (lysed), and the lysate is centrifuged cold to remove cellular debris. At low temperature the model protein is soluble, but then it phase separates on heating to body tempera-

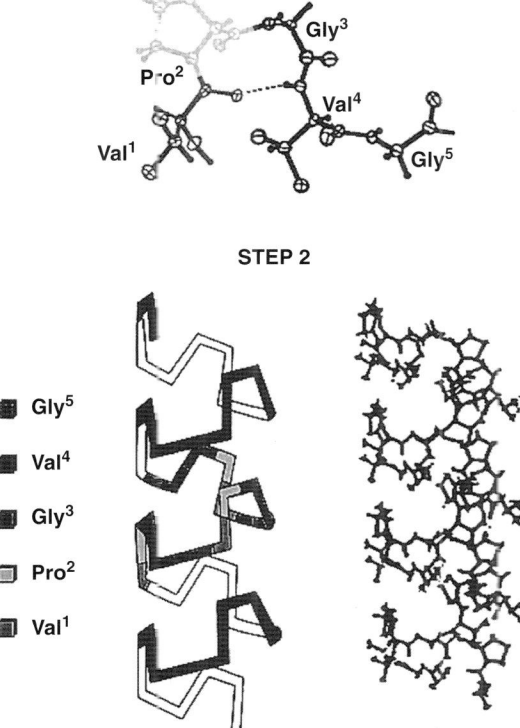

FIGURE 2.2. The folding pattern of the repeating five residue sequence of elastin, which for the secondary structural reason of the β-turn (step 1) is defined as (Val^1–Pro^2–Gly^3–Val^4–Gly^5) giving poly(VPGVG). For a large number of repeats, this should be recognized as equivalent to poly(GVGVP). In step 2, the chain of repeating pentamers folds into a helical structure called a β-*spiral*. For the complete description, several β-spirals associate to form a twisted filament, a cross-section of which is shown in Figure 2.1D. (Reproduced from Urry,[4] with permission of www.WorldandIJournal.com)

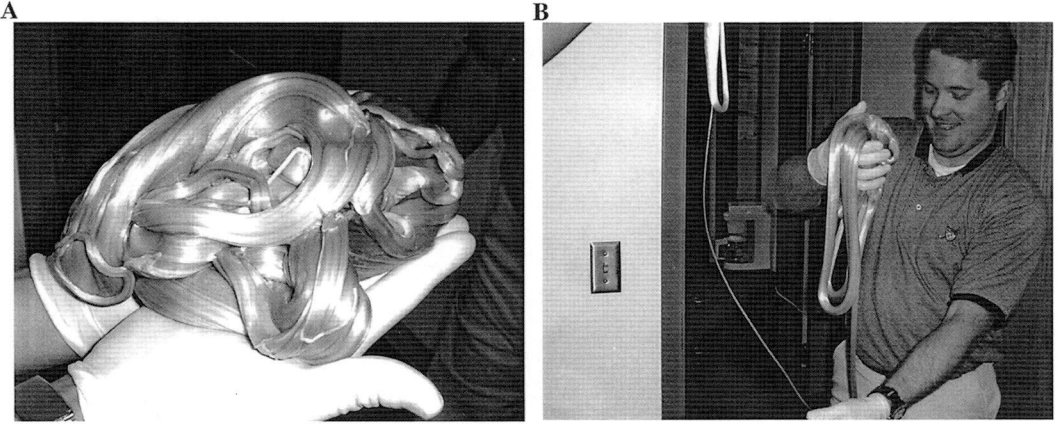

FIGURE 2.3. **A** Mass of the elastic-contractile model protein, $(GVGVP)_{251}$, from a single fermentation run of *E. coli*. **B** Suitably transformed to produce the protein material.

ture. Centrifugation of a warmed solution results in the mass of elastic-contractile model protein of Figure 2.3A. Under these conditions this model protein gives the appearance of glistening taffy. In a visual demonstration of the physical properties in Figure 2.3B, pulling on the viscoelastic model protein readily forms a long fine strand.

2.2 Our Window on What Sustains Life

2.2.1 Motion as an Index of Life

Albert Szent-Györgyi, perhaps the first person to know the excitement of seeing the motion of Life outside of the living organism, began a review of his remarkable studies with the statement, "Motion has always been looked upon as the index of life."[5] Szent-Györgyi, while prominent in the resistance in Hungary during World War II and while in sanctuary within the Swedish Embassy in Budapest, had removed the contractile proteins from skeletal muscle, fashioned them into a thick fiber, added the necessary chemicals, and saw the fiber contract.

2.2.2 Designed Model Proteins That Contract

By means of the design, and originally the chemical synthesis, of model proteins, our window on *what sustains Life* begins somewhat analogously. In the midst of a less lethal controversy on the nature of protein elasticity, we designed a series of elastic-contractile model proteins, fashioned them into fibers or bands, and on adding the appropriate source of energy visually observed the bands contract to produce motion and perform the mechanical work of lifting a weight. Not unlike the studies of Szent-Györgyi, an energy source produced motion; but, now man-made, the model proteins had produced the motion.

In addition to changes in concentrations of chemicals (chemical energy) to drive contraction of the Szent-Györgyi demonstration, we have now seen each of the sources of thermal energy, pressure-volume energy, chemical energy, electrical energy (reduction and oxidation), and light, acting on a particular *functional R-group*, reversibly drive contraction and relaxation, that is, result in demonstration of mechanical energy by a properly designed model protein. In all, six energies can be interconverted. In each case of contraction at constant temperature, the energy input alters the vinegar-like or oil-like character of the **R**-groups. These energies, the same that living organisms utilize, have been employed in graphic, visual demonstrations of pumping iron, that is, of the performance of work by synthetic model proteins formed into contracting and relaxing elastic bands that lift and lower weights (See Figure 2.4).[1,6,7]

2.2 Our Window on What Sustains Life

2.2.3 Model Proteins Convert Additional Energies That Sustain Living Organisms

The same molecular process of the separation of oil-like and vinegar-like groups that converted the five different energies into motion (pumping iron) can be used to convert any of those five energies from one into the other. Model proteins have been designed such that electrical energy of reduction (adding electrons) converts into the chemical energy of picking up acid (pumping protons). Furthermore, evidence from studies of the model proteins suggests that the chemical energy input of a change in acid concentration can convert into the chemical energy output of forming ATP.

These energy conversions represent fundamental energy transformations essential to the existence of plants and animals. Herein begins the foundation of a previously unrecognized and likely dominant mechanism for protein function.

2.2.4 *De Novo*–designed Model Protein-based Machines

The *de novo*-designed elastic-contractile model proteins—not modeled after, or copies of, contractile or other generally recognized energy-converting proteins—begin with a part of the mammalian elastic protein. Then functional groups, designed into the elastic model protein, allow responses to different energy inputs and

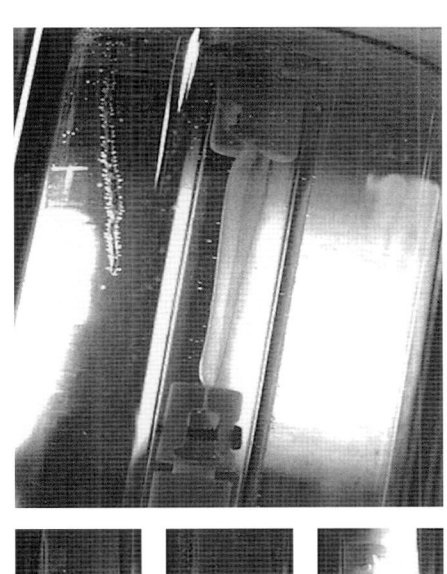

FIGURE 2.4. The lefthand side shows a representative sheet of γ-irradiation cross-linked elastic-contractile model protein, designed for the conversion of an input energy into the output of pumping iron, performing mechanical work of lifting a weight. The righthand side shows the synthetic protein machine lifting a weight in response to an energy input. (Reproduced from Urry.[1])

allow coupled pairs of different functional groups to be included. In this way the energy input, to which one functional group responds, converts, as the output energy, into the energy form to which the other functional group is responsive. For example, adding negative electrons to an attached positive group makes it more oil-like and forces a negatively charged, vinegar-like group to become more oil-like by doing the work of picking up a proton. In this case the electrical energy input of adding electrons results in the chemical work of pumping protons.

These experimental demonstrations of our capacity to design model proteins that perform key biological energy conversions constitute our access to the question of What Sustains Life.

2.3 Energy Sources Cause Proteins to Fold, Assemble, and Function

2.3.1 Certain Body Tissues and Proteins Lift Weights on Raising the Temperature

The underlying physical property, relevant to the above energy conversions, was evident as early as 1880 in a publication by Charles S. Roy.[8] Roy attached a 50-g weight to a 1 cm wide strip of human aorta, heated it from the temperature of a cool British laboratory to body temperature, and observed the strip of aorta contract and lift the weight. Repeatedly and reversibly, heating caused contraction (shortening) and cooling caused relaxation (extension). Roy recognized this to be an exception "to the general rule that heat causes expansion and cold contraction." The macromolecule in the aortic wall responsible for this counterintuitive property is the connective tissue protein elastin, but also involved is the actomyosin complex of smooth muscle that exhibits a related heat rigor (see Chapter 7).

The aorta, the major artery leaving the heart, experiences great demand to expand as the blood enters it on contraction of the heart's left ventricle. To sustain this demand, two thirds of the total protein in the aorta, as it courses through the chest, is elastin, the mammalian elastic protein. Part of the energy of the contracting heart is stored as mechanical energy in the stretched aortic wall. After the aortic valve closes to prevent flow back into the heart, this stored energy continues to propel the blood into distant smaller arteries just as a filled balloon expels its contents. The relationship between elastic storage of energy and Roy's observed contraction driven by heating is central to understanding both the basis of elasticity and the mechanism of muscle contraction.[9] This will be treated in Chapter 8.

2.3.2 Elastin is Cross-linked and Contains Repeating Sequences of Amino Acid Residues

Elastin is a chain molecule comprised of a string of nearly 800 linearly linked amino acid residues. The long chains are cross-linked into an extended network, which on spreading would appear much as the wires of a chain-link fence. Because of the cross-links, the chain molecules do not flow irreversibly passed each other during stretching, such that, after stretching, the elastic fibers can recoil to their original resting state.

The most striking and the longest sequence between cross-links of the elastin chain molecule of pigs[10] and cattle[11] contains a repeating sequence of five amino acid residues to produce a repeating pentamer, designated as poly(GVGVP) (see Fig. 2.2 and the associated text). As described in relation to Figure 2.2, the capital letter G stands for the glycine amino acid residue, $-HNCH_2CO-$; V stands for the oil-like valine residue, $-HNCHRCO-$, with an oily (hydrocarbon) **R**-group of $-CH(CH_3)_2$; and P stands for $-NCHRCO-$, with **R** being the hydrocarbon, $-CH_2CH_2CH_2-$, bridging over from the N to the C of the CH.

The synthetic elastic model protein, comprised of long chains of more than 200 repeating pentamers, constitutes the starting point for our exploration of how biologically accessible energy can be converted into the energy forms that sustain Life. When fashioned into a cross-linked fiber or sheet, as in Figure 2.4A, it exhibits a 10-fold greater shortening on heating

2.3 Energy Sources Cause Proteins to Fold, Assemble, and Function

than did the aortic wall of Roy's study. As discussed below, now more than a century after Roy's observations, the molecular process has been described and used not only with heating (thermal energy) but also with the many energies that biology utilizes in performing the work required to maintain viable organisms.[7]

2.3.3 Proteins Modeled After Elastin Convert Heat to Motion

The underlying mechanism for conversion of heat to motion derives from the interesting property exhibited by certain proteins in water whereby these chain molecules increase order on increasing the temperature. Heating causes the oil-like groups, dispersed along the chain, to come together, forcing the folding and assembly of chains, as shown in Figure 2.1C,D. Thus, an extended elastic model protein chain, conceptually anchored at one end and attached to a weight at the other, would fold and shorten to move or lift the weight as the temperature is raised over a particular range, say, as with Roy's study, from room temperature to body temperature.

A single chain of poly(GVGVP) in an extended state and in a contracted or folded state is shown symbolically lifting a weight in Figure 2.5. Observation of a single chain molecule, or several intertwined chain molecules contracting to lift a weight, however, has only recently been achieved.[12–14] Generally, however, γ-irradiation (from cobalt 60) cross-links a glue-like mix of poly(GVGVP) and water into an elastic-contractile band, which, as shown in Figure 2.4, can lift and lower a weight. Attaching a weight to this rubber-like band stretches it, and on heating from 0° to 20°C little happens, but on increasing the temperature from 20 to 40°C the band contracts and lifts the weight. On further raising the temperature above 40°C, again, little happens. For this composition of elastic-contractile model protein, the temperature interval from 20 to 40°C constitutes a *transition zone*. As shown graphically in Figure 2.6 and considered in greater detail below, additional energy inputs exhibit the same pattern of behavior when the model protein has been designed appropriately.

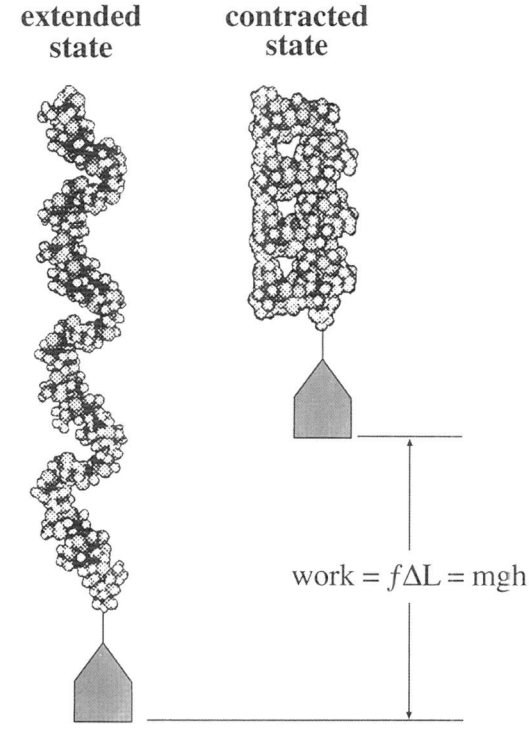

FIGURE 2.5. A space-filling model is shown of a single chain of poly(GVGVP) in the β-spiral structure extended to 130% by loading with a weight. With the appropriate energy input, oil-like groups between turns of the β-spiral structure associate and perform the work of lifting the weight. In reality, the fundamental structural unit is thought to be three β-spirals intertwined in the formation of a twisted filament. For simplicity of illustration, a single chain is used. By means of the proper design, for example, the inclusion of one or two functional groups per 100 residues, the energy input that makes the functionality more oil-like drives contraction and the performance of mechanical work of pumping iron, as indicated in Figure 2.6.

2.3.4 Graphical Representation of Localized Changes (Transitions) and of Consilience in Diverse Protein-based Machines

2.3.4.1 Using Thermal Energy to Drive Contraction

A dramatic change, like the above, localized shortening over a limited temperature interval

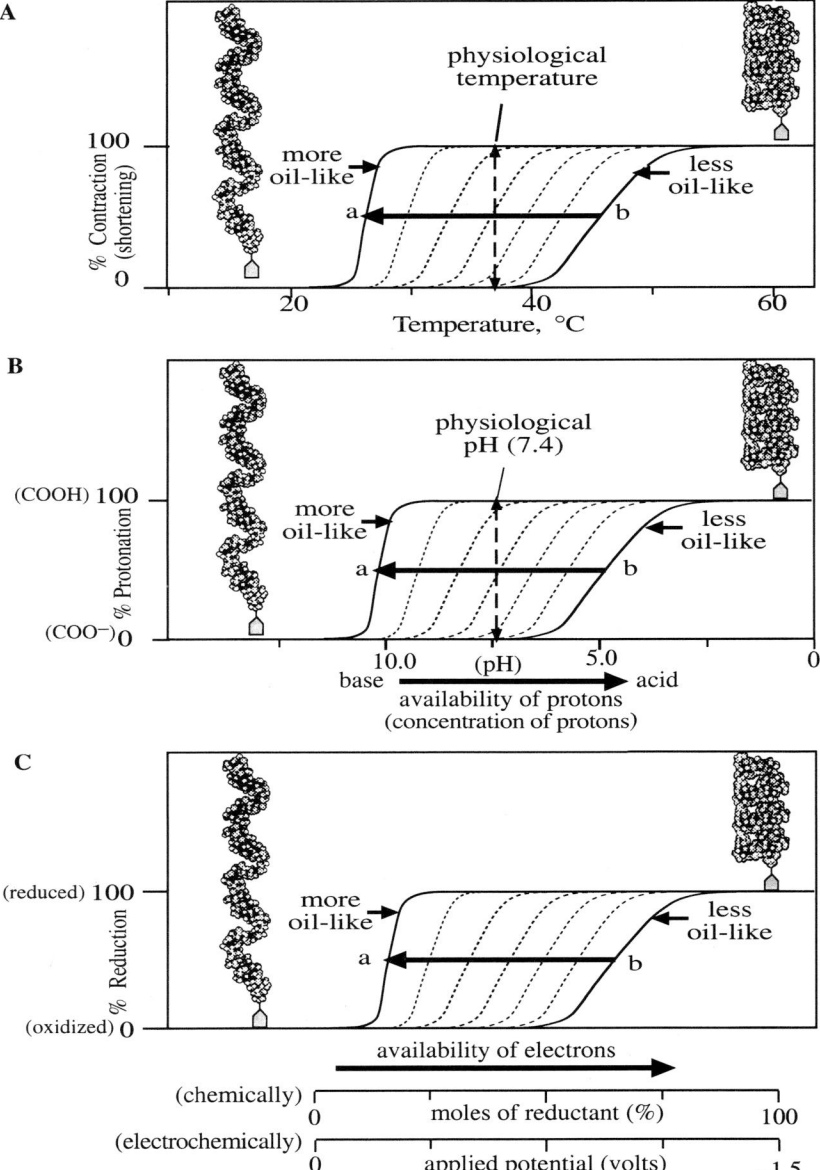

FIGURE 2.6. In general, the conversion from the extended state to the contracted state shown in Figure 2.5 is graphed here as a systematic family of sigmoid-shaped curves with a common dependence of oil-like character of the elastic-contractile model protein whether the energy input is thermal, chemical, or electrical. This is an expression of the consilient mechanism for protein-based machines and materials where the curves **A–C** for the different energy conversions exhibit the same dependencies. (Reproduced from Urry.[7])

is called a *transition*. A family of such transitions, that is, of how the length changes as the temperature is raised, is shown in the curves of Figure 2.6A for a series of cross-linked elastic-contractile model proteins of differing oil-like character. Part of the heat energy expended in increasing the temperature of the model protein and the surrounding water converts, by means of the protein in water, into the work of lifting the weight.

2.3 Energy Sources Cause Proteins to Fold, Assemble, and Function

This model protein chain intersperses oil-like, hydrophobic (water-fearing) **R**-groups and neutral **R**-groups along a chain of hydrophilic (water-loving) backbone peptide units (see Fig. 2.1A). At low temperature the protein dissolves in water, but, as the temperature is raised to the critical transition temperature for a given sequence of residues, the oil-like **R**-groups separate out of the water by folding to form contacts within and between chain molecules. Accordingly, *the contraction on heating occurs as oily units along the chain separate from water and force the chain to fold; the oily groups come together, minimize their contact with water, and result in association within and between chain molecules. In short, the oil-like groups simply become insoluble.*

The sigmoid-shaped curves of Figure 2.6A represent the shortening of contraction that occurs *on raising the temperature* through the relevant temperature interval for the particular extent of oil-like character of the model protein. Elastic-contractile model proteins of more oil-like composition contract at lower temperatures and over narrower temperature intervals.

2.3.4.2 Using Chemical Energy to Drive Contraction

The sigmoid-shaped curves of Figure 2.6B for model proteins containing negatively charged vinegar-like groups demonstrate contraction *on adding positively charged protons* (acid, H^+) over the appropriate concentration range of acid. The plot is percentage of protonation versus decreasing pH (i.e., increasing concentration of protons). Again, the negatively charged groups within model proteins of more oil-like composition neutralize at lower concentrations of acid and over a narrower concentration range, that is, with a larger Hill coefficient indicating a more efficient energy conversion. This is considered in Chapter 1, Figures 1.2 and 1.3 and more extensively in Chapter 5.

2.3.4.3 Using Electrical Energy to Drive Contraction

The sigmoid-shaped curves of Figure 2.6C for model proteins containing positively charged groups demonstrate contraction *on adding negatively charged electrons* (e^-) to again form a more oil-like group. The positively charged groups within model proteins of more oil-like composition neutralize at lower availability of electrons (a lower applied potential) and over a narrower change in applied potential. Again for the more oil-like compositions, there result larger equivalent Hill coefficients indicating a more efficient energy conversion in analogy to the positive cooperativity discussion in Chapter 1 and relevant to Figures 1.2 and 1.3, but treated mathematically in Chapter 5, sections 5.7 and 5.8.

Whether the energy input is heating with an increase in temperature or is at constant temperature with the chemical energy input of adding positively charged protons to negatively charged groups or with the electrical energy input of adding negatively charged electrons to positively charged groups, the dependence on the oil-like character of the model protein is the same. Chapter 5 presents these correlations with experimental data using a series of model proteins of systematically increased oil-like character.[7]

The correlations of Figure 2.6 represent one profound expression of the consilient mechanism for protein function in energy conversion; that is, the data of Figure 2.6 represent a "common groundwork of explanation"[15] for protein function.

2.3.5 Heating Reversibly Increases Protein Order, an Inverse Temperature Transition

In general, heating at the melting point of water causes a decrease in order, as occurs for the transition where crystalline water, ice (in which every atom exists in a fixed position), melts to form less-ordered liquid water in which molecules constantly shift associations from one cluster of molecules to another. Heating at the boiling point of water also causes a decrease in order in the transition in which water molecules in the liquid state, with a regular relationship dictated by direct contact between water molecules, become water vapor with only rare collisions between water molecules.

Quite the inverse occurs for water-dissolved protein of interest here; that is, by the consilient mechanism, heating from below to above the folding transition increases the order of the model protein. Because heating increases protein order, the transition is called an *inverse temperature transition*.

The inverse temperature transition is most unambiguously seen with the related cyclic minimodel-protein molecule, designated as cyclo(GVGVAPGVGVAP), where A stands for alanine with –CH$_3$ for the **R**-group, and G, V, and P are as defined above. At low temperature these cyclic molecules of 12 residues are randomly dispersed (disordered) in solution, but, on raising the temperature, they assemble into precisely ordered arrays constituting the crystals of Figure 2.7.[16] With lowered temperature, the crystals dissolve, and the molecules again become random, one molecule disordered with respect to another in solution.

These molecules become ordered when the temperature is raised and disordered when the temperature is lowered, just the inverse of the ice-to-liquid and the liquid-to-vapor transitions. The physical properties of the water around the hydrophobic (oil-like) valine **R**-groups, –CH(CH$_3$)$_2$, before association of the cyclic molecules, are responsible for this inverse behavior.

2.3.6 Pentagonal Arrangement of Water Around Oil-like Groups

Direct observation of water abutting oil-like (hydrophobic) groups comes from the work entitled " Zur Struktur der Gashydrate" published in 1951 by Stackelberg and Müller.[17] As shown in Figure 2.8,[1] water molecules surround a central molecular oil droplet, that is, a hydrophobic group of atoms such as methane (CH$_4$, the simplest hydrocarbon molecule) or propane (CH$_3$–CH$_2$–CH$_3$). The surrounding 20 water molecules organize at the corners of pentagons with relatively strong hydrogen bonds between water molecules, O-H •• O, and with 12 pentagons surrounding the hydrophobic group, but with weak interactions with the hydrophobic group.

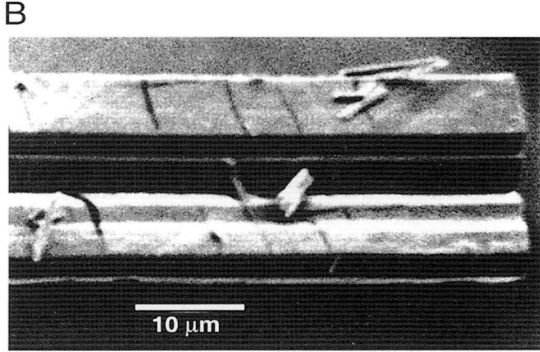

FIGURE 2.7. These crystals of cyclo(GVGVAPGVG-VAP) form when the temperature of aqueous solutions is raised and dissolve when the temperature is lowered. This finding represents an unambiguous demonstration that the model protein component of the aqueous solution becomes more ordered on higher temperature and is one of the reasons that the transition is called an *inverse temperature transition*. (Adapted with permission from Urry et al.[16])

These pentagonal arrangements of water molecules at the surface of oil-like groups enhance the water structure of additional layers. The addition of one CH$_2$ to the Val **R**-group, –CH(CH$_3$)$_2$, of the (GVGVP) repeating sequence to form (GVGIP) with **R**-group of I being –CH$_2$(CH$_3$)–CH$_2$–CH$_3$ causes addition of many more water molecules for each five residue repeat.[18] The increase in the number of ordered water molecules surrounding oil-like **R**-groups lowers the temperature of the transition. As apparent in Figure 2.6, an increase in the oil-like character of a model

2.3 Energy Sources Cause Proteins to Fold, Assemble, and Function

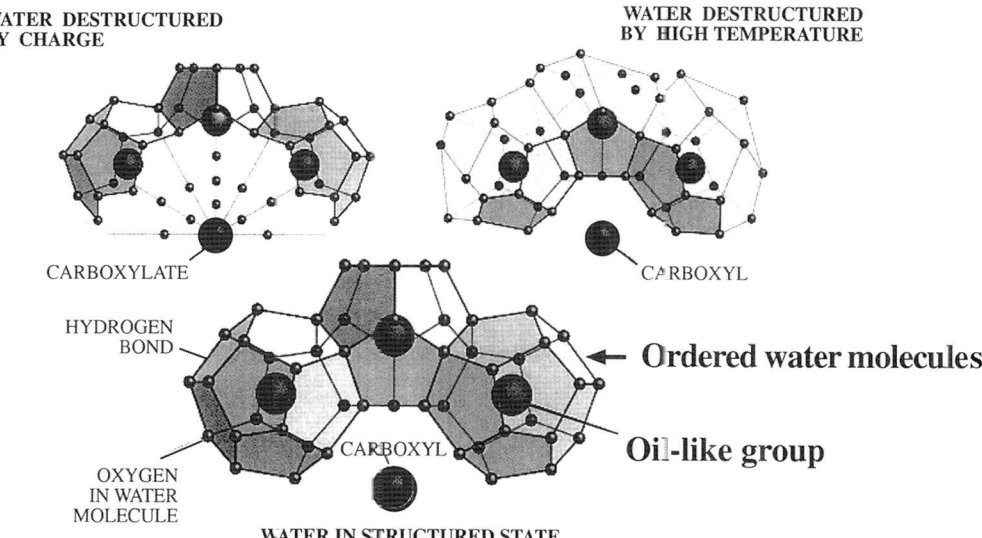

FIGURE 2.8. Ordered water molecules surround oil-like groups, as first shown by Stackelberg and Müller.[17] As oil-like groups associate, this structured water becomes less ordered liquid water; this is the large positive change in entropy responsible for the inverse temperature transition, the phase transition that is fundamental to protein function. As shown in the upper left, hydration of a proximal carboxylate destroys this ordered water and is responsible for the solubilization of proximal oil-like groups, but the thermal destructuring of the structured water shown at the upper right is not a correct concept (see Chapter 5 for a more complete discussion of this phenomenon). (Adapted with permission from Urry.[1])

protein could lower the transition temperature from above to below room temperature to result in folding. As the protein orders, these many, more-ordered water molecules around oil-like side chains become less-ordered bulk water with weak, more amorphous and more rapidly shifting hydrogen bonds.

Importantly, the increase in disorder of water molecules, on going from surrounding oil-like **R**-groups to becoming bulk water, is greater than the increase in order of the protein. The overall change for the transition is to a state of less order, as always observed by physicists and chemists, but a biologically very useful increase in order of the protein occurs. The overall increase in disorder for the transition is a measurable quantity referred to as an increase in entropy,[19] and the increase in order of the protein part, of the solution of protein and water, represents a decrease in entropy.

2.3.7 Inverse Temperature Transitions Provide Negative Entropy to Protein

The inverse temperature transition is a specific mechanism whereby thermal energy (heat) provides an increase in order of the protein part of the system. A decrease in entropy of this sort has been termed *negative entropy* by Schrödinger.[20] While the total entropy (disorder)[21] for the complete system of protein and water increases as the temperature is raised, the structural protein component, critical to the conversion of thermal energy to mechanical work, increases in negative entropy. The protein component increases in order by the folding that shortens length and by the assembly of oil-like domains that builds structures.

Living organisms, however, function at constant temperature. As would be necessary for warm-blooded animals and as is apparent in

Figure 2.6B,C, inverse temperature transitions can convert energy at constant temperature. Under these circumstances the energy inputs change the temperature at which the inverse temperature transition occurs, as depicted in Figure 1.1. This requires that the model proteins be designed with a composition such that the energy inputs change the transition temperature from above to below body temperature. In general, to lower the transition temperature, the energy input makes a vinegar-like **R**-group, or otherwise attached group, more oil-like.

2.3.8 Lowering the Temperature at Which Ordering Occurs Rather Than Increasing the Temperature

2.3.8.1 Increasing Oil-like Character Lowers Transition Temperature

Significantly, when the protein chain contains more hydrophobic (oil-like) groups, the transition for folding and assembly (contraction) occurs at a lower temperature, as in curve a of Figure 2.6A, and over a narrower temperature interval. Similarly, as shown in curve b of Figure 2.6A, a chain with less oil-like **R**-groups begins its folding at a higher temperature and requires a wider temperature interval. The dependence of the transition temperature on the degree of oil-like nature of the protein is pivotal to our view of how living organisms make use of available energy, that is, to the consilient mechanism of energy conversion and the negative entropy flow that sustains Life at constant temperature.

An example of a change to a less hydrophobic model protein would be the replacement of one of the valine (V) residues having the oil-like **R**-group, $-CH(CH_3)_2$, with an alanine (A) residue having the smaller oil-like R-group, $-CH_3$, as in poly(GAGVP). When γ-radiation cross-linked to form a rubber-like band, poly(GAGVP) contracts on raising the temperature, but it does so with a higher transition temperature as shown in curve b of Figure 2.6A. In general, any energy input that makes the protein of curve b more oil-like lowers the transition temperature, as in curve a, and drives contraction when the temperature is fixed at an intermediate value.

Therefore, instead of changing the temperature, one changes the temperature interval over which folding occurs. Depending on the composition of the designed model protein, different energy inputs increase the oil-like character of the protein and lower the temperature range over which folding occurs. Initially the temperature range for folding may be above body temperature, say 40° to 55°C, and an energy input lowers temperature range for folding to below body temperature, for example, to the interval of 25° to 37°C, and drives folding to move or lift a weight. In fact, any change that lowers the temperature for hydrophobic folding from above to below body temperature causes the chain, at a constant body temperature, to go from unfolded to folded. In this case, the *movable cusp of insolubility* of Figure 1.1 moves to lower temperatures.

2.3.8.2 Increasing Oil-like Character by Chemical Energy

An important example occurs when the protein chain contains charged, vinegar-like **R**-groups, for example, $-CH_2COO^-$. These charged groups raise the temperature interval for the inverse temperature transition. An increase in the chemical energy, due to adding acid (H^+), results in protonation of the vinegar-like carboxylate, COO^-, to form the uncharged, more oil-like $-CH_2COOH$ **R**-group. As shown in Figure 2.6B, the stepwise conversion of $-CH_2COO^-$ groups to $-CH_2COOH$ groups stepwise lowers the temperature of the inverse temperature transition and has the effect in Figure 2.6A of shifting the thermally driven contraction from curve b to curve a. *Protonation folds the protein!* Accordingly, the energy input of increasing the amount of acid, thereby, drives contraction. Energy inputs—such as changes in the concentration of certain chemicals that make the protein more oil-like—lower the temperature range over which the folding occurs. Such energy inputs drive contraction. They result in the performance of mechanical work.

2.3.8.3 Increasing Oil-like Character by Electrochemical Energy

As represented in Figure 2.6C, reduction of a more polar oxidized group attached to the model protein drives contraction. For more oil-like model proteins, the affinity for electrons is greater, as in curve a, and reduction requires less electrical energy due to the steeper curve. More oil-like model proteins require less electrical energy to drive contraction. Furthermore, reduction of a group attached to a model protein increases the oil-like character of the model protein, because reduction lowers the temperature at which contraction occurs, as shown in Figure 2.6A.

2.3.8.4 The ΔT_t-mechanism of Energy Conversion

A shorthand way of identifying this process for energy conversion becomes helpful. If the temperature at which the onset of the inverse temperature transition occurs is called T_t and the magnitude of the change in transition temperature is indicated by a delta, Δ, then the process of converting energy by changing the transition temperature can be called the ΔT_t-mechanism.

The insights of the ΔT_t-mechanism did not originate from studying a particular energy conversion of an existing biological process, such as muscle contraction. As described above, the understanding resulted from observing hydrophobic folding of elastic model proteins with increases in temperature. Then it was found, with the proper model protein design, that many energy inputs changed the transition temperature, that is, caused a ΔT_t.

Now we emphasize that the most effective chemical means for changing the temperature range for folding involves attachment of phosphate, for example, phosphorylation by ATP, or the binding of ATP itself.

2.3.8.5 Increasing Hydration Around Oil-like Groups Lowers T_t, That Is, Lowers the Cusp of Insolubility

The preceding discussion considered only phenomenology, but it is possible to state the mechanism whereby the contractions due to oil-like associations occur. Insight into the mechanism was apparent as early as 1937 in the work of Butler,[22] and it resides in the thermodynamics of the solubility of oil-like groups in water, as discussed in more detail in Chapter 5. The Gibbs free energy, ΔG, governs solubility, that is, $\Delta G(\text{solubility}) = \Delta H - T\Delta S$. ΔH is the heat of the transition (i.e., the area of a cusp of insolubility in Fig. 1.1) and is negative for favorable reactions that release heat, and ΔS, the change in entropy, is positive for an increase in disorder and negative for an increase in order.

In brief, the addition of oil-like groups in water results in the release of heat (ΔH is negative). *Somewhat surprisingly, the dissolution of oil-like groups in water is a favorable exothermic reaction.* Solubility of oil-like groups in water, however, is limited, because the second term of the Gibbs free energy for solubility, $-T\Delta S$, is positive for hydration of oil-like groups. The water that forms around oil-like groups, as in Figure 2.8, is structured water, obviously more ordered than bulk water. In this case, the ($-T\Delta S$) term is positive, because formation of structured (more ordered) water around oil-like groups (hydrophobic hydration) gives an inherently negative ΔS. Thus, as more oil-like hydration develops, an unfavorable positive ($-T\Delta S$) term begins to become larger than the inherently negative ΔH term; the $\Delta G(\text{solubility})$ becomes positive, and solubility is lost. *Accordingly, association of oil-like groups (insolubility) results when too much structured water develops around oil-like groups.* This constitutes a central insight into protein function in energy conversion. In summary, the development of too much hydrophobic hydration causes insolubility, as verified experimentally[18] and as discussed in Chapter 5.

The above analysis of the processes of Figure 2.6B,C allows for the simplistic representation shown in Figure 2.9. As depicted in reaction (i) of Figure 2.9, addition of proton, H^+, to a carboxylate, $-COO^-$, makes the model protein more oil-like and allows formation of more structured water around the model protein, and the more oil-like model protein becomes insoluble. The result is contraction due to association of oil-like groups, because the

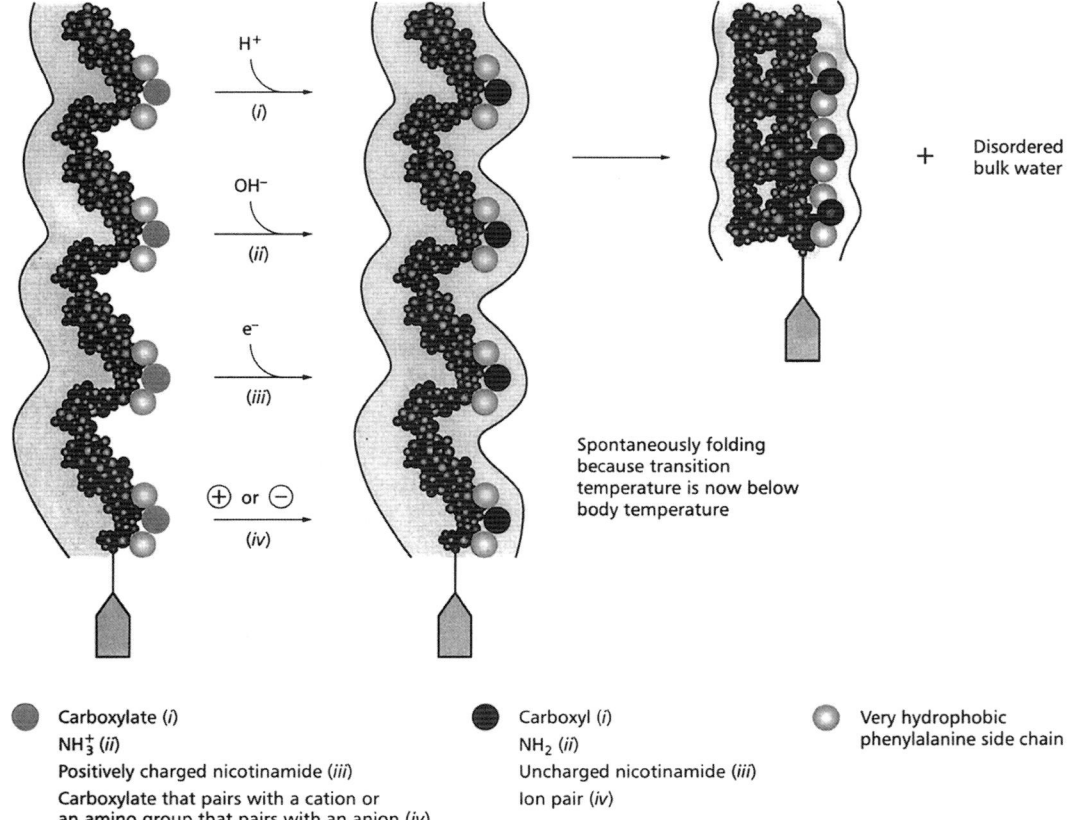

FIGURE 2.9. A set of reactions is shown, each of which causes the model protein to become more oil-like with the result of a contraction due to association of oil-like groups. See text for further discussion. (Reproduced with permission from Urry.[45])

positive ($-T\Delta S$) term dominates the smaller negative ΔH term, and ΔG(solubility) becomes positive. (Recall, a negative ΔG[solubility] means that solubility is favored, and a positive ΔG[solubility] means that solubility is lost.)

As depicted in reaction (ii) of Figure 2.9, the reaction of OH^- with $-NH_3^+$ groups of a model protein results in formation of H_2O and $-NH_2$ to give a more oil-like model protein. The effect is to form so much oil-like hydration, as shown in the central structure of Figure 2.9, that oil-like solubility is lost with the result of contraction.

As depicted in reaction (iii) of Figure 2.9, the addition of an electron to an oxidized group attached to the model protein, that is, the reduction of the model protein, causes the model protein to become more oil-like. The result is the formation of too much oil-like hydration, again as shown by the central structure of the figure, and insolubility ensues, the result of which is contraction.

As depicted in reaction (iv) of Figure 2.9, either the neutralization of a negative group (e.g., $-COO^-$) of the model protein by a positive ion (e.g., Na^+ or Ca^{++}) from solution or the neutralization of a positive group (e.g., $-NH_3^+$) of the model protein by a negative ion (e.g., Cl^-) from solution can cause the model protein to become too oil-like with the result of contraction due to too much oil-like hydration. Contraction occurs as the result of insolubilization of the oil-like model protein and is an example of the lowering of the cusp of insolubility as represented in Figure 1.1.

In general, *the key chemicals for changing the folding temperature in our elastic-contractile model proteins are prevalent triggers of function in biology.*

2.3.9 Inverse Temperature Transitions Extract Order (Negative Entropy) from Energy Sources!

Inverse temperature transitions provide the mechanism whereby order, that is, negative entropy, can be extracted from an energy source to sustain Life. This becomes a basis whereby the existing order of Life can be maintained. It requires only that the proteins of living organisms be of a composition that responds by a change in the transition temperature when acted on by an available energy source.[23] In our view, *the resulting hydrophobic association of oil-like domains of proteins constitutes the negative entropy flow that sustains the order of living organisms.*

Holding such a view requires substantial experimental verification. In our case the required evidence resides in the capacity to design model proteins that utilize inverse temperature transitions for converting energy from one form to another. This has been done for the conversion of thermal, pressure, chemical, electrical, and electromagnetic (e.g., light) energies into the mechanical work of lifting a weight simply by the mechanical aspect of association of oil-like domains of the model protein.

Quoting Schrödinger, "The living organism seems to be a macroscopic system which in part of its behaviour approaches to that purely mechanical (as contrasted with thermodynamical) conduct to which all systems tend, as the temperature approaches the absolute zero and *the molecular disorder is removed.*"[20]

While not approaching absolute zero of temperature, but rather due to raising the temperature, the central effect of the inverse temperature transition is, nonetheless, mechanical in origin with the analogous result of creating ordered structures. Even when energy conversion by this mechanism does not involve the performance of mechanical work, the model protein can go from dissociated oil-like domains to association of oil-like domains in the process of converting one form of energy to another. For example, as shown in more detail below, when the model protein contains both an attached positively charged oxidized group and an attached negatively charged carboxylate, the reduction of (the addition of negative electrons to) the positively charged, oxidized group lowers the transition temperature to below body temperature and causes the protein to fold. Significantly, in the process the energy input of reduction forces the uptake of a proton by a carboxylate, COO^-, at a distance along the protein chain from the reduced group. In this case, electrochemical energy converts to the chemical work of pumping protons, that is, of picking up protons during the mechanical (contractile) change of hydrophobic association.

2.4 Essential Energy Conversions That Sustain Life

2.4.1 Photosynthesis and Respiration: Summations of the Energy Conversions of Life

The overall statements for the energy conversions of biology have been known for more than a century. They are embodied in the expressions for photosynthesis and respiration, as currently written and balanced for the formation and utilization of a carbohydrate such as glucose with six carbon (C) atoms and the equivalent of a molecule of water, H_2O, for each carbon atom, consequently, the name *carbohydrate*.

Photosynthesis:

$$\text{carbon dioxide} + \text{water} + \text{light energy} \Rightarrow$$
$$6(CO_2) \quad 6(H_2O)$$
$$\text{carbohydrate} + \text{oxygen}$$
$$[C(H_2O)]_6 \quad 6(O_2)$$

Respiration:

$$\text{carbohydrate} + \text{oxygen} \Rightarrow \text{water} +$$
$$[C(H_2O)]_6 \quad 6(O_2) \quad 6(H_2O)$$
$$\text{carbon dioxide} + \text{chemical energy}$$
$$6(CO_2)$$

From the standpoint of atomic balance, respiration is essentially the reverse of photosynthesis, but there exists the fundamental distinction that the energy obtained from respiration is not light energy but rather the chemical energy utilized to construct and maintain living organisms. The combined effect of photosynthesis and respiration is to convert light energy (more than 50 photons to produce a glucose molecule) to chemical energy (some 36 ATPs per glucose molecule) that sustains Life. Again, we note that ATP, adenosine triphosphate, is biology's common denomination for readily accessible chemical energy. We may also note that an estimate of the efficiency of energy conversion from sunlight to ATP of direct use to animals is but a few percent (see Chapter 4).

The above equations for photosynthesis and respiration, exactly balanced with respect to CO_2, H_2O, $[C(H_2O)]_6$, and O_2, mask an extraordinarily intricate set of reactions where balance tends to be masked by blurring detail. The objective here, however, is to dissect out sufficient detail to expose the primary energy-converting steps common to both processes and to demonstrate that model proteins, utilizing inverse temperature transitions, emulate key elements of those energy-converting steps.[24]

2.4.1.1 Photosynthesis

For that purpose, photosynthesis is described in the three steps listed below and represented to a similar level of complexity in Figure 2.10, which shows that the energy conversions involve the thylakoid membrane of the chloroplast.

1. Inside the leaves of plants sunlight strikes the integrated proteins performing energy conversions (the protein machines) within the thylakoid membranes of the green chloroplasts. The light-energized protein machines split two water molecules ($2H_2O$) into two oxygen atoms that combine to form molecular oxygen, O_2, which is released to the atmosphere. The four hydrogen atoms (4H) become four positive protons (4 acid, H^+ groups) released to the thylakoid lumen, plus four negative electrons (4 e^-) that flow along the electron transport chain in the membrane and pump eight more protons into the thylakoid lumen. On boosting from another light reaction, the four electrons reduce two nicotinamide molecules.[25]

2. The protons, now at 1,000-fold higher concentration in the thylakoid lumen than in the surrounding stroma, return across the thylakoid membrane from lumen to stroma and produce the chemical energy ATP by means of the protein-based machine ATP synthase, briefly considered below and in some detail in Chapter 8.

3. The reduced nicotinamides and ATP convert carbon dioxide (CO_2) to carbohydrate, $[C(H_2O)]_6$, by means of the protein machines of the dark reactions, known as the Calvin cycle.

2.4.1.2 Respiration

Again for the purpose of identifying the primary energy converting steps, respiration is given in terms of the comfortingly analogous three steps, but in a somewhat inverted order as listed below. This is represented in Figure 2.11 to a like level of complexity, involving in this case the inner mitochondrial membrane[26]:

1. Oxidation of carbohydrate by means of the protein machines of the glycolysis pathway, a transition reaction, and the Krebs cycle regenerate carbon dioxide, CO_2, and produce reduced nicotinamides and flavins and some ATP.

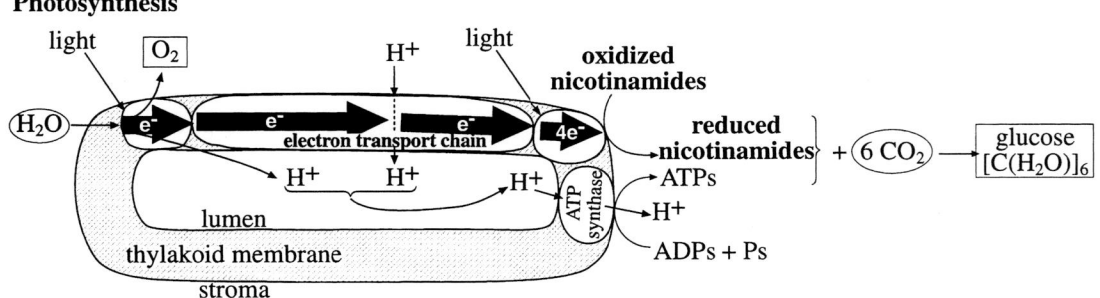

FIGURE 2.10. The energy conversion steps of photosynthesis are shown. See text for more discussion.

2.4 Essential Energy Conversions That Sustain Life

Respiration

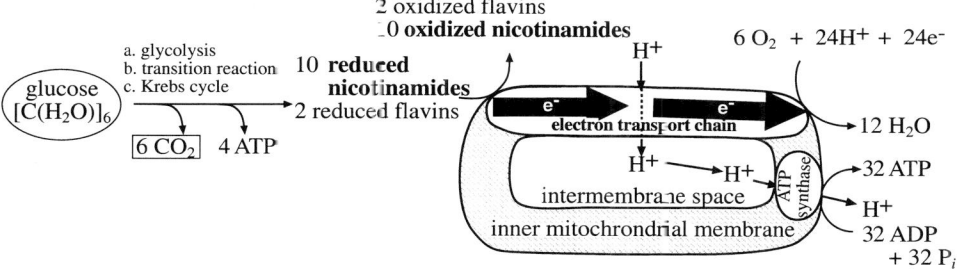

FIGURE 2.11. Simplified schematic representation of the energy conversion steps of respiration are shown. See text for more discussion.

2. Electrons, obtained by oxidation of reduced nicotinamides and flavins, flow through the electron transport chain (a series of proteins with associated groups that cyclically become oxidized and reduced) in the inner mitochondrial membrane ultimately to oxygen with regeneration of water, H_2O, while pumping protons (acid) across the inner mitochondrial membrane into the intermembrane space between the inner and outer mitochondrial membranes.

3. The 1,000-fold greater concentration of protons (acid) of the intermembrane space return across the inner mitochondrial membrane and produce chemical energy, in the form of ATP.[27] As in photosynthesis, the ATP results from the protein-based machine ATP synthase.

Key intermediate energy-converting steps common to both photosynthesis and respiration are (1) the cyclic reduction and oxidation of nicotinamides and flavins coupled to the pumping of acid (protons) from one side to the other of a membrane and (2) the return of acid (protons) across the membrane coupled to production of ATP.

2.4.2 Key Molecular Players Change Charge in Both Photosynthesis and Respiration

As seen above, the two primary molecular players of photosynthesis and respiration are nicotinamide and ATP. Both of these key molecules effect change in the charge of a protein machine during its function. In particular, oxidized nicotinamide is positively charged, and, on reduction (on addition of negative electrons), it becomes uncharged (more oil-like, more hydrophobic). ATP can add a negatively charged phosphate to a protein machine to make it less oil-like, and removal of the phosphate dramatically increases the oil-like character of the protein machine. The decrease of charge in model proteins, as in going from vinegar-like to more oil-like by means of reduction of nicotinamide or by means of dephosphorylation, can drive hydrophobic folding (association of oil-like groups).[28,29]

2.4.3 Key Molecular Players Perform Mechanical Work in Model Proteins

Reduction of oxidized nicotinamide attached to an elastic model protein lowers the transition temperature and drives hydrophobic folding, that is, reduction drives contraction with the lifting of a weight,[28] as depicted in Figures 2.5C and 2.9 reaction (*iii*). The process is reversible; re-oxidation of the attached nicotinamide returns the elastic model protein to its original unfolded state. ATP attaches one of its three charged phosphates to an elastic model protein and yields ADP. This phosphorylation, the addition of the super vinegar-like phosphate group, causes unfolding of the hydrophobically folded model protein.[29] Subsequent removal of the phosphate drives hydrophobic folding, that is, dephosphorylation can drive contraction with the lifting of a weight. Both molecular players of the key steps of photosynthesis and respiration have been used to perform mechanical work with properly designed model proteins. The cyclic reduction/oxidation and the cyclic dephosphorylation/phosphorylation achieve one contraction/relaxation cycle with the consumption of two electrons and a proton and with consumption of one phosphate of the ATP molecule, respectively.

2.4.4 Relationship of Key Molecular Players to Protons

Further direct analogy to the key intermediate steps of photosynthesis and respiration combines reduction of positively charged groups (to drive hydrophobic folding) with performing chemical work of changing the concentration of protons, of pumping protons. It further combines protonation of negatively charged groups (to drive hydrophobic folding) with performing the chemical work of energizing a phosphate plus ADP to form ATP.

The justification for this view of ours is in the demonstration that making a model protein more oil-like can increase the affinity of carboxylates for protons up to a millionfold,[30] whereas an increase of only some 3,000-fold is required to pump protons for the relevant steps of photosynthesis and respiration.

As noted above in Figures 2.6C and 2.9 reaction (*iii*), the reduction of a nicotinamide attached to a model protein lowers the folding temperature by increasing hydrophobic hydration and, therefore, drives folding at body temperature.[28] Importantly, however, when the model protein also contains a negatively charged carboxylate, –COO$^-$, increasing the hydrophobicity by reduction of the nicotinamide forces the functional vinegar-like, negatively charged carboxylate group to become uncharged and more oil-like by the uptake of a positive proton, H$^+$ (see Figure 2.12B).[31] Also directly, the association of proton, H$^+$, with a negatively charged **R**-group, for example, –CH$_2$–CH$_2$–COO$^-$, on a model protein lowers the folding temperature and drives hydrophobic folding at body temperature (See Figure 2.6B). Furthermore, protonation of a negatively charged –COO$^-$ group increases hydrophobicity and is expected, in our view, to energize an attached phosphate, P$_i$, such that it would have the capacity to phosphorylate ADP to produce ATP (see Figure 2.12A). ATP is, again, the basic source of energy that produces motion and performs other biological functions. This combining of two different functional groups within the same molecular machine becomes central to the use of designed elastic-contractile model proteins to provide insight into the mechanism for fundamental processes of photosynthesis and respiration. The point to retain is that the studies on designed elastic-contractile model protein demonstrate the correct energy considerations to perform the indicated energy conversions, but different structures may be used to harvest properly output energies other than mechanical.

2.4.5 Immediate Energy Sources for the Motion of Muscle Contraction

The source of energy for muscle contraction is ATP. During one cycle of contraction and relaxation, ATP is bound, split into the products of ADP and P$_i$ (or PO$_4^{2-}$), and the products are sequentially released.[32] The triggering mechanism for contraction, however, is the release of calcium ion (Ca^{2+}) from nearby stores into the muscle cell, where the calcium ions bind to carboxylates of the muscle proteins.

This sudden increase in Ca^{2+} concentration also constitutes an energy requirement for contraction. In fact, calcium pumps in the cell membrane use ATP to pump calcium ion out of the cell into nearby storage compartments to be held in readiness for release through transmembrane channels that are electrically triggered to open by a nerve impulse. Thus, the energy to drive contraction of the voluntary (or striated) muscle of the discus thrower comes directly from ATP and from an increase in calcium ion concentration, with the latter also being achieved through expenditure of ATP.

What can be demonstrated with model proteins functioning as contractile molecular machines is that two of the most effective means of lowering the temperature of an inverse temperature transition to drive contraction are positively charged calcium ions (Ca^{2+}) binding at paired negatively charged carboxylates (COO$^-$) to decrease net charge

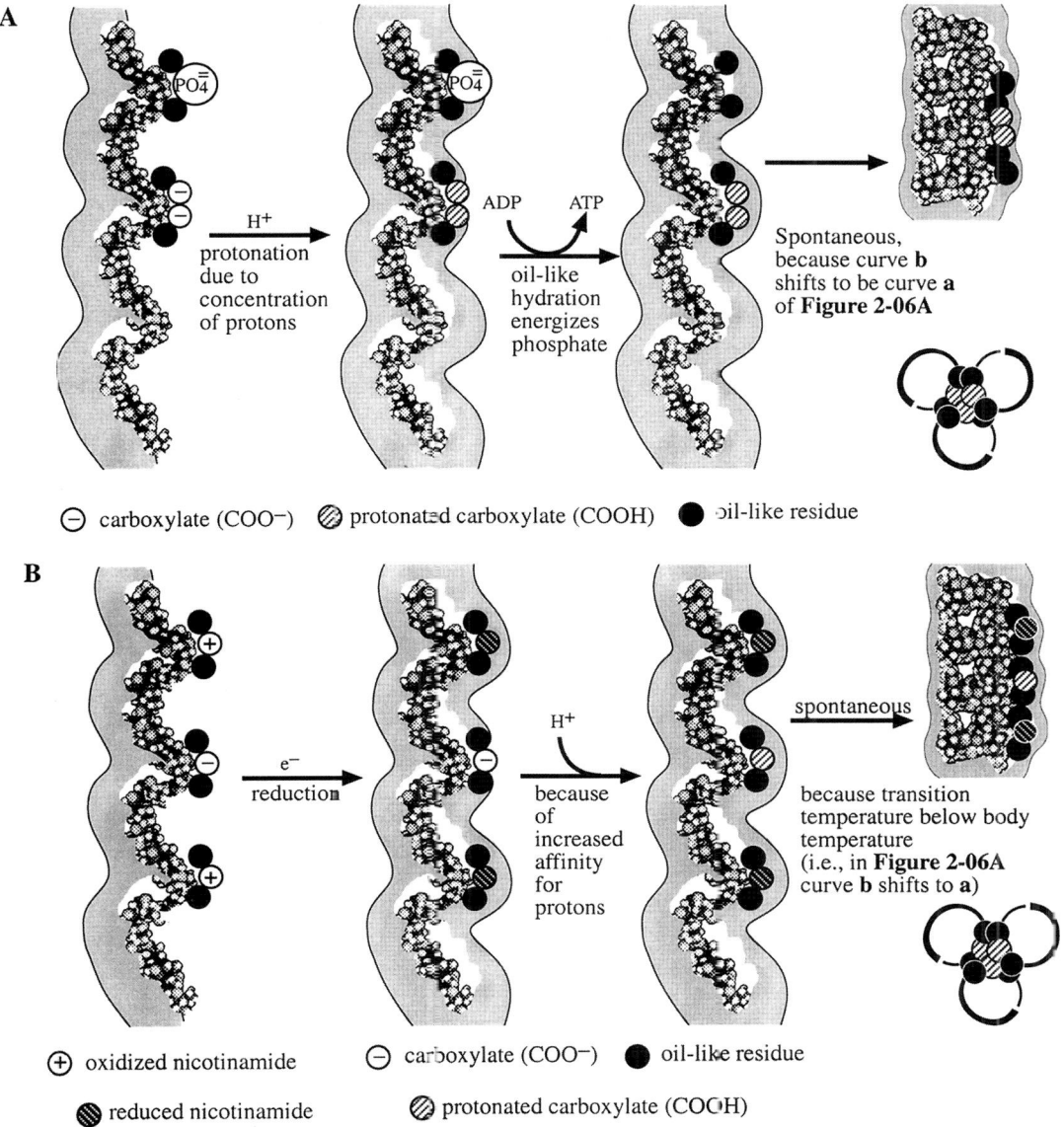

FIGURE 2.12. The thermodynamics of the special-structured pentagonally arranged water molecules around oil-like groups, as in Figure 2.8, is such that too much leads to loss of solubility with the result of hydrophobic association. This kind of water is indicated by the halo. On the side where there is no halo, charged groups have destructured the hydrophobic hydration in the process of achieving their required hydration. This is indicated by the absence of halo. **A** Both carboxylates and phosphates destructure water on one side. Protonation of the carboxyls allows restoration of so much hydrophobic hydration that the phosphate is at high energy due to lack of adequate hydration, just as inadequate hydration results in hydrophobic-induced pKa shifts that are a measure of the increase in free energy of carboxylates. The phosphate is so energized as to be able to add to ADP to form ATP with restoration of the complete halo of hydrophobic hydration with the result of insolubility and contraction. **B** The combination of carboxylates and positively charged oxidized nicotinamide destructure the hydrophobic hydration, the halo, on one side of the structure. Reduction of the nicotinamide converts a charged group to a hydrophobic group and restores so much hydrophobic hydration that the pKa of the carboxylate is higher than the pH of the solution. The carboxylate starved for hydration pulls a proton out of solution; the halo is restored, and insolubility and contraction result. This constitutes the pumping of protons driven by reduction of the nicotinamide.

(see Figures 5.15 and 5.27) and dephosphorylation to remove charge (see Table 5.2). These are the most immediate energy-converting events of muscle contraction.

2.5 Synthetic Model Protein Machines Emulate Energy Conversions of Photosynthesis, Respiration, and Motion

2.5.1 Model Protein Machines in the Coupling of Functions

Seen through the looking glass of energy conversion, a living organism constitutes the successful integration of many proteins functioning as molecular machines. By definition, a machine converts energy from one form to another, and an engine is the particular machine that performs useful mechanical work, for example, lifting a weight or throwing an object. Overall, therefore, the discus thrower functions as a complex molecular engine having converted the chemical energy derived from the intake of food into the useful mechanical work of throwing the discus. In doing so, the discus thrower has successfully utilized many coordinated molecular machines, proteins, of different energy-converting functions. In addition to the coordination of different machines, there is also the important coupling of functions within a single molecular machine.

Here we describe the coupling of functions. A function involves any chemical entity that can exist in either of two states, and by the consilient mechanism the two states differently affect hydrophobic hydration to a significant extent. Two functions become coupled when the more polar state of each decreases hydrophobic hydration while the more hydrophobic state of each increases the potential for hydrophobic hydration. The two functions can be a chemical couple such as –COO⁻/–COOH and a redox couple such as the interconvertible states of oxidized nicotinamide/reduced nicotinamide.

With both functional groups in the same model protein in their more polar states, that is, ionized and oxidized, an electrical or electrochemical energy input to produce the reduced state of the redox couple by making the model protein more hydrophobic brings about the chemical energy output of the increase in affinity of a carboxylate for its proton. The bases of the coupling of functions are the insights of the consilient mechanism arising from the coherence of the simpler contractile processes of Figure 2.6. In fact, the common form of the graphs of Figure 2.6, independent of the energy source, underscores the unifying element of controlling hydrophobic hydration as shown in Figure 2.9. Accordingly, *de novo*–designed elastic-contractile model proteins experimentally demonstrate the coupling of functions.[31]

The biomolecular machines of living organisms convert the chemical energies of food and molecular oxygen, O_2, through many different energy-conversion steps, to the mechanical work of motion to maintain Life and to the construction of the macromolecules and cellular structures that form living cells. The key steps, however, are the coupling of oxidation to proton release and reduction to proton uptake, and the coupling of proton transmembrane flow to phosphorylation of ADP to form ATP.

In the following we use the molecular structure of the elastic-contractile model protein, as exemplified in the simplified structural representations in Figures 2.5, 2.6, and 2.9, to demonstrate a principle that is transferable to protein machines of different configurations, such as linear and rotary motors. The fundamental principle arises out of the competition for hydration between apolar (oil-like) and polar (e.g., charged) groups, given the technical term of an *apolar–polar repulsive free energy of hydration* and represented as ΔG_{ap}.[7] Chapter 5 details the experimental foundation and analysis giving rise to ΔG_{ap}, which may also be thought of as a water-mediated repulsion between hydrophobic and charged species. Chapters 7 and 8 demonstrate the relevance of ΔG_{ap} to molecular machines of biology.

2.5.2 Synthetic Model Protein Machines Pump Protons on Reduction

Oxidized nicotinamide is more vinegar-like due to its positive charge; this is indicated in Figure 2.9 by reaction (*iii*) and in Figure 2.12B by the

2.5 Synthetic Model Protein Machines Emulate Energy Conversions of Photosynthesis

R-group with an open circle containing a plus sign in an elastic-contractile model protein that also contains carboxylates indicated by an open circle with a negative sign. Like the carboxylate in Figure 2.9 of reaction (*i*), charge destructures the pentagonally structured waters surrounding oil-like groups. Reduction of (addition of negative electrons to) oxidized nicotinamide makes it more oil-like and results in a more complete shell of hydrophobic hydration surrounding the second structure of Figure 2.12B. When oxidized nicotinamide, attached to a model protein with an inverse temperature transition above body temperature, is reduced, more low entropy water forms; the inverse temperature transition drops below body temperature and increases the apolar–polar repulsion between oil-like groups and carboxylates. The consequence is a shift in the pKa of the carboxylate, resulting in protonation, followed by completion of the shell of hydrophobic hydration and hydrophobic folding and contraction represented in Figure 2.12B[31] (see also Figure 2.9 and the graphical representations in Figure 2.6). The result of reduction is the uptake, that is, the pumping, of protons. *Thus, reduction pumps protons, i.e., causes the uptake of protons, and, although the structure of the elastic-contractile model protein is inadequate to localize a concentration buildup of protons, the same physical principle applies as demonstrated with the protein-based machines of the electron transport chain* (see Figures 8.12 through 8.25 of Chapter 8 and related text).

If reduction of the oxidized nicotinamide were to occur in an appropriately structured and carboxylate-containing protein within a membrane, however, the pickup of proton by the carboxylate could occur from one side of a membrane, thereby lowering the concentration on that side, and on oxidation the proton release could occur to a confined space on the other side, thereby increasing the proton concentration within the confined space on the other side. Our proposal is that the cyclic reduction and oxidation of redox couples within the thylakoid membrane of chloroplasts and within the inner mitochondrial membrane achieves proton pumping by means of ΔG_{ap}, the aqueous mediated repulsion between oil-like and charged groups that disrupts association of oil-like groups to open access on one side and then the other of these membranes (again, see Figures 8.12 to 8.25 of Chapter 8 and related text). The resulting proton concentration gradient, as described by Mitchell's chemiosmotic hypothesis,[33] provides the energy for phosphorylation of ADP to form ATP as noted below and discussed more extensively in the text associated with Figures 8.26 to 8.36 of Chapter 8.

2.5.3 Synthetic Elastic-contractile Model Protein Machines to Energize Phosphates

Both reduction of a positively charged nicotinamide and the protonation of a negatively charged carboxylate increase the oil-like nature of a protein. Both remove vinegar-like charge. Phosphate, however, is super vinegar; one phosphate attached to a model protein is equivalent to three or four carboxylates in its capacity to disrupt hydrophobic association by destructuring hydrophobic hydration. When the model protein, functioning as a biomolecular machine, is made more vinegar-like by binding of phosphate, the affinity of carboxylates for protons is less (so too is affinity of a negative for a positive charge as in ion-pair formation). The 1,000-fold higher concentration of proton on one side of the membrane could, however, still drive protonation from that side of the membrane.

Now, making the model protein more oil-like by protonation of several carboxylates allows much more pentagonally structured water to form, as schematically depicted in the Figure 2.12A. This, in our view, energizes the highly charged phosphate, as it competes for its own hydration by destructuring the waters hydrating the oil-like groups in an effort to obtain the hydration that it requires. With a properly designed active site, the phosphate could escape the energetically charged situation by adding to an ADP molecule to form ATP when faced off against a sufficiently oil-like surface.

This scenario, however, would not provide for the most efficient energy conversion because the phosphate, so activated, could have a significant probability of hydrolytic cleavage with release of heat. Too often the result could be the production of thermal energy instead of the desired chemical energy. As was noted in Chapter 1 and as will be treated in some detail in Chapter 8, biology has evolved a remarkably efficient rotary motor that spatially separates the protonation of carboxylates from the activation of phosphate. Nonetheless, from the perspective presented in Chapter 8, the same physical process, a counterintuitive competition for hydration between oil-like domains and vinegar-like species, obtains at both sites. First, however, Chapter 5 develops an extensive experimental and analytical basis for the water-mediated repulsion between oil-like and vinegar-like groups, the apolar–polar repulsive free energy of hydration, ΔG_{ap}.

2.5.4 Synthetic Model Protein Machines Pump Iron

One of the more dramatic ways to increase the oil-like character of a model protein functioning as a biomolecular machine is to neutralize the charge on a pair of carboxylates by ion pairing to a calcium ion with its double positive charge. Ion pairing of calcium ion with carboxylate ion lowers the transition from above to below body temperature, as shown in lowering the cusp of insolubility of Figure 1.1 and as is apparent in Figure 2.6A. As in Figures 5.15 and 5.27, this allows the waters of oil-like hydration to form with the consequence of lowering the transition temperature so much that it forces hydrophobic folding and drives contraction. Interestingly, the trigger for muscle contraction is calcium ion binding to a pair of carboxylates.

The very most dramatic way to increase the oil-like nature of a model protein is the removal of an attached phosphate. This is demonstrated in Figure 2.12A. Calcium ion binding to a pair of carboxylates is second only to protonation of a carboxylate in driving hydrophobic association (See Figures 5.27 and 5.34). Thus, the combination of calcium ion binding to a pair of carboxylates followed by phosphate removal, as occurs in muscle contraction, provides perhaps the most potent means of bringing about hydrophobic association and the associated contraction. It is our view that hydrophobic association caused through lowering the cusp of insolubility, as represented in Figure 1.1, provides the primary driving force for muscle contraction, as will be shown in part below in Figure 2.17 for the hydrophobic association that occurs in the cross-bridge of the myosin II motor in its rigor-like state and that is dissociated in the ATP bound state.

2.6 Progression to Biology's Machines from Model Protein Machines

2.6.1 Equivalence of Energy Conversion to Biology's Proteins yet Structural Limitations of Elastic-contractile Model Proteins

Designed elastic-contractile model proteins served to elucidate the consilient mechanism for energy conversion. With proper design, they remarkably perform mechanical work using five different energy inputs. Furthermore, as shown by Steinberg and coworkers,[34] there are designs whereby a continuous band of cross-linked elastic-contractile protein can provide a continuous driving force for rotary engines. The rotary engine contains two separated baths, one that drives contraction and the second that effects relaxation. In all, elastic-contractile model proteins functioning by the consilient mechanism are capable of demonstrating more than 15 classes of pairwise energy conversions.

It is important to realize that the principles so derived are independent of a specific molecular structure. The requirements of the molecular constructs are that they can exist in water, exhibit adequate elasticity or flexibility, and contain oil-like groups in variable spatial relationship to vinegar-like groups with the latter being capable of occurring in states of different degrees of vinegar-like character.

Much fundamental science came from the study of elastic-contractile model proteins. The elastic-contractile model proteins provided the molecular system for realizing the correct description of elasticity.[9] Furthermore, competition for hydration between oil-like and polar,

2.6 Progression to Biology's Machines from Model Protein Machines

for example, charged, groups came conclusively from numerous designed elastic-contractile model proteins and their characterization using a number of different physical methods. Chapter 5 presents this in some detail.

Despite these extraordinary virtues, elastic-contractile model proteins are structurally limited and thereby are themselves unable to take full advantage of the principles to which they gave rise. Biological protein-based machines, however, provide the fertile ground of examples wherein the consilient mechanism can be given generality. A progression of gedanken images from the elastic-contractile model proteins, based on a pentameric repeat with translational symmetry, demonstrate approaches to the very diverse structures of biology's protein-based machines. The remainder of this section begins a succession of steps, and Chapters 7 and 8 carry the argument directly to a number of the protein machines of biology.

2.6.2 ATP Synthase Common to Most Living Systems

As implied by Figure 2.10, which shows details of photosynthesis, ATP synthase plays the central role for the production of ATP in photosynthetic plants, which ATP is used in the synthesis of carbohydrate and all other organic molecules of plants and as the energy source to power the protein-based machines of plants. As similarly indicated in Figure 2.11 for details of respiration, the same ATP synthase provides nearly 90% of the ATP produced in the oxidation of glucose to carbon dioxide and water. Because of the central role of ATP in the synthesis of biomolecules and in powering biology's machines, it is not unreasonable to consider ATP synthase to be the most important protein-based machine of biology.

As introduced in Chapter 1, ATP synthase is in reality a pair of mechanically coupled rotary motors; the F_0-motor driven by the proton gradient provides the mechanical output of a rotating shaft, and the F_1-motor driven by the rotating shaft produces ATP from ADP and P_i (inorganic phosphate, PO_4^{-3}).

2.6.2.1 The F_0-motor of ATP Synthase

The F_0-motor presents a simplistic utilization of the consilient mechanism. The repulsion, between a vinegar-like carboxylate of a cleat of a wheel and the sea of oil of a cell membrane, becomes an attraction on protonation,[35] and the wheel rotates one cleat into the oily membrane as a second cleat rotates into a sufficiently polar, aqueous position for a carboxyl to release its proton to the other side of the membrane. The direction of rotation of the wheel and importantly of the axle (the γ-rotor) that extends from the wheel depends on the frequency with which protons enter and leave from each side of the membrane. This straightforward demonstration of the consilient mechanism will be considered in more detail in relation to Figures 8.26 through 8.30.

2.6.2.2 The F_1-motor: ATP Synthesis by Changing Repulsion Between the Oil-like Surface of the Rotor and the Charged Nucleotides at the Origin of Water-filled Clefts

The same repulsive interaction, ΔG_{ap}, between an oil-like surface and charges associated with the nucleotides occurs in the synthesis of ATP from ADP and P_i by the F_1-motor—the second, the extramembrane, component of ATP synthase. As noted above, the intramembrane component functions to rotate an axle-like projection, called the γ-rotor, from the center of an intramembrane rotary motor driven by protons.

In the extramembrane component the γ-rotor forms the stem and core of an orange-shaped structure comprised of six sections, three α-subunits and three β-subunits, arranged as threefold symmetrical (αβ) pairs, designated as $(αβ)_3$. The key element of the consilient mechanism applied to ATP synthase is that the γ-rotor exhibits three faces of very different hydrophobicity. In our view, rotation of the γ-rotor by the F_0-motor causes the very hydrophobic side of the rotor to be spatially opposed, through a water-filled cleft, to the catalytic site containing the most charged state,

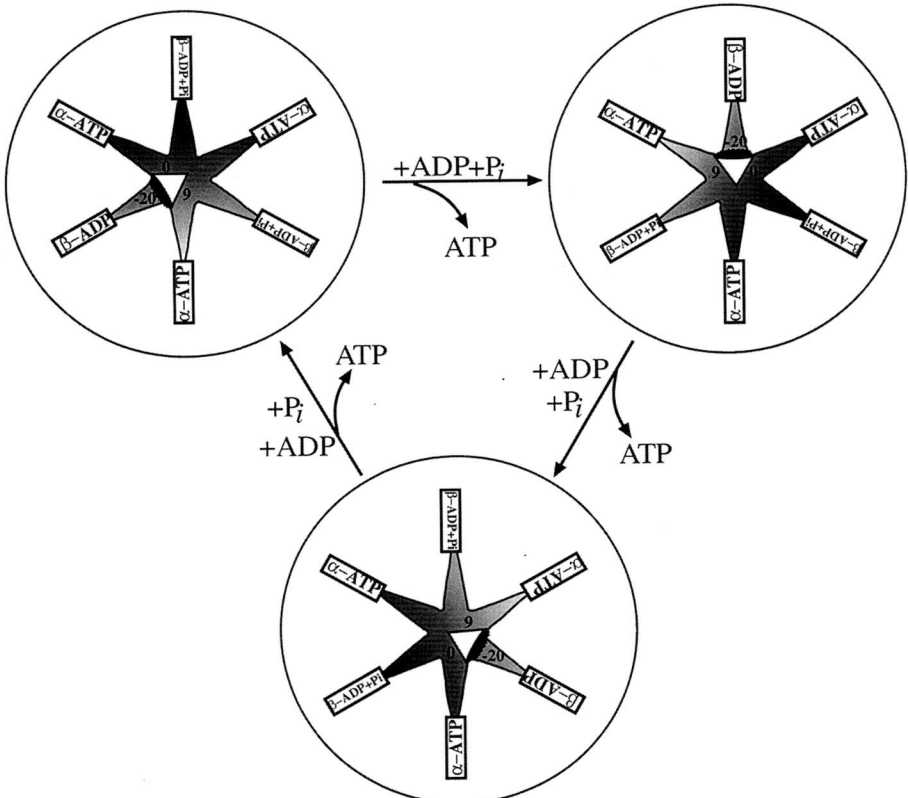

FIGURE 2.13. Shown is one complete cycle of the F_1-motor of ATP synthase on filling all catalytic sites with nucleotide and with a γ-rotor that has three faces of very different hydrophobicities, that is, of very different oil-like character. As discussed in Chapter 8, the relative oil-like character of the three faces compare as −20 kcal/mol-face for the most oil-like, +0 kcal/mol-face for an essentially neutral face, and +9 kcal/mol-face for the least oil-like face. The least oil-like face would allow ADP and P_i to enter the catalytic site. As the F_0-motor rotates the darkened, oil-like face of the rotor toward the catalytic site containing ADP plus P_i, the repulsion between the oil-like face and the filled site causes the ADP plus P_i to relax the repulsion by forming ATP. As rotation brings the oil-like face into complete opposition to the now ATP-filled site, the ATP molecule is expelled from the site. As the rotor continues on its way ADP again enters the catalytic site, followed by P_i. In this mechanism the noncatalytic ATP-containing site would have the function of repelling the most oil-like side of the rotor sufficiently to prevent hydrophobic associations that would produce a frictional drag on the rotor and decrease efficiency and rate of energy conversion.

ADP and P_i. This face-off raises the free energy of the reactants, that is, increases the magnitude of ΔG_{ap}, and results in formation of the less charged ATP. Figure 2.13 shows this cyclic process with all sites containing nucleotide and with the three sides of the γ-rotor, as described in more detail in the figure legend. Molecular details are discussed in the text that describes Figures 8.31 through 8.37.

2.6.3 Filament Assembly from a Clam-shaped Globular Protein

From the perspective of the consilient mechanism, the assembly of filaments as required for muscle contraction and the necessary movement of components within the cell involves hydrophobic association/dissociation between composite subunits. The actin thin filament of muscle contraction and the microtubules of intracellular transport present two relevant examples. Both pose the same challenge to the consilient mechanism. Assembly of the actin filament requires ATP and assembly of microtubules requires structurally and energetically equivalent GTP (guanosine triphosphate). Thus, how could binding a triphosphate to a soluble subunit unit bring about hydrophobic assembly, when, by the consilient mechanism, the triphosphate should be disrupting hydrophobic association? The issue is approached below by considering dimerization of a hydrophobically closed clam-shaped monomer.

2.6.3.1 Disrupting Intramolecular Hydrophobic Association to Form Intermolecular Hydrophobic Association

Phosphorylation initiated dimerization by hydrophobic association is illustrated in Figure 2.14. In Figure 2.14A, a closed clam-shaped globular protein, representing globular G-actin, is held closed by hydrophobic association; but phosphorylation near the mouth destroys a significant amount of the hydrophobic hydration and allows the clam-shaped globular protein to open[36] in analogy to the unfolding of our elastic model protein on phosphorylation.[29] The residual hydrophobic hydration of the open state, however, becomes the driving force for association, that is, for the dimerization shown in Figure 2.14B.

2.6.3.2 Treadmilling in Formation and Maintenance of Actin and Microtubular Tracts for Motion

Addition of a third dimension to the two-dimensional representation in Figure 2.14B provides an additional hydrophobic face for filament growth. Growth of the actin filament occurs at one end by the conversion of monomeric globular G-actin to filamentous F-actin on ATP binding. In this case ATP binding equates to the phosphorylation in Figure 2.14B. Dissociation to form a G-actin occurs on hydrolysis, giving ADP and P_i at the opposite end of the filament. Release of P_i, leaving ADP allows for recovery of the monomeric form with ADP release. Thus, there is a treadmilling with ATP-dependent growth at the positive end and hydrolysis-dependent dissociation at the negative end.

In the case of microtubules the unit that adds or leaves is an αβ heterodimer called tubulin.

2.6.4 Deriving Contraction from a Hydrophobically Closed Clam-shaped Globular Protein

2.6.4.1 Adapting the Clam-shaped Globular Protein Motif to Approach a Sliding Filament Element of Contraction

In Figure 2.15, the clam-shaped globular protein motif is adapted to demonstrate *one of many* possible ways that the control of hydrophobic folding could achieve motion by means of a sliding filament/cross-bridge model for muscle contraction. In this case the increase in energy of the attached and energized phosphate allows that water become the receptor molecule for the high-energy phosphate rather than a molecule of ADP.

We hasten to clarify the purpose of Figure 2.15, which is to exemplify the compliant nature of a mechanism based on the repulsion between oil-like and vinegar-like groups. It shows the ease of a conceptual progression from the more obvious phosphorylation control of opening and closing of a clam-shaped globular protein to a more involved swinging cross-bridge/sliding filament process.

2.6.5 A First Step in Putting an ATPase Motor to Work, Controlling Association with a Complementary Surface

The families of proteins that consume ATP while functioning as protein-based machines are called ATPases. Muscle contraction, just noted above, is a member of the family of linear (contractile) protein motors that also includes ATPases that walk along protein tubules and transport elements from one part of the cell to another. Another class of protein motors that uses ATP rotary and nonrotary ion pumps transports ions from one side to the other of the

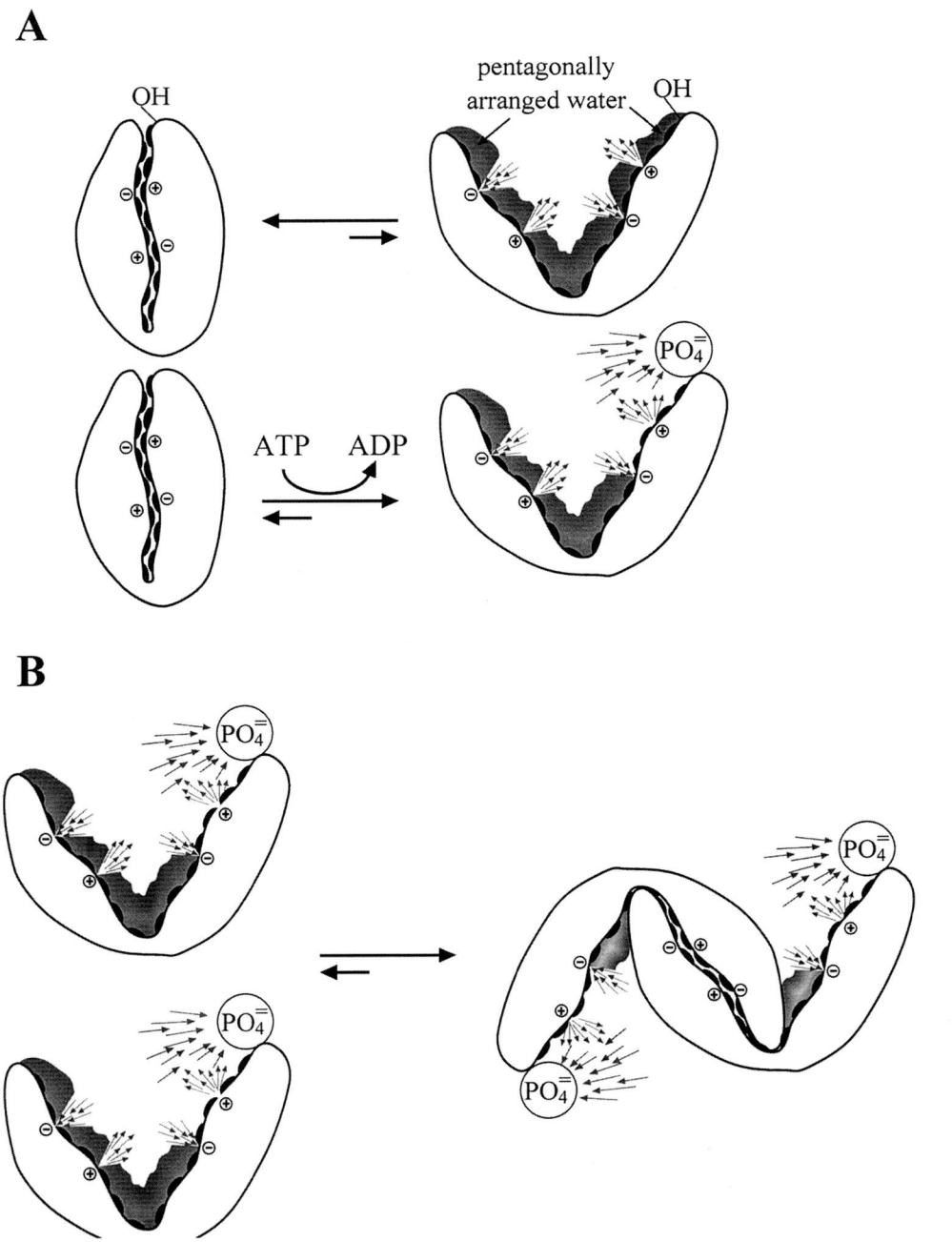

FIGURE 2.14. **A** Phosphorylation Driven Unfolding. A clam-shaped globular protein exhibits an opening fluctuation but closes due to the formation of too much hydrophobic hydration. Phosphorylation of the site near the mouth of the clam-shaped globular protein destructures so much hydrophobic hydration in the process of achieving its own hydration that the equilibrium is shifted to the open state. **B** Phosphorylation Driven Dimerization. On association of complementary surfaces containing hydrophobic hydration, an insolublilization of oil-like groups occurs such that dimerization results.

2.6 Progression to Biology's Machines from Model Protein Machines

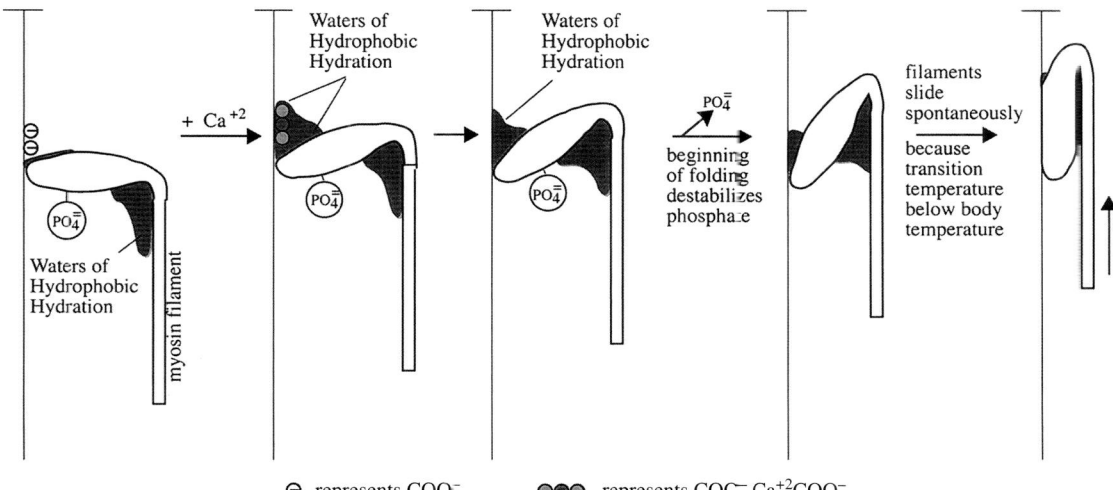

FIGURE 2.15. A primitive swinging cross-bridge/sliding filament model of muscle contraction is introduced for the purpose of providing a conceptual bridge from the simple oil-like dissociation/association for the opening/closing of a clam-shaped globular protein toward the more complex structural aspects of muscle contraction. Major limitations in the drawing include the attachment–detachment process and the required forward movement of the cross-bridge to achieve attachment at the next actin binding site. These issues are considered in Chapter 8, where a twisting and untwisting of the cross-bridge is implicated.

cell membrane. ATP-consuming protein-based machines travel along nucleic acids, correcting the structures in a number of ways or identifying a sight to be cleaved. ATP produces energetically equivalent GTP, and GTPases constitute a family of molecular switches. Thus, ATPases are recognized as the workhorse protein-based machines that perform the day-to-day chores of living organisms.

Putting the ATPase motor to work can begin with attachment to a surface, as shown in Figure 2.16. In this figure the simple attachment/detachment cycle begins with attachment. In the absence of ATP, an oil-like domain of the surface associates with an oil-like domain of the ATPase. Transient dissociations occur with opening of a cleft. The presence of complementary charges on each hydrophobic surface facilitates association as well as dissociation. More pKa-shifted charged groups, as in the series of model proteins in Figures 1.2 and 1.3 with increasing numbers of more oil-like phenylalanine (Phe, F) residues, favor association of oil-like domains with ion pairing. Limited hydration of the charged groups, due to competition with oil-like groups for hydration, facilitates ion-pair formation. Despite the presence of charges and because of their complementary positions, dissociation in the absence of ATP results in the formation of too much oil-like hydration such that the equilibrium is shifted toward association, as indicated by the length of the arrows connecting the two states in Figure 2.16A.

Bound ATP, with its highly charged tripeptide tail at the base of the cleft, destructures oil-like hydration by orienting water molecules for its own hydration, as shown in Figure 2.16B. The decreased amount of oil-like hydration becomes insufficient to maintain association, and the equilibrium shifts toward the dissociated state. The situation shown in Figure 2.16 parallels that of the protein-based motor of skeletal muscle with positive charges on the ATPase and complementary negative charges at the actin binding site. Again, to begin consideration of the more complex case, the lower part below the cleft recedes somewhat on dissociation to suggest further structural changes that can accompany association and dissocia-

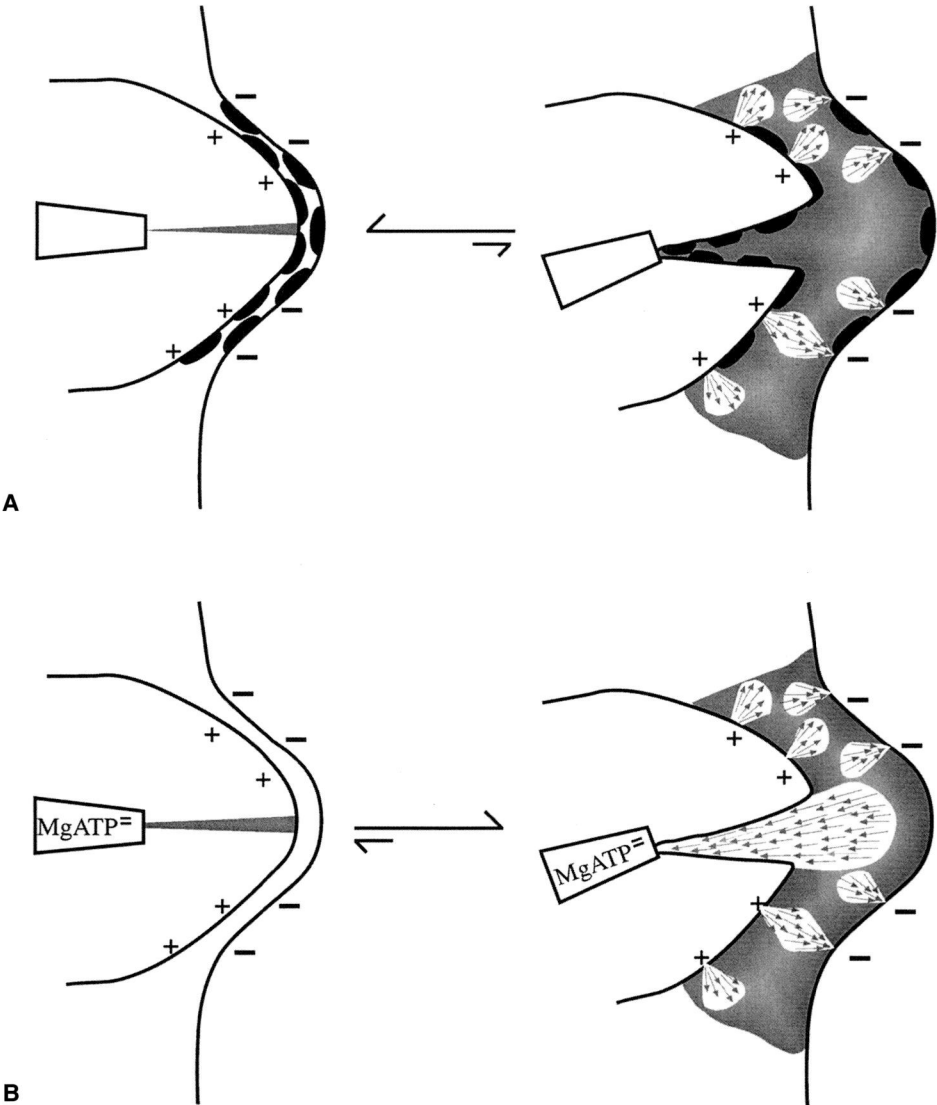

FIGURE 2.16. An initial step in converting the "idling" ATPase motor into a motor capable of performing work involves attachment and detachment as between components of actin and myosin involved in the sliding filaments of muscle contraction. **A** The cartoon depicts an equilibrium between states of oil-like association and dissociation with fluctuating cleft openings, but in the absence of ATP, the equilibrium markedly shifts toward association. **B** Binding of MgATP with its negative charges at the base of the cleft destroys the hydration around oil-like groups in the process of achieving its own hydration and in so doing shifts the equilibrium dramatically toward dissociation. This represents the first step of muscle contraction involving dissociation of the cross-bridge of the myosin II motor from the actin filament.

tion. There can also be structural changes within the ATPase that, as the result of phosphate release, develop important elastic forces on attachment to the surface and contraction at fixed length.

2.6.6 Scallop Muscle and the Clam-shaped Globular Protein Motif

The scallop adductor muscle, of course, is that tasty seafood delicacy that can be prepared

2.6 Progression to Biology's Machines from Model Protein Machines

fried, baked, or even broiled. For the scallop it functions to open and close the two halves of this beautiful classic bivalve shell. Importantly, the crystal structure data have been obtained of the cross-bridge of scallop muscle from the myosin filament to the site for actin binding, and the cross-bridge has been obtained in two different states, a near rigor state and a state with an ATP analogue.[37,38] Somewhat ironically, the structure of the cross-bridge provides an actual analogue of the clam-shaped globular protein in Figure 2.14A.

The crystal structure is of the S1 cross-bridge that contains the globular head, the α-helical lever arm, and its wrappings, the essential (ELC) and regulatory (RLC) light chains, which seem to contribute by strengthening the single helical rod. For our purposes here, the light chains will be removed in order to make apparent the changes that occur when going between contracted and ATP bound states. As we will see, ATP binding opens the hydrophobic association between key components of the cross-bridge, which bears analogy to the binding of a single phosphate to open the hydrophobically associated clam-shaped globular protein.

2.6.6.1 Binding of ATP Analogue to Cross-bridge "Opens" the Oil-like Associated State as Occurs with the Clam-shaped Globular Protein Motif

2.6.6.1.1 The Process of Muscle Contraction Itself Can Be Viewed as a Variation on the Clam-Shaped Globular Protein Theme

Figure 2.17 gives stereo views obtained from the crystal structure of the cross-bridge of the scallop myosin II motor in the contracted near-rigor state in **A** (from the side of the ATP binding site) and **C** (from the side opposite the ATP binding site) and in the bound ATP analogue state in **B** (from the side of the ATP binding site) and in **D** (from the side opposite the ATP binding site). Incredibly, as a direct analogy to the hydrophobically closed clam-shaped globular protein in Figure 2.14A, the binding of the negatively charged ATP with its triphosphate tail causes hydrophobic dissociation, that is, causes opening of the hydrophobically closed association.

2.6.6.1.2 Structural Reorientation on Dephosphorylation

The long limb in Figure 2.17 is the α-helical lever arm that in part connects the myosin filament to the globular part of the cross-bridge. The stump or pedicle-like structure is the N-terminal domain. In Figure 2.17A,C, the view is in a direction along the myosin filament, whereas in Figure 2.17B,D, the view is approximately perpendicular to the myosin filament as may be judged from the foot-like terminus that is taken as being directed parallel to the myosin filament. Thus, dephosphorylation results in a rotation of the cross-bridge around the head of the lever arm to tuck it under the stump of the N-terminal domain. As expected from the consilient mechanism, loss of the negatively charged phosphate results in association of oil-like domains. As considered in the text associated with Figures 8.47 through 8.57, this hydrophobic association provides the power stroke of muscle contraction.

2.6.6.2 The Clefts Within Which the ATP Resides Act as Conduits to Direct the Forces Emanating from the Thirsty Phosphates

The cleft at which the ATP resides runs in two directions, with the γ-phosphate positioned to direct its thirst for water toward the actin binding site and toward the open clam-like region between upper lever arm and pedicle of the N-terminal domain. Seen from the front, the actin binding side, the ATP site resides in a cleft under an overhang with its γ-phosphate in the depths at the aqueous edge of the cleft. The cleft runs in two directions. In one direction in Figure 2.17B the cleft runs outward and upward at about 10 o'clock to the junction of the myosin head with the actin filament, such that ATP binding disrupts attachment to the actin filament by the repulsion between oil-like and charged species. In the other direction the cleft runs to the oil-like association between the head of the lever arm and the N-terminal

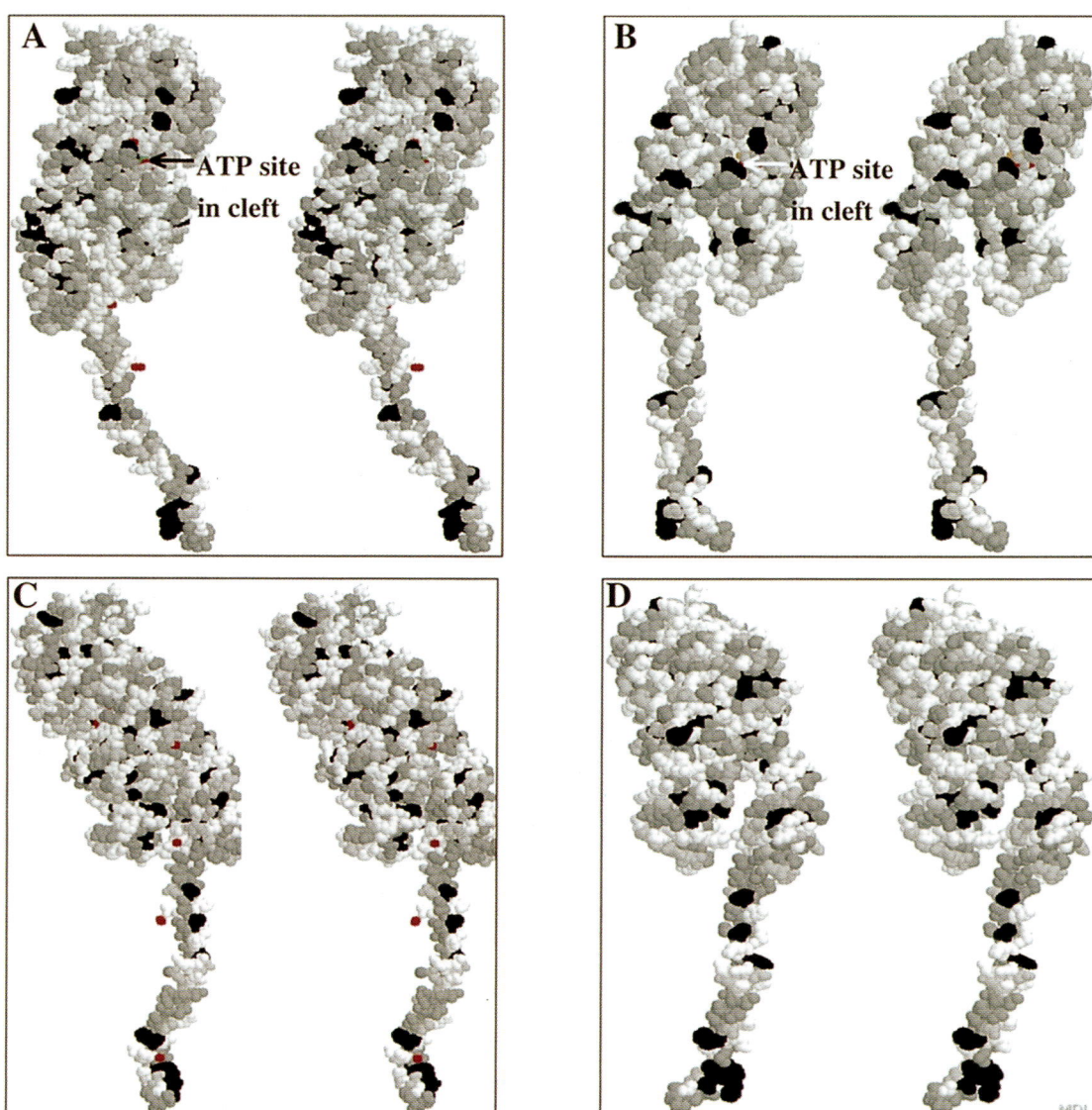

FIGURE 2.17. Cross-eye stereo views show a space-filling representation of scallop muscle cross-bridge (S1). The lever arm is present, but essential and regulatory light chains are removed with neutral residues light-gray, aromatics black, other hydrophobics gray, and charged residues white. **A** The near-rigor state shows a hydrophobic association between the N-terminal domain and the upper part of the lever arm with view looking approximately along the axis of the myosin filament (Protein Data Bank, Structure File 1KK7). **B** The ATP analogue state shows hydrophobic dissociation between the N-terminal domain and the upper part of the lever arm. The view is now approximately perpendicular to the myosin filament, that is, the lever arm rotates with an untwisting on hydrophobic dissociation. Protein Data Bank, Structure File 1KK8 **C,D** The backside of a cross-bridge, that is, the side opposite from ATP binding site is shown. These parts were prepared with the crystallographic results of Himmel et al.[37] as obtained from the Protein Data Bank, Structure Files 1KK7 and 1KK8.

domain to cause the association to "open", that is, to dissociate. The cleft acts like a conduit directing the thirst of the phosphate for hydration in both directions to disrupt hydrophobic associations. The cleft is shown from several different vantage points in Figures 8.58 and 8.59 and further discussed in their associated text.

In Chapter 8, more structural background and molecular details of contraction exhibited by the linear myosin II motor are considered after, in Chapter 5, the physical basis for the apolar (oil-like)–polar (vinegar-like) repulsive energy that controls hydrophobic association is experimentally and analytically developed. The crystal structures of the cross-bridge of scallop muscle provide remarkable examples of the consilient mechanism functioning in this protein-based machine!

2.7 Consequences of Protein Machines Based on the Inverse Temperature Transitions

We again draw comparison by quoting from Schrödinger's *What is Life?* as published in 1944: "What I wish to make clear in this last chapter is, in short, that from all that we have learnt about the structure of living matter, we must be prepared to find it working in a manner that cannot be reduced to the ordinary laws of physics."[20] In the present volume, our foundation is, of course, the somewhat counterintuitive inverse temperature transition. It is by means of the inverse temperature transition that the energies essential to sustain Life can, in fact, be accessed. This efficient consilient mechanism, however, requires no new laws of physics. In fact, the seeds of this mechanism are found in a 1937 report of Butler,[22] in which he analyzed the thermodynamic elements of the solubility of oil-like groups in water.

The essential aspect of the capacity of the inverse temperature transition to achieve diverse energy conversions resides within large chain molecules, which were just becoming known when the first edition of Schrödinger's book appeared. As we have sketched above, the functional properties of the model protein-based machines result from the interaction between hydrophobic (oil-like) and hydrophilic (vinegar-like) groups constrained to occur in unique sequence along the chain molecule, which combines with the capacity of vinegar-like functional groups to shift between being more vinegar-like and more oil-like.

The special effectiveness of proteins compared with all other known chain molecules arises from biology's capacity accurately to produce diverse sequences. As for accuracy, the error in placing the correct amino acid residue at the specified sequence position depends on the ability to delineate one **R**-group from another at the stage of attachment to t-RNA (see Chapter 4).[39] In general, the error is remarkably small. It is also very important that all of the residues be of the correct optical configuration,[40] as random mixing of mirror images of amino acid residues would destroy structure with the consequence of eliminating reliable function (see the paragraphs on Pasteur and mirror image molecules in section 3.1.6 of Chapter 3).

These features of protein polymers result in regular, nonrandom structures, which, as will be shown later, are the reason for efficient energy conversion. As for diversity, theoretically each position in the sequence can be any one of 20 different amino acid residues. Again very significantly, the protein sequence is specified by the nucleic acid sequence and not by the energetics within the sequence of a particular protein. In this way, a sequence that would normally be highly improbable becomes just as probable as any other sequence. (Improbabilities would otherwise arise by virtue of being a single sequential arrangement of amino acids among an inordinate number of possible arrangements of the same collection of amino acids and because of unfavorable interaction energies between the **R**-groups and thus being of high energy.) Of course, much energy has been spent in producing the "improbable" sequence, as discussed further in Chapter 4.

Production of unique and improbable sequences might have been one place where the far-from-equilibrium arguments of Prigogine could have come into play. Order is indeed achieved out of chaos, but it occurs at a high price in energy—in the energy required to copy the parent chain of DNA, in the energy required to transcribe the DNA sequence into RNA, and in the energy required to translate the nucleic acid sequence into a protein sequence. To the best of our present knowledge, however, each of the many steps in the procedure can be treated and studied in the test tube as an equilibrium process. Thus, once the machinery of the cell is in place, the individual steps within the cell appear to be reversible and explicable in terms of the thermodynamics of Boltzmann and Gibbs. However, as noted below and treated in more detail in Chapter 4, biology has employed a throw-away energy; the pyrophosphate product of each step of the assembly essential for reversibility is broken down by the interesting enzyme pyrophosphatase to ensure unidirectionality to both nucleic acid and protein synthesis.

Now, because it can cost just as much energy to produce an inefficient and more limited protein-based machine as it can to produce a more efficient and/or a new machine that can access a new energy source, obviously with natural selection the arrow of time for biology is toward greater complexity and diversity function (see Chapter 6).

2.7.1 Protein-based Machines as Catalysts for Energy Conversion

When an energy input causes an inverse temperature transition of hydrophobic folding, a protein converts part of the energy into its own folding and ordering; this constitutes one aspect of a decrease in entropy. On completing a cycle by returning to the unfolded and disordered state, the net entropy change, however, becomes zero for this cyclic folding and unfolding. When in the process of cycling—mechanical, chemical, electrical, or some other work—has been catalyzed, then there results a useful decrease in entropy resulting from the energy input. A productive negative entropy change does occur when, in the process of folding, the protein picks up protons and deposits them at a higher concentration, that is, pumps protons,

or the protein lifts a weight, that is, pumps iron, for some useful purpose, or the protein phosphorylates ADP to form ATP. In each case movement is away from disorder; a beneficial increase in order, that is, a decrease in entropy, has been achieved. Proteins that hydrophobically fold as the result of an energy input to result in mechanical work, chemical work, or some other energy output could reasonably be classified as negative entropy machines, but this is an incomplete characterization. An adequate characterization would be to view protein-based machines simply as catalysts of energy conversion.

It is now possible to design and prepare elastic-contractile model proteins capable of interconverting the energies that sustain Life. The set of six energies interconverted by living organisms are represented symbolically in Figure 2.18. Indicated by the bold interconnecting arrows are the pairwise energy conversions, directly demonstrated by designed model proteins. Additional energy conversions have been demonstrated less directly. These model proteins are biomolecular machines of further interest as advanced materials for medical and nonmedical applications to enhance quality of Life and to assist in sustaining an ever larger, more complex and more productive society.

2.7.2 Efficiency and the Definition of Chemical Energy

The efficiency, η, of a process can be described as the work achieved divided by the energy input, that is, η = work performed/energy input. The work performed (w) can be the lifting of a weight, which is the mass (m) of the object times the force of gravity (g) times the height (h) the object was raised, that is, w = mgh, as indicated in Figure 2.5. The energy input can be due to the increase in concentration of protons, which is proportional to the size of the step along the horizontal axis in Figure 2.6B. In Figure 2.6B, the size of step required to go completely from COO^- to COOH for a less oil-like protein with a broader sigmoid curve is greater than for the more oil-like protein with its narrower sigmoid curve. As the energy input comes

2.7 Consequences of Protein Machines Based on the Inverse Temperature Transitions

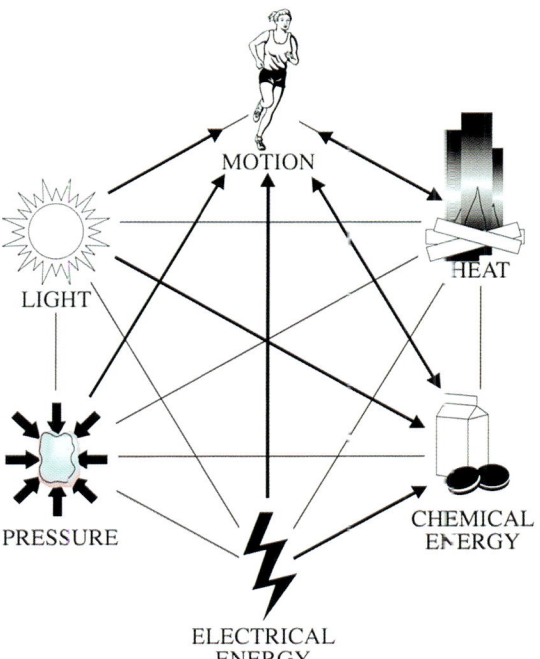

FIGURE 2.18. Energies are shown that can be interconverted by means of elastic-contractile model proteins capable of exhibiting inverse temperature transitions functioning by means of the competition for hydration between oil-like and charged groups called an apolar–polar repulsive free energy of hydration. See Chapter 5 for a more complete development of the phenomenology and physical basis and Chapter 8 for details of the molecular process.

in the denominator for the expression of efficiency, this means that energy conversion is more efficient for the more oil-like proteins. Another expression of relative efficiency, introduced in Chapter 1, is the Hill coefficient that quantifies positive cooperativity. As considered below, when it is possible to compare directly the apolar–polar repulsive free energy of hydration with the more commonly considered electrostatic mechanism of charge–charge repulsion in water, the former wins by several factors of ten in calculated efficiency (see Chapter 5 and, specifically, Figure 5.36). *The efficiency, with which proteins can convert available energy by means of the inverse temperature transitions to those energies required to sustain*

Life is one argument for the significance of this mechanism in biology.

2.7.3 Cooperativity and Survival of Efficient Mechanisms

The steeper acid–base titration curves (of Figures 1.2 and 2.6B, that result from more oil-like proteins) express positive cooperativity, which is to say that the binding of a first proton facilitates the binding of the second proton and the initial two cooperate to facilitate binding of the third, and so forth. (see Figures 1.4 and 5.31). Another mechanism for energy conversion results from the repulsion of like charges much as static electricity causes hair to stand out, preventing it from assembling on the head as one might like.

The charge–charge repulsion mechanism can be used to drive extension of a chain molecule. Then the chemical energy required to remove repulsion, relax extension, achieves the mechanical work of lifting a weight. Because the first charge repels the addition of the second charge, and so forth, however, the curves as represented in Figure 2.6B would all be much broader. As a consequence, much more energy input would be required to achieve the same amount of mechanical work. The electrostatic (charge–charge) repulsion mechanism is much less efficient than the mechanism based on water-mediated repulsion between apolar and polar groups that operates for inverse temperature transitions.

Calculation of the data in Figure 5.36 provides quantitative comparison of the contractile efficiencies of two cross-linked matrices where, in both cases, protonation of $-COO^-$ to form $-COOH$ drives contraction. The polymer using the charge–charge repulsion mechanism is poly(methacrylic acid), $[-CH_2-CH_3CCOOH-]_n$, and the model protein using the apolar–polar (oil-like/vinegar-like) repulsion mechanism of the inverse temperature transition is poly(GVGVP GVGFP GEGFP GVGVP GVGFP GVGFP), where E stands for the glutamic acid residue with the **R**-group, $-CH_2-CH_2-COOH$. Both matrices are loaded with a force, f = mg, of a 2-gram weight, and the contraction, that is, the decrease in length (ΔL) as the pH is lowered, occurs on addition of acid. Now efficiency can be written as $\eta = f\Delta L/(2.3RT\Delta pH)\Delta n$, where R is the gas constant (1.987 cal/mol-deg), T is the temperature in degrees K (centigrade degrees + 273), and Δn is the number of protons used. Based on the steepest slope for each curve in Figure 5.36, the ratio of the efficiencies of the inverse temperature transition mechanism, η_{ap}, to the charge–charge repulsion mechanism, η_{cc}, is given by $\eta_{ap}/\eta_{cc} = [\Delta pH\Delta n]_{PMA}/[\Delta pH\Delta n]_{E4F}$. On substitution of the experimental numbers, $\eta_{ap}/\eta_{cc} = [(2.24)(6.86 \times 10^{-5})]/[(0.59)(6.12 \times 10^{-6})] \approx 40$. The mechanism based on the inverse temperature transition of the designed elastic-contractile model proteins is more than 40 times more efficient. The electrostatic mechanism in water is only a few percent as effective as the mechanism that uses the inverse temperature transition.

Due to the struggle to survive under circumstances of limited food supply, organisms evolve to use the most efficient mechanism available to their composition. The most efficient mechanism available to the proteins that sustain Life would seem to be the apolar–polar repulsive free energy of hydration as observed for the inverse temperature transitions for hydrophobic association. The efficiency of designed elastic-contractile protein-based machines and a number of additional properties make designed protein-based materials of substantial promise for the marketplace of the future.

2.8 "What Sustains Life?" Becomes "What Sustains Society?" (Biomolecular Machines as Advanced Materials for the Future)

Advances in physics, chemistry, and biology spawned technologies that now support larger populations at higher standards of living; they have provided fossil fuel–powered, hydro–powered, nuclear-powered, and chemical-based energy sources (e.g., electricity, fuels, explosives); they have given us electronics and communications, agricultural enhancement methods and products, medicines, new materials such as plastics, and related medical devices. Now, fertile technologies, spawned from advances in molecular biology and molecular biophysics, promise opportunities to continue to sustain society at higher population levels with higher standards of living.

Here we utilize insights gained from the capacity of living organisms to access and transform energy and apply these protein-based machines to problems of consequence to an ever larger, more complex society. Three major concerns of our society are an unsustainable rise in medical care costs, the drug addiction problem that drives an increasing crime rate and supports terrorism, and pollution of the environment.

The preceding design of model proteins that emulate energy conversions of living organisms constitutes useful knowledge for the design of model proteins for medical and nonmedical use. In this brief overview, we mention three applications that relate to the above-noted problems in our society. Two are medical applications, tissue reconstruction and drug delivery, and the third is primarily a nonmedical application of (bio)degradable plastics. These are treated more extensively in Chapter 9.

The chosen three examples result from our newly developed understanding of protein-catalyzed energy conversion. This brings us to the modern-day situation so well-stated by J. Bronowski,[41] "In effect, the modern problem is no longer to design a structure from the materials, but to design the materials for a structure."

For medical applications the biocompatibility of the model proteins must, of course, be established; this has been done for three basic compositions, but verification of nontoxicity must be established for each specific composition designed for a particular human or veterinarian use.

The ultimate limitation to the utilization of model proteins as advanced materials for the future resides in low-cost, reliable production. Proteins are unique polymers in that they can be prepared by genetic engineering, allowing for fidelity and diversity of sequence not possible for any other known chain molecules. An overview would be incomplete without a

2.8 "What Sustains Life?" Becomes "What Sustains Society?"

few comments on this capacity as presently demonstrated.

2.8.1 Genetic Engineering of Model Proteins

On chemically synthesizing the first very long chains of poly(GVGVP) early in 1984 with the support from the Army Medical Research Office, our peptide chemist, K. U. Prasad, suggested that it would make a very nice chewing gum, an idea dismissed with a laugh for one stick of gum would have cost a few hundred dollars. As the result of an Accelerated Research Initiative for producing elastomers of repeating peptide sequences by genetic engineering, which was announced by the Office of Naval Research in 1985, the current anticipated cost for an edible stick of gum could become as little as a few pennies. This cost comparison between chemical synthesis in the research laboratory and microbial biosynthesis illustrates the expansion of applications possible with the recent success of recombinant DNA technology for producing $(GVGVP)_{251}$[42] and other protein-based polymers.

As discussed in subsequent chapters, the error-free preparation resulting from microbial biosynthesis is just as important for function as lowering production cost is for expansion of applications. The basic gene design and construction occur with essentially no limitations on sequence, and expression of the protein-based polymer proceeds with few constraints on sequence. The basic (monomer) gene is polymerized enzymatically to provide genes for the entire possible range of exact polymer chain lengths. To date, model proteins containing as many as 4,000 residues have been produced with remarkably high quantities prepared per volume of the fermentor. An example of the microbe *E. coli* producing $(GVGVP)_{121}$ is shown in Figure 2.19.[43] Later chapters contain further details of the genetic engineering approaches.

2.8.2 Tissue Reconstruction

Within a tissue such as skin, artery, or lung, cells attach to proteins of their extracellular matrix by means of short specific sequences of amino acid residues in extracellular proteins. Within the cells, cytoskeletal fibers, much like muscle fibers, span between attachment sites. When the tissue is stretched, so too are the intracellular cytoskeletal fibers. This mechanical stretching of the cytoskeletal fibers and their attachments results in chemical signals that instruct the nucleus to express those genes for producing the proteins and other extracellular macromolecules required to sustain the deforming forces. In recent years this has been called the *tensegrity principle*.[47]

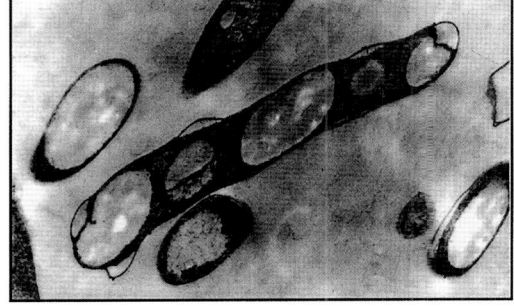
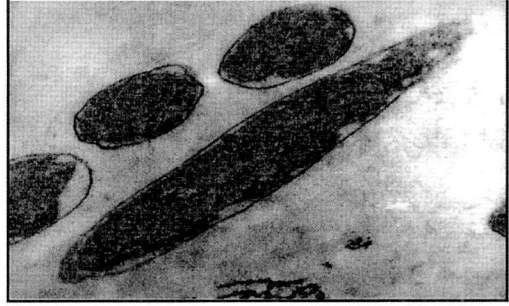

FIGURE 2.19. Example of the production of the elastic-contractile model protein, $(GVGVP)_{121}$, by genetic engineering of *E. coli*. **A** Transformed *E. coli* shows large inclusion bodies of product having phase separated intracellularly. **B** Nontransformed *E. coli* shows no such inclusion bodies. (Reproduced with permission from Urry et al.[43])

An equivalent conversion of mechanical energy into chemical energy occurs with the elastic model proteins. Stretching of properly designed and cross-linked elastic model proteins, such as the one in Figure 2.20B, increases affinity for protons and removes protons from the surrounding solution. Equivalently, we believe that stretching could energize an attached phosphate to phosphorylate another molecule as part of the chemical signal to the nucleus.

As a test of this perspective, we prepared elastic model proteins with and without cell attachment sequences and cross-linked them to form elastic sheets called *bioelastic matrices*. Cells do not adhere to these matrices without the cell attachment sequences; cells do adhere to the bioelastic matrices containing the cell attachment sequences, as depicted in Figure 2.20. Furthermore, the cells spread out and grew to completely cover the surface. Perhaps even more significantly, cyclic stretching of the

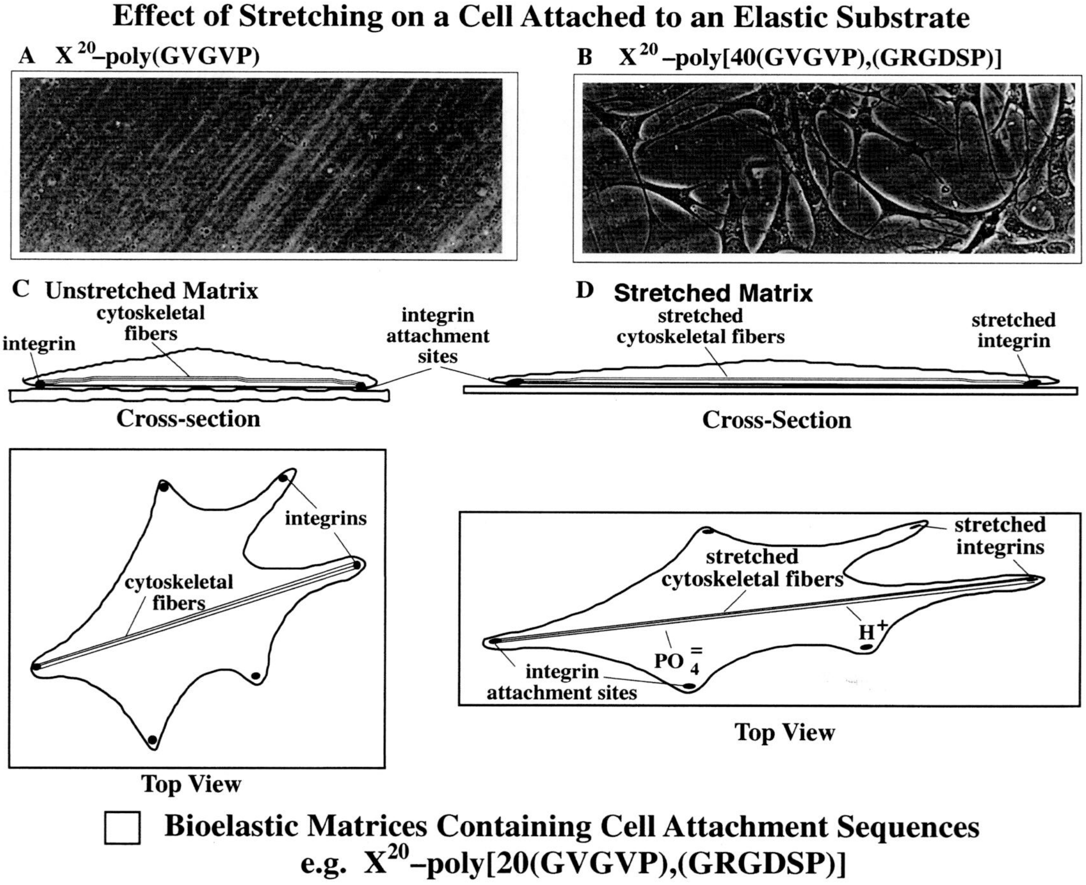

Effect of Stretching on a Cell Attached to an Elastic Substrate

A X^{20}-poly(GVGVP)

B X^{20}-poly[40(GVGVP),(GRGDSP)]

C Unstretched Matrix
cytoskeletal fibers, integrin, integrin attachment sites
Cross-section

D Stretched Matrix
stretched cytoskeletal fibers, stretched integrin
Cross-Section

Top View (integrins, cytoskeletal fibers)

Top View (stretched cytoskeletal fibers, stretched integrins, integrin attachment sites, $PO_4^=$, H^+)

☐ **Bioelastic Matrices Containing Cell Attachment Sequences**
e.g. X^{20}-poly[20(GVGVP),(GRGDSP)]

FIGURE 2.20. Elastic-contractile model proteins are used as temporary functional scaffolding for soft tissue restoration by means of an attached normal cell capable of mechanochemical transduction to restore a natural tissue. **A** Elastic matrices without cell attachment sequences show no attachment of human fibroblasts. **B** Elastic matrix with cell attachment sequences show human fibroblasts having attached, spread, and with capacity to grow to confluence. **C, D** Simultaneous stretching of matrix and cell causes cellular mechanochemical transduction. See text and Chapter 9 for further discussion. (Adapted from Urry.[46])

bioelastic matrix stimulates the appropriate attached cells to produce their own extracellular matrix, that is, to begin the reconstruction of a natural, functional tissue as shown in a simulated urinary bladder system.[44]

Thus, the potential exists for preparing bioelastic matrices that match the deformable properties of the tissue to be reconstructed, and the potential exists for the matrix to be a temporary functional scaffolding into which the natural cells migrate and attach. There the cells sense the forces that the tissue must sustain and begin reconstructing the appropriate natural tissue.

2.8.3 Drug Delivery

In one design for drug delivery, the drug itself provides the chemical energy to drive folding and contraction, that is, the drug provides the energy for its own packaging into the drug delivery vehicle. When the model protein contains negatively charged carboxylates under the influence of oil-like **R**-groups, the affinity for a positively charged drug increases just like the affinity increases for protons. Therefore, instead of adding proton, the positively charged drug is added to drive folding and contraction. The model protein separates out of solution holding the drug firmly to form the drug delivery vehicle. As the drug leaves the vehicle so too does the vehicle disperse. In effect, the drug provided the glue that held the delivery device together.

Many of the drugs that control pain, for example, the narcotics, are positively charged. The body has its own painkillers, referred to as *endorphins*; these are naturally released by the body after about 20 to 30 minutes of running and are considered responsible for the runner's high. An active part of the endorphins are the enkephalins. One, called Leu-enkephalin amide, is positively charged and has been loaded into a designed model protein with a strong oil-like sequence and with a carboxylate occurring every 15 amino acid residues, as shown in Figure 2.21. The release of this positively charged drug from a constant surface area in a test tube is shown in Figure 2.21 to maintain a near constant release for 3 months. Release extending out to a year has now been demonstrated.

With respect to control of drug addiction, the positively charged narcotic antagonist naltrex-

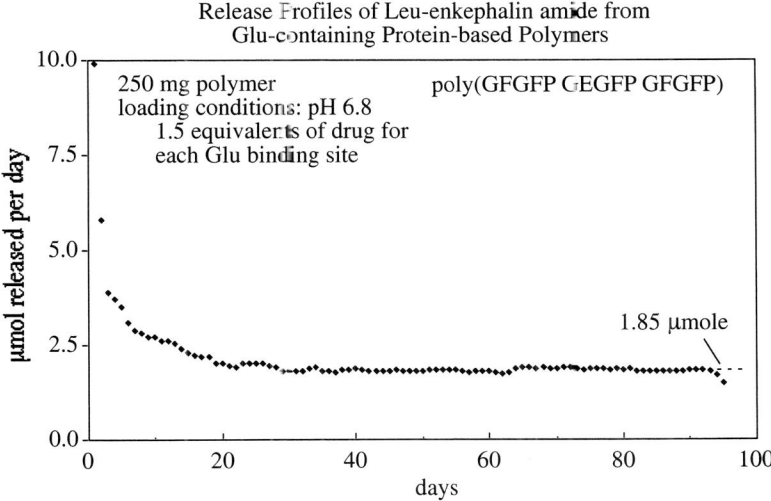

FIGURE 2.21. The designed elastic-contractile model proteins function by means of the apolar–polar repulsive free energy of hydration to achieve zero order release with simultaneous dispersal of delivery vehicle and to control the release level. See text and Chapter 9 for further discussion. (Reproduced from Urry et al.[48])

one, when in the blood, blocks the action of a millionfold higher dose of heroin. An addict with naltrexone in the bloodstream receives no high from heroin. Therefore, a controlled release of naltrexone from an implant in the body to the bloodstream of an addict, after withdrawal, would guard against return to dependency.

2.8.4 Smart Plastics

A smart plastic would harmlessly disintegrate once its useful Life were completed. Plastics made of plastic-contractile model proteins with controllable inverse temperature transitions can be designed as smart plastics. A smart protein-based plastic, having fulfilled its role, would swell and become a fragile, edible gelatin-like substance. Rather than foretell death for the fishes, a smart protein-based plastic could provide food for the fishes, once its useful Life as a plastic were complete.

The natural amino acids asparagine and glutamine provide chemical clocks for the disintegration of protein-based plastics; they have the **R**-groups $-CH_2CONH_2$ and $-CH_2CH_2CONH_2$, respectively. The carboxamide functional group, $-CONH_2$, hydrolyzes to negatively charged carboxylates, $-COO^-$, at predesigned rates. Half of the carboxamides will convert to carboxylates in as short a time as a few days or in as long a time as 10 years. The time can be set by the choice of the amino acid residues immediately preceding and following in the sequence and by the general oil-like character of the model protein. The desired half-life can be designed into the sequence of the protein-based plastic. As the carboxylates form from carboxamides at the surface, the plastic swells to a fragile, edible gel. Protein-based plastics could become food for marine life.

Examples of protein-based plastics with transition temperatures below 0°C are given in Figure 2.22 and Figure 1.10; they remain strong plastics in water until the chemical clocks, that is, the timed conversion of carboxamides to carboxylates, raise the transition temperature from below to above the temperature of the water. At this point, the plastics, having fulfilled their useful life, gracefully soften and disappear.

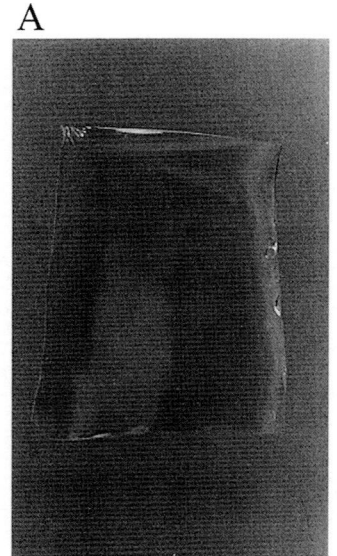
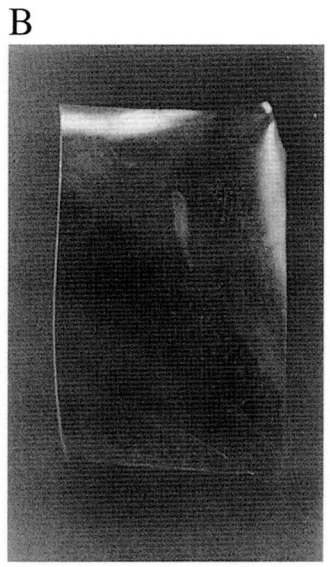
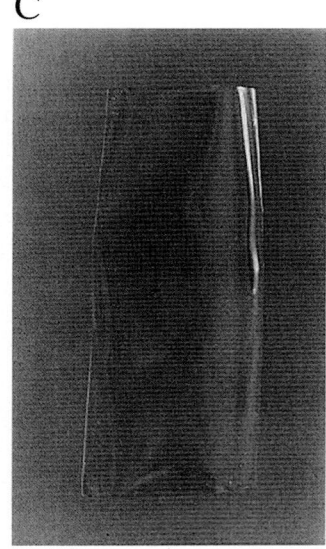

FIGURE 2.22. By changing one of the Gly residues of GVGVP to an L-amino acid residue, for example, to AVGVP, results in a new family of model proteins that become programmably biodegradable inverse thermoplastics. The three examples given are poly(FVGVP), poly(IVGVP), and poly(VVGVP). (D.W. Urry and D.C. Gowda, unpublished results.)

References

1. D.W. Urry, "Elastic Biomolecular Machines: Synthetic Chains of Amino Acids, Patterned After Those in Connective Tissue, Can Transform Heat and Chemical Energy into Motion." *Sci. Am.* January **1995**, 64–69.
2. Vinegar is a solution primarily of acetic acid, CH_3COOH, and water. As shown by Pasteur in 1864, vinegar is made from alcohol by certain bacteria under aerobic conditions.
3. C.B. Anfinsen, "Principles that Govern the Folding of Protein Chains." *Science*, **181**, 223–230, 1973.
4. D.W. Urry, "Protein Folding and the Movements of Life." *The World & I*, (Natural Science, At The Edge), **6**, 301–309, 1991.
5. A. Szent-Györgyi, "Studies on Muscle." *Acta Physiol. Scand.*, **9 (suppl XXV)**, 1–116, 1945.
6. D.W. Urry, "Molecular Machines: How Motion and Other Functions of Living Organisms Can Result from Reversible Chemical Changes." *Angew. Chem.* (German) **105**, 859–883, 1993; *Angew. Chem. Int. Ed. Engl.*, **32**, 819–841, 1993.
7. D.W. Urry, "Physical Chemistry of Biological Free Energy Transduction as Demonstrated by Elastic Protein-based Polymers," invited Feature Article, *J. Phys. Chem. B*, **101**, 11007–11028, 1997.
8. C.S. Roy, "The elastic properties of the arterial wall." *J. Physiol.*, **3**, 125–159, 1880.
9. D.W. Urry and T.M. Parker, "Mechanics of Elastin: Molecular Mechanism of Biological Elasticity and its Relevance to Contraction." *J. Muscle Res. Cell Motil.*, **23**, 541–557, 2002: Special Issue, *Mechanics of Elastic Biomolecules*. H. Granzier, M. Kellermayer, W. Linke, Eds.
10. L.B. Sandberg, N.T. Soskel, and J.B. Leslie, "Elastin Structure, Biosynthesis and Relation to Disease States." *N. Engl. J. Med.*, **304**, 566–579, 1981.
11. H. Yeh, N. Ornstein-Goldstein, Z. Indik, P. Sheppard, N. Anderson, J.C. Rosenbloom, G. Cicila, K. Yoon, and Rosenbloom, "Sequence Variation of Bovine Elastin mRNA Due to Alternative Splicing." *J. Collagen Rel. Res.*, **7**, 235–247, 1987.
12. It has now been possible to stretch a single elastic protein-based polymer chain and to obtain a uniformly increasing force versus extension curve.[13,14] This was done with a surprisingly simple device called an *atomic force microscope*, the development of which resulted in the Nobel prize for Paul Hansma. The performance of the mechanical work of lifting a weight, shown in Figure 2.4, utilized one-sixteenth of an inch thick and one-fourth of an inch wide elastic bands. We are now working to develop twisted filaments of some three chains about a millionth of an inch wide to perform similar energy conversions.
13. D.W. Urry, T. Hugel, M. Seitz, H. Gaub, L. Sheiba, J. Dea, J. Xu, and T. Parker, "Elastin: A Representative Ideal Protein Elastomer." *Philos. Trans. R. Soc. Lond. B*, **357**, 169–184, 2002.
14. D.W. Urry, T. Hugel, M. Seitz, H. Gaub, L. Sheiba, J. Dea, J. Xu, L. Hayes, F. Prochazka, and T. Parker, "Ideal Protein Elasticity: The Elastin Model." In *Elastomeric Proteins: Structures, Biomechanical Properties and Biological Roles*. P.R. Shewry, A.S. Tatham, and A.J. Bailey, Eds. Cambridge University Press, The Royal Society; Chapter Four, pages 54–93, 2003.
15. E.O. Wilson *Consilience: The Unity of Knowledge*. Alfred E. Knopf, New York, 1998, p. 8.
16. D.W. Urry, M.M. Long, and H. Sugano, "Cyclic Analog of Elastin Polyhexapeptide Exhibits an Inverse Temperature Transition Leading to Crystallization." *J. Biol. Chem.*, **253**, 6301–6302, 1978.
17. M.V. Stackelberg and H.R. Müller, "Zur Struktur der Gashydrate." *Naturwissenschaften*, **38**, 456–458, 1951.
18. D.W. Urry, S.Q. Peng, J. Xu, and D.T. McPherson, "Characterization of Waters of Hydrophobic Hydration by Microwave Dielectric Relaxation." *J. Am. Chem. Soc.*, **119**, 1161–1162, 1997.
19. Entropy measures the change in order; a positive change in entropy measures increased disorder, and a negative change in entropy measures increased order. For the transition where ice melts to form water at 0°C (273°K on the absolute scale where motion ceases at 0°K), the increase in entropy is the experimentally determined heat absorbed during the transition divided by the temperature for the transition, that is, 273°K (see Chapter 5 for a more complete discussion).
20. E. Schrödinger, *What is Life?* Cambridge University Press, Cambridge, England, first published in 1944, Canto edition with "Mind and Matter" and Autobiographical Sketches, Forward by R. Penrose, 1992.
21. It should be recognized that a change in entropy, ΔS, is a readily determined experimental quantity. The measured heat of the transition, ΔH, the heat in calories required to drive the folding represented in Figures 2.1 and 2.2, is simply divided by the temperature, T_t, at which the folding transition occurs, that is, $\Delta H/T_t = \Delta S$.

22. J.A.V. Butler, "The Energy and Entropy of Hydration of Organic Compounds." *Trans. Faraday Soc.*, **33**, 229–238, 1937.
23. Commonly for elastic contractile model proteins in aqueous solutions, the magnitude of the decrease in the transition temperature, T_t, is proportional to the negative entropy change for the transition. The increased hydrophobicity that lowers T_t translates into a greater exothermic heat change, ΔH_t, for the transition, that is, ΔH_t(after) − ΔH_t(before) is a negative quantity. The measure of the entropy change for the transition, ΔS_t, is simply $\Delta H_t/T_t$. Within this measured increase in entropy for the system of model protein plus water, however, is a negative entropy change due to the increase in order resulting from model protein folding and assembly. Thus, the energy sources that effect energy conversion by this mechanism do so by means of increasing negative entropy to the model protein part of the system.
24. For more details, see, for example, D. Voet, J.G. Voet, and C.W. Pratt, *Fundamentals of Biochemistry*. John Wiley & Sons, New York, 1999, pp. 529–561.
25. The structure of the functional component of the nicotinamides of photosynthesis and respiration that undergoes oxidation and reduction is given in Figure 3.6.
26. For more details, see, for example, D. Voet, J.G. Voet, and C.W. Pratt, *Fundamentals of Biochemistry*. John Wiley & Sons, New York, 1999, pp. 492–528.
27. As is now taught in middle school, living organisms oxidize glucose to produce the chemical energy of ATP, the universal energy currency of biology, whereas in our fireplaces the burning of wood, which is the oxidation of cellulose, a polymer of the same glucose molecule, provides thermal energy for heating our homes.
28. D.W. Urry, L.C. Hayes, and D. Channe Gowda, "Electromechanical Transduction: Reduction-driven Hydrophobic Folding Demonstrated in a Model Protein to Perform Mechanical Work." *Biochem. Biophys. Res. Commun.*, **204**, 230–237, 1994.
29. A. Pattanaik, D. Channe Gowda, and D.W. Urry, "Phosphorylation and Dephosphorylation Modulation of an Inverse Temperature Transition." *Biochem. Biophys. Res. Commun.*, **178**, 539–545, 1991.
30. D.W. Urry, D. Channe Gowda, S.Q. Peng, and T.M. Parker, "Non-linear Hydrophobic-induced pKa Shifts: Implications for Efficiency of Conversion to Chemical Energy." *Chem. Phys. Lett.*, **239**, 67–74, 1995. A millionfold increase in affinity is what would be required for pumping protons into the stomach.
31. D.W. Urry, L.C. Hayes, D. Channe Gowda, S.Q. Peng, and N. Jing, "Electrochemical Transduction in Elastic Protein-based Polymers." *Biochem. Biophys. Res. Commun.*, **210**, 1031–1039, 1995.
32. For more general details, see, for example, D. Voet, J.G. Voet, and C.W. Pratt, *Fundamentals of Biochemistry*. John Wiley & Sons, New York, 1999, pp. 180–186. Muscle contraction is also treated in some detail in Chapter 8.
33. P. Mitchell, "Keilin's Respiratory Chain Concept and its Chemiosmotic Consequences." *Science*, **206**, 1148–1159, 1979.
34. I.Z. Steinberg, A. Oplatka, and A. Katchalsky, "Mechanochemical Engines." *Nature*, **210**, 568–571, 1966.
35. Interestingly, on the basis of the differences in the ΔG_{HA} values for the Asp–COOH and Asp-COO$^-$ in Table 5.3, protonation causes Asp to become more oil-like and to seek out an oil-like site such as the lipid bilayer of the thylakoid and inner mitochondrial membranes. The change in ΔG_{HA} values times ten gives the calculated energy input for the transport of ten protons to drive one complete rotation of the F_0-motor, which would be 38 kcal/mol-rotation. As a single rotation of the F_0-motor can synthesize three ATP molecules from ADP plus P_i, the chemical energy output would be 24 kcal/mol-rotation, which gives a maximum possible efficiency of $100 \times 24/38$ or 63%.
36. D.W. Urry, "Free Energy Transduction in Polypeptides and Proteins Based on Inverse Temperature Transitions." *Prog. Biophys. Mol. Biol.*, **57**, 23–57, 1992.
37. D.M. Himmel, S. Gourinath, L. Reshetnikova, Y. Shen, A.G. Szent-Györgyi, and C. Cohen, "Crystallographic Findings on the Internally Uncoupled and Near Rigor States of Myosin: Further Insights into the Mechanics of the Motor." *Proc. Natl. Acad. Sci. U.S.A.*, **99**, 12645–12650, 2002.
38. It is a particular pleasure to point out in this context that Andrew G. Szent-Györgyi is the son of Albert Szent-Györgyi, the notable of section 2.2.1 who was perhaps the first person to see the motion of Life outside of the living organism, when he isolated actomyosin and added the chemicals to cause contraction.[5]
39. A.R. Fersht, "The Charging of tRNA," In *Accuracy in Molecular Processes: Its Control and Rel-*

References

evance to *Living Systems*, T.B.L. Kirkwood, R.F. Rosenberger, and D.J. Galas, Eds., Chapman and Hall, London, 1986, pp. 67–82.

40. D.W. Urry and H. Eyring, "Stereochemistry and Rate Theory in Protein Synthesis." *Arch. Biochem. Biophys.*, **Suppl. 1**, 52–62, 1962.

41. J. Bronowski, *The Ascent of Man*. Little, Brown and Company, Boston/Toronto, 1973, p. 110.

42. D.T. McPherson, J. Xu, and D.W. Urry, "Product Purification by Reversible Phase Transition Following *E. coli* Expression of Genes Encoding up to 251 Repeats of the Elastomeric Pentapeptide GVGVP." *Protein Expression Purification*, **7**, 51–57, 1996.

43. D.W. Urry, D.T. McPherson, J. Xu, H. Daniell, C. Guda, D.C. Gowda, N. Jing, T.M. Parker, "Protein-based Polymeric Materials: Syntheses and Properties." In *The Polymeric Materials Encyclopedia: Synthesis, Properties and Applications*, CRC Press, Boca Raton, pp. 7263–7279, 1996.

44. D.W. Urry and A. Pattanaik, "Elastic Protein-based Materials in Tissue Reconstruction." *Ann. N.Y. Acad Sci.*, **831**, 32–46, 1997.

45. D.W. Urry, "Engineers of Creation," *Chemistry in Britain*, **32**, 39–42, 1996.

46. Dan W. Urry, "Elastic Molecular Machines in Metabolism and Soft Tissue Restoration," *TIBTECH*, **17**, 249–257 (1999).

47. N. Wang, J.P. Butler, and D.E. Ingber, "Mechanotransduction across the cell surface and through the cytoskeleton." *Science*, **260**, 1124–1127, 1993.

48. D.W. Urry, A. Pattanaik, M.A. Accavitti, C-X. Luan, D.T. McPherson, J. Xu, D.C. Gowda, T.M. Parker, C.M. Harris, and N. Jing, "Transductional Elastic and Plastic Protein-based Polymers as Potential Medical Devices," in *Handbook of Biodegradable Polymers*, ed. by Domb, Kost, and Wiseman, Harwood Academic Publishers, Chur, Switzerland. pp. 367–386, 1997.

3
The Pilgrimage: Highlights in Our Understanding of the Products and Components of Living Organisms

3.1 Initial Glimpses

3.1.1 The First Ionian Enlightenment: Thales of Miletus

Historians point to a defined *period*, *place*, and *person* where began the ascent of man toward an empowering understanding of his world.[1] The period was the 6th century BC. The place was *Miletus* on the Aegean Sea in greater Greece, and the person was *Thales*. In the 6th century before the birth of Christ, Thales of Miletus forged the first Ionian revolution and began dispelling the illusions of knowledge embodied in Greek mythology. No longer would myth unchallenged provide an acceptable explanation of natural phenomena. No longer would Hesiod's accounts of the Greek Gods—Chaos, Gaea (Earth), and Eros (Desire); and of Gaea's giving birth to the Heavens (Uranus), the Mountains, and the Sea (Pontus)—satisfy the desire for understanding nature.

In its place Thales of Miletus launched the sense of inquiry with the simple question, "What is the world made of?" His simple rejoinder "water" has been given explanation; Thales of Miletus reasoned "that the principal is water … getting the notion perhaps from seeing that the nutriment of all things is moist and kept alive by it … and from the fact that the seeds of all things have a moist nature and that water is the nature of moist things."[2]

As regards living things and as advanced in this volume, water is the vital ingredient that uniquely enables efficient function of biomolecular machines. Water dissolves the more polar parts of chain biomolecules and, in a uniquely constrained way, transiently hydrates "water-insoluble" oil-like parts of biomolecules in a molecular interplay that powers protein-based machines. Biomolecular machines sustain Life, and, in an analogous manner, man's machines sustain society. The requirements for sustaining both Life and society necessitate access to adequate energy sources.

3.1.2 Staying the Prediction of Malthus

In 1798, Thomas Robert Malthus, in "An Essay on the Principle of Population" made two postulates: "First, That food is necessary to the existence of man. Secondly, That the passion between the sexes is necessary and will remain nearly in its present state."[3] Malthus then expanded upon his postulates: "I say that the power of population is indefinitely greater than the power in the earth to produce subsistence for man. Population, when unchecked, increases in a geometrical ratio. Subsistence increases only in an arithmetic ratio."

Malthus further argued, "The race of plants and the race of animals shrink under this great restrictive law. And the race of man cannot, by any efforts of reason, escape from it. Among plants and animals its effects are waste of seed, sickness, and premature death. Among mankind, misery and vice." How then has the world for 200 years stayed the prediction of Malthus?

3.1 Initial Glimpses

How does the earth sustain populations greater and standards of living higher than conceivable to Malthus? Machines and the energies that power them sustain Life and sustain society. Technological developments hold back the most disastrous devastations and derive sustenance from the earth and sun to provide standards of living beyond comprehension 200 years ago. By way of illustration, western technological advancements, descendent from that first Ionian revolution, find the oil, pump the oil out of the earth's crust, refine the oil, design the oil-powered machines, and thereby provide the energy that power the machines and increase the earth's capacity to sustain society.

By acquiring sufficient energy resources to fuel the molecular machines of biology (to provide food sources) and to fuel the man-made machines of society, technology has stayed the Malthusian catastrophe. Popular energy resources are not limitless, however, and associated pollutions increasingly challenge our biosphere. In this volume, we examine the biomolecular machines of Life and then, in an as yet very preliminary way, explore specific means whereby efficient, environmentally friendly biomolecular machines may help sustain society.

3.1.3 A Process for Obtaining Meaningful Answers

Beyond asking a meaningful question follows a process for determining a meaningful answer. That process has become known as the *scientific method*. At a fundamental level the Periodic Table of the Elements answers the question posed 26 centuries earlier by Thales of Miletus. Mendeleev formulated the periodic law for the elements circa 1870 (see Figure 3.1), yet decades passed before this foundation of chemical theory became generally accepted. Successful prediction and subsequent verification of additional elements not known in 1870 resulted in general acceptance and the establishment of the periodic law as represented in the Periodic Table of the Elements of Nature

TABELLE II

REIHEN	GRUPPE I. — R^2O	GRUPPE II. — RO	GRUPPE III. — R^2O^3	GRUPPE IV. RH^4 RO^2	GRUPPE V. RH^3 R^2O^5	GRUPPE VI. RH^2 RO^3	GRUPPE VII. RH R^2O^7	GRUPPE VIII. — RO^4
1	H=1							
2	Li=7	Be=9,4	B=11	C=12	N=14	O=16	F=19	
3	Na=23	Mg=24	Al=27,3	Si=28	P=31	S=32	Cl=35,5	
4	K=39	Ca=40	—=44	Ti=48	V=51	Cr=52	Mn=55	Fe=56, Co=59, Ni=59, Cu=63.
5	(Cu=63)	Zn=65	—=68	—=72	As=75	Se=78	Br=80	
6	Rb=85	Sr=87	?Yt=88	Zr=90	Nb=94	Mo=96	—=100	Ru=104, Rh=104, Pd=106, Ag=108.
7	(Ag=108)	Cd=112	In=115	Sn=118	Sb=122	Te=125	J=127	
8	Cs=133	Ba=137	?Di=138	?Ce=140	—	—	—	
9	(—)	—	—	—	—	—	—	
10	—	—	?Er=178	?La=180	Ta=182	W=184	—	Os=195, Ir=197, Pt=198, Au=199
11	(Au=199)	Hg=200	Tl=204	Pb=207	Bi=208	—	—	
12	—	—	—	Th=231	—	U=240	—	

- 25 centuries passed before Mendeleev answered, in part, the question of Thales of Miletus.
- Meaningful questions giving rise to experiments that lead to meaningful answers with verifiable predictions become the foundation of scientific inquiry.
- Verifiable predictions become empowering results on which science and societies advance.

FIGURE 3.1. The Periodic Table of the Elements: Mendeleev's last version as proposed in 1871–1872 with spaces left for elements that he deduced existed but that were unknown at the time. This provides a substantive answer to the question posed by Thales of Miletus and in doing so constitutes the beginning of one of the most fundamental systematizations of science. (Obtained from http://www.periodic.lanl.gov/mendeleev.htm)

(Figure 3.2). Meaningful prediction becomes the standard within the scientific method for establishing meaningful understanding. Our purpose here is not to elaborate on the scientific method, but to demonstrate the scientific method in action in the process of developing and substantiating a new basis for the function of biomolecular machines. As for the question of Thales of Miletus, the fundamental answer is that the world is made of more than 100 atomic elements. Once the periodic law of the elements was accurately established, it was determined forever; it never has to be redone. *The laws of nature are constant.* Quite unlike the laws emanating from our legislative bodies, once established, the determined laws of nature provide empowering understanding, but they remain so only as long as the historical record remains accessible.

For biology, the most common elements are carbon (C), hydrogen (H), oxygen (O), nitrogen (N), and, to a more limited extent, sulfur (S). These lighter atomic elements combine to form the molecules of biology, including those of its protein-based machines. Our task is to understand how the molecules, especially the macromolecules of biology, function to sustain Life. Before presenting an unfolding understanding of biomolecular machines, however, the mainstay of our pilgrimage recounts steps in the development of our knowledge of the molecules of biology.

3.1.4 Historical Proscription

There was a time, indeed quite recent in the ascent of man,[4] when products and components

Periodic Table of the Elements of Nature

1 H																	2 He
3 Li	4 Be											5 B	6 C	7 N	8 O	9 F	10 Ne
11 Na	12 Mg											13 Al	14 Si	15 P	16 S	17 Cl	18 Ar
19 K	20 Ca	21 Sc	22 Ti	23 V	24 Cr	25 Mn	26 Fe	27 Co	28 Ni	29 Cu	30 Zn	31 Ga	32 Ge	33 As	34 Se	35 Br	36 Kr
37 Rb	38 Sr	39 Y	40 Zr	41 Nb	42 Mo	43 Tc	44 Ru	45 Rh	46 Pd	47 Ag	48 Cd	49 In	50 Sn	51 Sb	52 Te	53 I	54 Xe
55 Cs	56 Ba	57 La	72 Hf	73 Ta	74 W	75 Re	76 Os	77 Ir	78 Pt	79 Au	80 Hg	81 Tl	82 Pb	83 Bi	84 Po	85 At	86 Rn
87 Fr	88 Ra	89 Ac	104 Rf	105 Db	106 Sg	107 Bh	108 Hs	109 Mt	110 110	111 111	112 112						
		58 Ce	59 Pr	60 Nd	61 Pm	62 Sm	63 Eu	64 Gd	65 Tb	66 Dy	67 Ho	68 Er	69 Tm	70 Yb	71 Lu		
		90 Th	91 Pa	92 U	93 Np	94 Pu	95 Am	96 Cm	97 Bk	98 Cf	99 Es	100 Fm	101 Md	102 No	103 Lr		

Other nonmetals	Alkali earth metals	Other metals	Noble gases
Alkali metals	Transition metals	Rare earth metals	Halogens

FIGURE 3.2. The Periodic Table of the Elements of Nature. This table constitutes today's most complete answer at the level of the elements, the atoms, to the question posed by Thales of Miletus 26 centuries ago. The spaces left by Mendeleev have been filled, and many more elements have been added.

3.1 Initial Glimpses

of living organisms were considered beyond man's competence to comprehend and beyond man's potential to produce. This boundary beyond which man's capacity was not to extend prevailed long after the building of the great gothic cathedrals. This barrier continued to stand even after many of the profound advances in science and medicine had withstood the test of centuries—even after the rousing advances of Vesalius, Copernicus, Galileo, Harvey, Malpighi, and Newton at renowned Universities like Padua, Bologna, and Cambridge.

In 1543, nearly three centuries before what was to be the accidental rout of this particular restraint to progress, two historic works were published. In the first great book of modern medical science, *Humana Corpora Fabrica*, Vesalius ignored the constraints against human dissection, accurately described human anatomy, and thereby overturned 13 centuries of Galenist dogma.[5] In an antithetical view of the cosmos entitled *On the Revolutions of the Celestial Spheres*, Copernicus removed the earth from the center of the universe.

Exactly two centuries before eventual biological enlightenment, in 1628, Harvey[6] published *De Motus Cordis* showing the heart to be not the seat of emotion but rather a mechanical pump. Harvey further predicted the existence of capillaries to connect the outward flow from the heart through the arteries to the return flow through the veins, as subsequently confirmed microscopically by Malpighi in 1661.

The Copernican view of the heavens had been championed by Galileo, refined by Kepler, and remarkably extended to a set of axioms by Newton with the three laws of mechanics, the resulting formulation of a universal law of gravitation, and the theory of the tides. Newton, who is viewed as the father of modern physics, presented much of this in 1687 in his book *Principia*, often considered "one of the most important works of science ever written," thus capping the Scientific Revolution of the seventeenth century.

Despite all this, more than another century was yet to pass before a product or component of a living organism was to be within man's capacity to produce and to comprehend.

3.1.5 Accidental Synthesis of Urea

The turning point, the first synthesis of a product of living organisms, occurred in 1828 when Frederick Wöhler,[7] a chemist working in Berlin, accidentally synthesized urea. Wöhler heated ammonium cyanate, NH_4OCN, a well-known inorganic compound comprised of the NH_4^+ and OCN^- ions. He recognized the rearrangement product of heating as urea, H_2NCONH_2, the natural non-ionic urinary excretory product of mammals, now known to be the primary nitrogen-containing product of protein metabolism.

This first organic[8] synthesis began an extraordinary odyssey, emerging highlights of which continue to be greeted regularly with excitement by the press, principally because of their potential significance to health care, to the economy, and to the environment. This proscribed yet singular, accidental synthesis of urea has transformed into the crescendo that is the Biotechnological Revolution of today.

Now man has the capacity to make, by chemical synthesis at the laboratory bench, the major products and components of living organisms. Man has come to understand much of the physical and chemical functioning of living organisms. With this understanding, he can design and prepare model proteins to emulate the set of energy conversions that living organisms require for survival. Furthermore, as a hallmark of the Biotechnological Revolution, man can transform living organisms into chemical factories to work in society's behalf, for example, to produce natural as well as newly designed products of use to society.

From the latter comes the fated promise to replace the finite petroleum reserves and to prepare products in an environmentally friendly manner to sustain an increasingly complex society. It is now possible, by genetic engineering, to produce proteins that evolution has never called upon nature to prepare and to produce protein-based materials for addressing primary problems required for sustaining our society.

3.1.6 Pasteur and Molecules with Mirror Images

From within the reddish sediment of wine casks and the genius of Pasteur began our understanding that carbon-containing molecules can exist as mirror images, just as the left hand is the mirror image of the right hand (see Figure 3.3). Pasteur found that the natural tartaric acid[9] molecule, from the red sediment of wine ferment, rotated the plane of polarized light in a rightward manner (dextro-rotatory), but that

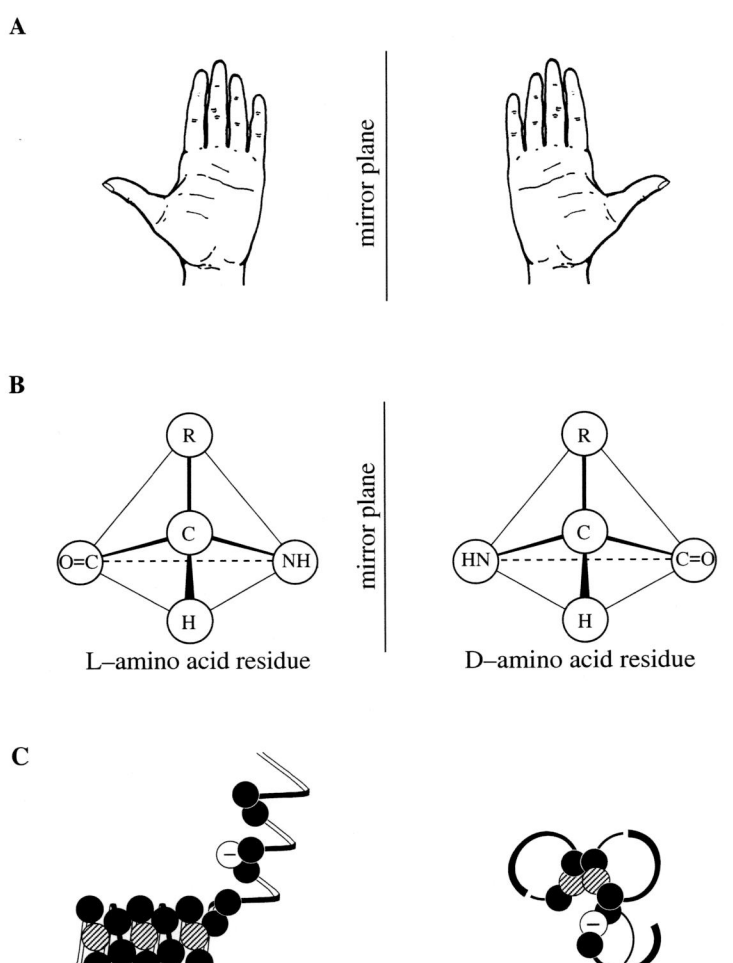

FIGURE 3.3. **A** Demonstration that the mirror image converts a left hand into a right hand, but the hands are otherwise identical. When a molecule has more than one asymmetrical center, however, as occurs commonly in carbon-containing molecules of biology, for each asymmetrical center, x, there are 2^x different structures as may be recognized by the Fischer projection of glucose in Figure 3.4. **B** The two optical (mirror image) isomers of the tetrahedral carbon atom of amino acid residues, designated as an L-amino acid residue on the left (the one that occurs in biology's genetically produced protein) and as a D-amino acid residue on the right (an optical isomer that does occur in biology, but in those peptides not encoded for by the genetic code). **C** The effect of insertion of a D-amino acid residue in an otherwise L-amino acid residue protein in the β-spiral structure of the elastic-contractile model protein of our focus would be to disrupt the regular structure. This is difficult to avoid completely in chemical synthesis, and it increases the temperature at which occurs the inverse temperature transition and decreases the heat of the transition due to less optimal association of oil-like groupings.

it no longer rotated the plane of polarized light after heating for several hours.[10] He learned this after having found that crystals of the inactive (paratartaric or racemic acid) product of heating existed as left-handed and right-handed mirror images and that solutions of the separated mirror image crystals rotated the plane of polarized light in opposite directions.

Most significantly for understanding the products and composition of living organisms, when Pasteur set the racemic mixture of mirror image crystals to ferment, the living organisms of the ferment consumed only the right-handed (dextro-rotatory) acid molecules leaving behind the left-handed (levo-rotatory) acid molecules. In every other physical and chemical property these molecules were identical, yet living organisms were able to distinguish between them, consuming one and leaving the other.

In the words of Pasteur, "Thus we find introduced into physiological principles and investigations the idea of the influence of the molecular asymmetry of natural organic products, of this great character which establishes perhaps the only well marked line of demarcation that can at present be drawn between the chemistry of dead matter and the chemistry of living matter."[8] In fact, we now know that for each asymmetrical carbon atom (a carbon atom with four different chemical groups attached) there are two optical isomers. Thus, for molecules such as glucose, with four asymmetrical carbon atoms, there would be 2^4 optical isomers. Biological molecules are comprised of but one of the possible optical isomers. This fact is fundamental to the success and efficiency with which proteins function as molecular machines accessing available energy and converting it to biologically essential functions.

Thus, in little more than three decades after Wöhler's accidental synthesis of urea, it was learned that living organisms exhibited a unique capacity to distinguish a molecule from its mirror image. So profoundly indicative of Life is the handedness of carbon-containing molecules that a search for the existence of Life in outer space once centered on finding such optically active molecules in matter-like meteorites. Although hydrocarbons have been reported in meteorites, the occurrence of molecules in meteorites that rotate the plane of polarized light has yet to be proven. With the heating that a meteorite sustains on entering the earth's atmosphere, it is not unlikely that racemization would occur as it did for the tartaric acid of Pasteur's original observations.

3.1.7 The Carbon Atom Key to Mirror Image Molecules

The mirror images made possible by the carbon atom result from four different substituents bound to a single carbon atom. As shown in Figure 3.3B, the carbon atom is at the center of a tetrahedron with a different substituent at each of the four apices. One of the two backbone carbon atoms of the amino acid residues shown in Figure 2.1A has four different substituents, the H atom, the C=O, the NH, and the R-group. Therefore, amino acid residues occur with mirror images. The substituents are "placed at the summits of an irregular tetrahedron" as had been anticipated by Pasteur[11] and as shown in Figure 3.3B.

As the surface of a protein interacts with an L-amino acid residue in Figure 3.3B, for example, the correct orientation of chemical contacts must occur on the protein surface in order to identify and associate with any three points of contact with the residue. Regardless of the side of approach to the tetrahedron *with four different substituents*, the three approachable substituents are arranged in a clockwise order in one case and a counterclockwise order for the mirror image. Therefore, different surfaces are required for proper fitting much as the handshake between two right hands fits whereas that between a left hand and a right hand does not.

In a protein sequence, the selection of only one of the two possible mirror image amino acid residues at each position is responsible for the regular folding and assembly of protein, such as the regular formation and association of helices shown in Figure 2.1C,D. Utilization of but one mirror image molecule makes possible the regular folding and assembly of proteins, which is responsible for the effective energy conversions to be discussed in subsequent chapters. The effect on folding and assembly of the inclusion of a single incorrect mirror image amino acid residue is depicted in Figure 3.3C, which disrupts the regularly folded and assembled structure depicted in Figure 2.1C,D. The random inclusion of very few mirror image monomers disrupts the formation of regular structures on which the efficiency of Life's energy conversions depends.

3.2 Determination of Biological Structures and Their Syntheses

Since the time of Wöhler (1828), the usual approach has been experimentally to determine the structure of a biological product by combination of chemical and physical methods and then to synthesize the product chemically as proof of structure. For the last half century, however, proof of structure has usually been achieved by determination of the crystal structure using X-ray diffraction methods, whereby a complete three-dimensional image of the molecule can be obtained.

3.2.1 Structure and Synthesis of Carbohydrate

The structure of the most common carbohydrate, the simple sugar D-glucose, was determined by the renowned chemist, Emil Fischer, in 1891 to be as shown in Figure 3.4. Viewing the six carbon atoms in the linear representation of glucose, four of the carbons contain four different substituents so that there are 16 (2^4) different optical isomers. On cyclizing to form the pyranoses, a fifth optically active carbon results with 64% forming the α-anomer and 36% the β-anomer for D-glucose. Fischer[12] synthesized each of the 16 optical isomers of glucose and demonstrated the biological form to be D-glucose, the one shown in Figure 3.4.

On combining into chain molecules with different attachments between the repeating units, D-glucose forms several well-known carbohydrates of remarkably different structural properties. There is *cellulose*, the primary component of wood with an estimated 10 trillion tons being produced annually through photosynthesis in trees. There are the *plant starches* (amylose and amylopectin) of cooking and nutrition note, and the *animal starch* glycogen, the stored source of energy in muscle.

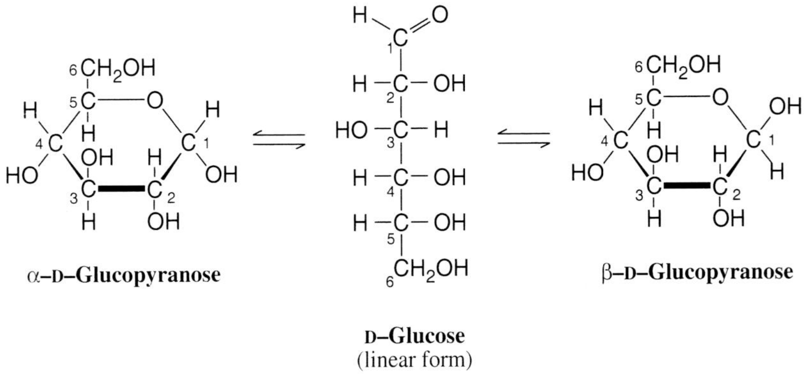

α-D-Glucopyranose   D-Glucose (linear form)   β-D-Glucopyranose

FIGURE 3.4. Structural representations of glucose, the major sugar and the primary source of energy of higher animals that gives rise to 36 ATP molecules during its oxidation to carbon dioxide and water. In the Fischer projection, the central structure, carbon atoms 2, 3, 4, and 5 each contains four different substituents, making them asymmetrical centers. Because each asymmetrical center can exist as either one of two mirror images, this means that there are $2^4 = 16$ distinguishable optical isomers. The single structure shown here is the structure of biological origin. Interactions with the active sites of enzymes distinguish this structure from the other 15.

3.2 Determination of Biological Structures and Their Syntheses

Each glucose monomer derived from glycogen gives rise to 36 ATP molecules[13] to provide the immediate energy source for functions such as muscle contraction, nucleic acid syntheses, protein synthesis, and maintaining the proper intracellular environment and extracellular matrix.

3.2.2 Limiting Optical Isomers (Mirror Images) Essential to Effective Function

Because D-glucose is for living organisms the pivotal molecular source of energy, its concentration in the organism must be carefully regulated. The result of inadequate regulation of glucose in humans is the disease diabetes mellitus. The diagnosis of diabetes mellitus—once nearly a certain death sentence arising from complications of cardiovascular disease, kidney disease, blindness, and frequent infection— transformed into the prognosis of a long and healthy life with the discovery of insulin.

The small protein insulin maintains the correct blood level of D-glucose, and the accurate recognition of a sugar such as D-glucose requires an intimate association of the sugar with a protein at each of its many sites of action. The intimacy of this interaction is such that a left-hand structure is distinguishable from a right-hand structure, again as apparent in a handshake. If there were not strict control of the optical isomers, one could imagine that control of glucose levels and the avoidance of disease would require 16 different systems; such a situation would be very cumbersome, inefficient, and more prone to disease by the factor of 16 raised to the power of the number of sites of interaction. It is difficult to imagine a successful organism under such circumstances.

3.2.3 Structures of Fats, Oils, and Membranes

The chemical syntheses of biological fats and oils do not lend themselves to such identifiable historic moments, but their preparation and reconstitution into functional membranes constitute achievements central to an understanding of cell function.

Commonly categorized as lipids, biological fats and oils are comprised of components such as fatty acids, glycerol, triglycerides, phospholipids, sterols, and steroids. They are of interest in energy conversion for two main reasons. First, nutritionally their metabolism provides two and one-half times more energy (calories) per gram than do carbohydrates and proteins. Second, they form the membranes that surround the living cell and that delimit internal organelles.

All cellular nutrients, excretory products, and components of the extracellular matrix must pass through the cell membrane. In addition, ions are pumped, certain ions into the cell and certain other ions out of the cell, to provide the required internal environment and to set the stage for response to stimuli. The ion pumps within the cell membrane constitute a very demanding aspect of energy consumption required to sustain the cell.

3.2.3.1 Fatty Acids

Fatty acids are simply the acetic acid molecules of vinegar, CH_3COOH, with long oil-like tails formed by the addition of hydrocarbons such as repeating saturated $-CH_2-CH_2-$ and unsaturated $-HC=CH-$ groups. A general example of saturated fats is $CH_3(CH_2)_nCOOH$, where n can vary from 10 to 22 (even values of n only). When n is 16, it is the primary fatty acid of red meats, called *stearic acid*. A common example of an unsaturated fatty acid is oleic acid, $CH_3(CH_2)_7HC=CH(CH_2)_7COOH$, the primary and beneficial component of olive oil.

3.2.3.2 Triglycerides

The attachment of three fatty acids to the three OH groups of glycerol, $CH_2OH-CHOH-CH_2OH$ as shown in Figure 3.5A, constitutes a triglyceride. This is the primary storage molecule of fat cells, and it is one of the lipids monitored during blood lipid chemistry determinations to assess risk of heart disease.

3.2.3.3 Phospholipids and Membranes

In phospholipids, two of the hydroxyls of the glycerol have attached fatty acids, whereas attached to the third hydroxyl is a hydrophilic, polar, or charged group. A common phospholipid, lecithin, shown in Figure 3.5B, forms the primary component of cell membranes. These molecules form so-called lipid bilayer membranes that constitute the membrane surrounding the living cell. They do so by the same separation of oil from vinegar and of oil from water described for proteins in the Chapter 2 overview and given a physical basis in Chapter 5. Once the lipid tail is long enough, there forms so much hydration around the oil-like moieties that the transition for assembly is below body temperature and the separation from water occurs with the geometry of the lipid bilayer membrane as the result. The relative size of the polar head group and the lipid tail determines whether the resulting structure is a near planar membrane or a sphere with an oily core and polar surface called a *micelle*.

The polar and charged phosphocholine part tries to stay in a watery environment, and the oil-like hydrocarbon chains of the fatty acids try to remain distant from the polar part and separated from water. The result is the assembly of the lipid bilayer membrane as an integral part of the cell membrane represented in Figure 3.6 with the oily tails of the lipids buried within and forming the oily membrane and the polar charged groups at the surface interacting with water. This involves the same forces of separation and segregation of oil-like groups from vinegar-like groups and oil from water that is central to the consilient mechanism of protein-

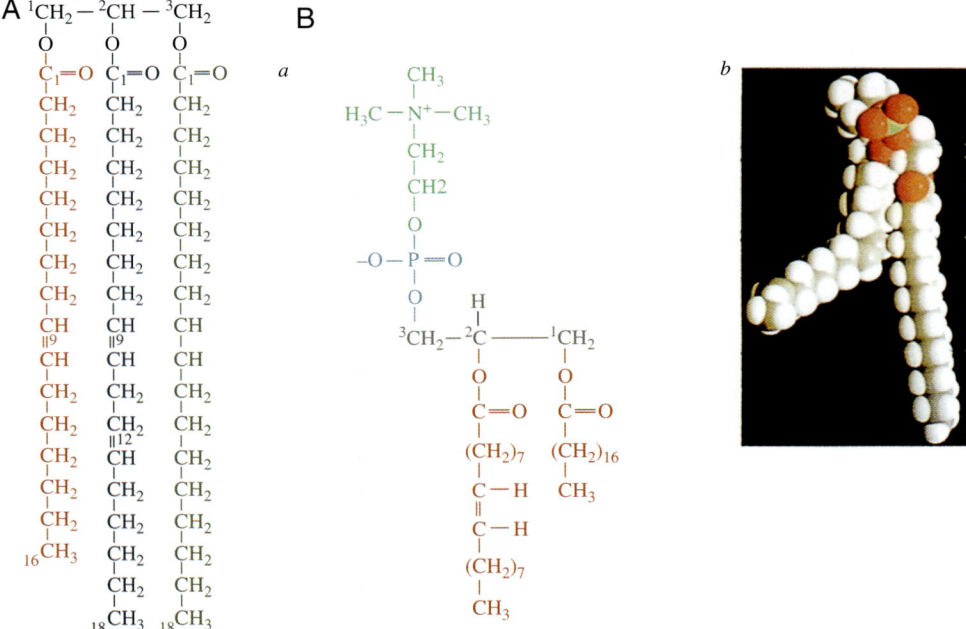

FIGURE 3.5. **A** A specific triglyceride comprised of a glycerol molecule, carbon atoms 1, 2, and 3, the hydroxyl functions of which are attached to three fatty acids. In this case it is three different fatty acids, two with different unsaturated, double bonds, and the third is the saturated fatty acid common in red meat. **Ba** A specific lecithin wherein the glycerol molecule has a charged phosphocholine group at carbon 3 and two different fatty acids, one saturated and the other unsaturated at carbons 1 and 2. **Bb** Space-filling model of the lecithin in (b) showing the effect of the double bond to be the introduction of a kink in an otherwise essentially linear hydrocarbon chain. (**B** Courtesy of Richard W. Pastor, FDA, Washington D.C. **A** and **B** from *Biochemistry*, Second Edition, D. Voet and J. Voet, Copyright © 1995, John Wiley & Sons, New York. Reprinted with permission of John Wiley & Sons, Inc.)

3.2 Determination of Biological Structures and Their Syntheses

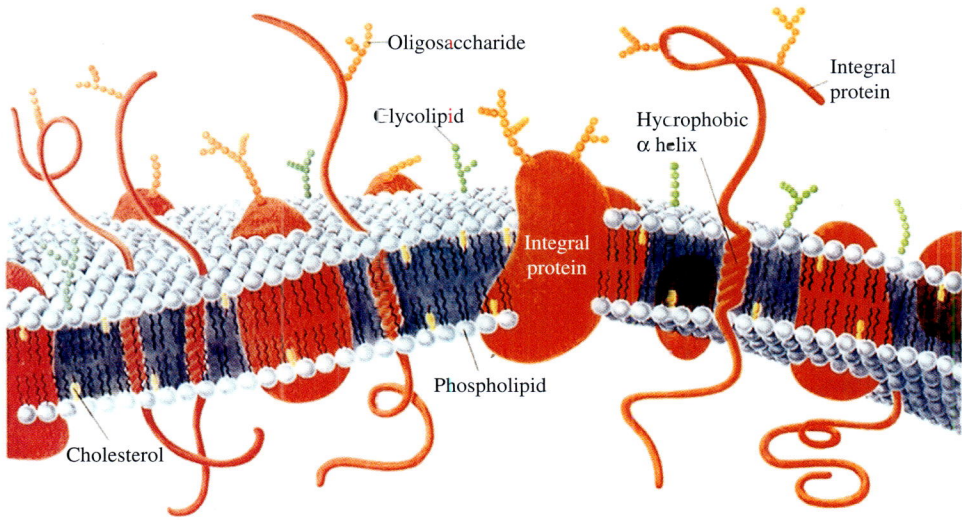

FIGURE 3.6. Cartoon of a cell membrane composed primarily of a lipid bilayer membrane, mostly containing the phospholipid lecithin, and decorated with occasional cholesterol molecules. Fundamental to the cell membrane function are integral membrane proteins that span from one side to the other of a cell membrane and commonly function in transmembrane transport and signaling, in cell recognition, and so forth. (From *Biochemistry*, Second Edition, D. Voet and J. Voet, Copyright © 1995, John Wiley & Sons, New York. Reprinted with permission of John Wiley & Sons, Inc.)

based machines and, as argued here, the primary driving forces for changes in protein structure that determine function.

3.2.4 The Essential Role of Protein Machines Within the Cell Membrane

As with proteins, the distribution of oil-like and vinegar-like groups dictates membrane structure. Furthermore, it is structure, and changes in structure, that dictate function, because charged species like protons, sodium ions, and potassium ions cannot pass through an intact oily lipid bilayer membrane of the cell without the cell membrane being specially modified by proteins called *integral membrane proteins* (see Figure 3.6). Integral membrane protein machines, therefore, control the passage of charged and other polar species into and out of the cell. The F_0-motor of ATP synthase represents perhaps the most important protein machine integral to the cell membrane, as it uses positively charged protons to produce some 90% of the ATP whereby living organisms function.

3.2.4.1 Protein Machines Pump Ions into and out of the Cell

With ATP as the source of energy, a particular protein in the cell membrane pumps two potassium ions, K^+, into the cell and three sodium ions, Na^+, out of the cell; another protein pumps calcium ions, Ca^{2+}, away from the contractile apparatus of muscle cells. The result is a polarized membrane with more negative charge inside than outside of the cell.

3.2.4.2 Protein Channels in Cell Membranes Open and Close to Allow Excitatory Flows of Ions

Still other integral membrane proteins, most common in the cell membrane of nerve cells, contain transmembrane channels that open with the proper stimuli, letting sodium ions rush in (depolarizing the membrane) as an excitatory action that activates neural networks required for thought processes. Additionally, nerve cell processes impinging on muscle cells depolarize and in turn activate channel openings in muscle cell membranes and ultimately trigger the rush of calcium ions into the muscle cell to initiate muscle contraction.

3.2.4.3 Protein-based Machines in the Thylakoid and Inner Mitochondrial Membranes

A flow of electrons in a series of cell membrane–associated oxidation and reduction cycles provides the energy source in the energy-converting thylakoid membranes of plants and inner mitochondrial membranes of plants and animals. In the process of this electron flow, membrane proteins pump protons across a cell membrane to increase the concentration of protons on one side of the membrane (to be considered in Chapter 8). Another protein-based machine uses the high concentration of protons on one side of a cell membrane to drive the formation of ATP as the protons pass to the low concentration side of the membrane (also to be considered in Chapter 8).

While defining the cell's outer boundary, the viable membrane, that is, one that contains the required complement of proteins functioning in the lipid bilayer membrane surrounding the cell, represents an essential component of the fundamental unit of Life. Life begins with the cell, and the cell no longer lives when the cell membrane has been rendered irreparably nonfunctional.

There is significance in the knowledge that man has synthesized the biological fats and oils; he has made the biological fats and oils into membranes, and he has incorporated proteins into the membranes with reconstitution of integral membrane protein function. As discussed below, man has also chemically synthesized protein. This completes the conceptual capacity to reconstitute the critical functional membrane component of the viable cell.

3.2.5 Syntheses of Sterols and Steroids

The total chemical synthesis of cholesterol (see Figure 3.7), the chief sterol of vertebrates and the essential precursor of the steroid hormones, was achieved in 1951 by R.B. Woodward, along

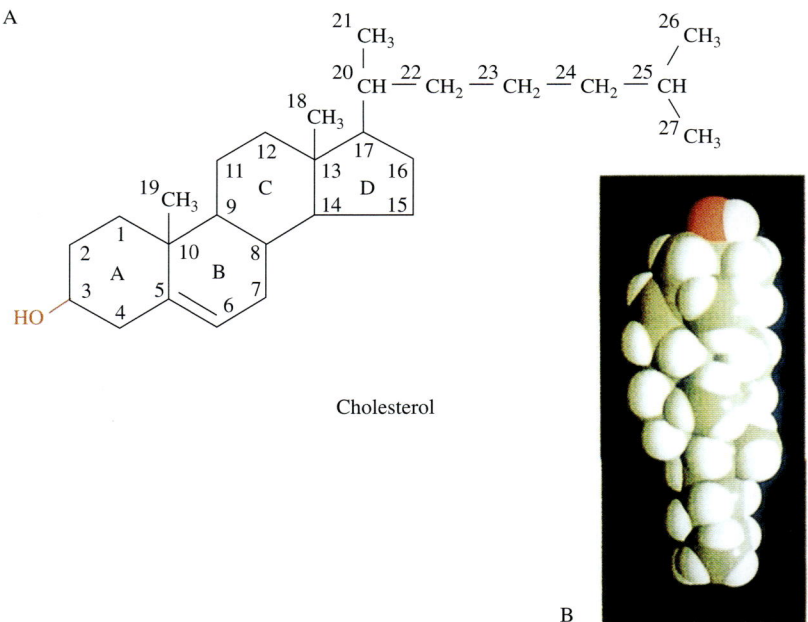

FIGURE 3.7. The structure of cholesterol of blood lipid chemistry noted for its relevance to heart disease **A** bond and **B** spacefilling representations of cholesterol, noted for its relevance to heart disease and for its precursor role in the biosynthesis of steroid hormones. (**B** Courtesy of Richard W. Pastor, FDA, Washington D.C. **A** and **B** from *Biochemistry*, Second Edition, D. Voet and J. Voet, Copyright © 1995, John Wiley & Sons, New York. Reprinted with permission of John Wiley & Sons, Inc.)

with the synthesis of cortisone, the anti-inflammatory steroid often obtained in creams to prevent itch, decrease redness, and enhance healing. In addition, steroids and their analogues are well known for their uses as oral contraceptives (estrogens and gestogens for birth control), as hormone replacement therapy for postmenopausal women, and as bodybuilding and athletic prowess enhancement factors (anabolic steroids, the androgens). Additionally, of course, blood levels of cholesterol signify predisposition to heart disease.

Clearly, the molecules that we can now produce by modern chemistry and the chemical energy that they represent have profound effects on humans and other living organisms of all levels of complexity. This is no longer an issue for debate as it was as recently as several decades ago. Instead, the debate is how to ensure the appropriate use of this capacity. For example, the pressure for athletes to use performance-enhancing steroid and peptide hormones and other drugs poses serious concerns for the International Olympic Committee, other sports regulatory efforts, and the overall long-term health of the athlete.

3.2.6 Animal and Plant Waxes

The wool of sheep repels water due to a coating of a wax called lanolin; it is a combination of sterols with long chain fatty acids and is often utilized as an ointment base in cosmetic and skin care formulations. Beeswax, perhaps the most commonly known animal wax, derives from the honeycombs of bees; it is the combination of straight long chain alcohols esterified to straight chain fatty acids. For example, the straight chain alcohol, $H_3C(CH_2)_{34}CH_2OH$ and the long chain fatty acid $HOOC(CH_2)_{34}CH_3$ combine with the loss of water, H_2O, to form the ester $H_3C(CH_2)_{34}COOC(CH_2)_{34}CH_3$, which is the longest chain molecule of beeswax; it is used in candles, cosmetics, modeling artificial fruits and flowers, and shoe polish.

A familiar plant wax is the wax coating of polished apples; it is a cuticle wax seen most often on the upper surfaces of leaves. Well-known cuticle waxes are carnauba wax from the fronds of a Brazilian palm tree and candelilla wax from the wild candelilla plants of Mexico and Texas. Carnauba is a hard, high-polish wax for automobile floor, and high quality shoe polishes. These plant and animal waxes should be distinguished from the similarly utilized paraffin waxes and petrolatum, which originate from petroleum. If biological origins of petroleum are considered, however, that distinction loses much of its significance.

3.2.7 Vitamins

It was appreciated in the early part of the twentieth century that a diet containing the correct amounts of carbohydrate, fats, and protein was insufficient for good health. Other "accessory food factors" were required. The initially identified "accessory substances" were amines, and so these "vital amines" were called *vitamines*. When it became clear that not all accessory substances were amines, the name became *vitamins*.

For humans, there are 13 essential vitamins, broadly classified as water soluble (or vinegar-like) and fat soluble (or oil-like). They often function as coenzymes or cofactors of metabolic reactions where they commonly change their oil-like or vinegar-like character, or they bring about a change in a protein to a more oil-like or vinegar-like state. For example, vitamin B_3 (niacin, the ring structure used to form nicotinamide as shown in Figure 3.8A) and vitamin B_2 (riboflavin, which becomes the flavins as shown in Figure 3.8B) are the principle coenzymes that receive electrons to form their more oil-like reduced state or give up electrons to form a more vinegar-like oxidized state. A change in the oxidative state of nicotinamide and flavin causes functional changes in the propensity of the protein, to which they are attached, to associate hydrophobically. Though beyond the scope of this effort, it would be of interest to assess, in general, the extent to which all vitamins, by effecting a change in oil-like character, drive folding and unfolding in accomplishing protein function.

In the early part of this century many vitamin deficiency diseases were identified and cured. Vitamin D cured rickets. Vitamin C (ascorbic

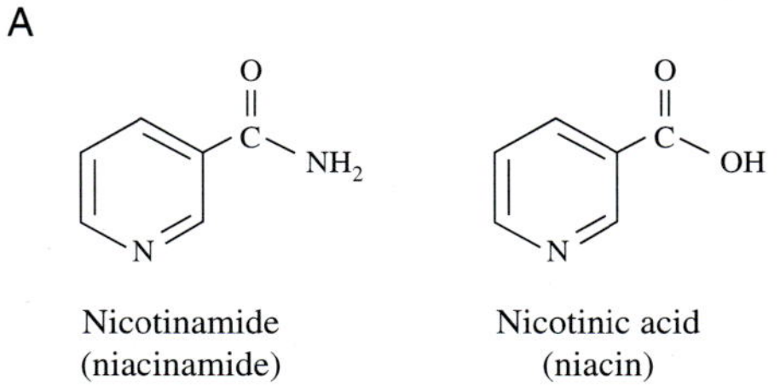

A

Nicotinamide (niacinamide)

Nicotinic acid (niacin)

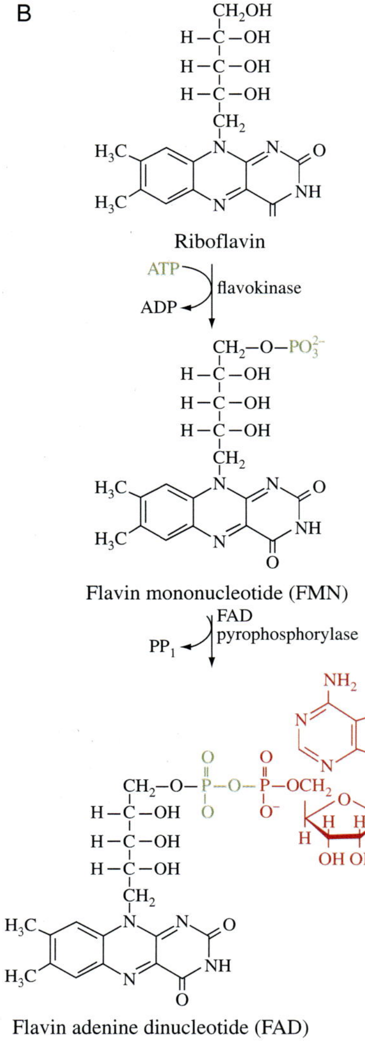

B

Riboflavin

Flavin mononucleotide (FMN)

Flavin adenine dinucleotide (FAD)

FIGURE 3.8. Structures of vitamins or vitamin-derived molecules that function in oxidation-reduction reactions. The oxidation of these redox groups in the inner mitochondrial membrane contributes to the electron transport chain that carries electrons from the oxidation of glucose to oxygen and in the process pumps protons from one side to the other of the inner mitochondrial membrane (see Chapter 8 for details). The proton gradient thus formed is used to phosphorylate ADP to form 32 of the 36 ATPs resulting from the oxidation of one glucose molecule to six CO_2 and six H_2O molecules. **A** Vitamin B_3, also called *niacin* or *nicotinic acid*, becomes converted to the amide (nicotinamide) and dressed up with a ribose sugar. Then, in a manner like that of riboflavin in B becomes phosphorylated to form nicotinamide mononucleotide (NMN) or further reacted with the addition of adenosine monophosphate (AMP) to form nicotinamide adenine dinucleotide (NAD). **B** Vitamin B_2, also known as *riboflavin*, is shown converted to the forms involved in redox reactions such as those of the electron transport chain. (From *Biochemistry*, Second Edition, D. Voet and J. Voet, Copyright © 1995, John Wiley & Sons, New York. Reprinted with permission of John Wiley & Sons, Inc.)

acid, discovered by the same Albert Szent-Györgyi of the early muscle contraction studies) cured scurvy. Vitamin B_1 cured beriberi. Pernicious anemia, which like diabetes mellitus once meant certain death, was cured by vitamin B_{12}, cyanocobalamine.

A large collaborative team of chemists headed by the same Woodward who had earlier synthesized cholesterol and cortisone, chemically synthesized vitamin B_{12}. In fact, all of the 13 essential vitamins for humans have been chemically synthesized. Thus continues the pilgrimage from proscription to successful preparation.

3.2.8 Structure and Chemical Synthesis of Natural Proteins

Emil Fischer achieved the first chemical synthesis of a protein-like fragment in 1907, when he assembled by peptide linkage a chain of 18 amino acids and showed that the chain was degraded by proteases, proteins functioning as enzymes to degrade protein.

Not surprisingly, the history of protein structure determination and synthesis involved proteins with key roles in energy conversion. Insulin, essential to the proper metabolism of glucose, was the first protein sequenced.

3.2 Determination of Biological Structures and Their Syntheses

Determination of the amino acid sequence of insulin by Sanger in 1953 provided the first demonstration that proteins do indeed have a specified sequence. Figure 3.9A gives the chain structure of preproinsulin that becomes proinsulin (Figure 3.10) on secretion from the cell with cleavage of the amino terminal signal sequence. The signal peptide is required for secretion of the protein through the oily lipid bilayer of the cell membrane (note in Figure 3.6) and in the process is cleaved off to give proinsulin. Then proinsulin becomes active insulin on excision of the C chain. Figure 3.9B gives the crystal structure, and Figure 3.9C gives the band or ribbon representation of the insulin molecule. The amino acid sequence of proinsulin is given in Figure 3.10, and the single-residue and mean-residue hydrophobicity plots for proinsulin are seen in Figure 3.11, which is a type of plot to be described and used in Chapter 7 for noting changes in composition and for identification of oil-like domains of special relevance to protein structure and function.

The first chemical synthesis of a natural protein was also that of insulin. Furthermore, the first three-dimensional protein structures determined were those of myoglobin, the oxygen storage/transport protein of muscle, and hemoglobin, the structurally related oxygen transport protein of blood. The structure and function of hemoglobins are considered in Chapter 7.

Chemical synthesis of insulin[14] occurred in the laboratories of three different research groups in the early 1960s. The protein hormone insulin is made up of two chains, one of approx-

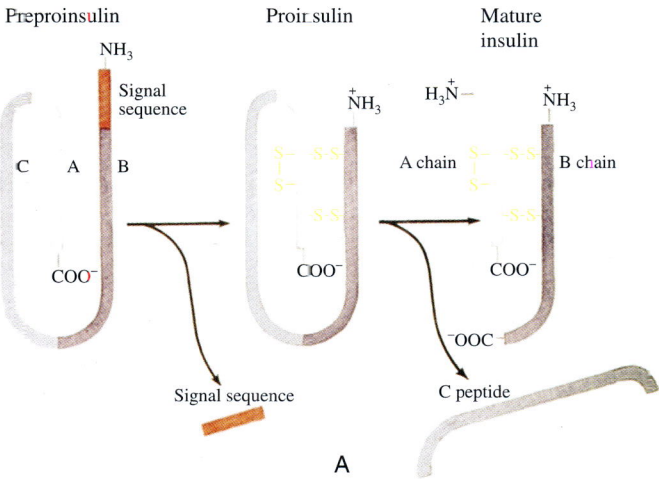

FIGURE 3.9. **A** Steps in the conversion of the gene product for insulin from preproinsulin to active (mature) insulin. **B** Space-filling representation of the active insulin molecule, as obtained from the crystal structure. **C** The ribbon model of active insulin showing the helical and extended regions of the A and B chains and the disulfide bridges that hold the A and B chains together after removal of the C peptide. (Reproduced with permission from Lehninger, Nelson, and Cox, *Principles of Biochemistry*, Second Edition, Worth Publishers.)

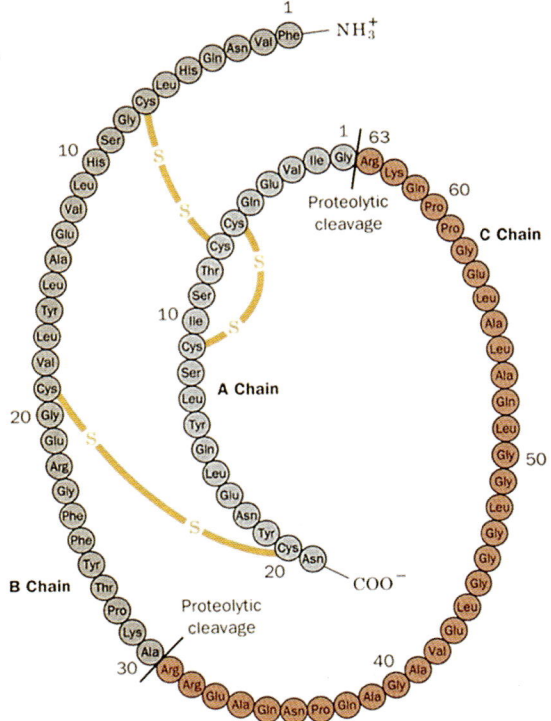

FIGURE 3.10. Primary structure (sequence of amino acid residues) of proinsulin, including location of the disulfide bridges with identification of the A, B, and C chains and locations of the cleavage sites for conversion to active (mature) insulin. (Reprinted with permission from R.E. Chance, R.M. Ellis and W.W. Brommer, *Science* **161**, 165–166, 1968. Copyright © 1968 AAAS.)

imately 20 and the second of approximately 30 amino acid residues. Johannes Meienhofer at the Technischen Hochschule in Aachen, Ching-i Niu at the Institute of Biochemistry in Shanghai, and P. G. Katsoyannis at the University of Pittsburgh with their colleagues each synthesized insulin. They determined the products of the different chemical syntheses to be active by determining biological activity and to be crystallizable with the formation of the same crystals as the natural product.

Classic chemical synthesis in solution, however, was never a practical way to produce commercial quantities of insulin for therapeutic use to control diabetes. On the other hand, Merrifield developed an alternative approach in the 1960s and 1970s. In this chemical method the protein chain was grown on a solid support, one residue at a time. The process used an automated machine to proceed through the many chemical steps required to add each amino acid residue. This solid phase chemical synthesis has been used to synthesize small proteins like insulin and interferon for commercial use. Now the most effective means of producing proteins of therapeutic value is to use the biological synthetic apparatus itself, by means of recombinant deoxyribonucleic acid (DNA) technology, also referred to as *genetic engineering*. Human insulin, marketed as Humulin by Eli Lilly, became the first commercial protein pharmaceutical produced by recombinant (DNA) technology.

3.2.9 Chemical Synthesis and Structure of Elastic-contractile Model Proteins

Prepared initially by chemical syntheses, *de novo*–designed model proteins were found to contract and produce motion and, in general, to perform mechanical and other kinds of work in response to chemical and other stimuli. In fact, the chemically synthesized elastic-contractile model proteins were found, just as their design had intended, to perform the kinds of energy conversions that sustain Life. The demonstrations of these energy conversions (Chapter 5), of the physical basis for the energy conversions (Chapter 5), of the relevance of the mechanism to naturally occurring proteins and protein-based machines (Chapters 7 and 8), and of their potential as materials for medical and nonmedical applications (Chapter 9) constitute the central messages of this volume.

A brief description of the chemical synthesis of elastic-contractile proteins immediately follows in the context of the general message of the chemical synthesis of natural biological products.

3.2.9.1 Chemical Synthesis of Elastic-contractile Model Proteins

Both classical solution and the Merrifield solid phase synthetic methods were used to prepare

3.2 Determination of Biological Structures and Their Syntheses

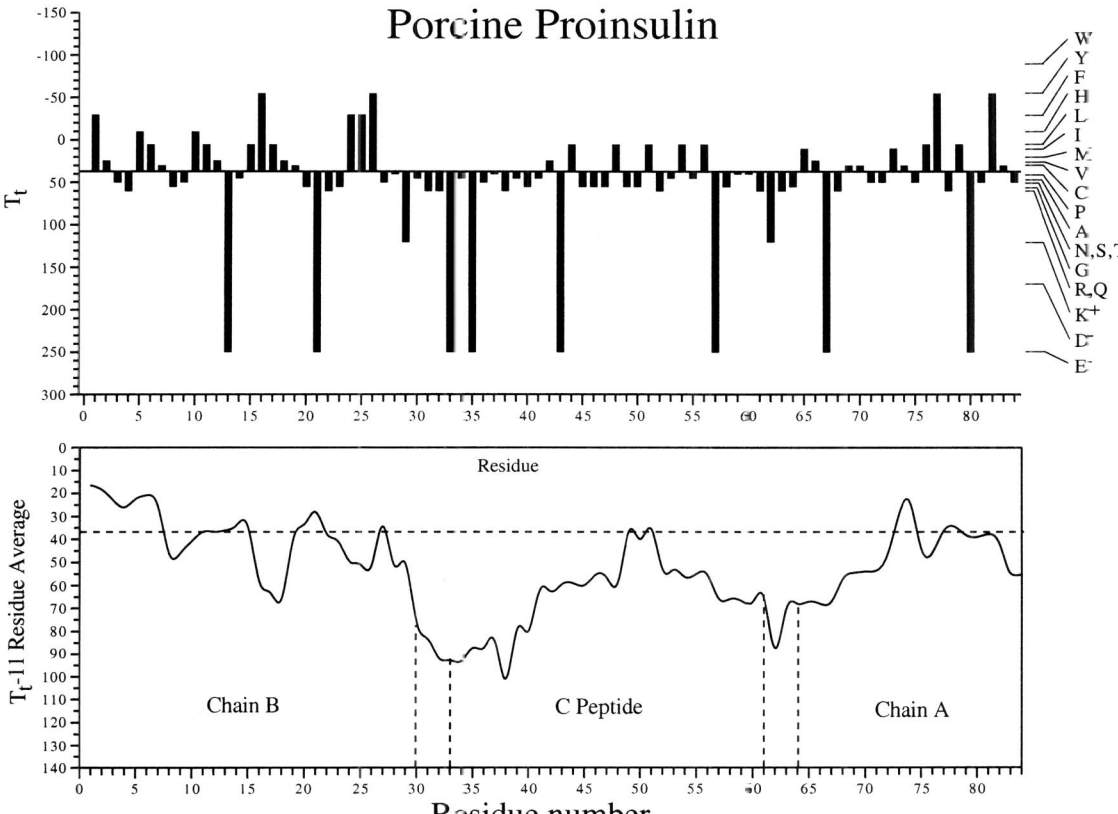

FIGURE 3.11. **Top.** Single-residue hydrophobicity plot of porcine proinsulin. At each residue position the hydrophobicity of the residue is given as derived from the T_t-based hydrophobicity scale. **Bottom.** Mean residue hydrophobicity plot (11-residue average) for proinsulin. The T_t-based hydrophobicity scale is derived in Chapter 5 and utilized in Chapters 7 and 8 to understand the hydrophobic associations attending function for selected protein systems.

the repeating sequences of mammalian elastin and variations thereof. Because of the extreme requirement for only L-amino acids in the chain molecule, classic chemical synthesis in solution proved most successful. The approach was to add V to P to form VP and to purify the dimer by repeated formation and dissolution of crystals. Then, VG was formed, crystallized, and G was added to form GVG, which was also crystallized. Finally, GVG was added to VP to form GVGVP. This sequence of five amino acid residues, called a pentamer, was vigorously purified, activated at the P end, and polymerized to form the polypentapeptide poly(GVGVP).

On forming a cross-linked matrix this chemically synthesized material contracts on raising the temperature much more strikingly than demonstrated for aorta long ago by C. S. Roy in 1880,[15] as noted in Chapter 2. The primary macromolecule in the aortic wall responsible for this property is the connective tissue protein elastin, which contains our starting sequence, poly(GVGVP), and related sequences. This inverse behavior as regards limited heating and cooling reflects what we now call the *inverse temperature transition*; it is central to the consilient mechanism and, we believe, to protein function.

By the occasional inclusion of amino acids with vinegar-like R-groups or other chemical groups that can be in two different states, more and less vinegar-like, energy sources other than heating have been shown to drive contraction.

These are detailed in Chapter 5. When these model proteins were subsequently made by living organisms, using recombinant DNA technology as described in the Chapter 2 overview, the biosynthesized model proteins converted energies just as had the chemically synthesized model proteins. Again, here disappears the delineation of things biological in origin from those of nonbiological origin so prevalent when this pilgrimage began with the accidental synthesis of urea in 1828 by Wöhler.

3.2.9.2 Structure of Elastic-contractile Model Proteins

The family of dynamic elastic-contractile model proteins that form the basis for the assertions of the central message of this volume do not lend themselves to the precise spatial descriptions of proteins that form crystals. Nonetheless, important structural description is possible for the poly(GVGVP) family. Indeed, the experimental and computational elucidation of structure will be noted in Chapter 5. Here we briefly provide the information in the context of the pilgrimage.

The most defined structural element is the β-turn shown repeating schematically in Figure 3.12A and in detail from the crystal structure of cyclo(GVGVP)$_3$ in Figure 3.12B. Thus there occurs a string of these β-turns in poly(GVGVP). The driving force for folding and assembly into a regular three-dimensional structure is the separation of the oil-like side chains from water. The question becomes what structure optimizes the separation of oil-like groups from water. Because the intramolecular and intermolecular oil-like contacts can be variable, the structural repetition need not be too precise. Based on a series of experimental studies and correlated computations, a single chain begins to fold into a helix as it associates with other chains. Representations are stepwise shown in Figure 3.12C–F. The helical structure, called a β-*spiral*, is shown in increasing detail in Figure 3.12C–E. It appears most likely that three β-spirals associate to form the twisted fil-

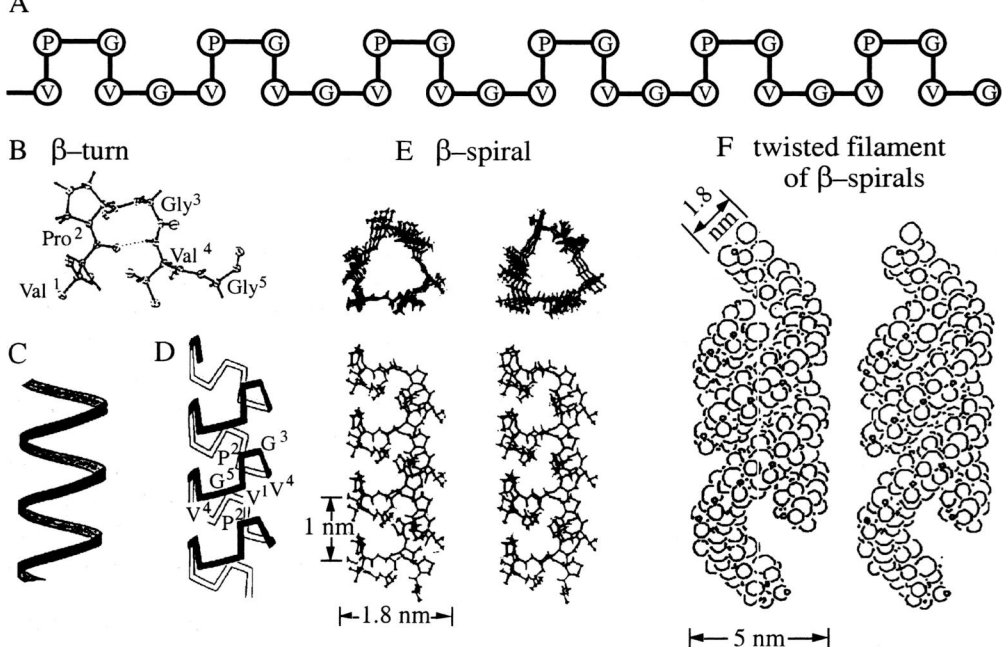

FIGURE 3.12. Molecular structure of the parent elastic-contractile model protein as described in the text. See Chapter 5 for discussion of experimental basis of the structure. (Parts **A**, **B**, **C**, **D**, **E**, and **F** reproduced with permission from Urry et al.,[100] Cook et al.,[24] Urry,[119] Urry,[121] and Urry et al.,[46] respectively.)

3.2 Determination of Biological Structures and Their Syntheses

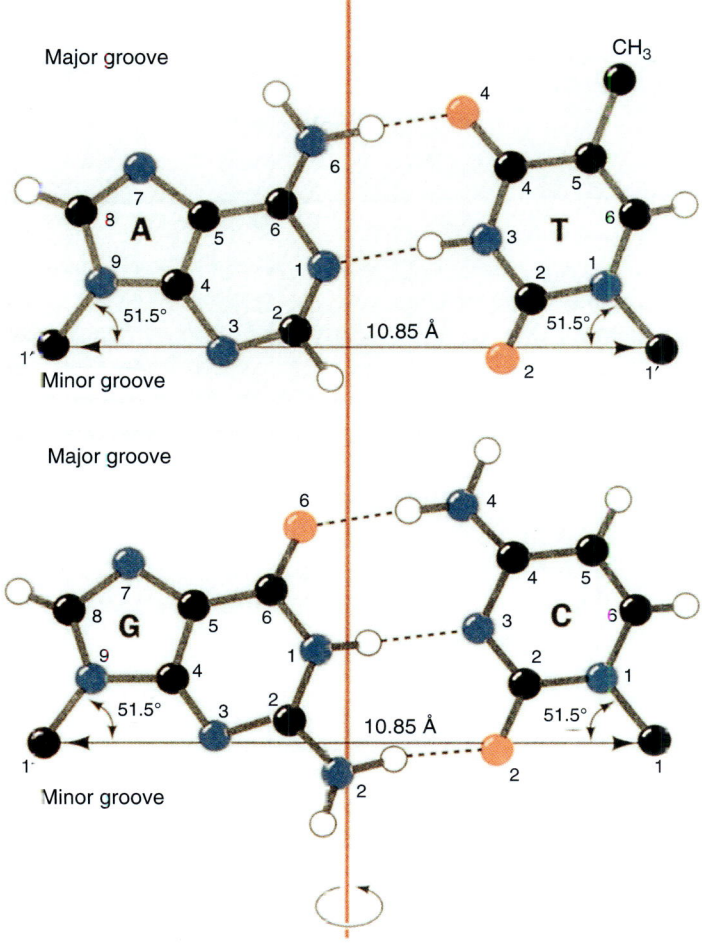

FIGURE 3.13. The hydrogen bonds exhibited by the A–T and G–C base pairs as they occur in the structure of DNA. This is the type of base pairing responsible for accurate replication of DNA, for transcription of DNA to form RNA, and for translation of RNA to form protein, as discussed in Chapter 4 in relation to the energetics of protein biosynthesis. (Reproduced with permission from S. Arnott, S.D. Dover, and A.J. Wonacott, *Acta Cryst.* **B25**, 2196, 1969. Copyright ©1969 IUCr journals.)

aments, as depicted in Figure 3.12F. For useful simplicity, Figures 2.5, 2.6, 2.9, and 2.12 show a single β-spiral in extended and contracted states to depict schematically the structural transition attending contraction/relaxation, but the twisted filament of Figure 3.12F represents the more accurate description.

3.2.10 Compositions, Syntheses, and Structures of Nucleic Acids

3.2.10.1 *The Structure, Stoichiometry, and Pairing of the Bases*

In the last decade of the 19th century, Kossel described the five bases that comprise nucleic acids. The four bases of DNA, deoxyribonucleic acid, are adenine (A), thymine (T), guanine (G), and cytosine (C). F for RNA, ribonucleic acid, A, G, and C occur, but uracil (U) replaces thymine. In the late 1940s, Chargaff found for DNA that the concentration of G equaled that of C, and the concentration of A was the same as that for T. This is because G selectively pairs with C by formation of three hydrogen bonds, and A selectively forms two hydrogen bonds with T in the case of DNA or U in the case of RNA. Figure 3.13 shows the hydrogen-bonded base pairs of DNA. This simple selective pairing by hydrogen bonding in combination with the genetic code is the basis for inheritance and for the translation of nucleic acid sequence into protein.

3.2.10.2 Sugar–Phosphate Backbone of the Nucleic Acids

As shown in Figure 3.14, poly(sugar–phosphate) forms the backbone of the nucleic acid polymers, and the four bases are side chains much as the 20 R-groups of protein are side chains. In addition to T of DNA being replaced by U of RNA, DNA differs from RNA by having the deoxyribose sugar where the 2' OH in Figure 3.14 for RNA is replaced by 2' H, hence the name deoxy. Consistent with the apolar–polar repulsive free energy of hydration developed and demonstrated in Chapter 5, the polar (more vinegar-like) sugar–phosphate backbone of the DNA double-stranded helix is outside and as distant as possible from the apolar (more oil-like) bases on the inside of the helix (see Figure 3.15). Thus, the structure of nucleic acids reflects the repulsion of oil-like and charged groups, called the apolar (hydrophobic)–polar (e.g., charged) repulsive free energy of hydration, as developed in Chapter 5.

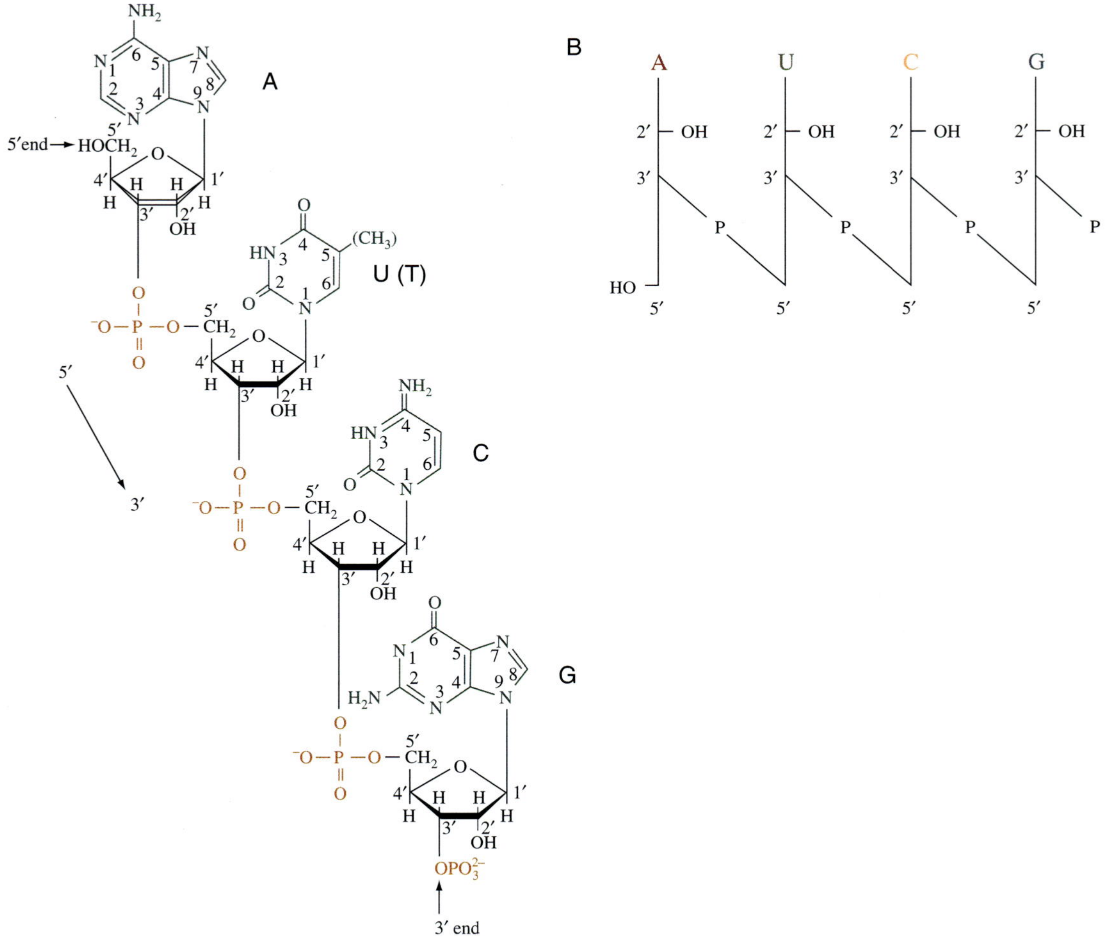

FIGURE 3.14. A sequence of RNA showing the attachment of one nucleotide monophosphate to the next. The structure is a poly(phosphoribose) with a base attached to each ribose group. (From *Biochemistry*, Second Edition, D. Voet and J. Voet, Copyright © 1995, John Wiley & Sons, New York. Reprinted with permission of John Wiley & Sons, Inc.)

3.2 Determination of Biological Structures and Their Syntheses

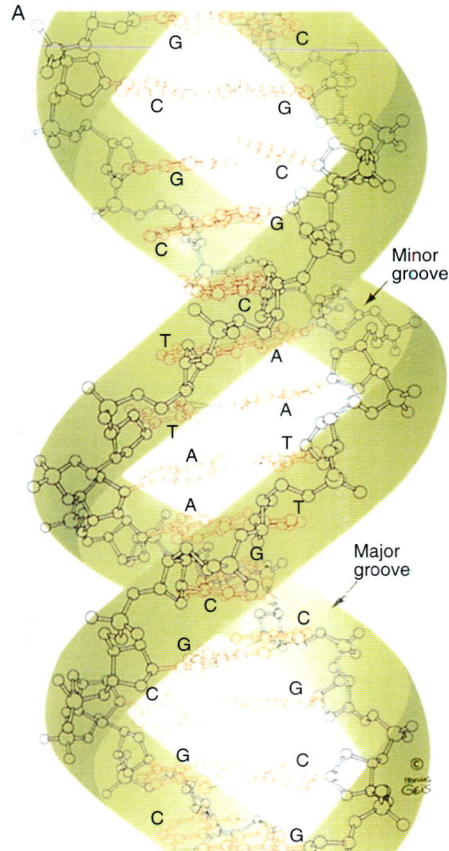
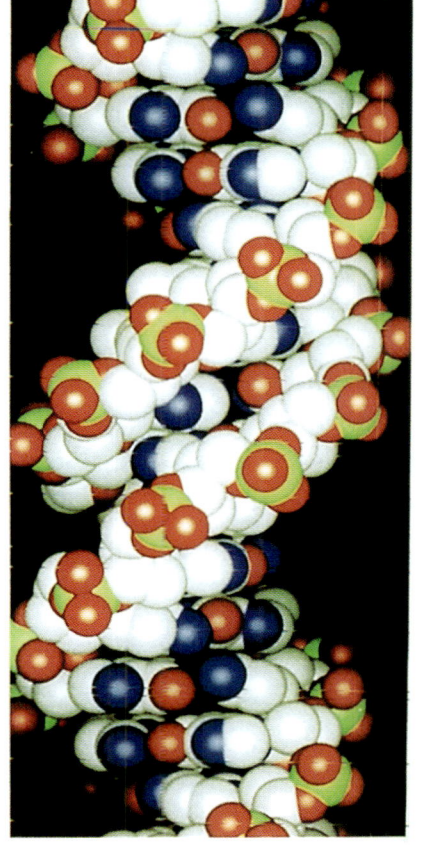

FIGURE 3.15. Two representations of the famous DNA double helix due to Watson and Crick (and Franklin). (**A** Reproduced with permission from The Irving Geis Archives, Rights owned by Howard Hughes Medical Institute, and **A** and **B** from *Biochemistry*, Second Edition, D. Voet and J. Voet, Copyright © 1995, John Wiley & Sons, New York. Reprinted with permission of John Wiley & Sons, Inc.)

3.2.10.3 The Watson-Crick DNA Double-stranded Helix

Perhaps standing as the most commanding symbol of the developments in biology and of the biotechnological revolution is the double-stranded helix of DNA shown in Figure 3.15, the original structure for which was due to Watson and Crick (Franklin).[16] This structure stands as the basis of heredity by its own replication where G pairs with C and A pairs with T. When being transcribed into RNA, A pairs with U, and the RNA sequence of bases is otherwise a copy of the DNA sequence of bases. Finally, by the genetic code explicitly considered in Chapter 6, a given RNA sequence of three bases encodes for a particular amino acid in the translation process to produce a specific protein. The probability and energy costs for protein biosynthesis are considered in Chapter 4.

3.2.10.4 Chemical and Enzymatic Synthesis of DNA

Continuing our pilgrimage to bring the products and components of Life within man's comprehension, in the 1960s, Gobind Khorana chemically synthesized nucleic acid, which was competent to specify synthesis of the appropriate protein, and in 1970 he enzymatically synthesized the first natural gene, a gene of yeast. Thus it would appear that all of the products and components of living organisms are within man's capacity to comprehend and to a more limited extent within man's capacity to produce. Now the most ambitious project yet is unfolding, the determination of the complete sequence of the human genome with what are expected to be truly extraordinary implications to health and to understanding man's relationship to other life forms.

3.3 Determination of the Human Genome

3.3.1 The Human Genome: "The Periodic Table of Life"

As noted above, the question posed by Thales of Miletus 26 centuries ago, "*What is the world made of?*" was answered at a fundamental level by completion of the *Periodic Table of the Elements*. The exact way in which the atomic elements combined to form the minerals of the earth's crust, the water of the oceans, the gases of the atmosphere, and the living organisms of biology, is not immediately obvious from the Periodic Table of the Elements, yet the very groupings within the table dictate the way in which the atomic elements combine to form complex combinations of atoms found on earth.

In a somewhat related way, the question of "What is Life?" would be answered at a certain level by the elucidation of the genomes of all living organisms. As the human genome contains many genes that are clearly related to lower forms of Life and as we pride ourselves on being the most advanced of creatures, it would not be too great an exaggeration to refer, for the sake of brevity, to the human genome as "The Periodic Table of Life."[17] An important distinction exists, however. Perhaps a more accurate description of the genome is *The Blueprint of Life*, for it is the proteins for which the genes encode that perform the essential functions in answering the question of "What Sustains Life?"

3.3.2 The Difference Between Lower Animals and Vertebrates

At this stage of analysis of the human genome data, the evidence indicates that evolution of vertebrates from lower animals occurred not so much by the addition of new genes encoding for new protein domains but rather from the novel ways that existing protein domains have evolved into combinations of domains to introduce a new element to function. In particular

Initial examination of the predicted protein complement of the human genome indicates that vertebrates have not evolved primarily by addition of new protein domains but through novel ways of putting these modules together to make protein. It is mostly the architecture rather than the building blocks that distinguishes us from other organisms.[18]

Perhaps the myoglobin protein domain provides an example. While retaining the same basic structure of myoglobin, homologous mutations led to the α-chain and β-chain variants that combine to give hemoglobin with new functional capacity. Our analysis of the amino acid mutations of myoglobin that gave rise to the α-chains and β-chains of hemoglobin provides insight. As is discussed in Chapter 7, it is hydrophobic association between domains to form the $\alpha_2\beta_2$-tetramer of hemoglobin that gives rise to new functions of survival value to vertebrates. It is the formation of less polar, more hydrophobic structures that results in the assembly of these domains to make a protein with a functional capacity required for a large animal that must have oxygen transported from lungs to the tissues. When we refer to an oil-like or hydrophobic domain, this should be recognized as a limited region of a protein domain that is involved in hydrophobic association between larger protein domains.

3.3.3 Genomics and Motility

The number of genes involved in the cytoskeletal systems (actin filaments, microtubules, and intermediate filaments) required for motility, as identified in the pre-human genome, is of the order of 100.[19] Some additional 14 genes have been observed from the human genome sequence, but two actin genes appear to have been missed. An equivalently large number of myosin, kinesin, and dynein motor domains have also been defined by traditional biochemical and genetic methods. Because of the size of the two former genes, the very large introns and the fragmentary nature of the identifications, some identified gene segments may result in the same gene being identified more than once. Nonetheless, it will continue to be vitalizing as the gaps are filled and the complete and accurate results emerge. In particular, there emerges the opportunity to identify and relate the consilient mechanism among the related motor domains and their interactions with cytoskeletal components.

3.3.4 Evolutionary Genomics

Evolutionary genomics[20] involves the relationship between protein domains (structural or functional units in a protein) and analyzes the relationships between lower organisms and man. Some 50% of the human genome is thought to result from the sharing or shuffling of protein domains of lower organisms. For example, Li et al.[20] count "1,865, 1,218, 1,183 and 973 domain types in human, fruitfly, nematode and yeast, respectively."

Based on the myoglobin/hemoglobin example to be discussed in Chapter 7, it is our perspective that changes in the oil-like surface of homologous domains gives rise to new aggregations of the protein domain resulting in new properties, and a common feature of the new property would be an increased positive cooperativity, that is, increased efficiency of function.

3.3.5 Genomics and Human Disease

The theories that all diseases have a genetic component and that every individual has genetic anomalies provide a basis for discussing the expected impact of the human genome in the practice of medicine.[21,22] Based on the former, targeting of newly designed drugs to active sites of a diseased protein, preparation by recombinant DNA technology of peptide and proteins to target appropriate sites, and gene delivery become obvious therapeutic approaches.

Based on the latter, individuals can exhibit different reactions to medical treatments, procedures, or prostheses. There are now known to be more than 1,100 genes with disease-related mutations (to follow this developing information, see www.ncbi.nlm.nih.gov/omim).

It has been estimated that 100,000 people die each year in reaction to medical treatments.[23] In the past, when a small number of individuals experience a severe reaction, the medical treatment has been discontinued even though the vast majority of the population may benefit. In the future, the people likely to have adverse reactions will be identified beforehand by their genetic makeup and be advised against the treatment, and those who can benefit from the procedure should be allowed to do so.

As the blueprint of human biology emerges, our focus turns to the molecular biophysics of the protein-based machines of biology to emphasize, as Bacon and Descartes would have us do, the development of insight into protein function by means of axioms and graphical representations (Chapter 5). Furthermore, medical applications of protein-based materials can provide means for delivering on some of the promise of medical implications that emerging knowledge of the human genome forecasts. Also as a point of clarification in Chapters 4 and 6 from a physical biochemical perspective, we attempt to alleviate some of the mystery that has shrouded the production and evolution of protein-based machines.

3.4 Design of Living Organisms as Chemical Factories

So commanding has become our understanding of the products, components, and functions of living organisms that we can now employ living organisms as factories to produce complex molecules of our own design.

Man has demonstrated the capacity to synthesize products and components of living organisms and, in key ways, to assemble those components and reconstitute natural function. He can chemically synthesize sequences of nucleic acid encoding for a specific protein sequence of interest.

Holding great promise for the future, man has learned (1) to produce new genes by chemical synthesis of nucleic acids with base pairing overlaps, (2) to use the renowned enzymatic polymerase chain reaction (PCR) to extend and form double-stranded DNA, and (3) to use the ligase enzyme to splice together large synthetic gene segments. The synthetic genes can be introduced into organisms in such a way as to induce the organism to produce the specified protein. Man can genetically engineer living organisms to produce not only proteins natural to other organisms but also new proteins of man's own design for functions that evolution never called upon living organisms to perform.

A few examples are discussed in Chapter 9 wherein the future will see the biosynthesis of biodegradable protein-based thermoplastics, materials to prevent postsurgical adhesions, temporary functional scaffoldings to direct tissue reconstruction, and drug delivery devices for new drug release regimens.

Indeed, the period from Wöhler's accidental synthesis of urea in 1828 to the Biotechnological Revolution of today presents a remarkable pilgrimage with the potential for dramatic and constructive consequences for individual health and societal development.

References

1. Our pilgrimage revisits highlights in the developing understanding of products and components of living organisms and thereby provides historical backdrop for the development of biomolecular machines and materials.
2. D.J. Boorstin, *The Seekers: The Story of Man's Continuing Quest to Understand His World.* Random House, New York, 1998, p. 22.
3. T.R. Malthus, *An Essay on the Principle of Population*, London, 1798; rev. ed., 1803. Quotation excerpted from *"Darwin,"* A Norton Critical Edition, Third Edition, Selected and Edited by Philip Appleman, W.W. Norton, New York, 2001, p. 39.
4. J. Bronowski, *The Ascent of Man*. Little, Brown and Company, Boston/Toronto, 1973.
5. So attacked was Vesalius for departing from the Galenist doctrine that he gave up his research on anatomy for the practice of medicine only to attempt a return in the year of his death at age 50.
6. Harvey counted Kings (James I and Charles I) and distinguished contemporaries such as Sir Francis Bacon among his patients. As a result of publishing his unorthodox views of the circulation of the blood, his prestigious practice is reported to have suffered.
7. Ionic ammonium cyanate, $NH_4^+ \bullet\bullet\bullet OCN^-$, contains the four most common elements of living matter, carbon (C), hydrogen (H), oxygen (O), and nitrogen (N). On heating, there is simply a rearrangement of the atoms to result in urea, H_2NCONH_2. This finding played a significant role in the debate between mechanists and vitalist, a philosophical issue that little interested Wöhler.
8. The term *organic* as used here indicates "of biological origin." More currently the term *organic chemistry* has come to mean the chemistry of the carbon atom.
9. The chemical structure of tartaric acid is HOOC–HCOH–HOCH–COOH, which has two carbon atoms with four different substituents, that is, two centers of asymmetry. The natural tartaric acid is dextrorotatory.
10. A familiar use of the capacity to distinguish plane-polarized light is found in the effectiveness of Polaroid sunglasses. Sunlight scattered from a road's surface, for example, is polarized horizontally, and Polaroid sunglasses are designed such that only vertically polarized light passes through them and therefore the bright light scattered from the road's surface does not enter the eyes. If a dextro-rotatory solution of tartaric acid or of sugar is placed in a tube and the tube is placed in front of the polaroid glasses, the scattered light from the surface of the road that passes through the tube will also pass through the sun glasses unless the sun glasses are rotated clockwise. At the correct amount of clockwise rotation, the tube becomes dark, and more of the scattered light from the surface of the road passes through the portion of the sunglasses that surround the tube.
11. L. Pasteur, "On the Molecular Asymmetry of Natural Organic Products." In *Lecons de Chimie professees en 1860*. Chemical Society of Paris, Paris, 1861. George Stewart and Co., printers, Edinburgh and London.
12. Because of his extraordinary knowledge of chemistry, Fischer was placed in charge of organizing chemical production in Germany during the First World War. Among the many misfortunes of that war were the combat deaths of his two sons. So distraught was Fischer by his circumstances that he ended his own life shortly thereafter—in 1919.

References

13. D. Voet and J. Voet, "Electron Transport and Oxidative Phosphorylation." In *Biochemistry*, Second Edition. John Wiley & Sons, New York, 1995.
14. In the usual manner the Katsoyannis publication of their work (*J. Am. Chem. Soc*, **85,** 2863, 1963) acknowledged at the end of the publication the support of the U. S. Public Health Service, National Institute of Arthritis and Metabolic Diseases. Interestingly, Niu and coworkers gave their acknowledgments in the Introductory Remarks with the statements "by holding aloft the great red banner of Mao Tse-tungs's thinking" and

    The total chemical synthesis of insulin is a piece of work which stems directly from the big leap forward movement. In 1958, when the movement had attained the stage of unbounded enthusiasm for achievements, we took upon ourselves the task of making synthetic insulin as the first protein ever to be synthesized, an achievement which we hope to accomplish in a short time to serve as a contribution to the scientific progress of our mother country and also as an expression of revolutionary initiative (*Kexue Tongbao*, **17**, 241, 1966)

15. C.S. Roy, "The Elastic Properties of the Arterial Wall." *J. Physiol.*, **3**, 125–159, 1880.
16. J.D. Watson and F.H.C. Crick, "Molecular Structure of Nucleic Acids." *Nature*, **171**, 737–738, 1953.
17. S. Pääbo, "The Human Genome and Our View of Ourselves." *Science*, **291** (February 16), 1219–1220, 2001.
18. Editorial Comments, *Science*, **291**, 1218, 2001.
19. T.D. Pollard, "Genomics, the Cytoskeleton and Motility." *Nature*, **409**, 843–845, 2001.
20. W-H. Li, Z. Gu, H. Wang, and A. Nekutenko, "Evolutionary Analysis of the Human Genome." *Nature*, **409**, 847–849, 2001.
21. L. Peltonen and V.A. McKusick, "Dissecting Human Disease in the Postgenomic Era." *Science*, **291**. 1224–1229, 2001.
22. G. Jimenez-Sanchez, B. Childs, and D. Valle, "Human Disease Genes." *Nature*, **409**, 853–855, 2001.
23. D. Drell and A. Adamson, "Fast Forward to 2020: What to Expect in Molecular Medicine." In *The Human Genome Project Information: Medicine*. June, 13, 2002 (from online magazine *TNTY Futures*).

4
Likelihood of Life's Protein Machines:
Extravagant in Construction Yet *Efficient in Function*

4.1 Introduction

A protein with its specified sequence, where each position is filled by but 1 of 20 different amino acid residues, represents an extremely improbable construct. Perhaps in part because of this improbability, searches for understanding have reasonably considered far-from-equilibrium conditions and addressed circumstances whereby order may appear out of chaos. On the other hand, biochemists with painstaking commitment have worked their way down to the component processes of DNA replication, of transcription of DNA into RNA, and of translation of RNA into protein sequence. They dissected each constituent process into its scores of energy-requiring reactions, assembled these reactions to provide a compelling description of each component, and emerged with a stunning tale of heredity and protein biosynthesis. Biology's story of protein biosynthesis is one of such elegant detail that, no matter how brilliant the mind, it could not have been derived otherwise. Accordingly, there is now little mystery in how a protein of specified sequence can come into being; it simply requires adequate energy. In fact, the details of protein biosynthesis reveal an extravagant process requiring extraordinary amounts of energy. We must yet, of course, admit mystery in how all of those integrated and interdependent molecular machines came into being in formation of the first cell with the capacity to synthesize protein of specified sequence.

In this chapter we briefly note the improbability of a protein comprised of 20 different amino acids in specified sequence. Then we step through the details whereby a protein of specified sequence comes into being, and, like the accountant, sum up the cost in terms of the biological energy currency, adenosine triphosphate (ATP), that is, in terms of the number of ATP molecules consumed in the production of a protein of unique sequence. Such an essentially ordinary exercise requires inclusion in this book in order to remove unnecessary mystery of protein production, to provide an example whereby molecular machines utilize energy to produce biological structure, and to place in perspective the ease with which new protein-based machines evolve when functioning by the consilient mechanism.

4.2 Improbability of a Unique Protein Sequence

4.2.1 Consideration of a Small 100 Residue Protein

Twenty different amino acid residues become incorporated into protein. The question may be stated as, what is the probability of occurrence of even a small protein of 100 amino acid residues of specified sequence? The calculation can be quite simple. For equal availability of all 20 amino acids, the probability of any 1 of the 20 amino acid residues occurring in a single position is (1/20). The probability for a completely specified sequence of 100 residues would be $(1/20)^{100}$, assuming an equal probability for the incorporation of each amino acid. This is one chance in 10^{130}. On this basis, the chance that a mixture of amino acids would assemble spontaneously into a specified protein sequence, even with catalyst and abundant energy to activate the assembly reaction, is essentially nonexistent. This is in part why arguments arise in relation to biology for extraordinary circumstances beyond those of common experience to chemists and physicists. In order to convey some sense of the magnitude of this improbability, we next consider mass.

4.2.2 Probability in Terms of Mass of 100 Residue Protein Molecule

For a mean residue weight of 100 grams per mole, 1 mole of a 100 residue protein would weigh about 10kg. With the number of molecules in a mole being Avogadro's number of $6.023 \times 10^{23} \approx 10^{24}$ molecules/mole, one molecule of a 100 residue protein would weigh about 10^{-20} grams. A quantity of 10^{130} distinct protein molecules of 100 residues with an average residue weight of 100 would weigh 10^{110} grams. This is the mass that would result for one molecule each of the possible 10^{130} distinct sequences of a 100 residue protein.

To illustrate the magnitude, this quantity of 10^{110} grams can be compared with the total mass of the earth. The mean density of the earth is given as 5.5 grams/cm^3, and the volume of the earth is approximately 10^{30} cm^3. The total mass of the earth becomes approximately 10^{31} grams. Thus, if the total mass of the earth were made up only of proteins of 100 residues of different sequences, the chance of finding a specified sequence would be no more than 1 chance in 10^{79}. When any 1 of 20 residues can be in each position in a 100 residue protein, the occurrence of a single molecule of specified sequence is so improbable that finding the specified sequence would still be unlikely, even if the known universe were comprised of nothing but 100 residue protein sequences.

How, then, can there be tens of thousands of unique and usually much longer protein sequences in living organisms? The answer, of course, comes from an understanding of the process and energetics of protein biosynthesis.

4.2.3 Reaching the Extraordinary Potential of such Enormous Structural Diversity

Proteins have no rival in the world of polymers. The difference is not small; rather, comparisons involve the enormous numbers of improbability noted above. Diversity in petroleum-based polymers utilizes a number of different repeating units, but compositional variation within a single polymer has been limited largely to so-called block copolymers.

Taking the most meaningful advantage of the enormous structural diversity of proteins requires some sense of the physical basis of function. Finding the way by a random selection of sequences poses no option, because the possibilities are too many. Knowledge of the laws that control structure and function, however, does provide access to the potential of such structural diversity. As presented in Chapter 5, the consilient mechanism offers the required insight, that is, the emerging comprehensive hydrophobic effect, and more specifically the apolar–polar repulsive free energy of hydration, systematizes progress that has been made in designing diverse and relatively efficient elastic-contractile model protein-based machines.

4.3 Protein Sequence as a Biosynthetic Construct

With detailed knowledge of the biological process of protein synthesis,[1] it becomes possible to estimate a minimal amount of energy required for biosynthesis of a specific protein sequence. We have just seen that a specific protein sequence is an extremely unlikely structure, because each position in the sequence could be occupied by any 1 of 20 of the naturally occurring amino acid residues. Surprisingly, as shown below, not only is the product of the biological synthesis very probable, but energetically biological synthesis of protein appears to be a wasteful process. The required fidelity of a precise protein sequence comes at an extraordinary cost in energy.

Biosynthesis of protein can be seen in three steps: replication, the production of the DNA for a daughter cell; transcription, the conversion of the DNA into the equivalent sequence of RNA; and translation, the conversion of the ribonucleic acid sequence into the specified protein sequence, that is,

1. DNA → replication → DNA
2. DNA → transcription → RNA
3. RNA → translation → protein

4.3.1 DNA Replication

4.3.1.1 Energy Cost to Produce DNA of the Required Length

Production of a DNA sequence that encodes for a 100 amino acid residue protein requires 300 bases (three for each triplet codon specifying a particular amino acid at a specific position) plus 3 bases for a start codon. (The table of the genetic code specifying the sequence of three bases [the triplet codon] that encodes for each amino acid residue is given in Chapter 6.) The bases of the nucleic acids begin as the triphosphates—adenosine triphosphates (ATP), guanosine triphosphates (GTP), thymidine triphosphates (TTP), and cytosine triphosphates (CTP). The structures of these nucleoside triphosphates (NTPs) may be gleaned from Figures 3.13 and 3.14. The equation for the conversion of nucleoside triphosphates into the DNA polymer can be written as follows:

$$pATP + qGTP + rTTP + sCTP = DNA + (p+q+r+s)PP \quad (4.1a)$$

This utilizes nATP equivalents where n = (p + q + r + s). As written, Equation (4.1a) is a reversible reaction. The key element of irreversibility results from Equation (4.1b), the removal of the side product, PP, which is an inorganic diphosphate called *pyrophosphate*. The enzyme pyrophosphatase catalyses the following very important reaction:

$$(p+q+r+s)PP \rightarrow \text{pyrophosphatase} \rightarrow 2(p+q+r+s)P \quad (4.1b)$$

Thus, another nATP equivalents are required where n = (p + q + r + s). One equivalent of ATP is required to convert a nucleoside monophosphate (NMP) into a nucleoside diphosphate (NDP) and a second is required to convert the nucleoside diphosphate into a nucleoside triphosphate (NTP). Accordingly, two equivalents of ATP are required for the addition of each base to the growing DNA polymer.

If the interest is in producing a 100 residue protein, then a 300 base DNA would be required, that is, n = 300 + 2 × 3, where one three is due to a triplet stop codon and the second three is due to the triplet start codon (AUG), which encodes for methionine but which is usually processed off. Therefore, 2n or *612 ATP equivalents* are required to duplicate a 306 base strand of DNA to encode for a 100 residue protein.

A key feature of the above equations is that the reactions in Equation (4.1a) are reversible, but those represented in Equation (4.1b) are irreversible. This represents the very significant means whereby synthesis of the gene, the poly(deoxyribonucleic acid) encoding for protein, is an irreversible process. Another important expense of energy is that only intermittent parts, called *exons*, of the DNA sequence (the gene) actually encode for protein. For the DNA that encodes elastin, less than 20% is exon, the remainder, called *introns*, does not encode for the protein. In addition, there is much "junk" DNA carried along during replication that appears to serve no useful purpose.

4.3.1.2 Improbability of the Specified 306 Base Sequence of DNA

It is in fact more improbable to have a 306 base sequence of DNA, poly(deoxyribonucleic acid), of an exactly specified sequence of bases than it is to have a 100 residue protein of specified sequence. In the case of DNA, any one of four bases can exist in each of the 306 base positions. The chance of any single position being of one of the four bases would be 1/4. Thus the chance that each position contains the specified base would be $(1/4)^{306}$, which is 10^{-184}. Recall that the number for the 100 residue protein was

4.3 Protein Sequence as a Biosynthetic Construct

10^{-130}. This result, however, requires modification due to redundancy where several codons can encode for a given amino acid residue. This comes from the fact that, given four different bases and a triplet codon, there are 64 different triplet codons to encode for 20 amino acid residues. (Table 6.2 shows how the 64 codons are used.)

4.3.1.3 Probability of DNA Sequence in Terms of an Equilibrium Constant

Writing the probability in terms of an equilibrium constant, K, uses the change in Gibbs free energy, ΔG, in the form $K = e^{-\Delta G/RT}$, where R is the gas constant, 1.98 cal/mole-degree, and T is the temperature in degrees Kelvin, for example, 310 for body temperature. Because the equilibrium constant is expressed as the ratio of reactants over products, it can be seen as a statement of probability, that is, an equilibrium constant of 1 indicates that reactant and products are equally likely. Now, K may be written for the combined reactions represented by Equations (4.1a) and (4.1b) as follows:

$$pATP + qGTP + rTTP + sCTP = DNA + 2(p+q+r+s)P \quad (4.1')$$

Because the equilibrium constant for Equation (4.1a) is approximately 1, that is, $\Delta G = 0$, and we know that the ΔG for hydrolysis of PP is $-8,000$ cal/mole,[2] each base addition of the overall reaction of Equation (4.1') would have an equilibrium constant of $K = e^{8,000/RT} = e^{13} = 4.4 \times 10^5 = 10^{5.6}$. Because there are 306 base additions, the overall equilibrium constant would be the product of 306 such reactions, that is, $(10^{5.6})^{306} = 10^{1,714}$. The reaction for DNA formation is irreversible, not because of far-from-equilibrium conditions, but rather because energy is thrown away in 8 kcal/mole steps. The result is that DNA is produced irreversibly at an extraordinarily high cost in energy.

4.3.2 Transcription: DNA → RNA

4.3.2.1 Energy Cost to Produce RNA of the Required Length

The following sets of analogous reactions transcribe the DNA, deoxyribonucleic acid sequence, into RNA, a ribonucleic acid sequence. In RNA the DNA base of thymine (T) is replaced by uracil (U), and, of course, the ribose of the RNA replaces the deoxyribose of DNA.

$$pATP + qGTP + rUTP + sCTP = RNA + (p+q+r+s)PP; \text{ nATP equivalents} \quad (4.2a)$$

$$(p+q+r+s)PP \rightarrow \text{pyrophosphatase} \rightarrow 2(p+q+r+s)P; \text{ nATP equivalents} \quad (4.2b)$$

where $n = (p - q + r + s)$. Again, for a 300 (n) base sequence of mRNA, and for the three base start codon, which becomes translated into a methionine residue, $2(300 + 3)$ equivalents of ATP are required, that is, another *606 ATP equivalents* are utilized.

4.3.2.2 Improbability of the Specified 303 Base Sequence of RNA

The same argument applies for RNA as for DNA except that there are 303 bases in the sequence. Again, any one of four bases may occur in each position of the 303 base sequence. Thus we would write, $(1/4)^{303} = 10^{-183}$. Of course, this again is more improbable than the 100 residue protein into which the RNA sequence is translated, but this number is again not quite so small due to codon redundancy.

4.3.2.3 Probability of RNA Sequence in Terms of an Equilibrium Constant

Now, by adding Equations (4.2a) and (4.2b), we have the equivalent overall reaction to that of Equation (4.1') given above, that is,

$$pATP + qGTP + rUTP + sCTP = RNA + 2(p+q+r+s)P \quad (4.2')$$

Again by analogy to the argument used above for DNA, we have for each base addition of Equation (4.2') that $K = e^{8,000/RT} = 10^{5.6}$. Now with the slightly fewer, 303, base additions, the overall equilibrium constant would be $(10^{5.6})^{303} = 10^{1,697}$. The reaction for RNA formation is irreversible, again because each addition is paid for by a relatively small 8 kcal/mole packet of energy. Thus, RNA is also produced, not by far-from-equilibrium conditions, but, rather, from the summation of stepwise base-by-base additions, due to an extraordinarily high overall expenditure of energy.

4.3.3 Translation: RNA → Protein

4.3.3.1 Energy Cost of Loading Amino Acids on tRNA

Before aligning the amino acids along the messenger ribonucleic acid (mRNA) template, the amino acid (AA) is activated by reaction to form AMP-AA and transferred to a small nucleic acid called tRNA, transfer ribonucleic acid, that is tRNA-AA, which contains the triplet codon for base pairing along the mRNA sequence. This two-step reaction is shown as Equations (4.3′a) and (4.3′b) below.

$$\eta ATP + \eta AA = \eta AMP\text{-}AA + \eta PP \quad (4.3'a)$$

$$\eta AMP\text{-}AA + tRNA = \eta AMP + tRNA\text{-}AA \quad (4.3'b)$$

where AMP is adenosine monophosphate and η is 100 for the 100 amino acids to be incorporated into the 100 residue protein. Reactions indicated by Equations (4.3′a) and (4.3′b) are reversible with the equilibrium constant for their sum being equal to 1. It is again the removal of the PP byproduct of the reactions that makes the activation and attachment of the amino acids to tRNA an irreversible process.

$$\eta PP \rightarrow \text{pyrophosphatase} \rightarrow 2\eta P \quad (4.3'c)$$

Accordingly, 2η ATP equivalents are required, which adds another *200 ATP equivalents*.

4.3.3.2 Energy Cost of Binding and Translocation of tRNA-AAs to Form Protein

The final steps in achieving the growth of a protein chain of specified sequence involve the binding of tRNA-AAs to position 1 of the ribosome and aligning along mRNA followed by translocation to position 2 of ribosome with incorporation into the growing peptide chain. These reactions can be indicated by the following equations:

$$\eta tRNA\text{-}AA + \eta \text{ ribosome(position 1)} + \eta GTP$$
$$\rightarrow \eta tRNA\text{-}AA\text{-}ribosome(\text{position 1})$$
$$+ \eta GDP + \eta P \quad (4.3a)$$

$$\eta tRNA\text{-}AA\text{-}ribosome(\text{position 1}) + \eta \text{ ribosome(position 2)} + \eta GTP \rightarrow \eta tRNA\text{-}AA\text{-}ribosome(\text{position 2}) + \eta \text{ ribosome}$$
$$(\text{position 1}) + \eta GDP + \eta P \quad (4.3b)$$

$$\eta(tRNA\text{-}AA\text{-}ribosome(\text{position 2}) \rightarrow$$
$$\eta tRNA + \eta AA \text{ in protein} \quad (4.3c)$$

Because GTP is energetically equivalent to ATP, this requires η (100) ATP equivalents for Equation (4.3a) and another η (100) ATP equivalents for Equation (4.3b) to give another *200 ATP equivalents*.

4.3.3.3 Probability of Protein Sequence Arising from Translation Alone

Adding the reactions indicated by Equations (4.3′a), (4.3′b), and (4.3′c) again gives the opportunity to calculate a statement of probability from the expression of the overall equilibrium constant for addition of an amino acid to its tRNA in terms of the change in Gibbs free energy, $K = e^{8,000/RT} = 10^{5.6}$; the ratio of product to reactant favors product by a factor of some 100,000. Thus, the activation of each amino acid for selective alignment at the ribosome is essentially an irreversible process.

The processes at the ribosome, required to add a single amino acid of tRNA-AA to the growing protein chain, results in the conversion of two molecules of GTP to two molecules of GDP plus two molecules of P. The standard free energy of hydrolysis of ATP to ADP and P is given as $-7,290\,\text{cal/mole}$,[2] and this same value can be used for the hydrolysis of GTP to GDP plus P. Thus for the equilibrium constant we can estimate $K = (e^{7,290/RT})^2 = 10^{11.8}$. Of course again the reaction goes to completion, but irreversibility always occurs as a result of a step by step loss of energy packets of about 8 kcal/mole.

4.3.3.4 The Cost of Producing Scores of tRNA Molecules

The tRNA that contains the anticodon for alanine (Ala, A), tRNA^{Ala}, has 76 bases such that 75 bonds must be formed to assemble the tRNA molecule with each bond utilizing 2 ATP equivalents. This requires 150 ATP equivalents. Furthermore, there are as many as 64 different tRNA molecules. If we assume a codon redundancy in a given species sufficient to require 40 different anticodons, the result would be 40 × 150 or 6,000 ATP equivalents.

4.3.3.4 The Cost of Producing the Ribosome

The total mass of the *Escherichia coli* ribosome is 2,520,000 Da. Considering only the protein that constitutes only one third of the ribosome, the molecular weight of the protein is 857,000 Da. Assuming the mean residue weight to be 100, there are some 8,570 residues. The formation of 8,570 peptide bonds alone would cost 17,000 ATP equivalents. Thus, the assembly of only the tRNA and protein components required for translation at the ribosome number requires much more than 23,000 ATP equivalents. Of course, a ribosome can be used for the synthesis of many copies of protein, but, even with synthesis of a large number of proteins before the ribosome would have to be replaced, the number of ATP equivalents is significant.

4.3.4 Summary of Minimal Energy Expenditure to Produce Protein

The biosynthesis of the DNA sequence, replication, to encode for a 100 residue protein requires *612 ATP* equivalents. Transcription, the biosynthesis of the RNA sequence to encode for a 100 residue protein requires an additional *606 ATP* equivalents. The activation of the amino acids and their attachment to the correct tRNA requires another *200 ATP* equivalents, and, finally, translation into the specified 100 residue protein requires yet another *200 ATP* equivalents. The total number of ATPs required is *1,618*. On the one hand, this overall number does not recognize that a single strand of DNA may produce many strands of RNA and that a single strand of RNA may be used to produce many protein chains. On the other hand, this sum does not include the ATP expense for producing the unused introns, the junk DNA, the complex nuclear and ribosomal structures, and all of the enzymes, for example, pyrophosphatase and RNA polymerase, that are necessary. In addition, there are other expenses, for example, the many housekeeping chores such as controlling the supercoiling of nucleic acids by type II topoisomerases and refolding of improperly folded protein by molecular chaperones. Thus, even by considering individually the replication of DNA, or the transcription to RNA, or the production of a protein chain, the production of biology's polymers occurs at an enormous cost in energy.

4.4 Protein Machines: Improbable or Extravagant Constructs

4.4.1 Comparison of Improbabilities and Minimal Energy Costs

The probabilities of product formation arising out of the individual processes of replication, transcription, and translation were considered separately above. Nonetheless, there remains a curiosity to obtain some overall number to compare with the factor of 10^{-130} arising out of the probability of a specific sequence for a protein when there are any of 20 residues possible for each position of a 100 residue protein. A sense of the sort of number might be obtained if the total energy were put in an equilibrium expression relevant to the overall process. Thus, the total energy consumed was 1,618 ATP equivalents. Thus we might write, $-8\,\text{kcal/mole-ATP} \times 1{,}618\,\text{ATP} = -12{,}944{,}000$ cal/mole and $K \approx e^{12,944,000/RT} \approx 10^{9,250}$. Of course, such numbers are so large as to have almost no meaning, except perhaps to justify the statement that the first copy of a 100 residue protein of specified sequence was paid for by an enormously extravagant expenditure of energy.

It is perhaps easier to discuss such biological overkill in the drive toward synthesis in terms of the addition of a single selected amino acid residue at a given position in the growing protein chain. The factor of 1/20 equals $10^{-1.3}$. Putting this in terms of $K = 10^{-\Delta G/2.3RT}$, $\Delta G =$

1.8 kcal/mole, yet the ΔG expenditure was approximately 30 kcal/mole. Favoring the selective addition by a factor of 1 million, seemingly an ample amount to ensure synthesis, would require 8.4 kcal/mole. Yet there are another 22 kcal/mole expended for this individual step.

Therefore, this "creation of structure" did not require "far-from-equilibrium" conditions; rather, it simply required an extravagant use of the energy coin of biology. "Order Out of Chaos" in terms of the production of the biological molecular machines occurs not under classic irreversible conditions but rather simply by the removal of a side product at the expense of an extraordinary amount of free energy.

4.4.2 Comparison of Energy Required to Form a 100 Residue Protein with that Released on Return of the Residues to Free Amino Acids

There is another way to compare the two probabilities. This is to compare the energy required to combine amino acids into a protein with that recovered when the protein of 99 peptide bonds is hydrolyzed to return to amino acid residues. In short, 1,618 ATP equivalents were required to form 99 peptide bonds of the 100 residue protein, which is just over 16 ATPs per peptide bond. This is a free energy cost of $8 \times 16 = 128$ kcal/mole. As hydrolysis of a peptide bond yields no more than 3 kcal/mole, the efficiency of energy conversion would be about 2%. Protein synthesis without considering production of the required DNA and RNA, but only the positioning of a single amino acid in the correct position, could properly be called extravagant as well.

4.4.3 Sun's Energy Required to Produce a Small Protein

We now continue further consideration of the biological machinery for production of protein. Because more than 50 photons are required to produce one glucose molecule, and one glucose molecule results in 36 ATPs, and 1,618 ATP equivalents (44.9 glucose molecules) are required to produce the 100 residue protein, some 2,250 photons would be required to produce this small protein. This is about 23 photons per peptide bond. Based on a value of 70 kcal/mole of photons, this means much more than 156,000 kcal is required to produce 1 mole of a 100 amino acid residue protein worth about 300 kcal on hydrolysis to free amino acids. In terms of energy recovered, this is an efficiency of 0.2%. The quantity, 156,000 kcal, is sufficient energy to bring to a boil from room temperature more than 500 gallons of water. Overall the biological process for creating a single small protein from the energy available is enormously extravagant.

This calculation assumes that the complex apparatus of the enzymes and nucleic acids of the ribosome appeared at no cost. The actual cost should take into consideration the number of times the individual components are used before replacement is required, that is, the so-called turnover number for each required component of the ribosome, of the t RNA, of the enzymes to carry out all of the syntheses of the nucleic acids, to break down the pyrophosphates, to catalyze the polymerizations, to correct errors, and so forth.

Despite this enormous expense, because the photons from the sun are free and the products such as glucose made with the sun's energy are so inexpensive, utilizing the biological mechanism to produce protein is orders of magnitude more efficient than man's capacity to produce protein by chemical synthesis. Based on large-scale chemical synthesis, $20/gram would be an optimistic cost for the synthesis of poly(GVGVP), whereas less than $20/kg has been indicated as possible with microbial biosynthesis. It becomes quite obvious how much we depend on the sun's energy reaching the earth's surface and the photosynthetic apparatus to harvest the sun's energy. This simply reinforces our recognition of the essential role of the photosynthesis of the world's forests and grasslands.

4.4.4 Implied Importance in Control of Protein Sequence

Surely the occurrence and maintenance of such a profuse process for protein biosynthesis demonstrate an unequivocal need for complete control of sequence and the continuing mandatory requirement of twenty diverse amino acids. Why should this be the case? In the non-biological world of petroleum-derived polymers with limited residues available and without control of sequence, function is much more limited. One presumes that diverse function of protein machines, by whatever means, requires complete control of such diverse sequence.

From the viewpoint of required energy, the biosynthetic process is extravagant indeed, yet protein machines can be remarkably efficient in spite of diverse function. For example the efficiency of the F_1-ATPase in rotating an actin filament has been estimated to approach 100%.[3] Why are these twenty amino acid residues required and why, for example, are leucine and valine residues functionally different from an isoleucine residue? Such questions raise the issue as to whether there might be a common groundwork with which to understand protein function using these twenty amino acid residues. Might there be a consilient mechanism wherein an understanding of otherwise subtle differences among certain of the twenty amino acids would become recognized as essential for diverse function. These issues are addressed in subsequent chapters.

References

1. See for example D. Voet and J. Voet, *Biochemistry*. Second Edition, John Wiley & Sons, Inc., New York, 1995, Chapter V. The Expression and Transmission of Genetic Information, pp 830–1019.
2. See for example D. Voet and J. Voet, *Biochemistry*. Second Edition, John Wiley & Sons, Inc., New York, 1995, standard free energies of phosphate hydrolysis, Table 15-3, page 430.
3. K. Kinosita, Jr., R. Yasuda, and H. Noji, "F_1-ATPase: a highly efficient rotary ATP machine." In *"Essays in Biochemistry: Molecular Motors"*, 35, 3–18, 2000, Edited by G. Banting and S.J. Higgins, Portland Press Ltd 59 Portland Place, London, W1N 3AJ, U.K.

5
Consilient Mechanisms for Diverse Protein-based Machines: The Efficient Comprehensive Hydrophobic Effect

5.1 Introduction

The machines of biology are of a mechanism more elemental than those of man's design. Consider, for example, the intermittent internal combustion (e.g., gasoline-reciprocating piston) engine that performs the work of putting a vehicle in motion. Fuel is injected in a timely manner into a series of chambers; the fuel vapors are ignited in each chamber; a series of explosions of hot expanding gasses occur in properly timed sequence; the hot expanding gasses supply the energy that drives the pistons, that by means of connecting rods rotates the crankshaft, that through a clutch assembly and speed-changing gears rotates the drive shaft, that by means of a differential gear box rotates the wheels, that puts the vehicle in motion.

The fundamental process is simpler for protein-based engines of biology. An energy input drives spatially separated oil-like domains of the protein into association and thereby produces the motion of a contraction. In our view, the protein-based engines of biology function by shifting oil-like domains between being water soluble and water insoluble. This perspective forms "a common groundwork of explanation" for diverse functions of protein machines; therefore, the perspective is that of a consilient mechanism.[1,2]

Due to the diverse molecular compositions of protein, different energy inputs cause the oil-like regions to come together or to separate. Because the many different energy inputs function by the same process of changing water solubility of the oil-like regions, energy outputs in addition to motion occur, and, because the association/dissociation of oil-like domains is generally reversible, energy inputs and energy outputs can exchange roles. How a given energy input changes the energy of the hydrophobically associated state becomes central to the consilient mechanism of protein-based machines.

In this chapter, we (1) note the structure and elasticity of our unadorned model protein-based machine, (2) catalog diverse energy resources available to this consilient mechanism as a function of model protein composition, (3) describe visual demonstration of contraction and relaxation of elastic protein bands functioning as engines that produce motion with diverse energy inputs, (4) demonstrate energy conversions not involving the work of motion by means of the coupling of functions through shared interaction with hydrophobic hydration, (5) review the kinds of work performed by these protein-based machines, (6) establish the physical basis for protein-based machines functioning by the consilient mechanism, (7) establish a general basis for positive cooperativity (Monod's "second secret of life"), and (8) relate 6 and 7 to provide an understanding of the remarkable efficiency made possible by the consilient mechanism.

For many coworkers, and myself, this chapter unfolds an ongoing journey of personal enlightenment, even of personal Ionian enchantment,[3]

5.1 Introduction

of longer than three decades. Developments presented here generally follow the sequence in which they occurred. The concluding paragraph of the most complete scientific review to date of the basic science aspects of our work, chronologically lists the sequence of developments.[4] That paragraph is quoted here separately.[5]

In this section, we introduce key thoughts and arguments in preparation for more complete discussions in sections 5.2 through 5.9. For even more detailed description and for different vantage points of examination of what we here call the *consilient mechanism*, please refer to original reviews at more[4,6] and less[7] technical levels and to the references listed therein and herein. First reported in this chapter, however, are thermodynamic derivations[8,9] and analyses (included in sections 5.1.3.4, 5.3, and 5.9). A developing general formalism for energy conversion by the consilient mechanism is begun in section 5.8.4. Those with a more general background in this area need not belabor these details in order to grasp the essential point of this chapter. The elemental message becomes a general sense of how one achieves diverse energy conversions by the design of model elastic-contractile protein-based machines wherein the key becomes control of hydrophobic association.

5.1.1 The Consilient Mechanism for Controlling Association of Oil-like Domains

5.1.1.1 Changes in Folding and Association of Oil-like Domains of Protein Chains Give Rise to Protein-catalyzed Energy Conversion

5.1.1.1.1 Energy Conversions Sustain Life

The living organism itself accesses energies available in its environment and converts those energies into the energies needed for the full range of biological functions. Machines convert energy from one form to another, and proteins, functioning as molecular machines, catalyze the energy conversions of biology.

5.1.1.1.2 Oil-like and Vinegar-like Parts of Proteins

In our view, an essential element of the many different protein-catalyzed energy conversions of biology resides in the old adage "oil and vinegar don't mix." We apply the adage to oil-like and vinegar-like parts[10] arising from different amino acid residues constrained by sequence to coexist along the protein chain. Key facets reside in the repulsive forces that cause the oil-like parts and vinegar-like parts of proteins to distance themselves one from the other and in the forces that cause oil-like parts of protein chain molecules to separate from water. These forces, between oil-like and changeable vinegar-like side chains (also referred to as **R**-groups as in Figure 2.1 and related text) distributed along the protein chain, control whether or not oil-like regions separate from water by association.

5.1.1.1.3 Sequence of Oil-like and Vinegar-like Parts Control Folding and Assembly of Proteins

Figure 2.1 provides example of the sequence-dependent interplay between oil-like and vinegar-like residues of our elastic-contractile model proteins that directs chain folding and association of folded chains. The relative position of oil-like and vinegar-like **R**-groups along a sequence and the state of the vinegar-like **R**-groups[11] determine when and in what arrangement the oil-like **R**-groups separate from water. In this chapter, we design diverse molecular machines of elastic-contractile model proteins. Repulsive interactions between oil-like and changeable vinegar-like **R**-groups attached to the model proteins build up on becoming more polar and relax on becoming less polar, and thereby they reversibly catalyze the many energy conversions exhibited by living organisms.

5.1.1.2 A Consilient Mechanism Provides a "Common Groundwork of Explanation" for the Diverse Energy Conversions of Biology

The mechanism whereby oil-like groups separate from vinegar-like groups provides a "common groundwork of explanation"[1] for diverse energy conversions that sustain Life; it is a consilient mechanism for hydrophobic association.[12] Fundamental to this process is the interplay of oil-like groups between being surrounded by water to being separated from water and self-associated, that is, from being soluble to being insoluble in water.

5.1.1.2.1 Phase Separation and Its Temperature Interval

Our model proteins provide a visible, that is, macroscopic, demonstration of oil-like regions going from being soluble to insoluble in water by means of a dramatic phase change. A particular energy input causes the single phase of a clear solution to become two visibly distinct phases. In Figure 5.1A, the particular energy input is heat, thermal energy, and the phase separation occurs as the temperature is raised from room temperature, 25°C (77°F), to body temperature, 37°C (98.6°F). For the parent elastic-contractile model protein, poly(GVGVP) or (GVGVP)$_n$,[13] in water, the interval from 25° to 37°C is its *temperature interval* for the phase transition from solubility to insolubility (see Figure 5.1C).

5.1.1.2.2 Phase Separations of Protein Polymers Produce Motion

The fascinating phase transitions exhibited by these elastic-contractile model proteins are analyzed below. It is the phase transition of these elastic model proteins that makes them contractile model proteins and that brings in other energy sources in addition to thermal by virtue of the capacity of other energy sources to change the temperature at which the phase separation occurs. The phase diagram perspective, therefore, provides a unifying means for describing the many energy conversions possible by the consilient mechanism. *It is the basic process of oil-like domains gaining dominance and then associating that drives the phase separation and constitutes a contraction that can be used to move an object.*

To produce motion is to perform mechanical work. In this case, mechanical work is the energy output resulting from the particular energy input that drove the contraction. In the most fundamental demonstration of the consilient mechanism (see Section 5.1.4), thermal energy is the input and mechanical work is the output. First, however, we consider familiar and not so familiar phase changes. Analysis of the resulting phase diagrams opens the door to the full range of energy conversions that occur in biology, that is, opens the door to the consilient mechanism of protein-based machines. This is a mechanism whereby the set of energy conversions of living organisms can occur.

5.1.2 Four Phase Changes (Transitions) in Model Protein–Water Systems

5.1.2.1 Familiar Phase Changes: Melting of Ice and Vaporization of Water

5.1.2.1.1 Heat and Entropy Changes of Phase Transitions

In schematic fashion, Figure 5.2 contains peaks representing our most familiar reversible phase transitions: the melting of ice near 0°C (32°F or 273°K) and the boiling of water near 100°C (212°F or 373°K). These phase transitions in the state of water represent changes from more order to less order on raising the temperature. The amount of heat required to melt ice to form less ordered water is 80 cal/gram (or, for the molar heat of the transition, ΔH_t[melting] = 80 cal/gram × 18 gram/mole = +1,440 cal/mole, where the subscript t denotes transition). The heat required to convert liquid water to vapor is a much larger 540 cal/gram, or ΔH_t (vaporization) = 540 cal/gram × 18 gram/mole = +9,720 cal/mole.

One way to look at this is to heat both ice-cubes in a pan on the stove and to the same amount of water in a second pan on a second burner on the stove. It will take about seven times more heat (produced by burning gas or passing an electric current through a heating element on a stove) to boil away the quantity of water converting it from water to vapor than to melt completely the same amount of water in ice cubes at 0°C to form liquid water at 0°C.

5.1 Introduction

FIGURE 5.1. Characterizations of a phase separation, that of the inverse temperature transition, for the elastic-contractile model protein, (GVGVP)$_{251}$. (**A**) A series of tubes show a clear solution of (GVGVP)$_{251}$ in water at temperatures below 25°C (left), a solution that became cloudy on raising the temperature above 25°C (middle), and the phase separation that occurs on standing (right). (**B**) Temperature profile for the onset of turbidity. Starting below 25°C with a clear solution containing at least 40 mg/ml of (GVGVP)$_{251}$ in water placed in a spectrophotometer, the temperature dependence for the onset of turbidity is accurately measured. The temperature at which half-maximal turbidity occurs is defined as T_t. (**C**) Differential scanning calorimetry curve of 40 mg/ml of (GVGVP)$_{251}$ in water. The trace shows the heat absorbed by the sample as the temperature is raised through the temperature interval of the inverse temperature transition. The area contained within the peak gives the heat of the transition, and the heat divided by the temperature, integrated over the transition, gives the entropy of the transition. (Adapted with permission from Urry.[4])

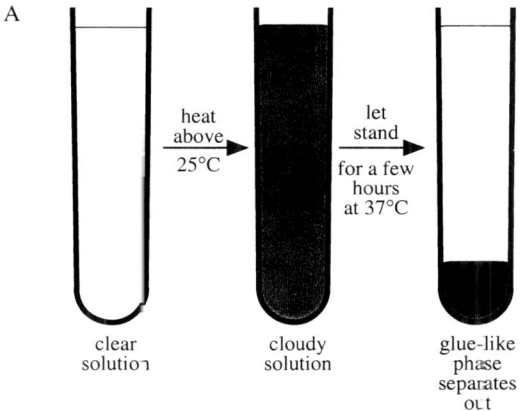

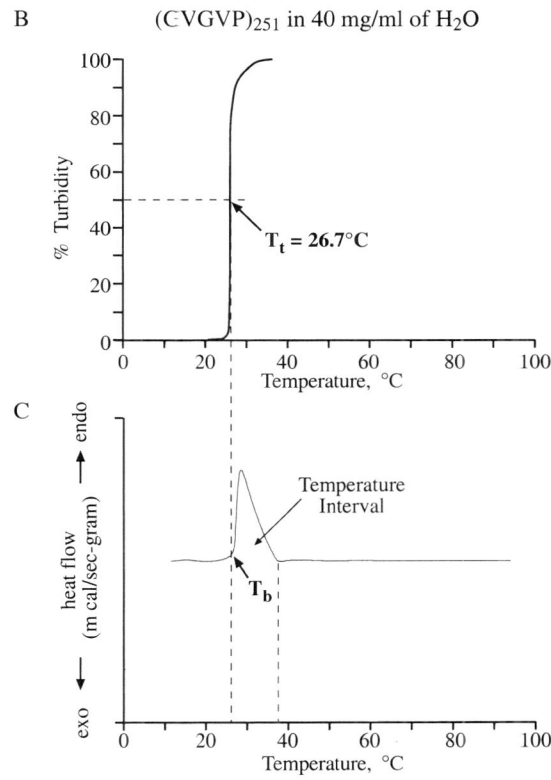

A quantitative measure of the decrease in order for each of these phase transitions is the entropy for the transition, ΔS_t, which is simply the heat of the transition, ΔH, divided by the temperature in degrees Kelvin (K) at which the transition occurred. For the melting of ice ΔS_t(ice → water) = 1,440 cal/mole/273° K = +5.3 cal/mole-degree (EU), and for the vaporization of water ΔS_t(water → vapor) = 9,720 cal/mole/373° K = +26 cal/mole-degree = +26 EU where 1 cal/mole-degree K is 1 entropy unit (EU).

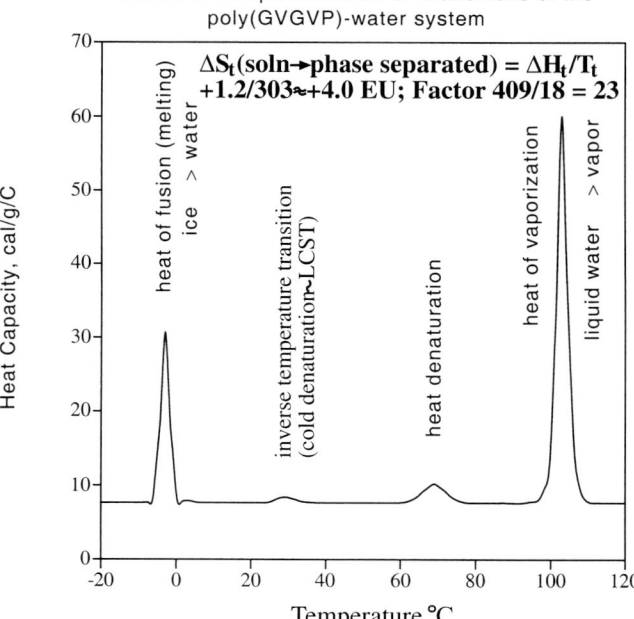

FIGURE 5.2. The four phase transitions of the model protein $(GVGVP)_{251}$ in water over the temperature range from $-20°$ to $120°C$. The familiar transition of the melting of ice and the vaporization of water are shown with the relative magnitudes of the heats of these transitions to those of protein heat denaturation and to the innocuous looking inverse temperature transition near $30°C$ that we believe to be the basis of the function of protein-based machines of Life. See text for discussion.

5.1.2.1.2 An Increase in Entropy Is a Decrease in Order

During the melting of ice, water molecules go from being precisely positioned, one molecule with respect to the other, in the well-known crystal structure of ice, to still being geometrically constrained to associate but able to shift positions in the liquid state. Melting of ice constitutes a decrease in order that is measured quantitatively as an increase in entropy, that is, +5.3 cal/mole-degree.

The decrease in order is about five times greater, however, for vaporization than for melting; $26/5.3 \approx 5$. This is understandable because, in water, though quite mobile and able to change positions, the molecules are still constrained to associate using geometries dictated by their shapes and hydrogen bonding of one water molecule to surrounding water molecules. On the other hand, once vaporized and in the gas phase, individual water molecules are no longer in contact. Instead, they are free to fly around completely independent of each other. These changes from states of more order to states of less order (ice to liquid water and liquid water to water molecules in the gas phase) are positive entropy changes (increases in entropy). A negative entropy change (a decrease in entropy) on going from one state to another indicates a change from a state of less order to a state of more order. Thus, the reverse processes (of water molecules in the air condensing on a cold window pane, or, if it is really cold outside, of water vapor forming frost [ice crystals] on the window pane) represent decreases in entropy.

These phase changes, very familiar to our daily experience, make real the thermodynamic quantities, ΔH and ΔS, that are fundamental to understanding how protein-based machines function. With knowledge of heat changes, ΔH, and entropy changes, ΔS, we unmask how model protein-based machines access energies in the environment and convert them to those energies essential for biological function. The phase transitions of the melting of ice and the vaporization of liquid water in terms of the thermodynamic quantities of heat change (also called change in heat content or enthalpy) and entropy change bring to life the energy conversions that sustain Life.

5.1 Introduction

5.1.2.2 Phase Transitions of Proteins

5.1.2.2.1 Heat Denaturation of Protein

Along the temperature scale shown in Figure 5.2, sandwiched between commonplace phase changes of the melting of ice at low temperature and the vaporization of liquid water at high temperature, are transitions exhibited by proteins in water. The higher temperature transition, shown near 70°C represents the denaturation of proteins. Below the denaturation transition temperature, the protein has a regular structure, where each amino acid residue of the protein chain occurs in a defined space. Above the denaturation transition temperature, the protein can be described as a disordered, irregular, and randomized tangle of chains. The location in space of a given amino acid residue in one chain with respect to other amino acids in the chain is different for each protein chain. During cooking (heating) of an egg, the proteins of the clear gel, surrounding the yellow yolk, become disordered, more solidified, and white on heat denaturation.

5.1.2.2.2 Low Heat of the Intermediate Transition at Lower Temperature

The lower temperature transition, occurring near 30°C for our basic elastic-contractile model protein, is difficult to detect, as plotted, because of the relatively small amount of heat involved in this transition. The representations in Figure 5.2 provide a visual bridge from the more familiar transitions of water, considered above, to the less familiar transitions central to the function of our model protein-based machines. Figure 5.2 is a composite of the transitions for pure or nearly pure water with the curve for poly(GVGVP) with the relative areas of the curves representing the circumstance for 1 gram of protein dissolved in 1 gram of water. The relative areas are plotted for equal weights of water and model protein, yet the differences in the areas of the peaks, which measure the relative heats of the transitions, are dramatic. *This seemingly innocuous transition, shown here peaking near 30°C for poly(GVGVP), is the source of the consilient mechanism for energy conversion. We believe this inconspicuous transition represents the primary physical process enabling the energy conversions that sustain Life.*

5.1.2.2.3 Heat Naturation of Protein

For our model contractile protein, poly(GVGVP), $\Delta H_t = +1.2$ kcal/mole-pentamer and ΔS_t(solution → phase separated) = +1.2/303 kcal/mole-degree K ≈ +4.0 cal/mole-pentamer-degree (EU).[4,14] Because 1 mole of (GVGVP) is 409 grams and 1 mole of water is 18 grams, on a per gram basis (4.0/409 = 0.0098 and 5.3/18 = 0.29) the entropy change for the melting of ice is 30 times greater (0.29/0.0098) than that of the transition for poly(GVGVP).

Despite the absorption of heat for the transition and the overall increase in entropy of +4.0 EU for the water plus protein, the protein component actually increases in order on raising the temperature. As unambiguously demonstrated by crystallization of a cyclic analog (see Figure 2.7), in this case the protein component of the water plus protein system becomes more ordered as the temperature is raised. For this and additional reasons, noted below in section 5.1.3, we call this transition exhibited by our model protein, poly(GVGVP), an *inverse temperature transition*.

5.1.2.2.4 The Inconspicuous Inverse Temperature Transition Is Equivalent to Cold Denaturation of Protein

In the early 1900s, it was recognized that lowering the temperature could cause enzymes to lose their catalytic activity due to an unfolding and/or disassembly of proteins and their subunits.[15] This is called *cold denaturation*.[16] Similarly with our model protein, lowering the temperature from above to below the *temperature interval* of the inverse temperature transition causes the model protein to unfold and disassemble, that is, to dissolve. In the case of our model protein, the reverse process of raising the temperature from below to above the temperature of the inverse temperature transition causes the massive, visible changes in solubility represented in Figure 5.1A. The phase diagram, discussed immediately below, provides systematic characterization of this phase separation. The phase separation analysis, however, accommodates many other energy inputs in addition to thermal energy, because many other energy inputs drive phase separation.

5.1.3 Phase Diagram for Inverse Temperature Transitions and Related Analyses

5.1.3.1 The Inverse Temperature Transition of Model Proteins

The model proteins of interest here are polymers of a distinctive family of repeating peptide sequences where the repeating unit may be as few as two residues or as many as several hundred residues. These elastic-contractile model proteins exhibit remarkably functional and distinctive reversible phase transitional behavior to increased polypeptide order on raising the temperature and on this basis was called an *inverse temperature transition*.[17] This inverse order–disorder transition demonstrable by elastic-contractile model proteins is the basis for many different energy conversions, including, we believe, the energy conversions existing in biology.[4,6]

5.1.3.2 Inverted Phase Transitional Behavior of Inverse Temperature Transitions

In general, solubility in a solvent, of whatever molecular species, increases with heating. The solubility of petroleum-based polymers in organic solvents and most petroleum-based polymers in water represent such cases. In general, solubility of the polymers increases on raising the temperature. For the case of the amount of solute within a solvent, phase diagrams graphically represent the change in solubility as a function of temperature. In the case of the phase diagram, the boundary between insolubility and solubility is called the *binodal* or *coexistence line*.[18] At the temperature of the coexistence line soluble and insoluble states coexist. In general, insolubility occurs at lower temperatures, and solubility takes place on increasing the temperature.

5.1.3.2.1 Our Model Proteins Become Insoluble on Increasing the Temperature

As demonstrated in Figure 5.3 for several model proteins, essentially unlimited solubility occurs at low temperature, and phase separation (insolubility) occurs as the temperature is raised. Also, for our model protein compositions,[19,20] the curvature of the coexistence line is inverted, having the shape of a valley instead of a smooth mountain peak. Because of this we call the phase transition, exhibited by elastic-contractile model proteins, an *inverse temperature transition*. Even more compelling reasons exist for the *inverse temperature transition* label.

5.1.3.2.2 Our Model Proteins Increase Order on Raising the Temperature

A vital property of these model proteins is that they are more ordered above the transition temperature defined by the binodal or coexistence line in Figure 5.3. The polymer component of this water–polypeptide system becomes more ordered or structured on increased temperature from below to above the transition. This behavior is the inverse of that observed for most systems, as discussed above. In particular, we developed the term *inverse temperature transition* when the precursor protein and chemical fragmentation products of the mammalian elastic fiber changed from a dissolved state, and therefore when molecules were randomly dispersed in solution, to a state of parallel-aligned twisted filaments as the temperature was raised from below to above the phase transition.[17,21,22]

Chemically synthesized high molecular weight polymers of repeating sequences found within the elastic fiber similarly form parallel-aligned twisted filaments (see Figure 5.4C) when the temperature is raised from below to above the coexistence line. Most definitively, cyclic analogues of the repeating peptide sequences crystallize when the temperature is raised above the transition temperature (see, e.g., Figure 2.5 and Figure 5.4A,B) and dissolve when the temperature is lowered below the transition.[23,24]

At first thought, this appears to contradict the second law of thermodynamics, which states

5.1 Introduction

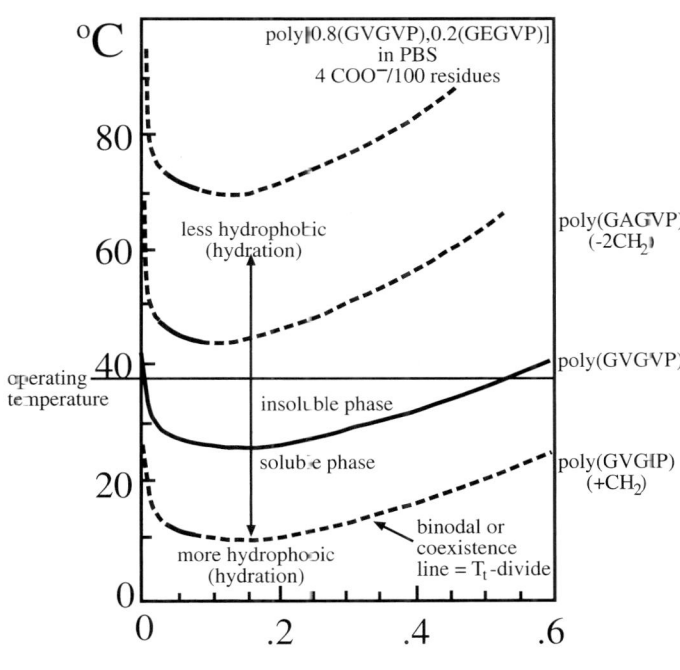

FIGURE 5.3. Phase diagram for several elastic-contractile model proteins, showing an inverted curvature to the binodal or coexistence line (when compared with petroleum-based polymers) that is equivalent to the T_t-divide, with the value of T_t determined as noted in Figure 5.1B. Solubility is also inverted with insolubility above and solubility below the binodal line, that is, solubility is lost on raising the temperature whereas solubility is achieved by raising the temperature of most petroleum-based polymers in their solvents. Note that addition of a CH_2 group lowers the T_t-divide and removal of the CH_2 group raises the T_t-divide. For these and the additional reason of increased ordering on increasing the temperature, the phase transitions of elastic-contractile model proteins are called *inverse temperature transitions*. (The curve for poly[GVGVP] is adapted with permission from Manno et al.[19] and Sciortino et al.[20]).

that the order of a system must decrease as the temperature is as raised, with the melting and vaporization transitions of water shown in Figure 5.2. This seeming violation of the second law of thermodynamics disappears on recognizing the presence of ordered water that surrounds oil-like side chains of the dissolved, low temperature state.[25,26] When the temperature is raised, this more-ordered water surrounding oil-like groups becomes less-ordered liquid or bulk water, and thereby provides the measured increase in entropy (see section 5.1.2.2) that drives the thermally induced phase separation.

5.1.3.3 A Glimpse at the Underlying Process Involving Water around Oil-like Groups

5.1.3.3.1 Effect of Adding CH_2 Units on Solubility in Water

Oil is composed mostly of $-CH_2$ and $-CH_3$ groups. When the small single oil-like group is attached to an OH, as in methanol, CH_3-OH, the methanol is infinitely soluble in water. For the series of CH_3-OH (methanol), CH_3-CH_2-OH (ethanol, the alcohol of the beverage variety), $CH_3-CH_2-CH_2-OH$ (1-propanol), $CH_3-CH_2-CH_2-CH_2-OH$ (1-butanol), $CH_3-CH_2-CH_2-CH_2-CH_2-OH$ (1-pentanol), the amount of the alcohol dissolvable in water decreases as each oil-like CH_2 unit is added. In fact, 1-octanol, $CH_3-CH_2-CH_2-CH_2-CH_2-CH_2-CH_2-CH_2-OH$, is insoluble in water, and the long chain, $CH_3-CH_2-CH_2-CH_2-CH_2-CH_2-CH_2-CH_2-$, is properly called *water fearing* or *hydrophobic*. Until the loss of solubility comes about, however, there is water surrounding the oil-like groups. For example, water surrounds the soluble oil-like butyl grouping, $CH_3-CH_2-CH_2-CH_2-$ when it is in water as 1-butanol. Although it may seem to be a contradiction in terms, this water that surrounds oil-like groups dissolved in water is called *hydrophobic hydration*. Significantly, the thermodynamic quantities of ΔH and ΔS describe changes in hydrophobic hydration. As discussed below, the formation of hydrophobic hydration occurs

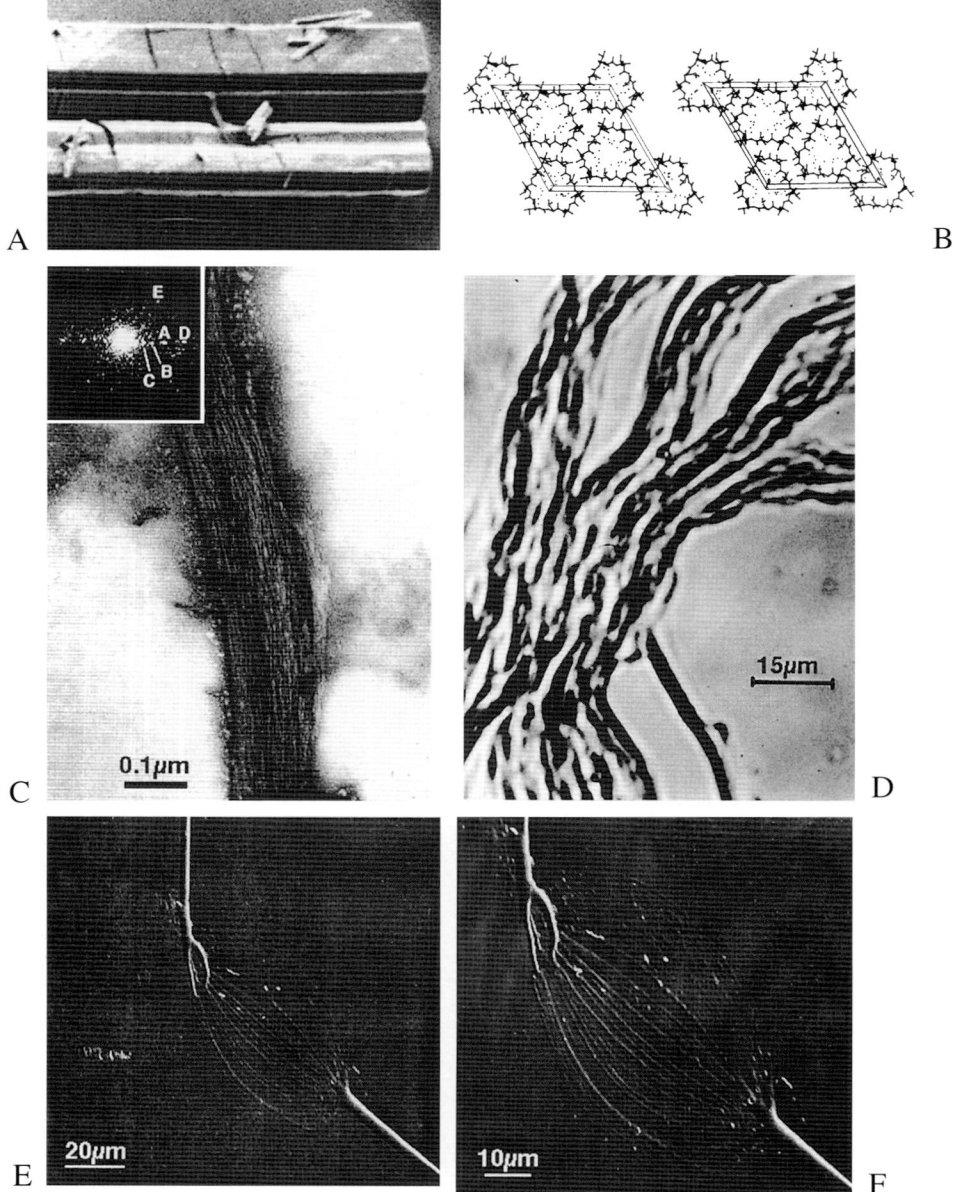

FIGURE 5.4. The products of the inverse temperature transition of elastic-contractile model proteins and their analogues are self-assembled (ordered) structures. (**A**) The cyclic analogue cyclo(GVGVAP)$_2$ crystallizes on raising the temperature through the inverse temperature and dissolves on lowering the temperature below the transition. (Adapted with permission from Urry et al.[23]) (**B**) The crystal structure of the cyclic analogue cyclo(GVGVP)$_3$, which also aggregates on raising the temperature and dissolves on lowering the temperature. (Reproduced with permission from Cook et al.[24]) (**C**) Electron micrograph of incipient aggregates of poly(GVGVP) after negative staining with uranyl acetate/oxalic acid showing parallel aligned, 5 nm diameter, twisted filaments that give the optical diffraction pattern shown in the inset. The twisted filaments are associated into a fibril of about 0.15 μm in cross section. (Reproduced with permission from Volpin et al.[54]) (**D**) Light micrograph, without any staining or fixative, showing aggregated state of poly(GVGVP) that has been chemically cross-linked above the temperature of the inverse temperature traansition due to the occasional inclusion of Lys and Glu residues. Fibers are seen to be composed of fibrils of a cross section similar to the fibril of part C. (Reproduced with permission from Urry.[119]) (**E,F**) Scanning electron micrographs at two different magnifications of a single fiber splaying out into parallel aligned fibrils and re-coalesced back into the same-sized fiber. (Reproduced with permission from Urry et al.[120])

5.1 Introduction

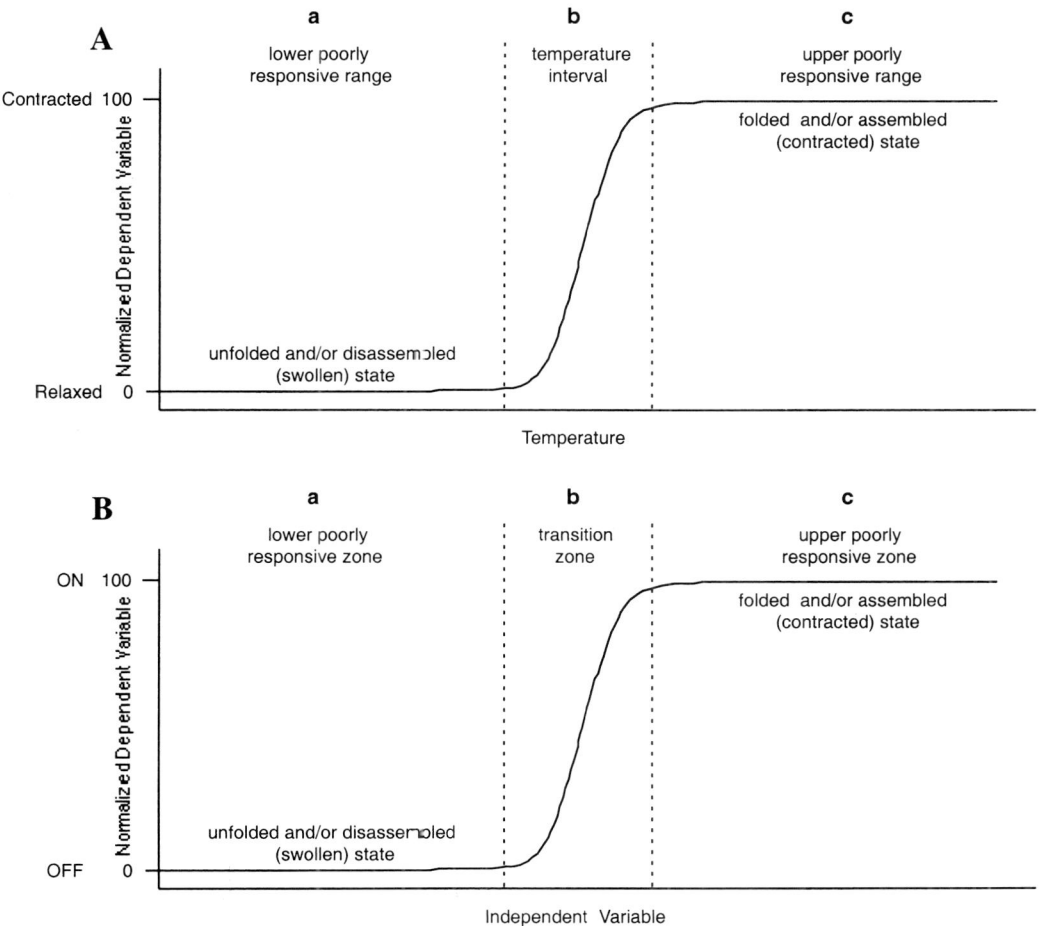

FIGURE 5.5. Transitions, plotted as independent variable versus dependent variable, showing a response limited to a particular range of independent variable. (**A**) Representation of the thermally driven contraction for an elastic-contractile model protein, such as the cross-linked poly(GVGVP), plotted as the percent contraction (dependent variable) versus temperature (independent variable). The plot shows a poorly responsive range below the onset of the transition, the *temperature interval* of the inverse temperature transition for hydrophobic association, and another poorly responsive region above the temperature range of the transition. (**B**) Generalized representation of transitions where any of the intensive (independent) variables of energies interconverted by designed elastic-contractile model proteins, such as temperature for thermal energy, chemical potential for chemical energy, electrochemical potential for electrical energy, and mechanical force for mechanical work, are plotted versus the normalized dependent variable. Again, there are the responsive *transition zone* and poorly responsive ranges above and below the transition zone.

with a negative ΔH and a negative ΔS. The signs of ΔH and ΔS for the formation of hydrophobic hydration and the way they enter into the expression governing solubility, of course, determine solubility or insolubility of oil-like groups and, thereby, provide the basis for energy conversions by the consilient mechanism.

5.1.3.3.2 Thermodynamic Expression for Solubility: The Gibbs Free Energy ($\Delta G = \Delta H - T\Delta S$)

The thermodynamic quantity that determines solubility is the Gibbs free energy, ΔG. It is made up of two terms, ΔH, the change in heat on dissolution, and $-T\Delta S$ where T is the temperature in degrees Kelvin (K) and ΔS quantifies the change in order attending dissolution. (These are the same thermodynamic quantities discussed above in relation to the phase changes exhibited by water and by proteins in water.) Accordingly, ΔG(solubility) = $\Delta H - T\Delta S$. When ΔG(solubility) for dissolution is negative, solubility occurs. As ΔG(solubility) becomes positive, solubility decreases.

In general, substances that form ions on dissolving in water evolve much heat; ΔH is negative. An example is caustic soda, NaOH. When water is carefully added to a test tube containing pellets of NaOH, they dissolve as Na^+ and OH^- ions with the production of much heat. The test tube becomes hot to the touch. In general, hydration of ions is an exothermic reaction. The same happens for glacial acetic acid (the major component of vinegar, CH_3–COOH, at a very high concentration). Glacial acetic acid must be added slowly and very carefully to water, because it dissolves as H^+ and CH_3–COO^- with the release of so much heat. ΔH for dissolution is large and negative. The full hydration of ions in water is a very exothermic, that is, very favorable, process.

5.1.3.3.3 The Thermodynamic Properties of Hydrophobic Hydration

Remarkably, in the above series of alcohols, addition of an oil-like CH_2 unit increases the release of heat when dissolved in water. Dissolving CH_3–CH_2–CH_2–CH_2–OH in water releases more heat per mole added than does dissolving CH_3–CH_2–CH_2–OH in water. *The dissolution of these oil-like CH_2 units in water is exothermic. Therefore, formation of hydrophobic hydration is exothermic.* Nonetheless, solubility decreases as the number of oil-like CH_2 units increases, despite a continuing favorable release of heat on dissolution. Now we must remember that ΔG(solubility) = $\Delta H - T\Delta S$ is the expression that governs solubility; and solubility occurs when ΔG is negative and decreases as ΔG becomes positive. Therefore, the loss of solubility, as the number of oil-like CH_2 units increases, is due to the sign and magnitude of the $(-T\Delta S)$ term. Based on the data of Butler,[25] the average change in the ΔH and $(-T\Delta S)$ terms for each of the four CH_2 units added on going from CH_3–OH to CH_3–CH_2–CH_2–CH_2–CH_2–OH is a favorable -1.4 kcal/mole-CH_2 for ΔH and an unfavorable $+1.7$ kcal/mole-CH_2 for $(-T\Delta S)$. Thus, insolubility ultimately happens when enough CH_2 units are added due to the larger and positive $(-T\Delta S)$ term. It measures the contribution to the Gibbs free energy due to an increase in order of water as it goes from being bulk water to being hydrophobic hydration. This ordering effect is so pronounced that Frank and Evans[26] referred to hydrophobic hydration as "icebergs" surrounding oil-like groups.

5.1.3.3.4 The Inverse Temperature Transition: A Hydrophobic Folding and Assembly Transition

The protein-based polymer is soluble in water at temperatures below its coexistence line where the hydrophobic residues are surrounded by hydrophobic hydration. As the positive $(-T\Delta S)$ term due to hydrophobic hydration becomes larger than the negative ΔH term, simply due to increasing the value of T, solubility of a protein-based polymer is lost, and it hydrophobically folds and assembles. The *inverse temperature transition* is a *hydrophobic association* transition.

5.1.3.4 Capacity to Move the T_t-divide as a Measure of Energy Available to the Consilient Mechanism (That Is, Moving the Temperature Interval)

5.1.3.4.1 Defining the T_t-divide

At the coexistence line of a phase transition, such as that plotted in Figure 5.3, soluble and

5.1 Introduction

insoluble states coexist. Heat added at this temperature, for an ideal case like ice dissolving in water, converts one state to the other rather than increases the temperature. At the temperature of the phase transition when there are two phases, the Gibbs free energy per mole of the molecules in the two phases are the same, that is, the chemical potential, $\mu = \Delta G/\text{mole}$, is the same for the molecules in solution and for the molecules in the phase-separated state. For the inverse temperature transition, the chemical potential for molecules in solution, that is, for molecules in the hydrophobically dissociated state, μ_{HD}, is the same as the chemical potential of the molecules in the separated, hydrophobically associated phase, μ_{HA}, that is, $\mu_{HD} = \mu_{HA}$. This means for the transition that $\Delta G_t = 0 = \Delta H_t - T_t \Delta S_t$, where the subscript **t** stands for the phase transition. Therefore, $\Delta H_t = T_t \Delta S_t$, and $T_t = \Delta H_t / \Delta S_t$, where T_t is the temperature for the transition. *Accordingly, the coexistence line may also be called the T_t-divide between soluble and insoluble states* (see Figure 5.3).[27] Importantly for the transition, $\Delta G_t = 0 = \Delta H_t - T_t \Delta S_t$ at the T_t-divide and $\Delta H_t = T_t \Delta S_t$.

5.1.3.4.2 Dependence of T_t on Chain Length and on Concentration

Meaningful comparison of the different means whereby the value of T_t may be moved requires definition of a common polymer chain length and concentration to function as a reference state.[19,20,28,29] Definition of the reference conditions occurred nearly two decades ago with chemically synthesized polymers,[28] before the following systematic comparisons began. In particular, chemically synthesized polymers dialyzed against 50,000 Da cut-off dialysis membranes[30] at 4°C (i.e., a temperature substantially below T_t) exhibit a lowering of T_t on increasing the concentration until a concentration of 40 mg/ml was reached. Increasing the concentration further had little effect on the value of T_t such that this became the concentration for comparison. This was extensively substantiated in the process of developing the phase diagrams for poly(VPGVG),[19,20,31] as the phase diagram is a plot of temperature versus concentration.

The chain length dependence was also determined on chemically synthesized polymers using dialysis membranes.[30] Subsequently, protein-based polymers using recombinant DNA technology were prepared with 41, 141, and 251 repeats of the pentamer.[29] Again, above 100,000 Da or over 200 repeats of the pentamer, the chain length dependence of T_t becomes limited. Accordingly using polymers of approximately 100,000 Da at concentrations of 40 mg/ml have become the reference conditions for the comparisons noted below.

5.1.3.4.3 Moving the T_t-divide by Adding a CH_2 Unit, That Is, Replacing Val by Ile (an Approximation to the Change in Gibbs Free Energy for Hydrophobic Association, ΔG_{HA})

From the data of Butler,[25] we can predict how the T_t-divide will change as CH_2 units are added or removed from the polymers plotted in Figure 5.3. If we add a CH_2 group to the **R**-group of the second Val residue of (GVGVP)$_n$, it becomes (GVGIP)$_n$, that is, the **R**-group for the V residue, $-CH(CH_3)_2$, becomes $-CH(CH_3)-CH_2-CH_3$ for an I residue. Because $\Delta H_t = T_t \Delta S_t$ for both compositions at their respective T_t-divides, the relevant equations become

$$\Delta H_t(\text{GVGVP}) = T_t(\text{GVGVP}) \Delta S_t(\text{GVGVP}) \quad (5.1)$$

and

$$\Delta H_t(\text{GVGIP}) = T_t(\text{GVGIP}) \Delta S_t(\text{GVGIP}). \quad (5.2)$$

Due to the experimental work of Butler and for reasons of the additive nature of the components within ΔH and of the components that make up ΔS, we separate the effect of the addition of a single CH_2 group and write that

$$\Delta H_t(\text{GVGIP}) = \Delta H_t(\text{GVGVP}) + \Delta H_t(CH_2) \quad (5.3)$$

and

$$T_t(GVGIP)\Delta S_t(GVGIP) \approx$$
$$T_t(GVGIP)[\Delta S_t(GVGVP) + \Delta S_t(CH_2)] \quad (5.4)$$

Substitution of Equations (5.3) and (5.4) into Equation (5.2) and subtracting Equation (5.1) gives

$$\Delta H_t(CH_2) - T_t(GVGIP)\Delta S_t(CH_2) =$$
$$[T_t(GVGIP) - T_t(GVGVP)]\Delta S_t(GVGVP) \quad (5.5)$$

Again, if the data of Butler[25] for the effect of addition of a CH_2 group on solubility are applicable, the hydrophobic association of the phase separation would be such that the magnitude of the $T_t(GVGIP)\Delta S_t(CH_2)$ term would be larger than that of the $\Delta H_t(CH_2)$ term, and for hydrophobic association, that is, for insolubilization, $\Delta S_t(GVGVP)$ is positive as more ordered hydrophobic hydration becomes less ordered bulk water on hydrophobic association. This means that $T_t(GVGIP)\Delta S_t(CH_2) > \Delta H_t(CH_2)$, such that

$$T_t(GVGVP) > T_t(GVGIP) \quad (5.6)$$

This is found experimentally for the phase separation of these model proteins. Addition of a CH_2 group lowers the value of T_t, that is, lowers the T_t-divide for $(GVGIP)_n$ as shown in Figure 5.3. The experimental estimates of T_t are nominally 25°C for $(GVGVP)_n$ and 10°C for $(GVGIP)_n$ in keeping with Equation (5.6). Also, removal of two CH_2 units, by replacing the first V residue of (GVGVP) with an alanine residue, A, having an **R**-group of $-CH_3$, results in a T_t value of 45°C. As expected from Equation (5.6) and from the data of Butler, $T_t(GVGVP) < T_t(GAGVP)$. This simple relationship between hydrophobicity and T_t provides insight into the T_t-based hydrophobicity scale for protein function and engineering, developed in section 5.3 below.

Returning to Equation (5.5), we define[32] $-\Delta G_{HA}(CH_2) \equiv \Delta H_t(CH_2) - T_t(GVGIP)\Delta S_t(CH_2)$ and rewrite Equation (5.5) to give

$$\Delta G_{HA}(CH_2) \approx [T_t(GVGVP) - T_t(GVGIP)] \Delta S_t(GVGVP) \quad (5.7)$$

In general, therefore,

$$\Delta G_{HA}(\chi) \approx [T_t(ref) - T_t(\chi)]\Delta S_t(ref) \quad (5.8)$$

or

$$\Delta G_{HA}(\chi) \approx -\Delta T_t(\chi)\Delta S_t(ref) \quad (5.9)$$

where χ is any variable that moves the T_t-divide from that of the reference (**ref**) model protein, and $\Delta G_{HA}(\chi)$ is the change in Gibbs free energy for forming the hydrophobically associated state of the model protein resulting from that variable. (Note that the change in Gibbs free energy is written here for hydrophobic association, that is, for insolubilization, which has the opposite sign of ΔG(solubility), as the latter considers the reverse process of solubilization or solubility discussed in section 5.1.3.3.) The approximation of Equation (5.9) is demonstrated below in Figure 5.10 in terms of the diagonal straight line through the sigmoid of data points. Within the approximation of Equation (5.9) $\Delta S_t(ref)$ would be 30 cal/mole-(GXGVP)-deg. That is the approximation of the practical T_t-based hydrophobicity scale.

Equation (5.9) provides an initial insight into the relationship between moving the T_t-divide and changes in free energy of hydrophobic association; it is of use when the value of $\Delta S_t(\chi)$, or $\Delta H_t(\chi)$ from which it is derived, is not available. Nonetheless, the limitations of Equation (5.9) become apparent when both $\Delta S_t(\chi)$ and $\Delta S_t(ref)$ are known, because reversal of roles of the reference and altered state should simply reverse the sign of $\Delta G_{HA}(\chi)$, and this is not the case due to the approximation of Equations (5.3) and (5.4).

As will be shown below on analysis of the data in Figure 5.10, a plot of T_t versus ΔG_{HA} has a sigmoid appearance. This means that the derivative dT_t/dG_{HA} gives a bell-shaped curve equal to $(dS_t)^{-1}$ and not a straight line as implied by Equation (5.9) and assumed by the T_t-based hydrophobicity scale. Equation (5.9), however, does appear to be meaningful when staying within the proper family of molecular systems, for example, as defined in Figure 5.10 by lines **b** (for the aliphatic side chains), **c** (for the aromatic side chains), and **d** (for the effect of charged species).

5.1 Introduction

Generally applicable statements result on simply subtracting Equation (5.2) from Equation (5.1) to give[9]

$$\Delta G_{HA}(\chi) = [T_t(\text{ref})\Delta S_t(\text{ref}) - T_t(\chi)\Delta S_t(\chi)] \quad (5.10a)$$

and as more recently derived[8]

$$\Delta G_{HA}(\chi) \approx [\Delta H_t(\text{ref}) - \Delta H_t(\chi)] \quad (5.10b)$$

One should realize that Equation (5.10b) includes the exothermic heat of van der Waals contacts due to the folding and assembly of the model protein during the transition. The association of model protein during the phase separation would be exothermic. The endothermic heats obtained from the inverse temperature (phase) transition arise from the conversion of hydration of oil-like groups to bulk water. Accordingly, the $\Delta G_{HA}(\chi)$ values calculated directly from the heat of the inverse temperature transition, as in Equations (5.10a) and (5.10b), would be minimal values when considered as the change in Gibbs free energy for conversion of hydrophobic hydration to bulk water (See Figure 8.1 and associated text).

5.1.3.4.4 Moving the T_t-divide by Replacing V by Other Naturally Occurring Amino Acids

Whatever amino acid replaces the valine (Val, V) residue, there results a change in the value of T_t. This dependency of T_t becomes a measure of the relative hydrophobicity of the different naturally occurring amino acid residues (see Table 5.1), called the T_t-based hydrophobicity scale.[4] Design of protein function by use of the T_t-based hydrophobicity scale requires knowledge of T_t only. Changing T_t provides a simple on–off switch for achieving function. Lowering T_t from above to below the operating temperature drives hydrophobic association, and raising T_t from below to above the operating temperature results in hydrophobic dissociation.

5.1.3.4.5 Moving the T_t-divide Without Changing the Amino Acid Composition

There are many ways, in addition to changing the amino acid composition of the model protein, to change the location of the T_t-divide. They fall into two main classes. One class is to have functional groups attached to the protein that can exist in two different states, for example, charged or uncharged, oxidized or reduced, phosphorylated or dephosphorylated, and so forth. The second class is to change the composition of the solvent in ways that alter the solvation of the oil-like groups without acting by changing the state of a functional side chain. Each of these means for moving the T_t-divide constitute energy inputs, and $\Delta G_{HA}(\chi)$, as identified above, becomes the energy output. Each energy input resulting in a $\Delta G_{HA}(\chi)$, depending on the sign of $\Delta G_{HA}(\chi)$, drives hydrophobic association (when negative) or dissociation (when positive). For cross-linked matrices, this is equivalent to contraction or relaxation, respectively, to produce motion. A negative $\Delta G_{HA}(\chi)$ produces contraction, and a positive $\Delta G_{HA}(\chi)$ results in relaxation. Examples of many energy inputs, as evidenced by changing T_t, are very briefly noted immediately below.

5.1.3.4.6 Moving the T_t-divide by Adding Protons to Carboxylates

Protonation, adding a proton, to the carboxylate group of the glutamic acid residue, or an aspartic acid residue, lowers the value of T_t and represents a chemical energy input that is one of the most effective ways to drive association of oil-like groups, that is, to result in the mechanical energy output resulting from contraction. Protonation of a carboxylate, $-COO^-$ + H^+ → $-COOH$, represents an energy input similar to reduction of a nicotinamide, and these are exceeded only by removal of a phosphate group in their effectiveness in lowering the value of T_t and consequently in producing a negative change in $\Delta G_{HA}(\chi)$, with which to drive hydrophobic association and its equivalent of contraction (see sections 5.14 and 5.15).

5.1.3.4.7 Moving the T_t-divide by Adding Electrons to Bound Oxidized Group

Reduction of an oxidized nicotinamide drives association of oil-like groups and is similar in effectiveness to protonation of a carboxylate in lowering the value of T_t. In oxidative

TABLE 5.1. T_t-based hydrophobicity scale for protein engineering of naturally occurring amino acid residues (in order of more hydrophobic [low T_t] to more polar [high T_t])[a].

Residue	R-group	Abbreviation	Letter	T_t^{ab}	ΔH_t (kcal/mole[c] ± 0.05)	ΔS_t (kcal/mole[c] ± 0.05)
Tryptophan	(indole CH$_2$-)	Trp	W	−90°C	2.10	7.37
Tyrosine	(−CH$_2$−C$_6$H$_4$−OH)	Tyr	Y	−55°C	1.87	6.32
Phenylalanine	(−CH$_2$−C$_6$H$_5$)	Phe	F	−30°C	1.93	6.61
Histidine	(−CH$_2$−imidazole)	His	H	−10°C		
Proline(calc.)[bd]	−CH$_2$CH$_2$CH$_2$−	Pro	P	(−8°C)		
Leucine	−CH$_2$CH(CH$_3$)$_2$	Leu	L	5°C	1.51	5.03
Isoleucine	−CH(CH$_3$)CH$_2$CH$_3$	Ile	I	10°C	1.43	4.60
Methionine	−CH$_2$CH$_2$SCH$_3$	Met	M	20°C	1.00	3.29
Valine	−CH(CH$_3$)$_2$	Val	V	24°C	1.20	3.90
Histidine	(−CH$_2$−imidazole H$^+$)	His$^+$	H$^+$	30°C		
Glutamic acid	−CH$_2$CH$_2$COOH	Glu	E	30°C	0.96	3.14
Cysteine	−CH$_2$SH	Cys	C	30°C		
Lysine	−CH$_2$CH$_2$CH$_2$CH$_2$NH$_2$	Lys°	K°	35°C	0.71	2.26
Proline(exptl)[e]	−CH$_2$CH$_2$CH$_2$−	Pro	P	40°C	0.92	2.98
Alanine	−CH$_3$	Ala	A	45°C	0.85	2.64
Aspartic acid	−CH$_2$COOH	Asp	D	45°C	0.78	2.57
Threonine	−CH(OH)CH$_3$	Thr	T	50°C	0.82	2.60
Asparagine	−CH$_2$CONH$_2$	Asn	N	50°C	0.71	2.29
Serine	−CH$_2$OH	Ser	S	50°C	0.59	1.86
Glycine	−H	Gly	G	55°C	0.70	2.25
Arginine	−CH$_2$CH$_2$CH$_2$NHC(NH)NH$_2$	Arg	R	60°C		
Glutamine	−CH$_2$CH$_2$CONH$_2$	Gln	Q	60°C	0.55	1.76
Lysine	−CH$_2$CH$_2$CH$_2$CH$_2$NH$_3^+$	Lys	K	120°C		
Tyrosinate	(−CH$_2$−C$_6$H$_4$−O$^-$)	Tyr$^-$	Y$^-$	120°C	0.31	0.94
Aspartate	−CH$_2$COO$^-$	Asp$^-$	D$^-$	120°C		
Glutamate	−CH$_2$CH$_2$COO$^-$	Glu$^-$	E$^-$	250°C		

[a] Scale uses poly[f_v(GVGVP),f_x(GXGVP)]. Poly(GVGVP) becomes reference for ΔT.
[b] T_t is the onset temperature for the hydrophobic folding and assembly transition, that is, inverse temperature transition, in pbs (0.15 N NaCl, 0.01 M phosphate) as determined by light scattering. The values are linearly extrapolated to $f_x = 1$ and rounded to a number divisible by 5. ΔH and ΔS are the values at $f_x = 0.2$ on the curve for a linear fit of the DSC derived endothermic heats and entropies of the transitions for the polymers in water.
[c] Per mole of pentamer.
[d] The calculated T_t value for Pro comes from poly(GVGVP) when the experimental values of Val and Gly are used. This hydrophobicity value of −8°C is unique to the β-spiral structure, where there is hydrophobic contact between the Val$_i^1$ γCH$_3$ and the adjacent Pro$_i^2$ δCH$_2$ and the interturn Pro$_{i+3}^2$ βCH$_2$ moieties.
[e] The experimental value determined from poly[f_V(GVGVP), f_P(GVGPP)].

phosphorylation of the inner mitochondrial membrane, the relationship between the change of state of a redox group and proton transport is central to the process of developing the proton gradient used to drive phosphorylation. As discussed in Chapter 8, section 8.3.4, during analysis of the structure and function of Complex III: ubiquinone: cytochrome c oxidoreductase (the cytochrome bc_1 complex) of the electron transport chain of the inner mitochondrial membrane, the oxidation of ubiquinol to produce a positively charged species and the reduction of ubiquinone to produce a negatively charged species both disrupt hydrophobic association to effect the release of two protons from the former and the uptake of two protons to the latter in effecting proton transport. In this manner and in the context of T_t, both the formation of positive charge and the formation of negative charge raise the value of T_t to disrupt hydrophobic association and thereby to allow, respectively, proton egress into the cytosol and ingress from the matrix for net transport across the membrane. In a different context, irreversible oxidation of protein components represents a means of inactivating protein function by irreversibly driving hydrophobic dissociation.

5.1.3.4.8 Moving the T_t-divide by Phosphorylation/Dephosphorylation

Phosphorylation, the covalent attachment of a phosphate to an OH group, has been to date the most effective way to raise the temperature of the T_t-divide. Thus, dephosphorylation, removal of phosphate, has been the most effective way to lower the temperature of the T_t-divide and thereby the most effective way to drive hydrophobic association and its equivalent of contraction. This is similar to a primary event in muscle contraction (see Chapters 7 and 8). The shift in the T_t-divide on binding of ATP can be as great as or greater than simple phosphorylation, depending on the interactions of ATP at the binding site. As discussed in Chapter 8, section 8.5, ATP binding drives hydrophobic dissociation, whereas loss of phosphate drives hydrophobic association both for the attachment to actin and for the power stroke to produce motion by the myosin II motor of muscle contraction.

5.1.3.4.9 Moving the T_t-divide by Ion Pairing with Oppositely Charged Vinegar-like R-groups

The ion pairing of sodium ion, Na^+, with negative carboxylate, COO^-, that is, $COO^- \bullet \bullet \bullet Na^+$, and the ion pairing of the ammonium ion functional group of lysine (Lys, K) with the chloride ion, that is, $-NH_3^+ \bullet \bullet \bullet Cl^-$, markedly lower the T_t-divide. The effectiveness of ion pairing in lowering the T_t-divide increases as the hydrophobicity (oil-like character) of the domain increases with which the charged group is associated. Interestingly, chloride ion (Cl^-) bridges between the $Val^1(\alpha-NH_3^+)$ and the Arg^{141}(guanidinium$^+$) in the ion-pairing network between subunits of deoxyhemoglobin, and the hydrophobic association of the deoxy state is further stabilized by the replacement of the $\alpha-NH_3^+ \bullet \bullet \bullet Cl^-$ with the carbamate, $R-NH-COO^-$, resulting from CO_2 pick up in the tissues (see Chapter 7). The most effective ion pairing involves association, within a hydrophobic domain, of calcium ion with a pair of carboxylates, that is, $-COO^- \bullet \bullet \bullet Ca^{2+} \bullet \bullet \bullet ^-OOC-$. Again, this has a familiar ring as the triggering event for muscle contraction (see Chapters 7 and 8).

5.1.3.4.10 Moving the T_t-divide by Changing the Solvent, That Is, Adding Salt to the Solution

Even when there are no vinegar-like (functional) groups in the model protein, as is the case for poly(GVGVP), adding salt, NaCl, lowers the T_t-divide. This effect is very anemic, however, requiring an order of magnitude (some 10 times) more salt than when there is a functional group with which to ion pair in order to drive hydrophobic association.

5.1.3.4.11 Moving the T_t-divide by Changing the Solvent, That is, Adding Organic Solutes to the Solution

The most effective organic solute in raising the T_t-divide is sodium dodecyl sulfate (SDS).

Much less effective is guanidinium hydrochloride, followed by urea. Glycerol and ethylene glycol have only a small effect in lowering the T_t-divide, and trifluoroethanol is somewhat more effective in lowering the T_t-divide. On this basis, the effectiveness of SDS polyacrylamide (SDS-PAGE) gels in achieving protein separations based on size becomes apparent.

5.1.3.4.12 Moving the T_t-divide by Increasing the Pressure

Especially when there are amino acid residues with aromatic side chains, such as tryptophan (Trp, W), phenylalanine (Phe, F), and tyrosine (Tyr, Y), increasing the pressure raises the temperature of the T_t-divide and favors protein unfolding (see the structures of the **R**-groups in Table 5.1).

5.1.3.4.13 Moving the T_t-divide by a Bound Chromophore That Changes Its Oil-like Character on Absorbing Light

Any chromophore attached to the protein that changes its hydrophobicity on absorbing a photon of light will change the temperature of the T_t-divide. The absorption of light is unique among the energy conversions of the consilient mechanism in that it is irreversible, for example, stretching does not generally cause the emission of light. To the best of our knowledge, there is no simple reversal of a consilient process driven by the absorption of light that can result in the emission of the same frequency of electromagnetic energy. The energy of a relevant photon is some 70 kcal/mole, whereas the energy output, $\Delta G_{HA}(\chi)$, will be an order of magnitude less, that is, the energy changes of biology are generally within ±8 kcal/mole.[33]

5.1.3.5 What Causes Loss of Solubility on Raising the Temperature from Below to Above the T_t-divide?

By definition of the change in Gibbs free energy, ΔG, solubility occurs when ΔG(solubility) is negative, and solubility is lost as ΔG(solubility) becomes positive. Here we restate that shown very early by Butler.[25] ΔH is negative and $-T\Delta S$ is positive when oil-like groups dissolve in water, that is, when hydrophobic hydration forms. Specifically, the addition of the four CH_2 moieties in the soluble alcohol series from methanol to n-pentanol results in an average $\Delta H/CH_2$ of -1.4 kcal/mole-CH_2, that is, formation of hydrophobic hydration results in a favorable release of heat. But $(-T\Delta S/CH_2)$ is $+1.7$ kcal/mole-CH_2. This is why solubility in water is ultimately lost when enough CH_2 moieties have been added such that ΔG has become positive.

Accordingly, for a fixed amount of hydrophobic hydration, simply raising the temperature linearly increases the magnitude of the positive $(-T\Delta S)$ term until it becomes greater than the negative ΔH term, that is, until ΔG becomes positive and solubility is lost. The result is the association of hydrophobic groups as an integral part of the *inverse temperature transition* of the hydrophobic folding and assembly transition. *Solubility is lost, which is readily seen in our model proteins as a phase separation, simply because increasing the temperature increases the magnitude of the positive $(-T\Delta S)$ term until it becomes greater than the negative ΔH term. Contrary to common description, solubility is not lost because of thermal destructuring, melting, or disordering of the hydrophobic hydration.*[34] As shown below in section 5.7.6 and Figure 5.27, the T_t-divide occurs at a higher temperature when the amount of hydrophobic hydration for a model protein composition becomes less.

5.1.3.6 An Experimental Method for Following Changes in Hydrophobic Hydration

Improved insight into the role of hydrophobicity in protein folding, assembly, and function requires an experimental means with which to estimate and follow changes in hydrophobic hydration as a function of different variables. Since the work of Butler[25] and of Frank and Evans,[26] the search has been for an experimental method with which to characterize changes in hydrophobic hydration and particularly to do so during changes in functional state of the protein. This goal was reached with the

5.1 Introduction

model proteins of focus here in part because conditions are possible where most of the water in the system can be hydrophobic hydration.[35]

The experimental method is microwave dielectric relaxation, which uses the same energy as that of a kitchen microwave oven. Hydrophobic hydration absorbs energy in the microwave frequency range at a slightly lower frequency than liquid water absorbs energy in microwave ovens to heat foods. It is now possible to follow the loss of hydrophobic hydration as insolubility takes place, that is, as hydrophobic association occurs, which includes intramolecular hydrophobic folding and/or intermolecular assembly. Most significantly, in section 5.7.3 and Figures 5.24 and 5.25, we can see the loss of hydrophobic hydration as hydrophobic association occurs and as vinegar-like groups ionize and take limited hydration away from hydrophobic groups in the process of obtaining their own hydration.

5.1.3.7 In Summary, Hydrophobic Hydration Disappears for Two Different Reasons with Opposite Consequences of Insolubility and Solubility

5.1.3.7.1 Hydrophobic Hydration Increases with Increased Number of Oil-like Groups Until Solubility Is Lost and Then Hydrophobic Hydration Disappears

Hydrophobic hydration increases as the number of oil-like groups increases, and then it abruptly disappears when insolubility is reached. As shown above with the expression $\Delta G(\text{solubility}) = \Delta H - T\Delta S$, insolubility occurs as the positive $(-T\Delta S)$ term becomes greater than the negative ΔH term. This is one reason for the disappearance of hydrophobic hydration. Knowledge of the source of insolubility due to hydrophobic association provides one of the two primary keys to understanding changes in protein folding during function. *De novo* design of elastic-contractile model proteins for diverse energy conversions using the T_t-based hydrophobicity scale demonstrates the usefulness of this view.

5.1.3.7.2 Hydrophobic Hydration is Destroyed as Proximal Vinegar-like Species Achieve Hydration on Becoming Charged

Formation of charged species, by the very act of requiring more hydration when ionized, destroys proximal hydrophobic hydration in the process of becoming charged. Consider a spatially localized association of oil-like domains in water. As the clusters of oil-like groups representing each domain begin to dissociate, hydrophobic hydration forms. As too much hydrophobic hydration forms, the positive $(-T\Delta S)$ term becomes larger than the negative ΔH term, resulting in a positive ΔG, and the association reforms. If, however, a proximal charged species should form concurrently, the nascent hydrophobic hydration is destroyed as the ionizing vinegar-like species gains its hydration. The result is a negative ΔG and the dissociation stands. that is, solubility occurs. To be sure, water molecules can continue to be adjacent to the exposed hydrophobic groups, but in this case the water molecules may no longer have the structure represented in Figure 2.8. In our view, this is the second of the two primary keys to understanding changes in protein folding in the process of function. Experimental studies on model proteins, designed to perform diverse energy conversions, as reviewed in this chapter, exemplify these explanations.

5.1.4 Heating/Cooling Pumps Iron!

5.1.4.1 Raising the Temperature from Below to Above the Temperature Interval of the Phase Transition Produces Folding and Association with the Result of Motion

The phase separated state of the parent elastic protein-based polymer, $(\text{GVGVP})_n$,[13] at 37°C is about 50% polymer and 50% water by weight. This was the composition for the transitions represented in Figure 5.2. On exposure of this state to 20 Mrad[36] of γ-irradiation, a nearly ideal elastic band forms[37] with properties similar to those of the major elastic arteries. On lowering the temperature to 25°C or below, the elastic band swells and increases its volume 10-fold. A return to body temperature, 37°C (98.6°F), causes the band to contract by hydrophobic association, which is to deswell. The γ-irradiation cross-linked (GVGVP)$_n$ forms an elastic-contractile band.

We then hang a weight on the elastic band at 25°C and it stretches until the elastic force due to deformation matches the attached weight. We then raise the temperature to 37°C; the band contacts and lifts the weight. Insignificant contraction occurs below 25°C, and insignificant contraction occurs above 37°C. In the *temperature interval* from 25° to 37°C, thermal energy converts to the mechanical work of lifting the weight (see Figures 2.4, 2.5, and 2.6A). This is *pumping iron*, which we all recognize as work. When thermal energy *pumps iron*, the technical term becomes *thermomechanical transduction*. Thermal energy, by way of this model protein-based machine, converts into mechanical work.

5.1.4.2 Graphic Representation of a Transition from One State to Another as Temperature Is Raised from Below to Above a Temperature Interval

Figure 5.5A plots the independent variable, temperature, and the resultant change in the dependent variable, contraction. As noted above, little change occurs below and little change occurs above the *temperature interval*. Also, as noted below, adding chemical energy in the form of increased concentration of salt lowers the temperature range of the *temperature interval*. This represents in graphic form the depiction of thermally driven contraction in Figure 2.4.

5.1.5 Pumping Iron without Changing Temperature

5.1.5.1 Using Salt with an Uncharged Polymer to Move the Temperature Interval Over Which Thermally Driven Contraction Occurs

The addition of 58 grams of NaCl, table salt, to 1 liter of water containing the elastic band of cross-linked (GVGVP)$_n$ lowers the *temperature interval* for the phase separation by 15°C. (This is approximately equal to dissolving the entire contents of the common grocery store container of table salt in 2 gallons of water.) Now, raising the temperature from 10° to 25°C in this 1 N salt solution causes the elastic strip to contract and lift a weight. Under these circumstances no significant contraction occurred below 10°C, and no significant contraction occurs above 25°C. In the *temperature interval* from 10° to 25°C, thermal energy again performs mechanical work. Now the work of lifting the weight is complete at 25°C in 1 N NaCl rather than at 37°C in salt-free water.[38]

When we again hang the weight on the swollen elastic band at 25°C in pure water and then add the above noted quantity of salt, the elastic band contracts and lifts the weight without heating and without a change in temperature. Further addition of salt at 25°C has only a limited effect. Addition of salt constitutes an input of chemical energy, and, under these circumstances, the addition of chemical energy results in the performance of mechanical work. The technical term is *chemomechanical transduction*. Chemical energy, by means of our model protein-based machine, produces mechanical work without a change in temperature. In the absence of a significant vinegar-like **R**-group, an energy other than heat can be used to produce mechanical work. In this case, however, performance of a modest amount of mechanical work required much chemical energy. With the correct vinegar-like **R**-group and chemical energy input, more efficient chemo-mechanical transduction becomes possible. As shown in Figure 5.5B, the independent variable can be chemical potential, the change in Gibbs free energy per mole of chemical added.

5.1.5.2 Using a Chemical Couple to Lower the Temperature Interval from Above to Below a Given Temperature Rather Than Raising the Temperature

Living organisms function without the changes in temperature required for thermomechanical transduction by the consilient mechanism. What makes the mechanism consilient or pervasive, however, is that many different energy inputs acting on many different vinegar-like **R**-groups change the temperature interval over which contraction occurs, and the resulting energy conversions can be very efficient. Now, as noted with the salt-driven contraction above, it is emphasized that energy conversions occur without a change in temperature.

In general, a change in polymer composition provides a particular vinegar-like group with which to access other energies. In this general case, the other energy input changes the vinegar-like group from a less to a more oil-like state. The functional group of the major component of vinegar provides a very effective chemical couple, namely, $-COOH/-COO^-$. The addition of acid, H^+, to $-COO^-$ converts the very vinegar-like carboxylate, $-COO^-$, to the more oil-like $-COOH$. This protonation of a single carboxylate group in the 100 residues of poly(GVGVP) can lower the temperature interval by tens of degrees centigrade and can require far less chemical energy to pump iron than is required in the NaCl example above for the uncharged polymer. Protonation/deprotonation very efficiently pumps iron.

Alternatively, the amino/ammonium chemical couple, $-NH_2/-NH_3^+$, functions similarly, although the change from $-NH_2$ to $-NH_3^+$ is only about half as effective in changing the value of T_t as is the $-COOH/-COO^-$ chemical couple (see Table 5.1).

5.1.5.3 Electrons, Added to a Bound Redox Couple, Lower the Temperature Interval from Above to Below a Given Temperature

Certain vitamins like B_2 (riboflavin) and B_3 (niacin) become chemically dressed up for attachment to proteins. As such these attached vitamins become redox couples that can accept electrons (become reduced) and give up electrons (become oxidized). A change in the redox state changes the temperature interval for the phase transition, that is, moves the T_t-divide. Just as protonation of a carboxylate lowers the temperature interval, so too does adding electrons to the oxidized state of a redox couple. Lowering the temperature interval from above to below the operating temperature by reduction drives contraction and performs mechanical work. The technical term is *electromechanical transduction*. In short, reduction/oxidation pumps iron.

5.1.5.4 Graphic Representation of the Generalized Transition Zone for Contraction Due to a Generalized Independent (Intensive) Variable

As mentioned above in reference to Figure 5.5A, as the temperature is raised, contraction of a band composed of elastic-contractile model protein occurs. Contraction occurs as the temperature is raised through a temperature interval. Crossing over the T_t-divide, defined in Figure 5.3, is to pass through the temperature interval over which contraction occurs; it is the result of the phase separation, specifically of the inverse temperature transition. Furthermore, the temperature interval for contraction occurs at a lower temperature when the model protein is more hydrophobic and at a higher temperature when the model protein is less hydrophobic.

As noted above, the temperature interval shifts on changing the concentration of a chemical, that is, on changing the chemical potential of a chemical for which the model protein is responsive, as in protonation of a carboxylate. Similarly, the temperature interval shifts on changing the availability of electrons, that is, on changing the electrochemical potential, such that an oxidized component of a redox couple becomes reduced.

Now, instead of plotting temperature, the chemical potential or electrochemical potential can be plotted, and contraction occurs as the concentration of a chemical passes through the critical range for reaction or as the availability of electrons passes through a critical range to achieve reduction. The range over which the contraction occurs is the *transition zone*, and the plot looks the same as in Figure 5.5A for temperature. Thus, the graphic representation in Figure 5.5B is common for all of the independent variables, such as temperature, pressure, and chemical potential, that can drive contraction, in which case contraction becomes the dependent variable.

In this perspective, the independent variables are the intensive variables of the free energy. The input energy to drive contraction, for example, is the product of two quantities, the intensive variable times the extensive variable. In the case of protonation of a carboxylate, the intensive variable of chemical potential is proportional to the change in concentration of the protons over which the contraction occurs, and the extensive variable is the number of protons utilized in driving the contraction. In the case of the intensive variable of temperature, the extensive variable is the entropy of the transition (ΔS_t), that is, the amount of heat (calories) divided by the temperature, and the input energy, of course, is simply the heat of the transition, ΔH_t (= $\mathbf{T_t}\Delta S_t$) at the $\mathbf{T_t}$-divide.

5.1.6 Energy Conversions in Addition to Pumping Iron

5.1.6.1 Energy Conversions Not Involving Mechanical Work

There are at least six kinds of free energy that interconvert by the consilient mechanism. They are mechanical, thermal, pressure-volume, chemical, electrical, and electromagnetic energy, for example, light, and the corresponding intensive variables are mechanical force, temperature, pressure, chemical potential, electrochemical potential, and electromagnetic radiation frequency. Above we noted only three energies (thermal, chemical, and electrical) of the five that can provide input energy for the performance of mechanical work. In fact any pair of the five, leaving mechanical energy aside, interconvert one into the other by the consilient mechanism, as treated more fully below in section 5.5. Immediately below we note interconversion of the pair of energies, chemical and electrical. In fact, some 18 classes of pairwise energy conversions occur by the consilient mechanism (see section 5.6).

5.1.6.2 Conversion of Electrical Energy into Chemical Work and Vice Versa

When a properly designed model protein contains both protonation/deprotonation chemical couples and reduction/oxidation redox couples, reduction drives contraction and protonation drives contraction. Most importantly, however, reduction (electrons) drives proton (acid) uptake, and protonation drives electron uptake. The former is *electro-chemical transduction,* and the latter is *chemo-electrical transduction.* This is the nature of the consilient mechanism as further described below. Two distinct energy sources, each of which performs mechanical work, can interconvert by effecting the changes in hydrophobic folding and assembly of the consilient mechanism.

5.1.7 The Consilient Mechanism in a Nutshell: The Comprehensive Hydrophobic Effect

5.1.7.1 Hydrophobic Replaces Oil-like

Hydrophobic, meaning *water fearing*, is the more technical term for *oil-like*. Instead of speaking of water surrounding oil-like groupings of atoms, or oil-like hydration, the term *hydrophobic hydration* is generally used. Even in the scientific community, however, those not directly active in the field find the term *hydrophobic hydration* contradictory and troublesome. Nonetheless, hydration of hydrophobic groups does occur, as shown in the pentagonal arrangements of water molecules in Figure 2.8 surrounding an oil-like group, and understanding the energetics of hydrophobic hydration is central to the consilient mechanism for function of protein-based machines. Several key aspects of hydrophobic hydration that constitute the comprehensive hydrophobic effect are briefly noted in this section and then expanded upon in section 5.7.

5.1.7.2 The Comprehensive Hydrophobic Effect Results from the Actual or Potential Hydrophobic Hydration That Can Occur for a Pair of Associable Hydrophobic Surfaces or Domains

If a pair of hydrophobic domains capable of association and dissociation would have too much hydrophobic hydration for a given temperature when fully exposed to water, the result is insolubility of the domains, that is, hydrophobic association of the domains with loss of the hydrophobic hydration. In our analysis we use the change in Gibbs free energy, ΔG, which governs solubility under common experimental conditions of constant temperature and pressure. To reiterate, ΔG involves two terms. The first term is the heat change (ΔH) that occurs, for example, as the molecule dissolves in water. When heat is released on dissolution, ΔH is negative. Release of heat on addition to water indicates solubility. The second term is −TΔS, where ΔS is the entropy change on dissolution. In particular, ΔG(solubility) = ΔH − TΔS. For the result of insolubility of hydrophobic domains, the (−TΔS) term is positive and of greater magnitude than the negative ΔH term.

5.1.7.3 Competition for Hydration Between Apolar (Hydrophobic) and Polar (e.g., Charged) Groups Controls Amount of Hydrophobic Hydration

If a charged species appears proximal to the associated pair of hydrophobic domains (within a few nanometers) as the pair of domains exhibit an opening fluctuation, the ionizing species destructures the nascent hydrophobic hydration in achieving its own hydration.[4,35] The decrease in hydrophobic hydration (in the number of pentagonally arranged water molecules) results in a smaller positive (−TΔS) term, the negative ΔH term dominates, and the hydrophobic domains dissociate. Also, of course, association occurs at a higher temperature when the positive (−TΔS) term becomes larger than the negative ΔH term, that is, **T$_t$** is at a higher temperature.

5.1.7.4 The Competition for Hydration Also Is Responsible for Large Hydrophobic-induced pKa Shifts with Positive Cooperativity That Results in Efficient Protein-based Machines

Competition for hydration between polar (e.g., charged) and hydrophobic (apolar) groups,[39] called the *apolar–polar repulsive free energy of hydration*, ΔG_{ap}, controls the amount of hydrophobic hydration.[4,35] Competition for limited hydration can result in large pKa shifts, because the vinegar-like group must do the work of destroying hydrophobic hydration in the process of obtaining its own adequate hydration.[40,41] In particular, carboxyls occur as part of hydrophobically associated domains that undergo opening and closing fluctuations. During an opening fluctuation, a carboxyl has the opportunity to ionize, but if the hydrophobic hydration is too great it recovers its proton. This continues until the pH is so high that the chance of proton recovery is too small, and a carboxylate forms, destructuring hydrophobic hydration to do so. The first carboxylate to form reached out a significant distance in gathering its hydration shell. The next carboxyl that has the opportunity to form a carboxylate during an opening fluctuation finds it has less hydrophobic hydration to destructure, and as a consequence its probability of remaining ionized increases. Thus the probability for the third carboxyl to remain ionized is even greater, and so on. *This is positive cooperativity described at the molecular level*. The positive cooperativity arises out of competition for hydration between hydrophobic and charged groups, and the mechanism requires that the energetics of hydrophobic-induced pKa shifts parallel the energetics reflected in the Hill coefficient that describes cooperativity. As shown in Figures 1.2 and 1.3 for **Model proteins i, ii, iii, iv**, and **v** and as will be further considered in section 5.8, nonlinear increases in pKa resulting from linear increases in hydrophobicity parallel nonlinear increases in Hill coefficients.

5.1.7.5 Competition by Positively and Negatively Charged Species for Hydration that Is Too Limited by Strongly Held Hydrophobic Hydration is Responsible for Ion Pair Formation in Protein Folding and Assembly

Consider an ion pair-containing hydrophobic association induced to open, due to the emergence of a new proximal polar (e.g., charged) species. These newly emergent charged species (formerly ion paired in the associated hydrophobic domains) propagate destruction of hydrophobic hydration, causing further removed hydrophobic domains to dissociate. In this manner, hydrophobic disassociation with emergence of separated positive and negative groups propagates hydrophobic dissociation outward many nanometers in a domino-like effect. As a consequence, hydrophobic dissociations occur in a positively cooperative manner with the result of efficient free energy transduction.[42]

Thus, common in the consilient mechanism is formation of ion pairs, the association of negatively charged with positively charged vinegar-like groups, each of which are part of an oil-like domain. The association of oppositely charged groups from two separate oil-like domains occurs as the oil-like property of the domains gains dominance, but then dissociation occurs as the pair of oil-like domains lose dominance on introduction of a new more polar group, for example, as a charged vinegar-like component is introduced into the vicinity of ion-pair-containing hydrophobic domains.

5.1.7.6 Multidimensional Representations of Protein-based Polymer Function as Molecular Machines

5.1.7.6.1 Seven-dimensional Phase Transitional Space for Protein-based Polymer Function as Molecular Machines

The position of the T_t-divide that separates soluble from insoluble (hydrophobically associated) states in the phase diagram depends on seven variables: on the six intensive variables[43] of temperature, chemical potential, electrochemical potential, mechanical force, pressure, and electromagnetic radiation, and on polymer volume fraction or concentration. Therefore, diverse protein-catalyzed energy conversions by the consilient mechanism result from designs that control the location of the T_t-divide in this seven-dimensional phase transitional space. Complete mathematical description has yet to be written for representation of the T_t-divide in seven-dimensional phase transitional space, but it may prove to be more relevant to molecular machines to develop representation of energy conversion in multidimensional free energy space.

5.1.7.6.2 Multidimensional Free Energy Space for Protein-based Polymer Function as Molecular Machines

On inclusion of the six energies interconverted by the consilient mechanism (see Figure 5.2, below) with the concentration dependence results in a seven-dimensional free energy space. In this connection a relationship between the magnitude of the shift of the T_t-divide and the change in Gibbs free energy of the protein-based polymer is given in Equation (5.9). Knowledge of the relevant ΔS_t(reference) in Equation (5.9) for a given variable, χ, would allow calculation of the complete χ-dimensional, energy conversion landscape of relevance to a specific protein-based machine. More comprehensive yet would be to break down each of the six energies into their intensive and extensive variables. In this case, the general description for energy conversion becomes a 13-dimensional, free energy landscape for energy conversion by diverse protein-based machines. In practice, graphic representations usually plot two dimensions or at most three dimensions. Nonetheless, complete description of this multidimensional dependency is required for the desired level of understanding of protein function and in order most effectively to engineer diverse protein-based machines.

5.2 Molecular Structure and Elasticity of Model Protein

Early reviews are available on the development of the molecular conformation of the tetrapeptide, pentapeptide, and hexapeptide repeating sequences of elastin[44,45,] and on the mechanism of elasticity of the basic elastic-contractile model protein.[46,47] These may be sought for historical background. Recent reviews of the mechanism of elasticity[48,49] and its relation to contractility[9] may be examined for a more

5.2.1 The Sequence of Amino Acid Residues: The Primary Structure

This journey of personal enchantment leading to the consilient mechanism of protein-based energy conversion began with the parent model protein (Gly-Val-Gly-Val-Pro)$_n$.[13] This five amino acid residue repeat, referred to as a *pentamer*, recurs 11 times in the elastic fibers of the pig and cow, that is, the subscript n is 11.[50,51] In our designed model proteins the pentamer generally repeats some 200 times or more, that is, the subscript n is often greater than 200. These protein-based polymers may also be called *polypentapeptides*. The general way to indicate this large repeating pentamer is poly(GVGVP) or (GVGVP)$_n$, where the value of n is specified for a given preparation. When made by genetic engineering, polymers of a precise number of repeats are obtainable. In Figure 5.1, the data represent those of a polypentapeptide comprised of 251 repeats, that is, (GVGVP)$_{251}$.

5.2.2 Secondary, Tertiary, and Quaternary Structures

5.2.2.1 Hydrogen Bonds Between Backbone Peptide Groups: The Secondary Structure

The molecular structure of poly(GVGVP) is one of a series of β-turns, one in each pentamer, inserted by the Pro-Gly sequence as schematically indicated in Figure 5.6A. Because of the location of the β-turn, which is a 10-atom hydrogen bonded ring involving the Val^{1}C-O–HNVal^4, early representations of the polypentapeptide were written as poly(VPGVG) or (Val^1-Pro^2-Gly^3-Val^4-Gly^5)$_n$.[52] The detailed β-turn shown in Figure 5.6B was obtained from the crystal structure of the cyclic molecule comprised of three pentamers[24] and was developed independently by the physical methods of nuclear magnetic resonance and circular dichroism and by the computational methods of molecular mechanics and dynamics.

The β-turn, the 10-atom hydrogen bonded ring involving the Val^1-CO to the Val^4-NH shown in Figure 5.6B, is the single secondary structural feature, the only hydrogen bond, in the repeating pentamer. Studies using Raman scattering indicate the presence of the β-turn below T_t, before the phase transition, and indicate no change in secondary structure, that is, no change in hydrogen bonding during the hydrophobic folding and assembly process attending the phase change.[53] This means during the phase transition that the significant changes within and between the chain molecules are changes in hydrophobic association.

5.2.2.2 Hydrophobic Association Represents Tertiary and Quaternary Structure

On raising the temperature above 25°C, the series of β-turns optimizes hydrophobic contacts within a single chain by wrapping up into a helical structure, called a *β-spiral*, as shown in detail by stereo pair in Figure 5.6E. The formation of intramolecular oil-like contacts, for example, the interaction between the ProβCH_2^i of pentamer i and the ValγCH_3^{i+3} of pentamer i + 3 in the sequence, may be considered changes in tertiary structure attending the inverse temperature transition. Finally, hydrophobic association between chains results in a twisted filament structure as indicated in Figure 5.5F. These are quaternary structural changes as developed from electron micrographs of negatively stained incipient aggregates formed on heating, from optical diffraction of the micrographs,[54] and from molecular mechanics and dynamics calculations.[55,56] In reality, the hydrophobic contacts within and between chains develop simultaneously each assisting the other, that is, in a cooperative manner.

For simplicity, we often display a single chain in the β-spiral structure; in reality, however, we do not expect such to form readily in isolation, but rather as part of a twisted filament as in Figure 5.6F. Accordingly, the schematic representations of the β-spiral structure of Figures 5.6C–E are regularly used to simplify but adequately make the intended point.

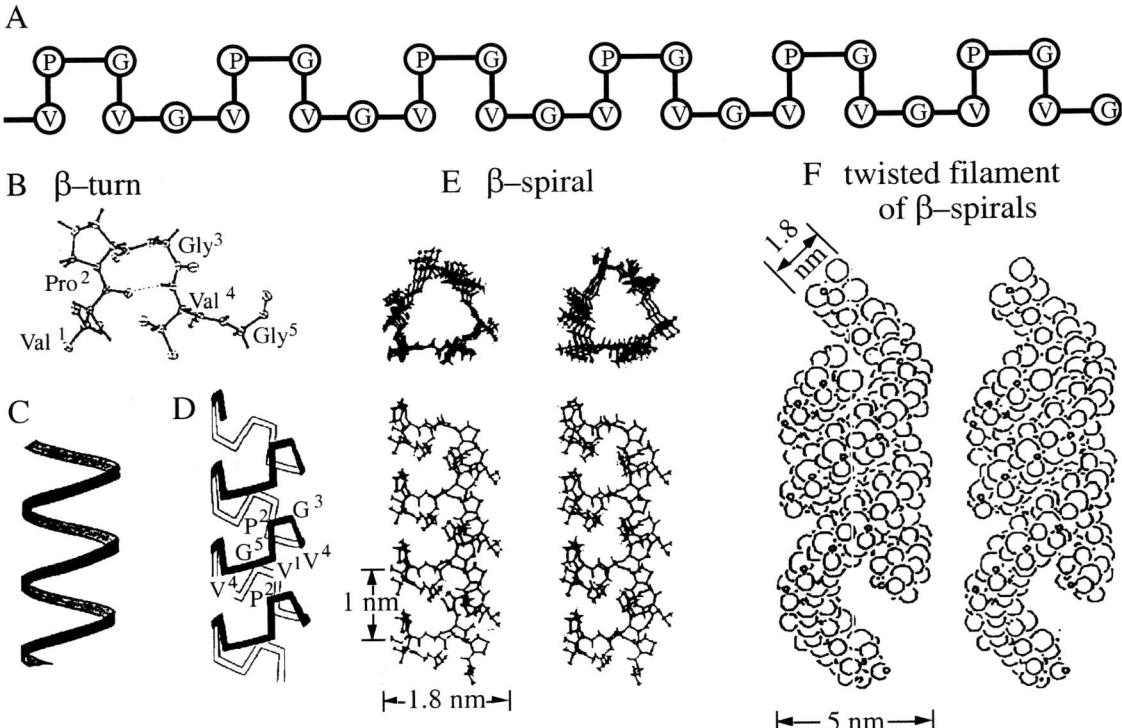

FIGURE 5.6. Description of the molecular structure of the parent model protein (GVGVP)$_n$, which should be recognized as equivalent to (VPGVG)$_n$. (**A**) A series of β-turns involving residues VPGV. (**B**) The β-turn obtained from the crystal structure of the cyclic conformational correlate cyclo(GVGVP)$_3$ shown in Figure 5.4B. (**C,D**) Schematic helical representations without β-turns (in **C**) and showing β-turns as spacers between turns of a helix, called the β-spiral. (**E**) Cross-eye stereo view of the poly(GVGVP) β-spiral in side view below and in axis view above. (**F**) Cross-eye stereo view of the twisted filament composed of three β-spirals. Before passing through the inverse temperature transition to increased order, as shown in Figure 5.1 and further characterized in Figures 5.2 through 5.5, this model protein contains the structural element of the β-turn, but otherwise is a disordered random chain. In the process of passing through the ordering transition, the molecular structure hydrophobically folds and assembles into a twisted filament composed of dynamic β-spirals, as represented in parts **E** and **F**. See section 5.2.2 for description and relevant references. (Parts **A**, **B**, **C**, **D**, **E**, and **F** reproduced with permission from Urry et al.,[100] Cook et al.,[24] Urry,[119] Urry,[121] and Urry et al.,[46] respectively.)

5.2.3 The Nature of the Elasticity of Cross-linked Poly(GVGVP)

5.2.3.1 Ideal, Dominantly Entropic, Elasticity

Stretching an elastic band constitutes an expenditure of mechanical energy. For the most efficient molecular machines, the mechanical energy expended on stretching the elastic band should be completely recovered on relaxation of the stretching force, that is, the force versus length curve on extension would exactly overlap with the force versus length curve obtained on relaxation. Such bands may be said to exhibit ideal elasticity. Bands of cross-linked poly(GVGVP), as well as the parent mammalian elastic fibers wherein the repeating pentamer sequence was initially observed,[50,51] have been found to exhibit such ideal (dominantly entropic) elasticity.[9,46,48]

Ideal elasticity would not occur when stretching involves friction-like interactions between the load-bearing chain and surrounding nonload-bearing chains. In such cases some of the mechanical energy expended during stretching is not recovered on removal of the stretching force. Chains having complex secondary (hydrogen bonded), tertiary, and quaternary structures provide mechanisms for the energy expended on extension to be dissipated away from the load-bearing chain segments.

5.2.3.2 The Decrease of Internal Chain Motions on Stretching

Stretching of the molecular structure described in Figure 5.6 does not involve the disruption of complex secondary, tertiary, and quaternary structures nor does significant frictional drag appear to occur between the chain bearing the load and adjacent nonload-bearing chains. In the structure most easily seen in stereo in Figure 5.6E, suspended segments, involving the VGV sequence, occur between the β-turns within a single chain. The two peptide groups, –CONH–, within VGV sequence are quite free to rotate or rock about the nearly colinear C–CONH–C bonds in the structure in Figure 5.6E. This motion of the backbone constitutes chain entropy. On stretching, the amplitude of these rocking motions decreases, and this causes the decrease in entropy that provides the restoring force after stretching. This mechanism of elasticity has been called the *damping of internal chain dynamics on extension*.[56] When the energy of stretching the elastomer primarily decreases internal chain motions, the elastomer becomes an entropic elastomer. This type of elasticity allows for more efficient energy conversions.

This mechanism of elasticity provides a "common groundwork of explanation" for the elasticity of all chain molecules regardless of composition and structure as long as there exists an internal chain motion that becomes decreased on deformation. For this reason, it too is a consilient mechanism. To delineate this consilient mechanism for elasticity from the consilient mechanism for hydrophobic association, as treated extensively in this volume, it will be referred to as the *elastic consilient mechanism*.

5.2.3.3 Importance of the Elastic Consilient Mechanism to Efficiency of Energy Conversion

An important element of an ideal elastomer is that the energy of deformation of an otherwise kinetically free chain must remain entirely in the backbone modes where it can be recovered on relaxation. Should the energy of deformation find its way into side chain motional modes, this can be dissipated into the solvent and into adjacent nonforce sustaining molecules and, therefore, be lost to the recovery during relaxation. The experimental result of this dissipation is hysteresis. Cross-linked poly(GVGVP) exhibits near ideal elasticity; the extension and relaxation curves overlap well. Single-chain force-extension curves of (GVGVP)$_{502}$ can exhibit extension and relaxation curves that overlap perfectly within instrumental limits.[48] On the other hand, cross-linked poly(GVGIP) always exhibits hysteresis; the relaxation curve falls well below the extension curve. Because the work of extension is the area under the force-extension curve, when the force-relaxation curve falls below the force-extension curve, the energy represented by the difference between extension and relaxation is lost. Expectedly, therefore, single-chain force-extension curves of (GVGIP)$_{rx260}$ do exhibit marked hysteresis; however, under conditions of extreme dilution occasional curves of (GVGIP)$_{nx260}$ can be obtained wherein the extension and relaxation curves superimpose.[57] (GVGIP)$_{nx260}$ with the greater tendency for hydrophobic association between chains, due to the bulkier more hydrophobic isoleucyl residue in each pentamer resulting from an added CH_2 moiety, provides an example of energy loss through side chain motions and interactions.

Hanging a weight on an elastic band of cross-linked poly(GVGVP), as described in section 5.1.4, constitutes a extending force; the band increases in length until the area under the force-extension curve becomes equivalent to the mechanical energy of the lifted weight. For elastic-contractile bands operating under the consilient mechanism, heating the band through the temperature interval or adding salt to move through the transition zone of Figure 5.5 lifts a weight a certain distance and results in the performance of a given amount of mechanical work. For bands that exhibited hysteresis, the distance a weight is lifted, on raising the temperature from below to above the phase transition for hydrophobic association, would be less than for the bands that exhibited a near ideal elasticity. The former bands would not be very good molecular machines for biological energy conversion. All of the energy conversions of the consilient mechanism involve changes in hydrophobic association, whether or not mechanical energy is the input or output. Because of this, all of the energy conversions are more efficient when the protein-based machine is a dominantly entropic elastomer.

Significantly, for elastin the majority of the heat and entropy change during chain extension results from changes in hydrophobic hydration.[58] Reversibility of an ideal elastomer, recovery to the same structure on relaxation, allows that the same change in hydrophobic hydration occur. To understand energy conversion adequately by the consilient mechanism, that is, by means of inverse temperature transitions, solvent entropy changes attending changes in hydrophobic association require delineation from changes in chain entropy for the development of elastic force.

5.2.4 Relationship Between Nature of Elasticity and Contractility

Historically, the question of mechanism of elasticity has been one of evaluating the relative contributions of three different proposed mechanisms: (1) the random chain network (classic rubber elasticity) theory,[18,58] (2) the solvent entropy theory,[59–61] and (3) the damping of internal chain dynamics on extension.[9,48,55,56] The first is due to the Flory school; the second was initiated by Weis-Fogh and Andersen, and the third is due to the present author and coworkers of the last quarter century.

The atomic force microscopy (AFM) single-chain force-extension studies on the elastic model proteins[9,48] demonstrate entropic elasticity for single chains without the presence of random chain networks with Gaussian distributions of end-to-end chain lengths as would have been required for the random chain network theory.[62] This leaves the proper delineation of solvent entropy and internal chain dynamics. The solvent entropy theory of elasticity requires consideration in the case of our elastic model proteins, because changes in solvent entropy attend changes in hydrophobic association during stretching and because most recent adherents of this view base their perspective on calculations of $(GVGVP)_n$.[60,61] In addition, a clear understanding of energy conversion by means of the inverse temperature transitions exhibited by elastic model proteins requires accurate delineation of the roles of chain and solvent entropy. Experimental delineation immediately follows.

5.2.4.1 Experiment to Determine Fraction of Ideal (Entropic) Elasticity

The classic experiment to estimate the amount of total elastic force, f, derived from the internal energy component of force, f_E, and the entropic component of force, f_S, where $f = f_E + f_S$, examines the temperature dependence of force at constant length. On plotting $\ln[f/T]$, with T in Kelvin, as a function of temperature while maintaining the elastomer element at fixed length, the slope of the plot multiplied by $(-T\ \mathrm{K})$ provides the f_E/f ratio. Such data are given in Figure 5.7A, where the f_E/f ratio above 40°C is found to be 0.9.[63] Whether in ethylene glycol:water (30:70), curve A, or pure water, curve B, the elastic band of cross-linked poly(GVGVP) exhibits an elastic force that is 90% entropic. When the solvent entropy change through the temperature interval of the inverse temperature transition is reduced to near zero, as occurs in ethylene glycol:water (30:70),[63] the entropic elastic force actually increases. This is contrary to the solvent entropy theory of elasticity. As analyzed below, however, the increase in entropic elastic force on going from 25° to 40°C eliminates solvent entropy change as a direct source of the estimated entropic component elastic force.

5.2 Molecular Structure and Elasticity of Model Protein

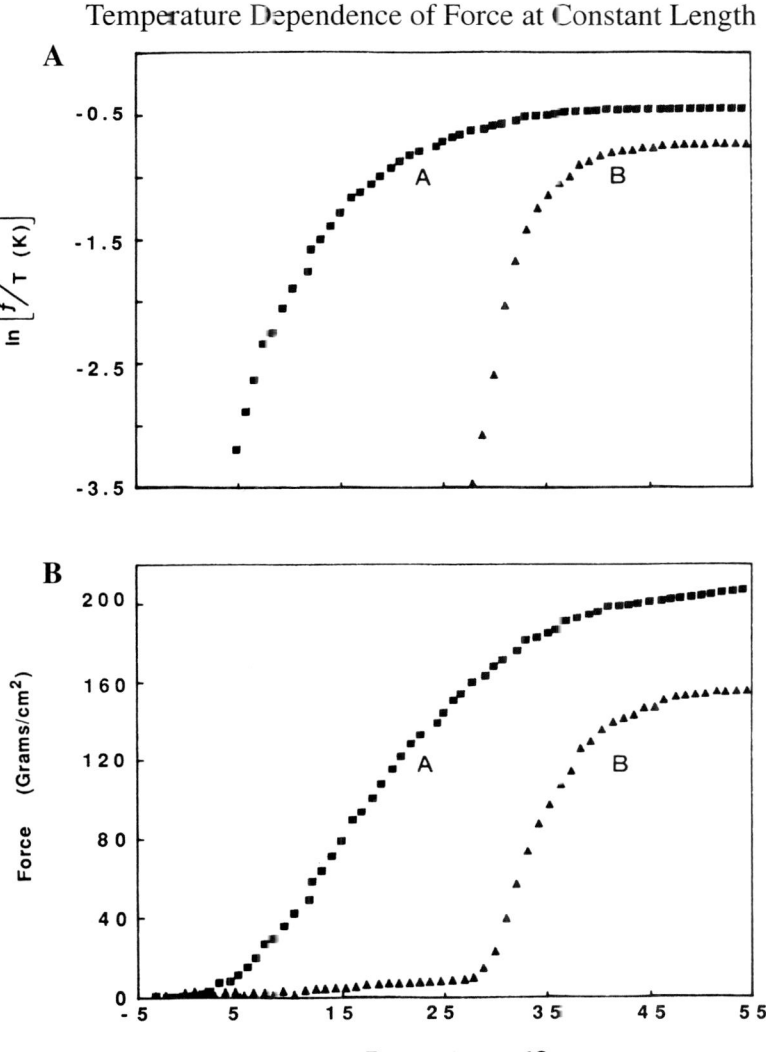

FIGURE 5.7. Thermoelasticity experiments to estimate the entropic component of elastic force in pure water (curves **B**) and in the solvent mixture of 30% ethylene glycol: 70% water (curves **A**). On increasing ethylene to 30%, the heat of the transition approaches zero, which means that the solvent entropy change approaches zero. The purpose of the experiment is to see if solvent entropy change contributes to the force developed on raising the temperature. Interestingly, the 90% entropic elastic force increases rather than decreases on removing solvent entropy change. The conclusion is that there is no evidence for the contribution of solvent entropy. Analysis of the fact of development of force under the experimental conditions of fixed length excludes solvent entropy as a contribution to the entropic elastic force. See section 5.2.4 for more complete discussion. (Reproduced with permission from Luan et al.[63]).

5.2.4.2 Implications of Force Changes at Constant Length

For poly(GVGVP) the temperature interval for hydrophobic association is 25° to 40°C. It is over this temperature range that hydrophobic hydration becomes higher entropy bulk water as the hydrophobic groups associate. For this temperature interval the ΔS for solvent is clearly positive. As shown in Figure 5.7, an elastic force develops over this temperature interval that is 90% entropic. Now, because $f_S = -T(\partial S/\partial L)_{V,T,n}$, entropy change for entropic elastic force development is negative. This means that the solvent entropy change is not contributing to the observed increase in the entropic component of elastic force.

The same argument applies for isometric contraction (force development at fixed length) driven by protonation of carboxylates (see Figure 1.4).[9,64] In this case, protonation of carboxylates in the cross-linked elastic matrix drives hydrophobic association with an increase in entropic elastic force due to a negative change in elastomer entropy. Under these conditions, however, hydrophobic hydration converts to bulk water; the solvent entropy change is positive, exactly the wrong sign to contribute to the entropic elastic force. How this can occur is depicted in Figure 5.8.[65]

5.2.5 Consequences of the Nature of Elasticity to Biocompatibility

5.2.5.1 Low Frequency Mechanical Resonances

Low frequency mechanical resonances within elastic protein-based polymers occur near 3 kHz (within the frequency range of sound absorption) and 5 MHz (near radio frequencies).[9,49,66] The physical methods of dielectric relaxation (loss permittivity), loss shear modulus, and acoustic absorption have been used to demonstrate these mechanical resonances in the 3 kHz range, and dielectric relaxation and nuclear magnetic relaxation data provided evidence for the 5 MHz resonance. These motions constitute an entropic contribution to the stability of the structure in Figure 5.6. By means of statistical mechanics, the harmonic oscillator partition function provides a gross sense of the magnitude of the favorable free energy of stabilization at these mechanical resonances; the value is about 9 kcal/mole-pentamer for the 5 MHz resonance and some 14 kcal/mole-pentamer for the acoustic resonance.[9,67]

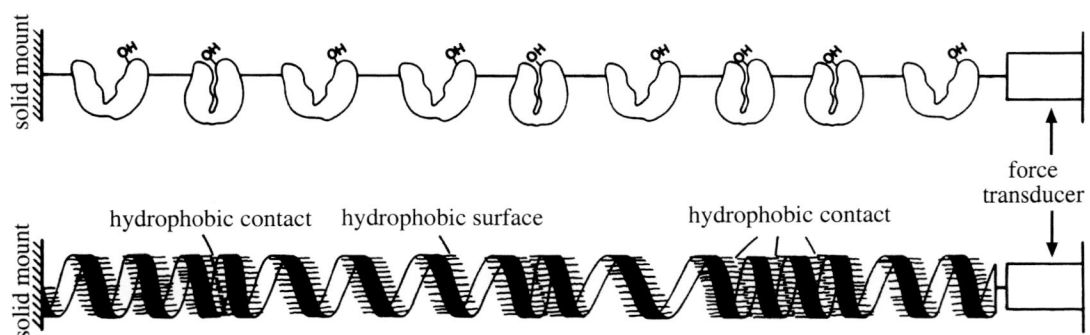

FIGURE 5.8. Hydrophobic association to produce increased elastic force at fixed length, as experimentally shown in Figure 5.7. (Top) A series of clam-shaped globular proteins, strung together by an elastic band attached near the mouth, are at equilibrium between open and closed states due to hydrophobic association. As the distribution shifts toward more closed states, the elastic force increases. (Bottom) A β-spiral with an equilibrium between hydrophobic association and dissociation between turns of the β-spiral. As the equilibrium shifts toward more hydrophobic association, turns of the β-spiral not involved in hydrophobic association become more extended. The decrease in internal chain motions in the more extended chain segments represents a decrease in entropy that is the source of an increase in entropic elastic force. (Reproduced with permission from Urry.[65])

5.2.5.2 Interaction, as in Antigen–Antibody Complexation, Would Increase Free Energy

For a molecular species to be antigenic, the antibody must bind to the antigen in the formation of a complex that would identify the molecular species as foreign. This necessitates suppression of the motions of the elastomer, of the mechanical resonances, at a cost of increasing the free energy of the complex. If the complexation free energy increase were less than one-half of that calculated by the harmonic oscillator partition function, the probability of interaction would be decreased by a factor of 10 million, making its formation insignificant. Because of this, the elastic protein-based polymers are expected, as they have been found (see Chapter 9) to be remarkably biocompatible. This property contributes to the usefulness of these elastic protein-based polymers over a large range of medical applications.

5.3 Cataloging the Energy Resources Available to the Consilient Mechanism of Energy Conversion

5.3.1 Each Means of Changing T_t Represents an Energy Source for Protein-based Machines

5.3.1.1 Point of View of Greatest Simplicity

"One of the principal objects of theoretical research in any department of knowledge is to find the *point of view* from which the subject appears in its *greatest simplicity*."[68] This statement was written by J. Willard Gibbs in 1881, the same Gibbs that gave us the Gibbs free energy, ΔG, so central to the above analyses. The significance of the quotation is compounded here. We propose the T_t-perspective to be a *point of view* of *greatest simplicity*, and in section 5.1.3.4 we derived approximations to the change in Gibbs free energy for hydrophobic association, $\Delta G_{HA}(\chi)$. From this expression, we recognize the change in T_t from a reference T_t to a new value of T_t caused by an energy input represented by χ to provide a measure of the change in Gibbs free energy for hydrophobic association of the protein-based polymer. Therefore, T_t, the onset temperature for the inverse temperature transition, represents an intrinsic property of the hydrophobic consilient mechanism of energy conversion.

5.3.1.2 Simplicity of the Observations

For our model proteins, determination of T_t is straightforward. The approximate value of T_t is evident to the unaided eye. A glass tube containing the parent model protein, poly(GVGVP), dissolved in water and at a comfortable room temperature, forms a clear solution (see the clear tube in Figure 5.1A). On grasping the test tube in the palm of one's hand for a few moments, the clear solution becomes cloudy, as represented in the middle tube in Figure 5.1A. On standing at palm temperature, two separate phases appear. Simply warming the tube from room temperature to body temperature causes the model protein to fold, assemble, and phase separate due to the association of oil-like groups.

Now, cross-linking the elastic model protein in the phase-separated state results in elastic bands. Similarly warming the band, swollen at room temperature (just below T_t), to body temperature (some 15 degrees above T_t) causes the band to contract with the performance of mechanical work. The band pumps iron on raising the temperature from below to above T_t. As scientific accounts go, the T_t perspective exemplifies simplicity.

5.3.1.3 Inputs That Move the T_t-divide Represent Energy Resources for the Consilient Mechanism of Energy Conversion Without Changing Temperature

Any change in T_t indicates a change in free energy for the folded (hydrophobically associated) state of the polymer. Therefore, in the first approximation the effectiveness of the energy input is obvious from the magnitude of the change in T_t. As a particularly relevant example, the most dramatic increase in T_t occurs on phosphorylation. Accordingly, phosphate binding provides the most effective energy input to drive unfolding (hydrophobic disassociation), and phosphate removal imparts the greatest output for contraction by favoring hydrophobic association. A list of the T_t-values resulting from the full range of energy inputs to a reference model protein becomes the catalog of energy resources available to the consilient mechanism for protein-based machines. Changes in the value of T_t, therefore, constitute the nuts and bolts of energy conversion by the consilient mechanism. *Most significantly, the following T_t-based cataloging of energy resources available to the consilient mechanism coincides with the catalog of the energy resources that run biology's machines.*

5.3.1.4 Impact of the Relative Oil-like Character of Functional Groups of Proteins on the T_t-divide

The capacity to exist in two different states defines a functional group. Here we divide the two different states into more and less hydrophobic or, from the opposite perspective, more and less polar. The clearest examples of functional groups are those groups that can be either uncharged or charged, that is, either neutral or ionized. The most obvious examples of ionizable functional groups of protein are the carboxyl/carboxylate, $-COOH/-COO^-$, and the amino/ammonium, $-NH_2/-NH_3^+$, chemical couples. A change in the amount of acid, H^+, present in the solution can change the state of the chemical couple from one into the other.

Importantly, the interaction between oil-like side chains and these functional groups changes the amount of acid required to convert from one state to the other (see Figures 1.2 and 5.20B and section 5.7.7). In other words, the extent of the oil-like character of the domain in which the functional group occurs can change the state of a functional group. From this perspective, the oil-like residues, normally considered nonfunctional, exhibit functionality by their capacity to change the state of a functional group. Oil-like groups can also be in different states; they can be surrounded by hydrophobic hydration, or by water that has been oriented for hydration of charged groups, or by no water as when in an insoluble state, that is, hydrophobically associated.

This section develops the foundation with which to utilize these different states in engineering protein function and in understanding the function of proteins as molecular machines. It does so by developing a scale that determines the relative effectiveness of chemical groups associated with protein to change the temperature at which the phase transition of poly(GVGVP) takes place, that is, to move the T_t-divide in Figure 5.3. That relative effectiveness in changing the location of the T_t-divide, as noted above, provides a measure of the relative effectiveness of the energy input that caused the ΔT_t.

5.3.2 The T_t-based Hydrophobicity Scale for Amino Acid Residues

5.3.2.1 The Preferred Site of Substitution

For our model protein, poly(GVGVP),[69] replacement of either of the Gly residues or the Pro residue markedly alters the structure and properties of the resulting protein-based polymer. For example, substitution of the Gly and Pro residues can destroy the favorable elasticity of the polymers. These positions will not be used for the comparisons of interest here. Substitution of the Val preceding the Pro is possible by most amino acid residues, but not all. Fortunately, any one of the 20 naturally occurring amino acid residues can replace the V of GVG without significant change to the basic structure and function.

5.3.2.2 An Expression for Substitution in Chemically Synthesized Model Proteins

The approach, therefore, is to introduce a guest residue in place of the V residue in the GVG sequence by preparing the pentamer GXGVP where X is any of the naturally occurring amino acid residues or a chemical modification of biological interest. (See Table 5.1 for identification of each of the **R**-groups of the 20 natural amino acid residues.) The pentamer with the guest

5.3 Cataloging the Energy Resources Available to the Consilient Mechanism of Energy Conversion   133

residue, X, is mixed in the reaction solution with the parent pentamer, GVGVP, at different ratios in different reaction vessels, and each mixture is polymerized. The result is a series of model proteins with different amounts of the guest residue.

The general expression for the composition of this series of model proteins becomes poly[f_x(GXGVP),f_v(GVGVP)], where f_x and f_v are called mole fractions wherein $f_x + f_v = 1$. By way of example, if f_x were 0.2 and, therefore, f_v were 0.8 (i.e., 1 − 0.2), there would be four GVGVP pentamers for each GXGVP pentamer, which means that on the average there would be 1 substitution, that is, one X residue, for every 25 residues or 4 substituted residues in every 100 residues.

5.3.2.3 The Experimental Observation

As long as the model proteins are not too oil-like, they are soluble at low temperature in a biological salt solution (physiological saline, 0.15 N NaCl), that is, the experiment begins at low temperature with a clear solution. The solution is heated; the temperature increases, and, at a critical temperature unique to the composition of the model protein, the sample becomes cloudy. This is shown in Figure 5.1B, where aggregation of the model protein causes cloudiness. A graph representing clear solutions of a number of different model protein compositions that turn cloudy at different temperatures on increasing the temperature is shown in Figure 5.9A. The temperature for the onset of

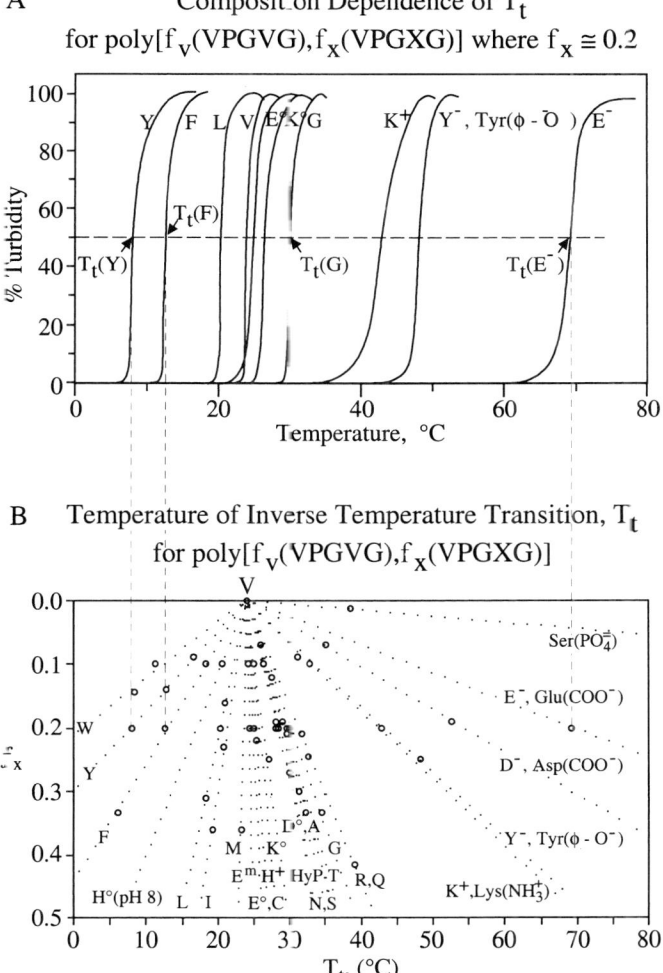

FIGURE 5.9. Experimental data for development of the T_t-based hydrophobicity scale. The general composition for the protein-based polymer is poly[f_x(GXGVP),f_v(GVGVP)], where X is the guest amino acid residue to be evaluated and f_x and f_v are mole fractions wherein $f_x + f_v = 1$. Part A contains the raw data for a number of guest residues substituted at a mole fraction of 0.2, which means 4 substituted residues per 100 residues of poly(GVGVP). The experimental conditions were 40 mg/ml of polymer of a molecular weight of about 100,000 Da in 0.15 N NaCl and 0.01 M phosphate at pH 7.4. Experimental T_t-values were obtained as shown in part A for $f_x = 0.2$, and additional polymers were characterized with different f_x values such that a plot of f_x versus T_t could be constructed as in part B. Extrapolation of the linear plots in part B to $f_x = 1$ gave the T_t-values that became the basis for the T_t-based hydrophobicity scale given in Table 5.1. (Adapted with permission from Urry.[6])

aggregation (of cloudiness) changes remarkably with substitution of only 4 guest residues in 100 residues. As established above in section 5.1.3.4, an increased oil-like character lowers the temperature for the onset of oil-like aggregation, and an increased vinegar-like character raises the temperature for onset of oil-like aggregation.

5.3.2.4 A Scale for Measuring Relative Oil-like Character

Establishing a scale requires the choice of reference conditions. We choose one guest residue per pentamer as the reference. Now, for the most oil-like residues the temperature for onset of aggregation for one guest residue per pentamer is below the freezing point of water, and for the very vinegar-like residues the temperature for aggregation for one vinegar-like residue per pentamer would be above the boiling point of water. In such cases, fewer guest residues are used, for example, one in five pentamers and one in ten pentamers, and the onset temperatures for aggregation are plotted. The plots give straight lines as shown in Figure 5.9B, as the fraction of guest residue increases. The lines can be extended to the value expected for one guest residue per pentamer, that is, for $f_x = 1$, and this value of T_t becomes the measure of relative oil-like character.

The relative oil-like character of each amino acid residue is given in terms of the T_t-based hydrophobicity scale in Table 5.1. T_t is measured by plotting as indicated in Figure 5.9; and the technical term for oil-like, hydrophobic (meaning water fearing), is used. The more commonly used technical term, *hydrophobicity*, replaces the equivalent statement of *oil-like character*.

5.3.2.5 Effect of Making **R**-groups More Oil-like Axiom I

Most simply stated, oil is a string of CH_2 groups. Adding a CH_2 group, or more generally only hydrocarbons (carbons, C, and hydrogens, H), to the **R**-group of a protein clearly represents an increase in oil-like character. Changing the **R**-group from the $-CH_3$ of alanine (Ala, A) to the $-CH(CH_3)_2$ of valine (Val, V) and to the $-CH_2CH(CH_3)_2$ of leucine (Leu, L) progressively creates more oil-like model proteins. Replacing the X in poly(GXGVP) by A, V, or L progressively lowers the temperature for onset of aggregation of the model proteins from 45°C to 24°C and to 5°C, respectively, as shown by the data in Figure 5.9 and listed in Table 5.1. Accordingly, a stepwise increase in oil-like character of the **R**-group stepwise *lowers* the temperature for aggregation!

Henceforth, any **R**-group substitution capable of lowering the temperature for aggregation is recognized as more oil-like than the **R**-group it replaced. A perspective so self-evident warrants statement as an axiom. In the following axiom, the transition is defined by the *temperature interval*, where the onset temperatures, noted above, identify the low temperature sides of the respective *temperature intervals*, as shown in Figures 5.1C and 5.5.

Axiom 1: The change in temperature interval, over which occurs the oil-like folding and assembly transition of a host model protein on introduction of different guest substituents, becomes a functional measure of relative oil-like character of the substituents, that is, of their relative hydrophobicity, and it provides a measure of the change in free energy of the hydrophobically associated state.

5.3.2.6 Effect of Changing the State of Vinegar-like Side Chains

Above, the increase in oil-like character lowered the value of T_t and, by Equation (5.9) for $\Delta G_{HA}(\chi)$, lowered the free energy of the hydrophobically associated state. The reverse occurs for a series of side chains that increase the vinegar-like character. Addition of **R**-groups containing charged amino groups, $-NH_3^+$, like that of lysine (Lys, K) or containing charged carboxylates, $-COO^-$, like that of glutamic acid (Glu, E) into poly(GXGVP) increases the temperature for aggregation to extrapolated values of 120°C for X = K and 250°C for X = E under the conditions described in more detail below. Accordingly, a stepwise increase in the vinegar-like character of the **R**-group stepwise *raises* the temperature for aggregation as the result of increasing the free energy of the hydrophobically associated state!

Therefore, within the practical approach of the T_t-based hydrophobicity scale, we argue that any **R**-group capable of increasing the temperature for aggregation is more vinegar-like than the **R**-group it replaced. Therefore, we evaluate each of the naturally occurring **R**-groups by these criteria. Recall that oil-like groups lower the *temperature interval* for aggregation, in our view, due to increasing the amount of hydrophobic hydration surrounding them. (Refer to the discussion in sections 5.1.3.3, 5.1.3.4, and 5.7.7) In our interpretation below, charged groups raise the temperature interval by destructuring the pentagonally arranged hydrophobic hydration surrounding oil-like groups in the unfolded and disassociated model protein, that is, by decreasing the total number of molecules of hydrophobic hydration possible in the disassociated model protein.

5.3.2.7 Systematic Comparison of All 20 R-groups of Proteins

Meaningful comparison of the relative oil-like character of all of the 20 naturally occurring **R**-groups of Table 5.1, including the different states of functional **R**-groups, requires adherence to stringent criteria and conditions. These experimental conditions, discussed above, are given by footnote here.[70] The criteria and conditions are remarkably met by the elastic repeating peptide sequence that originated in the mammalian elastic fiber, $(GVGVP)_n$.

Axiom 1, given above and the first of five axioms listed below in section 5.6.3, becomes the basis for consideration of protein function and for the design of protein-based materials for the future (see Chapter 9). As simply interpreted below, when there are more water molecules surrounding oil-like groups, the phase separation of the oil-like groups from water occurs at a lower temperature.

5.3.2.8 The Continuous Nature of the T_t-based Hydrophobicity Scale

The terms *oil-like* and *vinegar-like* provide good insight into the basic issues, but they are limited when addressing particular issues. To overcome preconceptions and to limit the need for clarifications of what is oil-like and what is vinegar-like when discussing the practical T_t-based hydrophobicity scale, more general terms prove helpful. Thus, the term *apolar* often replaces oil-like, and the term *polar* often replaces vinegar-like. Although the T_t-based hydrophobicity scale warrants much discussion, only a few issues are noted here.

5.3.2.8.1 A Scale from Apolar to Polar

As shown in Table 5.1, tryptophan (Trp, W) has the most apolar **R**-group, and glutamic acid (Glu, E), when ionized, has the most polar **R**-group. Between these two extremes exists an essentially continuous set of values. The side chain of glutamine, $-CH_2-CH_2-CO-NH_2$, though uncharged is quite polar, the most polar uncharged side chain. This indicates that the peptide moiety itself, $-CO-NH-$, is polar and suggests that some half-dozen peptide moieties could sum to be as effective as a single carboxylate. It would seem that the hydrogen bonding of peptide moieties with water could limit the amount of hydrophobic hydration and raise the value of T_t. Also when the oil-like residues dominate, even though the CO or NH may be at the interface with water, they would be at a higher free energy due to limited hydrogen bonding with water. This would lower the free energy for formation of secondary structure, that is, intramolecular and intermolecular hydrogen bonding becomes favored by the presence of more hydrophobic residues. In our view, this factor contributes to the growth of amyloid deposits of Alzheimer's disease and assists in driving prions into insoluble aggregates (see Chapter 7).

5.3.2.8.2 Comparison of Negative and Positive Ions

The T_t-based hydrophobicity scale shows the side chain of aspartic acid, $-CH_2-COO^-$, to be significantly less polar than that of glutamic acid, $-CH_2-CH_2-COO^-$. This is not as expected from the number of CH_2 groups in the side chains and likely arises from steric-dependent hydrogen bonding interactions with nearby peptide groups. Significant also is the difference between the effectiveness of $-NH_3^+$, and $-COO^-$, in raising the value of T_t. These delineations are relevant to protein function and dysfunction in Chapters 7 and 8 and are explored further below once estimates for $\Delta G_{HA}(E^-)$ and $\Delta G_{HA}(K^+)$ have been obtained.

5.3.3 The T_t-based Hydrophobicity Scale for Additional Biologically Relevant Chemical Groups

5.3.3.1 The Special Position of the Phosphate Group, $-OPO_3^=$

Many chemical groups in addition to the amino acid side chains are important in biology because of their attachment to proteins. Most notable is the phosphate group, $-OPO_3^=$, the most important chemical entity in biological energy conversion. The phosphate group is also the most important chemical entity in the consilient mechanism for the function of protein-based machines. The most polar chemical entity of all those examined for inclusion in the T_t-based hydrophobicity scale is the phosphate group. As shown in Table 5.2, phosphate attached to a Ser OH constitutes the most polar, the most supra-vinegar-like chemical group attachable to our model proteins; it is the least oil-like. The removal of a phosphate most dramatically makes the model protein more oil-like and most emphatically drives hydrophobic association.

5.3.3.2 Effects on T_t of Oxidation and Reduction of Redox Functional Groups

Nicotinamide plays a central role in both photosynthesis and respiration, as introduced in Chapter 2. In photosynthesis, light effectively splits water molecules into electrons, protons, and oxygen, O_2. The protons, H^+, drive the rotary ATP synthase motor to produce ATP from ADP and Pi. The electrons reduce nicotinamide, which then provides for reduction of CO_2 to result in glucose. In respiration, nicotinamides and flavins become reduced in the process of glucose oxidation to form CO_2. This occurs by the reactions of glycolysis, the transition reaction, and the Krebs cycle of intermediary metabolism. Subsequent oxidation of nicotinamides and flavins in the inner mitochondrial membrane results in the pumping of protons across the inner mitochondrial membrane to the intermembrane space between the inner and outer mitochondrial membranes (see Chapter 8 and specifically section 8.3). In this process oxidation of nicotinamide results in 50% more protons pumped across the membrane than the oxidation of flavin. The higher concentration of protons in the inner membrane space then drives the ATP synthase that combines ADP and Pi to produce 32 of the 36 ATP molecules for each molecule of glucose oxidized.

The change of the glutamic acid side chain from $-COO^-$ to $-COOH$ results in the largest change in T_t ($\Delta T_t = -220$) of all of the vinegar-like side chains of the natural amino acid residues (see Table 5.1). As reported in Table 5.2, the model nicotinamide, N-methyl nicotinamide (NMeN) used for its relative stability, exhibits a ΔT_t of $-250°C$ on reduction to the dihydro-NMeN. The ΔT_t decreases to $-105°C$ on subsequent reaction that adds a water molecule to give the more polar 6-OH tetrahydro-NMeN. Reduction of nicotinamide adenine dinucleotide (NAD$^+$ → NADH) results in a ΔT_t of $-150°C$, and that of flavin adenine dinucleotide (FAD → FADH$_2$) results in a ΔT_t of $-95°C$. The relative ΔT_t values of $-150°C$ for NAD$^+$ and $-95°C$ for FAD are consistent with their relative contributions to the pumping of protons across the inner mitochondrial membrane.

5.3.3.3 Nitration and Sulfation of the Aromatic Side Chain of Tyrosine

From Table 5.1 the T_t of tyrosine is $-55°C$ and in Table 5.2 that of sulfated tyrosine is $140°C$. This gives a large ΔT_t of $-195°C$. The ΔT_t due to nitration of tyrosine is an even larger $275°C$. These are large effects, but only one-third to one-fourth of the magnitude due to phosphorylation. It is no surprise, therefore, that phosphorylation/dephosphorylation became biology's primary process for energy conversion, biology's universal energy currency.

5.3 Cataloging the Energy Resources Available to the Consilient Mechanism of Energy Conversion 137

TABLE 5.2. Hydrophobicity scale (preliminary T_t and ΔG_{HA} values) for chemical modifications and prosthetic groups of proteins[a].

Residue X	ΔG_{HA} from Figure 5.10 (kcal/mol)[b]	T_t, linearly extrapolated to $f_X = 1$ (°C)
Lys(dihydro NMeN)[c]	−7.0	−130
Glu(NADH)[d]	−5.5	−30
Lys(6-OH tetrahydro NMeN)[c]	−3.0	15
Glu(FADH$_2$)	−2.5	25
Glu(AMP)	+1.0	70
Ser(—O—SO$_3$H)	+1.5	80
Thr(—O—SO$_3$H)	+2.0	100
Glu(NAD)[d]	+2.0	120
Lys(NMeN, oxidized)[c]	+2.0	120
Glu(FAD)	+2.0	120
Tyr(—O—SO$_3$H)[e]	+2.5	140
Tyr(—O—NO$_2^-$)[f]	+3.5	220
Ser(PO$_4^-$)	+8.0	860

[a] The usual conditions are for 40 mg/ml polymer, 0.15 N NaCl and 0.01 M phosphate at pH 7.4. T_t = Temperature of inverse temperature transition for poly[f_V(VPGVG),f_X(VPGXG)].
[b] Gross estimates of ΔG_{HA} using the T_t-values in the right column in combination with the T_b versus ΔG_{HA} values from Figure 5.10.
[c] NMeN is for an N-methyl nicotinamide pendant on a lysyl side chain, that is, N-methyl-nicotinate attached by amide linkage to the ε-NH$_2$ of Lys. The most hydrophobic reduced state is N-methyl-1,5-dihydronicotinamide (dihydro NMeN), and the second reduced state is N-methyl-6-OH 1,4,5,6-tetrahydronicotinamide or (6-OH tetrahydro NMeN). For the oxidized and reduced N-methyl nicotinamide, the conditions were 0.5 mg/ml polymer, 0.1 M potassium bicarbonate buffer at pH 9.5, and 0.1 M potassium chloride.
[d] For the oxidized and reduced nicotinamide adenine dinucleotides, the conditions were 2.5 mg/ml polymer, 0.2 M sodium bicarbonate buffer at pH 9.2.
[e] The pK$_a$ of polymer bound —O—SO$_3$H is 8.2.
[f] The pK$_a$ of Tyr(—O—NO$_2$) is 7.2.
Source: Adapted with permission from Urry et al.[41]

5.3.3.4 Cinnamide

Cinnamide undergoes a light-driven geometrical isomerization, which is a simple conversion from a *trans* to a *cis* configuration. This raises the value of T_t such that the effect of the absorption of light is a modest shift toward hydrophobic unfolding.[71]

5.3.4 Energy Resources of the Consilient Mechanism: Changing the Free Energy (ΔG_{HA}) for Hydrophobic Association

5.3.4.1 Different Estimates of Transition Temperature Used in Calculating the Gibbs Free Energy for Hydrophobic Association, ΔG_{HA}, by Equation (5.10a)

When estimating the value of ΔG_{HA}, as derived in section 5.1.3.4 and defined in Equation (5.10a), $\Delta G_{HA}(\chi) = [T_t(\text{ref})\Delta S_t(\text{ref}) - T_t(\chi)\Delta S_t(\chi)]$, three different ways are noted to represent the temperature of the transition, and they lead to different levels of accuracy for the calculated value. The first of course is T_t, the simplest measurement, which actually is the temperature of the onset of turbidity representing the initial aggregation event of the phase transition (see Figure 5.1B). When calculating ΔG_{HA} using T_t, the error will be greater as the widths of the transitions of the two states differ more significantly.

The second representation of the transition temperature, T_b, utilizes the onset of the transition as estimated from the differential scanning calorimetry (DSC) curve in Figure 5.1C, as estimated by the intersection of the initial baseline and the initial rise of the curve.

The third representation of the transition temperature utilizes the *extremum* (in this case the maximal value) of the DSC curve in Figure 5.1C. This estimate will be designated as T_m. Use of T_m would normally result in a more accurate ΔG_{HA}. This, however, requires accurate DSC data on both states. Such data are not as easily obtained because of the small heats of the transitions, the long time required for completing the association of the transition, and the higher value of T_m.

5.3.4.2 ΔG_{HA}: The Change in Gibbs Free Energy for Hydrophobic Association Due to Change in Amino Acid Composition With $\Delta G_{HA}(Gly) = 0$

5.3.4.2.1 Calculation of ΔG_{HA} and Choice of Reference State

Equation (5.10b), $\Delta G_{HA}(\chi) = [\Delta H_t(\text{ref}) - \Delta H_t(\chi)]$, utilizes the ΔH_t values for the phase transition. Because for the inverse temperature transition, $\Delta H_t = T_t \Delta S_t$, the heat of this transition to the hydrophobically associated state becomes a measure of the change in Gibbs free energy for the transition, ΔG_{HA}. It now becomes necessary to define an appropriate reference state to achieve meaningful comparisons. This involves both the fraction of composition as well as the specific amino acid composition to be referenced. For the former, the ΔH_t values reported in Table 5.1 for $f_X = 0.2$ are extrapolated to $f_X = 1$ and for the latter the Gly (G) residue is taken as the amino acid of reference, because its side chain, a single hydrogen atom, is neither oil-like nor vinegar-like. This change in Gibbs free energy for hydrophobic association on substitution of Gly (G) in GGGVP by the guest amino acid residue X may be written as $\Delta G_{HA}(GGGVP \rightarrow GXGVP)$, but is indicated as a reference value by the superscript.

5.3.4.2.2 Estimated Values for $\Delta G_{HA}(GGGVP \rightarrow GXGVP)$

For the following estimates, T_b values are available for guest amino acid residues without functional groups.[72] With Equation (5.10b), the calculated values for the residues more hydrophobic than glycine (Gly, G), that is, $\Delta G_{HA}(GGGVP \rightarrow GXGVP)$ substitutions, in units of kcal/mol-pentamer become

$\Delta G_{HA}(GGGVP \rightarrow GWGVP) = -7.00$,
$\Delta G_{HA}(GGGVP \rightarrow GYGVP) = -5.85$,
$\Delta G_{HA}(GGGVP \rightarrow GFGVP) = -6.15$,
$\Delta G_{HA}(GGGVP \rightarrow GLGVP) = -4.05$,
$\Delta G_{HA}(GGGVP \rightarrow GIGVP) = -3.65$,
$\Delta G_{HA}(GGGVP \rightarrow GVGVP) = -2.50$,
$\Delta G_{HA}(GGGVP \rightarrow GMGVP) = -1.50$,
$\Delta G_{HA}(GGGVP \rightarrow GE\ GVP) = -1.30$,
$\Delta G_{HA}(GGGVP \rightarrow GPGVP) = -1.10$,
$\Delta G_{HA}(GGGVP \rightarrow GAGVP) = -0.75$,
$\Delta G_{HA}(GGGVP \rightarrow GTGVP) = -0.60$,
$\Delta G_{HA}(GGGVP \rightarrow GD°GVP) = -0.40$,
$\Delta G_{HA}(GGGVP \rightarrow GK°GVP) = -0.05$,
$\Delta G_{HA}(GGGVP \rightarrow GNGVP) = -0.05$,

and $\Delta G_{HA}(GGGVP \rightarrow GGGVP) = 0.00$, Thus, we have a measure of the favorable decrease in free energy of the hydrophobically associated state for those residues more hydrophobic than G.

For residues less hydrophobic, or more polar, than glycine, $\Delta G_{HA}(GGGVP \rightarrow GSGVP) = +0.55$, $\Delta G_{HA}(GGGVP \rightarrow GQGVP) = +0.75$, $\Delta G_{HA}(GGGVP \rightarrow GY^-GVP) = +1.95$, $\Delta G_{HA}(GGGVP \rightarrow GD^-GVP) \approx +2.6$, $\Delta G_{HA}(GGGVP \rightarrow GK^+GVP) = +2.94$, and $\Delta G_{HA}(GGGVP \rightarrow GE^-GVP) = +3.72$. A positive ΔG_{HA} disfavors the hydrophobically associated state, and a negative ΔG_{HA} favors hydrophobic association. The values are plotted in Figure 5.10 and listed in Table 5.3.

In Figure 5.10 a sigmoid curve is drawn through the data points, which on the hydrophobic side follows closely the hydrophobic residues W, F, L, I, V, P, and A to the neutral G residue. Those residues with mixed, but uncharged, hydroxyl, carboxyl, amino, and so forth, functional substituents fall off of the sigmoid. On the polar side the sigmoid curves around until it takes the slope defined by the $K^+ - E^-$ data points. Interestingly, continuing the $K^+ - E^-$ slope to 860°C, the T_t-value resulting from a phosphate attached to a serine (Ser, S) residue, gives a value for $\Delta G_{HA}(PO_4^-)$ of approximately 8 kcal/mole-phosphate.

Another interesting point about Figure 5.10 is provided by the diagonal straight line though the data points. This line illustrates the approximation of the $T_b(T_t)$-based hydrophobicity scale, which provides the most directly usable

5.3 Cataloging the Energy Resources Available to the Consilient Mechanism of Energy Conversion

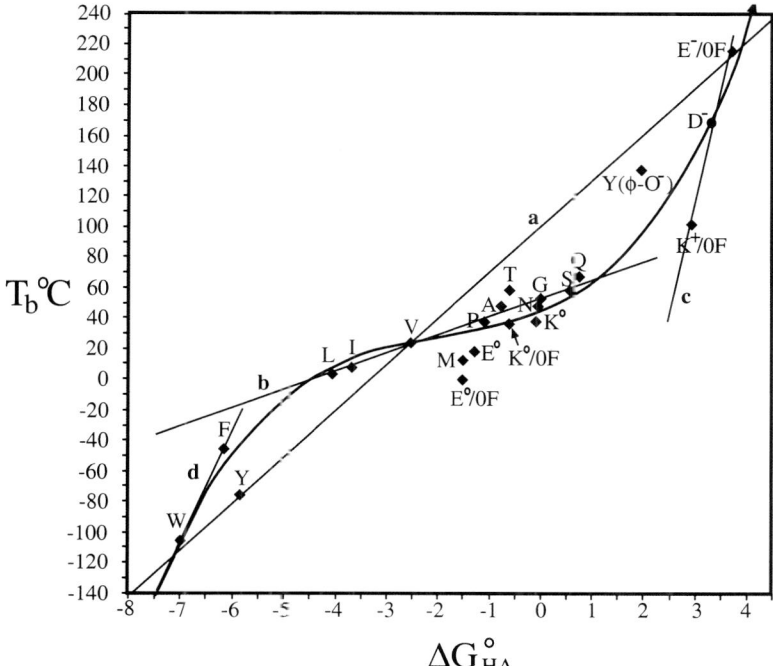

FIGURE 5.10. An embodiment of the comprehensive hydrophobic effect in terms of a plot of the temperature for the onset of phase separation for hydrophobic association, T_b, versus ΔG_{HA}, the Gibbs free energy of hydrophobic association for the amino acid residues, calculated by means of Equation (5.10b) using the heats of the phase (inverse temperature) transition (ΔH_t). Values were taken from Table 5.3. T_b and T_t were determined from the onset of the phase separation as defined in Figure 5.1C,B, respectively. The estimates of ΔG_{HA} utilized the ΔH_t data listed in Table 5.1 for $f_X = 0.2$ but extrapolated to $f_X = 1$, and the Gly (G) residue was taken as the zero reference. Identification of the delineated slopes: (**a**) The diagonal straight line gives the approximation of the $T_b(T_t)$-based hydrophobicity scale, as developed from the data in Figure 5.9 and listed in Table 5.1. (**b**) The slope for aliphatic hydrocarbons defined by Leu (L), Ile (I), Val (V), Pro (P), Ala (A), and Gly (G). (**c**) The slope, as defined by charged Lys (K^-) and Glu (E^-), for charged species in competition for hydration with the hydrophobic residues of the host model protein. (**d**) An approximate slope for the aromatic hydrocarbons defined by Phe (F) and Trp (W). (Adapted from Urry.[8])

information in terms of achieving function by turning on and off hydrophobic association. Equation (5.9), $\Delta G_{HA}(\chi) \approx -\Delta T_t(\chi)\Delta S_t(\text{ref})$, expresses the approximation of the $T_b(T_t)$-based hydrophobicity scale wherein the mean value of $\Delta S_t(\text{ref})$, as obtained from the diagonal line of Figure 5.10, would be approximately 30 cal/mole-pentamer-deg.

Perhaps most significant in Figure 5.10 is the delineation of the sigmoid into three slopes that group in terms of the classes of side chains, slope b for those amino acid residues containing only aliphatic groups, slope d for those amino acid residues containing aromatic groups, and slope c for those amino acid residues containing charged groups.

It should also be noted that the data in Table 5.3 in parentheses were obtained on microbially prepared **Model Proteins** i and x' in Table 5.5 (p. 153) and that the values for $\Delta G_{HA}(D^-)$ and $\Delta G_{HA}(PO_4^=)$ were obtained by using the experimentally derived T_t-values and reading the corresponding ΔG_{HA} values from the sigmoid of Figure 5.10 or extensions thereof.

TABLE 5.3. Hydrophobicity Scale in terms of ΔG_{HA}, the change in Gibbs free energy for hydrophobic association, for amino acid residue (X) of chemically synthesized poly[f_v(GVGVP), f_x(GXGVP)], 40 mg/ml, mw ≈ 100 kDa in 0.15 N NaCl, 0.01 M phosphate, using the net heat of the inverse temperature transition, $\Delta G_{HA} \approx [\Delta H_t(GGGVP) - \Delta H_t(GXGVP)]$ for the $f_x = 0.2$ data extrapolated to $f_x = 1$.

Residue X	T_b°C	$\Delta G_{HA}(\chi)$/kcal/mol pentamer
Trp	−105	−7.00
Phe	−45	−6.15
Tyr	−75	−5.85
His	−10 (T_t)	−4.80 (from graph)
Leu	5	−4.05
Ile	10	−3.65
Val	26	−2.50
Met	15	−1.50
His⁺	30 (T_t)	−1.90 (from graph)
Cys	30 (T_t)	−1.90 (from graph)
Glu(COOH)	20 (2)	−1.30 (−1.50)
Pro	40	−1.10
Ala	50	−0.75
Thr	60	−0.60
Asp(COOH)	40	−0.40
Lys(NH$_2$)	40 (38)	−0.05 (−0.60)
Asn	50	−0.05
Gly	55	0.00
Ser	60	+0.55
Arg	60 (T_t)	+0.80 (from graph)
Gln	70	+0.75
Tyr(φ-O⁻)	140	+1.95
Asp(COO⁻)	170 (T_t)	≈ +3.4 (from graph)
Lys(NH$_3^+$)	(104)	(+2.94)
Glu(COO⁻)	(218)	(+3.72)
Ser(PO$_4^=$)	860 (T_t)	≈ +8.0 (from graph)

Data within parentheses utilized microbial preparations of poly(30-mers), e.g., (GVGVP GVGVP GXGVP GVGVP GVGVP GVGVP)$_n$, with n ≈ 40.

The notation (from graph) indicates that the value of T_t from Table 5.1 was used with the sigmoid curve of Figure 5.10 to estimate $\Delta G_{HA}(\chi)$.

Source: Adapted from Urry.[8]

5.3.4.2.3 Considerations of the Experimental Conditions for the Evaluation of ΔG_{HA} and Their Relevance to the Use of the Values in Table 5.3

Appropriate use of the values in Table 5.3 begins with an understanding of the conditions for obtaining the experimental data. At a temperature below the onset of the inverse temperature transition, the side chains are fully exposed to and surrounded by water; they have their complete complement of hydration. At a temperature above the transition interval, the packing of the folded state determines the extent of loss of the hydration shell. Because the barrier to mobility of the model protein, poly(GVGVP), is a very low 1 to 1.5 kcal, it seems reasonable to expect that the transition to hydrophobic association would be essentially complete, as packing would not be limited by the energetics of association of more rigid interfaces. Fortunately, the reference values are not significantly compromised by such considerations.

Use of the values in Table 5.3 to estimate the hydrophobicity of given surfaces, as will be done in Chapters 7 and 8, requires an estimate of the surface exposure of a particular side chain to water at the interface of interest. Crude estimates of surface exposure can be made for hydrophobic residues, and future calculations should be able to improve those estimates. As regards charged residues, the key issue is whether or not they are ion paired or involved in "salt bridges," which is the term commonly used by crystallographers. Experimental results on the effect of ion pairing on the heat of the inverse temperature transition suggest a reduction in the heat to about one-half. Accordingly, when the structure suggests ion pairing, the value of ΔG_{HA} for a particular residue will be divided by two. Interestingly, for those surfaces where there is opportunity for complete exposure to water, the residues with charged side chains are fully extended and jutting out into the water. It is as though the charged component was trying to stay as far away as possible from the hydrophobic chain on which it is often perched. This is especially apparent for the side chains of lysine (–CH$_2$–CH$_2$–CH$_2$–CH$_2$–NH$_3^+$), arginine (–CH$_2$–CH$_2$–CH$_2$–NH–C[NH]–NH$_2^+$), and glutamic acid (–CH$_2$–CH$_2$–COO⁻). We believe this to be another reflection of the component of ΔG_{HA} referred to as an apolar–polar repulsive free energy of hydration, ΔG_{ap} (see Equation [5.13] of section 5.7.9.2 below). Thus, when the side chain is fully erect, as seen for certain residues in the crystal structures of the molecular chaperone, GroEL/GroES, and especially of the γ-rotor of ATP synthase, the full value of ΔG_{HA} should be used for that residue.

5.3 Cataloging the Energy Resources Available to the Consilient Mechanism of Energy Conversion

5.3.4.3 Determination of ΔG_{HA} for a Change in Functional State of Naturally Occurring Vinegar-like Side Chain

5.3.4.3.1 The ΔG_{HA}(Glu-COOH → Glu-COO$^-$)

Using Equation (5.10b) and differential calorimetry data on **Model protein i** of Table 5.5 as listed in Table 5.3 with a T_b(Glu-COOH) of 2°C, a T_b(Glu-COO$^-$) of 218°C, a ΔG_{HA}(Glu-COOH) of −1.50 kcal/mole, and a ΔG_{HA}(Glu-COO$^-$) of +3.72 kcal/mole, the calculated increase in Gibbs free energy of hydrophobic association on ionization of the carboxyl function, ΔG_{HA}(Glu-COOH → Glu-COO$^-$) = [+3.72 − (−1.50)] = +5.22 kcal/mole-pentamer. This quantity is of interest in connection with that calculated below for ΔG_{HA}(Lys-NH$_2$ → Lys-NH$_3^+$) for the purpose of estimating effects of ion pairing between model proteins.

Additional questions can be addressed. For example, how does ΔG_{HA}(Glu-COOH → Glu-COO$^-$) compare with the ΔG_{HA} of other functional groups in the same model protein, for example, functional groups that are coupled with proton pumping in oxidative phosphorylation of the mitochondria, such as nicotinamides? This question is answered in section 5.3.4.4. Another question is, how does the magnitude of ΔG_{HA}(Glu-COOH → Glu-COO$^-$) compare with that of an opposite signed vinegar-like functional group, for example, the quantity ΔG_{HA}(Lys-NH$_2$ → Lys-NH$_3^+$)? The latter is given immediately below.

5.3.4.3.2 The ΔG_{HA}(Lys-NH$_2$ → Lys-NH$_3^+$)

Using Equation (5.10b) and differential calorimetry data at pH 7.5 on **Model Protein x′** of Table 5.5 with a T_b(Lys-NH$_2$) of 38°C, a T_b(Lys-NH$_3^+$) of 104°C, a ΔG_{HA}(Lys-NH$_2$) of −0.60 kcal/mole-(GKGVP), and a ΔG_{HA}(Lys-NH$_3^+$) of +2.94 kcal/mole-(GK$^+$GVP), the calculated increase in Gibbs free energy for hydrophobic association on protonation of the amine function, ΔG_{HA}(Lys-NH$_2$ → Lys-NH$_3^+$) = **+3.54 kcal/mole-pentamer**. The effect on ΔG_{HA} of forming the charged Lys-NH$_3^+$ is less than on forming the charged Glu-COO$^-$, that is, ΔG_{HA}(Glu-COOH → Glu-COO$^-$) of **+5.22 kcal/mole-pentamer**. Thus, one might understand when using ion pairing to lower T_t and ΔG_{HA} that a thermodynamic equivalence could require more positively charged groups than negatively charged groups. In this regard, data on the allosteric protein hemoglobin are relevant. The allosteric binding of biphosphoglycerate with five negative charges to eight positively charged groups at the diad axis of hemoglobin, enhances oxygen release. This ion pairing favors the more hydrophobically associated deoxyhemoglobin state, with the charge stoichiometry such that there are to be anticipated more positive charges than negative charges (see Chapter 7).

5.3.4.3.3 Calculation of the Effect of Ion Pairing Between Chains, ΔG_{HA}[(Glu-COO$^-$; Lys-NH$_3^+$) → (COO$^-$ NH$_3^+$)]

Using Equation (5.10b) and differential calorimetry data on the COO$^-$ state of **Model Protein i** of Table 5.5 with a T_b(Glu-COO$^-$) of 218°C yields a value for ΔG_{HA}(Glu-COO$^-$) of 3.72 kcal/mole-(GE$^-$GVP). Similarly for **Model Protein x′** of Table 5.5 with a T_b(Lys-NH$_3^+$) of 104°C yields a value for ΔG_{HA}(Lys-NH$_3^+$) of +2.94 kcal/mole-(GK$^+$GVP). The ΔG_{HA}(COO$^-$ NH$_3^+$) value, obtained for a pH 7.5 solution containing equimolar (Glu-COO$^-$) and (Lys-NH$_3^+$) functional groups, is **0.36 kcal/mole**. This means that the decrease in Gibbs free energy for hydrophobic association due to the ion-pair formation between **Model proteins i** and **x′** is approximately **3 kcal/mole-pentamer**, that is, ΔG_{HA}[(Glu-COO$^-$; Lys-NH$_3^+$) → (COO$^-$ NH$_3^+$)], becomes [0.36 − (3.72 + 2.94)/2] ≈ **−3.0 kcal/mole-pentamer**. This exemplifies the power of ion pairing in lowering the Gibbs free energy for hydrophobic association, which increases with increasing hydrophobicity.

5.3.4.4 Calculation of ΔG_{HA} for a Change in the State of Bound Biologically Relevant Functional Groups

5.3.4.4.1 The ΔG_{HA} Due to the Reduction of Oxidized N-methyl Nicotinamide (NMeN$^+$)

As T_t is equivalent to T_b, a direct means of estimating values of ΔG_{HA}[GK(NMeN$^+$)GVP], of ΔG_{HA}[GK(dihydroNMeN)GVP], and of

$\Delta G_{HA}[GK(NMeN)GVP]$ is to use the values of T_t in Table 5.2 and the plot of T_b versus ΔG_{HA} of Figure 5.10. The Gibbs free energy of hydrophobic hydration, obtained in this way, is given in Table 5.2 to be +2.0, −3.0, and −7.0, respectively. Therefore, an estimate of the change in Gibbs free energy for hydrophobic association on reduction of the N-methyl nicotinamide, $\Delta G_{HA}[NMeN^+ \rightarrow NMeN]$ becomes [−7.0 − [+2.0]] = **−9.0 kcal/mole-pentamer**. Comparison of the change in free energy for hydrophobic association on reduction of this nicotinamide, $\Delta G_{HA}[NMeN^+ \rightarrow NMeN]$, to the decrease in free energy of hydrophobic association due to protonation of the carboxylate of Glu, $\Delta G_{HA}[Glu\text{-}COO^- \rightarrow Glu\text{-}COOH]$ of **−5.22 kcal/mole**, gives insight into the maximal efficiency for the coupling of these functional groups. Accordingly, the coupling of the two reactions, wherein reduction drives the uptake of a proton by a Glu-COO⁻, could be expected to occur in the particular model protein with a 60% efficiency of energy conversion. A similar number for efficiency is obtainable on calculating the carboxyl/carboxylate pKa shift due to reduction in the same polymer containing both functional groups by comparing the free energy change on reduction with the observed pKa shift (see section 5.9, below).[73]

5.3.4.4.2 The ΔG_{HA} Due to the Reduction of Oxidized Nicotinamide Adenine Dinucleotide (NAD)

Following the same process as used above for N-methyl nicotinamide, involving the data in Table 5.2 and the plot in Figure 5.10, an estimate of the change in Gibbs free energy for hydrophobic association on reduction of nicotinamide adenine dinucleotide, $\Delta G_{HA}(NAD \rightarrow NADH)$ becomes [−5.5 − (+2.0)] = **−7.5 kcal/mole-pentamer**.

5.3.4.4.3 The ΔG_{HA} Due to the Reduction of Oxidized Flavin Adenine Dinucleotide (FAD)

Continuing as immediately above using the data in Table 5.2 and in Figure 5.10, an estimate of the change in Gibbs free energy for hydrophobic association on reduction of flavin adenine dinucleotide, $\Delta G_{HA}(FAD \rightarrow FADH_2)$ becomes [−2.5 − (+2.0)] = **−4.5 kcal/mole-pentamer**. The qualitative relationship between $\Delta G_{HA}(NAD \rightarrow NADH)$ of **−7.5 kcal/mole-pentamer** and $\Delta G_{HA}(FAD \rightarrow FADH_2)$ of **−4.5 kcal/mole-pentamer** is consistent with the resulting relative proton translocation due to these molecules on oxidation in the electron transport chain of the inner mitochondrial membrane where the proton transport ratio for FADH₂:NADH is 2:3.

5.3.4.4.4 The ΔG_{HA} Due to Phosphorylation

Phosphorylation by the cardiac cyclic AMP-dependent protein kinase at the serine (Ser, S) residue of the polymer poly[30(GVGIP), (RGYSLG)] resulted in an estimated T_t-value in Table 5.2 of 860° C.[74,75] The equilibrium constant for the reaction

$$ATP + poly[30(GVGIP),(RGYSLG)] = ADP + poly(30[GVGIP],[RGYS\{-OPO_3^=\}LG]) \quad (5.11)$$

was essentially 1. In particular, the average phosphorylation from many phosphorylation attempts was 47%.[9,75] This means that the change in free energy of the phosphate on going from the terminal position of ATP to poly(30[GVGIP],[RGYS{−OPO₃⁼}LG]) is effectively zero. The conclusion is that the phosphate in poly(30[GVGIP],[RGYS{−OPO₃⁼} LG]) is at the same high free energy state as it is in the γ-position of ATP.

Because hydrophobic association does not occur until the temperature is above T_t for the phosphorylated state, the phosphate cannot be folded into a low dielectric constant site, but instead would be fully exposed to the solvent. It is fundamental to the consilient mechanism of protein-based machines to understand how this can be! As discussed in section 5.7, the high-energy state of the phosphate is due to competition for hydration between the phosphate and the hydrophobic groups, as reflected in the large change in T_t.

Following the above approach of combining the data in Table 5.2 with the relationship between T_b and ΔG_{HA} of Figure 5.10 and the realization that $T_b \approx T_t$, it becomes possible to achieve an estimate of $\Delta G_{HA}(-OPO_3^=)$. The charged species, K^+ and E^-, determine a slope with which to extrapolate from 100° C to the value of 860° C determined for the phosphorylation discussed above. This gross means of estimation results in the remarkable value of $\Delta G_{HA}(-OPO_3^=)$ = **+8 kcal/mole**.

This free energy of interaction occurs in the unfolded state and serves to raise the free energy for formation of the hydrophobically associated state. Because the high-energy state is the unfolded polymer, it cannot be due to charge–charge repulsion; there is only 1 phosphate in about 300 residues. The high-energy state cannot be due to the phosphate being in the low dielectric region of a folded protein. In our view, it represents a clear measure of the apolar–polar repulsive free energy of hydration arising out of the competition for hydration between the oil-like side chains of the isoleucine (Ile, I) and valine (Val, V) residues and the phosphate in the unfolded model protein.

Phosphate removal drives hydrophobic association, that is, contraction, more effectively than any other chemical change measured thus far. The binding of ATP, adenosine triphosphate, the universal energy currency of biology, could even more effectively increase the value of T_t, but this can be modified by the associations of the ATP at the binding site. Also, as in Table 5.2, the estimate of ΔG_{HA}[Glu(AMP)] is a modest +1 kcal/mole, and the addition of each phosphate would add approximately 8 kcal/mole. Furthermore, we expect that the state of greatest apolar–polar repulsion would result from the presence of ADP plus P_i.

5.3.4.5 Using ΔT_t Due to Addition of Salts to Calculate Change in Free Energy (ΔG_{HA}) for Hydrophobic Association of Neutral Poly(GVGVP)

The dependence of T_t for poly(GVGVP) on the addition of a series of salts (the Hoffmeister series) is shown in Figure 5.11A. With the exceptions of I^- and SCN^-, the addition of salts to poly(GVGVP) lowers the value of T_t. With the ΔH_t data from Table I of Luan et al.[76] for use in Equation (5.10b), calculation of ΔG_{HA} (0 N → 1.0 N NaCl) yields a value of **−0.57 kcal/mole**. Calorimetry (ΔH_t) data are not yet available for the other salts. If we assume the values to be proportional to ΔT_t with the NaCl data providing the proportionality constant, however, estimates of ΔG_{HA}(0 N → 1.0 N salt) can be obtained for a series of salts. These are listed below and in Table 5.4A. The decrease in T_t(GVGVP) due to the addition of 1 mole of salt is found in Table 3c of Luan and Urry.[72] In what follows, after identification of the polymer composition and salt, listed are the numbers for ΔT_t on increasing the amount of salt from 0 to 1.0 N and the calculated value of ΔG_{HA} in **kcal/mole** based on a proportionality to the NaCl data: Na_3PO_4(pH > 8), −140° C, **−5.7**; $(NH_4)_2SO_4$, −69° C, **−2.8**; Na_2CO_3(pH > 8), −28° C, **−1.1**; NaCl, −13.9° C, **−0.57**; $CaCl_2$, −6.6° C, **−0.27**; NaBr, −3.5° C, **−0.14**; NaI and NaSCN, +3.5° C, **+0.14**.

5.3.4.6 Using ΔT_t Due to Ion-Pairing of Salts with Functional Groups to Estimate Change in Free Energy (ΔG_{HA}) for Hydrophobic Association

As shown by comparison of the data in Figure 5.12A with those in Figure 5.11A, the effect of salts in lowering the free energy of the hydrophobically associated state of the protein-based polymer is much greater when the polymer contains a charged side chain. An estimate of ΔG_{HA}(Glu-COO⁻ ••• Na⁺; steepest ion-pair slope), that is, the change in Gibbs free energy for ion pairing over the change in salt concentration with the steepest slope, may be obtained using the slopes, $\Delta T_t/\Delta G_{HA}$ established in Figure 5.10 for the ionization of the carboxyl (E → E⁻) to be 380°/9 kcal/mole. Table 3d Luan and Urry[72] provides the data points used in Figure 5.12A from which the steepest slope (0.15 N to 0.25 N) is obtained to give the value of ΔT_t/N NaCl to be −395° C/N. The result is ΔG_{HA}(Glu-COO⁻ ••• Na⁺; steepest ion-pair slope) = **−9.4 kcal/mole-pentamer/N**. The effect of ion pairing of Na⁺ with COO⁻ lowers the free energy of the hydrophobically associated state by more than an order of magnitude when compared with the NaCl interaction with the neutral polymer, that is, ΔG_{HA}[poly(GVGVP); 0 N → 1 N NaCl] = **−0.57 kcal/mole**. Furthermore, once the effect of ion pairing is completed above 1 N NaCl, the effect of the salt for the step from 1 N to 2 N reduces to **−0.7 kcal/mole**. This estimate of the powerful effect of ion pairing on lowering the free energy of hydrophobic association is greater for Ca^{2+} (−15 kcal/mole) and greater yet as the polymer becomes more hydrophobic by replacement of Val by Phe and Ile (see Figure 5.15).

For calculation of the lowering of ΔG_{HA} due to the ion pairing of $-NH_3^+$ with Cl^-, within poly[0.76(GVGVP),0.24(GK^+GVP)] the data in Table 3d of Luan and Urry[72] provide the steepest slope (0.05 N to 0.20 N) to give the value of $\Delta T_t/N$ NaCl, that is, −177° C/N. From Figure 5.10 the $\Delta T_t/\Delta G_{HA}$ value established for the ionization of the amino (K → K^+) is 230°/12.5 kcal/mole. Accordingly, ΔG_{HA}(Lys-NH_3^+ ••• Cl^-; steepest ion-pairing slope) ≈ **−9.6 kcal/mole-pentamer/N**. After the effect of ion pairing is complete, that is, for the slope above 1 N NaCl, this reduces to ΔG_{HA}(Lys-NH_3^+ ••• Cl^-; 0.5 N → 1.5 N salt) = **−36**. Again, the effect of ion pairing to stabilize the hydrophobically associated state of the model protein is seen to be large.

Figure 5.12B demonstrates the effect of adding a CH_2 group, as for the more hydrophobic (GVGIP) repeat, on increasing the magnitude of the favorable change in the Gibbs free energy for hydrophobic association as the result of ion pairing. Also shown in Figure 5.12B is the greater capacity of calcium ion over sodium ion in lowering the free energy for hydrophobic association. This effect will be calculated below in relation to the data in Figure 5.15.

5.3.4.7 Limited Relevance of ΔT_t Due to Addition of Organic Solvents and Solutes to Calculate Change in Free Energy (ΔG_{HA}) for Hydrophobic Association of Neutral Poly(GVGVP)

The dependence of T_t for poly(GVGVP) on the addition of a series of organic solutes and solvents is shown in Figure 5.11B. Additional data are given by Luan and Urry,[72] who also include the thermodynamic values used below in combination with Equation (5.10b) to estimate ΔG_{HA}, as indicated in the footnotes of Table 5.4B. The solute with the steepest initial slope for increasing the value of T_t on adding the organic solute is SDS. One expects, therefore, that it would be most effective in raising the free energy of the hydrophobically folded state. There is no calorimetry data available for SDS,

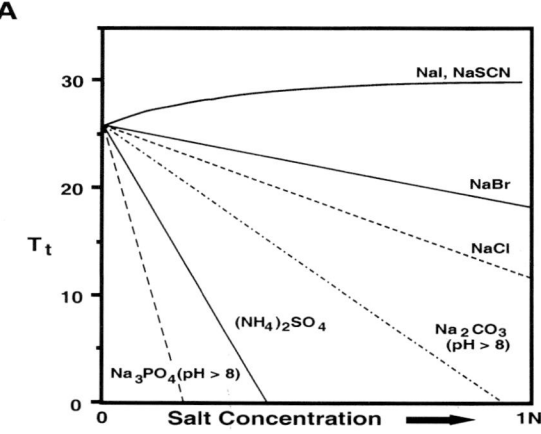

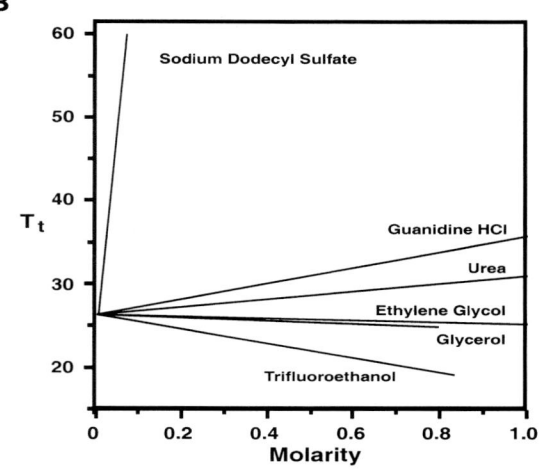

FIGURE 5.11. Dependences of T_t-values for poly(GVGVP) as a function of salt concentration, (**A**) and of organic solvents and solutes (**B**). These data are used in Table 5.4 on a per mole (normality) basis to calculate the ΔG_{HA}(solute) in terms of kcal/mole solute. (Adapted with permission from Urry.[6])

5.3 Cataloging the Energy Resources Available to the Consilient Mechanism of Energy Conversion 145

TABLE 5.4. $\Delta G_{HA}(\chi)$ for hydrophobic association of a given composition of model protein per mole for the steepest slope of ΔT_t per change in salt concentration.

A. Model protein and salt	ΔG_{HA}(salt) (kcal/mol/N)	Steepest slope and concentration range
Poly(GVGVP) & NaBr	−0.14	−3.5° C/N & 0N → 1N
Poly(GVGVP) & CaCl$_2$	−0.27	−6.6° C/N & 0N → 1N
Poly(GVGVP) & NaCl	−0.57	−14° C/N & 0N → 1N
Poly(GVGVP) & Na$_2$CO$_3$ (pH > 8)	−1.10	−28° C/N & 0N → 1N
Poly(GVGVP) & (NH$_4$)$_2$SO$_4$	−2.8	−69° C/N & 0N → 1N
Poly(GVGVP) & Na$_3$PO$_4$ (pH > 8)	−5.7	−140° C/N & 0N → 1N
Poly[0.8(GVGVP),0.2(GE$^-$GVP)] & NaCl	−9.4	−395° C/N & 0.15N → 0.25N*
Poly[0.76(GVGVP),0.24(GK$^+$GVP)] & NaCl	−9.6	−177° C/N & 0.05N → 0.20N*
Poly[0.8(GVGVP),0.2(GE$^-$GVP)] & CaCl$_2$	−15	−618° C/N & 0.125N → 0.05N*

*Utilizes solpes defined in Figure 5.10. See text for discussion.

B. Model protein and organic solute or solvent	ΔG_{HA}(solute) (kcal/mol/M)	Steepest slope and concentration range
Poly(GVGVP) & sodium dodecyl sulfate	+18[a]	+360° C/M & 0M → .25M
Poly(GVGVP) & dimethyl urea	+0.93[b]	+12.2° C/M & 0M → 1M
Poly(GVGVP) & methanol	+0.93[d]	+→−biphasic & 0M → 1M
Poly(GVGVP) & ethanol	+0.79[d]	+→−biphasic & 0M → 1M
Poly(GVGVP) & ethylene glycol	+0.79[d]	−0.6° C/M & 0M → 1M
Poly(GVGVP) & dioxane	+0.71[c]	+4.9° C/M & 0M → 1M
Poly(GVGVP) & dimethyl sulfoxide	+0.71[c]	+3.3° C/M & 0M → 1M
Poly(GVGVP) & acetone	+0.63[c]	+5.3° C/M & 0M → 1M
Poly(GVGVP) & glycerol	+0.57[d]	−1.8° C/M & 0M → 1M
Poly(GVGVP) & guanidine-HCl	+0.41[b]	+10.3° C/M & 0M → 1M
Poly(GVGVP) & urea	+0.23[b]	+5.3° C/M & 0M → 1M
Poly(GVGVP) & trifluoroethanol	−0.37[a]	−8.7° C/M & 0M → 1M
Poly(GVGVP) & guanidine$_2$-H$_2$SO$_4$	−0.43[b]	−11.4° C/M & 0M → 1M
Poly(GVGVP) in deuterium oxide (D$_2$O)	−0.26[e]	

[a] Calculated using guanidine-HCl ΔT_t/M data to calibrate slope for estimating ΔG_{HA}.
[b] Calculated using Eqn (10b) and calorimetry ($\Sigma \Delta Q_t$) data within Table 3f of Luan and Urry.[72]
[c] Calculated using Eqn (10b) and calorimetry (ΔH_t) data within Table 3g of Luan and Urry.[72]
[d] Calculated using Eqn (10b) and calorimetry (ΔH_t) data within Table 3k of Luan and Urry.[72]
[e] Calculated using Eqn (10b) and calorimetry (ΔH_t) data within Table 3j of Luan and Urry.[72]

but if we assume that SDS has the same slope, $\Delta T_t/\Delta G_{HA}$, as guanidine-HCl, one obtains an estimate of $\Delta G_{HA}(0 \to 0.1$ M SDS$) \approx$ **18 kcal/mole-pentamer/mole-solute**. Reasonably, SDS is the most effective agent identified to date in causing the polymer to hydrophobically unfold. This is to be compared with the commonly used denaturants of guanidineHCl of **+0.43 kcal/mole** and of urea of **+23 kcal/mole** (See Table 5.4B).

The correlation between ΔT_t and ΔG_{HA}, so well demonstrated by the results in Tables 5.3 and 5.4 and in Figure 5.10, breaks down when it comes to consideration of the effects of certain organic solvents. This is apparent with ethanol and methanol where the $\Delta T_t/M$ is initially positive and then shifts negative, whereas the experimental ΔH_t from which ΔG_{HA} is obtained by means of Equation (5.10b) simply decreases as M increases. Two possible explanations for this discrepancy would be the direct interaction of organic solute or solvent with the polymer and mixed effects on the structure of hydrophobic hydration. Elements of the latter are seen with amino acid side chains in Figure 5.10, where the −OH in Tyr (Y) and Thr (T), the −COOH of Glu (E°) and Asp (D°), the −NH$_2$ of Lys (K°), and so on are displaced from the diagonal line for the strict correlation. These functional groups are expected to interact with the hydrophobic hydration in ways that could either destabilize or stabilize it, that is, to modify the thermodynamics of hydrophobic hydration such that the con-

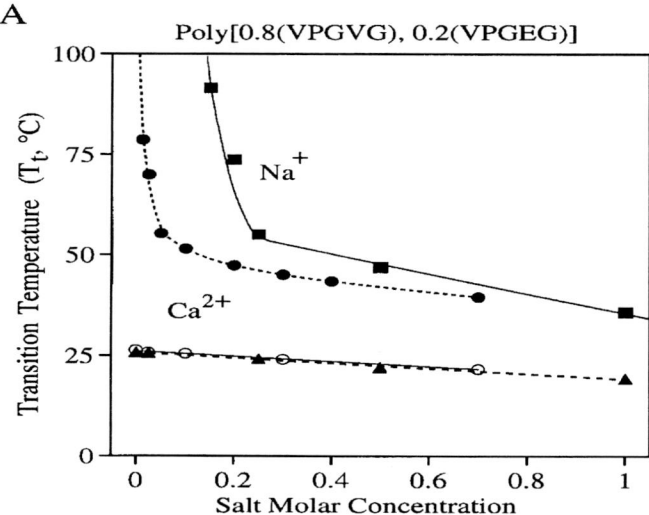
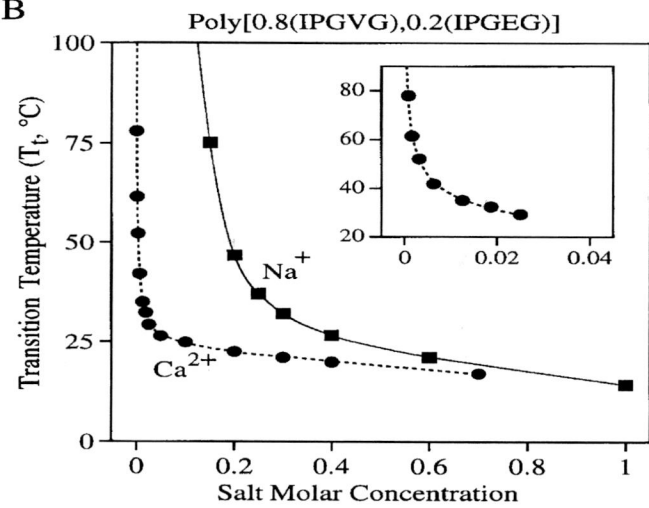

FIGURE 5.12. Dependences of T_t-values for poly[0.2(GE⁻GVP),0.8(GVGVP)] on sodium ion and calcium ion concentration. (**A**) At low ion concentrations the slopes are very steep, being close to 30 times steeper for Na^+ than for poly(GVGVP) and more than 90 times steeper for Ca^{2+}. This emphasizes the importance of ion pairing in lowering the free energy for hydrophobic association, that is, in inducing hydrophobic association. (**B**) Where there is one additional CH_2 group per pentamer, the steepness increases by almost a factor of two for Na^+ (some 50 times greater than for poly[GVGVP]) and a factor of 10 for Ca^{2+} (close to 1,000 times greater than for poly[GVGVP]. *Thus, ion pairing provides a powerful force for inducing hydrophobic association that becomes greater as the hydrophobic character of the model protein increases.* (**A** is plotted from Tables 3d and 3e of Luan and Urry,[72] and part B is from C.-H. Luan and D.W. Urry, unpublished results.)

tributions to ΔH_t are no longer dominated by a change in the amount of hydrophobic hydration. After all, the special thermodynamic properties of hydrophobic hydration determine ΔG_{HA}. Despite these thermodynamic complications, there remains the practical utilization of ΔT_t as a simple sliding on–off switch for controlling association and dissociation of our model proteins.

5.3.5 What the T_t-based Hydrophobicity Scale Implies About Physical Basis

If lowering T_t is the result of increased hydrophobic hydration, as we derived from the data of Butler[25] and of Frank and Evans,[26] then the increase in the value of T_t due to the presence of 1 carboxylate per 100 residues of poly(GVGVP) would seem to arise from the loss of hydrophobic hydration; this suggests that the negatively charged ion destructures hydrophobic hydration in the process of achieving its own hydration shell. Such insight, supported by a series of less direct arguments, became confirmed by the direct observation of hydrophobic hydration (see in section 5.7). Before definitively addressing the physical basis, however, demonstrations are given of many of the diverse energy conversion possible by the consilient mechanism.

5.4 Visual Observation of Elastic Protein-based Machines at Work

5.4.1 Straightforward Development of Elastic Protein-based Machines for Pumping Iron

Development of our understanding of energy conversion by the consilient mechanism of hydrophobic association began with the expected thermally driven contraction of cross-linked bands composed of model protein.[77,78] Next, it was anticipated and found that protonation of a carboxylate within the model protein lowered the temperature of the transition and that protonation of the carboxylate within the cross-linked elastic band drove contraction.[64] This was immediately followed by the predicted demonstration that addition of salt could drive contraction by lowering the temperature of the transition of the simple cross-linked model protein, poly(GVGVP).[79] Subsequently and expectedly came pressure-driven relaxation with aromatic residues[80] and reduction-driven contraction using a chemically attached redox couple.[81] These energy conversions with the energy output being the mechanical work of lifting a weight were visually demonstrated. Elastic bands, with the appropriate model protein composition and a suspended weight, contract and lift weights, that is, *pump iron*, when designed to be responsive to different energy inputs. Despite many different energy inputs, all of the contractions utilize the common mechanism of hydrophobic association to lift the weights. A video provides visual recordings of these *reversible* contractions and relaxations.

5.4.1.1 Simplicity of the Fundamental Observations

5.4.1.1.1 Onset of Aggregation on Raising the Temperature of Model Proteins in Solution

Again, the initial experimental observations are very simple. For the model proteins in solution, a clear solution becomes cloudy as the temperature is raised. The onset of cloudiness signals aggregation, or association, of the model proteins, as shown in Figure 5.1. During this association, oil-like **R**-groups of the model protein chains separate from water and become insoluble. Specifically, the differential calorimetry curve of Figure 5.1C provides graphic representation of the "cusp of insolubility." Moving the cusp to lower temperatures than the working temperature gives insolubility, and moving the "cusp" to higher temperatures than the operating temperature gives solubility. The temperature range over which aggregation (insolubilization) occurs is called the *temperature interval*. How each of 20 different **R**-groups changes the temperature interval for aggregation becomes the primary experimental database (see Figures 5.5 and 5.10 and associated discussion). The database, called a T_t-based hydrophobicity scale, in a most practical form reports the relative oil-like character of the 20 naturally occurring **R**-groups (see Table 5.1 and associated discussions). Figure 5.10 provides the relationship between T_t and the fundamental Gibbs free energy of hydrophobic association, ΔG_{HA}, required for a more complete analysis of energy conversion by protein-based machines, and relevant values are given in Tables 5.2, 5.3 and 5.4.

5.4.1.1.2 Temperature Dependence of Solubility of Oil-like Groups and Formation of Insoluble Structures

In relation to the above discussion of the Butler[25] alcohol series, which considered solubility of oil-like groups in terms of the equation $\Delta G = \Delta H - T\Delta S$, raising the temperature increases the magnitude of the (−TΔS) term until it dominates over the ΔH term and solubility is lost. In the basic model protein of interest here, poly(GVGVP), the aggregated state still contains 60% water by weight.[28] Because of this, the separated oil-like groups become spatially localized hydrophobic patches on the surface of helically folded and coiled model protein. This may be thought of in two steps: (1), on folding of the helices (β-spirals) the localized hydrophobic patches tend to form helical ribbons of oil-like groups on the surface of the β-spirals; and (2), the helically folded structures associate by means of the oil-like helical ribbons on the surface to form twisted filaments (see Figure 5.6E,F). In reality the hydrophobic folding of a single chain and the association (coiling) of chains to form twisted filaments are cooperative and generally inseparable.

5.4.1.2 Constructing Macroscopic Elastic Protein-based Machines

The simple process of constructing the contractile elastic bands follows. It involves forming the phase-separated state comprised of interpenetrating twisted filaments as represented in Figure 5.6F, shaping the phase-separated state into thin sheets and cross-linking the sheets into elastic matrices, most simply, by exposure to γ-irradiation.

5.4.1.2.1 Formation and Shaping of a Glue-like Substance

As shown in Figure 5.13 with the parent model protein, poly(GVGVP), the polymer is soluble in water below 25° C. On heating to 37° C, hydrophobic association of the model protein occurs, and, on standing, separation occurs to form a glue-like phase. The overlying solution is removed, a pestle is inserted, and the glue-like phase fills the space between the pestle and tube.

5.4.1.2.2 Cross-linking of the Glue-like Substance to Form Contractile Elastic Protein-based Sheets and Bands

The glue-like phase between the concentric cylinders of pestle and tube are maintained in the phase-separated state and exposed to 20 Mrad of γ-irradiation from a cobalt-60 source. (Alternatively, chemical cross-linking can occur with a suitable reactive chemical either without or with functional side chains, or cross-linking can occur enzymatically with introduction of appropriate sequences.) One of the advantages of γ-irradiation is sterilization and biocompatibility for medical applications.

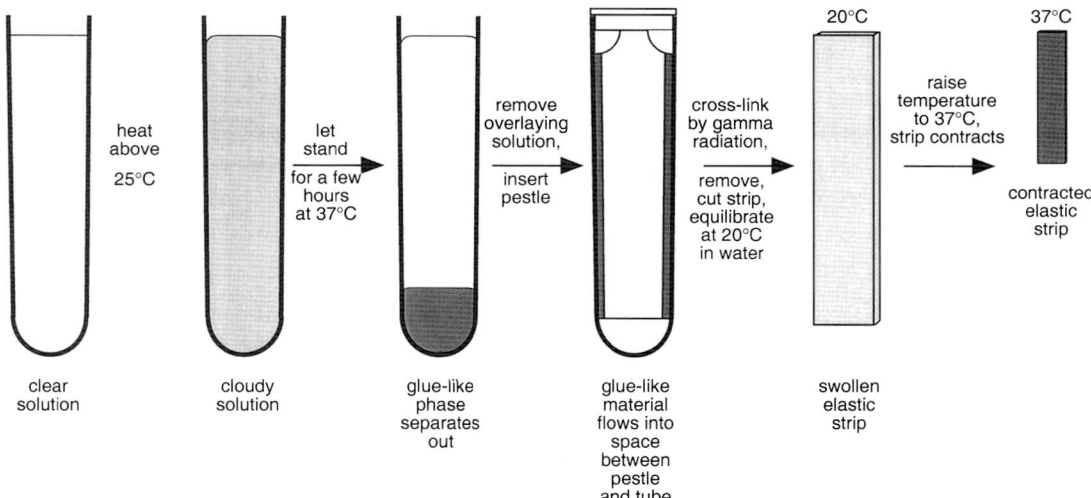

FIGURE 5.13. A procedure for the formation of elastic model proteins into γ-irradiation cross-linked sheets for characterization as elastic-contractile systems for performance of mechanical work of lifting a weight.

5.4 Visual Observation of Elastic Protein-based Machines at Work 149

Informative anecdotes relate to effective γ-irradiation cross-linking. Adequate cross-linking does not occur unless the temperature has been raised through the temperature interval to drive the hydrophobic folding and assembly. Solutions of polymer at concentrations lower than that of the usual phase-separated state, obtainable only by being at temperatures lower than the temperature interval, do not cross-link. Irreversible denaturation of poly(GVGVP), as represented by the denaturation peak of Figure 5.2, results in a phase-separated state that is twice as concentrated, 70% polymer and 30% water.[28] Even though at twice the concentration, the denatured polymer does not effectively cross-link by γ-irradiation. In each case the absence of the state of interpenetrating folded and regularly associated chains interferes with formation of cross-links between chains.

The elastic strip or sheet is removed from the tubes, and it is found to swell on lowering the temperature and to contract on raising the temperature through the temperature interval from 20° to 37° C, as depicted at the right in Figure 5.13. A sheet of γ-irradiation cross-linked (GVGVP)$_{251}$ is shown in Figure 5.14.

5.4.1.3 Hydrophobic Association in Solution Becomes Contraction in the Elastic Matrix

The same process of hydrophobic association, observed in solution and characterized by the phase diagram in Figure 5.3, occurs in the elastic matrix where hydrophobic association displays as a visible contraction. The parallel process of cloudy formation of polymers in solution transforms in the elastic matrix into a contraction capable of performing mechanical work.

Contraction of elastic matrices occurs as the temperature is raised through the temperature interval of Figure 5.5A; this constitutes the input of thermal energy (TE), measurable as the heat of the transition seen in Figure 5.1C. Visible contraction also occurs as any of the independent variables move through the *transition zone* in Figure 5.5B. Depending on the composition of the model protein that consti-

FIGURE 5.14. Sheet of elastic model protein of γ-irradiation cross-linked (GVGVP)$_{251}$ as demonstrated in Figure 5.13 directly usable in thermally driven contraction and in chemically driven contraction.

tutes the elastic matrix, the independent variable can be one of a number of energy inputs such as chemical energy (CE), electrochemical energy (ECE) pressure-volume energy (PVE), and electromagnetic radiation energy (EMRE) such as light. Thus, one can design elastic matrices capable of contracting and performing the mechanical work of lifting weights, and with the unaided eye all can see that the design is successful.

As discussed below, some designs of model proteins require less of a given energy input to perform a given amount of mechanical work. The effectiveness of the given energy input in performing mechanical work depends on the change in the Gibbs free energy for hydrophobic association, ΔG_{HA}, resulting from the energy input acting on the particular model protein composition.

5.4.1.4 Solvent Entropy Changes in Elastic Protein Band on Loading (Stretching)

5.4.1.4.1 Below T_t, Solvent Entropy Changes due to Stretching of a Fully Swollen Elastic Band

Carefully hanging a weight on the fully swollen elastic band causes elastic extension sufficient to support a light load. The elastic force, f, within the band that supports the load is comprised of two components, an internal energy component, f_E, and an entropy component, f_S, that is,

$$f = f_E + f_S \quad (5.12)$$

Stretching at a temperature below T_t of the fully swollen elastic band results in no change in hydrophobic hydration. *Therefore, any elastic force developed below T_t is not expected to arise due to solvent entropy changes.* The entropy component, f_S, of the elastomeric force arises from the decrease in the internal chain motion such as rotation about bonds, while the internal energy component, f_E, results from straining of bond lengths and deforming bond angles that could lead to chain rupture.

5.4.1.4.2 Above T_t, Solvent Entropy Changes Due to Stretching of a Contracted Elastic Band

The fully contracted elastic band is in a state of maximized hydrophobic association. Because of this, stretching of the contracted band exposes hydrophobic groups to water. Recall that when an oil-like group dissolves in water, formation of hydrophobic hydration occurs with the release of heat. Thus, hanging a weight on the contracted band exposes hydrophobic groups to water and releases heat as hydrophobic hydration forms. Associated with the exothermic hydrophobic hydration is a decrease in entropy of the water that goes from bulk water to hydrophobic hydration, such that the change in the $(-T\Delta S)$ term of the Gibbs free energy is positive. For elastic bands composed of poly(GVGVP), 90% of the force developed on stretching the hydrophobically associated state is due to f_S (see Figure 5.7).[63]

5.4.1.4.3 Effects on Solvent Entropy of Raising the Temperature of Loaded Elastic Band Through the Temperature Interval for Contraction

Obviously, the effect of raising the temperature of a loaded, cross-linked elastic band composed of elastic protein-based polymers of the poly(GVGVP) family is to drive hydrophobic association with the consequence of lifting of the attached weight. How does this combine with the above understanding of elasticity to perform mechanical work?

Hydrophobic association has the corollary of a positive change in entropy of water as hydrophobic hydration becomes bulk water. This occurs whether the experiment is the change in length at constant force (load) of pumping iron and the work done is $f\Delta L$ or whether the experiment is the increase in force at constant length and the energy output is $L\Delta f$. In the latter case there is an increase in force as thermally driven hydrophobic association occurs, and the increase in force occurs when the solvent entropy change is positive.

Thus, for the $L\Delta f$ case the increase in force cannot arise from a decrease in solvent entropy, because, in fact, the process is accompanied by an increase in solvent entropy. This brings us to the following perspective. *In both cases, the entropic elastic force is maintained or <u>increased by greater extension resulting in an increased damping of internal chain dynamics within a chain segment of fewer repeats</u> as the chains within a substantial part of the elastic band undergo hydrophobic association, as represented in Figure 5.8.*[65]

5.4.2 Thermal Energy Produces Mechanical Work of Lifting a Weight (Heating Through the Temperature Interval Drives Contraction)

5.4.2.1 Using Cross-linked Model Proteins as Thermally Driven Contractile Elastic Bands

For the performance of mechanical work by model proteins cross-linked to form elastic bands, a weight is attached to the rubber-like band and the temperature raised. Over a particular temperature interval, the band contracts and lifts the weight. Relatively little contraction occurs below the temperature interval, and relatively little contraction occurs above the temperature interval for the particular composition (see Figure 5.5A). The presence of more oil-like **R**-groups lowers the temperature interval over which contraction occurs, and the presence of less oil-like or more polar groups **R**-groups raises the temperature interval over which contraction occurs. Thus, the temperature dependence of contraction of the cross-linked elastic matrix follows the dependence of the T_t-divide on hydrophobicity of the protein-based polymers in solution. The relative hydrophobicity is reconfirmed in this simplistic and visual observation of the conversion of thermal energy (heat) into mechanical work (see Figures 2.4 and 2.5). An obvious elementary statement of the experimental findings is given as Axiom 2.

Axiom 2: Heating to raise the temperature from below to above the temperature interval for hydrophobic association of cross-linked elastic model protein chains drives contraction with the performance of mechanical work.

5.4.2.2 Thermal Energy (TE), Measurable as the Heat of the Transition of the Band Under Load, ΔH_l, Produces Mechanical Work (MW) of Lifting a Weight, That Is, $TE \rightarrow MW$

At any temperature within the temperature interval for the thermally driven transition, there exists an equilibrium between hydrophobic association and dissociation. Obviously, at the low temperature side of the temperature interval, limited hydrophobic association occurs. At the high temperature side of the temperature interval, hydrophobic association is limited only by the extending load. The relationship is such that hydrophobic association is inversely proportional to the load.

If we knew the heat of the transition for the elastic band under load, ΔH_l, we could calculate the ratio of the mechanical work performed to the heat input expended in lifting the load,

that is, the efficiency of the energy conversion. Stretching of these hydrophobically associated elastomers, however, is exothermic due to the forced exposure and hydration of hydrophobic groups. Accordingly, ΔH_l is less than ΔH_t, but the experimental determination of ΔH_l is difficult due to the nature of the differential scanning calorimeter required to determine the heats of the transitions. Fortunately, the chemical energy expended in moving the loaded elastomer through the transition zone for the performance of mechanical work can be determined, as discussed below.

5.4.3 Chemical Energy Produces Mechanical Work of Lifting Weight (Changing Concentration of Chemical *Through the Transition Zone* Drives Contraction)

5.4.3.1 *Equivalence of Moving the Temperature Interval and Moving Through the Transition Zone*

5.4.3.1.1 Moving the Temperature Interval

The temperature interval for aggregation of poly(GVGVP) begins just above room temperature in water, that is, above 25° C, and ends at about 37° C (see Figure 5.1C and Figure 5.5A). Addition of salt (NaCl) at 25° C lowers the temperature for aggregation and causes the solution to go cloudy. As discussed below, salt drives contraction by lowering the temperature interval for heat-driven contraction.

5.4.3.1.2 Relationship Between Temperature Interval and Transition Zone for Salt-driven Contraction in the Absence of Functional Side Chains

The decrease in temperature for the onset of the transition due to the addition of salt is 14° C for the addition of 58 grams in 1 liter, that is, 1 N NaCl.[79] Accordingly, the addition of 1 N NaCl moves the temperature interval originally 25° to 37° C approximately one interval width to 11° to 23° C and drives contraction of the elastic matrix Increasing the concentration of salt from 1 N to 2 N, however, does little to further the contraction. Thus the transition zone for the NaCl–cross-linked-poly(GVGVP) system is 0 to 1 N NaCl (see Figure 5.5). When observing the cross-linked elastic band attached to a weight, added salt causes the rubber-like band to contract and perform the work of lifting a weight. The cross-linked elastic-contractile model protein transforms chemical energy of increasing salt concentration into mechanical work of pumping iron.

5.4.3.1.3 Relationship Between Changing the Temperature Interval and Moving Through Transition Zone for Contraction by Changing the State of Functional Side Chains

When the polymer contains vinegar-like functional groups, the appropriate counter-ion of the salt ion pairs with the functional group. Ion pairing markedly lowers the temperature interval for contraction. Ion pairing places the transition zone at a much lower salt concentration, for example, between 0.01 and 0.15 N NaCl.

In particular, when the counter-ion for a carboxylate, COO^-, is the special case of a proton, H^+, the transition zone can occur at very low concentrations, for example, between a concentration-1 (C_1) of 0.000001 and concentration-2 (C_2) of 0.00001 N HCl. Another way to write this small concentration difference is to use powers of 10, that is, $C_1 = 10^{-6}$ N HCl and $C_2 = 10^{-5}$ N HCl. Yet another way to write this difference is to use logarithms to express the increase in concentration, for example, $\log(C_2) - \log(C_1) = (-5) - (-6) = 1$. The change in chemical energy, ΔE, required to drive a process is the product of the change in chemical potential, $\Delta \mu$, times the change in number of moles, Δn, that is, $\Delta E = \Delta\mu\Delta n$. For simplicity, the present argument focuses on chemical potential such that $\Delta E/\Delta n = 2.3RT[\log(C_2) - \log(C_1)] = $ **1420 cal/mole**, where R = 1.987 cal/mole-deg and at physiological temperature (e.g., 37° C such that T = 310 K). *Pumping iron, performing mechanical work, by the mechanism of charge neutralization, for example, ion pairing, requires less chemical energy to do the same amount of mechanical work than if the model protein had no charged functional group or if the charged group of the model protein were already neutralized. This comparison is graphically demonstrated below.*

5.4.3.2 Relationship Between Temperature Interval and Transition Zone for Chemical Energy Input

5.4.3.2.1 Relationship Between Changing the Temperature Interval and Moving Through the Transition Zone for Contraction Using Carboxylate (–COO⁻) Functional Side Chains and Calcium ion (Ca^{2+})

The temperature interval for poly(GVGVP) begins at 25° C and ends at 37° C. The same temperature interval is considered for two different microbially prepared analogues of poly(GVGVP), namely, **Model protein I**: $(GVGIP\ GFGEP\ GEGFP\ GVGVP\ GFGFP\ GFGIP)_{26}$ and **Model Protein ii**: $(GVGVP\ GVGFP\ GEGFP\ GVGVP\ GVGVP\ GVGVP)_{40}$ of Table 5.5. Figure 5.15 contains the dependence on the logarithm of the calcium ion concentration, $\log[Ca^{2+}]$, of the value of T_t for both model proteins in water. For **Model Protein ii** aggregation of the model protein in solution and contraction of the cross-linked band in the absence of a load begin at 37° C, as the calcium chloride reaches a concentration of 0.034 mole/liter ($\log[Ca^{2+}] = -1.47$), and hydrophobic association is complete as the calcium chloride concentration increases to 0.446 mole/liter ($\log[Ca^{2+}] = -0.351$). Thus, the temperature interval of 25° to 37° C corresponds to a transition zone for the independent variable of calcium ion concentration of 0.034 to 0.446 mole/liter. This is a change in concentration of 0.412 mole/liter, but, more to the point, on the log scale, the difference is 1.12. The chemical energy divided by the number of moles of calcium ion consumed in the process of driving contraction of **Model Protein ii** in the limit of zero load is $\Delta E/\Delta n \approx 2.3RT[1.12] \approx$ **1,580 cal/mole**. For **Model Protein I** the transition zone is narrower with a difference on the log scale of only 0.116. The chemical energy per mole required to drive the contraction of **Model Protein I** is $\Delta E/\Delta n \approx 2.3RT[0.116] \approx$ **164 cal/mole**. The two carboxyls of **Model Protein I** would reasonably bind a single calcium ion in a bidentate manner, and this is less likely for **Model Protein ii** where the carboxyls are separated by 30 residues instead of two residues.

5.4 Visual Observation of Elastic Protein-based Machines at Work

TABLE 5.5. Model Proteins and Protein-based Polymers.

		A. Microbially-produced Model Proteins	
Model Protein	I:	(GVGIP GFGEP GEGFP GVGVP GFGFP GFGIP)$_{26}$(GVGVP)	2E/5F/2I
Model Protein	i:	(GVGVP GVGVP GEGVP GVGVP GVGVP GVGVP)$_{36}$(GVGVP)	E/0F
Model Protein	ii:	(GVGVP GVGFP GEGFP GVGVP GVGVP GVGVP)$_{40}$(GVGVP)	E/2F
Model Protein	iii:	(GVGVP GVGVP GEGVP GVGVP GVGFP GFGFP)$_{39}$(GVGVP)	E/3F
Model Protein	iv:	(GVGVP GVGFP GEGFP GVGVP GVGFP GVGFP)$_{15}$(GVGVP)	E/4F
Model Protein	v:	(GVGVP GVGFP GEGFP GVGVP GVGFP GFGFP)$_{42}$(GVGVP)	E/5F
Model Protein	ii':	(GVGVP GVGFP GK{NMeN}GFP GVGVP GVGVP GVGVP)$_{22}$(GVGVP)	K/2F
Model Protein	iii':	(GVGVP GVGVP GK{NMeN}GVP GVGVP GVGFP GFGFP)$_{22}$(GVGVP)	K/3F
Model Protein	iv':	(GVGVP GVGFP GK{NMeN}GFP GVGVP GVGFP GVGFP)$_{21}$(GVGVP)	K/4F
Model Protein	v':	(GVGVP GVGFP GK{NMeN}GFP GVGVP GVGFP GFGFP)$_{21}$(GVGVP)	K/5F
Model Protein	vi':	(GVGVP GVGFP GEGFP GVGVP GVGVP GK{CnAm}GVP)$_{22}$(GVGVP)	E-K'2F
Model Protein	vii':	(GVGVP GVGFP GEGFP GVGVP GVGFP GK{CnAm}GVP)$_{21}$(GVGVP)	E-K'2F
Model Protein	viii':	(GVGVP GVGK{CnAm}P GEGFP GVGVP GVGVP GFGVP)$_{22}$(GVGVP)	E-K'2F
Model Protein	ix':	(GVGVP GVGK{CnAm}P GEGFP GVGVP GVGFP GFGVP)$_{21}$(GVGVP)	E-K'3F
Model Protein	x':	(GVGVP GVGVP GKGVP GVGVP GVGVP GVGVP)$_{22}$(GVGVP)	K/0F
Model Protein	xi':	(GVGVP GVGFF GKGFP GVGVP GVGVP GVGVP)$_{22}$(GVGVP)	K/2F
Model Protein	xii':	(GVGVP GVGVP GKGVP GVGVP GVGFP GFGFP)$_{22}$(GVGVP)	K/3F
Model Protein	xiii':	(GVGVP GVGFF GKGFP GVGVP GVGFP GVGFP)$_{21}$(GVGVP)	K/4F
Model Protein	xiv':	(GVGVP GVGFF GKGFP GVGVP GVGFP GFGFP)$_{21}$(GVGVP)	K/5F
		B. Chemically-synthesized Protein-based Polymers (Mean molecular weights approximately 100 kDa)	
Polymer	i:	Poly(G**D**GFP GVGVP GVGVP GFGVP GVGVP GVGK{NMeN}P)	DK/1F
Polymer	I:	Poly(GVGVP GVGVP G**D**GVP GVGVP GVGVP GVGVP)	D/0F
Polymer	II:	Poly(GVGVP GVGFP G**D**GFP GVGVP GVGVP GVGVP)	D/2F
Polymer	III:	Poly(GVGVP GVGVP G**D**GVP GVGVP GVGFP GFGFP)	D/3F
Polymer	IV:	Poly(GVGVP GVGFP G**D**GFP GVGVP GVGFP GVGFP)	D/4F
Polymer	V:	Poly(GVGVP GVGFP G**D**GFP GVGVP GVGFP GFGFP)	D/5F
Polymer	VI:	Poly(GVGVP GVGVP G**E**GVP GVGVP GVGVP GVGVP)	E/0F
Polymer	VII:	Poly(GVGVP GVGFP G**E**GFP GVGVP GVGVP GVGVP)	E/2F
Polymer	VIII:	Poly(GVGVP GVGVP G**E**GVP GVGVP GVGFP GFGFP)	E/3F
Polymer	IX:	Poly(GVGVP GVGFP G**E**GFP GVGVP GVGFP GVGFP)	E/4F
Polymer	X:	Poly(GVGVP GVGFP G**E**GFP GVGVP GVGFP GVGFP)	E/5F
Polymer	XI:	Poly(GFGFP GVGVP G**D**GVP GFGFP GFGVP GVGVP)	D/5F'
Polymer	XII:	Poly(GFGFP GVGVP G**E**GVP GFGFP GFGVP GVGVP)	E/5F'

Accordingly, the values for Δn are likely to differ by significantly less than a factor of two for the two model proteins. Thus, simply by changing the composition from that of **Model Protein ii** to that of **Model Protein I**, the contraction with which to perform mechanical work could possibly occur using one-tenth the amount of the chemical energy and certainly with no more than one-fifth the amount of the chemical energy. (See section 5.9.5.1 for further discussion of the relative efficiencies of **Model Proteins I** and **ii** in Table 5.5.)

5.4.3.2.2 Relationship Between Width of Transition Zone (Plotted Using the Log Scale), Efficiency of Energy Conversion, Energy Required for Biological Production of the Model Proteins, and Evolution of More Efficient Machines

The log scale shown in Figure 5.15 was chosen because when concentration is plotted in this way the width of the transition zone is proportional to chemical energy. Thus simply by looking at Figure 5.15, it is apparent that **Model**

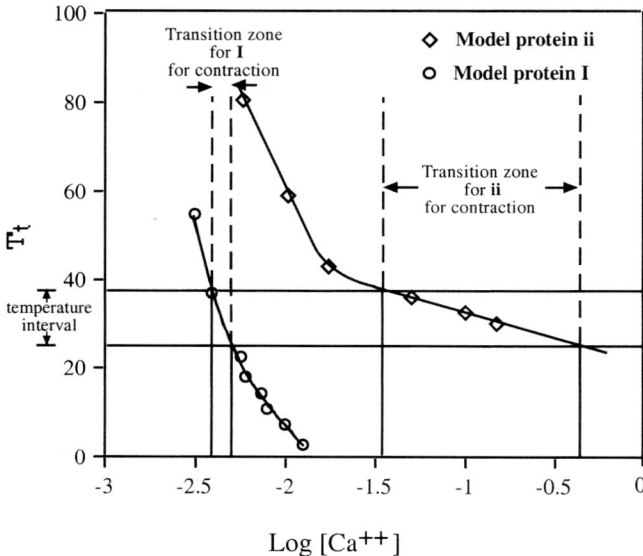

FIGURE 5.15. Plot of T_t as a function of log(calcium ion concentration), log[Ca^{2+}], which is proportional to chemical energy. Considering the biological transition zone of lowering T_t from 37°C to 25°C, it is possible visually to gain a sense of how the amount of chemical energy differs as the composition of the model protein changes. Clearly, as the hydrophobicity increases per carboxylate, the chemical energy required to drive hydrophobic association decreases, and it further decreases by having the carboxylates of the Glu (E) residues proximal for bidentate interaction with the doubly charged calcium ion. (D. Urry and C.-H. Luan, unpublished results.)

Protein I is on the order of 10 times more efficient in converting the chemical energy of calcium ion into mechanical work. Yet, as was discussed in Chapter 4 for proteins of the same chain length, it cost the organism the same amount of energy to produce the more efficient molecular machine as to produce the less efficient machine. As argued in Chapter 6, the evolution of **Model Protein ii** into the more efficient machine of **Model Protein I** can occur by single base mutations, and each mutation would step by step increase the efficiency of the protein for calcium driven energy conversion. *Given the consilient mechanism of energy conversion by single base mutations it is apparent that more complex and more efficient molecular machines would evolve due to natural selection.*

5.4.3.3 Contraction by Moving Through the Transition Zone and Its Equivalent of Moving the Temperature Interval for Heat-driven Contraction

5.4.3.3.1 Addition of Acid can Drive Contraction by Lowering Temperature Interval for Heat-driven Contraction, Which Moves the Elastomer Through the Transition Zone

Adding acid (proton, H^+) converts the negatively charged vinegar-like **R**-group (–CH_2–CH_2–COO^-) of the glutamic acid residue to –CH_2–CH_2–$COOH$. In the model protein, poly[4(GVGVP),(GEGVP)] with a ratio of four GVGVP pentamers to one GEGVP pentamer, there is one glutamic acid residue, designated by the letter E, in 25 residues. This model protein remains soluble in neutral or alkaline water; it will not aggregate no matter how high the temperature, and the rubber-like band formed by cross-linking this model protein will not contract. The temperature interval for this composition is above 100° C.

If acid is added to the solution of polymer, then aggregation begins at 28° C and is complete by 40° C, that is, the temperature interval is from 28° to 40° C. Similarly, for the cross-linked elastomer, contraction occurs as the temperature is raised through the temperature interval. Protonation of the charged (–CH_2–CH_2–COO^-) **R**-group results in a more oil-like **R**-group so that the separation of oil-like **R**-groups from water occurs at a lower temperature. With this composition of model protein, the chemical energy of increasing acid concentration transforms into the mechanical energy of lifting a weight. By increasing the concentration of acid through the range of the transition zone contraction occurs and results in the performance of mechanical work with lifting of a weight.

5.4 Visual Observation of Elastic Protein-based Machines at Work

Using the simplified representation of a single chain of model protein in Figure 5.16,[82] reaction (i) depicts the addition of proton to a carboxylate. This restores so much hydrophobic hydration in an intermediate for contraction such that a larger positive ($-T\Delta S$) term results in the insolubility of the oil-like groups and the chain folds in a contraction. In reality, of course, the fundamental structure is a multistranded twisted filament, and the result is a more complex hydrophobic association involving a concerted hydrophobic folding and assembly. Also, the relation between increased hydrophobic hydration and contraction (hydrophobic folding and assembly) is given below in section 5.7.

5.4.3.3.2 Addition of Base Can Drive Contraction by Lowering Temperature Interval for Heat-driven Contraction

Considering reaction (ii) of Figure 5.16, addition of base (hydroxyl ion, OH^-) converts the positively charged vinegar-like **R**-group ($-CH_2-CH_2-CH_2-CH_2-NH_3^+$) of the lysine residue to $-CH_2-CH_2-CH_2-CH_2-NH_2$, that is, $-CH_2-CH_2-CH_2-CH_2-NH_3^+ + OH^- \rightarrow -CH_2-CH_2-CH_2-CH_2-NH_2 + H_2O$.

In the model protein, poly[4(GVGVP),(GKGVP)] with a ratio of four GVGVP pentamers to one GKGVP pentamer, there is one lysine residue, designated by the letter K, in 25 residues. This model protein remains soluble in

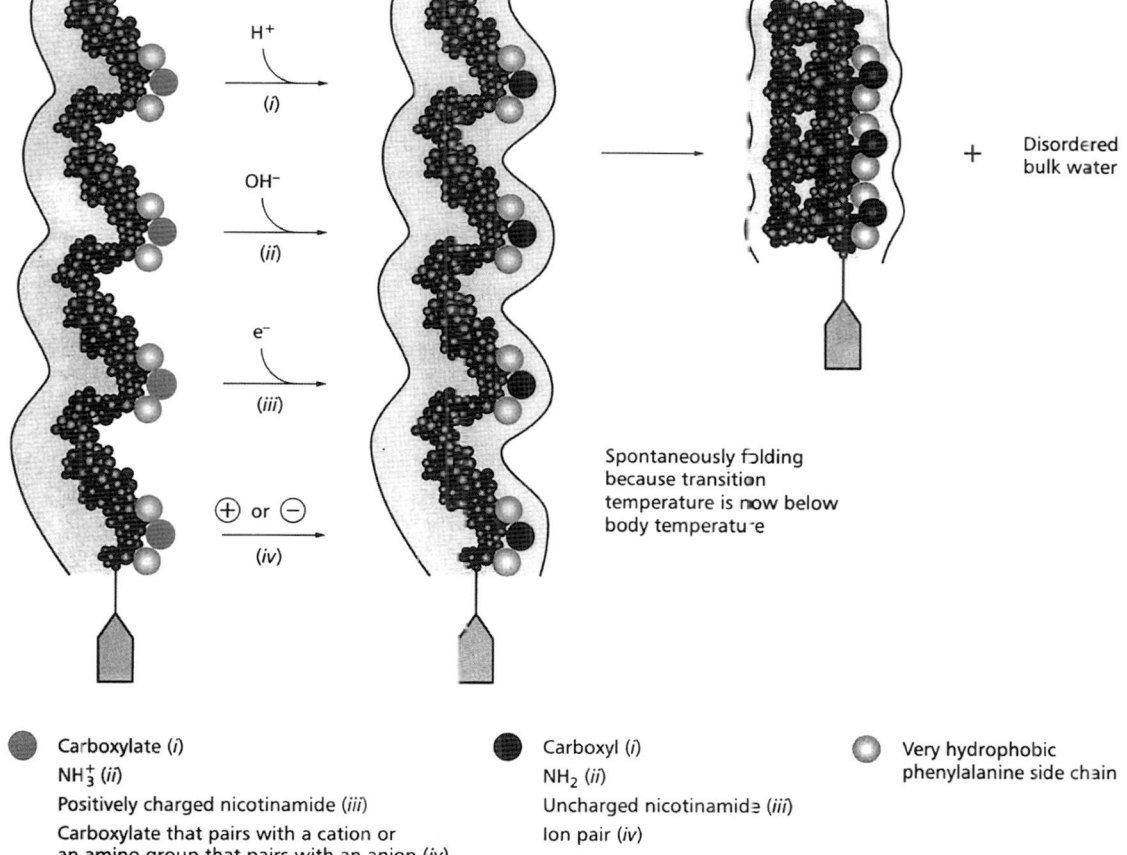

FIGURE 5.16. Contraction of a β-spiral as the result of four different types of reactions that cause the structure to become more hydrophobic. The fully hydrated intermediate state is only to indicate restored hydration potential if hydrophobic association had not occurred. (Reproduced with permission from Urry.[82])

neutral or acidic water; it will not begin aggregation in physiological saline until the temperature is greater than 45° C, and the rubber-like band formed by cross-linking this model protein will not begin to contract unless the temperature is greater than 45° C. The temperature interval begins just above 45° C. If hydroxyl ion is added to the polymer solution sufficient to convert all side chains to $-CH_2-CH_2-CH_2-CH_2-NH_2$, then aggregation begins at 28° C and is complete by 45° C, that is, the temperature interval is now from 28° to 45° C. Similarly, for the cross-linked elastomer, contraction occurs over the temperature interval from 28° C to 45° C. Removal of the charged proton to form the $-CH_2-CH_2-CH_2-CH_2-NH_2$ results in a more oil-like **R**-group so that the separation of oil-like **R**-groups from water occurs at a lower temperature. With this composition of model protein, the chemical energy of increasing hydroxyl ion concentration transforms into the mechanical energy of pumping iron.

Again using the simplified representation of a single chain of model protein in Figure 5.16, reaction (ii) depicts the removal of a proton from an ammonium, NH_3^+. Forming the uncharged NH_2 restores so much hydrophobic hydration that a more positive $(-T\Delta S)$ term causes ΔG(solubility) to become positive, resulting in insolubility of the oil-like groups, and the chain folds in a contraction due to hydrophobic association.

5.4.3.3.3 Ion Pairing Drives Contraction by Lowering Temperature Interval for Heat-driven Contraction

Whether using a model protein with a negatively charged functional group or one with a positively charged functional group, pairing with an oppositely charged ion lowers the temperature interval for heat-driven hydrophobic association. Correspondingly, ion pairing in the cross-linked elastic matrix moves the system through the transition zone and drives contraction.

Once more using the simplified representation of a single chain of the model protein in Figure 5.16, reaction (iv) depicts the addition of a counter-ion to the oppositely charged vinegar-like group. Partial neutralization of the charged functional group provides yet another means of restoring sufficient hydrophobic hydration that the $(-T\Delta S)$ term becomes even more positive. This results in ΔG(solubility) becoming positive and the oil-like groups becoming insolubile, and again the chain hydrophobically folds in a contraction.

5.4.4 Electrical Energy Produces Mechanical Work of Lifting Weight (Increasing the Electrochemical Potential *Through the Transition Zone* to Reduce Oxidized Redox Groups Drives Contraction)

5.4.4.1 Reduction (Adding Electrons) Drives Contraction by Lowering Temperature Interval for Heat-driven Contraction

Reduction of the oxidized nicotinamide (N-methyl nicotinamide, $NMeN^+$) in poly[0.73(GVGVP),0.27(GK{NMeN}GVP)] lowers the temperature for the onset of the temperature interval from 49° to 9° C, for a 60% reduction.[83] When forming the cross-linked elastic matrix, as shown in Figure 5.17 at room temperature and as available in video, reduction drives contraction and oxidation drives relaxation.[81]

Again with the simplified representation of a single chain of model protein in Figure 5.16, reaction (iii) depicts the addition of a negative electron to a positively charged component of a redox couple. Forming the reduced, uncharged component of the redox couple, again, restores so much hydrophobic hydration that a positive $(-T\Delta S)$ term causes ΔG(solubility) to become positive, resulting in insolubility of the oil-like groups, and the chain folds due to hydrophobic association with the result of a contraction.

5.4.5 Pressure-Volume Energy Produces Mechanical Work (MW) (Lowering Pressure *Through the Transition Zone* Drives Contraction)

When aromatic groups are present, the application of pressure increases the value of T_t. Pressure increases the temperature range over which the temperature interval occurs. The order of responsiveness to increases in pressure

5.4 Visual Observation of Elastic Protein-based Machines at Work

A Flow Through Electro-Mechanical Transduction Cell

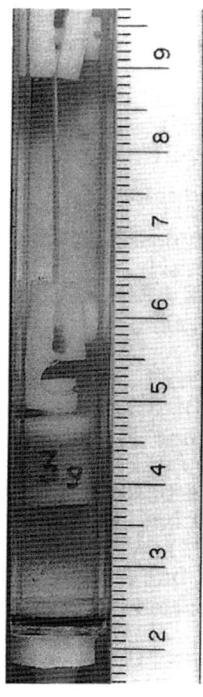

B X^{20}-Poly[0.73(GVGVP),0.27(GK{NMeN}GVP)] Redox Video Data

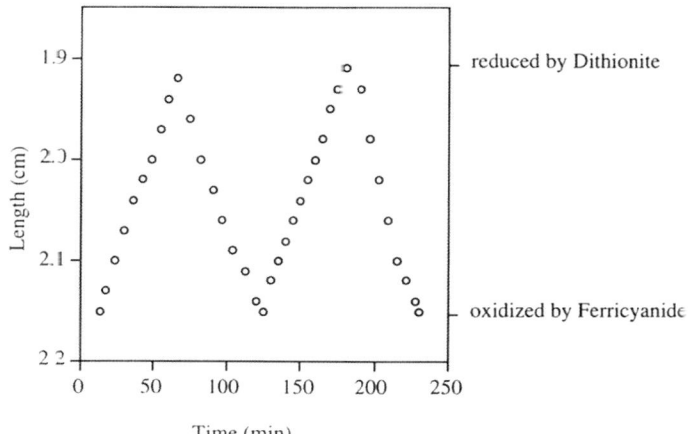

FIGURE 5.17. Reduction-driven contraction of model protein as an elastic sheet containing a redox couple. (**A**) Flow through cell for reductant and oxidant containing elastic band suspending a weight that is lifted and lowered. (**B**) Plot from video showing length changes attending reduction (contraction) and oxidation (relaxation). (Reproduced with permission from Urry et al.[81])

is tryptophan (Trp, W) > phenylalanine (Phe, F) > tyrosine (Tyr, Y).[84] Thus, the application of pressure can drive unfolding, and the release of pressure can be used to drive contraction.[80] This has also been captured by video using a model protein containing aromatic phenylalanine residues where application of pressure lowers the weight and removal of pressure causes contraction.

5.4.6 Electromagnetic (Light) Energy Produces Mechanical Work (Electromagnetic Radiation That Moves T_t Through the Transition Zone Drives Contraction/Relaxation)

Any interaction of light that causes a change in the oil-like character of the model protein will change the value of T_t and hence will change the location of the temperature interval with the result of a change in the state of folding. A simple result of the absorption of light by a chromophoric group is the *trans* to *cis* geometrical isomer transformation. In two light-absorbing molecules, one of biological origin, cinnamide and the other, azobenzene, the absorption of 300 nm light converts the cinnamide molecule from a *trans* to a *cis* geometrical isomer. In both cases this light-driven conversion raises the value of T_t and can be used to drive unfolding of a model protein to which it is attached.[71,85] Either a different wavelength of light or thermal energy can reverse the isomerization and result in the hydrophobic association of contraction.

The results of sections 5.4.3, 5.4.4, 5.4.5, and 5.4.6 form the experimental observations that give rise to Axiom 3.

Axiom 3: At constant temperature, an energy input that changes the temperature interval for thermally driven hydrophobic association in a model protein can drive contraction, that is, oil-like folding and assembly, with the performance of mechanical work; in other words, the energy input moves the system through the transition zone for contraction due to hydrophobic association.

5.4.7 Summary of Consilient Protein-based Machines (Engines and Motors) That Pump Iron, That Is, That Produce Motion

The term *machine* is a general designation for any device that converts energy from one form to another. The particular type of machine that performs mechanical work or produces motion is commonly called either a *motor* or an *engine*. Here, we summarize the capacity of protein-based machines to perform mechanical work (pump iron) by the consilient mechanism of hydrophobic association by utilizing a number of different energy inputs.

5.4.7.1 Pumping Iron by Raising and Lowering the Temperature Through the Temperature Interval

Figure 5.18A uses the representation of a single β-spiral in the extended and contracted states. The sigmoid curve at a represents the folding of, and defines the temperature interval for, a more oil-like model protein chain and that at b represents the folding of, and defines the temperature interval for, a less oil-like model protein. Any energy input that makes more oil-like the model protein with the temperature interval at b lowers the temperature interval of b toward that of a and defines the transition zone c for that energy input. Figure 5.18B, 1), represents the contraction of a band of elastic model protein when heating through the temperature interval for hydrophobic folding and assembly of the model protein.

5.4.7.1.1 Pumping Iron by Protonation of a Carboxylate ($COO^- \to COOH$) to Make the Model Protein More Oil-like

Increasing the concentration of acid, H^+, constitutes a particular chemical energy input. The addition of acid (H^+), over the concentration range that protonates the carboxylate, lowers the temperature interval for the model protein, drives contraction, and defines the transition zone for this energy input indicated as c in Figure 5.18A. Figure 5.18B, 2a), represents the contraction and relaxation by reversible protonation and deprotonation of the carboxyl functional group, i.e., *chemo-mechanical transduction*.

5.4.7.1.2 Pumping Iron by Adding Salt to a Neutral Model Protein or to a Model Protein with a Charged Functional Group

Simply by the addition of NaCl, for example, to a model protein with no ionized functional groups can lower the temperature interval, define a transition zone, and drive contraction. The change in salt concentration required to drive contraction of an uncharged model protein can be 10 times more than if a charged functional group is present. Driving contraction when there is a charged group to which a counter-ion from the salt can ion pair requires a much smaller change in concentration of salt, that is, a smaller chemical energy input. The energy conversion achieved by ion pairing is much more efficient. Both uses of salts are represented in Figure 5.18B, 2b).

5.4.7.2 Pumping Iron by Reduction of an Oxidized Functional Group Attached to the Model Protein

Figure 5.17 shows direct experimental results, which give rise to the generalized representation in Figure 5.18, 3). In the case of Figure 5.17, for every 100 residues of model protein, there are 5.4 lysine residues to each of which is attached an N-methyl nicotinamide (NMeN) group. When oxidized, the circumstance is as indicated by temperature interval b of Figure

5.4 Visual Observation of Elastic Protein-based Machines at Work 159

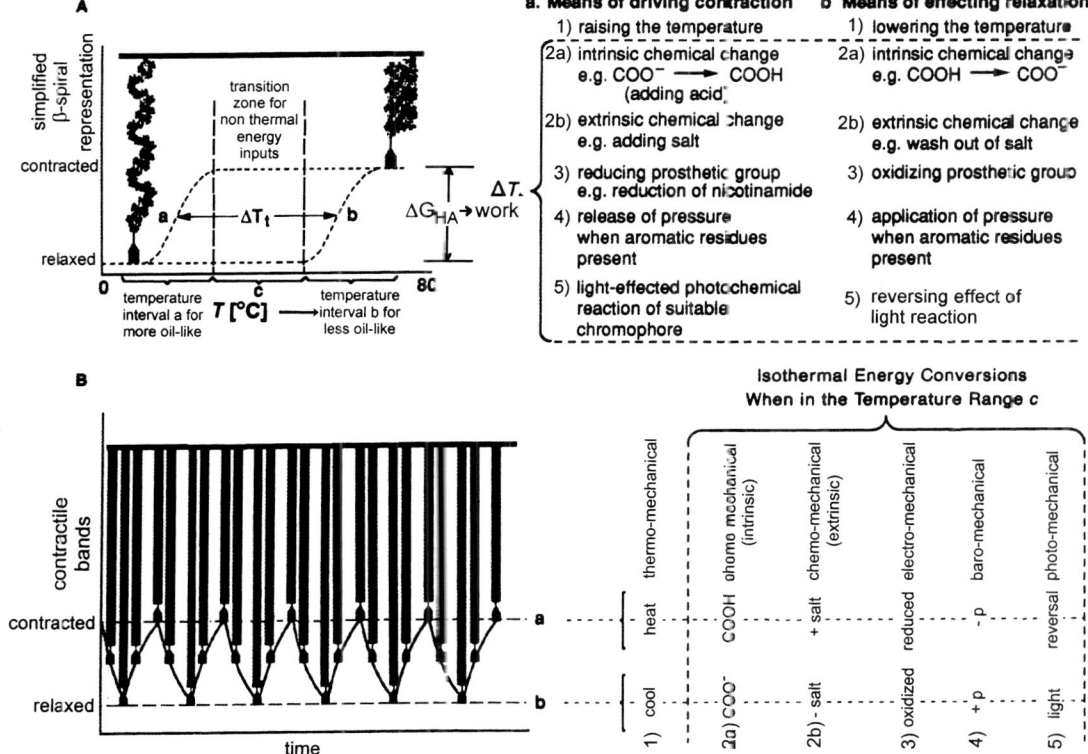

FIGURE 5.18. Performance of mechanical work by designed protein-based engines: Different energy inputs "pump iron" by lowering $\Delta G_{HA}(\chi)$ to drive hydrophobic association. See text for discussion. (Reproduced with permission from Urry.[6])

5.18A. On reduction by the chemical dithionite, the temperature interval for thermally driven contraction lowers to a temperature interval a and thereby moves the elastic band through the transition zone c. Contraction results when the operating temperature is in zone c. Oxidation by the chemical ferricyanide relaxes the elastic band and completes the reduction–oxidation cycle to return the band and the attached weight to its original state. With the proper mediator molecules to diffuse between electrode and redox couple in the cross-linked matrix, the reduction and oxidation can be carried out electrically; thus this may correctly be called *electro-mechanical transduction*.

5.4.7.3 Pumping Iron by Changing the Pressure Using Model Proteins with Aromatic (Large Oil-like Ring Structures) Side Chains

The application of pressure, when there are aromatic amino acids in the polymer such as tryptophan (Trp, W), phenylalanine (Phe, F), and tyrosine (Tyr, Y) as defined in Table 5.1, raises the temperature interval from a to b of Figure 5.18A. Accordingly, as indicated in Figure 5.18B, 4), the application of pressure can cause relaxation, and the release of pressure can result in contraction with the lifting of a weight.[80] This is *baro-mechanical transduction*.

5.4.7.4 Pumping Iron by Using Light to Change the Oil-like Character of a Chromophore Attached to the Model Protein

To obtain a change in hydrophobic association in response to light, the natural product cinnamic acid was attached to a lysine (Lys, K) residue of the polymer to form the cinnamide. With a protein-based polymer having 1K residue per 30 residues and generally 21 30-mers forming the high molecular weight protein-based polymer, the structure can be represented as (H_5C_6–CH=CH–CO–NH–Lys)–polymer. The orientation about the –CH=CH– bond can be either *trans* or *cis*. To begin with, the configuration is 100% *trans*, and, when it absorbs 300 nm ultraviolet light, it converts to about 90% *cis*. Conversion of *cis* back to *trans* occurs with 270 nm ultraviolet light. The temperature interval for the *cis* isomer occurs at a higher temperature, like interval b of Figure 5.18A, while that of the *trans* isomer occurs over the lower temperature interval, a. The *trans* isomer is more oil-like than the *cis* isomer. The change in T_t on an $f_X = 1$ basis is about 60° C.[71] Figure 5.18B, 5) represents *photo-mechanical transduction*.

5.4.7.5 The Set of Energies and Intensive Variables (Bold Faced) Involved in Using Model Proteins as Molecular Engines and Motors

Mechanical work = **mechanical force** × distance lifted = **f**ΔL = mgh
Pressure-volume = **pressure** × volume = **P** × V
Chemical energy = **chemical potential** × number of moles used = $\Delta\mu \times \Delta n$
Thermal energy = heat of the transition = ΔH_t = **temperature** × entropy = **T**ΔS_t
Electrical energy = **electrochemical potential** × number of electrons used = zFΔ**E** × Δn
Electromagnetic radiation = **photon energy(hν)** × number of photons used = **hν** N

5.5 Energy Conversions Due to Coupling of Functional Groups

Here we discuss the coupling of functional groups as part of the general phenomena of action and reaction and most specifically as an integral part of the consilient mechanism of energy conversion. The argument may be given using two statements that derive from experimental observations to be presented below.

Statement 1: Within the context of the hydrophobic consilient mechanism, a functional group is a chemical entity that can exist in one of two or more interconvertible states, each with a different hydrophobicity, as measured by the T_t-based and ΔG_{HA}-based hydrophobicity scales (see Tables 5.1, 5.2 and 5.3).

Examples of functional groups, each with a strong dependence on hydrophobicity, are:

1) Ionizable functional groups that are a function of pH (the change in chemical potential of proton, $\Delta\mu_H$) and exhibit hydrophobicity dependent pKa values.
2) Redox functional groups with hydrophobicity dependent reduction potentials for conversion between oxidized and reduced states.
3) Groups that change their state on absorption of light, e.g., convert from *trans* to *cis*.
4) A functional group such as a hydroxyl that can be reversibly, e.g., enzymatically, phosphorylated and dephosphorylated).

Statement 2: Because the uncharged states of ionizable functional groups and the reduced states of oxidized functional groups are each more oil-like or hydrophobic than the charged and oxidized states, the formation of the more hydrophobic state of one functional group can cause the more polar state of a second kind of attached and responsive functional group to convert to its more hydrophobic state. This is the coupling of functions by the consilient mechanism!

In particular, the reduction of NMeN$^+$ approximates the replacement of several valine residues by phenylalanine residues. Just as

5.5 Energy Conversions Due to Coupling of Functional Groups

increasing the hydrophobicity of a polymer causes the protonation of a carboxylate to occur at a lower acid concentration, it is to be expected, therefore, that the reduction of NMeN$^+$ would lower the acid concentration required to protonate a carboxylate attached to the same hydrophobic domain.

By the consilient mechanism, energy conversions due to coupling of functional groups result from a common dependence of the different functional groups on hydrophobicity of the model protein. This common dependence of the essential property of different functional groups (e.g., pKa, reduction potential) on hydrophobicity makes possible energy conversion by the coupling of functions using the consilient mechanism as described in more detail below.

5.5.1 Action Begets Reaction

5.5.1.1 For Protein-based Machines Input Energy Constitutes the Action and Output Energy Constitutes the Reaction

In physics and chemistry a number of laws, relations, and principles are statements that some action (commonly expressible as an energy input) causes a reaction (commonly expressible as an energy output). Examples are Guldberg and Waage's law of mass action for chemical reactions, Onsager's reciprocal relations for transport processes, and Newton's third law of motion where a force acting on an object results in a matching reaction force.

Another example is the Principle of Le Châtelier, which may be stated as follows: For any system at rest (at equilibrium) the introduction of a stress (in our case an input energy) causes the system to react in such a way as to relieve the stress (in our case by an output energy). This principle reasonably describes protein-catalyzed energy conversion, that is, the function of protein-based machines. Under prescribed conditions, properly designed model protein-based machines exhibit a behavior where for each action there is a reaction. In section 5.4, regardless of the action, which was any one of several different input energies, the performance of mechanical work was the reaction (there was a common output energy). In this section the energy inputs, which individually caused contraction, convert one into the other by their common dependence on the change in free energy of the hydrophobically associated (contracted) state. This provides definition of the subset of protein-based machines that, despite utilizing the hydrophobic consilient mechanism of hydrophobic association, are not characterized as having mechanical work as their output energy.

5.5.1.2 For Protein-based Machines, the Input Energy Must Occur Through the Transition Zone

Energy input must occur through the transition zone, which is determined by the design of polymer composition, to operate under the conditions in which the machine is to function. The change in temperature, in concentration of solute, in availability of electrons, and so forth, must be such that the change in the independent (intensive) variable traverses the transition zone of the model protein. When the variable is temperature, that is, when heat is the energy input, the transition zone would be the temperature interval (see Figures 5.1 and 5.5A) and the energy output can be contraction. When the variable is an increase in the amount of acid, H$^+$, in solution, that is, an increase in chemical energy per mole of proton, the transition zone (see Figure 5.5B) defines the range for the change in concentration of acid (chemical potential) that converts, for example, –COO$^-$ to –COOH, and contraction again can be the energy output. When the variable is the availability of electrons, that is, electrical energy is the input, the transition zone defines the change in electrochemical potential required to change an oxidized component of a redox couple to the reduced component, and again contraction can be the energy output.

5.5.1.3 A Functional Group Can Occur in Either One of at Least Two Different States

Recall that a functional group, for example, a vinegar-like group, can occur in either one of two or more different states. In general, one state of a functional group is more oil-like, and another is more vinegar-like. To say one is less polar and the other is more polar is an equivalent statement. The carboxyl/carboxylate, COOH/COO$^-$, pair represents one example of a functional group, where COOH is more oil-like (less polar) and COO$^-$ is less oil-like (more polar). The organic phosphate group can occur in three different states, $-OPO_3^=$, $-OPO_3H^-$, and $-OPO_3H_2$, in order of decreasing polarity and increasing oil-like character. In these cases, the proton, H$^+$, is the chemical entity that is added or released. Another functional group is the redox couple, NMeN$^+$ and NMeN, where the net addition of two electron, 2e$^-$, and a proton converts the more polar NMeN$^+$ to the more oil-like (less polar) NMeNH.

5.5.1.4 Each State of a Functional Group Couples to the States of Another Functional Group by Contributing to and Being Affected by the Oil-like Character of the Polymer

Recall also from section 5.3 that the value of T_t provides one measure of the oil-like character or hydrophobicity of the model protein. A lower value of T_t means a more oil-like character for the model protein. Accordingly, the coupling of one functional group to another, that is, the conversion of energy represented by one functional group to the form represented by another functional group, occurs because each energy change has the common action of effecting and being affected by the change in free energy of hydrophobic association, ΔG_{HA}.

5.5.2 Hydrophobicity (Degree of Oil-like Character) Positions Transition Zone

5.5.2.1 Changing the Oil-like Character by Progressive Addition of More Oil-like Phe Groups Moves the Temperature Interval (Transition Zone) to Lower Temperatures

When the transition zone is specifically the temperature interval, any change in oil-like character of the model protein-based machine moves the temperature range over which the temperature interval occurs. Specifically, the replacement of a less hydrophobic valine (Val, V) residue by the more hydrophobic phenylalanine (Phe, F) residue lowers and narrows the temperature interval over which hydrophobic association occurs. This is schematically represented in Figure 5.19A,[4] which shows what happens to a curve like that of Figure 5.5A, when the polymer involved is subjected to a systematic change in its oil-like character. As discussed below, any change in oil-like character can move the transition zone of any process that can itself change the temperature interval (i.e., move the T_t-divide; see sections 5.1.3.4 and 5.3).

5.5.2.2 Changing Oil-like Character by Progressive Addition of More Oil-like Phe Groups Moves the Transition Zone, Increases Affinity of the COOH/COO$^-$ Chemical Couple for Protons, and Decreases the Change in Chemical Energy Required to Protonate

Figure 5.19B represents the series of curves that result when the polymer series containing the COOH/COO$^-$ chemical couple is designed with systematic changes in hydrophobicity.[4] In practice the polymer may be from the family (GVGVP GVGXP GEGXP GVGVP GVGXP GXGXP)$_n$, where E is the glutamic acid residue with its carboxyl function and X may be either the less hydrophobic V or the more hydrophobic F residue. The less oil-like polymer would have V at all positions indicated by X. Then, as more positions indicated by X have their V residue replaced by the more oil-like F residue, the curves shift to indicate that lower concentrations of acid are able to protonate the carboxyl, and the protonation occurs over a narrower transition zone. The same family of curves is obtained for (GVGVP GVGXP GKGXP GVGVP GVGXP GXGXP)$_n$, where K is the lysine residue with its amine chemical couple ($-NH_2/-NH_3^+$) but where the independent variable involves the concentration of base (the hydroxyl ion, OH$^-$). Both families of curves take on the same form as found when the energy input was thermal in Figure 5.19A.

5.5 Energy Conversions Due to Coupling of Functional Groups 163

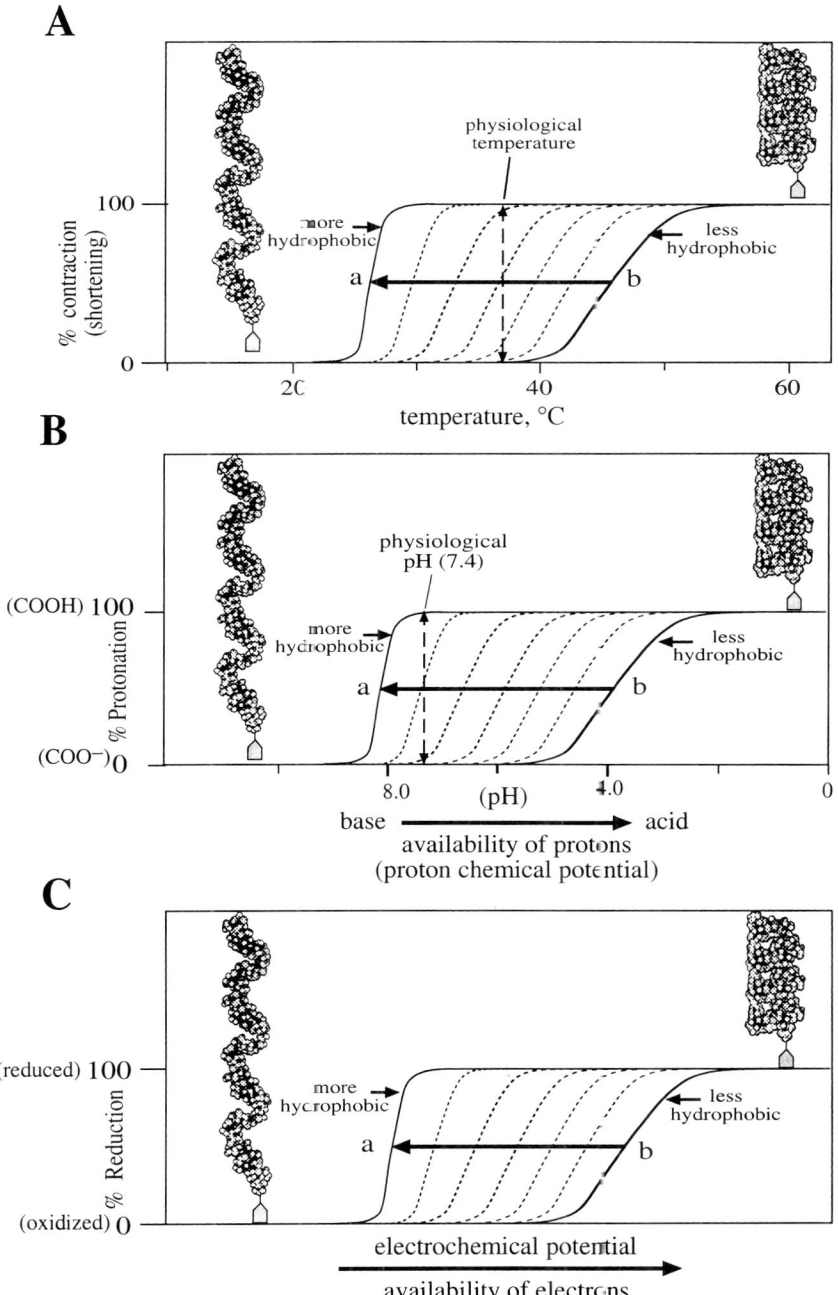

FIGURE 5.19. Demonstration of the similar form of the hydrophobic dependence of contractions independent of the form of energy used as long as the model protein is properly designed for the respective energy input. (**A**) Thermally driven contraction due to hydrophobic association. (**B**) Protonation-driven contraction due to hydrophobic association when acting on a protonatable functional group within the model protein. (**C**) Reduction-driven contraction due to hydrophobic association when acting on a redox functional group within the model protein. (Adapted with permission from Urry.[4])

5.5.2.3 Changing the Oil-like Character by Progressive Addition of More Oil-like Phe Groups Moves the Transition Zone, Increases Affinity of the NMeN/NMeN⁺ Redox Couple for Electrons, and Decreases the Electrical Energy Required to Reduce

Figure 5.19C represents the series of curves that result when the polymer series containing the NMeN/NMeN⁺ redox couple is also designed with systematic changes in hydrophobicity.[4] The same family of polymers can be designed as used for the lysine residue, but in this case the lysine residue is utilized for attachment of the N-methyl nicotinamide (NMeN), that is, (GVGVP GVGXP GK[NMeN]GXP GVGVP GVGXP GXGXP)$_n$. Again, as less oil-like Val (V) residues are systematically replaced by more oil-like Phe (F) residues, a lower concentration of electrons (a lower electrochemical potential) reduces the oxidized state of the redox couple. Again, the family of curves take on the same form as occurs when chemical energy was the input energy and when thermal energy was the input energy, as in Figure 5.19A,B.

5.5.2.4 Changing the Oil-like Character of a Model Protein by Any Energy Input, χ, That Moves the Transition Zone Can Be Used to Change the Energy of a Functional Group That Is Also Capable of Changing the Oil-like Character of the Model Protein

Additional input energies can be light energy interacting with a chromophore to change its hydrophobicity and pressure-volume energy acting most effectively on aromatic groups to change their expression of hydrophobicity.

Many other interaction energies come from the electromagnetic frequency spectrum. They can come from outside of the ultraviolet, visible, and infrared frequency ranges. All that is required is an element of the model protein in water that is able to take up the energy. The dipole moment of the peptide group with its positive end at the NH and its negative end at the oxygen of the CO provides one site of interaction. The acoustic and radio frequency ranges of the electromagnetic radiation spectrum represent two frequency ranges that are fundamental to our elastic model proteins capable of exhibiting inverse temperature transitions. When subjected to oscillating electric fields near 3 kHz (acoustic frequency) and 5 MHz (radio frequency), the peptide group of the protein backbone reorients its dipole moment and oscillates with the applied oscillating electric fields in these frequency ranges. The uptake of these energies by the elastic model proteins change the internal chain dynamics and are expected to alter the force exerted by a slightly extended polymer. Of more direct concern to a change in free energy of hydrophobic association dependent on the amount or potential amount of hydrophobic hydration is the 5 GHz absorption (microwave frequency) exhibited by hydrophobic hydration itself (see section 5.7). Energy absorbed directly by hydrophobic hydration is expected to affect the TΔS controlling solubility (i.e., hydrophobic association)

The innumerable ways in which the value of T_t can be changed constitute the number of energy inputs, χ, that can drive protein-based machines.

5.5.3 Coupling of Functional Groups by Means of the Consilient Mechanism and by Means of Their Common Dependence on the Gibbs Free Energy of Hydrophobic Association, ΔG$_{HA}$

*The energy input to a first functional group causes a change in the free energy of the hydrophobically associated state, **ΔG$_{HA}$**. That ΔG$_{HA}$ then causes a change in the free energy of a second functional group. This is the coupling of functions by the consilient mechanism. It represents the most explicit expression of the comprehensive hydrophobic effect.*

Section 5.4.3 and Tables 5.2 to 5.4 list many calculated estimates of ΔG$_{HA}$ arising from many different energy inputs. The estimated values of ΔG$_{HA}$ provide a quantitative measure of how a given energy input changes the energy of the hydrophobically associated state. This provides

5.5 Energy Conversions Due to Coupling of Functional Groups

experience for the many ways in which the energy of the hydrophobically associated state can be changed and the effectiveness of each way to do so. Specifically, section 5.3.4.4 provides estimates of the ΔG_{HA} due to the reduction of oxidized N-methyl nicotinamide (NMeN$^+$) in poly(GVGVP), about **−9.0 kcal/mole**, and section 5.3.4.3, estimates ΔG_{HA}(Glu-COOH → Glu-COO$^-$) to be about **+5.22 kcal/mole**. In this section, experimental results demonstrate the coupling of these two different functional groups.

5.5.3.1 Change in the Hydrophobicity (Extent of Oil-like Character) Due to Action on One Functional Group, for Example, Reduction of Oxidized N-methyl Nicotinamide, Moves the Transition Zone for a Second Functional Group, for Example, the Carboxyl/Carboxylate Chemical Couple: Electro ↔ Chemical Transduction

Figure 5.20A shows the shift in the transition zone for the carboxyl function in reaction to the reduction of the **K(NMeN$^+$)** redox function.[73] Specifically, the chemically synthesized model protein designed to show this coupling of functions was **Polymer i: Poly**(GDGFP GVGVP GVGVP GFGVP GVGVP GVGK[**NMeN**]P); D-K/1F of Table 5.5, where **D** is for the aspartic acid residue with the side chain, –CH$_2$–COOH, that accompanies the N-methyl nicotinamide functional group attached to a lysine side chain. Reduction of NMeN$^+$ to form NMeN shifts the transition zone for protonation of the carboxyl function by 2.5 pH units. This shows electrochemical transduction, the conversion of electrical energy into chemical energy. In particular, the reduction reaction is carried out near pH 9, where the reduced NMeN is most stable.

Under these circumstances, when the NMeN is in the oxidized state and the aspartic acid residue is charged, that is, –CH$_2$–COO$^-$, reduction of NMeN$^+$ near pH 9 causes the carboxylate function to pick up a proton and become the uncharged carboxyl function. Above we have referred to the performance of mechanical work as "pumping iron." The action of this protein-based machine could be referred as "pumping protons" or "pumping acid" as occurs in the stomach and for which we take Zantac® or PepcidAC® when our proton pumps are too active. Of course, the major place in biology where electro-chemical transduction occurs is in the energy-converting mitochondria, and the players are similar,

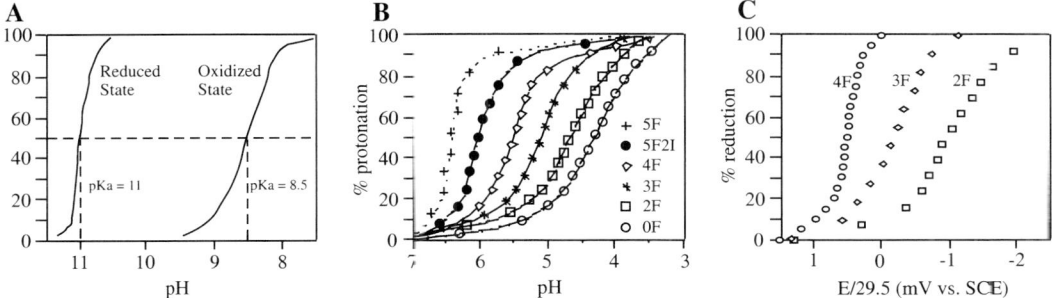

FIGURE 5.20. Experimental results that demonstrate the same form of hydrophobicity dependence as shown in Figure 5.19. (**A**) Conversion of electrical into chemical energy when the model protein contains both redox couple and protonatable chemical couple. (Adapted with permission from Urry.[4]) (**B**) Hydrophobicity dependence for the conversion of chemical energy into mechanical work when the model protein contains a protonatable function within a series of increasingly hydrophobic model proteins. (**C**) Hydrophobicity dependence for the conversion of electrical energy into mechanical work when the model protein contains a function capable of reduction and oxidation within a series of increasingly hydrophobic model proteins. Adapted with permission from Hayes.[103]) Regardless of energy conversion, the qualitative dependence on hydrophobicity is the same.

nicotinamide redox couples and carboxyl/carboxylate chemical couples (see Chapter 8).

On reduction of the nicotinamide, the transition zone in Figure 5.20A for the carboxyl/carboxylate chemical couple shifts to lower acid concentrations and becomes narrower, just as is demonstrated by data for the series of model proteins in Figure 5.20B (and schematically represented in Figure 5.19B) on increasing the oil-like character by replacing V by F. The series of model proteins for Figure 5.20B is the set of **Model Proteins i, ii, iii, iv,** and **v** of Table 5.5. Clearly, the increase in oil-like character due to systematic replacement of V by F for **Model Proteins i** through **v** shifts the pKa, the pH at 50% protonation, to higher pH values. The pKa defines the center of the transition zone. This unambiguous stepwise increase in hydrophobicity stepwise lowers proton concentrations and narrows the width of the transition zone, that is, it decreases the chemical energy required to drive the hydrophobic association. Figure 5.20B shows the experimental data on which the schematic representation in Figure 5.19B is based. The important point to note is that the reduction of nicotinamide has exactly the same effect on the carboxyl/carboxylate chemical couple as increasing the oil-like character by replacing less oil-like V by more oil-like F.

In an exactly analogous manner, equivalent behavior is seen with the series of **Model Proteins ii′, iii′,** and **iv′** of Table 5.5 containing the nicotinamide functional group attached to a lysine (Lys, K) residue by an amide linkage (See Figure 5.20C). The systematic replacement of V by F shifts the reduction potential of the nicotinamide, that is, shifts the transition zone to lower electron concentrations, and results in a narrower transition zone. Figure 5.20C represents an explicit experimental example of the schematic representation in Figure 5.19C.

In conclusion, coupling of different functional groups occurs when both functional groups affect and are affected by the oil-like domain of which they are a part. When one group brings about a change in the oil-like character of a common protein domain, that change affects the second functional group. This coupling of functions through the oil-like character of the protein-based machine occurs by means of the mutual dependence of the Gibbs free energy of hydrophobic association, ΔG_{HA}, on the state of the functional groups.

5.5.3.2 Change in the Hydrophobicity (Extent of Oil-like Character) Due to Action on One Functional Group, for Example, Absorption of Light by the Cinnamide Chromophore, Moves the Transition Zone for a Second Functional Group, for Example, the Carboxyl/Carboxylate Chemical Couple: Photo ↔ Chemical Transduction

In this demonstration of photo-mechanical and photo-chemical transduction, cinnamic acid, C_6H_5–CH=CH–COOH, was attached by amide linkage to the side chain of a lysine (Lys, K) residue, abbreviated as K{CnAm}. The studies utilized the designed **Model Proteins vi′, vii′, viii′** and **ix′** of Table 5.5. Light of 300 nm drove the *trans* geometrical isomer of cinnamide into the *cis* isomer and raised the value of T_t by about 10°C, which effects a small increase in ΔG_{HA}, that is, a shift to the hydrophobically dissociated state (see Figure 5.21A). When the acid–base titrations were carried out a small decrease of a few tenths of a pH unit was reported as shown in Figure 5.21C.[71] The pH shift is in line with what is expected from the small ΔT_t in Figure 5.21B with the resulting small increase in ΔG_{HA}.

5.5.3.3 The Statement for Energy Conversion by the Coupling of Functions with the Consilient Mechanism Becomes Axiom 4

Axiom 4: Two or more different functional groups of a model protein, each of which can be acted upon by a different energy input that changes the temperature interval for oil-like folding and assembly, become coupled one to the other by being part of the same oil-like folding and assembly domain, that is, the energy input acting on one functional constituent alters the oil-like character of the model protein and by doing so acts as an energy output on the other functional constituent through its dependency on the oil-like character of the model protein.

5.5 Energy Conversions Due to Coupling of Functional Groups

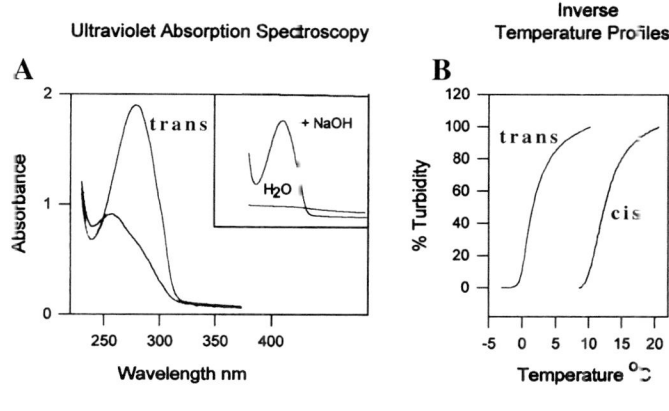
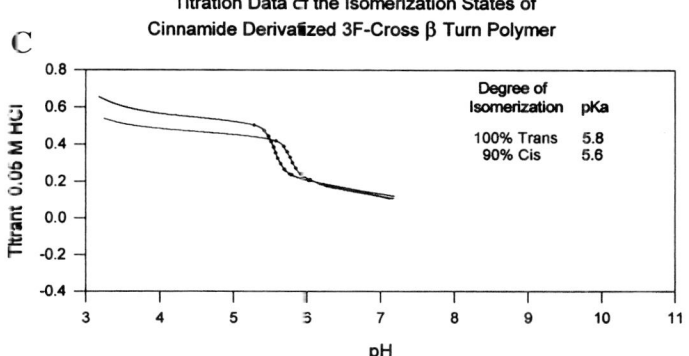

FIGURE 5.21. Photo-mechanical and photo-chemical transduction in a model protein that contains a protonatable glutamic acid residue and a chromophore capable of being driven from *trans* to *cis* on absorption of 300 nm light, both existing within the same hydrophobic association domain. (**A**) Ultraviolet absorption spectra showing the changes in the spectrum on going from 100% *trans* to 90% *cis* on illumination with 300 nm light. (**B**) Light-driven shift of isomer from *trans* to *cis* increases the value of T_t. (**C**) Light-driven shift of isomer from *trans* to *cis* lowers the pKa of Glu carboxylate. (From Heimbach.[71])

For example, the action of adding vinegar-like residues to a generally oil-like model protein results in the reaction of raising the temperature for oil-like separation such that the entire model protein acts less oil-like. Conversely, the inverse action of adding more oil-like groups to the model protein with some vinegar-like, charged residues results in the reaction of increasing the affinity of, for example, carboxylates, –COO⁻, for protons. The result becomes one of the vinegar-like residues changing functional states to become more oil-like.

5.5.3.4 *The Correlated Observation of the Narrowing of the Transition Zone on Becoming More Oil-like Gives Rise to the Efficiency Statement of Axiom 5*

Axiom 5: More hydrophobic domains result in more efficient energy conversions involving constituents undergoing conversion between different (more and less hydrophobic) functional states.

Furthermore, the action of introducing heat drives contraction, and the inverse action of stretching releases heat. On the other hand, the action of increasing proton concentration drives contraction, but, when there are carboxylates still present in the partially contracted matrix, the inverse action of stretching is more involved in that stretching causes proton uptake. These are not strictly parallel processes; to be so, the latter would have resulted in proton release on stretching, as occurs on stretching polymers, which relieves charge–charge repulsion between many closely spaced carboxylates. The underlying mechanism must provide an adequate explanation for the difference.

5.5.4 Coupling of Reactions: A Key Step Toward More Complex Biological Interactions

With the above sense of action and reaction, vinegar-like groups become coupled in model proteins that are sufficiently oil-like. For example, if a sufficiently oil-like model protein contains two different changeable (ionizable and reducible) **R**-groups, then the conversion of one to a more oil-like group by eliminating its charge causes the second vinegar-like **R**-group to adjust its property such that it too can become more oil-like. This is quite analogous to the principle of Le Châtelier; the stress caused by the increase in oil-like character on reduction of a positively charged group becomes relieved by the protonation of a negatively charged carboxylate, as discussed above in section 5.5.3.1.

5.6 Systematic Classification of Energy Conversions by Consilient Protein-based Machines

The change in Gibbs free energy of hydrophobic association, ΔG_{HA}, affects and is affected by virtually every energy accessible to, and convertible by, living organisms. It is at the foundation of the consilient mechanism; it is the result of inverse temperature transitions, and it is a representation of the comprehensive hydrophobic effect. In general, it would seem to provide for the most efficient mechanism whereby protein-based machines can perform various energy conversions in aqueous systems.

This section categorizes consilient protein-based machines. In preparation for doing so, however, we seek out appropriate definitions with which to delineate the several different kinds of machines possible by means of inverse temperature transitions.

5.6.1 Clarification of the Terminology for Energy Converting Machines

5.6.1.1 Machine Is the More General Term for a Device That Performs Energy Conversion

Based on *Webster's Third New International Dictionary*, a machine is "an assemblage of parts that transmit forces, motion, and energy one to another in some predetermined manner and to some desired end." Also drawing on the definition in the *McGraw-Hill Concise Encyclopedia of Science and Technology*, 2nd Edition, 1989, a machine is "a combination of rigid or resistant bodies having definite motions and capable of performing useful work." The protein-based machines of focus here are elastic bodies capable of contraction/relaxation by the motions arising out of hydrophobic association/dissociation.

5.6.1.2 Protein-based Machines Convert Mumerous Sources of Energy from One Form or Location to Another

The above definitions of machines do not limit the kind of useful work that is the output of the machine, nor do they limit the kind of energies that the machine interconverts, that is, that it changes one into another. In particular, protein-based machines are amphiphilic polymers, containing oil-like and vinegar-like substituents arranged along the chain molecule, that are capable of converting many different sources of energy from one form or location to another. The consideration of relocating an energy source from one site to another is to include, for example, the relocation of the chemical energy represented by the oxygen molecule from the lungs to the tissues.

5.6.1.3 Engines and Motors Are Machines That Produce Motion

According to the *McGraw-Hill Concise Encyclopedia of Science and Technology*, Second Edition, 1989, an engine is "a machine designed for the conversion of energy into useful mechanical motion." According to *Webster's Third New International Dictionary*, an engine is "a machine for converting energy (in such forms as heat, chemical energy, nuclear energy, radiation energy, and the potential energy of elevated water) into mechanical force and motion."

On the other hand, a motor, as defined by *Webster's*, is "one that imparts motion a source of mechanical power." In addition, *McGraw-Hill Concise Encyclopedia of Science and Technology*, Second Edition, gives a somewhat more restricting definition of a motor as "an electric rotating machine that converts electric energy into mechanical energy." The term *mechanical energy* includes both the kinetic energy of producing motion whereby an applied mechanical force, **f**, moves an object through a distance, ∆L, and the potential energy in terms of the development of mechanical force at fixed length, L∆f. Furthermore, there may be recognized both linear and rotary motors, and these may be driven by energies other than electrical.

5.6.1.4 Protein-based Engines and Motors Convert Numerous Sources of Energy into the Performance of Useful Mechanical Energy

Consistent with use in biology, we will not differentiate between engines and motors and will use the terms interchangeably to refer to machines that produce mechanical energy. To be more specific, designed protein-based engines and motors convert a number of energies, namely, thermal, pressure-volume, chemical, electrical, and electromagnetic, into motion. Also, as certain protein-based machines exhibit nearly ideal elastic behavior, the resulting elastic-contractile protein-based motors can be reversible. These elastic protein-based machines can be stretched, allowing that this mechanical energy input can, with the proper model protein composition, result in outputs of thermal, pressure-volume, chemical, electrical, and lower frequency electromagnetic radiation energies. The one clear exception to simple reversibility is photo-mechanical transduction because the light energy input is on the order of 70 kcal/mole, whereas the changes in Gibbs free energy of hydrophobic association would generally be no larger than ±10 kcal/mole. This means that in a single step of stretching there would not develop sufficient energy for the emission of visible or ultraviolet light, although built-up chromophores could conceivably exist within associated hydrophobic domains that would emit light on stretch-effected exposure of their hydrophobic domain to more polar water.

5.6.2 Summary Hexagon of Energy Conversions by the Consilient Mechanism of ΔG_{HA}

Figure 5.22 summarizes the set of different forms of energy that can be interconverted by the consilient mechanism of changes in the Gibbs free energy of hydrophobic association, ΔG_{HA}. As evident in Equation (5.9), the quantity, ΔT_t, is central to the definition of ΔG_{HA}. Because of this, we have referred to this emerging understanding as the ΔT_t hydrophobic paradigm for protein folding and function. Macromolecules that exhibit the phenomenon of hydrophobic association on raising the temperature, that is, the property of an inverse temperature transition of increase in polymer order on increasing the temperature, function as molecular machines by the consilient mechanism. Accordingly, we speak of molecular machines of the T_t-type.

Figure 5.22 represents the pairwise interconversion of six different energies. This gives rise to 15 classes of pairwise energy conversions. Furthermore, because the chemical couples of phosphorylated/dephosphorylated and carboxyl/carboxylate each independently change the free energy for hydrophobic association, they can participate in a chemo-chemical transduction. Consider, for example, a model protein capable of hydrophobic association/dissociation that contained both a carboxyl/carboxylate chemical couple and a phosphorylation site. In this case the model protein could be designed such that dephosphorylation would cause protonation of the carboxylate. This would be proton pumping by means of phosphorylation/dephosphorylation. Alternatively, protonation of carboxylates could raise the free energy of (i.e., activate) a bound phosphate.

Similarly, the model protein capable of hydrophobic association/dissociation could

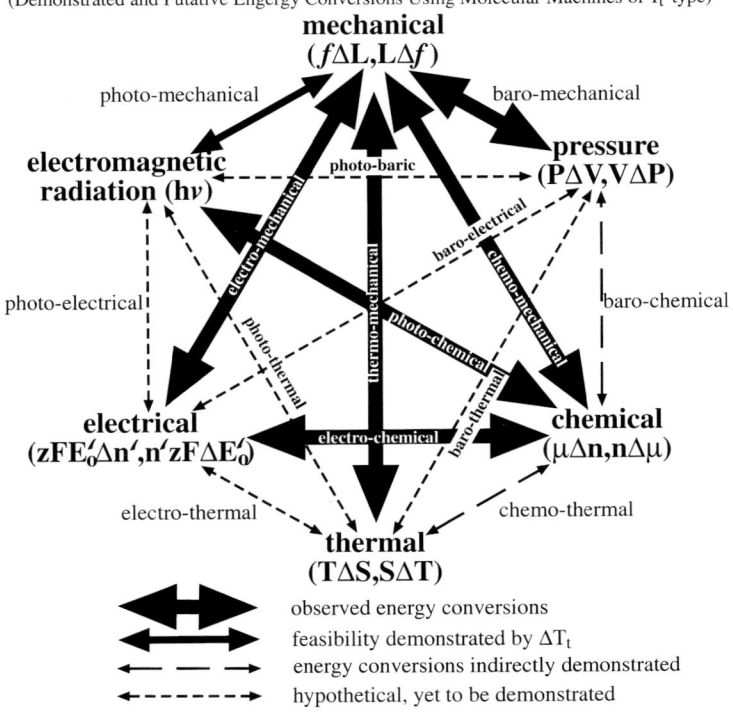

FIGURE 5.22. T_t-based molecular machines utilize the hydrophobic association transition to achieve 18 classes of pairwise energy conversions involving the six intensive variables of the free energy—pressure, mechanical force, temperature, chemical potential, electrochemical potential, and electromagnetic radiation (light and dielectric relaxation). See text for further discussion. (Reproduced with permission from Urry.[4])

contain two different redox couples, each with a different reduction potential. In this case, reduction of one redox couple would lower ΔG_{HA} of the model protein, and the lowered ΔG_{HA} would change the reduction potential of the second redox couple. In this case reduction of one could force the reduction of the second in what might be designated as an electro(1)-electrical(2) transduction. Additionally, model proteins can be designed to be simultaneously responsive to different frequencies and/or forms of electromagnetic radiation. Energy input at one frequency could cause a change in the absorption of a second frequency as an energy output. In sum, this makes 18 classes of pairwise energy conversions possible by ΔG_{HA}-based machines. This is why we have called this a *consilient* mechanism; it provides a "common groundwork of explanation" for very diverse energy conversions.

5.6.3 Five Axioms with Reversible Energy Conversions Represented

By the design of many elastic-contractile model proteins and the observations of the series of experiments on those designs noted above in this chapter, a set of apparent axiomatic statements emerged. Here the axioms are collected in one place and where relevant the types of energy conversion are listed with double-headed arrows indicating the reversibility of the energy conversion.

The following set of axioms provides the basis for the engineering of diverse protein-based machines and of the design of molecular machines from amphiphilic polymers in general, as considered in Chapter 9. They also open the door to new insights into the function of biology's protein-based machines, as discussed in Chapters 7 and 8.

Axiom 1: The change in temperature interval, over which occurs the oil-like folding and assembly transition of a host model protein on introduction of different guest substituents, becomes a functional measure of relative oil-like character of the substituents, that is, of their relative hydrophobicity, and it provides a measure of the change in free energy of the resulting hydrophobically associated state.

This axiom provides the database and insight for the design and function of protein-based molecular machines.

Axiom 2: Heating to raise the temperature from below to above the temperature interval for hydrophobic association of cross-linked elastic model protein chains drives contraction with the performance of mechanical work.

This is the thermally driven contraction inherent in inverse temperature transitions. Example: Thermo ↔ mechanical transduction.

Axiom 3: At constant temperature, an energy input that changes the temperature interval for thermally driven hydrophobic association in a model protein can drive contraction, that is, oil-like folding and assembly with the performance of mechanical work; in other words, the energy input moves the system through the transition zone for contraction due to hydrophobic association.

This requires macromolecules that can achieve equilibrium as the functional states of constituents fractionally vary; it provides a unifying basis not only for different means of producing motion of living organisms, but for all types of biological energy conversions. The double-headed arrows indicate the reversibility of the energy conversion in the following examples:

Chemo ↔ mechanical transduction
Electro ↔ mechanical transduction
Baro ↔ mechanical transduction
Photo → mechanical transduction

Axiom 4: Two or more different functional groups of an amphiphilic macromolecule, each of which can be acted upon by a different energy input that changes the temperature interval for oil-like folding and assembly, become coupled one to the other by being part of the same oil-like folding and assembly domain, that is, the energy input acting on one functional constituent by altering the oil-like character acts as an energy output on the other functional constituent by the latter's dependency on oil-like character.

This axiom is the fundamental *energy coupling* axiom; it utilizes the hydrophobic association transition common to all energy conversions by the consilient mechanism Importantly, in doing so, it involves the performance of kinds of work in addition to mechanical work. Examples include the following

Electro ↔ chemical transduction
Electro ↔ thermal transduction

Baro ↔ electrical transduction
Photo → voltaic transduction

Thermo ↔ chemical transduction
Photo → thermal transduction

Baro ↔ thermal transduction
Baro ↔ chemical transduction

Photo → baric transduction
Photo → chemical transduction

Chemo ↔ chemical transduction
Electro ↔ electrical transduction

Electromagnetic radiation(1) → electromagnetic radiation(2) transduction

Axiom 5: More hydrophobic domains make more efficient the energy conversions involving polar constituents undergoing conversion between more and less hydrophobic functional states.

This is the efficiency axiom (poising or biasing).

5.6.4 Protein–(T_t)-based Molecular Machines of the First Kind (Molecular Engines and Motors for the Performance of Mechanical Work)

Protein-based molecular machines of the first kind directly use the change in free energy of hydrophobic association, ΔG_{HA}, as a contraction for the performance of mechanical work.

As shown in the hexagonal array in Figure 5.22, five different energy inputs can perform mechanical work by the consilient mechanism. The set of elastic-contractile model proteins capable of direct utilization of hydrophobic association for contraction are called *protein-based molecular machines of the first kind*. These are enumerated below with brief consideration of the reversibility of these machines.

5.6.4.1 Thermo ↔ Mechanical Transduction

5.6.4.1.1 Thermo → Mechanical Transduction (Heat-Driven Contraction)

The input of thermal energy over the temperature interval for hydrophobic association (see Figure 5.5) drives contraction, as represented in Figures 5.18B and 5.19A.[77,78] This was first demonstrated with elastic protein-based polymers dissolved in water in which the temperature ranges for heat-induced aggregation of the polymers $(GVGVP)_n$, $(GVGIP)_n$, and $(GGVP)_n$ correlated with the temperature range for contraction of the cross-linked matrices.

5.6.4.1.2 Mechano → Thermal Transduction (Stretch-Induced Release of Heat)

As is expected for a reversible process, the reverse process of stretching, applying a mechanical force to extend the elastic protein band, is an exothermic process, primarily due to hydration of oil-like groups that become exposed on stretching.[58,59] As emphasized previously, Butler[25] in 1937 was the first to appreciate that hydration of oil-like groups was exothermic.

5.6.4.2 Chemo ↔ Mechanical Transduction

5.6.4.2.1 Chemo → Mechanical Transduction (Proton-Driven and Salt-Driven Contraction)

Chemo → mechanical transduction can be divided into two types: one (polymer based) utilizes a functional group of the polymer, and the other (solvent based) involves changing the activity of water or directly altering the hydrophobic hydration. The most apparent examples of polymer-based chemo → mechanical transduction involve changing the degree of ionization of a side chain by protonation/deprotonation[64] or the extent of ion pairing to the charged functional group. Examples of solvent-based chemo → mechanical transduction are best given with a polymer having no functional side chains, such as is the case with poly(GVGVP) and poly(GVGIP). The data in Figure 5.11 provide many examples of changes in solvent that change the value of T_t, and thereby those chemical energy sources can be used to drive contraction. Changes in the chemical energy (concentration) of NaCl has been explicitly demonstrated to drive contraction and relaxation.[79]

5.6.4.2.2 Mechano → Chemical Transduction (Stretch-Induced Proton Uptake)

The conversion of the mechanical energy of stretching into chemical energy recognized as a change in the concentration of a chemical species, mechano → chemical transduction, has been explicitly demonstrated by means of stretch-induced pKa shifts (see Figure 5.23) exhibited by the carboxyl function of a glutamic acid residue. As shown in Figure 5.23, stretching increases the affinity of the carboxyl of glutamic acid for its proton. In this case stretching induces proton uptake, and it could similarly increase affinity for ion pairing and effect formation of an ion pair.[86,87]

In the absence of a functional group, however, stretching can be used to effect a change in the concentration of a chemical in the surrounding solution. In particular, these protein-based machines functioning as mechano → chemical transducers could be used in desalination. On stretching there occurs an exothermic hydration of hydrophobic groups, that is, an uptake of pure water and formation of an environment where ions would be at high energy. Relaxation after removal from the overlying saline solution results in release of partially desalinated water.

5.6 Systematic Classification of Energy Conversions by Consilient Protein-based Machines

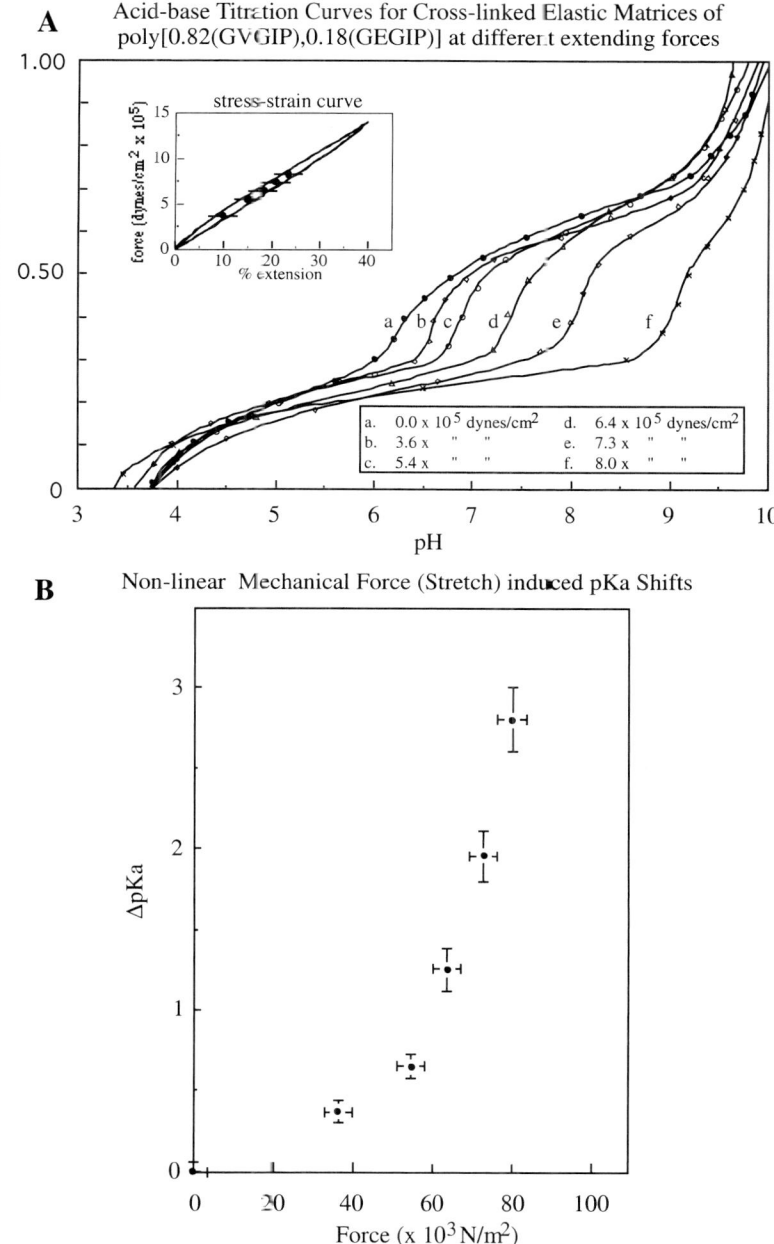

FIGURE 5.23. Mechano-chemical transduction exhibited by carboxyl-containing cross-linked elastic model protein (**A**) Stretch gives nonlinear increases in pKa values despite the linear stress-strain curve (inset). (**B**) Linear increase in mechanical force gives supra linear increase in the pKa of carboxyl function. (Reproduced with permission from Urry and Peng.[87])

5.6.4.3 Electro ↔ Mechanical Transduction

5.6.4.3.1 Electro → Mechanical Transduction (Reduction-Driven Contraction)

Synthesis of poly[0.73(GVGVP),0.27(GK{NMeN}GVP)] allowed determination of the dependence of the temperature interval for heat-induced hydrophobic association on reduction of the redox couple N-methyl nicotinamide (NMeN), which resulted in inclusion in the T_t-based hydrophobicity scale with the T_t-values given in Table 5.2.[83] On cross-linking to form the elastic matrix, reduction was found to drive contraction, and oxidation resulted in relaxation.[81]

5.6.4.3.2 Mechano → Electrochemical (Electrical) Transduction (Stretch-Induced Electron Uptake)

This experiment has yet to be done, but the expectation is clear and substantiated experimentally by hydrophobic-induced shifts in reduction potential as shown in Figure 5.20C. Stretching causes exposure of hydrophobic groups that would result in predictable stretch-induced changes in reduction potential.

5.6.4.4 Baro ↔ Mechanical Transduction

5.6.4.4.1 Baro → Mechanical Transduction (Pressure Release Drives Contraction)

Using the series of protein-based polymers poly[f_X(GXGVP),f_V(GVGVP)], where X may be tryptophan (Trp, W), phenylalanine (Phe, F), tyrosine (Tyr, Y), methionine (Met, M), or valine (Val, V) and where f_X values ranged from 0.14 to 1, the application of pressure was found to increase the value of T_t. The application of pressure leads to hydrophobic dissociation with an increase in the formation of hydrophobic hydration. Plots of ΔT_t versus log(P), with the pressure, P, ranging from 1.29 to 115 atmospheres, gave linear slopes from which decreases in volume due to application of pressure could be calculated.[84] Cross-linked elastic matrices, suspending a weight, relax and lower the weight on increasing pressure and contract to lift the weight on release of pressure.[80]

5.6.4.4.2 Mechano → Baric Transduction

Stretching is expected to decrease volume without change in composition. This would be due to the increase in water of hydrophobic hydration, which yields a state of lesser volume. Under conditions of fixed volume, stretching is expected to yield a decrease in pressure. With the perspective that the inverse temperature transition occurs when $\Delta H_t = T_t \Delta S_t$, an adequate explanation of the pressure effect requires a separate determination of the pressure dependence of ΔH.

5.6.4.5 Electromagneto ↔ Mechanical Transduction

In our studies and characterization of the elastic protein-based polymers, they have been found and/or designed to be responsive to energy inputs over a very wide frequency range of the electromagnetic spectrum from absorption of energy in the ultraviolet range of greater than 10^{15} Hz to the lowest dielectric relaxation for which a prominent absorption maximum has been found near 10^3 Hz. Additional important dielectric relaxations occur about 5 MHz (radio frequency region)[88,89] and 5 GHz (microwave frequency region).[35] Thus, input of energy into the polymer at specific frequencies over this electromagnetic spectrum could be expected to drive contraction/relaxation. Accordingly, energy inputs over the electromagnetic frequency range from the high-energy side of greater than 10^{15} Hz to the low-energy side of 10^3 Hz should be able to effect what might, for want of a better term, be called *electromagneto-mechanical transduction*.

The absorption in the microwave region at 5 GHz, as will be discussed below, is due to hydrophobic hydration. Putting in energy of this frequency ought to be the most effective way to heat the elastic protein-based polymer and drive contraction. Another intense absorption occurs near 5 MHz that is due to the internal chain motions of the protein backbone. It is the damping of these motions on stretching that is the source of the entropic component of the

5.6 Systematic Classification of Energy Conversions by Consilient Protein-based Machines

elastic force. Yet another interesting absorption in the dielectric relaxation spectrum occurs in the middle of the acoustic frequency range. All of these absorptions are expected to be relevant to electromagneto-mechanical transduction, and they would be expected to exhibit some reversibility. The most popular aspect of electromagneto-mechanical transduction, photomechanical transduction, as relevant to the consilient mechanism of energy conversion would be irreversible. A mechanical input alone would not reversibly result in emission of a photon of light. While stretching an elastic-contractile model protein could conceivably trigger emission of a photon, the energy recovered on relaxation could not reconstitute a chromophore capable of photon emission.

5.6.4.5.1 Photo → Mechanical Transduction

The examples demonstrated to date using elastic protein-based polymers have involved the light-driven *trans* to *cis* geometrical isomerization that raises the temperature of the inverse temperature transition. This has been seen with azobenzene[85] and with the more biologically relevant cinnamide, discussed above.[71] This capacity to effect relaxation can be reversed by heating or by a different frequency of light to regain the contracted state. Although a practical reversibility, this should not be confused with the reversibility in the sense that a mechanical energy input would result in the emission of a photon of light.

5.6.4.5.2 Mechano ↔ Electromagnetic Transduction

Stretching is expected to increase the dielectric absorption near 5 GHz and to decrease the dielectric absorptions near 3 kHz and 5 MHz. These effects could be used in the design of a number of transducers. Perhaps these effects could be referred to as *mechano ↔ dielectric transduction*. Furthermore, the 3 kHz dielectric relaxation exhibits the counterpart of an acoustic absorption.[48,66] This means that the mechanical oscillation of a sound wave could be detected as a change in the 3 kHz dielectric relaxation.

5.6.4.6 Device Reduction of a Linear Contracting Engine to a Rotary Engine

5.6.4.6.1 Collagen-Driven Rotary Engine of Katchalsky

The reduction of the elementary elastic strip that contracts and relaxes to a rotary motor has been demonstrated by Steinberg et al.[90] In that case, a circular elastic band was constructed. In a continuous manner it was wrapped around a set of pulleys and through a pair of baths, one bath providing the correct chemical energy for contraction and the other for relaxation. In this way a chemomechanical rotary motor was constructed.

5.6.4.6.2 Generalization to Input Energies Other Than Chemical

Similarly with one bath as the reducing system and the other as an oxidizing system, an electromechanical rotary motor could be constructed. Other means of driving contraction and relaxation represented in Figures 5.18 and 5.22 could be used in a similar manner to construct rotary motors.

5.6.5 Protein–(T_t)-Based Molecular Machines of the Second Kind (Molecular Machines for the Performance of Work Other Than Mechanical)

Discussion of each of the protein–(T_t)-based molecular machines of the second kind is beyond the scope of this volume. Above were discussed the set of five pairwise energy conversions that constituted protein–(T_t)-based molecular machines of the first kind. This leaves another 13 pairwise energy conversions of the second kind, which are listed under Axiom 4 in section 5.6.3. Most of the protein–(T_t)-based molecular machines of the second kind are of a more academic interest. Because they contain all of the coupling-of-functions, however, many are of central interest to biology. These have been discussed in section 5.5. Basically, when a hydrophobic domain contains two different vinegar-like **R**-groups, changing one to be more oil-like can induce the other to become more oil-like; this is one statement of the Principle of Le Chatelier. A favorable Gibbs free energy for hydrophobic association, $\Delta \mathbf{G_{HA}}$, is used to perform work other than mechanical work, and it does so in a way that gives rise to positive cooperativity.

The sort of energy conversions that constitute protein–($\mathbf{T_t}$)-based molecular machines of the second kind have been referred to as *pumping protons* and *pumping electrons*. There are two aspects to chemo ↔ chemical transduction. An example would be where a particular chemical energy input caused the elastic-contractile model protein to become more oil-like and thereby to energize another chemical group such as a phosphate. Another aspect of chemo-chemical transduction could include the classic consideration of enzyme catalysis, in which case there occurs a decrease in the chemical energy of reactants resulting in an increase in the chemical energy of products.

Our focus now turns to the physical basis whereby the energy conversions of the hydrophobic consilient mechanism occur, and, of course, it becomes an issue of what controls the inverse temperature transition of hydrophobic association.

5.7 What Is the Physical Basis for the Consilient Mechanism of Energy Conversion?

5.7.1 Answer: The Solvent-Mediated Interaction Between Oil-like and Vinegar-like Groups Constrained to Coexist Along and Between Protein Chain Molecules

Biology's machines are different from man's machines, primarily because they evolved to function efficiently in water. Many of man's machines commonly operate at high temperatures, above the boiling temperature for water. Such conditions are incompatible with life. Other manmade machines, electric motors, and generators and the many energy conversions of the electronic world do not function in water. As shown in Figure 2.16, even the chemically driven polymeric machines that do function in water, but that utilize the interactions of charges in water, are very inefficient compared with the protein-based machines available to biology by means of inverse temperature transition of the hydrophobic consilient mechanism.

As discussed in this section, the interplay between oil-like groups and vinegar-like groups in an aqueous medium is the basis of the energy conversion by the consilient mechanism. If there were no oil-like components and no through solvent interaction between oil-like and vinegar-like groups, energy conversion would have to rely on inefficient charge–charge interactions that are shielded by the high dielectric constant of water. Present arguments of "conformational energy" forcing vinegar-like groups into energetically unfavorable circumstances would also play a greater role. The interactions between apolar and polar components constrained by positions in the protein chain molecule make for efficient protein-based machines in water. To understand this better, we need to understand the nature and dynamics of hydrophobic hydration during the function of such systems.

5.7.2 The Structure of Water Surrounding Oil-like **R**-groups

Water molecules arrange in a special way around oil-like groups. Stackelberg and Müller[91] showed this with the crystal structure of hydrates of hydrocarbon gases (see Figure 2.8). The waters arrange at the apices of pentagons, and 12 pentagons arrange around a small oil-like molecule. When structural detail is sufficient, these pentagonal arrangements of water molecules have been seen at the surface of oil-like groups in protein. Martha Teeter[92] first observed hydrophobic hydration in protein, positioned adjacent to an oil-like side chain. Pentagonal arrangements of water molecules were observed adjacent to the side chain of a leucine (Leu, L) residue with its oil-like, $-CH_2CH(CH_3)_2$, **R**-group.

More oil-like **R**-groups in our model protein studies resulted in lower temperatures for the onset of the inverse temperature transition of hydrophobic folding and assembly (see section 5.3.2). We argued that more oil-like **R**-groups in

5.7 What Is the Physical Basis for the Consilient Mechanism of Energy Conversion?

polymers on the soluble side of the T_t-divide means more hydrophobic hydration and that more hydrophobic hydration means a lower temperature for the T_t-divide between solubility and insolubility. In fact, we proposed that the amount of hydrophobic hydration determines the temperature at which the T_t-divide occurred (section 5.3.5). In particular, as the argument went, any process that increased the amount of pentagonally arranged water lowered, and any process that decreased the amount of pentagonally arranged water necessarily raised, the *temperature interval* for hydrophobic association. Subsequently, the definitive test of this perspective became possible—the capacity to measure directly the amount of hydrophobic hydration and to determine the effect of increasing and decreasing the amount of hydrophobic hydration on the value of T_t. The most current studies and the most definitive studies are described immediately below in section 5.7.3. Then, the earlier studies are described in section 5.7.9. These latter studies preceded and, in fact, provided the impetus to obtain the equipment required for the definitive studies.

5.7.3 Direct Measurement of Hydrophobic Hydration

Two decades ago, during studies of the backbone motions of these elastic model proteins for the purpose of understanding the nature of the elasticity, an associated observation suggested that energy of the type produced by microwave ovens was increasingly absorbed as the amount of hydrophobic hydration increased.[88,89] The observation made possible a more complete understanding of the physical basis for the consilient mechanism of energy conversion that we call the *comprehensive hydrophobic effect* and that included thermodynamic characterization in terms of a Gibbs free energy of hydrophobic association, ΔG_{HA} (see sections 5.1.3 and 5.1.7.).

5.7.3.1 Identification of Hydrophobic Hydration

Microwave absorption by pure water as a function of frequency is shown in Figure 5.24A.[35] A solution of approximately half water and half model protein, (poly(GVGIP), by weight demonstrates the presence of an absorption at a lower frequency than that of water alone (see Figure 5.24B). Subtracting out the peak due to absorption by pure water leaves a peak for a kind of water that must be interacting with the model protein. Because the sum of both peaks represents a reasonable estimate of the total amount of water, and because we know the total amount of water present, the amount of interacting water can be calculated. Watching the behavior of the interacting water in response to several experimental variables provides for its identification.

How is the water interacting with the model protein? In Figure 5.25A, the number of interacting water molecules is plotted as a function of temperature for both poly(GVGIP) and poly(GVGVP). First, the more oil-like polymer, poly(GVGIP) with an added CH_2 group per pentamer, has more of the interacting water than the less hydrophobic poly(GVGVP), as would be expected for hydrophobic hydration. Most significantly, however, increasing the temperature through the transition temperature interval of hydrophobic folding and assembly, that is, from one side to the other of the T_t-divide, for each model protein causes the interacting water to disappear. We therefore conclude that *the interacting water with its absorption maximum near 5 GHz is hydrophobic hydration*. Now, we address the question, Does the previously deduced competition for hydration between oil-like and vinegar-like groups really occur?

5.7.3.2 Direct Observation of Competition for Hydration

We now examine the hydrophobic hydration of model proteins containing the vinegar-like glutamic acid residue with the **R**-group, $–CH_2–CH_2–COOH$. On decreasing the acid, H^+, concentration, the **R**-group ionizes to form $–CH_2–CH_2–COO^-$. The common measure for a decrease in acid (proton) concentration is an increase in pH. Initially, the model protein has the composition, poly[5(GVGVP),(GEGVP)], but with a precise sequence, (GVGVP GVGVP

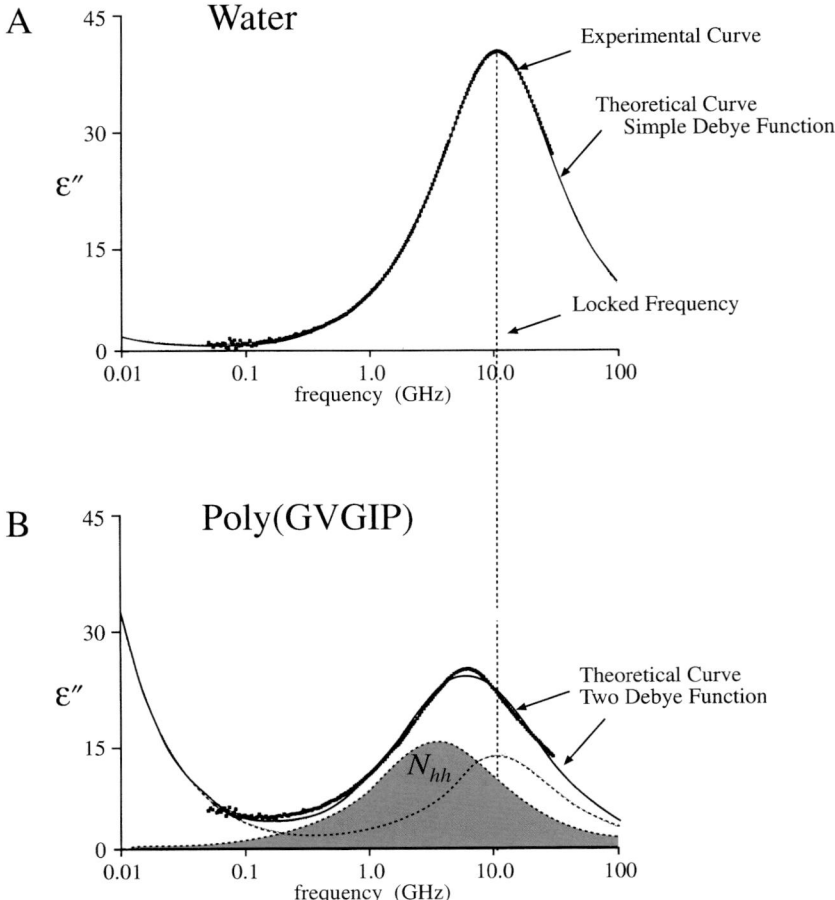

FIGURE 5.24. Microwave dielectric relaxation spectra utilized for the direct observation of hydrophobic hydration. (**A**) For pure water, showing a single Debye relaxation indicative of a single molecular species species. (**B**) For equal mass of model protein and water, the complex absorption resolves into two peaks, one for pure water and the other designated as *interacting water* (N_{hh}). With the reasonable approximation that the total absorption area represents essentially all of the water present, the relative areas of resolved curves allow calculation of amount of interacting water, N_{hh}, which is identified in Figure 5.25. (Reproduced with permission from Urry et al.[35])

GEGVP GVGVP GVGVP GVGVP)$_{36}$. This is **Model Protein i** of Table 5.5. As shown in Figure 5.25B, when protonated at low pH, this model protein has a similar amount of hydrophobic hydration as poly(GVGVP) of Figure 5.25A, This is consistent with the similar values of T_t in Table 5.1. On increasing the pH (as plotted in Figure 5.25B), hydrophobic hydration decreases as $-CH_2-CH_2-COO^-$ forms.

As expected, the more hydrophobic **Model Protein ii** of Table 5.5, with two Val (V) residues replaced by two more hydrophobic Phe (F) residues every six pentamers, is shown in Figure 5.25B to exhibit more hydrophobic hydration. Again, as $-CH_2-CH_2-COO^-$ forms on increasing the pH, the amount of hydrophobic hydration decreases. We conclude that *the formation of carboxylate groups destructures hydrophobic hydration*. Furthermore, the pH at which half of

5.7 What Is the Physical Basis for the Consilient Mechanism of Energy Conversion?

the carboxyls have converted to carboxylate, that is, the pKa where $[-COOH]/[-COO^-] = 1$, occurs at a higher pH, that is, at a lower proton concentration for the more hydrophobic polymer. This is a measure of the extra work required for carboxylate to form in the presence of more hydrophobic hydration.

As further amplified below, the data in Figure 5.25B indicate the two-way street of the competition for hydration. *Formation of carboxylates destroys hydrophobic hydration, but the carboxylate thus hydrated is at a higher Gibbs free energy as reflected by the higher pKa, the higher pH required when there is more hydrophobic hydration.*

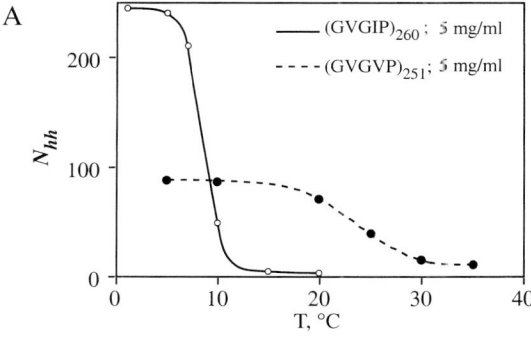
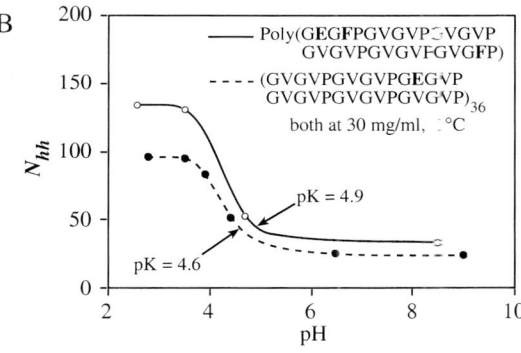

FIGURE 5.25. Temperature and pH dependences of the *interacting water*, N_{hh}. (**A**) Plot of interacting water, N_{hh}, as a function of temperature where the *interacting water* is seen to disappear as the temperature is raised through the *temperature interval* of the inverse temperature transition for hydrophobic association. Therefore, one concludes that the *interacting water*, N_{hh}, is hydrophobic hydration. (**B**) Plot of hydrophobic hydration, N_{hh}, as a function of decrease in pH where the carboxyl functions of the two model proteins ionize to form the charged carboxylate groups. By the time there is less than one charged carboxylate per 100 residues, approximately two-thirds of the hydrophobic hydration has been lost, and the point at which this occurs is at a higher pH as the model protein becomes more hydrophobic. From these data we conclude that there exists a competition for hydration between apolar (hydrophobic) and polar (e.g., charged) groups. Thus, there exists an apolar-repulsive free energy for hydration! (Reproduced with permission from Urry et al.[35])

5.7.4 Competition for Hydration Controls the Amount of Pentagonally Arranged Water

Salts dissolve very readily in water with release of heat because the positively and negatively charged groups (ions) of the salt prefer to have water surrounding them than to be ion paired in the salt crystal. The water around charged groups strongly orients with respect to the charge. We have just seen that the attraction between a charged (polar) group and water is so strong that the charged group, limited from abundant water by the polymer sequence of which it is a part, can dismantle the hydration around oil-like (apolar) groups. Figure 5.26 depicts this competition for hydration with changing proximity of a polar (e.g., charged) group to an oil-like (apolar) group. Thus, when both oil-like groups and charged groups are constrained to coexist along a protein sequence, they compete for hydration. The competition is seen as a loss of hydrophobic hydration, N_{hh}, with the consequences of an increase in T_t and an increase in the pKa, that is, a ΔpKa, which is a measure of the repulsion between apolar and polar groups.

This competition for hydration between charged and hydrophobic groups constrained by sequence to coexist occurs even within a side chain of a single residue composed of charged groups and hydrophobic groups. Amino acid residues such as glutamic acid, arginine, and lysine contain the charged function at the end of a string of hydrophobic CH_2 groups. It is interesting that protein crystal structure data tend to show these side chains, when standing alone, to be fully extended, as, for example, for

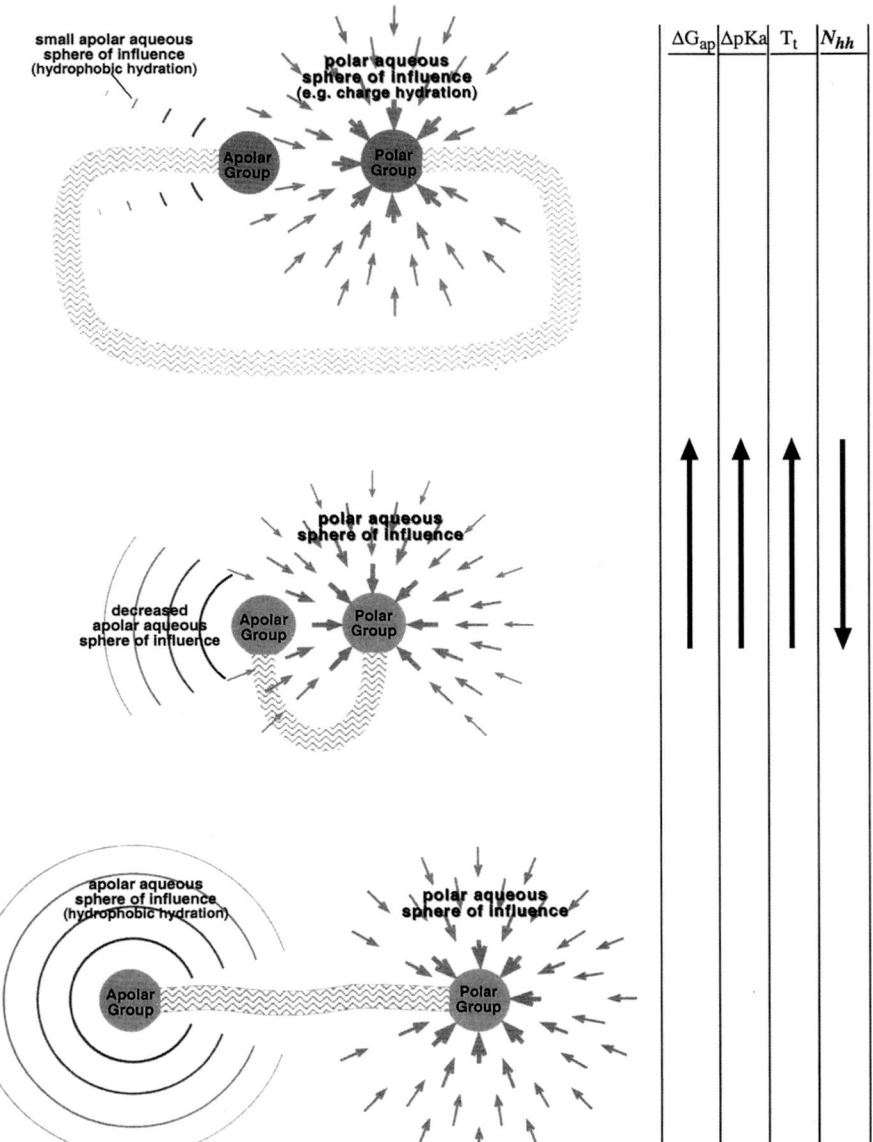

FIGURE 5.26. The apolar–polar repulsive free energy of hydration, where competition for limited hydration results in less hydrophobic hydration, N_{hh}, an increase in T_t, an increase in ΔG_{ap}, that is, an increase in the Gibbs free energy of the ion, $\Delta G(ion)$, and an increase in the magnitude of the ΔpKa. (Reproduced with permission from Urry.[122])

residues E261 and E264 of the γ-rotor of ATP synthase (see Chapter 8, Figures 8.31 and 8.33) and the R58 residue that marks the ATP binding site in the chaperonin structures in Chapter 7, Figures 7.43 through 7.46. To have the amino group of lysine folded back beside the string of four CH_2 groups would, by the apolar–polar repulsive free energy of hydration, limit the accessible hydration for both and increase the magnitude of ΔG_{ap}.

The forces between water and charged groups are much stronger than the forces that structure water adjacent to oil-like groups. A few charged residues can destructure much of the pentagonally arranged water. The formation of just four carboxylates in 100 residues in poly[4(GVGVP),(GEGVP)] raises the transition temperature for oil-like separation from 24°C for –COOH to 69°C for –COO⁻ as shown in the upper part of Figure 5.3.

There are two sides, however, to the coin of competition for hydration. The carboxylate must pay a price to wrench hydration away from hydrophobic groups and in having to do so it never achieves the level of hydration that it would as a dilute carboxylate in bulk water. The price appears in the amount of hydroxyl ion, OH⁻, necessary in the solution before the proton, H⁺, leaves the –COOH to form a carboxylate, that is, OH⁻ + –COOH = –COO⁻ + H_2O. In the absence of the oil-like residues, the amount of hydroxyl required to remove the proton and form COO⁻ is much less. When there are only 2 carboxyls per 100 residues in poly[9(GVGIP),(GEGIP)], it is so difficult for the COO⁻ to form that a pH of 6.4 is required,[93] which is a 250-fold increase in the amount of hydroxyl ion required. When nearly half of the oil-like Val residues are replaced by more oil-like Phe residues, as in **Polymer V** of Table 5.5, a million-fold increase in the amount of hydroxyl ion was required to form the carboxylate ion of the aspartic acid residues.[40]

5.7.5 We Now Know What Controls Insolubility and Solubility of Hydrophobic Domains

5.7.5.1 Loss of Solubility Due to Too Much Hydrophobic Hydration

In general, the solubility of a model protein such as poly(GVGVP) in water comes from the presence of the polar peptide group, –CONH–. The hydrogens of water, HOH, hydrogen bond to the oxygen of the CO, that is, CO ··· HOH, and the oxygen of water hydrogen bonds to the NH, that is, NH ··· OH_2. This hydrogen bonding gives rise to solubility. As oil-like groups are added to the model protein in water at a particular temperature, solubility is ultimately lost.

From Butler,[25] but also from the T_t-based hydrophobicity scale in Table 5.1, we learn the unique way in which this happens. Formation of hydration surrounding oil-like groups is a favorable exothermic reaction; heat is released; the heat charge for the hydration reaction, ΔH, is negative.

Despite this and as contradictory as it may initially sound, however, too much hydration of oil-like groups results in loss of solubility. This is due to the entropy change, ΔS, that accompanies hydrophobic hydration. Recall that solubility is governed by the Gibbs free energy, ΔG(solubility) = $\Delta H - T\Delta S$, where T is the temperature in degrees Kelvin, K, where K = C + 273. The water molecules of hydrophobic hydration, the pentagonally arranged water in Figure 2.8, are more structured than the water molecules in liquid or bulk water. The entropy change for formation of hydrophobic hydration has the opposite sign from the entropy changes for the melting of ice and the vaporization of liquid water in Figure 5.2. The latter changes are to states of less order on raising the temperature; so too is the loss of pentagonally arranged water to form liquid water on raising the temperature through the transition zone for the inverse temperature transition. Therefore, the entropy change, ΔS, is negative for formation of hydrophobic hydration on dissolution of oil-like groups in water, such that the $(-T\Delta S)$ term is positive. Now as each oil-like group is added, the positive increment in the $(-T\Delta S)$ term is greater than the negative increment in ΔH. Accordingly, the oil-like additions can occur with retention of solubility until at a given temperature ΔG(solubility) becomes positive and solubility is lost.

Whether a particular hydrophobic region or domain of a model protein or of a natural protein associates with a second hydrophobic domain in the same molecule or a separate molecule, the same process of loss of solubility occurs. If the two hydrophobic domains can associate and if together they have so much hydrophobic hydration that their T_t is below the temperature of the environment, they associate; they are insoluble; ΔG(solubility) is positive.

5.7.5.2 Solubility of Oil-like Groups Occurs Because Polar Groups Destructure Hydrophobic Hydration

If a charged vinegar-like group with inadequate hydration appears proximal to an opening fluctuation of associating hydrophobic domains, it will use the nascent hydrophobic hydration of the opening fluctuation for its own hydration. With less hydrophobic hydration, the magnitude of the positive ($-T\Delta S$) term becomes less than the magnitude of the negative ΔH term; ΔG(solubility) is negative, and the two domains gain solubility.

5.7.6 The Formation of Ion Pairs Makes Model Proteins More Hydrophobic

During the operation of many of biology's protein-based machines, the formation and separation of ion pairs is central to function. Very commonly the ion paired state forms as two oil-like domains associate, one oil-like domain containing the positive charge and the other the negative charge. Examples of this are seen in the binding and release of oxygen by hemoglobin in a manner that enables the transport of oxygen by hemoglobin from the lungs to the tissues. Alternatively, the ion for pairing may come from the solution. In this case, ion pairing at one potentially oil-like domain makes it more hydrophobic such that the more hydrophobic domain can then associate with another hydrophobic domain. Often in biology, calcium ion is released to the solution where it can ion pair with carboxylates of a potential hydrophobic domain. Often, such ion pairing provides the trigger for function, as in muscle contraction.

Here we report the effect of calcium ion binding to carboxylates that exhibit hydrophobic-induced pKa shifts. The pKa shifts result from the work required to destructure hydrophobic hydration in the process of ionization to form the carboxylate. In Figure 5.27, calcium ion interaction with carboxylates makes two different model proteins more hydrophobic to different extents, depending on their initial oil-like composition. Figure 5.27A considers **Model Protein ii** of Table 5.5, which for every 30 residues (six pentamers) has one carboxylate ($-COO^-$)-containing glutamic acid residue (Glu, E) and two, very oil-like phenylalanine (Phe, F) residues with the hydrocarbon **R**-group of $-CH_2-C_6H_5$ having replaced two valine (Val, V) residues with side chain $-CH_2-(CH_3)_2$. As calcium ion is added to 10 times equivalence, the T_t-value decreases, initially rapidly to about equal amounts of calcium ion and carboxylate, then much more slowly. In a near mirror image fashion, as calcium ion is added, the number of waters of hydrophobic hydration increases. Clearly, calcium ion pairing with carboxylates increases hydrophobicity, as indicated by the decrease in the value of T_t. Clearly, the decrease in T_t correlates with an increase in hydrophobic hydration. For this composition, however, not until a twofold excess of calcium has been added does hydrophobic association begin at body temperature.

Figure 5.27B shows the same data for **Model Protein I** of Table 5.5 with two proximal carboxylate (COO^-)-containing glutamic acid residues (Glu, E), with five very hydrophobic F residues and two more hydrophobic I residues with the side chain $-CH_2(CH_3)CH_2-CH_3$, having replaced nine of the valine (Val, V) residues with side chain $-CH_2-(CH_3)_2$, in every 30 residues. In this more hydrophobic model protein, before there is one calcium ion per carboxylate the temperature for hydrophobic association drops below $0°C$. These polymers, of course, are the **Model Proteins ii** and **I** considered in section 5.5.3.1 and included in the plots in Figure 5.20B.

5.7.7 The Amount of Hydrophobic Hydration Determines the Value of T_t

The data in Figure 5.27 experimentally establish the relationship between the temperature, T_t, for the onset of the inverse temperature transition and the number of molecules of hydrophobic hydration, N_{hh}. Without going in to the details of the experiments, it is possible to appreciate the relationship by visual inspection.

Microwave absorption by hydrophobic hydration as a function of frequency provides

5.7 What Is the Physical Basis for the Consilient Mechanism of Energy Conversion?

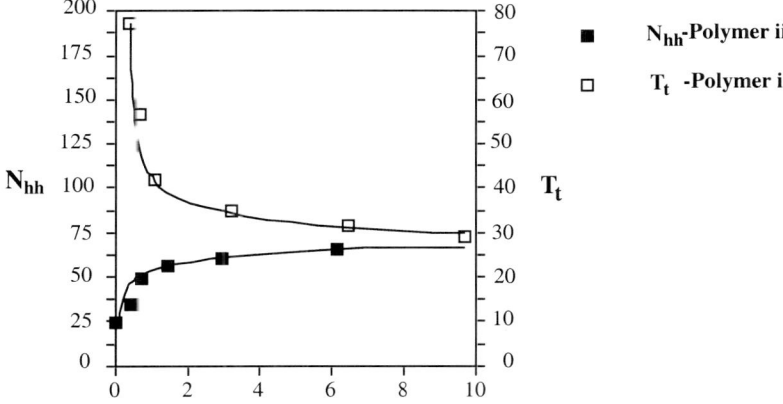

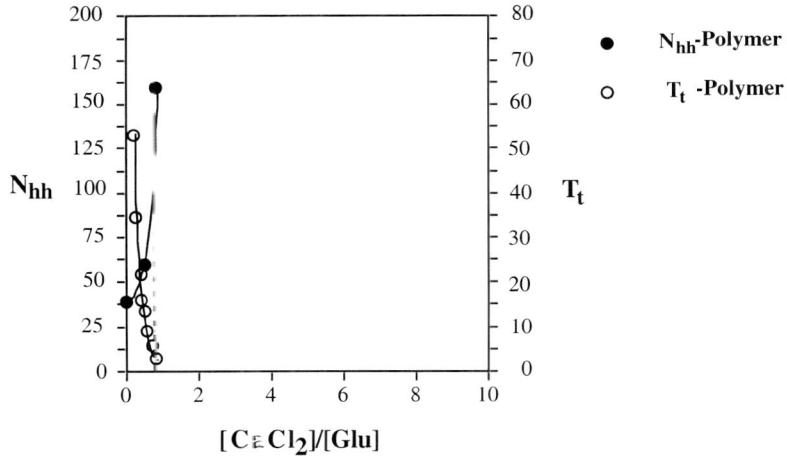

FIGURE 5.27. Development of hydrophobic hydration, N_{hh}, on ion pairing shows a mirrored decrease in, or lowering of, T_t. (Top) For the less hydrophobic **Model Protein ii**, the increase in N_{hh} is somewhat modest as is the decrease in the value of T_t. (Bottom) For the more hydrophobic **Model Protein I**, the increase in N_{hh} is very abrupt and the lowering of the value of T_t is correspondingly abrupt. Conclusions: (1) Amount of hydrophobic hydration, N_{hh} positions the T_t-divide! (2) Ion pairing markedly restores hydrophobic hydration, N_{hh}, and, once too much hydrophobic hydration has formed, the result is hydrophobic association as the $(-T\Delta S)$ term of the Gibbs free energy for solubility becomes too positive and dominates the free energy and insolubility of hydrophobic association results! (C.-H. Luan and D.W. Urry, unpublished results.)

an estimate of the number of molecules of hydrophobic hydration, N_{hh}, for the two model proteins. The model proteins both contain the vinegar-like group in its carboxylate state, $-COO^-$, and in both the Val residues have been replaced with more hydrophobic residues. Their dependence of T_t on concentration of calcium ion, however, differs remarkably.

The same difference in T_t exhibited by the two different polymer compositions appears in the inversely related number of water molecules of hydrophobic hydration, N_{hh}. This striking mirroring of N_{hh} and T_t occurs for each polymer composition, even though the calcium ion dependence is itself strikingly different between the less hydrophobic and the more hydrophobic polymers. Clearly T_t, the temperature at which the onset of hydrophobic association occurs, depends on the number of waters of hydrophobic hydration. As the amount of hydrophobic hydration increases, T_t decreases in an exactly mirrored manner. Accordingly, *the amount of hydrophobic hydration determines the value of T_t.*

5.7.8 Earlier Studies Indicated the Fundamental Process to be Competition for Hydration

Above, the special structural organization of these model proteins provided the unique opportunity to identify and observe the behavior of hydrophobic hydration. Under relevant circumstances hydrophobic hydration was more prevalent than bulk water. Key variables showed that the amount of hydrophobic hydration, N_{hh}, determined T_t. Numerous systematic studies, however, preceded the direct observation of hydrophobic hydration by microwave dielectric relaxation. The earlier studies built the foundation on which direct observation of N_{hh} became the capstone. Below, interpretations, implications, and/or conclusions follow brief descriptions that lay bare the power of the inverse temperature transition and add to the panoply of data constituting the comprehensive hydrophobic effect.

5.7.8.1 Dependence of Heat of Transition, ΔH_t, on Degree of Ionization

With the protein-based polymer poly[0.8GVGVP),0.2GEGVP)], at low pH when all 4 of the Glu (E) residues/100 residues are as COOH, the transition temperature is near 25°C and the heat of the transition, ΔH_t = 0.97 kcal/mole-pentamer (see Figure 5.28). On raising the pH to the point of less than two COO$^-$/100 residues, the heat of the transition has been reduced, and ΔH_t = 0.27 kcal/mole pentamers.[94] The preferred interpretation over a decade ago was (1) *that the formation of 2 COO$^-$/100 residues destructured almost three-fourths of the thermodynamically measured waters of hydrophobic hydration, and (2) that there exists a competition for hydration between apolar and polar residues, referred to as an apolar–polar repulsive free energy of hydration.*

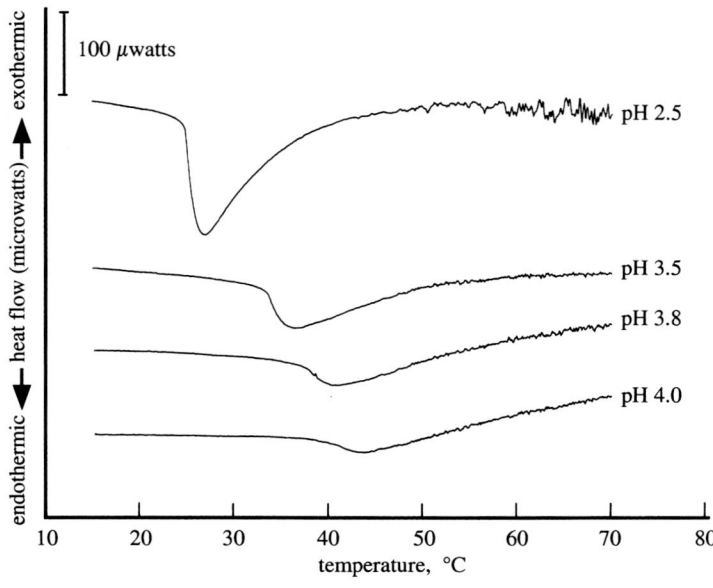

FIGURE 5.28. Differential scanning calorimetry of model protein, plotted (with heat absorption peak down) as function of increasing degree of ionization of carboxyl function. Ionization to form less than 2 carboxylates per 100 residues reduces the heat of the transition to about one-fourth. Charge destroys hydrophobic hydration, raises T_t, and gives rise to solubility. (Reproduced with permission from Urry et al.[94])

These interpretations were reinforced by the observation in the same time period of stretch-induced pKa shifts (discussed immediately below) by insights of the T_t-based hydrophobicity scale (discussed above) and then by hydrophobic-induced pKa shifts (as discussed further below).

5.7.8.2 Stretch-induced pKa Shifts

In 1937, using series of simple organic molecules, Butler[25] first observed that exposure of CH_2 groups to water resulted in an exothermic reaction with formation of a unique water around such oil-like groups. In 1970, Weis-Fogh and Andersen[59] reported the exothermic reaction on stretching elastin and interpreted it in terms of the exposure of hydrophobic groups with formation of hydrophobic hydration. Hoeve and Flory[58] further affirmed this in their 1974 publication. Accordingly, stretching of elastin-like protein forces exposure of hydrophobic groups to the surrounding water with the result of an exothermic reaction of hydrophobic hydration.

Consideration of the effects of stretching begins with hydrophobically associated and cross-linked elastomeric matrix composed of the same protein-based polymer as used in the calorimetry studies with poly[0.8(GVGVP), 0.2(GEGVP)]. As shown in Figure 5.29,[86] the acid–base titration curve of unstretched matrix of cross-linked poly[0.8(GVGVP), 0.2(GEGVP)] gives a pKa of 3.99, given adequate time for equilibration of medium with matrix. Repeating the acid–base titration after stretching on applying a load of 1 gram shows that the pKa shifted to 4.84. More extensive data from another polymer composition appears in Figure 5.23.[87] The chemical potential of base required to remove the proton from COOH increases on stretching. This means that the free energy for formation of the carboxylate, COO^-, has increased. Such a result is quite striking, because there is an increase in water in the matrix on stretching, which one might expect to facilitate ion formation. It would seem that the kind of water that enters the matrix on stretching is not suitable for hydration of a charged species. The type of water that

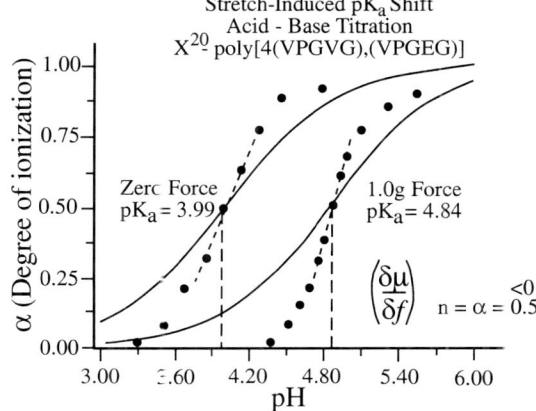

FIGURE 5.29. Stretch-induced pKa shift within hydrophobically associated elastic model protein implies competition for hydration between carboxylate and hydrophobic groups. Despite water uptake into the model elastic protein on stretching, the COO^- experiences less accessible hydration. The implication is that hydrophobic hydration, formed due to exposed hydrophobic groups on extension, is unsuited for COO^- hydration. (Reproduced with permission from Urry et al.[86])

enters the matrix on stretching is primarily hydrophobic hydration.

This brings us to the following analyses and conclusions:

1. *Even though there is an increase in the amount of water in the elastomer on stretching, much of which must be hydrophobic hydration, the formation of carboxylates has become disfavored. Therefore, water of hydrophobic hydration must not be suitable for hydration of carboxylates.*

2. $(\partial \mu / \partial f)_{n=\alpha=0.5} < 0$ *for the ΔT_t-mechanism*: The chemical potential, μ, is defined as $RT \ln \mathbf{a}$, where $\mathbf{a}$ is activity of proton in this case. In the pH range of 4 to 5, proton concentration, $[H^+]$, can be used in place of $\mathbf{a}$, such that $\mu = RT \ln[H^+]$. Because the choice for acid–base titrations is logarithm to the base 10, log, it can be stated that $\mu = 2.3 RT \log[H^+]$, and because pH $= -\log[H^+]$, then $\mu = -2.3 RT$ pH. At 50% ionization, that is, $\alpha = 0.5$, $\mu = -2.3 RT$ pKa, such that $\Delta\mu$(proton) $= -2.3 RT$ ΔpKa. From Figure 5.29 it is clear that an increase in load from zero

to 1.0 gram represents an increase in force; therefore, Δf is positive. On the other hand, the pH at the pKa increased from 3.99 to 4.84; therefore, Δμ(proton) is a negative quantity. Thus, it can be written that $(\partial \mu/\partial f)_{n=\alpha=0.5} < 0$, that is, stretching results in an uptake of protons, where n = α = 0.5 indicates evaluation at the degree of ionization of 0.5 which is the pKa, the point of 50% ionization.

3. $(\partial \mu/\partial f)_{n=\alpha=0.5} > 0$ *for the charge–charge repulsion mechanism*: Studying polyelectrolytes with the example of poly(methacrylic acid), [–CH–CCH₃COOH–]ₙ, Katchalsky et al.[95] in 1960 presented the argument that stretching resulted in the release of protons, that is, $(\partial \mu/\partial f)_{n=\alpha=0.5} > 0$. The description of the process for cross-linked poly(methacrylic acid), with a carboxyl function on every other backbone atom, follows: (a) At 60% ionization, charge–charge repulsion gives complete extension into stiff rods. (b) At 0 to 10% ionization, charge–charge repulsion is no longer present and the chain randomizes to give full contraction. (c) When the chemical potential of proton in the bathing medium allows 20% ionization, contraction will have occurred to the extent allowed by charge–charge repulsion. On stretching of the matrix at this point, there is an increase in distance between charges, and for the unchanged proton chemical potential in the bathing medium there can be a release of protons into the medium until the chemical potential of the fiber again matches that of the medium. (d) Therefore, stretching results in a release of protons for the charge–charge repulsion mechanism, which is to say that $(\partial \mu/\partial f)_{n=\alpha=0.5} > 0$ for the charge–charge repulsion mechanism, whereas for the ΔT$_t$-mechanism it is exactly the opposite, $(\partial \mu/\partial f)_{n=\alpha=0.5} < 0$. (e) Conclusion: *A mechanism other than charge–charge repulsion is responsible for the stretch-induced pKa shift exhibited by T$_t$-based molecular machines.*

4. *Relationship of cooperativity to ΔpKa*: Furthermore, the increase in Gibbs free energy of hydroxyl ion required to remove the proton from COOH, indicated by the pKa shift of 0.85, is $\Delta G_{2,1} = \Delta\mu(\text{hydroxyl ion}) = 2.3\, RT\, \Delta pKa = 1.16$ kcal/mole. In association with the pKa shift, there is an increase in steepness of the acid–base titration curve on stretching, indicating an increase in positive cooperativity. For small pKa shifts this may be estimated using the Wyman equation[96] $\Delta G = RT(1 - 1/n)/\alpha(1 - \alpha)$, where n (the Hill coefficient)[97] is the slope of the curve at the pKa and α is the degree of ionization, which at the pKa is 0.5. Under zero load, n was 1.07; on loading to 1 gram, n became 2.21. For the zero load case, ΔG_1 is 0.15 kcal/mole, whereas for the 1 gram load, ΔG_2 is 1.30, such that the difference, $\Delta G_{1,2}$, is 1.15 kcal/mole. Thus, both the pKa shift and the increase in positive cooperativity represent different expressions of the same energy of interaction. Conclusion: *Positive cooperativity is another expression of the competition for hydration between apolar and polar groups that gave rise to the stretch-induced pKa shift.*

5. *Nonlinear stretch-induced pKa shift*: The stress/strain curve of the cross-linked protein-based polymer X^{20}-poly[0.82(GVGIP), 0.18(GEGIP)] is shown in the inset of Figure 5.23A to be linear such that the work done on the elastomeric band is directly proportional to the applied force. On the other hand, the pKa shifts resulting from linear increases in load are very nonlinear (Figure 5.23B).[87] This carries significant implication with regard to the efficiency of the conversion of mechanical energy into the chemical work of pumping protons.

In addition, the acid–base titration curve becomes steeper, that is, exhibits an increase in positive cooperativity as the elastomeric band is further extended (Figure 5.23B). This effect was explicitly calculated above. Because the change in chemical potential times the change in number of moles involved to go from the relaxed state to the contracted state is the required chemical energy, the positive cooperativity means less chemical energy will be required to perform a given amount of mechanical work. Not only is there a different mechanism from that of the charge–charge repulsion mechanism, but the new mechanism is more efficient, as was briefly discussed in Chapter 2 and shown in Figure 2.16. This will be discussed more extensively below.

5.7.8.3 Hydrophobic-induced pKa Shifts

Four studies based on different model protein systems and experimental variables have been used to vary hydrophobicity systematically and thereby to demonstrate the basis of hydrophobic-induced pKa shifts. The four studies are (1) dilution of a recurrent ionizable residue in a poly(5-mer) from 1 ionizable group every 5 residues to 1.2 per 100 residues by stepwise replacement of a polar residue with an apolar (hydrophobic) Val (V) residue, (2) maintenance of the composition of a poly(30-mer) but rearranging the proximity of more hydrophobic residues to the ionizable residue in a poly(30-mer), (3) increasing systematically the hydrophobic residues within a poly(30-mer) while holding the ionizable residue of the poly(30-mer) constant at 1 per 30 residues, and (as discussed in section 5.7.8.2) (4) stretching incrementally a hydrophobically folded cross-linked elastic matrix containing ionizable residues (in this case the polymer composition was kept constant but the hydrophobicity was increased by application of an extending force that increasingly exposed hydrophobic groups to water in sufficient proximity to the ionizable function.

5.7.8.3.1 Increasing Hydrophobicity by Stepwise Replacement of the Functional Group with a More Hydrophobic Residue

The model system, poly[f_x(GVGIP), f_x(GXGIP)], is used where f_x and f_v are mole fractions and X is defined as follows for each of three studies: (1) X = Glu(E), f_x = 1.0 to 0.06; (2) X = Asp(D), f_x = 1.0 to 0.06; and (3) X = Lys(K), f_x = 1.0 to 0.06. In each of these studies, the starting polymer is poly(GXGIP), and then the residue X is stepwise replaced by the more hydrophobic valine (Val, V) until the X residue is very dilute in the polymer, 1.2 residues per 100 residues. As shown in Figure 5.30A,B for starting compositions of poly(GEGIP)[93] and poly(GKGIP),[98] respectively, the charge density is high enough such that charge–charge repulsion is observed in the absence of salt. This charge–charge repulsion is entirely removed by 0.15 N NaCl for X = E and D (data not shown)[99] and significantly relieved for X = K, and the essentially normal pKa is obtained for these ionizable functions. In the absence of salt, the charge–charge repulsion disappears for f_E = f_D = 0.7 and for f_K = 0.75. The remarkable feature of these studies is that, as the dilution of charged species continues, the pKa shift resumes with a vengeance, instead of decreasing as it must if charge-charge repulsion were relevant. The largest pKa shifts are found at the highest dilutions examined with 1.2 ionizable residues per 100 residues of polymer.

5.7.8.3.1.1 Poly[f_V(GVGIP),f_X(GXGIP)] with X = Glu(E) and f_X Varied from 1.0 to 0.06.[93]

Specifically for the Glu(E) residue, 10 polypeptides were synthesized, each with a different value of f_E. The value for f_E = 1 in pure water is 4.7, and on addition of 0.15 N NaCl the value drops to 4.35. The magnitude of the charge–charge repulsion is about 0.49 kcal/mole as estimated from the ΔpKa of 0.35. A pKa of 4.35 is the lowest value obtained in the absence of salt at f_E = 0.7. On further dilution, that is, for f_E < 0.7, the increase in pKa resumes until at f_E = 0.06 the pKa becomes 6.6, which relaxes to 6.1 on the addition of physiological saline. Thus, the mechanism operative in these T_t-type protein-based polymers, which is not charge–charge repulsion, has brought to bear a repulsive free energy of interaction of over 3 kcal/mole. As will be discussed later, this new mechanism can give rise to repulsive free energies of interaction of 8 to 10 kcal/mole.

Calculated curves using the charge–charge repulsion argument are included in Figure 5.30A, and the means of calculating is given in the following section on poly[f_V(GVGIP), f_K(GKGIP)]. The minimal experimental dielectric constant for poly(GVGIP) at 36° is obtainable from Table 5.6, where the high-frequency component (ε^∞ of 23) and the 5 MHz component (Δε of 42) sum to give a minimal value of 65. With this value, only a few tenths of a pH unit shift is calculated (see Figure 5.30A). Even when taking the unrealistic value of 5 for the dielectric constant in the absence of any ionizable residue, the experimental curves are not reasonably calculated by the Glu(E) residue. With the charge–charge repulsion approach, less than 40% of the shift observed for $f_E = 0.06$ is calculated. The same calculation is carried out below in more detail for the lysine series.

5.7.8.3.1.2 Poly[f_V(GVGIP),f_X(GXGIP)] with X = Asp(D) and f_D Varied from 1.0 to 0.06.[99] Specifically for the Asp(D) residue, 10 polypeptides were synthesized, each with a different value of f_D. The detailed information of the synthesized compositions and the pKa shifts with and without salt should be sought in Urry et al.[99] The magnitude of the charge–charge repulsion seen in the ΔpKa for an f_D of 1 was 0.8 pH units, representing a charge–charge repulsion of 1.1 kcal/mole with a charged aspartate every fifth residue. The repulsion was entirely removed by the addition of 0.15 N NaCl. Thus,

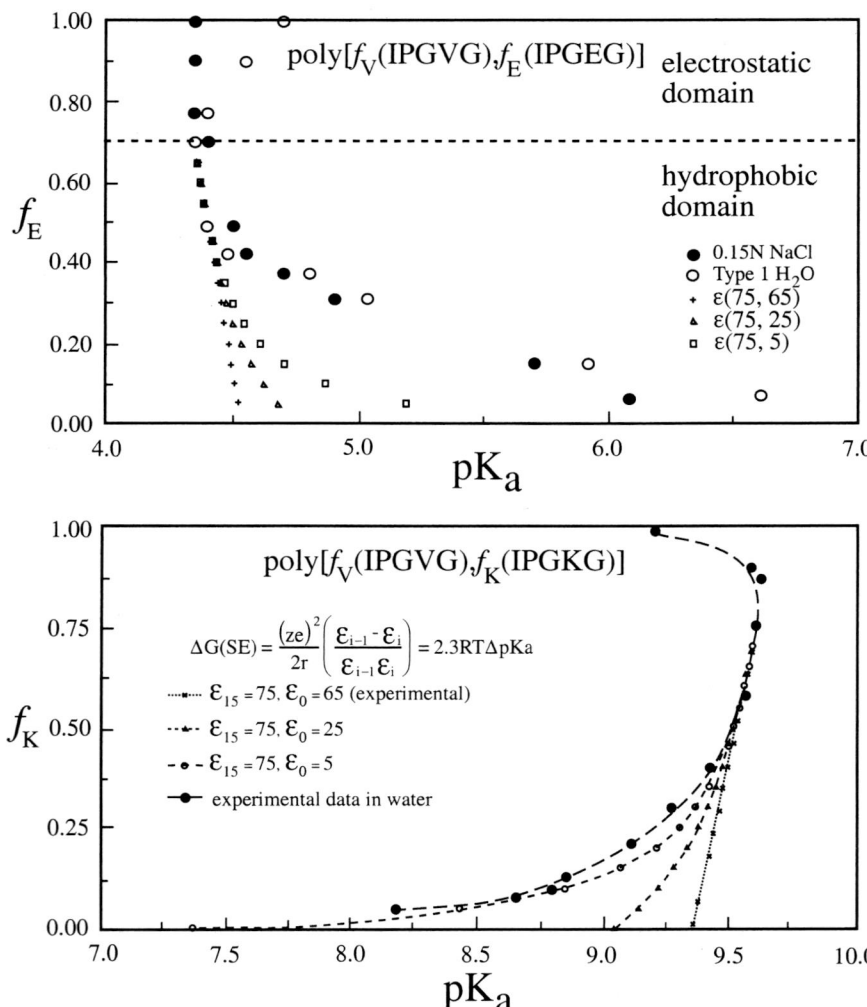

FIGURE 5.30. (**A**) Hydrophobic-induced pKa shifts as Glu is replaced by Val in the family of model proteins, poly[f_V(GVGIP),f_E(GEGIP)], demonstrate a limitation of the electrostatic argument. (Reproduced with permission from Urry et al.[93]) (**B**) Using the protein-based polymer poly[f_V(GVGIP), f_K(GKGIP)], hydrophobic-induced pKa shifts as Lys is replaced by Val, as with the data in **A**, also demonstrates limitation of electrostatic argument. (Reproduced with permission from Urry et al.[98])

TABLE 5.6. Dielectric relaxation of poly(GVGIP) in water as a function of temperature.[a]

Temp (°C)	Δε	τ (ns)	α	σ (mS/m)	ε" at 100 MHz ± 0.5
0[b]	26 ± 3	139 ± 6	0.36 ± 0.08	1.2 ± 0.2	32.3
6[b]	26 ± 4	87 ± 10	0.3 ± 0.1	1.8 ± 0.6	30.8
12	37 ± 2	49 ± 3	0.11 ± 0.03	0.5 ± 0.1	26.3
18	34 ± 2	44 ± 3	0.08 ± 0.02	0.6 ± 0.1	24.7
24	37 ± 3	44 ± 3	0.08 ± 0.07	0.6 ± 0.1	23.7
30	38 ± 2	38 ± 2	0.08 ± 0.02	0.6 ± 0.1	23.2
36	42 ± 3	36 ± 2	0.09 ± 0.02	0.6 ± 0.1	23
42	43 ± 3	35 ± 2	0.09 ± 0.02	0.7 ± 0.1	22.7
48	48 ± 3	39 ± 2	0.10 ± 0.02	0.7 ± 0.2	22.6
54	50 ± 3	35 ± 3	0.10 ± 0.02	0.9 ± 0.2	22.6
60	55 ± 4	31 ± 2	0.11 ± 0.03	1.0 ± 0.2	22.4

[a] The minimal dielectric constant at 37° C is 65, that is, 23 for the high frequency (100 MHz) limit plus 42 for the 5 MHz band. Values of 5 are not correct.
[b] The dielectric parameters of the samples at 0° and 6° C are only suggestive because the relevant dielectric relaxation is not yet very well resolved. (Data from Table 1 of Buchet et al.[89])

under physiological conditions with an aspartic carboxylate every fifth residue, there is no charge–charge repulsion in 0.15 N NaCl. On the other hand, on dilution to 1.2 residues/100 residues, the ΔpKa was 2.1 pH units in water and 1.5 pH units in 0.15 N NaCl. Significantly, even in this domain where charge–charge repulsion is ruled out, there is yet an effect of salt to relax the repulsion between hydrophobic and charged groups. This region of the hydrophobic domain will be calculated explicitly below for the lysine NH_3^+ example.

5.7.8.3.1.3 Poly[f_V(GVGIP),f_X(GXGIP)] with X = Lys(K) and f_K Varied from 1.0 to 0.05.[98]

For carboxylates, the high pH species is the charged $-COO^-$ species, whereas for amino functions the low pH species is the charged, $-NH_3^+$, group. Accordingly, the pKa values of lysine (Lys, K) residues will reverse direction and shift to lower values. Specifically, to examine the pKa shifts of the Lys(K) residue, 12 polypeptides were synthesized, each with a different value of f_K. The data in Figure 5.30B are for water. The highest pKa value is 9.6, and the value for f_K = 1 is 9.2 in Type I (low conductivity) water and 9.4 in the presence of normal saline. The relaxation of charge–charge repulsion appears to be less effective when the Cl^- anion is the shielding species rather than Na^+ cation. As the value of f_K approaches 0, the pKa decreases in pure water to 8.2 and to 8.6 in normal saline. Interestingly, the shielding by the Cl^- counter-ion appears to be more effective in the hydrophobic domain rather than in the charge–charge repulsion (electrostatic) domain.

We now focus on an attempt to use the electrostatic mechanism to calculate the large pKa shifts in the range for lower values of f_K (<0.75). The calculation uses the expression of solvation energy (SE) due to Born, SE = $[(ze)^2/2r][1 - 1/\varepsilon]$, where z is the charge on the ion, e is the unit electron charge, ε is the dielectric constant of the surrounding medium, and r is the radius of the ion. For the change in solvation energy, ΔSE, on going from ε_i to ε_{i-1} would be

$$\Delta SE = Q[\varepsilon_{i-1} - \varepsilon_i]/\varepsilon_{i-1}\varepsilon_i]$$

where Q is substituted for the coefficient $[(ze)^2/2r]$.

The four steps of the calculation are (1) the choice of dielectric constants for boundary conditions of f_K = 0.75 (at the highest pKa value) and 0, when there is no K present; (2) the change in dielectric constant for each replacement of a polar K residue by the more hydrophobic V residue; (3) use of an experimental value of ΔpKa to evaluate Q; and (4) calculation of the ΔpKa for each ionizable residue replaced by Val per 100 residues. The results for Lys(K) are given in Figure 5.30B and for Glu(E) in Figure 5.30A. At the temperature of the experiment, the dielectric constant of pure water is 75; this is the highest value possible. The dielectric constant for f_K = 0 has been experimentally determined to be 65 (See the 36°C data in Table 5.6).[89] The use of these values for the boundary conditions results in a maximal ΔpKa of only 0.3 pH units, whereas the true ΔpKa is five times larger. Only when a fictitious dielectric constant of 5 is used for the phase-separated state can the electrostatic calculations begin to approach the experimentally observed shifts.

One might argue that the lysine side chain is buried in a local dielectric constant of 5 even though the macroscopic dielectric constant is 65. This is not reasonable because the barrier to backbone mobility of poly(GVGIP) is only 1.1 kcal/mole.[89] A barrier many times greater would be required to hold a lysine side chain at a local dielectric constant of 5 for a viable electrostatic argument. In fact the magnitude of the pKa shifts can be as large as 8 to 10 kcal/mole for an Asp residue in a more hydrophobic T_t-type protein-based polymer.[100]

Conclusion: Neither charge–charge repulsion nor being buried in a medium of low dielectric constant provides a satisfactory explanation for the increasing pKa shift attending dilution of ionizable side chains in a moderately hydrophobic protein-based polymer.

5.7.8.3.2 Carboxylate-Containing Polytricosapeptides of Constant Composition but with Rearrangement of More Hydrophobic Residues: The Power of Sequence!

5.7.8.3.2.1 Effect of Different Pentamer Sequence Arrangements of the Same 30-mer Composition. Chemically synthesized polytricosapeptides, poly (30-mers), were prepared with compositions of 1 aspartic acid residue (Asp, D) and 5 more-hydrophobic phenylalanine (Phe, F) residues replacing valine (Val, V) residues per repeat of 30 residues, but with different relative locations of D and F residues. These compositions are written:

Polymer XI: Poly(GFGFP GVGVP G<u>D</u>G VP GFGFP GFGVP GVGVP)
Polymer V: Poly(GVGVP GVGFP G<u>D</u>G FP GVGVP GVGFP GFGFP)

The pKa for **Polymer V** was found to be 6.7, whereas that for **Polymer XI** under identical conditions was determined to be 10 (see Figure 5.32, below).[40] The normal pKa for aspartic in this polymer without any F residues having replaced V residues (see **Polymer I** in Table 5.5) is 3.8. The polymers were designed with the β-spiral structure as the basis. (This structure was developed in section 5.2.) Given the β-spiral structure as the design basis, **Polymer XI** had the more-hydrophobic F residues more distal from the D residue, whereas in **Polymer V** the more-hydrophobic F residues were more proximal to the D residue, as shown in Figure 5.32, below). Accordingly, without changing the mean composition of the polytricosapeptide, but by designing structures with different proximities to the ionizable function, the pKa shift changed from 6.7 − 3.8 = 2.9 to 10 − 3.8 = 6.2.[40] The Varied hydrophobic proximity of the functional groups, resulting from the primary structure, causes differences in hydrophobic-induced pKa shifts.

5.7.8.3.2.2 Comparison of a Fixed Sequence of Six Pentamers with a Random Mix of the Same Pentamers.[101] The importance of sequence control is also understood through two additional comparisons. The polytricosapeptide **Polymers X** and **XII** in Table 5.5, with specified sequences, exhibited pKa values of 8.1 and 7.8, respectively, whereas a random mixture of the same combination of pentamers, namely, poly[(GEGFP),2(GVGVP),2(GVGFP), (GFGFP)], gave a pKa of 5.2. Similarly, the specified sequence poly(GEGFP GVGVP GVGVP GVGVP GFGFP GFGFP) gave a pKa of 7.7, whereas the same pentamers in random mixture, namely, poly[(GEGFP, 3(GVGVP),2(GFGFP)] gave a pKa of 4.7 (see Table 1 in Urry et al.).[101] The pKa shift with specified sequence is four times larger than for random polymers of the same mean pentamer composition. In terms of energies, the average pKa shift of the polymers composed of random pentamers, 0.65, represents a repulsive free energy of less than 1 kcal/mole, whereas the average pKa shift of the fixed sequence protein-based polymers (3.6) represents a repulsive free energy of about 5 kcal/mole. For energy conversions required by living organisms, sequence matters. In part, this is why biology puts so much energy into synthesis of protein, as discussed in Chapter 4.

5.7.8.3.3 Increasing Hydrophobicity While the Composition of the Functional Group Remains Constant

In the polytricosapeptides of this study the aspartic acid residue in one series and the glutamic acid residue in a second series are kept constant at 1 per 30 residues, and the number of valine residues replaced by the more hydrophobic phenylalanine residues systematically increases from 0 to 2 to 3 to 4 and to 5, as follows:

Polymer	**I**:	**Poly**(GVGVP GVGVP G**D**GVP GVGVP GVGVP GVGVP)	D/0F
Polymer	**II**:	**Poly**(GVGVP GVGFP G**D**GFP GVGVP GVGVP GVGVP)	D/2F
Polymer	**III**:	**Poly**(GVGVP GVGVP G**D**GVP GVGVP GVGFP GFGFP)	D/3F
Polymer	**IV**:	**Poly**(GVGVP GVGFP G**D**GFP GVGVP GVGFP GVGFP)	D/4F
Polymer	**V**:	**Poly**(GVGVP GVGFP G**D**GFP GVGVP GVGFP GFGFP)	D/5F
Polymer	**VI**:	**Poly**(GVGVP GVGVP G**E**GVP GVGVP GVGVP GVGVP)	E/0F
Polymer	**VII**:	**Poly**(GVGVP GVGFP G**E**GFP GVGVP GVGVP GVGVP)	E/2F
Polymer	**VIII**:	**Poly**(GVGVP GVGVP G**E**GVP GVGVP GVGFP GFGFP)	E/3F
Polymer	**IX**:	**Poly**(GVGVP GVGFP G**E**GFP GVGVP GVGFP GVGFP)	E/4F
Polymer	**X**:	**Poly**(GVGVP GVGFP G**E**GFP GVGVP GVGFP GFGFP)	E/5F

The magnitude of the pKa shifts are given under the representations of the structures in Figure 5.32 (see below) with the reference value being 4.0 for aspartic acid and 4.4 for glutamic acid in the polymers without F residues.[40] Observed is a remarkably nonlinear pKa shift with increasing numbers of more hydrophobic Phe residues, which is analogous to that of the stretch-induced pKa shift discussed above and even similar to the form of the pKa shifts observed on replacing a V by E, D, or K, discussed immediately above.

Conclusion: There is a dramatic nonlinear pKa shift with increasing number of hydrophobic Phe residues and an associated increase in positive cooperativity that is quite analogous to the stretch-induced pKa shift (see Figure 5.23B) and with other means of systematically increasing hydrophobicity to achieve pKa shifts.

A similar series of polytricosapeptide model proteins containing the lysine residue has been prepared biosynthetically and characterized,[67,102,103] namely:

Model Protein i′:	(GVGVP GVGVP G**K**GVP GVGVP GVGVP GVGVP)$_{22}$(GVGVP)	K/0F
Model Protein ii′:	(GVGVP GVGFP G**K**GFP GVGVP GVGVP GVGVP)$_{22}$(GVGVP)	K/2F
Model Protein iii′:	(GVGVP GVGVP G**K**GVP GVGVP GVGFP GFGFP)$_{22}$(GVGVP)	K/3F
Model Protein iv′:	(GVGVP GVGFP G**K**GFP GVGVP GVGFP GVGFP)$_{21}$(GVGVP)	K/4F
Model Protein v′:	(GVGVP GVGFP G**K**GFP GVGVP GVGFP GFGFP)$_{21}$(GVGVP)	K/5F

Lys (K) residues exhibit similar nonlinear pKa shifts, in this case to lower pKa values, on increasing hydrophobicity by increasing the number of phenylalanine residues per 30-mer, but the magnitudes of the shifts are less than for glutamic and aspartic acids.[40,41]

The above profusion of studies demonstrate the breadth and depth of the foundation for the physical basis of the consilient mechanism of the inverse temperature transition. The studies discussed in section 5.7.8 preceded the microwave dielectric relaxation studies and were quite conclusive in their own right, but nothing is so convincing as identifying the hydrophobic hydration and directly observing the loss of hydrophobic hydration as the charged species form. They also point to more explicit descriptions and subtleties such as the basis for efficiency and cooperativity. We begin exploring these below.

5.7.9 Primary Source of pKa Shifts in Model Proteins i Through v: Competition for Hydration Between Hydrophobic and Charged Groups

In section 5.1.3.4, we derived expressions for the Gibbs free energy of hydrophobic association, ΔG_{HA}. In this section 5.7, hydrophobic-induced pKa shifts were shown to arise from a competition for hydration between the polar (ionized) and apolar (hydrophobic) groups. Accordingly, the pKa shifts provide an opportunity to express the Gibbs free energy represented by this competition for hydration, ΔG_{ap} and to compare it to the ΔG_{HA} determined under similar circumstances and to Monod's "second secret of life" embodied in allostery-based positive cooperativity.[104] In our case, the allostery involves the two conformational states, hydrophobically associated and hydrophobically dissociated.

5.7.9.1 Generally Considered Physical Bases for pKa Shifts in Polymers

The three generally considered sources for pKa shifts are (1) the electrostatic-based charge–charge repulsion, (2) the ionizable group being forced into an environment of low dielectric constant where it cannot ionize, and (3) the perspective developed by the systematic series of model proteins considered here of a competition for hydration between hydrophobic and charged groups constrained by sequence to coexist along a chain molecule such as a protein.

5.7.9.1.1 Electrostatic Charge–Charge Repulsion

As seen for the series poly[f_V(GVGIP), f_E(GEGIP)] in Figure 5.30, when $f_E = 1$, a small charge–charge repulsion occurs with four residues separating the ionizable carboxylate-containing side chain of the Glu (E) residues. The addition of 0.15N NaCl removes that repulsion. When $f_E = 0.17$, as in **Model Proteins i** through **v** of Table 5.5, with 29 residues separating charged residues, charge–charge repulsion is no longer a tenable explanation. This is verified by the observed cooperativity, which is negative for charge–charge repulsion, as shown in Figure 5.35 below for poly(methacrylic acid) with a carboxyl on every other backbone atom. On the other hand, **Model Proteins i** through **v** exhibit a positive cooperativity that increases with increased replacement of less hydrophobic Val (V) residues with more hydrophobic Phe (F) residues.

5.7.9.1.2 Local Environment of Low Dielectric Constant

The classic argument for pKa shifts in proteins is the claim that the energy of a folded state is so favorable that it can force the ionizable species into the energetically unfavorable circumstance of a local environment of low dielectric constant. This issue was considered in some detail and rejected by Urry et al.[98] in connection with the series poly[f_V(GVGIP), f_K(GKGIP)] for which the data are given in Figure 5.30.

5.7.9.1.3 Competition for Hydration Between Charged and Hydrophobic Groups

The conclusion that this is the mechanism for the hydrophobic-induced pKa shifts of the model proteins of consideration here was extensively developed earlier in this section 5.7. In what follows, this mechanism is explored further.

5.7.9.2 Derivation of ΔG_{ap}, the Change in Gibbs Free Energy Due to the Apolar–Polar Repulsive Free Energy of Hydration

Derivation of ΔG_{ap}, the change in Gibbs free energy due to an apolar–polar repulsive free energy of hydration, begins with the expression for chemical potential of acid (proton, H$^+$), μ_H = RTln(**a**), where **a** is the activity of proton. Under the normal low concentrations for acid, the activity coefficient is 1, and **a** can be replaced by the concentration of the acid, [H$^+$], to give μ_H = RTln[H$^+$]. Because acid–base titrations are plotted using a log scale, μ_H = 2.3RTlog[H$^+$] and as pH = $-$log[H$^+$], the chemical potential of proton becomes, μ_H = $-$2.3RTpH. The chemical potential for the value at which the pH = pKa is simply μ_H = $-$2.3RTpKa. Now the change in

5.7 What Is the Physical Basis for the Consilient Mechanism of Energy Conversion?

chemical potential, $\Delta\mu$, resulting from the hydrophobic-induced pKa shift can be written as $\Delta\mu_{2,1} = \mu_2 - \mu_1 = -2.3RT(pKa_2 - pKa_1) = -2.3RT\Delta pKa$. Finally, we note that the ΔpKa arises from the competition for hydration between charged (polar) and hydrophobic (apolar) groups, that is, competition resulting in repulsion arises on formation of the carboxylate, $-COO^-$, which parallels formation of hydroxyl ion. As we would like to express the change in Gibbs free energy per mole resulting from the apolar–polar repulsion as each species reaches out for water unperturbed by the other, this has the effect of sign reversal, such that we can write $\Delta\mu_H = -\Delta\mu_{OH} = -\Delta G_{ap}$, that is,

$$\Delta G_{ap} = 2.3RT\, \Delta pKa \qquad (5.13)$$

where the subscript ap stands for apolar–polar repulsion. It will be of interest to see how this quantity relates to ΔG_{HA} and to positive cooperativity and charge–charge repulsion in the treatment of acid–base titration data that consider both interactions in section 5.8. Before doing so, however, we note another experimental demonstration of ΔG_{ap}, but in this case there occurs delineation between the component that results from sequence without the capacity to form a conformation to relieve apolar–polar repulsion and that component resulting from lowering repulsion by hydrophobic association and charge neutralization.

5.7.9.3 Hill Plots Delineate ΔG_{ap} Due Solely to Sequence and That Associated with Allosteric Effects (Due to Formation of a Hydrophobically Associated State)

Analysis of the experimental data in Figure 5.31 allows fundamental delineation between hydrophobic-induced pKa shifts in the absence of the ordering due to hydrophobic association and hydrophobic-induced pKa shifts associated with positive cooperativity and the presence of a more-structured hydrophobically associated state. The analysis is consistent throughout the series of **Model Protein I** through **v** in Table 5.5. Sufficient for our analysis here is consideration of the acid–base titrations of two model proteins, namely,

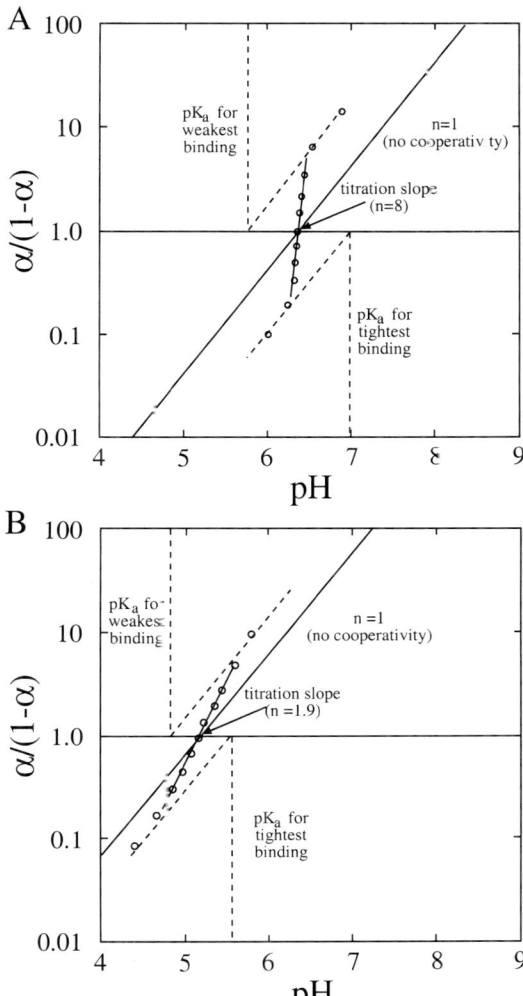

FIGURE 5.31. Hill plots, $\log[\alpha(1-\alpha)]$ vs. pH, for **Model Proteins iii** and **v**. The normal pK for E(COOH) is ~4. **Model Proteins iii** and **v** exhibit their first COOH pKa values at 5.53 and 7.0, respectively, and their last COOH pKa values at 4.84 and 5.7 for the 39th and 42nd COOH pKa values, respectively. The pKa values of completely unfolded model proteins exhibit residual shifts due to competition for hydration between oil-like and vinegar-like groups. Repulsions of 1.2 (=0.84 × 2.3RT) and 2.4 (=1.7 × 2.3RT) kcal/mole carboxyl remain in the completely hydrophobically unfolded model protein.

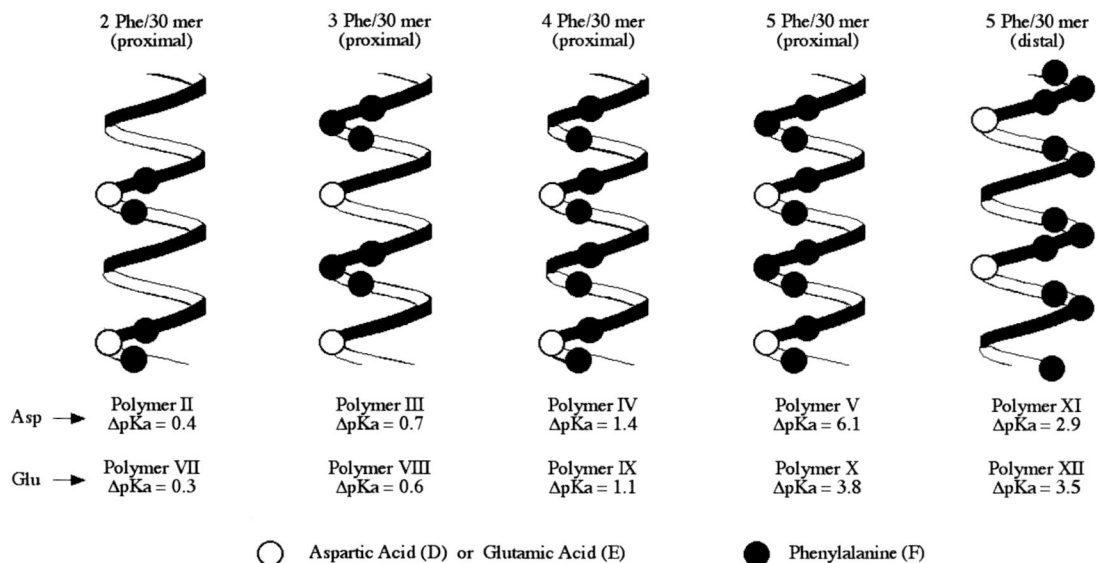

FIGURE 5.32. As indicated below the schematic β-spiral representations of the structures, stepwise increases in hydrophobicity as Phe replaces Val result in supralinear increases in carboxyl pKa values in studies of chemically synthesized polymers in analogy to the hydrophobic-induced pKa shifts in Figures 5.29, 5.30, and 5.31. (Reproduced with permission from Urry et al.[40]).

Model Protein iii: (GVGVP GVGVP G<u>E</u>GVP GVGVP GVGFP GFGFP)$_{39}$(GVGVP)   E/3F
Model Protein v: (GVGVP GVGFP G<u>E</u>GFP GVGVP GVGFP GFGFP)$_{32}$(GVGVP)   E/5F

The Hill plots in Figure 5.31 provide remarkable insight into the hydrophobic-induced pKa shifts. Starting in the hydrophobically associated state for **Model Protein v**, on decreasing acid concentration, that is, raising the pH, the pKa of the first carboxyl to form carboxylate is 7.0. In Figure 5.31A, this is the pKa for the most tightly bound proton. As more carboxylates form, further ionization becomes easier, and finally the last carboxyl to ionize does so with a pKa of 5.7. Remarkably, the last carboxyl to ionize does so with a pKa shifted 1.7 pH units from that of the unperturbed Glu pKa of about 4. Even in the completely unfolded state there remains an apolar–polar repulsion of 2.4 (= 1.7 × 2.3RT) kcal/mole-Glu.[105]

Those who cannot abandon the concept of the carboxyls being bound within a local environment of low dielectric constant may still wish to argue that the carboxyl is held within a low dielectric constant provided by the most proximal Phe (F) side chains. Thus it becomes of interest to remove the two most sequence proximal Phe (F) residues. This is achieved with **Model Protein iii**, and the experimental results are shown in Figure 5.31B. Although the magnitude of the pKa shift is reduced, the conundrum remains. The last carboxyl to ionize does so with a pKa shifted 0.84 pH units from that of the unperturbed Glu pKa of about 4. Here there remains a repulsion of 1.2 (= 0.84 × 2.3RT) kcal/mole-Glu even when the more hydrophobic Phe (F) residues are at a distance of some 11 residues along the sequence. It seems inescapable that apolar–polar repulsion, the effort of the charged and hydrophobic residues to reach water unperturbed by the other, reaches out several nanometers. More detailed scenarios follow.

5.7.9.4 Scenario for pKa Shifts Coupled with Positive Cooperativity: Starting from the Hydrophobically Associated State

During an opening fluctuation of a pair of associated hydrophobic domains, the formation of too much hydrophobic hydration, at a given temperature above T_t, causes the domains immediately to reassociate. Should an ionizable group, say, a carboxyl (–COOH), within the dissociating hydrophobic domain ionize to form the carboxylate (–COO$^-$), it must obtain its required hydration by destructuring nascent hydrophobic hydration. The result is an increase in free energy of the charged hydrophobic domain. Now there are two possibilities. Either there is sufficient proton in solution that the carboxylate recovers its proton, or a second carboxylate forms and cooperates with the first in destroying hydrophobic hydration such that the free energy of each of the two carboxylates is less than that of the lone carboxylate. Now formation of each new carboxylate does so with a lower free energy. Accordingly, as shown in Figure 5.30, the first carboxylate forms with a higher pKa, and each subsequent carboxyl ionizes with a lower pKa. The last carboxyl to form carboxylate completes the positive cooperativity for the model protein.

5.7.9.5 Basis for the pKa Shift Remaining After Complete Hydrophobic Dissociation

When considering the residual pKa shifts for the model protein series above, we need to be clear that even **Model Protein i** with no phenylalanine (Phe, F) residues exhibits a mean pKa shift to 4.5 so that the residual shift due solely to the three Phe (F) residues of **Model Protein iii** is only 0.3 units or 0.43 kcal/mole-Glu. On the other hand, the total ΔG_{ap} for **Model Protein v** with a ΔpKa of 2.87 would be 4.1 kcal/mole-Glu when at a temperature sufficiently above the T_t-value. Another relevant point is that in Figure 5.25B a residual hydrophobic hydration remains for **Model Proteins i** and **ii** with less than two carboxylates per 100 residues, which residual hydrophobic hydration remains as ionization continues to 3.4 carboxylates per 100 residues. The understanding is that the hydrophobic residues tend to arrange on the backside of the model protein from the carboxylate groups and are thereby shielded by the polymer chain from the otherwise overpowering carboxylates. Thus derives the image of an apolar–polar repulsion. It should be emphasized that the carboxylates, even though immersed in water, do not achieve full hydration when their pKa is greater than 4.

Accordingly, there results a residual pKa shift, after complete hydrophobic dissociation, as the result of the sequence-imposed proximity of the charged and hydrophobic side chains.

5.8 Integration of Cooperativities Due to Apolar–Polar and Charge–Charge Repulsion into Acid–Base Titration Theory

5.8.1 Hydrophobically Associated and Hydrophobically Dissociated: Two States for Diverse Allostery

Over the last century, one of the more enigmatic, yet fundamental properties exhibited by protein-based machines of biology has been positive cooperativity. So impressed was Monod by this phenomena that he is reported to have considered positive cooperativity "the second secret of life," with the first secret being the structure of DNA.[104] The initial example goes back to the first decade of the twentieth century, the binding of oxygen by hemoglobin.[97,106] Many remarkable researchers have made exceptional contributions over many years—Wyman,[96] Adair,[107] of course Perutz,[104] and then in the 1960s Monod and coworkers,[108,109] as well as Koshland, Nemethy, and Filmer.[110]

By means of mathematical formalism steeped in symmetry of multisubunit globular proteins, Monod and coworkers described the cooperative binding of oxygen by hemoglobin, which effort warranted Nobel recognition.[111] In the general perspective, each of the identical or near identical globular subunits could exist in two different states of order, hence Monod's use of the term *allostery*. The physical basis for the process was not specifically addressed, however, other than to credit a "conformational change." In fact, the place wherein we have found a mechanism had been discounted, with the perspective that "One may set aside the simple problem of fibrous proteins. Being used as scaffolding, shrouds and halyards, they fulfill these requirements by adopting relatively simple types of translational symmetries."[111] Thus, what we describe in this volume, using model proteins of inherent translational symmetry, had not been anticipated.

Another element, taught by the consilient mechanism that stands out as not having been anticipated prior to the current work but constitutes the scientific core of this volume, is contained in the words of Gregorio Weber in one of the currently outstanding treatments of protein interactions, "A complete description of the energetics of hemoglobin, or any other oligomeric protein, is well-nigh impossible. It would involve not only the determination of the energetic couplings of any number of ligands with each other and with the subunit interactions but also the variations of these quantities with pH, temperature and pressure."[112] It is here that the consilient mechanism, a name for chronicling the comprehensive hydrophobic effect, has something to contribute. Not only the variables of chemical potential (e.g., ligand binding and pH), temperature, and pressure, but also the variables of applied potential, mechanical force, and electromagnetic frequencies all sum (+ and −) to give the resultant change in Gibbs free energy for hydrophobic association/dissociation, $\Sigma_i \Delta G^i_{HA}$, as depicted in Figure 5.33. This constitutes a statement for a ΔG_{HA} additivity principle.

Here, by introducing ΔG_{HA}, which by comparing ionized and neutral states, can be equivalent to $\Delta G_{ap}(=2.3\ RT\ \Delta pKa)$ in the absence of charge–charge repulsion, we bring positive cooperativity into the formalism for acid–base titrations. As applied to our model proteins, this represents a problem in allostery quite equivalent to that of hemoglobin. In each repeating unit of hemoglobin there are two states—oxygen bound and oxygen free. For our model proteins with an ionizable functional group in each repeat, there are ionized and nonionized states. In the case of hemoglobin, the principle variable is the fraction of ligand bound Y, and it enters in the form Y/(1 − Y). This is equivalent in the acid–base titration theory to the degree of ionization, α, and it enters as $\alpha/(1 - \alpha)$. For another example, for model proteins containing redox couples, there is the degree of reduction, α', and it enters into the formalism as $\alpha'/(1 - \alpha')$. Thus, the following formalism is relevant, independent of whether one treats ligand binding to allosteric proteins, the ionization process in an acid–base titration curve, or the reduction process in a potentiometric titration. In our view of positive cooperativity, whether in the model protein-based machines of our design or in the protein machines of biology, in all cases two states—the hydrophobically associated state and the hydrophobically dissociated state—provide the common allosteric denominator with the fundamental fraction being the degree of dissociated and associated states.

5.8.2 The Henderson-Hasselbalch Equation for Dilute Weak Acids

For an isolated polypeptide containing one ionizable residue such as an aspartic or glutamic acid residue without significant hydrophobicity in the remaining residues, the Henderson-Hasselbalch equation applies. The derivation begins with the statement of an equilibrium constant, K_o, that describes the relationship between two states, in our case –COOH and –COO⁻.

$$K_o = [COO^-][H^+]/[COOH] = e^{-\Delta G_o/RT} = 10^{-\Delta G_o/2.3RT} \quad (5.14)$$

where ΔG_o is the change in Gibbs free energy on going from reactants and products. By taking the logarithm to the base 10 and introducing the definitions that pH = −log[H⁺] and pK_o = −logK_o, Equation (5.14) becomes

$$pH = pK_o + \log\{[COO^-]/[COOH]\} \quad (5.15)$$

With the definitions α = [COO⁻]/{[COOH] + [COO⁻]} for the fraction of ionized species and $(1 - \alpha)$ = [COOH]/{[COOH] + [COO⁻]} for the fraction of nonionized species, substitution into Equation (5.15) results in the well-known Henderson-Hasselbalch equation, that is,

$$pH = pK_o + \log[\alpha/(1 - \alpha)] \quad (5.16)$$

5.8 Integration of Cooperativities Due to Apolar–Polar and Charge–Charge Repulsion

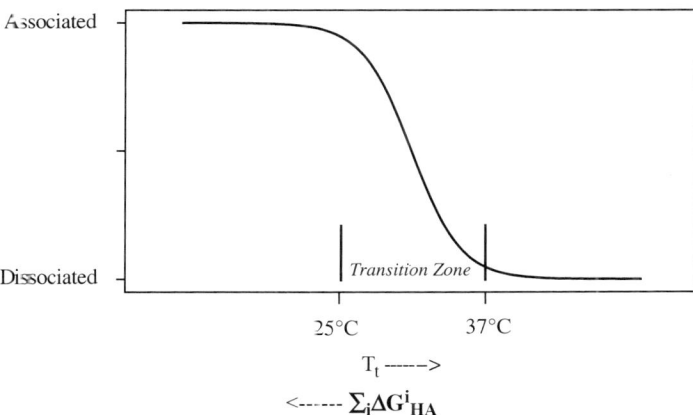

FIGURE 5.33. Representation of lowering the onset temperature, $T_{t(b)}$, of the inverse temperature (phase) transition for hydrophobic association from the dissociated state at 37°C to 25°C, that is, to a temperature just sufficient to achieve essentially complete hydrophobic association. Hydrophobic association results from the summation of all of the variables (+ and –) that contribute to ΔG_{HA}, i.e., $\Sigma_i \Delta G^i_{HA}$. This summation represents the ΔG_{HA} additivity principle.

5.8.3 Comparison of the Henderson-Hasselbalch Equation to Titration Curves of Polyelectrolytes and Model Proteins

Experience shows marked deviation from the Henderson-Hasselbalch equation. There are examples of acid–base titration curves with curves broader than predicted by the Henderson-Hasselbalch equation, an effect called *negative cooperativity*, and there are examples of acid–base titration curves with curves sharper than predicted by the Henderson-Hasselbalch equation, an effect called *positive cooperativity*. These deviations from Equation (5.16) can be demonstrated by cooperative interactions involving the charged functional group.

5.8.3.1 Negative Cooperativity of Poly(methacrylic acid) and Charge–Charge Repulsion

When the Gibbs free energy increases as each subsequent charged forms, negative cooperativity occurs as shown for polymethacrylic acid in Figure 5.35A (see page 209).[113] With polymethacrylic acid there occurs a carboxyl function on every other backbone atom. As carboxylates form to the extent of greater than 50% ionized, charge–charge repulsion becomes very significant. Even in water with its high bulk dielectric constant, there is insufficient space between negative charges to pack in water molecules with their large dipole moment to shield between charges. The broadened curve of Figure 5.35A results, where the experimental data are compared with the prediction in Equation (5.16), given as the solid line for the Henderson-Hasselbalch equation. As discussed in the next section, the chemical energy required to go from the COOH state to the COO⁻ state is proportional to the width of the curve in pH units and also to the shift in pK.

5.8.3.2 Positive Cooperativity of Model Proteins and Apolar–Polar Repulsion

Positive cooperativity results when a repulsive Gibbs free energy exists on formation of the first COO⁻, a repulsion that the presence of the first COO⁻ relieves to some extent for the formation of subsequent carboxylates. Then the second and subsequent carboxylates form with a steadily decreasing repulsive free energy. The result is a steepened and narrower acid–base titration curve. The titration curves of two model proteins in Figure 5.35B (page 209) are much steeper and narrower than the curve of Equation (5.16). Importantly, the curves become progressively steeper as the hydrophobicity of the model protein becomes greater. The result, in fact, is that less of a change in chemical energy is required to go from the COOH state to the COO⁻ state for the model proteins than for the idealized dilute weak acid case of the Henderson-Hasselbalch derivation.[114] Furthermore, the amount of chemical energy required decreases as more phenylalanine (Phe, F) residues, with the more-hydrophobic **R**-group $-CH_2-C_5H_6$, replace less-hydrophobic valine (Val, V) residues, with the **R**-group $-CH_2(CH_3)_2$.

The series of **Model Proteins i, ii, iii, iv**, and **v** in Table 5.5 with 0, 2, 3, 4, and 5 F residues per 30-mer exhibits a systematic nonlinear increase in steepness, that is, in positive cooperativity, and an associated nonlinear increased pKa shift, as plotted in Figure 5.34. The energy required to convert from the COOH state to the COO⁻ state systematically in a supralinear way becomes less and less, as more Phe residues replace Val residues. The energy required to convert from the hydrophobically dissociated state of COO⁻ to the hydrophobically associated (contracted) state of COOH becomes less, as the model protein becomes more hydrophobic. The elastic-contractile protein-based machine becomes more efficient as it becomes more hydrophobic. The cooperativity of **Model Protein iv** with a Hill coefficient of 2.6 is similar to that of hemoglobin with a Hill coefficient of 2.8. Remarkably, the Hill coefficient of 8 exhibited by **Model Protein v** exceeds that of most known allosteric proteins.

The formalism for introducing negative and positive cooperativity into the Henderson-Hasselbalch equation, Equation (5.16), follows.

5.8.4 Introduction of Cooperativity into the Henderson-Hasselbalch Equation for Polymers of Linear Repeats

The formalisms for describing the electrostatic interactions between charged groups follow those of Harris and Rice,[115] Overbeek,[116] and Katchalsky,[113] and the repulsive apolar–polar interactions are introduced in an analogous way, but with a distinguishing feature between the two. Interestingly, the pKa shifts for charge–charge repulsion and apolar–polar

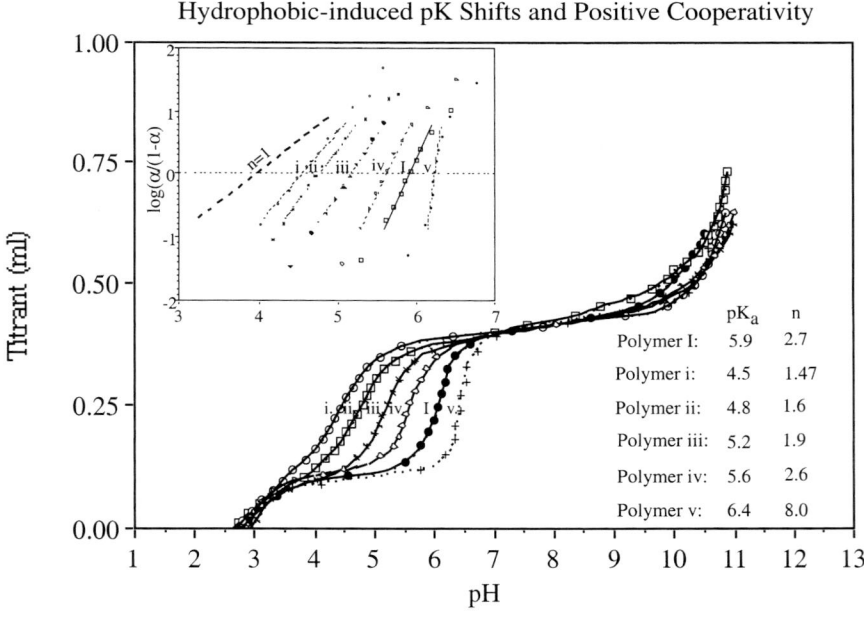

FIGURE 5.34. Acid–base titration curves of the series of elastic **Model Proteins I** and **i** through **v** of Table 5.5 that exhibit systematic increases in hydrophobic-induced pK shifts and positive cooperativity resulting from competition for hydration between apolar and polar groups. (Inset) Slope of the Henderson-Hasselbalch equation with n = 1, and the slopes for the Hill coefficients, n, of the positively cooperative **Model proteins i. ii, iii, iv, I**, and **v**, respectively. It may be noted for comparison that the negatively cooperative poly(methacrylic acid) (PMA) has a slope, n, of 0.5. The respective pKa and n values for the model proteins are listed within the figure. See text for further discussion.

5.8 Integration of Cooperativities Due to Apolar–Polar and Charge–Charge Repulsion

repulsion are additive. The cooperativities, on the other hand, are of opposite sign and can cancel. Thus, when the free energy of interaction represented by the pK shift and the free energy of interaction represented by the change in cooperativity are the same, as is the case in Figure 5.29, the apolar–polar repulsive free energy for hydration dominates, that is, there is no significant charge–charge repulsion.

5.8.4.1 Multiple Acid–base Equilibria in Protein-based Polymers with Widely Sequence Displaced Functional Groups

For the set **Model Proteins i** through **v** in Table 5.5, one glutamic acid residue (Glu, E) repeats every 30 residues. This means that there are 95 backbone and side chain atoms between the ionizable functions, and for **Model Protein i** there are 36 such repeats. There then exist 36 equilibrium constants, K_1 through K_{36}, one for the ionization of each carboxyl function. An overall equilibrium constant can be written, K^{36}, which is the product of all 36. This can be written in general as

$$K^n = K_1 K_2 \ldots K_n = [(COO^-)_n][H^+]^n / [(COOH)_n] \quad (5.17)$$

Taking the logarithm to the base 10,

$$n \log K = n \log[H^+] + \log\{[(COO^-)_n]/[(COOH)_n]\} \quad (5.18)$$

and again defining $pH = -\log[H^+]$, $pK = -\log K$, dividing by n, and rearranging gives

$$pH = pK + (1/n)\log\{[(COO^-)_n]/[(COOH)_n]\} \quad (5.19)$$

5.8.4.2 Introduction of Effect of Substantial Hydrophobic Hydration

As shown in Figure 5.34, in the interactions among carboxylates while under the influence of increasing numbers of hydrophobic residues, $pK \neq pK_o$, but rather $pK = pK_o + \Delta pK$, that is, the pKa shifts of section 5.7.8. Furthermore, it is assumed that when one chain starts to ionize, due to the positive cooperativity, it completely ionizes before the next chain starts such that, to an adequate approximation, the experimentally measurable degree of ionization, $\alpha \approx$ $[(COO^-)_n]/[(COOH)_n]$ and n is replaced by a phenomenological n to correspond with this assumption. Equation (5.19) can now be restated as,

$$pH = pK_o + \Delta pK + (1/n)\log[\alpha/(1-\alpha)] \quad (5.20)$$

Equation (5.20) retains the form of the Henderson-Hasselbalch equation, but it has been generalized to introduce pK shifts and cooperativity into the expression. This phenomenological introduction of cooperativity is analogous to the treatments describing the oxygenation of hemoglobin. The quantity n is equivalent to a Hill coefficient.[97] Furthermore, in point 4 of section 5.7.8.2, we noted for pK shifts of about 1 that the Wyman equation,[96] $\Delta\Delta G_o = RT(1 - 1/n)/\alpha(1 - \alpha)$, when evaluated at the pK at $\alpha = 0.5$ and the relevant Hill coefficients (n), gave the same value for the energy of interaction as that calculated for the pK shift. In particular, by Equation (5.13), $\Delta G_{ap} = 2.3RT\Delta pK$, and when evaluating at $\alpha = 0.5$, $\Delta\Delta G_o = 4RT(1 - 1/n)$. Thus, as occurred in the data in Figure 5.29 where $\Delta G_{ap} \approx \Delta\Delta G_o$, the relationship between ΔpK and Hill coefficient, n, becomes $\Delta pK = 1.74(1 - 1/n)$. This suggests that positive cooperativity and hydrophobic-induced pK shifts can be different representations of the same interaction energy.

Larger pK shifts, however, have been observed than are calculable by the Wyman equation. For example, a pK shift of 6 has been observed for aspartic acid in a very hydrophobic polymer (see Figure 5.32), whereas from the Wyman equation, $\Delta pK = 1.74(1 - 1/n)$, the ΔpK could not exceed 1.74. Also with a reference pK value of 4.0 for the Glu carboxyl, a pK shift of 2.4 is seen in Figure 5.34 with a corresponding Hill coefficient of 8. For larger Hill coefficients and pK shifts, the limitation of the correlation is likely due to limiting approximations of the Wyman equation that have been overcome, for example, by the Monod treatment.

Additionally, as shown in Figure 5.31A, pK shifts separate into an apolar–polar repulsive component that can be relaxed by a change in hydrophobic association and an apolar–polar repulsive component that is dictated by sequence and thereby retained even in the most unfolded and disassembled state. Expectedly, the pKa shift, relaxed by a disruption of all hydrophobic association, is the component that results in positive cooperativity and in Hill coefficients of greater than 1.

To continue with the derivation, because $K = e^{-\Delta G/RT} = 10^{-\Delta G/2.3RT}$, $\log K_o = -\Delta G_o/2.3RT = -pK_o$, and Equation (5.20) can be rewritten in terms of changes in Gibbs free energy, ΔG, to give

$$pH = \Delta G_o/2.3RT + \Delta\Delta G_o/2.3RT \\ + (1/n)\log[\alpha/(1-\alpha)] \quad (5.21)$$

Derivation of Equation (5.21) considers hydrophobic-induced pK shifts and hydrophobic-induced increases in positive cooperativity. In Figure 5.30, however, there is experimental delineation between the hydrophobic domain where the above considerations dominate and the electrostatic domain where charge–charge repulsion dominates with a negative cooperativity. Thus we require an expression that would properly include both effects.

5.8.5 Toward a Generalized Acid–Base Titration Expression

5.8.5.1 Introduction of the Harris and Rice Treatment of Cooperativity

Following Harris and Rice,[115] the last term of Equation (5.8) can be effectively restated as $\log[\alpha/(1-\alpha)] + (\partial\Delta G/\partial\alpha)_T/2.3RT$, where $(\partial\Delta G/\partial\alpha)_T = 0$ at $\alpha = 0.5$, to give

$$pH = pK_o + \Delta pK + \log[\alpha/(1-\alpha)] \\ + (\partial\Delta G/\partial\alpha)_T/2.3RT \quad (5.22)$$

that is, the last term of Equation (5.22) replaces the Hill coefficient with a term that provides for the change in Gibbs free energy as the degree of ionization changes, $(\partial\Delta G/\partial\alpha)_T$, which changes the steepness of the curve from that given by the simple Henderson-Hasselbalch equation.

5.8.5.2 Introduction of the Effects of Both the Charge–Charge Repulsion, ΔG_{cc}, and Apolar–Polar Repulsive Free Energy of Hydration, ΔG_{ap}

As discussed above, there are two primary mechanisms in aqueous dominated media for pK shifts and changes in the steepness of the experimental titration curve. These are referred to as *electrostatic-induced*, principally charge–charge repulsion (cc) and as *hydrophobic-induced*, that is, the apolar–polar repulsive free energy of hydration (ap). The general expression can now be written specifically to include these two primary interactions as

$$pH = pK_o + \Delta pK_{cc} + \Delta pK_{ap} + \\ \log[\alpha/(1-\alpha)] + \{[(\partial\Delta G/\partial\alpha)_T]_{cc} \\ + [(\partial\Delta G/\partial\alpha)_T]_{ap}\}/2.3RT \quad (5.23)$$

Although the ΔpK_{cc} and ΔpK_{ap} values are of the same sign and add, the partial derivatives are of the opposite sign and counter each other. In particular, $[(\partial\Delta G/\partial\alpha)_T]_{c-c}$ is the result of a negative cooperativity with a curve that is broader than given by the Henderson-Hasselbalch equation, and $[(\partial\Delta G/\partial\alpha)_T]_{a-p}$ is the result of a positive cooperativity with curve that is steeper than given by the Henderson-Hasselbalch equation.

5.8.5.3 Experimental Delineation of ΔG_{cc} and ΔG_{ap} in Series of Model Proteins

The experimental titration data for the series of **Model Proteins I, i, ii, iii, iv**, and **v** are listed in Table 5.5 and shown in Figure 5.34, where the differential effects of charge–charge and apolar–polar repulsion become apparent. For **Model Proteins i, ii, iii, iv**, and **v**, for each systematic shift in pKa there occurs a corresponding increase in Hill coefficient. **Model Protein I**, however, does not fall within the series, because it exhibits a pKa shift without an increase in magnitude of the Hill coefficient. **Model Protein I** and **iv** exhibit essentially the same Hill coefficient of 2.7, whereas the pKa values differ by 0.3 pH units.

On inspection of the sequence of **Model Protein I**, it is more hydrophobic with five F plus two I residues having replaced seven less hydrophobic V residues in the repeating 30 residue sequence. If this were the only determinant it would be expected to have a greater pKa and a larger Hill coefficient than **Model Protein v**. **Model Protein I**, however, has two proximal E residues with carboxylates separated by only two residues, that is, –EPGE–. From the data in

5.8 Integration of Cooperativities Due to Apolar–Polar and Charge–Charge Repulsion

Figure 5.30, in poly(GEGIP) the Glu residues are separated by four residues, –EGIPGE–, and the pKa shift due to charge–charge repulsion is ≥0.35 pH units. The charge–charge repulsion component of the pKa shift for –EPGE– is expected to be greater than for –EGIPGE–. Accordingly, in the titration curve of **Model Protein I**, when compared with those of **Model Proteins i, ii, iii, iv**, and **v**, the differential effects of charge–charge and apolar–polar repulsion become apparent, and an iterative fitting process using Equation (5.23) should allow determination of ΔG_{cc} and ΔG_{ap} in this mixed case. In doing so the recognition of the residual pKa shift after complete unfolding, shown in Figure 5.31, should also be delineated.

5.8.5.4 Positive Cooperativity as a Fundamental Property of the Competition for Hydration Between Apolar (Hydrophobic) and Polar (e.g., Charged) Species

From the analysis of the acid–base titration data in Figures 5.30 through 5.34, positive cooperativity results from the apolar–polar repulsive free energy of hydration, that is, from the competition for hydration between apolar (hydrophobic) and polar (e.g., charged) species. The general statement can be that the appearance on the scene of the first polar, for example, charged, species must do the work of destructuring hydrophobic hydration in order to achieve adequate hydration for itself.

To put this into perspective, consider again the Gibbs free energy for solubility, ΔG(solubility) = $\Delta H - T\Delta S$, and recall the discussion of Butler's findings (see section 5.1.3.3), where insolubility results from the formation of too much hydrophobic hydration. Too much hydrophobic hydration causes the positive $(-T\Delta S)$ term to become larger than the negative (exothermic) ΔH term, that is, when ΔG(solubility) becomes positive and solubility is lost. The insolubility comes in the form of the association of a pair of hydrophobic domains. However, the pair of domains undergoes dissociation and association fluctuations. Now, a polar species can emerge (e.g., COOH → COO$^-$) proximal to an opening fluctuation, if it can achieve adequate hydration by destructuring the hydrophobic hydration. In doing so it lowers the magnitude of the positive $(-T\Delta S)$ term sufficiently to allow the dissociation to stand. Even though there may yet be substantial hydrophobic hydration covering the surfaces of the previously paired hydrophobic domains, there is additional water, and the emergence of the second polar species achieves its hydration without having to destructure quite so much hydrophobic hydration, and its equilibrium constant is more favorable. This is the description of positive cooperativity at the molecular level arising from an apolar–polar repulsive free energy of hydration.

To the best of my knowledge, the hemoglobin oxygenation curve is historically the first example of a biologically essential positive cooperativity. Because of this, it becomes an important objective to explore the phenomenology of hemoglobin's positive cooperativity and compare it with that of the consilient mechanism due to an apolar–polar repulsive free energy of hydration (as is done in Chapter 7) and, in fact, to do so for a number of protein-based machines that exhibit positive cooperativity.

5.8.5.5 Evaluation of $[(\partial \Delta G/\partial \alpha)_T]_{ap}$ Within the Single Polymer, **Model Protein v**

From the plot in Figure 5.31A of the titration of **Model Protein v**, the pKa of the first carboxyls to ionize may be taken as 7.0, whereas that of the last carboxyls to ionize is about 5.7. Therefore, in the process of driving hydrophobic disassociation of a relaxation or, inversely, in the process of driving hydrophobic association of a contraction, the pKa shifts by some 1.3 pH units. By Equation (5.13), ΔG_{ap} = 2.3 RT ΔpKa ≈ 2.3 × 1.987 × 310 × 1.3 = 1.8 kcal/mole-carboxyl. Now, because $\Delta G_{ap} = \int [(\partial \Delta G/\partial \alpha)_T]_{ap} d\alpha$, we have that $\int [(\partial \Delta G/\partial \alpha)_T]_{ap} d\alpha ≈ 1.8$ kcal/mole-carboxyl. This is an example wherein only the apolar–polar repulsive free energy of hydration is relevant, as the presence of a Glu (E) residue every 30 residues does not give rise to charge–charge repulsion in the unfolded state. Also, when totally unfolded the carboxyl pKa is still shifted from the unperturbed pH value of near 4.0. Even with the model protein unfolded, there still exists an apolar–polar repulsive free energy ΔG_{ap} of 2.3 RT ΔpKa = 2.3 × 1.987 × 310(5.7 – 4.0) = 2.4 kcal/mole-carboxyl. Even when disassembled and unfolded, the sequence of **Model Protein v** does not allow the carboxylate complete access to water unencumbered by hydrophobic groups. Furthermore, the maximum pKa shift experienced by the carboxyls in this hydrophobic model protein is 7.0 – 4.0 = 3.0, giving a ΔG_{ap} of 4.2 kcal/mole-carboxyl (see Figure 5.31A).

For **Model Protein I**, there occur 2 glutamic acid residues (Glu, E) per 30 residues, and importantly these are separated by only two residues. In this case there exist significant values for both ΔG_{ap} and ΔG_{cc}. Recall that the free energies add for the pKa shift but are of the opposite sign for cooperativity. This being the case, it is possible to write two equations with the two unknowns, ΔG_{ap} and ΔG_{cc}, and to solve the values. In this case, approximate values for ΔG_{ap} of 2.0 kcal/mole-carboxylate and ΔG_{cc} of 0.7 kcal/mole-carboxylate pair can be estimated. Thus, when laid out in the form of Equation (5.23), it becomes possible to begin separation of different contributions to the Gibbs free energy of interaction.

5.8.5.6 Relationship of the Consilient Mechanism to Cold Denaturation and to Hydrophobic Dissociation on Being Made More Polar

Hydrophobic association on raising the temperature is the most fundamental aspect of the consilient mechanism, arising as it does from the inverse temperature transition. An equivalent statement would be that hydrophobic dissociation on lowering the temperature is fundamental to the consilient mechanism. Historically, this has been called *cold denaturation of enzymes*.[16] In our view, those protein systems that associate on heating to physiological temperatures in order to achieve a functional state should be considered in terms of the consilient mechanism.

In fact, any protein function that involves cycling between hydrophobically associated and dissociated states by whatever energy input, for example, whether chemical or electrochemical, should be considered in terms of the consilient mechanism of the apolar–polar repulsive free energy of hydration.

5.8.6 Toward a Generalized Theory for Achieving Function of Protein-based Machines Founded on Hydrophobic Association

Equation (5.23) provides a general expression for fitting the family of curves schematically shown in Figure 5.19B and experimentally demonstrated in Figure 5.20A–C. The relevant energy conversions can be considered as electro-chemical (**A**), chemo-mechanical (**B**), and electro-mechanical (**C**). In each case, positive cooperativity is the result of increasing hydrophobicity, and it would appear to apply to the many energy conversions described in section 5.6. Thus, it would seem necessary only to cloak the expression in the correct energy terms, recognizing as in Equation (5.21) that pH, for example, is a free energy divided by 2.3 RT. It would seem, whatever set of energies drive function, that a complete description could be achieved by properly including the required terms in such an equation.

Hopefully, the above proposed integration of energy conversion, involving the recognition of a commonality of protein function that includes an apolar–polar repulsive free energy of hydration, might bring us one step closer to the daunting challenge recognized by Weber[112] of achieving a more "complete description of the energetics" of protein-based machines.

5.9 On the Efficiency of Consilient Protein-based Machines

5.9.1 Introductory Remarks

5.9.1.1 The Central Role of the Gibbs Free Energy for Hydrophobic Association in Diverse Energy Conversions

The change in Gibbs free energy for hydrophobic association, ΔG_{HA}, affects and is affected by the forms of energy utilized by living organ-

5.9 On the Efficiency of Consilient Protein-based Machines

isms. It originates from inverse temperature transitions; it establishes the comprehensive hydrophobic effect, and, because of the pervasiveness of its role in catalyzing energy conversion, the comprehensive hydrophobic effect provides the basis for the consilient mechanism. From our developing perspective and as demonstrated in part below, it appears that ΔG_{HA} provides the foundation for the most efficient mechanism of diverse energy conversion available in aqueous systems.

5.9.1.2 The Design of Poised Elastic-Contractile Protein-based Machines

The design of efficient protein-based machines requires judicious choice of model protein composition based on the intended energy source and operating conditions of temperature, medium, and so on. Proper design poises the system immediately to initiate function on addition of the driving force, the input energy. Of course, in recognizing this, protein function by modulating hydrophobic association simply depends in a complementary way on the many variables of the Weber concern,[112] for example, ligand binding, subunit association, pH, temperature, and pressure, as well as phosphorylation–dephosphorylation, oxidation–reduction, absorption of light, and so forth.

From a practical standpoint, for example, T_t would be set just above the transition zone, as defined in Figure 5.5, such that the energy input for which the system is designed can most efficiently lower T_t through the transition zone. For biological conditions T_t would be poised at 37°C, and the added energy would lower T_t the width of the transition zone to about 25°C. The reaction would be represented by an equilibrium constant for hydrophobic association, K_{HA}, expressible as $\log K_{HA} = -\Delta G_{HA}/2.3RT$, where $\Delta G_{HA}/2.3RT$ would be plotted versus Y″, the fraction of hydrophobic association in analogy to Equation (5.23), and modified to include cooperativity as shown for a protonation/deprotonation reaction also treated in Equation (5.23). The chemical couple COOH/COO⁻, is a specific example, but the reaction can involve any of the energy conversions represented in the Hexagon in Figure 5.22. Of course, Axiom 5 (see section 5.5.3) provides a guiding principle for increasing efficiency of energy conversion, which has the effect of increasing positive cooperativity.

5.9.1.3 ΔG_{HA} Provides an Efficiency Limit and Becomes the Coupling Process in Energy Conversion by the Consilient Mechanism

Interestingly, for each model protein composition and energy source the consilient mechanism sets an efficiency limit, $\hat{e}_{CM}$. It is simply the change in Gibbs free energy for hydrophobic association, the output energy, divided by the input energy, that is, $\hat{e}_{CM} = \Delta G_{HA}/\text{input energy}$. For the coupling of functions, as, for example, occurs in electro ↔ chemical transduction, reduction becomes the energy input that results in an output energy of ΔG_{HA}, a decrease in the Gibbs free energy of hydrophobic association due to an increase in hydrophobicity; the ΔG_{HA} thus produced then becomes the energy input for the output of a change in chemical energy. In general, these energy conversions can be reversible, and ΔG_{HA} emerges as the coupling process in energy conversion by the consilient mechanism.

To elaborate on an electro ↔ chemical transduction example, the chemical group to be reduced can be a biosynthetic modification of vitamin B_3 (niacin, shown in Figure 3.8A) or of vitamin B_2 (riboflavin, shown in Figure 3.8B). Electrical energy is expended to reduce the oxidized state of a modified vitamin B_3 or B_2 attached to the protein, making the protein more oil-like and lowering the Gibbs free energy for hydrophobic association, ΔG_{HA}, sufficient to drive hydrophobic association. This decrease in ΔG_{HA} changes the free energy of a carboxylate and causes it to take up a proton (see Figure 5.20A and associated discussion).

Analyses of these elements are considered below after a brief general definition of efficiency.

5.9.2 General Statement for Efficiency of Energy Conversion

The usual symbol for efficiency is η. The definition of η is the output energy divided by the input energy times 100 so that the efficiency is given as a percentage:

$$\eta = (\text{output energy}/\text{input energy})\,100 \quad (5.24)$$

5.9.3 Consilient Mechanism's Efficiency Limit, $\hat{e}_{CM}$, for Model Protein Composition and Energy Source

The Gibbs free energy for hydrophobic association, ΔG_{HA}, may be viewed as an output energy that results from a particular input energy acting on a specific composition of protein-based machine. Thus, we define an efficiency limit for energy conversion by the consilient mechanism, $\hat{e}_{CM}$, for a particular energy input and model protein composition.

$$\hat{e}_{CM} = \Delta G_{HA}/\text{input energy} \quad (5.25)$$

The same input energy can result in very different efficiency limits depending on the protein composition. Furthermore, the same energy input, acting on the same polymer composition, can vary depending on the range of the intensive variable. For example, for the composition $(GVGVP)_n$, raising the temperature from 5° to 20°C results in no ΔG_{HA} (see Figure 5.5A). On the other hand, raising the temperature from 25° to 40°C results in hydrophobic association that is useful in many ways including the output of mechanical work. Alternatively, an energy input acting to lower the value of T_t would have little effect until T_t begins to enter the transition zone, which for most mammals would begin at about 37°C and be complete by about 25°C (see Figure 5.5B). In relation to Figure 5.5, T_t on the x-axis could be plotted versus Y'', the fraction of hydrophobically associated state. With T_t above 37°C, Y'' would be zero; as T_t is lowered through the transition zone, there would be a sigmoidal increase in Y'' until a value of 1 were reached at about 25°C. Graphically this is simply the sigmoid of Figure 5.33.

A prevalent example in biology involves a chemical energy and its intensive variable of chemical potential, μ, the change in Gibbs free energy per mole of chemical. For the polymer composition in Figure 5.12A, at low salt concentration the slope $\Delta T_t(\chi)/\Delta\mu$, apparent as $\Delta T_t(\chi)/\Delta N$, is steep, whereas at higher concentrations the slope is much less. By Equation (5.9), ΔG_{HA} is larger for the low salt concentration range than the high concentration range. Thus, $\hat{e}_{CM}$ varies significantly depending on the effectiveness of the energy input in achieving a ΔG_{HA}. The latter quantity is calculable by employing Equation (5.10b), $\Delta G_{HA}(\chi) = [\Delta H_t(\text{ref}) - \Delta H_t(\chi)]$, and using relevant experimental data, much of which are found in Luan and Urry's review[72] entitled "Elastic, Plastic, and Hydrogel-Forming Protein-based Polymers" in the *Polymer Data Handbook*. Key results are listed in Table 5.4. Accordingly, in a plot of T_t versus concentration it is the slope at any point along the curve, and the calculated ΔG_{HA} per mole, that is, μ_{HA}, would be for that concentration. A good example of the importance of model protein composition is shown in Figure 5.15 and the associated discussions of relative effectiveness in section 5.4.3.2.

5.9.3.1 *CaCl₂ and NaCl as Chemical Energy Inputs to an Uncharged Model Protein to Drive Hydrophobic Association*

We begin with NaCl, for which calorimetry data, ΔH_t, are available from Table I of Luan et al.[117] for this salt interacting with poly(GVGVP) and used in Equation (5.10b). Chemical energy is defined as $\Delta\mu\Delta n$, where the chemical potential, $\Delta\mu$, is $2.3RT \log(a'/a'')$. The a' and a'' are the activities of the ions at two different activities, a' and a'', which equal concentration at sufficiently low ion concentrations, [C]. This assumption will be made here for simplicity, but the activity coefficients required for accurate calculations with salts are available.[118] This situation is further complicated by the determination of Δn when there is no functional group being titrated. Accordingly the numbers calculated are taken for a Δn of 1. These complications are not significant for comparison of the effect of different chloride salts on uncharged polymers, but they do become relevant when making comparisons with data from charged model proteins.

For the concentration step from 0.1N to 1.0N, ΔT_t is 12.5 and $\Delta G_{HA} = \Delta H_t(0.1\,N - 1.0\,N) = 0.5\,\text{kcal/mole-pentamer}$. Note for the concentration step used (0.1 N – 1.0 N), these numbers are slightly different from the values in Table 5.4 for the concentration step of (0 N – 1.0 N). The concentration step of a factor of 10 (0.1 N – 1.0 N) is chosen so that the chemical energy input at physiological temperature, 37°C (310°K), would be 2.3RT log(10) = 1.42 kcal/mole, for utilization in the denominator of Equation (5.25). Therefore,

$$\hat{e}_{CM}[\text{poly(GVGVP), NAC1}] = 0.50 \text{ kcal/mole}/1.42 \text{ kcal/mole} = 0.35 \quad (5.26)$$

For $CaCl_2$ there is as yet no experimental value for $\Delta H_t(0.1 N - 1.0 N)$ determined from calorimetry. Thus, to have an estimate, the same slope will be assumed as for NaCl, that is, $\Delta G_H[\text{poly(GVGVP)}, CaCl_2(0.1 N - 1.0 N)] = (0.50/12.5)(6.6 \times 0.9) = 0.24$. Under this approximation,

$$\hat{e}_{CM} = [\text{poly(GVGVP)}, CaCl_2] = 0.24 \text{ cal/mole}/1.42 \text{ cal/mole} = 0.17 \quad (5.27)$$

In this approximate way, the efficiency limit when using the increment of (0.1 N – 1.0 N) with a chemical energy per mole of 1.42 kcal at 37°C is 0.35 for NaCl and an even smaller value of 0.17 for $CaCl_2$. The effectiveness of salt in driving hydrophobic association increases dramatically when there are charged side chains with which to ion pair.

5.9.3.2 *CaCl₂ and NaCl as Chemical Energy Inputs to a Charged Model Protein: Impact of Ion Pairing on Driving Hydrophobic Association*

In the absence of titration data for accurate $\Delta\mu$ and Δn values, in the absence of corrections to activities (see above), and without direct calorimetry data to estimate ΔH_t for the specific salt and model protein systems of interest, gross comparison derives from the ΔG_{HA}(salt) values listed in Table 5.4A, as determined in section 5.3.4.

5.9.3.2.1 For **Model Protein i** Using $CaCl_2$ as the Chemical Energy Input

For **Model Protein i** in Table 5.5 and from data in Figure 5.10 and Table 5.3, the protonation of –COO⁻ (E⁻/0F) to give –COOH (E⁻/0F) gives a slope of (9 kcal/mole)/(380°C). As ion pairing constitutes partial neutralization, we will assume this slope. For $CaCl_2$, in the 0.0125 N to 0.05 N concentration interval and as shown in Figure 5.12A and listed in Table 3e of Luan and Urry,[72] at low ion concentrations the steepest slope is estimated to have a $\Delta T_t/N$ value of –618°C/N, which gives ΔG_{HA} = **–14.6 kcal/mole-pentamer/N-calcium ion**.

It should be emphasized that this value is relevant in only a very narrow concentration range, causing only partial folding and unfolding of the model protein. Nonetheless, a sense of the increased effectiveness of ion pairing in driving hydrophobic association derives on division of 14.6 by 0.27 from Table 5.4. *By this estimate, calcium ion pairing with polymer carboxylate as compared with the calcium chloride salt effect on an uncharged polymer approximates to be some 50 times more efficient in the use of chemical energy to produce a change in the Gibbs free energy for hydrophobic association*, ΔG_{HA}.

5.9.3.2.2 For **Model Protein i** Using NaCl as the Chemical Energy Input

For NaCl in the 0.15 N to 0.25 N concentration interval and also as shown in Figure 5.12A and listed in Table 3e of Luan and Urry,[72] at low ion concentrations the steepest slope is estimated to have a $\Delta T_t/N$ value of –395°C/N. This gives ΔG_{HA} = **–9.35 kcal/mole-pentamer/N-sodium ion**. Division by 0.57 for NaCl with poly(GVGVP) of Table 5.4 provides an estimate of better than *a 15-fold increase in the effectiveness of Na⁺ ion pairing with –COO⁻, that is, Glu-COO⁻ … Na⁺, in driving hydrophobic association than when there is no site for ion pairing*.

5.9.3.2.3 For **Model Protein x'** and Poly[0.76(GVGVP),0.24(GKGVP)] Using the NaCl Energy Input

First an estimate of the lowering of the Gibbs free energy of hydrophobic association, ΔG_{HA}, obtained for poly[0.76(GVGVP),0.24(GK⁺GVP)], derives from the data in Table 3d of Luan and Urry.[72] The steepest slope, $\Delta T_t/N$ (0.05 N to 0.20 N) arising from ion pairing of –NH₃⁺ with Cl⁻, gives a value for NaCl of

−177° C/N. Next, the value for $\Delta T_t/\Delta G_{HA}$ of 12.5 kcal/mole/230°C derives from Figure 5.10 using **Model Protein x′** of Table 5.5. On multiplication, ΔG_{HA}(Lys-NH$_3^+$ ... Cl$^-$; steepest ion-pairing slope) ≈ **− 9.6 kcal/mole-pentamer/N-chloride ion**. Division of this number by 0.57 of Table 5.4 for NaCl with poly(GVGVP) gives *a relative effectiveness of approximately 17*. Again, we see the dramatic effect of ion pairing on driving hydrophobic association.

5.9.3.3 Acid as Energy Input for Model Protein with Titrable Functional Groups

The acid–base titration curve is sigmoid in shape as is commonly the case for equilibrium processes wherein fraction of completion of the reaction is plotted versus log[C]. In the case of the acid–base titration, pH = −log[H$^+$]. Because of this, some basis must be decided upon for the measurement of the relative proton chemical energy required to drive from 0% to 100% completion. Here we utilize the steepest part of the sigmoid and talk of the efficiency for operating in this linear range, which would give the maximal efficiency for the process. The curve for a dilute weak acid, described by the Henderson-Hasselbalch equation, exhibits a steepest slope of 0.9 pH units for going from degree of ionization, α, of 0 to 1, that is, from the protonated to the charged state.

To make comparisons with the above salt effects, however, accurate values of Δn from acid–base titrations and correction of concentrations to activities should be considered. At this time, however, several different graphical representations of relative efficiencies are possible. These include comparison of the relative effectiveness of changes in chemical potential, $\Delta\mu$, to drive T_t from just above to below the operating temperature, comparison of the relative $\Delta\mu\Delta$n areas determined from acid–base titration curves, and comparison of the significance of different degrees of positive cooperativity, that is, the impact of changes in the Hill coefficient.

5.9.4 Graphical Insights into Relative Efficiencies for Chemical Energy Input

Three different graphical representations provide insight into relative efficiency in response to a chemical energy input. The first shows the relative amount of chemical energy required to lower the onset temperature for hydrophobic association from just above the operating temperature to a temperature just sufficient to cross the width of the transition zone of Figure 5.5. For (GVGVP)$_n$ that would be to lower T_t from the physiological temperature of 37°C where the polymer is hydrophobically dissociated to 25°C where the model protein would be almost completely hydrophobically associated, as depicted in Figure 5.33.

A second representation comes from the perspective of acid–base titration curves. The relative efficiency, for example, of two different proton-driven molecular machines becomes apparent by the relative areas of $\Delta\mu\times\Delta$n, where $\Delta\mu$ (= −2.3RT ΔpH) is measured by a step along the pH axis and Δn by a step along the y-axis, in terms of the change in number of moles required to convert from one state to the other. This is demonstrated in Figure 5.34.

A third measure of relative efficiency derives from comparison of the Hill coefficients where the larger Hill coefficient indicates a more efficient conversion of chemical energy into the hydrophobically associated state. This is demonstrated in the inset in Figure 5.34.

5.9.4.1 Relative Efficiency Limits for Chemical Energy Inputs Using the Consilient Mechanism

The efficiency limit of the consilient mechanism for a chemical energy input is written as

$$\hat{e}_{CM} = \Delta G_{HA}/\Delta\mu\Delta n \qquad (5.28)$$

$\Delta\mu\Delta$n is the chemical energy where $\Delta\mu$ is the change in chemical potential and is the Gibbs free energy per mole of the chemical used to drive hydrophobic association, given by 2.3RT log{[C$_2$]} − log{[C$_1$]} or equivalently 2.3RT log{[C$_2$]/[C$_1$]} where the approximation of concentration is used for activity and the change in chemical potential is positive when [C$_2$] > [C$_1$]. Δn is the change in number of moles of the chemical used to achieve the hydrophobic association, for example, the number of moles of functional groups titrated.

5.9 On the Efficiency of Consilient Protein-based Machines

The relative efficiency of two systems, $\eta(\text{system-1})/\eta(\text{system-2}) = \eta_1/\eta_2$, can be expressed as the ratio of efficiency limits for the two systems when being compared using the same chemical energy.

$$\eta_1/\eta_2 = \eta(\text{system-1})/\eta(\text{system-2}) = \hat{e}_{CM}(\text{system-1})/\hat{e}_{CM}(\text{system-2})$$

$$\eta_1/\eta_2 = (\Delta G_{HA}/\Delta\mu\Delta n)_1/(\Delta G_{HA}/\Delta\mu\Delta n)_2 \quad (5.29)$$

5.9.4.2 Relative Efficiencies by Chemical Energy Required to Move the Cusp of Insolubility Across Biology's Transition Zone

When concerned with the physiological operating temperature of 37°C, a model protein with its $T_{t(b)}$-value at 37°C will not have appreciably hydrophobically associated. Any variable that lowers the $T_{t(b)}$-value to 25°C, which is the width of the transition zone for $(GVGVP)_{251}$, will have essentially completed the transition to the hydrophobically associated state, as depicted in Figure 5.33. The variable will have moved the cusp of insolubility across the transition zone for biology. In particular, the interest is in the variable of the chemical energy per mole required for a ΔT_t just sufficient to traverse the transition zone from hydrophobically dissociated at 37°C to hydrophobically associated at 25°C.

Two examples are represented in Figure 5.15 for **Model Proteins I** and **ii** of Table 5.5, where the T_t-value is plotted as a function of an increase in chemical energy, that is, in this case an increase in the concentration of calcium ion. Section 5.4.3.2 considered the relative efficiencies on the basis of graphical comparisons With respect to Equation (5.29), that comparison assumed the values of $(\Delta G_{HA})_1$ and $(\Delta G_{HA})_2$ and of Δn_1 and Δn_2 to be equivalent. The approximation becomes $\eta_1/\eta_2 \approx \Delta\mu_2/\Delta\mu_1$, which gives the ratio of $0.116/1.12 \approx 0.1$. With adequate differential calorimetry and titration data, the relative efficiencies in terms of relative efficiency limits can be determined accurately. Actual efficiency ratios for converting chemical energy into mechanical work using cross-linked matrices are calculated below in section 5.9.5.3.

5.9.4.3 Relative Efficiencies by $\Delta\mu\Delta n$ Areas of Acid–base Titration Curves

Another ready visual means of approximating relative efficiencies is given in Figure 5.34 for the series of **Model Proteins i** through **v**. In particular, **Model Proteins i** and **v** are compared by the boxes indicated. When the titrations are carried out with the same concentration of carboxyls for each model protein and the same concentration of titrant, the vertical displacement is proportional to Δn and the horizontal displacement is proportional to $\Delta\mu$. Accordingly, comparison of the area, $\Delta\mu_i\Delta n_i$, as indicated for **Model Protein i** with the area, $\Delta\mu_v\Delta n_v$, approximated for **Model Protein v** gives the relative efficiencies more accurately than was done above for the data in Figure 5.15. Improved accuracy here would require choosing accurately the same interval for the degree of ionization, α, for example, the interval of 0.1 to 0.9.

5.9.4.4 Relative Efficiencies by Comparison of Hill Coefficients

The most efficient operational design would be for the machine to operate over the range of the acid–base titration curve with the steepest $\Delta n/\Delta\mu$ slope. Because the Hill coefficient, n, as defined in Equation (5.20) is a measure of the slope, it provides for ready comparison of efficiencies. The Hill coefficients for **Model Proteins i** through **v** are listed in Figure 5.34, and the slopes are plotted in the inset. Accordingly, the comparison of the efficiencies of **Model Proteins i** and **v** simply becomes $\eta_i/\eta_v = 1.5/8.0 = 0.19$. Thus, by increasing the hydrophobicity by the replacement of five Val (V) residues by five Phe (F) residues, as indicated in Table 5.5, increases the efficiency of the protein-based machine by just over fivefold.

5.9.5 Relative Chemomechanical Transductional Efficiencies of the Electrostatic Charge–Charge Repulsion and Consilient Mechanisms by Experimental Determination of $\Delta\mu$, Δn, and $f\Delta L$

The comparison of relative efficiencies above considered only those polymers functioning by the consilient mechanism. This involved the inverse temperature transition for hydrophobic association. Using the same functional group, the carboxyl, and its interconversion between –COO⁻ to –COOH, it is now possible to compare directly the chemomechanical transductional efficiencies of poly(methacrylic acid), [–CH$_2$–(CH$_3$)C(COOH)–]$_n$, abbreviated as PMA, to that of **Model Protein iv** in Table 5.5. PMA utilizes the electrostatic charge–charge repulsion mechanism, and **Model Protein iv** utilizes the apolar–polar repulsive free energy of hydration of the inverse temperature transition developed in this chapter.

The same graphical insights utilized above can be used to obtain insight into relative efficiency. These can then be compared with the direct experimental determination of the efficiency of each polymeric system and of the relative efficiencies. The relative $\Delta\mu\Delta n$ areas can be observed for the polymers as well as their Hill coefficients. Finally, the acid–base titration curves are given for the cross-linked elastic matrices held at fixed force and the change in length determined.

5.9.5.1 Relative Efficiencies by $\Delta\mu\Delta n$ Areas and Hill Coefficients of Polymer Acid–Base Titration Curves

5.9.5.1.1 Comparison of the $\Delta\mu\Delta n$ Areas

The acid–base titration curve for PMA is given in Figure 5.35A, where the change in degree of ionization and change in pH are given as the box or $\Delta\mu\Delta n$ area over which the polymer functions effectively in contraction/relaxation. In Figure 5.35B occurs the acid–base titration for **Model Protein v** and the $\Delta\mu\Delta n$ box for its contraction/relaxation cycle. The $\Delta\mu\Delta n$ box for **Model Protein v** in Figure 5.35B is transposed as a small shaded box onto the box for PMA in Figure 5.35A, with a gross correction for the difference in Δn. Even so, the difference in efficiencies can be visually appreciated. This visual comparison would more accurately carry over to relative efficiencies if the change in length at the same fixed force resulted in similar $f\Delta L$ terms for the mechanical work performed.

5.9.5.1.2 Comparison of the Hill Coefficients

As shown in the inset of Figure 5.34, the Hill coefficient for PMA is 0.5, whereas that of **Model Protein v** is 2.7. Thus, without consideration of differences in Δn, the relative efficiency would be 0.5/8.0 = 0.06 or one-sixteenth as efficient as the model protein. After the statement of efficiency for chemomechanical transduction immediately below, the relative efficiency of the two mechanisms will be given using **Model Protein iv** as the representative of the consilient mechanism. The ratio of Hill coefficients for the latter model protein is 0.5/2.7 or differing only by a factor of 5 rather than 16. The experimentally determined relative efficiency is still very large.

5.9.5.2 Statement of Efficiency for Chemomechanical Transduction

The proper expression for the efficiency of a chemomechanical engine is

$$\eta = f\Delta L / \Delta\mu\Delta n \qquad (5.30)$$

where f is the fixed force or weight being lifted and ΔL is the distance that the weight is raised. The quantities $\Delta\mu$ and Δn, as considered above, are the change in chemical potential and the change in number of moles of proton required during the energy conversion. In what follows a stress–strain/acid–base titration apparatus is used to obtain the information required for Equation (5.30).

5.9.5.3 Relative Chemomechanical Efficiencies of Electrostatic Repulsion and Consilient (Inverse Temperature Transition) Mechanisms

The quantity that we wish to calculate is given by Equation (5.31) where the subscript ap stands for the apolar–polar repulsive free energy of hydration mechanism and cc stands for the charge–charge repulsion mechanism:

$$\eta_{ap}/\eta_{cc} = [f\Delta L(ap)/\Delta\mu\Delta n(ap)]/ \\ [f\Delta L(cc)/\Delta\mu\Delta n(cc)] \qquad (5.31)$$

The experimental data for ΔL and $\Delta\mu$ are obtained from the graph in Figure 5.36 for a con-

5.9 On the Efficiency of Consilient Protein-based Machines

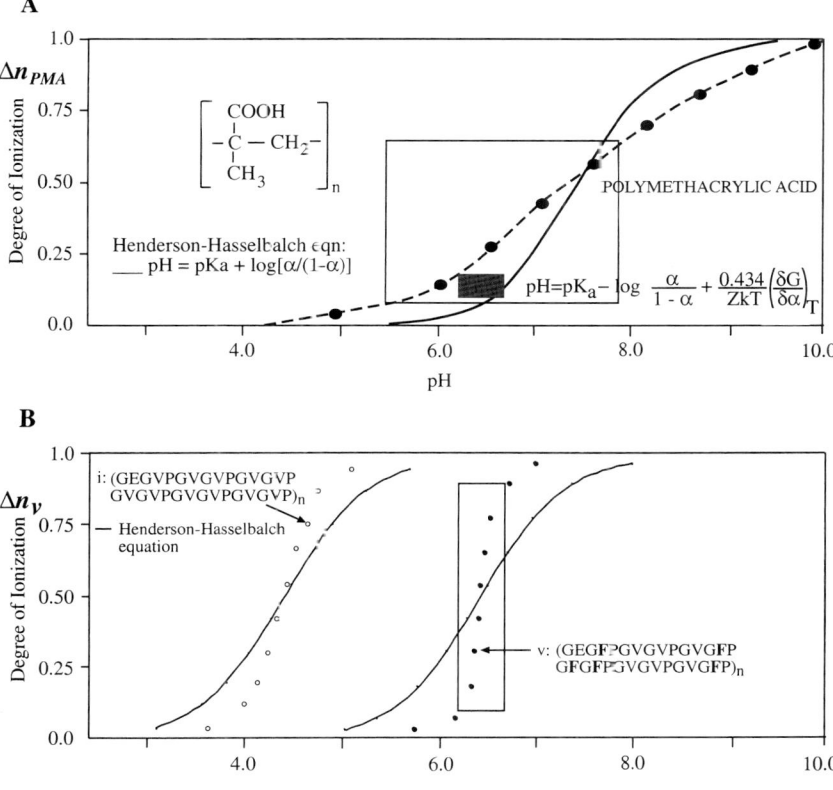

FIGURE 5.35. Visualization of the relative efficiencies of the electrostatic charge–charge repulsion and the apolar–polar repulsion mechanisms by comparison of acid–base titration curves for poly(methacrylic acid), which exhibits charge–charge repulsion (negative cooperativity), and for the **Model Proteins i** and **v**, which exhibit apolar–polar repulsion (positive cooperativity). The polymers are each compared with the Henderson-Hasselbalch curve as reference. Chemical energy is $\Delta\mu \cdot \Delta n$, where $\Delta\mu = 2.3RT\Delta pH$ for the change in pH to go from one state to the other, and Δn is the number of moles to go from a degree of ionization, α, of zero to one. Accordingly, the area of an acid–base titration curve as indicated in **A** and in **B**, corrected for the difference in number of moles required to go from one state to another, gives the relative amounts of input energy used. These areas are indicated separately in **A** and in **B**, and then with the correction for Δn, the small shaded area in the larger area of part **A** demonstrates how much less energy is required for the apolar–polar repulsive free energy of hydration with its positive cooperativity to perform a similar amount of work.

stant force of 2 grams. For the calculation the steepest slope is used for both systems. Because PMA exhibits a lower slope over a considerable range of the ΔL versus pH plot, this gives the best working range for PMA and uses its most favorable efficiency range. As the slopes are taken for the common ΔL of 9 mm length changes, the differences are readily seen in the relative ΔpH required to achieve the common length change with a ΔpH_{PMA} of 2.24 and a $\Delta pH_{E/4F}$ of 0.59. Also the Δn values are 6.86×10^{-5} for PMA and 6.12×10^{-6} for **Model Protein iv**, also indicated in the subscript as E/4F. Accordingly,

$$\eta_{ap}/\eta_{cc} = [(2.24)(6.86 \times 10^{-5})]/[(0.59)(6.12 \times 10^{-6})] \approx 40 \quad (5.32)$$

The result is dramatic! The efficiency of the charge–charge repulsion mechanism is very limited indeed.

For years the common discussion of mechanism even in proteins has centered on the inter-

210 5. Consilient Mechanisms for Diverse Protein-based Machines

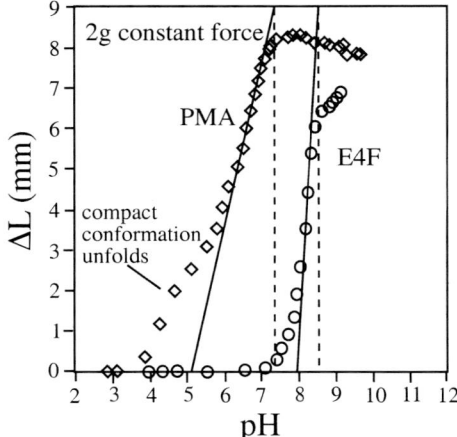

FIGURE 5.36. Experimental plots of cross-linked matrices, each loaded with a 2 gram weight, that contract on lowering pH to lift the weight through the distances, ΔL, as indicated. These data provide a direct experimental comparison with the relative efficiencies for chemomechanical transduction of η_{cc} (charge–charge repulsion mechanism) using poly(methacrylic acid) and of η_{ap} (apolar–polar repulsion mechanism) using elastic **Model Protein iv** of Table 5.5. The Δn values are 6.86×10^{-5} mole for PMA and 6.12×10^{-6} mole for E/4F, **Model Protein iv**. The calculation given in the lower part demonstrates the apolar–polar repulsion mechanism of the model protein to be about 40 times more efficient than the charge–charge repulsion mechanism of PMA, poly(methacrylic acid). See text for further discussion. (Unpublished data of D.W. Urry, L. Hayes, and J. Lee.)

action of charges, the attraction of opposite charges and the repulsion of like charges. In dominantly aqueous systems, however, salts dissolve to become ions because the free energy of separated hydrated ions is more favorable than the ion-pair associations in the crystal. The phase separation process of inverse temperature transitions occurs in the absence of any charge at all, but charge under the influence of a hydrophobic domain regains its affinity for association even in a dominantly aqueous system and becomes a most potent means of controlling phase separation.

It seems quite evident that the function of biological macromolecules, and proteins in particular with their ease to evolve the most efficient mechanism for accessing and utilizing its energy sources biological systems as discussed in Chapter 6, would utilize the inverse temperature transition mechanism developed and demonstrated in this chapter.

Clearly, it would seem not unreasonable to propose the consilient mechanism as the dominant mechanism in protein structure formation and function. The comprehensive hydrophobic effect should be the foundation from which to engineer protein materials for medical and non-medical uses.

References

1. E.O. Wilson, *Consilience, The Unity of Knowledge*. Alfred E. Knopf, New York, 1998, page 8, gives a definition of the word consilience as providing a "common groundwork of explanation."
2. Following E.O. Wilson, we speak of a consilient mechanism as a "common groundwork of explanation" for each of two interlinked physical processes in the function of protein-based machines of biology. The first physical process is that of hydrophobic association, herein developed as the comprehensive hydrophobic effect (CHE). CHE derives its consilience from being the dominant mechanism capable of achieving all of the diverse energy conversions for which biology is renown and more general yet from being relevant to all amphiphilic polymeric systems, that is, for all polymers that contain both oil-like (hydrophobic) and polar (e.g., charged) constituents. The second physical process is that of entropic elastic force development, which is essential to efficient energy conversion, especially when motion is involved; this force development is most evident as an increase in elastic force under isometric conditions. The elastic consilient mechanism derives its consilience from being relevant to all polymeric systems, of whatever composition, whenever deformation results in a decrease in chain backbone mobility, for example, the damping of internal chain dynamics on extension. This second physical process, which constitutes a consilient mechanism for ideal or entropic elastic, will be delineated as the elastic consilient mechanism.
3. E.O. Wilson, *Consilience: The Unity of Knowledge*. Alfred E. Knopf, New York, 1998, pp. 4–5.
4. D.W. Urry, "Physical Chemistry of Biological Free Energy Transduction as Demonstrated by Elastic Protein-based Polymers." *J. Phys. Chem. B*, **101**, 11007–11028, 1997.
5. *"Concepts Introduced During Development of Elastic Protein-based Polymers for Free*

References

Energy Transduction: The conclusions of this article can also be given in terms of the following chronological listing of the concepts introduced during the development of elastic protein-based polymers for free energy transduction: (1) the concept of the damping of internal chain dynamics on extension as the source of entropic elastomeric force, called the *librational entropy mechanism of elasticity*; (2) the concept of T_t, the temperature of the hydrophobic folding and assembly transition, being used as the fundamental measure of hydrophobicity and providing a practical on–off switching capacity; (3) the generally obvious concept that raising the temperature from below to above T_t is a means of performing mechanical work by cross-linked elastic protein-based polymers; (4) the concept of the ΔT_t-mechanism wherein the value of T_t is changed, rather than the temperature, as a means of achieving free energy transduction; (5) the concept of energy conversion by means of the coupling of different functional moieties by being part of the same hydrophobic folding and assembly domain arising out of, for example, (a) hydrophobic-induced pKa shifts, (b) hydrophobic-induced shift in redox potential, and (c) demonstrated coupling of carboxyl and redox functions to result in electrochemical transduction; (6) the concept of the competition for hydration between apolar (hydrophobic) and polar (e.g., charged) moieties to give rise to pKa shifts and positive cooperativity; (7) the concept of 'poising' for achieving higher efficiencies; (8) essential equivalence of the inverse temperature transition of a phase separation and the intramolecular phase separation of the hydrophobic folding of a globular protein or assembly of the protomer subunits to form a multisubunit globular protein such as phosphofructose kinase, that is, the extension to globular proteins; and (9) extension to all polymers where, however, the degree of expression of the above effects is limited due to the lack of the many advantages of protein-based polymers of Table I." (From Urry.[4])

6. D.W. Urry, "Molecular Machines: How Motion and Other Functions of Living Organisms Can Result from Reversible Chemical Changes." *Angew. Chem.* [German], **105**, 859–883. 1993; *Angew. Chem. Int. Ed. Engl.*, **32**, 819–841. 1993.

7. D.W. Urry, "Elastic Biomolecular Machines: Synthetic Chains of Amino Acids, Patterned After Those in Connective Tissue, can Transform Heat and Chemical Energy into Motion." *Sci. Am.* January 1995, 64–69.

8. D.W. Urry, "The Change in Gibbs Free Energy for Hydrophobic Association: Derivation and Evaluation by means of Inverse Temperature Transitions." *Chem. Phys. Letters*, **399**, 177–183, 2004.

9. D.W. Urry and T.M. Parker, "Mechanics of Elastin: Molecular Mechanism of Biological Elasticity and its Relevance to Contraction." *J. Muscle Res. Cell Motil.*, **23**, issue 5–6 (2002); Special Issue: *Mechanics of Elastic Biomolecules*, H. Granzier, M. Kellermayer, W. Linke, Eds.

10. Recall that oil-like groups contain hydrocarbon; the classic example is the CH_2 unit, the most common chemical grouping of oil. The principal ingredient of vinegar is acetic acid, which when negatively charged is written as $CH_3–COO^-$, but our use of the term *vinegar-like* also includes positively charged species like amine functions, for example, $–NH_3^+$, and it can even include very polar groups like oxygen and the peptide group itself. Of particular interest in protein function are the vinegar-like groups that can exist in either of two or more states, such as charged or uncharged.

11. The vinegar-like side chains can exist in either of two different states, for example, the carboxyl/carboxylate, $COOH/COO^-$, chemical couple and the amino/ammonium, $-NH_2/NH_3^+$, chemical couple. As demonstrated below, the uncharged state of the couple favors association of oil-like domains, whereas the charged state of the couple disrupts association of oil-like domains by having destroyed the special hydration of oil-like groups in the process of achieving its own hydration.

12. The word *consilient* is used as the adjective form of the noun *consilience*. As listed in *Webster's Third New International Dictionary*, consilience contains the sense that there exist pervasive, fundamental laws of nature underlying related disciplines that provide a common groundwork of understanding. Here we refer to the consilient mechanism related to hydrophobic association as a water-dependent, pervasive, and fundamental process by which the macromolecules of living organisms function in providing the essential diverse energy conversions of Life. In general, if we speak of the consilient mechanism without further delineation, it will be the consilient mechanism of hydrophobic association. When referring to elastic, reference will be to the consilient mechanism for entropic (ideal) elasticity or simply to the elastic consilient mechanism.

13. (GVGVP)$_n$ uses the single letter amino acid residue abbreviation for the repeating pentapeptide sequence (glycyl-valyl-glycyl-valyl-prolyl)$_n$. Also, the three letter abbreviation would be (Gly-Val-Gly-Val-Pro)$_n$. The peptide residue is given in general as –NH–CHR–CO– and the **R**-group side chain is –H for G, –CH–(CH$_3$)$_2$ for V, and –CH$_2$–CH$_2$–CH$_2$– for P, with the three CH$_2$ moieties bridging from the N (replacing the H) to the central atom of the backbone of the peptide residue, C. Thus the V and P residues contain the oil-like groupings of atoms, which are in fact the hydrocarbon moieties of oil. The interconnecting peptide group itself, –CO–NH–, formed between the defined residues represents a vinegar-like grouping, even though it is without a net charge. It is considered polar, as there are substantial partial charges on the atoms, but the sum of the partial charges of the four atoms in the peptide moiety is essentially zero. These structural considerations were introduced in Chapter 2 and are treated in more detail below in section 5.2.
14. For a comparison per gram with the transitions of melting and vaporization of water, these numbers per pentamer would need to be divided by 23, which is (409 grams/mole-pentamer divided by 18 grams/mole-water).
15. F. Franks, "Protein Destabilization at Low Temperatures." *Adv. Protein Chem.*, **46**, 107–139, 1995.
16. P.L. Privalov, "Cold Inactivation of Enzymes." *Crit. Rev. Biochem. Mol. Biol.*, **25**, 281–305, 1990.
17. The term *inverse transition* was first used in connection with the increase in order of the antibiotic stendomycin on raising the temperature (D.W. Urry and A. Ruiter, "Conformation of Polypeptide Antibiotics. VI. Circular Dichroism of Stendomycin." *Biochem. Biophys. Res. Commun.*, **38**, 800–806, 1970). The term became specifically *inverse temperature transition* in relation to coacervation of elastin fragments that exhibited a phase separation with increased order on raising the temperature (B.C. Starcher, G. Saccomani, and D.W. Urry, "Coacervation and Ion-Binding Studies on Aortic Elastin." *Biochim. Biophys. Acta*, **310**, 481–486, 1973, and D.W. Urry, B. Starcher, and S.M. Partridge, "Coacervation of Solubilized Elastin Effects a Notable Conformational Change." *Nature*, **222**, 795–796, 1969).
18. P.J. Flory, *Principles of Polymer Chemistry*. Cornell University Press, Ithaca, New York, 1953, Figure 121.
19. M. Manno, A. Emanuele, V. Martorana, P.L. San Biagio, D. Bulone, M.B. Palma-Vitorelli, D.T. McPherson, J. Xu, T.M. Parker, and D.W. Urry, "Interaction of Processes on Different/time scales in a bioelastomer capable of performing energy conversion." *Biopolymers*, **59**, 51–64, 2001.
20. F. Sciortino, K.U. Prasad, D.W. Urry, and M.U. Palma, "Self-Assembly of Bioelastomeric Structures From Solutions: Mean Field Critical Behavior and Flory-Huggins Free-Energy of Interaction." *Biopolymers*, **33**, 743–52, 1993.
21. B.A. Cox, B.C. Starcher, and D.W. Urry, "Coacervation of α-Elastin Results in Fiber Formation." *Biochim. Biophys. Acta.*, **317**, 209–213, 1973.
22. B.A. Cox, B.C. Starcher, and D.W. Urry, "Coacervation of Tropoelastin Results in Fiber Formation." *J. Biol. Chem.*, **249**, 997–998, 1974.
23. D.W. Urry, M.M. Long, and H. Sugano, "Cyclic Analog of Elastin Polyhexapeptide Exhibits an Inverse Temperature Transition Leading to Crystallization." *J. Biol. Chem.*, **253**, 6301–6302, 1978.
24. W.J. Cook, H.M. Einspahr, T.L. Trapane, D.W. Urry, and C.E. Bugg, "Crystal Structure and Conformation of the Cyclic Trimer of a Repeat Pentapeptide of Elastin, Cyclo-(L-Valyl-L-prolylglycyl-L-valylglycyl)3." *J. Am. Chem. Soc.*, **102**, 5502–5505, 1980.
25. J.A.V. Butler, "The energy and entropy of hydration of organic compounds." *Transaction Faraday Society*, **33**, 229–238, 1937.
26. H.S. Frank and M.E. Evans, "Free Volume and Entropy in Condensed Systems: III. Entropy in Binary Liquid Mixtures; Partial Molal Entropy in Dilute Solutions; Structure and Thermodynamics in Aqueous Electrolytes." *J. Chem. Phys.*, **13**, 507–532, 1945.
27. A convenient estimate of the value of T_t is shown in Figure 5.1B, where an increase in turbidity of the solution is plotted with an increase in temperature and the 50% point for maximal turbidity measures T_t. As shown in Figure 5.1C, this measure of T_t occurs near the onset of the transition as measured by differential scanning calorimetry (DSC), where it is referred to as T_b and the shaded area of the curve gives the exothermic heat of the transition, ΔH_t, for (GVGVP)$_{251}$. It should be further noted that a more accurate, but not yet precise measure of ΔS_t uses a mean temperature for the transition, which we may denote as T_m. The accurate measure of ΔS_t sums over the DSC curve in small increments of heat released divided by the mean temperature for the increment. For

convenience, we refer to T_t, but the differences should be kept in mind when carrying out calculations as in 5.1.3.4.

28. D.W. Urry, T.L. Trapane, and K.U Prasad, "Phase-Structure Transitions of the Elastin Polypentapeptide-Water System Within the Framework of Composition-Temperature Studies." *Biopolymers*, **24**, 2345–2356, 1985.

29. D.W. Urry, D.T. McPherson, J. Xu, H. Daniell, C. Guda, D.C. Gowda, N. Jing, and T.M. Parker, "Protein-Based Polymeric Materials: Syntheses and Properties." In *The Polymeric Materials Encyclopedia: Synthesis, Properties and Applications*, CRC Press, Boca Raton, pp. 7253–7279, 1996. See Figure 6.

30. As dialysis membranes are calibrated against globular proteins and as below T_t these protein-based polymers are unfolded and disassembled, it was expected that the polymer that could pass through the membrane by a process of reptition (reptilian motion) would be much larger than 50,000 Da. After preparing the polymers by recombinant DNA technology, where the chain length was precisely known, for example, just over 100,000 Da for $(GVGVP)_{251}$, it was apparent that the chemically synthesized protein-based polymers retained by the 50,000 cut-off membranes were over 100,000 Da.

31. F. Sciortino, M.U. Palma, D.W. Urry, and K.U. Prasad, "Nucleation and Accretion of Bioelastomeric Fibers at Biological Temperatures and Low Concentrations," *Biochem. Biophys. Res. Commun.* **157**, 1061–1066, 1988.

32. Several points require consideration on identification of $\Delta H_t(CH_2) - T_t(GVGIP)\Delta S_t(CH_2)$ as $-\Delta G_{HA}(CH_2)$. The points include the separability assumption of Equations (5.3) and (5.4), the relevance of the model protein to such identification, and the choice of reference state in order that the nonlinearity of hydrophobic-induced pKa shifts be included. From the data of Butler,[25] the separability is reasonable for a simple CH_2 group, but examination of the calculated result is required to be satisfied whether or not extension to more complex substituents is warranted. As the inverse temperature transition of $(GVGVP)_n$ has been experimentally shown to involve no Raman detectable changes in secondary structure,[53] the elastic-contractile model proteins of focus here reasonably represent the best known model available for such an effort. It should be noted, however, that NMR studies on the temperature and solvent dependence of peptide NH and ^{13}CO chemical shifts suggest minor changes in the exposure of peptide NH and CO groups to solvent during the transition[45,46] To include the nonlinear effects, such as the hydrophobic-induced pKa shifts, where $pKa = \Delta G/2.3RT$, the choice of reference state becomes important. For situations where these nonlinear effects are significant, $\Delta S_t(GXGVP)$ is the preferred reference state. This requires extrapolation of $\Delta S_t(GXGVP)$ given for $f_X = 0.2$ in Table 5.1 to $f_X = 1$, as was done for T_t. Also this requires DSC data for the modified composition that is not always easy to obtain. When DSC data is not available for the perturbed state, Equation (5.8) can yet be used with $\Delta S_t(GVGVP) = \Delta S_t(ref)$, for certain comparison such as ratios of interest. The complete derivation of Equation (5.10b) and its evaluation by means of differential scanning calorimetry (DSC) data is given in reference 8, which should be sought for a more in depth treatment.

33. Bioluminescence presents an interesting case where a chemiluminescence occurs in which the build up of the chemical required many stepwise energy inputs. Also there could possibly be a stretch-induced component, for example, as seen at night in the wake of ships at sea.

34. If this were not the case we would be faced with the puzzling question. Since the value of T_t is higher when there are fewer CH_2 units in the polymer, and therefore less hydrophobic hydration, why should a higher temperature be required to melt the lesser amount of more poorly structured hydrophobic hydration? The thermally-induced inverse temperature transition does not result from a thermal destructuring of hydrophobic hydration, as we, with others, have indicated in the past (See for example the upper right illustration in Figure 2.8).[7]

35. D.W. Urry, S-Q. Peng, J. Xu, and D.T. McPherson, "Characterization of Waters of Hydrophobic Hydration by Microwave Dielectric Relaxation." *J. Amer. Chem. Soc.*, **119**, 1161–1162, 1997.

36. 20 Mrad means a radiation-absorbed dose of 20 million Roentgens.

37. To be an ideal elastic band means that the energy expended during stretching of the band is entirely recovered on relaxation of the deforming force.

38. This aspect of the consilient mechanism has been called the ΔT_t-mechanism, because the energy conversion occurs not by a change in temperature but rather by a change in the transition temperature.

39. Here we note more general terms: *polar* for vinegar-like or for charged or ionized and *apolar* for oil-like or hydrophobic, but recognize that there exist a continuum of groups from polar to apolar.
40. D.W. Urry, D.C. Gowda, S.-Q. Peng, and T.M. Parker, "Non-linear Hydrophobic-induced pKa Shifts: Implications for Efficiency of Conversion to Chemical Energy." *Chem. Phys. Lett.*, **239**, 67–74, 1995.
41. D.W. Urry, S.Q. Peng, L.C. Hayes, D.T. McPherson, Jie Xu, T.C. Woods, D.C. Gowda, and A. Pattanaik, "Engineering Protein-based Machines to Emulate Key Steps of Metabolism (Biological Energy Conversion)." *Biotechnol. Bioeng.*, **58**, 175–190, 1998.
42. D.W. Urry, L. Hayes, C.X. Luan, D.C. Gowda, D. McPherson, J. Xu, and T. Parker, "ΔT_t-Mechanism in the Design of Self-Assembling Structures," In *Self-assembling Peptide Systems in Biology, Medicine and Engineering.* A. Aggeli, N. Boden, S. Zhang, Eds., Kluwer Academic Publishers, Dordrecht, The Netherlands, 2001, pp. 323–340.
43. Energy is the product of two quantities, the change in the intensive variable times the change in the corresponding extensive variable. For mechanical energy, the intensive variable is the applied force, f, and the extensive variable is the distance over which the force is applied, ΔL. The amount of mechanical energy involved in a process becomes the product of the force to produce the change in motion times the change in length or position, that is, $f\Delta L$. Chemical energy is the product of the extensive variable, the number of a molecular species, n, such as that of a particular ion utilized in the process and the intensive variable, called chemical potential (μ = RT lna where **a** is the activity, which becomes concentration at low concentrations that limit effects of ion-ion interactions).
44. D.W. Urry and M.M. Long, "Conformations of the Repeat Peptides of Elastin in Solution: An Application of Proton and Carbon-13 Magnetic Resonance to the Determination of Polypeptide Secondary Structure." *CRC Crit. Rev. Biochemistry*, **4**, 1–45, 1976.
45. D.W. Urry, "Characterization of Soluble Peptides of Elastin by Physical Techniques." In *Methods in Enzymology*, **82**, 673–716, 1982, (L.W. Cunningham and D.W. Frederiksen, Eds.) Academic Press, Inc., New York, New York.
46. D.W. Urry, C.M. Venkatachalam, M.M. Long, and K.U. Prasad, "Dynamic β-Spirals and A Librational Entropy Mechanism of Elasticity." In *Conformation in Biol.* (R. Srinivasan and R.H. Sarma, Eds.) G.N. Ramachandran Festschrift Volume, Adenine Press, USA, 11–27, 1982.
47. D.W. Urry, "Thermally Driven Self-assembly, Molecular Structuring and Entropic Mechanisms in Elastomeric Polypeptides." In *Mol. Conformation and Biol. Interactions* (P. Balaram and S. Ramaseshan, Eds.) Indian Acad. of Sci., Bangalore, India, pp. 555–583, 1991.
48. D.W. Urry, T. Hugel, M. Seitz, H. Gaub, L. Sheiba, J. Dea, J. Xu, and T. Parker, "Elastin: A Representative Ideal Protein Elastomer." *Phil. Trans. R. Soc. Lond.*, B **357**, 169–184, 2002.
49. D.W. Urry, T. Hugel, M. Seitz, H. Gaub, L. Sheiba, J. Dea, J. Xu, L. Hayes, F. Prochazka, and T. Parker, In *Ideal Protein Elasticity: The Elastin Model*, P. Shewry and A. Bailey, Eds., Cambridge University Press, (in press) 2003.
50. L.B. Sandberg, J.G. Leslie, C.T. Leach, V.L. Torres, A.R. Smith, and D.W. Smith, "Elastin Covalent Structure as Determined by Solid State Amino Acid Sequencing." *Pathol. Biol.*, **33**, 266–274, 1985.
51. H. Yeh, N. Ornstein-Goldstein, Z. Indik, P. Sheppard, N. Anderson, J.C. Rosenbloom, G. Cicila, K. Yoon, and J. Rosenbloom, "Sequence Variation of Bovine Elastin mRNA due to Alternative Splicing." *J. Collagen Rel. Res.*, **7**, 235–247, 1987.
52. When the number of repeats is large, poly(GVGVP) is equivalent to poly(VPGVG), because the polymers differ only by the particular 2 or 3 residues that begin or terminate the polymer. For the remainder of the 1,000 or so residues, the polymers are identical.
53. G.J. Thomas, Jr., B. Prescott, and D.W. Urry, "Raman Amide Bands of Type-II β-Turns in Cyclo-(VPGVG)$_3$ and Poly(VPGVG), and Implications for Protein Secondary Structure Analysis." *Biopolymers*, **26**, 921–934, 1987.
54. D. Volpin, D.W. Urry, I. Pasquali-Ronchetti, and L. Gotte, "Studies by Electron Microscopy on the Structure of Coacervates of Synthetic Polypeptides of Tropoelastin." *Micron*, 7, 193–198, 1976.
55. D.W. Urry, C.M. Venkatachalam, M.M. Long, and K.U. Prasad, "Dynamic β-Spirals and a Librational Entropy Mechanism of Elasticity." In *Conformation in Biology*, R. Srinivasan and R.H. Sarma, Eds., G.N. Ramachandran Festschrift Volume, Adenine Press, USA, 11–27, 1982.
56. D.K. Chang and D.W. Urry, "Polypentapeptide of Elastin: Damping of Internal Chain Dynamics on Extension." *J. Computational Chem.*, **10**, 850–855, 1989.

References

57. T. Hugel, M. Seitz, H. Gaub, and D. Urry, unpublished results.
58. C.A.J. Hoeve and P.J. Flory, "Elastic Properties of Elastin." *Biopolymers*, **13**, 677–686, 1974.
59. T. Weis-Fogh and S.O. Andersen, "New Molecular Model for the Long-range Elasticity of Elastin." *Nature*, **227**, 718–721, 1970.
60. L.B. Alonso, B.J. Bennion, and V. Daggett, "Hydrophobic Hydration is an Important Source of Elasticity in Elastin-based Polymers." *J. Am. Chem. Soc.*, **123**, 11991–11998, 2001.
61. Z.R. Wasserman and F.R. Salemme, "A Molecular Dynamics Investigation of the Elastomeric Restoring Force in Elastin." *Biopolymers*, **29**, 1613–1631, 1990.
62. P.J. Flory, "Molecular Interpretation of Rubber Elasticity." *Rubber Chem. Tech.*, **41**, G41–G48, 1968.
63. C.-H. Luan, J. Jaggard, R.D. Harris, and D.W. Urry, "On the Source of Entropic Elastomeric Force in Polypeptides and Proteins: Backbone Configurational vs. Side Chain Solvational Entropy." *Int. J. Quant. Chem. Quant. Biol. Symp.*, **16**, 235–244, 1989.
64. D.W. Urry, B. Haynes, H. Zhang, R.D. Harris, and K.U. Prasad, "Mechanochemical Coupling in Synthetic Polypeptides by Modulation of an Inverse Temperature Transition." *Proc. Natl. Acad. Sci. USA*, **85**, 3407–3411, 1988.
65. D.W. Urry, "Protein Folding and Assembly: An Hydration-Mediated Free Energy Driving Force." In *Protein Folding: Deciphering the Second Half of the Genetic Code*. (Lila Gierasch and Johnathan King, Eds.), Am. Assoc. for the Advancement of Sci., Washington, D. C. 63–71, 1990.
66. D.W. Urry, J. Xu, W. Wang, L. Hayes, F. Prochazka, and T.M. Parker, "Development of Elastic Protein-based Polymers as Materials for Acoustic Absorption." *Mat. Res. Soc. Symp. Proc.: Materials Inspired by Biology*, **774**, 81–92, 2003.
67. D.W. Urry, T.C. Woods, L.C. Hayes, J. Xu, D.T. McPherson, M. Iwama, M. Furuta, T. Hayashi, M. Murata, and T. M. Parker, "Elastic Protein-Based Biomaterials: Elements of Basic Science, Controlled Release and Biocompatibility." In: *Tissue Engineering and Novel Delivery Systems*, Marcel Dekker, Inc., New York, Chapter 2, pp. 31–54, 2004.
68. As reported in L.P. Wheeler, *Josiah Willard Gibbs, the History of a Great Mind*; Yale University Press: New Haven, CT, and London, 1952; pp 88–89, this statement was penned by J. Willard Gibbs in an 1881 letter to the American Academy of Arts and Sciences.

69. Poly(GVGVP) stands for the repeating sequence (glycyl-valyl-glycyl-valyl-prolyl)$_n$, also abbreviated as (Gly-Val-Gly-Val-Pro)$_n$, or even more simply as (GVGVP)$_n$ where n is the number of times that the sequence of five amino acid residues repeats. When given as poly(GVGVP), the value of n is an unspecified average but for our use below is quite large. For the comparisons of interest here for development of the T_t-based hydrophobicity scale the average value of n is of the order of 200.
70. Comparisons of this type require a standardized set of conditions of protein chain length, of concentration of model protein, of solution conditions and of a chosen model protein within which to make systematic replacements. The chain length should approach 1,000 residues; the required concentration has been determined to be 40 grams per liter or more; the solution should have the same concentration of salt as occurs in biology, the so-called physiological saline solution of 8.7 grams NaCl in one liter; and the chosen model protein should be elastic, should have an intermediate temperature range for aggregation, should have a favorable location for substitution that minimally affects the nature of the folding and assembly, and should be easily perturbed by selected substitution in the same favorable location in the sequence.
71. C.J. Heimbach, *Photochemical Transduction by Hydrophobically Poised Bioelastic Proteins*. Ph.D. Dissertation, The University of Alabama, Birmingham, 1998.
72. C.-H. Luan and D.W. Urry, "Elastic, Plastic, and Hydrogel Protein-based Polymers." In *Polymer Data Handbook*, J.E. Mark, Ed., 1999, Oxford University Press, New York, pp. 78–89, Tables 1 and 3a.
73. D.W. Urry, L.C. Hayes, D.C. Gowda, S.-Q. Peng, and N. Jing, "Electro-chemical Transduction in Elastic Protein-based Polymers." *Biochem. Biophys. Res. Commun.*, **210**, 1031–1039, 1995.
74. A. Pattanaik, D.C. Gowda, and D.W. Urry, "Phosphorylation and Dephosphorylation Modulation of an Inverse Temperature Transition." *Biochem. Biophys. Res. Comm.*, **178**, 539–545, 1991.
75. Asima Pattanaik, private communication. The value of the percentage phosphorylation of Pattanaik et al.[74] of 23.5% has been corrected to 47% as an average of five runs. This gives a ΔT_t of 840°C (860°C–20°C).
76. C.-H. Luan, T. Parker, K.U. Prasad, and D.W. Urry, "DSC Studies of NaCl Effect on the

Inverse Temperature Transition of Some Elastin-based Polytetra-, Polypenta-, and Polynonapeptides." *Biopolymers*, **31**, 465–475, 1991.
77. D.W. Urry, M.M. Long, R.D. Harris, and K.U. Prasad, "Temperature Correlated Force and Structure Development in Elastomeric Polypeptides: The Ile1 Analog of the Polypentapeptide of Elastin." *Biopolymers*, **25**, 1939–1953, 1986.
78. D.W. Urry, R.D. Harris, M.M. Long, and K.U. Prasad, "Polytetrapeptide of Elastin: Temperature Correlated Elastomeric Force and Structure Development." *Int. J. Pept. Protein Res.*, **28**, 649–660, 1986.
79. D.W. Urry, R.D. Harris, and K.U. Prasad, "Chemical Potential Driven Contraction and Relaxation by Ionic Strength Modulation of an Inverse Temperature Transition." *J. Am. Chem. Soc.*, **110**, 3303–3305, 1988.
80. D.W. Urry, L.C. Hayes, T.M. Parker, and R.D. Harris, "Baromechanical Transduction in a Model Protein by the ΔT_t Mechanism." *Chem. Phys. Lett.*, **201**, 336–340, 1993.
81. D.W. Urry, L.C. Hayes, and D.C. Gowda, "Electromechanical Transduction: Reduction-driven Hydrophobic Folding Demonstrated in a Model Protein to Perform Mechanical Work." *Biochem. Biophys. Res. Commun.*, **204**, 230–237, 1994.
82. D.W. Urry, "Engineers of Creation." *Chemistry in Britain*, 39–42, 1996.
83. D.W. Urry, L.C. Hayes, D.C. Gowda, C.M. Harris, and R.D. Harris, "Reduction-driven Polypeptide Folding by the ΔT_t Mechanism." *Biochem. Biophys. Res. Commun.*, **188**, 611–617, 1992.
84. D.W. Urry, L.C. Hayes, D.C. Gowda, and T.M. Parker, "Pressure Effect on Inverse Temperature Transitions: Biological Implications." *Chem. Phys. Lett.*, **182**, 101–106, 1991.
85. L.A. Strzegowski, M.B. Martinez, D.C. Gowda, D.W. Urry, and D.A. Tirrell, "Photomodulation of the Inverse Temperature Transition of a Modified Elastin Poly(pentapeptide)." *J. Am. Chem. Soc.*, **116**, 813–814, 1994.
86. D.W. Urry, S.-Q. Peng, L. Hayes, J. Jaggard, and R.D. Harris, "A New Mechanism of Mechanochemical Coupling: Stretch-induced Increase in Carboxyl pK_a as a Diagnostic." *Biopolymers*, **30**, 215–218, 1990.
87. D.W. Urry and S.-Q. Peng, "Non-linear Mechanical Force-induced pK_a Shifts: Implications for Efficiency of Conversion to Chemical Energy." *J. Am. Chem. Soc.*, **117**, 8478–8479, 1995.
88. R. Henze and D.W. Urry, "Dielectric Relaxation Studies Demonstrate a Peptide Librational Mode in the Polypentapeptide of Elastin." *J. Am. Chem. Soc.*, **107**, 2991–2993, 1985.
89. R. Buchet, C.-H. Luan, K.U. Prasad, R.D. Harris, and D.W. Urry, "Dielectric Relaxation Studies on Analogs of the Polypentapeptide of Elastin." *J. Phys. Chem.*, **92**, 511–517, 1988.
90. I.Z. Steinberg, A. Oplatka, and A. Katchalsky, "Mechanochemical Engines." *Nature*, **210**, 568–571, 1966.
91. M.V. Stackelberg and H.R. Müller, "Zur Struktur der Gashydrate." *Naturwissenschaften*, **38**, 456, 1951; M.V. Stackelberg and H.R. Müller, "Feste Gashydrate II: Struktur und Raumchemie." *Z. Elektochem.*, **54**, 25–39, 1954.
92. M.M. Teeter, "Hydrophobic Protein at Atomic Resolution: Pentagonal Rings of Water Molecules in Crystals of Crambin." *Proc. Natl. Acad. Sci. U.S.A.*, **81**, 6014–6018, 1984.
93. D.W. Urry, S.-Q. Peng, and T.M. Parker, "Delineation of Electrostatic- and Hydrophobic-Induced pKa Shifts in Polypentapeptides: The Glutamic Acid Residue." *J. Am. Chem. Soc.*, **115**, 7509–7510, 1993.
94. D.W. Urry, C.-H. Luan, R.D. Harris, and K.U. Prasad, "Aqueous Interfacial Driving Forces in the Folding and Assembly of Protein (Elastin)-Based Polymers: Differential Scanning Calorimetry Studies." *Polym. Preprints, Div. Polym. Chem., Am. Chem. Soc.*, **31**, 188–189, 1990.
95. A. Katchalsky, S. Lifson, I. Michaeli, and M. Zwick, "Elementary Mechanochemical Processes." In *Size & Shape of Contractile Polymers: Conversion of Chemical & Mechanical Energy*. Pergamon Press, New York, 1960, pp. 1–40, see page 11.
96. J. Wyman, "Allosteric Effects in Hemoglobin." *Cold Spring Harbor Symp. Quant. Biol.*, **28**, 483–489, 1963.
97. A.V. Hill, "The Possible Effect of the Aggregation of Hemoglobin on its Dissociation Curves." Proceedings of the Physiological Society, *J. Physiol.*, **40**, iv–vii, 1910, and A.V. Hill *J. Biochem.*, **7**, 471–480, 1913.
98. D.W. Urry, S.-Q. Peng, D.C. Gowda, T.M. Parker, and R.D. Harris, "Comparison of Electrostatic- and Hydrophobic-induced pKa Shifts in Polypentapeptides: The Lysine Residue." *Chem. Phys. Lett.*, **225**, 97–103, 1994.
99. D.W. Urry, S.-Q. Peng, T.M. Parker, D.C. Gowda, and R.D. Harris, "Relative Significance of Electrostatic- and Hydrophobic-Induced pK_a Shifts in a Model Protein: The Aspartic

Acid Residue." *Angew. Chem.* [German], **105**, 1523–1525, 1993; *Angew. Chem. Int. Ed. Engl.*, **32**, 1440–1442, 1993.

100. D.W. Urry, D.C. Gowda, S.-Q. Peng, T.M. Parker, N. Jing, and R.D. Harris, "Nanometric Design of Extraordinary Hydrophobicity-induced pKa Shifts for Aspartic Acid: Relevance to Protein Mechanisms." *Biopolymers*, **34**, 889–896, 1994, Figures 4 and 5B.

101. D.W. Urry, D.T. McPherson, J. Xu, H. Daniell, C. Guda, D.C. Gowda, N. Jing, and T.M. Parker, "Protein-based Polymeric Materials: Syntheses and Properties." In *The Polymeric Materials Encyclopedia: Synthesis, Properties and Applications*. CRC Press, Boca Raton, FL, pp. 7263–7279, 1996.

102. T. Cooper Woods, *Protein-based polymers as Delivery Vehicles for Antisense Oligonucleotides*. Ph.D. Dissertation, University of Alabama, Birmingham, 1998.

103. L. Hayes, *Effect of Hydrophobicity of Elastic Protein-based Polymers on Redox Potential*. Ph.D. Dissertation, University of Alabama, Birmingham, 1998.

104. M.F. Perutz, "Mechanisms of Cooperativity and Allosteric Regulation in Proteins." *Q. Rev. Biophys.*, **22**, 139–236, 1989.

105. The presence of positive cooperativity, explained by the interconversion between different conformational states, also demonstrates that these entropic elastic model proteins are not properly described as random chain networks, as the adherents of the relevance of the classical theory of rubber elasticity to protein elasticity are compelled to argue.

106. C. Bohr, K.A. Hasselbalch, and A. Krogh, "Uber einem in biologischen Beziehung wichtigen Einfluss, den die Kohlensaurespannung des Blutes auf dessen Sauerstoffbinding übt." *Skand. Arch. Physiol.*, **16**, 401–412, 1904.

107. G. S. Adair, "The Hemoglobin System. VI. The Oxygen Dissociation Curve of Hemoglobin." *J. Biol. Chem.*, **63**, 529–545, 1925.

108. J. Monod, J.-P. Changeux, and F. Jacob, "Allosteric proteins and molecular control systems." *J. Mol. Biol.*, **6**, 306–329, 1963.

109. J. Monod, J. Wyman, and J.-P. Changeux, "On the nature of allosteric transitions: a plausible model." *J. Mol. Biol.*, **12**, 88–118, 1965.

110. D.E. Koshland, G. Nemethy, and D. Filmer, "Comparison of experimental binding data and theoretical models in proteins containing subunits." *Biochemistry*, **5**, 365–385, 1966.

111. J. Monod, "On Symmetry and Function in Biological Systems." In *Nobel Symposium 11: Symmetry and Function of Biological Systems at the Macromolecular Level*. (A. Engstrom and B. Strandberg, eds.), Almqvist & Wiksell Forlag AB, Stockholm, 1968, page 1527. Also reprinted in *Selected Papers in Molecular Biology* by Jacques Monod, A. Lwoff and A. Ullmann. Eds Academic Press, 1978, page 708.

112. G. Weber *Protein Interactions*. Chapman and Hall, New York, 1992, page 104.

113. A. Katchalsky, "Solutions of Polyelectrolytes and Mechanochemical Systems." *J. Polymer Sci.*, **7**, 393–412, 1951.

114. The change in chemical energy per mole ($\Delta G/\Delta n$), also called the change in chemical potential ($\Delta \mu$), is proportional to the change in pH required to go from the COOH state to the COO$^-$ state. Obviously for the narrower, steeper curves a smaller change in chemical energy would be required. The result would be a more efficient energy conversion, if, for example, the change resulted in the performance of mechanical work, as treated below in section **5.9.5**.

115. F.E. Harris and S.A. Rice, "A Chain Model of Polyelectrolytes. I." *J. Phys. Chem.*, **581**, 725–732, 1954.

116. J. Th. G. Overbeek, "The Dissociation and Titration Constants of Polybasic Acids." *Bull. Soc. Chim. Belg.*, **57**, 252–261, 1948.

117. C.-H. Luan, T. Parker, K.U. Prasad, and D.W. Urry, "DSC Studies of NaCl Effect on the Inverse Temperature Transition of Some Elastin-based Polytetra-, Polypenta-, and Polynonapeptides." *Biopolymers*, **31**, 465–475, 1991.

118. H. S. Harned and B. B. Owen, *The Physical Chemistry of Electrolyte Solutions*. 3rd Ed. Rheinhold, New York, 1967.

119. D.W. Urry, "What is Elastin; What is Not." *Ultrastruct. Pathol.*, **4**, 227–251, 1983.

120. D.W. Urry, K. Okamoto, R.D. Harris, C.F. Hendrix, and M.M. Long, "Synthetic, Cross-Linked Polypentapeptide of Tropoelastin: An Anisotropic, Fibrillar Elastomer." *Biochemistry*, **15**, 4083–4089, 1976.

121. D. W. Urry, "Free Energy Transduction in Polypeptides and Proteins Based on Inverse Temperature Transitions." *Prog. Biophys. Molec. Biol.*, **57**, 23–57, 1992.

122. D. W. Urry, "Five Axioms for the Functional Design of Peptide-Based Polymers as Molecular Machines and Materials: Principle for Macromolecular Assemblies." *Biopolymers (Peptide Science)*, **47**, 167–178 (1998).

6
On the Evolution of Protein-based Machines (Toward Complexity of Structure and Diversity of Function)

"There is grandeur in this view of life, with its several powers, having been originally breathed into a few forms or into one; and that...from so simple a beginning endless forms most beautiful and most wonderful have been, and are being, evolved."

Charles Darwin, 1859
On the Origin of the Species[1]

6.1 Introduction

This volume presents a consilient mechanism, "a common groundwork of explanation," for energy conversions whereby protein-based machines interconvert the set of six energies interconverted by living organisms. This chapter examines the simple stepwise process whereby protein compositional changes can access each of the six energy sources and can improve the efficiency of each of a resulting 18 pairwise energy conversions.

How complex are the changes in model protein composition required to obtain new molecular machines capable of new energy conversions? In fact, with knowledge of the consilient mechanism, they are, by experimental demonstration, almost trivial! How complex are the compositional changes required to make a given molecular machine more efficient? Again, by means of the rules emergent from the fundamental mechanism, the requisite compositional changes in the model proteins are extraordinarily simple!

Given the common genetic code for all Life, we now ask, how complex are the mutations required to make the compositional changes that access new energies and that make protein-based machines more efficient? Remarkably, arriving at the necessary mutations is uncomplicated! A single base mutation is all that is required to access a new energy source or to make a given energy conversion more efficient. As we will discuss, by single base mutations, we evolve a single molecular machine, a thermally driven engine, step after naturally selected step into diverse and increasingly efficient protein-based machines capable of interconverting the energies essential for Life. Interestingly, the organization of the genetic code itself delineates oil-like and vinegar-like residues and allows evolution to be straightforward.

Accordingly, this chapter addresses the simplicity of molecular evolution for biological energy conversion. In presenting a scenario for the evolution of protein machines, we accept the challenge of Behe, which also underscores the quintessential role of molecular machines in Life's processes. In Behe's words, "if you search the scientific literature on evolution, and if you focus your search on the question of how molecular machines—the basis of life—developed, you find an eerie and complete silence. The complexity of life's foundation has paralyzed science's attempt to account for it; molecular machines present an as-yet-impenetrable barrier to Darwinism's universal reach."[2]

Armed with the consilient mechanism for energy conversion, the "as-yet-impenetrable barrier" of Behe transforms into a springboard for insight into early evolution of the molecular machines that sustain Life. Each step to access an additional energy and each step to achieve greater efficiency for a given energy conversion represents a step toward greater complexity of structure. Clearly, each such single step underscores an improved capacity to survive. If production of the molecular machines of Life is such that each new and/or more efficient naturally preferred machine is achievable by a simple process, then, given the Blueprint of Life seen in the genomes of humans and lower species,[3,4] there remains little bewilderment in the natural flow toward increased diversity and greater complexity.

Alvin Toffler, in his Forward for Prigogine and Stengers' book *Order out of chaos*,[5] accepts the message of Darwinian evolution with this statement:

Imagine the problems introduced by Darwin and his followers! For evolution, far from pointing toward reduced organization and diversity, points in the opposite direction. Evolution proceeds from simple to complex, from "lower" to "higher" forms of Life, from undifferentiated to differentiated structures. And, from a human point of view, all is quite optimistic. The (biological) universe gets "better" organized as it ages, continually advancing to a higher level as time sweeps by.

For the universe as a whole, time's arrow points to increasing disorder and uniformity, yet for biology time's arrow points to increasing diversity and complexity of structure and function. We will see how this can be for the structural evolution of the molecular machines of Life. Then we examine how the molecular machines, themselves functioning by the consilient mechanism, facilitate the creation of order.

Life represents a perpetuating, diversifying energy-collecting system of stepwise increasing complexity. Is the physical mechanism for this energy-driven march from disorder to order itself efficient? As shown later in this chapter, the answer here also becomes a simple yes. Because of the way in which it accesses and employs energy, Life represents an efficient energy-driven assembling, ordering, negative entropy machine.

The physical basis for this march toward greater order with time is the special energy-collecting, phase transition of biology, referred to previously in this volume as an *inverse temperature transition*. The separation of oil from water and the repulsion of oil-like from vinegar-like groups constrained to coexist along a protein chain molecule, so readily positioned by the genetic code for the biosynthesis of protein, become the physical basis for biology's energy-driven march away from disorder.

6.2 Accessing Available Energies to Achieve Useful Function

We recall the Szent-Györgyi assertion, "Motion has always been looked upon as the index of Life."[6] Here, we begin with the historical sequence of energy sources used in our laboratory to produce motion by means of elastic–contractile model proteins specifically designed to be responsive to those differing energy sources. Thermal energy (heat) and then chemical energy in the form of increased concentration of table salt (NaCl) and in the form of increased concentration of acid were considered in turn and found to cause elastic strips, made from properly designed model proteins, to contract and to produce motion. Also, pressure energy produced motion by acting on natural aromatic amino acids. The energy conversions were unmistakably demonstrated by the simple visual observation of lifting weights, that is, of pumping iron. This was not after-the-fact rationalization of experimental results, but rather successful designs of protein-based machines intended to interconvert the energy conversions subsequently demonstrated.

The preceding energy sources—thermal, chemical, and pressure—acted directly on the model proteins with natural amino acid residues available through use of the genetic code. Accessing light and electrical energies requires attachment of vitamin or vitamin-like biological molecules to the elastic model proteins after including yet another natural amino acid. In all, five different input energies produce motion, that is, result in the energy output described as the performance of mechanical work, which is easily observed as the lifting of a weight. The following describes the simplicity of changes in model proteins required to access these energies to produce motion. In addition, an appropriate pair of functional groups, by being part of the same oil-like domain on phase separation, can be utilized to interconvert any two of the five different energy inputs that independently produce motion.

6.2.1 Using Heat (Thermal Energy) to Produce Motion

We begin, as the story for these protein-based polymers began, by using temperature to drive the phase separation mechanism as demonstrated by a synthetic high polymer based on a naturally occurring repeat of five residues. Cross-linking the polymer composed of the repeating pentamer, $(GVGVP)_{251}$, also given below as **Polymer I**, produces elastic sheets or strips (see Figure 5.14). Raising the temperature of the elastic-contractile sheet from room temperature to body temperature, that is, to above the temperature for the onset of the phase separation, causes the oil-like groups distributed along the extended protein chain molecules that were surrounded by water to be separated from water. As a result, the polymer chains fold, producing a contractile event and the lifting of a weight (see Figure 6.1).[7–9]

6.2.2 Using Salt (Chemical Energy) to Produce Motion

With the same elastic sheets as above, addition of salt (NaCl) lowers the temperature of the phase separation from above to below the operating temperature and drives contraction with the moving of a weight, but *without changing the temperature*. A point to be made here is the amount of salt required for achieving contraction under these conditions. Replacing tap water with seawater would result in but a small contraction, but doing so with waters of the Dead Sea or of the Great Salt Lake would cause a more significant contraction.[10]

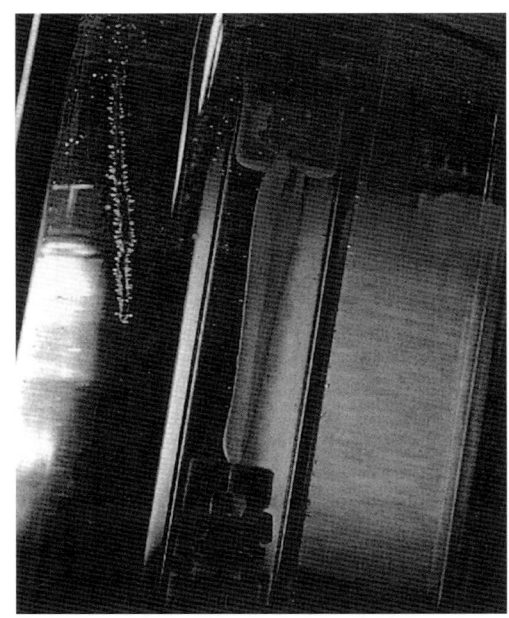
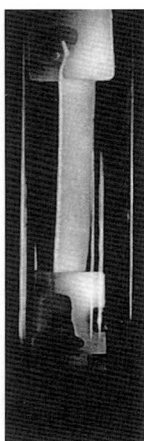

FIGURE 6.1. Elastic-contractile strip of cross-linked poly(GVGVP) in water contracting to lift a weight (time to half contraction: 5 minutes for millimeter thickness, seconds for 50μ thick fibers). (Reproduced from Urry.[8])

6.2.3 More Efficient Conversion of Increased Salt into Motion

To this point thermal and chemical energies have driven contraction without a change in amino acid composition of the basic model protein. An elastic sequence from the mammalian elastic fiber has been used as naturally found, albeit a sequence separate from and syn-

6.2 Accessing Available Energies to Achieve Useful Function

thesized to form a 20- to 30-fold longer chain. Now we depart from what was directly available from a natural sequence and design a change in composition of the basic model protein to achieve more efficient energy conversion.

6.2.3.1 Replacing Val by Glu

The infrequent replacement of valine (Val, V) residues by glutamic acid residues (Glu, E), as in **Polymer II** in Table 6.1, replaces the oil-like group ($-CH(CH_3)_2$) by a vinegar-like functional group ($-CH_2CH_2-COOH$). This introduces at physiological pH the negatively charged carboxylate ($-COO^-$) group. Addition of salt (NaCl) to water results in positively charged sodium ions (Na^+) and negatively charged chloride ions (Cl^-). The positively charged sodium ions (Na^+) ion pair with the negatively charged carboxylate ($-COO^-$) group. Formation of the ion pairs ($-COO^- \ldots Na^+$) lowers the temperature of the phase separation from above to below the operating temperature and drives contraction with the moving of a weight.

As is apparent from Figure 5.12, at room temperature the amount of salt required to effect the phase separation and drive contraction by ion pairing is less than one-tenth that amount required when there are no negatively charged carboxylate ($-COO^-$) groups. This means a 10-fold increase in efficiency of the function of this salt-driven molecular machine. The simple replacement of Val by Glu has increased the efficiency of salt-driven contraction by more than 10-fold (Figure 5.12).

6.2.3.2 Replacing Other Val Residues by Phe Residues Further Increases Efficiency

Polymers III through **VI** represent a systematic increase in the number of Val (V) residues replaced by more oil-like Phe (F) residues. Each step increase in oil-like character of the model protein, on going from 0 to 2 to 3 to 4 and to 5 Phe residues for every 30 residues, stepwise increases the affinity of Na^+ for $-COO^-$. Each step increase in oil-like character means that less of an increase in salt is required to drive contraction. **Polymer VI**, when cross-linked into elastic sheets, provides the most efficient molecular machine of the set. This molecular machine requires less chemical energy to produce a given amount of motion, that is, to perform a given amount of mechanical work.

6.2.4 Accessing Acid (Another Form of Chemical Energy) to Produce Motion

An even more effective way to neutralize the affect of the vinegar-like carboxylate is to add acid, (H^+). This is the association of a negative carboxylate, ($-COO^-$) with the positive proton (H^+). The association of proton with carboxylate represents a kind of ion pairing and forms a more stable carboxyl, that is, $-COO^- + H^+ = -COOH$. Using **Polymer II**, the amount of

TABLE 6.1. The polymers (model proteins) as considered in this chapter.

Polymer	I:	(GVGVP GVGVP GVGVP GVGVP GVGVP GVGVP)$_{41}$(GVGVP)$_5$	
Polymer	II:	(GVGVP GVGVP GEGVP GVGVP GVGVP GVGVP)$_n$(GVGVP);	E/0F
Polymer	III:	(GVGVP GVGFP GEGFP GVGVP GVGVP GVGVP)$_n$(GVGVP);	E/2F
Polymer	IV:	(GVGVP GVGVP GEGVP GVGVP GVGFP GFGFP)$_n$(GVGVP);	E/3F
Polymer	V:	(GVGVP GVGFP GEGFP GVGVP GVGFP GVGFP)$_n$(GVGVP);	E/4F
Polymer	VI:	(GVGVP GVGFP GEGFP GVGVP GVGFP GFGFP)$_n$(GVGVP);	E/5F
Polymer	VII:	(GVGVP GVGVP GKGVP GVGVP GVGVP GVGVP)$_n$(GVGVP);	K/0F
Polymer	VIII:	(GVGVP GVGVP GKGFP GVGVP GVGVP GVGVP)$_n$(GVGVP);	K/2F
Polymer	IX:	(GVGVP GVGVP GKGVP GVGVP GVGFP GFGFP)$_n$(GVGVP);	K/3F
Polymer	X:	(GVGVP GVGFP GKGFP GVGVP GVGFP GVGFP)$_n$(GVGVP);	K/4F
Polymer	XI:	(GVGVP GVGFP GKGFP GVGVP GVGFP GFGFP)$_n$(GVGVP);	K/5F
Polymer	XII:	(GVGIP GVGIP GVGIP GVGIP GVGIP GVGIP)$_{43}$(GVGIP)$_2$(GVGVP)	

The subscript n indicates the number of repeats of the 30 residues grounded as 6 pentamers, and n can range from 2 to 40 and more.

added acid that drives contraction is one-thousandth the amount of salt required in the most efficient salt-driven contraction noted above.[11] As more Phe residues replace Val residues, the amount of added acid required to drive contraction becomes less and less. Finally, with **Polymer VI**, a several hundred- to a thousand-fold decrease in the amount of acid required to drive contraction occurs.[9] The efficiency of the molecular machine has increased by another several hundred to a thousand-fold. Now, a small change in acidity from physiological conditions can be used to drive contraction (see Figures 1.2, 1.3, and 5.19B). Just the common uptake of carbon dioxide from the air, which combines with water to form carbonic acid, or addition of a small amount of the weak acid, for example, vinegar, becomes sufficient to drive contraction of elastic strips of **Polymer VI**.

Thus, compositional changes achieved by the simple substitution of Val by Glu and then of Val by Phe create successively more efficient molecular machines with which to produce motion. In fact, protonation of carboxylates provides the driving force for the propeller-like motion of the bacterial flagella,[12] which provides the mobility whereby the bacterium obtains food. The structure of the flagellar rotary motor has evolved to become much more complex than our linearly contracting model proteins. A simpler rotary motor, ATP synthase,[13] the primary energy-converting motor of biology, uses protonation of carboxylates to drive a rotor that is utilized within the F_1-ATPase component to result in the synthesis of ATP from ADP and P_i.[14] Whether it is the conversion of chemical energy into motion or into another form of chemical energy, we believe the underlying mechanism to be the same.

In our view, cyclic protonation/deprotonation of carboxylates drives the cyclic association/dissociation of oil-like groups responsible for flagellar rotary motion and for rotation of an asymmetric oil-like rotor to produce ATP by ATP synthase. In the process, protons move from the high concentration to the low concentration side of a cell membrane in which the membrane component of the rotary motor resides. In fact, the proton concentration difference from one side to the other of the membrane would not need to be particularly high to drive the rotary motor. The efficiency of the flagellar motor is reported to be about 50% and that of the ATPase component of ATP synthase operating in reverse has been calculated to approach 80% to 100%[15] (see Chapter 8).

6.2.5 Producing Motion by Changes in Pressure

With the above change of the aliphatic hydrocarbon side chain of the Val (V) residue to the aromatic hydrocarbon side chain of the Phe (F) residue also comes the capacity for pressure to produce motion.[16] With the correct balance of oil-like and vinegar-like residues to set the temperature at which oil-like separation from water occurs, the application of pressure causes the unfolding and setting down of a weight, and release of pressure results in the lifting of a weight. Remarkably, with only the conversion of an occasional Val to Glu or an occasional Val to Phe, both chemical energy and pressure energy have been accessed by these protein-based molecular machines.

6.2.6 Accessing Electrons (Electrical Energy) to Produce Motion

6.2.6.1 Reduction of Lysine-Attached Nicotinamide Drives Contraction

Production of motion by the addition of electrons requires attachment of vitamin-like molecules such as B_2 and B_3 (see Chapters 3 and 5), because the natural amino acids do not easily take up and release electrons. In our experimental demonstration, an N-methyl nicotinamide was attached to a lysine residue by the amide moiety, and indeed reduction drove contraction.[17] Furthermore, increasing the oil-like character on replacing Val (V) by Phe (F), as with **Polymers XIII**, **IX**, and **X** in Table 6.1, increases the affinity of an oxidized group for electrons, as discussed in Chapter 5 and represented in Figure 5.19C. In the circumstance of natural proteins, the reduction–oxidation (redox) vitamin-like molecules bind to oppositely charged sites by ion pairing and by oil-like associations, as discussed below.

6.2.6.2 Adding Positively Charged Residues

In living organisms, the vitamin-like molecules that accept and release electrons are commonly negatively charged due to the presence of phosphate groups. Examples are phosphorylated nicotinamide (e.g., nicotinamide mononucleotide, which may be written nicotinamide-ribose-phosphate) and flavin (e.g., flavin mononucleotide, which may be written flavin-pentose-phosphate). Introduction of positively charged lysine (Lys, K) and arginine (Arg, R) residues provides sites for binding of negatively charged phosphate of the vitamin-derived molecules that accept and release electrons. **Polymers VII** through **XI** in Table 6.1 demonstrate the effect of an increase in the more hydrophobic Phe (F) residues. The increase in Phe residues increases both the binding affinity of the negatively charged vitamin-like nucleotide to the positively charged protein sites and the affinity of the vitamin-like molecules for electrons. The latter result in more efficient electron-driven molecular machines.

As was discussed in Chapter 5, for Figure 5.17, addition of electrons to a positively charged redox group increases oil-like character and drives model protein folding, which result in contraction and the performance of mechanical work. The increase in affinity for electrons of the vitamin-like molecule that occurs on replacement of Val by Phe (see Figure 5.20C) makes for a more efficient electron-driven contraction. Thus, a genetic code that would allow easy mutational steps to become more oil-like would, here again, provide for evolution of more efficient protein-based machines.

6.2.6.3 Increasing Affinity for Electrons Using Negatively Charged Residues

If the chemical group that can take up and release electrons is positively charged, the same changes in composition that increased affinity for Na^+ or H^+ above become operative. In particular, if the chemical group in its electron-deficient state were a positively charged vitamin B_3, niacin, the series of **Polymers II** through **VI** would demonstrate a systematic increase in affinity for the binding of positively charged B_3. Correspondingly, the stepwise increase in oil-like character on progressing from **Polymer II** to **VI** would also result in a systematic increase in affinity for electrons (see Figure 5.20C). Development of binding capacity to a vitamin-like molecule with both oil-like and charged components depends on the charge and the oil-like character of the polymer.

Alternatively, the covalent attachment of a redox group like niacin to the lysine side chain rather than ion-pair formation, as with **Polymers VII** through **XI** in Table 6.1 and as shown in Figure 5.20C, results in nicotinamides that exhibit increased affinity for electrons as the polymer contains more hydrophobic Phe residues. *Thus, stepwise increases in the oil-like character of the polymer result in stepwise increases in electron affinity and correspondingly steeper curves indicating greater efficiencies.*

6.2.7 Accessing Light Energy to Produce Motion

For light to be accessed to produce motion by the consilient mechanism, there must be bound to the model protein a group that changes its oil-like character on absorption of a photon. There are many ways in which this can happen. Given the correct conditions of the solution, even the nicotinamide considered above can have its oil-like character changed by the absorption of light. Another biological molecule, cinnamic acid, can be attached to a Lys side chain to form a cinnamide. The cinnamice is normally a linear group, but on absorption of a photon it becomes kinked. Absorption of a different photon can reverse the process and straighten out the kink in the group. The linear molecule is more oil-like than the kinked molecule. Therefore, with the proper balance of oil-like and vinegar-like side chains of the model protein, absorption of the photon that straightens the group can cause contraction and the lifting of a weight, and absorption of the photon that introduces the kink would effect the lowering of the weight. The principle of a light-induced kink in the carotinoid of the visual process is an important element of the energy conversion leading to vision, as in rhodopsin of the vertebrate eye and bacteriorhodopsin.

6.2.8 Accessing Energies to Perform Work Other Than Motion

For electrical energy to be converted to chemical energy, one need only utilize the amino acid substitutions already considered. The presence of a positively charged Lys residue for attachment of a phosphorylated nicotinamide-ribose group and the presence of a negatively charged Glu or Asp with the carboxylate function in the side chain are sufficient to interconvert electrical and chemical energy. Reduction of the nicotinamide drives oil-like folding and forces uptake of a proton by the carboxylate group (see Figures 5.20A and 2.12B). This represents the conversion of electrical energy to chemical energy using the same groups as are involved in mitochondria, the energy-converting factories of the cell. Using an elastic model protein with an adequate oil-like character and containing both a lysine-attached N-methyl nicotinamide and a carboxylate group, reduction of the oxidized N-methyl nicotinamide causes the uptake of a proton.[18] Electrical energy is converted to the chemical energy of pumping protons.

6.2.9 Summary of Minimal Amino Acid Changes to Achieve the Diverse Energy Conversions of Biology

Starting from the elementary elastic protein sequence $(GVGVP)_n$, by simple substitution, the set of protein-based machines sufficient to provide the set of biological energy conversions can be achieved. It is only necessary to substitute the Val (V) residue by Glu (E) and Phe (F) and Ile (I) by Lys (K). Substitution by Glu accesses chemical energy. Substitution by Phe accesses pressure energy and increases efficiency of energy conversions by the consilient mechanism. Substitution by Lys accesses electrical and light energies. The question to be addressed now is just how difficult such substitutions would be by means of evolution under the constraints of the universal genetic code.

6.3 Mutations and Evolution of Protein-based Machines

6.3.1 A Single Genetic Code for All Life

All known Life is based on but a single genetic code (see Table 6.2).[19] Because of this, all Life would seem to be descended from a single primitive cell. The momentous event or series of events of Life's origin with a single genetic code are not addressed here. From the viewpoint of energy conversion, however, it would have culminated in a moment when the minimal set of molecular machines assembled to form the first primitive and most competitive self-replicating cell.

Four bases occur in the macromolecule deoxyribonucleic acid that ultimately encode protein sequences. They are the bases U (uracil), C (cytosine), A (adenine), or G (guanine), which encode for 20 amino acid residues. They do so by means of specified sequences of three bases called *triplets*. Given four possible bases but a sequence of three being required to specify a given residue, there are $4^3 = 4 \times 4 \times 4 = 64$ possible triplets. Three triplets specify the end of a protein sequence; these are called *stop codons*. This leaves the other 61 triplets to encode for 20 amino acids, and all triplets are used. As a result, degeneracy or redundancy occurs, that is, several different triplets can encode for the same residue.

With all that we currently know, no compelling physicochemical basis appears to exist to explain why a particular triplet codon must encode for a particular amino acid residue throughout all of biology. Furthermore, why do six different triplet codons encode for leucine, serine, and arginine and four different codons encode each for valine, glycine, proline, alanine, and threonine? On the other hand, for the standard genetic code, one triplet encodes for tryptophan and one for methionine; two encode for aspartic acid, glutamic acid, tyrosine, phenylalanine, histidine, glutamine, asparagine, lysine, and cysteine, and three triplets encode for isoleucine. From this perspective, there seems to be little coherence in the genetic code.

6.3 Mutations and Evolution of Protein-based Machines

TABLE 6.2. The genetic code.

First position (5' end)	Second position				Third position (3' end)
	U	C	A	G	
U	UUU Phe	UCU Ser	UAU Tyr	UGU Cys	U
	UUC	UCC	UAC	UGC	C
	UUA Leu	UCA	UAA STOP	UGA STOP	A
	UUG	UCG	UAG STOP	UGG Trp	G
C	CUU	CCU	CAU His	CGU	U
	CUC Leu	CCC Pro	CAC	CGC Arg	C
	CUA	CCA	CAA Gln	CGA	A
	CUG	CCG	CAG	CGG	G
A	AUU	ACU	AAU Asn	AGU Ser	U
	AUC Ile	ACC Thr	AAC	AGC	C
	AUA	ACA	AAA Lys	AGA Arg	A
	AUG Met[b]	ACG	AAG	AGG	G
G	GUU	GCU	GAU Asp	GGU	U
	GUC Val	GCC Ala	GAC	GGC Gly	C
	GUA	GCA	GAA Glu	GGA	A
	GUG	GCG	GAG	GGG	G

[a] Nonpolar amino acid residues are tan, basic residues are blue, acidic residues are red, and polar uncharged residues are purple.
[b] AUG forms part of the initiation signal as well as coding for internal Met residues

Source: D. voet, J. Voet, and C. Pratl, Fundamentals of Biochemistry, John Wiley and Sons, New York, 1990.

The set of codons that encode for the same residue, however, do not occur at random. Instead, a set of four codons that may encode for the same amino acid residue is such that the third base in the triplet codon can change without changing the amino acid residue. Of the four codons for valine, for example, the first two bases are always GU and the third base can be any one of the four bases, U (uracil), C (cytosine), A (adenine), or G (guanine). Thus, this residue has a factor in favor of its occurrence, and because of this changes in the third base of the triplet do not result in a mutation. These are fundamental issues relating to the origin and evolution of self-replicating Life.

As reviewed above and demonstrated in Chapter 5, substitutions of the Val residue by another residue allowed access to a new energy source and/or a more efficient use of a given energy source. *Here we simply look at the genetic code and observe whether the critical process for this particular evolution of molecular machines is difficult or easy to achieve by means of the changes dictated by the genetic code.* As argued by Behe, evolution of biology's molecular machines presents a bewildering enigma to the gradualism required of Darwinian evolution.[2] As we will note below, quite the opposite obtains. When considering the consilient mechanism, functional relationship emerges between the genetic code and achieving diverse and more efficient protein-based machines.

Clearly, the capacity to access energy available in the environment and to convert that energy to useful function was key to survival of primitive Life, once established. In fact, perhaps the most remarkable and central characteristic of Life is its evolutionary capacity to access virtually every available energy source and to adapt to utilize each energy source more efficiently. Accordingly, it is central to evolution to understand how simple or complex the process of accessing a new energy source may be on the basis of what we have learned about protein-based machines functioning by the considered hydrophobic consilient mechanism.

6.3.2 General Relationship of the Genetic Code to the Consilient Mechanism

Using the genetic code for all living organisms given in Table 6.2, we now address whether a coherence or a consilience can be identified within the genetic code. In approaching this issue, the genetic code may be looked upon as families of amino acid residues. A family would be that set of amino acid residues that has the same base for the second position of the triplet codon, and a member of the family would have any one of the other four bases in the first position. Now we ask if a family exists for which there is good coherence in the property of the **R**-groups of the family. By this we mean that a change in the first and third base of the triplet codons for the family would have only a qualitative effect on the resulting protein sequence. In other words, does a family exist for which any mutation, any change, involving the first and/or third base of the triplet codon would not fundamentally change or destroy function of the resulting protein. If only one such family exists, then it might be called the *primary family* of the genetic code.

By considering the T_t-based hydrophobicity scale noted in Chapter 2 and developed in Chapter 5 (see Table 5.1), the primary family is readily recognized; it is valine (Val, V), methionine (Met, M), isoleucine (Ile, I), leucine (Leu, L), and phenylalanine (Phe, F). These are all hydrophobic residues without any other functional capacity. The residues valine and methionine exhibit a similar degree of oil-like character. Substitution of one by the other would hardly change the temperature of the inverse temperature transition at all. Conversion from Val to Ile or Leu results in the simple addition of a CH_2 group, which constitutes a modest increase in oil-like character. Conversion of Val to Phe does involve a substantial increase in oil-like character, but adds no other physical property, only an increase in oil-like character.

It is important to note that tyrosine (Tyr, Y) and tryptophan (Trp, W) are more hydrophobic than valine in the T_t-based hydrophobicity scale. Therefore, Y and W would be candidates for the primary, the hydrophobic, family of the genetic code, except that these residues add additional physical properties. Both have large dipole moments and exhibit chemical reactivities that can dramatically change their oil-like character. Most notably, tyrosine can be a site for phosphorylation, which converts it to a supervinegar-like residue and dramatically disrupts hydrophobic folding, assembly, and associated functions. On our scale, based directly on the hydrophobic association event, tryptophan is the most hydrophobic residue, whereas other scales that utilize less direct means of assessing functional hydrophobicity place trytophan at much less hydrophobic positions. Thus, these residues, tyrosine and tryptophan, would not

6.3 Mutations and Evolution of Protein-based Machines

constitute part of a family that is unambiguously oil-like in its function.

The other three families of the genetic code encode for less homogeneous sets of amino acid residues. Perhaps the second most coherent family would be that with C (cytosine) in the second position of the triplet. These residues occupy a neutral range in the hydrophobicity scale. These are alanine (Ala, A), serine (Ser, S), proline (Pro, P), and threonine (Thr, T), but S and T can be phosphorylated to become supervinegar-like residues. The most diverse family would appear to be that with A (adenine) in position two of the triplet codon. These residues span from the most polar glutamic acid residue with a negative charge, to lysine with the opposite signed charge, to the very hydrophobic tyrosine with a site for phosphorylation, to histidine, which, due to its structure, provides key functionality in active sites of many enzymes, and finally to glutamine and asparagine, which are relatively unstable and with time can break down to glutamic and aspartic acids. The remaining family with G (guanine) in position two is also a mixed bag containing glycine, cysteine, tryptophan, arginine, and, again, serine. Clearly, the most homogeneous and coherent family is the unambiguously hydrophobic family with U (uracil) in position two.

Starting with the primary (hydrophobic) family of amino acid residues with U as the second base of the triplet codon, only a change in the second base provides the opportunity to change to a less oil-like, more vinegar-like residue. Among the more vinegar-like residues are the negatively charged and positively charged residues that provide direct access to new energy sources, as noted above and seen in Chapter 5.

6.3.3 Mutations That Create New and More Efficient Molecular Machines

The capacity to access a new energy source constitutes the cardinal step in the evolution of an organism. As we will discuss, it represents but a trivial step in structural modification at the DNA level. The capacity to access a new energy source or to more efficiently use a source of energy renders an organism more fit to survive whenever an energy source becomes limiting or wherever a different energy source is available. This symbolizes natural selection and survival of the fittest in terms of molecular machines.

6.3.3.1 Changing Val to Glu (a Step Toward Diversity and Greater Complexity)

The mutation, the change in base sequence of DNA to insert a different amino acid residue in the resulting protein, required to access a new energy source is remarkably minimal. A single base change in the codon encoding for valine allows the change from a thermally driven protein-based machine to a more efficient chemically driven protein-based machine. The mutation from a Val residue to a carboxylate-containing Glu or Asp residue requires but a single base change in the triplet codon, and it can occur with any of the four codons for the Val residue.

Changing the second base of two of the four triplet codons for Val, namely, GUA and GUG to GAA and GAG, respectively, are two ways to convert Val to Glu. Changing the second base of the other two triplet codons for Val, for example, GUU to GAU and GUC to GAC, converts Val to Asp. Thus, the Val triplet codons are such that a single base change of the second base from U to A results in amino acid residues with carboxylate side chains. *A single mutation converts a thermally driven (and also a poor chemically driven) protein-based machine into a more efficient chemically driven protein-based machine. How trivial and likely the diversification of biology's molecular machines, especially because it costs no more energy (of biosynthesis) to produce the new or improved protein-based machine.*

6.3.3.2 Changing Val to Phe (Another Step Toward Diversity and Complexity)

The mutation from the Val residue to the more oil-like Phe residue again requires but a single base change in the triplet codon, and it can occur with either of two codons. Instead of changing the second base of GUU and GUC to A to get aspartic acid, the first base, G, is changed to U to give the very hydrophobic phenylalanine residue. The change of Val to Phe results in more efficient chemically driven protein-based machines as well as accesses pressure energy, as discussed above (also see Figures 1.2, 1.3, and 5.34). More modest increases in hydrophobicity and in resulting efficiency occur by changing the first base in the valine triplet codons from G to C, for example, to give leucine, and to A to give isoleucine from any three of valine's four triplet codons. For the fourth triplet codon of valine, GUG, conversion of the first base to A gives methionine with an insignificant change in oil-like character from that of valine.

6.3.3.3 Replacing Oil-Like Residues by Positively Charged Residues (Further Steps Toward Diversity and Greater Complexity)

As indicated above, isoleucine (Ile, I) is only slightly more hydrophobic than valine, and methionine (Met, M) is very similar to valine when compared using the T_t-based hydrophobicity scale. Thus these are functionally equivalent, or nearly so, to the valine residue. The AUA triplet codon for isoleucine and the AUG triplet codon for methionine convert to codons for lysine by changing the second base in these triplet codons to A. A single base change from these hydrophobic residues gives lysine (Lys, K). Changing the same isoleucine and methionine codons to G in the second position gives the positively charged arginine residue, as does the same change to four CUX leucine codons. As noted above, the presence of lysine provides entry to electromechanical and electrochemical transduction by the capacity of the positively charged lysyl side chain to bind the negatively charged nicotinamide and flavin mononucleotides, for example. This added structural complexity provides for electrically(redox)-driven protein-based molecular machines.

6.3.3.4 Providing Sites for Phosphorylation

It is remarkable that the above identified primary family of amino acids of the genetic code can, by single base mutations, give rise to the full range of protein-based machines for the conversion of one form of energy to another. Another fundamental energy conversion is from one form of chemical energy to another. The central energy conversion between chemical energies in biology involves phosphorylation. The required mutations to provide sites for attachment of phosphate are those that result in amino acid residues with the –OH group available. The residues with the OH functional group are tyrosine, serine, and threonine. Converting the second, the central, U of phenylalanine to A accesses tyrosine. On the other hand, conversion of the central U of phenylalanine to C gives serine. Finally, threonine derives from conversion of the central U of the triplet codon of isoleucine and methionine to C.

6.3.4 Single Base Mutations Access New and/or More Efficient Machines

In short, the primary (hydrophobic) family of the genetic code, which provides for the hydrophobic phase separation mechanism, also provides the capacity to access all classes of protein-based machines, that is, all energy sources, by a single base change. This is the nature of the genetic code. *In this regard, the genetic code becomes understood in terms of T_t-type protein-based machines. Might protein machines based on inverse temperature transitional behavior be fundamental for Life to exist?*

6.4 Energy Inputs Create Order Out of Chaos (Biology's Reversal of Time's Arrow for the Universe)

From the Preface of *Order Out of Chaos*[5]:

Our scientific heritage includes two basic questions which till now no answer was provided. One is the relation between order and disorder. The famous law of increase of entropy describes the world as evolving from order to disorder; still, biological or social evolution shows us the complex emerging from the simple. How is this possible? How can structure arise from disorder? Great progress has been realized in this question. We know now that nonequilibrium,

6.4 Energy Inputs Create Order Out of Chaos

the flow of matter and energy, may be a source of order.

More to the point, we now see biology's access to energy by means of the consilient mechanism of energy conversion, combined with readily available mutations to improve protein-based machines, as the source of increased structural order and functional diversity.

Again drawing from Toffler's Forward for *Order Out of Chaos*[5]:

In classical or mechanistic science, events begin with "initial conditions," and their atoms or particles follow "world lines" or trajectories. These can be traced either backward into the past or forward into the future. This is just the opposite of certain chemical reactions, for example, in which two liquids poured into the same pot diffuse until the mixture is uniform or homogenous. These liquids do not de-diffuse themselves. At each moment of time the mixture is different, the entire process is "time oriented."

The time orientation is relentlessly in the direction of maximal disorder.

While indeed the two liquids do not of themselves de-diffuse, we are familiar with energy inputs that drive de-mixing. During Prohibition in the United States, the so-called revenoors were bent on shutting down the stills in the backwoods. In this process known as distillation, the two liquids are water and ethanol. With sufficient input of thermal energy, the alcohol could be largely separated from water and other components of the fermentation process to make "moonshine." The demand for "moonshine" more than paid for the heat energy required to distill the ferment. Thus, given sufficient energy the chaos of the mixture is reversed.

The phase separation mechanism of biology, the separation of oil-like groups of protein from water, is such that it has the capacity to take in energy efficiently and thereby to reverse the flow toward disorder. Given appropriate protein machines and energy sources, each being accessible by a simple single base mutation, the phase separation mechanism of biology can, with sufficiently reliable inputs of energy, be used to create ever-increasing order and complexity.

6.4.1 Input of Energy Reverses Chaos of a Mixture

6.4.1.1 Heating (Thermal Energy) Achieves De-mixing and Self-assembly

Two protein-based polymers with different temperatures for their inverse temperature transitions, for example, **Polymer I**, $(GVGVP)_{251}$ with a T_t of $25°C$, and **Polymer XII**, $(GVGIP)_{260}(GVGVP)$ with a T_t of $10°C$, can be thoroughly mixed and dissolved in a solution below $10°C$. The protein-based polymers of whichever composition are completely disordered one with respect to the other. Chains of **Polymer XII** are completely randomly dispersed with respect to themselves and with respect to the chains of **Polymer I**. There is complete chaos in the relationship among the polymers in solution.

Remember that the difference between these polymers is quite modest. They differ by only one CH_2 in each five residues. However, because they exhibit inverse temperature transitions at different onset temperatures, one polymer can be separated from the other and self-assembled into an ordered structure with a minimal input of energy.

For example, the completely disordered and mixed solution of **Polymers I** and **XII** can be heated from below $10°C$ to $20°C$, and the model protein with the lower transition temperature will de-mix completely; it will self-separate and self-assemble (see Figure 6.2A).[20] Because of the nature of the phase separation mechanism (the inverse temperature transition), the energy required to de-mix and self-assemble is small. On a per gram basis it is very much less than the energy required to distil spirits, that is, to separate alcohol from a fermentation mixture. Remember also the simpler cyclic analogue that can be dissolved in solution, but which on raising the temperature reversibly forms crystals (see Figure 2.7). This is the ultimate example of an increase in order, and it occurs on raising the temperature. Of course, the essential player in the self-separation process is the structured water surrounding oil-like groups when dissolved in solution.

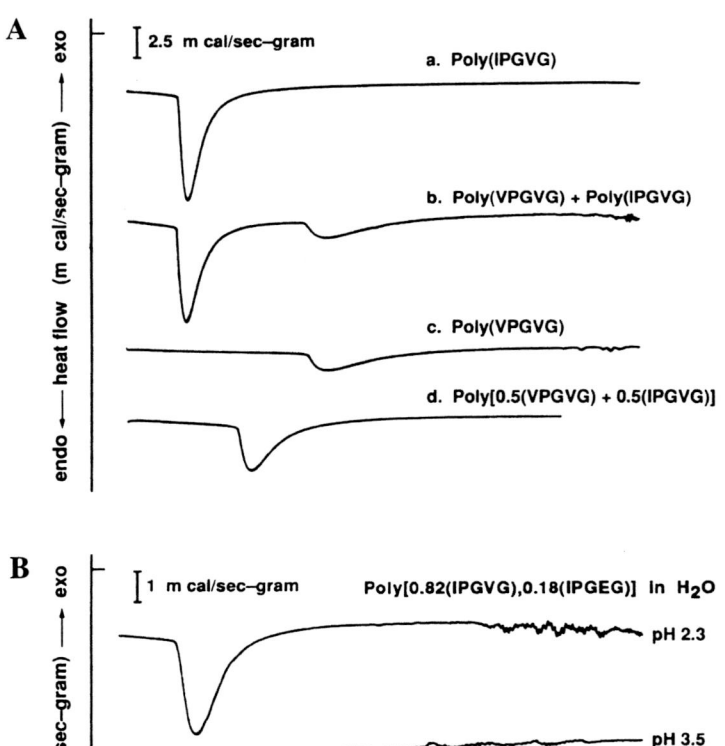

FIGURE 6.2. Differential scanning calorimetry data of elastic-contractile model proteins. (A) Phase separation transition for **Polymers I** and **XII**, alone in solution (curves a and c) and when mixed in the same solution (curve b). Even when mixed, the individual polymers separate from each other; they de-mix due to the input of thermal energy during a slow increase in temperature. Also a polymer was synthesized that contained equal amounts of the two pentamers, and its phase transition is found at an intermediate temperature (curve d). (B) With a composition having a carboxylate function, the input of chemical energy of protons (addition of acid to lower the pH) drives the phase separation to lower temperatures. See text for discussion. (Reproduced with permission from Urry et al.[20])

6.4.1.2 Chemical Energy Achieves De-mixing and Self-assembly

Continuing with the mixed solution of **Polymers I** and **XII** as defined in Table 6.1, separation and self-assembly can also be achieved with the addition of chemical energy at constant temperature. The addition of a chemical energy, such as the addition of salt at 25°C to a solution of poly(GVGVP) plus poly(GVGIP), just sufficient to lower the temperature of the polymer with the lower transition temperature, **Polymer XII**, to below the operating temperature, can cause that polymer to self-separate.[21] Again, the energy input leads to the decrease in entropy represented by de-mixing and by the ordering, the self-assembly, of **Polymer XII**, in analogy to Figure 6.2A, but the input energy now is increasing salt concentration rather than increasing temperature.

Addition of more salt can cause the **Polymer I** with the higher transition temperature also to separate out of solution. Even with one phase-separated polymer layered over the other, these molecules do not again become significantly rescrambled. Replacement of the overlying salt

6.4 Energy Inputs Create Order Out of Chaos

solution with pure water results in the slow dissolution and re-mixing of the two polymer compositions. Thus, this is a completely reversible open system when there is an appropriate energy change.

The amount of energy required for de-mixing and assembly sets biological macromolecular systems apart from other molecular systems. De-mixing of water and alcohol is a much more difficult problem. Distillation, for example, is itself an energy-demanding process, and the separation of alcohol from water is not complete because the alcohol distillate still contains a fixed amount of water. This alcohol and water distil as an azeotrope, with a set ratio of water to alcohol. Further energy input is yet required to achieve complete separation. On the other hand, the complete de-mixing of our two protein-based polymers requires but a small amount of energy. Furthermore, we should note that each separates from the other with their own predetermined water composition

6.4.1.3 Heating Assembles Different But Complementary Model Proteins from the Chaos of Solution

Polymer V, (GVGVP GVGFP G$\underline{E}$GFP GVGVP GVGFP GVGFP)$_n$(GVGVP). with 1 Glu (E) and 4 Phe (F) per 30 residues, and **Polymer VII**, (GVGVP GVGVP G$\underline{K}$GVP GVGVP GVGVP GVGVP)$_n$(GVGVP), with 1 Lys (K) and no Phe (F) per 30 residues, will each dissolve in solution at room temperature and physiological pH to become completely mixed in solution. As with **Polymers I and XII** above, they become completely and randomly dispersed, each polymer chain exhibiting no relation with respect to all others. If the temperature is then raised above a critical temperature of 13°C, the positive charge of **Polymer VII** pairs with the negative charge of **Polymer V** in a pairwise self-assembly and separation from solution, as shown in Figure 6.3B.[22] The polymers communicate with each other to form a more complex structure, shown in Figure 6.4. Because in each case the charge has destructured a significant part of the ordered water around the oil-like groups of each polymer in solution, the amount of energy required to achieve this striking organization is even less than that required for the association of uncharged polymers. *Heating causes these complementary model proteins to proceed efficiently from the chaos of solution to form a specific organized structure.* The time's arrow for this system, solution of water plus two hydrated model proteins, however, as a whole continues toward disorder. Under the circumstances described, heating provides for formation of an ordered structure containing the two complementary model proteins in Figure 6.4, but the more structured water surrounding the oil-like Val and Phe groups of the dissolved model proteins becomes less ordered bulk water, such that the overall change is toward disorder.

6.4.2 Poised Complementary Model Proteins Each Contribute Energy for Creating Order Out of Chaos

Let us again consider **Polymers V** and **VII** as above, but in this example they are each dissolved in separate solutions with **Polymer I**. At room temperature and physiological pH, **Polymer I** is completely mixed in solution with **Polymer V** in one vial, and in a second vial **Polymer I** is mixed in solution with **Polymer VII**. With regard to the polymers, chaos reigns in both solutions. When the solutions of the two vials are mixed, however, instead of increasing the chaos further, **Polymer V** assembles with **Polymer VII** to form the organized structure in Figure 6.4, leaving **Polymer I** alone in solution. This association to form the structuring in Figure 6.4 occurred without the addition of any external energy. These two different solutions poured into the same pot do not "diffuse until the mixture is uniform or homogenous." Energy, arising out of the sequence, resides within each of the dissolved polymers, **Polymer V** and **VII**. *Each model protein has effectively contributed energy to the other polymer for the mutual assembly of the specifically complementary polymers.* By virtue of complementary structures, these molecules communicate with one another to create order out of chaos! The energy input for this ordering process was the energy expended during protein synthesis. All of the polymers, however, cost the same amount

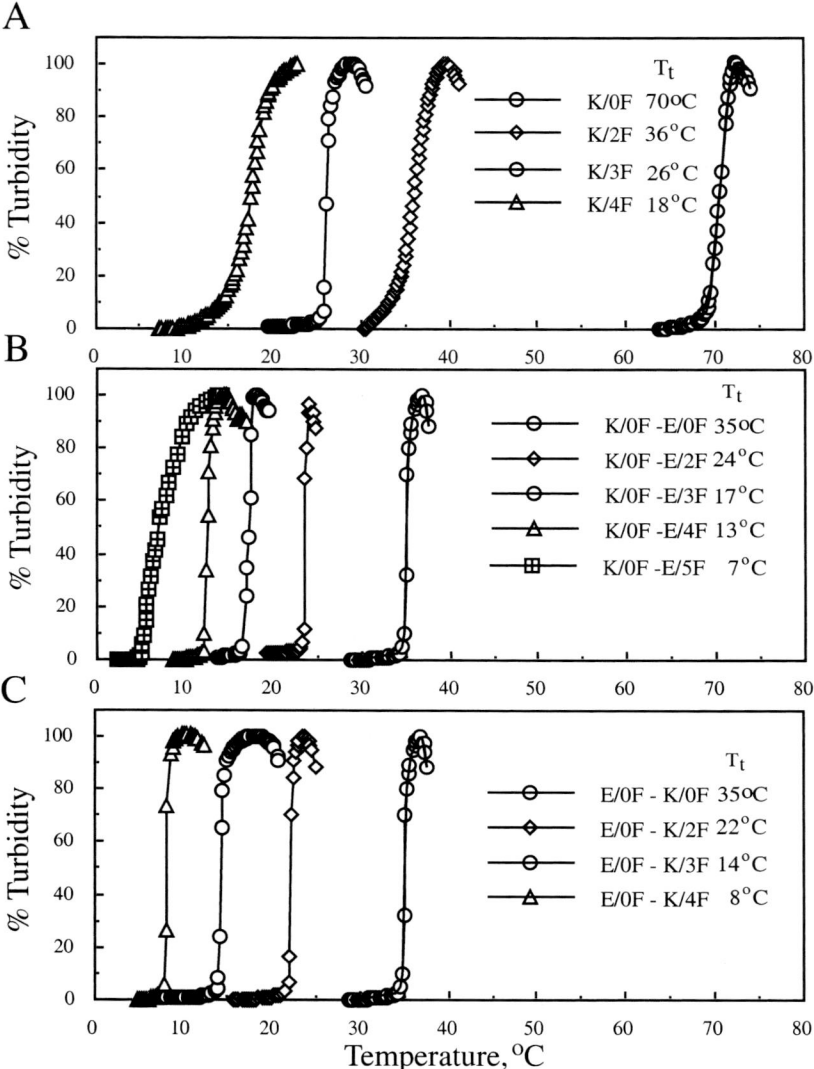

FIGURE 6.3. Temperature profiles for assembly of a series of elastic-contractile model proteins containing one charged residue in a 30 residue repeat, either a Lys (K) residue with the $-(CH_2)_4NH_3^+$ charged side chain or a Glu (E) residue with the $-(CH_2)_2COO^-$ charged side chain, and containing increasing hydrophobicity by Val residue replacement with more oil-like Phe (F) residues. These polymers are defined in Table 6.1 as **Polymers II** though **XI**. Conditions are 40 mg/ml polymer at pH 7.5 in 0.01 M phosphate. (A) Temperature-elicited self-assembly of **Polymers VII** through **X**, identified as K/0F, K/2F, K/3F and K/4F. (B) Temperature-driven association of oppositely charged (complementary) polymers with a constant Lys (K)–containing polymer, K/0F, and varying hydrophobicity of the Glu (E)–containing polymers, E/0F, E/2F, E/3F, E/4F, and E/5F. (C) Temperature-driven association of oppositely charged (complementary) polymers with a constant Glu (E)–containing polymer, E/0F, and varying hydrophobicity of the Lys (K)-containing polymers, K/0F, K/2F, K/3F, and K/4F. Under the conditions used, the self-assembly of E/0F is greater than 80°C (not shown) and that of K/0F occurs at 70°C (shown in A) such that individually at 37°C both polymers form clear solutions. On combining the two solutions above 13°C, they associate. Each complementary polymer provides the chemical energy to the other for structure formation. See text for implications. (Reproduced with permission from Urry et al.[22])

6.4 Energy Inputs Create Order Out of Chaos

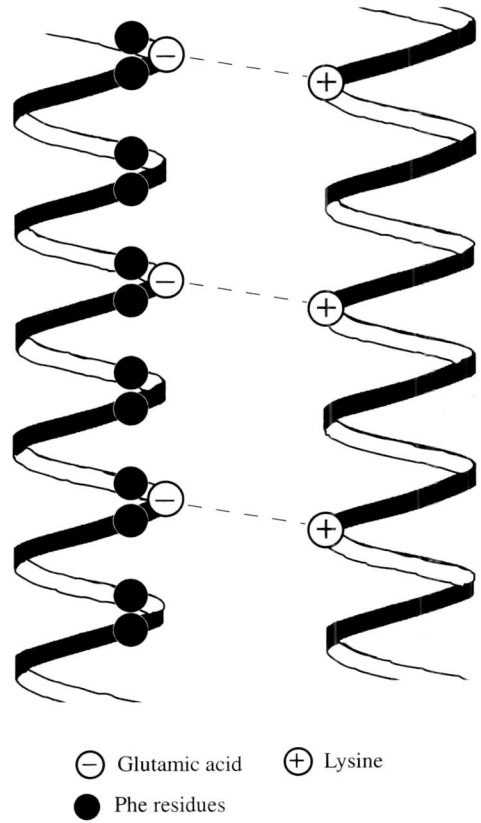

Self-assembly of protein-based polymers by ion-pairing of hydrophobically poised charges

⊖ Glutamic acid   ⊕ Lysine
● Phe residues

FIGURE 6.4. Ion-paired complementary structure formed on combining solutions of two model proteins, E/4F and K/0F, that is, **Polymers V** and **VII**, respectively, as defined in Table 6.1. The data are given in Figure 6.3B. See text for discussion.

of energy to produce by biosynthesis, but complementary sequences caused two to combine and create order out of chaos.

6.4.2.1 Within Each Model Protein There Exists a Repulsive Free Energy Between Charged Groups and Oil-like Groups as Each Reaches for Water of Hydration Unperturbed by the Other

Conventional wisdom would say that the addition of a charged group to a generally oil-like model protein would increase its solubility, and this is correct. Conventional wisdom would not recognize, however, that the charged groups of these fully dissolved and generally unstructured model proteins before mixing would be at increased free energy. After all, in the conventional view, increases in free energy of a charged species arise either from charge–charge repulsion or by charge being constrained in some way to exist in a medium of low dielectric constant, as being forced by a "conformational change." Neither of these conventional arguments is applicable to this case of 1 charge per 30 residues and generally disordered dissolved model protein.

As developed in Chapter 5, the interaction energy that is applicable is a repulsive free energy of hydration between the charged groups and the oil-like groups arising out of a competition for hydration between the charged and oil-like groups. This we have called an *apolar–polar repulsive* free energy of hydration, ΔG_{ap}, as approximated by hydrophobic-induced pKa shifts (see Figures 1.2 and 5.20). Furthermore, when there is no charge–charge repulsion, $\Delta G_{ap} = -\delta \Delta G_{HA}$, the Gibbs free energy of hydrophobic association (see Chapter 5 for the derivations).

For **Polymers II** through **XI**, ΔG_{ap} measures the increase in free energy of the carboxylate, $-COO^-$, or of the charged amino, $-NH_3^+$, which is reflected in systematic increases in pKa shifts (see Figures 1.2 and 5.20). This results from an increase in oil-like hydration due to replacement of Val (V) by Phe (F), which, by competition for hydration, limits hydration of the charged species In an entirely analogous way, we believe that the free energy of a phosphate, $-OPO_3^=$, can be increased, that is, energized, due to a face-off between extremely polar phosphate and oil-like domains. As argued in Chapter 8, we believe this to be relevant to the function of ATP synthase and to the myosin II motor of muscle contraction.

6.4.2.2 By Protein Biosynthesis, **Polymers I, V, and VII** Each Requires the Same Energy to Produce Chains of the Same Length

As was discussed in Chapter 4, the energy required to produce a specific protein sequence is independent of the sequence. Therefore, it requires no more energy to produce a 1,000 residue sequence of **Polymer V** or of **Polymer VII** than it does to produce a 1,000 residue sequence of **Polymer I**. Yet **Polymers V** and **VII** are at higher energy due to the repulsive free energy of interaction between the oil-like and charged residues. The higher energies are directly measurable from the magnitudes of their hydrophobic-induced pKa shifts exhibited by their charged $-COO^-$ and $-NH_3^+$ groups. Because of their energized complementary sequences, **Polymers V** and **VII** are poised for hydrophobic association induced by ion-pair formation.[23] They are energized for structure formation due to the large apolar-polar repulsion, ΔG_{ap}, within the separated chains that relaxes on ion pairing with hydrophobic association.

In an analogous manner, mutations in homologous protein domains result in new associations of the protein domains with the result of new structure and new function. From the emerging data on the human genome, it appears that as many as half of the human genes were derived in this way from the genes of lower species to result in diversity of structure and improved function required of higher species.

6.4.2.3 *"Order Out of Chaos" by Combining Complementary Protein Domains with Each Separately Being at Equilibrium*

To provide broader implications of this association of complementary model proteins, we turn to Prigogine and Stengers' *Order Out of Chaos*[24]:

On the other hand, far from equilibrium there appears a variety of mechanisms corresponding to the possibility of occurrence of various types of dissipative structures. For example, far from equilibrium we ... may also have processes of self-organization leading to nonhomogeneous structures to nonequilibrium crystals.... We can speak of a new coherence, of a mechanism of "communication" among molecules. But this type of communication can arise only in far from equilibrium conditions. It is quite interesting that such communication seems to be the rule in the world of biology. It may in fact be taken as the very basis of the definition of a biological system.

In our example of communication among molecules, the source of the communication resides within the energetics of each dissolved polymer sequence. We considered in Chapter 4 the biosynthesis of protein of specified sequence. The biosynthesis of protein requires a great deal of energy, and thereby a protein sequence might be considered a "dissipative structure." More correct, however, would be to consider a protein sequence as a unique storage device with the distinctive propensity for organization of structure and specialized function. Nor would the many small energy steps, taken during biosynthesis, be reasonably representative of far-from-equilibrium conditions. Nor would the individual **Polymer V** or **VII** separately in solution be reasonably considered as existing under conditions far from equilibrium, because their ΔG_{ap} values would generally be less than 8 to 10 kcal/mole. Therefore, it would seem that proteins of unique sequence, even when energized by an apolar-polar repulsive free energy and capable of forming ordered structures of complex function by virtue of their sequence, are not well-represented by terms such as *dissipative* and *far from equilibrium*.

Importantly, for polymers of the same length, the energy required for biosynthesis of **Polymers V** and **VII** would be no different from the energy to produce **Polymer I**, and yet only **Polymers V** and **VII** exhibited this particular propensity to self-organize. Directing time's arrow relentlessly toward synthesis utilizes, in sum, the simple doubling of the energy required for each step by the trick of removing the pyrophosphate reaction product (see Chapter 4). Thus, biological creation of structures would not seem to require far-from-equilibrium conditions.

Instead, biology creates new structures and functions by employing relatively small packets of energy, using approximately 8 kcal/mole steps, derived ultimately from the energy of sunlight. In particular, the energy of the sun as trapped by biology's photosynthetic process creates the glucose structure; the glucose structure through the energy conversions of biology results in the universal biological energy currency ATP (adenosine triphosphate), and ATP and its equivalents provide the energy for protein synthesis of specified sequences by small 8 kcal/mole steps.

6.4.3 Mutations Accessing New, More Efficient Machines Make Inevitable the Flow Toward Greater Diversity and Complexity

For the universe as a whole, viewed in terms of the second law of thermodynamics, useful energy continually decreases; the universe is becoming less organized; time's arrow for the universe points toward disorder and energy death. On the other hand, that component of the universe, occupied by living systems, behaves inversely. Time's arrow for living systems points toward greater organization and diversity and improved function. Not surprisingly, as seen above, Life does so by continually consuming energy. Life comprises a set of energy-accessing machines.

In terms of the biological origins of protein-based machines, we have been addressing above what Prigogine and Stengers in *Order Out of Chaos* consider one of the two basic questions of our scientific heritage[25]:

Our scientific heritage includes two basic questions to which till now no answer was provided. One is the relation between disorder and order. The famous law of increase of entropy describes the world as evolving from order to disorder; still, biological or social evolution shows us the complex emerging from the simple. How is this possible? How can structure arise from disorder? Great progress has been realized in this question. We know now that nonequilibrium, the flow of energy and matter, may be a source of order.

Three key properties of living systems combine to provide the recipe for an inevitable flow toward greater diversity and complexity: (1) a means of producing protein whereby any sequence is equally likely, that is, at its beginnings a protein sequence capable of efficient energy conversion is no more costly to produce than an inefficient protein-based machine; (2) a common groundwork of explanation, a consilient mechanism, for the efficient energy conversions catalyzed by protein, wherein only small excursions from equilibrium provide function; and (3) abundant and enduring energy sources. One of the necessary ingredients does not appear to be "far from equilibrium conditions." Of course, in a universe where time's arrow points to increasing disorder and decreasing energy, at some point in time the energy sources utilizable by biology will be neither abundant nor enduring.

6.5 The Mechanist Versus Vitalist Debate

Webster's *Third New International Dictionary* defines vitalism as "a doctrine that the functions of a living organism are due to a vital principle distinct from physiochemical forces." The relevant definition of mechanism reads "a doctrine that holds that natural processes, esp. the processes of Life, are mechanically determined and capable of complete explanation by the laws of physics and chemistry." In the pre-Wohler era (see Chapter 3), the vitalist position held sway that an understanding of the molecules of Life was beyond man's grasp. This position began its fall in 1828 when Wohler accidentally synthesized urea, the primary nonionic urinary excretory product of mammals, which arises from the natural breakdown of protein.

For much of twentieth century the mechanist vs. vitalist debate centered on the fossil record and the issue of a "missing link" in the evolution of man from lower animals. Recently, the debate has shifted focus. If there ever were a critical missing link in the fossil record, there certainly is no missing link in the exquisite biochemical record. The common genetic code solidifies man's relationship to other living organisms and indeed to the common origins of all Life. The small detailed differences in sequences of man's proteins to those of other Life forms provides a detailed map of those relationships.

The recent developments in the determination of the human genome and its comparison to that of lower species allows not only that all Life has the same genetic code, but in fact that a large part of the human genome directly derives from lower animals. The genes of humans come, in substantial part, from small modifications of the protein domains of lower organisms that combine to form new associations of protein domains giving rise to new structures and functions. Chapter 7 gives the myoglobin–hemoglobin example for such modifications of a protein domain to arrive at new structure and function.

In his book, *Darwin's Black Box: The Biochemical Challenge to Evolution*, Behe[2] seems to accept the biochemical argument arising from a common genetic code and the demonstration that protein sequences map man's kinship to all other Life. In its place the vitalist stand of Behe notes the complexity and interdependency of components of the molecular machines of Life and argues that the gradualism of Darwin cannot work, rather that there must have been a creator responsible for the intelligent designs of molecular machines. Overall the position of Behe represents a useful step in the mechanist vs. vitalist debate, because it gives a tangible position that can be examined.

This chapter demonstrates that the gradualism of Darwin is demonstrable in the evolution of *de novo*–designed protein-based machines toward diverse function. It has attempted to show how readily the origins of the diverse molecular machines of Life can derive from details of protein biosynthesis, including the genetic code, and the consilient mechanism of energy conversion. Furthermore, ever-increasing complexity is the natural result of the consilient mechanism coupled to mutation during protein synthesis with the results being acted upon by natural selection for the establishment of improved molecular machines. At the more advanced level, the gradualism of Darwin would seem to be evident in the relationship between the human genome and that of lower animals.

6.5.1 The Challenge to the Mechanist—The Challenge to the Vitalist

The suggestion that the profound interdependence and intricate detail of complex molecular machines is such that gradualism, arising from a stepwise series of mutations, could not have given rise to such complexity calls to mind the exceptional Preface of the 1923 text by G.N. Lewis and M. Randall[26]:

There are ancient cathedrals which, apart from their consecrated purpose, inspire solemnity and awe. Even the curious visitor speaks of serious things, with hushed voice, and as each whisper reverberates through the vaulted nave, the returning echo seems to bear a message of mystery. The labor of generations of architects and artisans has been forgotten, the scaffolding erected for their toil has long since been removed, their mistakes have been erased, or have become hidden by the dust of centuries. Seeing only the perfection of the completed whole, we are impressed as by some superhuman agency. But sometimes we enter such an edifice that is still partly under construction; then the sound of hammers, the reek of tobacco, the trivial jests bandied from workman to workman, enable us to realize that these great structures are but the result of giving to ordinary human effort a direction and a purpose.

6.5.2 "Seeing Only the Perfection of the Completed Whole, We Are Impressed as by Some Superhuman Agency"

The mechanist has yet to define the early simpler self-sufficient Life forms. In evolution the mechanist has yet to complete an understanding of the temporary scaffoldings that allowed transition from prebiotic to biotic and from minimally self-sufficient more primitive Life forms to result in the more complex completed edifice that we all celebrate. When plausible transitional scaffoldings from prebiotic to biotic evolution have been reconstructed, the mechanist will yet be left with a seemingly unanswerable query. Why are there immutable laws that man can identify and that are often so well-described with a mathematics that man has derived? It is hard to imagine a state of knowledge when there would not remain an unknown without which a "vital force" would be disallowed.

The future of the mechanist holds many daunting challenges. To date, however, results of the scientific method seem to have provided limited guidance in maintaining a healthy social structure. Therefore, going forward, one could hope for a constructive role for the modern vitalist in utilizing scientific advancements to support a more complex and humane society. Informed and wisdom-providing guidance from the vitalist would seem appropriate to assist in bringing the benefits of the Biotechnological Revolution to relieve suffering, to provide continual improvement in quality of Life, and generally to raise standard of living while maintaining respect for the remarkable edifice that each defined being represents.

References

1. P. Appleman, Ed., *Darwin*, Third Edition, W. W. Norton, New York, 2001, p. 1.
2. M.J. Behe, *Darwin's Black Box: The Biochemical Challenge to Evolution*. Simon and Schuster, New York, 1996, p. 5.
3. W.-H. Li, Z. Gu, H. Wang, and A. Nekutenko, "Evolutionary Analysis of the Human Genome." Nature, **409**, 847–849, 2001.
4. Editorial Comments. *Science*, **291**, 1218, 2001.
5. A. Toffler, Forward, to I. Prigogine and I. Stengers, *Order Out of Chaos: Man's New Dialogue with Nature*. Bantam Books, New York, 1984, p. xx.
6. A. Szent-Györgyi, "Studies on Muscle." *Acta Physiol. Scand.* **9** (Suppl. XXV), 1–116, 1945.
7. D.W. Urry, "Molecular Machines: How Motion and Other Functions of Living Organisms Can Result from Reversible Chemical Changes." *Angew. Chem.* [German], **105**, 859–883, 1993; *Angew. Chem. Int. Ed. Engl.*, **32**, 819–841, 1993.
8. D.W. Urry, "Elastic Biomolecular Machines: Synthetic Chains of Amino Acids, Patterned After Those in Connective Tissue, can Transform Heat and Chemical Energy into Motion." *Sci. Am.*, January, 1995, 64–69.
9. D.W. Urry, "Physical Chemistry of Biological Free Energy Transduction as Demonstrated by Elastic Protein-based Polymers." *J. Phys. Chem. B*, **101**, 11007–11028, 1997.
10. D.W. Urry, R.D. Harris, and K.U. Prasad, "Chemical Potential Driven Contraction and Relaxation by Ionic Strength Modulation of an Inverse Temperature Transition." *J. Am. Chem. Soc.*, **110**, 3303–3305, 1988.
11. D.W. Urry, B. Haynes, H. Zhang, R.D. Harris, and K.U. Prasad, "Mechanochemical Coupling in Synthetic Polypeptides by Modulation of an Inverse Temperature Transition." *Proc. Natl. Acad. Sci. U.S.A.*, **85**, 3407–3411, 1988.
12. S.C. Schuster and S. Khan, "The Bacterial Flagellar Motor" *Annu. Rev. Biophys. Biomol. Struct.*, **23**, 509–539, 1994.
13. P.D. Boyer,' The Binding Change Mechanism for ATP Synthase—Some probabilities and possibilities." *Biochim. Biophys. Acta*, **1140**, 215–250, 1993.
14. J.P. Abrahams, A.G.W. Leslie, R. Lutter, and J.E. Walker, "Structure at 2.8 Å Resolution of F_1-ATPase from Bovine Heart Mitochondria." *Nature*, **370**, 621–628, 1994.
15. K. Kinosita, Jr., R. Yasuda, H. Noji, and K. Adachi, 'A Rotary Molecular Motor that can Work at Near 100% Efficiency." *Philos. Trans. R. Soc. Lond. B Biol. Sci.*, **355**, 473–489, 2000.
16. D.W. Urry, L.C. Hayes, T.M. Parker, and R.D. Harris, "Baromechanical Transduction in a Model Protein by the ΔT_t Mechanism." *Chem. Phys. Lett.*, **201**, 336–340, 1993.
17. D.W. Urry, L.C. Hayes, and D. C. Gowda, "Electromechanical Transduction: Reduction-driven Hydrophobic Folding Demonstrated in a Model Protein to Perform Mechanical Work." *Biochem. Biophys. Res. Commun.*, **204**, 230–237, 1994.
18. D.W. Urry, L.C. Hayes, D.C. Gowda, S.-Q. Peng, and N. Jing, "Electro-chemical Transduction in Elastic Protein-based Polymers." *Biochem. Biophys. Res. Commun.*, **210**, 1031–1039, 1995.
19. D. Voet and J. Voet, *Biochemistry*, Second Edition, John Wiley & Sons, New York, 1995, p. 966, Table 30-2.
20. D.W. Urry, C.-H. Luan, S.Q. Peng, T.M. Parker, and D.C. Gowda, "Hierarchical and Modulable Hydrophobic Folding and Self-assembly in Elastic Protein-based Polymers: Implications for Signal Transduction." *Mater. Res. Soc. Symp. Proc.*, **255**, 411–422, 1992.
21. C.-H. Luan, T. Parker, K.U. Prasad, and D.W. Urry, "DSC Studies of NaCl Effect on the Inverse Temperature Transition of Some Elastin-based Polytetra-, Polypenta-, and Polynonapeptides." *Biopolymers*, **31**, 465–475, 1991.
22. D.W. Urry, L. Hayes, C.-X. Luan, D.C. Gowda, D. McPherson, J. Xu, and T. Parker, "ΔT_t-Mechanism in the Design of Self-Assembling Structures." In *Self-assembling Peptide Systems in*

Biology, Medicine and Engineering, A. Aggeli, N. Boden, S. Zhang, Eds., Kluwer Academic Publishers, Dordrecht, The Netherlands, 2001, p. 323–340.

23. In point of fact, this design capacity to energize for structure association has been used to improve polymer alignment for cross-linking to produce stronger materials useful for a number of applications (see Chapter 9).

24. I. Prigogine and I. Stengers, *Order Out of Chaos: Man's New Dialogue with Nature*. Bantam Books, New York, 1984, p. 13.

25. Preface to I. Prigogine and I. Stengers, *Order Out of Chaos: Man's New Dialogue with Nature*. Bantam Books, New York, 1984, p. xxix.

26. Preface to G.N. Lewis and M. Randall, *Thermodynamics and the Free Energy of Chemical Substances*. McGraw-Hill, New York, 1923.

7
Biology Thrives Near a Movable Cusp of Insolubility

7.1 Introduction

Biology thrives near a movable cusp of insolubility, yet excursions too far, either direction, into the realms of insolubility or solubility spell disease and death.

Even to the casual observer, the past decades have witnessed remarkable and unparalleled progress in the understanding and utilizing of biological processes. Because of this, the proposal of a new pervasive theme such as the consilient mechanism for hydrophobic association introduced here and its representation by the sweeping statement that "biology thrives near a movable cusp of insolubility" should not be undertaken casually. This is particularly the case, for example, for such thoroughly researched areas as muscle contraction, complexes of the electron transport chain and ATP synthase of the inner mitochondrial membrane, oxygen transport by hemoglobin, blood clotting, chaperones, and so forth.

Nonetheless, the extensive experimental results detailed in Chapter 5 on elastic-contractile model proteins in our view are compelling. Furthermore, present daring derives from the emerging acceptance of our proposed mechanism of protein elasticity of two decades ago; limited encouragement and general disparagement have, in the last few years, transformed into more general acceptance.[1,2] The present work, in fact, gains from our new insight into entropic (ideal) protein elasticity, which has earned the right to be considered a consilient mechanism of elasticity. This is because the mechanism applies to all chain molecules independent of composition and structure as long as the chain molecule exhibits internal motions that decrease in magnitude on deformation. The belief that regular, nonrandom, dynamic structures could exhibit near-ideal elasticity allowed the design of elastic model proteins capable of the diverse energy conversions of biology with remarkable positive cooperativity required for efficient energy conversion. Reviews of the basic model protein data[3,4] underlying the consilient mechanism for hydrophobic association have been well received. On the other hand, suggestions of relevance to specific biological systems, such as those presented in this and the next chapter, have met with resistance familiar from the elasticity experience and not uncommon in scientific efforts.[5]

Most compelling in our view, however, is the *de novo* design of diverse protein-based machines that utilize the unifying principle presented in Chapter 5. Successful design has simply been too facile and too comprehensive not to persevere. Furthermore, when one begins to apply the single underlying principle to what is known of natural biological systems, for example, Life's machines, in our view the phenomenological correlations appear too coherent to be unsound.

In particular, in this chapter we examine a number of protein-based biological systems that can present excesses of insolubility or solubility. Some of these, for example, muscle contraction and hemoglobin transport of oxygen have been extensively studied, as noted above. The example of blood clotting (thrombus formation) has great medical significance because, in our view, good health and life itself require proper balance near the movable cusp of insolubility. Shifts toward excess insolubility precipitate heart attacks (coronary occlusion, thrombus in coronary arteries), strokes (apoplexy, thrombus in arteries of the brain), and phlebitis (thrombus in veins). On the other hand, shifts toward excess solubility result in the life-threatening bleeding disease of hemophilia and the degradation of elastic tissue in the lungs with the result of pulmonary emphysema.

Other examples have emerged more recently. Prion proteins induce insolubility and cause the ravages of Alzheimer's and mad cow diseases. Then there are chaperones that reverse inappropriate insolubilities. In these latter cases considered mechanisms are not so deeply ingrained. In none of these, however, has the sense of an apolar–polar repulsive free energy of hydration, ΔG_{ap}, emerged. In none of these has there been a suggestion of the competition for hydration between hydrophobic and polar species that is the basis for repeated experimental demonstrations of large hydrophobic-induced pKa shifts.

In general, mechanisms have often focused on the energetics arising out of electrostatics such as charge–charge repulsion and ion-pair formation under the assumption of a uniform medium of low dielectric constant. As for the rationalization of pKa shifts exhibited by carboxylates of glutamic acid and aspartic acid and by the ε-amino group of lysine, the argument seems to be that the polar species is forced by a "conformational change" into unfavorable low dielectric constant environments in order to account for states of high free energy. Instead of such a high-energy state resulting from a "conformational change," in our view these high-energy states, such as hydrophobic-induced pKa shifts, result from competition for hydration. To restate, in our view, a change in competition for hydration due to a change in the state of a functional group drives conformational change, not the reverse.

We again note the simple message contained within the familiar adage that "oil and vinegar don't mix" and combine it with the effects of oil-like and vinegar-like groups being constrained to coexist along the chain molecules made possible by the unique process, reviewed in Chapter 4, of protein biosynthesis. Experimental examination of our systematically designed model proteins reveals the force that exists between oil-like and vinegar-like groups. Thermodynamic analysis translates that force between oil-like and vinegar-like groups into a change in Gibbs free energy for hydrophobic association, ΔG_{HA} (see Chapter 5), a key functional component of which we call the apolar–polar repulsive free energy of hydration, ΔG_{ap}. We believe ΔG_{HA} and its functional component, ΔG_{ap}, to be a dominant free energy of interaction that underpins protein function and disfunction.

Electrostatic interactions such as ion-pair formation are indeed important, but in our view their role in protein function resides in their effect on the value of ΔG_{HA}. In particular, the electrostatic interactions of ion-pairing and hydrogen-bonding lower ΔG_{HA} (make hydrophobic association more favorable) by relieving the apolar–polar repulsive free energies of hydration, ΔG_{ap}. Ion pairs and hydrogen bonds form in the aqueous environment within and between protein domains because of a limited hydration of the separated polar groups. The limited hydration of polar groups, although seemingly bathed in water, arises from competition for hydration with hydrophobic groups. *ΔG_{ap} results from spatial, through water, interactions between oil-like (hydrophobic) and vinegar-like (polar, e.g. charged) residues constrained by protein sequence and relieved by protein folding and assembly. Changes in ΔG_{ap}, as the direct result of changes in the polar state of a functional group, drive function by means of effecting changes in folding and assembly.*

Particularly in this and the next chapter, we regret that time does not permit development of the desired level of expertise in each of the protein systems to be considered. Nonetheless, it is thought that the pervasiveness of the con-

7.2 The Movable Cusp of Insolubility

silient mechanism becomes persuasive with the examples as given, and this allows for a more timely completion of this volume.

7.2 The Movable Cusp of Insolubility

The differential calorimetry curve in Figure 5.1C graphically represents a "cusp of insolubility." By definition, a cusp is "a fixed point on a mathematical curve (graph) at which a point tracing the curve would exactly reverse its direction, that is, *a point at which traversing in opposite directions has opposite consequences*."[6] In particular, from a point located at the peak of the curve in Figure 5.1C and for any one of the cusps in Figure 7.1, tracing the curve to lower temperature finds solubility whereas tracing the curve to higher temperature gives insolubility. Alternatively, as represented in Figure 7.1, at a constant temperature an energy input to which the protein composition is receptive can move the curve, the cusp of insolubility, to lower temperature to give insolubility or move the curve to higher temperature to give solubility.

Moving the cusp of insolubility occurs by the energy inputs that give rise to a change in the Gibbs free energy of hydrophobic association, ΔG_{HA}, as developed in Chapter 5. An effective energy input either makes the protein more oil-like and thereby lowers the temperature interval of the cusp of insolubility (lowers ΔG_{HA}) or makes the protein more polar and thereby raises the temperature for the onset of the cusp of insolubility (raises ΔG_{HA}). T_t (determined from light-scattering measurements) or T_b (determined from calorimetry measurements as the leading (low-temperature) edge of the temperature interval shown in Figure 7.1) experimentally defines the onset temperature for the cusp of insolubility. Both T_t and T_b become convenient approximations to the divide between solubility and insolubility. In fact, this boundary between solubility and insolubility, as represented in Figure 5.3, we call the T_t-divide.

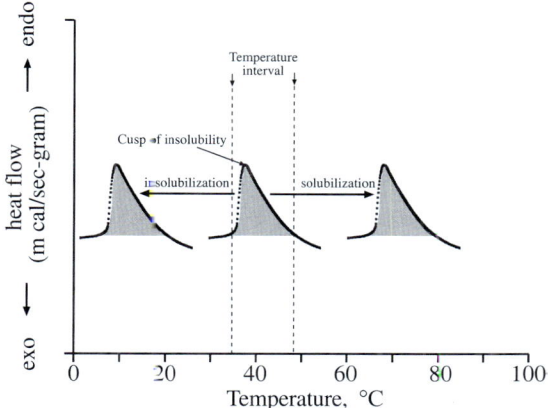

FIGURE 7.1. The movable "cusp of insolubility" represented as the calorimetry curve for the inverse temperature transition of $(GVGVP)_{251}$ due to hydrophobic association. Hydrophobic association occurs either by raising the temperature from below to above the temperature interval of the transition or by introducing an energy that lowers the cusp of insolubility, that is, lowers the temperature at which the transition occurs. Hydrophobic dissociation occurs either by lowering the temperature from above to below physiological temperature or by moving the cusp to higher temperatures by the introduction of an energy that increases the temperature at which the transition occurs from below to above physiological temperature.

As demonstrated in this chapter, moving the cusp of insolubility too far in one direction or the other is harmful. *Thus, excursions too far, either direction, into the realms of insolubility or solubility spell disease and death.* Also as noted below and more extensively in Chapter 8, living organisms thrive on relatively small energy excursions, generally less than ±10 kcal/mole (42 kJ/mole). Furthermore, as discussed in broader context in the Epilogue, life thrives never very far from equilibrium, that is, never very far from the T_t-divide, never very far from the cusp of insolubility. The distinct, stepwise energy changes that sustain Life are very many and generally small, but in sum they represent a substantial quantity of energy.

7.2.1 Biological Function by Association/Dissociation of Pairs of Spatially Assembled Hydrophobic Domains

7.2.1.1 Hydrophobic Hydration Disappears for Two Different Reasons with Diametrically Opposite Consequences of Either Insolubility or Solubility

7.2.1.1.1 Hydrophobic Hydration Disappears When a Pair of Hydrophobic Domains Associate, That Is, When the Pair of Domains Becomes Insoluble

If the pair of domains has too much hydrophobic hydration for a given temperature, when fully exposed to water, hydrophobic association with loss of the hydrophobic hydration results. As discussed extensively in Chapter 5, the reason lies in the statement of the Gibbs free energy governing solubility, which is ΔG(solubility) = $\Delta H - T\Delta S$. When there is too much hydrophobic hydration, the $(-T\Delta S)$ term is positive and larger than the negative ΔH term such that insolubility ensues. This is because systems move in the direction of minimizing the free energy, ΔG, that is, systems move toward lower values of ΔG. Accordingly, the pair of hydrophobic domains proceeds to solubility when the ΔG(solubility) is negative, and insolubility occurs when ΔG(solubility) is positive.

The change in Gibbs free energy for hydrophobic association, ΔG(hydrophobic association) = ΔG_{HA}, arose in Chapter 5 to describe those interactions of model proteins that affect the state of hydrophobic association. ΔG_{HA} is calculated from the heat of the inverse temperature transition, that is, from the area of the cusp of insolubility in Figure 5.1C and Figure 7.1. ΔG_{HA} may also be estimated from a shift of the cusp of insolubility along the temperature axis, resulting from introduction of the perturbation, times the entropy change for the transition (heat of the transition divided by the transition temperature). The change in T_t measures the movement of the cusp of insolubility along the temperature axis. When the change is in the solvent and does not represent a change in the chemical structure of the model protein, as in the solute effect on a neutral polymer, the estimate of the entropy change refers to the (reference) state before the perturbation (see Equations [5.8] and [5.9] in Chapter 5). When the chemical nature of the polymer varies, then the entropy change for the chemically altered structure is used, as in Equation (5.10a) of Chapter 5.

Obviously, ΔG_{HA} bears the opposite sign from ΔG(solubility). ΔG_{HA} refers to hydrophobic association, that is, the loss of solubility of hydrophobic groups, whereas ΔG(solubility) refers to the opposite process of dissolution. When considering polymers such as proteins with both oil-like and vinegar-like side chains, an important distinction between the two recognizes, for example, that a soluble globular protein contains hydrophobic association within the folded structure of the single macromolecule.

7.2.1.1.2 Some Hydrophobic Hydration Disappears When a Water-exposed Hydrophobic Domain Occurs Proximal to an Emergent Polar (e.g., Charged) Species

Consider a pair of hydrophobic domains exactly at the cusp of insolubility. In this case the equilibrium constant for association/dissociation is essentially one; half of the pairs of hydrophobic domains are dissociated and covered with hydrophobic hydration, and the other half are associated and devoid of hydrophobic hydration. If an ionizable species becomes proximal to the transiently dissociated pair of hydrophobic domains, the ionizing species destructures hydrophobic hydration in achieving its own hydration. In terms of ΔG(solubility), the loss of hydrophobic hydration for the open state (with its inherently positive ΔS)[7] causes the $(-T\Delta S)$ term to become less positive; the negative ΔH term for solubility then dominates, and the shift to the dissociated state occurs. In terms of ΔG_{HA}, the introduction of a polar interaction moves the cusp of insolubility to a higher temperature, such that ΔG_{HA} is positive and hydrophobic association is lost, that is, solubility results. The cusp of insolubility moved to a higher temperature, placing the pair of hydrophobic domains on the soluble side of the cusp. Under these circumstances, the water at the surface of the

7.2 The Movable Cusp of Insolubility

hydrophobic domain is not structured hydrophobic hydration, but instead orients toward the charged species and contributes to its hydration shell.

As discussed in Chapter 5, another expression of the change in Gibbs free energy for hydrophobic association results from the experimental observation of hydrophobic-induced pKa shifts. This change in Gibbs free energy, referred to as an apolar–polar repulsive free energy of hydration, ΔG_{ap}, is calculable as ΔG_{ap} = 2.3RTΔpKa (see the derivation of Equation [5.13] in Chapter 5). The ΔpKa arises from competition for hydration between polar and hydrophobic groups, and the competition determines the amount of hydrophobic hydration. As this dictates whether the hydrophobic domains are associated or not, it should be equivalent to ΔG_{HA} as long as the pKa shift arises solely due to the competition for hydration between apolar and polar groups and does not include charge–charge repulsion as part of the reason for the pKa shift, that is, for increasing the free energy of the ionized state (See Equation [5.23] of Chapter 5). Because of this, ΔG_{ap} is also calculable from the heats of the inverse temperature transition via Equation (5.10b) of Chapter 5, i.e., $\Delta G_{ap} = \Delta G_{HA}[E/0F \rightarrow E^-/0F]$ when properly written per E residue of the same model protein. This increase in the Gibbs free energy of a negatively charged (anionic) state favors neutralization either by protonation or by ion-pair formation. Ion-pair (salt-bridge) formation, therefore, can allow so much hydrophobic hydration to reoccur that a return to the hydrophobically associated state results. For this reason, ion-pair formation and even hydrogen binding are often an integral part of hydrophobic association in protein folding, assembly, and function. This is very relevant to the function of hemoglobin as a machine for oxygen transport.

7.2.1.2 Identification of Hydrophobic Association with Rigor in Muscle and with the Tense, T-State, of Allosteric Proteins Exhibiting Positive Cooperativity

Rigor is a state of being rigid or stiff. The term is commonly used in relation to muscle stiffness, but relevance exists, in general, to the taut or tense (contracted, insoluble) T-state as opposed to the relaxed (soluble) R-state of allosteric proteins such as hemoglobin that exhibit enhanced effectiveness for oxygen transport due to the property of positive cooperativity. From our perspective, this rigid or stiff state equates to insolubility due to association of hydrophobic protein domains, and the relaxed state equates to the removal or partial removal of hydrophobic association.

The hydrophobically associated state represents the insolubility side of the T_t-(solubility/insolubility)divide. As discussed in Chapter 5, contraction of the model elastic-contractile model proteins capable of inverse temperature transitions arises due to hydrophobic association. Hydrophobic association occurs, most fundamentally, on raising the temperature, on adding acid (H^+) to protonate and neutralize carboxylates ($-COO^-$), and on adding calcium ion to bind to and neutralize carboxylates. Most dramatically, hydrophobic association occurs on dephosphorylation of (i.e., phosphate release from) protein, and it commonly occurs with formation of ion pairs or "salt bridges" between associated hydrophobic domains.

Changing the state of functional groups associated with hydrophobic domains occurs with remarkable positive cooperativity. Also, as most emphatically shown in Figure 5.34 of Chapter 5, more hydrophobic model proteins exhibit greater positive cooperativity and correspondingly larger pKa shifts. Importantly, a major part of the free energy change represented by the shift in pKa is the same as that represented by the change in positive cooperativity (the change in Hill coefficient). In our view, this is because the pKa shift and the extent of positive cooperativity result from the same physical process, the competition for hydration between hydrophobic (oil-like) and polar (vinegar-like, e.g., charged) groups constrained by sequence to coexist in proteins.

It is possible, however, to have a residual pKa shift after the allosteric interchange between two states has resulted in the state of complete solubility. As shown in Figure 5.31A, when ionization is complete and only the unfolded state is left, there remains a residual pKa shift. The carboxylates and hydrophobic groups continue to behave as though repulsed, as they try unsuccessfully to access water completely undisturbed by the other.

As discussed below, these considerations are relevant to formation of the rigor state in muscle, to the tense or taut T-state of proteins that exhibit positive cooperativity, to the formation of the oxygenated and deoxygenated states of hemoglobin, to the proper balance of blood clotting, to the formation of insoluble protein-based fibers giving rise to Alzheimer's and mad cow diseases, to the unfolding and proper refolding of incorrectly folded protein by chaperonins, and to the unfolding and degradation of oxidized or otherwise solubilized protein states.

7.2.2 Rigor: Hydrophobic Association in Muscle at the Gross Anatomical Level

A satisfying affirmation of the fundamental significance and pervasiveness of hydrophobic association of the consilient mechanism is found, even at the anatomical level, in Dorland's *Illustrated Medical Dictionary* in the following definitions, given as alphabetically listed, under the term *rigor*.[8]

7.2.2.1 Acid Rigor, "Coagulation of Protein of Muscle Produced by Acids"

Webster's *Third New International Dictionary* defines coagulation as "the process by which such change of state takes place consisting of the alteration of a soluble substance, usually protein, into an insoluble form...."[9] Acids produce insolubility of muscle protein. At the physiological pH of 7.4, all but the most pKa-shifted carboxyl groups of protein occur as ionized carboxylates. Acids protonate carboxylates to form the uncharged carboxyl, that is, $-COO^- + H^+ = -COOH$.

By the consilient mechanism, protonation of carboxylates restores hydrophobic hydration to hydrophobic domains competing for hydration. As discussed in Chapter 5 and noted above, introduction of too much hydrophobic hydration causes insolubility. Recall the expression for the Gibbs free energy for solubility, that is, $\Delta G(\text{solubility}) = \Delta H - T\Delta S$. When there is too much hydrophobic hydration, the magnitude of the positive $(-T\Delta S)$ term is larger than the negative ΔH term. Accordingly, addition of acid to muscle protein results in an unfavorable positive $\Delta G(\text{solubility})$ with the result of hydrophobic association. In other words, by removal of charged groups, acids move the cusp of insolubility below physiological temperature to cause the stiffness called *rigor*.

7.2.2.2 Calcium Rigor, "Systolic Cardiac Arrest Caused by an Excess of Calcium"

From the definitions of *systole* and *systolic* in Dorland's *Illustrated Medical Dictionary*, systolic cardiac arrest is recognized as the heart muscle having become fixed in the contracted state. We know in the process of muscle contraction that release of calcium ion triggers contraction by binding at carboxylates, and we have seen in Figures 5.15 and 5.27 that, with the appropriate protein composition, a small increase in calcium ion can drive the hydrophobic association of contraction by ion pairing with pKa-shifted carboxylates. Thus, even on a macroscopic anatomical scale, it is possible to see the consequences of the consilient mechanism of hydrophobic association whereby calcium ion pairing with carboxylates lowers the T_t-(solubility/insolubility) divide with the consequence of hydrophobic association fixing muscle protein in the contracted state.

7.2.2.3 Heat Rigor "Rigidity of Muscles Induced by Heat"

Increasing the temperature to result in insolubility is the hallmark of hydrophobic association by inverse temperature transitions. As temperature is a factor in the positive $(-T\Delta S)$ term for formation of hydrophobic hydration, an increase in temperature obviously increases the magnitude of the positive $(-T\Delta S)$. An increase in the magnitude of the positive $(-T\Delta S)$ term means that $\Delta G(\text{solubility}) = \Delta H - T\Delta S$ becomes positive, and solubility is lost. The cusp of insolubility, the T_t-(solubility/insolubility) divide, is surmounted by heating. On

7.2 The Movable Cusp of Insolubility

heating, the system passes from soluble to insoluble. On heating, the protein of muscle forms the contracted, the rigid, the stiff state as the result of hydrophobic association.

More than a century ago Roy attached a 50-gram weight to a 1 cm wide strip of human aorta[10] and heated the strip repeatedly back and forth from room temperature to physiological temperature. The aortic vascular tissue primarily composed of smooth muscle and elastic fibers reversibly contracted on heating and relaxed on cooling, thereby performing the mechanical work of raising and lowering the weight. Approximately half of the force developed on heating results from smooth muscle contraction and the other half from the inverse temperature transition of the elastic fibers.

7.2.2.4 Rigor Mortis, "The Stiffening of a Dead Body, Accompanying the Depletion of Adenosine Triphosphate in the Muscle Fibers"

Death is the ultimate manifestation of excursion excess into the realm of insolubility. ATP Adenosine Triphosphate (ATP), the universal biological energy currency, is the ultimate solubilizer of protein. In our view, the negative hypercharged phosphate functions as a supercarboxylate to destroy hydrophobic hydration in the process of satisfying its own thirst for hydration. Thus, as paired associated hydrophobic surfaces undergo an opening fluctuation, hydrophobic hydration that would form is immediately recruited for hydration of added phosphate. As substantial hydrophobic hydration is required for the positive ($-T\Delta S$) to dominate and to result in insolubility, the result of phosphorylation is solubility. In this way bound phosphate provides for protein solubilization by separation of hydrophobically associated domains.

Organic phosphates, for example, phosphorylated protein and the phosphates of adenosine, are generally unstable and break down over time to water-soluble inorganic phosphate. Accordingly, ATP must be continually re-supplied by the metabolism of a living organism. At death, respiration stops; oxidative phosphorylation stops; the re-supply of ATP terminates; and the pool of ATP decreases with time. As is well known, the development of rigor mortis provides a measure of the time of death. In our view and by the consilient mechanism, based on the inverse temperature transition of hydrophobic association, depletion of ATP in muscle necessarily results in the hydrophobic association of muscle protein, that is, in the formation of the stiff, rigid, contracted state of muscle.

Interestingly, death is not a case of the disordering, dissolving, or melting away of structure maintained by the energy conversions of the living organism. Quite the contrary, death results in the rigidification of structures ordered by hydrophobic association. The ultimate disintegration of a dead body is due to active processes such as degradation by proteolytic enzymes (primarily of other organisms), toxic chemicals, excess heat, and so forth.

These examples of rigor at the gross anatomical level in muscle are manifestations of hydrophobic association at the molecular level; they reflect the fundamental role of hydrophobic association/dissociation in the function of muscle. These occurrences of rigor in muscle do not support the view of electrostatic interactions, of direct charge–charge interactions independent of hydrophobic domains being predominant in muscle function. These anatomical manifestations that correlate muscle contraction with hydrophobic association continue below the physiological level.

7.2.3 Hydrophobic Association in Muscle Contraction and in Contractile Model Proteins: Phenomenological Correlations

The phenomena that drive muscle contraction—thermal activation, pH activation, calcium ion activation, stretch activation in insect flight muscle, and dephosphorylation itself—have all been shown to drive contraction by hydrophobic association in the elastic-contractile model proteins discussed in Chapter 5. As concerns a pair of hydrophobic domains, all of these processes surmount the T_t-divide (the cusp of insolubility in Figure 7.1) to go from a soluble state to an insoluble state either by raising the temperature from below to above the T_t-divide or by lowering the temperature of the T_t-divide (moving the cusp of insolubility) from above to below physiological temperature.

Thus, as is discussed below in terms of phenomenological correlations, we recognize hydrophobic association as the fundamental molecular process of muscle contraction. In general, the perspective is one of controlling the association/dissociation of localized hydrophobic domains in the process of contraction/relaxation. In Chapter 8, molecular details of muscle contraction are examined in this context. Here at the physiological level, phenomenological correlations are considered in terms of solubility/insolubility of hydrophobic domains, or, as included in the title of this chapter, in terms of the relationship of the operating temperature to the location along the temperature axis of the cusp of insolubility, the temperature at which occurs the inverse temperature transition of hydrophobic association.

7.2.3.1 Thermal Activation of Muscle Contraction

7.2.3.1.1 The Inverse Temperature Transition and Thermal Activation in Muscle

The phenomenon of hydrophobic association on raising the temperature, as noted above and treated in detail in Chapter 5, derives from the thermodynamics of structured water surrounding hydrophobic moieties. Hydrophobic hydration disappears, due to an unfavorable Gibbs free energy for solubility, as the temperature is raised from below to above the transition temperature that reaches the cusp of insolubility represented in Figure 7.1. This causes the hydrophobic domains to separate from water by means of intra- and intermolecular hydrophobic association.

Under the conditions at which the coexistence line in the phase diagram in Figure 5.3 occurs, associated and dissociated states of a pair of hydrophobic domains coexist, and the change in the Gibbs free energy is zero, that is, $\Delta G_t = \Delta H_t - T_t \Delta S_t = 0$. This means for the transition that $T_t \approx \Delta H_t/\Delta S_t$. The temperature, T_t, at which the hydrophobic association transition occurs, depends on the quantity of hydrophobic hydration. When there is more water of hydrophobic hydration, the transition temperature is lower, that is, the value of T_t is lower because $\Delta H_t/\Delta S_t$ is a smaller quantity. When there is less water of hydrophobic hydration, the value of T_t is higher because $\Delta H_t/\Delta S_t$ is a larger quantity.

7.2.3.1.2 Muscle and the Mammalian Elastic Fiber Are Exothermic on Stretching

Stretching of the mammalian elastic fiber in water (as well as that of cross-linked poly[GVGVP], the most prominent repeating sequence of bovine and pig elastic fibers)[11] is an exothermic process[12] due to the hydration of hydrophobic groups that become exposed to water as extension forces disassociation of hydrophobic groups.[13] Stretching of muscle also gives off heat, as would occur on forced separation of hydrophobically associated domains of protein within muscle.[14] For the mammalian elastic fiber, the exothermic heat of hydration on extension is several times (about four times) greater than the elastic energy stored in the entropic elastic fiber on extension.[13]

7.2.3.1.3 Muscle and Cross-linked Elastic Protein-based Polymer, (GVGVP)$_n$, Contract on Raising the Temperature over the Same Temperature Range

Just as stretching releases heat, heating causes contraction. For cross-linked poly(GVGVP) under isometric conditions, as shown in Figure 5.7B, raising the temperature from near 0° C to 25° C results in a small linear increase in force, but raising the temperature from 25° C to 40° C results in a supralinear development of tension under isometric conditions or in the lifting of a weight under isotonic conditions. The marked force development above 25° C is clearly the result of the inverse temperature transition of hydrophobic association.

In 1970, Hill described the same properties for intact frog sartorius muscle.[15] Hill reported a linear increase in isometric tension on raising the temperature from 0° to 23° C, but on raising the temperature higher there occurred a more

7.2 The Movable Cusp of Insolubility

pronounced increase in tension, which Hill ascribed to an "active" process. Given the above considerations and those of isometric force development represented in Figures 5.7 and 5.8 and associated discussion in section 5.2.4, it is reasonable to view the "active" process in terms of a thermally driven inverse temperature transition of hydrophobic association.

More recently, Ranatunga described analogous properties[16] for chemically skinned, glycerinated, rabbit psoas muscle fibers immersed in relaxing solution (pH ≈ 7.2, pCa ≈ 8, ionic strength 200 mM) where membrane active processes would be less likely. It is not unreasonable, therefore, to suggest that a hydrophobic association transition is the source of the pronounced thermally induced tension development in mammalian skeletal muscle, not unlike that of our elastic protein-based polymer model system.

Mammalian muscle acts as though it is poised for hydrophobic association at room temperature much as is the $(GVGVP)_n$ elastic-contractile protein-based polymer on which are based the visually demonstrated contractions reviewed in Chapter 5.

7.2.3.2 pH Activation of Muscle Contraction

As discussed in Chapter 5, a single Glu (E) in a sequence of 30 residues, or 1 Asp (D) per 30 mer, exhibits very large, very nonlinear pKa shifts, as Val residues are stepwise replaced by more hydrophobic Phe residues.[17] This demonstrates, in yet another way, hydrophobic-induced pKa shifts resulting from the competition for hydration between hydrophobic and charged residues. For our particular purposes here, it demonstrates that the pKa of a carboxyl group can be hydrophobically shifted over the range from 4 to 10.[17] Furthermore, the data demonstrate a nonlinear increase in positive cooperativity (quantified by the Hill coefficient)[18] that parallels the nonlinear shift in pKa. This means that a carboxylate moiety can readily have a pKa just below physiological pH and that a small change in pH can result in the complete protonation of the hydrophobically perturbed carboxylate.

As noted in Chapter 5, as listed in Table 5.5 and as given in Figure 5.34, the pKa of **Model Protein ii**, (GVGVP GVGFP GEGFP GVGVP GVGVP GVGVP)$_{40}$(GVGVP), is 4.8 (Hill coefficient of 1.6), whereas the pKa for **Model Protein I**, (GVGIP GFGEP GEGFP GVGVP GFGFP GFGIP)$_{26}$(GVGVP), is 5.9 (Hill coefficient of 2.7) and that of **Model Protein v**, (GVGVP GVGFP GEGFP GVGVP GVGFP GFGFP)$_{42}$(GVGVP), is 6.6 with a Hill coefficient of 8. Because in ischemia the pH can drop to 6.5 or lower,[19] a carboxylate such as in **Model Protein v** could be protonated in ischemia to drive muscle contraction. In particular, pH-induced contraction has been reported in the aorta of rats, and increasing blood pressure enhances such pH-induced contraction.[20] The latter effect would, with an elevation in blood pressure, be aided by a stretch-induced increase in pKa (see below and Figure 5.23) and would make the effect of ischemia more profound.

7.2.3.3 Calcium Ion Activation of Muscle Contraction

Data regarding the dependence of T_t and N_{hh} on calcium ion concentration for **Model Proteins I** and **ii** in Figure 5.27, and the above self-consistent analysis of muscle contraction being one of controlling the hydrophobic association of an inverse temperature transition, combine to support the hypothesis that calcium ion binding at carboxylates increases the water of hydrophobic hydration. This would, in our view of competition for hydration between hydrophobic domains and phosphate, increase the free energy of bound phosphate leading to an increase in the rate of phosphate release. The loss of the very polar phosphate, whether as a covalently bound phosphate, as a free phosphate, or as bound nucleotide phosphate, favors the hydrophobic association of contraction that was deduced above to occur as part of muscle contraction. This point of view is considered in more detail in Chapter 8, after discussion has progressed from the anatomical and physiological levels considered in this chapter to the molecular level for the myosin II motor to be considered in that chapter.

7.2.3.4 Stretch Activation of Muscle Contraction

With the elastic model system, γ-irradiation cross-linked poly[0.82(GVGIP),0.18(GEGIP)], extension increases the pKa in a supralinear fashion, as previously emphasized in Figure 5.23 of Chapter 5.[21] This demonstrates what could be called a *stretch activation* of the carboxylate moiety; the free energy of the carboxylate increases nonlinearly as the extension increases. As stretching exposes hydrophobic groups of the hydrophobically associated matrix, the increased waters of hydrophobic hydration resulting from stretch exposure of the hydrophobic moieties must necessarily be considered as unsuited for hydration of the carboxylate moiety; the result is the observed increase in free energy of the carboxylate.

As noted in Chapter 5, phosphate is a supercarboxylate, being equivalent to several carboxylates in its capacity to shift the value of T_t, that is, in its capacity to destructure hydrophobic hydration. Thus, we anticipate that stretching of a muscle fiber with exposure of hydrophobic moieties would, in an entirely analogous manner, increase the free energy of a phosphate. Such an increase in the free energy of a phosphate would increase the rate of phosphate release. This would naturally be interpreted by the biochemist as an increase in ATPase activity.

Stretch activation of muscle is a well-described phenomenon; it was the subject of The Croonian Lecture (1977) given by Pringle,[22] and it has been extensively researched and reported in the literature over the ensuing decades. For example, the basic description becomes "When active insect flight muscle is stretched, its ATPase rate increases. . . ."[23] This we take as yet another demonstration of a fundamental process whereby a phosphate present in a protein can be activated, "energized," as the result of an increase in hydrophobicity. It is an example of the competition for hydration between apolar and polar species, that is, an example of the apolar–polar repulsive free energy of hydration active in muscle contraction.

7.2.3.5 Dephosphorylation Initiates Hydrophobic Associations of Muscle Contraction

Phosphate attached to a model protein is three to four times more effective on a mole fraction basis than carboxylate in raising the T_t-divide for hydrophobic association, which we have shown is due to a decrease in hydrophobic hydration (see Figures 5.25 and 5.27). Dephosphorylation, therefore, would re-establish hydrophobic hydration and dramatically lower the T_t-divide, which is to lower the temperature range of the cusp of insolubility to below physiological temperature. The result would be an insolubilization of hydrophobic domains (a hydrophobic association) that we consider to be the power stroke of muscle contraction.

A consistent feature of the myosin II motor in its many facets is that binding ATP causes dissociation. Again, a spatially localized pair of associated hydrophobic domains remains as such as long as during an opening fluctuation there develops too much hydrophobic hydration, and they reassociate. The binding of a water-thirsty ATP molecule in the proximity of an opening fluctuation would recruit the nascent hydrophobic hydration for its own hydration and leave the pair of hydrophobic domains dissociated, as depicted in Figure 2.16. *Thus, the most natural feature of ATP binding as seen by the consilient mechanism would be dissociation of hydrophobic domains.*

The relationship between the states of the myosin II motor domain, subfragment 1 (S1), and waters of hydrophobic hydration has been directly observed by the same microwave dielectric relaxation measurements as used with the data in Figures 5.24 and 5.25. Suzuki and coworkers[24] have observed the ADP-bound state of S1 to have more water of hydrophobic hydration than the ADP•P_i state, that is, than the most polar state. Thus, the effects of the additional negative charges are to destroy water of hydrophobic hydration and to raise the value of T_t above physiological temperatures. The removal of the phosphate would result in an increase in N_{hh} and a lowering of T_t below the physiological temperature with the consequence of the association of paired hydrophobic domains in the process of muscle contraction.

7.2.3.6 Scenario of Muscle Contraction by the Inverse Temperature Transition of Hydrophobic Association

Whether concern is for an explanation of the cause of rigor at the gross anatomical level or whether it is a more detailed consideration of the physiology of muscle contraction there occurs a correlation of phenomena whereby contraction is well described by the hydrophobic association of an inverse temperature transition. In Chapter 8, detailed molecular interactions attending the contraction/relaxation cycle of the myosin II motor are considered in terms of the consilient mechanism of the control of association/dissociation of hydrophobic domains. In particular, detailed comparisons of the crystal structures of scallop muscle cross-bridge in the near rigor, and bound ATP analogue states demonstrate a dramatic hydrophobic dissociation between the head of the lever arm and the N-terminal domain for the ATP analogue state that becomes a hydrophobic association for the near rigor state. The latter markedly changes the orientation of the lever arm.

7.2.4 Hemoglobin Transport of Oxygen by Positioning the "Cusp of Insolubility" Through Control of Intersubunit Hydrophobic Association/Dissociation

The globin protein provides an example of particular interest to the present discussion. As stated by Perutz,[25] "Ionized residues are excluded from the interior of globin chains, which is filled largely by hydrocarbon side chains, but some serines and threonines also occur there." This perspective immediately emerged from the first few protein structures determined by X-ray diffraction methods. As many more protein structures have been so determined, the only essential modification of this quotation is that whenever charge is found buried within the hydrophobic associations, it is always found ion paired (also referred to as a salt bridge). This is, of course, as required by the consilient mechanism of protein structure and function. Reemergence of the separated charges contributes to the destruction of hydrophobic hydration on dissociation and makes facile the modulation of hydrophobic association/dissociation when assisted by introduction of a proximal modest charge or other polarity.

Armed only with the information that two states exist, determination of the involvement of the consilient mechanism of an inverse temperature transition becomes straightforward. *Raising the temperature and decreasing net charge under conditions that change the state both bring about formation of the hydrophobically associated state.* Hydrophobic association occurs with loss of water, and hydrophobic dissociation occurs with an uptake of water. This too can be diagnostic. Application of these simple tests to the diverse data on hemoglobin below makes it quite clear that the two states, deoxygenated and oxygenated, are related by an inverse temperature transition, with the deoxygenated state being the hydrophobically associated state. This is discussed below. What is less clear from analysis of structural data in the literature is exactly how the binding of the polar group of oxygen, O_2, triggers the conversion from hydrophobically associated to hydrophobically dissociated. Can it be a direct effect of the introduction of the polar oxygen molecule? Or does the binding of oxygen cause conformational changes such that a mismatch of salt bridges and hydrogen bonds develops within and/or between the hydrophobically associated domains? The latter could expose these polar species and thereby disrupt a particular hydrophobic association of a pair of hydrophobic domains.

7.2.4.1 Hemoglobin: Hydrophobic Association of Four Myoglobin-Like Proteins

7.2.4.1.1 The Relationship Between the Oxygen Binding Proteins of Hemoglobin and Myoglobin

Hemoglobin is a tetramer composed of two α-chains and two β-chains, with paired αβ dimers delineated as $\alpha^1\beta^1$ and $\alpha^2\beta^2$. (In the nomenclature used here, the superscript identifies the subunit and the subscript indicates the number of such subunits.) Myoglobin and the α-chain of hemoglobin arose evolutionarily from the same primordial globin gene, and the β-chain of hemoglobin evolved from the α-chain. Each chain binds a heme group, which is the site of oxygen binding.

The sequences of the α- and β-chains of adult hemoglobin A (HbA) and of myoglobin are represented in Figure 7.2 in terms of single residue hydrophobicity plots. This is based on the T_t-based hydrophobicity scale reported in Chapter 5, section 5.3.2. The T_t-values for each amino acid residue are given in Table 5.1 and were derived from the data in Figure 5.9. As shown in Table 5.1, the most vinegar-like residue is glutamic acid (Glu, E); it is the most effective residue for disrupting or preventing hydrophobic association. The Glu (E) residue is the most polar residue, as represented in Figure 7.2, with the largest downward deflection. The carboxylate, $-COO^-$, of the Glu (E) residue is the most effective amino acid side chain in destructuring hydrophobic hydration. Because of the special thermodynamic properties of hydrophobic hydration discussed in Chapter 5, an increase in the potential for hydrophobic hydration leads to hydrophobic association, whereas a decrease in the potential for hydrophobic hydration disrupts hydrophobic association.

7.2.4.1.2 Distribution of Charged and Oil-like Residues in the Globin Chain

In general, from the perspective of the consilient mechanism, the distribution of charged or polar residues (shown by the downward deflections from the 37° C line in Figure 7.2) and the oil-like or hydrophobic residues (shown by the upward deflection from the 37° C line in Figure 7.2), constitutes the primary contribution to the common tertiary structure for myoglobin in Figure 7.3 and to that of the α- and β-chains of hemoglobin in Figure 7.4.

7.2.4.1.3 Insights into Association of Chains from the Single Residue T_t-based Hydrophobicity Plots

From the viewpoint of hydrophobic association between chains, therefore, the primary difference in amino acid sequence between myoglobin and the α- and β-chains of hemoglobin resides in the number and distribution of glutamic acid (Glu, E) residues. As shown in Figure 7.2, myoglobin contains 14 Glu (E) residues, whereas the α-chain contains only 4 and the β-chain only 8. Of all the amino acid residues, the presence of Glu (E) residues is the most effective means of disrupting subunit interaction by hydrophobic association. Primarily because of this, myoglobin exists as a monomer and exhibits little or no tendency for intermolecular hydrophobic association, whereas the propensity for the α- and β-chains to associate hydrophobically is much greater, resulting in the $\alpha_2\beta_2$ tetramer under physiological conditions.

7.2.4.1.4 The Relative Oxygen Binding Properties of Myoglobin and Hemoglobin

Myoglobin exhibits a hyperbolic curve for the binding of oxygen as the result of a simple binding process wherein a linear increase in bound oxygen results from a linear increase in the partial pressure of oxygen, pO_2, until saturation is approached. On the other hand, hemoglobin exhibits a sigmoidal oxygen-binding curve, in which hemoglobin initially exhibits a lower affinity for oxygen (Figure 7.5). When hemoglobin completes its binding of oxygen at a higher chemical potential, that is, at a higher partial pressure of oxygen, it has done so with a sigmoid binding curve in which the binding of the fourth oxygen molecule to the tetramer occurs with a 200-fold greater affinity, that is, oxygen binding occurs with positive cooperativity (Figure 7.6). The initial bound oxygen facilitates binding of the second oxygen. This means that hemoglobin can change its degree of saturation markedly with a small change in partial pressures of oxygen with near saturation in the lungs and near complete release on going to the tissues. Myoglobin, on the other hand, shows little change in degree of saturation for the same change in partial pressure of oxygen. For hemoglobin, this, of course, is the required property for transport of oxygen from the lungs to the tissues.

7.2 The Movable Cusp of Insolubility   251

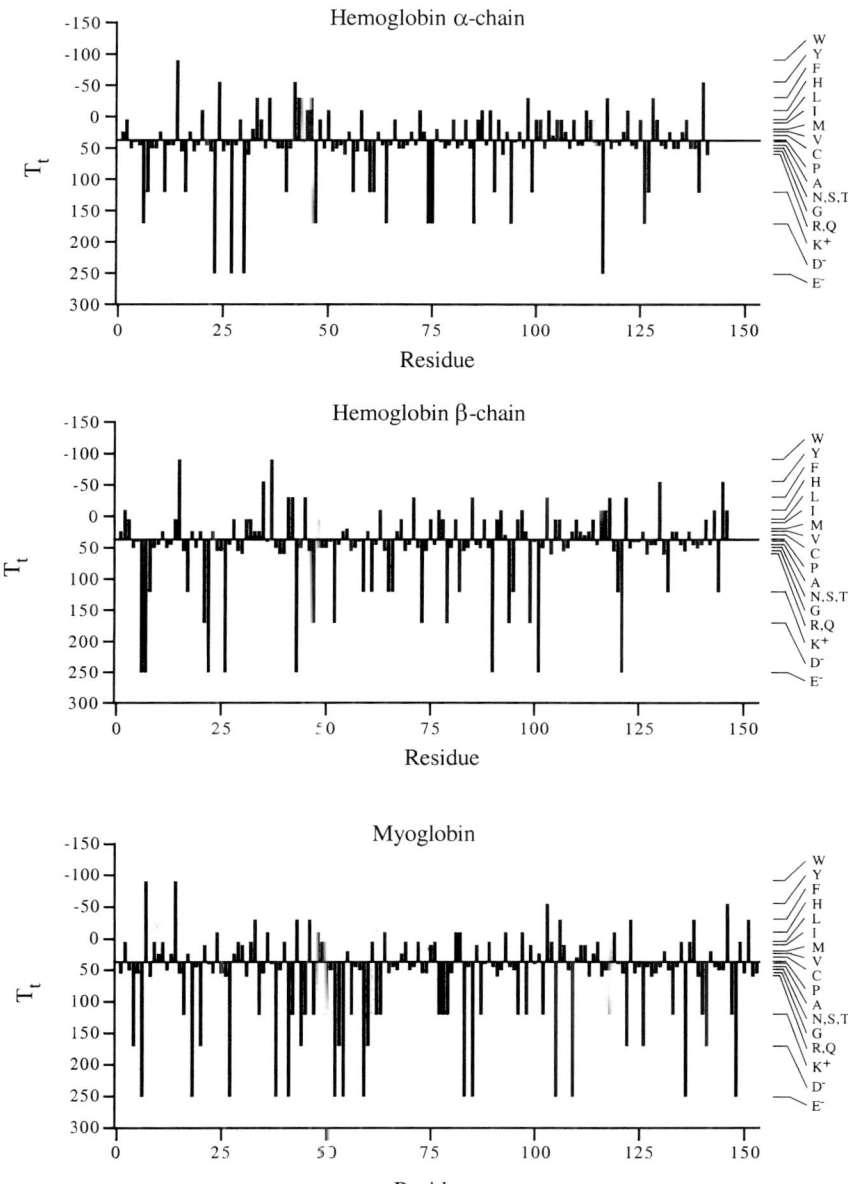

FIGURE 7.2. Single residue hydrophobicity plots for the α- and β-chains of hemoglobin and for myoglobin. Note the marked decrease in polar residues (the negative deflections) in the plots for the α- and β- chains of hemoglobin compared with the plot for myoglobin; this is responsible for hydrophobic association between α- and β-chains in the formation of the tetramer of chains that is hemoglobin.

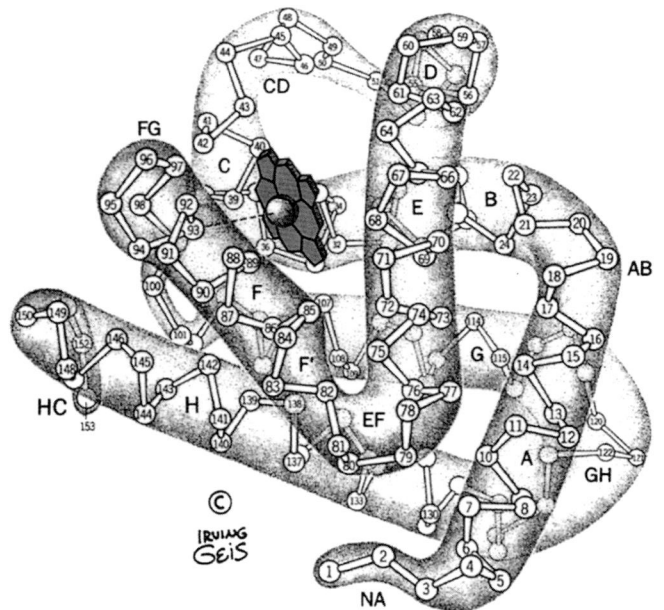

FIGURE 7.3. Structure of myoglobin (Kendrew sausage representation), showing location of the oxygen binding heme group and indicating its coordination to the imidazole of His^{90}. (Reproduced with permission. Illustration, Irving Geis. Rights owned by Howard Hughes Medical Institute.)

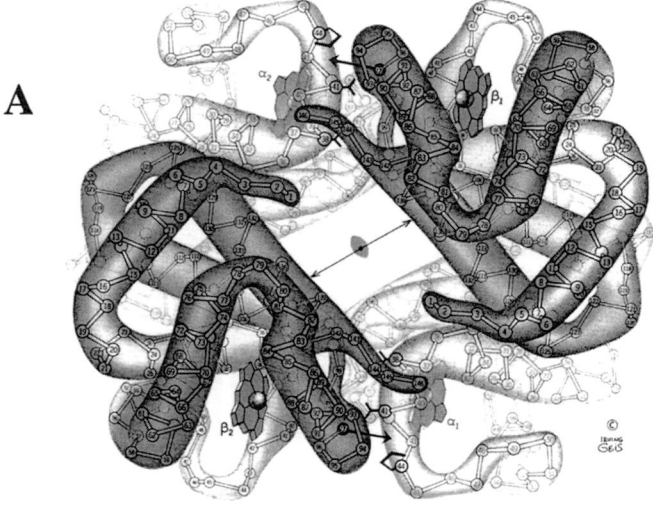

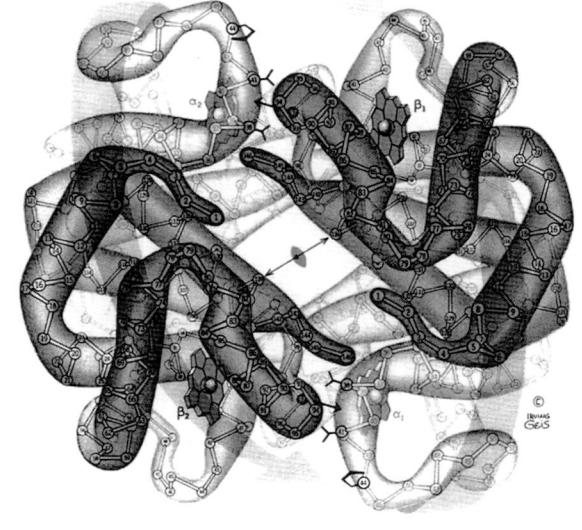

FIGURE 7.4. Structure of hemoglobin (Perutz sausage model representation). **A** Deoxyhemoglobin. **B** oxyhemoglobin. Dyad axis view showing four subunits, constituted by two α-chains and two β-chains with the same folding of individual chains as in myoglobin. (Reproduced with permission. Illustration, Irving Geis. Rights owned by Howard Hughes Medical Institute.)

7.2 The Movable Cusp of Insolubility

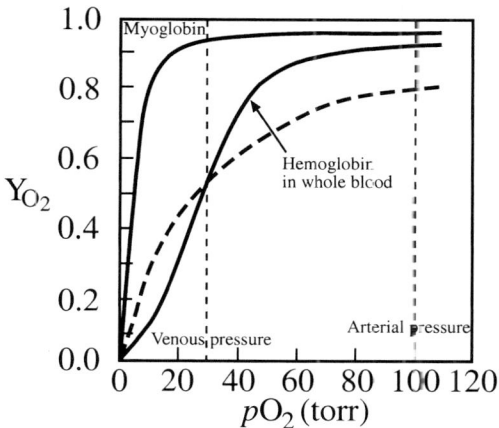

FIGURE 7.5. Plots of the oxygen binding curves for myoglobin and hemoglobin in whole blood, with a dashed curve for the binding that hemoglobin would exhibit with the same average binding affinity but without positive cooperativity. Y_{O_2} is the fraction of sites occupied by oxygen (i.e., the fraction of saturation), and pO_2 is the partial pressure of oxygen. The arterial pO_2 reflects the pressure of oxygen in the lungs.

7.2.4.1.5 Positive Cooperativity Arises from the Subunit Interactions

The isolated heme bound α-chain and the tetramer of β-chains, like myoglobin, bind oxygen with a simple hyperbolic curve; they do not exhibit positive cooperativity. Furthermore, a wide change in neither pH nor CO_2 significantly affects the binding of oxygen by myoglobin, whereas these have significant and physiologically relevant effects on the oxygen transport binding properties of hemoglobin.

FIGURE 7.6. Hill plots, $\log[Y_{O_2}/(1 - Y_{O_2})]$ to obtain the coefficients that show oxygen binding affinity to hemoglobin from the start to the finish of the titration. As measured from the intercepts with the 0.0 line, the binding affinity to hemoglobin increases 100-fold from binding of the first oxygen to the fourth oxygen; this is positive cooperativity. In contrast, the oxygen binding to myoglobin shows a slope of one, that is, no cooperativity. (Adapted with permission from Voet and Voet.[103])

Therefore, the interactions between subunits must contain the source of the positive cooperativity that provides for effective transport function of hemoglobin.

7.2.4.1.6 A Small Difference in the Geometry of the Heme Binding Site in the Oxygenated and Deoxygenated States of Myoglobin and Hemoglobin

In deoxymyoglobin and in deoxyhemoglobin, the heme iron is out of the heme (porphyrin) plane by 0.55 and 0.60 Å, respectively, whereas in oxymyoglobin and in oxyhemoglobin the heme iron is out of the porphyrin plane by 0.22 and 0.00 Å, respectively. In all cases of displacement, the heme iron is displaced toward the coordinated imidazole side chain of a histidine residue in formation of a tetragonal pyramid bond structure, and oxygen binds to the opposite side of the heme, restoring or partially returning the heme iron to the plane defined by the heme.

This difference in heme iron displacement from the porphyrin plane has been proposed by Perutz[25] to be the basis for the positive cooperativity, that is, for the sigmoid oxygen binding curve essential for the function of hemoglobin in its oxygen transport role. Our purpose here is to bring function into focus in terms of the consilient mechanism of inverse temperature transitions of hydrophobic association/dissociation of defined pairs of hydrophobic domains within the hemoglobin tetramer. More to the point is the question, could the consilient mechanism involving changes in the Gibbs free

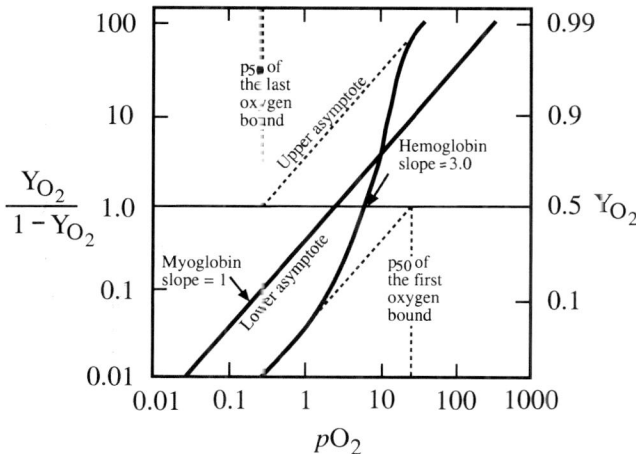

energy of hydrophobic association contribute to the sigmoid oxygen binding property and thereby contribute to hemoglobin function?

7.2.4.2 Occurrence of a Temperature Effect Diagnostic of the Consilient Mechanism Despite Absence of an Essential Functional Role

The consilient mechanism is most fundamentally identified by an inverse temperature transition that results in hydrophobic association on increasing the temperature through the transition zone, that is, through the responsive temperature interval. The observation of a temperature-driven release of oxygen from tetrameric hemoglobin has no recognized physiological role, because there is no significant increase in temperature on going from the lungs to the tissues.

Also, it is important to note that thermally driven release of oxygen from tetrameric hemoglobin is greater than the temperature-effected release from the related monomer myoglobin. The state obtained on raising the temperature for a molecular system functioning by the consilient mechanism of an inverse temperature transition of hydrophobic association would necessarily be a state of greater hydrophobic association. By the consilient mechanism, therefore, thermally driven oxygen dissociation that is greater for hemoglobin than for myoglobin defines deoxyhemoglobin as the more hydrophobically associated state of hemoglobin.

A modest increase in temperature moves the hemoglobin tetramer across the T_t-divide, surmounting the cusp of insolubility, to arrive at the more insoluble side wherein pairs of hydrophobic domains have moved from having greater exposure to water to exhibiting more hydrophobic association. This prediction of the consilient mechanism is borne out by the crystal structures of the oxygenated and deoxygenated states of hemoglobin. Associated pairs of hydrophobic domains characterize deoxyhemoglobin in the crystal structure, and certain components of these domains become exposed to water in oxyhemoglobin.

7.2.4.3 Positive Cooperativity on O_2 Binding Analogous to Formation of Carboxylates Within Increasingly Hydrophobic Domains

7.2.4.3.1 Positive Cooperativity of Carboxyl Ionization in Elastic Model Proteins Increases with Increasing Hydrophobicity

The above prediction of the consilient mechanism that deoxyhemoglobin is the state of greater hydrophobic association would allow that binding of an initial polar oxygen molecule could be analogous to the formation of an initial carboxylate group, as occurs in our stepwise increasingly hydrophobic model elastic protein-based polymeric machines described in Chapter 5. Specifically, as shown in Figure 5.34, ionization of the initial carboxyl to form the polar carboxylate group is increasingly delayed, that is, occurs at increasingly higher pH values, as the hydrophobicity of the model protein increases. Associated with the supralinear increase in pKa shifts with a linear increase in hydrophobicity, there occur corresponding supralinear increases in positive cooperativity. The Hill coefficients,[18] (n), increase from 1.5 to 8.0 as the result of stepwise increases in hydrophobicity. Figure 7.7A contains plots of the Hill coefficients. The set of model proteins from Table 5.5 and their Hill coefficients follow:

Model Protein i: (GVGVP GVGVP GEGVP GVGVP GVGVP GVGVP)$_{36}$(GVGVP), E/0F ⇒ 1.5
Model Protein ii: (GVGVP GVGFP GEGFP GVGVP GVGVP GVGVP)$_{40}$(GVGVP), E/2F ⇒ 1.6
Model Protein iii: (GVGVP GVGVP GEGVP GVGVP GVGFP GFGFP)$_{39}$(GVGVP), E/3F ⇒ 1.9
Model Protein iv: (GVGVP GVGFP GEGFP GVGVP GVGFP GVGFP)$_{15}$(GVGVP), E/4F ⇒ 2.7
Model Protein v: (GVGVP GVGFP GEGFP GVGVP GVGFP GFGFP)$_{42}$(GVGVP), E/5F ⇒ 8.0

7.2 The Movable Cusp of Insolubility 255

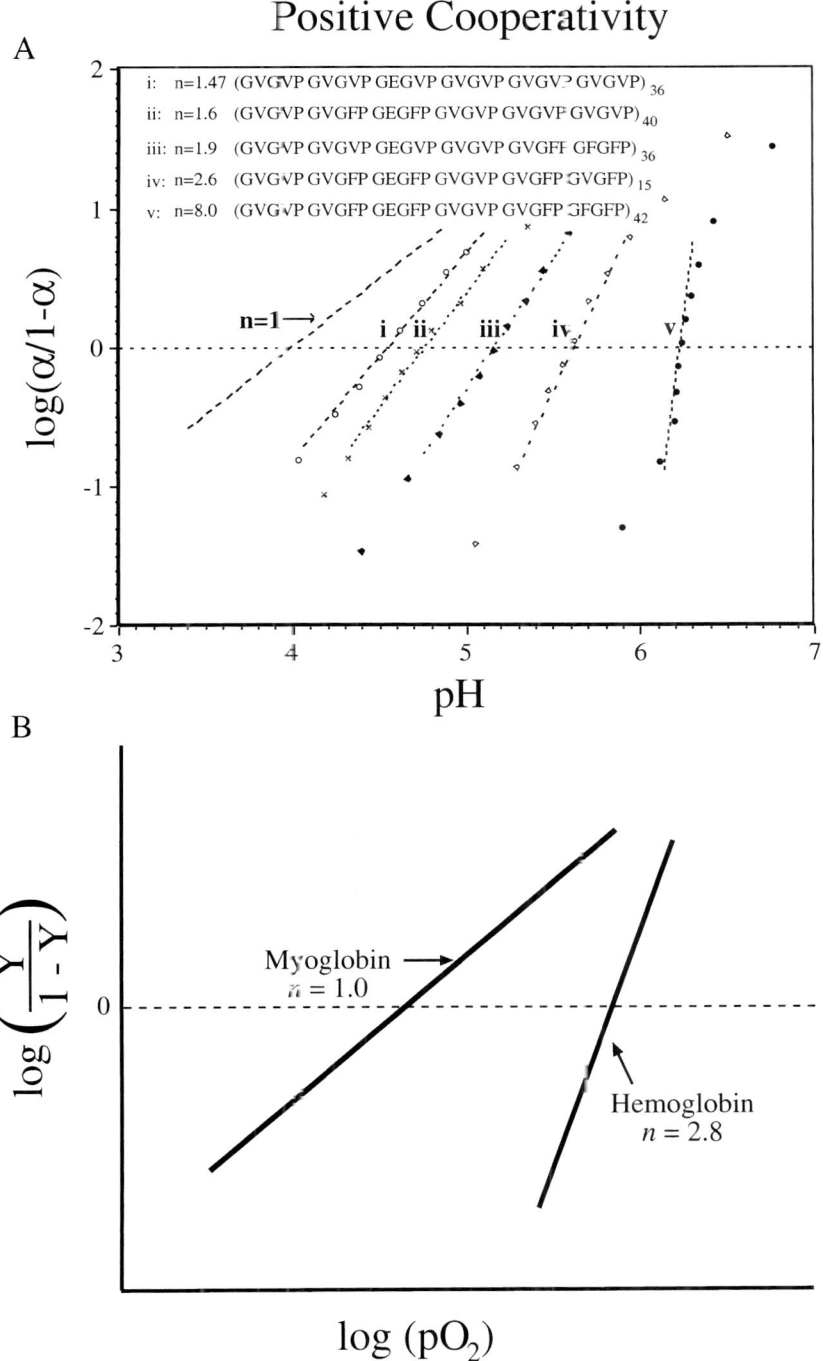

FIGURE 7.7. Hill plots to obtain Hill coefficients using log[α/(1 − α)] versus activity of reactant. **(A)** For the acid–base titration curves of the **Model Proteins i** through **v** of Table 5.5, showing a supralinear increase in Hill coefficient from n = 1.5, 1.6, 1.9, 2.7 to 8.0 as the hydrophobicity increases linearly. **(B)** For the oxygenation of myoglobin (n = 1) and for the oxygenation of hemoglobin (n = 2.8). A more dramatic increase in positive cooperativity can be observed with the designed model proteins.

In these elastic model proteins development of such dramatic positive cooperativity occurs without significant conformational strain, but rather simply results from competition for hydration between hydrophobic and charged residues. Obviously, this positive cooperativity occurs without an electrostatic component to the conformational energy, such as repulsion between charges. The charge–charge repulsion of poly(methacrylic acid) (PMA) results in a negative cooperativity, which, when represented by a Hill coefficient, gives an *n* of 0.5 with 1/2 the slope of myoglobin in Figure 7.7B. Recall the data for the elastic model proteins discussed in Chapter 5, section 5.7. Increased positive cooperativity results from competition for hydration between the polar carboxylate and the hydrated hydrophobic domains. Competition for hydration with hydrophobic groups increases the Gibbs free energy of the carboxylate. Not only can the elastic model proteins be designed to exhibit the magnitude of positive cooperativity exhibited by Hb, they can also be designed with positive cooperativity that substantially surpasses the vaunted hemoglobin example.

7.2.4.3.2 Analogy to Hemoglobin Binding of Oxygen

In the case of hemoglobin, an analogy exists between the higher pH, that is, the higher hydroxyl ion concentration, required to form carboxylate and the higher oxygen concentration, the higher pO_2 values, required for the first oxygen molecule to bind one of the four hemes in hemoglobin. Hill plots of the carboxyl-containing elastic protein-based polymers in Figure 7.7A can be compared with those of myoglobin and hemoglobin in Figure 7.7B. A structural analysis of oxygen binding will be required to consider whether the oxygen acts as a polar species to destroy hydrophobic hydration and enable ion pairs of hydrophobic domains to separate or in some other manner to cause the associated hydrophobic domains to separate.

The cases are quite analogous; both exhibit delayed formation of the more polar species. For the tetrameric hemoglobin, binding of the more polar oxygen molecule requires a higher pO_2, a higher chemical potential of oxygen, than for the related monomeric myoglobin. For the model proteins with increased hydrophobicity, deprotonation of carboxyl to form the more polar carboxylate requires a higher chemical potential of hydroxyl ion. Thus, ionization is delayed on raising the pH as the hydrophobicity of the model protein increases. The ionization is delayed because the carboxylate must do the work of destructuring hydrophobic hydration in order to form. Once one carboxylate forms, having destructured substantial hydrophobic hydration, subsequent carboxylates form more easily because they have less hydrophobic hydration to destructure. The magnitude of the Hill coefficient (n), that is, the positive cooperativity, increases with each replacement of a Val (V) residue by a more hydrophobic Phe (F) residue, because the task of the initial charge formation in obtaining adequate hydration increases as the elastic model protein becomes more hydrophobic.

7.2.4.3.3 Dissociation of the $\alpha_2\beta_2$ Tetramer on Oxygenation at Low Concentrations ($< 3\,\mu M$ Heme)

At concentrations of less than $3\,\mu M$ heme, the deoxygenated state occurs as the $\alpha_2\beta_2$ tetramer, and oxygen binding exhibits positive cooperativity. On oxygenation the tetramer splits into $\alpha^1\beta^1$ and $\alpha^2\beta^2$ dimers. As discussed below, separation at low concentration of the tetramer into dimers on oxygen binding results from the separation of ion pairs that had been brought into salt-bridge formation due to the dominance of hydrophobic domains. In other words, the association between the $\alpha^1\beta^1$ and $\alpha^2\beta^2$ dimers is dominantly hydrophobic.

7.2.4.3.4 Is the Consilient Mechanism Relevant to Oxygen Binding by Hemoglobin?

Should the structural data show the $\alpha^1\beta^1$–$\alpha^2\beta^2$ interface to be one of hydrophobic association that is disrupted on oxygenation, then in our view the answer is clearly yes, the cooperative binding of oxygen to hemoglobin results from the consilient mechanism. In our view the presence of hydrophobic association at the $\alpha^1\beta^1$–$\alpha^2\beta^2$ interface would indicate that positive cooperativity, essential to hemoglobin function in oxygen transport, arises due to the consilient

7.2 The Movable Cusp of Insolubility

mechanism with the underlying physical process being an apolar–polar repulsive free energy of hydration. Once the first oxygen molecule in the process of binding effects some hydrophobic dissociation of the $\alpha^1\beta^1$ and $\alpha^2\beta^2$ dimers, by destructuring hydrophobic hydration that would occur on partial dissociation at the interface between the two dimers, the second oxygen can bind more readily to the second subunit because its intersubunit hydrophobic hydration has already been partially destructured. Thus, there would result a sigmoid binding curve, quantified in terms of a Hill coefficient of greater than 1.

Yet much more phenomenological data, showing the coherence of phenomena between hemoglobin function and the consilient mechanism of the inverse temperature transition for hydrophobic association, continues below, before direct examination of the molecular structures, as generally presented. Subsequently, in section 7.3, the molecular structures are examined to look for the specific interactions most significant to the consilient mechanism.

7.2.4.4 Existence of Two Conformational States, T and R, of Hemoglobin

7.2.4.4.1 Definition of the T State: The Tense or Taut Form Exhibited by Deoxyhemoglobin

The T state is like a rigor or contracted state and is analogous to dephosphorylated, low pH (protonated), and ion-paired states apparent in the contracted state of elastic model proteins and of muscle. In our perspective, the T state is the state with more hydrophobic association between subunits. Within the associated hydrophobic domains of the T state can be anticipated ion pairs and hydrogen bonds within and between the $\alpha^1\beta^1$ and $\alpha^2\beta^2$ dimers that would facilitate shifting the cusp of insolubility. The cluster of ion pairs and hydrogen bonds that change at the $\alpha^1\beta^1$–$\alpha^2\beta^2$ interface on oxygenation are shown in Figure 7.8. What is not apparent in Figure 7.8 is that these interactions form because these polar groups are under the influence of hydrophobic domains. The spatially localized hydrophobic contacts between subunits represent movable regions that begin on the insolubility side of the cusp of

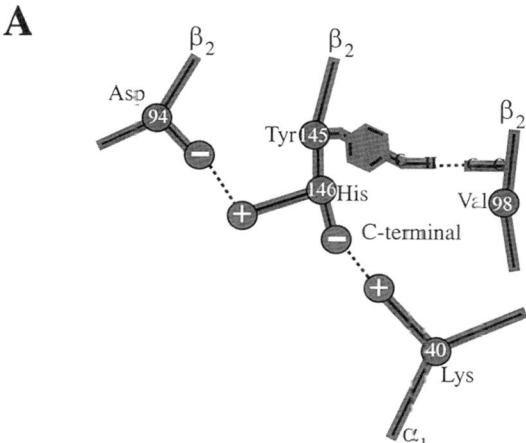

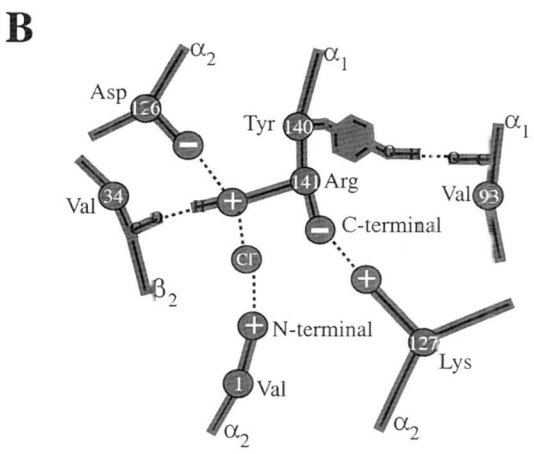

FIGURE 7.8. Networks of ion pairs and hydrogen bonds of the "switch region" (A) and the "joint-associated region" (B) at the interface between the $\alpha^1\beta^1$ and the $\alpha^2\beta^2$ dimers that comprise the hemoglobin tetramer. These polar interactions result from the hydrophobic association between the dimers that is maximal in deoxyhemoglobin and partially relaxed on oxygenation. The key ion pairs that orient the $\alpha^1\beta^1$ and the $\alpha^2\beta^2$ dimers are indicated by the circled 1 and 2 and are identified as such on the interface in Figure 7.18. See text for discussion. (Adapted with permission from Voet and Voet.[103])

insolubility, but binding of the polar oxygen molecule moves the cusp of insolubility to higher temperature such that now the system shifts toward the more soluble side of the cusp of insolubility. The T state with no bound oxygen contains hydrophobic domains thermodynamically poised on the insolubility side of the cusp of insolubility.

7.2.4.4.2 Definition of the R State: The Relaxed Form Exhibited by Oxyhemoglobin

The R state is the more polar, more hydrated, relaxed state of hemoglobin. With respect to the separable paired hydrophobic domains, that is, the separated $\alpha^1\beta^1$ and $\alpha^2\beta^2$ dimers seen at low concentration, the R state resides on the solubility side of the movable cusp of insolubility. It is the hydrophobically dissociated state resulting from introduction of a polar species, namely, oxygen. By the consilient mechanism, the spatially localized hydrophobic domains poised on the insolubility side of the T_t-divide in the T state become disrupted on introduction of a polar species and are now found on the solubility side of the T_t-divide. This disrupts the intersubunit hydrophobic associations between the $\alpha^1\beta^1$ and $\alpha^2\beta^2$ dimers that had been made possible by the now disrupted ion pairs and hydrogen bonds. In other words, in the case of hemoglobin, as the result of oxygen binding, the cusp of insolubility moves to a higher temperature along the temperature axis, as represented by the rightmost cuspid in Figure 7.1. Again, with regard to the dissociable pairs of hydrophobic domains, the R state of hemoglobin at physiological concentrations with four bound oxygen molecules has shifted toward the solubility side of the cusp of insolubility.

7.2.4.5 *The T → R State Transition Involves Uptake of Water Despite a Decrease in Volume of Aqueous Channel of Oxyhemoglobin*

7.2.4.5.1 General Considerations of Transient Openings of Hydrophobically Associated Domains of the T State

Recall from Figure 5.27 and associated discussion that, as the amount of hydrophobic hydration increases, the value of T_t decreases. The cusp of insolubility moves to lower temperatures, from above to below physiological temperature. Accordingly, hydrophobic association between two surfaces occurs because the surfaces are sufficiently hydrophobic that during a transient opening so much hydrophobic hydration forms that the T_t-divide is below the operating temperature. This causes the transiently opened paired hydrophobic domains to reassociate. If a sufficiently polar species should be introduced into the protein during a transient opening, then, due to competition for hydration, the polar species results in a decrease in the amount of hydrophobic hydration. As a result, the thermodynamic driving force for association can be lost, that is, the T_t-divide can be raised above the operating temperature. Direct competition for hydration between the added polar species and the hydrophobic groups of the hydrophobic domain may be the most obvious way to achieve the loss of hydrophobic hydration, but it should not be considered the only way to do so. *Any interaction that causes the associated hydrophobic domains to be less hydrophobic, that is, to have less hydrophobic hydration on separation, constitutes the central feature of the consilient mechanism.*

7.2.4.5.2 The Uptake of Water on Transition from the T to the R State

Bulone et al.[26,27] report the average number of water molecules that attends the T to R state transition in hemoglobin is 75. Colombo and coworkers[28] independently report the uptake of a similar number of water molecules, 60, on conversion from the hydrophobically associated to the partially dissociated state. At normal concentrations, this addition of water can be identified with the partial hydrophobic dissociation at the interfaces between the paired $\alpha\beta$ dimers ($\alpha^1\beta^1$ and $\alpha^2\beta^2$). As noted above, at low concentrations (less than $3\mu M$ heme), the dissociation of the functional units, the paired $\alpha\beta$ dimers ($\alpha^1\beta^1$ and $\alpha^2\beta^2$), is complete on the binding of oxygen.[25]

7.2.4.6 *Hydrophobic Interfaces Between Subunits Identified by Mean Residue Hydrophobicity Plots*

7.2.4.6.1 Mean Residue Hydrophobicity Plots for the Hemoglobin α- And β-Chains and for Myoglobin

Figure 7.9 provides a comparison of the T_t-based mean residue hydrophobicity plots of the α- and β-chains and myoglobin as a function of

7.2 The Movable Cusp of Insolubility

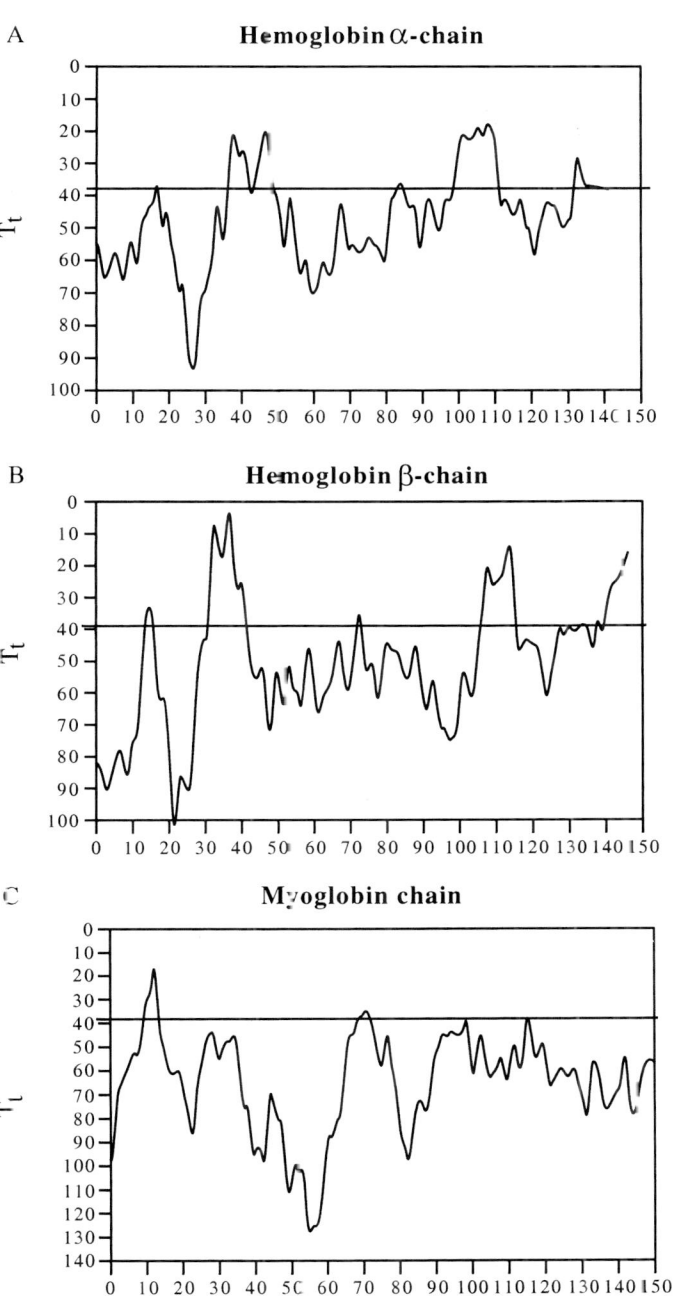

FIGURE 7.9. Mean residue hydrophobicity plots of the α-chain (A) and the β-chain (B) of hemoglobin and of myoglobin (C). The strong hydrophobicity peaks just above 100 residues are responsible for stable formation of the $\alpha^1\beta^1$ dimer. The strong hydrophobicity peaks near 40 residues may be considered part of the "switch region." These hydrophobic residues provide an arc for hydrophobic association at the $\alpha^1\beta^1$–$\alpha^2\beta^2$ interface. By the consilient mechanism, decreased heme hydrophobicity on oxygen binding relaxes this inter-dimer hydrophobic association and allows the T → R transition, as shown in Figure 7.10, to occur. This is further visualized in Figure 7.18 as discussed in the text. See text for further discussion.

sequence. A sliding window of the average value for 11 residues is used with the data point given for the central residue. Upward deflecting peaks from the 37° C baseline identify the more hydrophobic sequences. The plot for myoglobin exhibits only minor upward deflections that are recognized as contributing to the intramolecular (tertiary) hydrophobic folding (see Figure 7.9C).

The prominent peaks of the mean residue hydrophobicity plots for the α- and β-chains in Figure 7.9A,B, immediately upon inspection, provide interesting insight into the hydrophobic quaternary structural interactions that hold the subunits together. The effect of the changes in amino acid residues from myoglobin to the α- and β-chains of hemoglobin is to create hydrophobic domains. These hydrophobic domains become the basis for the hydrophobic association between subunits to form the stable α–β functional dimeric unit and the hydrophobic association at the interface between the $\alpha^1\beta^1$ and $\alpha^2\beta^2$ functional units.

7.2.4.6.2 Hydrophobic Association Holds Together the Fundamental Functional Unit, the αβ Dimer

The mean residue hydrophobicity plots for the α- and β-chains of hemoglobin exhibit two prominent upward deflections. These prominent hydrophobic sequences, not present in myoglobin, are responsible for the hydrophobic associations between chains. Those residues responsible for two of the four peaks in the mean hydrophobicity plots of the α-chain and β-chains in Figure 7.9A,B, specifically due to residues 100 to 110 of the α-chain (LLSHCLLVTLA) and 105 to 115 of the β-chain (LLGNVLVCVLA), form the primary hydrophobic association that holds together the basic structural unit, the αβ dimers ($\alpha^1\beta^1$ and $\alpha^2\beta^2$). Except for the hydrophobic histidine residue, H^{130}, these sequences contain no ionizable functions. Because of this they form the nondissociating αβ functional unit. Hydrophobic association within each basic αβ dimer is represented in Figure 7.10 as the association of G-helices. This hydrophobic association remains unchanged on oxygenation.

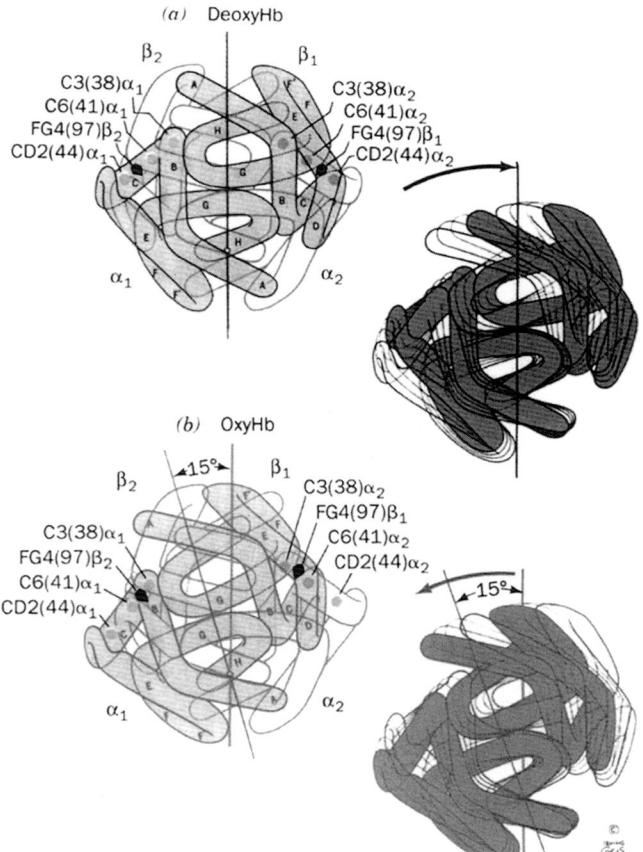

FIGURE 7.10. Perspective of hemoglobin to demonstrate the change on oxygenation to be the rotation of the $\alpha^1\beta^1$ dimer with respect to the $\alpha^2\beta^2$ dimer. The fulcrum of the rotation is the "joint-associated" region, and the shift at the "switch region" is noted. Also shown are the contacts between the α- and β-chains of the $\alpha^1\beta^1$ dimer. See text for discussion here and Figure 7.18 and associated text there for insight into the rotation that occurs at the α^1–β^1 and α^2–β^2 interface. (**A** Reproduced with permission from The Irving Geis Archives. Rights owned by Howard Hughes Medical Institute, reproduced from D. Voet and J. Voet, *Biochemistry*, Second Edition, John Wiley & Sons, Inc. New York, 1995.)

7.2 The Movable Cusp of Insolubility

7.2.4.6.3 Positive Cooperativity Due to Changeable Hydrophobic Interface Between αβ Dimers

The additional identified hydrophobic sequences of Figures 7.9A,B undergo changes on oxygenation. Residues 36 to 49 of the α-chain, FPTT**K**TYFPHF**D**LS, contribute to the so-called switch region, whereas the more hydrophobic residues 31 to 41 of the β-chain, LLVVYPWTQ**R**F, contribute to the so-called joint region (Figure 7.10A). Also shown in Figure 7.9B, the $β_2$ carboxyl-terminal sequence LAHKYH provides another prominent hydrophobic sequence that contributes to the switch region.

In particular as shown in Figure 7.8a, the Y145 and the terminal H146 contribute to two ion pairs and a hydrogen bond as key polar interactions of the hydrophobic switch region while being quite hydrophobic in their own right. In an analogous way, as shown in Figure 7.8b, the $α_1$-carboxyl-terminal Y140 and R141 with the aid of a chloride ion (Cl⁻) stitch together three ion pairs and two hydrogen bonds while in association and influenced by the hydrophobicity of the joint region. In our view, the neutralization of these polar groups by ion pairing and hydrogen bonding lowers the cusp of insolubility for an enlarged hydrophobic domain at the interface between αβ dimers that includes residues 31 through 41 of the β-chain, identified in the corresponding hydrophobic peak in Figure 7.9B. The shift between ion pair enhanced and enlarged hydrophobic association of the T state to the more limited hydrophobic association on dissociation of the ion pairs of the R state provides for the pivotal 15° motion of the joint region at the interface between the pairs of αβ dimers, as depicted in Figure 7.10.

7.2.4.6.4 Changes in Tertiary Structure on Oxygen Binding Disrupt Geometry Required for Ion-pair Formation

Drawing on the consilient mechanism for a working hypothesis, changes in tertiary structure on oxygen binding alter the positioning of the polar species in Figure 7.8, particularly of the charged species, such that they can no longer be sufficiently proximal to function as ion pairs. The separated ion pairs now function more as individual charged species and derive more hydration from their vicinity. Accordingly, fluctuations that separate the proximal hydrophobic domains allow the newly emerged charged species to destructure nascent hydrophobic hydration and thereby decrease the driving force for the associations of these hydrophobic domains. The result would be that hydrophobic associations of the T state open up with an uptake of water to give the R state. The fundamental question becomes, what is the principal reason for the disruption of the ion-pair and hydrogen-bonding interactions. This will be discussed further below in relation to the Perutz and consilient mechanisms.

7.2.4.6.5 More Detailed Identification of the Paired Hydrophobic Domains That Alter Association During the T to R State Transition

The principal sites considered responsible for the change in hydration, noted above, involve networks of ion pairs and hydrogen bonds (see Figure 7.8) between paired hydrophobic domains that are disrupted during oxygen binding. They occur at the $α_1$–$β_2$ interface and are referred to as the *switch region*. Changes occur at another site of hydrophobic association, the interface between the $α_1$ and $α_2$ chains, but this joint region only partially disassociates for increased exposure to water. These features are visited further in section 7.3 with additional representations of the crystal structures that focus on these domains.

7.2.4.7 Role of Ion-pair Formation in Hydrophobic Association of Elastic-contractile Model Proteins

7.2.4.7.1 Moving the Spatially Localized Hydrophobic Associations of Hemoglobin Back and Forth Across the T_t-divide (Moving the Cusp of Insolubility)

7.2.4.7.1.1 Polar Species Act Cooperatively. When separated, negatively and positively charged species individually act in a cooperative manner to destructure hydrophobic hydration. When ion paired, this capacity of the separated ions to destroy hydrophobic hydration becomes muted. In the case of hemoglobin, the negative species would be carboxylates, carbamates, biphosphoglycerates, and the electronegative oxygen molecule itself, and positive species, though inherently less effective than negative species, would be positively charged amino, guanidinium, and imidazolium groups. Such polar species when part of, or sufficiently proximal to, a pair of hydrophobic domains, shift the system to the solubility side of the divide, which is the R state. Charged species, whether positive or negative, shift the cusp of insolubility toward higher temperatures than the operating temperature, that is, toward solubility.

7.2.4.7.1.2 Ion-pair Formation Decreases the Effect of Charge. Because polar species cooperate to shift a pair of hydrophobic domains toward solubility (hydrophobic dissociation), ion pairing and to a lesser extent hydrogen bonding simultaneously neutralize pairs of polar species and markedly shift the system toward insolubility (hydrophobic association). It is important in understanding the state of hemoglobin to recognize that *ion pairing dramatically decreases the effect of charge to disrupt hydrophobic association.* Ion-pair formation is about half as effective as protonation of a carboxylate in shifting a spatially localized site toward the insolubility side of the divide as demonstrated in Figure 5.34 with the model proteins in Table 5.5.

7.2.4.7.2 Ion Pairing Between Chains of Model Proteins

The striking effectiveness of ion-pair formation in lowering the free energy of hydrophobic association, ΔG_{HA}, between moderately hydrophobic chains of elastic model proteins is demonstrated in Figure 6.3. The elastic **Model Protein x′** in Table 5.5, but with an *n* of 36 instead of 22, that is, (GVGVP GVGVP GKGVP GVGVP GVGVP GVGVP)$_{36}$(GVGVP), does not begin hydrophobic association at pH 7.5 in water until 70°C (158°F) and the Glu (E)-containing elastic model protein, **Model Protein v**, (GVGVP GVGFP GEGFP GVGVP GVGFP GFGFP)$_{42}$(GVGVP), even with 5 Phe (F) residues per 30 mer and under the same solvent conditions, does not begin hydrophobic association until 55°C (131°F). Remarkably, when these model protein chains are combined in the same solution, aggregation begins at 5°C (41°F).

Even when the paired model proteins have no Phe (F) residues, the effects of hydrophobic association are notable. In solution alone, the K-containing chain exhibits a T_t of 70°C (158°F), as above, and separately in solution the E-containing chain, (GVGVP GVGVP GEGVP GVGVP GVGVP GVGVP)$_{36}$ (GVGVP), exhibits a T_t-value that extrapolates to over 100°C (212°F). On combining in a 1:1 ratio, the T_t-value is 35°C (95°F). The formation of ion pairs between these chains lowers the free energy for hydrophobic association on a per GEGVP and GKGVP (ion pair) basis by some 3.5 kcal/mole-pentamer. This exemplifies the importance of ion-pair formation, even between modestly hydrophobic chains, in promoting hydrophobic association. As more hydrophobic residues—such as Ile (I), Leu (L), Phe (F), Tyr (Y), and Trp (W)—participate in the hydrophobic domain, the free energy for hydrophobic association becomes even more favorable.

Specific effects described below are known for shifting the equilibrium in the direction of either the T or the R state. These effects are explicable in terms of the ΔT_t-mechanism for moving the T_t-divide and the approximately equivalent Gibbs free energy of hydrophobic association, ΔG_{HA}, and its component the apolar–polar repulsive free energy of hydration, ΔG_{ap}. In all cases ion-pair formation associated with hydrophobic domains drives hydrophobic association.

7.2.4.8 Further Correlation of Phenomena

7.2.4.8.1 The Bohr Effect

7.2.4.8.1.1 The Bohr Effect is the Release of H^+ on Binding of O_2. Hemoglobin releases about 0.6 protons for each O_2 bound. These protons derive from partial proton release from the α_2-chain Val^1(α-NH$_3^+$) and the β_2-chain H146(imidazolium). These charged species are part of the ion-paired clusters shown in Figure 7.8. On oxygenation the polarity sufficiently increases to relieve the hydrophobic hydration driving force for maintaining these ion-paired networks.[29,30]

7.2.4.8.1.2 Reversible Combining of CO_2 to the N-terminal Amino Groups.

Associated with the Bohr effect is the reversible combining of CO_2 to the N-terminal amino groups such as the V1(α-NH_3^+) to result in carbamates. The reaction for carbamate formation is written, R–NH_2 + CO_2 = R–NH–COO^- + H^+. This –COO^- provides a more effective negative charge to replace the Cl^- (shown in Figure 7.8B) that bridged between the V1(α-NH_3^+) and the R141(guanidinium) in the ion-pairing network of the T state that is deoxyhemoglobin.

7.2.4.8.1.3 Analogy of Cl^- in the Associated Region of Deoxyhemoglobin and the Effect of NaCl on Lys-containing Elastic Model Proteins.

For the Lys-containing elastic model protein poly[0.76(GVGVP),0.24(GKGVP)] in water, the T_t-value is greater than 100°C. On addition of 0.05 N NaCl, the value of T_t occurs at 70°C, and at 0.25 N NaCl the value of T_t occurs at 35°C. On a per mole NaCl basis, this would be a ΔT_t of −175°C. The value of ΔS_t for poly(GK$^+$GVP), when ionized, is 3.9 cal/mole-pentamer °K. Using these values in Equation (5.9) of Chapter 5, $\Delta G_{HA}(\chi) = \Delta T_t(\chi)\Delta S_t(\text{ref})$, gives a ΔG_{HA}(Cl^- ion pairing) of −0.68 cal/mole-pentamer.

This estimate of a favorable change in Gibbs free energy for Cl^- ion pairing with positively charged Lys(K^+) would be greater for more hydrophobic domains, as appears to be the case for the relevant paired hydrophobic domain of deoxyhemoglobin. When the model protein is the more hydrophobic (GVGIP GVGIP GK$^+$GIP GVGIP GVGIP GVGIP)$_{42}$, the favorable decrease in ΔG_{HA}(Cl^- ion pairing) becomes −1,480 cal/mole-pentamer. Even the more modest ion pairing of Cl^- with Lys(K^+) contributes significantly to enhance hydrophobic association.

7.2.4.8.2 2,3-Biphosphoglycerate Binding: [CH_2($PO_4^=$)–CH($PO_4^=$)–COO^-]

The interaction of 2,3-biphosphoglycerate (BPG) represents the most dramatic use of ion pairing in the oxygen transport process of hemoglobin. A single molecule of BPG interacts with a total of eight positive charges. Four positive charges derive from the first two residues of the two β-chains, V1α-NH_3^+ and the His^2 imidazolium and an additional four positive charges come from two K82 ε-amino groups of the β-chains and two additional H143 imidazolium residues of the β-chains. By the consilient mechanism, such dramatic ion-pair formation and its proximity to the switch region decreases polarity and shifts the equilibrium from the R state toward the T state, that is, it induces dissociation of oxygen. In terms of the ΔT_t-mechanism, BPG binding lowers the T_t-divide. The cusp of insolubility moves to lower temperatures and shifts the interface between the $\alpha\beta$ dimers toward the greater hydrophobic association of deoxyhemoglobin.

7.2.4.8.3 Fetal Hemoglobin.

The subunit composition of fetal hemoglobin is $\alpha_2\gamma_2$ instead of the $\alpha_2\beta_2$ of adult hemoglobin A. The γ-chain differs from the β-chain by replacement of the positively charged side chain of the β-chain H143 with an uncharged Ser residue. The two H143 residues, one on each of the two adult β-chains, are part of the BPG binding site of the maternal hemoglobin. Replacement of the two β-chain H143 residues by Ser residues in fetal hemoglobin reduces the number of charges from eight to six and reduces BPG affinity to fetal hemoglobin. The result is retention of a higher affinity for oxygen due to a decrease in the shift to the deoxygenated T state. This facilitates transfer of oxygen from maternal to fetal hemoglobin *in utero*.

7.3 Hemoglobin Structures Demonstrate the Consilient Mechanism

7.3.1 Extending the Structural Insights

In the foregoing phenomenological considerations of muscle and hemoglobin function, a clear pattern emerged. It was one of a correlation of phenomena exhibited by muscle and hemoglobin on the one hand and elastic-contractile protein-based polymers on the other. At the core of elastic protein-based polymer function resides the phenomenon of an inverse temperature transition of hydrophobic association. Control of hydrophobic association became the basis for the *de novo* design of a family of efficient protein-based machines capable of performing the energy conversions extant in biology.

As developed in Chapter 5, control of hydrophobic association derived from control of hydrophobic hydration, and the physical basis for controlling hydrophobic hydration to achieve the diverse functions became described as an apolar–polar repulsive free energy of hydration, ΔG_{ap}. More directly stated, ΔG_{ap} derives from a competition for hydration between apolar (hydrophobic) and polar (e.g., charged) groups constrained by structure to coexist in a limited, shared space. As charged species gain dominance in the competition, they reduce the amount of hydrophobic hydration and effect hydrophobic dissociation. As hydrophobic residues gain dominance in the competition, they gather so much hydrophobic hydration that they become insoluble. Yes, too much hydrophobic hydration for a given temperature results in an inverse temperature transition to hydrophobic association with loss of water. Butler's 1937 report[31] contains within it this insight into the thermodynamics of the thermodynamics of solubility of hydrophobic groups. We grew to understand this reality while designing diverse protein-based machines. The demonstrably empowering mechanism was given the name of the *consilient mechanism*, because it provides a "common groundwork of explanation," as noted initially in Chapter 1.

Despite the above noted correlation of phenomena, current descriptions of molecular structure and resulting function of hemoglobin and myoglobin (as well as of muscle contraction to be addressed at the molecular level in Chapter 8) proceed without consideration of the consilient mechanism. With the consilient mechanism in mind, however, a distinctive way of looking at protein structure and function materializes. The availability of so many protein crystal structures from The Protein Data Bank[32] and, as employed in our case, the capacity to examine them by the software FrontDoor to Protein Explorer[33] and to illustrate them by Adobe® Photoshop® 5.5 provides the opportunity to consider the function of biology's proteins specifically in terms of the consilient mechanism. Immediately below, we begin with the hemoglobin molecule, the first of several proteins thus considered in this volume.

7.3.2 Resulting Functional Insights

Positive cooperativity (demonstrated by the sigmoid oxygen binding curve of hemoglobin) simply results when the binding of oxygen at one heme increases the affinity for oxygen binding at a second heme. An earlier proposal of direct heme–heme interaction in hemoglobin due to a *theoretically calculated* dramatic change in heme polarizability on oxygen binding was shown to be incorrect based on analysis of spectroscopic data of model heme–heme interacting systems.[34] Further consideration of the direct interaction between heme groups in hemoglobin has generally been discounted because the heme centers of the α^1–β^1-subunits are separated by 40 Å and the heme centers of the β^1–α^2-subunits are separated by 24 Å.

Nonetheless, as discussed below, detailed comparisons of the crystal structures of T state deoxyhemoglobin,[35] deoxyHb, and T state oxyhemoglobin,[36,37] oxyHb(T), demonstrate significant changes due to oxygen binding that occur in a crevasse separating the hemes of the β^1- and α^2-subunits and the equivalent crevasse between the hemes of the β^2- and α^1-subunits. It is our view, drawn from the developments in Chapter 5, that those changes reflect a decrease in the apolar–polar repulsive free energy of hydration, ΔG_{ap}, emanating from the heme as the result of oxygen binding. ΔG_{ap} acts through water and propagates as a function of distance as charged groups, associated with the crevasse, change their state.

In particular, the decrease in heme hydrophobicity on oxygen binding eases the competition for hydration with proximal charged groups. The charged groups, once held due to lack of hydration as ion pairs or with their charge turned away (repulsed) from the hydrophobic dominance, separate or reorient in a way to minimize ΔG_{ap}. The separated and reoriented charged groups now reach out for more hydration, such that the oxygen-binding event at one heme decreases the hydrophobicity (the amount of hydrophobic hydration) of the second heme at the other end of the crevasse and facilitates the binding oxygen to the second heme. The decrease in hydrophobicity of the second heme due to oxygen binding to the first heme uniquely becomes the consilient mechanism's explanation for the sigmoid binding curve, that is, for the effective oxygen transport function of hemoglobin.

In short, we propose that first polar oxygen, in its approach to the heme binding site of hemoglobin, experiences repulsion due to the hydrophobicity of the heme and its environment. The barrier results from the loss of hydration available to the oxygen molecule as it approaches the hydrophobic hydration of the exposed heme and its surrounding hydrophobic groups. Once the first polar oxygen molecule binds, however, the resulting decrease in hydrophobicity transmits through the water-filled crevasse, by reoriented, more competitive charged groups, to the second heme and facilitates binding of the second oxygen. Structural justifications for this view follow.

7.3.3 Examination of the Structural Changes Resulting from Oxygen Binding

7.3.3.1 General Appearance of the Crystal Structures of Deoxyhemoglobin and Oxyhemoglobin

7.3.3.1.1 The Size of the Aqueous Core of the T State of Hemoglobin

A space-filling representation, Figure 7.11A,B presents stereo views of deoxyHb and oxyHb(T) from the perspective of the twofold (diad) axis. A first point to note with this perspective is that it gives the most minimal size of the aqueous core of the hemoglobin molecule. Looking by stereo view into the aqueous core at several slight angles from the diad axis, one sees a much larger water-filled core. Thus, water is not only accessible from the outside of the molecule to the areas of association between subunits, but water is also available to the quaternary structure from the inside. In fact, as measured from slabs taken perpendicular to the twofold axis, the water-filled core, which waters would constitute part of the "waters of Thales," approaches 40% of the total diameter. The thickness of the wall on one side can be less than the width of the aqueous channel.

7.3.3.1.2 The Limited External Differences Between deoxyHb and oxyHb(T)

From the perspective of Figure 7.11, small differences are seen between the β-chains and between the α-chains within the same tetramer. Surprisingly, these small intramolecular differences appear to be as great as the intermolecular differences between deoxyHb and oxyHb(T). Attempts to oxygenate crystalline deoxyHb at room temperature destroy crystalline order. However, crystals of deoxyHb formed at 4°C, carefully changed to a 40% polyethylene glycol–water solution, and oxygenated at 4°C survive with oxygen bound at all four hemes.[36] The result is T-state oxyhemoglobin, that is, oxyHb(T). The protocol for forming oxyHb(T) is a prescription for reigning in the consilient mechanism certainly at the level of changing quaternary structure. Nonetheless, we choose to utilize oxyHb(T) because, if oxygen binding does decrease the hydrophobicity of the heme group, evidence of this should be in the heme environment as long as there does remain adequate water in the crystal to evidence the competition for hydration.

7.3.3.1.3 Looking to a Molecular Basis for the Coherence of Phenomena

The phenomenological observations detailed above in section 7.2.4 clearly exhibit a coherence of phenomena with the consilient mechanism. Our objective, therefore, is to determine the particular expressions of the consilient mechanism at the molecular level. As a starting point, we look from the aqueous medium to the heme group in its protein setting and make gross comparison between the heme environment in myoglobin and the environments of the hemes of hemoglobin. In our view, the first clue

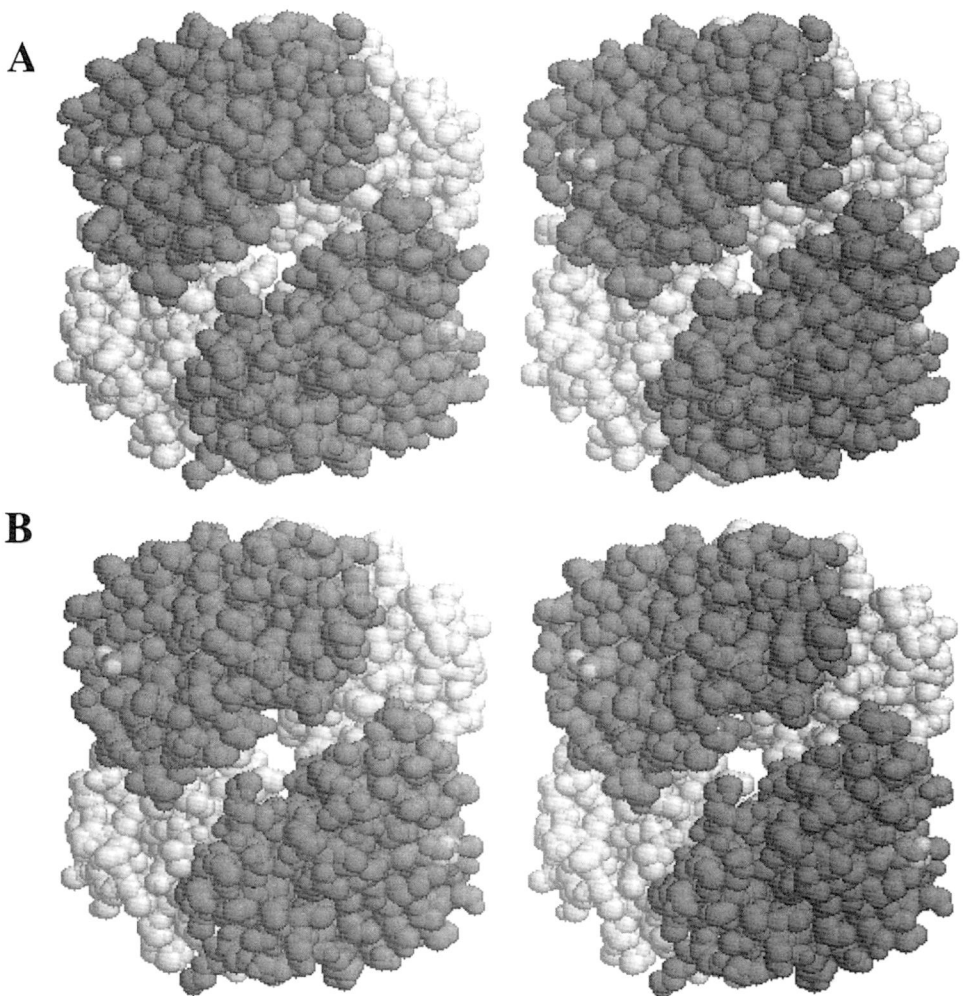

FIGURE 7.11. Stereo view of the human hemoglobin tetramer from the perspective of the twofold (diad) axis using space-filling representation with α-chains in white and β-chains in gray and with the light dots for the propionate oxygens of heme ligands within the upper right and lower left quadrants. **(A)** Deoxyhemoglobin (deoxyHb).[35] (Protein Data Bank, Structure File, 1A3N.) **(B)** Oxyhemoglobin in the T state, oxyHb(T).[36] (Protein Data Bank, Structure File, 1GZX.)

to the molecular basis for differences in oxygen binding, represented in Figures 7.5, 7.6, and 7.7, gains from such a gross perspective.

7.3.3.2 Structure and Disposition of Heme Groups Before and After Oxygen Binding

7.3.3.2.1 Gaining Perspective of the Heme Group

A key aspect of our consilient view of oxygen transport by hemoglobin, derived as it has been from the inverse temperature transition for hydrophobic association, places the heme group in the unique position of being the most hydrophobic component of hemoglobin. Furthermore, our general experience has been that the addition of a polar group like the oxygen molecule would significantly decrease heme hydrophobicity. A sense of the heme group begins with Figure 7.12, which presents in ball and stick representation a stereo view of the structures of the heme (ferroprotoporphyrin

7.3 Hemoglobin Structures Demonstrate the Consilient Mechanism

IX) group, as found in the binding sites of the α-chain for deoxyHb in Figure 7.12A and oxyHb(T) in Figure 7.12B.

Figure 7.13 provides some sense of the relative size and disposition of the heme group *in situ* within the β^1-chain. The structures without and with bound oxygen are shown in stereo view in a space-filling representation for the hemes and stick representation for the β^1-chain. Figure 7.13A presents the heme for deoxyHb from the side where histidine binds the central iron (upper part) and from the side to which oxygen binds (lower part). Figure 7.13B similarly presents the heme for oxyHb(T) where the bound oxygen molecule appears in its lower part.

With the β^1-chain in stick representation, the heme groups can still be seen and yet retain

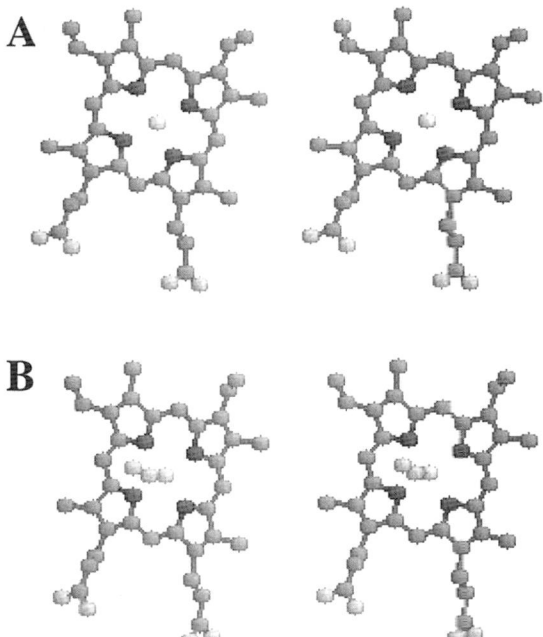

FIGURE 7.12. Stereo view of β-chain heme group of human hemoglobin tetramer in ball and stick representation with an iron atom in the center surrounded by protoporphyrin IX. See text for further description. **(A)** Deoxyhemoglobin (deoxyHb).[35] (Protein Data Bank, Structure File, 1A3N.) **(B)** Oxyhemoglobin in the T state, oxyHb(T), showing the diatomic oxygen atom bound to the central heme iron.[36] (Protein Data Bank, Structure File, 1GZX.)

some (albeit a somewhat exaggerated) sense of the relative size of heme to that of the protein and its aromatic side chains. The heme is basically a very large hydrophobic (aromatic) group moderated by two attached polar propionate groups, $-CH-CH-COO^-$, that are equivalent to the side chain of glutamic acid (Glu, E). Also the iron at the heme center presents a binding site, one side for the aromatic His ligand and the other side for the very polar oxygen molecule. The heme ring structure extends further due to the substitution at its edge of two vinyl groups, $-CH=CH_2$, and four methyl groups, $-CH_3$, which further increase its hydrophobicity.

7.3.3.2.2 Views from Aqueous Solution of the Heme in Myoglobin and the Hemes in Hemoglobin

Figure 7.14 presents from solution a stereo view of the crystal structure of myoglobin[38] in space-filling representation with the protein chain in white, except for the most hydrophobic aromatic residues shown in black, and with the heme in gray and its propionate oxygens as light atoms on gray heme. Figure 7.14A gives the structure of deoxymyoglobin, deoxyMb, and Figure 7.14B presents the structure of oxymyoglobin, oxyMb. Also included in both myoglobin crystal structures are two sulfate ions at the upper right hand edge.

In similar fashion, Figure 7.15A presents the deoxyhemoglobin $\alpha^2\beta^1$ dimer and Figure 7.15B the oxyhemoglobin $\alpha^2\beta^1$ dimer with the view in both cases being taken from the β^1-heme to the α^2-heme. Strikingly, the hemes of hemoglobin reside in sites that are much more hydrophobic than those of the heme of myoglobin, as readily seen by the black nonhistidinemost hydrophobic aromatic residues. In our view, this feature alone ensures that the affinity for oxygen would be lower for Hb than for Mb. By the apolar–polar repulsive free energy of hydration, ΔG_{ap}, a polar oxygen molecule would experience greater repulsion on approaching the heme binding sites in hemoglobin than the heme site in myoglobin. The question now advances to the issue of

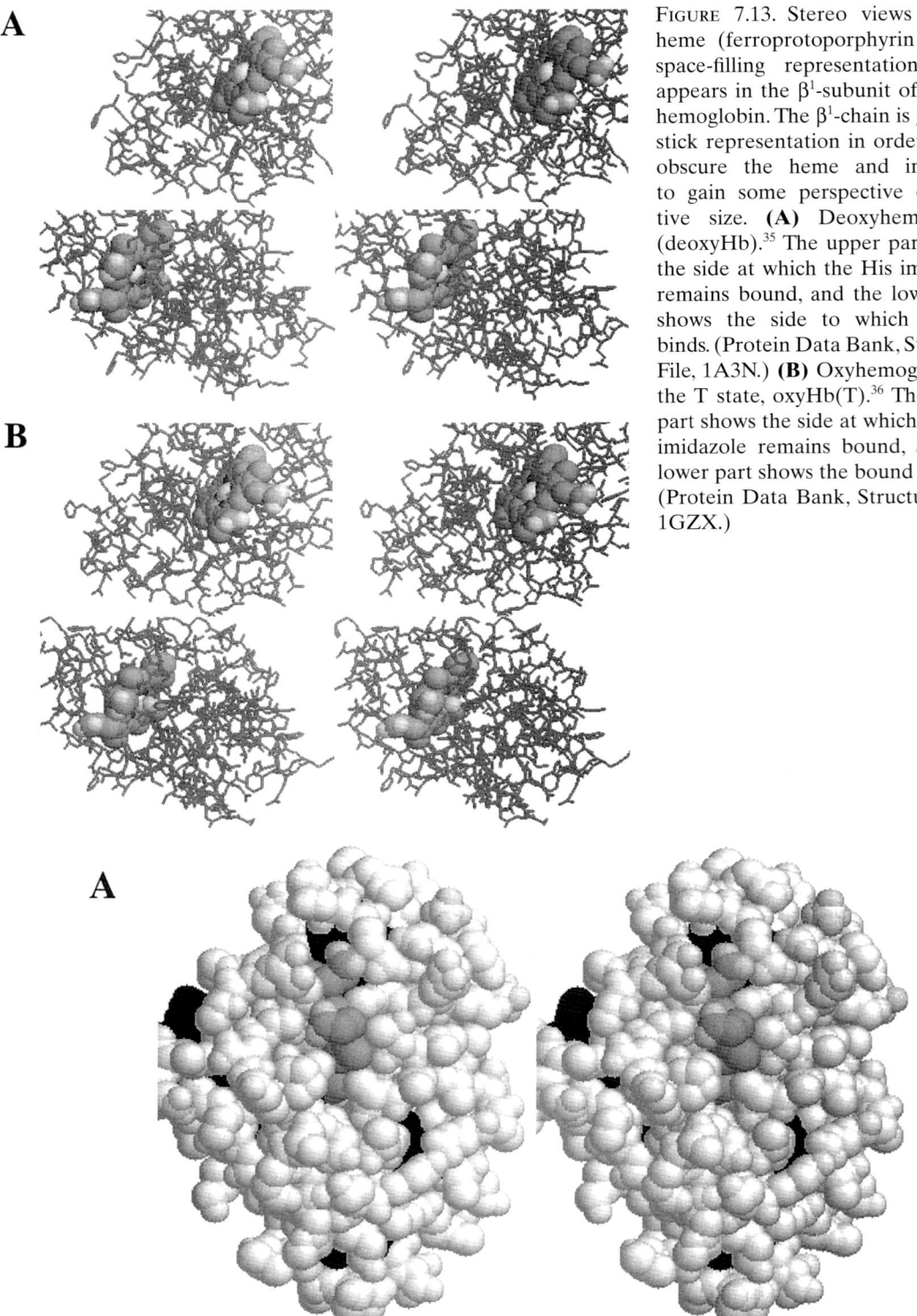

FIGURE 7.13. Stereo views of the heme (ferroprotoporphyrin IX) in space-filling representation as it appears in the β^1-subunit of human hemoglobin. The β^1-chain is given in stick representation in order not to obscure the heme and in order to gain some perspective of relative size. **(A)** Deoxyhemoglobin (deoxyHb).[35] The upper part shows the side at which the His imidazole remains bound, and the lower part shows the side to which oxygen binds. (Protein Data Bank, Structure File, 1A3N.) **(B)** Oxyhemoglobin in the T state, oxyHb(T).[36] The upper part shows the side at which the His imidazole remains bound, and the lower part shows the bound oxygen. (Protein Data Bank, Structure File, 1GZX.)

FIGURE 7.14. Stereo view of the crystal structure of myoglobin[38] in space-filling representation with a white protein chain, except for black non-His aromatic residues, and gray heme ligand with propionate oxygens shown as light atoms on gray heme at upper center of structure. **(A)** Deoxymyoglobin (Protein Data Bank, Structure File, 1A6N). **(B)** Oxymyoglobin (Protein Data Bank, Structure File, 1A6M). The crystal structure for both states contains two identically positioned sulfate ions on the surface. Note the limited exposure of aromatic residues to the surface, particularly in the area occupied by the heme group, and compare it with that observed in hemoglobin in Figures 7.15 and 7.16.

7.3 Hemoglobin Structures Demonstrate the Consilient Mechanism

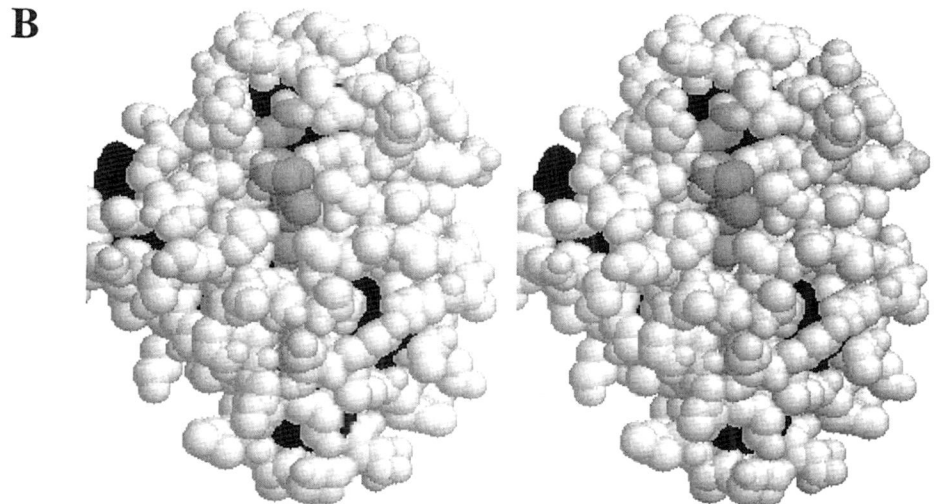

FIGURE 7.14. *Continued*

how the binding of oxygen at a first heme enhances oxygen affinity at a second heme. Based on the view in Figure 7.16 from the α^2-heme to the β^1-heme, the approach to the answer gains from an expanded view of the $\alpha^2\beta^1$ heme pair.

7.3.3.2.3 The Presence of a Hydrophobic Crevasse on the Protein Surface That Connects the $\beta^1\alpha^2$-heme Pair

Figure 7.17 presents an expanded stereo view of Figure 7.16 and discloses details of a crevasse

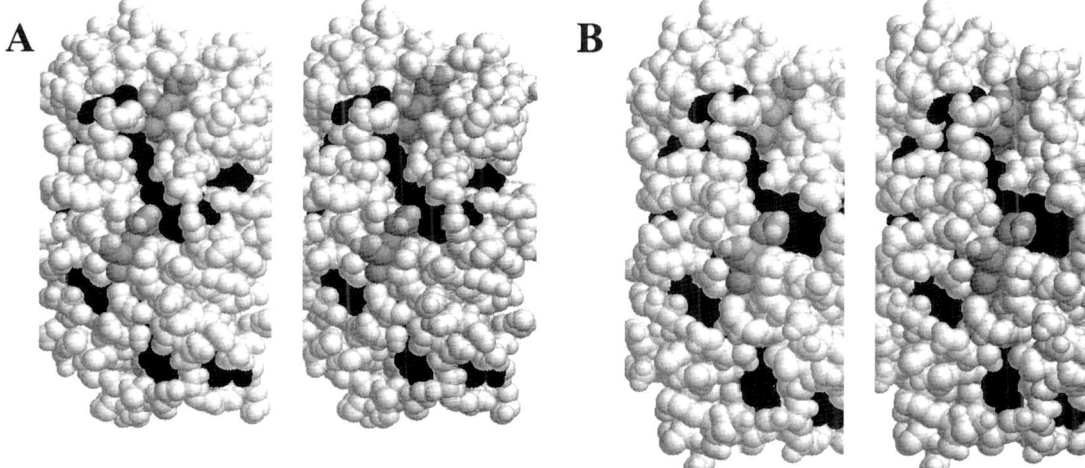

FIGURE 7.15. Stereo view of the hemoglobin $\alpha^2\beta^1$ dimer in space-filling representation with the protein chain in white, except for non-His aromatics in black, and with the light gray heme showing propionate oxygens as light atoms on two gray hemes starting at center left and upper center and oriented toward upper right of the structure. The view is taken from the β^1-heme to the α^2-heme. Note the most hydrophobic aromatic residues connecting the two hemes. **(A)** Deoxyhemoglobin (deoxyHb).[35] (Protein Data Bank, Structure File, 1A3N.) **(B)** Oxyhemoglobin in the T state, oxyHb(T).[36] (Protein Data Bank, Structure File, 1GZX.)

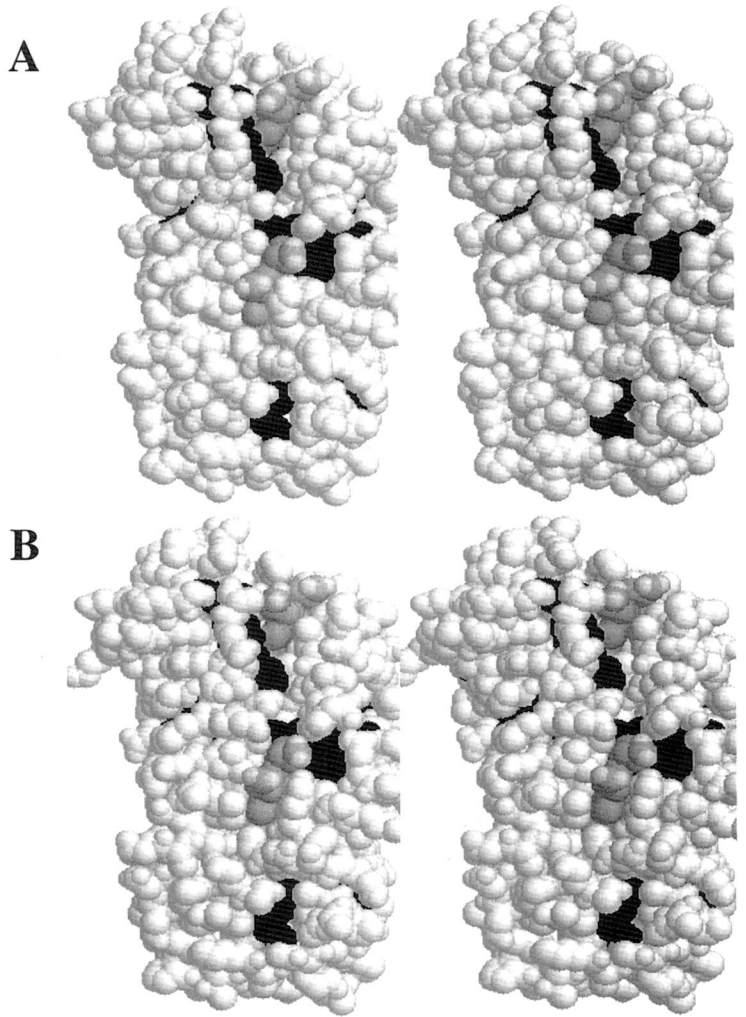

FIGURE 7.16. Stereo view of the hemoglobin $\alpha^2\beta^1$ dimer in space-filling representation with the protein chain in white, except for non-His aromatics in black, and with the light-gray heme showing propionate oxygens as light atoms on two gray hemes starting from center and upper center and tilted toward the upper right of the structure. This view is taken from the α^2-heme to the β^1-heme. Also note the presence of the very hydrophobic aromatic residues between the two hemes. **(A)** Deoxyhemoglobin (deoxyHb).[35] (Protein Data Bank, Structure File, 1A3N.) **(B)** Oxyhemoglobin in the T state, oxyHb(T).[36] (Protein Data Bank, Structure File, 1GZX.)

connecting the $\alpha^2\beta^1$ heme pair. Whether in the deoxyHb or the oxyHb(T) state, the exposed hydrophobic hemes form the ends of the crevasse that is lined with very hydrophobic (black) groups. In particular, contributed from the β^1-chain, there are the hydrophobic side chains of F45, F42, F41, H92, L88, L91, L96, and H97, and contributed from the α^2-chain there are Y72, H45, F46, H58, L91, L86, and L83.

In addition, when looking from the α^2-heme to the base of the β^1-heme, at the very bottom of the crevasse, abutting the β^1-heme in Figure 7.17, resides the V98 residue, and an enlarged view of the crevasse in Figure 7.15 shows the similarly positioned V93 residue at the base of the α^2-heme. As shown in Figure 7.8, these residues participate in hydrogen bonds with residues Y145 and Y140, respectively, that are disrupted on oxygen binding. Most significantly, a widened view of Figure 7.17, but taken from directly above the crevasse, is included below in Figures 7.19 and 7.20. It shows the key residues Y145 and H146 of the β^1-chain and Y140 and R141 of the α^2-chain. Therefore, all of the critical interactions in Figure 7.8 that change on oxygen binding reside within the area of influence of these hemes. Thus, the happenings within the interheme crevasse become a means of communication not only between the hemes of the particular crevasse but also provide communication between the $\alpha^2\beta^1$-heme pair and the $\alpha^1\beta^2$-heme pair.

Returning to the interheme crevasse, the unique grouping of hydrophobic residues associated with the $\alpha^2\beta^1$-heme pair and with the equivalent $\alpha^1\beta^2$-heme pair constitutes two remarkably hydrophobic domains for exposure at an aqueous surface. It is true, however, that much of the hydrophobicity lies in the depths of the water-filled crevasse. These are circumstances under which the consilient mechanism dictates an acute competition for hydration between the hydrophobic residues of each crevasse and its associated surface charged groups. Such a competition gives rise to a repulsion between hydrophobic and charged groups as each requires water unperturbed by the other to lower their free energy. Based on the studies reviewed in Chapter 5, this, of course, is what we have called an *apolar–polar repulsive free energy of hydration* and have indicated by ΔG_{ap}. As the hydrophobic groups are held as a more rigid part of the protein structure, it would seem to be left to the charged side chains at the ends of their aliphatic chains with their greater range of motion to reflect the repulsion.

7.3.3.2.4 The Disposition of Charged Side Chains Associated with the Crevasse Connecting the β^1- and α^2-Heme Groups for deoxyHb

There are basically two ways in which charged side chains of amino acid residues can lower their free energy when confronted with a hydrophobic domain. One way is to form ion pairs. The second way, when possible, is to move toward locations more remote from the most hydrophobic surface. Both examples are shown in the crevasse that connects the β^1 and α^2 heme groups as well as in the equivalent crevasse connecting the $\alpha^1\beta^2$-heme groups.

As shown in Figure 7.17A, there are seven charged residues in the immediate vicinity of the $\beta^1\alpha^2$ crevasse. Starting at six o'clock and reading clockwise, they are K61, K90, R92, R40, E43, K66, and K95. Interestingly, this means six positively charged and one negatively charged side chains. However, the two heme groups add four more negatively charged groups associated with each crevasse. Those residues in position for ion pairing are K61, R92, E43, and K66. Despite literally being bathed in water, ion pairs form. The positively charged side chains of K90, R40, and K95 are expected to have moved to their positions of lowest accessible free energy, that is, to minimize repulsion coming from the apolar crevasse. Thus these charges are expected to be pointing in the direction of lower hydrophobicity. This finding indicates that the region of the α-heme is less hydrophobic than that of the β-heme. This observation is consistent with a number of reports that the T state exhibits a fivefold greater binding at the α-heme than at the β-heme.[39–42]

7.3.3.2.5 The Effect of Oxygen Binding on the Disposition of Charged Side Chains Proximal to the Crevasse Connecting the β^1- and α^2-Heme Groups

As shown in Figure 7.17B, oxygen binding effects a remarkable change in the disposition of the seven charged side chains. The four side chains positioned for ion paring separate from their ion paired positions, and the charges of the three positively charged, non-ion-paired residues, K90 R40, and K95, reorient their charges. Yet the forces that effect these changes *cannot* have come through the protein structure. In our view, the driving force for these oxygenation-effected side chain reorientations came through the "waters of Thales" of the hydrophobic crevasse. We believe this demonstrates that oxygen binding, even though it occurs on the other side of the hemes, decreases the hydrophobicity of the hydrophobic crevasse sufficiently to allow improved hydration of these charged species. Thus, we arrive at a specific set of structural changes on oxygenation consistent with the consilient mechanism.

7.3.4 Consideration of Communication Between the $\alpha^2\beta^1$- and $\alpha^1\beta^2$-Heme Pairs

The consilient mechanism provides a means of communication between the pair of hemes of

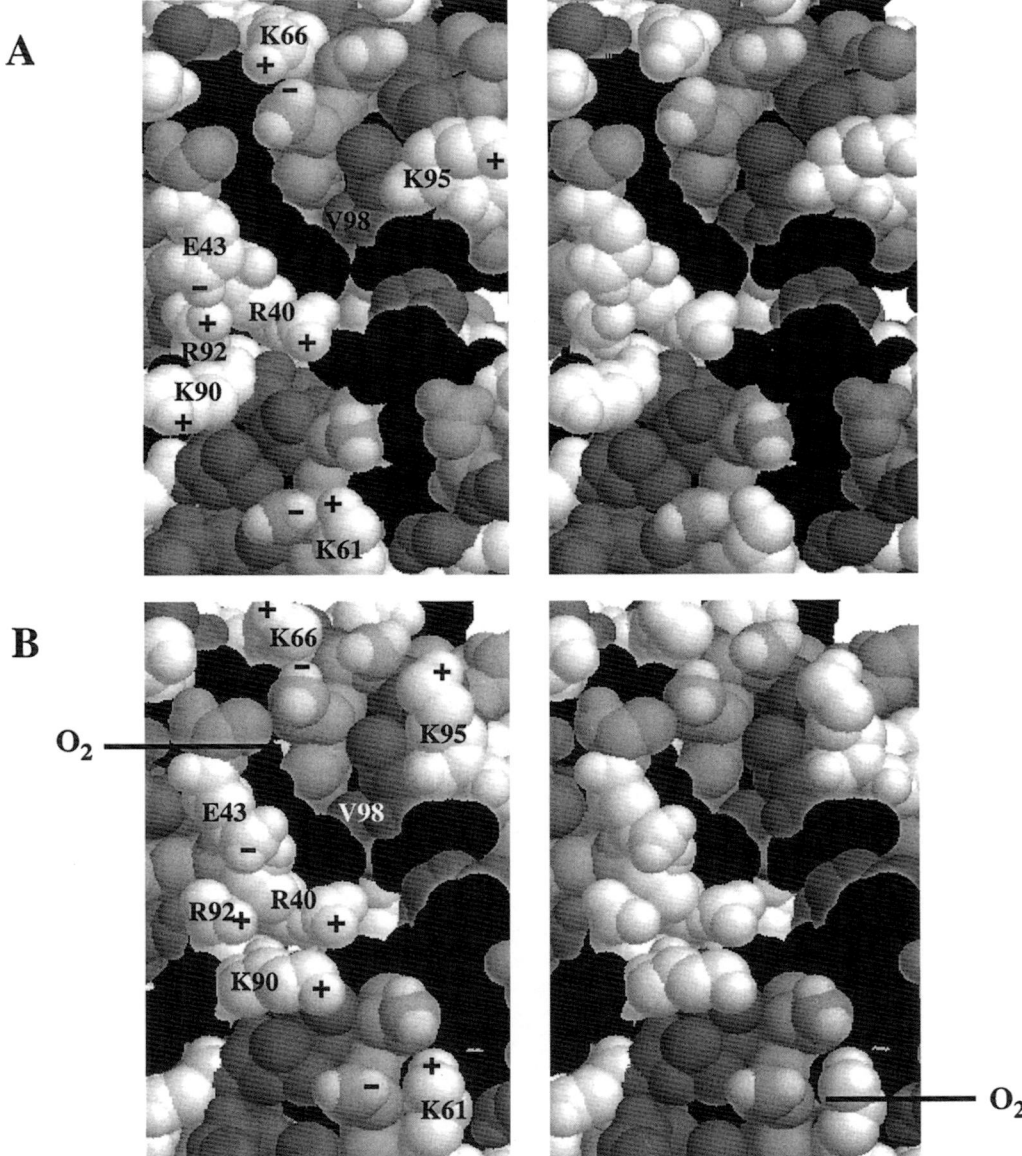

FIGURE 7.17. Crevasse on hemoglobin surface connecting α^2- and β^1-hemes: Stereo view showing a space-filling representation in which the residues are shaded with aromatics in black, other hydrophobics in gray, neutrals in light gray, charged in white, and with the heme in a lighter gray with propionate oxygens shown as light atoms. This is a magnified view taken from the α^2-heme to the β^1-heme as in Figure 7.16 in order to define the disposition of the charged groups. (A) Deoxyhemoglobin (deoxyHb).[35] (Protein Data Bank, Structure File, 1A3N.) (B) Oxyhemoglobin in the T state, oxyHb(T).[36] On the binding of oxygen, the charged residues reorient in a manner indicating a decrease in hydrophobicity in the inter-heme crevasse. See text for discussion. (Protein Data Bank, Structure File, 1GZX.)

7.3 Hemoglobin Structures Demonstrate the Consilient Mechanism

the $\alpha^2\beta^1$ crevasse and equivalently of the $\alpha^1\beta^2$ crevasse. However, can the consilient mechanism be directly responsible for the communication between the $\alpha^2\beta^1$- and $\alpha^1\beta^2$-heme pairs? We begin by consideration of the $\alpha^1\beta^1$ interface, which is equivalent to the $\alpha^2\beta^2$ interface.

7.3.4.1 Changes in the Interface Between the $\alpha^1\beta^1$ and $\alpha^2\beta^2$ Dimers Due to Oxygen Binding

There is general appreciation that the principal quaternary structural changes giving rise to cooperative oxygen binding occur at the $\alpha^1\beta^2$ (and $\alpha^2\beta^1$) interfaces. For example, "*The quaternary structural change preserves hemoglobin's exact two-fold symmetry and takes place entirely across its α^1–β^2 (and α^2–β^1) interface. The α^1–β^1 (and α^2–β^2) contact is unchanged.* ..."[43] The hydrophobic crevasse discussed above constitutes part of that interface.

7.3.4.1.1 Estimates of the Hydrophobicities of the deoxyHb $\alpha^1\beta^1$ Interface

With the representation of the interface given in Figure 7.18, it becomes possible to approximate, ΔG_{HA}, the Gibbs free energy for hydrophobic association of the $\alpha^1\beta^1$–$\alpha^2\beta^2$ interfaces. The calculation uses the hydrophobicity scale of Table 5.3, estimates the area of the residue exposed to the interface, and reduces the effect of charge by one-half when the charged residue is ion paired, as was the experience from ion pairing between complementary chains of elastic protein-based polymers. The estimated value of ΔG_{HA} was –9 kcal/mole-face for the α-face and –24 kcal/mole-face for the β-face. These appear to be reasonable approximations, but do not take into consideration the fact that some of the residues, such as K99 and R104, lie on the aqueous core of the twofold axis shown in Figure 7.11.

7.3.4.1.2 Locating the Ion-pair Interactions Between Interfaces and Visualizing the Overlay of the $\alpha^1\beta^1$ and $\alpha^2\beta^2$ Dimers and the Change on Oxygen Binding

From Figure 7.8, important ion pair interactions in deoxyHb that orient the $\alpha^1\beta^1$ dimer with

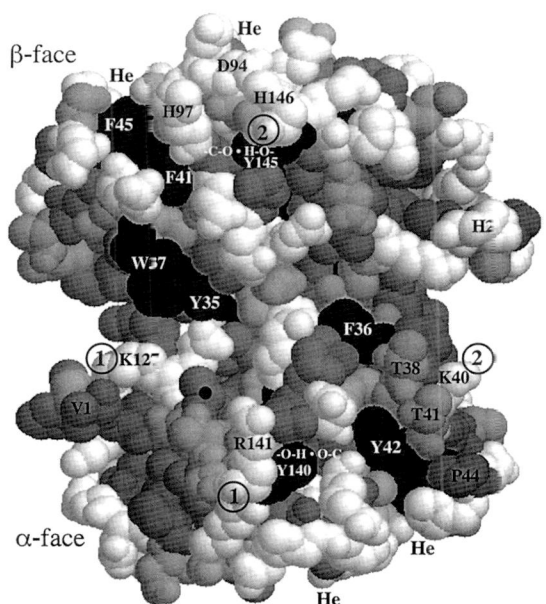

FIGURE 7.18. Human deoxyhemoglobin[35] showing, in space-filling representation, the interface between αβ dimers in which the residues are shaded with aromatics in black, other hydrophobics in gray, neutrals in light gray, charged in white, and with the two hemes (indicated by **He** at each propionate) in a lighter gray with attached propionate oxygens seen as light atoms. Key residues are indicated. The circled numbers indicate ion-pairing sites between interfaces, and the black dot in the lower left quadrant gives the point of rotation between interfaces. See text for discussion. (Protein Data Bank, Structure File, 1A3N.)

respect to the $\alpha^2\beta^2$ dimer are indicated by the circled numbers 1 and 2, as defined in Figure 7.18, which represents the face of the $\alpha^1\beta^1$ dimer. A view of the interaction of the two dimers becomes possible by making a transparency of Figure 7.18. The transparent copy is placed exactly over the original, turned over, rotated by 90 degrees clockwise, and adjusted until the two circles with 2 in them exactly overlay. The transparency now becomes the $\alpha^2\beta^2$ dimer. This arrangement approximates the association between the two faces for deoxyhemoglobin. Note that H97 of the β-face aligns between residues P44 and T41 of the α-face of the $\alpha^1\beta^1$ dimer. On going from the T state of deoxyhemoglobin to the R state of oxyhemoglobin, a rotation occurs, centered at the black dot, such that H97 now aligns between residues T41 and T38. This rotation, otherwise indicated in Figure 7.10, represents the quaternary structural change between the $\alpha^1\beta^1$ and $\alpha^2\beta^2$ dimers that results from oxygen binding.

7.3.4.1.3 Visualizing the Relationship Between the $\alpha^2\beta^1$- and $\alpha^1\beta^2$-heme Pairs Before and After Oxygen Binding

The He labels on Figure 7.18, located at heme propionates, indicate the widths of the α^1- and β^1-hemes. With the transparency, representing the $\alpha^2\beta^2$ dimer properly placed over the $\alpha^1\beta^1$ dimer as indicated above, the relationship between the α^2- and β^1-hemes is shown at the upperleft and the relationship between the α^1- and β^2-hemes is shown at the lower right. Performing the above indicated rotation increases the distance between the $\alpha^2\beta^1$-hemes and the nearly equivalent $\alpha^1\beta^2$-hemes. The relationship between the $\alpha^2\beta^1$ hemes is shown in Figures 7.15A and 7.16A.

7.3.4.2 Looking Directly into the $\alpha^2\beta^1$ ($\alpha^1\beta^2$) Interheme Crevasse also to See the Classic Intersubunit Interactions That Change on Oxygen Binding

Having gained a sense of the three-dimensional relationship between the hemes and realizing that the quaternary structural changes are principally due to disruption of the ion pairs involving the H146 and R141 C-terminal carboxylates, we now look for the relationship of these functional groups to the $\alpha^2\beta^1$ ($\alpha^1\beta^2$) interheme hydrophobic crevasse. For that purpose, Figure 7.19 provides a look directly into the $\alpha^2\beta^1$ crevasse for the full width of that interfacial interaction with a few of the critical residues labeled.

7.3.4.2.1 Are the Critical Ion Pairs for Quaternary Structural Change Within Reach of the Heme-based Consilient Mechanism?

Importantly, the H146–K40 ion pair that occurs between the $\alpha^2\beta^1$ interface, located by the circled '2's in Figure 7.18, is readily seen in the expanded view of the $\alpha^2\beta^1$ interheme crevasse of Figure 7.19A. Also, the C-terminal carboxylate of the α^2 R141 is visible that contributes its negative charge to the ion pair with α^1 K127 (the circled '1's interactions in Figure 7.18). The question now becomes, can oxygenation of a heme in the $\alpha^2\beta^1$ (or the $\alpha^1\beta^2$) interheme hydrophobic crevasse weaken these ion pairs that are thought to trigger the quaternary structural changes considered responsible for the interaction between $\alpha^2\beta^1$- and $\alpha^1\beta^2$-heme pairs? To do so by the consilient mechanism would seem to require intervening "waters of Thales."

7.3.4.2.2 Demonstrating the Presence of the "Waters of Thales"

Figure 7.19B shows the observed "waters of Thales" when the chains are given in space-filling representation. The observed water molecules in a crystallographic structure represent a minimal number of water molecules actually present, as many more waters are too mobile to be positioned by X-ray diffraction. Conversion of the chains in Figure 7.19 to the stick representation in Figure 7.20 allows delineation of the unseen residues in Figure 7.19. Furthermore, the stick representation allows for three-dimensional visualization of all of the diffraction-detected waters in this region. From Figure 7.20B with the detected water molecules, it is not unthinkable that the consilient mechanism might play a role in the relaxation of these critical ion pairs and thereby contribute to the communication between $\alpha^2\beta^1$- and $\alpha^1\beta^2$-heme pairs that facilitates oxygen binding. After all, formation of oxyhemoglobin occurs with the uptake of water, in our view, as ion pairs and hydrophobic regions dissociate. Figure 7.21 gives an expanded view of the part of the β^1-chain that undergoes disruption of ion pairs and hydrogen bond. Figure 7.21 can also facilitate identification of residues in Figure 7.20 without the distraction of so many labels.

7.3 Hemoglobin Structures Demonstrate the Consilient Mechanism 275

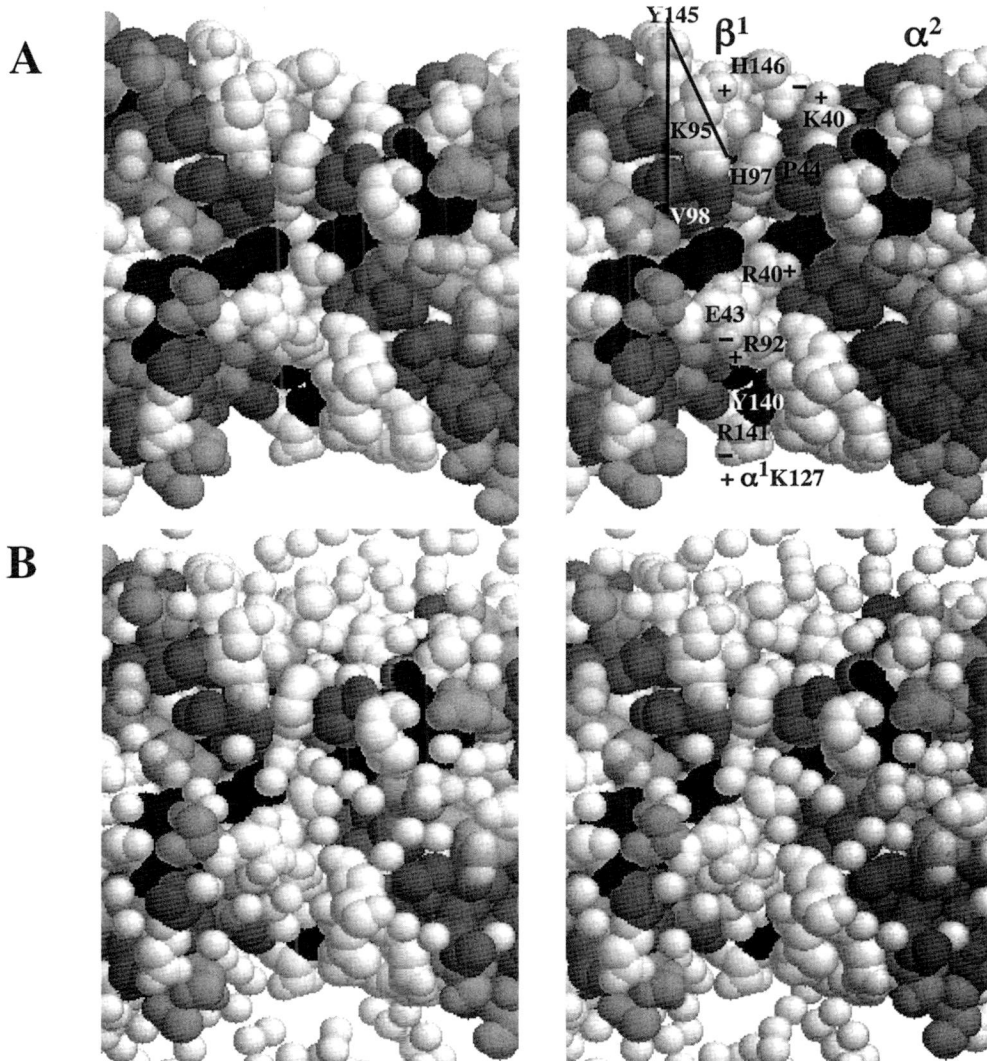

FIGURE 7.19. **(A)** Human deoxyhemoglobin[35]: Stereo view of the $\beta^1\alpha^2$ interface axis looking directly into the inter-heme crevasse; space-filling representation with an α^2-chain on the right and a β^1-chain on the left, and shadings of groups as indicated in Figure 7.18. Key residues are labeled. See text for discussion. **(B)** The water molecules, detectable by diffraction, have been added and are seen as light shaded small balls. (Protein Data Bank, Structure File, 1A3N.)

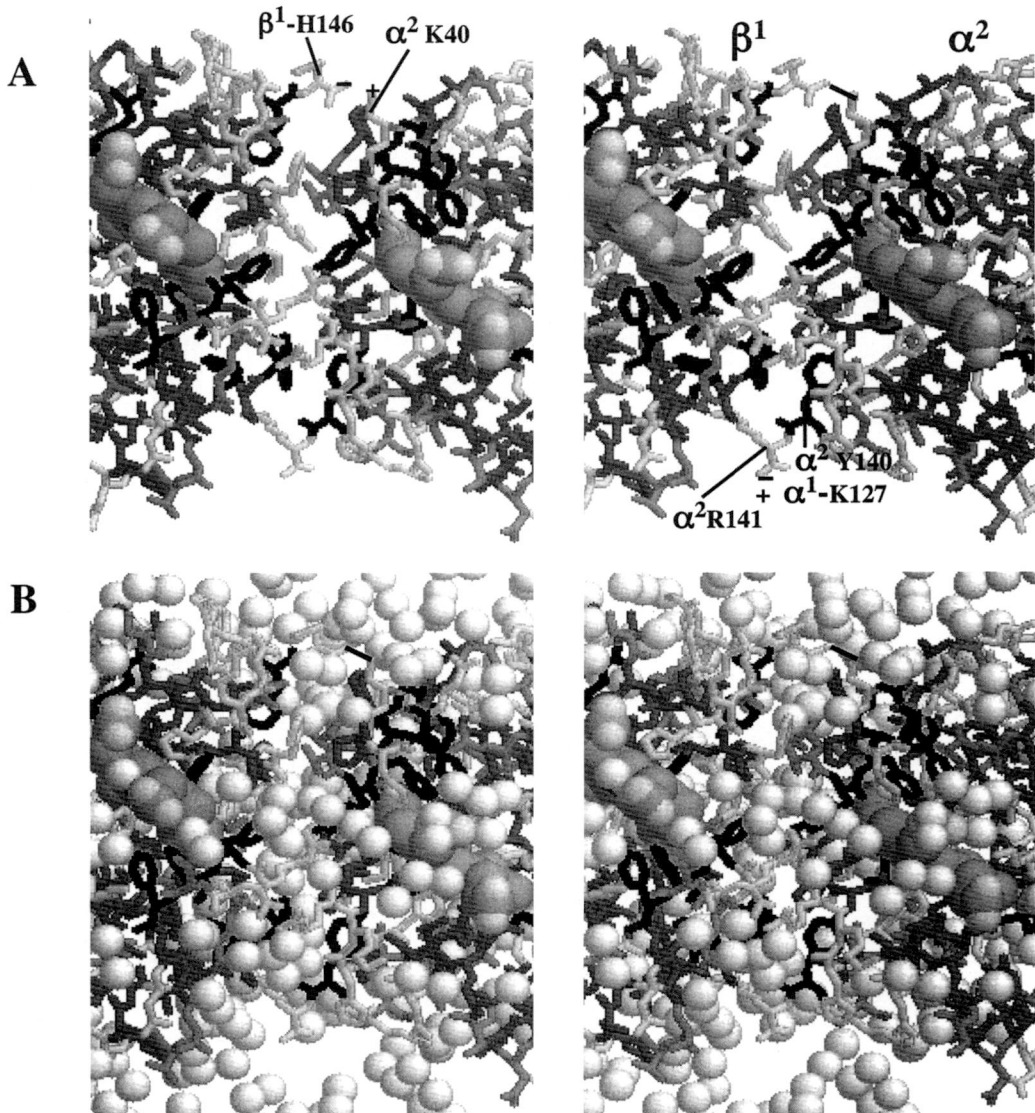

FIGURE 7.20. **(A)** Human deoxyhemoglobin[35]: Same stereo view as in Figure 7.19 of the $\beta^1\alpha^2$ interface axis looking directly into the inter-heme crevasse with an α^2-chain on the right and a β^1-chain on the left, and shadings of groups as indicated in Figure 7.19, but in this case using the stick representation. Key residues are labeled. **(B)** Same as in A but the water molecules (light shaded balls), detectable by diffraction, can be seen permeating through the interface where they could be expected to function as "waters of Thales" in facilitating the quaternary structural changes by the consilient mechanism. See text for discussion. (Protein Data Bank, Structure File, 1A3N.)

7.3 Hemoglobin Structures Demonstrate the Consilient Mechanism

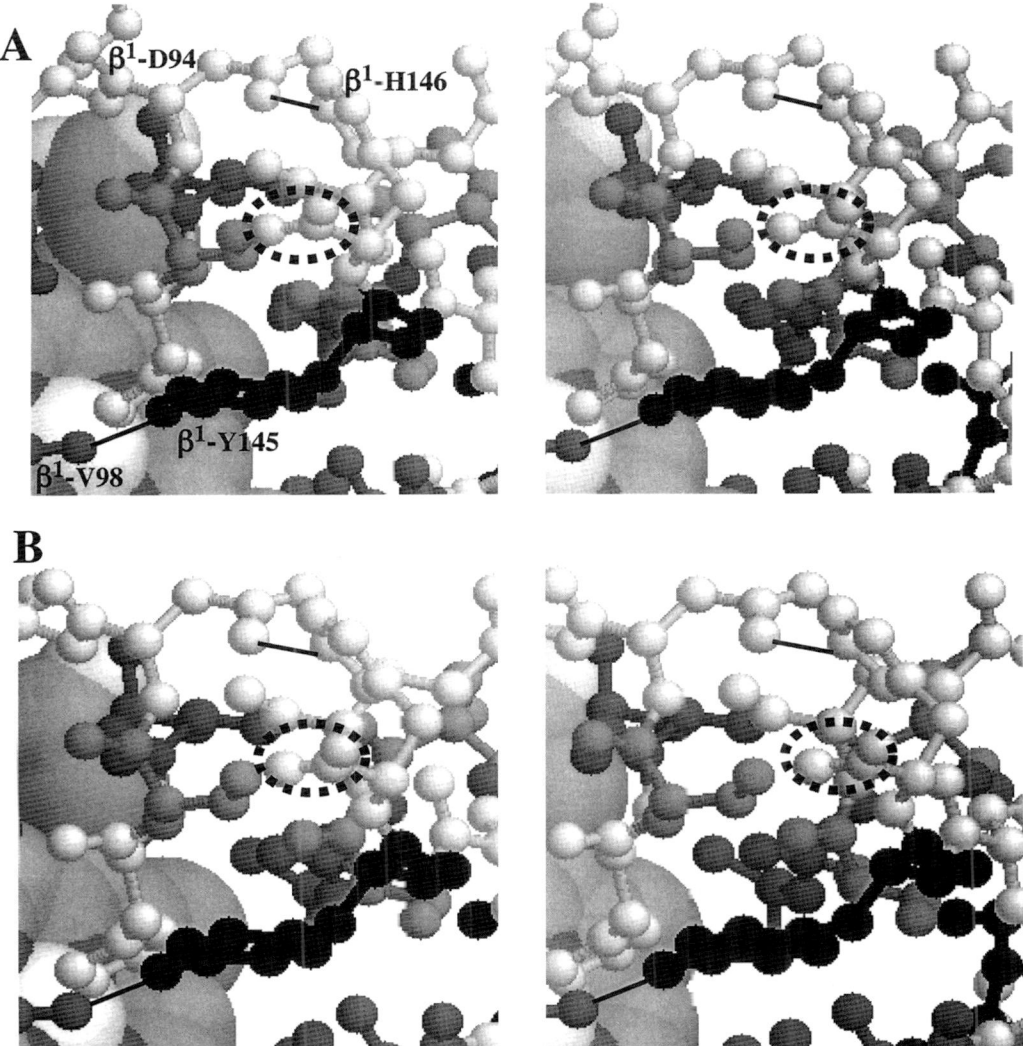

FIGURE 7.21. Stereo view of the human hemoglobin β^1-chain in ball and stick representation with shadings of groups as indicated in Figure 7.18. Key residues are labeled that change their interactions on oxygen binding that effects the T → R transition. The lines indicate ion pairs or hydrogen bonds within the β^1-chain, and the ellipse indicates the carboxyl-terminal carboxylate of β^1H146 that ion pairs with the charged side chain of α^2K40 in the T state, as directly observed in Figures 7.19 and 7.20 and as may be further discerned by using Figure 7.18 with a transparent overlay as discussed in the text. **(A)** Deoxyhemoglobin (deoxyHb).[35] (Protein Data Bank, Structure File, 1A3N.) **(B)** Oxyhemoglobin in the T state, oxyHb(T).[36] (Protein Data Bank, Structure File, 1GZX.)

7.3.5 The Consilient Mechanism Summarized as the Source of the Sigmoid Oxygen Binding of Hemoglobin

7.3.5.1 Positive Cooperativity in Hemoglobin by the Consilient Mechanism

7.3.5.1.1 The Interheme Hydrophobic Crevasse, Filled with Hydrophobic Hydration, Repulses Charge and the Approach of Polar Oxygen

From the perspective of the consilient mechanism, the approach of an oxygen molecule to a heme of deoxyhemoglobin faces loss of hydration as it enters the sphere of hydrophobic hydration emanating from the hydrophobic heme group and its hydrophobic surroundings, including that of the interheme crevasse. This constitutes ΔG_{ap} and kinetic barriers.

7.3.5.1.2 Binding of Initial Oxygen Is Poor Due to Repulsive ΔG_{ap}

The work performed by the now-bound oxygen molecule in disrupting hydrophobic hydration also results in a lowered binding affinity because of the expended ΔG_{ap}. An important part of that work involves the disruption of ion pairs and reorientation of the charged side chains of the crevasse on oxygen binding.

7.3.5.1.3 Oxygen Binding Increases Heme Polarity and Results in Hydration of Dissociating Polar Groups and of Dissociating Hydrophobic Groups with the Consequence of a Generalized Water Uptake

The increased polarity due to oxygen binding relaxes ion pairs and effects dissociation of hydrophobic groups. It does so, because it allows polar species to gain in the competition for hydration with hydrophobic groups. Some of the hydrophobic hydration that previously formed on hydrophobic dissociation can now be recruited by the emerging polar groups, and the dissociations of both stand. The result is swelling, as reflected in the experimentally measured uptake of water and in the shattering of crystalline deoxyHb on efforts to bind oxygen. (These are generalized and fundamental properties of increasing the polarity of heme groups that can be expected to extend to oxidation of cytochromes of the electron transport chain.)

7.3.5.1.4 Each Oxygen That Binds Facilitates Binding of the Subsequent Oxygen

Communication between hemes readily occurs through the interheme crevasse. The decrease in hydrophobicity of the oxygen-bound heme allows separation of ion pairs and reorientation of charges associated with the interheme crevasse. This increase in polarity propagates to the second heme. Now the work to be performed by oxygen binding to the second heme decreases and affinity increases.

With the proximity of the β^1 H146–α^2 K40 ion pair to the $\beta^1\alpha^2$ interheme crevasse, shown in Figure 7.19A, it is not unreasonable to suppose that oxygen binding to one or both hemes of the $\beta^1\alpha^2$ heme pair would facilitate this ion pair separation that is one of the two keys to the quarternary structure change on oxygen binding. Furthermore, it is not unreasonable to suggest that the α^2 R141–α^1 K127 ion pair, also shown in Figure 7.19A, might be influenced by oxygen binding at the $\beta^1\alpha^2$-heme pair. Similar relationships occur at the $\alpha^1\beta^2$-heme pair where the key ion pairs would be β^2 H146(COO$^-$)–α^1 K40 and the α^1 R141(COO$^-$)–α^1 K127 with similar proximities to those of the $\beta^1\alpha^2$-heme pair in Figure 7.19A. These relationships can be visualized by a combination of Figure 7.19A and the transparency overlay in Figure 7.18.

7.3.5.2 Melding of the Perutz and Consilient Mechanisms

Should oxygen binding alter the tertiary structure of an individual globin chain such that the above-noted set of four ion pairs at the interface between the $\alpha\beta$ dimers ($\alpha^1\beta^1$ and $\alpha^2\beta^2$) could no longer sterically form, then the separated ions could be expected to disrupt hydrophobic hydration, and the free energy of hydrophobic association would become less favorable. The small change in His-Fe distance and in angles at the heme, transmitted by the "gears and levers" as proposed in the Perutz mechanism, could readily effect separation of

7.3 Hemoglobin Structures Demonstrate the Consilient Mechanism

the ion pairs and hydrogen bonds in the switch and joint regions. Given the relatively rigid structures of the α- and the β-chains, the effect of the small conformational change would be to reduce the complementarity between the $\alpha^1\beta^1$ and $\alpha^2\beta^2$ interfaces.

This would constitute merging of the consilient mechanism with the Perutz mechanism.[25] The competition for hydration between polar (e.g., charged) and hydrophobic groups responsible for positive cooperativity would remain at these sites, but in the Perutz mechanism it would have been the result of the tertiary structural change rather than arising directly from changes in the balance of the competition for hydration at the $\beta^1\alpha^2$- and $\alpha^1\beta^2$-heme pairs that result from oxygen binding.

7.3.6 Sickle Cell Anemia: Lowers the Temperature Interval of the "Cusp of Insolubility" with the Consequences of Disease and Death

The movable "cusp of insolubility" depicted in Figure 7.1 provides meaningful visualization of the disease, sickle cell anemia. A single mutation in the β-chain of the normal adult human hemoglobin, hemoglobin A, to produce hemoglobin S lowers the "cusp of insolubility" for deoxyHbS from above to below physiological temperatures. Under conditions of low oxygenation, the hemoglobin S aggregates and causes the red blood cell to sickle, as shown in Figure 7.22. Without treatment, homozygous individuals, in which all β-chains contain the mutation, suffer from acute onset of abdominal pain and ulcerations of the lower extremities and have a life expectancy of about 45 years.[44] Heterozygous individuals, in which half of their β-chains contain the mutation and the other half are normal, escape such dire symptoms.

7.3.6.1 Hemoglobin S

7.3.6.1.1 Hemoglobin S Results from the Mutation of Glu^6 (E6) to Val^6 (V6) in the β-chain

Hemoglobin S is the result of a single base mutation in the DNA sequence for the β-chain that replaces the most polar residue, the charged carboxylate-containing E6 residue, by a modestly hydrophobic V6 residue (see the Genetic Code in Table 6.2 for the single base change required for the replacement of Glu by Val). This substitution shifts the hemoglobin tetramer toward hydrophobic association, that is, lowers the cusp of insolubility from above to below physiological temperature, when in the deoxygenated state. Such a simple substitution lowers the T_t-(solubility/insolubility)divide for the aggregation of the more hydrophobic tetrameric deoxyhemoglobin S molecules into fibers shown in Figure 7.23. The T_t-divide, however, remains above physiological temperatures as long as there is a high enough level of oxygen to maintain the more polar oxyhemoglobin state.

7.3.6.1.2 Hydrophobic Association of Hemoglobin S Tetramers (Insolubilization)

The critical intertetramer contact that lowers the T_t-divide for aggregation directly involves the mutated hydrophobic residue, V6, of a β_2-chain. As shown in Figure 7.24, the oil-like Val^6 side chain, $-CH-(CH_3)_2$ of a β_2-chain associates with the very oil-like F8 side chain, $-CH_2-C_6H_5$ and L89 side chain, $-CH_2-CH-(CH_3)_2$, of the β_1-chain in an adjacent tetramer. This hydrophobic interaction lowers the cusp of insolubility for association of deoxyhemoglobin S to below physiological temperatures. The hydrophobic association between tetramers propagates fiber formation, illustrated in Figure 7.23, and distorts the hemoglobin transporting cell, the red blood cell, as shown in Figure 7.22.

7.3.6.1.3 Manifestation of the Oxygenated State that Prevents Hydrophobic Association of Hemoglobin S Tetramers

Qualitatively, oxygen binding to hemoglobin forms a more polar state. The more polar state in its thirst for hydration destructures the hydrophobic hydration that would otherwise form around the oil-like V6 side chain, $-CH-(CH_3)_2$ of the β_2-chain and the very oil-like Phe^{85} side chain, $-CH_2-C_6H_5$ and L89 side chain, $-CH_2-CH-(CH_3)_2$, of the β_1-chain of an

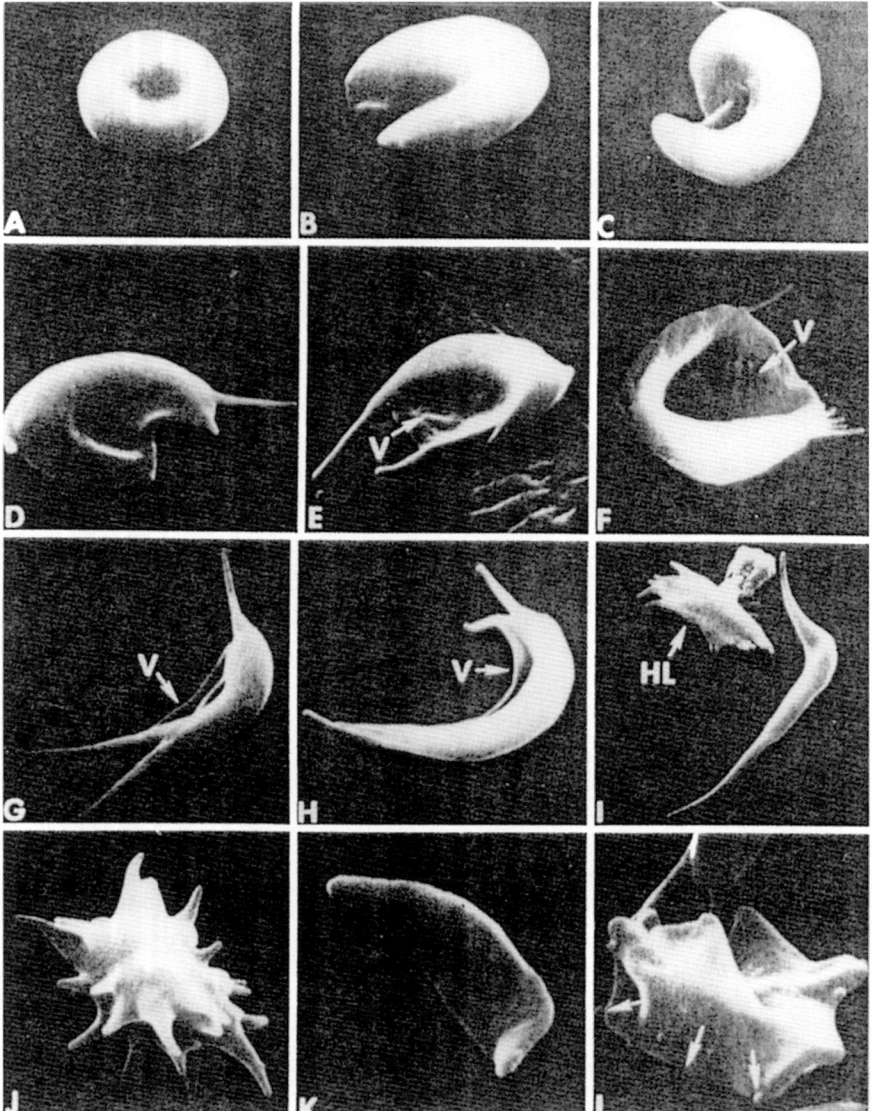

FIGURE 7.22. Sickled (hemoglobin S–containing) red blood cells as shown by scanning electron microscopy. (**A**) Oxygenated, showing normal shape. (**B**) Deoxygenated, showing varying degrees of sickling and crenation. (Archives Internal Medicine, 1974, **133**, 545–562, *Copyright © 1974, American Medical Association*. All rights reserved. With permission from White.[104])

adjacent tetramer. Oxygen binding causes the separation of the ion pair networks of Figure 7.8. These individual charged species now each reach out and gather their own hydration shell and cooperatively assist other polar species to do the same. This emergence of polar species, as noted above, brings in some 70 additional water molecules. In doing so they destructure substantial hydrophobic hydration such that there is no longer sufficient hydrophobic hydration to drive hydrophobic association.

Recall from Chapter 5, section 5.1.3.3, that insolubility (association) of hydrophobic groups occurs when there has developed too much hydrophobic hydration. In particular, $\Delta G(\text{solubility}) = \Delta H - T\Delta S$, where ΔH is nega-

7.3 Hemoglobin Structures Demonstrate the Consilient Mechanism

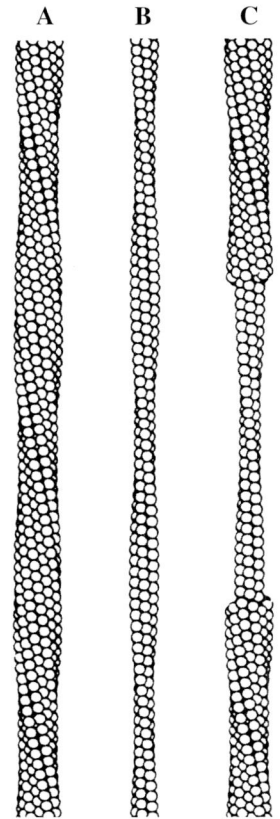

FIGURE 7.23 Model of hemoglobin S fibers as reconstructed from electron micrographs with each sphere representing a molecule of tetrameric hemoglobin. (A) An outer layer of 10 strands of hemoglobin molecules. (B) An inner core of four strands of hemoglobin molecules. (C) The combined representation of strands. (Reprinted from Journal of Molecular Biology, Vol. **130**, G.W. Dykes, R.H. Crepeau, and S.J. Edelstein, Three-dimensional Reconstruction of 14-Filament Fibers of Hemoglobin S, 451–472, 1979,[105] with permission from Elsevier.)

tive and favorable for formation of hydrophobic hydration and $(-T\Delta S)$ is positive and unfavorable for formation of hydrophobic hydration. Because of this, solubilization of oil-like groups occurs by lowering the amount of hydrophobic hydration, and insolubility occurs as too much hydrophobic hydration develops. Thus the "cusp of insolubility" for hydrophobic association shifts to lower temperatures as the amount of hydrophobic hydration increases and shifts to higher temperatures as the amount of hydrophobic hydration decreases, as occurs

by means of competition for hydration between charged and oil-like groups.

7.3.6.1.4 Persistence of This Mutation Due to the Protective Effect of a Heterozygous Condition Against the Malarial Parasite

Twenty-five percent of African blacks are heterozygous for this trait. The persistence of this mutation at this high level is due to the effectiveness of the heterozygous condition as protection against malaria. "When the red cells of a person with sickle-cell trait are invaded by the malarial parasite, the red cells adhere to blood

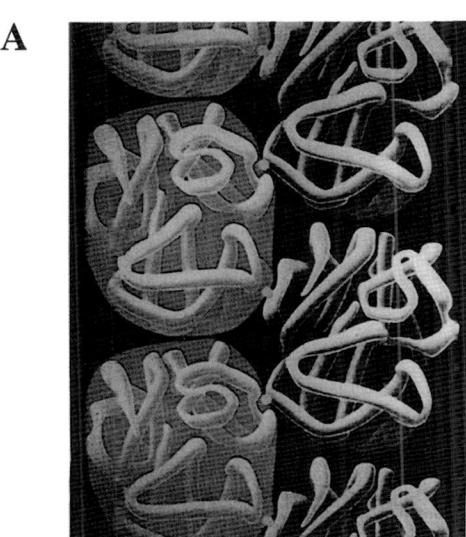

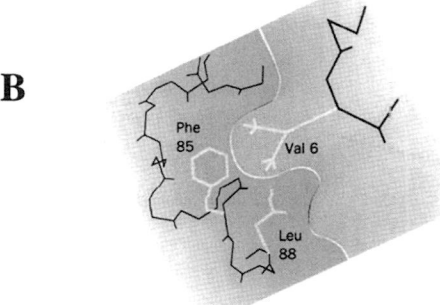

FIGURE 7.24. (A) Two strands of deoxyhemoglobin S tetramers showing the point of hydrophobic association responsible for fiber formation. (B) The mutation of a Glu^6 to Val^6 in the β_2-chain lowers the cusp of insolubility by means of association with the hydrophobic Phe^{85} and Leu^{88} residues of the β_1-chain of an adjacent tetramer. (Copyright © by Irving Geis.)

vessel walls, become deoxygenated, assume the sickled shape, and then are destroyed, the parasite being destroyed with them."[44]

7.3.6.2 Phase Diagrams of Hemoglobins S and A

7.3.6.2.1 Temperature Effect Diagnostic of Consilient Mechanism

The most fundamental characteristic of the consilient mechanism, based as it is on proteins that exhibit inverse temperature transitions, is the property of aggregation on raising the temperature and dissolution on lowering the temperature. That the consilient mechanism applies to hemoglobin S can be appreciated in the statement of Eaton and Hofrichter[45] that "the most important characteristic of hemoglobin S polymerization is that a gel can be prepared by heating a liquid solution at the appropriate concentration and 'melted' by cooling." Of course, the gel is the state of hydrophobically associated fibers, and melting is the process of fiber dissolution.

7.3.6.2.2 The Phase Diagram of the Elastic Model Protein (GVGVP)$_{251}$

The phase diagrams of the several model proteins given in Figure 5.3 show the binodal or coexistence lines for each that we here call the T_t-divide. As shown specifically for (GVGVP)$_{251}$ in Figure 7.25, there is a second line,[46,47] the spinodal line (T_{sp}), and it becomes coincident with the T_t-divide for this model protein at modest concentrations. The spinodal line is obtained by extrapolation of data from lower temperatures as indicated in the insert. The spinodal line exhibits the particular advantage that it can be determined for aggregations that would occur at higher temperatures were it not for other effects such as thermal denaturation. As demonstrated by the insightful work of San Biagio and Palma,[48,49] this provides substantial advantage in the comparison of hemoglobins S and A.

7.3.6.2.3 Inverted Phase Diagrams of Hemoglobins A and S: A Further Diagnostic of Inverse Temperature Transition Using the Spinodal Line

As shown in Figure 7.26, San Biagio and Palma[48,49] determined the spinodal lines for the aggregation of hemoglobins S and A and demonstrated the occurrence of inverse temperature transitions. Rather than being domed (concave with respect to the horizontal axis), as occurs for the most common petroleum-based polymers, the transition line is shaped valley-like (convex with regard to the horizontal axis). This indicates that intermolecular hydrophobic association dominates the aggregation, as shown in Figure 7.24 and as occurs for (GVGVP)$_n$ and the other polymers in Figure 5.3.

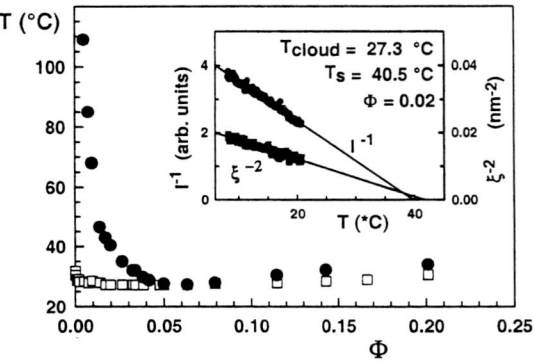

FIGURE 7.25. Phase diagram of (GVGVP)$_{251}$ showing the spinodal line in addition to the usual binodal (coexistence) line that we call the T_t-divide. **Inset** shows experimental determination of the spinodal line by extrapolation to the x-axis intercept of data on the temperature dependence of concentration fluctuations obtained at lower temperature. This means that critical data can be obtained for phase separations that would occur at elevated temperatures if denaturation did not occur. Note that spinodal and binodal lines overlap for part of the volume fraction axis. (Reproduced with permission from Manno et al.[46])

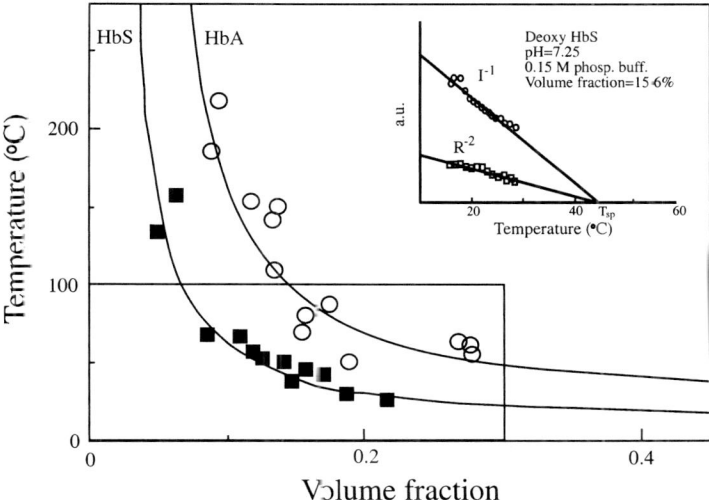

FIGURE 7.26. Spinodal lines for hemoglobin S (■), which does aggregate at physiological temperatures, and hemoglobin A (○), which has the potential to aggregate (hydrophobically associate) if the temperature could reach above 60°C. (Adapted with permission from San Biagio and Palma.[48])

7.3.6.2.4 Estimate of ΔG_{HA} (HbS → HbA)

By use of Equation (5.8) of Chapter 5, $\Delta G_{HA}(\chi) = [T_t(\chi) - T_t(\text{ref})] \Delta S_t(\text{ref})$, and the data in Figure 7.26, it becomes possible to estimate the change in Gibbs free energy for hydrophobic association due to replacement of the E6 to V6 in the β-chain. Deoxyhemoglobin S becomes the reference state, because the calorimetry data are available for hemoglobin S. Drawing from the extensive review of Eaton and Hofrichter,[45] the fitted value for the heat capacity of −234 cal/mole K is used for ΔS_t (deoxyHbS aggregation). From the phase diagram of San Biagio and Palma, T_{sp}(deoxyHbS) is 303K and T_{sp}(deoxyHbA) is 341K. Accordingly, ΔG_{HA}(deoxyHbS → deoxyHbA) = (303K − 341K)(−234 cal/mole K) = 8.9 kcal/mole, and, of course, ΔG_{HA}(deoxyHbA → deoxyHbS) = −8.9 kcal/mole. An excursion by 8.9 kcal/mole into the realm of insolubility means disease and early death for the homozygous condition.

Accordingly, the important oxygen transport capacity of hemoglobin and the disease state of sickle cell anemia are explicable in terms of spatially localized and moveable cusps of insolubility, that is, water solubility–insolubility divides, in terms of the consilient mechanism.

7.4 Blood Clotting: Poised at the Cusp of Insolubility

7.4.1 Poised Between Bleeding and Blocking Blood Flow

Formation of a blood clot involves a complex cascade of enzymatically controlled reactions the penultimate step of which is the formation of thrombin. Thrombin then cleaves peptides from fibrinogen to form fibrin monomers that associate to form the fibrin clot. Fibrin clot formation is a carefully poised process whereby a relatively minor injury will not allow excessive bleeding and result in death and whereby excessive clot formation will not block blood flow and result in death. As will be seen by examination of the relevant molecular structures, regardless of the balance struck, the key process of clot formation is the hydrophobic association of fibrin monomers. In demonstrating this perspective, the same T_t-based mean residue hydrophobicity plot will be used as was used above in Figure 7.9 for understanding the hydrophobic association of hemoglobin subunits.

7.4.1.1 Thrombus Formation Leading to Heart Attack, Stroke, and Phlebitis

Excess clot (thrombus) formation in the blood vessels results in disease and death. When thrombus forms in the coronary arteries that service the heart muscle, occlusion of a coronary artery can result in a heart attack with death of a region of heart muscle, a myocardial infarction. When the blockage destroys a large enough or critical portion of the heart, pumping function is lost and death results.

When thrombus forms in an artery of the brain, referred to as *apoplexy* or a *cerebrovascular accident*, a portion of brain tissue dies. Again, the size and location of the artery blocked determines the amount of lost brain tissue and whether the lost function results in partial paralysis, loss of speech, or death.

When blood clot forms in a vein, without inflammation and called *phlebothrombosis* or in response to inflammation and termed *thrombophlebitis*, localized pain, swelling, redness, and heat develop with resulting loss of function. From the perspective of an inverse temperature transition as the basis for clot formation, the natural increase in temperature would aggravate the problem.

7.4.1.2 Hemophilia (Bleeding Diseases)

Hemophilia arises from any of a number of factors (commonly due to sex-linked inheritance) required for thrombin formation such that proper fibrin clot does not form. In infancy anemia and death can occur. Bruises occur when the source of the bruise was so innocuous as not to have been noticed. Internal bleeding occurs in the mouth, nose, gastrointestinal tract, and in the joints with swelling and impairment of function. As we will discuss, all of this occurs because the hydrophobic association required for fibrin clot formation cannot happen. All of this is the result of loss of control of the movable "cusp of insolubility."

7.4.2 Control of Fibrin Clot (Thrombus) Formation

7.4.2.1 Structure of Fibrinogen, the Precursor Protein

7.4.2.1.1 The Subunit Structure of Fibrinogen

The fibrinogen molecule is comprised of three chains—Aα, Bβ, and γ—that are cross-linked by disulfide bridges among themselves and to a second set of three chains to give a multiply disulfide-bridged structure, $(A\alpha)_2(B\beta)_2(\gamma)_2$, as schematically represented in Figure 7.27A. The A stands for 16 residues at the amino end of the α-chain, and B represents 14 residues at the amino end of the β-chain. Note that all six chains are associated amino end to amino end and held in that position by numerous disulfide cross-links, indicated by the interconnecting lines. The A and B sequences are so delineated because their removal initiates fibrin clot formation. The removal of the A peptide for activation represents the removal of two A peptides at the amino-end–amino-end junction that then hydrophobically associate with a pair of end-to-end associated globular γ domains.

7.4.2.1.2 The Primary Structure of the α, β, and γ Chains of Human Fibrinogen

The amino acid sequences of the Aα (610 residues), Bβ (461 residues), and γ (411 residues) chains of human fibrinogen are given in Table 7.1.[50] Having the amino acid sequences directly available will be helpful in discussing the peptide cleavages and the interchain interactions relevant to fibrin formation.

7.4.2.1.3 The Crystal Structure of Chicken Fibrinogen

The crystal structure of approximately 70% of the residues of chicken fibrinogen has become available from the laboratory of Doolittle and coworkers[51] as shown in Figure 7.27B. The $(A\alpha)_2(B\beta)_2(\gamma)_2$ subunit structure combines to form a flattened sigmoidal or S-like shape, which with the missing sequences will be indicated as $(\alpha\beta\gamma\alpha'\beta'\gamma')$. The amino termini of all six chains, the two chains—$\alpha\alpha'$, $\beta\beta'$, and $\gamma\gamma'$—all meet in the central part of the structure,

7.4 Blood Clotting: Poised at the Cusp of Insolubility

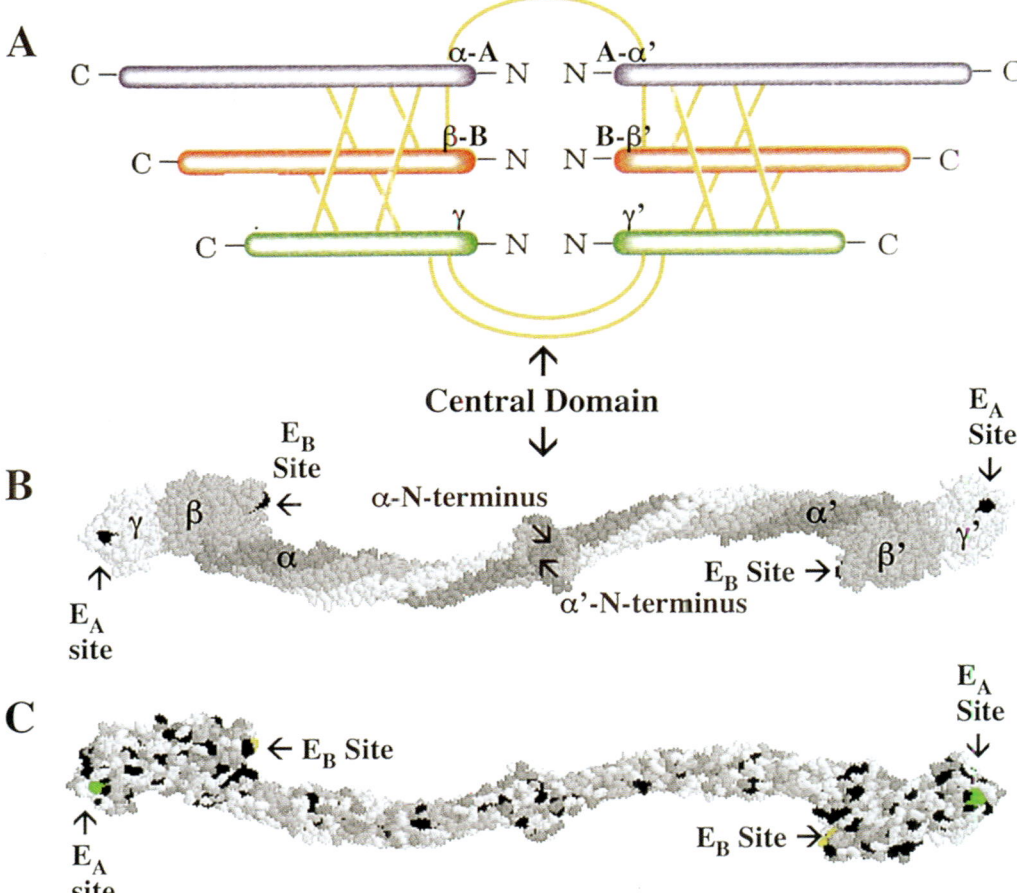

FIGURE 7.27. **(A)** Schematic representation of the structure of one molecule of fibrinogen, showing disulfide bridges and N-end-to-N-end arrangement of chains. (Adapted with permission from Voet and Voet.[103]) **(B)** Crystal structure of one side of the native chicken fibrinogen molecule with gray, light gray, and white for delineation of the α-, β-, and γ-chains, respectively, where β- and γ-chains are shown to have globular components. Soft fibrin clot forms on removal of a 16 residue A peptide at the central domain exposing a new amino-terminal α-chain GPRP sequence that initiates association at the hydrophobic γ-globule E_A site (shown in stereo in Figure 7.32A), and on removal of a 14 residue B peptide at the central domain to expose the β-chain GHRP sequence for association with the hydrophobic β-globule E_B site (shown in stereo in Figure 7.32B). **(C)** Crystal structure of the same side of the native chicken fibrinogen molecule for the purpose of delineating hydrophobicity with the most hydrophobic (nonhistidine) aromatic residues in black, other hydrophobic residues in gray, neutral residues in light gray, and the charged residues, acidic and basic (including histidine), as white. It is obvious even at this gross level of examination that the carboxyl-terminal globular components of the β- and γ-chains are much more hydrophobic than the remainder of the molecule. (B and C prepared using the crystallographic results of Yang et al.[51] as obtained from the Protein Data Bank, Structure File 1M1J.)

TABLE 7.1. Human fibrinogen: sequences of the Aα, Bβ, and γ chains[a].

Aα-Fibrinogen

		E_A-site		
1	ADSGEGDFLA	EGGGVR↓*GPRV*	VERHQSACKD	SDWPFCSDED	WNYKCPSGCR	MKGLIDEVNQ
61	DFTNRINKLK	NSLFEYQKNN	KDSHSLTTNI	MEILRGDFSS	ANNRDNTYNR	VSEDLRSRIE
121	VLKRKVIEKV	QHIQLLQKNV	RAQLVDMKRL	EVDIDIKIRS	CRGSCSRALA	REVDLKDYED
181	QQKQLEQVIA	KDLLPSRDRQ	HLPLIKMKPV	PDLVPGNFKS	QLQKVPPEWK	ALTDMPQMRM
241	ELERPGGNEI	TRGGSTSYGT	GSETESPRNP	SSAGSWNSGS	SGPGSTGNRN	PGSSGTGGTA
301	TWKPGSSGPG	STGSWNSGSS	GTGSTGNQNP	GSPRPGSTGT	WNPGSSERGS	AGHWTSESSV
361	SGSTGQWHSE	SGSFRPDSPG	SGNARPNNPD	WGTFEEVSGN	VSPGTRREYH	TEKLVTSKGD
421	KELRTGKEKV	TSGSTTTTRR	SCSKTVTKTV	IGPDGHKEVT	KEVVTSEDGS	DCPEAMDLGT
481	LSGIGTLDGF	RHRHPDEAAF	FDTASTGKTF	PGFFSPMLGE	FVSETESRGS	ESGIFTNTKE
541	SSSHHPGIAE	FPSRGKSSSY	SKQFTSSTSY	NRGDSTFESK	SYKMADEAGS	EADHEGTHST
601	KRGHAKSRPV					

Bβ-Fibrinogen

		E_B-site & E_A polymerization site				
1	QGVNDNEEGF	FSAR↓*GHRPLD*	*PAPPPISGGG*	*YRARPAKAAA*	TQKKVERKAP	
61	DAGGCLHADP	DLGVLCPTGC	RPIRNSVDEL	NNNVEAVSQT	SSSSFQYMYL	
121	LKDLWQKRQK	QVKDNENVVN	LYIDETVNSN	IPTNLRVLRS	ILENLRSKIQ	
181	KLESDVSAQM	EYCRITPCTVS	CEEIIRKGGE	TSEMYLIQPD	SSVKPYRVYC	
241	DMNTENGGWT	VIQNRQDGSV	QGFGNVATNT	DGKNYCGLPG	EYWLGNDKIS	
301	QLTRMGPTEL	LIEMEDWKGD	VQNEANKYQI	SVNKYRGTAG	NALMDGASQL	
361	MGENRTMTIH	NGMFFSTYDR	RKQCSKEDGG	GWWYNRCHAA	NPNGRYYWGG	
421	*QYTWDMAKHG*	TDDGVVWMNW	SMKIRPFFPQ	Q:	••Proposed	

••Experimental hydrophobic association at E_B sites
hydrophobic association at E_A sites

γ-Fibrinogen

1	YVATRDNCCI	LDERFGSYCP	TTCGIADFLS	TYQTKVDKDL	QSLEDILHQV	ENKTSEVKQL
61	IKAIQLTYNP	DESSKPNMID	AATLKSRIML	EEIMKYEASI	LTHDSSIRYL	QEIYNSNNQK
121	IVNLKEKVAQ	LEAQCQEPCK	DTVQIHDITG	KDCQDIANKG	AKQSGLYFIK	PLKANQQFLV
181	YCEIDGSGNG	WTVFQKRLDG	SVDFKKNWIQ	YKEGFGHLSP	TGTTEFWLGN	EKIHLISTQS
241	AIPYALRVEL	EDWNGRTSTA	DYAMFKVGPE	ADKYRLTYAY	FAGGDAGDAF	DGFDFGDDPS
301	DKFFTSHNGM	QFSTWDNDND	KFEGNCAEQD	GSGWWMNKCH	AGHLNGVYYQ	GGTYSKASTP
361	*NGYDNGIIWA*	TWKTRWYSMK	KTTMKIIPFN	RLTIGEGQQH	HLGGAKQAGD	V:

[a] Cleavage points for the A and B peptides (indicated by arrows) open the E_A and E_B sites, respectively. Underlined near the carboxyl terminus of the γ is the identified hydrophobic sequence that associates with the E_A site, and near the carboxyl terminus of the β chain is the proposed sequence for hydrophobic association to the E_B site.

7.4 Blood Clotting: Poised at the Cusp of Insolubility

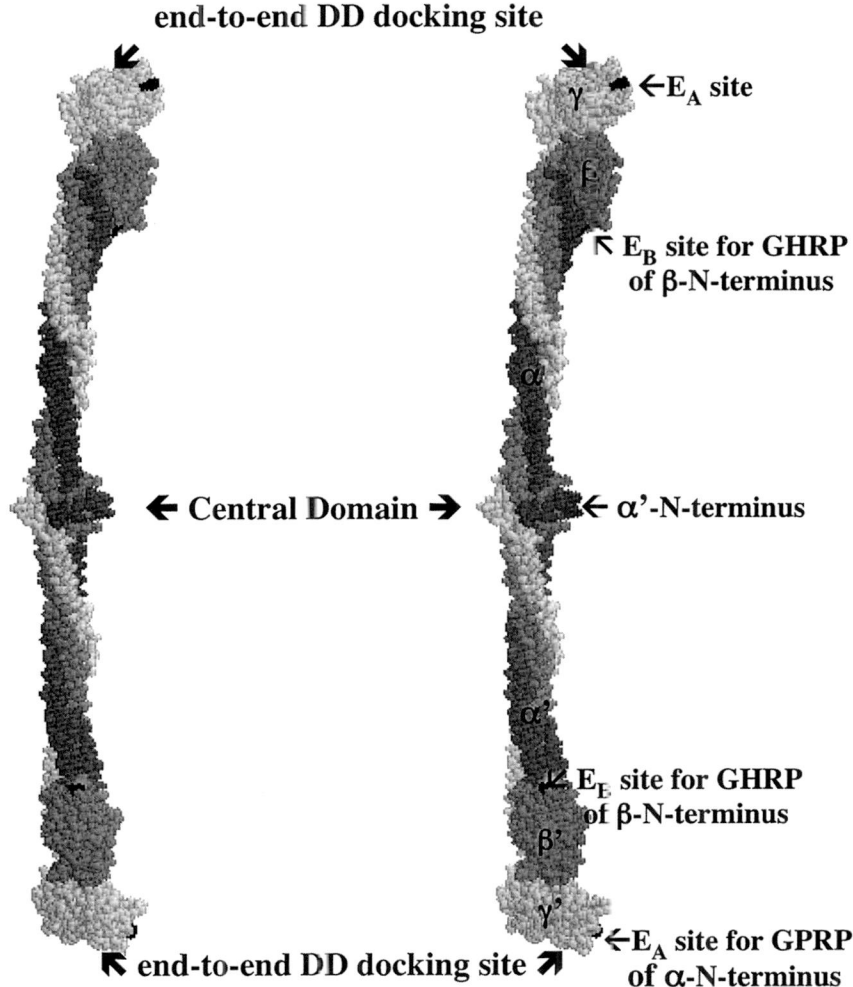

FIGURE 7.28. Stereo view of fibrinogen, $(\alpha\beta\gamma)_2$, showing three sites for intermolecular contacts resulting in formation of the fibrin clot: (1) the end-to-end docking site for linear fiber growth that works in cooperation with association at the E_A site; (2) the E_A site of the carboxyl-terminal γ-globule hydrophobic domain, which provides for fibrin formation in a second dimension due to association with the GPRP of an α-amino terminus (at the central domain) that was exposed on removal of the 16 residue A peptide; and (3) the E_B site of the carboxyl-terminal β-globule hydrophobic domain, which provides for fibrin formation in the third dimension by association with GHRP of a β-amino terminus (at the central domain) that becomes exposed on removal of the 14 residue B peptide. These, as well as the end-to-end polymerization, are all sites of hydrophobic association as shown in Figures 7.31 and 7.32. (Prepared using the crystallographic results of Yang et al.[51] as obtained from the Protein Data Bank, Structure File 1M1J)

indicated as the central domain. Front and back sides of the fibrinogen molecule are shown in Figure 7.27B as obtained from a crystal structure that is 65% water. Interestingly, the phase-separated state of poly(GVGVP) is 63% water by weight.[52] Of the 1,364 amino acid residues of the half molecule of chicken fibrinogen, 272 residues of the carboxyl-terminal sequence of the α-chain are missing; at the amino termini of the three chains 96 residues are missing, and 16 residues are not seen at the carboxyl terminus of the γ-chain. These missing sequences are so disordered as to provide no diffraction pattern.

The absence of the residues noted does not appear to block an understanding of the key elements of fibrin formation as much as might have been initially expected. Binding of the key residue sequences that become exposed on removal of the A and B peptides identifies the E_A and E_B sites for intermolecular association. The sequence GPRP, which binds at the E_A site, and the sequence GHRP, which binds at the E_B site, are found in the expected sites as labeled in Figure 7.27B, where they look much like black dots in this whole-molecule view.

The primary message of this consideration of blood clotting in relation to the movable cusp of insolubility becomes apparent in this whole-molecule view of fibrinogen in Figure 7.27C. Even at this level of inspection with the most hydrophobic amino acid residues given in black in Figure 7.27C, the greater hydrophobic character of the carboxyl-terminal globular units of the γ-chain and β-chains is apparent. This aspect is pursued in more detail below when considering the double fragment D from human fibrin itself, which is the end-to-end association of the γ-chain globular units.

7.4.2.2 T_t-based Hydrophobicity Plots of Human Fibrinogen Chains

The single residue T_t-based hydrophobicity plot and the mean residue hydrophobicity plot reported for the chains of fibrinogen provide interesting insights. A glimpse of the single residue plot provides immediate sequence insights that are easier to glean from such a plot than from the sequences as listed in Table 7.1. The mean residue hydrophobicity plot provides an opportunity to visualize hydrophobic weightings of sequences and to recognize pattern similarities between chains that may be associated with function, as seen with the myoglobin and hemoglobin chains in Figures 7.2 and 7.9. Recall that the mean residue hydrophobicity plot averages the value for an 11 residue sliding window with the numbered residue as the central residue and with 37°C being the value used entering and leaving the sequence. For example, the mean value for residue 1 uses T_t-values of 37°C for positions 1 to 5 and the averaged value of residues 1 through 6; and the mean T_t-value for residue 6 is the average value for the first 11 residues.

7.4.2.2.1 γ-Fibrinogen—A Striking Hydrophobic Sequence Near the Carboxyl End

As shown in Figure 7.29, from the perspective of hydrophobic sequences, residues 337 through 379 of the γ-fibrinogen chain constitutes the most hydrophobic sequence, and the mean residue hydrophobicity plot bears an interesting pattern of hydrophobicity peaks that bears a striking resemblance in shape and intensity to the 401 to 448 sequence of the Bβ-fibrinogen chain (Figure 7.30B). There are additional mean residue hydrophobicity peaks of lesser prominence centered near residues 240, 278, and 310. This provides an interesting opportunity to look for correlations with points of intermolecular association leading to fibrin clot formation as represented by the crystal structure data.

7.4.2.2.2 Aα-Fibrinogen—A Striking Hydrophobic Sequence (~80 Residues) at Midsequence

The most striking feature of the Aα-fibrinogen mean residue hydrophobicity plot (plot not included), a sequence of 610 residues, is the central 81 residue sequence (residues 284 to 365) that is devoid of all but a single charged residue, which is the moderate lysine (Lys, K) residue and which contains four tryptophan (Trp, W) residues, the most hydrophobic residue. The mean residue plot for Aα-fibrinogen (plot not shown), however, is devoid of significant hydrophobic peaks in contrast to those observed in Figure 7.9 for the α- and β-chains of hemoglobin A and for those of the γ-fibrinogen chain in Figure 7.29. The presence of the repetitive central sequence in humans, but not in the chicken sequence, would provide a striking opportunity for hydrophobic associa-

7.4 Blood Clotting: Poised at the Cusp of Insolubility

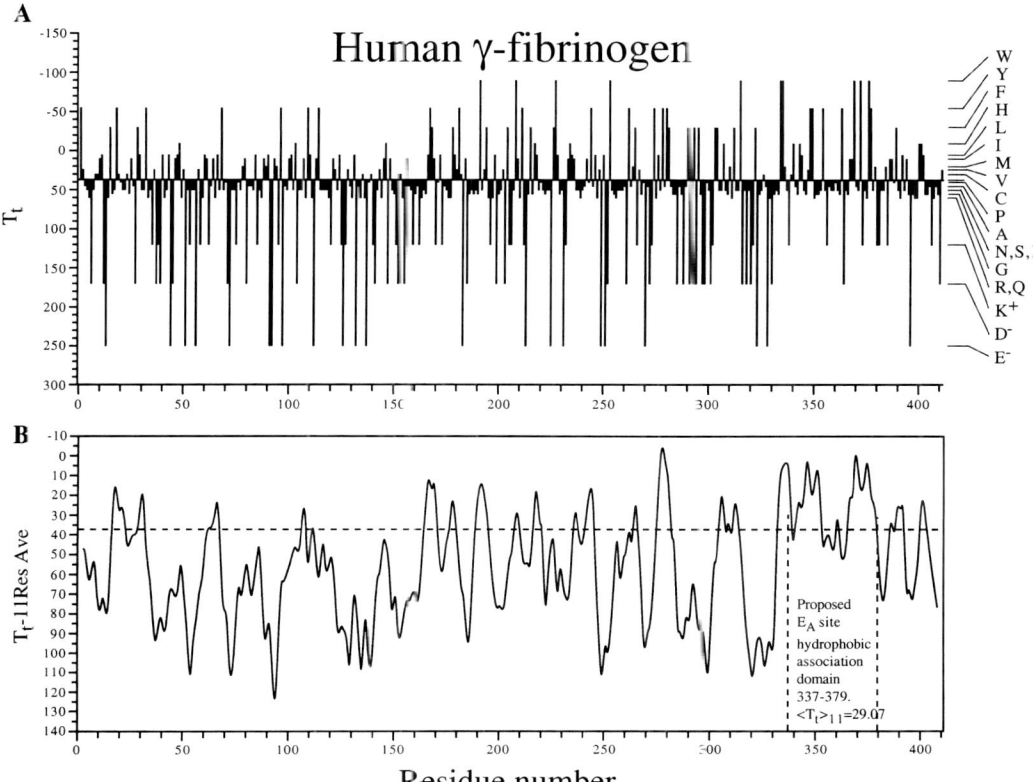

FIGURE 7.29. Single residue (**A**) and mean residue (**B**) hydrophobicity plots of human γ-fibrinogen. The most striking feature of the mean residue plot is the hydrophobic sequence, residues 337 to 379, experimentally shown to provide the E_A site for interaction with the α-chain amino-terminus on removal of the A peptide and exposure of GPRP and its surrounding hydrophobic residues for association with the hydrophobic domain of the E_A site shown in Figure 7.32A. A similar situation obtains for the β chain E_B site, as shown in Figure 7.32B. In our view, A and B peptide removals lower the cusp of insolubility for fibrin clot formation by making possible these hydrophobic associations. Accordingly, blood clotting occurs by means of a movable cusp of insolubility, the central point being that hydrophobic association, lowering the cusp of insolubility, is central to aggregation of molecules for blood clotting.

tion. This feature does not come into play in the structures in Figures 7.31 and 7.32, nor do the missing 372 amino-terminal residues of the chicken sequence. These differences with humans do not seem to interfere with carrying the insight of the chicken structure to the fragment DoubleD of human fibrin. Fortunately, the human fibrin-DD was obtained with the same bound GPRP and GHRP peptides, noted above, the exposures of which trigger fibrin clot formation.

7.4.2.2.3 Bβ-Fibrinogen—A Striking Hydrophobic Sequence (~50 Residues) Near the Carboxyl End

The mean residue hydrophobicity plot of Bβ-fibrinogen in Figure 7.30B exhibits two minor hydrophobic sequences near residues 115 and 375 that could serve hydrophobic folding and assembly within the fibrinogen molecule, but would be less likely to participate in the hydrophobic association between molecules.

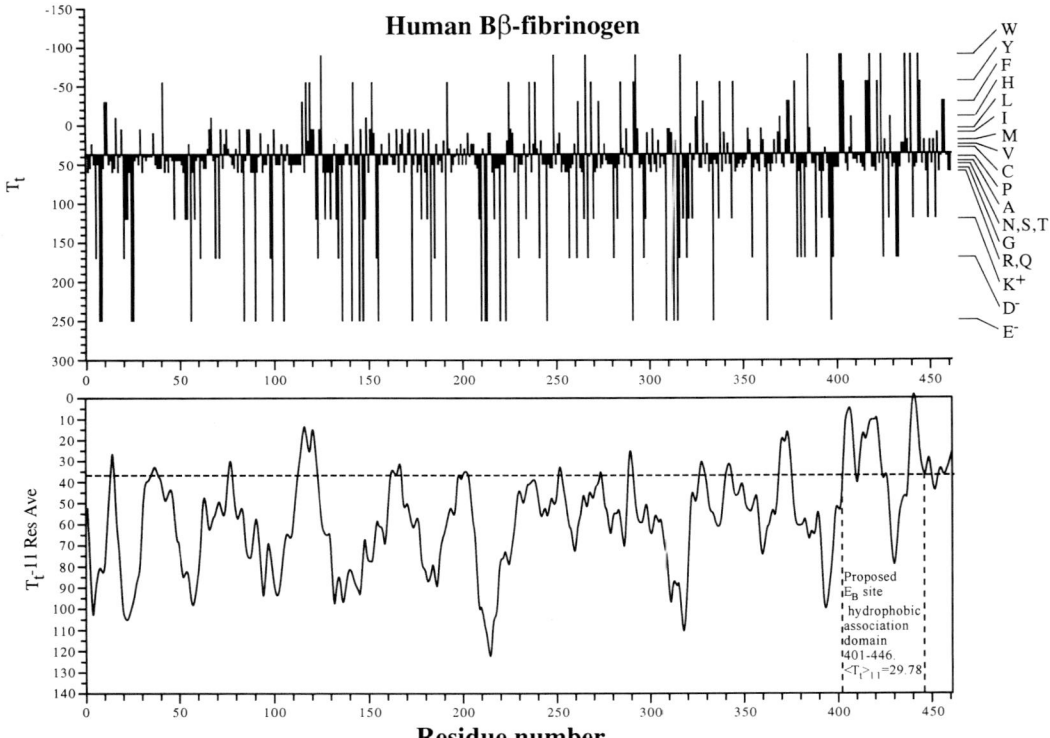

FIGURE 7.30. Single residue (A) and mean residue (B) hydrophobicity plots of human Bβ-fibrinogen. The most striking feature of the mean residue hydrophobic plot of this sequence is due to residues 401 to 446. The pattern of mean hydrophobicity arising from residues 401 to 446 bears striking resemblance to that of residues 337 to 379 of γ fibrinogen (see Figure 7.29B) that have been defined as forming the E_A site. Based on the pattern similarity with the E_A site, residues 401 to 446 become recognized as the E_B site for hydrophobic association with site exposed on removal of the B peptide at the central domain. Accordingly, removal of the B peptide exposes GHRP and its hydrophobic surroundings for hydrophobic association with the β-chain E_B site shown in Figure 7.32B.

The most striking hydrophobic sequence of Bβ-fibrinogen runs from residues 401 through 448, and the pattern of the mean hydrophobicity bears a striking resemblance to that of residues 337 through 379 of the γ-fibrinogen chain. The latter became a definition of the E_A site such that the pattern of mean hydrophobicity of residues 401 through 448 of β-fibrinogen becomes a definition of the E_B site. The actual sites have now been determined by X-ray diffraction of a human fibrin fragment, and direct observation of the sites is given in Figures 7.31 and 7.32.

7.4.2.3 Thrombin Activation of Fibrinogen Provides for Hydrophobic Association to Form the Soft Fibrin Clot

7.4.2.3.1 Removal of Peptide A (16 Residues) Initiates Intermolecular Hydrophobic Association

Removal of the first 16 residues of Aα-fibrinogen activates fibrinogen for fibrin formation by removing the repulsive effect of four negative charges of carboxylates, $-COO^-$, from two glutamic acid (Glu, E) and two aspartic acid (Asp, D) residues. This also decreases the apolar–polar repulsive free energy of interaction expressed in Equation (5.13) of Chapter 5, section 5.7.9, and allows hydrophobic association to proceed. The result is the exposure of a GPRP sequence from each α-chain of the central domain for association with E_A sites of γ-fibrinogen.

The E_A site provides for association with another fibrinogen molecule, that is, with the $(A\alpha)_2(B\beta)_2(\gamma)_2$ molecule. The experimentally defined sequence that effects polymerization of

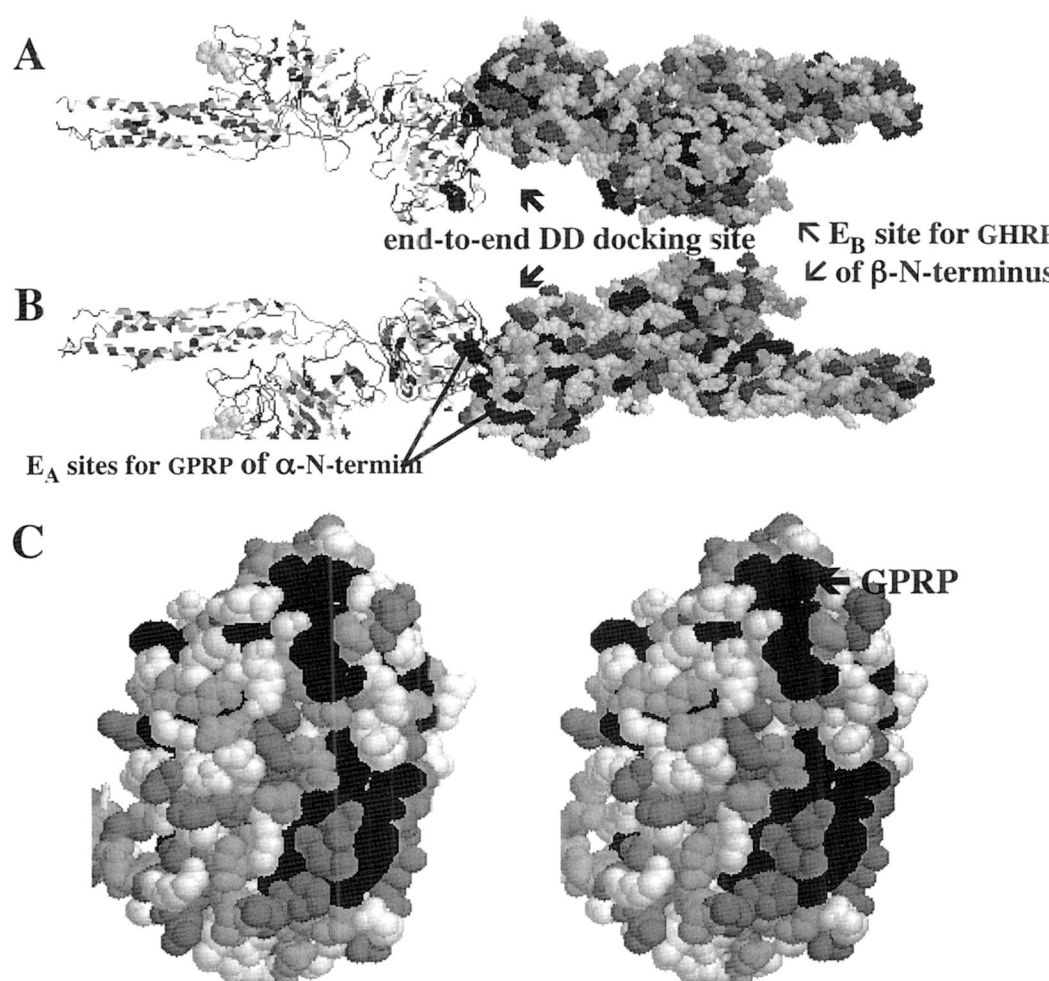

FIGURE 7.31. Human fibrin–DD association sites with residues given as aromatics (black), other hydrophobics (gray), neutrals (light gray), and acidics and basics (white). **(A)** One side of end-to-end docking of D fragments, with D on the left side in ribbon representation and D on the right side in space-filling representation. **(B)** Opposite side of end-to-end docking of D fragments, with D on the left side in ribbon representation and D on the right side in space-filling representation. The latter shows both E_A sites that would interact with the two GPRP sequences exposed at the central domain on removal of the A peptides by thrombin to reveal E_A site, identified here by bound GPRP tetrapeptide. **(C)** Stereo view of one end of end-to-end DD docking site. Only about one-half of the area shown is actually contact area, and it is of modest hydrophobicity, $\Delta G_{HA} \approx -8$ kcal/mole-contact surface. (Prepared using the crystallographic results of Everse et al.[55] as obtained from the Protein Data Bank, Structure File 1FZC.)

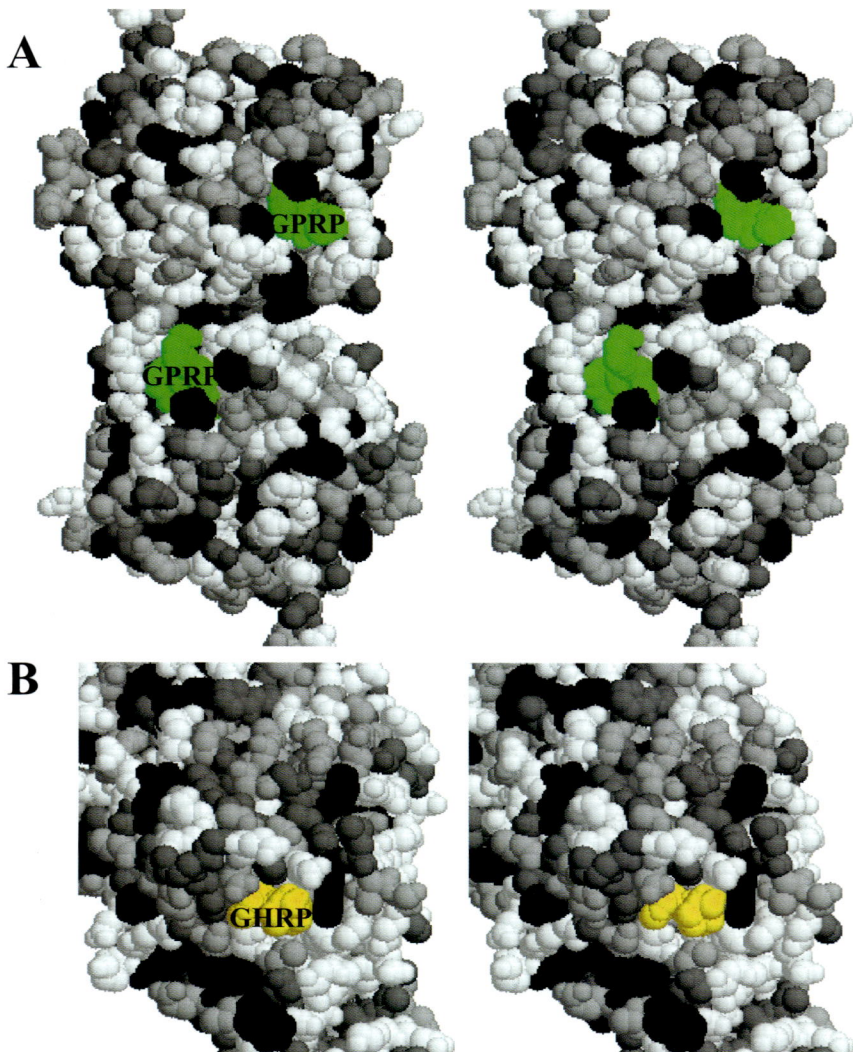

FIGURE 7.32. Stereo views of human fibrin–DD association sites with residues given as aromatics (black), other hydrophobics (gray), neutrals (light gray), and acidics and basics (white). **(A)** E_A site (with bound GPRP tetrapeptide), at γ-globular prominence with an estimated ΔG_{HA} of –16 kcal/mole-surface, indicating a compelling globular prominence for hydrophobic association, especially when doubled as occurs in the end-to-end docked structure. **(B)** E_B site (with bound GHRP tetrapeptide), at β-globular prominence with an estimated ΔG_{HA} of –8 kcal/mole of an approximated contact surface area indicating a modestly hydrophobic globular prominence for association. (Prepared using the crystallographic results of Everse et al.[55] as obtained from the Protein Data Bank, Structure File 1FZC.)

fibrin by association with the E_A site is sequence 337 to 379 of γ-fibrinogen.[53] The polymerization sites are delineated in Table 7.1, which gives the sequences. *As shown in Figure 7.29B and as noted above, this is the most striking hydrophobic sequence of γ-fibrinogen.* Accordingly, by simple inspection of the mean hydrophobicity plot for γ-fibrinogen, we see the activation of fibrinogen as the emergence of a site for hydrophobic association with the most hydrophobic sequence in a second molecule.

Furthermore, as seen in Figure 7.28, the E_A site is approximately at right angles to and shares a common edge with the end-to-end DoubleD association site such that these sites could cooperate in initiation of the hydrophobic associations for soft fibrin clot formation.

7.4.2.3.2 Removal of Peptide B (14 Residues) Allows for Further Intermolecular Hydrophobic Association

Removal of the first 14 residues of Bβ-fibrinogen activates fibrinogen for fibrin formation by removing the repulsive effect of three negative charges of carboxylates, $-COO^-$, from two glutamic acid (Glu, E) residues and one aspartic acid (Asp, D) residue. Removal of the B peptide creates or exposes a GHRP site near the central domain defined in Figure 7.27 that associates with the carboxyl-terminal end of the β-fibrinogen, the E_B site, chain of another fibrinogen molecule.[53] This sequence has not been specifically delineated as above for γ-fibrinogen. Nonetheless, based on the structural homology between γ-fibrinogen and β-fibrinogen, as demonstrated by the β-chain and γ-chain topologies given in Figure 3a of Spraggon et al.,[54] and particularly based on the mean residue plot of the γ-chain (Figure 7.29B) with its striking similarity to the mean residue plot of the β-chain (Figure 7.30B), the corresponding sequence for association with the E_B site would involve residues within the sequence from residues 401 through 448. The mean residue hydrophobicity plot for residues 401 through 448 in Bβ-fibrinogen demonstrates hydrophobic homology to sequences 337 to 379 of γ-fibrinogen (Figure 7.29B). Accordingly, by simple inspection of a mean hydrophobicity plot, this time for β-fibrinogen, we again see a dominantly hydrophobic sequence and recognize hydrophobic association as the basis for soft fibrin clot formation.

7.4.2.4 Hydrophobic Associations of Fragment DoubleD from Human Fibrin

When the fibrin clot is cleaved by the protease plasmin, it fragments into two parts, a D fragment and a smaller E fragment. Although no crystal structures have been obtained for the E fragment, Doolittle and coworkers have obtained crystal structures for a factor XIII (thrombin-activated transglutaminase) cross-linked DoubleD fragment in which D fragments are attached end to end.[54] Fortunately, these crystal have been obtained and the structures solved with bound GPRP and GHRP peptides such that the E_A and E_B sites have been identified.[55] With these accomplishments of the Doolittle group, it becomes possible to characterize the associations of chains that result in blood clotting. The objective is to assess the hydrophobicity of the association sites to challenge further our perspective that "biology thrives at the cusp of insolubility," which cusp of insolubility represents the process of hydrophobic association.

7.4.2.4.1 Hydrophobicity of the Points of Association

Three key points of association become clear. They are the E_A and E_B sites identified on the fibrinogen molecule of Figures 7.27B and 7.28 and demonstrated in the human fragment DoubleD of Figure 7.31A,B and the end-to-end docking site found in the human fragment DoubleD. This provides the opportunity to examine directly the hydrophobicity of these three sites of aggregation for soft fibrin clot formation and to estimate hydrophobicity in terms of estimates of relative values for the Gibbs free energy of hydrophobic association, ΔG_{HA}, of the contact faces or prominences.

7.4.2.4.2 End-to-End Docking Uses the Carboxyl-terminal Globular Component of the γ-chain

With the crystal structure, the end-to-end docking prominence becomes defined. The docking site is shown in side views in Figure 7.31A,B and end on in Figure 7.31C. This is a site of modest hydrophobicity that involves using only about half of the end observed in Figure 7.31C. One estimates the Gibbs free energy of hydrophobic association, ΔG_{HA}, to be $-8\,\text{kcal/mole}$ of contact surface. It should be appreciated that this is a gross estimate that should be improved once appropriate programs have been prepared. The calculation utilizes the values, ΔG_{HA}, in Table 5.3 for the amino acid residues and subjectively reduces those values with crude estimates of the amount of exposure of the side chain to the surface concerned and a further reduction of the contributions of charged residues by half when ion paired.

7.4.2.4.3 Hydrophobicity of the E_A Site and a Presumed Cooperative Binding with End-to-End Docking

As is apparent in Figure 7.31B, the E_A site occurs as a double site at the junction of two molecules. Coarse calculations of the free energy for hydrophobic association at each of the E_A sites in Figure 7.32A indicate an especially hydrophobic surface with an estimate for ΔG_{HA} of −16 kcal/mole-E_A site, which would be twice that number for the double site. It seems reasonable to expect that two sites of the fibrinogen molecule exposed at the central domain on removal of the A peptides from the amino terminus of each α-chain would themselves cooperatively bind with both E_A sites. The process of assembly of the two E_A sites on different fibrinogen molecules with the two thrombin exposed GPRP-containing sites in proper relationship at the central domain could be expected to cooperatively facilitate the end-to-end docking. In this way thrombin activation of fibrinogen would initiate fibrin clot formation. Initiation of fibrin clot formation becomes the stitching together of hydrophobic domains. "Biology thrives at the movable cusp of insolubility. . . ."

7.4.2.4.4 Hydrophobicity of the E_B Site

The E_B site gains its potential for hydrophobic association primarily by the unusual clustering of the most hydrophobic residue, tryptophan, specifically residues W424, W440, W444, and W385, as well as the additional very hydrophobic aromatic residues F374, F457, Y345, and Y445. Five of those residues contribute to the mean residue hydrophobicity pattern in Figure 7.30B, and small adjustments in surface exposure on association could create the most hydrophobic single site for fibrin assembly.

Perhaps the 272 missing carboxyl-terminal residues of the α-chains serve to cover the hydrophobic E_B and E_A sites until pushed aside by the much greater hydrophobic association potential of the sites exposed on removal of the A and B peptides.

7.4.2.4.5 The Limited Information on the GPRP and GHRP Sites at the Central Domain

The missing 96 residues at the amino termini of the three chains, in the vicinity of the central domain, preclude definition of the hydrophobic domains that pair with the E_A and E_B sites, but one predicts the capacity of these missing residues to arrange in formation of hydrophobic domains with which to pair with the total of four E_A and E_B sites.

A description of the general conclusion of blood clotting as driven by hydrophobic association follows.

7.4.2.5 Cross-linking to Form the Hard Fibrin Clot

The hydrophobic associations with the E_A site of the hydrophobic 337 to 379 sequence of the γ-chain and with the E_B site of the hydrophobic 409 to 451 sequence of the β-chain result in antiparallel alignment of the γ-chains from different fibrin molecules near the end-to-end γ-chain–γ-chain docking site. In particular, by means of a transamidase, glutamine-399 of one γ-chain cross-links to the lysine-406 of the abutting second γ-chain, and glutamine-399 of the second γ-chain cross-links with the lysine-406 of the first γ-chain to form two ε-(γ-glutamyl)lysine cross-links. These chemical cross-links transform the soft fibrin clot irreversibly into the hard fibrin clot, which can only be disassembled by subsequent proteolytic enzymes.

Thus, the key process of fibrin clot formation is one of cleaving peptide sequences, the result of which is to lower the cusp of insolubility, representing the hydrophobic association between molecules, from above to below physiological temperature.

7.5 Grave Medical Consequences of Protein Insolubilities

7.5.1 Introductory Remarks

7.5.1.1 Excursions Too Far into the Realm of Protein Insolubility

Excursions too far into the realm of protein insolubility present a variety of tragic and ultimately fatal neurodegenerative diseases. In humans there are two causative proteins: the prion protein of transmissible spongiform encephalopathies and the amyloid precursor protein of Alzheimer's disease. In general, there are some 15 proteins[56] that by mutations or processing errors exhibit similar protein amyloid deposits[57] that result in organ damage. Such protein insolubilities occur in mammals, birds, and even yeast.[58]

7.5.1.2 The Etiology Is Common to the Formation of all Amyloid Deposits

Very stable protein aggregates, originating within the host or introduced into the host, act as "condensation nuclei"[59] on which the natural host protein grows until the resultant aggregates destroy their host. Thus, the protein aggregates constitute proteinaceous-only infectious agents, named *prions* by Prusiner,[50,61] that destroy tissue and ultimately the host.

A common theme runs through all of these injurious protein aggregates. They contain a β-structure composed of a relatively hydrophobic protein sequence.

7.5.1.3 General Comments on Our Proposed Physical Basis for Amyloid Formation

We believe that an understanding of the development of amyloid deposits resides within the development in Chapter 5 of the physical process underlying hydrophobic association dissociation. The simplest statement of the physical process is the presence of a competition for water between oil-like (hydrophobic) and polar (e.g., charged) groups of protein. The competition was expressed by Equation (5.13) in Chapter 5 as an apolar–polar repulsive free energy of hydration, ΔG_{ap}, on the basis of pKa shifts. The physical process of hydrophobic association/dissociation was also given expression in Equation (5.8) of Chapter 5 as a Gibbs free energy of hydrophobic association, ΔG_{HA}, which can be related to the shifts in the temperature at which the inverse temperature transition occurs.

The factors underlying protein insolubility diseases can be understood in terms of the same factors that control hydrophobic association. Loss of charged groups favors hydrophobic association, that is, favors protein insolubility. Obviously an increase in hydrophobicity is a shift toward insolubility. However, increased hydrophobicity makes unfavorable the presence of polar groups, that is, raises the free energy of even the peptide groups of the protein backbone.

An increase in hydrophobicity provides a driving force for hydrogen-bonded association of protein chains. In our view, this can occur under conditions where one would have otherwise thought that the peptide groups were already quite satisfied by their exposure to water. The first clear insight into this perspective came from the stretch-induced pKa shifts seen in our hydrophobically associated elastic model proteins (see Chapter 5, section 5.7.8.2). In this case even though stretching with increased exposure of hydrophobic groups increased the amount of water within the elastic matrix, carboxylates, COO^-, experienced less available water; they became water thirsty even though more water was in the elastic matrix. Accordingly, the water surrounding hydrophobic groups must be the wrong kind of water for hydrating polar groups such as carboxylates and, to a lesser but significant extent, peptide groups.

7.5.1.4 Segue to the Magnitude of the Medical Problem

Before pondering specific structural examples, however, a brief discussion of the medical significance of protein insolubility sharpens interest because of the increasingly common nature of the problem. Perhaps even more critically, relevance of the issue to our everyday lives becomes apparent on equating protein insolubility with the growing problem of Alzheimer's disease and current well-known victims, notably former U.S. President Ronald Reagan and most recently respected actor Charlton Heston. The fearsome side of protein insolubility also equates to the recent scares of mad cow disease and reports of the insoluble protein infecting humans who have eaten tainted beef. Then there are tragic inherited diseases in humans and in other animals and organisms that enhance the process of insoluble protein formation. It is our belief that the science developed in Chapter 5 provides the basis whereby these disease processes can be explained.

7.5.2 Magnitude of the Medical Problem of Excessive Protein Insolubility

7.5.2.1 Alzheimer's Disease

Currently in the United about 4 million people suffer from Alzheimer's disease. As the population demographics continue to shift with the percentage of people in the older groups growing most rapidly, the medical care problem becomes more demanding. It is estimated that nearly 50% of people over 85 years of age have Alzheimer's disease. Furthermore, in the western world Alzheimer's disease is considered the fourth leading cause of death.

The time course of the disease from diagnosis to death is from several years to two decades. Over that period, Alzheimer's disease presents a steady progression toward decreased mental faculties. Limited mental function commonly becomes apparent in the loss of ability to think through even simple mathematical problems. The debilitation then progresses to loss of recall of common numbers, like those of telephone numbers, dates, and addresses. The progression continues with loss of recognition of friends and family, with loss of coordination resulting in incapacity for personal care and speech, and with loss of awareness of surroundings.

This progressive dementia results from growing protein deposits, insoluble protein plaques (amyloid plaques), and insoluble protein neurofibrillary tangles in the brain that cause death of brain cells and loss of brain tissue.

7.5.2.2 Prion Diseases

Two decades ago Prusiner proposed protein particles as the infectious agent of scrapie,[62] the neurodegenerative disease of sheep.[60] He called the infectious protein a *prion*, standing for protein infectious only. The infectious agent was protein without involvement of pathogens, such as bacteria and viruses, and devoid of any nucleic acid. During the same period and working among the Stone Age Fore Tribe in Papua New Guinea, Gajdusek identified a similar neurodegenerative disease process in humans, called *kuru*.[63] The ritual of eating the brain tissue of deceased tribal members provided for transmission of the disease. This identified the occurrence of an infectious disease in humans with insoluble protein as the causative agent for transmission.

Collectively known as *transmissible spongiform encephalopathies* (TSEs),[64] the recognized transmissible prion diseases now include Creutzfeld-Jakob disease (CJD), Gerstmann-Sträussler-Scheinker disease (GSS), fatal familial insomnia (FFI), familial thalamic dementia, and kuru. Furthermore, animal TSEs include scrapies in sheep and goats, mad cow disease (bovine spongiform encephalopathy, BSE), feline spongiform encephalopathy (FSE), transmissible mink encephalopathy, and chronic wasting disease (CWD) in deer and elk.

These transmissible diseases, with insoluble protein as the infectious agent, would seem to have originated as mutations, and within this context they became inherited prion diseases. Creutzfeld-Jakob disease, GSS, and FFI number among those now recognized as inherited prion diseases in humans.

7.5.3 Phenomenology of Amyloid (Insoluble Protein) Deposits from Mad Cow Disease to Alzheimer's Disease

7.5.3.1 How Can a Natural Protein Become a Transmissible Infectious Agent?

A prion protein, designated as **PrP**, has a natural monomeric form, designated **PrP^C**, and an insoluble, phase-separated or associated form, designated **PrP^{Sc}**, where the superscript **C** stands for the normal cellular form and the superscript **Sc** stands for the infectious scrapie form. **PrP^C** is degradable by proteolytic enzymes, whereas **PrP^{Sc}**, the smaller component of **PrP^C**, forms a proteolytic resistant core.

The prion protein, therefore, exhibits a phase transition from a monomeric protein to an aggregated protein that bears analogy in fundamental respects to the phase transition exhibited by the model proteins discussed in Chapter 5. For the phase transition of our model proteins, association is relatively fast and dissociation or dissolution is quite slow. The difference with the prion phase transition is only qualitative in that the association step of the phase transition of prion protein is extremely slow, and dissociation is so slow as to be irreversible. The slow relentless growth of insoluble prion protein fibers continues until it destroys the cell or tissue with which it may be associated. This calls to mind sickle cell anemia due to an inverse temperature transition of hemoglobin S wherein intracellular fiber growth distorts and sickles red blood cells and leads to their destruction (see Figure 7.22).

A fundamental question of particular interest here becomes, is the prion aggregation step controlled by the same forces that control hydrophobic association? In other words, is the prion protein solubility–insolubility phase transition an inverse temperature transition? To find the answer, we again look for a coherence of phenomena characteristic of inverse temperature transitions.

7.5.3.2 How Infectious Can a Prion Protein Be?

Infectiousness of the insoluble fiber form of prion protein, **PrP^{Sc}**, derives from its capacity to induce the normal **PrP^C** form to change its three-dimensional shape and grow on the **PrP^{Sc}** template. Part of the severity of the medical problem is that autoclaving and other possible sterilization procedures are insufficient to destroy the fibrous **PrP^{Sc}** form.[65,66] The stability of the **PrP^{Sc}** form has resulted in transmission of CJD from contaminated, autoclaved surgical instruments, from donor dura mater, from donor corneas, and from human growth hormone—all materials associated in some way with neural tissue.

Transmission of prion diseases from one person to another through use of diseased human tissues becomes a particularly difficult problem, because the growth of the insoluble **PrP^{Sc}** form can require years before symptoms of the disease are apparent. The gestation period for prion diseases can be years, yet the **PrP^{Sc}** would seem to be infectious during that period.[67]

The epidemic of "mad cow disease" that resulted in the loss of upwards of 200,000 cattle apparently arose due to feeding cattle a meal that contained contaminated neural tissue. Animal feed containing nervous tissue is considered to be the source of prion diseases occurring in sheep and goats. There also appears to be cross-infectivity between species, as scores of humans have contracted a variant form of Creutzfeld-Jakob disease (vCJD) from eating contaminated beef.[68–70]

Having qualitatively noted the phenomenology of prion-related diseases arising from excursions into the realm of excess insolubility, we now look for correlations of phenomena that point to a consilient mechanism.

7.5.4 Mechanism of Amyloid Fiber Formation and Relentless Fiber Growth

7.5.4.1 Scenario of Hydrophobic Associations Based on Inverse Temperature Transitions and on the Apolar–Polar Repulsive Free Energy of Hydration, ΔG_{ap}

Whenever an opening fluctuation toward hydrophobic dissociation occurs, progression to dissociation continues only as long as too much hydrophobic hydration *does not* result. Should there be polar groups sufficiently proximal to the site of hydrophobic dissociation, such as charged groups able to compete for hydration and in doing so to destructure enough nascent hydrophobic hydration, then dissociation results. If instead too much hydrophobic hydration results during an opening fluctuation, the T_t-divide drops too far below the operating temperature, and hydrophobic association holds firm. This we believe is the origin of the insoluble **PrP**^{Sc} form that gives rise to formation of insoluble protein of the fibrous aggregates. The fibrous aggregates utilize a particular structural element in the aggregation process. It is a structural element different from that of hemoglobin S fibers. In our view and as a consequence of the consilient mechanism, the structural element is responsible for the extremely slow but relentless growth.

7.5.4.2 β-Structures: The Common Theme of Prion Protein Amyloid Fibers

7.5.4.2.1 The Infective Form of Prion Protein Is Aggregated β-structure

The experimental findings are that **PrP**^{Sc} forms insoluble aggregates with more β-structure and less α-helix than occurs in the **PrP**^{C} monomeric form of the protein. As shown in Figure 7.33, in the α-helical structure all peptide groups, –CONH–, within a single chain hydrogen bond within that chain.[71] This is the dominant conformation, in this case the primary structure, of myoglobin and hemoglobin shown in Figures 7.3, 7.4, and 7.10. The largely α-helical hemoglobin subunits interact by hydrophobic association and by hydrophobic association facilitated by ion-pair and other polar interactions, as also occurs during formation of hemoglobin S fibers of sickle cell anemia (see Figures 7.13 and 7.14).

As shown in Figure 7.34,[71] the association of parallel and antiparallel β-chains in the formation of β-sheets is exactly the inverse; all peptide groups form hydrogen bonds between chains. The consequences of this are significant when the protein chains have a composition within which exists substantial competition for hydration between hydrophobic (apolar) and polar (e.g., charged or peptide) groups. Of the three chains shown in the parallel and antiparallel β-sheets in Figure 7.34, all of the CO and NH components of the peptide groups in the central chain are hydrogen bonded. On the other hand, either the CO or the NH component of each peptide group of a β-chain at the growing edge of the sheet is not hydrogen bonded to another chain. It is generally thought, however, that interpeptide hydrogen bonding is energetically equivalent to peptide hydrogen bonding to water. With this perspective, therefore, from the standpoint of hydrogen bonding, there would be little to distinguish an edge chain from an inner chain. However, when the apolar–polar repulsive free energy of hydration described and developed in Chapter

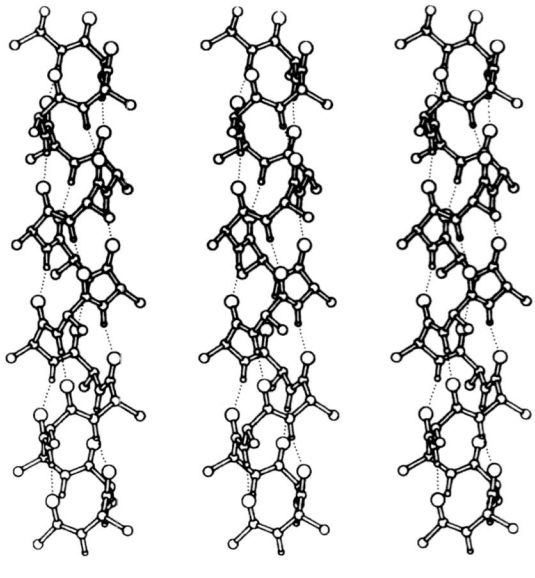

FIGURE 7.33. Stereo perspectives of the right-handed α-helical structure, with the lefthand pair for cross-eye viewing and the righthand pair for walleye viewing. Note that all peptide NH and CO groups are intrachain hydrogen bonded, except for three CO groups at the carboxyl end and three NH groups at the amino end of the structure. (Reproduced with permission from Urry and Luan.[71])

7.5 Grave Medical Consequences of Protein Insolubilities

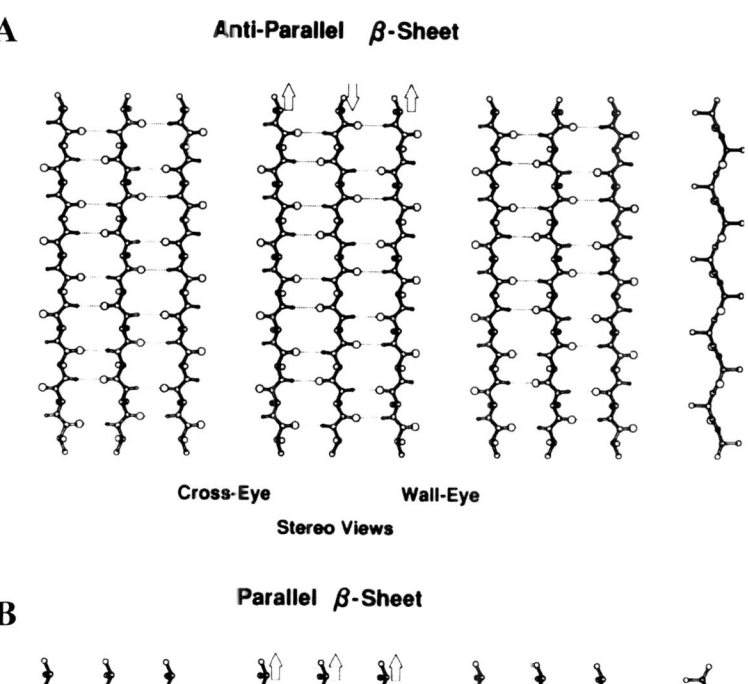

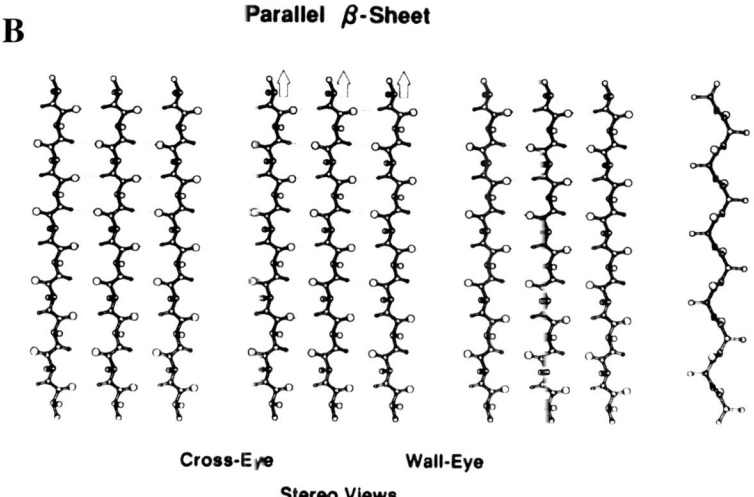

FIGURE 7.34. Stereo perspectives of antiparallel **(A)** and parallel **(B)** β-pleated sheet structures (three chains each), with the lefthand pair for cross-eye viewing and the righthand pair for walleye viewing. Open arrows indicate chain direction, antiparallel above and parallel below. Note that all peptide NH and CO groups are interchain hydrogen bonded, except for the CO and NH groups of growing edge chain that are directed into water. At right, single chain shown in side view. (Reproduced with permission from Urry and Luan.[71])

5 is considered, the circumstance can be quite different, as explained below.

7.5.4.2.2 The Peptide Unit Is a Polar Group in its Own Right

As shown in Table 5.1, the side chains of a protein are spread over the entire T_t-based hydrophobicity scale from the most polar carboxylate, COO^-, of the glutamate side chain, $-CH_2-CH_2-COO^-$, with a T_t of 250°C to the most hydrophobic side chain of tryptophan with a T_t of −90°C. On the polar side of the scale at 60°C is glutamine, with the side chain $-CH_2-CH_2-CONH_2$. Now, it is not unreasonable to assume that the peptide backbone group, –CONH–, would be in the same range of polarity as the glutamine side chain. Using the reference state of valine with a T_t-value of 25°C, the difference in T_t-values, ΔT_t, for glutamine is 35°C (60°C–25°C), and the ΔT_t for the most polar glutamate is 225°C. In this case one might estimate that a half dozen (225/35) peptide groups would sum to be as polar as a single carboxylate of glutamic acid.[72]

With either the CO or NH of each peptide group (but not both) of β-chain hydrogen bonded on one side to a second β-chain, about 12 peptides of a β-chain at the edge of a β-sheet would be equivalent to a single carboxylate. Regardless of the exact numerical equivalency, in our view there exists a competition for hydration between hydrophobic groups and a nonhydrogen-bonded CO or NH component of a peptide group. Thus, as further explained below, the nonhydrogen-bonded portion of a peptide group at the growing edge of a hydrophobic β-sheet could be a polar group starving for hydrogen bonding even though directed into what appears to be a sea of available water molecules.

7.5.4.3 How Can a Natural Protein in a β-sheet–containing Amyloid Fiber Function as a Seed Crystal to Induce Fiber Propagation That Is Ultimately Injurious to the Host Site?

We have seen hydrophobic-induced pKa shifts for carboxylates that were produced by systematically increasing hydrophobicity and due to stretching of hydrophobic elastic matrices containing carboxylates (see Chapter 5, section 5.6, and Figures 5.20B, 5.23, 5.29, 5.30, 5.31, and 5.34). The pKa shift arises due to inadequate hydration for the carboxylate—due to a competition for hydration between hydrophobic groups and charged carboxylate for water (see Figures 5.24 and 5.25 of Chapter 5). This has been called an *apolar–polar repulsive free energy of hydration*, ΔG_{ap} (see Equation [5.13] of Chapter 5).

The peptide group also is a polar group, and although it does not have an ionizable function to quantify its lack of adequate hydration, when part of a hydrophobic peptide, it is nonetheless at an unfavorable increase in free energy.[72] For an edge chain of a β-sheet, the high energy is relieved only by hydrogen bonding to an additional β-chain. *In this way the dominantly hydrophobic environs of the edge chain of a natural protein in a β-sheet–based fiber acts as a seed site for fiber propagation.*

7.5.4.4 Propensity of Prion Protein for Fiber Propagation: Correlation with the Consilient Mechanism

If the above mechanism, based as it is on an inverse temperature transition, is indeed responsible for the formation of infectious prion protein aggregates, then the variables that lower the value of T_t would be expected to favor fiber formation and infectivity. Increase in hydrophobicity and decrease in charge either by neutralization or by mutation would be expected to favor fiber formation and growth. Also as described above, because the free energy that favors the association of β-chains is described as an apolar–polar repulsive free energy of hydration, the presence of limiting hydration would lead to an increase in β-sheet propensity.

7.5.4.5 The T_t-based Single Residue Plot of the Human Prion Protein

7.5.4.5.1 Increase in Hydrophobicity by Replacement with a more Oil-like Residue, P → L, A → V

The effect of more hydrophobic mutants demonstrates an increased propensity for the insolubility of fiber formation. For example, "The substitution of amino acids in the mutants, e.g., 3A → V (at positions 113, 115, and 118) and P101L, enhances the folding of the peptide into compact structural units, significantly enhancing the formation of the extensive β-sheet fibrils."[73] The Pro (P) to Leu (L) mutant[74] both enhances hydrophobicity and removes the kink due to a proline residue that would interfere with a continuous β-chain structure. In particular, consideration of the 55 residue

7.5 Grave Medical Consequences of Protein Insolubilities

neurodegenerative core of mouse PrP, namely PrP(89–143), sequence and its P101L mutant

GQ GGGTHNQWNK PSKPKTNLKH VAG AAAAGAV VGGLGGYMLG SAMSRP MIHF GSD
GQ GGGTHNQWNK LSKPKTNLKH VAG AAAAGAV VGGLGGYMLG SAMSRP MIHF GSD

shows that the P101L mutant "is completely converted into β-sheet, suggesting that formation of a specific β-sheet structure may be required for the peptide to induce disease."[75] The equivalent sequence in the human prion protein is indicated by the bar between arrows in Figure 7.35, which also shows the sites of amino end and carboxyl end post-translational cleavage (the open arrows).[76]

Thus the β-sheet structure may be identified as the infective, self-propagating, structural element, and we believe the argument to be compelling that the driving force for self-propagation derives from the special coupling of this structural feature with apolar–polar repulsive free energy of hydration, ΔG_{ap}, that increases with increased hydrophobicity.

7.5.4.5.2 Observations on the Special Role of Water

In our view the special role of water in the inverse temperature transition sets the stage for the fiber growth and stability. It might be anticipated that removal of water by lyophilization, which achieves total water removal, would destabilize and fragment β-sheet structures more so than drying under ambient conditions that would leave much bound water in place. Interesting experimental data have been reported on this feature. "Samples of SHa106–122 that formed assemblies while drying under ambient conditions showed X-ray patterns indicative of 33 Å thick slab-like structures having extensive H-bonding and inter-sheet stacking. By contrast, lyophilized peptide that was equilibrated against 100% relative humidity showed assemblies with only a few layers of β-sheet."[73]

7.5.4.5.3 The Human Prion Protein with Related Mutations Causative in Gerstmann-Sträussler-Scheinker Disease

An analogous mutation in the human prion protein, P102L, to the above P101L causes GSS

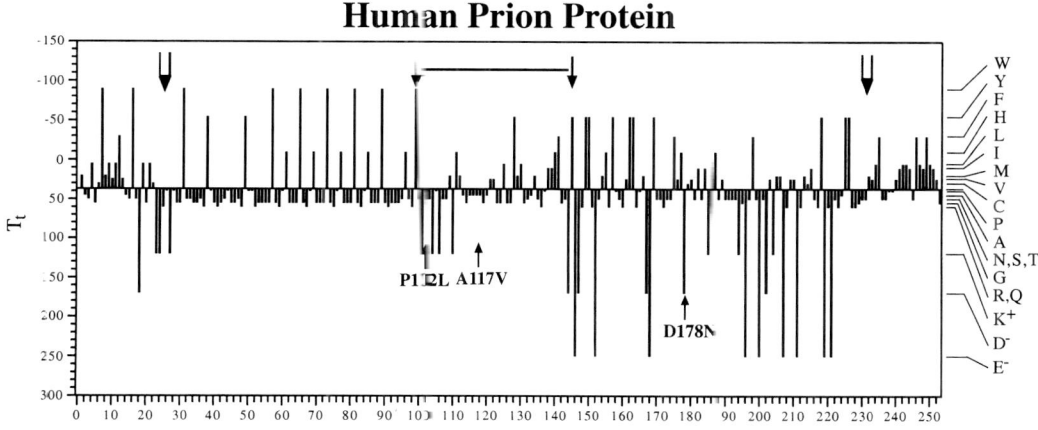

FIGURE 7.35. Human prion protein: single residue hydrophobicity plot. Arrows separated by bar indicate 55 residue sequence that completely forms a β-sheet. Open arrows indicate normal post-translational cleavage sites. Lower arrows indicate disease-causing mutations that replace more polar residues with more hydrophobic residues. See text for discussion. (Sequence obtained from Harper and Lansbury.[106])

disease as does the A117V mutation.[70,73] The positions of these mutations are indicated by the upward-directed arrows in Figure 7.35 for the human prion protein.

7.5.4.5.4 A Mutation Causative for Fatal Familial Insomnia

The mutation responsible for the human prion disease FFI, decreases charge on the human prion protein in a sensitive sequence. The mutation D178N converts a charged aspartic acid residue to an uncharged asparagine. This too is indicated in Figure 7.35. Again, the removal of the D178 stand-alone carboxylate opens up the second longest apolar sequence to add to β-sheet formation. If the mutation were instead D202N, we would expect no significant effect on amyloid formation. The remainder of polar residues in the 196 to 221 sequence would dominate and prevent assembly of hydrophobic β-chains into a β-sheet.

7.5.4.5.5 Low pH and Ion Pairing Enhance Insolubilization of Fiber Formation

Lowering the pH of the **PrP**^{Sc} form in the range of 4.4 to 6 favors amyloid formation.[77,78] Low pH, however, appears to have no effect on the conformation of the carboxyl-terminal portion of the **PrP**^{Sc} form, but low pH does facilitate incorporation of the amino-terminal portion into amyloid. The few residues with carboxylates in the 140 to 170 relatively hydrophobic sequence could experience hydrophobic-induced pKa shifts as demonstrated in Figures 5.30 and 5.34 of Chapter 5 and become uncharged carboxyls.

7.5.4.5.6 Insight from Electron Crystallography for a Working β-structure of Scrapie Prion Protein

Inadequate crystallinity in the amyloid deposits of the **PrP**^{Sc} form of human prion protein or of any of its mutated and truncated forms limits the details of the β-structure available from X-ray crystallography. A recent report of electron crystallography of two-dimensional crystals of the proteinase K–treated **PrP**^{Sc} form giving essentially the sequence from residues 90 to 231, however, is available.[79] The model focusing on the β-structure appears in top view in Figure 7.36A and in side view in Figure 7.36B. The

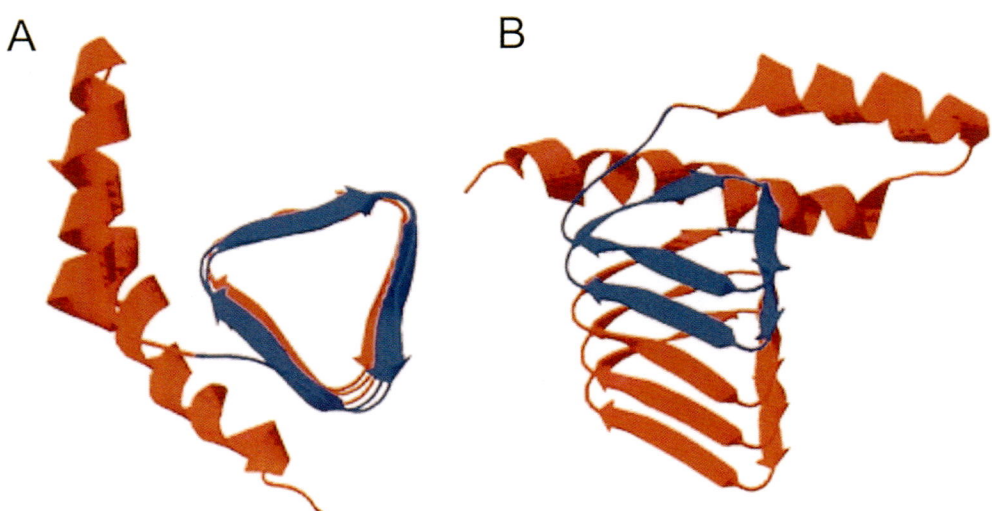

FIGURE 7.36. Possible structural model of scrapie prion protein from electron crystallography studies; partial view shows triangular arrangement of parallel β-pleated sheets. **(A)** Top view. **(B)** Side view. Complete top view perspective is shown in Figure 7.37. (Reproduced with permission from Wille et al.[79] Copyright 2002 National Academy of Sciences, U.S.A.)

7.5 Grave Medical Consequences of Protein Insolubilities

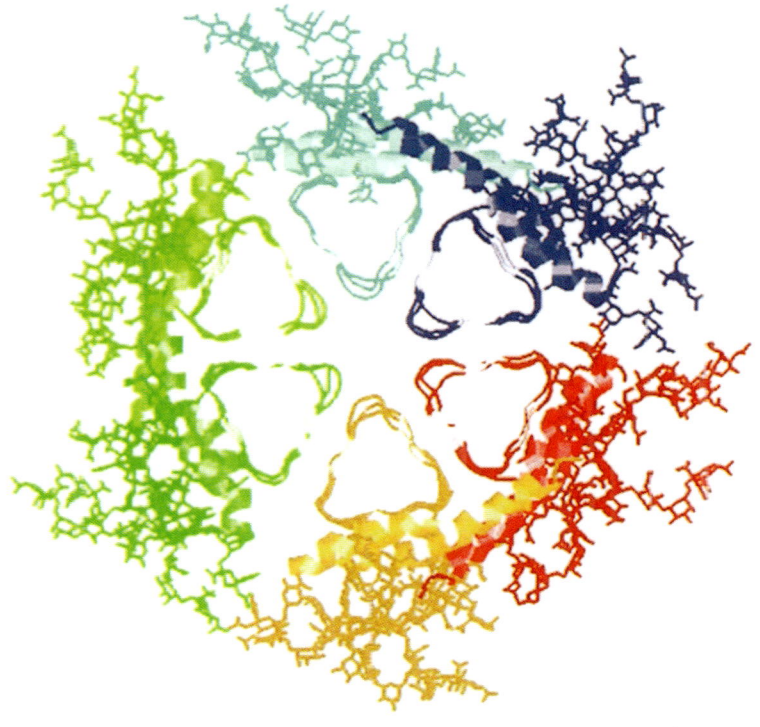

FIGURE 7.37. Possible structural model of scrapie prion protein from electron crystallography studies, showing full sixfold symmetric top view with triangular arrangement of parallel β-pleated sheets at the center surrounded by additional partially α-helical sequences. (Reproduced with permission from Wille et al.[79] Copyright 2002 National Academy of Sciences, U.S.A.)

broad line or band ending in an arrow indicates a β-chain in three segments forming a triangular arrangement with intermolecular hydrogen bonding between parallel-aligned β-chain segments. Thus one side of the triangular structure would represent a parallel β-sheet as laid out in Figure 7.34B.

In Figure 7.37 six such structures are packed in hexagonal fashion. The triangular-shaped β-structures reside in a proteolytic resistant core with α-helical and additional sequences decorating the outside. One might wonder how such a structure could be consistent with the reported extraordinary thermal stability.[65] Surely the hexagonal packing would not withstand such high temperatures. Given such a structure, fiber growth could occur by adding individual units in a direction perpendicular to the plane in Figure 7.37. One possibility could be that an individual fibril might form with a 120° rotation on addition of each unit.

7.5.4.6 The Seminal Amyloid-forming Sequence of Alzheimer's Disease

The single residue T_t-based hydrophobicity plot for the amyloid precursor protein (APP) is shown in Figure 7.38. APP is a cell surface protein of unknown function that becomes degraded to β-amyloid peptide. The representative amyloid plaque forming sequence is indicated as Aβ-amyloid (1–42), and it is located in the APP sequence by the labeled bar in the vicinity of residue 700. From what we have discussed earlier in this chapter for identifying sequences having a propensity for hydrophobic association, recognition of the most prominent apolar sequence in APP is straightforward; it is the sequence in the range of residues 690 to 730. This is the sequence that we would identify *a priori* as the most likely sequence to form β-amyloid plaque; it is indeed the Aβ-amyloid. Again, it would seem that the T_t-based

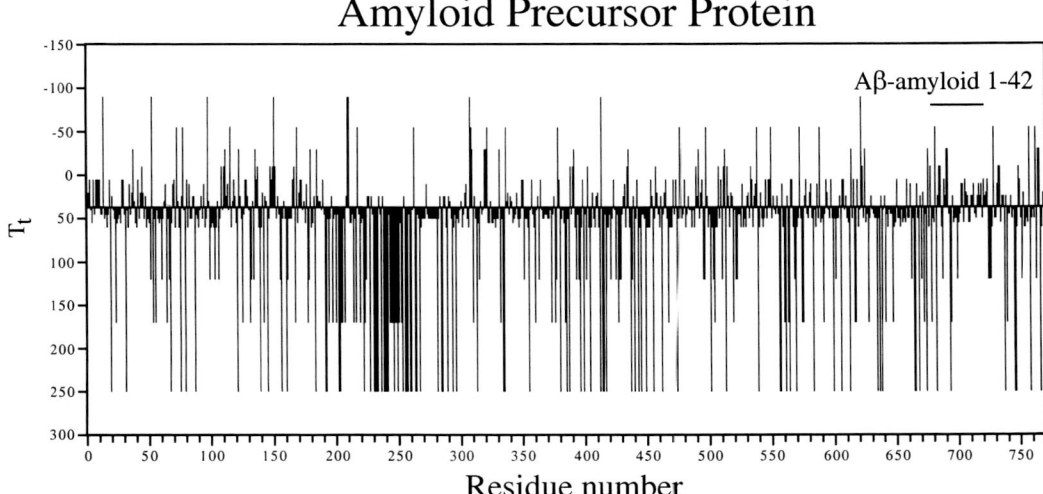

FIGURE 7.38. Amyloid precursor protein: Single residue T_t-based hydrophobicity plot. Bar indicates Aβ-amyloid (1–42), an amyloid fiber–producing sequence, the longest, most apolar β-structure–producing sequence of the protein.

hydrophobicity plots provide insight into the propensity of protein sequences for hydrophobic association.

7.5.4.7 Amyloidosis

These considerations apply to the broader set of proteins with the tendency to form β-structure–containing aggregates, a process designated in general as *amyloidosis*. A single example is briefly noted.

7.5.4.8 Lysozyme Amyloidosis

Presence of the αβ **fold** occurs in lysozyme, characterized by hydrophobic association between an amphiphilic helix and a hydrophobic side of a β-sheet. The β-sheets are thought to associate and become the proteolytic resistant core as the α-helical regions become cleaved by proteolysis.[80]

7.6 Molecular Chaperones: Biology's Effort to Overcome Improper Insolubility

Molecular chaperones perform general housekeeping functions in the cell, which include catalysis of the unfolding and correct refolding of misfolded proteins with external hydrophobic patches that should be internalized, of protein transport, and of disaggregation of protein conglomerates. These molecular chaperone machines commonly involve two or more proteins that work together to achieve their housekeeping roles.

Improperly hydrophobically folded proteins occur whenever spatially localized hydrophobic folds have formed wherein thermal fluctuations leading to opening of the fold result in an overwhelming amount of hydrophobic hydration that prevents dissociation. In short, the T_t-divide of Figure 5.3 or the cusp of insolubility of Figure 7.1 is too far below the operating temperature. The localized hydrophobic association cannot on its own unfold and refold correctly. However, if a hydrophobically misfolded protein could be placed within a cage lined with charges, by the consilient mechanism the protein would unfold. For the particular case of chaperonin, seven of the ultimate in charged groups, ATPs, bind in a positively cooperative interaction in the process of repulsing hydrophobic domains from the interior of the cage and in the process expose other polar species. The result becomes interior walls lined with charges that destructure the hydrophobic hydration of the fluctuating misfolded protein and result in an unfolding. Within the cage surrounded by polar species, the T_t-divide, or the cusp of insolubility for the hydrophobic associations for correct protein folding, is substantially above the operating temperature. Then the cusp of insolubility must be lowered in a carefully controlled manner to sort out the correct hydrophobically folded structure. Thus, by the consilient mechanism, *ATP binding to the chaperonin raises the cusp of insolubility for protein folding above physiological temperature to give solubility, and the subsequent stepwise removal of P_i lowers the cusp of insolubility step by step, allowing for a controlled refolding of the spatially localized hydrophobic domains as ΔG_{ap}, the apolar–polar repulsive free energy of hydration between wall and protein substrate, abates.*

7.6.1 Molecular Structure of Chaperonins

7.6.1.1 Structure of Escherichia coli GroEL: A Homo-oligomer with 14 Subunits of 57,000 Da Each

The 14 subunits arrange as two heptameric rings, that is, with 7 units having a C7 symmetry axis, related to the second 7 units by a twofold dyad axis. This results in two water-filled cavities of 85,000 Å^3 each.[81] Each of the 14 subunits consists of three domains—an equatorial domain, an intermediate domain, and an apical domain. Seven ATP molecules bind to the 7 equatorial domains at a position near the intermediate domain and, in our view by means of ΔG_{ap}, propagate dissociations of hydrophobic domains (including their intrinsic ion pairs) and domain rotations in the intermediate and apical domains due to apolar–polar repulsions. The result is a water-filled cavity with its size doubled to 175,000 Å^3.[81]

7.6.1.2 Structure of E. coli GroES

Seven GroES, 10,000 Da protein molecules, bind coincident with ATP binding at one end of the D7 GroEL structure to cap the structure and enclose a water-filled cavity of 175,000 Å^3 with the capacity to contain protein or peptide substrate inside of a size up to 70 kDa.

7.6.1.3 Crystal Structure of One State of the GroEL/GroES Chaperonin

Figure 7.39A shows a space-filling representation of the crystal structure of one state of the GroEL/GroES chaperonin with a ring of 7 subunits of apo-GroEL (without ligand) at the bottom, a second ring of 7 subunits of (ADP)$_7$–GroEL in the middle, plus a cap of 7 subunits of GroES.[82] This single structure exemplifies the change in structure that attends nucleotide binding (the middle ring) as the bottom ring of 7 GroEL subunits is without any bound nucleotide. The purpose of the added Z-shaped lines is discussed below.

7.6.2 Substrate Considerations

7.6.2.1 Substrates for Chaperonin

Substrates of improperly folded protein are characterized by having exposed hydrophobic surfaces that bind in the groove between the H and I helices of the apical portion of GroEL. GroES subunits displace such snagged substrates inward in a concerted process with ATP binding. At this entrance to the internal chamber of apo-GroEL, there are nine residues, eight hydrophobic and one Ser available for hydrophobic association with the non-native polypeptide. In general, any protein state with significant exposure of hydrophobic surface could be expected to bind to chaperonin at this apical site. A common substrate for GroEL is protein containing an αβ fold where a hydrophobic side of an α-helix associates with a hydrophobic side of a β-sheet. This is reminiscent of the types of protein structures that on partial hydrolysis to remove α-helix give rise to amyloidosis, a β-sheet of hydrophobically associated protein chains common to prions, as discussed above.

7.6.2.2 A Structure for Snagging the Substrate

The asymmetrical structure in Figure 7.39A provides an opportunity to visualize how a ring of seven nonliganded GroEL, that is, an apo-GroEL, subunits snags its substrate. This

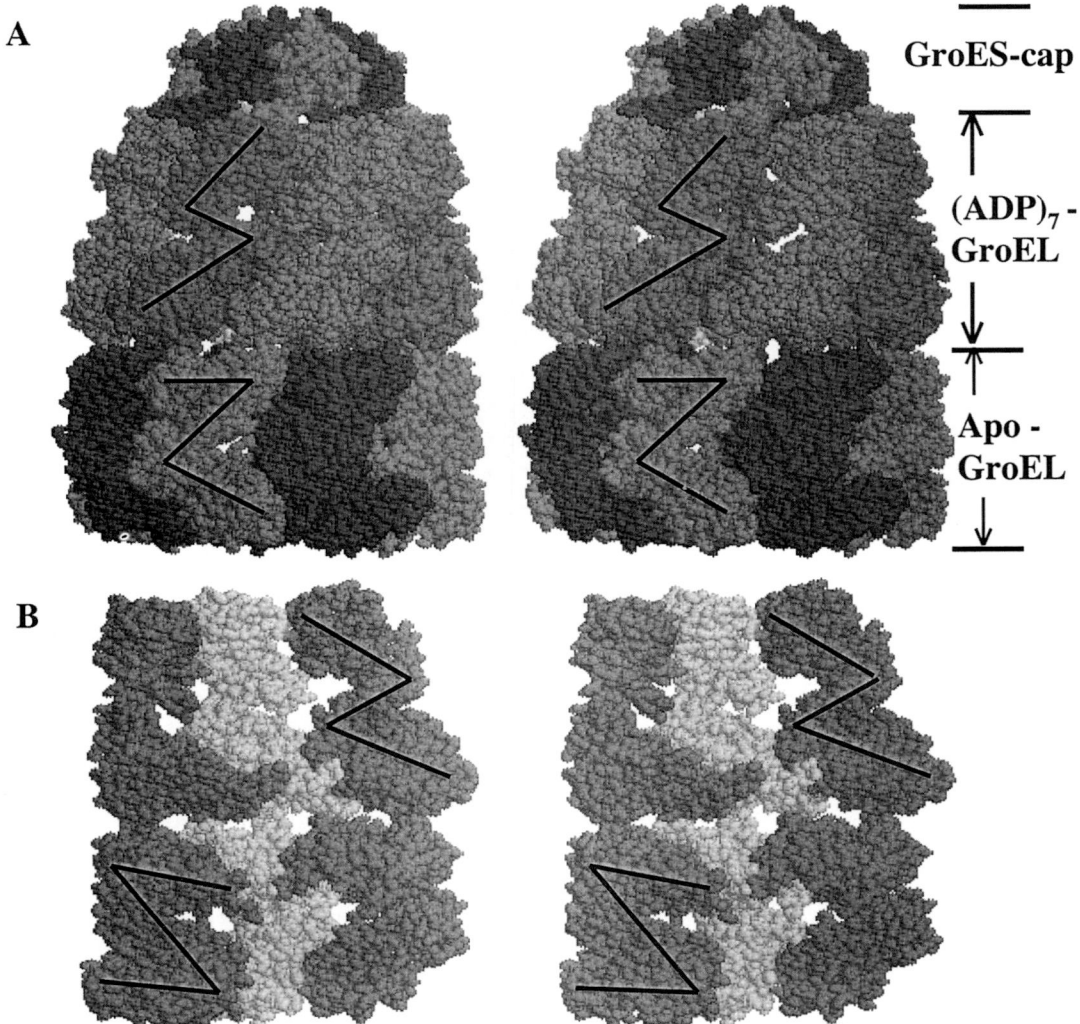

FIGURE 7.39. Stereo view of the crystal structure of the asymmetric GroEL–GroES-(ADP)$_7$ chaperonin complex. **(A)** Exterior view of all 21 subunits—7 GroES subunits at the top, 7 *cis* ADP-GroEL subunits in between, and 7 *trans* GroEL at the bottom. **(B)** Internal view of 3 *cis* ADP-GroEL and 3 *trans* GroEL subunits. The difference between ADP-GroEL and GroEL subunits is shown by the extension of a three-dimensional Z shape on ligand binding. (Prepared using the crystallographic results of Xu et al.[82] as obtained from the Protein Data Bank, Structure File 1AON.)

becomes possible by taking a slab, by cutting away the front half of the asymmetrical structure, and looking inside. In the slab stereo view of the asymmetrical structure in Figure 7.40A, the aromatic and the remaining hydrophobic residues are given as dark gray to gray, and the neutral and charged residues are white. The entrance to the bottom ring, the apo-GroEL ring without bound adenosine di- or triphosphate, is seen to be dominantly hydrophobic. It is this hydrophobic entryway that snags the misfolded substrate with its exposed hydrophobic surface elements. A more detailed look at this surface is considered later in Figures 7.43 and 7.44.

7.6.2.3 Changes in the Interior Surface on Ligand Binding and Capping

Importantly, the interior surface of the apo-GroEL ring contains more hydrophobic residues than the interior surface of the $(ADP)_7$–GroEL ring. The change from a somewhat hydrophobic surface to a highly charged surface on binding nucleotide phosphate is shown more effectively in Figure 7.40B, where the charged residues are now shown as the red residues and the hydrophobic and neutral residues are shown as white. Although there are charged surface residues on the interior sides of both GroEL rings, in this gross view it becomes very clear that a much greater proportion of the

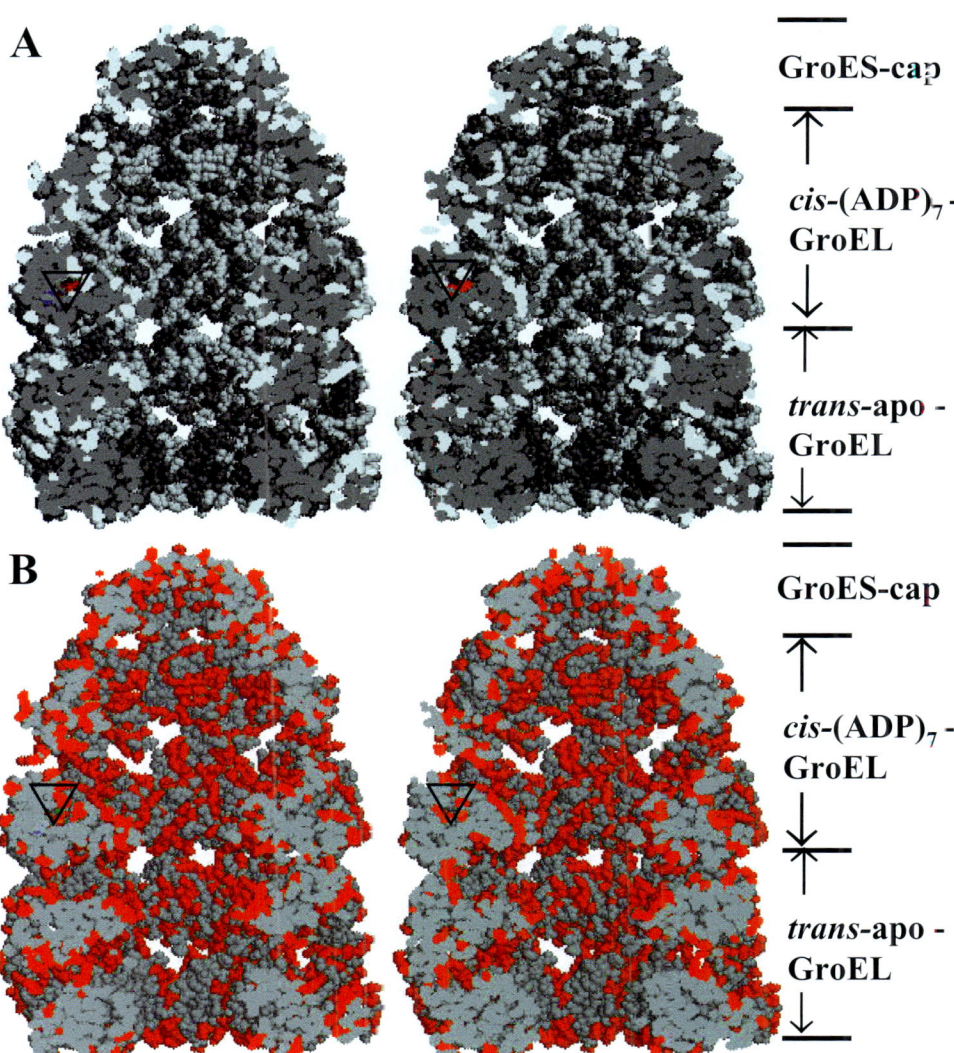

FIGURE 7.40. Slab stereo views with the front half of the structure cut away to show the interior half of the crystal structure of the asymmetric GroEL–GroES-$(ADP)_7$ chaperonin complex. Triangles locate ADP at the base of an aqueous cleft, with the phosphate tail directed toward the aqueous chamber. (A) Hydrophobic residue distribution: gray (hydrophobic), dark gray (aromatic), and neutral and charged residues white. (B) Charged residues red and all remaining residues light. (Prepared using the crystallographic results of Xu et al.[82] as obtained from the Protein Data Bank, Structure File 1AON.)

surface contains charged residues once bound to nucleotide phosphate and capped with GroES. After reviewing the mechanical steps of this two-cycle unfolding–refolding machine immediately below, the physical basis as seen through the perspective of the consilient mechanism will be considered at further molecular detail.

7.6.2.4 Summary of Domain Mechanics During an ATP-mediated Unfolding–Folding Cycle (see Figure 7.41)

Step 1. Substrate of misfolded protein binds at the *cis* apical region by hydrophobic association to the H and I helices and the loop region.

Step 2. ATP and GroES bind with positive cooperativity to the *cis* side of the ring containing the misfolded protein, displacing it from the apical binding site and releasing it into an aqueous cage that, during the process, increases in size from $85,000 \text{ Å}^3$ to $175,000 \text{ Å}^3$ (see Figure 7.41).[81] In the resulting large polar cavity, called the *Anfinsen cage*,[83] the caged protein unfolds, and the nucleotide is trapped in a way that ensures complete hydrolysis of ATP.

Step 3. The seven *cis* ATP molecules are hydrolyzed to ADP and inorganic phosphate, P_i with stepwise release of P_i.

Step 4. The unfolded polypeptide refolds to the native or a more nearly native state during stepwise release of P_i, but does not always complete proper refolding in a single cycle.

Step 5. Misfolded polypeptide attaches to the *trans* side and is joined by ATP and GroES as in step 2 above, but this ATP binding exhibits negative cooperativity and causes the release of seven ADP, GroES, and properly folded polypeptide on the cis side. The *trans* side then goes through steps 2 through 4. If proper refolding has not occurred on the *cis* side, the exposed hydrophobic groups of the as yet improperly folded protein will again be snagged at the apical region on the *cis* side for a repeat cycle. If proper folding has occurred, the native protein will escape, and a new misfolded protein with exposed hydrophobic patches will again bind on the *cis* side followed by steps 2 through 4 on the *cis* side.

7.6.2.5 Recap of Cooperativity

The binding of ATP to the first side of the D7 (C7 + C2 symmetry) symmetrical structure of GroEL exhibits positive cooperativity, that is, the first ATP binds weakly and then each subsequent ATP binds more tightly. The binding of ATP to the *trans* side, while there remains ADP in the seven sites and GroES on the *cis* side, shows negative cooperativity, that is, the first ATP binds more tightly, and each subsequently bound ATP binds more weakly in the process of displacing the ADP and GroES of the *cis* side.

7.6.3 Insights Provided by the Consilient Mechanism into Chaperonin-Directed Hydrophobic Unfolding and Refolding

The consilient mechanism was born out of controlling the hydrophobic association–dissociation of elastic-contractile model proteins to achieve the possibility of some 18 classes of pairwise energy conversions (see Chapter 5, section 5.6). In the process a set of five Axioms became the phenomenology out of which the consilient mechanism arose. For the first time "a common groundwork of explanation"[84] was able to perform the diverse energy conversions of biology.

Application of the consilient mechanism to chaperonins, therefore, seems most fitting, because the role of chaperonin is to hydrophobically unfold and refold proteins that have hydrophobically misfolded. The problem divides into two parts functioning under the same dominant interaction, changes in hydrophobic association within the chaperonin itself and the hydrophobic unfolding and refolding of the substrate. By the consilient mechanism, phosphate is the most effective chemical grouping known for disrupting hydrophobic association. In the GroEL chaperonin seven phosphate groupings, as seven ATPs, ring the protein to be unfolded and

7.6 Molecular Chaperones: Biology's Effort to Overcome Improper Insolubility

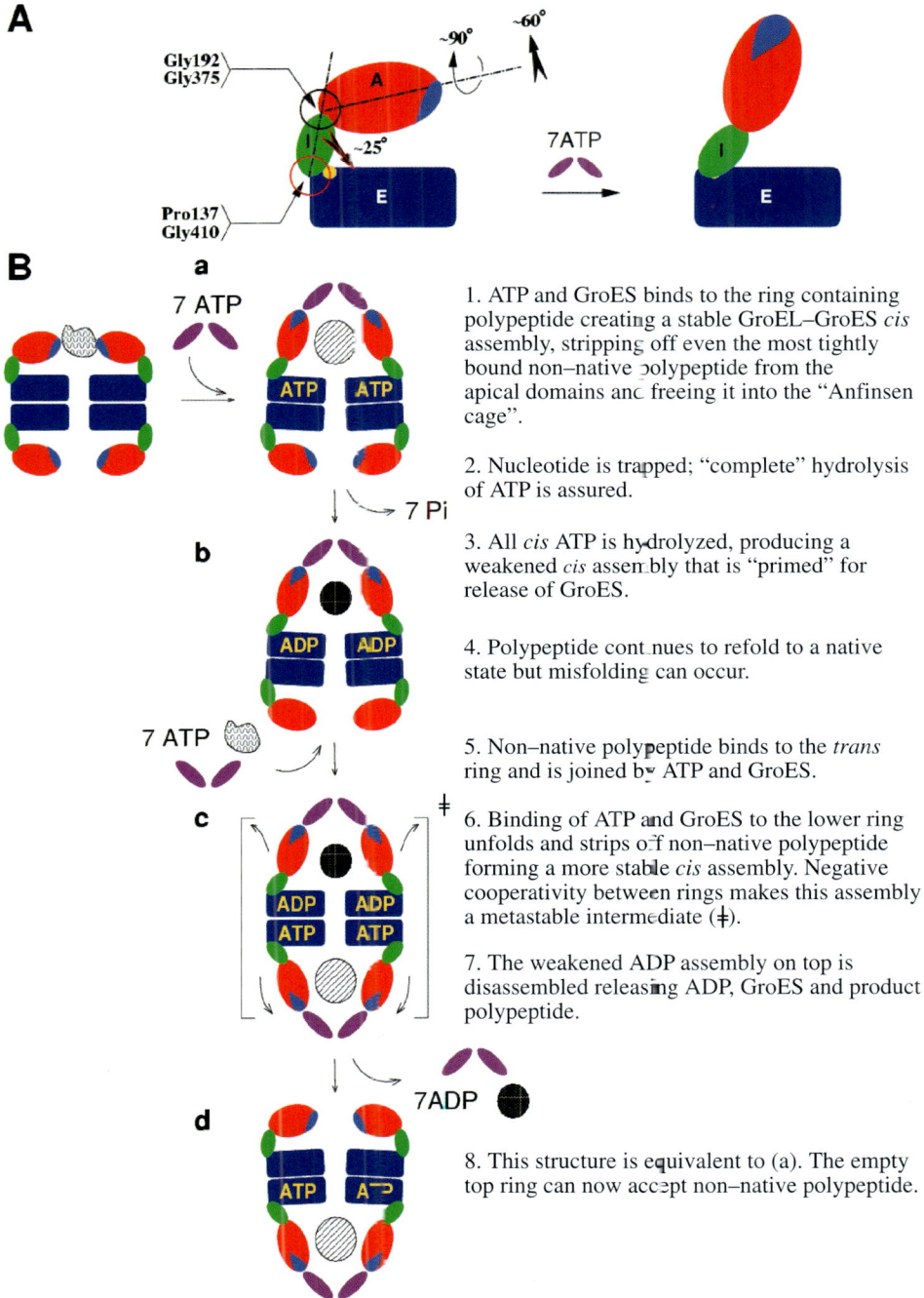

FIGURE 7.41. Schematic representation of the chaperonin refolding machine composed of a twofold GroEL (body)/GroES (cap). Misfolded protein unfolds in polar cavity due to seven ATPs and properly refolds as seven P_i are released stepwise. See text for discussion. (Reprinted from Journal of Structural Biology, Vol. **124**, G. Xu, P.B. Zigler, GroEL/GroES: Structure and Function of a Two-stroke Folding Machine, 129–141, 1998,[81] with permission from Elsevier.)

subsequently refolded. Obviously, the stage is set. The apolar–polar repulsive free energy of hydration, ΔG_{ap}, becomes the specific physical process of focus (see Chapter 5, sections 5.1.7.4, 5.3.3, and 5.7.9). However, do the molecular details support the contention that ΔG_{ap} drives the changes in GroEL structure and dominates the interaction of the GroEL/GroES structure with hydrophobically misfolded substrate?

7.6.3.1 ATP Binds with its Triphosphate Tail at an Internal Aqueous Cleft

As has become classic for ATPases, which are the proteins that hydrolyze ATP during function, ATP binds at the base of an aqueous cleft. Even more to the point, ATP binds with its triphosphate tail specifically directed toward the protein–aqueous interface. As shown in Figure 7.42A, the triphosphate tail can be seen when looking at the wall from within the aqueous chamber.[85] The triphosphate tail in the equatorial domain of GroEL is positioned to obtain hydration from within the aqueous chamber. In so doing the polar phosphate tail orients the water molecules in the cleft and beyond in a manner unsuited for hydrophobic hydration. Accordingly, while many water molecules are seen in the crystal structure (see Figure 7.42B) and may be called "waters of Thales," most of the waters involved in function are those within the enclosed large aqueous chamber of some $175,000 \text{Å}^3$. The water molecules of the large aqueous chamber, however, are too mobile to be located by crystallographic means.

By the apolar–polar repulsive free energy of hydration, ΔG_{ap}, which results from the competition for hydration between hydrophobic and charged species, the water molecules within the chamber, nonetheless, take on a preferred orientation in which they are properly oriented toward the charges that develop over the interior surface during the concerted process of nucleotide binding and capping with GroES. As explained below on the basis of the consilient mechanism, such interior water does not readily form hydrophobic hydration as the trapped and misfolded protein substrate begins to unfold. The result is that hydrophobic dissociation within the substrate dominates as long as the nucleotide binding sites are filled with ATP and the chamber is capped with GroES.

7.6.3.2 Recap of the Thermodynamic Basis for Hydrophobic Association/Dissociation in Water

The key aspect to keep in mind for hydrophobic association in water is that it is equivalent to insolubility of hydrophobic groups in water. Remarkably, the 1937 report of Butler[31] contains the fundamental insight. Methanol (CH_3–OH) is miscible with water in all proportions, and the process of dissolving methanol in water is exothermic. Heat is released, and the combination of water and methanol results in a favorable state of lower free energy. Dissolution of ethanol (CH_3–CH_2–OH) is even more exothermic, releasing approximately 1.4 kcal more heat per mole on dissolution than does methanol. So too for n-propanol (CH_3–CH_2–CH_2–OH); its dissolution releases some 1.4 kcal more heat per mole on being dissolved in water than does ethanol. The same holds for each CH_2 group added, according to the Butler data, up to n-pentanol (CH_3–CH_2–CH_2–CH_2–CH_2–OH). For the methanol to pentanol series as each CH_2 group is added, the average change in heat (ΔH) is -1.4 kcal/mole-CH_2. (The negative sign indicates the release of heat.) Therefore, as far as the release of heat is concerned, formation of hydrophobic hydration is a favorable reaction. Indeed, hydrophobic hydration occurs, it is a fact and has been seen in crystal structures (see, for example, Figure 2.8). However, we also know that n-octanol, CH_3–CH_2–CH_2–CH_2–CH_2 CH_2–CH_2–CH_2–OH, is insoluble in water, despite the fact that the heat released increases for each CH_2 group added.

The governing thermodynamic expression for solubility is $\Delta G(\text{solubility}) = \Delta H - T\Delta S$, and as long as $\Delta G(\text{solubility})$ is negative, solubility occurs. Therefore, the reason for decreased solubility and ultimate loss of solubility as the number of CH_2 groups increases must be due to a positive ($-T\Delta S$) term. The formation of the ordered hydrophobic hydration from less ordered bulk water results in a negative ΔS due to formation of water more ordered than bulk

7.6 Molecular Chaperones: Biology's Effort to Overcome Improper Insolubility

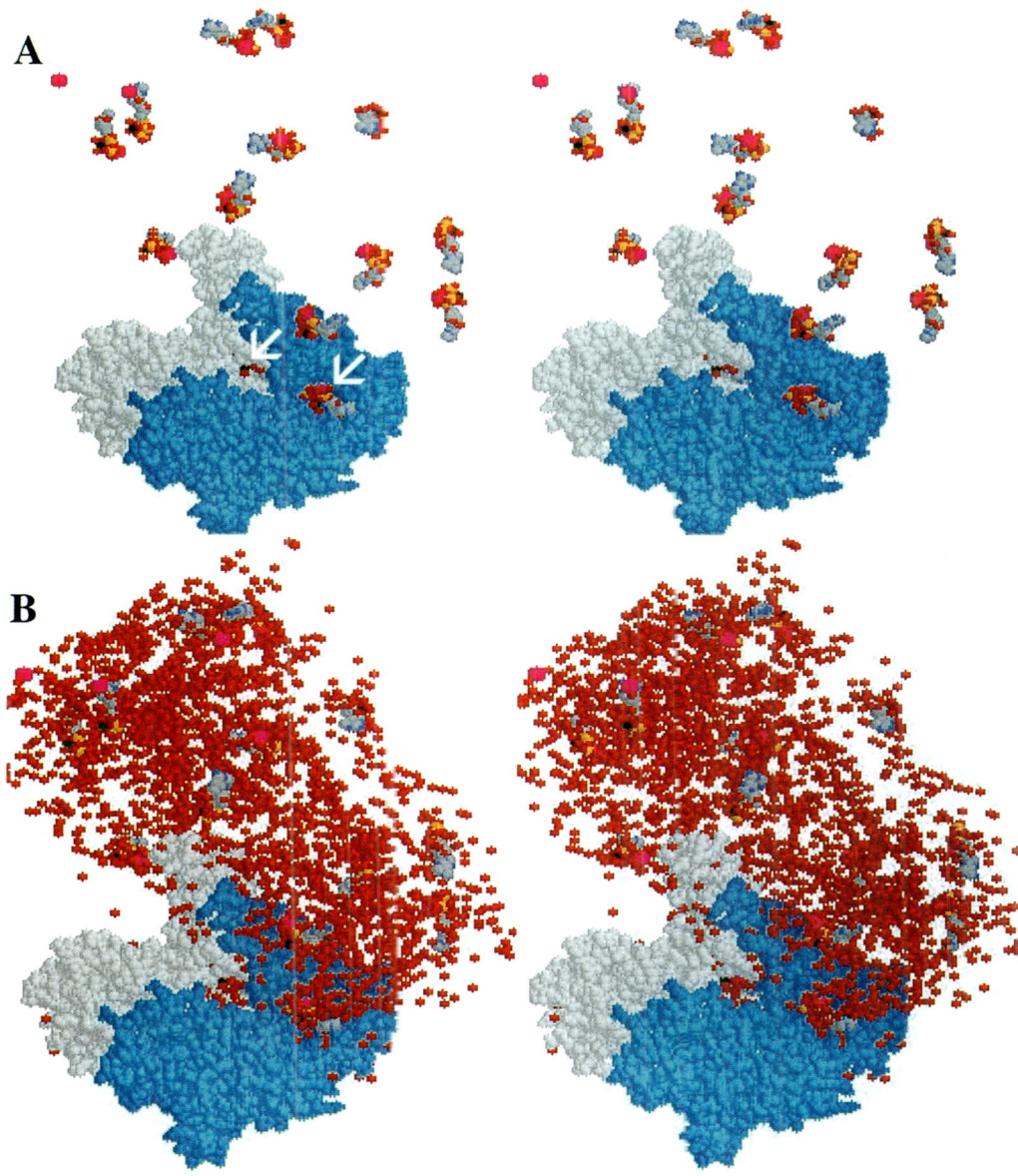

FIGURE 7.42. **(A)** Stereo view of chaperonin structure, [(ATP)$_7$GroEL)]$_2$, with all 14 ligands (KMgATP), but no GroES cap exists in the structure. It specifically shows two adjacent subunits from an angle within the cage in order to see the exposed triphosphate tails (indicated by the white arrows) and the inside of the large aqueous cage. As in general with other ATPases (proteins that hydrolyze ATP), the triphosphate tail occurs at an aqueous cleft that opens out into the large aqueous chamber. **(B)** Crystallographically reported "waters of Thales" within and between the adjacent equatorial domains of the two C7 donuts. The overwhelming "waters of Thales" reside within the large aqueous chamber. With all seven ATP molecules in place within a single donut and the dissociated ion pairs they effect, the charged groups all cooperate in disrupting emergent hydrophobic hydration as they orient water for their own hydration and thereby unfold the hydrophobically misfolded protein. (Prepared using the crystallographic results of Wang and Boisvert,[85] as obtained from the Protein Data Bank, Structure File 1KP8.)

water, such that the $(-T\Delta S)$ term is indeed positive. Now, for a pair of associated hydrophobic domains, a transient opening fluctuation causes the build up of hydrophobic hydration, but when so much hydrophobic hydration has occurred that the positive $(-T\Delta S)$ term becomes larger than the negative ΔH term, ΔG(solubility) becomes positive, which means insolubility. During a transient dissociation of a pair of hydrophobic domains hydrophobic association recurs, when too much hydrophobic hydration forms.

On the other hand, by the consilient mechanism, if a proximal charged group should appear during the opening fluctuation, then it recruits the water of hydrophobic hydration for its own hydration. With the development of insufficient hydrophobic hydration due to the presence of proximal charge during the opening fluctuation, hydrophobic dissociation stands. Thus the task of the chaperonin is to develop sufficient proximal charge to allow dissolution of the hydrophobically misfolded protein substrate. Then proper hydrophobic refolding would take place during careful withdrawal of the charge, as would occur on hydrolysis of ATP to form ADP and P_i and release of the highly charged inorganic phosphate. Does the structure of the GroEL/GroES chaperonin accommodate such a mechanism?

7.6.3.3 In the Absence of Ligand Binding, Charges Can Be Buried as Ion Pairs Within Complementary Hydrophobic Domains

Some of the unseen charges in the interior of the lower ring, the *trans*-apo-GroEL ring of Figure 7.40B, occur as ion-pairs. Due to the influence of proximal hydrophobic domains the separated ions lack adequate hydration and lower their free energy by forming ion pairs. Wang and Boisvert[85] have noted some ion-pair separations, namely,

Upon binding of ATP, the hydrophobic property of the apical domain surface is reduced. In the apo structure[86,87] many charged residues (K207, K226, R231, K272, E216, E252, E255, and E257) are involved in inter-subunit interactions and are interlocked at the apical–apical domain interface. In this structure, many of these interactions are substantially weakened.... This change weakens the electrostatic interactions among a large number of charged residues (K207, K226, R231, K272, E216, E252, E255, and E257) at the interface.

The ion pairs often constitute complementary hydrophobic surfaces, where one hydrophobic surface contains a negative charge and the other contains a positive charge, and the hydrophobic surfaces structurally fit such that the ions pair on association. When a proximal charge causes dissociation of the complementary hydrophobic surfaces, the separated charges that emerge from the previous ion pairs now contribute as separated charged species to propagate hydrophobic dissociation to greater distances from the initial source of charge at the ATP binding site. *In this way, ATP binding propagates ion-pair separation, extending the reach of ΔG_{ap} to greater distances.*

7.6.3.4 Repulsion Between Bound ATP and the Hydrophobic Domain Responsible for Positive Cooperativity and for Apical Domain Rotation

A more direct use of ΔG_{ap} brings in abundant charge. ΔG_{ap} results from the competition for hydration between hydrophobic and charged groups. Therefore, charged groups and hydrophobic groups achieve the state of lowest free energy when they each access water undisturbed by the other. This translates into an apolar(hydrophobic)–polar(charge) repulsive free energy of hydration. Thus, if ATP were to bind at a site where the ATP must compete with a hydrophobic domain for hydration, there would occur a repulsion that makes binding less favorable and that applies a force for moving the domain.[88] Because the location of the ATP binding site occurs at one side of its subunit, as shown in Figures 7.42A and 7.43B, binding of the first ATP must also contend with the hydrophobic domain of the neighboring subunit. Binding of the second ATP in that neighboring subunit would contend with less repulsion so that it binds with a higher affinity. The work of binding the second ATP has already been done, in part, by the binding of the

7.6 Molecular Chaperones: Biology's Effort to Overcome Improper Insolubility 313

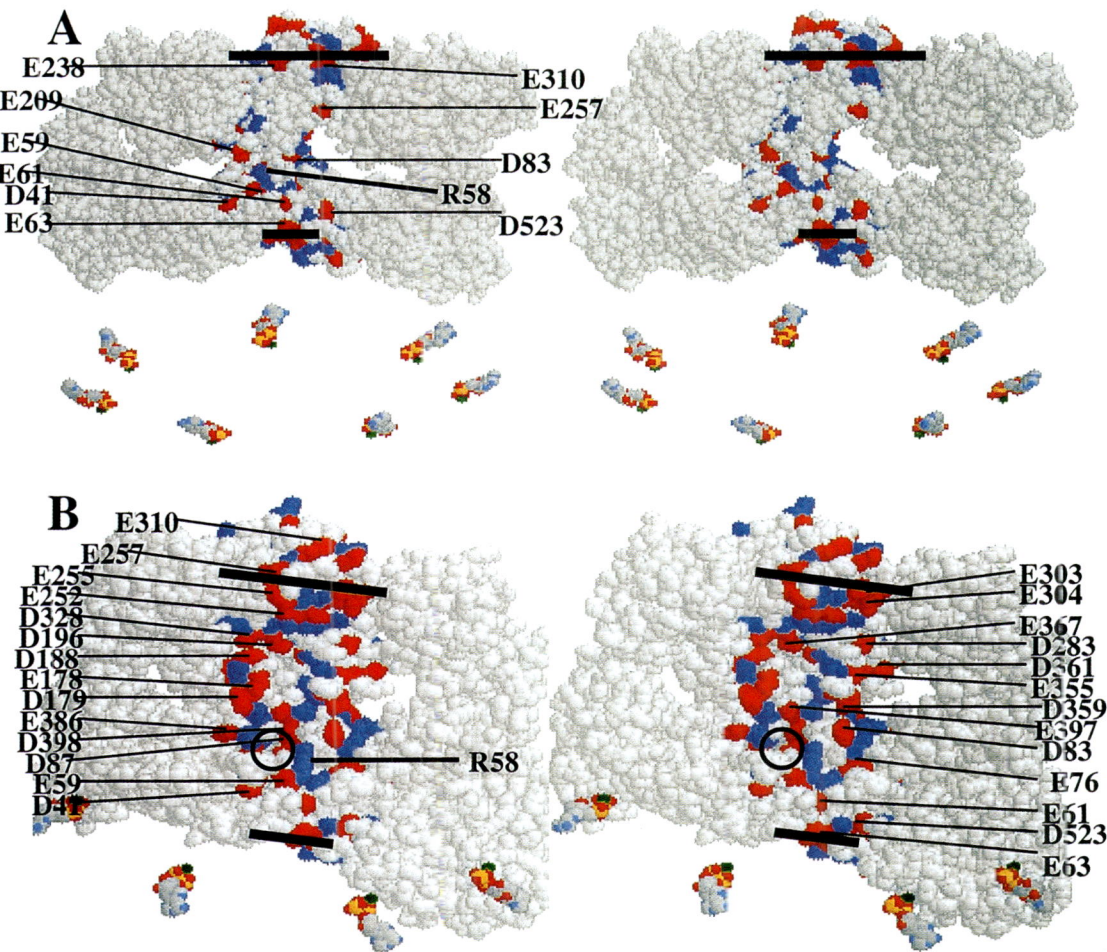

FIGURE 7.43. Stereo views of the interior sides of three subunits of the asymmetric GroEL structure showing charged residues for the central subunit. The red residues are the glutamic acid (E) and aspartic acid (D) residues with negative carboxylates, and the blue residues are for lysine and arginine residues that contain positive charges. The heavy bars set the limits of the interior chamber. R58 provides a general landmark for locating the binding site in subsequent figures, whereas D87 and D 398, when observable as in part B, provide carboxyl oxygen atoms for coordination of Mg^{2+} with oxygen atoms of the nucleotide phosphate tail. (A) GroEL in the absence of nucleotide showing small interior chamber with fewer charged residues. Eleven residues with negative carboxylates could be considered to line the interior wall that contains many hydrophobic residues as shown in Figure 7.40A. This figure is the same as that in the lower part of Figure 7.40 and indicated as $trans$-Apo-GroEL, but has been turned upside down for a more ready comparison with the cis-$(ADP)_7$-GroEL section in part B. (B) The cis-$(ADP)_7$-GroEL section of Figure 7.40 with ADP located at the circle where D87 and D398 carboxylates share Mg coordination with α- and β-phosphates of ADP. Approximately 24 carboxylate-containing residues line the interior chamber. Due to space limitations, only the negatively charged residues are labeled. (Prepared using the crystallographic results of Xu et al.[82] as obtained from the Protein Data Bank, Structure File 1AON.)

first ATP. This provides the physical basis for positive cooperativity.

Repulsion between hydrophobic surfaces of the apical domain and ATP molecules in their binding sites within each subunit also provides a driving force for rotation of the apical domains. The repulsion between ATP in the binding site of its subunit and the hydrophobic domain of the adjacent subunit would drive the rotation in a clockwise rotation as seen looking from the end of the apical domain toward the intermediate and equatorial domains. The rotation moves the hydrophobic residues into position for intersubunit hydrophobic association and, at the same time, brings charged residues from the sides of aqueous pores into the chamber formed by seven GroEL subunits. From the structure and the perspective of the consilient mechanism, the ΔG_{ap} developed on ATP binding would appear to contribute to both the apical and intermediate domain rotations.

Thus, the apolar–polar repulsive free energy of hydration drives both domain rotation and the separation of ion pairs during the separation of complementary hydrophobic surfaces. The result becomes apparent on comparison of the two interior surfaces in Figure 7.43. The most effective residues contributing to ΔG_{ap} are the carboxylates. Because of this and in order to keep the comparisons in Figure 7.43 from being too cluttered, only the carboxylate-containing residues are labeled. In short, there are more than twice as many carboxylates attributable to the surface of the aqueous chamber after the nucleotide and GroES have bound. Also, the magnitude of the rotation of the apical domain becomes apparent on noting the positions of E257 and E310 in Figure 7.43A before rotation with their locations in Figure 7.43B after rotation.

7.6.3.5 Estimate of Change in ΔG_{HA} of Inner Wall on Nucleotide Binding and GroES Capping

Differential scanning calorimetry data of the inverse temperature transition for hydrophobic association of a series of model proteins, reported in Chapter 5, allowed calculation of the relative values of the Gibbs free energy of hydrophobic association, ΔG_{HA}, for the side chain of each amino acid residue. Table 5.3 lists the values of ΔG_{HA} with the glycine (Gly, G) residue taken as zero. When the residues seen from the central aqueous chamber of the apo-GroEL ring are listed, as labeled in Figure 7.44, the summation over ΔG_{HA} for the labeled residues yields a relatively hydrophobic value of approximately −10 kcal/mole-subunit of internal surface as seen from the aqueous chamber. On the other hand, when the summation is taken over the residues labeled in Figure 7.45, which is the surface of the large aqueous chamber after nucleotide binding and GroES capping, that is, the interior surface of $(ADP)_7$–GroEL/GroES, the summation over ΔG_{HA} reports a very polar surface of greater than +50 kcal/mole-subunit of internal surface as seen from the aqueous chamber. Clearly, nucleotide binding and GroES capping create a very polar surface.

7.6.3.6 View of the ATP Binding Site Along Inner Wall from the Top of the Apical Domain

Figure 7.46 demonstrates continuity from bound nucleotide to the full length of the polar surface. Looking from the top of the apical domain downward along the inner face of the chamber exposes the α- and β-phosphates of ADP at the base of the classic ATPase cleft. Thus, ATP binds the equatorial domain at the aqueous interface of an aqueous cleft. From this position, by means of the apolar–polar repulsive free energy of hydration, ΔG_{ap}, seven ATP molecules working in a positively cooperative manner transform the uniquely poised structure to create a remarkably polar face that reaches far into the large chamber for adequate hydration. How then can such a structure perform the task of unfolding and properly refolding hydrophobically misfolded proteins?

7.6.4 Chaperonin Design Exemplifies the Consilient Mechanism

The design of the GroEL/GroES molecular chaperone presents a clear statement of the

7.6 Molecular Chaperones: Biology's Effort to Overcome Improper Insolubility

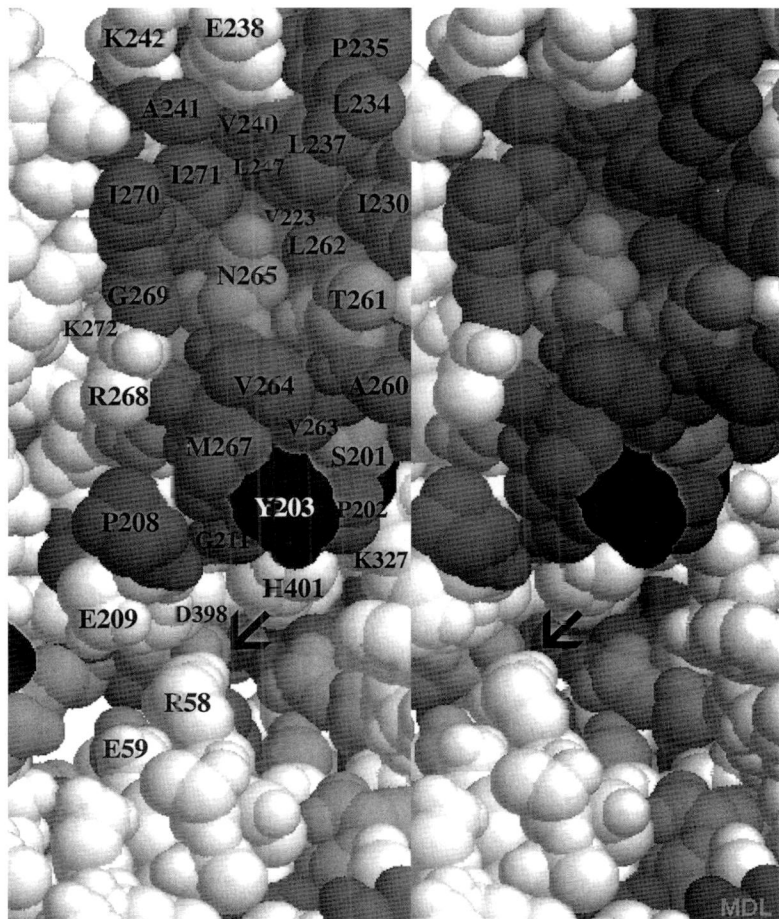

FIGURE 7.44. Stereo view of a subunit of Apo-GroEL with the black arrow pointing to the location on the equatorial domain where the ATP triphosphate tail binds. Because of the large presence of hydrophobic residues, ATP would be expected to bind with a substantial apolar–polar repulsive free energy of hydration, ΔG_{ap}. This could drive clockwise rotation of the apical domain that on combination with ATP binding at other subunits would give rise to positive cooperativity. (Prepared using the crystallographic results of Xu et al.[82] as obtained from the Protein Data Bank, Structure File 1AON.)

consilient mechanism. The binding of seven ATP molecules to a chaperonin surrounds the hydrophobically misfolded protein with a cage of charge. Why? The structural change on ATP binding states simply that sufficient proximal charge solubilizes hydrophobic groups, that is, charge disrupts hydrophobic association. To the best of our knowledge, only the consilient mechanism, with a fundamental element being the competition for hydration between charged and hydrophobic groups, explains the phenomenon of what has been called the *Anfinsen cage*. Only the experimentally calculable apolar–polar repulsive free energy of hydration, ΔG_{ap}, embodies this concept. ΔG_{ap} contributes to the Gibbs free energy of hydrophobic association, ΔG_{HA}. Independent experimental determinations show that under the proper experimental variable, a change in ΔG_{HA}, determined from the change in heat of an inverse temperature transition due to an increase in hydrophobicity, and a change in ΔG_{ap}, due to a change in pKa resulting from the same change in hydrophobicity, yield the

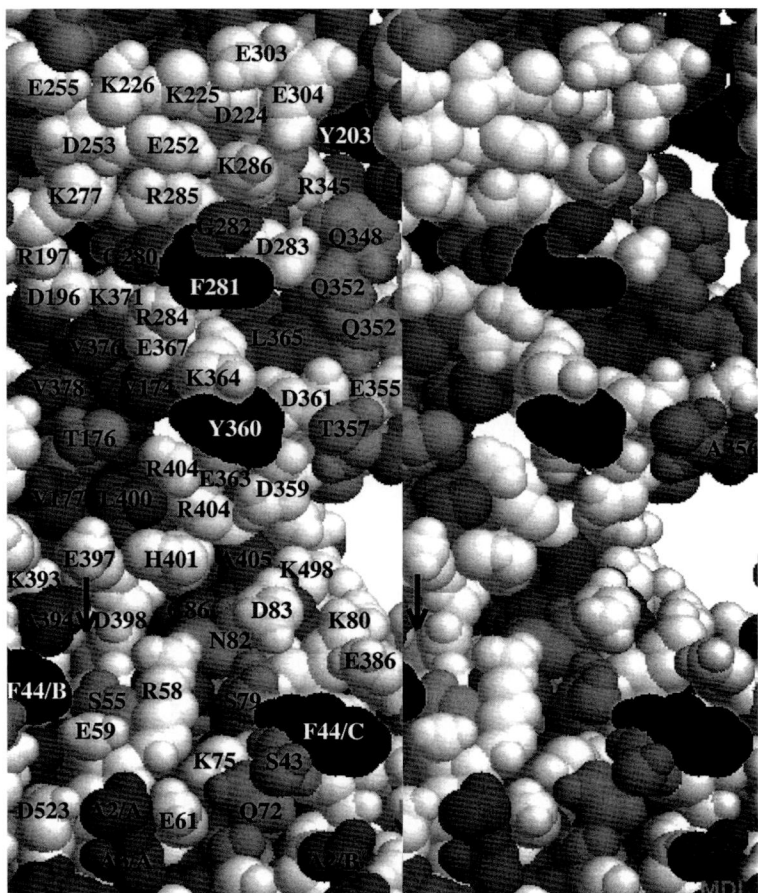

FIGURE 7.45. Stereo view of the inside of an asymmetric cis-GroEL–GroES-(ADP)$_7$. Residues D87 and D398 coordinate Mg^{2+} (along with α- and β-phosphates of ADP), and residue R58 combines to serve as a reference point; the black arrow points the direction to the ADP binding site. Viewed from within the large aqueous cavity, approximately at right angles to the view in Figure 7.46. Figure prepared using the crystallographic results of Xu et al.[82] as obtained from the Protein Data Bank, Structure File 1AON.

same change in kcal/mole (see Chapter 8, section 8.1.8.1.9).

7.6.4.1 Explicit Mechanism for the Function of an Anfinsen Cage

In 1973, Anfinsen[83] stated the principle that the sequence of a protein determined the resulting full three-dimensional structure. The rationale for calling the large aqueous cage that forms upon binding of seven ATP molecules to the GroEL/GroES molecular chaperone the *Anfinsen cage* was that this structure provided the near ideal environment whereby the Anfinsen principle would be unfailing. However, as noted above, there is a physical reason why an aqueous cavity surrounded by very polar walls causes the unfolding of hydrophobically misfolded protein and why decreasing the polarity by stepwise removal of the most polar group, phosphate, would facilitate proper hydrophobic folding. It is the physical process of competition for hydration between hydrophobic and polar (especially charged) groups that causes hydrophobic unfolding and the relaxation of that competition by a decrease in charge

7.6 Molecular Chaperones: Biology's Effort to Overcome Improper Insolubility 317

that allows for proper refolding. That physical process was demonstrated with care by a series of experimental results in Chapter 5, and it was called the *apolar–polar repulsive free energy of hydration* for the reasons given above.

Of course, a visualization of this GroEL/GroES Anfinsen cage is incomplete without a representation of the GroES cap of the cage.

The dome of the Anfinsen cage is shown in Figure 7.47, and, to us, it seems to be every bit as much of an architectural attraction as the domes of the ancient cathedrals and the modern sports arenas. The actual portion that constitutes the inside enclosure contributed by the dome represents a relatively small area of the inside wall, but additional chains fold over and into pores in the GroEL part of the

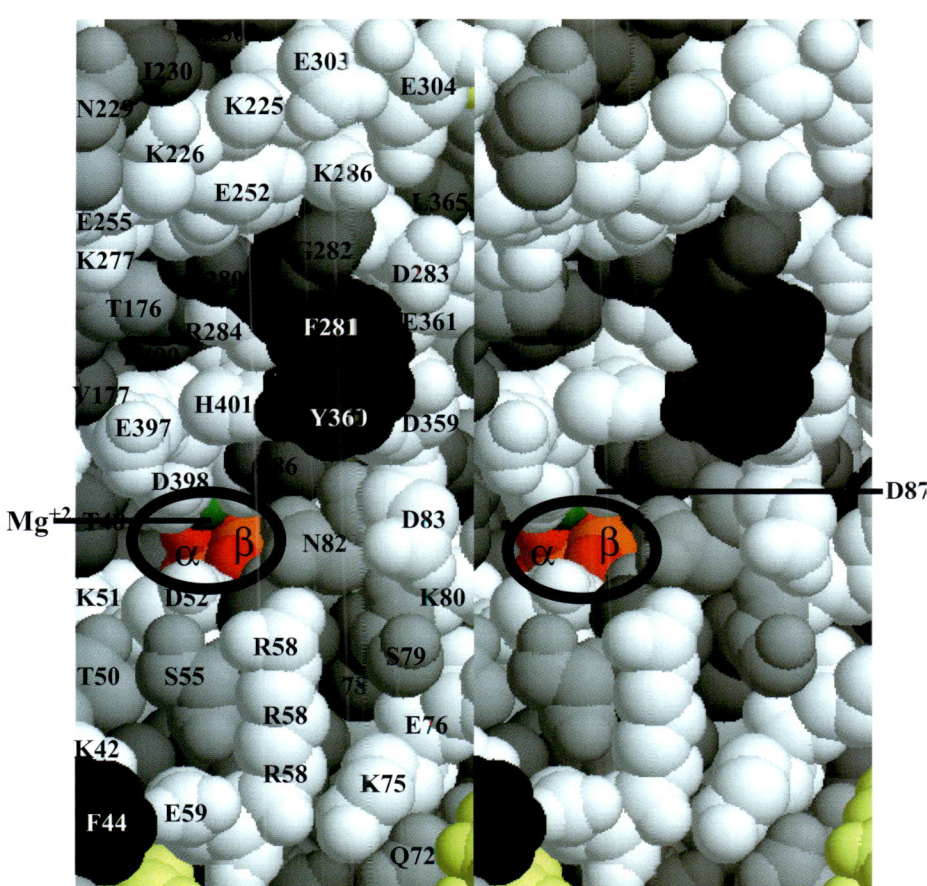

FIGURE 7.46. Stereo view of the ADP binding site of asymmetric (ADP)$_7$-GroEL showing the α- and β-phosphates of ADP classically positioned at the base of an aqueous cleft, that is, this figure provides a view of the ADP binding site looking down at the interior floor of the large aqueous cavity. The α- and β-phosphates within the circle and the D87 and D398, all of which provide coordination for the Mg ion, accurately define the nucleotide binding site. Note, therefore, that the carboxylates of the D87 and D398 residues are not in position to contribute as directly to the polar surface and are not included in the summation over ΔG_{HA} for the surface. Note also that inclusion of residue R58 provides easy reference to the location of the nucleotide binding site for the structures of Figures 7.43A, 7.44, 7.45, and 7.46. (Prepared using the crystallographic results of Xu et al.[82] as obtained from the Protein Data Bank, Structure File 1AON.)

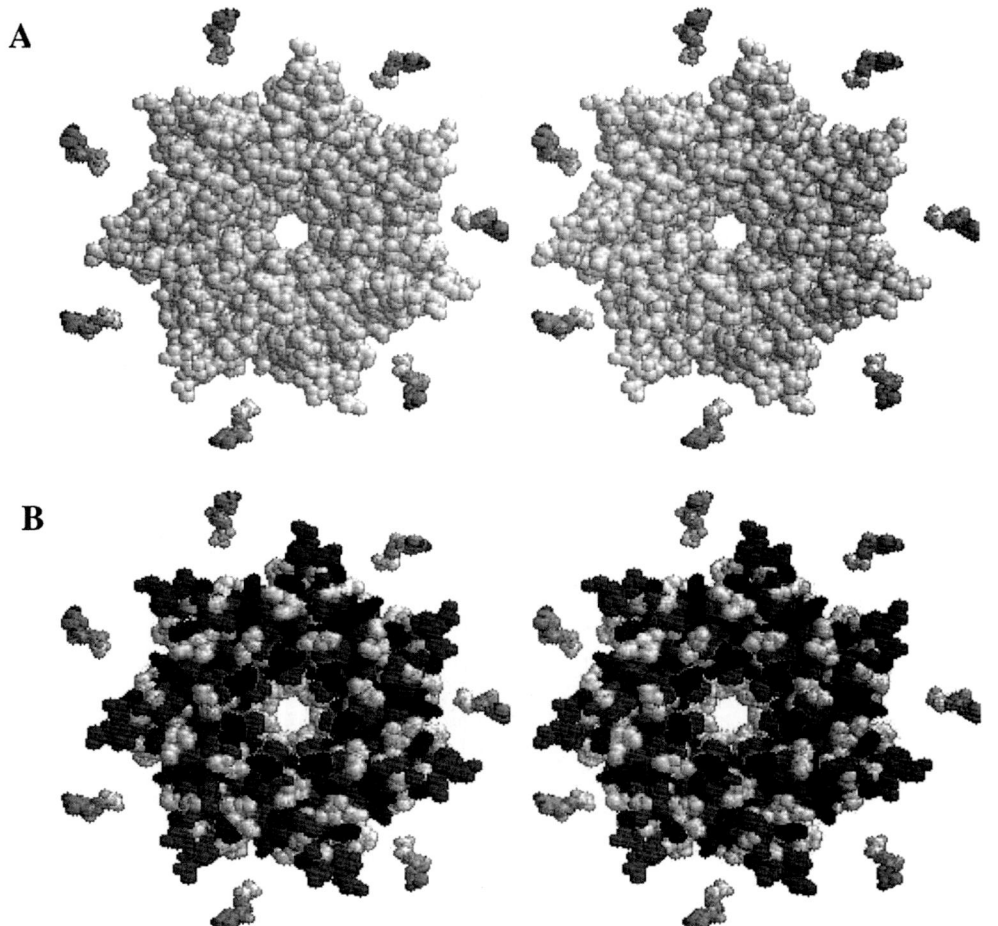

FIGURE 7.47. Stereo view of the aqueous chamber side of the (ADP)$_7$-GroEL–GroES cap, which may be called the *dome* of the Anfinsen cage for protein folding. ADP molecules were included to provide orientation. **(A)** Sevenfold symmetric structure in white relief. **(B)** Sevenfold symmetric structure with black, dark gray, gray, light gray, and white representing aromatic, acidic, basic, other hydrophobic, and neutral residues, respectively. The dome provides a very polar surface as demonstrated with the perspective in Figure 7.37. (Prepared using the crystallographic results of Xu et al.[82] as obtained from the Protein Data Bank, Structure File 1AON.)

chamber, as shown in Figure 7.48. Interestingly, a summation over ΔG_{HA} for a comparison of the relative hydrophobicity, as done for a subunit of the walls indicated in Figures 7.44 and 7.45, indicates a very polar cap topping the aqueous chamber that is dominated by five E, three D, and three K residues and sums to approach +30 kcal/mole-subunit for a relatively small surface area.[89]

7.6.4.2 Effect of ATP Binding and GroES Capping on the Substrate's Movable Cusp of Insolubility

ΔG_{HA} for the apo-GroEL wall of the chamber was approximated to be about –10 kcal/mole subunit inner surface, as estimated for the surface shown in Figure 7.44. ATP binding and the resulting emergence mostly by rotation of

7.6 Molecular Chaperones: Biology's Effort to Overcome Improper Insolubility

charge on the interior surface raises the temperature of the movable cusp of insolubility to hydrophobically unfold the substrate. ΔG_{HA} for the inner wall of the (ADP)$_7$-GroEl/GroES state approximates to be more than +50 kcal/mole subunit inner surface, as estimated by the surface shown in Figure 7.45. With an average radius to the inner wall of about 4 nm[82] and with a mean diameter of the misfolded protein of a 2 to 3 nm diameter, this means that the misfolded protein would come within a nanometer of the wall. This places the misfolded protein within the estimated reach, for the apolar–polar repulsive free energy of hydration, found for the elastic contractile model proteins of Chapter 5. The apolar–polar repulsive free energy of hydration, ΔG_{ap}, would tend to center the hydrophobically misfolded protein substrate, but Brownian motion and transient openings would facilitate escape from the metastable hydrophobically misfolded state.

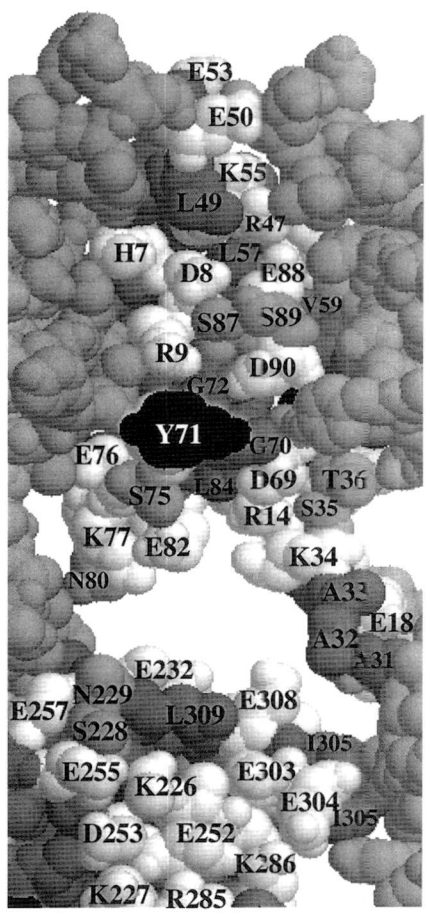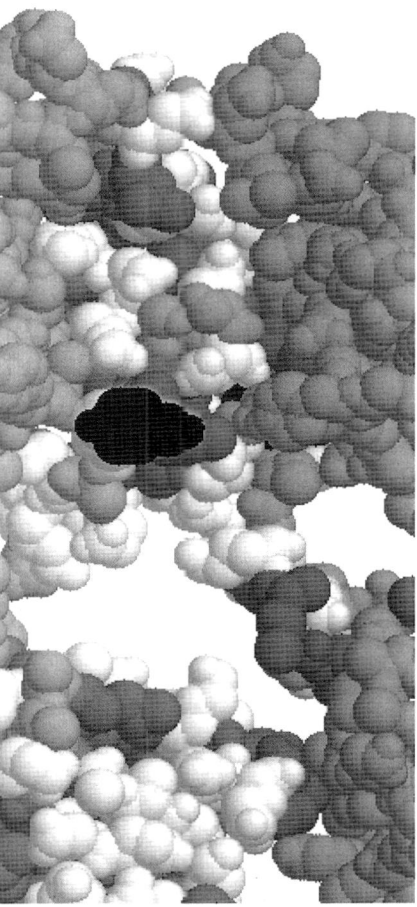

FIGURE 7.48. Stereo view of the aqueous chamber side of the GroES cap at the GroEL–GroES junction for the asymmetric (ADP)$_7$-GroEL structure. The purpose is to estimate surface polarity by obtaining a sum over ΔG_{HA} for the face and to compare the value to the inner faces of GroEL as represented in Figures 7.44 and 7.45. Double-digit numbers are for the GroES cap, and triple digit numbers indicate the top of the apical domain of GroEL. (Prepared using the crystallographic results of Xu et al.[82] as obtained from the Protein Data Bank, Structure File 1AON.)

7.6.4.3 Effect of Splitting ATP to Form the Most Polar State, ADP Plus P_i

Introduction of the potent charged species ATP prepares the Anfinsen cage. However, the most potent collection of charge occurs when ATP is hydrolyzed, that is, splits, to give ADP plus P_i. If the ADP plus P_i should stay in position for a few moments, it would present an extra jolt of charge, possibly amplified as in the original ATP binding, to result in the pulse of an even greater ΔG_{HA} in the form of ΔG_{ap}. It is, of course, the ΔG_{ap} contribution to ΔG_{HA} that drives hydrophobic unfolding.

7.6.4.4 Stepwise Release of Inorganic Phosphate Steadily Lowers the Temperature of the Substrate's Movable Cusp of Insolubility to Promote Systematic Hydrophobic Refolding

ΔG_{ap} peaks at the separation of ATP to form the most charged state of ADP plus P_i, which raises the movable cusp of insolubility to its highest temperature (see Figure 7.1) and in an equivalent representation raises the T_t-divide (also called the *coexistence or bimodal line*, as shown in Figure 5.3) to its highest value. The subsequent stepwise release of P_i steadily lowers the temperature of the movable cusp of insolubility to refold the substrate systematically and hydrophobically. This allows the substrate surfaces to begin sorting out the most favorable hydrophobic associations as the inorganic phosphates, $HPO_4^=$, peel off stepwise to lower systematically the T_t-divides for different cooperating hydrophobic domains toward and finally below the operating temperature.

7.7 Consequences of Excursions into Excess Solubility Due to Oxidative Processes

7.7.1 The Wide-ranging Impact of Oxidative Stress

From all that we have discussed in the foregoing process of developing the consilient mechanism for hydrophobic association, the proper function of biology's great macromolecules, proteins and nucleic acids, demands a graceful balance between hydrophobic association and dissociation. Disruption of that fine ballet of "the movable cusp of insolubility" brings disease and death. In general terms, unchecked oxidative processes, due to oxygen directly and molecular species derived from oxygen, are among the most troublesome means of shifting toward excess solubility. Hence, current medical research interests emphasize disentangling the role of numerous antioxidants in attenuating oxidative stress implicated in the full range of disease processes.

Oxidative stress has been implicated in disease states from cancers, heart disease, Lou Gherig's disease (amyotrophic lateral sclerosis), infections, environmental toxicities, to primary and secondary aging neurovascular disease processes. Oxidative stress occurs in general during critical illness in the intensive care setting, where the conclusion has been that "the oxidative damage to cells and tissues eventually contributes to organ failure."[90] Being such a fundamental problem, nature has devised a direct attack on the most offensive errant oxidants by means of superoxide dismutase, the enzyme that disarms superoxide, O_2^-, and hydrogen peroxide, H_2O_2. These and other oxidants are given off by cells that attack foreign bodies from coal dust particles to infectious agents

Our focus here continues along the lines of studies on the elastic-contractile model proteins. Examples will show that hydroxylation of the basic model protein, poly(GVGVP), indeed moves the cusp of insolubility to higher temperatures resulting in solubility that becomes relevant during wound repair. Furthermore, the argument will be presented that the oxidative bursts of macrophages in their attack on foreign agents in the lung lead to the development of pulmonary emphysema, because the oxidative bursts are not well targeted to the foreign agent alone. Defensive oxidative bursts also lead to the destruction of natural elastic tissue so essential to lung function.

7.7.2 Solubility: The Consequence of Oxidative Hydroxylation of Proline

7.7.2.1 Demonstration with the Parent Model Protein, $(GVGVP)_n$

Early in our studies it was expected that the post-translational modification of proline hydroxylation, so important to proper collagen structure and function, would raise the value of the temperature, T_t, for the onset of the inverse temperature transition for models of elastin. Accordingly, hydroxyproline (Hyp) was incorporated by chemical synthesis into the basic repeating sequence to give the protein-based polymers poly[f_{Val}(Val-Pro-Gly-Val-Gly), f_{Hyp}(Val-Hyp-Gly-Val-Gly)], where $f_{Val} + f_{Hyp} = 1$ and values of f_{Hyp} were 0, 0.01, and 0.1. The effect of prolyl hydroxylation is shown in Figure 7.49.[91] Replacement of proline by hydroxyproline markedly raises the temperature for hydrophobic association. Prolyl hydroxylation moves the "movable cusp of insolubility" to higher temperatures and shifts the molecular system toward solubility.

Poly(VPGVG) is a substrate for the natural enzyme prolyl hydroxylase,[92] which uses molecular oxygen and the cofactor vitamin C (ascorbic acid) for the reaction. Figure 7.49 also contains the temperature profile for fiber formation that results from hydroxylation by prolyl hydroxylase. When 100% hydroxylated as in poly(Val-Hyp-Gly-Val-Gly), T_t is about 65°C. An estimated 1% hydroxylation by prolyl hydroxylase results in a value of T_t of about 40°C. As shown in Figure 5.1C, T_t is just at the very onset of the aggregation, such that raising the value of T_t from about 30° to about 40°C would prevent assembly into fibers at body temperature. *Prolyl hydroxylation of the polypentapeptide model of elastin impairs fiber formation at physiological temperature.*

7.7.2.2 Prolyl Hydroxylation Blocks Elastic Fiber Formation In Vitro and In Vivo

7.7.2.2.1 In Cell Culture of Vascular Smooth Muscle Cells

The prediction in 1979 that hydroxylation of proline (Pro, P) would impair fiber formation[91] was borne out 6 years later in cell culture.[93] Using aortic smooth muscle cells in culture with substantial amount of the cofactor ascorbate, Barone et al.[95] found an overhydroxylation of the Pro residues in elastin. Furthermore, overhydroxylated the elastin remained in solution at 37°C at a higher than normal concentration with limited formation of insoluble elastin. Thus, as had been predicted from what is now called *the consilient mechanism for hydrophobic association,* making the elastin more polar by hydroxylation impairs fiber formation.

7.7.2.2.2 In Vivo Scar Tissue Formation of Wound Repair

In wound repair the activity of prolyl hydroxylase increases markedly to produce the necessary quantity and quality of collagen. Although this is the primary objective of hydroxylating the proline of collagen, a side effect of hydrox-

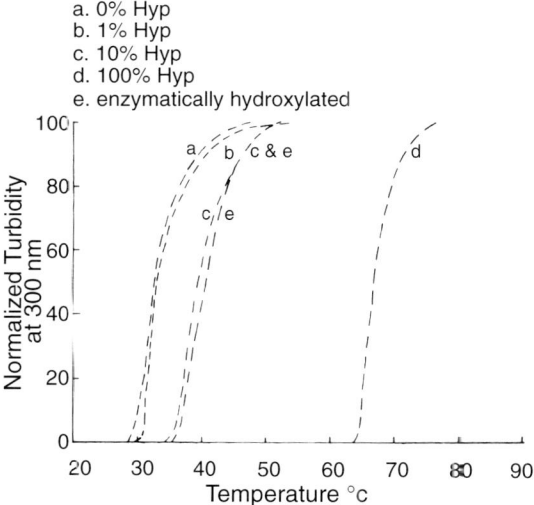

FIGURE 7.49. Effect of prolyl hydroxylation on hydrophobic association (insolubility). Temperature profiles show turbidity formation due to the aggregation attending the onset of the inverse temperature transition. The value for the onset temperature is taken at 50% of the maximal turbidity and is called T_t. Hydroxylation markedly shifts the polymer to solubility. (Reproduced with permission from Urry et al.[91])

ylation of elastin has the effect of impairing formation of elastic fibers. The result is stated well in the treatise on wound repair by Peacock,[94] "In a scar, where there are few elastic fibers and where collagen fibers become oriented primarily along the lines of tension, there is little 'give' and stretching and relaxation are not possible." Furthermore, "Failure to include new elastic fibers in repair tissue until long after collagen fibers are formed is another example of the inferiority of scar tissue to normal tissue. . . ."[94] One presumes that evolution toward rapid closure of a cut or tear, possibly to limit infection, has proven to be more important than ensuring quality of repair. Once again, this time *in vivo*, the effect of an oxidative process, that of hydroxylation, results in preventing the hydrophobic association required for fiber formation.

7.7.3 Loss of Elastic Recoil in the Elastic Fiber on Oxidation by a Superoxide and Hydrogen Peroxide Generating Natural Enzyme

Applying what is now the consilient mechanism for hydrophobic association to the elastic fiber, the prediction became that oxidation of the elastic fiber by a natural enzyme with the biological role of producing superoxide and hydrogen peroxide would cause hydrophobic dissociation evidenced by a swelling and a loss of elastic recoil. A xanthine oxidase superoxide generating system prepared by Bruce Freeman was used. The stress–strain curve was determined at time zero, and then the superoxide generating system was added and followed for 2 hours, when a new stress–strain curve was again determined. The superoxide generating system was replenished and run for another 2 hours, and the stress–strain curve was determined for a third time. The process was repeated for 12 hours. As shown in Figure 7.50,[95] each exposure to superoxide and hydrogen peroxide from the enzymatic generating system caused the elastic modulus to decrease and the extension, required before force development began, to increase. The interpretation is that the formation of polar species on the elastin chains by the processes of these highly oxidative molecular species moves the "movable cusp of insolubility" to higher temperatures resulting in hydrophobic dissociation. Thus, the elastic fiber is susceptible to the natural oxidative bursts emitted by macrophages, as they encounter a foreign agent.

7.7.4 Relevance to Environmentally Induced Lung Disease

7.7.4.1 *Oxidative Attack on Foreign Substances by the Body's Macrophages*

The body's natural defense mechanism involves macrophages (called *monocytes* when in the bloodstream) that give off oxidative bursts on encountering foreign agents and pathogens. As

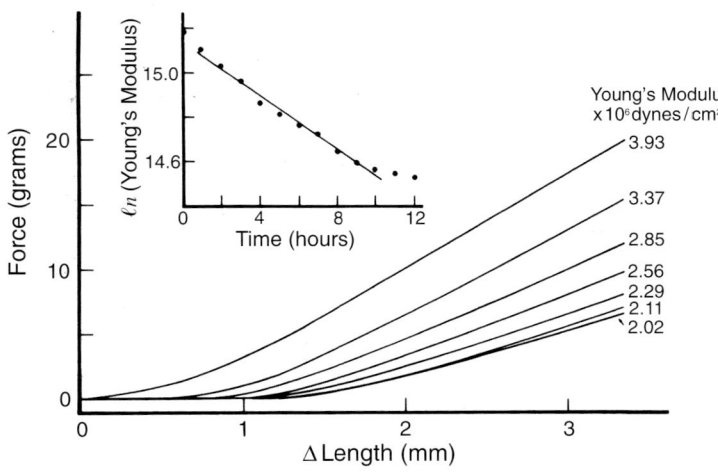

FIGURE 7.50. Effect of enzymatic superoxide generating system on the elastic properties of ligamentum nuchae elastin. With time of exposure of purified fibrous elastin to O_2^- and H_2O_2, the elastic modulus decreases, and larger and larger extensions are required before development of an elastic force occurs. This is the result of swelling due to hydrophobic dissociation. Once the fiber is hydrophobically dissociated, it becomes susceptible to proteolytic degradation. (Reproduced with permission from Urry et al.[95])

would be expected from the consilient mechanism for hydrophobic association within macromolecular species, a most effective approach to render the agent harmless would be to oxidize it, thereby, causing it to become susceptible to biodegradation, such as proteolysis, and even to direct dissolution by oxidation. The problem with such an oxidative attack is that it lacks the capacity to be directed solely at the eliciting agent. Because of this, natural constituents of the extracellular matrix inadvertently become targets of the oxidative bursts. Herein lies the introduction to problems of environmentally induced lung diseases such as pulmonary emphysema and pulmonary fibrosis.

7.7.4.2 Molecular Basis for Defective Elastic Tissue in Environmentally Induced Lung Disease: Pulmonary Emphysema

Pulmonary emphysema is a chronic progressive disorder wherein the elastic fibers of the lung become fragmented and dysfunctional with the result of a loss of elastic recoil.[96,97] One result is that the patient must consciously and forcibly exhale. A prominent proposal for the etiology of the disease has been the lack of proteinase inhibitors that would otherwise prevent proteolytic degradation of the fiber.[98–100] To help study the disease, a successful animal model for pulmonary emphysema was developed in which direct instillation of the proteolytic enzyme elastase into the lung became the causative agent.[101,102]

Another element of the story is the finding that the antiprotease, α1-proteinase inhibitor is damaged by superoxide. We, of course, argue that inactivation by oxidation of this inhibitor protein would be due to hydrophobic dissociation, that is, due to moving the "movable cusp of insolubility." Furthermore, from our studies to develop elastic model proteins as materials for medical applications (see Chapter 9), the finding is that these materials are essentially nonbiodegradable *in situ* as long as the value of T_t remains about 15°C below body temperature. Only after the value of T_t increases above body temperature, as when carboxamide chemical clocks convert to carboxylates, does degradation occur.

Thus our scenario for the pathogenesis of pulmonary emphysema becomes a primary event of macrophages encountering foreign agents inspired into the lungs and emitting oxidative bursts that inadvertently oxidize the elastic fiber. As shown in Figure 7.50, this causes the elastic fiber to hydrophobically dissociate, that is, to swell and lose its elastic recoil, and to become susceptible to proteolytic digestion. *Thus, underlying the pathogenesis of pulmonary emphysema resides a "movable cusp of insolubility" where excursion into the realm of increased solubility means disease and death.*

References

1. D.W. Urry, T. Hugel, M. Seitz, H. Gaub, L. Sheiba, J. Dea, J. Xu, and T. Parker, "Elastin: A Representative Ideal Protein Elastomer." *Pailos. Trans. R. Soc. Lond.*, B **357**, 169–184. 2002.
2. D.W. Urry and T.M. Parker, "Mechanics of Elastin: Molecular Mechanism of Biological Elasticity and its Relevance to Contraction." *J. Muscle Res. Cell Motil.*, Spec. Issue, **23**, 2002.
3. D.W. Urry, "Physical Chemistry of Biological Free Energy Transduction as Demonstrated by Elastic Protein-based Polymers." *J. Phys. Chem.* B, **101**, 11007–11028, 1997.
4. D.W. Urry, "Elastic Biomolecular Machines: Synthetic Chains of Amino Acids, Patterned After Those in Connective Tissue, can Transform Heat and Chemical Energy into Motion." *Sci. Am.*, January 1995, 64–67.
5. Two curiously complementary and impeding truisms face any newly proposed concept or outlook in science. They may be demonstrated by two quotations. The first, attributed to Maurice Maeterlinck, indicates the struggle of those who would make new proposals, "At every crossway on the road that leads to the future, tradition, has placed against each of us, 10,000 men to guard the past." The second suggests why it is so easy, even for those without a direct personal stake, to hold with tradition, "Keep in mind that new ideas are commonplace, and *almost always wrong*" (E.O. Wilson, *Consilience. The Unity of Knowledge*, Alfred E. Knopf, New York, 1998). Accordingly, to stand against a new idea is to be *almost always right*, with neither intellectual effort nor actual understanding required. Those who might practice default to pretense could seldom find a better opportunity.

There is another aspect noted by Francis Bacon in *The New Organon*, "The human understanding is no dry light, but receives an infusion from the will and affections; whence proceed sciences which may be called 'sciences as one would wish.' For what a man had rather were true he readily believes." It is particularly difficult for anyone to allow ready marginalization of a hard-won and once very empowering viewpoint.

Of course, the basic issue in science always remains the extent to which any concept continues to bear fruit. The wonderful thing about scientific pursuit is that there is a correct and unchanging truth and that progress is best achieved by hewing to that path as closely as possible. This was attempted, as reported in part in Chapter 5, by repeatedly designing new protein-based machines and testing those designs for intended function.

6. Webster's *Third New International Dictionary of the English Language*, Unabridged, Merriam-Webster, Inc., Springfield, MA, 1993, p. 557.
7. The formation of hydrophobic hydration from bulk water gives a negative change in ΔS, which makes the $(-T\Delta S)$ term positive for this process, whereas the conversion of hydrophobic hydration to bulk water results in a positive change in ΔS such that the $(-T\Delta S)$ term is negative for this process.
8. Dorland's *Illustrated Medical Dictionary*, 27th Edition. W.B. Saunders, Philadelphia, 1985, p. 1468.
9. Webster's *Third New International Dictionary of the English Language*, Unabridged. Merriam-Webster, Inc., Springfield, MA, 1993, p. 432.
10. C.S. Roy, "The Elastic Properties of the Arterial Wall." *J. Physiol.*, **3**, 125–159, 1880.
11. L.B. Sandberg, J.G. Leslie, C.T. Leach, V.L. Alvarez, A.R. Torres, and D.W. Smith, "Elastin Covalent Structure as Determined by Solid Phase Amino Acid Sequencing." *Pathol. Biol.*, **33**, 266–74, 1985.
12. T. Weis-Fogh and S.O. Andersen, "New Molecular Model for the Long-range Elasticity of Elastin." *Nature*, **227**, 718–721, 1970.
13. C.A. Hoeve and P.J. Flory, "The Elastic Properties of Elastin." *Biopolymers*, **13**, 677–686, 1974.
14. A.V. Hill, "A Discussion on the Thermodynamics of Elasticity in Physiological Tissues." *Proc. R. Soc. Lond.* B, **139**, 464–497, 1952.
15. D.K. Hill, "The Effect of Temperature in the Range 0–35 Degrees C on the Resting Tension of Frog's Muscle." *J. Physiol.*, **208**, 725–739, 1970.
16. K.W. Ranatunga, "Thermal Stress and Ca-independent Contractile Activation in Mammalian Skeletal Muscle Fibers at High Temperatures." *Biophys. J.*, **66**, 1531–1541, 1994.
17. D.W. Urry, D.C. Gowda, S.-Q. Peng, and T.M. Parker, "Non-linear Hydrophobic-induced pKa Shifts: Implications for Efficiency of Conversion to Chemical Energy." *Chem. Phys. Lett.*, **239**, 67–74, 1995.
18. A.V. Hill, "The Possible Effect of the Aggregation of Hemoglobin on its Dissociation Curves." Proceedings of the Physiological Society, *J. Physiol.*, **40**, iv–vii, 1910, and A.V. Hill. *J. Biochem.*, **7**, 471–480, 1913.
19. N.B. Butwell, R. Ramasamy, I. Lazar, A.D. Sherry, and C.R. Malloy, "Effect of Lidocaine on Contracture, Intracellular Sodium, and pH in Ischemic Rat Hearts." *Am. J. Physiol.*, **264**(6Pt2), H1884–H1889, 1993.
20. K.I. Furakawa, J. Komaba, H. Sakai, and Y. Ohizumi, "The Mechanism of Acidic pH-Induced Contraction in Aortae from SHR and WKY Rats Enhanced by Increasing Blood Pressure." *Br. J. Pharmacol.*, **118**, 485–492, 1996.
21. D.W. Urry and S.-Q. Peng, "Non-linear Mechanical Force-induced pKa Shifts: Implications for Efficiency of Conversion to Chemical Energy." *J. Am. Chem. Soc.*, **117**, 8478–8479, 1995.
22. J.W. Pringle, "The Croonian Lecture, 1977. Stretch Activation of Muscle: Function and Mechanism." *Proc. R. Soc. Lond. Ser. B Biol. Sci.*, **201**(1143), 107–130, 1978.
23. N. Thomas and R.A. Thornhill, "Stretch Activation and Nonlinear Elasticity of Muscle Crossbridges." *Biophys. J.*, **70**(6):2807–2818, 1996.
24. M. Suzuki, J. Shigematsu, Y. Fukunishi, Y. Harada, and T. Yanagida, T. Komada, "Coupling of Protein Surface Hydrophobicity Change to ATP Hydrolysis by Myosin Motor Domain." *Biophys. J.*, **72**, 18–23, 1997.
25. M.F. Perutz, "Mechanisms of Cooperativity and Allosteric Regulation in Proteins." *Q. Rev. Biophys.*, **22**, 139–236, 1989.
26. D. Bulone, M.B. Palma-Vittorelli, and M.U. Palma, "Enthalpic and Entropic Contributions of Water Molecules to Functional T → R Transition of Human Hemoglobin in Solution." *Int. J. Q. Chem.*, **42**, 1427–1437, 1992.
27. D. Bulone, P.L. San Biagio, M.B. Palma-Vittorelli, and M.U. Palma, "On the Role of Water in Hemoglobin Function and Stability, Response." *Science*, **259**, 1335–1336, 1993.

28. M.F. Colombo, D.C. Rau, and V.A. Parsegian, "Protein Solution in Allosteric Regulation: A Water Effect on Hemoglobin." *Science*, **256**, 655–659, 1992.
29. C. Bohr, K.A. Hasselbalch, and A. Krogh, "Uber einem in biologischen Beziehung wichtigen Einfluss, den die Kohlersaurespannung des Blutes auf dessen Sauerstoffbinding übt." *Skand. Arch. Physiol.*, **16**, 401–412, 1904.
30. Christian Bohr was the father of quantum physicist Niels Bohr.
31. J.A.V. Butler, "The Energy and Entropy of Hydration of Organic Compounds." *Trans. Faraday Soc.*, **33**, 229–238, 1937.
32. H.M. Berman, J. Westbrook, Z. Feng, G. Gilliland, T.N. Bhat, H. Weissig, I.N. Shindyalov, and P.E. Bourne, "The Protein Data Bank." *Nucleic Acids Res.*, **28**, 235–242, 2000.
33. E. Martz, "FrontDoor to Protein Explorer 1.982 Beta" Copyright © 2002 Website: proteinexplorer.org.
34. D.W. Urry and J.W. Pettegrew, "Model Systems for Interacting Heme Moieties. II. The Ferriheme Octapeptide of Cytochrome." *J. Am. Chem. Soc.*, **89**, 5276–5283, 1967.
35. J.R.H. Tame and B. Vallone, "The Structures of Deoxy Human Haemoglobin and the Mutant Tyrα42 His at 120K." *Acta Crystal. D*, **56**, 805–811, 2000.
36. M. Paoli, R. Liddington, J. Tame, A. Wilkinson, and G. Dodson, "Crystal Structure of T State Haemoglobin with Oxygen Bound at all Four Haems." *J. Mol. Biol.*, **256**, 775–792, 1996.
37. The choice of the crystal structure of Paoli et al.,[36] with each of the heme sites occupied by oxygen (while remaining in the T state), goes to the primary purpose, which is to determine the effect of binding the very polar oxygen molecule to the heme on the hydrophobicity of heme and to assess its influence on proximal groups through ΔG_{ap}. As such, this crystal provides a challenging test of the consilient mechanism to see if the effects of ΔG_{ap} are apparent even though changes in quaternary structure are limited. For our purposes, the structure will be delineated as oxyHb(T).
38. J. Vojtechovsky, J. Berendzen, K. Chu, L. Schlichting, and R.M. Sweet, "Implications for the Mechanism of Ligand Discrimination and Identification of Substrates Derived from Crystal Structures of Myoglobin–Ligand Complexes at Atomic Resolution." Protein Data Bank, Structure Files, 1A6N and 1A6M.
39. A. Brzowowski, Z. Derewenda, E. Dodson, G. Dodson, M.J. Grabowowski, R. Liddington, T. Skarzynski, and D. Valley, "Bonding of Molecular Oxygen to T State Human Haemoglobin." *Nature*, **307**, 74–76, 1984.
40. R.C. Liddington, Z. Derewenda, G. Dodson, and D. Harris, "Structure of the Liganded T State of Haemoglobin Identifies the Origins of Cooperative Oxygen Binding." *Nature*, **331**, 725–728, 1988.
41. R.C. Liddington, Z. Derewenda, E. Dodson, R. Hubbard, and G. Dodson, "High resolution crystal structures and comparisons of T state deoxy and two liganded T state hemoglobins: T (αoxy) Haemoglobin and T (met) Haemoglobin." *J. Mol. Biol.*, **228**, 551–579, 1992.
42. C. Rivetti, A. Mozarelli, G.L. Rossi, E.R. Henry, and W.A. Eaton, "Oxygen Binding by Single Crystals of Haemoglobin." *Biochemistry*, **32** 2888–2906, 1993.
43. D. Voet and J.G. Voet, *Biochemistry*, 2nd Edition, John Wiley & Sons, New York, 1995, p. 226.
44. See, for example, "Sickle Cell Anemia." *Encyclopedia Britannica*, 2002.
45. W.A. Eaton and J. Hofrichter, "Sickle Cell Hemoglobin Polymerization," *Adv. Prot. Chem.*, **40**, 63–279, 1990.
46. M. Manno, A. Emanuele, V. Martorana, P.L. San Biagio, D. Bulone, M.B. Palma-Vitorelli, D.T. McPherson, J. Xu, T.M. Parker, and D.W. Urry, "Interaction of Processes on Different Time Scales in a Bioelastomer Capable of Performing Energy Conversion." *Biopolymers*, **59**, 51–64, 2001.
47. F. Sciortino, K.U. Prasad, D.W. Urry, and M.U. Palma, "Self-assembly of Bioelastomeric Structures From Solutions: Mean Field Critical Behavior and Flory-Huggins Free-energy of Interaction." *Biopolymers*, **33**, 743–752, 1993.
48. P.L. San Biagio and M.U. Palma, "Spinodal Lines and Flory-Huggins Free-energies for Solutions of Human Hemoglobins HbS and HbA." *Biophys. J.*, **60**, 508–512, 1991.
49. P.L. San Biagio and M.U. Palma, "Solvent-induced Forces and Fluctuations: A Novel Comparison of Human Hemoglobin S and A." *Comments Theor. Biol.*, **2**, 453–470, 1992.
50. D.W. Chung, J.E. Harris, and E. Harris, "Nucleotide Sequences of the Three Genes Encoding for Human Fibrinogen." *Adv. Exp. Med. Biol.*, **281**, 39–48, 1990.
51. Z. Yang, J.M. Kollman, L. Pandi, and R.F. Doolittle, "Crystal Structure of Native Chicken

Fibrinogen at 2.7 Å Resolution." *Biochemistry*, **40**, 12515–12520, 2001.

52. D.W. Urry, T.L. Trapane, and K.U. Prasad, "Phase-structure Transitions of the Elastin Polypentapeptide-water system Within the Framework of Composition–Temperature Studies." *Biopolymers*, **24**, 2345–2356, 1985.

53. M.W. Mosesson, K.R. Siebenlist, and D.A. Meh, "The Structure and Biological Features of Fibrinogen and Fibrin." *Ann. N.Y. Acad. Sci.*, **936**, 11–30, 2001.

54. G. Spraggon, S.J. Everse, and R.S. Doolittle, "Crystal Structures of Fragment D from Human Fibrinogen and its Crosslinked Counterpart from Fibrin." *Nature*, **389**, 455–462, 1997.

55. S.J. Everse, G. Spraggon, L. Veerapandian, M. Riley, and R.F. Doolittle, "Crystal Structure of Fragment Double D from Human Fibrin with Two Different Bound Ligands." *Biochemistry*, **37**, 8637–8642, 1998.

56. L.C. Serpell, M. Sunde, and C.C.F. Blake, "The molecular basis of amyloidosis." *Cell. Mol. Life Sci.*, **53**, 871–887, 1997.

57. Amyloid deposits are fibrous protein structures of high β-sheet content that are identified microscopically from their staining characteristics, giving a green birefringence when stained with the dye Congo red.

58. R.B. Wickner, K.L. Taylor, H.K. Edskes, M.-L. Maddelein, H. Moriyama, and B.T. Roberts, "Prions in *Saccharomyces* and *Podospora* spp.: Protein-based Inheritance." *Microbiol. Mol. Biol. Revs.*, **63**, 844–861, 1999.

59. J.S. Griffiths, "Self-replication and Scrapie." *Nature*, **215**, 1043–1044, 1967.

60. S.B. Prusiner, "Novel Proteinaceous Particles Cause Scrapie." *Science*, **216**, 136–144, 1982.

61. S.B. Prusiner, "Scrapie Prions." *Annu. Rev. Microbiol.*, **43**, 345–374, 1989.

62. The word *scrapie* evolved from the habit of sheep with the disease to "scrape off" their coats by rubbing against available surfaces such as trees, apparently due to an itching symptom characteristic of the disease.

63. S.B. Prusiner, D.C. Gajdusek, and M.P. Alpers, "Kuru with Incubation Periods Exceeding Two Decades." *Ann. Neurol.*, **12**, 1–9, 1982.

64. The term *spongiform encephalopathy* arises from the appearance of the brain tissue wherein the brain is riddled with lacunae or vacuoles giving rise to a sponge-like appearance.

65. P. Brown, P.P. Liberski, A. Wolff, and D.C. Gajdusek, "Resistance to Steam Autoclaving After Formaldehyde Fixation and Limited Survival After Ashing at 360 Degrees C: Practical and Theoretical Limitations." *J. Infect. Dis.*, **161**, 467–472, 1990.

66. D.M. Taylor, H. Fraser, I. McConnell, D.A. Brown, K.A. Lanza, and G.R. Smith, "Decontamination Studies with the Agents of Bovine Spongiform Encephalopathy and Scrapie." *Arch. Virol.*, **139**, 313–326, 1994.

67. An advantage of using elastic protein-based polymers as scaffoldings for soft tissue restoration as discussed in Chapter 9 would be to overcome the problem of infectious prions.

68. J. Collinge, K.C. Sidle, J. Meads, J. Ironside, and A.F. Hill, "Molecular Analysis of Prion Strain Variation and the Aetiology of 'New Variant' CJD." *Nature*, **383**, 685–690, 1996.

69. Health Canada—Canada Comminicable Disease Report, Vol. 22–17. Website: http://www.hc-sc.gc.ca/hpb/lcdc/publicat/ccdr/96vol22/dr2217ec.html.

70. See the "Kimball's Biology Pages" Website: http://users.rcn.com/jkimball.ma.ultranet/BiologyPages/W/Welcome.html, specifically, http://users.rcn.com/jkimball.ma.ultranet/BiologyPages/P/Prions.html.

71. D.W. Urry and C.-H. Luan, "Proteins: Structure, Folding and Function." In *Bioelectrochemistry: Principles and Practice*, G. Lenaz, Ed., Birkhäuser Verlag AG, Basel, Switzerland, 1995, pp. 105–182.

72. When using the data for ΔG_{HA} of Table 5.3 with the values $\Delta G_{HA}(Gln) = +0.75$ kcal/mole-pentamer and $\Delta G_{HA}(Glu^-) = +3.72$ kcal/mole-pentamer, five peptide units would sum to give the an equivalent ΔG_{ap} of an ionized glutamate residue. If a sufficient concentration of edge chains of a β-sheet could be experimentally prepared and examined by NMR, the lack of adequate peptide hydrogen bonding might be observable by distinctive chemical shifts of the C*O* and N*H* peptide atoms of the edge β-chain.

73. H. Inouye, J. Bond, M.A. Baldwin, H.L. Ball, S.B. Prusiner, and D.A. Kirschner, "Structural Changes in a Hydrophobic Domain of the Prion Protein Induced by Hydration and by Ala → Val and Pro → Leu Substitutions." *J. Mol. Biol.*, **300**, 1283–1296, 2000.

74. The designation P101L indicates that the proline (Pro, P) residue 101 of the normal sequence has mutated to give a leucine (Leu, L) residue, and similarly the A → V designation

without enumeration of the residue number means a less hydrophobic alanine (Ala, A) residue has been mutated to a more oil-like valine (Val, V) residue.

75. D.D. Laws, H-M.L. Bitter, K. Liu, H.L. Ball, K. Kaneko, H. Wille, F.E. Cohen, S.B. Prusiner, A. Pines, and D.E. Wemmer, "Solid-state NMR Studies of the Secondary Structure of a Mutant Prion Protein Fragment of 55 Residues that Induces Neurodegeneration." *Proc. Natl. Acad. Sci. U.S.A.*, **98**, 11686–11690, 2001.

76. S.B. Prusiner, "Molecular Biology of Prion Diseases." *Science*, **252**, 1515–1522, 1991.

77. W. Swietnicki, R. Petersen, P. Gambetti, and W.K. Surewicz, "pH Dependent Stability and Conformation at the Recombinant Human Prion Protein PrP (90–231)." *J. Biol. Chem.*, **272**, 27517–27520, 1997.

78. S. Hornemann and R. Glockshuber, "A Scrapie-like Unfolding Intermediate at the Prion Protein Domain PrP (121–231) Induced by Acidic pH." *Proc. Natl. Acad. Sci. U.S.A.*, **95**, 6010–6014, 1998.

79. H. Wille, M.D. Michelitsch, V. Guénebaut, S. Supattapone, A. Serban, F.E. Cohen, D.A. Agard, and S.B. Prusiner, "Structural Studies of the Scrapie Prion Protein by Electron Crystallography." *Proc. Natl. Acad. Sci. U.S.A.*, **99**, 3563–3568, 2002.

80. D.R. Booth, M. Sunde, V. Bellotti, C.V. Robinson, W.L. Hutchinson, P.E. Frazer, P.N. Hawkins, C.M. Dobson, S.E. Radford, C.C.F. Blake, and M.B. Pepys, "Instability Unfolding and Aggregation of Human Lysozyme Variants Underlying Amyloid Fibrillogenesis." *Nature*, **385**, 787–793, 1997.

81. Z. Xu and P.B. Sigler, "GroEL/GroES: Structure and Function of a Two-stroke Folding Machine." *J. Struct. Biol.*, **124**, 129–141, 1998.

82. Z. Xu, A.L. Horwich, and P.B. Sigler, "The Crystal Structure of Asymmetric GroEL–GroEs-(ADP)$_7$ chaperonin complex." *Nature*, **388**, 741–750, 1997.

83. C.B. Anfinsen, "Principles that Govern the Folding of Protein Chains." *Science*, **181**, 223–230, 1973. The Anfinsen principle is that the sequence of a protein dictates the full three-dimensional structure that it would form, that is, sequence dictates protein folding and assembly. The need for molecular chaperones suggests that the correctly folded protein, the lowest energy structure, is not always the result and that the problem arises out of improper hydrophobic associations. Interestingly, the need for molecular chaperones is greatest when temperature is elevated. They were first identified as heat shock proteins, which are just the conditions where metastable hydrophobic associations would occur. The rise in temperature from that of the cusp of insolubility to the insoluble hydrophobically associated state occurred without opportunity for the protein to sort out the lowest energy, hydrophobically associated state. Thus, biology employs a polar cage that ensures hydrophobic dissociation and then systematically decreases polarity of the cage by hydrolysis of ATP with phosphate release to allow for correct hydrophobic re-association.

84. See references 4 and 5 of Chapter 1.

85. J. Wang and D.C. Boisvert, "Structural Basis for GroEL-assisted Protein Folding from the Crystal Structure of (GroEL-KMgATP)$_{14}$ at 2.0 Å Resolution." *J. Mol. Biol.*, **327**, 843–855, 2003.

86. K. Braig, Z. Otwinowski, R. Hedge, D.C. Boisvert, A. Joachimiak, A.L. Horwich, and P.B. Sigler, "The Crystal Structure of Bacterial Chaperonin GroEL at 2.8 Å." *Nature*, **371**, 578–586, 1994.

87. K. Braig, P. Adams, and A.T. Brunger, "Conformational Variability in the Refined Structure of the Chaperonin GroEL at 2.8 Å Resolution." *Nature Struct. Biol.*, **2**, 1083–1094, 1995.

88. Such an apolar–polar repulsion became apparent on analysis of the data in Figure 5.31. In that case, even though **Model Protein v**, (GVGVP GVGFP GEGFP GVGVP GVGFP GFGFP)$_{42}$ (GVGVP), was completely unfolded and the carboxylates were free to reach out for water unperturbed by the other, there still remained a measured repulsion of 2.4 kcal/mole-carboxylate. The freedom of motion of the otherwise random coil structure was constrained from a range of conformations by the 2.4 kcal/mole repulsion between carboxylate and the more hydrophobic phenylalanine (Phe, F) residues.

89. Calculations of the surface ΔG_{HA} values are necessarily very gross until they are given per nanometer squared and until a set of rules are developed to take into consideration the extent of residue exposure to the surface and especially the exposure of the charged or polar component of the functional side chain to the surface.

90. R. Lovat and J.C. Preiser, "Antioxidant Therapy in Intensive Care." *Curr. Opin. Crit. Care*, **9**, 266–270, 2003.

91. D.W. Urry, H. Sugano, K.U. Prasad, M.M. Long, and R.S. Bhatnagar, "Prolyl Hydroxylation of

the Polypentapeptide Model of Elastin Impairs Fiber Formation." *Biochem. Biophys. Res. Commun.*, **90**, 194–198, 1979.
92. R.S. Bhatnagar, R.S. Rapaka, and D.W. Urry, "Interaction of Polypeptide Models of Elastin with Prolyl Hydroxylase." *FEBS Lett.*, **95**, 61–64, 1978.
93. L.M. Barone, B. Faris, S.D. Chipman, P. Toselli, B.W. Oakes, and C. Franzblau, "Alteration of the Extracellular Matrix of Smooth Muscle Cells by Ascorbate Treatment." *Biochchim. Biophys. Acta*, **840**, 245–254, 1985.
94. E.R. Peacock, Jr., *Wound Repair*, W.B. Saunders, Philadelphia, 1984, pp. 56–101.
95. D.W. Urry, "Entropic Elastic Processes in Protein Mechanism. II. Simple (Passive) and Coupled (Active) Development of Elastic Forces." *J. Protein Chem.*, **7**, 81–114, 1988, especially Figure 12.
96. D.W. Urry, R.S. Bhatnagar, H. Sugano, K.U. Prasad, and R.S. Rapaka, "A Molecular Basis for Defective Elastic Tissue in Environmentally Induced Lung Disease." In *Molecular Basis of Environmental Toxicity*, R.S. Bhatnagar, Ed., Ann Arbor Scientific Publishers, Inc., Ann Arbor, MI, pp. 515–530, 1980.
97. J.G. Clark, C. Kuhn, and R.P. Meacham, "Lung Connective Tissue." *Int. Rev. Connect. Tissue Res.*, **10**, 249–331, 1983.
98. P.J. Stone, "The Elastase–Antielastase Hypothesis of the Pathogenesis of Emphysema." *Clin. Chest Med.*, **4**, 405–412, 1983.
99. G.L. Snider, "Two Decades of Research in the Pathogenesis of Emphysema." *Schweiz Med. Wochenschr.*, **114**, 898–906, 1984.
100. A. Janoff, "Elastases and Emphysema. Current Assessment of the Protease–Antiprotease Hypothesis." *Am. Rev. Respir. Dis.*, **132**, 417–433, 1985.
101. C. Kuhn, S.-Y. Yu, M. Chraplyvy, H.E. Linder, and R.M. Senior, "The Induction of Emphysema with Elastase. II. Changes in Connective Tissue." *Lab. Invest.*, **34**, 372–380, 1976.
102. M. Osman, S. Keller, Y. Hosannah, J.O. Cantor, G.M. Turino, and I. Mandl, "Impairment of Elastin Resynthesis in the Lungs of Hampsters with Experimental Emphysema Induced by Sequential Administration of Elastase and Trypsin." **105**, 254–258, 1985.
103. D. Voet and J.G. Voet, *Biochemistry*, 2nd Ed., John Wiley & Sons, New York, 1995.
104. J.G. White, "Ultrastructural Features of Erythrocyte and Hemoglobin Sickling." *Arch. Intern. Med.*, **133**, 545–562, 1974.
105. G.W. Dykes, R.H. Crepeau, and S.J. Edelstein, "Three-dimensional Reconstruction of 14-Filament Fibers of Hemoglobin S." *J. Mol. Biol.*, **130**, 451–472, 1979
106. J.D. Harper and P.T. Lansbury, Jr., "Models of Amyloid Seeding in Alzheimer's Disease and Scrapie: Mechanistic Truths and Physiological Consequences of the Time-dependent Solubility of Amyloid Proteins." *Annu. Rev. Biochem.*, **66**, 385–407, 1997.

8
Consilient Mechanisms for Protein-based Machines of Biology

8.1 Thesis

Biology thrives near a movable cusp of insolubility, and the <u>forces</u> that, in a positively cooperative manner, power the molecular machines of biology <u>drive spatially localized hydrophobic protein domains</u> back and forth <u>across</u> thermodynamically movable <u>water-solubility-insolubility divides</u>.

8.1.1 Further Considerations of the Volume Title

This chapter is the principal focal point for the question asked in the title, What Sustains Life? The obvious answer, of course, is that protein machines access energy available in the environment and convert it to those forms of energy required for Life to flourish. The specific rejoinder in the title speaks of "Consilient Mechanisms for...." In our usage, *consilient mechanism* means a particular "common groundwork of explanation,"[1] whereby protein-based machines access the whole range of available energies and convert those energies to the many energy forms required to build and operate the structures that constitute the functional living entity.

A more complete technical description of the consilient mechanisms of concern to biological energy conversion encompasses two distinct but interlinked physical processes of *hydrophobic association* and of *elastic force development*. The latter results from a generally applicable mechanism of entropic elastic force arising from the damping of internal chain dynamics on deformation of an interconnecting chain segment that began with significant kinetic freedom. The fundamentals of hydrophobic association are embodied within the comprehensive hydrophobic effect, expressed in terms of the Gibbs free energy of hydrophobic association, ΔG_{HA}. As developed in Chapter 5, the operative component of ΔG_{HA} is the apolar–polar repulsive free energy of hydration, ΔG_{ap}. Extensive studies on systematically varied elastic-contractile model proteins demonstrate that ΔG_{ap} derives from the competition for hydration between apolar (hydrophobic) and polar (e.g., charged) groups.

The rejoinder in the title continues, "Protein-based Machines...." Biology's protein-based machines group into three functional categories: (1) those of the electron transport chain that produce the proton concentration gradient by pumping protons from one side to the other of the inner mitochondrial membrane; (2) those that synthesize adenosine triphosphate (ATP), principally ATP synthase of the inner mitochondrial membrane that utilizes the proton gradient to produce ATP; and (3) those many ATPases that use ATP in carrying out the essential functions of the living cell. Arguably, the protein-based machine most vital to energy conversion in living systems is ATP synthase of the inner mitochondrial membrane. This proton-powered rotary motor produces nearly 90% of the ATP, the energy coin, of biology.

329

ATP and equivalent nucleotide triphosphates (NTPs) power essentially all subsequent protein-based machines of biology, either directly or indirectly.

The crucial protons that fuel ATP synthase result from the action of four complexes, four protein-based machines, of the electron transport chain within the inner mitochondrial membrane. Of the four, Complex III (ubiquinone: cytochrome c oxidoreductase[2]) provides a particularly revealing protein-based machine, wherein the power of ΔG_{ap} achieves ready recognition in the role of proton gatekeeper and wherein a single strand of protein chain, stretched by hydrophobic association, provides for the elastic force development essential for subsequent domain movement resulting in electron transfer. Accordingly, the hydrophobic and elastic consilient mechanisms, in our view as developed in Chapter 5 and given example below, provide salient insight into the function of Complex III and, as we shall also see, into the function of ATP synthase and the myosin II motor of muscle contraction.

For the casual observer, muscle contraction presents the most notable ATPase-based protein machine of biology. Specifically, the linear myosin II motor of striated muscle most signifies the motion of mammalian Life, as exemplified by the discus thrower mentioned in the introduction to Chapter 2. In particular, the powerstroke of the myosin II motor provides another stunning example of the hydrophobic consilient mechanism. In this case, binding of the very polar ATP molecule or an equivalent chemical analogue disrupts hydrophobic association that becomes re-established in the absence of ATP, for example, in the near-rigor state.[3] The myosin II motor in part illustrates the case of the hydrophobic consilient mechanism by providing insight into the physical basis for what Rayment and his colleagues described as a thermodynamic instability in their insightful paper, "Revised Model for the Molecular Basis of Muscle Contraction."[4] Therefore, a short restatement of the physical bases for the pervasive hydrophobic and elastic consilient mechanisms follows, initially in a qualitative sense and then in a more detailed sense in the subsequent parts of section 8.1.

8.1.2 Representation of the Movable Cusp of Insolubility

The experimental endothermic heat of the inverse temperature transition of hydrophobic association symbolizes the hydrophobic consilient mechanism. That symbol, the sign of the cusp of insolubility, is recognized in Figures 1.1 and 7.1. Different nonthermal energy inputs power biology's protein-based machines through the inverse temperature transition by moving the cusp of insolubility along the temperature axis.

8.1.3 Moving the Cusp of Insolubility

The physical basis of the hydrophobic consilient mechanism draws from the comprehensive hydrophobic effect in the form of a Gibbs free energy of hydrophobic association, ΔG_{HA}, and the size of the cusp of insolubility determines the magnitude of ΔG_{HA}. Hydrophobic hydration provides the dominant thermodynamic contribution to ΔG_{HA}. The foremost functional component of ΔG_{HA} is the change in apolar–polar repulsive free energy of hydration, ΔG_{ap}, which arises from the competition for hydration between apolar (hydrophobic) and polar (e.g., charged) groups. ΔG_{ap} dictates the interchange of structural states of protein domains from hydrophobically associated (insoluble) to hydrophobically dissociated (soluble). Thus, as developed in Chapter 5 with designed elastic-contractile model proteins, ΔG_{ap} is the thermodynamic quantity that moves the cusp of insolubility; it drives spatially localized hydrophobic domains of model proteins back and forth between solubility and insolubility, and, in doing so, it catalyzes the set of energy conversions extant in biology.

8.1.4 Positive Cooperativity

The ΔG_{ap}-driven interconversion between states of solubility and insolubility exhibits positive cooperativity and as such becomes the physical basis for the allosteric effect of Monod.[5] Thus, ΔG_{ap}, demonstrated with designed model proteins, results in a positively cooperative interchange between hydrophobic association (insolubility) and hydrophobic dissociation (solubility) that constitutes the movable cusp of insolubility. *Therefore, we ask, are the Gibbs free energy of hydrophobic association and its operative apolar–polar repulsive free energy of hydration relevant to the molecular machines of biology? In other words, does Monod's "second secret of Life"*[6,7] *reside within the particulars of ΔG_{ap}?*

As shown in Chapter 5, the molecular basis of positive cooperativity and the source of ΔG_{ap} arise from changes in the competition for hydration that result when converting a functional group from a more polar to a less polar state and vice versa. When the polar species becomes neutralized, for example, $H^+ + -COO^- \rightarrow$ **–COOH**, or $-NH_3^+ \rightarrow$ **–NH$_2$** $+ H^+$, or an oxidized state $\rightarrow$ a **reduced state**, or $MgATP^{-2} + H_2O \rightarrow$ **MgADP$^-$** $+ HPO_4^{-2} + H^+$, the remaining less polar (the less charged, bold-faced) product no longer destructures as much hydrophobic hydration. The resulting reconstitution of sufficient additional hydrophobic hydration drives hydrophobic association, because the $(-T\Delta S)$ term becomes more positive and overwhelms the inherently negative ΔH term. In short, ΔG(solubility), which equals $\Delta H - T\Delta S$, becomes positive and solubility is lost.

Limitations in the amount of polar hydration, arising due to the competition for hydration with hydrophobic groups, result in large shifts in pKa values and in reduction potentials and associated changes in positive cooperativity (see Figure 5.20). *These hydrophobic-induced shifts in the free energy of the more polar state of functional groups, that is, groups that can occur in two or more states of polarity, are central to the hydrophobic consilient mechanism for energy conversion by controlling hydrophobic association/dissociation.*

8.1.5 Contractility as the Coupling of Hydrophobic Association and Elastic Force Development

Hydrophobic association within a protein chain constitutes an element of contraction. From our design and study of elastic-contractile model proteins, a contraction comprises two distinct but interlinked physical processes. They are hydrophobic association between domains within the protein construct and the associated development of an elastic force within interconnecting chain segments.[8]

The experimental condition of isometric contraction, the development of elastic force at fixed extension, most clearly brings to light the role of the elastic element of contraction. At this stage one begins to recognize two consilient mechanisms—the hydrophobic consilient mechanism for hydrophobic association and the elastic consilient mechanism for the fundamental process of elastic force development. The coupled physical processes of hydrophobic association/dissociation and changes in elastic force constitute key aspects of function and efficiency of contractile protein-based machines. Working in combination, the hydrophobic and elastic consilient mechanisms are considered relevant to the function of the Rieske Iron Protein movement of Complex III and of the myosin II linear motor, most notably during isometric contractions (see sections 8.3.4 and 8.5.3).

8.1.6 On the Relevance of Hydrophobic and Elastic Consilient Mechanisms to Biology's Protein-based Machines

Before proceeding further with these introductory comments, the reader should be aware that professionals in the specialized areas have proposed mechanistic detail for a number of biology's protein-based machines. None of those descriptions, however, has considered what we describe as the comprehensive hydrophobic effect. Although an awareness of hydrophobic effects is pervasive with a long history going back most notably to Edsall in 1935[9] and Butler in 1937,[10] there remains an absence of the concept of a competition for hydration between apolar and polar groups for hydration, that is, of the ΔG_{ap} component of the comprehensive hydrophobic effect. Consequently, there is no appreciation of the dominance of ΔG_{ap} as a controlling element within ΔG_{HA}. As a result, the proposed fundamental role of ΔG_{ap} within protein mechanisms has been absent.

Similarly for the elastic consilient mechanism, elastic force development can be the result of hydrophobic association resulting in the damping of internal chain dynamics within interconnecting chain segments, as we believe occurs in Complex III of the electron transport chain and the myosin II motor. Ideal or reversible elastic force development does not require random chain networks and it does not arise directly from changes in solvent entropy. Furthermore, the apolar–polar repulsive free energy of hydration, ΔG_{ap}, can provide a repulsive force that elastically constrains a protein segment providing an elastic impulse that can result in efficient motion, as we believe occurs in ATP synthase.

Accordingly, the perspectives in Chapter 5, developed on elastic-contractile protein-based polymers, introduce new concepts into the functional description of biology's protein-based machines. As with the introductory comments in Chapter 7, the footnote relevant to reactions toward new concepts in science is repeated here in footnote form.[11]

8.1.7 Our Platform from Which to Explore Energy Conversion in Biology

Furthermore, to the best of our understanding, *de novo* design of the family of elastic model protein-based machines, described in Chapter 5, and the extensive exploration of mechanism, presented therein only in part, stand unparalleled in the development of an understanding of energy conversion by protein-based materials. As such these demonstrated energy conversions, using *de novo* designed molecular machines, present a firm and warranted foundation on which to explore the relevance of the hydrophobic and elastic consilient mechanisms to biology's protein-based machines.

This chapter discusses key protein-based machines of biology to demonstrate the relevance of the hydrophobic and elastic consilient mechanisms. The objective in this chapter, therefore, is to investigate selected examples of biology's protein-based machines and to look at the molecular level for a coherence of phenomena with the designed elastic model protein-based machines described in Chapter 5. Accordingly, the following constitutes the second of two chapters that address the third assertion of this book, *the assertion of biological relevance*, as introduced in Chapter 1. The order in which we consider protein-based machines of biology will follow the natural flow of energy conversions in respiration from the redox-driven proton pumps, to the proton-driven rotary protein motor that produces ATP, and to a protein motor that utilizes ATP as the source of energy with which to drive function.

8.1.8 The Comprehensive Hydrophobic Effect: Can This Be a Consilient Mechanism for the Energy Conversions of Biology?

Can the comprehensive hydrophobic effect, developed in Chapter 5, be a consilient mechanism, a "common groundwork of explanation,"[1] for biology's protein-based machines? We put forward the point of view that a "common groundwork of explanation" does indeed pertain to the function of the diverse protein-based machines that sustain Life. The "common groundwork of explanation" resides in the thesis that introduces this chapter. Simply restated, biology thrives by means of controlling the association (insolubility) and dissociation (solubility) of hydrophobic domains of protein-based machines to achieve the essential energy conversions that sustain Life. Surprisingly, the extensively studied myosin II motor of muscle contraction provides a previously unrecognized and clear demonstration whereby addition of polar phosphate disrupts hydrophobic association and phosphate removal drives the hydrophobic association that provides the powerstroke of contraction (see section 8.5.3).

8.1.8.1 Brief Considerations Relevant to the Comprehensive Hydrophobic Effect

To date the sigmoid curve in Figure 5.10 offers perhaps the most effective introduction to the comprehensive hydrophobic effect. It is a plot of the reference temperature, either T_b or T_t,[12] for the onset of the inverse temperature transition for hydrophobic association versus ΔG_{HA}, the Gibbs free energy of hydrophobic association. Calculation of ΔG_{HA} utilizes the heat of the inverse temperature transition obtained on a series of (GVGVP)-based polymers in which guest pentamers (GXGVP) are introduced where X is each one of the naturally occurring amino acid residues as well as biologically relevant chemical modifications thereof and attached relevant protein prosthetic groups (see Tables 5.2 and 5.3). For an inverse temperature transition for hydrophobic association, the heat of the transition itself, under experimental conditions in Figures 1.1 and 7.1, becomes the Gibbs free energy for hydrophobic association, which by Equation 5.10b is written. $\Delta G_{HA}(\chi) \approx \Delta H_t(\text{ref}) - \Delta H_t(\chi)$.

8.1.8.1.1 The Comprehensive Hydrophobic Effect Resolves Three Categories of Amino Acid Residues

Figure 5.10 delineates the dependencies of ΔG_{HA} on T_t, or its equivalent T_b, into slopes of three categories—that of the aliphatic hydrocarbons, that of the aromatic hydrocarbons, and that resulting from the interaction of charged species with host hydrophobic residues. The slopes define the entropy change attending hydrophobic association for each category. The first category, defining the central part of the sigmoid, contains the amino acid residues that differ only by the number of aliphatic hydrocarbons (e.g., $-CH_2-$, $-CH-$, $-CH_3$), namely, Leu, Ile, Val, Pro, Ala, and Gly. These purely aliphatic amino acid residues define a linear relationship between ΔG_{HA} and T_t (obtained from temperature profiles for aggregation) or T_b (obtained from differential scanning calorimetry data). The slope for peptides containing only aliphatic hydrocarbons and peptide groups, the first category, is $\Delta S_{HA} = 80$ cal/deg-mole(GXGVP). The aliphatic amino acid residues give a large change in ΔG_{HA} (or in ΔS_{HA}) for a small change in T_t or T_b.

The second category, shown at low temperature, involves the amino acid residues with aromatic side chains, namely, Phe and Trp, which define a steeper dependence of T_t versus ΔG_{HA}. Tyr with its phenolic hydrogen is within the cluster of aromatic residues but is displaced from the F–W line. This low temperature segment for aromatic residues exhibits a slope, when defined by the data points for F and W, of $\Delta S_{HA} = 14$ cal/deg-mole(GXGVP).

The third category in Figure 5.10 is at high values of T_t or T_t and is defined by the charged residues K$^+$ and E$^-$, with a slope of $\Delta S_{FA} = 7$ cal/deg-mole(GXGVP). This slope results from the competition for hydration between apolar and charged species. The effectiveness of the formation of charged species in destructuring hydrophobic hydration is most readily seen as large increases in $T_t(T_b)$-values that result from charge formation.

8.1.8.1.2 Source of the Different Slopes for Aliphatic and Aromatic Amino Acid Residues

For the inverse temperature transition of hydrophobic association, the temperature for the onset of the transition, T_t or T_b, is approximately $\Delta H_t/\Delta S_t$, where ΔH_t (the net endothermic heat of the transition) and ΔS_t (the entropy change for the transition) derive from the change in structure and change in amount of water attending the transition from hydrophobic hydration to bulk water. The pentagonal dodecahedral structure of hydrophobic hydration surrounding a simple aliphatic hydrocarbon gas, due to Stackelberg and Müller,[14] appears in the lower part of Figure 2.8. As long as the hydrophobic group is comprised of noncyclic saturated hydrocarbons, as in the series of aliphatic residues L, I, V, P, A, and G of the first category in Figure 5.10, the pentagonal dodecahedral hydration structure remains quite representative. In other words, as long as the series of amino acid residues is essentially described by a change of a $-CH_2-$ or $-CH_3$ hydrocarbon unit, as in the first category, it is reasonable that a common slope for a T_t versus ΔG_{HA} plot would be experimentally observed. The $\Delta G_{HA}/CH_2 \approx -1$ kcal/deg-mole(GXGVP).

For an aromatic hydrocarbon, as in the phenyl group, $-C_6H_5$, of phenylalanine (Phe, F), however, only the arrangement of water around the periphery of the aromatic ring could begin to approximate a portion of the pentagonal dodecahedral arrangement. The water molecules hydrating the top and bottom surfaces of the planar aromatic structure would be expected to associate with different values for ΔH_t and ΔS_t on formation of hydrophobic hydration. Thus, the slope in the T_b versus ΔG_{HA} plot for aromatic, unsaturated cyclic groups would be expected to differ from that of simpler aliphatic hydrocarbons. As shown in Figure 5.10, the slope for aromatic groups is steeper than that for aliphatic groups.

8.1.8.1.3 Source of the Slopes for Amino Acid Residues with Different Charged Functional Groups

The slope defined by the data points for $K^+/0F$ and $E^-/0F$, which is due to the charged states of **Model Protein x′**, (GVGVP GVGVP G$\underline{K}^+$GVP GVGVP GVGVP GVGVP)$_{22}$(GVGVP), and **Model Protein i**, (GVGVP GVGVP G$\underline{E}^-$GVP GVGVP GVGVP GVGVP)$_{36}$(GVGVP), results from the hydrophobic hydration remaining after formation of the $-NH_3^+$ and the $-COO^-$ states, respectively. When the host polymer is poly(GVGVP), the formation of the charged species, $-COO^-$, destroys more hydrophobic hydration than does the $-NH_3^+$ charged species. Thus the position of a charged species along this slope indicates, among other things, the relative effectiveness of that charged species to destructure hydrophobic hydration in order to achieve its own hydration.

If we continue along this slope in Figure 5.10 to 860°C, which is the T_t-value obtained on extrapolation to one $-Ser-O-PO_3 =$ per (GVGIP) on phosphorylation of the serine (S) in poly[30(GVGIP),(GRGDSP)], a ΔG_{HA} of 8 kcal/mole-phosphate is found. Accordingly, the phosphate group becomes recognized as a remarkably effective charged species in destructuring hydrophobic hydration. In fact, phosphate is the most effective species that we have found for disrupting hydrophobic association (see Table 5.2), and its capacity to change ΔG_{HA} approximates its free energy of hydrolysis. In our view, this element constitutes the primary process whereby the rotary motor, ATP synthase, and the linear motor, myosin II, function.

8.1.8.1.4 Net Heat Changes for the Inverse Temperature Transitions of the Comprehensive Hydrophobic Effect Are Endothermic Due to the Conversion of Hydrophobic Hydration to Bulk Water!

Solubility is governed by the Gibbs free energy for solubilization, $\Delta G(\text{solubility}) = \Delta H - T\Delta S$. The early work of Butler[10] established the remarkable and fundamental finding that formation of water around hydrophobic groups is exothermic, that is, the formation of hydrophobic hydration constitutes a favorable exothermic reaction. He further demonstrated that solubility of organic molecules containing hydrophobic groups in water becomes lost because the $(-T\Delta S)$ term is positive for the CH_2 group and grows faster with each added CH_2 than the favorable, negative ΔH term (see Chapter 5, section 5.1.3.3). During an inverse temperature transition, solubility of hydrophobic groups is lost on raising the temperature, that is, hydrophobic groups associate because raising the temperature increases the magnitude of the positive $(-T\Delta S)$ term. Accordingly, this reverse change of hydrophobic hydration to bulk water, as occurs during an inverse temperature transition, should be an endothermic transition, as it is.

Putting aside changes in hydration that are the fundamental defining feature of inverse temperature transitions, the physical association of the model protein molecules containing hydrophobic groups is expected to be of the opposite sign, that is, to be exothermic, arising from effects such as van der Waals interactions. In preparation of Figure 5.10, the net endothermic ΔH_t was used to calculate ΔG_{HA}. The presence of an exothermic component resulting from physical association of model protein chains was not explicitly considered. As it should be, ΔG_{HA}, calculated from the heat of the inverse temperature transition for hydrophobic association, is the net result of all changes in energy that attend the inverse temperature transition. Nonetheless, an understanding of the comprehensive hydrophobic effect is incomplete without some sense of the relative magnitudes of the dominant endothermic and the smaller exothermic components.

8.1.8.1.5 Relative Magnitude of the Endothermic and Exothermic Components of the Inverse Temperature Transition and Relevance to Biology's Protein-based Machines

Recent work by the Rodriguez-Cabello group in Valladolid, Spain, using temperature-modulated differential scanning calorimetry (TMDSC), provides insight into the magnitude of the exothermic component of the inverse temperature transition and how much it reduces the experimentally measured endothermic heat.[15] As shown in Figure 8.1,[16] the TMDSC data for (GVGVP)$_{251}$ suggests for this composition that the oppositely signed exothermic component due to the physical association of chains is about one-third the magnitude of the endothermic heat that effects loss of hydrophobic hydration. The results in Figure 8.1 should help to clarify the comprehensive hydrophobic effect to those in the community of physical biochemists and biophysicists who yet think that the hydrophobic effect arises dominantly from the exothermic component. The endothermic heat of the inverse temperature transition, in fact defines the inverse temperature transition and is the basis for the "hydrophobic effect" or, as we have called the more complete understanding, the *comprehensive hydrophobic effect*.

Thus, values of ΔG_{HA} should be recognized as the net result of a transition dominated by the endothermic conversion of hydrophobic hydration to bulk water attending hydrophobic association with a lesser contribution due to the exothermic association of the model protein molecules, resulting from van der Waals interactions usually calculated using the Lennard-

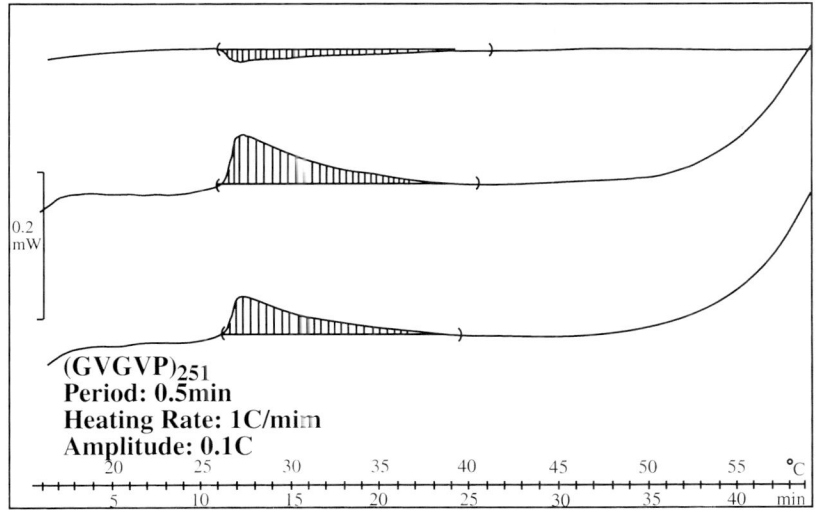

FIGURE 8.1. Component heats of hydrophobic association of an inverse temperature transition obtained by means of temperature-modulated differential scanning calorimetry (TMDSC). (Upper curve) An exothermic component of the inverse temperature transition due to the physical (van der Waals) interaction between associating molecules. (Middle curve) The endothermic component (due to disruption of hydrophobic hydration), which is the fundamental feature of an inverse temperature transition of hydrophobic association. (Lower curve) Net endothermic heat of an inverse temperature transition that one measures in the standard differential scanning calorimetry. This is the value used for determining the Gibbs free energy for hydrophobic association, ΔG_{HA}, arising out of an inverse temperature transition. Noting the expanded temperature scale, it is apparent that the width (the transition zone) of the curves determined here at a much higher concentration is similar to that of Figure 5.1C. (Preliminary communication from J. Carlos Rodriguez-Cabello · produced with permission of J. Carlos Rodriguez-Cabello.[15])

Jones 6–12 potential[17] or the Buckingham potential functions.[18] It will be interesting, in future work, to determine the relative magnitude of the endothermic and exothermic components for each of the amino acid residues and for other biologically relevant chemical modifications, as they contribute as guest residues to the inverse temperature transition of $(GVGVP)_n$ and of other informative host model proteins.

8.1.8.1.6 The "Waters of Thales" as a Requirement of the Consilient Mechanisms

As regards biology's protein-based machines, the comprehensive hydrophobic effect, dominated as it is by changes in hydrophobic hydration of the inverse temperature transition, would become less dominant as the presence of water in the more polar state of a protein-based machine became more limited, that is, as the "waters of Thales" decrease.[19,20] The long-recognized idea of cold denaturation of proteins (enzymes),[21] that is, the loss of structure required for function on lowering the temperature, constitutes the same phenomenon as the formation of structure on raising the temperature. On observation of the phenomenon in elastic-contractile model proteins, we developed the more general term of *inverse temperature transition*, as it can be observed under circumstances not relevant to cold denaturation, such as the dissolution of crystals of hydrophobically associated cyclic model proteins on lowering the temperature.[22] Of course, the comprehensive hydrophobic effect derives from the inverse temperature transition. The fundamental change underlying the inverse temperature transition, however considered, is one of changing the amount and nature of protein hydration.

One of the more challenging locations, therefore, for consideration of the comprehensive hydrophobic effect in the panoply of biological energy conversions is the electron transport chain embedded within the inner mitochondrial membrane. Essential parts of these protein-based machines insert into and function in very hydrophobic lipid bilayers. Here the ingress and egress of protons for development of the proton concentration gradient across the inner mitochondrial membrane becomes the output energy of an electron transfer process that is so crucial to living organisms. A central player, the ubiquinone coenzyme Q,[23] converts from very hydrophobic lipid diffusible species on the one hand to a positively charged species on oxidation at one site and on the other hand to a negatively charged species on receiving electrons at another site. Both occur as key elements of the function of Complex III of the electron transport chain.

By the hydrophobic consilient mechanism and specifically due to the apolar–polar repulsive free energy of hydration, ΔG_{ap}, formations of these charged coenzyme Q species contain the thermodynamic key to transforming "buried water molecules"[24,25] observed in the crystal structure into pathways for proton entry into and exit from the inner mitochondrial membrane that results in the development of the proton concentration gradient. In particular, formation by oxidation of the positively charged ubiquinol near one side of the membrane would disrupt hydrophobic association, allowing release of protons to that side of the membrane, whereas formation of the negatively charged ubiquinone by reduction near the other side of the membrane would disrupt hydrophobic association to open channels for the uptake of proton from that side of the membrane (see discussion below in section 8.4.4).

8.1.8.1.7 The Coupling of Hydrophobic Association with Development of Elastic Force

Complex III is an example of the consilient mechanism for elasticity that includes the coupling of hydrophobic association with development of an elastic force. In particular, the Rieske iron protein (RIP) of Complex III resides on the cytoplasmic side and contains a long hydrophobic α-helix that passes through the lipid bilayer from the cytoplasmic side to emerge on the matrix side with charged residues that combine to anchor the iron protein to the membrane. On the cytoplasmic side, a sequence of about 15 residues that is continuous with the transmembrane anchor tethers the globular component containing the FeS center.

By the hydrophobic and elastic consilient mechanisms the physical processes would proceed as follows: The most favorable hydrophobic association available to RIP is the hydrophobic association of the hydrophobic tip containing the FeS center with the Q_o site containing the ubiquinol within cytochrome **b**. For this most favorable hydrophobic association to occur, the tether becomes stretched. On changing the charge by addition of the negative electron to the FeS center and passing a second electron on to the heme of cytochrome b_L, the ubiquinol at the Q_o oxidation site takes on two positive charges that disrupt the hydrophobic association of this site. Hydrophobic dissociation of the very hydrophobic tip of the FeS center allows the stretched entropic elastic segment of the tether to lift the globular component of RIP using an aromatic side chain as a fulcrum and to flip the tip to hydrophobically associate at the site for reduction of cytochrome c_1. This represents an interlinking of hydrophobic association and elastic force development as repeatedly demonstrated with the elastic-contractile model proteins in Chapter 5. On oxidation of the FeS center at cytochrome c_1, the hydrophobic tip containing the FeS center returns to its most favorable hydrophobic association at the Q_o site containing a new molecule of ubiquinol to be positioned for the next cycle.

Thus, the challenge to the consilient mechanisms posed by the electron transport chain transforms into a showcase example that includes the two distinct but interlinked physical processes of the development of entropic elastic force during hydrophobic association to bring both consilient mechanisms to bear.

8.1.8.1.8 ΔT_t as a Simple On–Off Switch

Another point to note is the simplicity of looking at hydrophobic association/dissociation from the perspective of T_t. Should T_t be just above the operating temperature, say, the physiological temperature of 37°C, then reducing the expression of charged species sufficient to lower T_t the width of the temperature interval for the transition (see Figure 5.5) drives essentially complete hydrophobic association. Thus, the change in T_t, the ΔT_t, functions as a simple sliding on–off switch for driving hydrophobic association/dissociation. For the slope in Figure 5.10 resulting from charge formation, a ΔT_t adequate to turn on or off hydrophobic association requires a relatively small ΔG_{HA}.

8.1.8.1.9 Coherence Between ΔG_{ap} and ΔG_{HA} Calculated Using Different Experimental Data Demonstrates Presence and Dominance of the Consilient Mechanism in Model Proteins

For **Model Protein i** in Table 5.5, which is (GVGVP GVGVP GEGVP GVGVP GVGVP GVGVP)$_{36}$(GVGVP), abbreviated as E/0F, we can write ΔG_{HA}[E/0F → E$^-$/0F], where E stands for the –COOH state and E$^-$ for the –COO$^-$ state of the glutamic acid (Glu, E) residue From Figure 5.10 and Table 5.3, we have that ΔG_{HA}[E/0F → E$^-$/0F] is 5.22 kcal/mole-(GEGVP). Using the data in Figure 5.34, the pKa of **Model Protein i** is 4.5. Now the pKa of a carboxyl group in glutamic acid (Glu, E) or aspartic acid (Asp, D), absent significant electrostatic interactions and hydrophobic competition for hydration, is 3.8 to 4.0. Therefore, the pKa of **Model Protein i** is shifted by 0.5 to 0.7 pH units due to the presence of the valine (Val, V) residues from the situation without such hydrophobic residues and without significant proximity of other polymer charges.

From Equation (5.13) in Chapter 5, we have that ΔG_{ap} = 2.3 RT ΔpKa, such that, ΔG_{ap} for **Model Protein i** becomes 0.7 to 1.0 kcal/mole(E); this is for one E in six pentamers. As the value of ΔG_{HA}[E/0F → E$^-$/0F] is 5.22 kcal/mole-(GEGVP) for one E in each pentamer, 5.22 kcal/mole-(GEGVP) should be divided by six (5.22/6 = 0.87) for proper comparison. As 0.87 kcal/mole obtained from ΔG_{HA}[E/0F → E$^-$/0F] is within the range of 0.7 to 1.0 kcal/mole(E) obtained from ΔG_{ap} of acid-base titration data, we have coherence between the effect of ionization on change in the Gibbs free energy for hydrophobic association, ΔG_{HA}, which is derived from calorimetric data and the hydrophobic induced pKa shift, ΔG_{ap}, which is derived from acid–base titration data. Thus, an additional reason exists why $\Delta G_{HA}[E/0F \rightarrow E^-/0F]$ should be interpreted as resulting from an apolar–polar repulsive free energy of hydration.

Accordingly, the comprehensive hydrophobic effect provides an explanation for both pKa shifts and the control of hydrophobic association by changes in the polarity of functional groups such as quinones, flavins, nicotinamides, and hemes of cytochromes. Furthermore, the phosphorylation or dephosphorylation of a serine (Ser, S), of a threonine (Thr, T), or of a tyrosine (Tyr, Y) and the change from ADP to ATP and the most polar state of the localized presence of both ADP and inorganic phosphate (P_i) represent the most dramatic changes in polarity that occur routinely in biology. As discussed in section 8.1.11.2, a sound basis exists for the view that the high energy released on hydrolysis of phosphate compounds derives from a limitation in the availability of adequate hydration prior to hydrolysis. Therefore, controlling the availability of hydration for phosphate groups by varying the presence of competing hydrophobic groups provides the means whereby a protein can either energize a phosphate or utilize the energy of phosphate hydrolysis.

8.1.8.1.10 Does There Exist a Competition for Hydration Between Hydrophobic and Polar (e.g., Charged) Groups as a Key Part of the Functioning of Protein-based Machines of Biology?

In our view, ΔG_{ap} provides the basis whereby raising the free energy of ADP and P_i, by forced apposition of the very hydrophobic side of the γ-rotor in ATP synthase, results in synthesis of ATP. Also, in myosin II motor ΔG_{ap} provides the basis whereby this ATPase drives muscle contraction. In particular, in broad-brush strokes, ATP binds in a cleft directed in two directions, (1) toward the hydrophobic association of the cross-bridge to actin binding site and (2) toward the hydrophobic association between the head of the lever arm and the amino-terminal domain of the cross-bridge. Directing the ATP thirst for water in both directions by means of the cleft causes the cross-bridge to hydrophobically dissociate from the actin binding site and the head of the lever arm to hydrophobically dissociate from the amino-terminal domain of the globular component of the cross-bridge. Conversely, the splitting of ATP to form ADP plus P_i and release of P_i results in a decrease in ΔG_{ap} with the consequence of hydrophobic reattachment of cross-bridge to actin site and hydrophobic re-association of head of the lever arm to the amino-terminal domain to provide the power-stroke (the contraction) of the myosin II motor.

8.1.8.1.11 Positive Cooperativity Increases as Charged Species Compete for Water with Additional More-hydrophobic Residues

As developed in Chapter 5, positive cooperativity arises due to competition for hydration between apolar (hydrophobic) and polar (e.g., charged) residues constrained by location to coexist along a protein sequence and ultimately within the folded and assembled structure dictated by the sequence. By way of description of the molecular process, we begin with associated hydrophobic domains with a proximal carboxyl, –COOH. During occasional fluctuations, paired hydrophobic domains momentarily begin to dissociate. As too much hydrophobic hydration forms, re-association (insolubility of hydrophobic groups) recurs, because the free energy becomes unfavorable, that is, the $(-T\Delta S)$ term for solubility of the hydrophobic groups becomes too positive and overwhelms the inherently negative ΔH term.

During the occasional fluctuation of hydrophobic dissociation, a carboxyl ionizes to form the charged carboxylate, $-COO^-$. To form, the charged carboxylate, however, must achieve its hydration by destructuring hydrophobic hydration. This constitutes an increase in free energy for the system, and reclosure occurs. Should the pH be high enough to give the opportunity for formation of a second carboxylate during an opening fluctuation, the second carboxylate actually forms more readily than the first, because it has less hydrophobic hydration to disrupt. Because together the carboxylates destructure more hydrophobic hydration due to overlapping hydration spheres, the probability for remaining dissociated increases. The occurrence of two carboxylates increases the chance for formation of a third and fourth carboxylate, and so forth. As with the second carboxylate, the third and fourth carboxylates have less hydrophobic hydration to destructure, and therefore they have a greater probability of formation; they form with a lower pKa.

In fact, as seen in the Hill plots in Figure 5.31, ionizations of subsequent carboxyls occur at lower pKa values. As more valine (Val, V) residues are stepwise replaced by more hydrophobic phenylalanine (Phe, F) residues, this trend continues with an increase in the initial pKa value and greater decreases in the pKa values for subsequent ionizations during the hydrophobic dissociation process. By the time five Val (V) residues have been replaced by the more hydrophobic Phe (F) residues, as in **Model Protein v**, (GVGVP GVGFP GEGFP GVGVP GVGFP GFGFP)$_{42}$(GVGVP) in Figure 5.31A, the mean pKa has increased to 6.4. Furthermore, analysis of the Hill plot in Figure 5.31A shows that formation of the first carboxylate occurs with a pKa of 7.0 and that of the last carboxylate of the 42 repeating units of six pentamers occurs with a much lower pKa of 5.7.

8.1.8.1.12 Ionization of the Last Carboxyl at a pKa Greater than 3.8 to 4.0 Requires That a Residual Apolar–Polar Repulsive Free Energy of Hydration Exists Even in the Completely Unfolded, Fully Dissolved, Soluble State

Even when (GVGVP GVGFP GEGFP GVGVP GVGFP GFGFP)$_{42}$(GVGVP) is completely unfolded with F residues spatially distributed toward near-maximal distances from E$^-$ residues, to the extent permitted by the sequence, there remains an apolar–polar repulsive free energy of hydration. Even when the carboxylates are as completely surrounded by water as possible, they still cannot access water completely undisturbed by hydrophobic residues. The common explanation that pKa shifts result from the carboxyl being forced by some "conformational change" into a pocket of low dielectric constant within a protein is not tenable for the completely ionized, unfolded state of this model protein.

Even if the two Phe (F) residues most proximal in the sequence to the carboxylate are Val (V) residues as in (GVGVP GVGVP GEGVP GVGVP GVGFP GFGFP)$_{39}$(GVGVP), a residual pKa shift for the hydrophobically dissociated state remains after all carboxyls are ionized, as shown in Figure 5.31B. *These findings argue that treatments of the structure and function of protein-based machines are unsatisfactory until proper inclusion of the apolar–polar repulsive free energy of hydration occurs.*

8.1.8.2 The Amount of Hydrophobic Hydration on Potentially Complementary Surfaces Determines Whether Hydrophobic Association or Dissociation Occurs

Figure 8.2 provides an elementary schematic of the association of globular proteins with contoured surfaces for proper steric association, but each having different elements whereby complementarity could be achieved.[26] In part the following paragraphs demonstrate the relevance to globular proteins in the analyses of Chapter 5 that utilized model proteins of translational symmetry undergoing phase transitions. This section is preliminary to the more involved but parallel discussion of the hydrophobic association of myosin cross-bridge with actin (and of components within the cross-bridge) in the absence of ATP that dissociate in the presence of bound ATP, as schematically shown in Figure 2.16.

8.1.8.2.1 Complementary Surfaces Associate when They Have the Potential for Too Much Hydrophobic Hydration at a Given Temperature

Consider a situation with globular protein subunits that are soluble in solution at a sufficiently low temperature and that have simple complementary hydrophobic surfaces covered with much hydrophobic hydration, as in Figure 8.2A.[26] Raising the temperature from below to

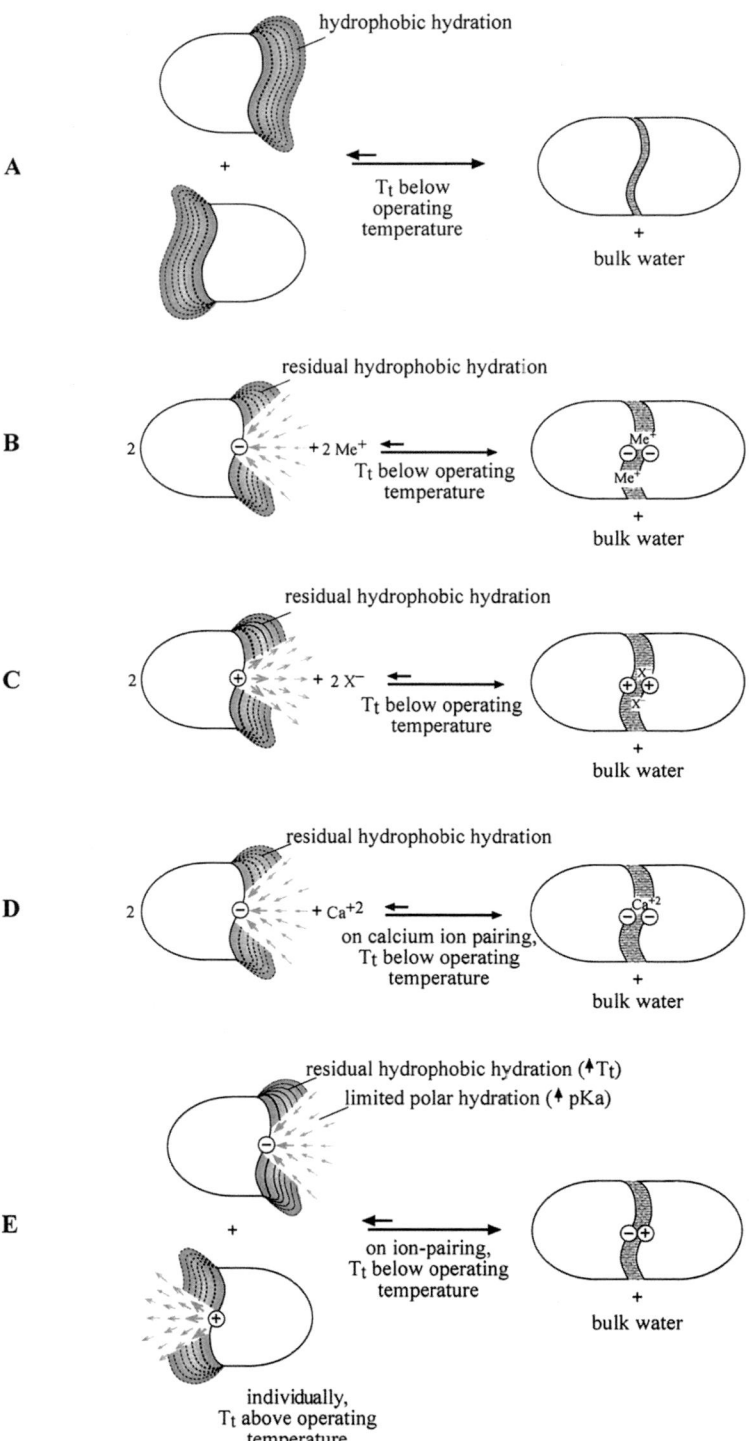

FIGURE 8.2. Elementary schematics (A–E) of the association of globular proteins with different elements to their complementary surfaces. (Reproduced with permission from Urry.[26])

above the temperature for the onset of the inverse temperature transition for hydrophobic association initiates dimer formation. Now, whether or not phase separation occurs depends simply on whether or not the dimer is soluble. If the dimer is soluble, there would be no phase separation. If the dimer were itself insoluble by means of extended association of dimers, there could be fiber formation with gelation and even phase separation. The former applies to hemoglobin S and to actin filament formation, where G(globular)-actin is soluble, but, when ATP binding initiates association, F(filamentous)-actin forms.

Independent of the question of dimer solubility, the dimerization process is one of hydrophobic association, and the change in Gibbs free energy for dimer formation by hydrophobic association, ΔG_{HA}(dimerization) = $\Delta H_D - T\Delta S_D$, obtains. In the case of hydrophobic association, the transition is endothermic; ΔH_D is positive. Therefore, as with the inverse temperature transition, association occurs because of a large negative $(-T\Delta S_D)$ term that results from the large positive change in ΔS_D as more ordered hydrophobic hydration becomes less ordered bulk water. Accordingly, dimerization occurs due to a sufficiently large and negative ΔG_{HA}, wherein changes in solvent entropy drive dimerization. Furthermore, the process can be discussed in terms of equilibria, where the sign and magnitude of the equilibrium constant $K_D = \exp(-\Delta G_{HA}/RT)$. As written, the dimerizations in Figure 8.2 are favorable with negative ΔG_{HA} values as crudely represented by the size and direction of the arrows.

8.1.8.2.2 Surfaces with Too Little Hydrophobic Hydration Do Not Hydrophobically Associate

When too few hydrophobic residues or too many charged species occur on the surface, too little hydrophobic hydration results. Dimerization will not occur, that is, the monomeric subunits retain their solubility in water. When the charged species exhibit hydrophobically shifted pKa values, then the surfaces are poised for association, particularly if the charges on opposing surfaces are of the opposite sign and are arranged in a complementary way, as in Figure 8.2E. A number of examples follow in the next paragraph.

8.1.8.2.3 Charged Species Destructure Hydrophobic Hydration But Experience an Increase in Free Energy on Doing So

As shown in Figure 8.2B, C, even when sufficiently hydrophobic surfaces have the same charge, the oppositely signed ion in solution can pair with the surface charge, and dimerization by hydrophobic association can result.[26] This was observed with elastic-contractile model proteins in Figure 5.12. The more hydrophobic poly[0.8(GVGIP),0.2(GEGIP)] associates at physiological temperature in the presence of 0.3 N or more NaCl, whereas the less hydrophobic poly[0.8(GVGVP),0.2(GEGVP)] does not (see Figure 5.12).

In Figure 8.2D calcium ion triggers hydrophobic association of negatively charged surfaces. As shown in Figures 5.15 and 5.27, calcium ion is particularly effective in lowering the value of T_t of **Model Protein I**, (GVGIP GFGEP GEGFP GVGVP GFGFP GFGIP)$_{40}$ (GVGVP); 2E/5F/2I, and **Model Protein ii**: (GVGVP GVGFP GEGFP GVGVP GVGVP GVGVP)$_{40}$(GVGVP); E/2F. When the model protein is more hydrophobic and contains two negative carboxylates in position for bidentate interaction with the divalent calcium ion, the decrease in T_t occurs at a much lower ion concentration for **Model Protein I**, which may be considered a poor man's E–F hand. In particular, Figure 5.27 shows the decrease in T_t to mirror the increase in hydrophobic hydration, which ties directly to Figure 8.2D.

When the globular subunits have oppositely signed charges on their surface, as in Figure 8.2E, each can exist alone as soluble monomers in solution, but, when combined in a single solution, they associate to form dimers. The analogy for the model proteins is shown in Figure 6.3. In particular, the two model proteins, **Model Protein x′**: (GVGVP GVGVP GKGVP GVGVP GVGVP GVGVP)$_{22}$(GVGVP;

K/0F and **Model Protein ii**: (GVGVP GVGFP GEGFP GVGVP GVGVP GVGVP)$_{40}$ (GVGVP); E/2F are each soluble in water at physiological temperatures. When the two solutions are combined, however, they hydrophobically associate as the temperature is raised above 24°C (see Figure 6.3B).

The consilient mechanism depicted in the elementary structures of Figure 8.2 are relevant to protein function arising out of association of globular components of molecular machines. The principles arise in almost textbook fashion in the function of the myosin II motor where the key charged species are ATP, ADP, and P$_i$, but in an important way also include the charged side chains (see section 8.5.3).

8.1.8.3 "Oil and Vinegar Don't Mix," But, When Oil-like and Vinegar-like Side Chains of Proteins Are Forced to Coexist by Virtue of Structure (Primary, Secondary, Tertiary, and Quaternary), They Interact in the Most Profound Way to Achieve Function

Thus, the old adage that "oil and vinegar don't mix" remains relevant when the oil-like and vinegar-like groups are constrained to coexist along a protein sequence and in proximity to a pair of hydrophobically associating/dissociating surfaces. Although structure causes these disparate groups to interact, they nonetheless attempt to separate as oil from vinegar. In fact, as substituents along a protein chain, oil-like and vinegar-like side chains effectively repulse each other as they each reach out for water undisturbed by the other. Another important element to recognize is that the apolar–polar repulsion can propagate along a water-filled cleft. Ion pairs, at a distance from the ATP binding site within the hydrophobically lined cleft, can be caused to separate by ATP binding, and then the separated ion pairs, even by small distances, assist to propagate polar dominance to a more distant pair of hydrophobic faces. As briefly considered immediately below, this driving force provides a more efficient mechanism for energy conversion in an aqueous medium than the more commonly considered electrostatic mechanisms.

8.1.9 On Efficiencies of Energy Conversion in the Aqueous Medium of Biology

This section considers efficiency of energy conversion by protein-based machines of biology by noting two points. The points arise from studies on elastic-contractile model proteins and draw from relatively new concepts concerned with control of hydrophobic association and with the nature of elasticity. The first point provides direct experimental evidence for a dramatically greater efficiency of the apolar–polar repulsion mechanism over the classic electrostatic charge–charge repulsion mechanism for chemomechanical transduction by polymers in water. The first point goes on to address electrochemical transduction efficiency with relevance, in principle, to the energy conversion of the electron transport chain.

The second point addresses the nature of elastic force development in relation to understanding efficient energy conversion. If the energy required for chain deformation during elastic force development becomes lost to other parts of the protein and to the surrounding water, then so too is efficient energy conversion lost. In other words, elastomeric force development on deformation in a protein-based machine followed by marked hysteresis on relaxation necessarily denotes an inefficient protein-based machine.

8.1.9.1 For Mechanochemical Transduction in Water Apolar–Polar Repulsion-based Is More Efficient than Charge–Charge Repulsion-based Molecular Machines

8.1.9.1.1 Simple Comparisons of the Chemomechanical Transductional Efficiencies, $\eta = f\Delta l/\Delta\mu \times \Delta n$, of the Charge–Charge Repulsion and the Apolar–Polar Repulsion, ΔG_{ap}, Mechanisms

The two molecular systems to be compared are PMA, poly(methacrylic acid), which is [–CH$_2$–(CH$_3$)C(COOH)–]$_n$, and **Model Protein iv** in Table 5.5, namely, (GVGVP GVGFP GEGFP GVGVP GVGFP GVGFP)$_n$ (GVGVP), abbreviated as E/4F. The two poly-

8.1 Thesis

mers were cross-linked to achieve similar elastic moduli, and both respond to changes in proton chemical energy, $\Delta\mu\Delta n = (-2.3RT\Delta pH)\Delta n$. In the following paragraphs, visual comparisons will be made with reference to Figures 5.34 (comparing Hill coefficients) and 5.35 (comparing the area of acid base titration curves).

The statement of efficiency, η, for proton-driven chemomechanical engines can be written as $\eta = f\Delta L/\Delta\mu\Delta n = f\Delta L/(-2.3RT\Delta pH)\Delta n$. Because the chemical energy, $\Delta\mu\Delta n$, is proportional to the change in pH, the slope of the acid–base titration curve provides an immediate comparison, where an increase in the magnitude of the Hill coefficient means a proportionately greater efficiency. The Hill coefficient for PMA is 0.5, whereas, as seen in Figure 5.34, that for E/4F is 2.7, suggesting on this basis an expectation of at least a fivefold more efficient energy conversion by E/4F.

The chemical energy also contains the number of moles of proton, Δn_H, consumed in the process. To consider both $\Delta\mu_H$ and Δn_H, the acid–base titration curve can be used. The y-axis contains the amount of acid consumed on protonating the carboxylates, and the x-axis provides the ΔpH required to do so. Accordingly, the area defined by the acid–base titration curve gives a measure of the chemical energy. Because PMA has one carboxyl for every two-backbone atoms, whereas E/4F has only one carboxyl for every 90-backbone atoms, many more protons will be consumed on driving the PMA chemomechanical engine, as is apparent from Figure 5.35. Missing in this comparison, however, is the amount of work performed by each engine. As discussed below, this is directly determined and compared in Figure 5.36.

8.1.9.1.2 Experimental Comparison Following pH Dependence of Length at Fixed Force Using Acid–Base Titration Curves, That Is, During Actual Performance of Mechanical Work

Figure 5.36 shows the change in length for the constant load of 2 grams, where the changes in length, despite very different consumptions of energy, are quite similar. In the comparison in Figure 5.36, the PMA engine required 6.86 × 10^{-5} mole of protons, whereas the E/4F engine requires 6.12× 10^{-6} mole of protons. If comparison is made for the best $\Delta L/\Delta pH$ slopes for each engine, as shown in Figure 5.36, the ratio of efficiencies, η_{ap}/η_{cc} is of the order of 40. *Clearly, in the aqueous milieu of biology the electrostatic mechanism of charge–charge repulsion is not a serious contender for the efficient machines available to biology.* As noted in section 8.1.8.3, ionizable functional amino acids and other polar prosthetic groups forced, through the process of biosynthesis (see Chapter 6), to coexist within a protein chain with substantial hydrophobic residues provides the ingredients for efficient protein function.

8.1.9.2 Efficiency of Electro-chemical Transduction in Water

8.1.9.2.1 Experimental Data on Electro-chemical Transduction Efficiency

The previous considerations of energy conversion resulted in the performance of mechanical work, and, for the considered model proteins, there is the obvious realization that hydrophobic association is tantamount to contraction. The control of hydrophobic association, however, can interconvert any two energies that can individually drive hydrophobic association. The example here draws from the demonstration that reduction of a redox couple can power contraction by driving hydrophobic association and that protonation of a carboxyl can do likewise. Both processes, protonation of a carboxylate and reduction of a redox group, effect a decrease in ΔG_{HA}, and both functional groups are affected by any ΔG_{HA}. As shown in Figures 5.20B, 5.23, 5.25, 5.29, 5.31, 5.32, 5.34, and 5.35B, Chapter 5 is replete with hydrophobic-induced pKa shifts. Equivalent hydrophobic-induced shifts in reduction potential for amino-methyl nicotinamide are shown in Figure 5.20C. Because pKa values and reduction potential are both coupled to hydrophobicity, it is to be expected that a change in redox state that changes hydrophobicity would bring about a change in pKa and vice versa. Here we demonstrate the efficiency of that conversion for a modest level of host polymer hydrophobicity.[27]

Reduction-induced pKa shift was demonstrated in the designed polytricosapeptide in Table 5.5, that is, **Polymer i:** Poly(G<u>D</u>GFP GVGVP GVGVP GFGVP GVGVP GVGK{NMeN}P), abbreviated as DK{NMeN}/2F, containing both redox (LysNMeN) and carboxyl (Asp, D) functional entities. From Table 5.2, the difference in ΔG_{HA} on going from the oxidized state, NMeN$^+$, to the relevant reduced state, 6-OH tetrahydro NMeN, is an approximate -4 kcal/mole-NMeN. On the other hand, for the pKa shift, the change in $\Delta G_{HA} = \Delta G_{ap} = 2.3RT\Delta pKa = -2.3(582$ cal/mole$)(2.5) = -3.35$ kcal/mole. From the preceding information, the efficiency can be written as η = (output energy/input energy) 100 = {$\Delta G_{ap}/\Delta G_{HA}$[LysNMeN$^+$ (oxidized) → LysNmethyl-6-OH-tetrahydronicotinamide (reduced)]}100 = (3.35/4)100 ≈ 83%. Thus, one estimates a high efficiency when the coupling of functional groups occurs by means of a common dependency on ΔG_{HA}.

8.1.9.2.2 Increase in Hydrophobicity Directly Shifts Reduction Potentials and Increases Efficiency, That Is, the Hill Coefficient and Positive Cooperativity, for NMeN

Attaching NMeN to the appropriate designed series of microbially prepared polymers gives the following polymers from Table 5.5:

Model Protein ii': (GVGVP GVGFP GK{NMeN}GFP GVGVP GVGVP GVGVP)$_{22}$ (GVGVP), K{NMeN}/2F
Model Protein iii': (GVGVP GVGVP GK{NMeN}GVP GVGVP GVGFP GFGFP)$_{22}$ (GVGVP), K{NMeN}/3F
Model Protein iv': (GVGVP GVGFP GK{NMeN}GFP GVGVP GVGFP GVGFP)$_{21}$ (GVGVP), K{NMeN}/4F

The indicated increases in hydrophobicity on replacing Val (V) by Phe (F) were found to cause systematic shifts in the reduction potential and corresponding increases in steepness (positive cooperativity) as shown in Figure 5.20C. In particular, the reduction potentials are K{NMeN}/0F (-145 mV), K{NMeN}/2F (-127 mV), K{NMeN}/3F (-107 mV), and K{NMeN}/4F (-90 mV).[28] Thus, for the redox couple NMeN/NMeN$^+$ the reduction potentials and efficiency depend on hydrophobicity in a systematic way, analogous to the dependence on hydrophobicity of pKa and efficiency for a chemical couple such as $-COOH/-COO^-$. On this basis, the dependence of reduction potential on the hydrophobicity of surrounding residues for players, such as heme b$_H$ and heme b$_L$, in the electron transport chain can be considered.[28]

8.1.9.3 Further Consideration of the Relevance of the Nature of Elastic Force to Efficient Mechanisms for Energy Conversion

8.1.9.3.1 The Case of Isometric Contractions

The development of force under conditions of fixed length, as in an isometric contraction, involves the elastic deformation of a chain or chains within the protein-based machine. On relaxation, ideal elastic elements return the total energy of deformation to the protein-based machine for the performance of mechanical work. Thus, the approach toward high efficiency for the function of a protein-based linear motor, or even for the RIP domain movement in Complex III, depends on how nearly the extension of an elastomeric chain segment approaches ideal elasticity.

It is important to realize that the model protein (GVGVP)$_{251}$ can exhibit ideal elastic behavior.[8] It is also significant to recognize that the source of ideal elastic force resides in the internal motions, the internal chain dynamics, of the elastic chain segment. To the extent that the energy of deformation becomes dispersed into molecules or chain segments not integral to the extended chain segment, energy is lost and efficiency decreases. Thus, the nature of the elastomeric force limits the possible efficiency of a protein-based machine. For an effective protein-based machine, deformations of chains should occur in such a way that the energies involved be recoverable.

8.1.9.3.2 Relevance of Nonideal Elasticity (Hysteresis) to Efficient Performance of Mechanical Work

As noted above for *an ideal elastomer*, the energy expended in deformation, for example,

8.1 Thesis

in extension, is completely recovered on release of the deforming force, that is, on relaxation. On the other hand, for *a nonideal elastomer*, all of the energy expended in deformation is not recovered on relaxation. In this case, the stress–strain curve is said to exhibit hysteresis.

8.1.9.3.3 Performance of Mechanical Work with an Isometric Contraction Followed by an Isotonic Contraction with Relaxation to Initial Force

During an isometric contraction, for example, due to hydrophobic association, force develops by the damping of internal chain dynamics of an interconnecting elastic chain segment (see Figure 5.8). To the extent that the chain segment exhibits hysteresis, on allowing the decrease in chain length to ensue (as in an isotonic contraction), only a part of the energy of deformation is recovered. Such a mechanochemical engine could be no more efficient than allowed by the nonideal (hysteretic) elastic chain segment.

In our analysis of hysteresis, appreciation was clear for an elastic segment that stores all of the deforming energy in the chain backbone. Thus, a mechanochemical motor, in which loss of deformation energy into associated chains occurred, would decrease efficiency for the conversion of chemical energy into mechanical work and vice versa. This begins to place constraints on the mechanism of efficient biological motors, and it begins to limit the utilization of levers, ratchets, and physical pushing derived from mechanical interactions between chains when searching for a more efficient mechanism for the performance of mechanical, chemical, or even electrical work.

8.1.10 The Electrostatic Argument of a Low Dielectric Constant Versus ΔG_{ap} to Explain pKa Shifts and Ion-pair Formation

8.1.10.1 Why Ion Pairs Form in Protein-based Polymers

Salts dissolve in water because the free energy of the separated hydrated ion is lower than when associated with other ions within the salt crystal. Ionized side chains of proteins remain in water as long as they retain adequate hydration. Therefore, any process that decreases effective hydration of ionic side chains of proteins becomes a driving force for ion pairing.

Commonly, hydration has been treated by assuming a uniform medium of a given dielectric constant, ε, where the force, f, between two charges, q and q', is given by the product, qq', divided by the distance, r, between charges squared times the dielectric constant, that is, $f = qq'/\varepsilon r^2$. When the charges are on a pair of oppositely charged plates, with water between them, the value of ε for water is experimentally determined to be about 80 at room temperature. Thus, by this argument, and consistent with limited water, ions pair only when ε becomes sufficiently small. In calculations of protein, an ε as low as 5 or less is used, and in calculations of small peptides where there are partial charges on atoms instead of a charge of one or more, ε has been commonly taken as unity.[29,30]

Qualitatively, the argument for using a low dielectric constant seems reasonable. The low dielectric constant comes into question, however, on closer scrutiny of experimental results on model proteins for which control of composition is possible, and it makes difficult what must be subjective judgments of a correct dielectric constant for water-filled clefts and crevices of various dimensions found in many protein-based machines.

8.1.10.2 The Usual Electrostatic Argument of a Low Dielectric Constant as Responsible for pKa Shifts

The usual electrostatic consideration of pKa shifts has been to assume that the ionizable function resides in a location within the protein that is of low dielectric constant. As the reasoning goes, because of the low dielectric locale, the carboxyl function does not ionize to form carboxylate until the pH is much higher, that is, the carboxyl exhibits a pKa shift. A simple means of relating pKa shift to dielectric constant uses the expression for solvation energy, SE, due to Born, $SE = [(ze)^2/2r](1 - 1/\varepsilon)$, where z is the charge on the ion, e is the unit electron charge, ε is the dielectric constant of the surrounding medium, and r now stands for the radius of the ion. As treated in Chapter 5, section 5.7.8.3, the interest is in fitting the series of experimental data points due to pKa shift as a function of mole fraction, f_K, of the lysine functional group in the polymer, poly[f_V(GVGIP),f_X(GXGIP)] with X = Lys(K) and f_K varied from 1.0 to 0.06.[31] The data are included in Figure 5.30 along with the data for poly[f_V(GVGIP),f_X(GXGIP)] with X = Glu(E), and f_x again varied from 1.0 to 0.06.[32]

Interest centers on the change in solvation energy, $\Delta SE = Q[(\varepsilon_{i-1} - \varepsilon_i)/\varepsilon_{i-1}\varepsilon_i]$, where Q is substituted for the coefficient $[(ze)^2/2r]$. To fit the data in Figure 5.30A,B, a dielectric constant of 5 or less had to be used. Yet from experimental determination of the polymer, poly(GVGIP), at the phase-separated concentration of 61% polymer and 39% water by weight, the minimal value for the dielectric constant was 65.[33] Because the polymer is a repeating pentapeptide sequence, because the barrier to backbone mobility is between 1 and 1.5 kcal/mole, and because any introduction of a charged functional group must raise the value of the dielectric constant, there is no way that the functional group can be held in a locale of low dielectric constant. There must be another physical basis for the hydrophobic-induced pKa shifts in Figure 5.30. As developed in Chapter 5 and reviewed briefly above, the physical basis is the competition for hydration between hydrophobic and charged groups, called **ΔG_{ap}**, the apolar–polar repulsive free energy of hydration.

Accordingly, one should maintain a healthy skepticism of the results of calculations of protein structure and function where low dielectric constants become the basis to drive ion-pair formation. This is especially the case when there are water-filled crevices and clefts on the surface of proteins and indeed coursing through the protein structure. The "waters of Thales" are there as an integral part of protein structure and function. To assume otherwise has been a useful approximation in the past, but it would seem no longer to be the most productive approach.

Another limitation of the assumption of low dielectric constants to explain pKa shifts is the absence of an explanation for the obligatory coupling of pKa shifts to positive cooperativity shown in Figures 5.20, 5.23, 5.31, and 5.34. Also, as argued above, positive cooperativity is a central aspect of efficient energy conversion by protein-based machines and naturally arises out of a competition for hydration between hydrophobic and charged groups.

Positive cooperativity arises from a competition for hydration between hydrophobic and charged residues just as do pKa shifts. On the other hand, the classic electrostatic argument would appear to have no obvious way to become integral with positive cooperativity: Furthermore, in Figure 1.4 a residual pKa shift of 1.7 pH units is found under circumstances where the model protein is completely unfolded. Under such circumstances one is compelled by the electrostatic analysis to choose a dielectric constant approaching 80 where there would be no significant energy of interaction. In our view, the classic electrostatic argument is simply left wanting, unable to describe the extensive data on elastic-contractile model proteins.

8.1.10.3 Electrostatics Become Significant in Protein Structure and Function When Charges Are Under the Influence of Hydrophobic Domains

Electrostatic interactions, considered here simply as the interactions of charged species, become significant in proteins in aqueous media when under the influence of hydrophobic domains (see Figures 5.30A,B), that is, when competing for adequate hydration with hydrophobic groups. In particular, pKa shifts and changes in reduction potential do not necessarily result from being buried in a localized region (medium) of low dielectric constant within the protein or from direct electrostatic interactions, but rather result from competition for hydration between hydrophobic and polar groups.

It is our belief that the presence of hydrated hydrophobic side chains limits the available hydration for ionic side chains and increases the probability of ion pairing. *Thus, while the formation of ion pairs may be considered an electrostatic event, it occurs in the aqueous milieu of protein folding, assembly, and function due to the dominance of the presence of, or potential for, hydrophobic hydration.*

8.1 Thesis

Above, in section 8.1.9.1, we considered a purely electrostatic mechanism for chemomechanical transduction. A high density of charged groups dominated a relatively minimal presence of hydrophobic groups in the cross-linked polymer PMA. Its efficiency for chemomechanical transduction was compared with the inverse circumstance of limited charge under the influence of a dominantly hydrophobic model protein. The latter proved to be much more efficient even though the functional group was the same, the –COOH/–COO⁻ chemical couple.

8.1.11 ATP: Biology's Energy Currency

The apolar–polar repulsive free energy of hydration, ΔG_{ap}, as reviewed above, results from a competition for hydration between charged species and hydrophobic groups. In the most extreme case reported thus far, this competition raised the pKa of a carboxyl from about 4 to about 11.[27] This constitutes an amount sufficient to raise the free energy of carboxylates by 8 to 10 kcal/mole. The basic issue to be addressed here is whether there is reason to believe that the free energy of the phosphate group would also be subject to competition for hydration with hydrophobic groups.

Simply stated, is the free energy of hydrolysis of ATP and other related phosphoanhydrides sensitive to limitations in hydration? If the answer is no, then the apolar–polar repulsive free energy of hydration, ΔG_{ap}, would not be relevant to ATP's role in energy conversion. If the answer is yes, then phosphates would indeed be sensitive to the competition for hydration with hydrophobic groups, as extensively documented experimentally with the carboxylate group. Again, if the answer is yes, then the development, within Chapter 5, most specifically section 5.7 and the associated data in Figures 5.23 through 5.32, of the concept of an apolar–polar repulsive free energy of hydration would be fundamentally congruent with the energizing and functioning of high-energy phosphates in biology.

In the process of building the case for this equivalence of energizing carboxylates and phosphates, a number of quotations will be drawn from the literature rather than relying in all cases on our interpretation of the literature. In doing so, the case for high-energy phosphates resulting from limited hydration will become very evident, and the energizing of phosphates, as with the energizing of carboxylates as evidenced by hydrophobic induced pKa shifts, becomes another illustration of consilience, that is, of a "common groundwork of explanation."

In terms of the molecular process whereby ATP forms and functions as biology's energy coin, there are two aspects—the protonation/deprotonation of carboxylates due to a proton concentration gradient that drives a rotor to form ATP from ADP and P_i, as in the ATP synthase, and the molecular process whereby the breakdown of ATP to ADP with release of P_i drives the protein-based machines of biology, as in the myosin II motor of muscle contraction.

8.1.11.1 Standard Free Energies of Hydrolysis of Phosphate Bonds

In this section the standard free energy of hydrolysis of ATP (and its equivalent nucleotide triphosphates, NTPs) to ADP and AMP and related biologically phosphorylated molecules are noted.[34] The standard free energy of hydrolysis of ATP to form ADP and P_i can be obtained from the equilibrium constant for the reaction

$$ATP \rightleftarrows ADP + P_i \quad (8.1)$$

to give the equilibrium constant, K_{eq},

$$K_{eq} = ([ADP][P_i])/[ATP] \quad (8.2)$$

The reference conditions for the reaction must be carefully defined and maintained, especially the availability of water and the presence of salts, when making comparisons of different high-energy compounds. From equilibrium theory, K_{eq} can be written in terms of the change in Gibbs free energy ΔG for the reaction

$$K_{eq} = e^{-\Delta G/RT} \quad (8.3)$$

TABLE 8.1. Standard Free Energies of Hydrolysis of Phosphate Bonds directly involved in Energy Conversion by Protein-based Machines.[1]

Biological compound	ΔG°′ (kJ/mol)	ΔG°′ (kcal/mol)
[2]Phosphocreatine	−43.1	−10.3
[3]PP$_i$ → 2P$_i$	−33.5	−8.0
[4]ATP → AMP + PP$_i$	−32.2	−7.7
[5]ATP → ADP + P$_i$	−30.5	−7.3

[1] Free energies obtained from W.P. Jencks in G.D. Fasman (Ed.) *Handbook of Biochemistry and Molecular Biology* (3rd ed), Physical and Chemical Data, Vol. I, 296–304, CRC Press, 1976.
[2] Energy reservoir in muscle and nerve cells used to regenerate ATP as it is depleted during cell function.
[3] The reaction catalyzed by the ubiquitous enzyme, pyrophosphatase, that shifts the equilibrium irreversibly toward chain growth in the synthesis of proteins and nucleic acids (See Chapter 4).
[4] The hydrolysis utilized in synthesis of nucleic acids and proteins that gives rise to pyrophosphate, **PP$_i$**, the splitting of which to inorganic phosphate makes chain growth irreversible (See Chapter 4).
[5] The basic reaction used to power the protein machines of biology.

Accordingly, for a favorable reaction where the equilibrium is very much toward products, the change in Gibbs free energy for the reaction ΔG(reaction) is a substantial negative quantity.

The standard free energies for a number of biological high-energy phosphates are given in Table 8.1,[34] in terms of both calories (cal or kcal) and Joules (J), where 1 cal = 4.184 J.

8.1.11.2 Considered Sources of the Standard Free Energies of Hydrolysis of Phosphoanhydride Bonds of ATP

Historically, three mechanisms have emerged to provide an understanding of the basis for so-called high-energy phosphate compounds. They are opposing resonance, electrostatic repulsion, and, more recently, limited hydration of the multivalent ions of di- and triphosphates.

8.1.11.2.1 Resonance Destabilization of the Bridging Oxygen Atom

In phosphoanhydrides, resonance structures may be written that shift the electron density of the P-O-P backbone in both directions toward the phosphate atoms and away from the bridge oxygen between phosphates. This is considered to destabilize and increase the free energy of the phosphoanhydride bond, a concept developed by Kalckar[35] and given the term *opposing resonance*. This shifting of density away from a bridge oxygen has been suggested to have possibly greater relevance for carboxyl phosphates with the C-O-P backbone in which resonance structures shift density toward the phosphate on the one hand and toward the carbonyl (C=O) carbon on the other.

8.1.11.2.2 Electrostatic Repulsion and Electron Distribution Along the P-O-P Backbone

When fully ionized, ATP carries four negative charges, four electronic charges, distributed among the nonbackbone oxygen atoms of the triphosphate string.[36–38] Packing these negative charges, over the oxygens within the volume the size of the triphosphate structure is expected to result in charge–charge repulsion and, therefore, would be described as a configuration of high energy. Even in this case hydration would be expected to relax this high-energy state, and protonation to remove charge could be expected to have a greater effect. Given this argument, it is somewhat surprising that protonation to neutralize the expected repulsion does not appreciably lower the energy released on hydrolysis.[39] Accordingly, this argument is left wanting.

8.1.11.2.3 Limited Hydration

As argued in Chapter 5 and briefly reviewed above, the case was built for limited hydration of carboxylates that arose out of competition for hydration between charged groups and hydrophobic groups. This was considered responsible for raising the free energy of carboxylates in experimentally demonstrated hydrophobic-induced pKa shifts. Similarly, the result of ion pairing in protein systems was argued as being driven by limited hydration arising again out of relaxing the competition for hydration between charged groups and hydrophobic groups.

In their 1970 landmark paper, George et al.[39] stated the case well in approaching the issue of high-energy phosphate compounds:

The very existence of ions in aqueous solution is due to their solvation energies compensating for the large amount of energy required to disrupt the crystal lattice in the dissolution of an ionic crystal, or to break the covalent bond to hydrogen and bring about electron transfer in the ionization of an acid. ... Hence even in the absence of values for the high energy phosphate compounds the conclusion is inescapable that at biological pH an important term contributing to their thermodynamic reactivity will be the difference between the solvation energies for the multi-charged reactant and product ions.

Drawing from their thermodynamic data and calculations on ATP, ADP and AMP hydrolyses, George et al.[39] concluded that the source of release of high energy on hydrolysis of ATP and ADP is a limitation of hydration in the reactants, for example, in ATP.

Almost two decades later Hayes et al.[40] concluded,

Although intramolecular (opposing resonance and electrostatic) effects play an important role in determining the energy of hydrolysis in some of these reactions, it is concluded that in those hydrolyses of most importance in energy storage and transduction (ATP → ADP + orthophosphate, and phosphocreatine + ADP → creatine + ATP), relative solvation energies of reactant and products are by far the most important factors in determining these energies.

From de Mies,[41]

However, from the values of Table IV, it can be inferred that a small change in the organization of solvent around the molecules of reactants and products might easily lead to a significant change in the thermodynamic parameters of a reaction.

This brings us directly to our argument that, from these data Hayes et al.[40] concluded that solvation energies of reactants and products are by far the most important factors in determining the energies of hydrolysis of phosphate compounds that are used for energy storage and transfer in the living cell. The same conclusion has been reached recently by Ewig and Van Wazer[42] who calculated the energy of hydrolysis of pyrophosphate in the gas phase. Furthermore, due to intramolecular competition for hydration, even greater effects of limited solvation energy can be expected for the phosphoanhydrides of biology such as ATP and ADP that also contain the adenine aromatic hydrophobic group than for the inorganic phosphate products calculated by Ewig and Van Wazer.[42]

8.1.11.3 Means of Energizing a Phosphate Sufficient for Addition to ADP

8.1.11.3.1 Our View: Competition for Hydration with Hydrophobic Moieties Energizes Polar Species

From an analysis of the hydrophobic-induced pKa and reduction potential shifts with the associated destruction of hydrophobic hydration on formation of carboxylates (see Figure 5.25B), we argue that the solvation limitation of phosphates can be made critical by the competition for hydration between polar and hydrophobic domains.

8.1.11.3.2 Analogy Between Formation of High-energy Phosphate Bonds and Hydrophobic-induced pKa Shifts: The General Argument

The magnitude of this apolar–polar repulsive free energy of hydration, when the polar species is carboxylate, has been demonstrated experimentally to approach the order of 40 kJ/mole or 8 to 10 kcal/mole. This would be sufficient to raise the free energy of an inorganic phosphate for addition to an ADP and in general to destructure hydrophobic hydration and open hydrophobic folds.

8.1.11.3.3 Question: How is the Free Energy of Phosphate Raised Sufficiently for Addition to ADP with Formation of ATP?

The free energy of phosphate is raised by the forced confrontation of the site containing ADP and P_i (or $R-P_i$) with a hydrophobically hydrated surface, as argued below for ATP synthase.

8.1.11.3.4 Question: How Can the Binding of ATP Drive a Conformational Change?

The binding of ATP to a site in sufficient proximity to a spatially localized hydrophobic association can result in the separation of associated hydrophobic domains. In the absence of ATP, the site exhibits transient partial openings due to normal thermal fluctuations. During such a fluctuation, the build up of too much hydrophobic hydration causes the hydrophobic association to re-form, because the T_t-divide is below the operating temperature for that spatially localized association. If during the fluctuation ATP should become proximal, the incipient hydrophobic hydration is destructured; there is no longer a driving force for closure, and the conformational change involving separation of associated hydrophobic domains occurs. The cross-bridge of the myosin II motor demonstrates such features in its interaction with actin and in the interaction of the head of the lever arm with the amino-terminal domain (see Figures 2.17 and 8.53).

8.1.11.3.5 Question: Over What Distances Can such Destruction of Hydrophobic Hydration Be Effective?

The destructuring of hydrophobic hydration by polar species is a cooperative process. This allows the opening of a hydrophobic association that includes an ion pairing to proceed with the result of adding the separated positive and negative charges for each to function in their own right as polar species for disruption of further removed hydrophobic associations in a sort of propagating domino or unzipping effect. Thus, the distance over which the destabilizing of hydrophobic associations can occur depends on the structure involved, but could readily become several nanometers, as is the case for the GroEL/GroES chaperonin (see Chapter 7).

The specific distance would depend on the details of the distances between the newly emerged charges on dissociation of the ion-pair-containing hydrophobic domains. If we consider 20 Mrad γ-irradiation cross-linked **Model Protein i** of Table 5.5, (GVGVP GVGVP GEGVP GVGVP GVGVP GVGVP)$_{39}$(GVGVP), in the contracted state, the density of carboxyls is such that the carboxyls are separated by more than 2 nm. By the time 50% ionization has occurred, the volume has increased 10-fold, the maximal allowed with 20 Mrad γ-irradiation cross-linking. This occurs with a mean distance between charges of more than 4 nm. Contraction due to hydrophobic association begins as carboxylates become less than one every 4 nm. Thus the effect of a charged carboxylate can be said to reach out a distance of more than 2 nm in its capacity to destructure hydrophobic hydration. The domino effect of separating ion pairs can be expected to increase the distance further.

The multivalent phosphates, already significantly limited in hydration, can be expected to reach out substantially further than carboxylates in their search for adequate hydration. This effect becomes enhanced when phosphate access to water is limited. When the phosphate occurs at the base of a cleft, the direction for access of water becomes severely limited and the thirst for hydration can be directed by the cleft to target sites of hydrophobic association. In other words, the cleft functions as a conduit to direct the thirst for hydration. By means of the cleft, the capacity for disrupting hydrophobic hydration to target sites can be boosted by effecting separation of ion pairs enroute, which boost the polar species capacity to disrupt hydrophobic hydration. Accordingly, the use of structure to direct the forces of apolar–polar repulsion becomes a useful design feature in certain ATP-driven protein-based machines, such as the myosin II motor.

8.1.11.3.6 The Distance Issue Can Also Be an Issue of Contact Area

In another context, the distance issue can become a matter of contact surface area of the hydrophobically associating hydrophobic domains that are held together by a negative ΔG_{HA}. Even if the surface area for hydrophobic contact were more than 10 nm^2, only sufficient hydrophobic hydration would have to be destroyed during a transient fluctuation to cause ΔG_{HA} to become positive. The change from a negative to a positive ΔG_{HA} is to shift from dominantly associated to primarily dissociated.

8.1.11.3.7 Estimate of $\Delta G_{HA}(PO_4^=)$ on Phosphorylation of Poly[30(GVGIP),(RGYSLG)]

As calculated and discussed in section 5.3.4.4.4, the $\Delta G_{HA}(PO_4^=)$ due to phosphorylation of poly[30(GVGIP)(GRGDSP)] at the Ser (S) residue raises the value of T_t to an extent that the calculated increase in free energy for hydrophobic association becomes about

- *Moving the T_t-divide by whatever means χ:*

$$\Delta G_{HA}(\chi) = \Delta H_t(GVGIP) - \Delta H_t(GVGVP) \quad (1)$$

where χ is the interaction that changes ΔH_t from that of the reference polymer and $\Delta G_{HA}(\chi)$ is the change in Gibbs free energy for hydrophobic assocciatiob due to the interaction.

- *$K \approx 1$ for the following phosphorylation by ATP*

$$ATP + poly[30GVGIP),(RGYSLG)] = \\ ADP + poly[30GVGIP)],(RGYS\{PO_4^=\}LG) \quad (2)$$

Using the slope, $\Delta T_t/\Delta G_{HA}$, defined by charged species $K^+/0F$-$E^-/0F$ of Figure 5-10 (6.8 cal/mol-pent/deg) and $\Delta T_t(PO_4^=)$ of 860°K allows an estimate of $\Delta G_{HA}(PO_4^=)$. Accordingly, $\Delta G_{HA}(PO_4^=) = 6.8 \times 860 + 2.3 \approx 8$ kcal/mol-pentamer

∴ **Phosphorylation raised the free energy of the insoluble (hydrophobically assocuated) state of the model protein by ~ 8 kcal/mol-$PO_4^=$, i.e., polymer-$PO_4^=$ is in an energized state in the soluble, unfolded model protein, as implied by $K \approx 1$.**

FIGURE 8.3. Independent estimates of the increase in free energy on phosphorylation of poly[30(GVGIP),(RGYSLG)].

8 kcal/mole. Also, the equilibrium constant for the enzymatic phosphorylation of poly[30(GVGIP),(RGYSLG)] using ATP was essentially one, which by Table 8.1 would be essentially 7 to 8 kcal/mole for hydrolysis of the terminal phosphate of ATP under standard conditions. This argument is laid out in Figure 8.3. Accordingly, these experimental results provided direct demonstration that the serine–phosphate bond can be fully energized by an increase in competition for hydration with hydrophobic groups. From this, we conclude that the free energy of hydrolysis can be a direct function of the presence of hydrophobic residues and that an increase in free energy of phosphates can arise from a competition for hydration.

8.1.11.4 Uses of ATP in Construction and Maintenance of the Living Organism

The following is a partial list of the uses of ATP or an energetically equivalent NTP: (1) to produce the macromolecules nucleic acids, proteins, and polysaccharides; (2) to produce the organelles and membranes (lipid biosynthesis); (3) to pump ions across cell membranes against concentration gradients to maintain the proper electrochemical gradients across cell membranes; (4) to transport nutrients into the cells and remove waste; (5) to drive trafficking of molecules, vesicles, and organelles (e.g., directional transport of Golgi, lysosomes, and chloroplasts) within the cells; (6) to use molecular chaperones to achieve proper protein folding and proteasomes and cellusomes to degrade spent protein; (7) to change supercoiling (Type II topoisomerase) and correct errors in nucleic acids; (8) to achieve motility as in cell crawling, muscle contraction, the beating of cilia and flagella, the actuation of hearing; (9) to provide additional energy requirements that include motor-dependent tasks in the cell such as "mitosis and cytokinesis, the morphogenesis of the endoplasmic reticulum, exo-, endo-, and pinocytosis, the polarization of the egg, gastrulation, and even thought itself [neuronal growth-cone motility and the axonal transport of neurotransmitter-containing vesicles]."[43]

8.1.12 ATPases, Biology's Workhorse Protein-based Machines, and ΔG_{ap}

8.1.12.1 The Disposition of ATP at Its ATPase Binding Site

From the perspective of the consilient mechanism and its functional component, ΔG_{ap}, the most striking feature of ATPases derives from the way in which ATP molecules attach to the protein. ATP binds near an aqueous interface of the protein. Even more telling is that the ATP molecule orients such that the triphosphate tail resides at the aqueous interface. This is where hydrolysis occurs. More revealing with regard to mechanism, ATP often binds with its phosphates at the edge of a water-filled cleft.

8.1.12.2 The Specific Case of the ATPase of the Myosin II Motor

In particular, as a step in the functional cycle of the myosin II motor, ATP binding occurs at a water-filled cleft.[4] In the scallop muscle, as shown below, the water-filled cleft can be seen to communicate in two directions. In one direction it communicates with the hydrophobic association of an actin cross-bridge attachment site. In the other direction the water-filled cleft communicates to the hydrophobic association involving principally the head of the lever arm and the amino-terminal domain of the cross-bridge. In both cases the hydrophobic associations become disrupted on ATP binding. In the first case binding results in detachment from the actin filament. In the second case, by disrupting the hydrophobic association between the amino-terminal domain and the head of the lever arm, ATP binding causes the cross-bridge to reach toward its next actin site of attachment. A credible mechanism requires that it provide insight into these fundamental structural features of ATP binding to ATPases, in general, and specifically to the ATPase that is the myosin II motor.

By the hydrophobic consilient mechanism for the myosin II motor and specifically by means of ΔG_{ap}, ATP binding effects both hydrophobic dissociation from the actin binding site and release of the hydrophobic association at the head of the lever arm, allowing the cross-bridge to move forward toward the next attachment site. Of course, loss of phosphate would reconstitute the hydrophobic associations, that is, would effect hydrophobic re-attachment to the actin binding site in concert with re-association of the head of the lever arm with the amino-terminal domain to result in the powerstroke. Section 8.5.4 presents crystal structure stereo views from which the above-noted perspective derives.

8.1.12.3 Three Aspects of Force Generation in ATPases

The concept of an apolar (hydrophobic)–polar (e.g., charge) repulsive free energy of hydration, ΔG_{ap}, contributes to understanding the mechanism whereby ATPases function in three distinguishing respects. First and second, ATP binding, and particularly on hydrolysis with formation of ADP plus P_i, has the potential to effect both "push" and "pull" components of force. Third, release of P_i results in development of an elastic "pull" component of force that is most in evidence during isometric contractions. These three elements of force development are discussed immediately below.

8.1.12.4 Binding ATP Energizes a Protein-based Machine by the Mechanical Result of Hydrophobic Dissociation Due to ΔG_{ap}

In reviewing their studies on the F_1 motor of ATP synthase, Oster and Wang[44] recently stated, "In many motors, the force generating step is associated with the binding of nucleotide to the catalytic site. *We propose that this is true of all ATPases*. In particular, for the F_1 motor this is the only way to reconcile all of the biochemical and mechanical measurements with its high mechanical efficiency." (The italics are the authors' own emphasis.) As presented here, the hydrophobic consilient mechanism makes a distinction between energizing and the use of that energy in force generation. In our view, this distinction is not so significant for the rotary F_1 motor of ATP synthase functioning as an ATPase. In this case by the hydrophobic consilient mechanism, ATP-driven rotation of the rotor results from repulsion due to an increase in ΔG_{ap}, the apolar–polar *repulsive* free energy of hydration.

As discussed below in section 8.4.4.11, ATP binding provides the major "push" component of force, but we expect the peak in ΔG_{ap} to occur at the moment of hydrolysis when the charge concentration is greatest with the momentary presence of both ADP and P_i. In the synthesis function of the F_1 motor of ATP synthase, we expect that the maximum repulsion occurs between the most hydrophobic side of the rotor and the ADP and P_i state and that this maximal repulsion decreases on ATP formation, which, in the consilient view, drives ATP formation. Accordingly, because repulsion is the force that drives the ATPase function of the F_1 motor and because repulsion drives rotation, ATP binding would provide near-maximal force generation, enhanced only at the moment of hydrolysis to form ADP plus P_i.

8.1.12.5 Development of an Elastic "Pull" Component of Force Resulting from an Apolar–Polar Repulsion on ATP Binding

On the basis of the consilient mechanism, and ΔG_{ap} in particular, one can also consider the development of elastic force on ATP binding to result from the damping of internal chain dynamics by repulsion of otherwise more kinetically free chains or loops. Insight into this aspect of the consilient mechanism surfaces with **Model Protein v** in Figures 1.4 and 5.31, as considered in the associated discussions in Chapters 1 and 5. Repulsion between the negatively charged carboxylates and hydrophobic phenylalanine (Phe, F) residues results in a repulsive free energy of 2.4 kcal/mole-carboxylate, as measured experimentally by the residual pKa shift, after dissolution is complete. In this case keeping the phenylalanine (Phe, F) residues distant from the carboxylates of the glutamate (Glu, E⁻) limits torsion angles that would otherwise be possible if the repulsion were not present. This constitutes a damping of internal chain dynamics. This repulsion limits the freedom of backbone motion; it restricts the backbone torsion angle combinations that would bring carboxylates and phenylalanine side chains too close. Such a restriction of accessible torsion angles constitutes a decrease in chain entropy that should be expressed as an increase in elastic force between the beginning and end points of the restricted chain segment. If the 30mer chain segment were held at the mean distance during repulsion, removal of the repulsion, in this case protonation to form –COOH, should relax the deforming force.

Based on the consilient interpretations of the model protein data, the presence of a negatively and multiply charged ATP separated by water molecules from a nearby chain or loop that contained hydrophobic residues would repulse the hydrophobic groups, causing them to reside at a greater distance and possibly in the shadow of the backbone. This constitutes a "deformation" that limits the freedom of motion of the chain or loop.

Similarly, apolar–polar repulsion between a highly charged group and very hydrophobic side chains of a chain or loop of protein would be expected to limit the freedom of rotation about backbone bonds of sufficiently kinetically free chain segment or loop. This increase in ΔG_{ap} would decrease the number of states accessible to the polymer and decrease the entropy of the chain. The resulting development of an elastic force in the chain segment or loop of protein could be used to perform mechanical work of changing protein structure. The mechanism of elasticity would now take on the description of force development due to "a damping of internal dynamics by repulsion" rather than by extension. In this specialized circumstance, ATP binding could provide a useful force-generating step in an ATPase, also not inconsistent with the suggestion of Oster and Wang.[44]

Accordingly, reasoning from the consilient mechanisms, release of phosphate has the potential to relax the apolar–polar repulsion and relax an entropic elastic deforming force. Thus, instead of an electrostatic repulsion, for example, the less efficient charge–charge repulsion, there would be the apolar–polar repulsion, which is argued above in section 8.1.9.1 and shown in Figure 5.36 to be more than 10 times more efficient as a mechanism for chemo-mechanical transduction. In our view, to be efficient, a machine should avoid the frictional losses inherent in latches, ratchets, and mechanically driven camshafts. Entropic elastic chains and through-solvent apolar–polar repulsive forces provide for efficient interconversion of chemical and mechanical energy. At present, however, we have no example of this development of elastic force in the myosin II motor. Therefore, it stands as a potential source of force development without example, as yet, in the protein machines of biology.

8.1.12.6 Release of P_i Discharges on Energized Protein-based Machine by the Third Mechanical Result of a "Pull" Component of Force Due to Re-established Hydrophobic Association

When considering ATPases that function as classic contractile linear motors, the distinction between energizing and force generation becomes much clearer. ATP binding energizes by disrupting hydrophobic association, perhaps usefully boosted at the moment of hydrolysis, but then force generation does not occur until release of P_i that allows hydrophobic association of the contracted state to recur. This could be thought of as a "pull" component of force generation that under isometric conditions expresses as the generation of an elastic force. Thus, ATP binding could be viewed as effecting relaxation, whereas loss of ATP yields the ultimate in contraction, approaching a rigor-like state, as noted in Chapter 7. Such would be the case unless or until a swelling pressure became relevant.

Release of the γ-phosphate from protein-bound ATP, leaving bound ADP, restores much hydrophobic association that existed in the protein before ATP binding. Release of the γ-phosphate permits reconstitution of sufficient hydrophobic hydration to drive hydrophobic association. In the broad view of the consilient mechanism, as regards ATPases, ATP binding causes hydrophobic dissociation, and P_i and ADP release re-establishes maximal hydrophobic association. In terms of the movable cusp of insolubility, binding of the polar ATP molecule raises the temperature of the movable cusp of insolubility to give solubility, and the decrease in polaritys on phosphate release lowers the movable cusp of insolubility to re-establish the insolubility of hydrophobic association

In the preceding discussion of ATPase types of protein-based machines, the hydrophobic and elastic consilient mechanisms, by means of a single operative component of the Gibbs free energy of hydrophobic association, namely, that of apolar–polar repulsion, have provided three distinct contexts for force development. Thus, we believe the consilient mechanisms will prove to be fundamental to the function of the ATPases of biology. As argued below, evidence emerges for the underlying competition for hydration between hydrophobic and polar (e.g., charged) groups to be a basic element in the functioning of key protein-based machines of biology.

8.2 Overview of Energy Conversion in Biological Systems

8.2.1 The Overall Energy Conversions: Photosynthesis and Respiration

The overall energy conversions of biology were briefly considered in Chapter 2, the overview chapter, and grossly represented in Figures 2.10 (for photosynthesis) and 2.11 (for respiration).

Here the statements of photosynthesis and respiration are given below for ready reference in order to place in context the component energy flows that sustain Life.[45,46]

Photosynthesis, in overview, uses light energy from the sun to combine water and carbon dioxide to produce carbohydrate and oxygen. Respiration, in overview, is exactly the reverse with the exception that, instead of an input of light energy, the output is chemical energy, as given in Figure 8.4 and considered below.

The light reactions themselves involve energy steps almost 10 times greater than the energy steps involved in the subsequent reactions of photosynthesis and of respiration. In general, there are many thousands of subsequent reactions, where the energy steps are 8 to 10 kcal/reaction or less. From this perspective, the task of gaining a sense of what sustains Life would seem to be overwhelming.

Fortunately, as will be discussed below for respiration, there are only a few tens of reactions in the oxidation of a glucose molecule to produce $6(CO_2)$ and $6(H_2O)$. Most significantly, there are but three products that constitute the most immediate resulting chemical energy. These are reduced nicotinamide adenine dinucleotide (NADH, 10 molecules), reduced flavin

<u>*Photosynthesis*</u>: **The first step in biology's reversal of the arrow of time**

carbon dioxide + water + light energy
$\quad 6(CO_2) \quad\quad\quad 6(H_2O)$

$\quad\quad\quad\quad\quad\quad\quad$ → **carbohydrate + oxygen**
$\quad$ (1) $\quad\quad\quad\quad\quad [C(H_2O)]_6 \quad\quad 6(O_2)$

<u>*Respiration*</u>: **Animals and plants in the dark use the chermical energy of respiration to create more diverse and complex life forms.**

carbohydrate + oxygen
$[C(H_2O)]_6 \quad\quad 6(O_2)$

$\quad\quad\quad\quad$ → **water + carbon dioxide + chemical energy (2)**
$\quad\quad\quad\quad\quad 6(H_2O) \quad\quad\quad 6(CO_2)$

But these remarkable chemical equations do not explain how!

FIGURE 8.4. Photosynthesis and respiration represent very broad expressions of the energy conversions that sustain Life.

adenine dinucleotide (FADH$_2$, 2 molecules), and adenosine triphosphate (ATP, 4 molecules). As ATP is the basic energy denomination of biology, the complete conversion of one glucose molecule to the chemical energy currency of biology comes down to the oxidation of NADH and FADH$_2$ in the production of an additional 32 molecules of ATP. This involves five protein-based machines that constitute what has been called *oxidative phosphorylation* and that occur in the mitochondria, the so-called energy factories of biology. Four of the protein-based machines (Complexes I through IV) effect the pumping of protons, and the fifth complex utilizes the return of the protons across the inner mitochondrial membrane to produce the 32 molecules of ATP.

In general, then, the energy conversions of biology reduce to the production of ATP and the uses of ATP, that is, the production of ATP by the five protein-based machines of the inner mitochondrial membrane and the thousands of subsequent protein-based machines that do the necessary work of the cell. This constitutes yet an enormous task that will fill hundreds of volumes in the future of protein-based machines. The intention of this volume, however, is to add a simplifying feature of a "common groundwork of explanation" for each of the hydrophobic and elastic consilient mechanisms. For the function of protein-based machines of biology, this perspective recovers an attractive element of simplification.

8.2.1.1 Photosynthesis for the Production of Carbohydrate (Glucose)

The following statement of photosynthesis is essentially a question of a balanced chemical equation where each atom among the reactants, on the lefthand side of the equation, is accounted for in the products on the right–hand side. For photosynthesis,

$$\text{Carbon Dioxide} + \text{Water} + \text{Light energy}$$
$$6(CO_2) \quad\quad 6(H_2O)$$
$$\Rightarrow \text{Carbohydrate} + \text{Oxygen} \quad\quad (8.4)$$
$$[C(H_2O)]_6 \quad\quad 6(O_2)$$

The remarkable simplicity of this statement, known for over a century, is stunning. It involves only the atoms of carbon (C), oxygen (O), and hydrogen (H) with truly simple stoichiometry. While there is great beauty in the simplicity of the equation, it obscures intricate detail of just what does happen in the process and, importantly, how it happens.

8.2.1.2 Respiration: Oxidation of Glucose with Ultimate Reduction of Oxygen to Result in Water and Carbon Dioxide

The chemical products of photosynthesis become the chemical reactants of respiration, and the chemical products of respiration become the chemical reactants of photosynthesis. Of course, absent the light energy for photosynthesis, there would be no products of carbohydrate and oxygen, and absent the chemical reactants of respiration, there would be no chemical energy with which to sustain Life. Thus, instead of oxidizing glucose by fire to give rise to heat, as occurs in fireplaces during the burning of wood (which is cellulose composed of strings of glucose molecules), living organisms evolved the capacity, step by step, to turn the oxidation of glucose into production of the energy contained within the ATP molecule. For respiration,

$$\text{Carbohydrate} + \text{Oxygen} \Rightarrow \text{Water}$$
$$[C(H_2O)]_6 \quad\quad 6(O_2) \quad\quad 6(H_2O)$$
$$+ \text{Carbon Dioxide} + \text{Chem Energy}$$
$$6(CO_2) \quad\quad\quad\quad\quad\quad\quad (8.5)$$

Again, the simplicity of the chemically balanced equation, this time representing respiration, belies what may seem an almost bewildering underlying set of complex reactions. Before addressing the key energy-converting steps of respiration, however, another pair of analytical expressions points in the direction of recognizing the separation of the hydrogen atom into its proton, H^-, and its electron, e^-, which represent elemental features laid bare without the blur of molecular detail.

Respiration can be seen as a pair of half-reactions, which provides initial insight into the underlying process carried out by the metabolic machinery of the cell. For oxidation of glucose, the oxidative half reaction is

$$C_6H_{12}O_6 + 6(H_2O) \Rightarrow 6(CO_2) + 24H^+ - 24e^-$$
$$(8.6)$$

For reduction of oxygen, the reductive half reaction is

$$6(O_2) + 24H^+ + 24e^- \rightarrow 12(H_2O) \quad (8.7)$$

Addition of the two half-reactions gives the expression for respiration on explicitly including the statement for the chemical energy obtained. The analytical simplicity of the half-reactions lays out the underlying essential result of the biological electron transport chain. Indeed, the electron transport chain of the mitochondrion achieves the separation into protons, H^+, and electrons, e^-. (For the structure of the mitochondrion, refer to Figure 8.5, below.)

The protons are released to one side of an otherwise generally proton-impermeable inner mitochondrial membrane to collect the protons in the space between the inner and outer membranes of the mitochondrion. The resulting proton concentration gradient then drives formation of ATP by the quintessential protein-based machine, ATP synthase, as the protons flow back through the inner mitochondrial membrane by means of another special path effecting proton permeability. Thus there are two fundamental questions. The first is, how does electron flow within the membrane achieve unidirectional proton flow across the membrane? The second is, how does the return flow of protons result in the formation of ATP, the energy coin of biology?

Incredibly, the marvelous work of many biochemists, biophysicists, and crystallographers has provided reaction stoichiometries, mechanistic details, and structures. It is a beautiful evolving story into which we would like to introduce the reasoning of the consilient mechanisms. As presented in Chapter 5, the hydrophobic and elastic consilient mechanisms developed from inverse temperature transitions of hydrophobic association exhibited by elastic-contractile model proteins functioning as protein-based machines capable of interconverting the same energies interconverted by living organisms. In particular, we introduce the insight of an apolar–polar repulsion into considerations of understanding molecular mechanism. First to be considered, however, are a number of intermediate molecular players and processes in the early oxidation stages of glucose in preparation for utilization of the reduced molecules in the biological electron transport chain.

8.2.1.3 Products of Glucose Oxidation by the Reactions of Intermediary Metabolism

Despite what at initial consideration may appear as maze-like detail, the many reactions that accomplish respiration, given as the oxidation of glucose to CO_2 and water, group as two cycles tied together by a single reaction. The first cycle is the *glycolysis cycle* that by way of the *transition reaction* feeds into the *citric acid cycle*, also called the Krebs cycle or the tricarboxylic acid cycle. As briefly enumerated below, the products of these reactions either are or become convertible to the universal energy currency of biology, ATP.

8.2.1.3.1 Glycolysis Cycle

By the glycolysis cycle, the oxidation of glucose yields two molecules of pyruvate, two molecules of reduced nicotinamide adenine dinucleotide (2 NADH), and two molecules of adenosine triphosphate (2 ATP).

8.2.1.3.2 Transition Reaction

In the transition reaction, the two pyruvate molecules are converted to two molecules of acetyl coenzyme A, two molecules of carbon dioxide (2 CO_2), and two molecules of reduced nicotinamide adenine dinucleotide (2 NADH).

8.2.1.3.3 Citric Acid (Krebs) Cycle

The Krebs cycle takes two molecules of acetyl coenzyme A from the transition cycle and converts them into four molecules of carbon dioxide (4 CO_2), six molecules of reduced nicotinamide adenine dinucleotide (6 NADH), two molecules of reduced flavin adenine dinucleotide (2 $FADH_2$), and two molecules of adenosine triphosphate (2 ATP).

8.2.1.3.4 Summary of the Oxidation of Glucose by Intermediary Metabolism

The products of intermediary metabolism for the oxidation of one molecule of glucose are 6 molecules of carbon dioxide (6 CO_2), 10 molecules of reduced nicotinamide adenine dinucleotide (10 NADH), 2 molecules of reduced

8.2 Overview of Energy Conversion in Biological Systems

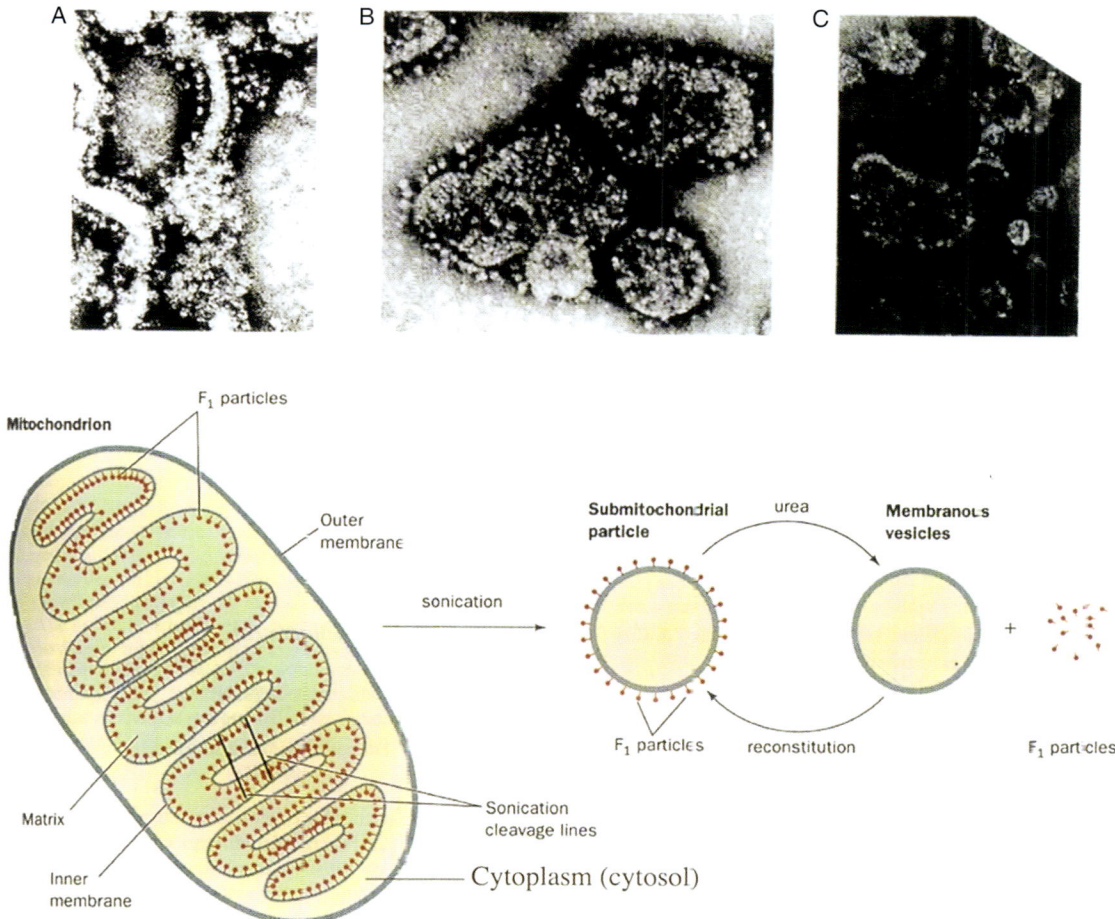

FIGURE 8.5. The mitochondrion, the energy factory of the cell. *(Top)* Electron micrographs **(A)**, **(B)**, and **(C)** of the inner mitochondrial membrane studded with stalks and headpieces that are the extramembrane components of ATP synthase with the remainder contained within the inner membrane. **(Bottom)** Drawing of a mitochondrion with an outer membrane and a folded inner mitochondrial membrane enclosing the matrix portion and separated from the outer membrane by the cytoplasm, also referred to as the *cytosol*. (**A.** Reprinted with permission from D.F. Parsons, Science **140**, 985 (1963). Copyright © 1963 AAAS. **B, C**, and lower part reprinted from *Biochemistry*, Second Edition, D. Voet and J. Voet, Copyright © 1995, John Wiley & Sons, New York. Reprinted with permission of John Wiley & Sons, Inc.)

flavin adenine dinucleotide (2 $FADH_2$), and 4 molecules of adenosine triphosphate (4 ATP).

8.2.2 Oxidative Phosphorylation in the Mitochondrion: The Primary Source of ATP

8.2.2.1 *Structure of the Mitochondrion*

A schematic representation of the mitochondrion and of vesicles derived from it, as well as electron micrographs showing the stalk and headpiece of ATP synthase extending from the membrane, is given in Figure 8.5.[47] Furthermore, the four complexes of the electron transport chain that produce the proton concentration gradient across the inner mitochondrial membrane and the fifth complex, ATP synthase, that utilizes the proton concentration gradient to produce ATP are schematically represented in Figure 8.6.[48]

Part of our challenge, to assess the relevance of the hydrophobic and elastic consilient mechanisms and specifically of apolar–polar

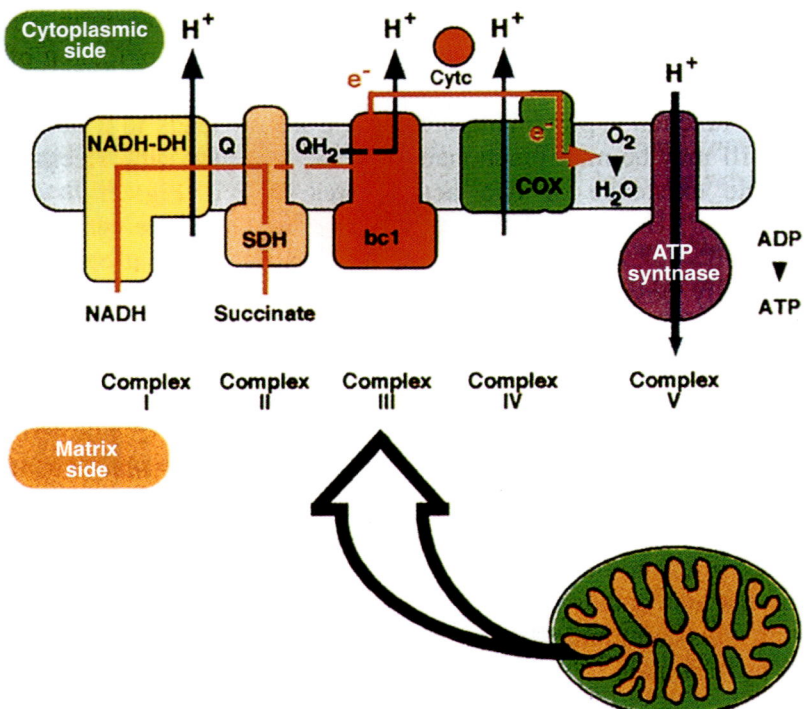

FIGURE 8.6. Oxidative phosphorylation of the mitochondria, showing the four complexes of the electron transport chain that pump protons across the inner mitochondrial membrane into the cytoplasmic side to produce a higher concentration of proton within the inner membrane cytosolic space and showing the ATP synthase (at right) that uses the proton flow, resulting from the higher concentration within the inner membrane space, from the cytoplasmic side to the matrix side to produce ATP from ADP and P_i (inorganic phosphate). (Reprinted with permission from M. Saraste, "Oxidative Phosphorylation at the fin de siecle." Science, 283, 1488–1493, 1999.[48] Copyright © 1999 AAAS.)

repulsive free energy of hydration to oxidative phosphorylation in the mitochondrion, resides in the diversity of the structural entities involved. In the abstract of an excellent and timely review on free energy transduction of mitochondrial respiratory enzymes, Schultz and Chan wrote that "Each of the respiratory enzymes uses a different strategy for performing proton pumping."[49]

Of the four complexes involved in the electron transport chain that achieve proton pumping, crystal structures are available for three. Fortunately, two known structures, Complexes III and IV, pump two-thirds of the protons that develop the proton concentration gradient, as is much of the crystal structure known for the fifth complex that produces the ATP. That complex, ATP synthase, utilizes the proton concentration gradient to produce almost 90% of the ATP obtained on oxidation of glucose. Thus there exists an opportunity to gain some perspective on the relevance of the consilient mechanisms. Section 8.3 considers Complexes III and IV in some detail, while section 8.4 presents ATP synthase. First, however, the overview and background perspectives continue.

8.2.2.2 Reactions of the Inner Mitochondrial Membrane

The complete oxidation of one glucose molecule to 6 molecules of carbon dioxide results in the reduction of 10 NAD and 2 FAD to produce 10 NADH and 2 $FADH_2$ molecules. Their oxidation by the inner mitochondrial membrane results in the flow of electrons through the four complexes of electron transport chain that pump protons into the intermembrane space and reduces oxygen molecules to water. The return of the protons through the fifth complex, ATP synthase, results in the production of 32 molecules of ATP. The chemical energy obtained on oxidation of the nucleotides, reduced on oxidation of one molecule of glucose, ultimately appears in the formation of 32 molecules of ATP.

This brings the total number of ATP molecules formed from ADP and P_i to 36 in the process of the oxidation of a single glucose molecule. The set of reactions in the mitochondrion that produces 32 of the 36 ATP molecules are collectively called *oxidative phosphorylation*. This occurs in two steps. Step 1 utilizes a series of four protein-based machines, Complexes I, II, III, and IV, collectively referred to as the *electron transport chain*. These complexes oxidize the 10 NADH and 2 FADH$_2$ molecules and pump protons across the inner mitochondrial membrane to produce a proton concentration gradient. Step 2 involves the most important energy converting protein-based machine of biology (ATP synthase) that utilizes the concentration gradient of protons across the inner membrane of the mitochondria to produce ATP from ADP and P_i.

8.2.2.3 Step 1: Oxidation of NADH and FADH$_2$, Via the Electron Transport Chain, Pumps Protons into the Intermembrane Space and Results in the Reduction of Oxygen to Water

Step 1 involves four enzyme complexes with reactions much as given by Schultz and Chan[49]:

Complex I: *Oxidizes* one molecule of NADH, and, by an as yet unknown mechanism, *pumps 4 protons into the intermembrane (cytosolic) space* and reduces *ubiquininone* (coenzyme Q) to produce *ubiquinol*, QH$_2$. Overall equation:

$$NADH + 5H^+(matrix) + Q \rightarrow NAD^+ + QH_2 + 4H^+(cytosol) \quad (8.8)$$

Complex II: FAD oxidizes succinate to produce fumarate and FADH$_2$, which reduces *ubiquininone* to produce *ubiquinol*, QH$_2$.

Overall equation:

$$Succinate + Q \rightarrow Fumarate + QH_2 \quad (8.9)$$

Complex III: The *cytochrome bc$_1$ complex oxidizes 2 molecules of ubiquinol, pumps 4 protons into the intermembrane (cytosolic) space, and reduces two molecules of cytochrome c*. Overall equation:

$$2QH_2 - 2cyt\ \mathbf{c}^{3+} + Q + 2H^+(matrix) \rightarrow \\ 2Q + 2cyt\ \mathbf{c}^{2+} + QH_2 + 4H^+(cytosol) \quad (8.10)$$

Complex IV: *Cytochrome c oxidase oxidizes cytochrome c pumps 2 to 4 protons into the intermembrane (cytosolic) space* and reduces oxygen to water. Overall equation:

$$4cyt\ \mathbf{c}^{2-} + O_2 + 4H^+(scalar) + 4H^+(matrix) \rightarrow \\ 4cyt\ \mathbf{c}^{3+} + 2H_2O + 4H^+(cytosol) \quad (8.11)$$

Complexes III and IV are considered in some structural detail in section 8.3.

Based on the above outline of key elements of the four complexes, 10 NADH molecules would contribute about 100 protons and 2 FADH$_2$ molecules would contribute another 12 protons. As indicated below, production of 32 ATPs by ATP synthase would utilize an equivalent number, nominally 106 protons.

8.2.2.4 Step 2: Proton Driven Phosphorylation Via ATP Synthase

The result of the oxidation of the reduced coenzyme molecules, 10 NADH and 2FADH$_2$, is the development of a proton concentration gradient across the inner mitochondrial membrane. The return of the protons back across the inner mitochondrial membrane results in the production of 32 ATP molecules from 32 molecules of ADP plus 32 molecules of P_i by means of the ATP synthase. The understanding of this source of energy for ATP synthesis derives from the now well accepted, but once highly controversial, Mitchell chemiosmotic hypothesis.[50]

8.2.3 Question: Why Do Ions Form and Why do Ions Pair?

Despite this issue having been addressed in several contexts above, it is again stated succinctly here to emphasize the importance of the perspective.

8.2.3.1 When There Is Adequate Water, Ions Dissolve, That Is, They Prefer to Be Fully Hydrated

Ions dissolve from salts, or whenever otherwise ion paired, due to lowering of the ion (self-energy) free energy on achieving full hydration. When there is inadequate hydration, the driving force is for charge neutralization. Neutralization of a negatively charged group could occur by protonation or by pairing with a cation, and neutralization of a positively charged group would occur by loss of a proton or by pairing with an anion. In either case the neutralizing ion could be either an inorganic or an organic ion, as in the case of the charged side chains of amino acid residues.

8.2.3.2 ΔG_{ap} Provides the Driving Force for Ion Pairing, Which Is an Expression of Limited Availability of Water for Hydration of the Ion

When a clustering of hydrophobic groups occurs, there develops a competition for hydration between charged and hydrophobic groups that raises the free energy of the charged species. This results in an apolar–polar repulsive free energy of hydration (ΔG_{ap}) that is seen as a shift in pKa. Functional groups with a shift in pKa are at an increase in free energy and would lower that free energy by ion pairing. Ion pairing within or between protein subunits is the result of this ΔG_{ap} and is caused by the proximity of hydrophobic groups. As concerns the protein-based machine, it is the sequence-imposed coexistence of polar and apolar groups along a protein chain that gives rise to association of hydrophobic domains, and ion-pair formation becomes an integral part of the energetics for hydrophobic association. An apolar–polar repulsive free energy of hydration also applies to oxidized and reduced states of prosthetic groups and cofactors and to their interactions.

8.2.4 Question: Can Free Energy Transduction Be Localized at Some Crucial Part of the Enzymatic Cycle?

By the hydrophobic consilient mechanism of energy transduction, the energizing of a protein (an enzyme) results from ΔG_{ap}, the change in apolar–polar repulsive free energy of hydration due to molecules or inorganic ions interacting with the enzyme or protein of interest.[51] Depending on the changes in the state of the molecules or inorganic ions interacting with the protein, the protein may even be energized further, as on the splitting of ATP to form the most energized state, ADP plus P_i. There then occur stepwise decreases in the energized state that generally results most dramatically on release of P_i. The energized state abates as the apolar–polar repulsions continue to relax. As the cycle progresses, the enzyme becomes fully de-energized only on return to its initial state. *Therefore, the answer to the posed question is no; the free energy transduction cannot be localized at any particular part of the cycle, because different amounts of energy reside within the enzyme at different steps of the cycle, that is, due primarily to apolar–polar repulsion, the energized state of the enzyme is different at each step of the cycle.*

8.2.5 A Glimpse of Selected Energy Conversions Going Forward

This section provides in a relatively limited detail a standard overview of energy conversion. For all but the photosynthetic organisms, however, energy conversion can be represented in terms of three stages. The first stage is the development of a proton concentration gradient across the inner membrane of the mitochondrion. The second stage is the use of that concentration gradient to produce ATP, the representative biological energy coin, again centered on the inner membrane of the mitochondrion. The third stage is the use of ATP and its equivalent nucleotide NTPs by ATPases or in general by NTPases.

In subsequent sections one representative protein-based machine is emphasized for each stage. For proton pumping across the inner mitochondrial membrane discussed in section 8.3, Complex III (ubiquinone:cytochrome c oxidoreductase), also called the cytochrome bc_1 complex, provides an amazing example of both aspects of the consilient mechanism, those controlling hydrophobic association, the apolar–polar repulsive free energy of hydration (ΔG_{ap}), and the development of elastic force due to a decrease in internal chain dynamics between two fixed points in a protein chain segment. The singular choice of ATP synthase also provides an opportunity in section 8.4 to emphasize the use of ΔG_{ap} to energize ADP plus P_i to form ATP. Among the myriad of ATPases, the myosin II motor of muscle contraction, a linear contractile motor, discussed in section 8.5 and as anticipated in Chapter 7, provides a remarkably apparent example of ΔG_{ap}, whereby binding ATP disrupts hydrophobic association of the amino-terminal domain with the head of the lever arm, and loss of phosphate re-establishes the hydrophobic association to provide the powerstroke.

8.3 The Electron Transport Chain: Protein Machines as Redox-driven Proton Pumps

8.3.1 Introductory Comments

8.3.1.1 Structural Status of the Four Protein Machines (Complexes I, II, III, and IV) of the Electron Transport Chain

8.3.1.1.1 Current Status of Structural Details Available for the Four Complexes of the Electron Transport Chain

Complex I is a very large protein machine with about 43 subunits. Because of this the structural information has only been achieved to date to 22 Å resolution.[52,53] The resolutions for Complexes II (2.9 Å), III (2.3 to 2.9 Å), and IV (2.3 to 3.0 Å) are nearly an order of magnitude better and provide details at a level that allows consideration of mechanism.

8.3.1.1.2 General Sources for More Background Information

Two excellent reviews are available for the protein-based machines of oxidative phosphorylation from which more background information may be obtained. One is a clear and succinct statement by Saraste,[48] and the other, by Schultz and Chan,[49] is an excellent, more extensive account of the state of information on the structures, strategies, and thermodynamics of the five protein-based machines of oxidative phosphorylation. Both of these reviews may be sought for additional insight and detail. In addition, the fifth edition of *Biochemistry* by Berg, Tymozcko, and Stryer[46] provides an outstanding account of energy conversion in the mitochondrion. Our objective in section 8.3 is to consider the structures and functions of the electron-transport/proton-pumping complexes in terms of the consilient mechanism. We begin by a summary statement of simple hypotheses whereby the apolar–polar repulsive free energy of hydration, ΔG_{ap}, would play the pivotal role of proton gatekeeper that is based on a structural analysis of Complex III and whereby elastic force generation plays a pivotal role in domain movement for selected electron transfer. Due to both time constraints and limitations in structural data, the other complexes are not considered in as much detail.

8.3.1.2 Relevance of the Consilient Mechanisms to the Coupling of Electron Transport to Proton Gating/Pumping

8.3.1.2.1 Hypothesis 1: ΔG_{ap} as the Gatekeeper for Transmembrane Proton Pumping by the Electron Transport Chain

By the operative component, ΔG_{ap}, of the Gibbs free energy of hydrophobic association, ΔG_{HA}, polar groups compete with hydrophobic groups for hydration; the result is that the formation of charged groups disrupts hydrophobic association. Recognition of two aspects of the competition becomes central to understanding mechanism. One is that hydrophobic association occurs whenever too much hydrophobic hydration begins to form during an opening fluctuation. The second is the formation of a charged group, *whether positive or negative*, that recruits emergent hydrophobic hydration for its own charged hydration and thereby effects hydrophobic dissociation and the opening of an aqueous channel.

Complex III, ubiquinone:cytochrome **c** oxidoreductase, provides this example remarkably well. Our proposal states that the oxidation of ubiquinol, QH_2, at the Q_o site on the intermembrane side of the inner mitochondrial membrane produces a positively charged molecule, for example QH_2^{2+}. Emergence of this positively charged species due to the electron flow of oxidation disrupts hydrophobic association of the hydrophobic tip of the RIP with the Q_o site and thereby allows two protons, $2H^+$, to pass into the intermembrane cytoplasmic space with the result of a molecule of ubiquinone. Furthermore, our proposal contends that the reduction of ubiquinone at the Q_i site produces a negatively charged molecule, for example Q^{2-}, on the matrix side of the inner mitochondrial membrane. Emergence of this negatively charged species due to the electron flow of reduction converts a string of hydrophobically enclosed water molecules into a water-filled channel for the entrance of two protons, $2H^+$, from the matrix space to produce ubiquinol. In this way by ΔG_{ap}, the formation of a positively charged species opens the gate for proton egress from the inner mitochondrial membrane, and the formation of a negatively charged species opens the gate for proton ingress into the inner mitochondrial membrane. Because of ΔG_{ap} and the structural locations of the redox-formed positively charged and negatively charged groups, the redox events and proton transport events are tightly coupled.

Thus, given the example of Complex III oxidation of ubiquinol at the Q_o site and reduction of ubiquinone at the Q_i site, we would like to generalize the proton pumping mechanism to the similar formation of a charged species of whatever molecular basis at appropriate locations in the inner mitochondrial membrane. *General hypothesis of proton gating/pumping: Proton translocation across the inner mitochondrial membrane occurs (1) by oxidative formation of a positively charged species that opens an aqueous passageway to the cytoplasmic side of the membrane and that becomes neutralized by proton release to the intermembrane (cytoplasmic) space and (2) by the reductive formation of a negatively charged group that opens an aqueous passageway to the matrix side of the membrane and that becomes neutralized by proton uptake from the matrix space.*

8.3.1.2.2 Hypothesis 2: Hydrophobic Association Within Complex III of the Hydrophobic (FeS) Tip of the Rieske Iron Protein with the Hydrophobic Ubiquinol-containing Qo Site Causes Extension and Damping of Internal Chain Dynamics in the Tether of the Iron Protein

Complex III exemplifies another aspect of the consilient mechanisms. The FeS center of the RIP resides at a very hydrophobic tip of a stylus-like structure with two sites for hydrophobic association, one at the Q_o site and a second at the cytochrome c_1 site. As part of our hypothesis, the relative affinity due to hydrophobic association at each site depends on the redox state of the FeS center and the charge at the Q_o site. Before receipt of the electron, ΔG_{HA} is more favorable at the Q_o site, whereas after receipt of the electron and oxidation of ubiquinol, the ΔG_{HA} becomes sufficiently less favorable at the Q_o site to allow the elastic force of the stretched tether to withdraw the FeS center from the Q_o site for translocation to the cytochrome c_1 site. On transfer of the electron to cytochrome c_1, the affinity at this site decreases and the FeS center returns to the very favorable hydrophobic association at the Q_o site, now occupied by a new molecule of hydrophobic ubiquinol. The affinity due to hydrophobic association is so favorable at the ubiquinol-occupied Q_o site that the interconnecting chain segment between membrane anchor and the globular component of the RIP becomes stretched. We suggest that this is yet another example of the elastic-contractile mechanism demonstrated in the model proteins discussed in Chapter 5 in the context of hydrophobic association causing extension of an interconnecting chain segment. Hydrophobic association coupled with damping of internal chain dynamics on extension of interconnecting chain segments results in development of elastic force.

Contraction by the muscle myosin II motor is another example of hydrophobic association

8.3 The Electron Transport Chain: Protein Machines as Redox-driven Proton Pumps

on decreased charge, that is, on loss of phosphate, and one where damping of internal chain motions on hydrophobic association must surely exist as evidenced by an increase in the elastomeric force under isometric conditions, but the chain or chains involved have not been as clearly identified.

8.3.2 Complex I: NADH:ubiquinone Oxidoreductase—Oxidation of NADH for Reduction of Ubiquinone to Ubiquinol While Translocating Four Protons from Matrix to Cytosol for Every NADH Oxidized

8.3.2.1 Structural and Redox Center Information for Complex I

The structure of Complex I of mammalian mitochondria comprises an unusually large complex of about 43 distinct protein subunits with a three-dimensional structure yet to be determined at atomic resolution. From electron microscopy, it has been observed with an L-shaped structure, as indicated in Figure 8.6 with a 20nm length in the membrane and an 8nm protrusion into the matrix. Equation (8.5), the overall reaction, NADH + 5 H$^+$ (matrix) + Q → NAD$^+$ + QH$_2$ + 4 H$^+$ (cytosol), belies a more involved set of electron carriers. The process of reducing ubiquinone, Q, to ubiquinol, QH$_2$, involves a substantial number of redox centers with the coupled result of the transmembrane transport of four protons for each NADH molecule oxidized.

The process whereby oxidation of NADH reduces ubiquinone to ubiquinol involves the redox groups of flavin mononucleotide (FMN) and six FeS centers. It appears, however, that the membrane spanning subunits of the L-shaped structure do not contain redox centers, but do contain the ubiquinone. Images at 2.2nm resolution[52] demonstrate a structure more contorted than the monolithic L-shaped block indicated for Complex I in Figure 8.6. Accordingly, when making suggestions of mechanism based on ΔG_{ap}, it is tempting to consider the possibility of redox groups being sufficiently proximal to the cytoplasmic side of the inner mitochondrial membrane to form species such as QH$_2^{2+}$ and the analogous oxidized state of flavin, FMNH$_2^{2+}$, to provide a source of proton release to the cytoplasmic side of the inner mitochondrial membrane. With regard to the matrix side of the membrane, the possibility would continue with the reduction of ubiquinone and FMN to form negative species, such as Q^{2-} and FMN^{2-}, as could happen on addition of two electrons to each of the oxidized ubiquinone and flavin rings, that would open channels for protons to enter from the matrix side of the membrane. Of course, the problem with this analogy to the quinol/quinone use of Complex III is the apparent absence of structural information placing flavin and ubiquinone appropriately at two sites in the membrane component of Complex I.

8.3.2.2 Further Structural Considerations

8.3.2.2.1 Additional Detail for L-shaped Structure

As shown in Figure 8.7 A for bovine Complex I at 22 Å resolution in ice, the monolithic block of Figure 8.6 becomes a structure of a greater defined shape with a bulbous matrix component and a much more convoluted membrane component. If the redox components, with the exception of ubiquinone itself, reside in the bulbous component, then the puzzle remains one of how proton translocation occurs.

8.3.2.2.2 A re-arranged Horseshoe-shaped Structure as a Candidate for the Native State

Further electron microscopic studies of an enzymatically active Complex I show the rearrangement from the L-shaped structure in Figure 8.7B to the horseshoe-shaped structure in Figure 8.7C.[53] This reversible rearrangement due to a decrease in ionic strength reversibly turned off (at high ionic strength) and turned on (at low ionic strength). Specifically, raising the concentration of NaCl to 0.2M abolished NADH:decylubiquinone reductase activity. This appears to be a general salt effect on hydrophobic association, as the same effect was reported for KCl, LiCl, MgCl$_2$, CaCl$_2$, Li$_2$SO$_4$, and NaNO$_3$, which is analogous to the results in Figures 5.11 and 5.12 for elastic-contractile model proteins. Accordingly, this gives the possibility that redox centers might occur within the inner mitochondrial membrane.

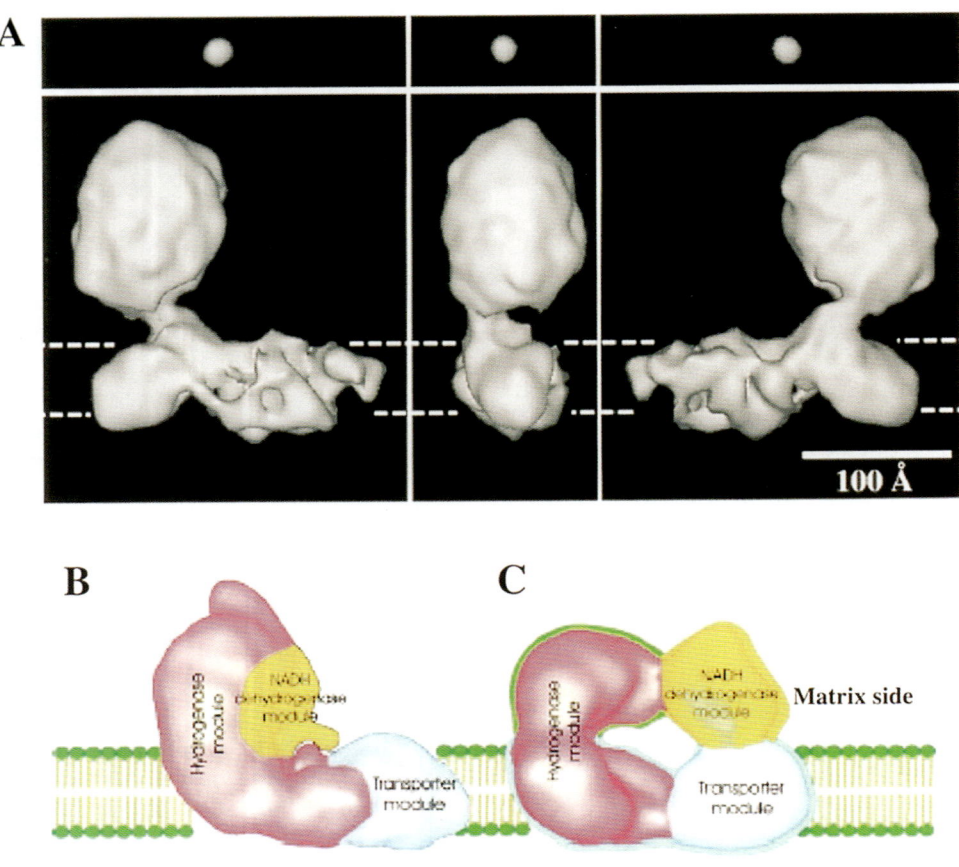

FIGURE 8.7. Reconstructions from electron micrographs of Complex I (NADH:ubiquinone oxidoreductase) of the electron transport chain of the inner membrane of the mitochondrion. **(A)** A more detailed L-shaped structure obtained at high ionic strength. **(B,C)** Structures obtained at high ionic strength (B) and at low ionic strength (C) for an enzymatically active state. (Reprinted from Journal of Molecular Biology, **277**, N. Grigorieff, Three-dimensional Structure of Bovine NADH: Ubiqunone Oxidoreductase (Complex I) at 22Å in Ice. 1033–1048, Copyright 1998.)

8.3.2.3 Potential Prosthetic Groups for Proton Uptake and for Proton Release

8.3.2.3.1 Proton Uptake from the Matrix into the Inner Mitochondrial Membrane

The isoalloxazine ring of FMN is quinone-like in that it can accept two electrons and take up two protons on reduction and give off two protons after oxidation that removes two electrons. This becomes the mechanism for proton translocation to be discussed below for Complex III as integral to the ubiquinone cycle. In particular, at the matrix side of the membrane two electrons could be added to ubiquinone to form Q^{2-} and analogously to FMN to give FMN^{2-}. By the apolar–polar repulsive free energy of hydration, ΔG_{ap}, of the consilient mechanism, these charged species transiently open channels for proton entry into the membrane for neutralization of the charge. One possible concern with using the isoalloxazine ring is that it is a system of three fused aromatic rings such that charge would not be as localized as it is for the ubiquinone system or for that matter for nicotinamide, both of which can form single aromatic rings and take up protons on reduction.

8.3.2.3.2 Proton Release to the Cytoplasmic Side of the Inner Mitochondrial Membrane

As redox systems are not known to be near the cytoplasmic side of the inner mitochondrial membrane, it is more difficult to consider oxidation sites where formation of the positively charged species, QH_2^{2+} and $FMNH_2^{2+}$, could occur. These are both species that, theoretically, could release protons to the cytoplasmic side of the membrane to complete proton transit in a tightly coupled way across the inner mitochondrial membrane. These positively charged species are expected to exhibit decreased lipid solubility, although the intramolecular ion pairing of negatively charged phosphate to the positively charged isoalloxazine ring of $FMNH_2^{2+}$ might improve the possibility of FMN use. Obviously, our limited knowledge of the molecular structure of Complex I does not allow sufficient insight to make a specific consilient-based proposal.

8.3.2.4 Dependence of Reduction Potential on Hydrophobicity of Binding Site?

As was shown for N-methyl nicotinamide in Figure 5.20C, the reduction potential of nicotinamide varies with hydrophobicity of binding site. This allows the possibility that ubiquinol could be oxidized at two different sites at the cytoplasmic side of the inner mitochondrial membrane, but the question becomes oxidized by what, as the redox centers have been considered to be on the matrix side of the membrane. These problems with understanding proton transport by Complex I are not present with Complex III, as discussed in section 8.3.4.

8.3.3 Complex II: Succinate:ubiquinone Reductase—Reduction of FAD by Succinate to Produce FADH₂, Which Ultimately Results in Reduction of Ubiquinone to Produce Ubiquinol for Subsequent Reactions of the Electron Transport Chain

Mammalian Complex II contains four subunits. The larger two subunits are extramembrane on the matrix side. They are a 79 kDa flavoprotein and a 29 kDa iron–sulfur protein with three iron sulfur centers: Fe_2S_2, Fe_4S_4, and Fe_3S_4. The two membrane bound subunits bind a single heme **b** and are reported to contain two or more ubiquinone/ubiquinol binding sites.

8.3.3.1 Function of Succinate Dehydrogenase, the E. coli Analogue of Complex II (Succinate:ubiquinone Reductase)

The Singular Purpose of Succinate Dehydrogenase is to Provide Reduced Ubiquinone, That is, Ubiquinol, to the Inner Mitochondrial Membrane as an Electron Carrier and as Chemical Energy for Proton Transport. The overall reaction is as follows:

$$^-OOCCH_2CH_2COO^- + Q = {}^-OOCCH=CHCOO^- + QH_2$$

$$\text{Succinate} + \text{Ubiquinone} = \text{Fumarate} + \text{Ubiquinol} \quad (8.12)$$

Equation (8.9) is a reversible reaction; it can run either direction by changes in the relative concentrations of succinate and fumarate. Even so, biology has different enzymes specializing in producing fumarate from succinate under aerobic conditions and for producing succinate from fumarate under anaerobic conditions. The two enzymes appear essentially identical with the same prosthetic groups, that is, FAD and the same three iron sulfur centers. There is, however, an important difference of side products of reactive oxygen species (ROS) that is greater in the reaction catalyzed by fumarate reductase enzyme (see section 8.3.3.4 below) than in the reaction catalyzed by succinate:ubiquinone reductase. It has to do with a small difference in the accessibility of the flavin prosthetic group to the aqueous phase. The flavin appears better shielded from dissolved oxygen in succinate:ubiquinone reductase.[54] Obviously, the electron transport chain of the inner membrane of the mitochondrion must function under aerobic conditions as the terminal reaction due to Complex IV (cytochrome **c** reductase) involves molecular oxygen (O_2) as a reactant, that is, the reduction of oxygen to water.

8.3.3.2 Succinate:ubiquinone Reductase in Three Distinct Reaction Steps

8.3.3.2.1 Step 1: Succinate Reduction of FAD to Produce FADH$_2$ and Fumarate

$$^-OOCCH_2CH_2COO^- + FAD =$$
$$^-OOCCH = CHCOO^- + FADH_2 \quad (8.12a)$$

In this reaction, the protons and electrons transfer together as hydrogen atoms.

8.3.3.2.2 Step 2: Oxidation of FADH$_2$ to FAD by Transfer of Electrons to FeS Centers with Release of Protons to Matrix

$$FADH_2 + 2FeS = FAD + 2FeS^- + 2H^+ \quad (8.12b)$$

This reaction of Complex II begins the separation of electrons from protons. However, why does Complex II not contribute to proton transport? If FADH$_2$ were oxidized on the cytoplasmic side of the lipid bilayer membrane, then conceivably the two protons could be released to the cytosol, as required for effective proton transport.

8.3.3.2.3 Step 3: Transfer of Electrons to Ubiquininone (Coenzyme Q) from FeS Centers to Produce Negatively Charged Ubiquinone, Q^{2-}, Which on Receipt of Protons from Matrix Produces QH$_2$

$$2FeS^- + Q + 2H^+ = 2FeS + QH_2 \quad (8.12c)$$

The reduction of ubiquinone by receipt of two electrons to produce Q^{2-} within the lipid bilayer membrane, but yet on the matrix side, would allow the pick up of two protons from the matrix. The process would result from formation of the negatively charged Q^{2-} whereby the competition for hydration due to the apolar–polar repulsive free energy of hydration, ΔG_{ap}, would open an aqueous channel to the matrix side of the membrane for entry of two protons, $2H^+$. With the above-preferred location for reaction (8.9b) combined with the just noted location of reaction (8.9c), there would be a net transport of two protons from the matrix side to the cytoplasmic side of the membrane. This would be in keeping with the direction of proton transport for Complexes I, III, and IV. As shown below in Figures 8.8 and 8.9, however, transport could not occur by this utilization of hypothesis 1 of the consilient mechanism for proton pumping (see section 8.3.1.2), because the arrangement of the redox centers is not as required to release two protons to the cytosol. In keeping with this perspective, of course, Complex II does not pump protons. Nonetheless, when ubiquinol is used by Complex III, recall that the two protons that added to ubiquinone to form ubiquinol came from the matrix side.

8.3.3.3 Structure of Succinate Dehydrogenase, the E. coli Analogue of Complex II (Succinate:ubiquinone Reductase)

The molecular structure of the analogue for Complex II, succinate dehydrogenase from *E. coli*, is shown in Figure 8.8[54] to exhibit the general form shown in Figure 8.6. Figure 8.8A uses a space-filling representation where neutral residues are light gray, aromatic residues are black, other hydrophobic residues are gray, and charged residues are white. With this means of representation, immediately apparent is a hydrophobic tip for insertion into the lipid layer of the membrane. Also apparent is a larger globular, more hydrophilic component that resides within the matrix side of the membrane.

With the ribbon representation in Figure 8.8B, a more transparent view of the structure allows visualization of the redox groups and the distribution of water molecules, shown as light dots. The FAD resides nearly centered within the globular component and is associated with very few water molecules. Then, in a stepwise manner, the three iron–sulfide centers fill in between FAD and the lipid bilayer, and just within the region of the lipid bilayer resides the heme **b** group. Reduction of ubiquinol occurs between the Fe$_3$S$_4$ center. On removal of the protein, this distribution of redox centers becomes more apparent in Figure 8.9A,B.

An expanded view of the FAD group is shown in Figure 8.9C along with the inhibitor, oxaloacetate, that resides in the substrate (succinate) site. FAD is shown to be the phosphate to phosphate coupling of adenosine monophosphate (AMP) with FMN. The key point of Figure 8.9C is the proximity of substrate to the

8.3 The Electron Transport Chain: Protein Machines as Redox-driven Proton Pumps

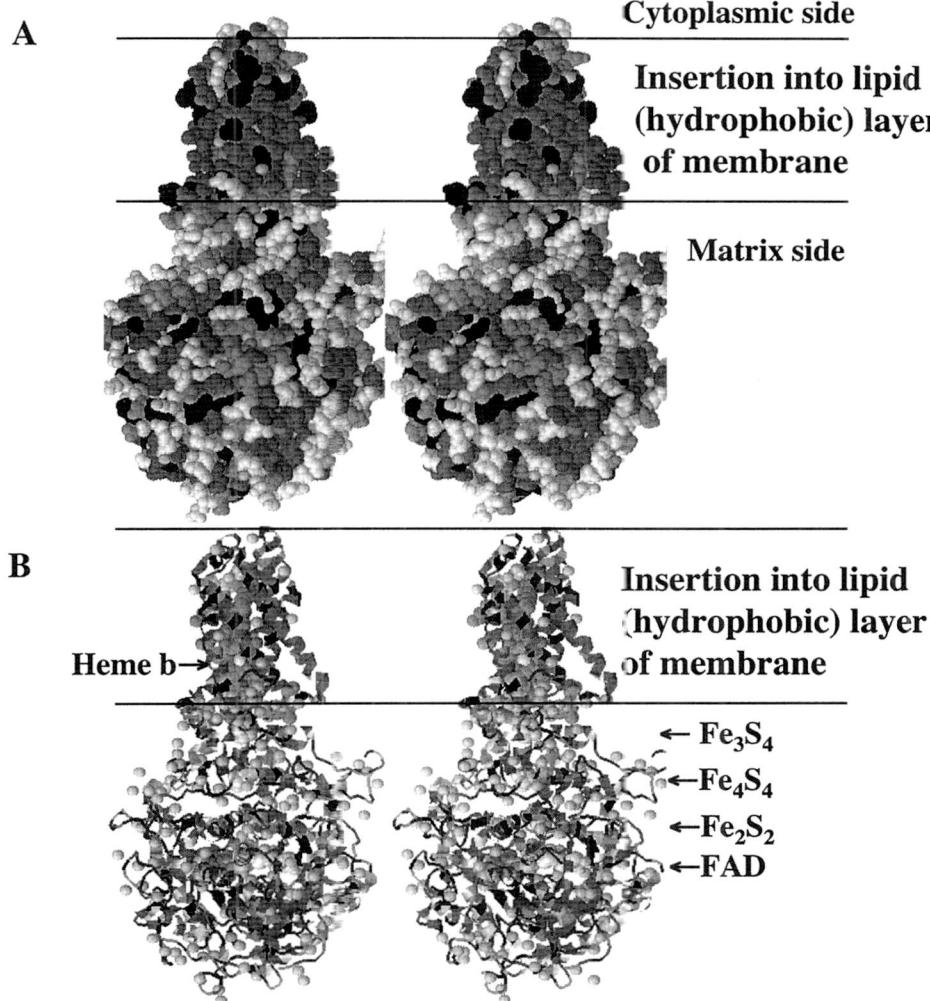

FIGURE 8.8. Stereo view of *E. coli* succinate: ubiquinone oxidoreductase, which is an analogue of Complex II of the electron transport chain of mitochondria, with neutral residues light gray, aromatics black, other hydrophobics gray, and charged residues white. **(A)** Space-filling representation with water shown as light dots. **(B)** Ribbon representation with ligands and "waters of Thales." (Prepared using the crystallographic results of Yankovskaya et al.[54] as obtained from the Protein Data Bank, Structure File, 1NEN.)

three-ring aromatic redox group isoalloxazine of the flavin component. This shows the reduction of FAD to form $FADH_2$ to occur by direct transfer of the two hydrogen atoms, that is, without physical separation of proton from electron. Thus there is no opportunity for formation of an anion, such as FAD^{2-}, that could then pick up two protons from the matrix side of the membrane as one part of the two-part process of proton transport from one side to the other of the lipid bilayer membrane

Reduction of ubiquinone by addition of two electrons on the matrix side of the membrane could be the basis for the pick up of two protons from the matrix side. However, the second part of the release of two protons to the cytoplasmic side of the membrane to achieve net transport of two protons is absent. Thus, from a structural standpoint, it is understandable that Complex II does not contribute to proton pumping. Complex II does make an important contribution of diffusible ubiquinol to the membrane

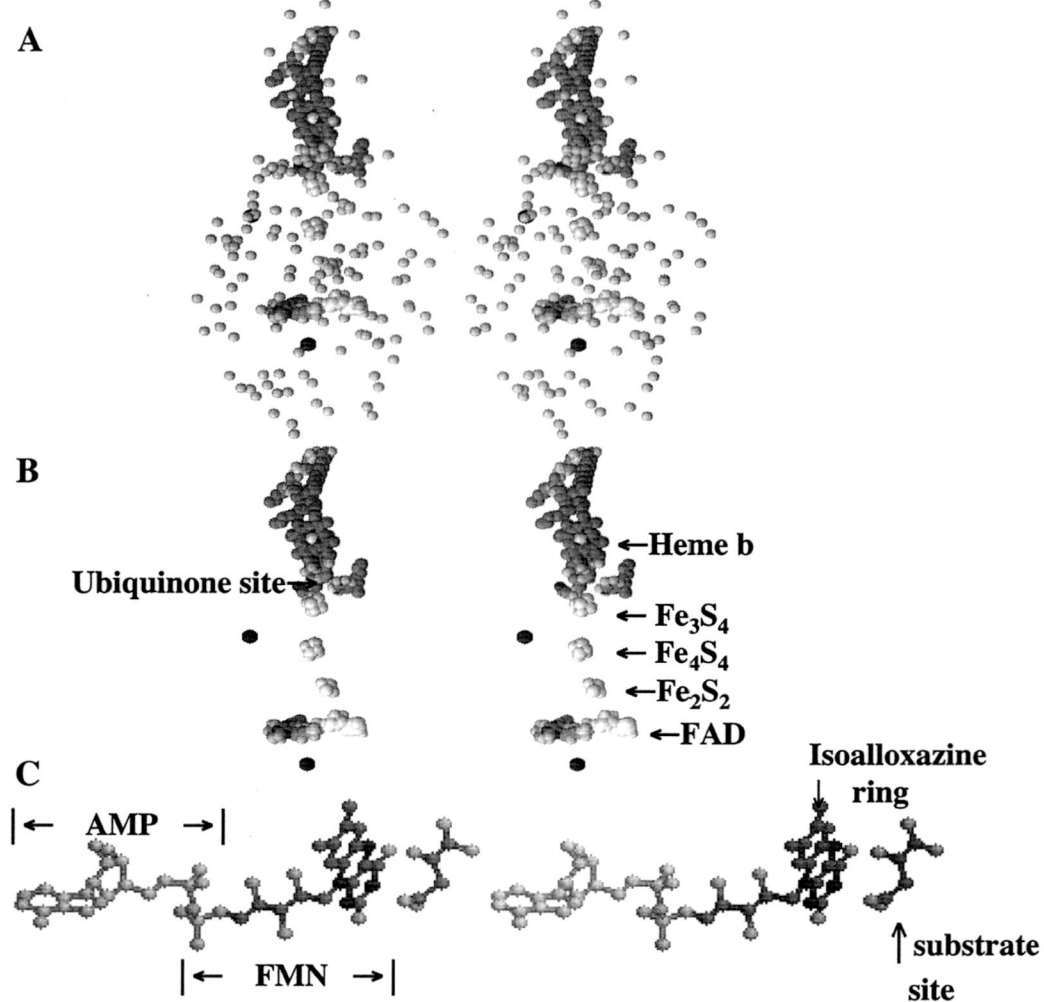

FIGURE 8.9. Stereo view of *E. coli* succinate: ubiquinone oxidoreductase, an analogue of Complex II of the electron transport chain of mitochondria. **(A)** Prosthetic groups (ligands) and "waters of Thales." **(B)** Ligands (FAD, FeS centers, heme **b**, etc.) **(C)** FAD with inhibitor oxaloacetate in the substrate reaction site. (Prepared using the crystallographic results of Yankovskaya et al.[54] as obtained from the Protein Data Bank, Structure File, 1NEN.)

that becomes oxidized by Complex III as part of the quinone cycle of proton pumping as part of electron transport.

8.3.3.4 Fumarate Reductase from Wolinella succinogenes, *Another Analogue for Complex II of the Electron Transport Chain of Mitochondria*

Before structural determination of succinate dehydrogenase, fumarate reductase provided insight into the structure and function of Complex II (succinate:ubiquinone reductase) of the electron transport chain.[49] As noted above, the succinate dehydrogenase reaction (8.9) succinate + ubiquinone = fumarate + ubiquinol, is reversible, and in *Wolinella succinogenes* the function of fumarate reductase is the reverse, to produce succinate from fumarate.[55]

Figures 8.10 and 8.11 present the structure of fumarate reductase in the same manner as used in Figures 8.8 and 8.9 for succinate dehydroge-

8.3 The Electron Transport Chain: Protein Machines as Redox-driven Proton Pumps

nase. Figure 8.10A shows in stereo view and space-filling representation the dimeric state of fumarate reductase with neutral amino acid residues in light gray, with aromatic residues in black, other hydrophobic residues in gray, and charged residues in white. Again, a hydrophobic band identifies the location of the lipid bilayer membrane. Figure 8.10B uses the ribbon representation of the protein, which allows easy visualization of the α-helices and β-sheet structural elements of the protein-based machine, as well as of some of the distribution of detectable water molecules.

Figure 8.11A gives the redox groups in ball and stick representation and the substantial number of detectable "waters of Thales" as

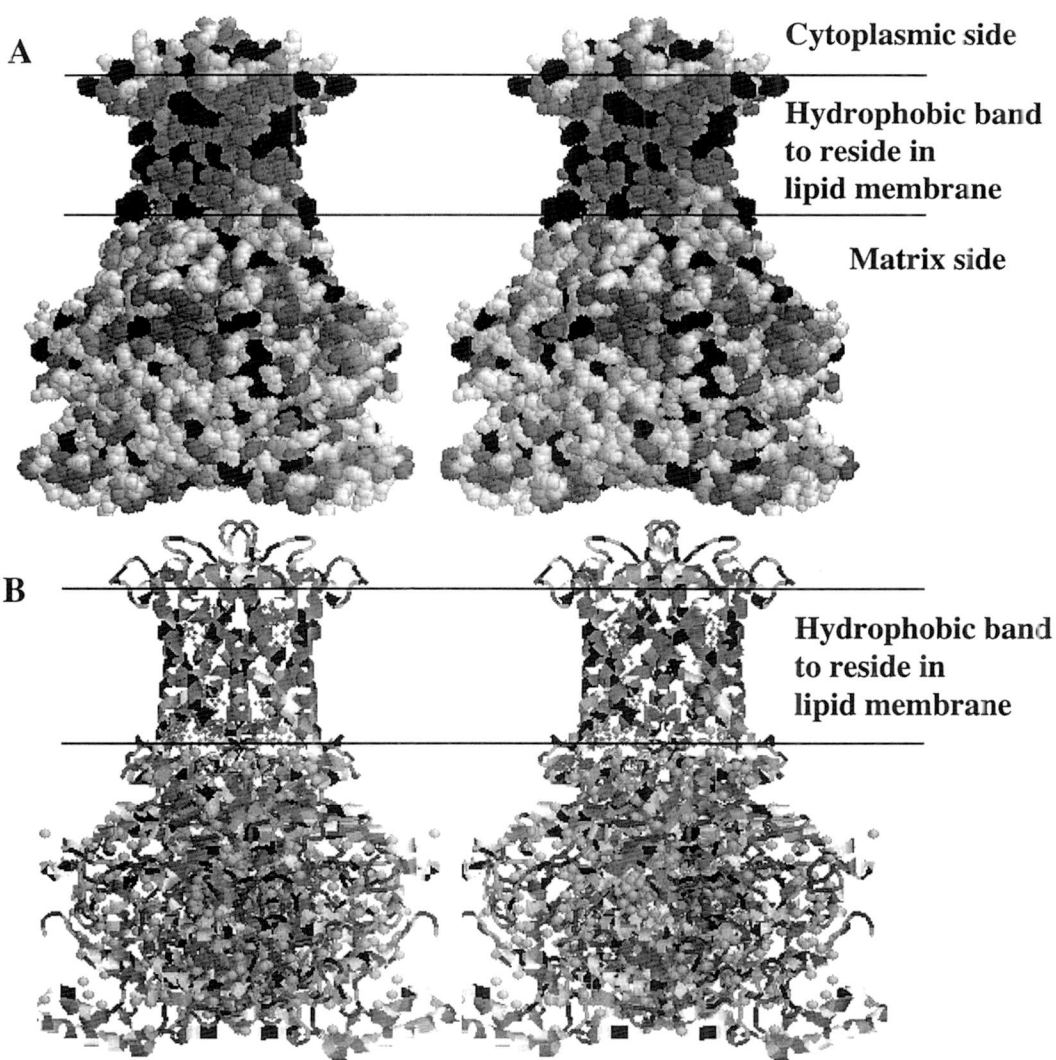

FIGURE 8.10. Stereo view of *Wolinella succinogenes* fumarate reductase dimer, an analogue of succinate:ubiquinone oxidoreductase (Complex II) of the electron transport chain of mitochondria, with neutral residues light gray, aromatic residues black, other hydrophobic residues gray, and charged residues white. (**A**) Space-filling representation with water shown as light dots. (**B**) Ribbon representation with ligands and "waters of Thales." (Prepared using the crystallographic results of Lancaster et al.[55] as obtained from the Protein Data Bank, Structure File, 1QLB.)

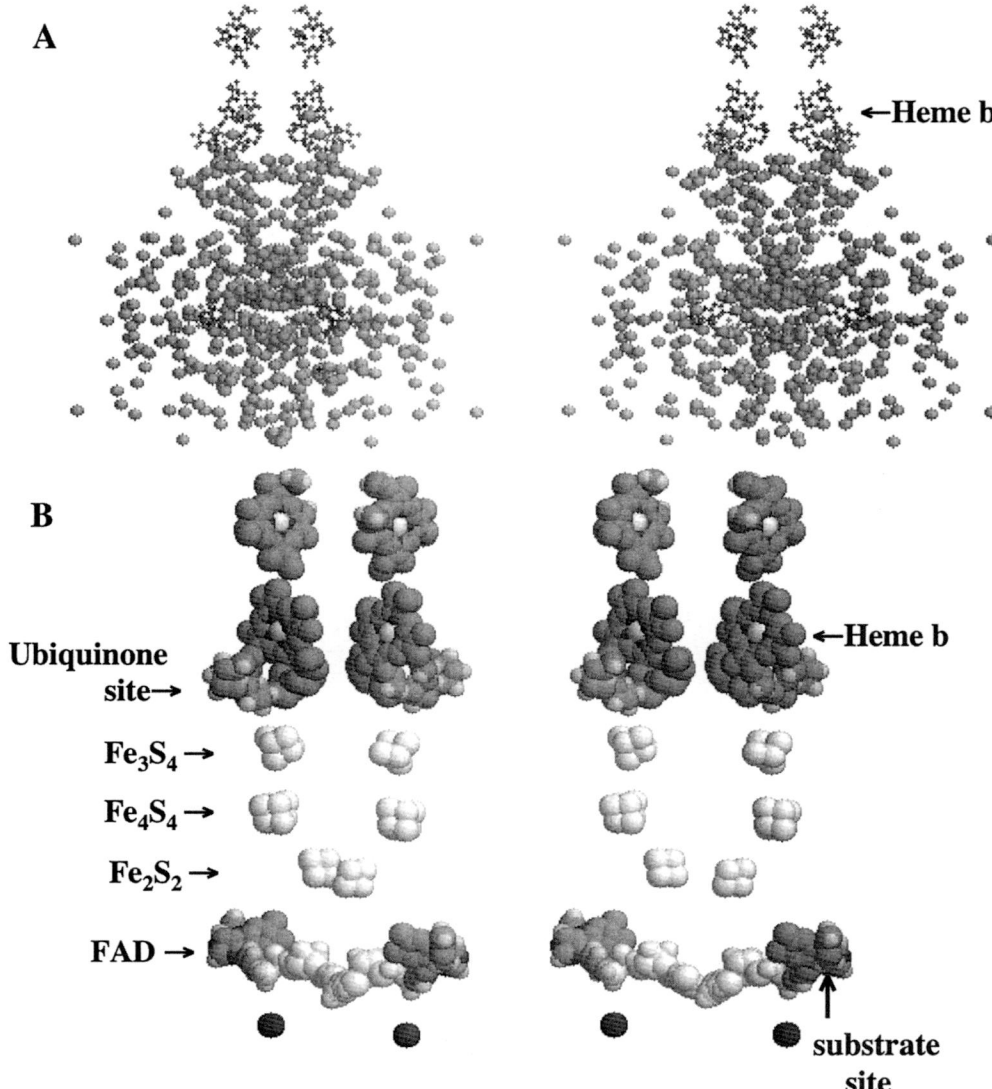

FIGURE 8.11. Stereo view of *Wolinella succinogenes* fumarate reductase dimer, an analogue of mitochondrial succinate:ubiquinone oxidoreductase (Complex II) of the electron transport chain. **(A)** Ligands in ball and stick representation and "waters of Thales" shown as light dots. **(B)** Ligands in space-filling representation, showing close association of substrate with the flavin isoalloxazine heteroaromatic three-ring system. (Prepared using the crystallographic results of Lancaster et al.[55] as obtained from the Protein Data Bank, Structure File, 1QLB.)

light dots while hiding the protein structure itself. In Figure 8.11B shown in space-filling representation are the redox groups and substrate and putative location of the ubiquinone site in analogy to those of Figure 8.9 for succinate dehydrogenase. In fact, the structural parallels between succinate dehydrogenase and fumarate reductase with regard to distribution of redox functions are essentially identical, but, of course, in duplicate for the latter. Seemingly, the only obvious difference in redox functions is the presence of a second heme **b** in the latter.

This, however, is without known functional consequence, as both proteins seem to function in the absence of the heme groups.[49,55] Also, very apparent is the greater number of detectable water molecules proximal to the flavin group. Although this organism functions anaerobically, molecular oxygen, when present, is an order of magnitude more likely to become ROS.[54] The disarming of ROS and their damaging effects is the reason that antioxidants are so beneficial. Of course, by the consilient mechanism, the action of ROS on proteins destroys function, because such polar additions to proteins limit the hydrophobic associations essential to function (see section 7.7 of Chapter 7).

Again, the structure provides an understanding of why this protein with all of the redox functions to effect proton transport does not pump protons. As shown below, the structure of Complex III (ubiquinone:cytochrome c oxidoreductase) provides remarkable insight into the coupling of electron flow to proton pumping, especially when viewed from the vantage point of the consilient mechanism with its functional element of an apolar–polar repulsive free energy of hydration, ΔG_{ap}.

8.3.4 Complex III: Ubiquinone:cytochrome c Oxidoreductase (The Cytochrome bc_1 Complex)

With the exciting achievement of molecular structures for Complex III,[24,25,56–59] combined with extensive studies on kinetics, thermodynamics, and stoichiometry,[49] came the first clear insight into the coupling of electron transport to transmembrane proton pumping. This process, so essential to sustain Life, is no longer shrouded in mystery. The following constitutes the first report whereby the proposed consilient mechanism extends our understanding of mechanistic detail for the function of Complex III, and both hydrophobic and elastic consilient mechanisms come into play.[59]

The hydrophobic and elastic consilient mechanisms, two distinct but interlinked physical processes of hydrophobic association/dissociation and elastic force development/relaxation, couple to achieve movement. In functional homodimer of Complex III, hydrophobic association of the RIP FeS center at the Q_o site in one monomer occurs with elastic force development in the tether that connects the globular component of the RIP to its membrane anchor in the other monomer. Formation of the positively charged ubiquinol, as electrons pass to the FeS center and to heme b_L, effects hydrophobic dissociation and allows relaxation of the elastic forces whereby the extended tether moves the FeS center to heme c_1 for transfer of its electron.

Proton gating represents another demonstration of the consilient mechanism in Complex III. The operative element of the change in Gibbs free energy for hydrophobic association, the apolar–polar repulsive free energy of hydrophobic hydration, ΔG_{ap}, uses the development of negative charge at the Q_i site to open an aqueous channel for uptake of proton from the matrix side of the membrane. As noted above, the formation of positive charge at the Q_o site displaces the hydrophobically associated RIP and enables release of two protons to the cytoplasmic side of the membrane. At both Q_o and Q_i sites the formation of charge disrupts hydrophobic association by means of ΔG_{ap} and opens the door to proton egress and ingress.

These roles of the consilient mechanism are demonstrated with structures of Complex III reported in the literature and are available for development and analysis from the Protein Data Bank using programs such as FrontDoor to Protein Explorer by Eric Martz, which is available at no cost at www.proteinexplorer.org.

8.3.4.1 General Structural Description of Complex III (Ubiquinol:cytochrome c Oxidoreductase)

8.3.4.1.1 Composition

Eleven protein subunits comprise the monomer of Complex III, but only three of these subunits contain the redox centers. They are cytochrome **b**, an integral membrane protein with two hemes (b_L and b_H), the RIP on the cytoplasmic side with a single α-helical membrane anchor, and cytochrome c_1 also located on the cytoplasmic side and also with a single α-helical membrane anchor. Although the membrane anchor for the RIP resides with cytochrome **b** of one monomer, its FeS center interacts at the Q_o and cytochrome c_1 sites of the other monomer. Therefore, this is referred to as a *functional homodimer*, wherein both monomers are required to achieve function.

8.3.4.1.2 Structure

Complex III, shown in stereo view and in the light-gray ribbon representation of Figure 8.12A, exhibits twofold symmetry, as there are two copies of each subunit related by a twofold rotational symmetry axis perpendicular to the

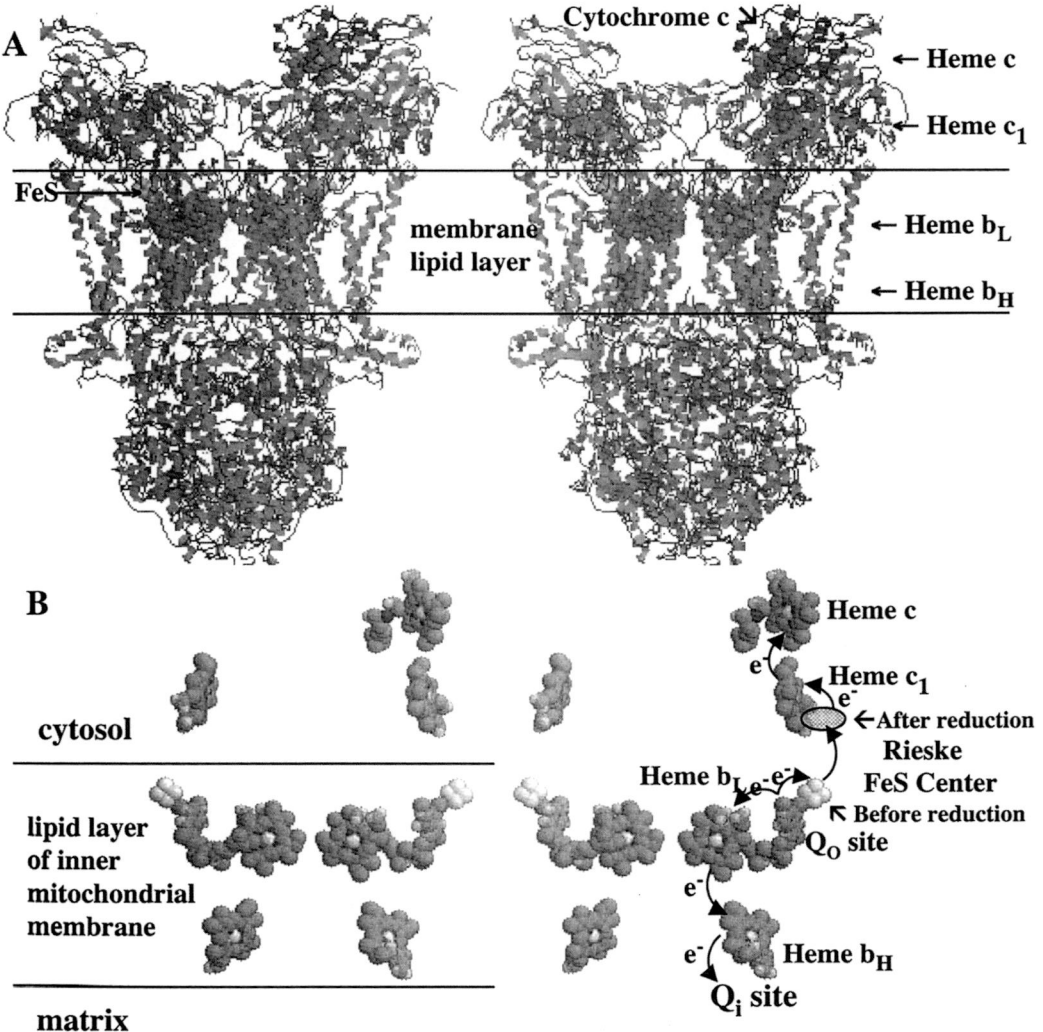

FIGURE 8.12. **(A)** Complete stereo view of the homodimeric structure (light gray) of yeast Complex III, cytochrome bc_1 complex (ubiquinol:cytochrome **c** oxidoreductase), plus cytochrome **c** (gray at upper right). The location of the heme groups and the FeS center are indicated in reference to the lipid bilayer membrane. **(B)** Stereo view of redox centers showing electron transfers, Q being for Coenzyme Q (ubiquinone), with Q_o identifying the oxidation site that becomes a positively charged site, for example, QH_2^{+2}, and Q_i identifying the reduction site that becomes a negatively charged site, that is, Q^{2-}. (Prepared using the crystallographic results of Lange and Hunte[59] as obtained from the Protein Data Bank, Structure File, 1KYO.)

membrane plane. Also shown in a darker gray ribbon representation in Figure 8.12A is the presence of a single molecule of cytochrome **c** at the top righthand side in position to receive an electron from the heme of cytochrome c_1 and to leave that site on the cytosolic side as the reduction product of the overall reaction. The prosthetic groups are shown again in stereo view and in twofold symmetry in Figure 8.12B plus the single cytochrome **c** heme center at the upper right.

The ribbon representation shows transmembrane α-helices traversing through the lipid bilayer membrane, containing heme **b** prosthetic groups, the Q_o and Q_i sites, and the FeS center at the tip of the globular component of the RIP and associated with the Q_o site.

8.3.4.2 Phenomenology of Ubiquinol Oxidation to Reduce Cytochrome c with Net Proton Transport

8.3.4.2.1 Overall Reaction

The overall equation for the reaction catalyzed by Complex III follows is given above in Equation (8.7). At first glance this balanced equation seems to have redundant reactants and products. Why does the equation have two QH_2 molecules as reactant and yet give one QH_2 as product? Inversely for Q, why is there one Q molecule as a reactant and two as a product? Also, why does the reaction read two protons from the matrix with four protons being released to the cytosol, and yet this reaction is generally credited with pumping four protons? Taking the last question first, the additional two protons came from the matrix by means of reduction of ubiquinone by Complex II. Thus, by supplying ubiquinol to the membrane pool, Complex II in reality contributes to the net transport of four protons by Complex III.

The answer to the first two questions is that Equation (8.7), in addition to stoichiometry, reveals elements of mechanism. A bifurcation of electron flow on oxidation of a single QH_2 molecule at the Q_o site occurs with the first electron going to the FeS center and the second passing through hemes b_L and b_H to add an electron to Q at the Q_i site. Now a second QH_2 molecule at the Q_o site repeats the bifurcation to complete the reduction of ubiquinone at the Q_i site, which Q^{2-} picks up two protons from the matrix side of the membrane to complete the stoichiometry of two QH_2 molecules used at the Q_o site with one recovered at the Q_i site. The net result is two QH_2 molecules used giving two Q molecules, but one molecule of Q was reduced, thereby regaining one molecule of QH_2.

8.3.4.2.2 Electron Transfer Steps

Focusing on the electron transfer steps indicated in Figure 8.12B, a single QH_2 molecule at the Q_o site passes one electron to the FeS center; the FeS center then moves to the heme c_1 and reduces it; the reduced heme c_1 transfers its electron to the heme of cytochrome c^{2+}. The result is the production of one molecule of cytochrome c^{2+} and QH_2^+ at the Q_o site. The QH_2^+ molecule at the Q_o site then transfers a second electron that passes through the **b** hemes to ubiquinone at the Q_i site. A second QH_2 molecule is similarly oxidized with the reduction of a second cytochrome c^{3+} to result in two molecules of cytochrome c^{2+}. Although the electron transfer phenomenology is well recognized and even the mechanism of electron tunneling over significant distances is an accepted mechanism, there remain issues of the mechanism whereby domains associate/dissociate and move where and when necessary. These points are discussed in terms of the hydrophobic and elastic consilient mechanisms, the principal message of this book, in section 8.3.4.3.

8.3.4.2.3 Proton Transport Sites

Occupied Q_o and Q_i sites involved in proton transport are shown in stereo view in association with the heme b_L and b_H ligands in Figure 8.13A in ball and stick representation and in Figure 8.13B in space-filling representation. This is a monomer subset shown in stereo of the ligands that were shown in the homodimer of Complex III in Figure 8.12B, which contained in additional the heme c_1 and **c** ligands. The added schematic representation of Figure 8.13A illustrates that the oxidation due to the removal of two electrons from a molecule of ubiquinol at the Q_o site sets the stage for transfer of two protons to the cytoplasmic side of the inner mitochondrial membrane. Also, the

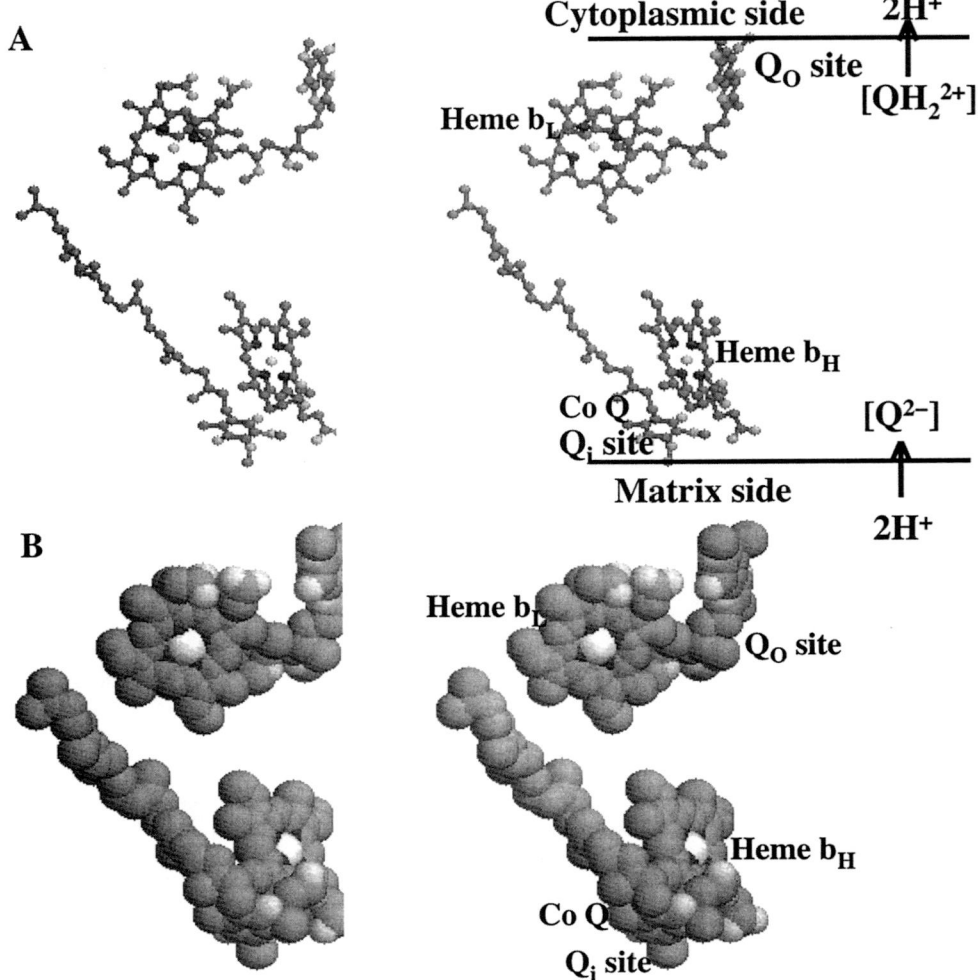

FIGURE 8.13. Yeast cytochrome bc_1 stereo view of the membrane-bound ligands—hemes b_L and b_H, stigmatellin occupying the Q_o (ubiquinol) site, and ubiquinone at the Q_i site. **(A)** Ball and stick representation to show molecular backbone structures. **(B)** Space-filling representation to give proper volume representation. By means of ΔG_{ap}, ubiquinol, on becoming positively charged by oxidation at the Q_o site, opens an aqueous channel to the cytoplasmic side of the inner mitochondrial membrane for proton release, and ubiquinone (coenzyme Q, CoQ), on becoming negatively charged by reduction at the Q_i site, opens an aqueous channel for proton uptake from the matrix side of the inner mitochondrial membrane. (Prepared using the crystallographic results of Hunte et al.[25] as obtained from the Protein Data Bank, Structure File, 1EZV.)

reduction of a molecule of ubiquinone due to the addition of two electrons results in the uptake of two protons from the matrix side of the inner mitochondrial membrane.

Ready visualization of this phenomenology derives from the remarkable structural and biochemical kinetic studies on Complex III. The issue addressed below, however, becomes one of understanding the intermolecular forces whereby such proton transport is coupled to electron transfer without significant leakage of proton across the inner mitochondrial mem-

8.3 The Electron Transport Chain: Protein Machines as Redox-driven Proton Pumps   375

brane. Introduction of the hydrophobic consilient mechanism to proton gating, whereby charge disrupts hydrophobic association and opens aqueous channels, occurs as the studies on elastic-contractile model protein reviewed in Chapter 5 become viewed in the light of the molecular structure of Complex III. This is discussed in section 8.3.4.4 with consideration of the Hunte et al.[25] structure of Complex III that includes diffraction-detectable water molecules.

8.3.4.3 Consilient Mechanisms for Electron Transfer/Proton Transport Reside in the Movement of the Rieske Iron Protein FeS Center

8.3.4.3.1 Molecular Structures That Show Relocation of the Globular Domain of the Rieske Iron Protein

Molecular structures of Complex III are often obtained with the FeS center of the RIP in position for docking at the Q_o site of the second monomer. This orientation, obtained using stigmatellin and antimycin inhibitors, is shown for the monomer in stereo view in Figure 8.14A with charged residues in white, neutral residues in light gray, aromatic residues in black, and other hydrophobic residues in gray. Thus, the darker regions indicate the more hydrophobic domains of the protein subunit. Zhang et al.[57] determined a crystal structure with the FeS center of the RIP at the position of the cytochrome c_1 site of the second monomer, as similarly shown in stereo view in Figure 8.14B. Thus, electron transfer is established in this unique case to be one of protein domain relocation subsequent to addition of the electron to the redox center. How does this dynamic mechanical event occur? As discussed immediately below, this utilizes both hydrophobic and elastic consilient mechanisms by interlinking hydrophobic association/dissociation and elastic extension/relaxation.

8.3.4.3.2 Consideration of Hydrophobic Association on Docking of the Rieske FeS Center at Q_o and Cytochrome c_1 Sites

The two docking sites for the FeS center of the RIP are shown in space-filling representation in Figure 8.15. By the use of white for charged residues and gray to black for increasingly hydrophobic residues, it becomes apparent by inspection of the involved surface areas, without calculation of the Gibbs free energy for hydrophobic association, ΔG_{HA}, that the Q_o site is much more hydrophobic than the cytochrome c_1 site. The hydrophobicity of the globular component of the RIP that associates with either the Q_o site or the cytochrome c_1 site is even more apparent. The very dark tip, shown in end view and in side view in Figures 8.16A,B, respectively, makes clear that the interaction of the FeS center with the Q_o site would clearly be dominated by hydrophobic association.

8.3.4.3.3 Stretched Tether Connecting the Membrane Anchor to the Globular Component of the Rieske Iron Protein When Hydrophobically Associated at the Q_o Site

The membrane anchor of the RIP associates at an angle with other transmembrane helices that define it as part of one monomer. It then crosses over by means of a tether from one monomer to interact with the other. The hydrophobic globular component of the RIP that contains the FeS center has its most favorable hydrophobic association at the Q_o site of the adjacent monomer. As shown in Figure 8.17 and also notable in Figure 8.14, to achieve this most favorable ΔG_{HA} a section of the tether becomes stretched to near full extension.

The two issues that now require examination are whether the visually extended tether is kinetically free and whether the chain does in fact contract (shorten) as the FeS center moves from the Q_o site of cytochrome b to the heme of cytochrome c_1.

8.3.4.3.4 Is the Tether of the Rieske Iron Protein a Free Standing (Kinetically Free) Chain When the FeS Center Is at the Q_o Site?

Useful elastic force develops at the fixed ends of a chain segment when the motion within the backbone of the chain segment decreases as the result of a deformation such as an increase in

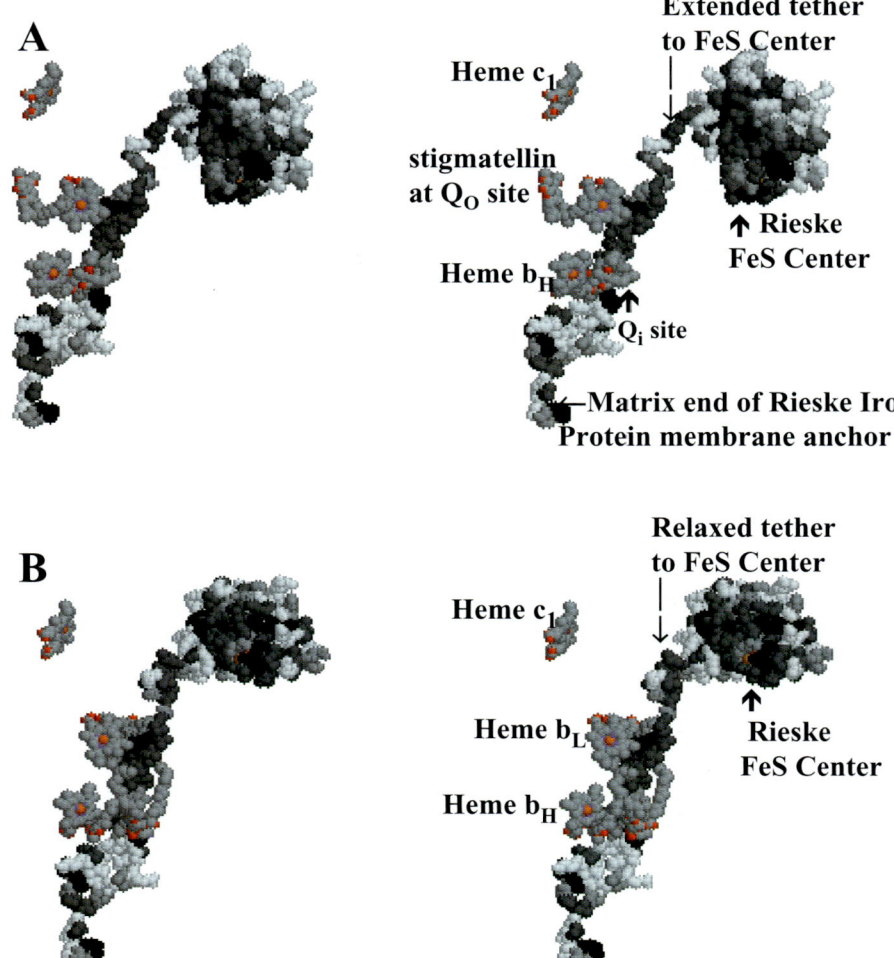

FIGURE 8.14. Stereo view of the monomer of Complex III: the cytochrome bc_1 complex of chicken. Shown are ligands plus the Rieske Iron Protein, where the inhibitors stigmatellin and antimycin are used to show FeS center relocation on reduction. The Rieske Iron Protein anchors in one monomer and then reaches across to accept an electron at its FeS center from the Q_o site of the second monomer. **(A)** With inhibitors, the FeS center is in position, with the tip of the globular component of the Rieske Iron Protein pointing directly downward at the level of the Q_o site, where it is to be reduced by the second component of the functional homodimer. (Prepared using the crystallographic results of Zhang et al.[57] as obtained from the Protein Data Bank, Structure File 3BCC.) **(B)** Without inhibitors, the FeS center is in position with the tip of the globular component of the Rieske Iron Protein directed out of the plane at the level of the heme c_1 to reduce cytochrome c_1 by the second unseen component of the functional homodimer. (Prepared using the crystallographic results of Zhang et al.[57] as obtained from the Protein Data Bank, Structure File 1BCC.) Note that the tether connecting the membrane anchor to the globular component of the Rieske Iron Protein subunit is extended in part **B** when the FeS center is at the Q_o site and relaxed in part **A** when the FeS site is at the level of the heme c_1 site.

8.3 The Electron Transport Chain: Protein Machines as Redox-driven Proton Pumps

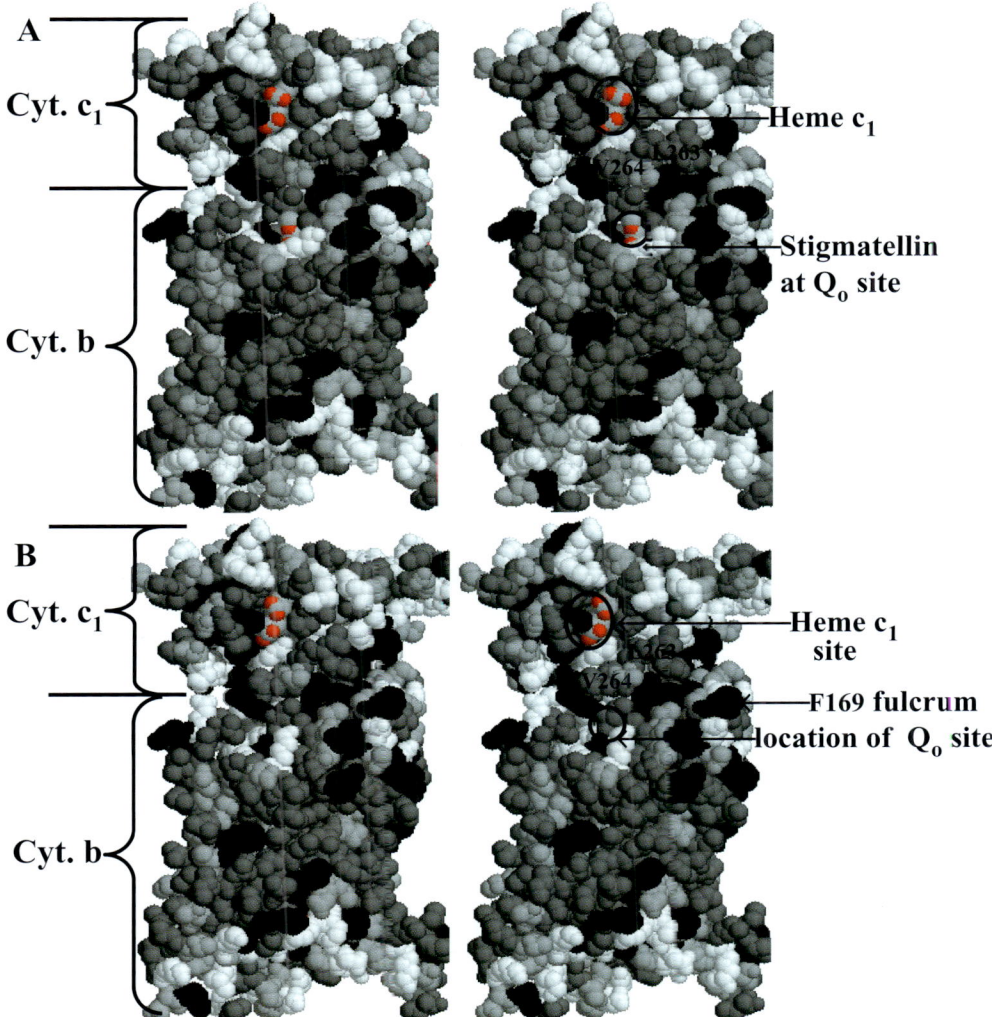

FIGURE 8.15. Stereo view of yeast cytochromes **b** and c_1 of Complex III showing Q_o and heme c_1 binding sites for the FeS center of the Rieske Iron Protein. Neutral residues are given in light gray, aromatic residues in black, other hydrophobic residues in gray, and charged residues in white in order to show visually the relative hydrophobicity of the two sites. Note at the Q_o site that ubiquinol would reside at the base of a hydrophobic pit into which the hydrophobic FeS tip of the Rieske Iron Protein fits (see Figure 8.16), whereas the heme c_1 site is not as hydrophobic. Also note that the hydrophobic residues L263 and V264 constitute a raised rim of the pit over which the FeS center must pass in order to dock at the heme c_1 site. **(A)** Sites in the presence of inhibitors where the FeS center resides at the Q_o site with the configuration shown in Figure 8.16. (Prepared using the crystallographic results of Zhang et al.[57] as obtained from the Protein Data Bank, Structure File 3BCC.) **(B)** Sites in the absence of inhibitors where the FeS center resides at the heme c_1 site (Prepared using the crystallographic results of Zhang et al.[57] as obtained from the Protein Data Bank, Structure File 1BCC.)

378 8. Consilient Mechanisms for Protein-based Machines of Biology

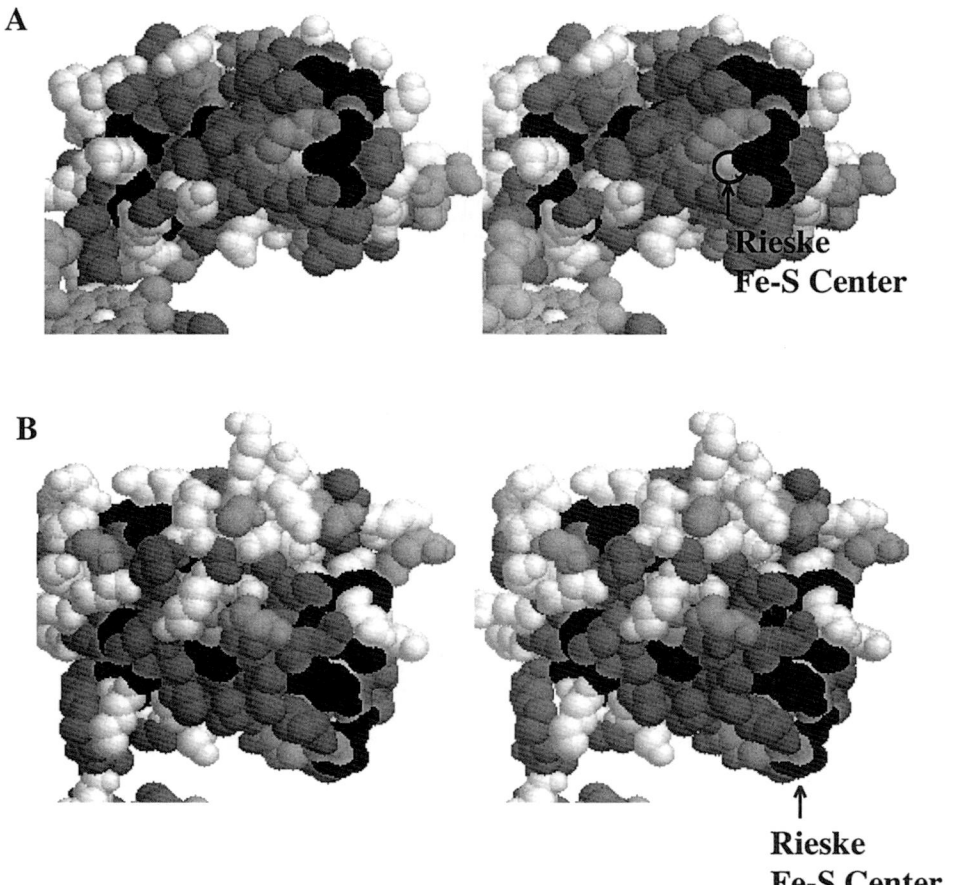

FIGURE 8.16. Stereo view of the Rieske Iron Protein (RIP) component of the monomer of Complex III: cytochrome bc_1 complex of chicken. Shown in space-filling representation is the RIP with neutral residues in light gray, hydrophobic residues in gray, aromatic residues in black, and charged residues in white. (Prepared using the crystallographic results of Zhang et al.[57] as obtained from the Protein Data Bank, Structure File 3BCC.) **(A)** Stereo view showing the very hydrophobic tip at the FeS center of the globular component of the RIP. **(B)** Stereo side view of the globular component of the RIP showing the hydrophobicity becoming greatest at the tip that would form a particularly favorable hydrophobic association in the pit of the Q_o site.

the distance between the fixed ends. This requires that the chain segment be sufficiently free from intermolecular interaction so that the backbone motions are responsive to the limitations posed by deformation. A kinetically free and sterically unrestrained chain is the best way to achieve this structural requirement. Thus an examination of the structure in space-filling representation with all chains present, that is, of the complete functional homodimer, provides the most direct opportunity to determine whether a chain segment may be applying a force at its ends.

By ribbon representation, Figure 8.18B provides a stereo view perpendicular to the twofold axis of an expanded section of dimer in which the tethers are seen crossing from one monomer to the other. These tethers that connect the RIP membrane anchor to the globular component for both monomers are shown totally free standing, without any interchain interactions. The cross-eye stereo view in space-filling repre-

8.3 The Electron Transport Chain: Protein Machines as Redox-driven Proton Pumps

sentation in Figure 8.18A of the same expanded section shows the tether in the foreground to be free standing without contact with other protein chain segments. Thus, the tether is kinetically free, being constrained in its motion only by the extending forces at its ends. This is exactly as required to provide an entropic elastic force by the mechanism of the damping of internal chain dynamics on extension.

8.3.4.3.5 Is the Tether Contracted When the FeS Center of the Rieske Iron Protein Resides at the Heme c_1 Site?

For the tether to be responsible for the motion of the FeS center of the globular component of the Rieske center from the Q_o site to the heme c_1 site it must have contracted; the length of the tether must have shortened. This question

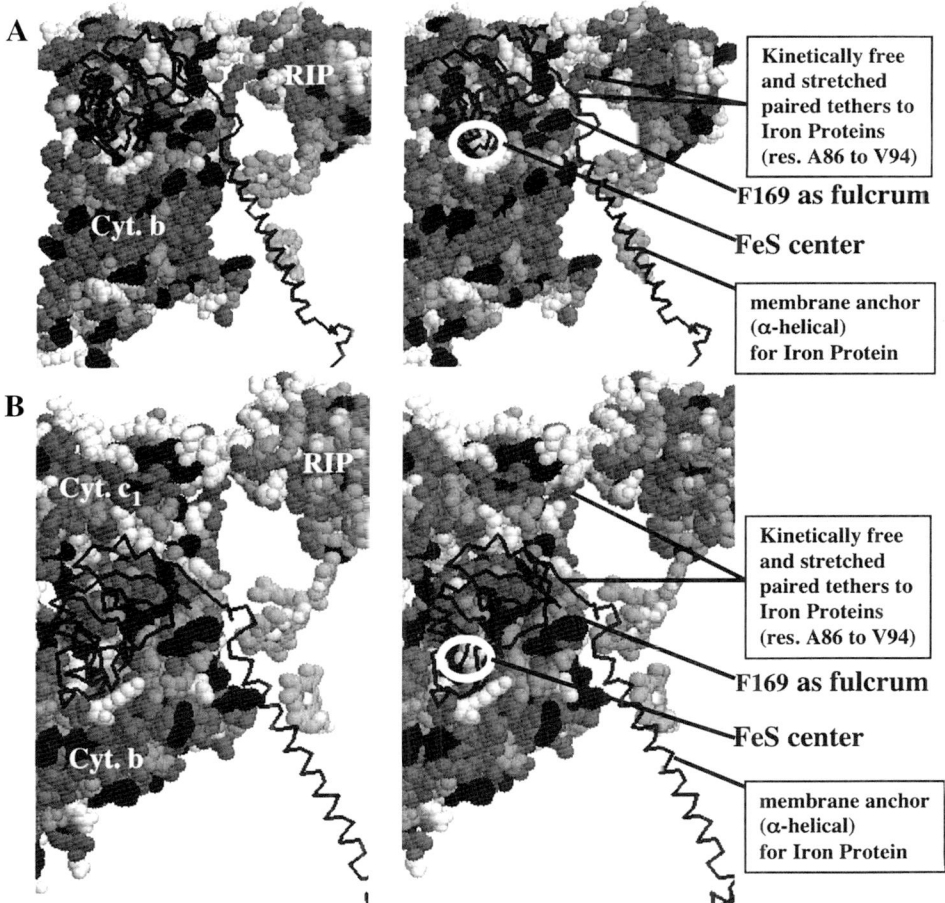

FIGURE 8.17. Stereo view of the components of the homodimer of yeast cytochrome bc_1:—one molecule of cytochrome **b** in space-filling representation, one molecule of cytochrome c_1 in space-filling representation, and both of the FeS-containing Rieske Iron Proteins present are shown with one in gray backbone representation and the other in space-filling representation. As usual for the space-filling representations, neutral residues are in light gray, hydrophobic residues in gray, aromatic residues in black and charged residues in white. The FeS centers of the Rieske Iron Proteins are at their Q_o sites on the cytoplasmic sides of the molecules of cytochrome **b**. (**A**) View perpendicular to twofold axis to show both tethers, one in backbone representation on the left and the other in space-filling representation on the right. (**B**) View rotated to be seen from above with heme **c** included. (Prepared using the crystallographic results of Lange and Hunte[59] as obtained from the Protein Data Bank, Structure File, 1KYO.)

can be addressed using the crystal structure of cytochrome bc_1 from chicken that is available for both states of Q_o site and heme c_1 site occupancy.[57] In the most appropriate backbone representation in Figure 8.19B, the tether is shown in its extended state by noting the distance from residues A66 to K77, whereas the A66 to K77 end-to-end distance is shortened in Figure 8.19A. This shortening is also apparent in Figure 8.14A,B. As shown in Figure 8.19, the tether is most dramatically shortened for the segment from A66 to I74. Thus, the second criterion for the tether to function as a contractile element to achieve domain movement has been met.

As with the elastic-contractile model proteins discussed in Chapter 5, favorable hydrophobic association (in this case of RIP globular protein tip with the Q_o site) stretches interconnecting chain segments. Thus, the answer to the question asked by Crofts et al.[60] in the title of their article, becomes clear: "Interactions of quinone with iron–sulfur protein of the bc_1 complex: Is the mechanism spring-

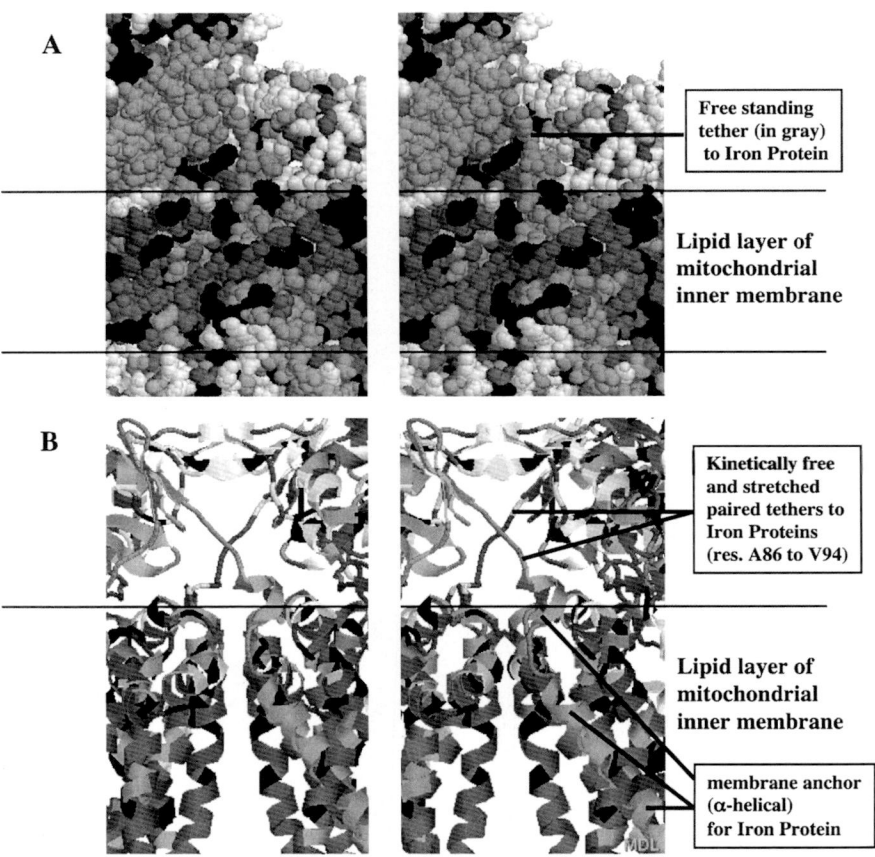

FIGURE 8.18. An expanded stereo view of the complete yeast Complex III: dimer of the cytochrome bc_1 complex. Identified are the tethers of Rieske Iron Protein connecting the membrane anchor to the globular component. The purpose is to demonstrate that the tether sequence from A66 to V94 is free standing in the structure and would, therefore, be capable of exhibiting the changes in entropy on extension required to produce an entropic elastic force. (A) Space-filling representation showing all of the Rieske Iron Protein in gray, but allowing only one tether to be seen. (B) A similar view to that in A, but using the ribbon representation that allows both tethers to be seen in symmetric relation in an open region of the structure. (Prepared using the crystallographic results of Lange and Hunte[59] as obtained from the Protein Data Bank, Structure File, 1KYO.)

8.3 The Electron Transport Chain: Protein Machines as Redox-driven Proton Pumps

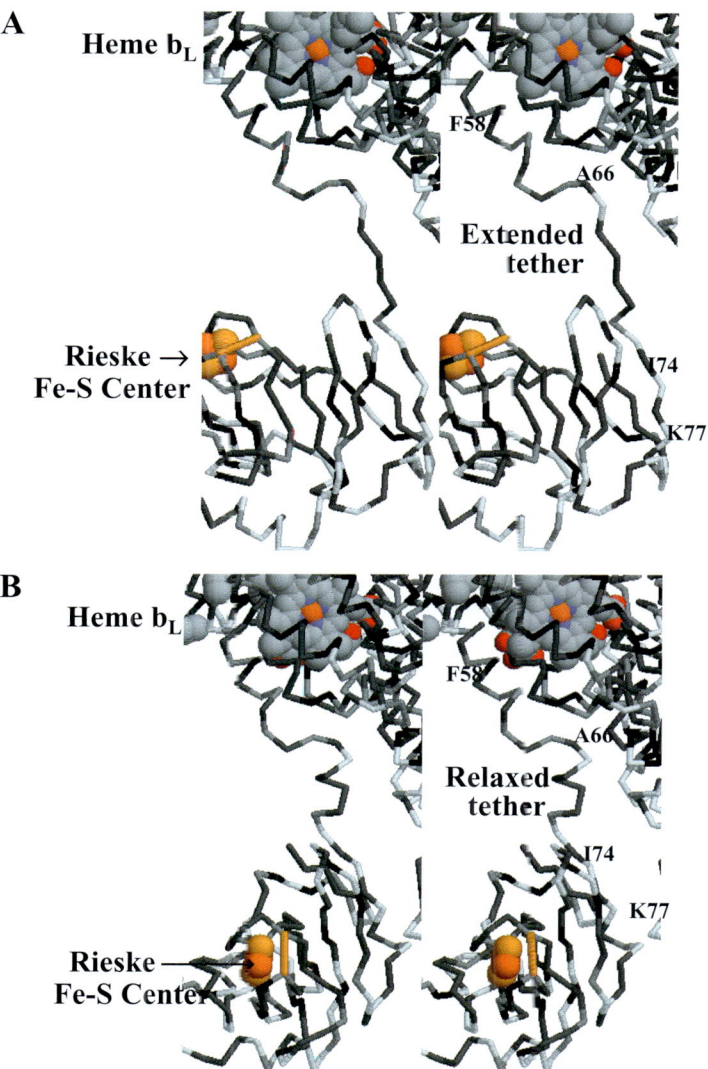

FIGURE 8.19. Stereo views of monomers of Complex III: cytochrome bc_1 complex of chicken with (A) and without (B) inhibitors. Protein in backbone representation with neutral residues in light gray, hydrophobics in gray, aromatics in black, and charged residues in white. (A) The tether is extended when the FeS is at Q_o site. (Prepared using the crystallographic results of Zhang et al.[57] as obtained from the Protein Data Bank, Structure File 3BCC.) (B) The tether is relaxed when the FeS center is at the heme c_1 site. Note the extension of the tether that is apparent by observing the distance between A66 and I74. (Prepared using the crystallographic results of Zhang et al.[57] as obtained from the Protein Data Bank, Structure File 1BCC.)

loaded?" The foregoing structural analysis does indeed indicate that hydrophobic association spring-loads the "iron–sulfur protein of the bc_1 complex." Now, if this is to provide a meaningful step forward in understanding the mechanism, an explanation must follow whereby the favorable hydrophobic association involving the ubiquinol at the Q_o site becomes lost so that the elastic retraction can bring about domain movement. The explanation comes from the change in the Gibbs free energy of hydrophobic association due to the contribution of apolar–polar repulsive free energy of hydration (ΔG_{ap}) that arises from charge formation.

8.3.4.3.6 Formation of QH_2^{2+} Would Disrupt Hydrophobic Association at the Q_o Site Due to ΔG_{ap}

Perhaps the most fundamental and certainly the most dramatic observation during the study of hydrophobic association using the elastic-contractile model proteins has been that the introduction of charge disrupts hydrophobic association. This finding and indeed the principal basis for controlling hydrophobic association, ΔG_{ap}, stands out in the release of hydrophobic association at the Q_o site. The transfer from ubiquinol, QH_2, of one electron to the FeS center and of a second electron to the heme b_L center leaves a positively charged ubiquinol, for example, QH_2^{2+}.

Before electron transfer, as fluctuations toward dissociation occur with occupancy by the hydrophobic QH_2 molecule, sufficient hydrophobic hydration would form to result in water insolubility such that reassociation results. If, on the other hand, fluctuations toward dissociation occur for QH_2^{2+} occupancy, as hydrophobic hydration formed it would be recruited to align for hydration of the positively charged QH_2^{2+}. Thus, increased hydrophobic hydration that ensured hydrophobic reassociation would not occur and hydrophobic dissociation would be the result. Without the favorable decrease in Gibbs free energy for association, the deforming force that caused extension of the tether would be gone, and elastic retraction would result. Additional mechanical details of the resulting domain movement follow below.

8.3.4.3.7 Mechanics Whereby the Stretched Tether Could Elastically Contract to Relocate the FeS Center from the Q_o Site to the Cytochrome c_1 Heme Site

The expected mechanics for relocation of the FeS center within the globular domain of the RIP come from the yeast Complex III depicted in Figure 8.17. One of the two RIP subunits occurs in backbone representation such that it is possible to see through to the underlying Q_o site for hydrophobic association. Immediately below the extended tether resides a raised side chain of residue F169. The bulky F169 side chain is positioned to function as a fulcrum that bears the force of the elastic contraction. This raises the stylus-like point of the FeS center (see Figure 8.16B) out of the pit of the Q_o site and over the L262 and V264 rim of the pit separating the Q_o and heme c_1 sites. In this manner the elastic contraction would lift the globular domain and rotate it into the less hydrophobic heme c_1 site for electron transfer. Transfer of the electron to the heme c_1 and replacement at the Q_o site of a hydrophobic ubiquinol molecule re-establishes the favorable Gibbs free energy for hydrophobic association at the Q_o site and positions the FeS center for the next electron transfer.

8.3.4.3.8 Proposed Calculation of Tether Dynamics to Assess Development of Entropic Elastic Force on Observed Extension

The observed structural changes attending the movement of the globular domain of the RIP are exactly as one would expect for the hydrophobic and elastic consilient mechanisms. Nonetheless, one should make and repeatedly test predictions based on the proposed mechanism, as was done over and over again when establishing the consilient mechanism for the diverse energy conversions of the designed elastic-contractile model proteins (see Chapter 5). In this way a proposed mechanism can achieve the required evaluation for a given system. The most meaningful tests are those for which there is a physical result obtained from instrumentation, not a result from a "computational experiment." Our ultimate goal is, of course, to reach the stage where the understanding of the forces involved is such that sufficiently reliable mathematical descriptions can be formulated. The present state of such calculations is not yet at that stage. In our view, using a low dielectric constant to make ion pairing favorable instead of an apolar–polar repulsive free energy of hydration demonstrates the present limitations. Even so, the ultimate goal remains. Accordingly, with extensive data on the versatile elastic-contractile model proteins of this book, a computational approach that served the model system well could become a meaningful step with which to clarify the mechanism of interest.

One computational test would be to perform a molecular dynamics calculation of the change in internal chain dynamics of the observed extension, that is, to calculate the change in chain entropy resulting from the extension observed in Figure 8.19. The two available tether sequences are residues 64 through 74 (A D V L A M S K I E I) of the chicken sequence and residues 84 through 94 (A D V L A M A K V E V) of the yeast sequence.

8.3.4.4 Consilient Mechanism for Proton Transport at Q_o and Q_i Sites

The focus has been on both aspects of the consilient mechanism (hydrophobic association/dissociation and elastic force development/relaxation) involved in the unique domain movement for electron transfer within Complex III. In what follows, the same aspect of hydrophobic association/dissociation of the consilient mechanism is proposed for facilitating proton gating.

The scenario often repeated throughout this book speaks of associating hydrophobic domains in which fluctuations in the direction of dissociation proceed until so much hydrophobic hydration has developed that hydrophobic association (insolubility) recurs. However, if a charged group should appear sufficiently proximal to the hydrophobically associating domains, now as the fluctuation towards dissociation occurs, the hydration that enters is immediately oriented for hydration of the charged group such that reassociation does not occur, because insufficient hydrophobic hydration formed to result in insolubility. Accordingly, charge disrupts hydrophobic association by disrupting hydrophobic hydration. This competition for hydration that blocks the build up of too much hydrophobic hydration can often be easily quantified. In many cases a change in hydrophobicity results in a very large change in pKa or in the affinity for other counterions. In such cases the change in Gibbs free energy for apolar–polar repulsion, ΔG_{ap}, becomes easily quantified. For example, $\Delta G_{ap} = 2.3 RT \Delta pKa$ (see Equation [5.13]).

In the context of proton gating, at both the Q_o and Q_i sites charge formation occurs as the result of electron transfer. Herein lies the coupling of electron transport to proton pumping. By means of ΔG_{ap}, the formation of charge opens aqueous access for proton egress and ingress. As shown in Figure 8.20, both sites exhibit "waters of Thales" that constitute incipient water channels through which protons could possibly move. The presence of hydrophobic groups, however, with the apolar–polar repulsive free energy of hydration, require that a greater aqueous pathways be formed, as considered below.

8.3.4.4.1 Proton Access to Cytosol from the Q_o Site

It has already been argued above that formation of the positively charged QH_2^{2+} triggers the displacement of the globular domain of the RIP from the Q_o site. Formation of the positive charge on electron transfer from QH_2 disrupts the hydrophobic association of a very hydrophobic tip of RIP with a substantially hydrophobic pit of the Q_o site. This is shown in Figures 8.15, 8.16, and 8.17. By this element of the consilient mechanism, competition for hydration between hydrophobic and charged residues gives rise to positive cooperativity, wherein the appearance of charge at one location facilitates charge formation, for example by ion-pair separation, at another location. Accordingly, charged groups cooperate to destructure hydrophobic hydration and thereby propagate an environment for charged groups.

Based on studies of the elastic-contractile model proteins reviewed in Chapter 5, during contraction by hydrophobic association proton diffuses in the contractile matrix at the rate of one-fifth that of the diffusion of water out of the matrix. Specifically, 20 Mrad γ-irradiation cross-linked poly(GVGVP), that is, X^{20}-poly(GVGVP), contracted by hydrophobic association and lifted a weight more than five times faster on addition of 0.15 N NaCl and 0.01 M sodium phosphate (PBS) at pH 7.4 and 25°C[61] than did X^{20}-poly[4(GVGVP), (GVGVP)] in PBS on lowering the pH from 7.4 to 2.1 at 37°C.[62] Despite there being more than 60% water in the matrix of the elastic-

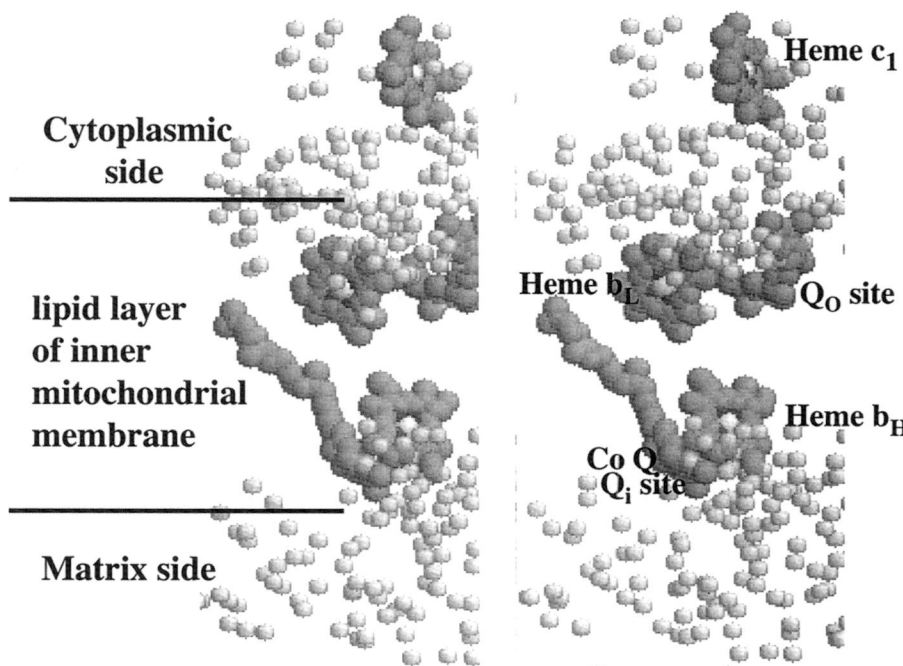

FIGURE 8.20. Yeast cytochrome bc_1 stereo view of hemes, the ubiquinol Q_o site and the ubiquinone Q_i site, and the "waters of Thales." By ΔG_{ap}, the development of a positive charge at the Q_o site on oxidation of ubiquinol disrupts the otherwise most favorable hydrophobic association of the FeS center of Reiske Iron Protein and allows for elastic retraction by the extended tethers. Removal of the FeS center of the Reiske Iron Protein from the Q_o site allows release of the protons of QH_2^{+2} to the cytoplasmic side of the inner mitochondrial membrane. Furthermore, development of a negative charge at the Q_i site on reduction of ubiquinone to form Q^{-2} opens an aqueous channel for the uptake of two protons from the matrix side to reconstitute ubiquinol, QH_2. (Prepared using the crystallographic results of Hunte et al.[25] as obtained from the Protein Data Bank, Structure File, 1EZV.)

contractile model protein, we interpret this to mean that water inside of the matrix is substantially hydrophobic hydration, which is unsuited for hydration of proton. Stretch-induced and hydrophobic-induced pKa shifts provide the same argument that hydrophobic hydration is unsuited for hydration of charged species (see Chapter 5).

8.3.4.4.2 ΔG_{ap} Control of Proton Ingress from the Matrix to the Q_i Site

As shown in Figure 8.20, the Q_i site, even before the presence of negatively charged ubiquinone, exhibits strings of water molecules emanating toward the matrix side of the inner mitochondrial membrane. As stated by Hunte when discussing associated ionizable residues of H181 and E272 of RIP, "These residues are connected with the bulk solvent, either directly or by a short array of hydrogen-bonded, buried water molecules..."[24] Adding to these potential charges, reduction of ubiquinone provides for the formation of as many as two additional negative charges, that is, Q^{2-}. By means of ΔG_{ap}, association of proximal hydrophobic groups would be disrupted on Q^{2-} formation. This would have the effect of expanding the strings of water molecules into aqueous channels capable of allowing the influx of protons at proper rates from the matrix to complete the conversion of ubiquinone to ubiquinol and thereby to couple electron transfer within the lipid layer of the inner mitochondrial mem-

8.3 The Electron Transport Chain: Protein Machines as Redox-driven Proton Pumps

brane to proton transport across the inner mitochondrial membrane.

At this point the facility of proton movement through water requires further consideration. There is the well-established perspective when bulk water is present that protons can be passed along a string of water molecules by a very fast proton jump mechanism whereby a proton may enter at one end of the string and a different proton be given off at the other end of the string. This was clearly demonstrated in the 1960s by the beautiful fast reaction studies of Manfred Eigen and others. As discussed above and in Chapter 5, when there are hydrophobic residues present and resulting hydrophobic hydration, proton transfer becomes lowered to about one-fifth that of the speed of diffusion of the water molecule itself. Accordingly, for the uptake of protons from the matrix at the Q_i site, as shown in Figure 8.21, the proximity of

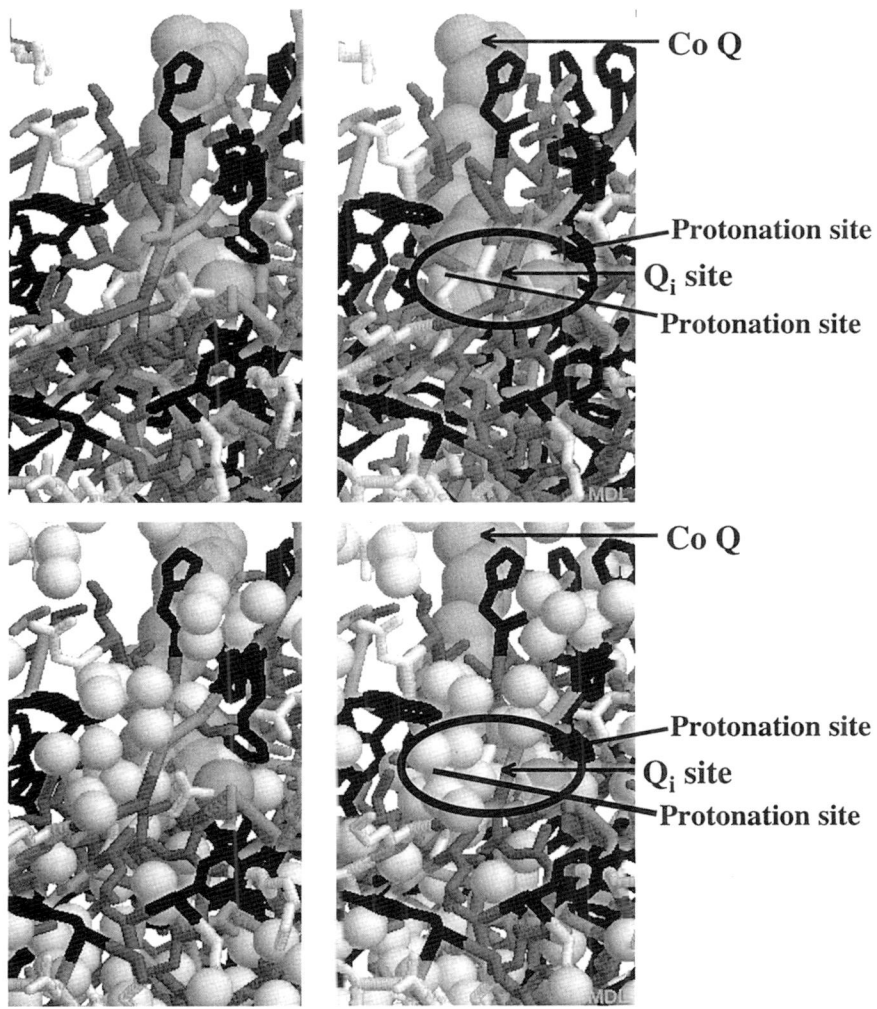

FIGURE 8.21. Yeast cytochrome bc_1 stereo view of the ubiquinone Q_i site from the bottom and "waters of Thales." ΔG_{ap} becoming positively charged on oxidation of ubiquinol at the Q_o site opens an aqueous access to the cytoplasmic side for proton release and becoming negatively charged on reduction of ubiquinone at the Q_i site opens an aqueous channel for proton uptake on the matrix side. The line to the protonation site in each of the four cases goes to an oxygen atom of O^{-2}. (Prepared using the crystallographic results of Hunte et al.[25] as obtained from the Protein Data Bank, Structure File, 1EZV.)

hydrophobic residues and the limited amount of water argues, from the elastic-contractile model protein studies, that the desired rate of proton ingress would require an enlarged aqueous channel of polar hydration. However, this required enlargement of the aqueous channel, the barrier presented by the lipid bilayer itself, and the absence of water molecules between hemes b_L and b_H in Figure 8.20 combine to ensure the stoichiometry for coupling electron transfer in Complex III to proton pumping across the inner mitochondrial membrane.

8.3.4.5 Complex III and the "Movable Cusp of Insolubility"

8.3.4.5.1 Movement of Rieske Iron Protein, Both Metaphorically and Literally, Represents a Movable Cusp of Insolubility

On oxidation of ubiquinol at the Q_o site, the resulting positive charges raise the temperature interval of the cusp of insolubility, as might be symbolically represented by shifting the central cuspid of Figure 7.1 to that of the rightmost cuspid. The result of hydrophobic dissociation, due to ΔG_{ap}, evidences the metaphorical movement of the cusp of insolubility. However, the cusp of insolubility of the tip of the RIP literally moves spatially from the Q_o site to reside at the cytochrome c_1 site as may be considered by examining Figures 8.14 through 8.17. Thus, as with the γ-rotor of ATP synthase considered below, the globular domain of RIP exhibits both literal and metaphorical movement of the cusp of insolubility, "and the forces that . . . power the molecular machines of biology drive spatially localized hydrophobic protein domains back and forth across thermodynamically movable water-solubility-insolubility divides."

8.3.4.5.2 Reduction of Ubiquinone at the Q_i Site Effects a Metaphorical Movement of the Cusp of Insolubility to Open Aqueous Channel for Proton Ingress

When reduction of ubiquinone occurs at the Q_i site, the negatively charged ubiquinone causes an increase in ΔG_{ap}. During a transient hydrophobic dissociation involving the strings of water molecules directed toward the matrix side of the inner mitochondrial membrane, the negatively charged site recruits the otherwise hydrophobic hydration for its own hydration. The result is an aqueous channel sufficient to allow proton ingress in a symbolical movement of the cusp of insolubility.

Our conclusion is that dissociation of the RIP from the Q_o site in a single step represents a unique intersection of electron transfer and proton translocation that simultaneously employs both the hydrophobic and elastic consilient mechanisms.

8.3.5 Complex IV: Cytochrome c Oxidase

Crystal structure data on Complex IV developed relatively early[63,64] compared with the other complexes of the electron transport chain. Nonetheless, the understanding of mechanism does not rise to the same level, for example, as that of Complex III. The following consideration of Complex IV is less extensive than that for Complex III above, but it continues to argue for the hydrophobic consilient mechanism by means of the apolar–polar repulsive free energy of hydration, ΔG_{ap}, and it also speculates on possible molecular players for proton egress and ingress.

Complex IV (cytochrome c oxidase) uses electrons from the product of Complex III, ferrocytochrome c, to reduce molecular oxygen (O_2) with the result of forming water. However, it is useful to realize that cytochrome c oxidase is actually part of a superfamily of heme–copper respiratory oxidases.[65] One member of the superfamily is a ubiquinol oxidase from *E. coli*,[66] In this protein-based machine ubiquinol, QH_2, replaces ferrocytochrome c (cyt c^{2+}) as an electron source. Furthermore, ubiquinol oxidase contains no binuclear Cu_A center that normally oxidizes ferrocytochrome c in Complex IV.

The central question in the comparison, however, becomes what molecular species in cytochrome c oxidase replaces QH_2 as the proton source. As discussed regarding Complex III, QH_2 was the source of both electrons to subsequent redox centers and protons for release to the cytoplasmic side of the inner mitochondrial membrane, and ubiquinone, Q, receives both electrons from heme b_H and protons from the matrix side of the membrane. Because QH_2 is lipid soluble, it diffuses through the lipid layer and functions simultaneously as proton and electron carrier to provide tight coupling of electron transport to proton pumping.

The following consideration of cytochrome **c** oxidase reviews (1) composition, (2) structure, (3) overall reaction, (4) the electron transfer steps of cytochrome **c** oxidase, (5) status of proton translocation and proposed aqueous D- and K-channels for proton ingress, (6) the redox Bohr effect and its correlation with electrochemical transduction in elastic-contractile model proteins, and (7) possible molecular sources of protons for translocation with an abundant and uniquely positioned functional side chain that exhibits interesting parallels to the states of QH_2 with, however, single rather than double proton changes and coordination to a metal ion required for electron transfer capability.

Although different molecular species and arrangements of redox centers occur in Complex III than in Complex IV, we proceed with the perspective that the underlying physical process is the same, that is, the operative apolar–polar repulsive free energy of hydration, ΔG_{ap}, aspect of the hydrophobic consilient mechanism. In this view, a change in charge due to electron transfer to or from a redox center changes that centers contribution to ΔG_{ap} in a way that can open or close aqueous channels for proton ingress into, and egress from, the inner mitochondrial membrane.

8.3.5.1 General Structural Description of Complex IV: Cytochrome c Oxidase

8.3.5.1.1 Composition

Bovine cytochrome **c** oxidase contains 13 protein subunits, summing to a molecular weight of 200 kDa. Three of the subunits, encoded in the mitochondrial DNA, appear common to all species, subunits I, II and III. In all cases the four redox centers are localized to subunits I and II. One redox center, a binuclear copper center, Cu_A, occurs in subunit II. The three other redox centers are found in subunit I; these are a heme **a**, a heme a_3, and a single copper ion, Cu_B, closely associated with heme a_3.

8.3.5.1.2 Molecular Structure

A cross-eye stereo view of the dimer of Complex IV (cytochrome **c** oxidase) from bovine heart mitochondria is shown in Figure 8.22. Figure 8.22A shows the molecular structure in space-filling representation with neutral residues in light gray, the most hydrophobic aromatic residues in black, other hydrophobic residues in gray, and the charged residues in white. With this shading, a dark hydrophobic band surrounding the middle of the structure positions the complex in the lipid (totally hydrophobic) layer of the inner mitochondrial membrane. The ribbon representation in Figure 8.22B allows visualization of the transmembrane helices within which are embedded the redox centers of heme **a**, heme a_3, and the copper center, Cu_B, associated with heme a_3. The fourth redox center, Cu_A, resides just outside of the lipid layer on the cytoplasmic side of the membrane, as labeled in Figure 8.22B.

8.3.5.2 Phenomenology of Cytochrome c Oxidation to Reduce Oxygen with Net Proton Transport: Structural Studies of Oxidase

8.3.5.2.1 Overall Reaction

As given, for example, by Schultz and Chan,[49] the overall reaction of cytochrome **c** oxidase is given in Equation (8.8) above. As the reduction of one molecule of oxygen, O_2, to water requires four electrons, these are shown coming from four molecules of ferrocytochrome **c** (4cyt c^{2+}), which become four molecules of ferricytochrome **c** (4cyt c^{3+}). Conversion of O_2 to two molecules of water, of course, also requires four protons; these are indicated as $4H^+$ (scalar). The fundamental purpose of the electron transport chain, of course, is to produce an increased concentration of protons in the cytosol. In some as yet not well-understood process, four protons

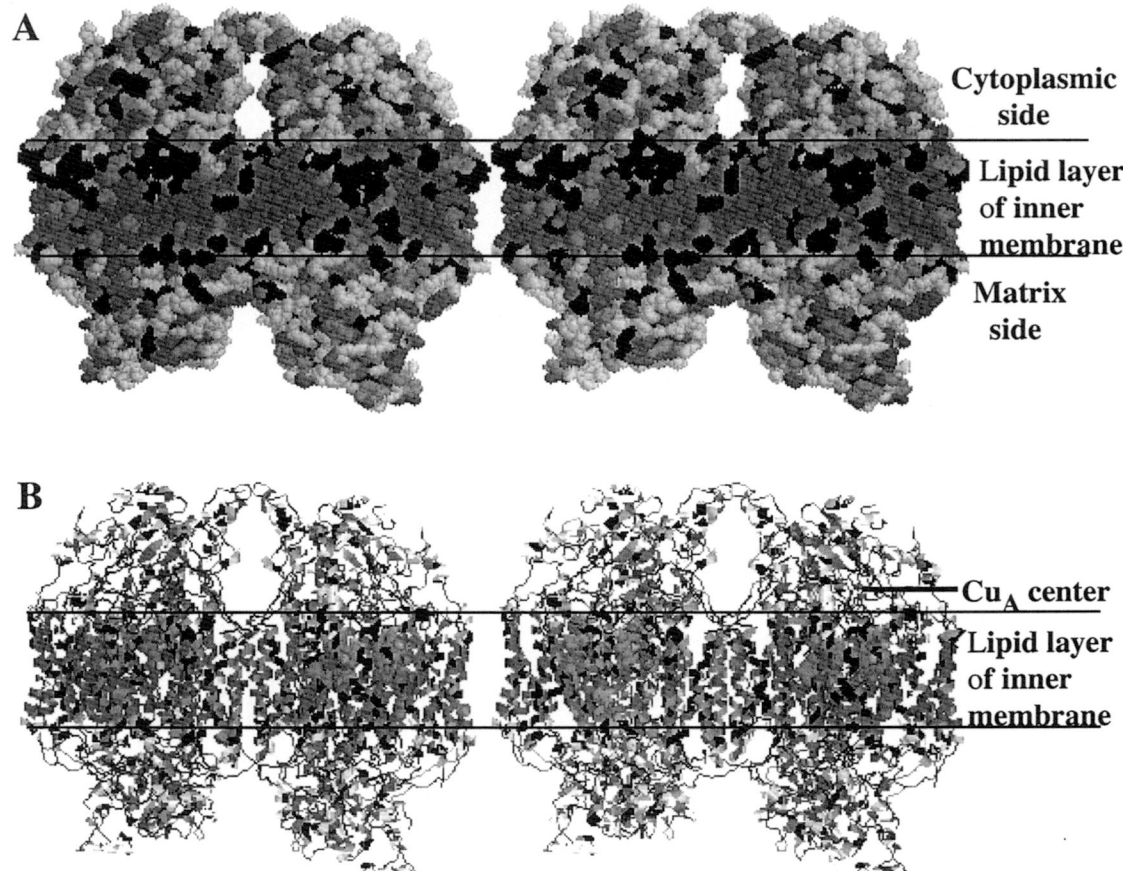

FIGURE 8.22. Stereo view of dimer of bovine heart Complex IV: cytochrome c oxidase, reduced without oxygen and with neutral residues in light gray, aromatic residues in black, other hydrophobic residues in gray, and charged residues in white. **(A)** Space-filling representation showing a hydrophobic band wherein the structure rests in the lipid bilayer. **(B)** Ribbon representation with the embedded redox centers discernible. (Prepared using the crystallographic results of Yoshikawa et al[67] as obtained from the Protein Data Bank, Structure File, 1OCR.)

from the matrix side of the membrane become transported to the cytoplasmic side of the membrane.

As argued for Complex III above, shifting the balance from hydrophobic to charged groups of whichever sign, positive or negative, opens aqueous access for protons into and out of the membrane region of Complex III. This results from an increase in ΔG_{ap}, the apolar–polar repulsive free energy for hydration. Many studies have been carried with Complex IV out to show changes in proton uptake on changes in redox state of the carriers, and many species have been considered for their contribution to proton translocation. Along with those, the unique locations and possible states of the imidazole side chain of histidine will again be noted because of states that parallel those of the ubiquinone system. Most fundamental to the hydrophobic consilient mechanism, however, will be the correlation of phenomena with the elastic-contractile model proteins that have been designed for analogous electrochemical transduction, wherein reduction of positively charged species causes proton uptake and increased hydrophobicity causes

8.3 The Electron Transport Chain: Protein Machines as Redox-driven Proton Pumps

increased electron affinity of positively charged species. First, however, the electron-transfer steps and the involved redox carriers are noted.

8.3.5.2.2 Electron-Transfer Steps: Ferrocytochrome c → Cu_A → Heme a → (Heme a_3 and Cu_B) → O_2

The series of single electron transfer steps of Complex IV are represented in Figure 8.23A. Ferrocytochrome c (cyt c^{2+}), the aqueous diffusible reduction product of Complex III, transfers a single electron to result in ferricytochrome c (cyt c^{3+}) and a reduced binuclear copper center, $(Cu_A)_2$. The single electron then transfers to the heme a and from there to the binuclear heme a_3—Cu_B center, which binds O_2 between the heme iron and Cu_B coordinated by the imidazole side chains of two histidine residues. This process repeats for four times to complete the reduction of O_2, which on addition of four protons completes the conversion to two molecules of water.

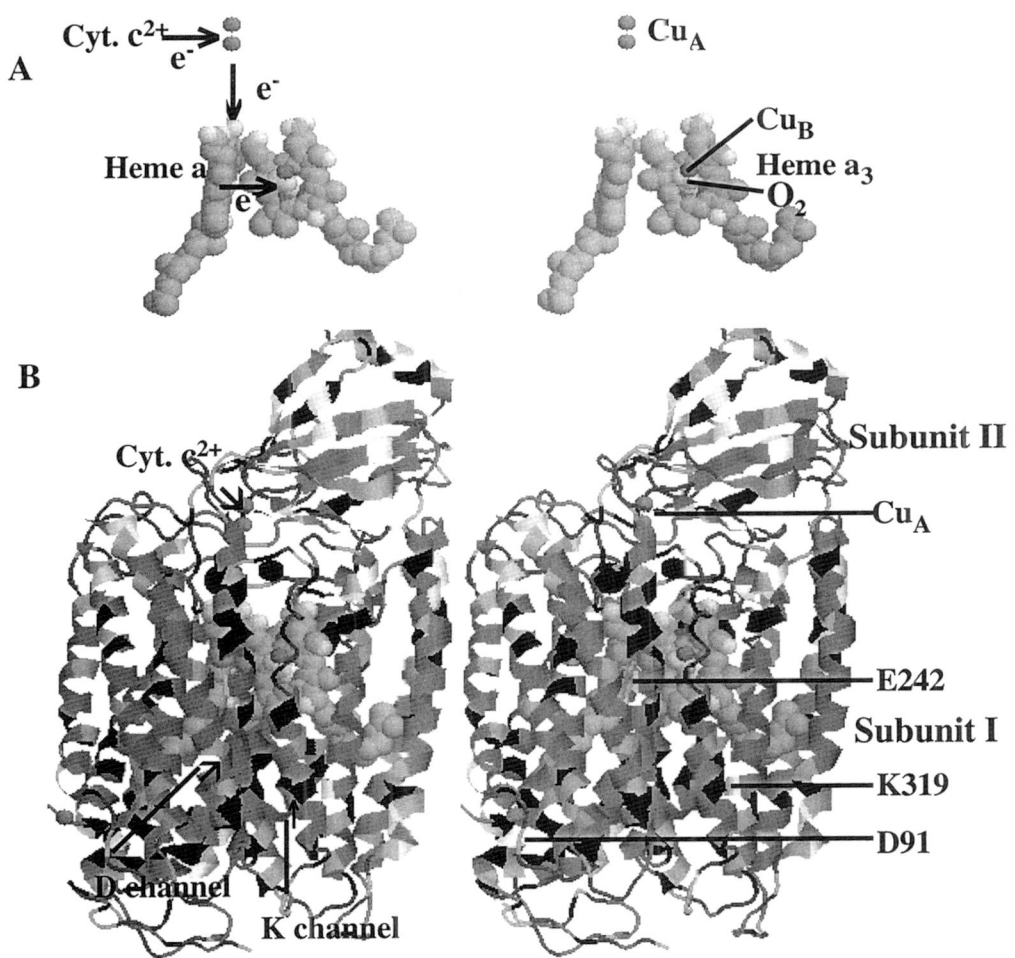

FIGURE 8.23. Stereo view of monomer of bovine heart Complex IV: cytochrome c oxidase with bound O_2. (A) Redox centers: Cu_A, heme a, heme a_3, and Cu_B with single electron transfers indicated by arrows at left. Same orientation as in B. (B) Subunits I and II in ribbon representation, showing in subunit II the locations of D91 and K319 residues with D and K channels indicated as well as the suggested coupled residue E242. (Prepared using the crystallographic results of Yoshikawa et al.[67] as obtained from the Protein Data Bank, Structure File 2OCC.)

8.3.5.2.3 Status of Proton Translocation

Understanding of the proton translocation has been complicated by the need to address a steady-state circumstance rather than to consider static states. For example, in the oxidized state in the absence of reductant, the heme–copper binuclear center relaxes to a state that lies outside of the catalytic cycle. With proper consideration of the catalytic cycle according to Wikström and Verkhovsky,[68] "Two protons each are pumped during the oxidative and reductive halves of the cycle." This important insight has yet to be carried to the molecular species involved, although an OH^- ligand for $Cu_B[II]$ has been considered, as well as earlier implication of a histidine liganded to Cu_B.[69]

8.3.5.2.4 Proposed Channels for Proton Translocation

Based on crystal structure data, two channels have been suggested for proton translocation.[64] One channel, the K-channel identified by residue K319, runs from the matrix side to the dioxygen center, where O_2 resides between the iron atom of heme a_3 and Cu_B, as shown in Figure 8.23B. A second channel, the D-channel, crosses the entire range of the lipid layer from the matrix side to the Cu_A site.

The current perspective for proton translocation by Complex IV has been summarized by Schultz and Chan[49]: "Rather, there is likely just a chain of proton carriers through the enzyme, and proton pumping consists of protons hopping roughly in concert along a chain of hydrogen bonds. The energy transduction consists of altering the orientation of the hydrogen bonds, changing the pK_a's of the amino acid residues, and transiently removing and restoring the osmotic barrier."

On the basis of the physical processes identified from the studies on free energy transduction by elastic-contractile model proteins functioning by inverse temperature transitions, the operative interaction energy that alters "the orientation of the hydrogen bonds" and changes "the pKa's of the amino acid residues" is the apolar–polar repulsive free energy of hydration, ΔG_{ap}, as developed in Chapter 5 and reviewed above.

8.3.5.2.5 Negative Dioxygen as a Super-Q_i Site (Negatively Charged Quinone) of Complex III

Figure 8.24B for Complex IV exhibits "waters of Thales" that are more nearly continuous across the membrane than for Complex III.[70] For example, a string of water molecules connects heme a_3 to the matrix side of the inner mitochondrial membrane. Such a limited string of water molecules, however, does not necessarily constitute a channel for proton transit, particularly if they are waters of hydrophobic hydration. Proton movement through the elastic-contractile model protein matrix during chemomechanical transduction by inverse temperature transition is very slow,[62] even when the contracted state is more than 60% water. Proton movement is very slow because so much of the water is hydrophobic hydration. The point to be made is that water molecules, arranged as hydrophobic hydration, are not suitable for hydrating a proton. This same effect results in the hydrophobic-induced pKa shifts, for example, of Figures 5.29 through 5.32 and 5.34. Accordingly, ΔG_{ap}, resulting from the formation of charge, disrupts hydrophobic hydration, including the incipient hydrophobic hydration of a transient hydrophobic dissociation. Thus, the formation of charge, by the hydrophobic consilient mechanism, is precisely suited for activating or opening aqueous channels in protein for proton translocation.

Accordingly, by the hydrophobic consilient mechanism, the receipt of electrons at the dioxygen sandwiched between heme a_3 and Cu_B, as shown in Figure 8.22B, creates a negatively charged site that turns the incipient channel apparent in Figure 8.23B into an aqueous channel suitable for proton ingress. The result, of course, is the conversion of O_2 into $2H_2O$. This would be the explanation for the four scalar protons noted in Equation (8.8).

8.3 The Electron Transport Chain: Protein Machines as Redox-driven Proton Pumps

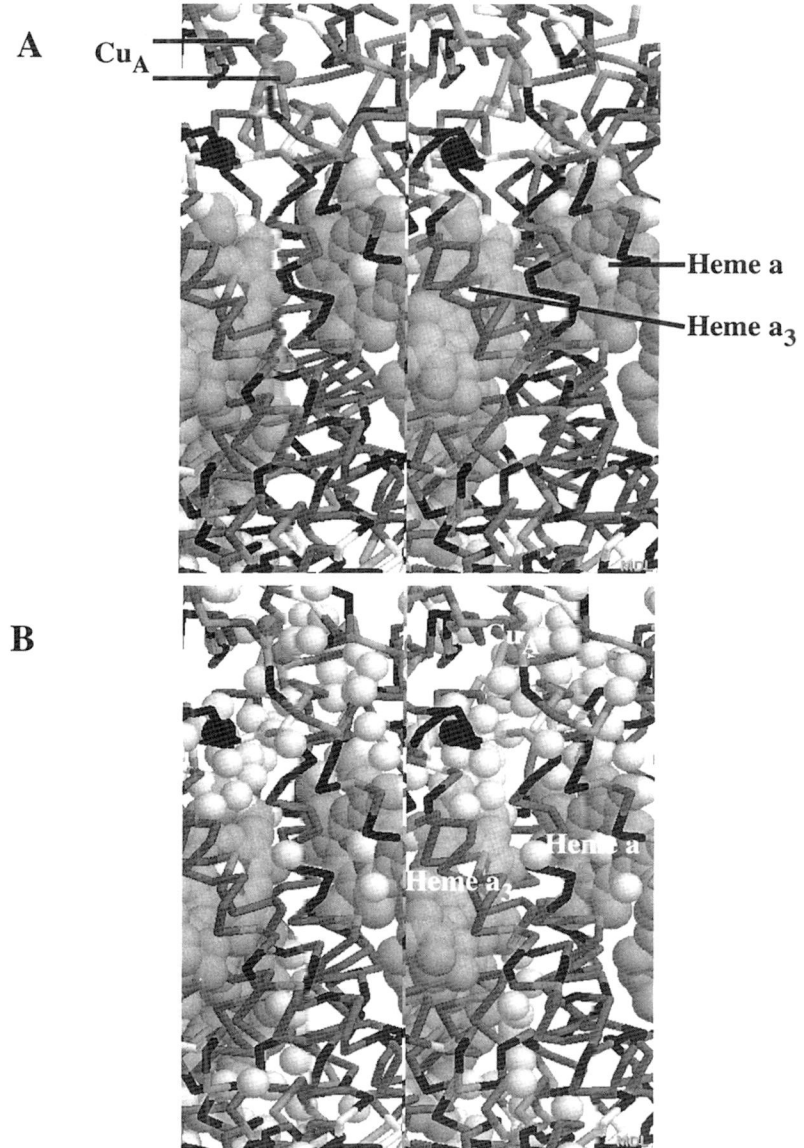

FIGURE 8.24. *Rhodobacter sphaeroides* Complex IV: cytochrome **c** oxidase. Stereo view in backbone representation of subunits I and II showing the greater permeation of water through the region of the lipid layer of Complex IV than for that of Complex III in Figure 8.20. **(A)** Heme redox centers at the heart of the lipid layer. **(B)** "Waters of Thales" distributed at almost all levels through the membrane region. (Prepared using the crystallographic results of Svensson-Ek et al.[70] as obtained from the Protein Data Bank, Structure File 1M56.)

8.3.5.3 The Redox Bohr Effect of Complex IV Represents the Hydrophobic Consilient Mechanism's ΔG_{ap} in Action

8.3.5.3.1 The Redox Bohr Effect in Complex IV

For more than a decade it has been known and repeatedly reaffirmed that reduction results in proton uptake.[68-73] In particular, two protons are taken up on reduction and two protons are released on oxidation. As stated by Wikström and Verkhovsky,[68] "Two protons each are pumped during the oxidative and reductive halves of the cycle, respectively," and "with an efficiency of two translocated charges per transferred electron in the steady state." Because four protons add to the negative dioxygen, it should be clear, however, that "for every electron that is transferred from cytochrome c to cytochrome c oxidase, one proton is pumped through the protein giving four translocated protons. . . ." These are the four matrix protons of Equation (8.8).

In relation to the hydrophobic consilient mechanism, it is relevant to note that the redox centers are positively charged metal ions that become less charged on the addition of the electron and more charged on oxidation. In general, the effect of reduction of the redox sites, for example, heme iron and copper centers, is to increase hydrophobicity. Because a change in the redox state results in proton translocation, the protein-based machine, Complex IV, performs electro-chemical transduction.

8.3.5.3.2 Coherence of Phenomena for Electro-chemical Transduction Demonstrated in Designed Elastic-contractile Model Protein and the Redox Bohr Effect in Complex IV

In the case of our elastic-contractile protein-based polymer poly(GDGFP GVGVP GV GVP GFGVP GVGVP GVGK{NmeN⁺}P), the positively charged N-methyl nicotinamide {NmeN⁺} becomes more hydrophobic on reduction. This causes an increase in the pKa of the carboxylate of the aspartic acid (D) residue by 2.5 pH units, as occurs on increasing the hydrophobicity or oil-like character of the polymer.[27] With the elastic-contractile protein-based polymer, reduction causes proton uptake and oxidation causes proton release. This is an example of an apolar–polar repulsive free energy, ΔG_{ap}, whereby reduction of a positively charged species (an increase in hydrophobicity) causes proton uptake. Thus, there exists a coherence of phenomena for electro-chemical transduction demonstrated in designed elastic-contractile model proteins and in the redox Bohr effect of Complex IV.

8.3.5.3.3 Equivalence of the Reduction/Oxidation (Redox) Bohr Effect in Complex IV and the Original Deoxygenation/Oxygenation Bohr Effect in Hemoglobin

As with hemoglobin, discussed in Chapter 7, a Bohr effect occurs with cytochrome c oxidase. Again from the viewpoint of the hydrophobic consilient mechanism, these phenomena are analogous. Formation of the less polar states on reduction of Complex IV and on forming deoxyhemoglobin result in proton uptake, whereas formation of the more polar oxidized state of Complex IV and the more polar oxygenated state of hemoglobin result in proton release. This is as expected from the ΔG_{ap} of the comprehensive hydrophobic effect, as discussed above.

8.3.5.4 Analogy Between States of the Metal–Ion-complexed Imidazole Side Chain of Histidine and the Ubiquinone/Ubiquinol System

During the transit of one electron from ferrocytochrome c to the (heme a_3–Cu$_B$) binuclear center, one proton is transported from the matrix side to the cytosolic side of the membrane. In analogy to Complex III one suspects that oxidation of a site with access to the cytosolic side would release a proton to that side, whereas on reduction the proton would be replenished from the matrix side. The problem could be stated as one of identifying the groups that would replace ubiquinol and ubiquinone of Complex III.

The striking feature of the functional side chains associated with the redox sites of cytochrome c oxidase is the preponderance of histidine residues, as shown in Figure 8.25. There are eight histidine residues in coordination at the redox sites. $(Cu_A)_2$ has two His residues (H116 and H204); heme **a** coordinates two histidines (H61 and H240) and the (heme a_3–Cu_B) binuclear center has four coordinated histidines (H240, H290, and H291 on the O_2 side and H376 on the backside).

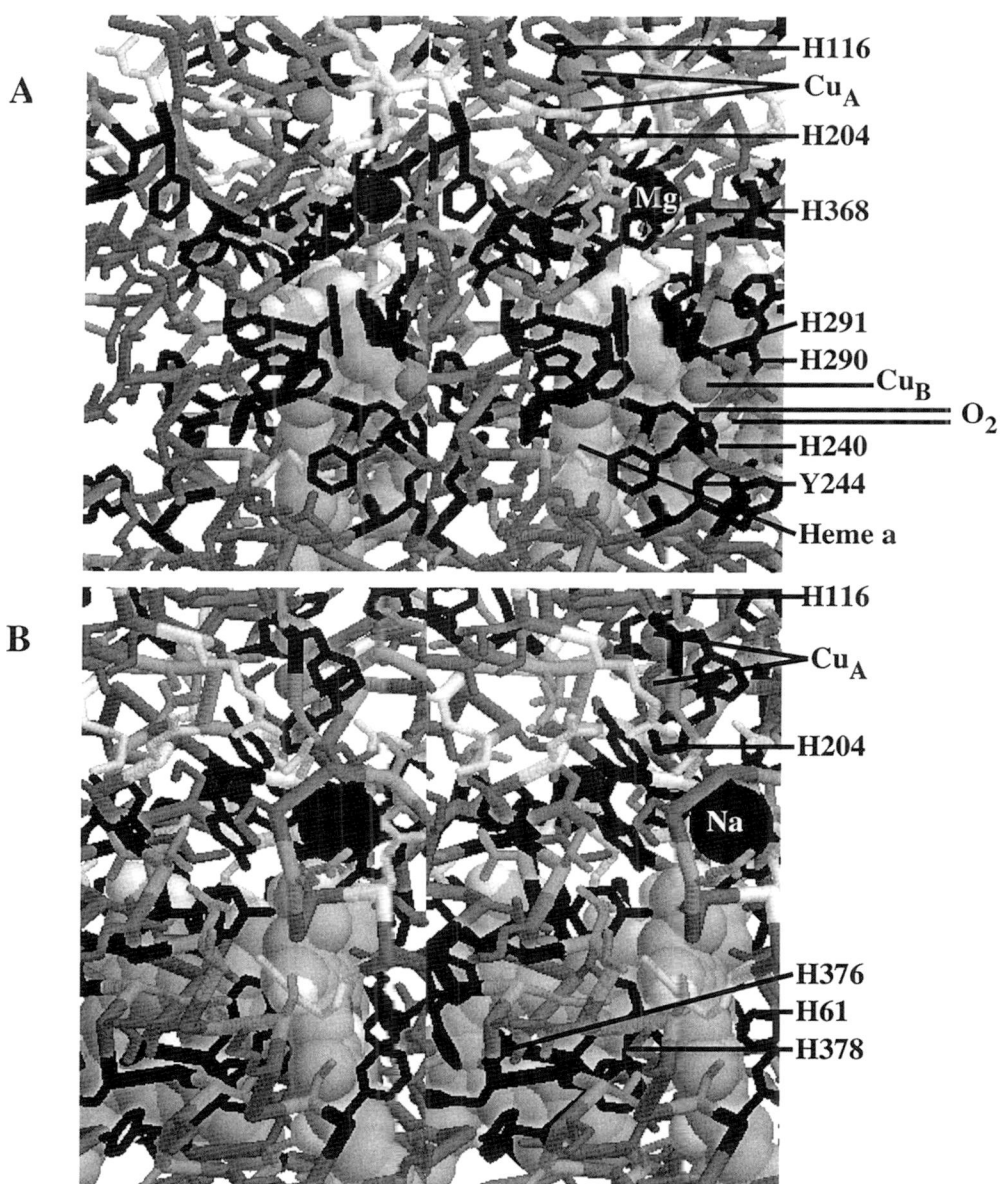

FIGURE 8.25. Complex IV: cytochrome c oxidase. Stereo view of subunits I and II given in vine representation to show a total of eight His residues coordinated to redox sites. **(A)** O_2 side of heme a_3–Cu_B binuclear center with three coordinated His residues and Cu_A with two coordinated His residues. **(B)** Backside of heme a_3 with one coordinated His residue, side view of heme **a** with two coordinated His residues, and again Cu_A with two coordinated His residues. (Prepared using the crystallographic results of Yoshikawa et al.[67] as obtained from the Protein Data Bank, Structure File 2OCC.)

Given the dominance of His coordination of the metal ions at the redox sites, it is difficult not to consider the interesting properties of histidine's imidazole (Im) side chain, identified in Table 5.1. Furthermore, imidazole can exist in three different states of protonation in analogy to the ubiquinol/ubiquinone system except that the three states differ by one proton for imidazole and two protons for ubiquinol/ubiquinone. In particular, the three states for imidazole are $HImH^+$ (imidazolium) $\leftrightarrow$ ImH (imidazole) $\leftrightarrow$ Im^- (imidazolate), and the equivalent three states are $QH_2^{2+} \leftrightarrow Q \leftrightarrow Q^{2-}$ for the ubiquinol/ubiquinone system. There occurs the important distinction, of course, that the ubiquinone system is a two-electron–two-proton system, whereas imidazole in association with a metal ion would vary as a one-electron–one-proton system as would be fitting for the single electron transfers from cytochrome c.

In a purely aqueous system, simple chelation of copper by a histidine residue shifts the pKa of the imidazole group by about 3 pH units.[74] On the other hand, in a more hydrophobic site, as in the inner mitochondrial membrane site of Complex IV, it would be interesting to see whether pKa shifts would be sufficient to access the imidazolate state in order to pick up and release proton as the redox state of the relevant center changed. Another point of consideration in this case relates to the superfamily of heme–copper respiratory oxidases, wherein there are ubiquinol oxidases in which ubiquinol replaces the binuclear copper center. This raises the question as to whether the binuclear $(Cu_A)_2$ center might be examined as replacement for ubiquinol as a site of proton release on oxidation.[75]

8.4 ATP Synthase: The Twofold Rotary Protein Motor of Oxidative Phosphorylation

8.4.1 Insights into Biology's Vital Protein-based Machine: ATP Synthase

When considering ATP synthase, this pivotal protein-based machine of biology, what do the *phenomenological and mechanistic assertions* arising out of the data and analyses in Chapter 5 bring to the *assertion of biological relevance*? It could be, perhaps, too much to expect that the answer would be as clear-cut as the detailed enlightenment attending the forgoing analysis of Complex III function. Nonetheless, the hydrophobic and elastic consilient mechanisms continue with demonstration, in this case, of their capacity to predict structural relationships and functional results. That the central rotor of ATP synthase, the so-called γ-rotor, be hydrophobically asymmetric constitutes the most fundamental prediction. The corollary to this prediction is the expectation that rotor orientation would be a predictable function of occupancy states of the catalytic sites. Most essential with regard to function, the apolar–polar repulsive free energy of hydration, ΔG_{ap}, applied to the structural data would foretell the direction of rotor rotation for both the ATPase and synthesis modes of action. In addition, the findings point to a number of specific experimental tests that can be carried out to assess this assertion of the importance to function of the hydrophobic and elastic consilient mechanisms.

8.4.1.1 Mechanistic Insights That the Assertions of This Volume Contribute to Understanding the Function of ATP Synthase

8.4.1.1.1 Breakdown of ATP Synthase into its Component and Overall Energy Conversions

ATP synthase represents the coupling of two protein machines, one that performs chemo-mechanical transduction and a second that performs mechano-chemical transduction. The net result of the coupling becomes chemo-chemical transduction in which the input of proton chemical energy produces the output chemical energy to yield ATP from ADP plus inorganic phosphate (HPO_4^{2-}, also indicated as P_i). The chemical energy output can be assessed by its relative magnitude but opposite sign to the heat released, approximately −8 kcal/mole, on hydrolysis of ATP to give ADP plus P_i.

8.4 ATP Synthase: The Twofold Rotary Protein Motor of Oxidative Phosphorylation

Both protein machines of ATP synthase are rotary motors, made possible by the mechanical coupling of chemo-mechanical transduction due to the F_0-motor to mechano-chemical transduction of the F_1-motor. A single rotor driven by the membranous F_0-motor in turn drives the extramembranous F_1-motor. The F_0-motor of yeast mitochondria, for example, exhibits a 10-fold rotational symmetry, providing a 10-step rotary engine, whereas the F_1-motor of all ATP synthases exhibits a threefold rotational symmetry, resulting in a threefold rotary engine.

8.4.1.1.2 Relevance of the Hydrophobic Consilient Mechanism to the F_0-motor

The relevance of the F_0-motor of ATP synthase to hydrophobicity is obvious in a way that has long been appreciated. It surprises no one that protonation of the $-COO^-$ of aspartate or glutamate to achieve the carboxyl $-COOH$ would facilitate solubility in a lipid bilayer. In fact, our design for using the inverse temperature transition in the first elastic-contractile model protein-based machine, driven by proton chemical energy, was based on the expectation that $-COOH$ would be more hydrophobic than $-COO^-$.[62] To argue that the hydrophobic consilient mechanism is relevant to function of the F_0-motor presents no strikingly new statement. However, it does become possible to use values for the Gibbs free energy of hydrophobic association obtained on analysis of the elastic-contractile model proteins containing each of the amino acid residues to provide reasonable thermodynamic and efficiency insights for ATP synthase.

8.4.1.1.3 Relevance of the Hydrophobic Elastic Consilient Mechanism to the F_1-motor Functioning as an ATPase: Analogy to the Internal Combustion Rotary Engine

Analogy may be drawn between the F_1-motor functioning in the ATPase mode, that is, a threefold chemo-mechanical rotary engine, and the threefold rotary engine of the Mazda automobile. For the latter, localized energy (pressure-volume) bursts produce motion by the expansion resulting from conversion of liquid to hot gasses on fuel ignition. Based on our view of the hydrophobic consilient mechanism and specifically the apolar–polar repulsive free energy of hydration, ΔG_{ap}, the analogy would be that hydrolysis of ATP to form ADP and P_i provides a burst of apolar–polar repulsion directed at the most hydrophobic side of a hydrophobically asymmetric rotor in a manner that drives rotation.

Such an analogy could be cautioned by the experimental finding that the ratio of ATP to $ADP + P_i$ at the catalytic site can be measured as close to unity. If this were incorrectly viewed as an equilibrium circumstance at the catalytic site, then one would consider the free energy change on going from ATP to $ADP + P_i$ to be zero. This is, however, erroneous. We know under standard conditions that the hydrolytic breakdown of ATP to $ADP + P_i$ releases about 8 kcal/mole of heat. Therefore, at the catalytic site the energy resulting from the conversion of ATP to $ADP + P_i$ must raise the free energy of the molecular structure. We propose that the development of an increase of 8 kcal/mole in ΔG_{ap}, the apolar–polar repulsive free energy of hydration, provides a rotational impulse by structural deformation as required to drive the γ-rotor in the ATPase mode.

8.4.1.1.4 Relevance of the Hydrophobic Elastic Consilient Mechanism to the F_1-motor Functioning as an ATPase: Analogy to a Three-pole DC Motor

Perhaps the better analogy, however, occurs with a threefold symmetric electrical motor, described by Kinosita et al.[76] as a three-pole DC motor. In their analogy "The rotor is a permanent magnet, a static component." Rather than being static, however, in our view the rotor and the housing would have the capacity to store elastic deformation. Instead of the repulsion between like poles of a magnet, the push component of force arises from the apolar–polar repulsion between the most polar state attained on formation of $ADP + P_i$ in the housing and the most hydrophobic face of the γ-rotor. Also, instead of acting through space, the push component of force requires an aqueous solvent through which to function.

Thus the fundamental predictions of the hydrophobic elastic consilient mechanism are that the rotor would exhibit asymmetric hydrophobicity, that different arrangements of nucleotide analogues representing different states of polarity at the catalytic sites would orient the rotor, and that hydrolysis of ATP in formation of the most polar state at a catalytic site of the involved protein subunit(s) would demonstrate a near-ideal elastic deformation of the γ-rotor and the protein subunit(s). Of course, such a mechanism would exhibit high efficiency and reversibility.

8.4.1.2 Predictions and the Demonstrated Relevance of the Consilient Mechanisms to the F_1-motor

8.4.1.2.1 Prediction of Hydrophobically Asymmetric and Elastically Deformable Rotor and Housing

Accordingly, in contrast to the obviousness of the role of the hydrophobic consilient mechanism to the F_0-motor of ATP synthase, the relevance of the hydrophobic consilient mechanism to the F_1-motor of ATP synthase calls for presentation of new and otherwise unexpected perspectives. For the hydrophobic consilient mechanism to be a dominant factor in the function of the F_1-motor, the primary predictions focus on the rotor that mechanically couples the two rotary engines. As noted above, the first prediction is that the γ-rotor must be hydrophobically asymmetric within the catalytic structure. A secondary prediction becomes that rotor hydrophobic asymmetry must be such that rotational orientation of the rotor responds to the occupancy states of the catalytic sites of the F_1-motor and that the rotor and associated catalytic subunit(s) be capable of nearly ideal reversible storage of deformation energy. Another important enabling prediction, regarding the F_1-motor functioning as an ATPase, is that elastic deformation arise out of the sudden development of significant ΔG_{ap} on hydrolysis of ATP to form ADP and P_i and that this deforming force defines the direction of rotation. Finally, the presence of the ΔG_{ap} on formation of the most polar state would be recognizable in the orientation and interactions, or lack thereof, of charged species between rotor and the subunit containing the most polar state.

8.4.1.2.2 Demonstration of an Asymmetrically Hydrophobic Rotor by Calculation of ΔG_{HA}, Gibbs Free Energy of Hydrophobic Association

Demonstrations of these predictions constitute the message of this section 8.4, and its success introduces the perspective of a conjoined hydrophobic elastic consilient mechanism. With the values in Table 5.3 and the crystal structure with three different states of occupancy, empty, ATP, and ADP, the three sides of the rotor can be identified and the respective Gibbs free energies of hydrophobic association, ΔG_{HA}, have been estimated to be −20, 0, and +9 kcal/mole. The most hydrophobic face associates with the empty site, the neutral face with the ATP bound site, and the most polar face with the ADP site which in the synthesis mode would be in position to add P_i. As expected from the magnitude of the resulting ΔG_{ap} for a series of crystal structures wherein the least polar occupancy state for the catalytic site could be defined, the most hydrophobic side of the rotor resides in apposition to the least polar site.

8.4.1.2.3 Prediction and Demonstration That the Most Hydrophobic Face of the Rotor Hydrophobically Associates with the Most Hydrophobic State of the Housing

On the basis of the hydrophobic consilient mechanism, predictions of the relative hydrophobicities of the housings at the location of the changeable catalytic sites in order of increasing hydrophobicity (decreasing polarity) are ADP plus P_i, ATP, ADP, P_i when the latter is not in position to have direct through-water interaction with the rotor, and the empty catalytic site. The empty catalytic site indeed hydrophobically associates with the side of the rotor found by calculation to have greatest hydrophobicity. Furthermore, for several different patterns of site occupancy available from crystal structures, the side of the rotor deemed most hydrophobic by the consilient hydropho-

8.4 ATP Synthase: The Twofold Rotary Protein Motor of Oxidative Phosphorylation

bic mechanism always aligns with the most hydrophobic side of the rotor.

8.4.1.2.4 Prediction of the Direction of Rotor Rotation for the ATPase and Synthesis Modes

Fortunately, a crystal structure has been reported in which a stable analogue, ADP plus SO_4^{2-}, of the most polar state, ADP plus HPO_4^{2-}, has been solved.[77] As expected for the most potent configuration, the SO_4^{2-} was nearly fully exposed through an intervening aqueous solvent to the γ-rotor, and the very polar sulfate was located just above the level at which the rotor changed from a double-stranded α-helical coiled coil at the amino-terminus of the γ-chain to a single-stranded α-helix that ultimately ended at the carboxyl terminus of the γ-chain at the base of the F_1-motor. At this position the apolar–polar repulsion occurring between the polar SO_4^{2-} and the hydrophobic side of the rotor applies most directly to the amino-terminal side of the double strand resulting in a torque that would provide a counterclockwise rotation for the γ-rotor during function as an ATPase. Synthesis would be achieved by a reversal of the direction of rotation, wherein the maximal apolar–polar repulsion would be applied to the ADP plus P_i state and would be relieved by formation of ATP.

8.4.1.2.5 Demonstration of an Elastically Deformable Rotor and Housing

In the crystal structure with a β-subunit containing the sulfate analogue of the most polar ADP plus P_i state, the γ-rotor and the most polar catalytic β-subunit are found displaced from each other by a mean distance of 2.9 Å distance and the γ-rotor is twisted up to 20° when compared with their relationship when the β-subunit is empty.[77] Quoting from Menz et al.,[77] "Note that interacting residues in the $β_E$-subunit and the γ-subunit move in opposite directions." This repulsion occurs for a configuration in which the most hydrophobic side of the γ-rotor is in apposition to a slightly less polar analogue, $ADP^{3-} + SO_4^{2-}$, of the most polar natural occupancy state, $ADP^{3-} + HPO_4^{2-}$, for a catalytic β-subunit. In our view, this displacement results from a very large near-maximal ΔG_{ap} available to the F_1-motor, that is, a near maximal apolar–polar repulsive free energy of hydration between the hydrophobic side of the γ-rotor and the very polar state, $ADP + SO_4^{2-}$, of a catalytic β-subunit. This near-maximal repulsion provides an elastic deformation of rotor and housing as required for efficient function of the ATPase in its performance of chemo-mechanical transduction.

8.4.1.2.6 Pattern of Charged Side Chain Orientations Reflects Presence of a Dominant ΔG_{aF}

An interesting orientation of side chains in the housing, on the inner surface of the $(αβ)_3$-subunit structure, becomes rational once the presence of an apolar–polar repulsion is recognized. In particular, the negatively charged aspartic acid residue, D315, is observed in the crystal structure between the analogue of the most repulsive state of the catalytic site and the γ-rotor. At this position the carboxylate of D315 is surrounded by water molecules and is bent away from the hydrophobic side of the rotor and toward the sulfate group with its two negative charges yet with space for only a few water molecules separating sulfate from carboxylate.

Why, with the capacity to position itself at greater distance, would the carboxylate of D315 accept a location of higher charge–charge repulsion? In our view, an apolar repulsion emanating from the hydrophobic side of the γ-rotor causes it to reside in such a configuration and as such the repulsion would effect an element of elastic deformation in the housing of the F_1-motor. On the other hand, residue D316, which is adjacent to the hydrophobic rotor, is bent flat in the opposite direction against the rotor where it exhibits ion pairing and hydrogen bonding. These side chain orientations reflect the ΔG_{ap} causing elastic deformations due to the proximal hydrophobic side of the γ-rotor.

If we had not already derived ΔG_{aF} from analysis of the data on elastic-contractile model proteins, as reviewed in Chapter 5, with the perspectives presented below we would have had to invent such a repulsive force to explain the structural information available for the F_1-motor of ATP synthase. Of course, ΔG_{ap} provides the force required to adapt the three-pole DC motor described in Figures 5, 7, and 8 of Kinosita et al.[76] to those forces available to protein-based machines functioning in an aqueous milieu.

8.4.2 Schematic Representation of ATP Synthase Structure: A Chemo-chemical Transducer

8.4.2.1 Subunits Divided into Five Structural Classes: Membranous Rotor, Extramembranous Rotor, Motor Housings for Each of the Two Rotors, and Stator Interlocking Motor Housings

As represented schematically in Figure 8.26,[78] ATP synthase contains a basic set of components that are common throughout biology. There are two rotary motors, the membranous F_0-motor and the extramembranous F_1-motor. An extramembranous rotor driven by the membranous F_0-motor couples rotations in each motor and drives ATP synthesis in the extramembranous F_1-motor. The effective housings of the two rotary motors are held fixed, one with respect to the other, by a stator element. In this way one complete rotation in the F_0-motor accomplishes one complete rotation in the F_1-motor.

Thus, utilizing Figure 8.26, the initial description of structure becomes enumeration of the constituent protein subunits and their number of repetitions given as a subscript for each of the four structural components: (1) The membranous protein subunits are the 10 repeats of protein subunit c, that is, c_{10}. (2) The motor housing for the membranous rotor utilizes the a protein subunit in combination with the lipid bilayer. (3) The extramembranous rotor is comprised of the γ and ε protein subunits. (4) The motor housing for the extramembranous rotor utilizes three α and three β protein subunits. (5) The stator component for interlocking the two motor housings utilizes two b protein subunits in combination with the δ protein subunit.

In addition, in animals mitochondrial ATP synthase contains another eight types of subunits, a number of which function in the coupling of the F_0-rotor to the γ-rotor.[79]

8.4.2.2 Structural and Functional Roles of Basic Protein Subunits in ATP Synthase

8.4.2.2.1 Rotor of the F_0-motor—The 10 c Protein Subunits

The number of c protein subunits constituting the F_0-motor depends on the species and has been reported to vary from 9 to 14. In yeast, it is 10, as in Figure 8.26, whereas in *E. coli* it is 12. The c subunit occurs as an α-helix in a hairpin configuration, with one side of the hairpin forming the inner wall and the second side of the α-helical hairpin interacting with the housing of the F_0-motor.

8.4.2.2.2 Housing of the F_0-motor—the a Protein Subunit and the Lipid Bilayer

As presently described, the housing of the F_0-motor becomes the lipid bilayer and five transmembrane helices of subunit a that combine to form two half-proton channels.[80] As shown in Figure 8.26, one half-channel spans from the cytosol to the middle of the membrane, and the second half-channel completes the transmembrane transit from middle of the lipid bilayer to the matrix side of the membrane.

8.4.2.2.3 Rotor Driven by the F_0-motor (Rotor of the F_1-motor)—the Single γ and ε Protein Subunits

The rotor that is driven by the F_0-motor comprises a single γ-subunit and a small ε-subunit attached to the γ-subunit at a point proximal to the base of the F_0-motor. This is called the γ-rotor. In the hydrophobic elastic consilient mechanism, the interactions of a hydrophobically asymmetric γ-rotor with the housing of the F_1-motor with different occupancy states of the catalytic sites constitute the basis for mechano-chemical transduction of the F_1-motor.

8.4.2.2.4 Housing of the F_1-motor—the Three α and Three β Protein Subunits

The housing of the F_1-motor contains the heart of the ATP synthase function (when the γ-rotor is driven clockwise by the F_0-motor) or ATPase

8.4 ATP Synthase: The Twofold Rotary Protein Motor of Oxidative Phosphorylation

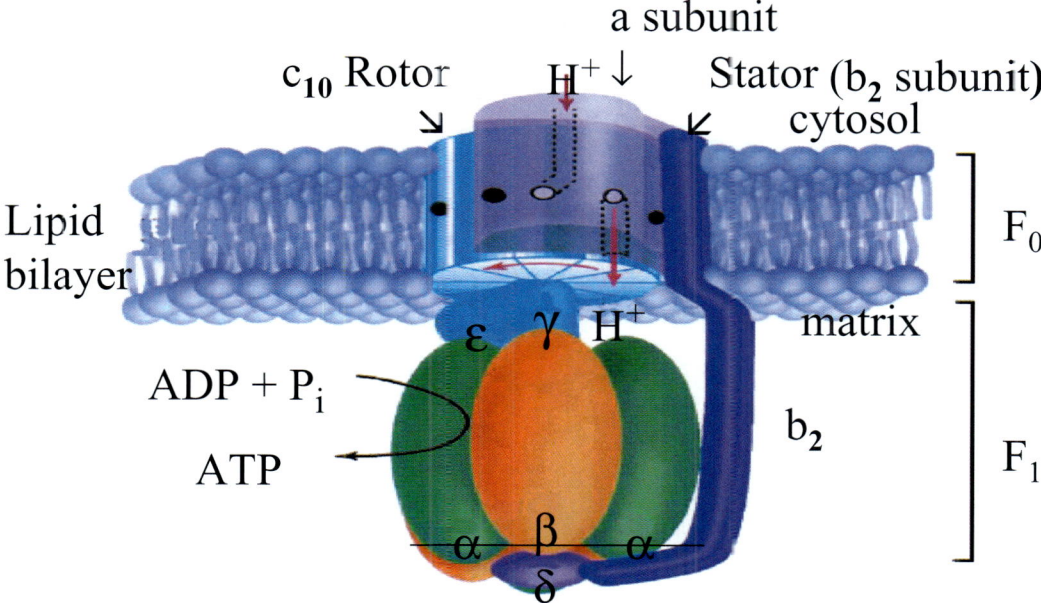

FIGURE 8.26. ATP synthase: proton flow from the cytosolic to the matrix side of inner mitochondrial membrane rotates upper intrinsic membrane F_0-motor with an attached axle, which thereby is made to rotate clockwise within the hexameric globule on the matrix side (the F_1-motor). The latter produces ATP from ADP and P_i to make 32 of the 36 ATPs (≈90%) resulting from glucose oxidation. The $(\alpha\beta)_3$ hexameric globule and $\epsilon\gamma$ stalk constitute the F_1-motor, which when functioning separately becomes the F_1-ATPase rotary motor. The separated extra-membrane hexameric globule and stalk constituting the F_1-ATPase rotary motor functions in the reverse (counterclockwise) direction with a high efficiency for the conversion of the energy of ATP hydrolysis, that is, on return of ATP to ADP and P_i, to the mechanical work of rotating a filament attached to the stalk as shown in Figure 8.42. (Adapted with permission from R.H. Fillingame, "Molecular Rotary Motors." *Science*, **286**, 1687–1688, 1999.[78] Copyright © 1999 AAAS.)

function (when driven counterclockwise by hydrolysis of ATP to form ADP and P_i). The housing of the F_1-motor is comprised of the three αβ dimers, that is, $(\alpha\beta)_3$. The three α-subunits are noncatalytic with sites occupied by ATP, that is, αATP. The β-subunits are catalytic and have variable occupancy of their catalytic sites.

8.4.2.2.5 The Stator of the F_0-motor—the δ and Two b Protein Subunits

For the rotation of the F_0-motor to be utilized by the F_1 motor, there must be a stator that interlocks the housings of the F_0- and F_1-motors. Two b protein subunits arranged as a double-stranded α-helical coiled coil fulfill this function by attachment at one end to the channel-containing housing of the F_0-motor and at the other end to the δ-subunit at the base of the threefold symmetric $(\alpha\beta)_3$ housing of the F_1-motor, as depicted in Figure 8.26. The δ-subunit would have the role of binding the six subunits together at the base of the housing of the F_1-motor. As such the δ-subunit constitutes part of the stator construction.

8.4.2.3 F_1-ATPase Assembles on Raising the Temperature by Hydrophobic Association of an Inverse Temperature Transition

The F_1-ATPase is composed of five different subunits comprising eight subunits in total, with the stoichiometry of $\alpha_3\beta_3\gamma\delta\epsilon$ and constituting a molecular weight of 371,100 kDa. The subunits of F$_1$-ATPase dissociate on lowering the temperature to 0°C, that is, they exhibit cold denaturation.[21,81,82] This, of course, indicates that under the appropriate conditions the subunits would assemble on raising the temperature. For enzymatic activity, this would necessarily include the γ-rotor with its double-stranded α-helical coiled coil within the threefold symmetric (αβ)$_3$ with approximate threefold symmetry but which looks much like an orange with six sections. Accordingly, the F$_1$-ATPase is of a nature to exhibit an inverse temperature transition of hydrophobic association/dissociation with a **T$_t$**-value somewhere between physiological temperature and 4°C.

Knowledge of the differential scanning calorimetry data and the temperature of the transition for assembly on raising the temperature would allow determination of the free energy of hydrophobic association, the **ΔG$_{HA}$**. It would be of interest to obtain such data with systematic increases in the number of sites occupied by ATP. Based on the hydrophobic consilient mechanism and in analogy to forming the more polar state by ionization of carboxyls to form carboxylates in the elastic-contractile model proteins, we would predict that the heat of the inverse temperature transition would decrease and the temperature of the transition would rise as occupancy of the nucleotide-binding sites progressed from being the most hydrophobic state of empty to becoming the most polar state with all sites containing ATP.

8.4.3 Initial Structural Information for ATP Synthase from Electron Density Maps and Thermodynamic Data

8.4.3.1 Overall Structure

By means of electron density maps on crystals of yeast mitochondrial ATP synthase, Stock et al.[83] obtained the associated F$_0$- and F$_1$-motors, as demonstrated in Figure 8.27A. The F$_0$-rotor was obtained in sufficient detail to identify 10 c subunits and to locate the critical D61 residue that undergoes protonation/deprotonation of its carboxylate functional group. The small white spherical dots crossing the middle of the F$_0$-rotor locate the D61 residue in each subunit.

Figure 8.27B shows only the rotating elements of ATP synthase with 10 α-helical hairpins making up the F$_0$-rotor and the double-stranded α-helical coiled coil (upper two-thirds) of γ-subunit decorated at its upper part with the ε-subunit shown jutting out to constitute the γ-rotor (the F$_1$-rotor). Also shown is the coupling component, which is incomplete.

8.4.3.2 Three Perspectives of the Structure of the F$_0$-rotor

Shown in cross-eye stereo side view in Figure 8.28 are three views of the F$_0$-rotor, that is, of the rotating wheel of the F$_0$-motor. At this level of resolution each residue is given as a sphere. Furthermore, each of the 10 subunits of the F$_0$-rotor is represented as a double-stranded α-helical hairpin with the ends at the top (cytosolic side) and the turn at the bottom (matrix side) and with one side of the hairpin in direct interaction with the lipid bilayer and the other side forming the inner wall of the rotating wheel. In Figure 8.28A the aspartic acid residue D61 is noted from outside. This residue exists with its side chain as a carboxyl (–COOH) when adjacent to the lipid bilayer, but it releases its proton to form the carboxylate (–COO$^-$) when it reaches the effective channel when adjacent to the a-subunit.

8.4.3.3 Proton Flow (Cytosol to Matrix) Drives Clockwise Rotation of the F$_0$-motor

A stereo view from the cytosolic (top) side of the F$_0$-rotor is given in Figure 8.28B. From this view the ends of the c subunits are seen. From the matrix (bottom) side in Figure 8.28C, the hairpin turn of the c subunits are seen. Briefly, following Fillingame et al.,[80] a cytosolic proton flows down a half-channel using the fourth of five transmembrane helices of subunit a and adds to the carboxylate of D61 that was ion paired with arginine-210 (R210) of the fourth transmembrane helix of subunit a. The more hydrophobic newly protonated helix of the c subunit rotates in a closkwise direction into the lipid layer as the subsequent c subunit releases its D61 proton by a second half-channel to the

8.4 ATP Synthase: The Twofold Rotary Protein Motor of Oxidative Phosphorylation

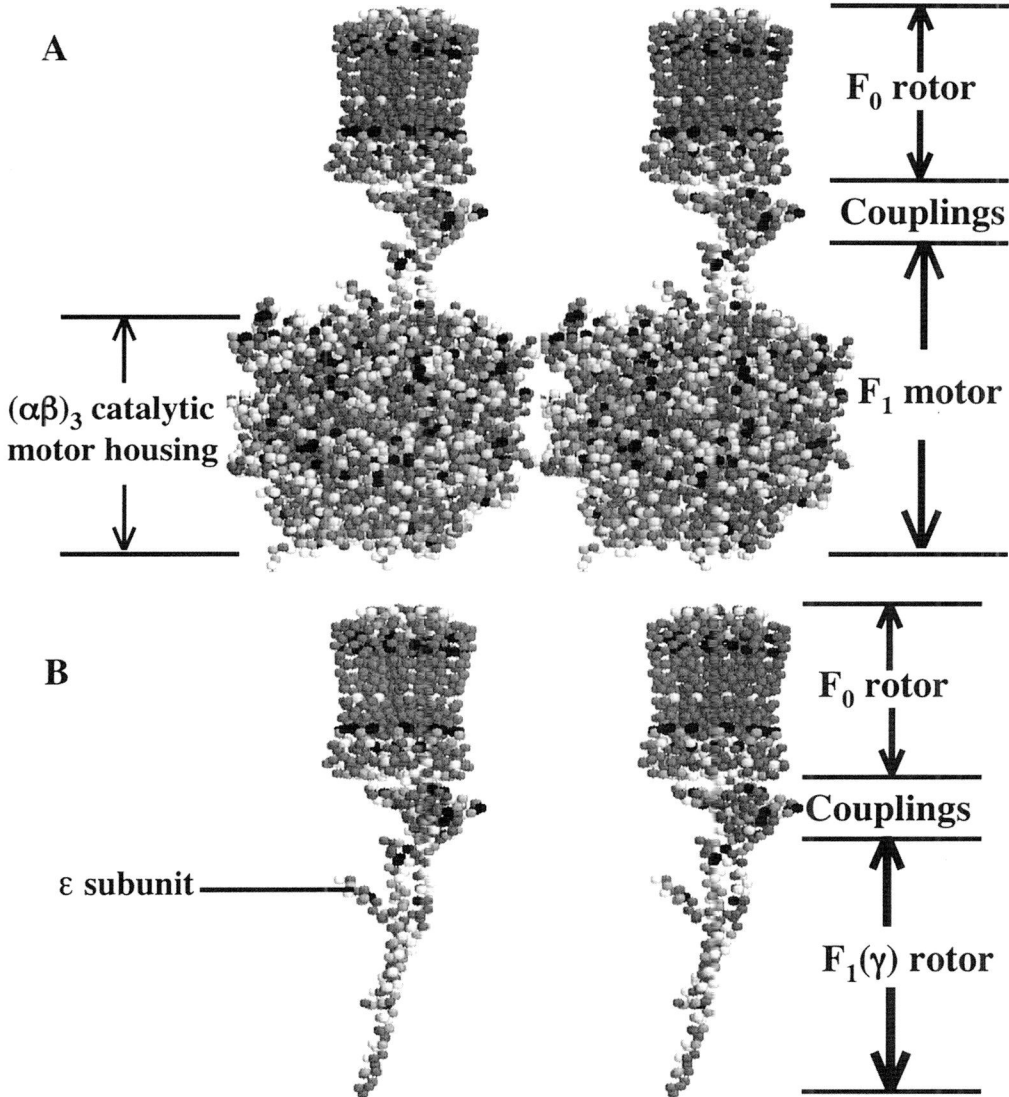

FIGURE 8.27. Stereo view of ATP synthase from the mitochondria of yeast, neutral residues in light gray, aromatics in black, other hydrophobics in gray, and charged residues in white. **(A)** A largely complete ATP synthase double rotary motor: F_0-motor, connecting rotor and F_1-motor, with only part of protein subunits observed that couple the F_0-motor to the rotor. **(B)** F_0-rotor, partial connector, and rotor assembly of F_1-motor. (Prepared using the crystallographic results of Stock et al.[83] as obtained from the Protein Data Bank, Structure File 1QO1.)

matrix and replaces the previous c subunit by ion pairing with the R210 at the fourth transmembrane helix. This completes the flow of a single proton from the cytosol to the matrix side of the inner mitochondrial membrane to give a 36° clockwise rotation in the F_0 rotor. With further stochastic motion the D61 carboxylate of the newly positioned c subunit picks up another proton from the cytosolic side of the membrane and so on to complete the transit of a second proton from one side to the other of the membrane to effect a second 36° rotation in

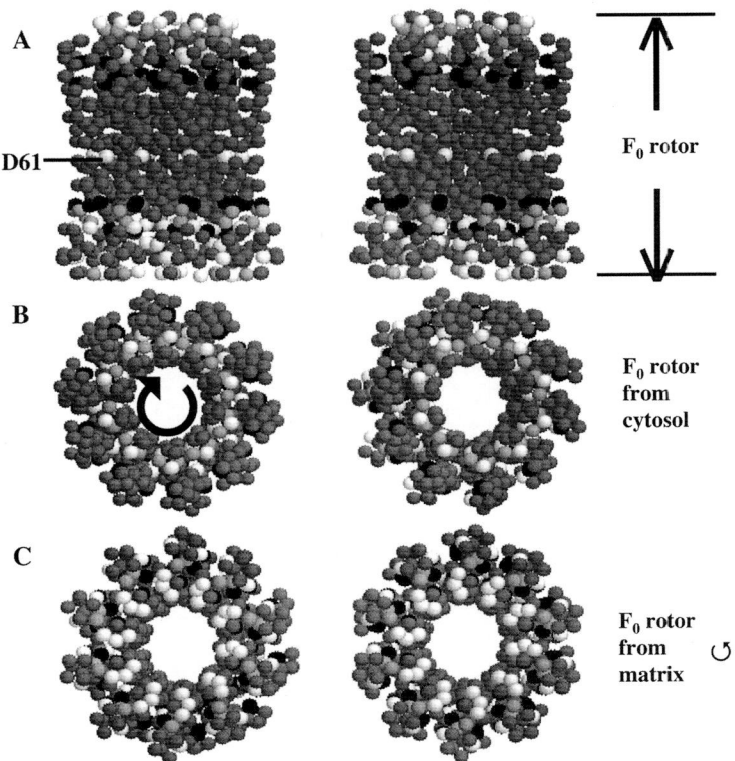

FIGURE 8.28. Stereo view of ATP synthase from mitochondria of yeast, neutral residues in light gray, aromatics in black, other hydrophobics in gray, and charged residues in white. **(A)** Rotating wheel of F_0-motor, side view. **(B)** Rotating wheel of F_0-motor, top view (cytoplasmic side). **(C)** Rotating wheel of F_0-motor, bottom view (matrix side). (Prepared using the crystallographic results of Stock et al.[83] as obtained from the Protein Data Bank, Structure File 1QO1.)

the F_0 rotor. The net direction of proton flow and of rotation of the F_0-rotor would depend on the frequency with which a proton enters from one side or the other of the membrane.

8.4.3.4 Barrier to Rotation of the F_0-motor Removed on Protonation

8.4.3.4.1 Hydrophobicity Plots, T_t and ΔG_{HA}, of the c Subunit of Bovine ATP Synthase

The single residue T_t hydrophobicity plot for the c subunit of bovine ATP synthase appears in Figures 8.29A. The hydrophobic residues are plotted with an upward deflection and the polar residues plotted with a downward direction. Residues D2 and K5 evidence the polar amino terminus of the c subunit that resides at the cytoplasmic (cytosolic) side of the inner mitochondrial membrane, and the polar residues at the matrix side of the membrane, most notably K42, provides for polar orientation at this polar site. With the single additional exception of residue E57, the remaining stretches of the c subunit would be very hydrophobic. Of greatest significance for proton flow, of course, is the single ionizable carboxyl function of residue E57.

As shown in the mean (7) residue plot of Figure 8.29B, in the carboxylate state the c subunit would present a barrier to movement into the region of the lipid bilayer, but this barrier is removed on protonation. The mean residue plot given in terms of the Gibbs free energy for hydrophobic association in Figure 8.29C shows that the equilibrium for moving into the region of the lipid bilayer would be favored by protonation of the carboxylate to form the carboxyl state.

As noted in the introduction of this section 8.4, this is as one would intuitively expect. Some quantification of this in terms of the comprehensive hydrophobic effect of the hydrophobic consilient mechanism follows immediately below for the *E. coli* example.

8.4 ATP Synthase: The Twofold Rotary Protein Motor of Oxidative Phosphorylation

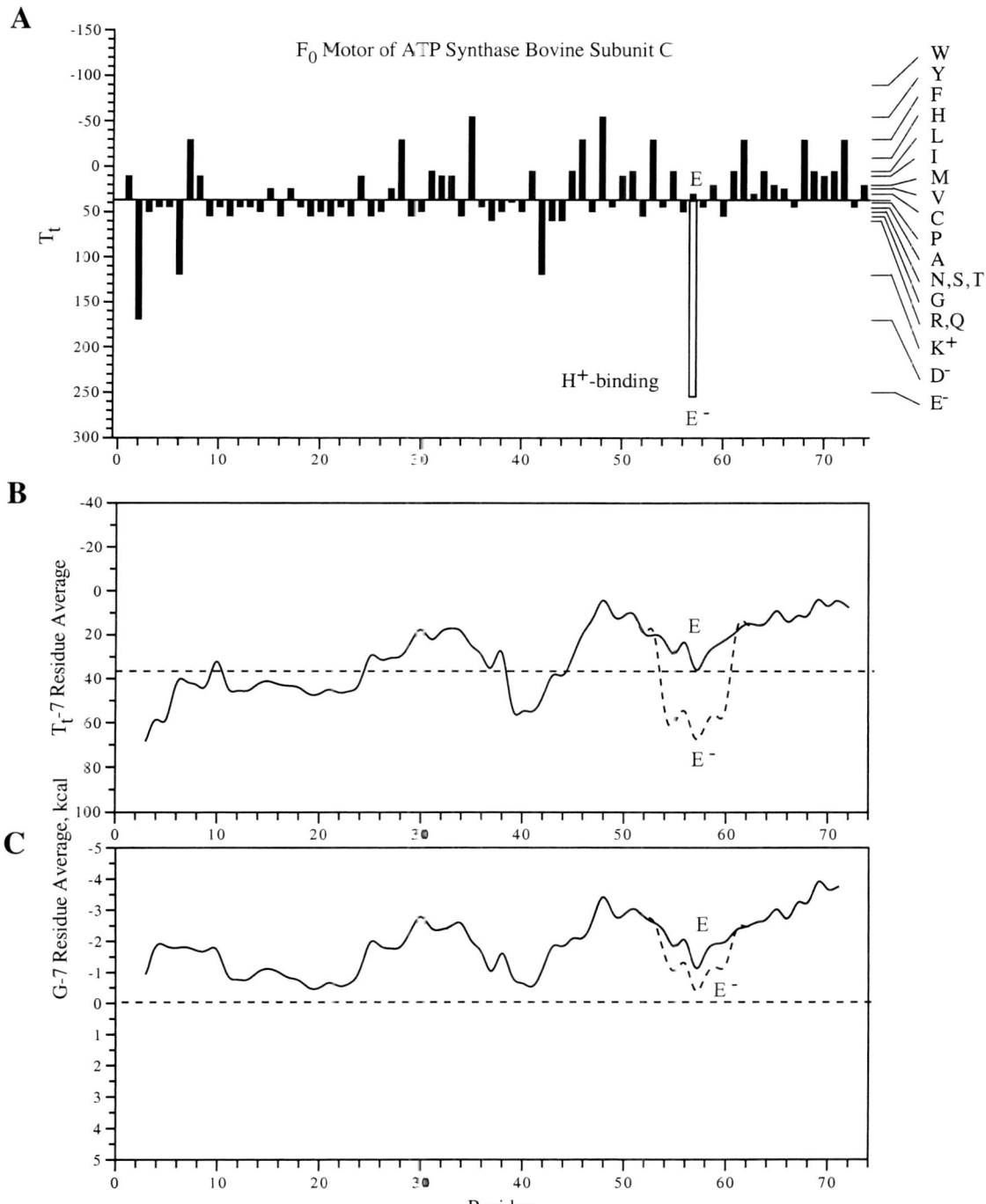

FIGURE 8.29. Hydrophobicity plots of 1 of 10 c-subunits of the F_0-motor of bovine ATP synthase. (A) Single residue T_t plot showing the hydrophobic sequence from residue 8 forward with the exception of very polar residue E57, which on protonation becomes a small upward deflection. (B) Mean (7) residue T_t plot. (C) Hydrophobicity plot in terms of the mean (7) residue plot of ΔG_{HA}, the Gibbs free energy of hydrophobic association.

8.4.3.4.2 Thermodynamics of Protonation and Efficiency of Energy Conversion: $\Delta G_{HA}[D^- \rightarrow D^0] \approx -3.8\,kcal/mole$

From Table 5.3 we find that the decrease in Gibbs free energy for hydrophobic association on protonation of the carboxylate (–COO⁻) of aspartate, D^-, to produce the carboxyl (–COOH) of aspartic acid, D^0, is approximately –3.8 kcal/mole. The protonation of 10 aspartic acid residues as required for one complete rotation of the F_0-motor could be expected to provide an energy of 38 kcal/rotation, which constitutes one complete rotation of the γ-rotor. Accordingly, by means of the F_1-motor, one complete rotation of the γ-rotor translates into the production of 3 mole of ATP from ADP plus P_i. Because the approximate heat released on hydrolysis of 1 mol of ATP to ADP plus P_i is 8 kcal/mole, one complete rotation of the γ-rotor would produce 24 kcal worth of ATP per rotation. An energy input of 30 kcal/mole and an output energy of 24 kcal per rotation gives an efficiency maximum of about 63%. Thus, the value of $\Delta G_{HA}[D^- \rightarrow D^0] \approx -3.8$ kcal/mole appears reasonable, even though it was a value obtained from the experimental determination of the temperature for onset of the inverse temperature transition using the sigmoid curve of Figure 5.10 rather than directly measured by differential scanning calorimetry using the difference in heat of the inverse temperature transition for the two states of the aspartic acid residue.

8.4.4 Crystal Structure and Function of the F_1-motor of ATP Synthase–F_1-ATPase

8.4.4.1 Boyer Caution About Crystal Structure Not Being That of Active Catalytic State

Our biochemical understanding of the mechanism of ATP synthase[84] comes in largest measure from the pioneering work of Boyer.[85–87] Because of this extensive expertise and insight, the Boyer perspective on crystal structure data reflects the judicious critic of the limitations of crystal structures when attempting to describe the dynamics of catalysis:

The contribution of Menz et al is welcomed as providing essential information for the difficult task still ahead of satisfactorily correlating structure with events of substrate binding, covalent catalysis, and product release. For this task, it needs to be recognized that forms revealed by X-ray analysis of static inhibited enzymes may not appear as such during active catalysis, even though they are likely representative of most intermediate forms.[85]

Even so, crystal structures provide the best snapshots of forces in action. Crystal structures provide an unparalleled opportunity to assess relevance to the major protein-based machines of biology of the free energy transduction so dominantly displayed by elastic-contractile model proteins (as developed in Chapter 5). If the apolar–polar repulsive free energy of hydration, ΔG_{ap}, the operative component of the Gibbs free energy of hydrophobic association, ΔG_{HA}, is active in ATP synthase, then it should become apparent in these snapshots.

8.4.4.2 Assembled Subunits, $(\alpha\beta)_3\,\gamma\delta$, of the F_1-ATPase

8.4.4.2.1 Structural Information

The F_1-ATPase contains five different polypeptide subunits, α, β, γ, δ and ε, with the stoichiometry of 3:3:1:1:1, that is, $\alpha_3\beta_3\gamma_1\delta_1\epsilon_1$. From crystal structure data the α- and β-subunits may be described as the trimer, $(\alpha\beta)_3$, with approximate threefold symmetry, but looking much like a peeled orange with six nearly identical sections. A cross section of the $(\alpha\beta)_3\gamma$ structure at the level of the β,γ-phosphates is shown in Figure 8.30, which, among other things, provides visualization of the approximate threefold symmetry despite the different occupancy sites and the obviously asymmetric position of the γ-rotor. The γ-subunit forms the core of the structure residing close to the threefold axis.

As shown in Figure 8.27B, the γ-subunit enters the upper two-thirds of the $(\alpha\beta)_3$ orange-

8.4 ATP Synthase: The Twofold Rotary Protein Motor of Oxidative Phosphorylation

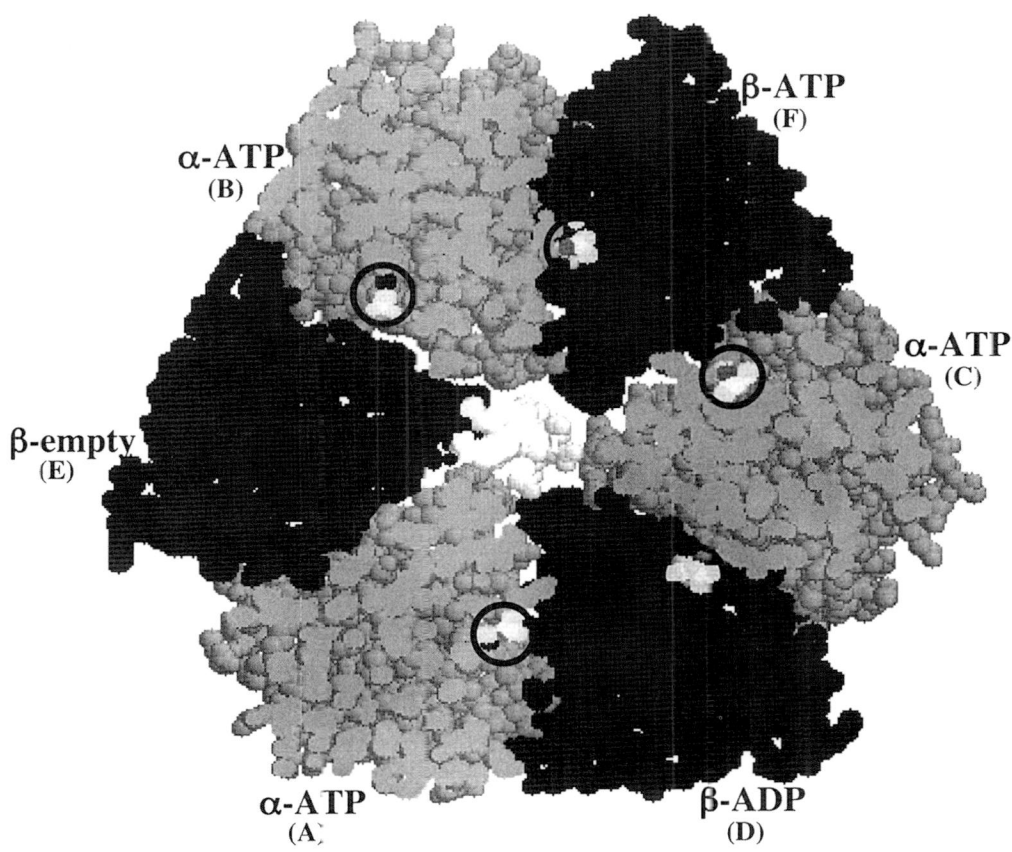

FIGURE 8.30. F_1-motor of ATP synthase, horizontal cross section of space-filling representation, showing three noncatalytic α-ATP subunits; the β-empty, β-ATP, and β-ADP catalytic subunits with sites circled and the γ-rotor. (Adapted from Urry.[109])

like structure as a generally hydrophobic double-stranded α-helical coiled coil and the lower third continues on as a single α-helix. The γ-subunit passes through a central aqueous cavity like a skewer, but with hydrophobic bearings. One hydrophobic bearing occurs as it passes double stranded into the orange-like structure, and most strikingly the primary hydrophobic bearing occurs as the single-stranded section of the γ-subunit terminates without quite passing to the outside. The ε-subunit is a small polypeptide that binds to the stem of the γ-subunit where it enters the $(αβ)_3$ structure. The $(αβ)_3$ structure functions as both a dynamic motor housing and a site of catalysis for the F_1-motor.

8.4.4.2.2 The Rotational Directions of the γ-rotor

The γ-subunit is driven to rotate in a clockwise direction by the F_0-motor and, when the F_1-motor is functioning as an ATPase, it rotates in a counterclockwise direction due to the energy derived from the hydrolysis of ATP to form ADP and P_i. This occurs in each of three active sites, one in each of the three β-subunits, arranged with near-threefold symmetry, as shown in Figure 8.30. The lack of exact threefold symmetry comes from the inherent asymmetry of the γ-rotor at the core of the $(αβ)_3$ structure and from different occupancy states of the three β-subunits.

FIGURE 8.31. β-(Empty) face of the γ-rotor of ATP synthase identified as the *very hydrophobic face* by the calculation shown of Table 8.2 to give $\Sigma\Delta G_{HA}$(empty face) ≈ −20 kcal/mole. Shown at right are the α-ATP and β-ADP ligands, at left the α-ATP and β-ATP ligands, and the α-ATP ligand behind. The neutral residues are light gray, the aromatic residues are black, the other hydrophobic residues are gray, and the charged residues are white. (Adapted from Urry.[109])

8.4.4.2.3 Early Crystal Structure Data

The early crystal structure data, utilized in Figure 8.30, proved to be particularly interesting as it occurred with noncatalytic ATP bound in each the three α-subunits, but most interestingly because the three β-subunits were solved with ATP bound at one unit, with ADP bound at the second unit, and with an empty third β-subunit. This structure provides many opportunities with which to begin analyses of mechanism. Among other things, this structure permits definition of three faces of the γ-rotor, as discussed immediately below.

8.4.4.3 The Three Faces of the γ-rotor

Our perspective of the hydrophobic consilient mechanism as it would apply to ATP synthase requires that there be an asymmetric hydrophobic rotor and that a variable apolar–polar repulsion occur between the different occupancy states of the catalytic β-subunits and the rotor. If the hydrophobic consilient mechanism is relevant, then the structure utilized in Figure 8.30 should occur with a distinctly and hydrophobically asymmetric γ-rotor.

8.4.4.3.1 The Hydrophobic Face Defined by β-Empty Subunit

In the static crystal structure, on the basis of the hydrophobic consilient mechanism one expects the side of the γ-rotor adjacent to the β-empty subunit to be the most hydrophobic. By the apolar–polar repulsive free energy of hydration, ΔG_{ap}, the polar nucleotide ligands should repulse the most hydrophobic side of the γ-rotor.

The ligands themselves allow correct identification of the orientation. By using the labeled cross section of Figure 8.30 as a guide, positioning the α-ATP and β-ADP at right, the α-ATP and β-ATP at the left, and an α-ATP behind allows direct observation of the β-empty face of the γ-rotor, as shown in Figure 8.31.

The hydrophobicity scale in Table 5.3 lists the contribution of each amino acid residue to the Gibbs free energy of hydrophobic association, ΔG_{HA}. Table 5.3 also provides the information required to calculate numbers for the relative hydrophobicities of the faces of the γ-rotor. The resulting numbers are tabulated and summed in Table 8.2, where the $\Sigma\Delta G_{HA}$(β-empty face) ≈ −20 kcal/mole. This is indeed a very hydrophobic value.

8.4.4.3.2 The Polar Face Defined by the β-ADP Subunit

Identification of the face defined by the β-ADP subunit comes from placing the α-ATP and β-ATP ligands at right, the α-ATP ligand and β-empty site at left, and the third α-ATP ligand behind, as shown in Figure 8.32. With K4, D5, K260, E261, and E264 contributing to this face, it is clearly a polar face. On tabulating the contribution of the visible amino acid residues for this face and summing, as shown in Table 8.2, $\Sigma\Delta G_{HA}$(β − ADP face) ≈ +9 kcal/mole. The calculation confirms a quite polar face.

8.4.4.3.3 The Neutral Face Defined by the β-ATP Subunit

Identification of the face defined by the β-ATP subunit comes from placing the α-ATP ligand and β-empty site at right, the α-ATP and β-ADP ligands at left, and the third α-ATP ligand behind, as shown in Figure 8.33. Interestingly, as shown in Table 8.2, tabulation of the amino

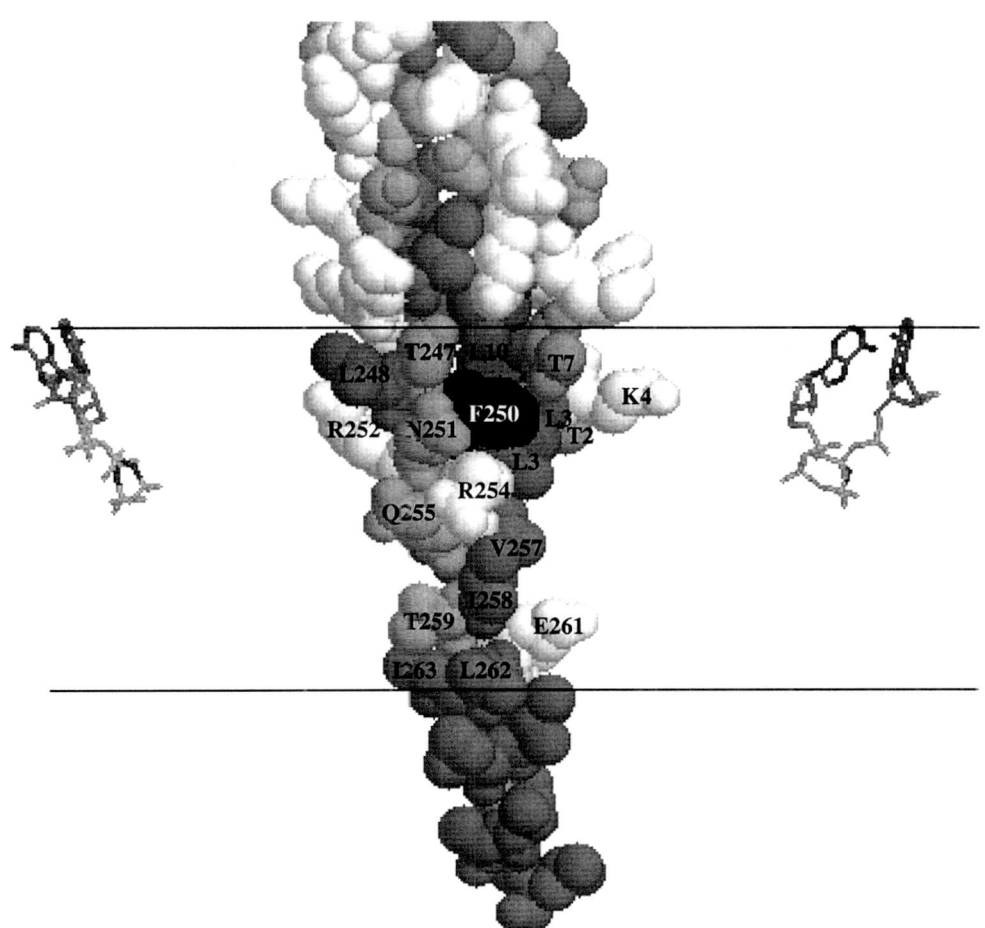

TABLE 8.2. Values for $\Sigma\Delta G_{HA}$(γ-rotor faces).[a]

β-empty face		β-ATP face		β-ADP face	
Res. No.	ΔG_{HA}	Res. No.	ΔG_{HA}	Res. No.	ΔG_{HA}
Thr 2/3	−0.20	Ala 1	−0.75	Ala 1	−0.75
Leu 3	−4.05	Thr 2/3	−0.20	Thr 2	−0.60
Lys 4	+2.94	Asp 5	+3.40	Leu 3/3	−1.30
Thr 7	−0.60	Ile 6/3	−1.20	Lys 4	+2.94
Leu 10/2	−2.00	Glu 264	+3.72	Asp 5	+3.40
Ile 263/3	−1.20	Ile 263/3	−1.20	Thr 7/2	−0.30
Leu 262/2	−2.00	Lys 260	+2.94	Glu 264	+3.72
Glu 261/2	+1.85	Thr 259	−0.60	Glu 261	+3.72
Thr 259	−0.60	Ile 258/3	−1.20	Lys 260	+2.94
Ile 258	−3.65	Ala 256	−0.75	Ile 258/2	−1.80
Val 257/2	−1.25	Gln 255	+0.75	Val 257	−2.50
Gln 255	+0.75	Val 257/3	−0.80	Ala 256/3	−0.25
Arg 254	+0.70	Thr 253	−0.60	Arg 254/2	+0.35
Arg 252/2	+0.35	Arg 252	+0.70	Thr 253/3	−0.20
Asn 251	−0.05	Asn 251	−0.05	Thr 249/3	−0.20
Phe 250	−6.15	Leu 248/2	−2.00		
Leu 248	−4.05	Thr 249	−0.60		
Thr 247	−0.60				
Sum	−19.8	Sum	+0.4	Sum	+9.2

[a] $\Sigma\Delta G_{HA}$(β-empty face) ≈ −20 kcal/mole; $\Sigma\Delta G_{HA}$(β-ATP face) ≈ +0 kcal/mole; $\Sigma\Delta G_{HA}$(β-ADP face) ≈ +9 kcal/mole.
Source: Adapted from Urry.[109]

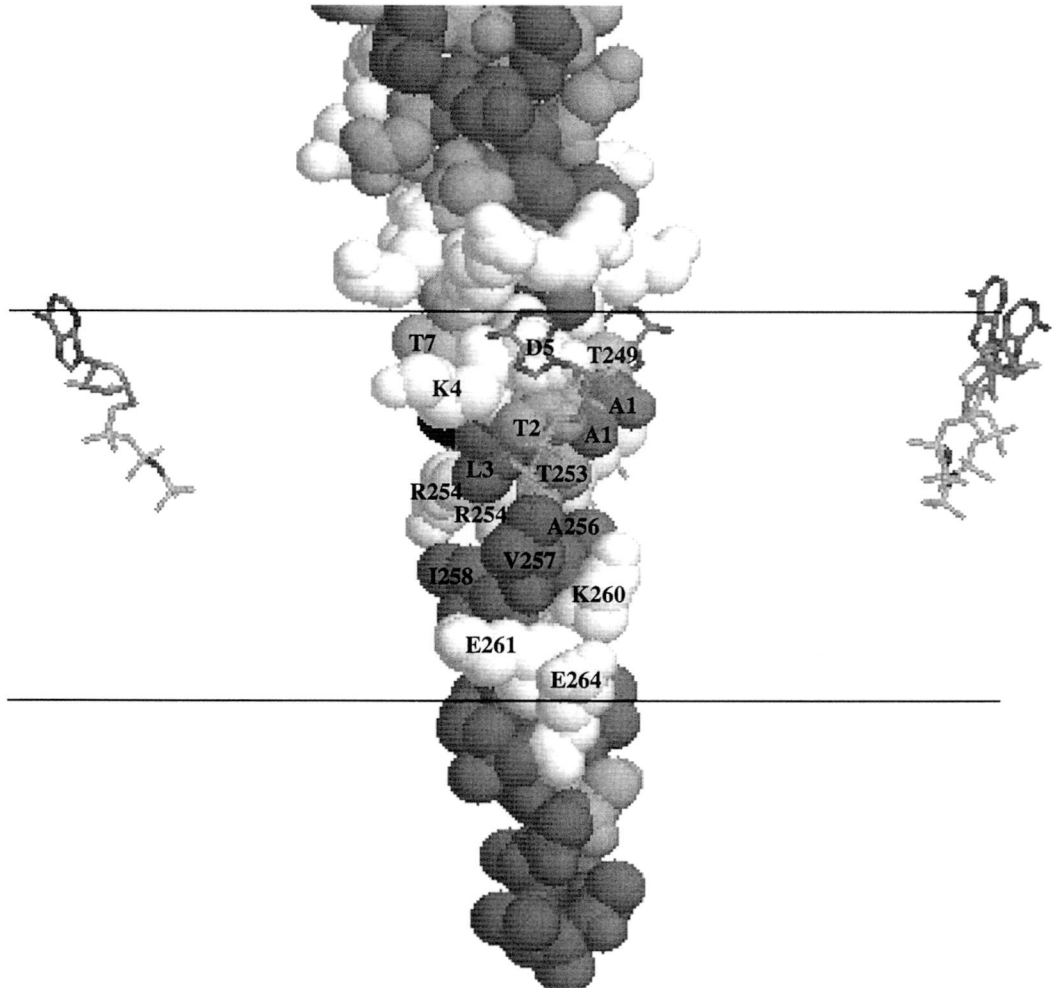

FIGURE 8.32. β-(ADP) face of the γ-rotor of ATP synthase seen from the side of the ADP site with ADP overlying the ADP face identified as the *polar face* by the calculation in Table 8.2 to give $\Sigma\Delta G_{HA}$(β-ADP face) ≈ +9 kcal/mole. Shown at right are α-ATP and β-ATP ligands, at left the α-ATP ligand and the β-(empty) site, and the α-ATP ligand behind. The neutral residues are light gray, the aromatic residues are black, the other hydrophobic residues are gray, and the charged residues are white. (Adapted from Urry.[109])

acid residues contributing to this face and summing gives $\Sigma\Delta G_{HA}$(β-ADP face) ≈ 0 kcal/mole.

8.4.4.3.4 An Additional Point About the Relative Hydrophobicities of the Faces of the γ-rotor

Thus, the calculated hydrophobicities of the structurally defined rotor delineate the three faces to give a polar face, a neutral face, and a hydrophobic face. Although the order of hydrophobicity allows striking delineation between empty and nucleotide-containing sites, the finer distinction of the decrease in hydrophobicity of the rotor faces with increase in polarity from the β-ADP site to the β-ATP site does not follow in proportion. If one considers the mechanism for the ATP synthase

8.4 ATP Synthase: The Twofold Rotary Protein Motor of Oxidative Phosphorylation

function, however, the relative order does make sense. In order to add P_i to an ADP-containing site, this site should be in apposition to the most polar face of the γ-rotor. Such a configuration would exhibit the lesser apolar–polar repulsive free energy of hydration, ΔG_{ap}, and would make formation of the most polar ADP plus P_i occupancy state more probable. Accordingly, the hydrophobically asymmetric γ-rotor is as expected for the hydrophobic consilient mechanism to be dominant in the function of ATP synthase.

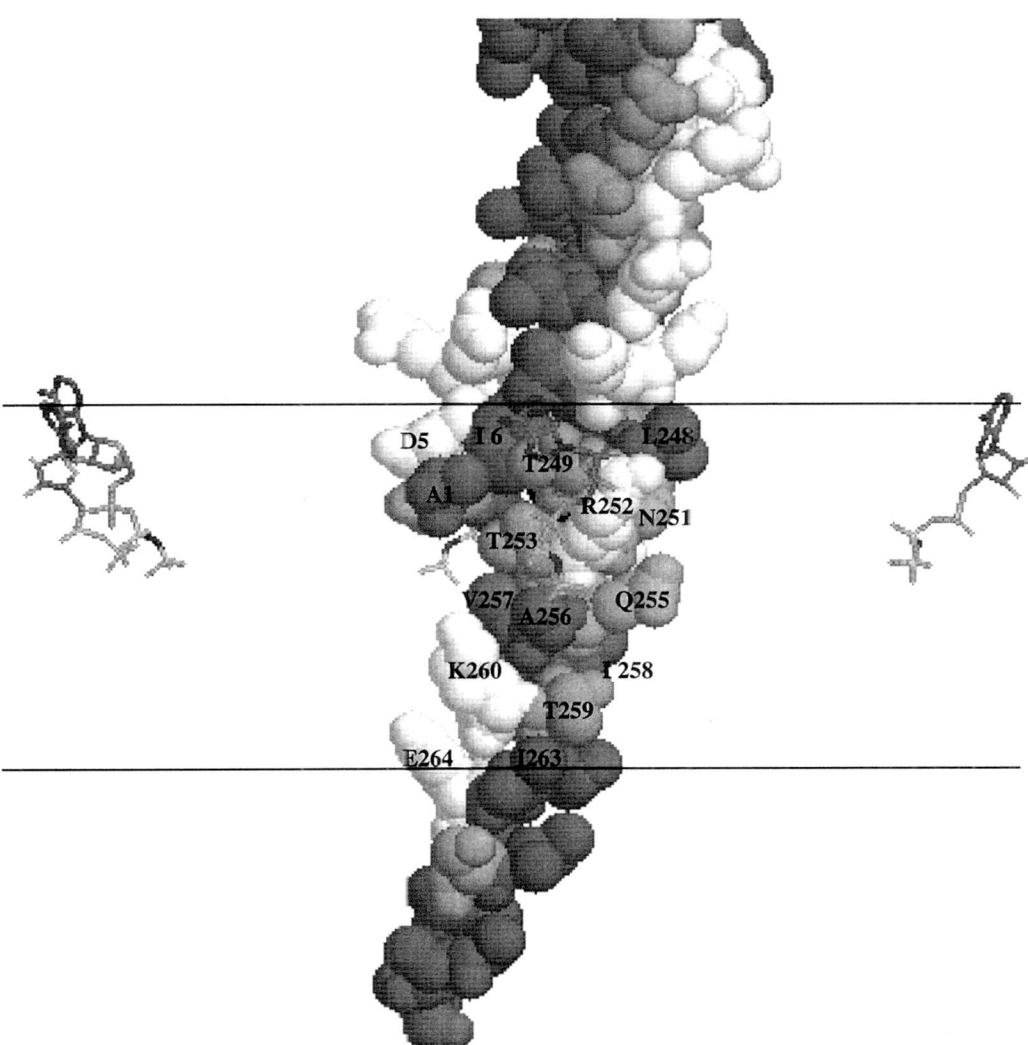

FIGURE 8.33. β-(ATP) face of the γ-rotor of ATP synthase seen from the side of the ATP site with ATP overlying the ATP face and identified as *the neutral face* by the calculation in Table 8.2 to give $\Sigma\Delta G_{HA}$(β-ATP face) ≈ +0 kcal/mole. Shown at right are the α-ATP ligand and the β-(empty) site, at left the α-ATP and β-ADP ligands, and the α-ATP ligand behind. The neutral residues are light gray, the aromatic residues are black, the other hydrophobic residues are gray, and the charged residues are white. (Adapted from Urry.[109])

8.4.4.4 Side Views (3) of the Interaction of the Paired Diametrically Opposed Subunits with the Asymmetric Rotor

Now the question becomes whether the different hydrophobicities of the faces of the γ-rotor affect the interaction between rotor and catalytic housing of the F_1-motor. Examination of side views of the three arrangements of diametrically opposed α–β-subunits in association with the rotor provides the answer. Using the chain designations of the crystal structure, included in Figure 8.30 with the γ-rotor also indicated by its chain designation G, the three arrangements may be given as β-empty(E)–γ-rotor(G)–α-ATP(C), β-ADP(D)–γ-rotor(G)–α-ATP(B), and β-ATP(F)–γ-rotor(G)–α-ATP(A).

To examine these interactions in Figure 8.34, cross-eye stereo views with the protein subunits in space-filling representation are used in which the neutral residues are light gray, the aromatic residues are black, the other hydrophobic residues are gray, and the charged residues are white. This choice of representation and residue shading allows immediate identification of dark regions as hydrophobic domains and the hydrophobic associations between subunits and rotor, if and when they occur.

8.4.4.4.1 β-Empty(E)–γ-Rotor(G)–α-ATP(C): The First of Three Arrangements of β-Catalytic Subunits with the Diametrically Opposed ATP-containing α-Subunits

This configuration, which for simplicity may be designated as EGC, makes a profound statement. As shown in Figure 8.34A, the association between the β-empty subunit and the γ-rotor is profoundly hydrophobic at the levels of both the nucleotide sites and the interaction as the γ-rotor enters the (αβ)₃ construct. The most hydrophobic face of the γ-rotor hydrophobically associates extensively with the β-empty subunit. Interestingly, except at the very tip of the γ-rotor, there is no interaction with the α-ATP(C) subunit; instead, there is an aqueous chasm separating G from C down to but not including the tip of the γ-rotor. It is difficult to imagine a more stark contrast between operative polar and apolar associations.

8.4.4.4.2 β-ADP(D)–γ-Rotor(G)–α-ATP(B): The Second of Three Arrangements of β-Catalytic Subunits with the Diametrically Opposed ATP-containing α-Subunits

In the DGB interaction shown in Figure 8.34B, the interaction of the γ-rotor with the β-ADP subunit is significantly hydrophobic, but in a limited way when compared with the interaction, EG, between β-empty subunit and the γ-rotor. For this DGB configuration, the α-ATP(B) interaction with the γ-rotor is partly re-established.

8.4.4.4.3 β-ATP(F)–γ-Rotor(G)–α-ATP(A): The Third of Three Arrangements of β-Catalytic Subunits with the Diametrically Opposed ATP-containing α-Subunits

For the FGA configuration shown in Figure 8.34C, both the α- and β-subunits contain ATP, and strikingly the γ-rotor lies centered between the two with more nearly similar interactions of the γ-rotor with the two subunits. This FGA association is very different from the off-center positioning of the γ-rotor for the EGC configuration in Figure 8.34A, where the association of the β-empty subunit with the γ-rotor gives the impression of being pulled over by the attraction of hydrophobic domains.

8.4.4.5 Association of γ-Rotor with F_1-motor Housing Looks Like the "Pull" of Hydrophobic Association, ΔG_{HA}, Coupled with the "Push" of Apolar–Polar Repulsion, ΔG_{ap}

The association of hydrophobic domains between the γ-rotor and the β-empty subunit in Figure 8.34A profoundly dominates the structural view of the EG interaction. Also striking is the seeming repulsion between the γ-rotor(G) and the α-ATP(C) subunit seen as an aqueous cleft between the G and C chains. The symmetry is partially restored when the diametrically opposed subunits contain β-ADP and α-ATP, as shown in Figure 8.34B. The

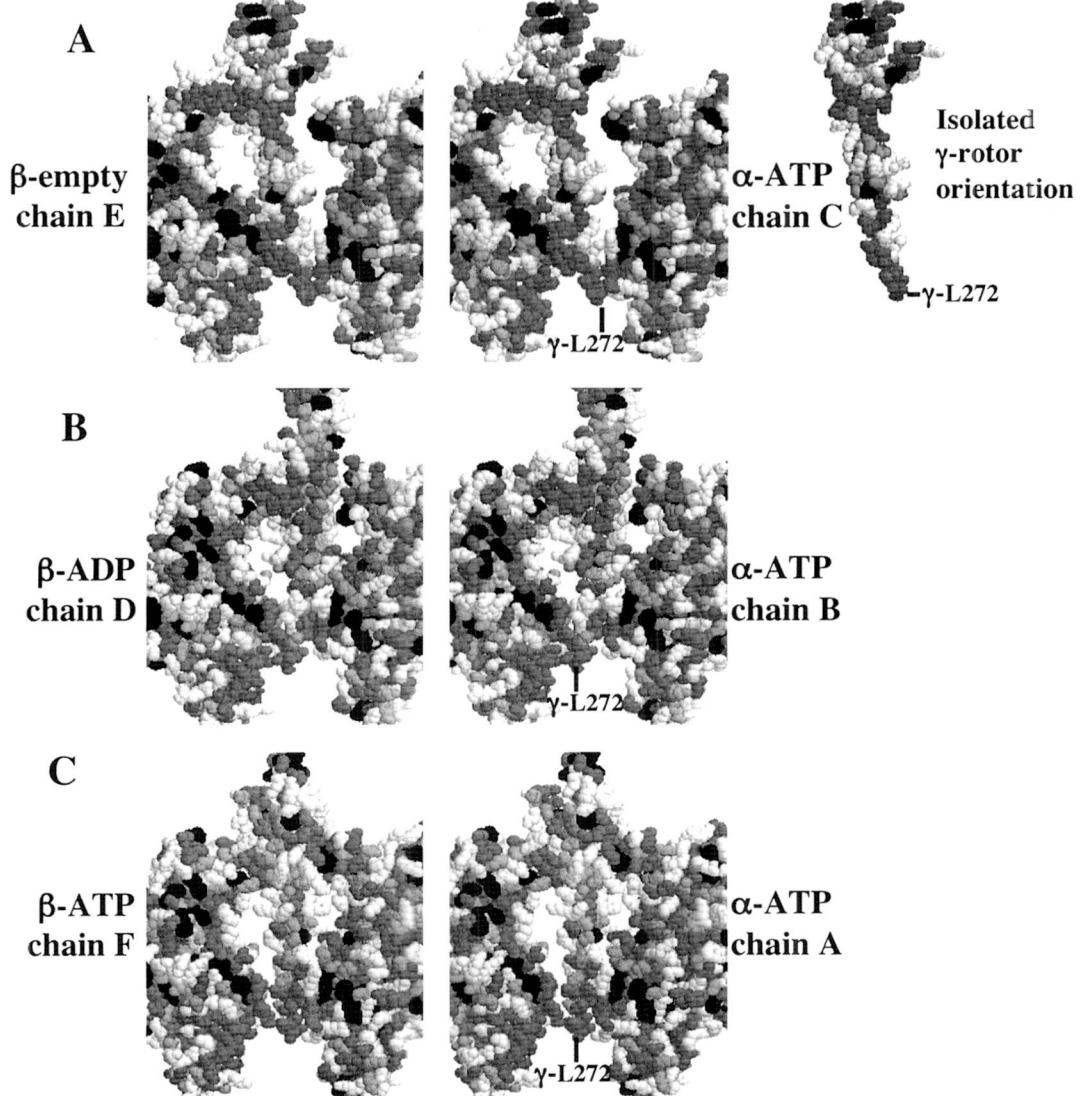

FIGURE 8.34. F_1-ATPase: vertical stereo views in space-filling representation showing the central γ-rotor with diametrically opposed subunits on each side with the neutral residues light gray, the aromatic residues black, the other hydrophobic residues gray, and the charged residues white. The β-subunits are on the left, and the diametrically opposed α-subunits are on the right. In terms of crystallographically defined chains and as defined in Figure 8.30: **(A)** the empty β-subunit (chain E) pairs with α-subunit (chain C) containing ATP; **(B)** the β-subunit containing ADP (chain D) pairs with the α-subunit (chain B) containing ATP; and **(C)** the β-subunit containing ATP (chain F) pairs with the α-subunit (chain A) containing ATP. By this graytone shading hydrophobic associations are seen as dark regions. In A, the γ-rotor hydrophobically associates so extensively with the empty β-subunit due to negative $\Sigma\Delta G_{HA}(\chi)$ that the two are not visually separable with the shading scheme. Therefore, the individual γ-rotor is shown separately in position to be seen in stereo. In B, The ADP-bound β-subunit hydrophobically associates to a lesser extent with the γ-rotor. In C, The ATP-bound β-subunit loses essentially all of the lower hydrophobic association with the γ-rotor and is very nearly centered between the two ATP-containing subunits. Very significant in part (**A**) is the remaining aqueous chamber between the β-empty subunit and the γ-rotor through which the apolar–polar repulsive forces can act. See text for further discussion. (Prepared using the crystallographic results of Abrahams et al.[84] as obtained from the Protein Data Bank, Structure File 1BMF.)

symmetry is nearly entirely restored when the diametrically opposed subunits both contain ATP, as in the configuration of β-ATP(F)–γ-rotor(G)–α-ATP(A) given in Figure 8.34C.

Seen grossly, these relationships are as one might expect if the hydrophobic consilient mechanism provided the dominant interaction energies. Additional crystal structures, examined below, allow this possibility to be explored further. The result is most encouraging for the consilient mechanisms that constitute the major message of this volume.

8.4.4.6 Stereo Top View of Assembled F_1-ATPase in Which All Sites Are Occupied by Nucleotides

As appreciated by Boyer,[85] a more recent structure of the F_1-ATPase due to Menz et al.[77] makes major strides in giving further insight into the function of ATP synthase. The top perspective of this structure is shown in stereo view in Figure 8.35 using space-filling representation. Included are the water molecules detectable by X-ray diffraction.

The few water molecule observed on the surface in Figure 8.35A tend to occur in crevices where the likelihood is greater that they would be sufficiently fixed in position to be detectable by the long time periods required for determination of structure by X-ray diffraction. On removal of the protein subunits in Figure 8.35B, it becomes possible to see the water molecules internal to the structure as well as the ligands. Almost surprising is the very large number of detectable water molecules. This water must contain a clue to function and, as such, we call them the *waters of Thales*. As water is essential for the hydrophobic consilient mechanism to be relevant, this observation bodes well for such a mechanism. However, this structure with its particular set of ligands becomes especially informative when viewed in the light of the ΔG_{ap} element of the hydrophobic consilient mechanism, as considered below.

8.4.4.7 Stereo View of γ-Rotor Surrounded with Nucleotides in All Sites (3 ADP in the α-Subunits, Plus One ADP-SO_4, Plus Two ADP-AlF_4 Bound in the β-Subunit Sites)

8.4.4.7.1 Side View of γ-Rotor Surrounded by Ligands

The cross-eye stereo perspective of the γ-rotor in side view and in space-filling representation is shown in Figure 8.36A surrounded by six nucleotides with two inhibitory nucleotides, ADP-AlF_4^-, labeled. Unlabeled are four smaller glycerol molecules around the γ-rotor. The labeled F250 residue identifies the hydrophobic side of the rotor, as may be seen by comparison with Figure 8.31.

8.4.4.7.2 Top View of γ-Rotor Surrounded by Ligands

Figure 8.36B gives the top view of the γ-rotor with surrounding ligands, labeled with respect to both ligand structure and crystallographically defined chain of residence. Three ADP ligands are in the α-subunits, chains A, B, and C. The β-catalytic subunits are filled with two nucleotides that inhibit catalysis, ADP-AlF_4^-, in chains D and F, and with ADP-SO_4^{2-} in the third catalytic subunit in chain E. Based on the argument of an apolar–polar repulsion between ligands in the nucleotide-binding sites and the γ-rotor, the observation of the hydrophobic face opposite the ADP-SO_4^{2-} occupied site would suggest that this site is less polar than the effect of the two opposing ADP-AlF_4^-–occupied sites. The Pauling electronegativity scale provides helpful insight with which to evaluate the relative polarities of these ligands.

8.4.4.7.3 Evaluation of Relative Polarity of Ligands Using the Pauling Electronegativity Scale

The Pauling electronegativity scale[88,89] places fluorine, F, with an electronegativity of 4.0 as the most electronegative atom, followed by

8.4 ATP Synthase: The Twofold Rotary Protein Motor of Oxidative Phosphorylation

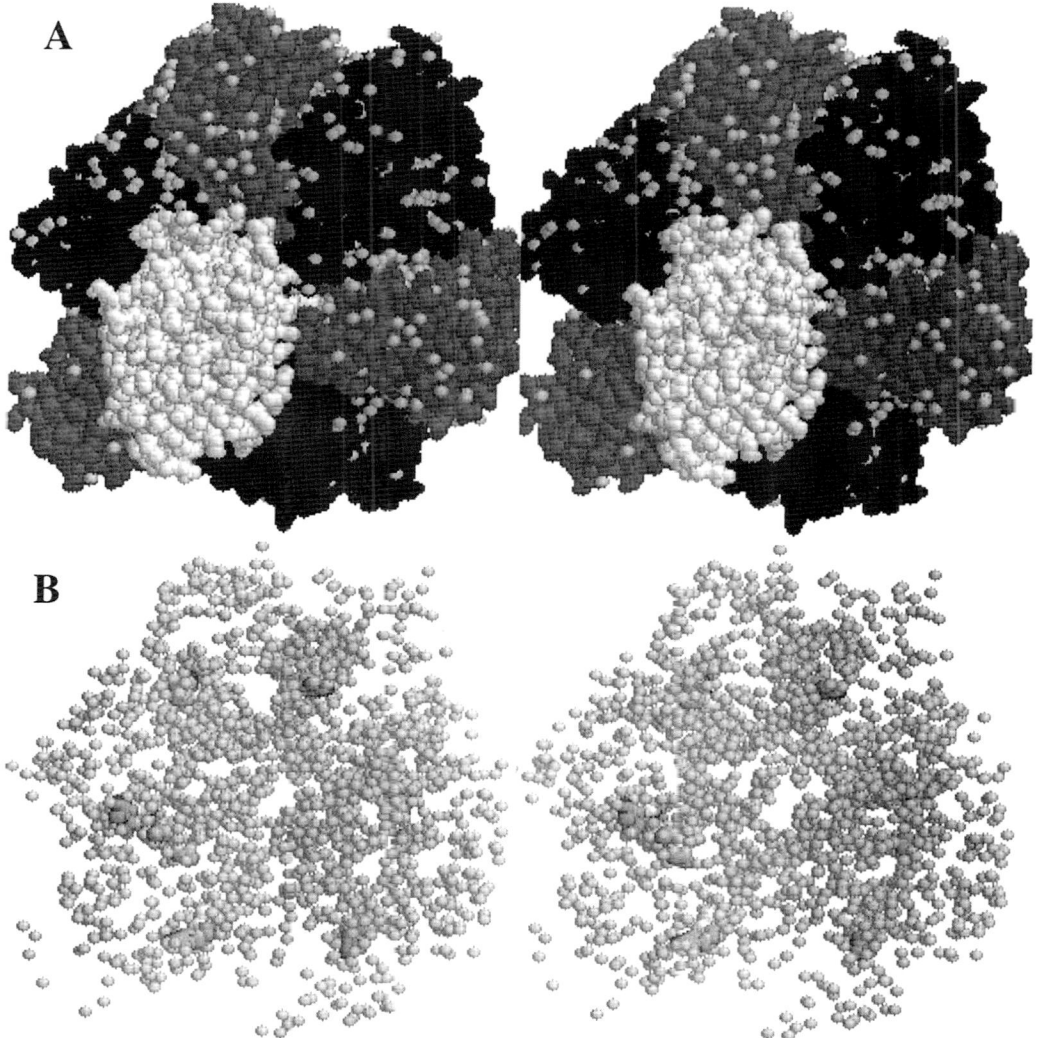

FIGURE 8.35. Crystal structure of the F_1-motor of ATP synthase in cross-eye stereo view from the top of the structure that has all sites filled with nucleotide, as indicated in Figure 8.36. **(A)** Space-filling model of $\gamma(\alpha\beta)_3$ showing a few detected water molecules on the surface with a white γ-rotor somewhat off center. **(B)** Space-filling view with protein subunits removed to show all of the water molecules detected by X-ray diffraction. These are the internal "waters of Thales" available to function by the consilient mechanisms for protein-based machines. (Prepared using the crystallographic results of Menz et al.[77] as obtained from the Protein Data Bank. Structure File 1H8E.)

oxygen, O, with an electronegativity of 3.5. The electronegativity of hydrogen, H, and phosphorus, P, are both 2.1, while that of sulfur, S, is 2.5 and that of aluminum, Al, is 1.5. Electronegativity is defined as "a measure of the power of an atom in a molecule to attract electrons to itself." We relate the electronegativity of an atomic grouping to its capacity to function as a polar entity expressed as a thirst for hydration of relevance to the hydrophobic consilient mechanism, but the formal charge the group presents yet an additional factor.

If we compare the atomic groupings of $-PO_3^{2-}$, SO_4^{2-}, AlF_4^-, and HPO_4^{2-}, the sum of elec-

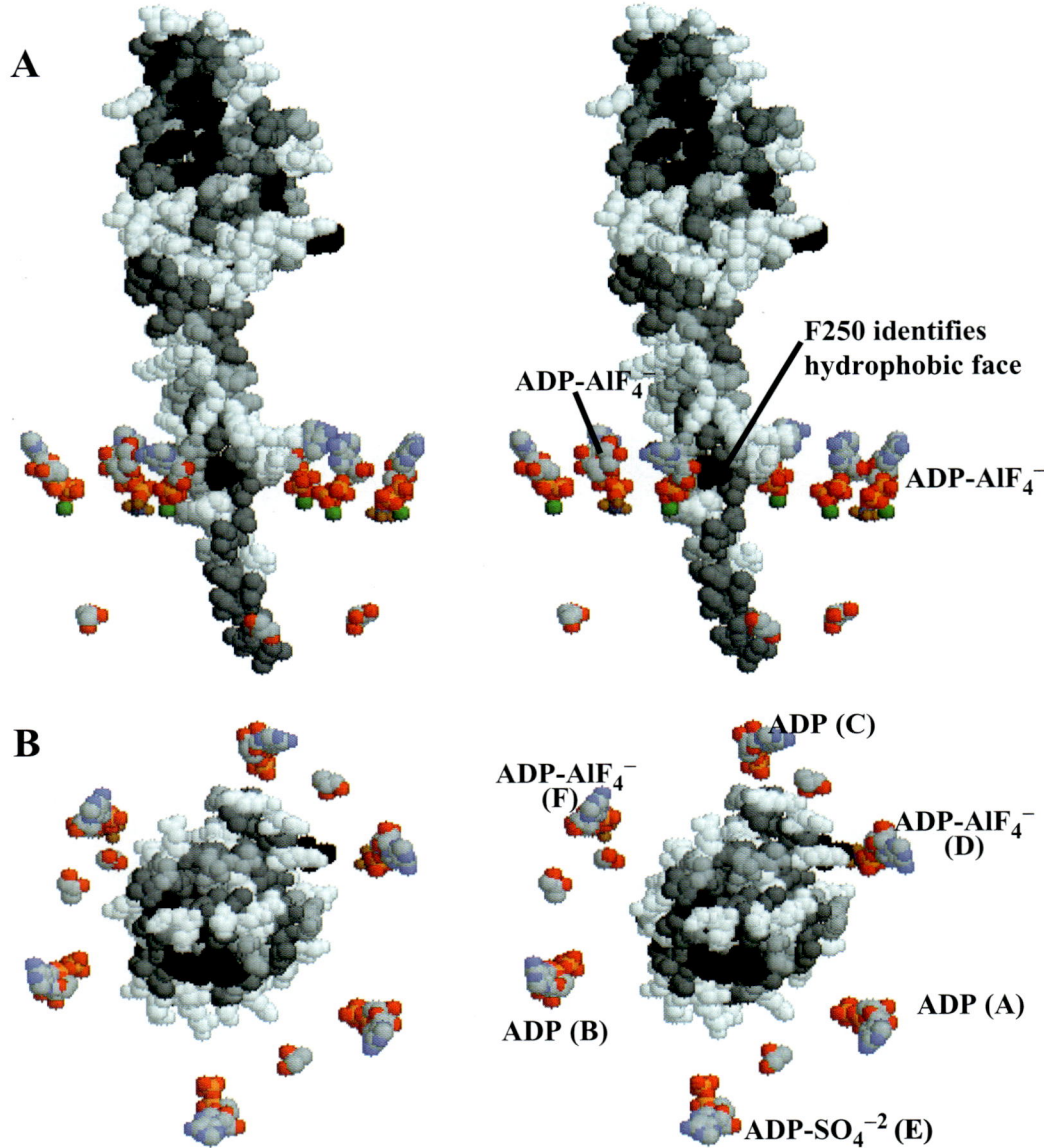

FIGURE 8.36. **(A)** Hydrophobic face of the γ-rotor of ATP synthase $\Sigma\Delta G_{HA}$(opposite pair of very polar ADP-AlF) ≈ −20 kcal/mole. With the most polar ADP-AlF$_4$ at rear sides and with one ADP at rear and three in front, ΔG_{ap} would place the hydrophobic face at front. F250 identifies the hydrophobic side of the γ-rotor, again to show the orientation effect of ΔG_{ap}. The neutral residues are light gray, the aromatic residues are black, the other hydrophobic residues are gray, and the charged residues are white. **(B)** Top view showing the γ-rotor and confirming orientation of ligands. One of five structures that confirm the hydrophobic face of the γ-rotor face away from the most polar sites. (Prepared using the crystallographic results of Menz et al.[77] as obtained from the Protein Data Bank, Structure File 1H8E.)

tronegativities are 12.6, 16.5, 17.5 and 18.2, respectively. The single charge on AlF$_4^-$ could be expected to lower its thirst for hydration in relation to the other groups. Even so, on this basis one expects the polarity of the molecular grouping of ADP-AlF$_4$ to be equivalent to and perhaps even more polar than that of ADP-SO$_4$. Thus, for the structures in Figures 8.35 and

8.36, on the basis of ΔG_{ap}, one would expect the γ-rotor to orient with its most hydrophobic face away from the two catalytic sites containing ADP-AlF$_4$ and therefore be directed toward the β$_E$ catalytic site containing ADP-SO$_4$. This is the orientation of the γ-rotor in the structure in Figure 8.36. Before proceeding to the most significant aspect of the structures in Figures 8.35 and 8.36, however, additional structures with different β-subunit occupancies can be examined to address the consistency of the expectation that the orientation of hydrophobic face of the γ-rotor would be toward the least polar occupancy site.

8.4.4.8 Consistency of Least Apolar–Polar Repulsion Positioning the Hydrophobic Face with Three Additional and Different Binding Site Occupancies

In addition to the structures in Figures 8.30[84] and 8.36,[77] three more crystal structures of the F$_1$-ATPase have been determined by the Walker group. The first of the three involve the occupancy states of ATP(A), ATP(B), ATP(C), ADP(D), PO$_4$(E), and ATP(F) for the protein structure listed in the Protein Data Bank as Structure File 1H8H.[90] The second of the three involve the occupancy states of ANP(A), ANP(B), ANP(C), ADP(D), PO$_4$(E), and ANP(F) where PNP may also be represented as AMPPNP, that is, a nitrogen replaces the bridge oxygen of ATP between the β- and γ-phosphates; this structure, shown in Figure 8.37, is listed in the Protein Data Bank as Structure File 1E1Q.[91] The third of the three additional structures involves occupancy states of ANP(A), ANP(B), ANP(C), ANP(D), empty(E), and ANP(F) and is listed in the Protein Data Bank as Structure File 1OHH.[92] In each case the hydrophobic side of the γ-rotor faces the occupancy state of chain E.

Certainly, the third structure holds the hydrophobic side of the γ-rotor most tenaciously at the β-empty catalytic site, even more so than the structures in Figures 8.30 through 8.34, because all five additional sites are occupied by the ATP equivalent, namely, ANP. The other two structures depend on the relative polarities of ADP(D) and PO$_4$(E) in their respective positions of exposure to the γ-rotor. That the β-PO$_4$(E) is least polar from the perspective of the γ-rotor becomes apparent from the display of the second structure of the three in Figure 8.37B, where the β-PO$_4$ of chain E is positioned so as to be sterically obscured from the γ-rotor. Accordingly, consistency holds for all five structures of the F$_1$-ATPase; the most hydrophobic side of the γ-rotor is always found opposite the least polar occupancy site, where the apolar–polar repulsive free energy of hydration, ΔG_{ap}, is the least.

8.4.4.9 Role of ATP in the α-ATP Subunits: Triangulation of Repulsive Forces

When the β-ATP and α-ATP sites are diametrically opposed, as in Figure 8.34C, the appearance read in terms of ΔG_{ap} is that the β-ATP site expresses slightly more apolar–polar repulsion than does the α-ATP site. From the point of view of the hydrophobic consilient mechanism, this suggests that the exposure of phosphates of the α-ATP sites through water to the γ-rotor is greater for the β-ATP sites. Figure 8.38 shows the exposure of the γ-rotor to α-ATP phosphates of chains A, B, and C of the structures utilized in Figures 8.30 through 8.34. In each of the α-ATP only a single oxygen of the γ-phosphate can be seen through a small peephole. While exposure of the α-ATP phosphates to the γ-rotor could be larger in the dynamic functional state, this situation is compared with a slightly greater exposure of the phosphates of the β-ATP site immediately below and to an analogue of the massively greater exposure on hydrolysis to form ADP plus P$_i$.

By the hydrophobic consilient mechanism, the α-ATP sites provide a critical role of establishing a triangulation of repulsive forces that serves to limit the hydrophobic associations of the γ-rotor. This triangulation of repulsive forces prevents the occurrence of a frictional drag on rotor rotation that would seriously limit motor efficiency.

416　　8. Consilient Mechanisms for Protein-based Machines of Biology

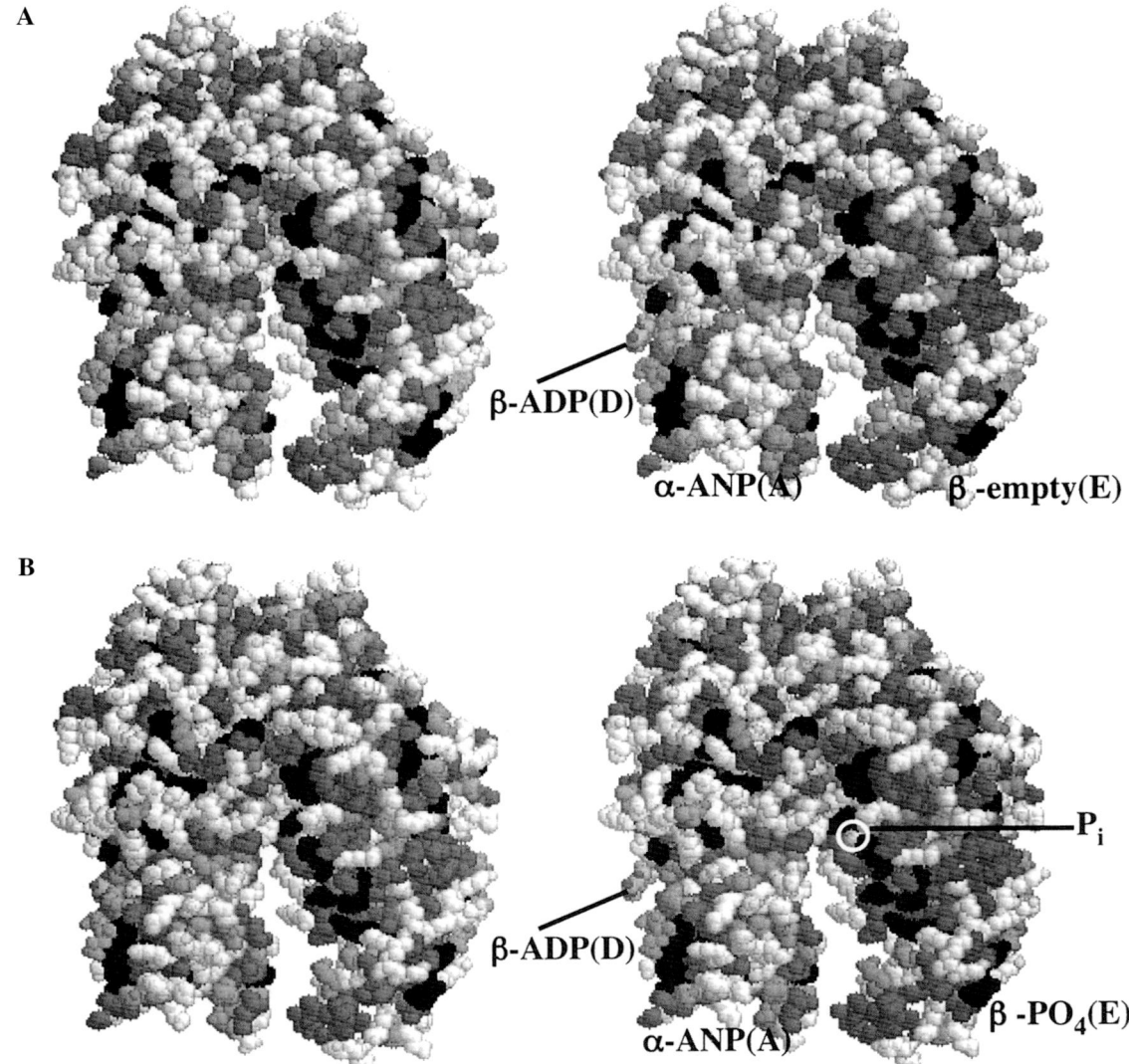

FIGURE 8.37. Stereo view of the F_1-motor of ATP synthase in space-filling representation with neutral residues light gray, aromatic residues black, other hydrophobic residues gray, and charged residues white. **(A)** Association of the β-empty subunit with the noncatalytic α-ATP subunit as seen from the outside. (Prepared using the crystallographic results of Abrahams et al.[84] as obtained from the Protein Data Bank, Structure File 1BMF.) **(B)** Interaction of P_i with the β-subunit in association with the non-catalytic α-ATP subunit as seen from the outside. (Prepared using the crystallographic results of Braig et al.[91] as obtained from the Protein Data Bank, Structure File 1E1Q.)

8.4.4.10 Obscured γ-Phosphate of β-ATP Becomes Full P_i Exposure After Hydrolysis to Produce Maximal ΔG_{ap} for Repulsive Interaction with the γ-Rotor

When looking toward the β-ATP site from the γ-rotor of the structure in Figure 8.30, the only part of the triphosphate observed is an oxygen on the β-phosphate, as shown in Figure 8.39. Also labeled are a number of reference residues and a location, indicated by the large X. The X approximates the location where the sulfate of ADP-SO$_4$, an analogue for ADP-PO$_4$, appears in the very interesting structure of

8.4 ATP Synthase: The Twofold Rotary Protein Motor of Oxidative Phosphorylation

Menz et al.,[77] shown in Figure 8.40. In this structure the SO_4 is a more stable analogue for the hydrolyzed γ-PO_4.

The very polar sulfate opens up the structure to achieve a direct through-water repulsive interaction with the γ-rotor. This observation provided by the structure in Figure 8.40 has the most fundamental implications with respect to the relevance of the hydrophobic consilient mechanism as describe in Chapter 1.

The most profound and fundamental prediction of the hydrophobic consilient mechanism, as regards F_1-ATPase function, is that hydrolysis to form ADP and P_i provides a water-mediated burst of apolar–polar repulsion that would drive the γ-rotor in the ATPase mode

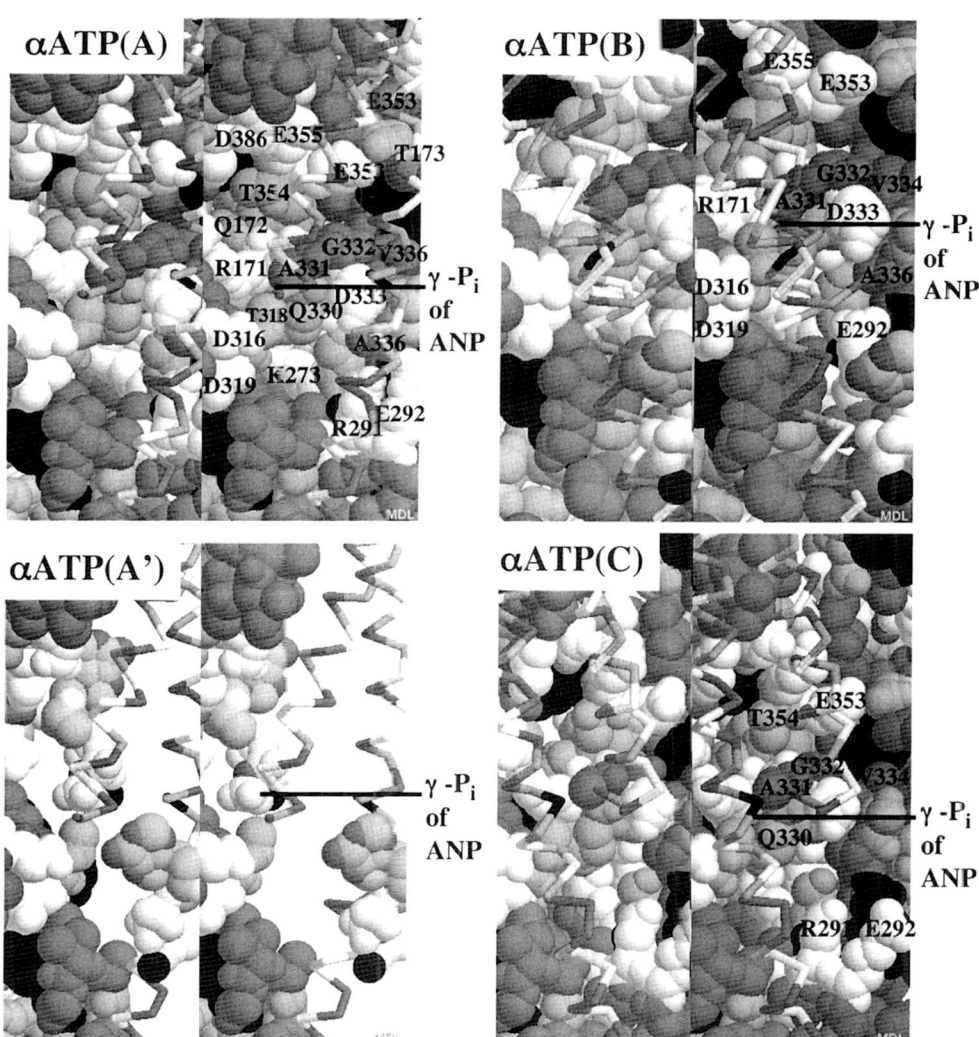

FIGURE 8.38. Stereo views of each of the α-ATP sites of the F_1-motor of ATP synthase in space-filling representation with neutral residues light gray, aromatic residues black, other hydrophobic residues gray, and charged residues white. View is from the γ-rotor of the peephole for oxygen of the γ-phosphate of the α-ATP sites for all three sites as defined in Figure 8.31. αATP(A') shows removal of chain A to expose all of the ATP analogue (AMPPNP). α-ATP(A), α-ATP(A'), α-ATP(B), and α-ATP(C) have the overlying γ-rotor in backbone representation. (Prepared using the crystallographic results of Abrahams et al.[84] as obtained from the Protein Data Bank, Structure File 1BMF.)

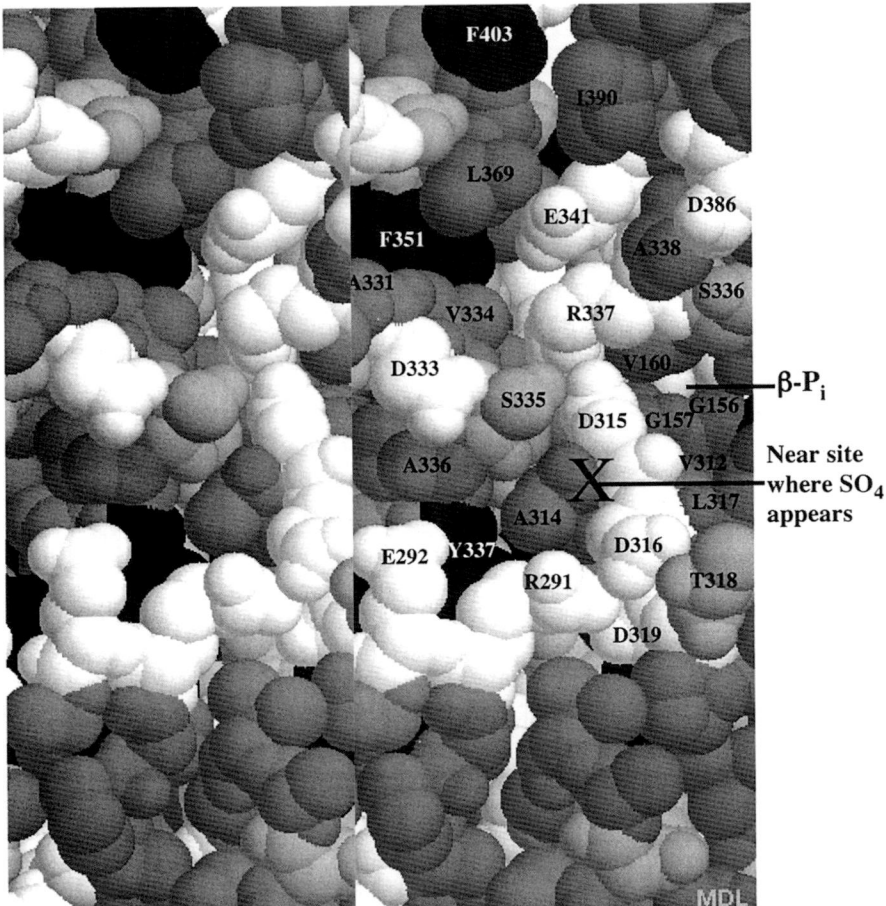

FIGURE 8.39. Stereo view of the F_1-motor of ATP synthase in space-filling representation with neutral residues light gray, aromatic residues black, other hydrophobic residues gray, and charged residues white. View is from the γ-rotor of the peephole for oxygen (O_1) of the β-phosphate of β-ATP(F), as defined in Figure 8.31. The X shows the approximate site where emerges the SO_4 of ADP-SO_4(E) of Figure 8.36, which is taken as the analogue for ADP-HPO_4(E). (Prepared using the crystallographic results of Abrahams et al.[84] as obtained from the Protein Data Bank, Structure File 1BMF.)

and as regards ATP synthase function that the state of highest repulsion in the synthesis mode comes from the face off between the most hydrophobic face of the γ-rotor and the most polar state of ADP and P_i. This severe repulsion is relieved by reducing the polarity on formation of ATP. A corollary of this prediction is that the P_i be as fully exposed as possible through water to the most hydrophobic side of the γ-rotor so that a maximal apolar–polar repulsion can occur.

If the γ-rotor were to be driven by an off-center application of torque at the point where the γ-rotor enters the $(\alpha\beta)_3$ structure, such an exposure of the hydrolyzed phosphate directly to the γ-rotor would seem to have little direct relevance. On the other hand, with regard to the hydrophobic consilient mechanism, it is central to function. This off-center application of torque occurs just before the γ-rotor goes from double stranded to single stranded and explains why the double-stranded rotor con-

8.4 ATP Synthase: The Twofold Rotary Protein Motor of Oxidative Phosphorylation

tinues as it does beyond entry into the $(\alpha\beta)_3$ motor housing.

8.4.4.11 Crystal Structure Evidence for a Significant ΔG_{ap} Between the ADP-SO_4-Occupied β_E-subunit and the γ-rotor

Above, the predicted hydrophobically asymmetric rotor was found. Furthermore, the hydrophobically asymmetric rotor associated hydrophobically with the $(\alpha\beta)_3$ housing of the F_1-motor and was appropriately responsive to the occupancy states of the nucleotide-binding sites. These findings all support the role of the Gibbs free energy of hydrophobic association, ΔG_{HA}, and of its operative apolar–polar repulsive free energy of hydration, ΔG_{ap}, in the function of the F_1-motor. Now by comparison of two crystal structures, one with the β_E-catalytic site empty and the second with the β_E-subunit con-

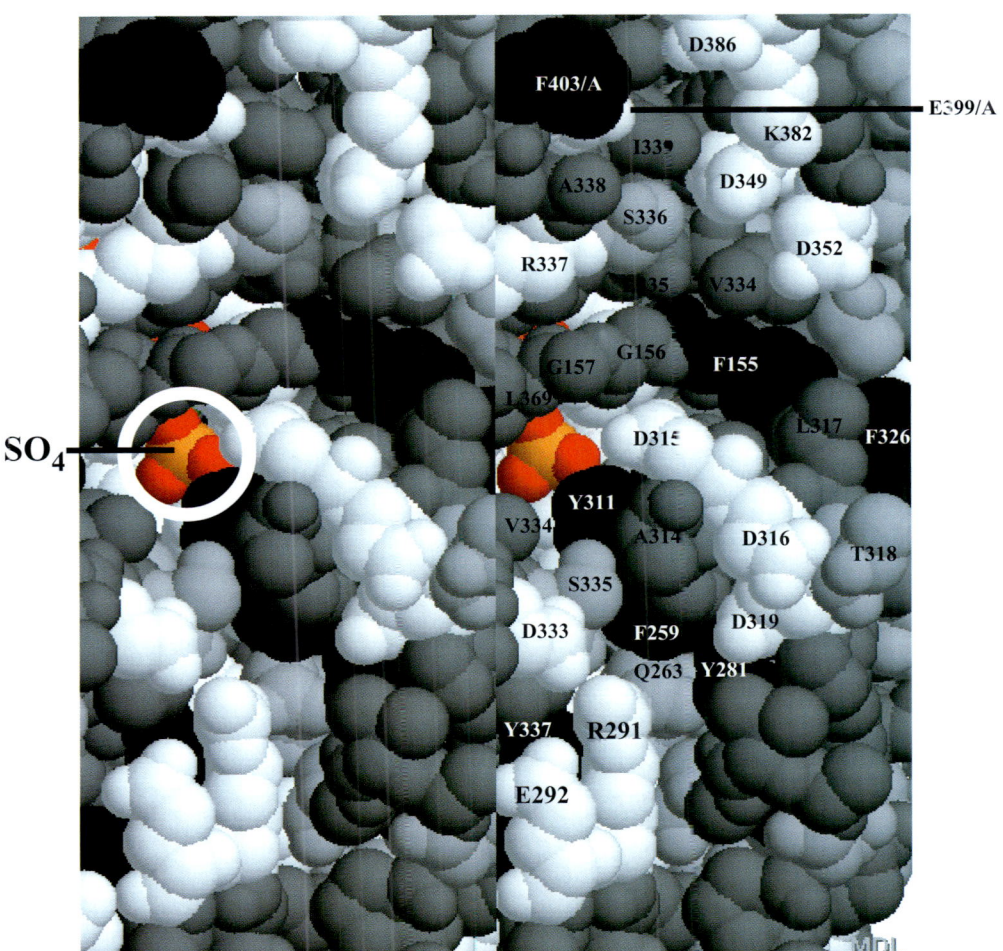

FIGURE 8.40. Stereo view of the F_1-motor of ATP synthase in space-filling representation with neutral residues light gray, aromatic residues black, other hydrophobic residues gray, and charged residues white. View is of the more stable SO_4 in place of the more polar PO_4 (the hydrolyzed γ-P of ATP) that occurs in chain E and is shown in Figure 8.41 with the overlying γ-rotor in gray ribbon representation. The polar SO_4 (encircled) opens up the protein structure of chain E (when compared with the β-ATP(F) site in Figure 8.39) to expose the SO_4 to the hydrophobic side of the γ-rotor and thereby to express a large ΔG_{ap} repulsive force through the intervening waters of Thales. (Prepared using the crystallographic results of Menz et al[77] as obtained from the Protein Data Bank, Structure File 1H8E.)

taining ADP-SO$_4$, a ΔG_{ap} repulsion between the ADP-SO$_4$–containing β$_E$-subunit and the γ-rotor is found.

To quote from the legend of Figure 5 of Menz et al.,[77] "Note that interacting residues in the β$_E$-subunit and the γ-subunit move in opposite directions." This movement in opposite directions of the very polar ADP-SO$_4$–containing β$_E$-subunit and the hydrophobic side of the γ-rotor, in our view, is a direct result of the ΔG_{ap} repulsion between the apolar and polar subunits. The separation of housing and rotor identifies a repulsion that introduces an elastic displacement and twist in the γ-rotor. Because of the relative stiffness of the rotor and housing, this elastic deformation (from otherwise preferred torsion angles) would have a large elastic modulus. Accordingly, the reported mean displacement of 2.9 Å and twist of up to 20 would reflect the application of a substantial torque to the γ-rotor. This occurrence again unites the hydrophobic and elastic consilient mechanisms in a way that would result in high efficiencies of energy conversion. The question now becomes whether or not this applied torque, due to the ΔG_{ap} repulsion between the exposed sulfate and the hydrophobic side of the γ-rotor, can predict the direction of rotation of the F$_1$-ATPase.

8.4.4.12 Predicted Direction of Rotation for F$_1$-ATPase

ΔG_{ap}[sulfate ↔ hydrophobic rotor] provides a water-mediated off-center "push" on the γ-rotor like an off-center push on a camshaft. The direction of the repulsion between exposed sulfate and the hydrophobic side of the γ-rotor, written as *ΔG_{ap}[sulfate ↔ hydrophobic rotor]*, can be deduced from the structural relationships of Figure 8.41. The apolar–polar repulsion between sulfate and γ-rotor applies toward the lower part of the double-stranded section of the γ-rotor above the amino-terminus of the γ-subunit. As evidenced by the structural relationships showing the overlay of the γ-rotor on the ADP-SO$_4$–containing β$_E$-subunit, the **ΔG_{ap}[sulfate ↔ hydrophobic rotor]** repulsive interaction would provide an impulse to rotate the γ-rotor in a counterclockwise direction.

8.4.4.13 Demonstrated Direction of Rotation of the γ-rotor

As seen from the landmark study of Noji et al.,[93,94] and represented in Figure 8.42, the direction of rotation for the F$_1$-ATPase is indeed in the counterclockwise direction. *Thus, the foundation for the hydrophobic and elastic consilient mechanisms in the function of ATP synthase and F$_1$-ATPase becomes increasingly compelling.*

8.4.5 Consideration of the Efficiency and Cooperativities Exhibited by F$_1$-ATPase

By analysis of crystal structures, we have just demonstrated the success of a significant number of predictions fundamental to the biophysics of the hydrophobic and elastic consilient mechanisms. Now we address the unique and fundamental properties of efficiency of energy conversion and of negative cooperativity of ATP binding and yet of positive cooperativity of the effect of ATP binding on catalytic rate. Can an understanding of these challenging phenomena exhibited by F$_1$-ATPase also fall to the comprehension flowing from the hydrophobic consilient mechanism?

8.4.5.1 High Efficiency for Performance of Mechanical Work

8.4.5.1.1 Near 100% Efficiency of the F$_1$-ATPase

With the capacity to attach an actin filament to the γ-rotor and to observe microscopically the rotation of the actin filament driven by the F$_1$-ATPase,[93,94] estimates of work performed on rotation of the actin filament could be estimated as could the energy consumed in the performance of that work in terms of the hydrolysis of ATP to form ADP and P$_i$. The resulting estimate was remarkably close to 100%.[95] Such a result stands as a severe challenge to any proposed mechanism.

Insight comes from considering elastic deformation as a machine for storing energy. By means of atomic force microscopy in the single-chain force extension mode, a single chain of (GVGVP)$_{502}$ could be observed to exhibit an

8.4 ATP Synthase: The Twofold Rotary Protein Motor of Oxidative Phosphorylation

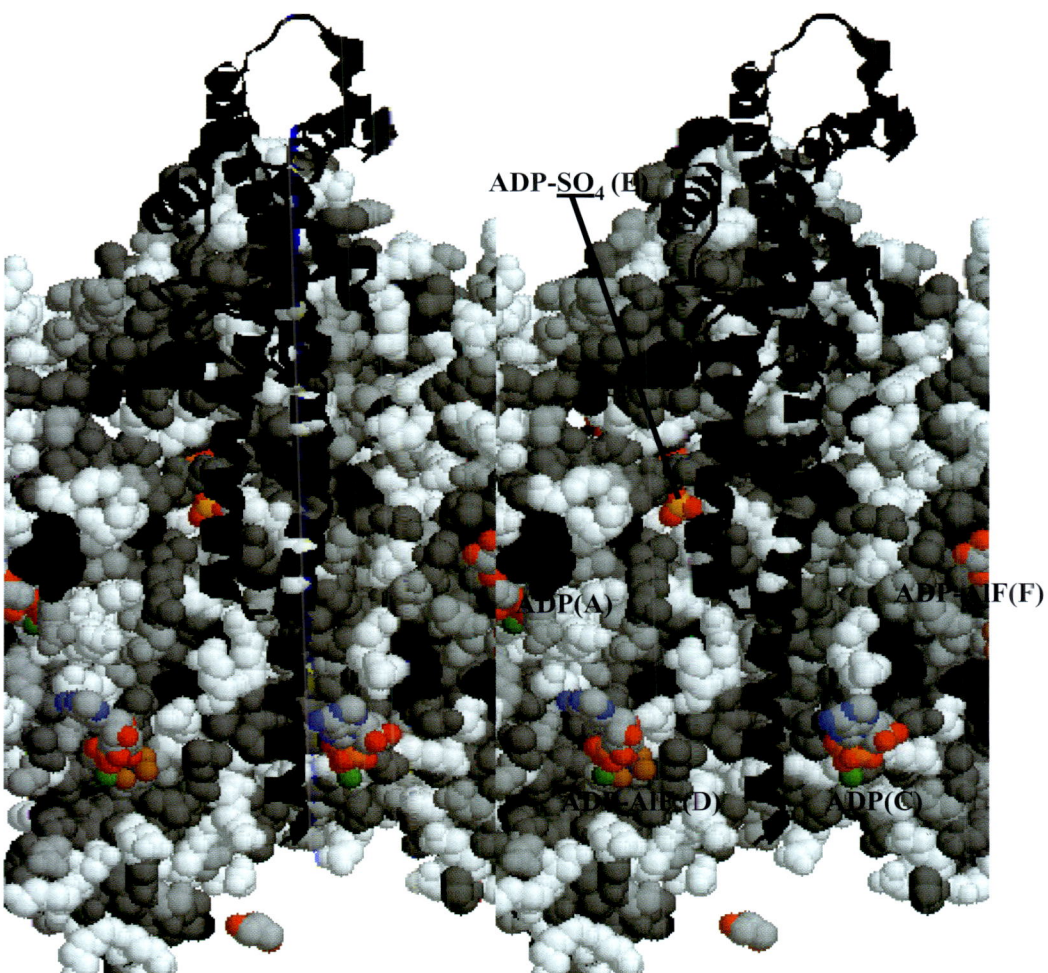

FIGURE 8.41. Inside stereo view of chains A, E, and B and their ligands of the F_1-motor of ATP synthase in space-filling representation with neutral residues light gray, aromatic residues black, other hydrophobic residues gray, and charged residues white and with the ligands for chains D, C, and F in the foreground. With this perspective, the overlying γ-rotor in gray ribbon representation properly overlays the catalytic site of the $β_E$-subunit containing the exposed sulfate. The large repulsive ΔG_{ap} force due to SO_4 applies a torque just above the amino-terminal sequence of the γ-rotor that would give a counterclockwise rotation when functioning as the ATPase. This has analogy to repulsion between like poles of two magnets that made such an effective analogue model for the F_1-motor, as presented by Kinosita et al.[76] The experimentally determined direction of rotation is counterclockwise, as shown in Figure 8.42. (Adapted from Urry.[109])

ideal elasticity in which all of the energy expended in deformation was recovered on relaxation (see Figure 5A of Urry and Parker[8]). On the other hand, when there are adherent chains that do not sustain the deformation as occurs with $(GVGIP)_{520}$, there occurs a marked hysteresis; the energy of deformation is not all recovered on relaxation, as shown in Figure 5B of Urry and Parker.[8] Accordingly, chains that take up energy during deformation but do not directly bear and store the force of deformation cannot return the energy on relaxation. When-

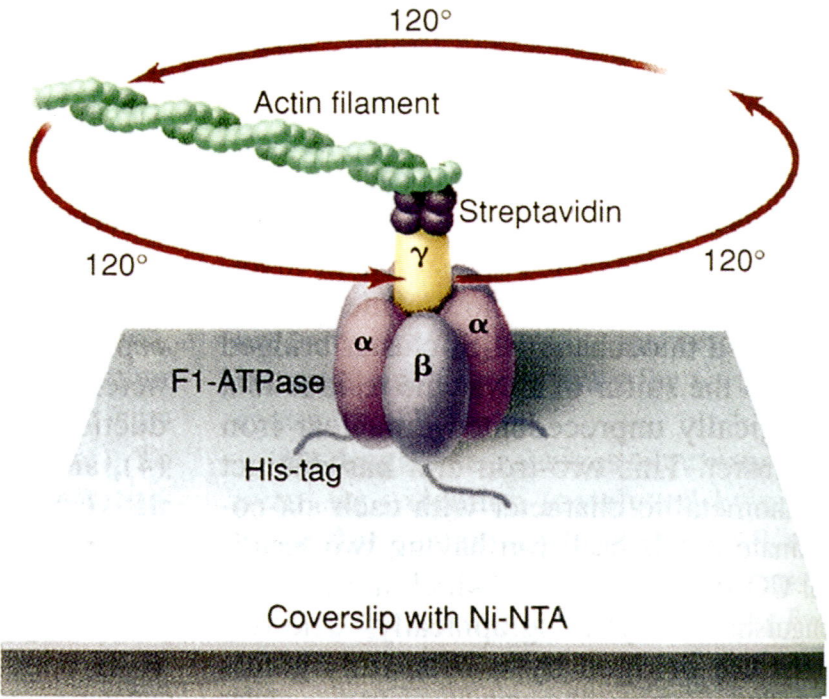

FIGURE 8.42. Actin filament attached to the γ-rotor by streptavidin and F_1-ATPase fixed on substrate. ATP drives counter clockwise rotation of the actin filament in 120° steps, which are directly observed by video microscope.[93,94] (Reprinted with permission from H. Noji, "Pharmacia Biotech & Science Prize: The Rotary Enzyme of the Cell: The Rotation of F_1-ATPase." Science, **282**, 1844–1845, 1998.[94] Copyright © 1998 AAAS.)

ever this occurs the efficiency of energy conversion becomes limited. Therefore, a mechanism wherein one protein subunit mechanically applies a torque to the rotor by physical contact would seem necessarily and irreversibly to lose energy into its three-dimensional lattice of chain segments in a way that cannot be recovered.

8.4.5.1.2 Efficient Use of a Sudden Burst of Repulsion, as Occurs on Hydrolysis of ATP to Produce ADP and P_i, Requires Storage in a Near-ideal Elastic Deformation

On the other hand, when elastic deformation occurs through a repulsive force mediated by the solvent, as would be the interpretation of the consilient mechanism for the elastic deformation reported by Menz et al.,[77] then dissipation through a lattice of chains would be limited and a high efficiency could be expected. The deformation in this case becomes the direct result of the repulsion and a return to the "equilibrium" state, which existed prior to the occurrence of the repulsive force, would be expected.

Accordingly, on hydrolysis to form ADP^{3-} Mg^{2+} plus HPO_4^{2-} the sudden burst of repulsion due to HPO_4^{2-} would result in an elastic deformation of the γ-rotor and of the associated β-subunit that would limit torsion angles in a storage of energy that would be nearly fully recovered on the relaxation of rotational motion.

8.4.5.2 Binding of ATP Exhibits Negative Cooperativity Due to Increases in ΔG_{ap} with Increasingly Hydrophobic Faces of the γ-Rotor

Negative cooperativity for binding of ATP would reasonably arise from increasing hydrophobicity of the three faces of the

8.4 ATP Synthase: The Twofold Rotary Protein Motor of Oxidative Phosphorylation

asymmetrically hydrophobic rotor, that is, an increase in the apolar–polar repulsive free energy of hydration, ΔG_{ap}, with each addition of ATP. The first ATP would be expected to bind adjacent to the most polar face ($\Delta G_{HA} \approx +9$ kcal/mole), the second to the essentially neutral face ($\Delta G_{HA} \approx 0$ kcal/mole), and the third opposite the most hydrophobic face of the γ-rotor ($\Delta G_{HA} \approx -20$ kcal/mole). The presence of the first ATP in the most favorable site would leave a less favorable site for the second ATP binding. Similarly the presence of the two ATP molecules bound to the more favorable sites would leave the site with the most repulsion. Thus, given the structure of the γ-rotor with its hydrophobic asymmetry, negative cooperativity is the necessary result. While a small contribution to the negative cooperativity could arise from direct charge–charge repulsion between multivalent ATP anions, acting through the aqueous compartment, this is expected to be small compared with the increasing apolar–polar repulsive free energy of hydration, ΔG_{ap}, that each subsequent ATP molecule would experience as the result of being confronted with a more hydrophobic side of the γ-rotor.

8.4.5.3 ATP Hydrolytic Rate Exhibits Positive Cooperativity as Each Additional Bound ATP Facilitates Hydrophobic Dissociation of γ-Rotor from the Inner Wall of $(\alpha\beta)_3$

On the basis of the hydrophobic consilient mechanism, the three ATP molecules in the noncatalytic α-subunits are expected to apply a triangular array of repulsive thrusts that would limit the viscous drag resulting from hydrophobic association between the γ-rotor and the $(\alpha\beta)_3$ housing of the F_1-motor. Even so, when there is an empty β-subunit, prominent hydrophobic association occurs. Thus, as each β-catalytic site is filled, the hydrophobic association would decrease until the binding of ATP to the third β-catalytic site would dramatically lower the viscous drag arising from hydrophobic association of the $(\alpha\beta)_3$ housing of the F_1-motor with the γ-rotor.

8.4.6 Review of Correlations Between the Hydrophobic Elastic Consilient Mechanism and the Properties of ATP Synthase/F_1-ATPase

8.4.6.1 Nine Major Points of Correlation

Below we list nine major points of correlation of the ATP synthase/F_1-ATPase. It is not a complete listing of the correlations, as there are many more details that should be included in a more exhausting consideration. Nonetheless, in our view the nine points are compelling.

1. As required for the hydrophobic consilient mechanism to be operative for an ATP synthase/F_1-ATPase protein-based machine, the three faces of the γ-rotor exhibit very different hydrophobicities, that is, different ΔG_{HA} values, namely, approximately -20, 0, and $+9$ kcal/mole.

2. Furthermore, the most hydrophobic face of the γ-rotor functions as such; it associates hydrophobically with the β-empty subunit, which becomes the most hydrophobic β-subunit due to the absence of polar (charged) nucleotides.

3. The F_1-ATPase exhibits extensive "waters of Thales"; as required for the existence of an apolar–polar repulsive free energy of hydration, water-filled clefts exist between the γ-rotor and polar nucleotide filled sites.

4. Negative cooperativity for binding of ATP occurs because each ATP addition is confronted through the "waters of Thales" with a progressively more hydrophobic face of the rotor; the consequence of increases in the apolar–polar repulsive free energy of hydration, ΔG_{ap}, constitute increasing barriers to the approach of polar ATP molecules and result in stepwise decreases in binding affinity.

5. The hydrophobic consilient mechanism gives a functional role to the α-ATP noncatalytic sites; these sites provide a fixed triangulation of repulsive ΔG_{ap} that helps to center the rotor and to prevent locking up due to the viscoelastic drag of hydrophobic association.

6. As each of the β-catalytic sites fills with ATP, the rate of hydrolysis would increase due to the decrease in viscoelastic drag that would

result in a progressively increased rotational rate of the γ-rotor, that is, there exists a positive cooperativity in the catalytic rate as all β-catalytic sites become occupied with ATP.

7. Because assembly of the F_1-ATPase is dominated by an inverse temperature transition of hydrophobic association, the structure disassembles on lowering the temperature, that is, it exhibits cold denaturation.

8. The maximal apolar–polar repulsive free energy of hydration, ΔG_{ap}, acting through the "waters of Thales," occurs between the most hydrophobic side of the γ-rotor and the most polar state of the β-catalytic site, ADP^{3-} Mg^{2+} plus hydrolyzed phosphate, HPO_4^{2-}. The structural relationship between the sulfate analogue of the hydrolyzed phosphate and the most hydrophobic side of the γ-rotor predicts the direction of rotation for the F_1-ATPase to be counterclockwise. Furthermore, this decisive repulsive thrust indicates why the second chain of the two-stranded α-helical coiled coil continues to the level of the γ-phosphate of ATP and is no longer and no shorter than it needs to be for this function.

9. The remarkably high efficiency of the F_1-ATPase in converting the chemical energy of ATP hydrolysis into the performance of mechanical work occurs because the repulsive force acts through solvent to produce a reversible elastic deformation of γ-rotor and the $(\alpha\beta)_3$ housing, rather than a mechanical mechanism where the steps of binding ATP and its hydrolysis could be expected to cause conformational changes in the apical portions of the β subunit that mechanically drove the eccentric cam-like strucure of the γ-rotor as it enters the $(\alpha\beta)_3$ structure as an off-center double-stranded helical coiled coil.

8.4.6.2 Restatement of the Critical Synthesis Step of ATP Synthase

In the eighth point of correlation of the hydrophobic elastic consilient mechanism given above, the maximal stage of apolar–polar repulsion occurred when the most polar occupancy state, ADP^{3-} Mg^{2+} plus HPO_4^{2-}, faced off against the most hydrophobic side of the γ-rotor to provide the thrust for a counterclockwise rotation. When the γ-rotor is driven in a clockwise direction by the F_0-motor of ATP synthase, the most hydrophobic face of the γ-rotor is driven into a face-off with the most polar occupancy state, ADP^{3-} Mg^{2+} plus HPO_4^{2-}, to result in a maximal apolar–polar repulsion that is relieved by the formation of ATP. This we believe is the mechanism whereby this pivotal protein-based machine of living organisms achieves its essential function of producing nearly 90% of biology's energy for performing the many functions required in the process of sustaining Life.

8.5 The Myosin II Motor of Muscle Contraction, a Representative ATPase

8.5.1 Hypothesis: Efficient Production of Motion by Muscle Contraction Derives from the Hydrophobic and Elastic Consilient Mechanisms, Whereby Dephosphorylation Results in Hydrophobic Association Coupled to Near-ideal Elastic Force Development

8.5.2 Coherence of Phenomena Between Contraction of Muscle and Contraction of Model Proteins Using the Inverse Temperature Transition

As reviewed in Chapter 7 with a focus on the issue of insolubility, extensive phenomenological correlations exist between muscle contraction and contraction by model proteins capable of inverse temperature transitions of hydrophobic association. As we proceed to examination of muscle contraction at the molecular level, a brief restatement of those correlations follows with observations of rigor at the gross anatomical level and with related physiological phenomena at the myofibril level. Each of the phenomena, seen in the elastic-contractile model proteins as an integral part of the comprehensive hydrophobic effect, reappear in the properties and behavior of muscle. More complete descriptions with references are given in Chapter 7, sections 7.2.2, and 7.2.3.

8.5.2.1 Rigor States Substantiate Hydrophobic Association in Muscle at the Gross Anatomical Level

As considered in Chapter 7, Dorland's *Medical Dictionary* has four entries for "rigor."[96] Alphabetically listed, they are (1) *acid rigor*, "coagulation of protein of muscle produced by acids," explained by the protonation of carboxylates to form the uncharged carboxyl, that is, $-COO^- + H^+ = -COOH$ with the result of contraction by hydrophobic association, (2) *calcium rigor*, "systolic cardiac arrest caused by an excess of calcium," whereby calcium ion pairing with paired carboxylates in the presence of hydrophobic residues also drives the contraction by hydrophobic association; (3) *heat rigor* "rigidity of muscles induced by heat," explained by the fundamental property of the consilient mechanism whereby raising the temperature drives contraction by hydrophobic association; and (4) *rigor mortis*, "the stiffening of a dead body, accompanying the depletion of adenosine triphosphate in the muscle fibers," whereby the suppleness of hydrophobic dissociation in the presence of the very polar ATP molecule disappears on breakdown of ATP following death that results in the stiffness of hydrophobic association.

8.5.2.2 Coherence of Phenomena Relating to Hydrophobic Association in the Myofibril and in Elastic-contractile Model Proteins

8.5.2.2.1 Thermal Activation of Muscle Contraction

Raising the temperature to drive contraction by hydrophobic association is the fundamental property of the consilient mechanism as demonstrated in Chapter 5 by means of designed elastic-contractile model proteins. Thermal activation of muscle contraction also correlates with contraction by hydrophobic association, but assisted in this case by the thermal instability of phosphoanhydride bonds associated with ATP, which on breakdown most dramatically drive hydrophobic association. In particular, both muscle and cross-linked elastic protein-based polymer, $(GVGVP)_n$ contract on raising the temperature over the same temperature range. Furthermore, there is the corollary of the release of heat on stretching both muscle and elastic-contractile model proteins. Both are exothermic on stretching due to exposure of hydrophobic groups to water, as may be argued from the original studies of Butler[10] in 1937.

8.5.2.2.2 Calcium Ion, pH, and Stretch Activation of Muscle Contraction

At the level of the myofibril, the addition of calcium ion, the lowering of pH, and stretching have each been shown to activate muscle contraction as well as to drive contraction of suitably designed elastic contractile protein-based polymers by hydrophobic association (see more extensive discussion in Chapter 7).

8.5.2.2.3 Dephosphorylation Drives Contraction Whereas ATP Binding Drives Relaxation

Because the energy for contraction comes from ATP, original expectations were that ATP and hydrolysis binding would cause contraction. This turned out not to be the case. Phosphorylation in its many forms causes relaxation by raising the temperature for the onset of an inverse temperature transition above physiological temperature, and dephosphorylation drives contraction exactly in parallel with the designed elastic-contractile model proteins. In the latter case, phosphorylation has been shown to cause relaxation by disrupting hydrophobic association, and dephosphorylation drives contraction by allowing re-establishment of hydrophobic association. The very same phenomenological correlations are discussed below at the molecular level for the myosin II motor.

8.5.2.2.4 Scenario of Muscle Contraction by the Inverse Temperature Transition of Hydrophobic Association

Whether at the anatomical level with the phenomenon of rigor or at the myofibril level with the variables of physiology, an extensive coherence of phenomena exists. Now we address the myosin II motor at the molecular level to determine if indeed muscle contraction may be explained in terms of the consilient mechanism of hydrophobic association under the control of a competition for hydration between polar (e.g., charged) and apolar (hydrophobic) groups. In this case the common functional groups capable of existing in different degrees of polarity would be ADP^{3-} Mg^{2+} HPO_4^{2-}; $ATP^{4-} \cdot Mg^{2+}$; HPO_4^{2-}; ADP^{3-} Mg^{2+}; $-COO^-$; $-COO^-$ Me^+; and $-COOH$, listed in approximate decreasing order of polarity.

8.5.3 Consideration of Muscle Contraction at the Molecular Level

8.5.3.1 Approach to the Molecular Level

In the context of relevance of the consilient mechanism to function of the myosin II motor, remarkable points are the location and orientation of ATP molecules bound to the cross-bridge and access to control hydrophobic associations/dissociations. As considered below in section 8.5.4.2, narrow clefts function as conduit through which forces arising from the polar phosphates are directed at a target site.

In this section on the myosin II motor, coherence of phenomena with that of the consilient mechanisms of energy conversion is addressed at the molecular level. Specifically, the importance of hydrophobic interactions is noted, as has been generally appreciated. More to the point, the presence of the apolar–polar repulsive free energy of hydration appears as a prominent factor in the contraction/relaxation cycle, and this has not been previously appreciated.

We begin with a brief orientation to the gross structural aspects of striated muscle and quickly move to the microscopic level where thick myosin and thin actin filaments are driven into greater overlap with each other during the cyclic process of myosin cross-bridge attachment to actin, contraction, and detachment from actin. Then, in a key surfacing of the consilient mechanism, we consider the molecular detail of ATP orientation in its binding site within the cross-bridge in relation to the means whereby ATP binding would bring about detachment of the cross-bridge from the actin binding site. Next, we consider geometric relationships of hydrophobic associations and dissociations relative to ATP binding, hydrolysis to form the most polar state of bound ADP plus P_i, phosphate release, and finally ADP release. An effort is made to integrate the role of the calcium ion trigger for muscle contraction in relation to changing hydrophobic associations attending the contraction/relaxation cycle. Finally, we consider the thermodynamic efficiency of the myosin II motor in relation to the development of elastic forces during contraction/relaxation and note the substantial energy requirement of the essential calcium ion trigger in the contraction/relaxation cycle.

8.5.3.2 Structure of Striated Muscle and the Sliding Filament Mechanism of Muscle Contraction

A striated muscle, such as the biceps, is comprised of bundles of muscle fibers. The fundamental unit of a muscle fiber is the myofibril composed of a series of repeating units called *sarcomeres* defined by the periodicity of Z lines (disks) at repeat distances of just over $2\mu m$ (e.g., $2.3\mu m$). The structural relationships proceeding from the anatomical level of the biceps to the microscopic level of the sarcomere are shown in Figures 8.43,[97] 8.44,[98] and 8.45.[97]

FIGURE 8.44. Drawing of a muscle fiber containing six myofibrils with each surrounded by the sarcoplasmic reticulum that releases calcium ion to trigger contraction and that pumps calcium ion back out to allow for the relaxation in preparation for the next contraction/relaxation cycle. Just inside the sarcolemma of the muscle fiber that surrounds the bundles of myofibrils are the mitochondria that supply the ATP required to convert the fiber from a contracted state to a relaxed but energized state in wait for the next release of calcium ion for triggering phosphate release for onset of the next contraction/relaxation cycle. Also shown is the location of the Z disk that separates sarcomeres (the fundamental unit of muscle contraction) and the arrangement of the A and I bands that are defined in detail in Figure 8.45.) (From *Fundamental of Biochemistry*, D. Voet, J. Voet & C.W. Pratt,[97] Copyright © 1999, John Wiley & Sons, New York. Copyright © 1999, John Wiley & Sons, New York. Reprinted with permission of John Wiley & Sons, Inc.)

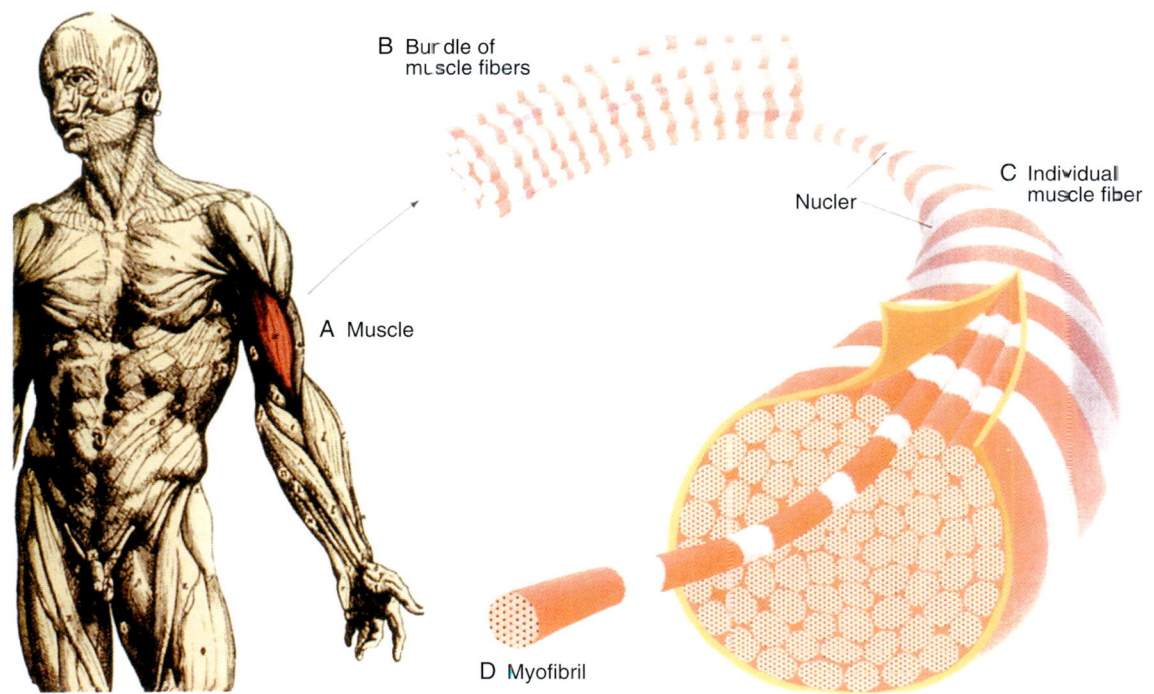

FIGURE 8.43. Musculature of a man that highlights the biceps muscle, perhaps best recognized for performing the mechanical work of lifting a weight, with a cutaway to an individual muscle fiber, presented as a bundle of myofibrils that contains the fundamental contractile element. (From *Fundamentals of Biochemistry*, D. Voet, J. Voet & C.W. Pratt,[97] Copyright © 1999, John Wiley & Sons, New York. Below: from *Biochemistry*, D. Voet and J. Voet,[98] Copyright © 1995, John Wiley & Sons, New York. Reprinted with permission of John Wiley & Sons, Inc.)

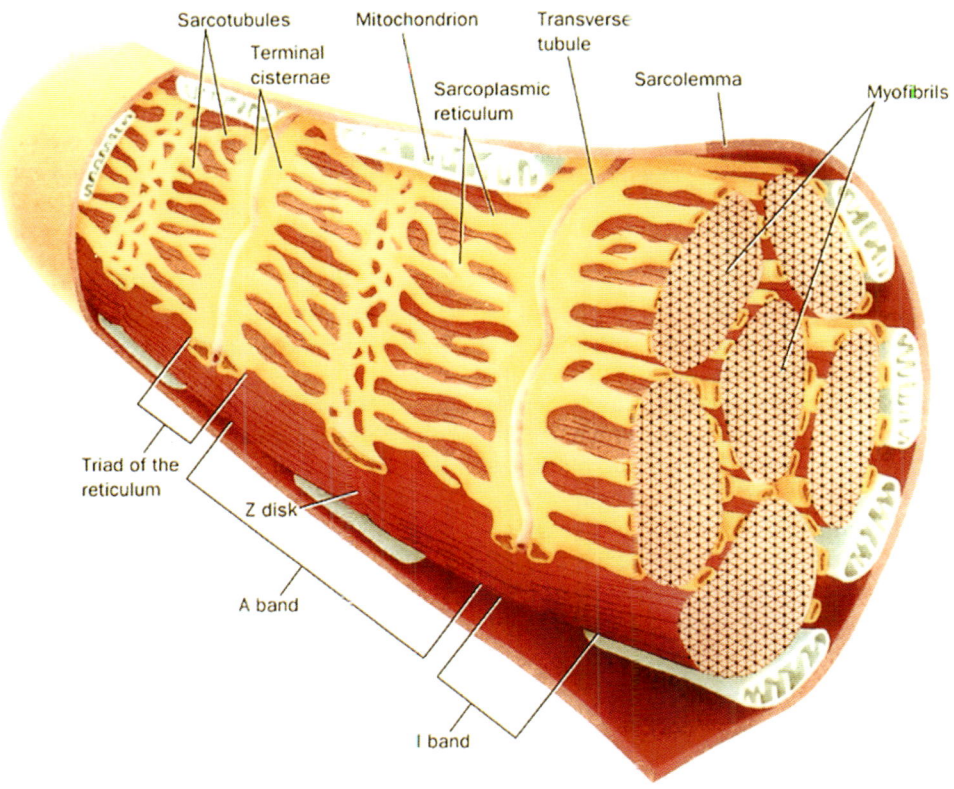

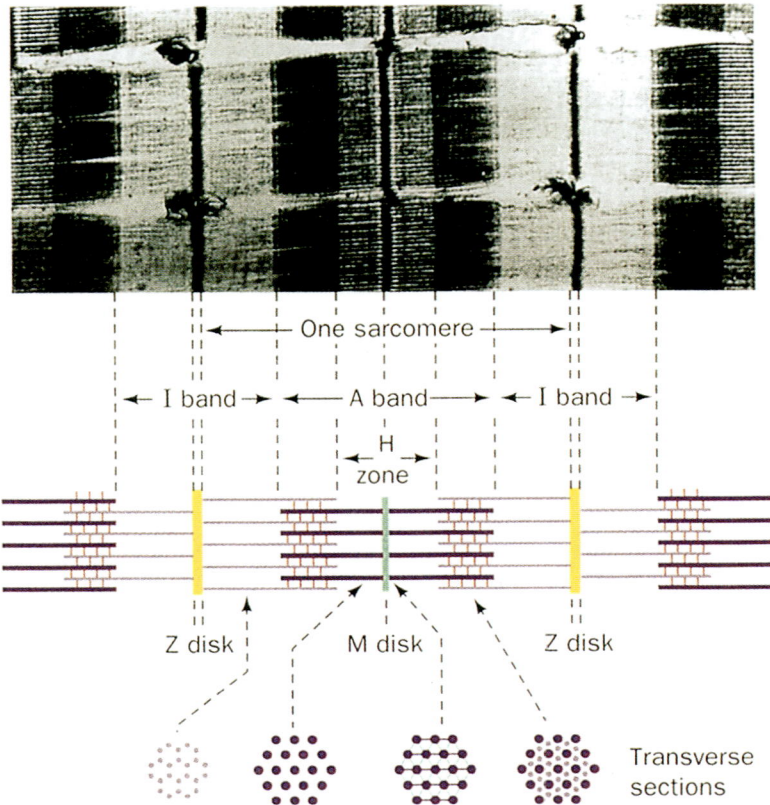

FIGURE 8.45. (Top) Electron micrograph of a myofibril delineating the A and I bands, the Z disk, and the H zone. (Bottom) Schematic representation of a microfibril showing thin actin filaments emanating from Z disks and overlapping with thick myosin filaments to form part of the A band and with myosin filaments occurring alone in the region of the H zone. Four transverse sections at indicated sites show packing of actin and myosin filaments. (Reproduced with permission from Voet et al.[97])

Emanating from the Z lines are thin actin filaments, and between the thin actin filaments, and generally overlapping only partially, are thick myosin filaments that contact the actin filaments by means of cross-bridges. Contraction results as the thick myosin filaments drive into greater overlap with the thin actin filaments due to the ATP-driven action of the cross-bridges.[99,100] This sliding filament by crossbridge action decreases the distance between Z lines and constitutes contraction by shortening of the muscle fibers. Preparations of the cross-bridges are variously referred to as the *myosin subfragment-1* (S1) and the *myosin head*.

8.5.3.3 Swinging Cross-bridge Representation of the Myosin II Contraction/Relaxation Cycle

A common perspective of the contraction/relaxation cycle comes from the text of Voet et al.,[101] as shown in Figure 8.46. An illustration at the top of Figure 8.46 shows the thick myosin filament between thin actin filaments with globular cross-bridges directed from the myosin filament toward the thin actin filaments. The sliding filament model has the cyclic attachment–contraction/detachment–relaxation action of the cross-bridges driving the thick

8.5 The Myosin II Motor of Muscle Contraction, a Representative ATPase

filament further into overlap with the thin filaments and resulting in the ends of the thin filaments moving toward each other. Again, this motion shortens the distance between Z lines that defines the fundamental unit of muscle contraction, the sarcomere, and results in the shortening of the muscle fiber that constitutes muscle contraction.

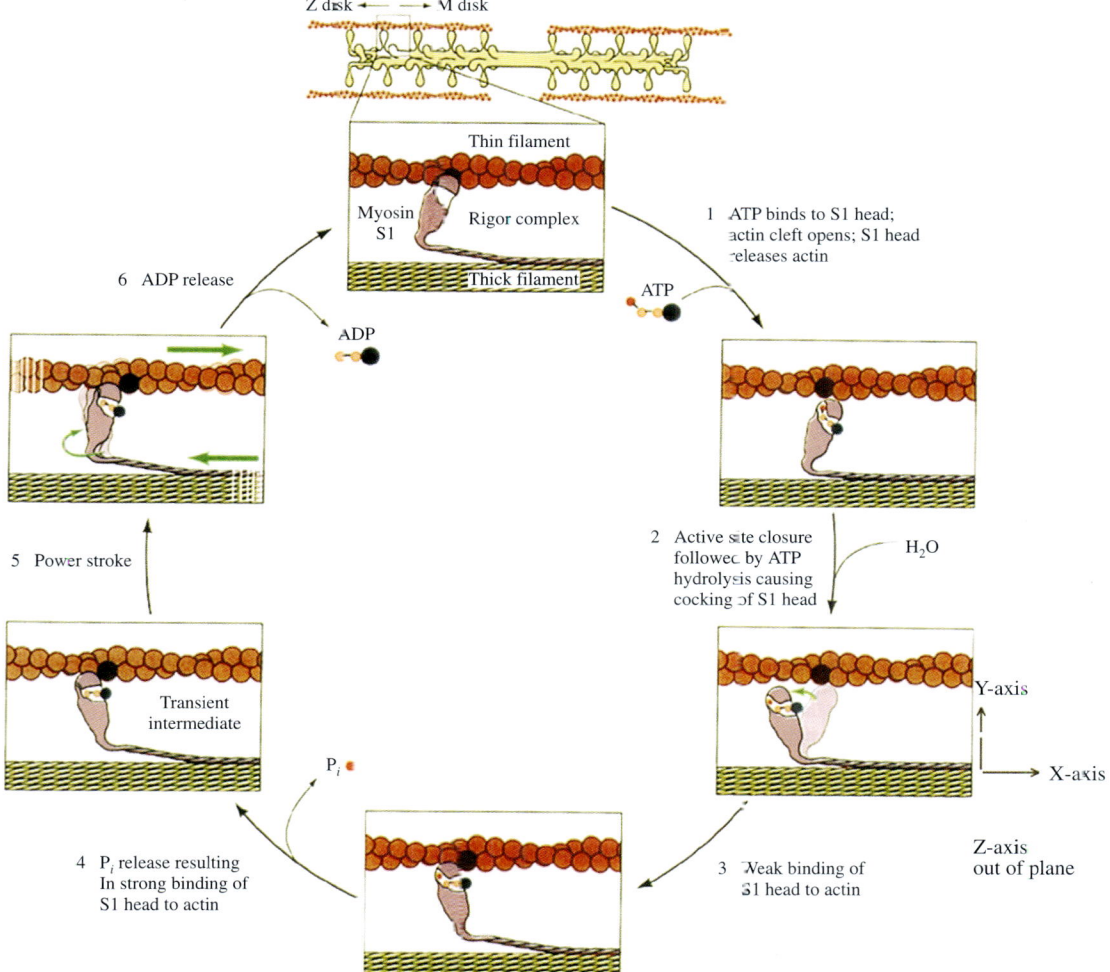

FIGURE 8.46. Contraction cycle in a swinging cross-bridge representation for the myosin II motor showing the cross-bridge performing a rowing-like motion. This rendering of the contraction/relaxation cycle begins with the cross-bridge attached to the actin filament. ATP binding induces detachment. Formation of ADP·P_i induces forward extension. P_i release yields firm re-attachment to the actin filament and results in the powerstroke for contraction. This is followed by ADP release in readiness for the next cycle to begin again with ATP binding. A coordinate system has been added at lower right in which the positive X-axis takes the direction of the myosin filament pointing away from the site where a cross-bridge emerges; the Y-axis approximates the direction of the α-helical lower segment of the lever arm in the direction of the actin filament, and the Z-axis is out of the plane. The cross-bridge structures examined in subsequent figures are viewed approximately with the three projections in the three planes, YZ, XY, and XZ, but in stereo projection in each case to give the third dimension. These three projections are given in Figures 8.47 through 8.52 using initially the space-filling representation followed by either backbone or ribbon representations. (From *Biochemistry*, D. Voet and J. Voet,[98] Copyright © (1995, John Wiley & Sons, New York). Reprinted with permission of John Wiley & Sons, Inc.)

In the expanded view of cross-bridge–actin interaction of Figure 8.46, step 1, the binding of ATP to the cross-bridge causes detachment of cross-bridge from the actin filament. Step 2 is the hydrolysis of bound ATP to yield bound ADP plus bound phosphate. This occurs with the forward movement of the cross-bridge (as in the moving forward of an oar) along the actin filament and with re-establishment of a weak interaction of the cross-bridge with actin as step 3. Release of phosphate as step 4 results in the combined strong binding of cross-bridge to actin immediately followed by the powerstroke. Step 5 represents the sliding of the thick filament further into the array of thin filaments. With release of ADP as step 6, the cross-bridge–actin interaction returns to the situation before ATP binding, thereby completing the cycle. In what follows, these steps will be considered in terms of detailed molecular interactions with emphasis on the consilient mechanism involving the apolar–polar repulsive free energy of interaction and its relationship to controlling hydrophobic hydration and hydrophobic association.

8.5.3.3.1 Analyses of Crystal Structures Below Do Not Support the Bending Motion of the Cross-bridge as Depicted in Figure 8.46

As shown below in Figures 8.47 through 8.52, which compare the different axial projections of the near-rigor and nucleotide bound states, there is no bending motion as illustrated in the results of steps 2 and 5. The contraction appears as a more complex motion involving a twisting within the lower part of the globular portion of the cross-bridge that changes the direction at which the single α-helix of the lever arm leaves the globular component. Rather than being strengthened by α-helical multistranding of the lever arm, it would appear that the essential and regulatory light chains fulfill such a re-enforcement role.

8.5.3.3.2 Absence in Figure 8.46 of Including Calcium Ion Release as the Event That Triggers Contraction

Another critical element in the contraction/relaxation cycle is to position the point of calcium ion release that provides the *in vivo* trigger for the contractile event. This will be considered in relation to details of an interplay of hydrophobic dissociations/associations that attend the contractile process.

8.5.4 Crystal Structure Analyses and Consilient Mechanisms in the Myosin II Motor

The crystal structure analyses utilize two crystal structures of the myosin II motor of scallop muscle available from the Protein Data Bank (http://www.rcsb.org/pdb) as Structure Files 1KK7 and 1KK8.[3] Structure 1KK7 is a near-rigor state with a sulfate at the active site, and structure 1KK8 is a nucleotide-containing state with ADP-BeF$_X$ (where BeF$_X$ is an analogue for the γ-phosphate) at the active site. The illustrations utilize FrontDoor to Protein Explorer 1.982 Beta (Copyright © 2002 by Eric Martz), which can be obtained at no cost at www.proteinexplorer.org. All structures are represented in shades of gray spanning from black for aromatic residues, gray for other hydrophobic residues, light gray for neutral residues, and white for charged residues, as fits with the delineation of polar (charged) as white and apolar varying from gray to black. In this way apolar regions can be distinguished at a glance from polar (charged) regions, with oil-like hydrophobic domains seen as dark regions and with polar domains seen as white regions. This allows immediate recognition of the location, size, and intensity of hydrophobic domains. The only exception to this coding scheme is in Figure 8.47, where the shades of gray are used to delineate the three different chains of the cross-bridge.

8.5.4.1 Structural Composition of the Myosin Cross-bridge

The structures used here contain that part of the amino terminus of myosin (residues 1 to 835) just sufficient to include the essential and regulatory light chains. Thus, the cross-bridge is composed of three chains—approximately 800 residues from the amino terminus of the myosin chain, the essential light chain, and the regula-

8.5 The Myosin II Motor of Muscle Contraction, a Representative ATPase

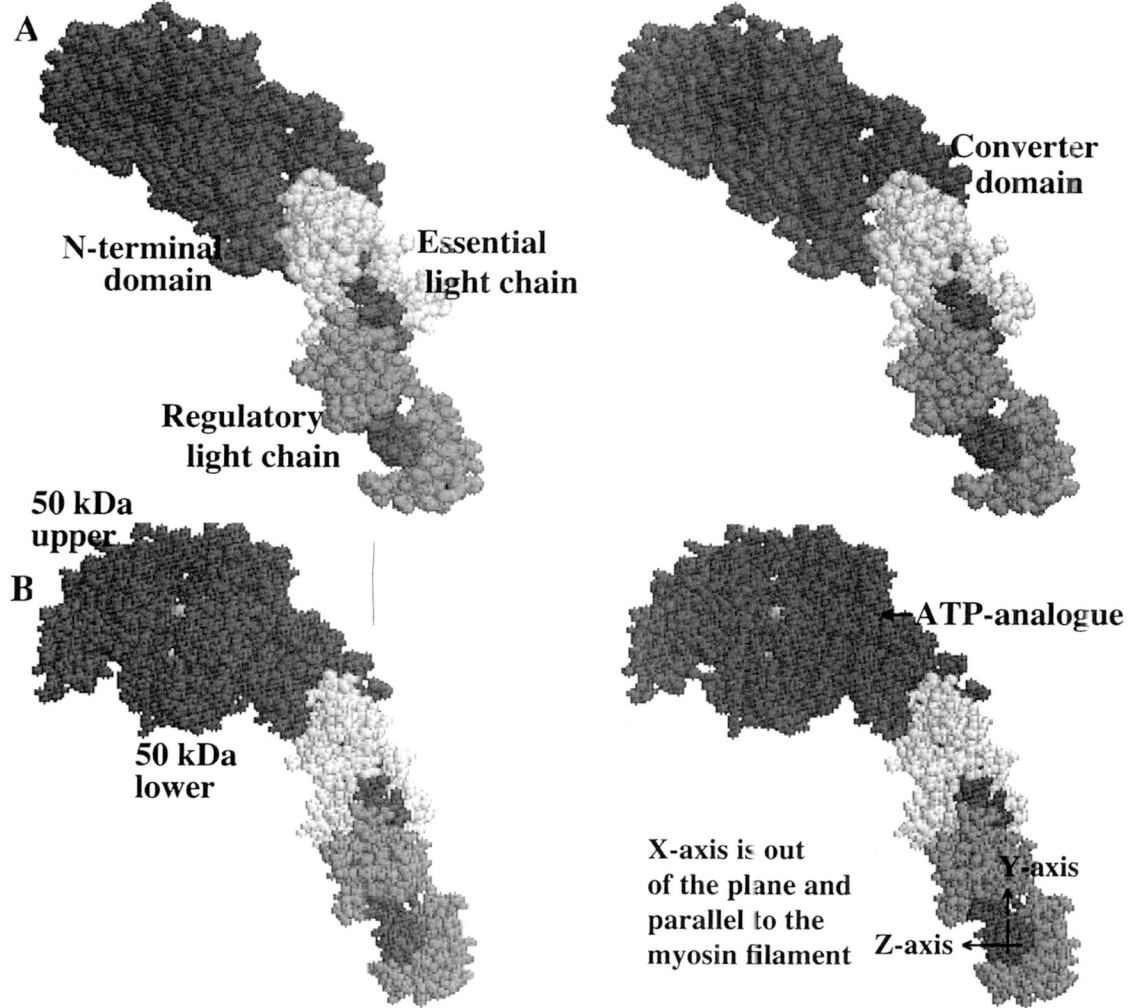

FIGURE 8.47. Stereo view (cross-eye) in space-filling representation of the scallop muscle cross-bridge (S1) in gray, with the essential light chain in white and the regulatory light chain in light gray. The coordinate system is chosen with the foot of the lever arm (obscured by the regulatory light chain) pointing in the direction of the myosin filament from which it arises. Accordingly, this is a projection on the YZ plane looking in the negative X-direction before the approximate 90° kink in the α-helical lever arm that connects the globular portion of the cross-bridge to the myosin filament. This alignment of the axes is better seen in Figures 8.48, 8.50, and 8.52. (A) Near-rigor state. (Prepared using the crystallographic results of Himmel et al.[3] as obtained from the Protein Data Bank, Structure File 1KK7.) (B) ATP analogue state (ADP-BeF$_X$). (Prepared using the crystallographic results of Himmel et al.[3] as obtained from the Protein Data Bank, Structure File 1KK8.)

tory light chain. The myosin sequence within the cross-bridge includes the lever arm that connects the globular portion of the cross-bridge to the myosin filament, which α-helical sequence is decorated at the filament end with the regulatory chain and at the globular end with the essential light chain. The globular portion of the cross-bridge is divided into a set of domains: the amino-terminal domain that we point to below as important in the hydrophobic association associated with the powerstroke, a 50 kDa segment divided into upper and lower domains separated by a cleft that runs from the nucleotide binding site to the actin binding site, and a converter domain that resides at the head of the lever arm that structurally ties to the lower domain of the 50 kDa segment.

8.5.4.2 Early Seminal Contributions of Rayment and Coworkers and Their Relationship to the Consilient Mechanisms

8.5.4.2.1 The Early Insights of Rayment and Coworkers

In 1993, Rayment and colleagues[102,103] published two seminal papers on the "Three-dimensional structure of the head portion of myosin, or subfragment-1, which contains both the actin and nucleotide-binding sites...." Thereby, these publications provided key structural aspects of the contraction/relaxation cycle. Shortly thereafter in a report of their structural studies of the myosin head from *Dictostelium* myosin II, Rayment and coworkers concluded that "The current structural results emphasize the importance of the narrow cleft that splits the 50-kDa segment in the molecular origin of myosin based motility. They suggest further that it functions not only in sensing the presence of the γ-phosphate of ATP but is also responsible for transducing the conformational change that results in the powerstroke."[4]

8.5.4.2.2 Extensions of the Early Insights of Rayment and Coworkers by Means of the Hydrophobic Consilient Mechanism

In what follows, we extend these insights in two significant ways. One way recognizes that the "narrow cleft" not only "splits the 50-kDa segment" in the direction of the binding site with actin, but also that the "narrow cleft" is directed toward the junction between the amino-terminal domain and the head of the lever arm to, in our view, effect "the conformational change that results in the powerstroke."

Our second extension is in the nature of the force emanating from the γ-phosphate in both directions. The force disrupts the hydrophobic association responsible for attachment to actin and the hydrophobic association between the amino-terminal domain and the head of the lever arm. That force given focus and direction by the "narrow cleft" derives from the apolar–polar repulsive free energy of hydration, ΔG_{ap}. ΔG_{ap} derives from a competition for hydration between hydrophobic and charged groups and as such constitutes a repulsive force that disrupts hydrophobic hydration and thereby disrupts hydrophobic association.

Hydrolysis of ATP to release the γ-phosphate as P_i that leaves the structure, therefore, has two consequences. Strong hydrophobic association is re-established between the cross-bridge and the actin binding site, and strong hydrophobic association is re-established between the amino-terminal domain and the head of the lever arm to provide the powerstroke. This perspective resides at the heart of the proposed contribution of the hydrophobic consilient mechanism to function of the myosin II motor. It is considered further below and most directly in section 8.5.4.7.

8.5.4.2.3 Efficient Energy Transduction Requires Coupling to Near-ideal Elastic Force Development

The above perspectives are natural consequences of both the hydrophobic and the elastic consilient mechanisms as applied to the structural data on the myosin II motor. Here we briefly explore the elastic element. An ideal elastomer exhibits exactly reversible stress–strain curves with complete recovery on relaxation of the energy of deformation. On the other hand, an elastomer that exhibits hysteresis does not recover all of the energy on relaxation that was expended on deformation. Accordingly, efficient muscle contraction should involve the deformation of near-ideal elastic segments to utilize more efficiently the energy expended in driving contraction. The mechanism of elasticity that can provide such near-ideal elasticity is the damping of internal chain dynamics on extension.

There are many aspects of the hydrophobic and elastic consilient mechanisms that warrant

8.5 The Myosin II Motor of Muscle Contraction, a Representative ATPase

discussion in relation to the mechanism of muscle contraction. Limitations of time and space, however, necessarily restrict consideration of many important aspects of muscle contraction that naturally flow from the insight of these consilient mechanisms. In what follows, we emphasize the role of the narrow cleft in terms of the presence and absence of nucleotide phosphates controlling hydrophobic associations/dissociations by means of the apolar–polar repulsive free energy of hydration, ΔG_{ap}. Specifically, the hydrophobic associations/dissociations involve the binding of the cross-bridge to the actin filament and the interaction of the amino-terminal domain with the head of the lever arm to achieve the powerstroke.

8.5.4.3 The Near-rigor and ATP (Analogue) Bound States of the Cross-bridge Compared in Three Planes Defined by Axes Set at the Myosin Filament

In setting up the coordinate system, the X-axis is taken parallel to the axis of the myosin filament; the Y-axis is taken perpendicular to the myosin filament in the direction of the actin filament to which the cross-bridge would attach, and the Z-axis is approximated by sighting through the center of the α-helical sequence of the lever arm involving residues 825 to 800 of scallop muscle.

8.5.4.3.1 View of Complete Cross-bridge in the YZ Plane Perpendicular to the Myosin Filament

A space-filling representation of the complete cross-bridge is shown in Figure 8.47, with the near-rigor state in **A** and the state containing the ATP analogue in **B**. The 50 kDa upper and lower domains and other domains of the myosin chain segment, for example, the amino-terminal domain and the converter domain, are in gray. Also shown and labeled are the essential light chain in white and the regulatory light chain in light gray. The globular head of the myosin chain, including in particular the amino-terminal domain, appears to be twisted in a clockwise direction as seen from the top.

The structural rearrangements appear more obvious in Figure 8.48, which is the same view given in backbone representation. The α-helical lever arm is shown in its entirety from foot to head with a knee bend at the junction of the essential and regulatory light chains and with an essentially unchanged orientation on going from the near-rigor state to an analogue representative of the ATP bound state. In this view the relocation of the amino-terminal domain is apparent, but becomes clearer in subsequent perspectives below in Figures 8.50 and 8.52.

8.5.4.3.2 View of the Cross-bridge in the XY Plane Demonstrates Absence of the Bending Motion at the Myosin Filament End of the Lever Arm

The cross-eye stereo view of the myosin cross-bridge of scallop muscle is given for the near-rigor state in Figure 8.49A and in the ATP analogue state in Figure 8.49B. In this perspective the amino-terminal domain is shown to have shifted from the left side of the head of the lever arm to the righthand side, while the essential light chain seems to have shifted very little. Again, the globular head of the myosin chain follows the same reorientation as the amino-terminal domain, exhibiting a clockwise rotation when seen from above.

Exactly the same perspectives of Figure 8.49 in space-filling representation are given in ribbon representation in Figure 8.50, which allows for a clearer view of any structural rearrangements that occur between near-rigor and ATP bound states. Again, it appears that the essential light chain changes little but that the amino-terminal domain, the leading edge of which is identified by residue G53, undergoes a large relocation on conversion to the near-rigor state. Although there may be a slight change in the bend at the knee of the α-helical lever arm, there is no detectable change in the approximately right angle turn on going from myosin filament segment to the lever arm. The bending motions indicated in steps 2 and 5 of Figure 8.46 do not occur.

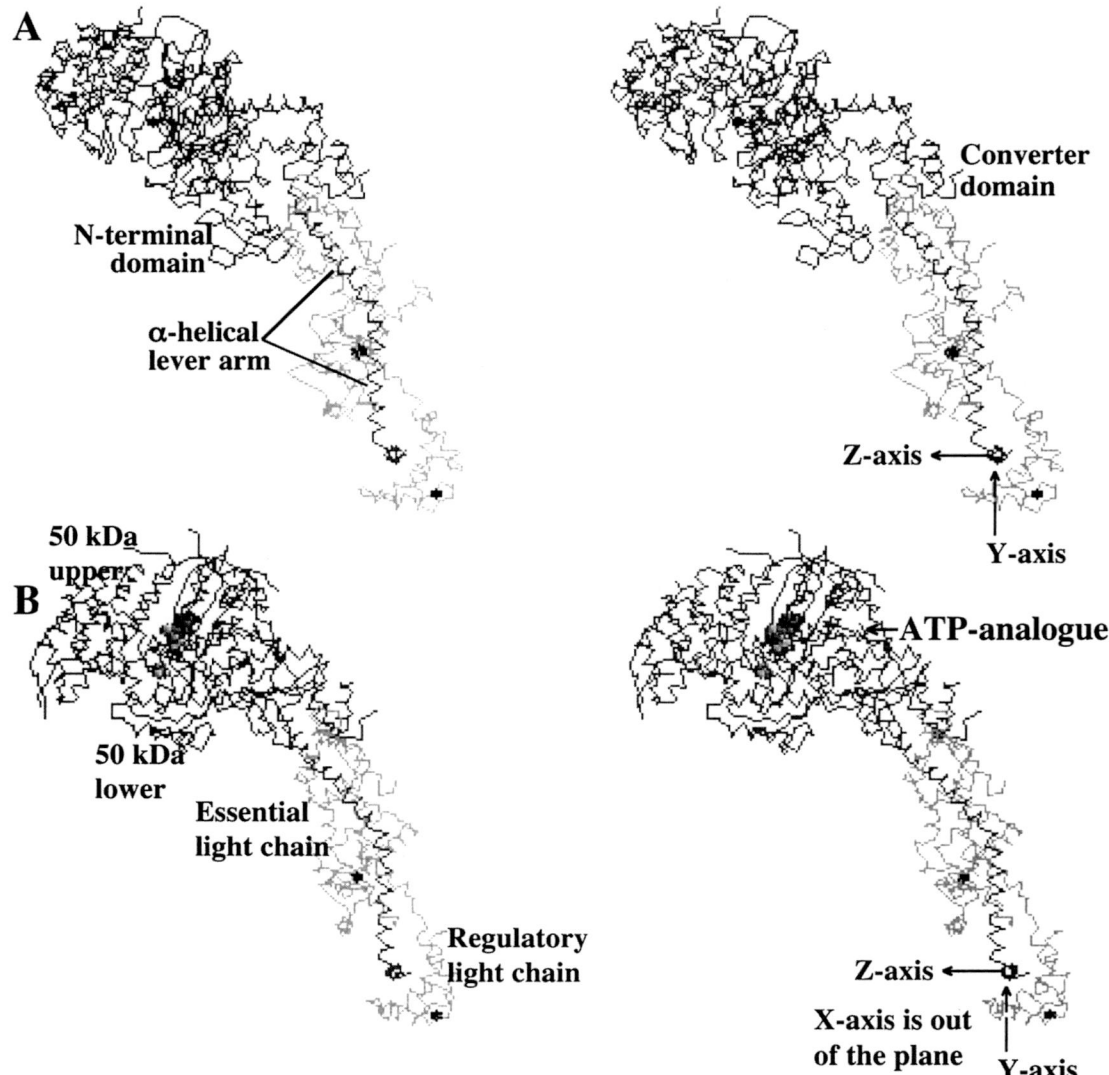

FIGURE 8.48. Stereo view (cross-eye) in backbone representation of scallop muscle cross-bridge (S1) in gray, with the essential and regulatory light chains in light gray. Projection on the YZ plane looking in the negative X-direction before the 90° kink, forming a foot-like structure where the α-helical lever arm connects to the myosin filament. The α-helical lever arm exhibits a knee-like bend at the function of the essential and regulatory light chains midway between a foot section and the head portion that nestles between the converter domain and the amino-terminal domain in A. **(A)** Near-rigor state that contains a single sulfate at the active site. (Prepared using the crystallographic results of Himmel et al.[3] as obtained from the Protein Data Bank, Structure File 1KK7.) **(B)** ATP state with ATP analogue (ADP-BeF$_X$) shown in space-filling representation. (Prepared using the crystallographic results of Himmel et al.[3] as obtained from the Protein Data Bank, Structure File 1KK8.)

8.5 The Myosin II Motor of Muscle Contraction, a Representative ATPase

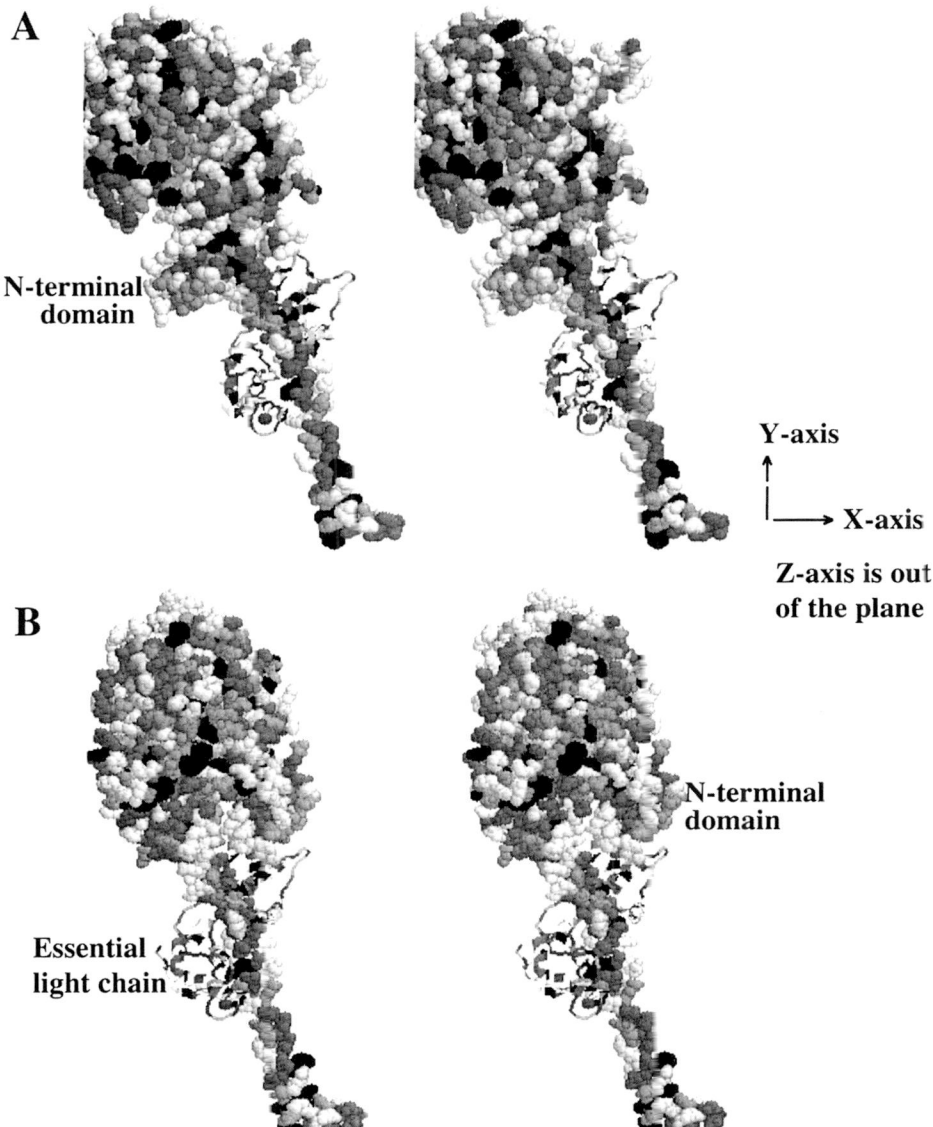

FIGURE 8.49. Stereo view (cross-eye) in space-filling representation of the scallop muscle cross-bridge (S1) with the essential light chain in ribbon representation, without the regulatory light chain shown and with neutral residues light gray, aromatic residues black, other hydrophobic residues gray, and charged residues white. The coordinate system places the foot of the lever arm pointing in the direction of the myosin filament from which it arises. This projection of the XY plane directly observes the near 90° kink as the α-helical lever arm connects the myosin filament to the globular portion of the cross-bridge. (A) Near-rigor state. (Prepared using the crystallographic results of Himmel et al.[3] as obtained from the Protein Data Bank, Structure File 1KK7.) (B) ATP analogue state (ADP-BeF$_X$). (Prepared using the crystallographic results of Himmel et al.[3] as obtained from the Protein Data Bank, Structure File 1KK8.)

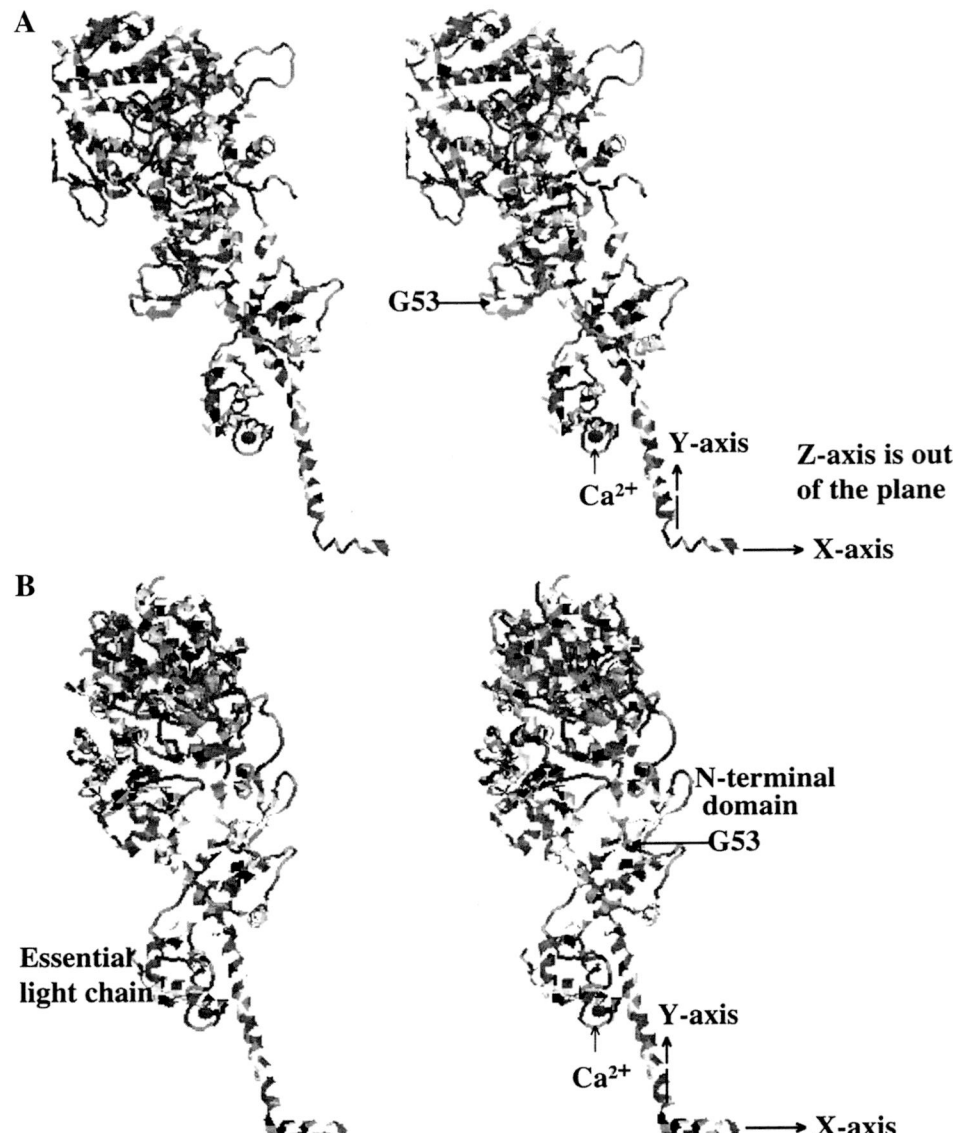

FIGURE 8.50. Same projection as in Figure 8.49 except that the cross-eye stereo view is of a ribbon representation of the scallop muscle cross-bridge (S1) with the essential light chain included but no regulatory light chain and with neutral residues light gray, aromatic residues black, other hydrophobic residues gray, and charged residues white. The coordinate system is chosen with the foot of the lever arm pointing in the direction of the myosin filament from which it arises. This projection at the XY plane directly shows, as unchanged between the two states, the 90° kink as the α-helical lever arm connects the myosin filament to the globular portion of the cross-bridge. See text for discussion. (A) Near-rigor state. (Prepared using the crystallographic results of Himmel et al.[3] as obtained from the Protein Data Bank, Structure File 1KK7.) (B) ATP state with the ATP analogue (ADP-BeF$_X$). (Prepared using the crystallographic results of Himmel et al.[3] as obtained from the Protein Data Bank, Structure File 1KK8.)

8.5 The Myosin II Motor of Muscle Contraction, a Representative ATPase

8.5.4.3.3 View of the Cross-bridge in the XZ Plane Demonstrates Major Change in Relationship Between the Head of the Lever Arm and the Amino-terminal Domain

In the space-filling representation in Figure 8.51, the perspective is essentially looking up the lever arm from the foot to its head. In this view of the projection in the XZ plane, the major discernible structural rearrangement involves the amino-terminal domain. The same perspective is given in backbone representation in Figure 8.52. To get the perspectives the same for both states, the Z-axis is sighted directly

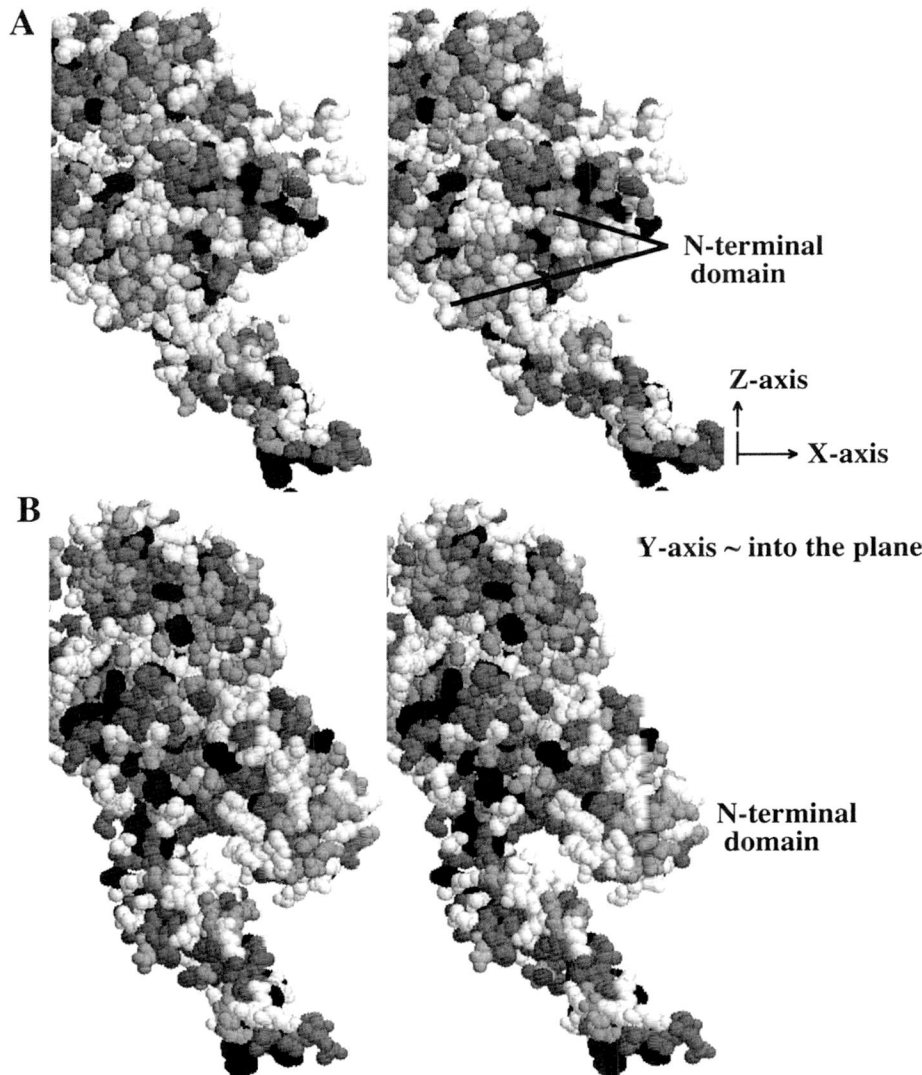

FIGURE 8.51. Stereo view in space-filling representation of scallop cross-bridge (S1) without light chains but with neutral residues in light gray, aromatics in black, other hydrophobics in gray, and charged residues in white. This perspective of the XZ plane, looking in the negative Z-direction, provides the best view of the movement of the amino-terminal domain that is positioned as a flap over the head of the lever arm in A and moves to be a free standing pedicle in B. The best view of this movement is shown in the same perspective but in backbone representation as in Figure 8.52. (**A**) Near-rigor state. (Prepared using the crystallographic results of Himmel et al.[3] as obtained from the Protein Data Bank, Structure File 1KK7.) (**B**) ATP analogue state (ADP-BeF$_x$). (Prepared using the crystallographic results of Himmel et al.[3] as obtained from the Protein Data Bank, Structure File 1KK8.)

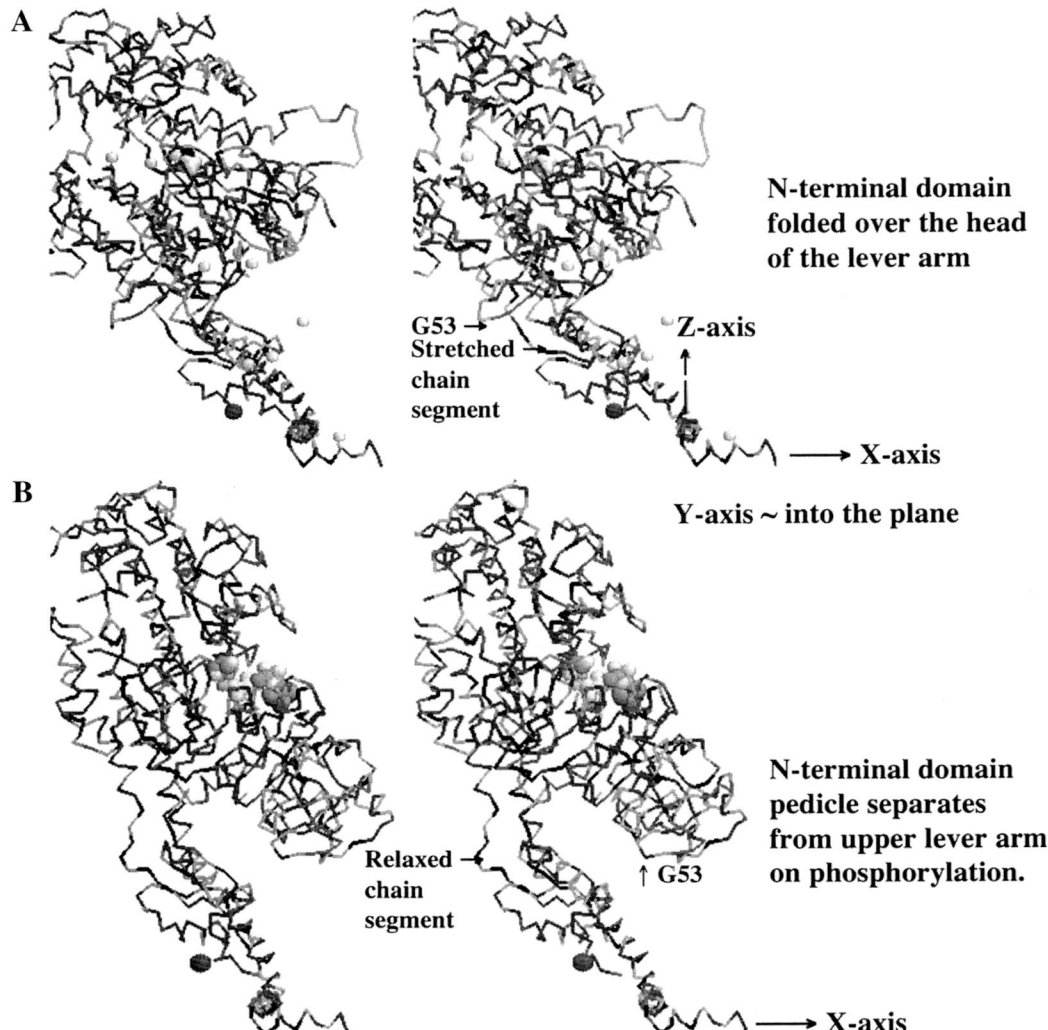

FIGURE 8.52. The same perspective for the stereo view (cross-eye) of scallop cross-bridge (S1) as shown in Figure 8.51, except that it is in backbone representation and thereby allows better delineation of domain movements. In A the amino-terminal domain is shown as a flap over the head of the lever arm in the near-rigor state, whereas it separates as a free-standing pedicle in B, the ATP state. The amino-terminal domain movement also becomes apparent by the changing G53 location at its leading edge. Another interesting feature in this backbone representation of the XZ plane perspective is the chain segment that appears stretched in A but relaxed in B. **(A)** Near-rigor state. (Prepared using the crystallographic results of Himmel et al.[3] as obtained from the Protein Data Bank, Structure File 1KK7.) **(B)** ATP analogue state (ADP-BeF$_X$). (Prepared using the crystallographic results of Himmel et al.[3] as obtained from the Protein Data Bank, Structure File 1KK8.)

along the α-helix of the lower part of the lever arm from residue 825 to residue 800. From this viewpoint the rotation of the 50 kDa upper domain is apparent from comparison of the reorientation of the major α-helix of the upper domain. This clockwise rotation of the 50 kDa upper domain, as seen from above, on going from the ATP state to the near-rigor state, appears to result from the amino-terminal domain movement. In what follows in Figures

8.53 and 8.54, the hydrophobic consilient mechanism provides an explanation of the physical basis for this movement.

Another feature of this perspective involves the apparent stretching of a chain segment, as labeled in Figure 8.52. The elastic force resulting from such an extension derives from the elastic consilient mechanism. As will be argued below, hydrophobic association gives rise to the development of a functional elastic force. Such a perspective may become a classic demonstration in protein-based machines, as it has been seen in Complex III of the electron transport chain and in the repulsive aspect of the comprehensive hydrophobic effect causing an elastic deformation during functioning of ATP synthase/F_1-ATPase.

8.5.4.4 Hydrophobic Association/Dissociation of the Head of the Lever Arm and the Amino-terminal Domain

Figure 8.53 shows stereo views of the scallop muscle cross-bridge, absent the light chains, that were oriented to better characterize the nature of the association (in the near-rigor state) and the dissociation (in the ATP bound state) involving the head of the lever arm and the amino-terminal domain. The stereo views help to show that the association in Figure 8.53A,C is hydrophobic, whereas the proximal surfaces of the head of the lever arm and amino-terminal domain in Figure 8.53B,D are dominated by charged residues.

Enlarged views of the relationship between the amino-terminal domain and the head of the lever arm, shown in stereo in Figure 8.53, are given in Figure 8.54 for the purpose of better seeing the interacting residues in the associated near-rigor state. On the left side of the figure the association is hydrophobic, with, as seen in the stereo view of Figures 8.53A, the lower charged part between the domains having twisted slightly out of co-planarity. On the other hand, the ATP analogue state is dissociated with a preponderance of charged (white) residues separating the two surfaces.

The explanation for the structural transition between the near-rigor and ATP bound states is given in section 8.5.4.7 (with the assistance of Figures 8.58 and 8.59). First, however, two issues require discussion before this pair of states can be considered models for producing the motion of muscle contraction. One issue involves a closer examination of the stretched/relaxed chain segment of Figure 8.52 in order to see if this chain segment can provide an example of elastic force development that results from hydrophobic association demonstrated in Figures 8.53 and 8.54. The pair of structures to be of greater significance should provide an understanding of force development, as required to explain isometric contraction and to understand energy-efficient production of motion. The second issue is how effectively the two structures demonstrate the displacement required for the sliding filament mechanism of muscle contraction.

8.5.4.5 Evidence for Elastic Force Development Resulting from Hydrophobic Association

8.5.4.5.1 Stretching the Relay Loop on Hydrophobic Association of the Head of the Lever Arm and the Amino-terminal Domain

The suggestion of a stretched chain segment is shown in Figure 8.52. An optimized perspective for considering this potential source of elastic force development is given in Figure 8.55. The estimate of length change uses the change of ratios of the relay loop segment I505–D511 to the relay helix segment Y500–F472 that accompanies the change of states at the nucleotide binding site. There appears to be an extension of as much as 40% accompanying the hydrophobic association resulting from loss of phosphate (or equivalent) from the nucleotide binding site.

8.5.4.5.2 Presence of a Number of Flexible Loops for Potential Storage of Elastic Force

By the elastic consilient mechanism the extension of single flexible loops causes an increase in the elastic force. It appears that one such loop has been identified in Figure 8.55. Rayment et al[104] note a number of flexible loops in the myosin cross-bridge. Each of these becomes a candidate for elastic force

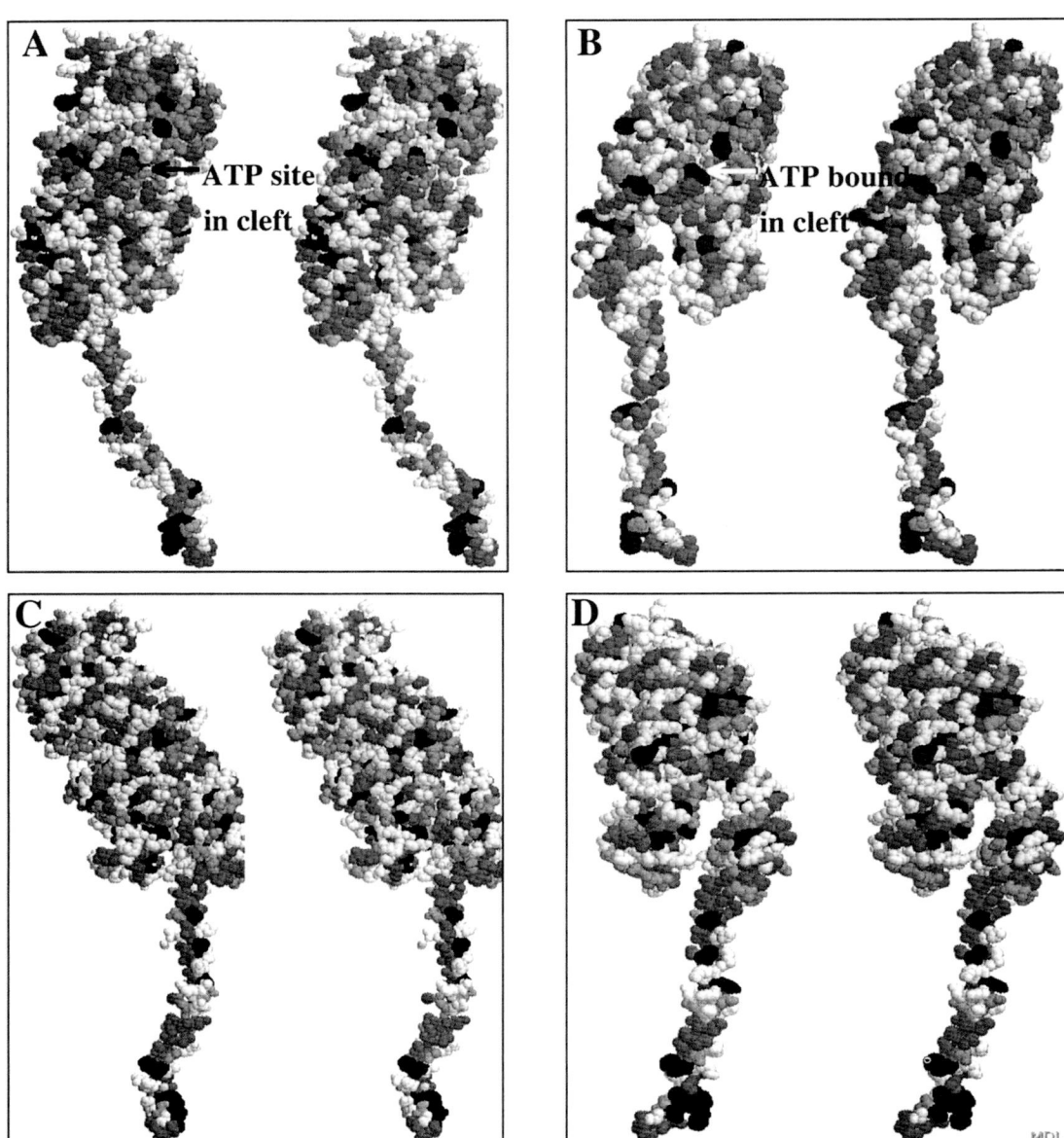

FIGURE 8.53. Cross-eye stereo views in space-filling representation of scallop muscle cross-bridge (S1), without essential and regulatory light chains but with cross-bridge oriented to better show the changing relationship between the lever arm and the amino-terminal domain. To facilitate observation of hydrophobic domains and their association, neutral residues are light gray, aromatic residues are black, other hydrophobic residues are gray, and charged residues are white. (A,C) Near-rigor state showing hydrophobic association between the amino-terminal domain and the upper part of the lever arm with the view taken approximately along the axis of the myosin filament. A shows the side of the nucleotide binding site referred to as the *front side*, whereas C shows the opposite (back) side. (Prepared using the crystallographic results of Himmel et al.[3] as obtained from the Protein Data Bank, Structure File 1KK7.) (B,D) ATP analogue state (ADP-BeF$_X$) showing hydrophobic dissociation of the amino-terminal domain and the upper part of lever arm. B shows the side of the nucleotide binding site referred to as the *front side*, whereas D shows the opposite (back) side. View now more nearly perpendicular to the myosin filament, that is, the lever arm rotates with an untwisting on hydrophobic dissociation. (Prepared using the crystallographic results of Himmel et al.[3] as obtained from the Protein Data Bank, Structure File 1KK8.)

8.5 The Myosin II Motor of Muscle Contraction, a Representative ATPase

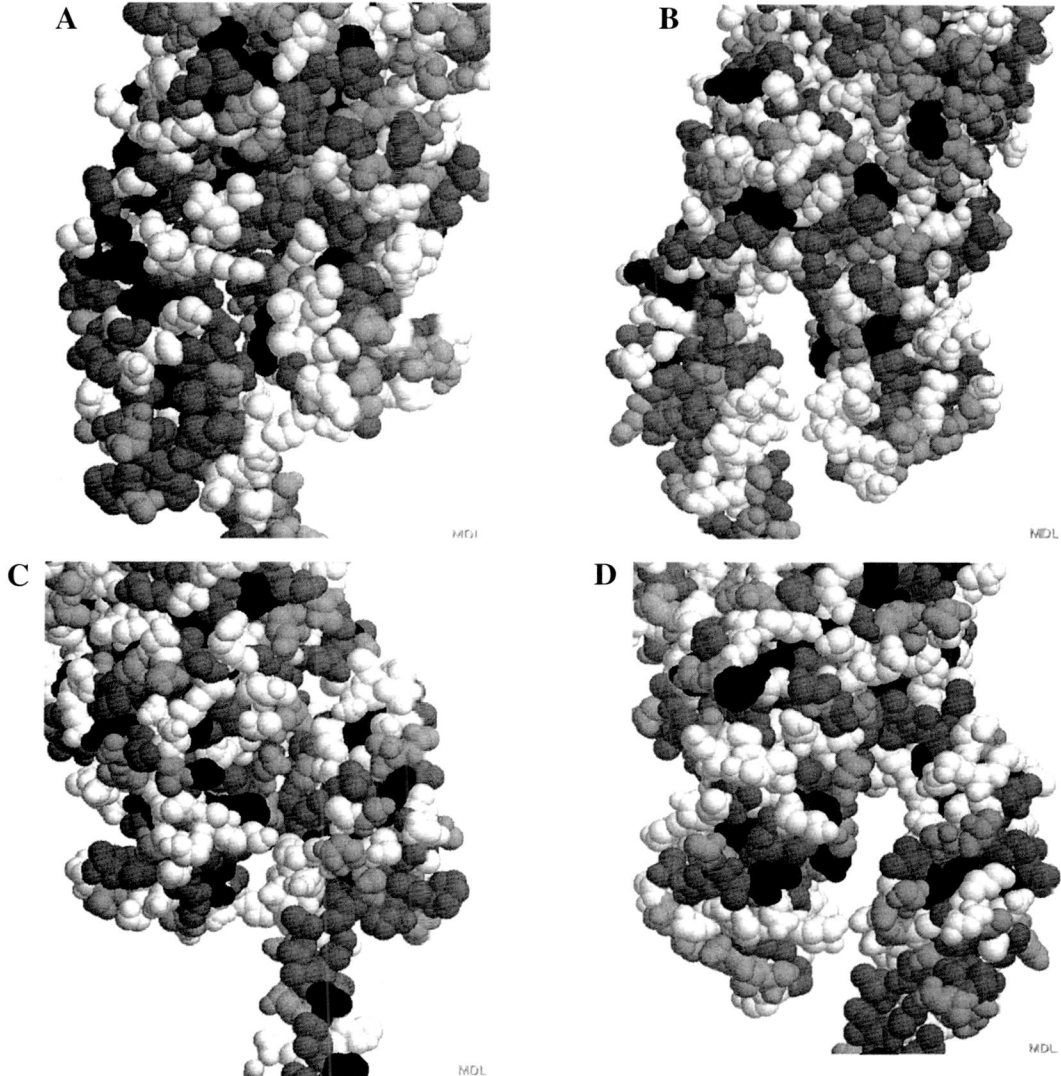

FIGURE 8.54. Scallop muscle cross-bridge (S1): Enlarged non-stereo views and orientations of those in Figure 8.53 to characterize the nature of the changing relationship between the amino-terminal domain and the head of the lever arm with charged residues in white, aromatic residues in black, other hydrophobic residues in gray, and neutral residues in light gray. **(A,C)** Near-rigor state with nucleotide binding site (front side) at the upper left and the opposite (back side) at the lower left. The shading of the associating residues demonstrates hydrophobic association. (Prepared using the crystallographic results of Himmel et al.[3] as obtained from the Protein Data Bank, Structure File 1KK7.) **(B,D)** The ATP analogue (ADP-BeF$_X$) bound state at the upper right (front side) and lower right (back side) showing hydrophobic dissociation due to the preponderance of charged residues separated by a wide aqueous corridor. (Prepared using the crystallographic results of Himmel et al.[3] as obtained from the Protein Data Bank, Structure File 1KK8.)

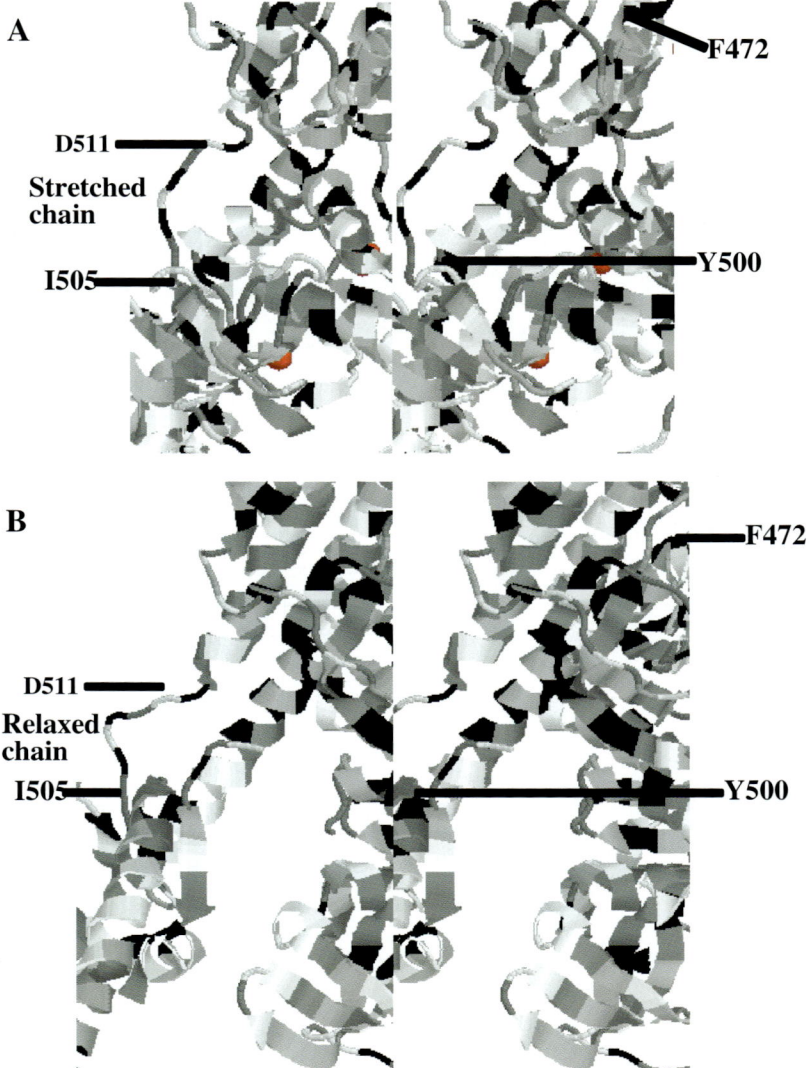

FIGURE 8.55. Expanded stereo view (cross-eye) of scallop muscle cross-bridge (S1) in ribbon representation to estimate length change of a portion of the relay loop. The ratio of the length of I505–D511 loop segment to that of the Y500–F474 segment of the relay helix, when comparing the two states, indicates as much as a 40% extension on hydrophobic association attending loss of phosphates. (A) Near-rigor state without nucleotide in the binding site. (Prepared using the crystallographic results of Himmel et al.[3] as obtained from the Protein Data Bank, Structure File 1KK7.) (B) ATP state using the ATP analogue ADP-BeF$_X$. (Prepared using the crystallographic results of Himmel et al.[3] as obtained from the Protein Data Bank, Structure File 1KK8.)

development on hydrophobic association attending dephosphorylation, as most readily evaluated during isometric contraction, that is, force development at fixed length. The increase in elastic force on extending a single chain, as determined from single-chain force-extension studies on elastic-contractile model protein, suggest that several chains would be stretched during the hydrophobic association of muscle contraction to achieve adequate force levels.

8.5.4.6 Use of Actin Binding Site on the Myosin Cross-bridge as a Reference to Evaluate Changes in Lever Arm Orientation

To consider a pair of states of the myosin cross-bridge (such as the near-rigor and ADP-BeF$_X$ states being analyzed above) as relevant to the mechanism of muscle contraction, there should be evidence of lever arm displacement. No significant displacement was in evidence in the perspectives utilized in Figures 8.47 through 8.53. The perspectives in Figures 8.56 and 8.57 use the actin binding site on the myosin cross-bridge as a reference for evaluating changes in orientation of the lever arm that would result from dephosphorylation.

The motion of the lever arm appears best described as a tucking under the amino-terminal domain due to hydrophobic association. The

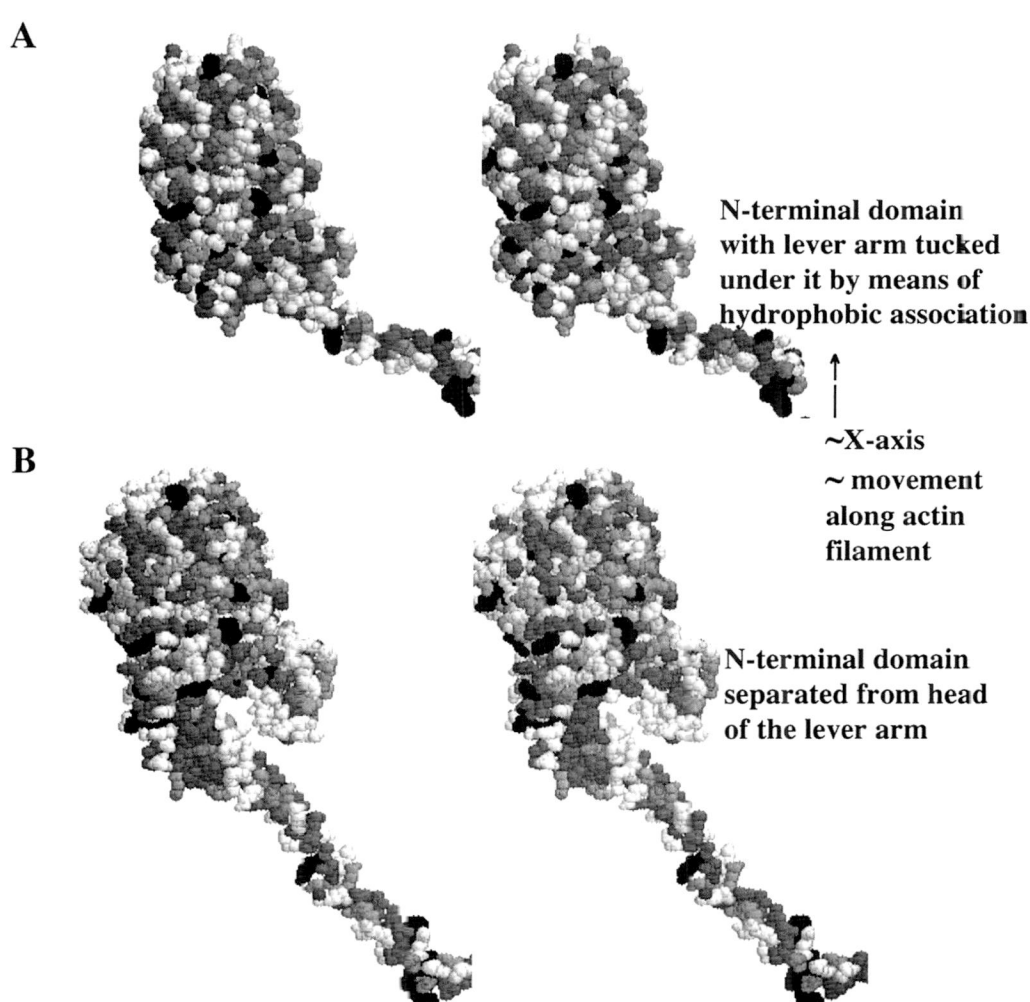

FIGURE 8.56. Stereo view in space-filling representation of a scallop cross-bridge (S1) view of the actin binding site on myosin cross-bridge, with continuity of the cleft from the ADP-BeF$_x$ binding site to the actin binding site that dissociates on ATP binding (A) Near-rigor state without nucleotide in the binding site. (Prepared using the crystallographic results of Himmel et al.[3] as obtained from the Protein Data Bank, Structure File 1KK7.) (B) ATP state using the ATP analogue ADP-BeF$_X$. (Prepared using the crystallographic results of Himmel et al.[3] as obtained from the Protein Data Bank, Structure File 1KK8.)

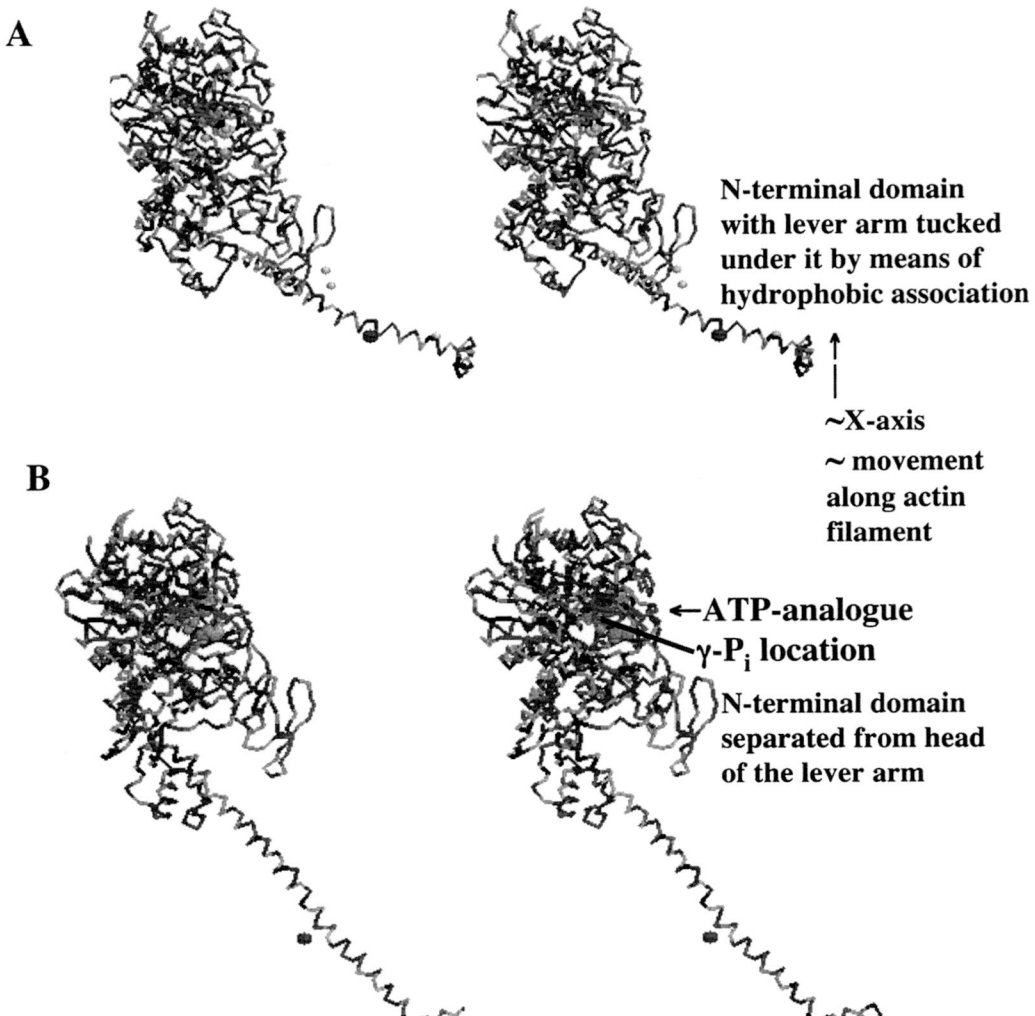

FIGURE 8.57. Stereo view of scallop muscle cross-bridge (S1) in backbone representation from the perspective of the actin binding site on the myosin cross-bridge. **(A)** Near-rigor state without nucleotide in the binding site. (Prepared using the crystallographic results of Himmel et al.[3] as obtained from the Protein Data Bank, Structure File 1KK7.) **(B)** ATP state using the ATP analogue ADP-BeF$_X$. (Prepared using the crystallographic results of Himmel et al.[3] as obtained from the Protein Data Bank, Structure File 1KK8.) On going to the near-rigor state, the amino-terminal domain moves slightly forward as the lever arm rotates about 90° and tucks its head under the amino-terminal domain by means of a hydrophobic association to provide for the major motion of the powerstroke. On ATP binding this hydrophobic association is lost due to apolar–polar repulsion from phosphate in the active site.

hydrophobic association would be most effectively disrupted by the ADP-P$_i$ occupancy state of the ATP binding site, and the hydrophobic association would most effectively reform on dephosphorylation. Importantly, the displacement of the lever arm is as required for movement along the actin filament. Also, the tucking of the head of the lever arm accompanied by a twist would stretch the relay loop of Figure 8.55 in the associated development of an elastic force. These are as required for achieving the motion of muscle contraction. What follows immediately below is a clarification of the force that would bring about the observed changes.

8.5 The Myosin II Motor of Muscle Contraction, a Representative ATPase

8.5.4.7 Role of the Narrow Cleft Emanating in Two Directions from the ATP Binding Site to the Actin Binding Site and to the Origin of the Powerstroke

As noted in section 8.5.4.2 and as foreseen by Rayment and coworkers,[102,103] the "narrow cleft" that "splits the 50-kDa segment" plays a central role in the function of the myosin II motor. In terms of the hydrophobic consilient mechanism, the narrow cleft becomes the structural conduit through which the force arising from the increased thirst for hydration experienced on hydrolysis of ATP to ADP plus P_i is directed to its principal sites of action. Stereo views that locate the nucleotide binding site at the base of, and at a turn of, a cleft are shown in Figure 8.58 for the near-rigor-like state in A

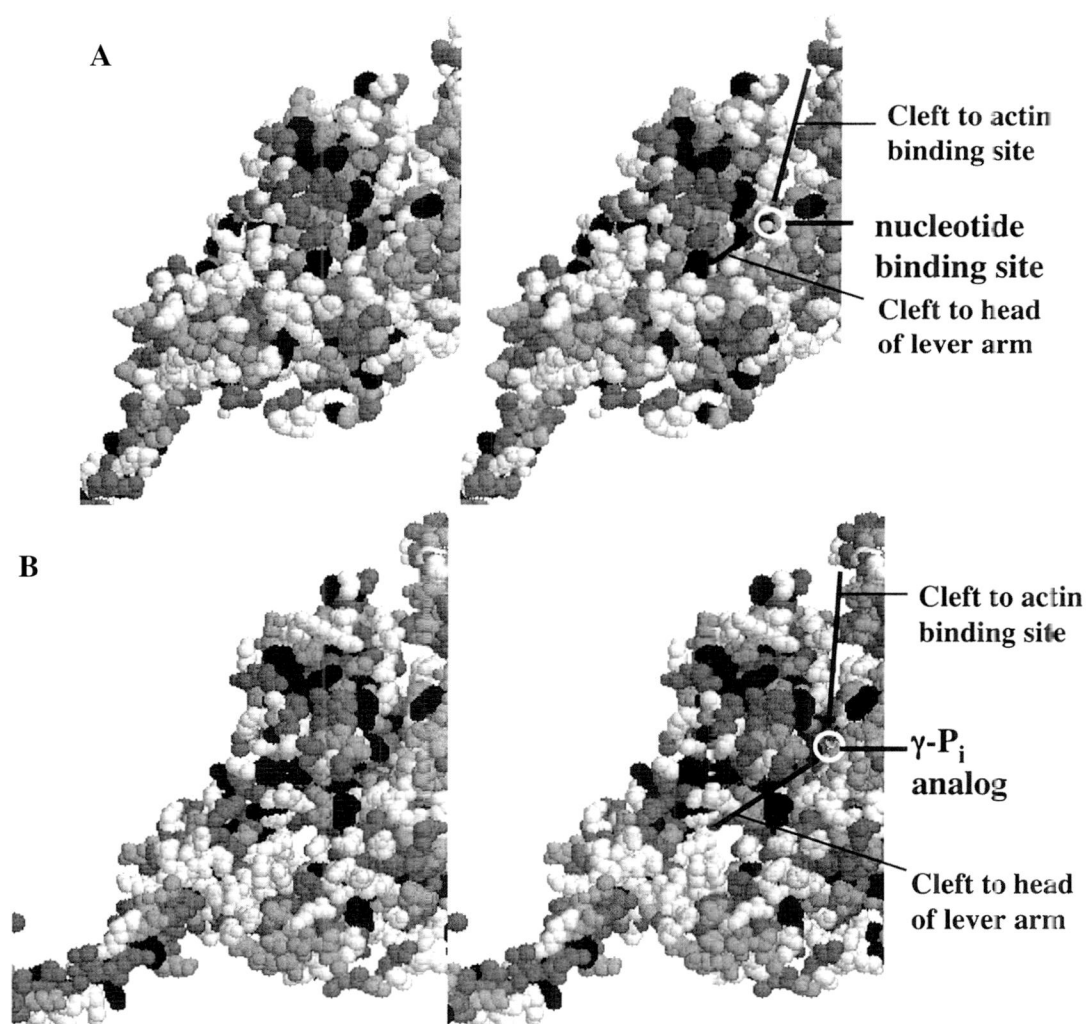

FIGURE 8.58. Stereo view in space-filling representation of scallop cross-bridge (S1) showing narrow clefts emanating in two directions from the nucleotide binding site, located by the white circle. One cleft runs upward to the actin binding site, and the second direction is toward the hydrophobic association between the amino-terminal domain and the head of the lever arm in A to effect the hydrophobic dissociation of head from lever arm as shown in B. (A) Near-rigor state. (Prepared using the crystallographic results of Himmel et al.[3] as obtained from the Protein Data Bank, Structure File 1KK7.) (B) ATP state due to ATP analogue ADP-BeF$_X$. (Prepared using the crystallographic results of Himmel et al.[3] as obtained from the Protein Data Bank, Structure File 1KK8.)

and for the state resulting from the ATP analogue, ADP-BeF$_X$, in **B**. As shown in **B**, the phosphate analogue is actually closer to the opened chasm between the head of the lever arm and the amino-terminal domain than to the actin binding site, and it is clearly located at a turn in the narrow cleft in order to direct its thirst for hydration in two directions, that is, to direct apolar–polar repulsion toward both hydrophobic association/dissociation sites.

As argued in Chapter 5 on the basis of extensive experimental data on elastic-contractile model proteins, on exposure to water hydrophobic groups hydrate in an exothermic reaction until too much hydrophobic hydration forms and solubility is lost. Charged groups, of course, dissolve in water due to the decrease in free energy derived from hydration on passing from the ionic interactions in a salt. In circumstances where hydrophobic groups and charged groups are forced by structure to compete for limited water, charged groups recruit emergent hydrophobic hydration during transient hydrophobic dissociation so that the driving force for hydrophobic association is lost.

The hydrophobic consilient mechanism describes this competition for hydration between hydrophobic and charged groups as an apolar–polar repulsive free energy of hydration, ΔG_{ap}, because, when motion is available to these groups, they reach out for water unperturbed by the other in order to lower their respective free energies of hydration. Thus, the competition for water is seen as a repulsion between charged and hydrophobic groups, as these groups attempt to distance themselves from each other in order to access water unperturbed by the other.

The occupancy state for a nucleotide binding site with the greatest thirst for water, and hence with the greatest repulsion between charged and hydrophobic groups, would be ADP^{3-} Mg^{2+} HPO_4^{2-}. Formation of this occupancy state would most dramatically drive the structural transition observed on going from the near-rigor state to the ATP analogue state that is represented in the preceding series of figures. The most effective way to direct the thirst for hydration is by means of the narrow clefts noted in Figures 8.58 and 8.59. Thus, we believe ΔG_{ap} to be the decisive force in the function of the myosin II motor.

8.5.5 The Hydrophobic Consilient Mechanism's Integration of the Calcium Ion Trigger into the Contraction/Relaxation Cycle

8.5.5.1 Putting the Near-rigor and ATP Analogue States of Scallop Muscle in Perspective

8.5.5.1.1 Correlation of the Consilient Mechanisms with Differences Between the Near-rigor and ATP Analogue States of Scallop Muscle

In Figures 8.47 through 8.58, two states of the myosin II motor are compared using specific reference conditions. In our view, these two states differ most dramatically by their extents of hydrophobic association. Clearly, the state with less polar occupancy of the nucleotide binding site favors greater hydrophobic association, and the state with more polar occupancy of the nucleotide binding site favors hydrophobic dissociation. The differences between the two states become explicable in terms of the hydrophobic and elastic consilient mechanisms, and the difference between the two states provides for displacement of the sort required for the motion of muscle contraction.

8.5.5.1.2 The Two States Truly Represent Neither the Desired Least Polar, Unoccupied Binding Site Nor the Most Polar State with ADP^{3-} Mg^{2+} HPO_4^{2-} in the Nucleotide Binding Site

The near-rigor state contains the polar magnesium sulfate salt, SO_4^{2-} Mg^{2+}, and, even though the charges of the magnesium sulfate ion pair sum to a net zero charge, as regards thirst for hydration, the polarity of the sulfate would be reduced to no more than one-half. Because of this, the analyzed near-rigor state should not be assumed to exhibit the maximal hydrophobic association of an unoccupied nucleotide binding site. Also, the ATP analogue state achieved on occupancy by ADP^{3-} Mg^{2+} BeF_3^{1-} would not be as polar as the state with ADP^{3-}

8.5 The Myosin II Motor of Muscle Contraction, a Representative ATPase

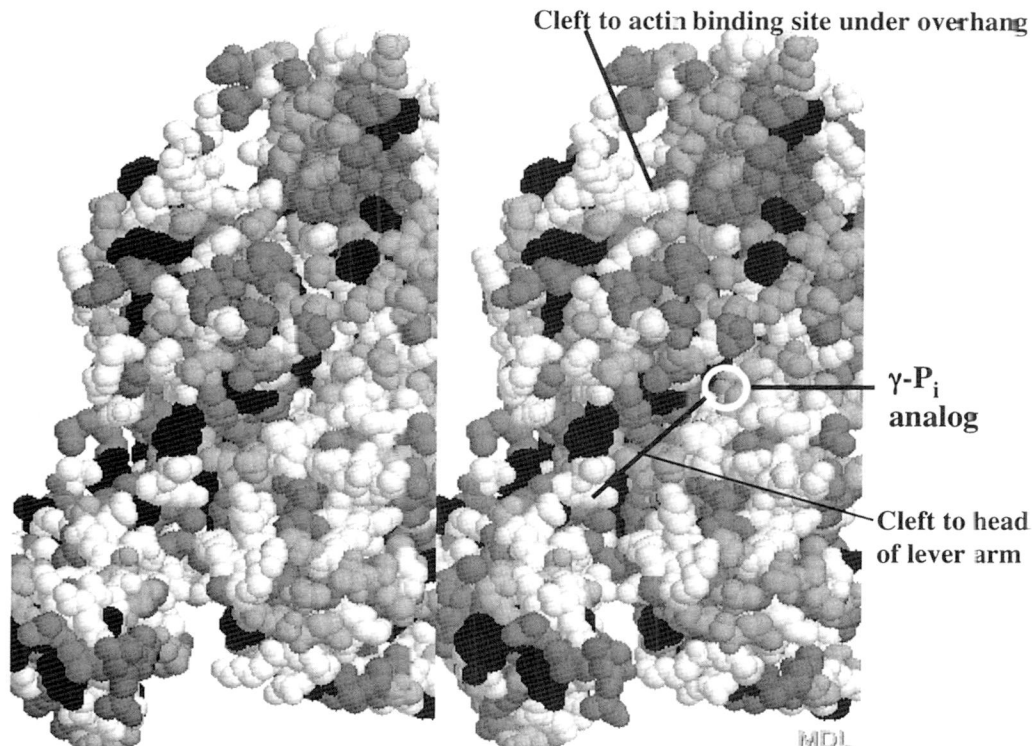

FIGURE 8.59. Stereo view in space filling representation of scallop muscle cross-bridge (S1) looking along the lever arm (lower left) from the foot to the nucleotide binding site that contains the ATP analogue ADP-BeF$_X$ and showing continuity of narrow clefts from the binding site in both directions—to the actin binding site from center to upward left and to effect hydrophobic dissociation of the head of lever arm from under the amino-terminal domain. (Prepared using the crystallographic results of Himmel et al.[3] as obtained from the Protein Data Bank, Structure File 1KK8.)

Mg^{2+} HPO_4^{2-} occupying the nucleotide binding site. Accordingly, this ATP analogue state may not convert the cross-bridge intramolecularly to a completely hydrophobically dissociated state. Thus, one should not be surprised to find states at even more extremes of hydrophobic dissociation and of hydrophobic association occurring in the myosin II motor. Nonetheless, this important work of Himmel et al.[3] demonstrates changes that, in our view, remarkably substantiate the hydrophobic elastic consilient mechanism. On the basis of the hydrophobic consilient mechanism and especially its component apolar–polar repulsive free energy of hydration, ΔG_{ap}, these two structures provide the opportunity with which to introduce the calcium ion trigger into the contraction/relaxation cycle.

8.5.5.2 Hydrophobic Consilient Mechanism Whereby Calcium Ion Can Trigger the Powerstroke

8.5.5.2.1 Calcium Ion Binding Displaces Tropomyosin to Open the Tight Binding Site on Actin for Hydrophobic Association with the Myosin Cross-bridge

The excitatory nerve impulse stimulates release of calcium from the sarcoplasmic reticulum. The calcium ion then binds to carboxylates of troponin C in an otherwise hydrophobic region located on the actin filament. This creates a new site for hydrophobic association that brings about the relocation of tropomyosin from its actin site of hydrophobic association to a new hydrophobic site associated with the actin filament. By relocation of tropomyosin, the tight binding site opens up on actin for hydrophobic association of the myosin cross-bridge. The association is expected to involve particularly complimentary actin and myosin surfaces with expression of high hydrophobicity due to reduction of polarity by well-positioned ion pairing between surfaces involving negative carboxylates and positively charged amino acid side chains, as discussed in relation to Figure 8.2.

8.5.5.2.2. Proximity of Hydrophobic Binding Site on Actin to the Nucleotide Binding Site of the Myosin Head Displaces Phosphate by Apolar–Polar Repulsion

This increase in hydrophobicity presented by the actin component of the complementary myosin-actin binding site directs an apolar–polar repulsion through the conduit of the narrow cleft to assist in expelling the phosphate. More simply stated, cross-bridge binding to the tight site on actin limits access of ADP plus P_i to hydration, and thereby raises the free energy of ADP plus P_i at the nucleotide-binding site with the result of the release of P_i.

8.5.5.2.3 Expulsion of Phosphate Sets Off Hydrophobic Association of the Head of the Lever Arm with the Amino-terminal Domain to Give the Powerstroke

The expulsion of phosphate removes the apolar–polar repulsion holding open the aqueous chasm between the head of the lever arm and the amino-terminal domain to result in the powerstroke. Then ADP releases more slowly, because ADP is repulsed to a lesser extent than phosphate and because ADP does not fit the nucleotide binding site as well as it fits ATP.

8.5.5.3 *ATP Binding and the State of ATP in Detached Cross-bridges*

8.5.5.3.1 ATP Binding Releases Cross-bridge Attachment to the Actin Filament

With the empty nucleotide binding site being exactly configured for occupancy by the ATP molecule,[104] ATP occupies the nucleotide binding site with a greater affinity than ADP. In doing so, the polar ATP molecule introduces an apolar–polar repulsion directed at the actin binding site and thereby effects detachment of the cross-bridge from the actin filament. Now, further insight into the contraction/relaxation cycle comes from knowing the enzymatic activity within the nucleotide binding site of the detached state.

8.5.5.3.2 The State of the Nucleotide in the Detached Cross-bridge

First we note that as few as 2% of the cross-bridges are attached to the actin filament at any one time.[105] This being the case, it becomes central to the mechanism to know the situation within the nucleotide binding site of the 98% of cross-bridges that are detached.[106] As described over three decades ago by Lymn and Taylor,[107] "actin binds with the myosin ADP P complex and displaces products of the hydrolysis reaction. It is concluded that this step is responsible for activation of myosin ATPase by actin." With 98% of the cross-bridges detached and with the rapid conversion of ATP to ADP + P_i (often noted as ADP P),[108] the most polar state of the cross-bridge would be the one readied for hydrophobic association with the complementary tropomyosin-vacated, hydrophobic binding site on actin, which could provide an apolar-polar boost for repulsion of phosphate from the cross-bridge.

Thereby, the hydrophobic consilient mechanism integrates the calcium ion trigger into the contraction/relaxation of the myosin II motor as well as provides the physical basis for the powerstroke.

References

1. E.O. Wilson, *Consilience: The Unity of Knowledge.* Alfred E. Knopf, New York, 1998, p. 8.
2. C. Lange and C. Hunte, "Crystal structure of the yeast cytochrome bc$_1$ complex with its bound substrate cytochrome c." *Proc. Nat. Acad. Sci. USA,* **99**, 2800–2805, 2002.
3. D.M. Himmel, S. Gourinath, L Reshetnikova, Y. Shen, A.G. Szent-Gyorgyi, and C. Cohen, "Crystallographic Findings on the Internally Uncou-

pled and Near-rigor States of Myosin: Further Insights into the Mechanics of the Motor." *Proc. Natl. Acad. Sci. U.S.A.*, **99**, 12645–12650, 2002.
4. A.J. Fisher, C.A. Smith, J. Thoden, R. Smith, K. Sutoh, H.M. Holden, and I. Rayment, "Structural Studies of Myosin:Nucleotide Complexes: A Revised Model for the Molecular Basis of Muscle Contraction." *Biophys. J.*, **68**, 19s–28s, 1995.
5. J. Monod, "On Symmetry and Function in Biological Systems." In *Nobel Symposium 11: Symmetry and Function of Biological Systems at the Macromolecular Level*, A Engstrom and B. Strandberg, Eds., Almqvist & Wiksell Forlag AB, Stockholm. 1968, p. 1527. Also reprinted in *Selected Papers in Molecular Biology by Jacques Monod*, A. Lwoff and A. Ullmann, Eds., Academic Press, New York. 1978, p. 708.
6. M.F. Perutz, "Mechanisms of Cooperativity and Allosteric Regulation in Proteins." *Q. Rev. Biophys.*, **22**, 139–236, 1989.
7. M.F. Perutz, *Mechanisms of Cooperativity and Allosteric Regulation in Proteins*. Cambridge University Press, Cambridge, 1990, page ••.
8. D.W. Urry and T.M. Parker, "Mechanics of Elastin: Molecular Mechanism of Biological Elasticity and its Relevance to Contraction." *J. Muscle Res. Cell Motility*, **23**, 541–547, 2002.
9. J.T. Edsall, "Apparent Molal Heat Capacities of Amino Acids and Other Organic Compounds." *J. Am. Chem. Soc.*, **57**, 1506–1507, 1935.
10. J.A.V. Butler, "The Energy and Entropy of Hydration of Organic Compounds." *Trans. Faraday Soc.*, **33**, 229–238, 1937.
11. Two curiously complementary and impeding truisms face any newly proposed concept or outlook in science. They are demonstrable by two quotations. The first, attributed to Maurice Maeterlinck, indicates the struggle faced by those who would make new proposals, "At every crossway on the road that leads to the future, tradition has placed against each of us, 10,000 men to guard the past." The second suggests why it is so easy for those holding with tradition, "Keep in mind that new ideas are commonplace, and *almost always wrong*" (E.O. Wilson, *Consilience: The Unity of Knowledge*, 1998). Accordingly, we have the corollary that to stand against a new idea is to be *almost always right*, with neither understanding nor thought required. This provides an irresistible opportunity for the more pretentious, but less so in the realm of scientific inquiry than in the machinations of society at large.

Then there is another aspect, noted by Francis Bacon in *The New Organon*: "The human understanding is no dry light, but receives an infusion from the will and affections; whence proceed sciences which may be called 'sciences as one would wish.' For what a man had rather were true he readily believes.' Clearly, it is particularly difficult for anyone to allow even partial marginalization of a hard-won and once thoroughly empowering viewpoint. Of course, the basic test of any point of view remains the extent to which it leads to gains in further understanding of the laws of nature.

In fact, the most satisfying aspect of scientific pursuit is that there is a correct and unchanging truth and that progress is most assured by hewing to the path of truth as closely as humanly possible. As reported in part in Chapter 5, the sequential *de novo* design of one class of protein-based machines after another and the immediately successful test of each design for its intended new energy conversion ensures the fundamental correctness of the hydrophobic consilient mechanism, now most directly described in technical terms as the comprehensive hydrophobic effect and its operative component, the apolar–polar repulsive free energy of hydration. Within polymers changes in hydrophobic association often couple to changes in elastic force.

12. T_b and T_t are different experimental measurements of the onset temperature for the transition to hydrophobic association that have been referenced by extrapolation to a common state of $(GXGVP)_n$, as shown in Figure 5.9B and described in section 5.3.2, where X is any of the naturally occurring amino acid residues or a chemical modification thereof. As shown in Figure 5.1, T_t is the onset temperature for aggregation as followed by light scattering, and T_b is the onset temperature for the transition for hydrophobic association, as obtained from calorimetry data of the inverse temperature transition. The bold-faced quantities represent the extrapolated reference values whereas the non-bold-faced quantities, T_b and T_t, represent the experimentally measured quantity on the particular composition of protein-based polymer. For the basic homopolypentapeptide, $(GVGVP)_n$ the bold-faced and non-bold-faced quantities are the same.
13. These approximations derive from the finite width of the inverse temperature transition, that is, from the substantial temperature inter-

val of the cusp of insolubility in Figure 7.1. Should the width of the transition be one degree or less, then the equalities would essentially be exact.
14. M.V. Stackelberg and H.R. Müller, "Zur Struktur der Gashydrate." *Naturwissenschaften*, **38**, 456, 1951.
15. Personal communication from J. Carlos Rodriguez-Cabello, Dpto. Fisica de la Materia Condensada, E.T.S.I.I., Universidad de Valladolid, Spain, 2003.
16. J.C. Rodríguez-Cabello, J. Reguera, M. Alonso, T.M. Parker, D.T. McPherson, and D.W. Urry, "Endothermic and Exothermic Components of an Inverse Temperature Transition for Hydrophobic Association by TMDSC." *Chem. Phys. Lett.*, **388**, 127–131, 2004.
17. J.E. Lennard-Jones, "The Equation of State of Gases and Critical Phenomena." *Physica*, **4**, 941–956, 1937.
18. R.A. Buckingham, "The Classical Equation of State of Gaseous Helium, Neon and Argon." *Proc. R. Soc.* A, **168**, 264–283, 1938.
19. As noted in the opening paragraph of the Prologue and in the initial paragraphs of Chapter 3, *Thales of Miletus* launched scientific investigation in the sixth century BC with the inquiry "What is the world made of?" His simple answer of "water" gave rise to subsequent derision. As his reasoning seems to have been "perhaps from seeing that the nutriment of all things is moist and kept alive by it,"[20] we believe Thales to have been essentially correct as regards function of living things. In recognition of Thales' initiation of our quest and because of the central role that water fills (by the consilient mechanisms) in function of the protein-based machines that sustain Life, we refer to this key water as the *waters of Thales*.
20. D.J. Boorstin, *The Seekers: The Story of Man's Continuing Quest to Understand His World*. Random House, New York, 1998, p. 22.
21. P.L. Privalov, "Cold Inactivation of Enzymes." *Crit. Rev. Biochem. Mol. Biol.*, **25**, 281–305, 1990.
22. D.W. Urry, M.M. Long, and H. Sugano, "Cyclic Analog of Elastin Polyhexapeptide Exhibits an Inverse Temperature Transition Leading to Crystallization." *J. Biol. Chem.* **253**, 6301–6302, 1978.
23. G. Lenaz, "A Critical Appraisal of the Mitochondrial Coenzyme Q Pool." *FEBS Lett.*, **509**, 151–155, 2001.
24. C. Hunte, "Insights from the Structure of the Yeast Cytochromes bc_1 Complex: Crystallization of Membrane Proteins with Antibody Fragments." *FEBS Lett.*, **504**, 126–132, 2001.
25. C. Hunte, J. Koepke, C. Lange, T. Rossmanith, and H. Michel, "Structure at 2.3 Å Resolution of Cytochrome bc_1 Complex from the Yeast *Saccharomyces cerevisiae* Co-crystallized with an Antibody Fv Fragment." *Structure*, **8**, 669–684, 2000.
26. D.W. Urry, "Five Axioms for the Functional Design of Peptide-Based Polymers as Molecular Machines and Materials: Principle for Macromolecular Assemblies." *Biopolymers (Peptide Science)*, **47**, 167–178, 1998.
27. D.W. Urry, L.C. Hayes, D.C. Gowda, S.-Q. Peng, and N. Jing, "Electro-chemical Transduction in Elastic Protein-based Polymers." *Biochem. Biophys. Res. Commun.*, **210**, 1031–1039, 1995.
28. L. Hayes, "Effect of Hydrophobicity of Elastic Protein-based Polymers on Redox Potential." Ph.D. Dissertation, The University of Alabama at Birmingham, 1998.
29. M.A. Khaled, V. Renugopalakrishnan, and D.W. Urry, "Proton Magnetic Resonance and Conformational Energy Calculations of Repeat Peptides of Tropoelastin: The Tetrapeptide." *J. Am. Chem. Soc.*, **98**, 7547–7553, 1976.
30. V. Renugopalakrishnan, M.A. Khaled, and D.W. Urry, "Proton Magnetic Resonance and Conformational Energy Calculations of Repeat Peptides of Tropoelastin: The Pentapeptide." *J. Chem. Soc.*, Perkin II, 111–119, 1978.
31. D.W. Urry, S.-Q. Peng, D.C. Gowda, T.M. Parker, and R.D. Harris, "Comparison of Electrostatic- and Hydrophobic-induced pKa Shifts in Polypentapeptides: The Lysine Residue." *Chem. Phys. Lett.*, **225**, 97–103, 1994.
32. D.W. Urry, S.-Q. Peng, and T.M. Parker, "Delineation of Electrostatic- and Hydrophobic-Induced pKa Shifts in Polypentapeptides: The Glutamic Acid Residue." *J. Am. Chem. Soc.*, **115**, 7509–7510, 1993.
33. R. Buchet, C.-H. Luan, K.U. Prasad, R.D. Harris, and D.W. Urry, "Dielectric Relaxation Studies on Analogs of the Polypentapeptide of Elastin." *J. Phys. Chem.* **92**, 511–517, 1988.
34. W.P. Jencks, "Free Energies of Hydrolysis and Decarboxylation." In *Handbook of Biochemistry and Molecular Biology*, Third Edition, G.D. Fasman, Ed., *Physical and Chemical Data*, Vol. I, CRC Press, Boca Raton, FL, 1976, pp. 296–304.

35. H.M. Kalckar, "The Nature of Energetic Coupling in Biological Synthesis." *Chem. Rev.*, **28**, 71–142, 1941.
36. T.L. Hill and M.H. Morales, "On 'High Energy Phosphate Bonds' of Biochemical Interest." *J. Am. Chem. Soc.*, **73**, 1656–1660, 1951.
37. A. Pullman and B. Pullman, *Quantum Biochemistry*, Interscience, New York, 1963.
38. D.B. Boyd and W.N. Lipscomb, "Electronic Structures for Energy-rich Phosphates." *J. Theor. Biol.*, **25**, 403–420, 1969.
39. P. George, R.J. Witonsky, M. Trachtman, C. Wu, W. Dorwart, L. Richman, W. Richman, F. Shurayh, and B. Lentz, "'Squiggle-H_2O' An Enquiry into the Importance of Solvation Effects in Phosphate Ester and Anhydride Reactions." *Biochim. Biophys. Acta*, **223**, 1–15, 1970.
40. M.D. Hayes, L.G. Kenyon, and P.A. Kollman, "Theoretical Calculations of the Hydrolysis Energies of Some 'High Energy' Molecules. 2. A Survey of Some Biologically Important Hydrolytic Reactions." *J. Chem. Soc.*, **100**, 4331–4340, 1978.
41. L. de Meis, "Role of Water in the Energy of Hydrolysis of Phosphate Compounds—Energy Transduction in Biological Membranes." *Biochim. Biophys. Acta*, **973**, 333–349, 1989.
42. C.S. Ewig and J.R. Van Wazer, "Ab Initio Structrures of Phosphoric Acids and Ester. 3. The P-O-P Bridged Compounds $H_4P_2O_{2n-1}$ for n = 1 to 4." *J. Am. Chem. Soc.*, **110**, 79–86, 1988.
43. R. Cross, "Introduction: The Mechanobiochemistry of Molecular Motors." *Essays Biochem. Mol. Motors*, **35**, 1–2, 2000.
44. G. Oster and H. Wang, "How Protein Motors Convert Chemical Energy into Mechanical Work." In *Molecular Motors*, M. Schliwa, Ed., Wiley-VCH GmbH and Co. KgaA, Weinheim, 2003, pp. 207–227.
45. Two excellent and current biochemistry texts may be sought for more background and additional detail: D. Voet, J.G. Voet, and C.W. Pratt, *Fundamentals of Biochemistry*, John Wiley & Sons, Inc., New York, 1999, and Berg et al.[46]
46. J.M. Berg, J.L. Tymoczko, and L. Stryer, *Biochemistry*, Fifth Edition, W.H. Freeman and Company, New York, 2002.
47. D. Voet and J.G. Voet, *Biochemistry*, Second Edition, John Wiley & Sons, 1995, p. 587.
48. M. Saraste, "Oxidative Phosphorylation as the fin de siecle." *Science*, **283**, 1488–1493, 1999.
49. B.E. Schultz and S.I. Chan, "Structures and Proton-pumping Strategies of Mitochondrial Respiratory Enzymes." *Annu. Rev. Biophys. Biomol. Struct.*, **30**, 23–65, 2001.
50. P. Mitchell, "Keilin's Respiratory Chain Concept and its Chemiosmotic Consequences." *Science*, **206**, 1148–1159, 1979.
51. T.L. Hill and E. Eisenberg, "Can Free Energy Transduction be Localized at Some Crucial Part of the Enzymatic Cycle?" *Quart. Rev. Biophys.*, **14**, 463–511, 1981.
52. N. Grigorieff, "Three-dimensional Structure of Bovine NADH:Ubiquinone Oxidoreductase (Complex I) at 22 Å in Ice." *J. Mol. Biol.*, **277**, 1033–1048, 1998.
53. B. Böttcher, D. Scheide, M. Hasterberg, L. Nagel-Stegerand, and T. Friedrich, "A Novel, Enzymatically Active Conformation of the *Escherichia coli* NADH:Ubiquinone Oxidoreductase (Complex I)." *J. Biol. Chem.*, **277**, 17970–17977, 2002.
54. V. Yankovskaya, R. Horsefield, S. Törnroth, C. Luna-Chavez, H. Miyoshi, C. Léger, B. Byrne, G. Cecchini, and S. Iwata, "Architecture of Succinate Dehydrogenase and Reactive Oxygen Species Generation." *Science*, **299**, 700–704, 2003.
55. C.R.D. Lancaster, A. Kröger, M. Auer, and H. Michel, "Structure of Fumarate Reductase from *Wolinella succinogenes* at 2.2 Å resolution." *Nature*, **402**, 377–385, 1999.
56. D. Xia, C.A. Yu, H. Kim, J.Z. Xia, A.M. Kachurin, L. Zhang, L. Yu, and J. Deisenhofer, "Crystal of the Cytochrome bc_1 Complex from Bovine Heart Mitochondria." *Science*, **277**, 60–66, 1997.
57. Z. Zhang, L. Huang, V.M. Shulmeister, Y.I. Chi, K.K. Kim, L.W. Hung, A.R. Crofts, E.A. Berry, and S.H. Kim, "Electron Transfer by Domain Movement in Cytochrome bc_1." *Nature*, **392**, 677–684, 1998.
58. S. Iwata, J.W. Lee, K. Okada, J.K. Lee, M. Iwata, B. Rasmussen, T.A. Link, S. Ramaswamy, and B.K. Jap, "Complete Structure of the 11-Subunit Bovine Mitochondrial Cytochrome bc_1 Complex." *Science*, **281**, 64–71, 1998.
59. C. Lange and C. Hunte, "Crystal Structure of the Yeast Cytochrome bc_1 Complex with its Bound Substrate Cytochrome c." *Proc. Natl. Acad. Sci. U.S.A.*, **99**, 2800–2805, 2002.
60. A.R. Crofts, V.P. Shinkarev, S.A. Dikanov, R.I. Samoilova, and D. Kolling, "Interactions of Quinone with Iron–sulfur Protein of the bc_1

Complex: Is the Mechanism Spring-loaded?" *Biochim. Biophys. Acta*, **1555**, 48–53, 2002.
61. D.W. Urry, R.D. Harris, and K.U. Prasad, "Chemical Potential Driven Contraction and Relaxation by Ionic Strength Modulation of an Inverse Temperature Transition," *J. Am. Chem. Soc.*, **110**, 3303–3305, 1988.
62. D.W. Urry, B. Haynes, H. Zhang, R.D. Harris, and K.U. Prasad, "Mechanochemical Coupling in Synthetic Polypeptides by Modulation of an Inverse Temperature Transition." *Proc. Natl. Acad. Sci. U.S.A.*, **85**, 3407–3411, 1988.
63. T. Tsukihara, H. Aoyama, E. Yamashita, T. Tomizaki, H. Yamaguchi, K. Shinzawa-Itoh, R. Nakashima, R. Yaono, and S. Yoshikawa, "Structures of Metal Sites of Oxidized Bovine Heart Cytochrome c Oxidase at 2.8 Å." *Science*, **269**, 1069–1074, 1995.
64. T. Tsukihara, H. Aoyama, E. Yamashita, T. Tomizaki, H. Yamaguchi, K. Shinzawa-Itoh, R. Nakashima, R. Yaono, and S. Yoshikawa, "The Whole Structure of the 13-Subunit Oxidixed Cytochrome c Oxidase at 2.8 Å." *Science*, **272**, 1136–1144, 1996.
65. J.A. García-Horsman, B. Barquera, J. Rumbly, J. Ma, and R.B. Gennis, "The Superfamily of Heme-copper Respiratory Oxidases." *J. Bacteriol.*, **176**, 5587–5600, 1994.
66. J. Abramson, S. Riistama, G. Larsson, A. Jasiatis, M. Svensson-Ek, L. Laakkonen, A. Puustinen, S. Iwata, and M. Wikstrom, "The Structure of the Ubiquinol Oxidase from *E. coli* and its Ubiquinone Binding Site." *Nature Struct. Biol.*, **7**, 910–917, 2000.
67. S. Yoshikawa, K. Shinzawa-Itoh, R. Nakashima, R. Yaono, E. Yamashita, N. Inoue, M. Yao, M.J. Fel, C.P. Libeu, T, Mizushima, H. Yamaguchi, T. Tomizaki, and T. Tsukihara, "Redox-coupled Crystal Structural Changes in Bovine Heart Cytochrome c Oxidase." *Science*, **280**, 1723–1729, 1998.
68. M. Wikström and M.I. Verkhovsky, "Proton Translocation by Cytochrome c Oxidase in Different Phases of the Catalytic Cycle." *Biochim. Biophys. Acta*, **1555**, 128–132, 2002.
69. M. Wikström, "Mechanism of Proton Translocation by Cytochrome c Oxidase: A New Four-stroke Histidine Cycle." *Biochim. Biophys. Acta*, **1458**, 188–198, 2000.
70. M. Svensson-Ek, J. Abramson, G. Larsson, S. Törnroth, P. Brzezinski, and S. Iwata, "The X-ray Crystal Structures of Wild-type and EQ (I-286) Mutant Cytochrome c Oxidases from *Rhodobacter sphaeroides*. *J. Mol. Biol.*, **321**, 329–339, 2002.
71. R. Mitchell, P. Mitchell, and P.R. Rich, "Protonation of the Catalytic Intermediates of Cytochrome c Oxidase." *Biochim. Biophys. Acta*, **1101**, 188–191, 1992.
72. R. Mitchell and P.R. Rich, "Proton Uptake by Cytochrome c Oxidase on Reduction and on Ligand Binding." *Biochim. Biophys. Acta*, **1186**, 19–26, 1994.
73. N. Capitanio, T.V. Vygodina, G. Capitanio, A.K. Konstantinov, P. Nichols, and S. Papa, "Redox-Linked Proteolytic Reactions in Soluble Cytochrome-c Oxidase from Beef Heart Mitochondria: Redox Bohr Effects." *Biochim. Biophys. Acta*, **1318**, 255–265, 1997.
74. D.W. Urry and H. Eyring, "Optical Rotatory Disperson Studies of L-Histidine Chelation." *J. Am. Chem. Soc.*, **86**, 4574–4580, 1964.
75. *Could the Cu_A center function in a manner equivalent to the Q_o site of Complex III?* Considerations: (1) The Cu_A center receives electrons from cytochrome **c**. (2) There are quinol oxidase enzymes that have the heme-heme-copper dioxygen reduction site but not the Cu_A center, that is, the Cu_A center would seem to replace the quinol oxidation site—the Q_o site. (3) Comparison of the oxidized and reduced states of cytochrome **c** oxidase show that the singular substantial conformational change includes histidine (H204) coordinating the Cu_A center. (4) The coordination geometry of the Cu_A center showed reduction to shift from mixed valence to the fully reduced form and a small increase in Cu-Cu distance (2.43 to 2.51 Å). This, we suggest, could correlate on subsequent oxidation with the release of a proton from H204. (5) The proton uptake by cytochrome **c** oxidase on reduction and on ligand binding[72]: Where 2 of 2.4 protons taken up on reduction were assigned to the binuclear center, the remaining fraction had the heme **a**/**Cu_A**. *Possible problem: the Cu_A center was not within the lipid layer, and there was no obvious obligatory source of protons from within the lipid layer except perhaps by means of the D-channel.*
76. K. Kinosita, R. Yasuda, and H. Noji, "F_1-ATPase: A Highly Efficient Rotary ATP Machine." *Essays Biochem. Mol. Motors*, **35**, 3–18, 2000.
77. R.I. Menz, J.E. Walker, and A.G.W. Leslie, "Structure of Bovine Mitochondrial F_1-ATPase with Nucleotide Bound to All Three Catalytic Sites: Implications for Mechanism of Rotary Catalysis." *Cell*, **106**, 331–341, 2001.

78. R.H. Fillingame, "Molecular Rotary Motors." *Science*, **286**, 1687–1688, 1999.
79. P.L. Pedersen, Y.H. Ko, and S. Hong, "ATP Synthases in the Year 2000: Evolving Views about the Structures of These Remarkable Enzyme Complexes." *J. Bioenerget. Biomembranes*, **32**, 325–332, 2000.
80. R.H. Fillingame, C.M. Angevine, and O.Y. Dmitriev, "Coupling Proton Movements to c-Ring Rotation in F1F0 ATP Synthase: Aqueous Access Channels and Helix Rotations at the a–c Interface." *Biochim. Biophys. Acta*, **1555**, 29–36, 2002.
81. A. Horak, H. Horak, and M. Packer, "Subunit Composition and Cold Stability of the Pea Cotyledon Mitochondrial F1-ATPase." *Biochim. Biophys. Acta*, **893**, 190–196, 1987.
82. M.E. Pullman, H.S. Penefsky, A. Datta, and E. Racker, "Partial Resolution of the Enzymes Catalyzing Oxidative Phosphorylation. I. Purification and Properties of Soluble, Dinitrophenol-stimulated Adenosine Triphosphatase." *J. Biol. Chem.*, **235**, 3322–3329, 1960.
83. D. Stock, A.G.W. Leslie, and J.E. Walker, "Molecular Architecture of the Rotary Motor in ATP Synthase." *Science*, **286**, 1700–1705, 1999.
84. J.P. Abrahams, A.G.W. Leslie, R. Lutter, and J.E. Walker, "Structure at 2.8 Å of F_1-ATPase from Bovine Heart Mitochondria." *Nature (Lond.)*, **370**, 621–628, 1994.
85. P.B. Boyer, "New Insights into One of Nature's Remarkable Catalysts, the ATP Synthase." *Mol. Cell Prev.*, **8**, 246–247, 2001.
86. P.D. Boyer, "The Binding Change Mechanism for ATP Synthase—Some Probabilities and Possibilities." *Biochim. Biophys. Acta*, **1140**, 215–250, 1993.
87. P.D. Boyer, "The ATP Synthase—A Splendid Molecular Machine." *Annu. Rev. Biochem.*, **66**, 717–749, 1997.
88. L. Pauling, "The Energy of Single Bonds and the Relative Electonegativities of Atoms." *J. Am. Chem. Soc.*, **54**, 3570–3582, 1932.
89. L. Pauling, *The Nature of the Chemical Bond and the Structure of Molecules and Crystals: An Introduction to Modern Structural Chemistry*, Third Edition, Cornell University Press, Ithaca, New York, 1960, p. 93.
90. R.I. Menz, A.G. Leslie, and J.E. Walker, "The Structure and Nucleotide Occupancy of Bovine Mitochondrial F_1-ATPase are not Influenced by Crystallization at High Concentrations of Nucleotide." *FEBS Lett.*, **494**, 11–14, 2001.
91. K. Braig, R.I. Menz, M.G. Montgomery, A.G.W. Leslie, and J.E. Walker, "Structure of Bovine Mitochondrial F_1-ATPase Inhibited by Mg(2+) ADP and Aluminum Fluoride." *Structure (Lond.)*, **8**, 567–573, 2000.
92. E. Cabezon, M.G. Montgomery, A.G.W. Leslie, and J.E. Walker, "The Structure of Bovine Mitochondrial F_1-ATPase in Complex with its Regulatory Protein If1." *Nature Struct. Biol.*, **10**, 744–750.
93. H. Noji, R. Yasuda, M. Yoshida and K. Kinosita, "Direct Observation of the Rotation of F_1-ATPase." *Nature (Lond.)*, **386**, 299–302, 1997.
94. H. Noji "Amersham Pharmacia Biotech & Science Prize: The Rotary Enzyme of the Cell: The Rotation of F_1-ATPase." *Science*, **282**, 1844–1845, 1998.
95. K. Kinosita Jr., R. Yasuda, H. Noji, and K. Adachi, "A Rotary Motor that can Work at Near 100% Efficiency." *Philos. Trans. R. Soc. Lond. B*, **355**, 473–489, 2000.
96. *Dorland's Illustrated Medical Dictionary* 27th Edition, W.B. Saunders, 1985, p. 1468.
97. D. Voet, J.G. Voet & C.W. Pratt, *Fundamentals of Biochemistry*, John Wiley & Sons, New York, 1999, Figures 7–22 and 7–23, p. 181.
98. D. Voet and J.G. Voet, *Biochemistry*, Second Edition, John Wiley & Sons, 1995, Figure 34–65, p. 1247, and 34–67, p. 1249.
99. A.F. Huxley and R. Niedergerke, "Interference Microscopy of Living Muscle Fibers." *Nature*, **173**, 971–973, 1954.
100. A.F. Huxley, "Muscle Structure and Theories of Contraction." *Prog. Biophys. Biophys. Chem.*, **7**, 255–318, 1957.
101. D. Voet, J.G. Voet, and C.W. Pratt, *Fundamentals of Biochemistry*, John Wiley & Sons, New York, 1999, Figure 7–29, p. 183.
102. I. Rayment, W.R. Rypniewski, K. Schmidt-Base, R. Smith, D.R. Tomchick, M.M. Benning, D.A. Winkelmar, G. Wesenberg, and H.M. Holden, "Three–Dimensional Structure of Myosin Subfragment-1 A Molecular Motor." *Science*, **261**, 50–58, 1993.
103. I. Rayment, H.M. Holden, M. Whittaker, C.B. Yohn, M. Lorenz, K.C. Holmes, and R.A. Milligan, "Structure of the Actin–Myosin Complex and its Implications for Muscle Contraction." *Science*, **261**, 58–65, 1993.
104. I. Rayment. C. Smith, and R.G. Yount, "The Active Site of Myosin." *Annu. Rev. Physiol.*, **58**, 671–702, 1996.
105. Y. Lecarpentier, D. Chemla, J.C. Pourny, and F.-X. Blanc, "Myosin Cross Bridges in Skeletal Muscles: 'Rower' Molecular Motors." *J. Appl. Physiol.*, **91**, 2479–2486, 2001.

106. One might raise the issue of whether the 98% of detached cross-bridges and the arrangement of the tropomyosin–troponin (T-C-I) complex on actin might influence availability of troponin C to the released calcium ion such that there could be a selection of those troponin C sites available for calcium ion binding that would have associated cross-bridges in position to take advantage of the emerging tight binding site.

107. R.W. Lymn and E.W. Taylor, "Mechanism of ATP Hydrolysis by Actomyosin." *Biochemistry*, **10**, 4617–4624, 1971.

108. T. Duke, "Cooperativity of Myosin Molecules Through Strain-dependent Chemistry." *Philos. Trans. R. Soc. Lond. B*, **355**, 529–538, 2000.

109. D.W. Urry, "Function of the F_1-motor (F_1-ATPase) of ATP synthase by Apolar-polar Repulsion through Internal Interfacial Water." *Cell Biology International*, **30**, (1), 44–55, 2006.

9
Advanced Materials for the Future: Protein-based Materials with Potential to Sustain Individual Health and Societal Development

The industrial revolution of the Nineteenth Century was born of the water-to-steam phase transition. As implicit in the last three chapters and explicit in the Epilogue, biology was born of the inverse temperature transition. As argued in this chapter, the advanced biomaterials renaissance of the Twenty-first Century will also have been born of the inverse temperature transition.

9.1 Introduction

9.1.1 The Appeal of New Discoveries and Their Utilization

"Nothing is more agreeable to men (those) devoted to a scientific career than to increase the number of discoveries, but when the result of their observation is demonstrated by practical utility their joy is complete."[1] Louis Pasteur penned these words a century and a half ago. So appropriate is this perspective today to those "devoted to a scientific career" that one might only add that further joy comes when "practical utility" enables further discoveries by providing the necessary funding. It is in this vein, fueled by the firm belief in the remarkable range of applications of protein-based machines and materials, that we have efforts underway to develop uses. The potential applications promise to help society attain a number of its goals, such as improving health and quality of life, reducing cost of health care, alleviating addiction, and relieving environmental pollution.

9.1.2 The Unrivaled Opportunity to Achieve Utility of Advanced Biomaterials

In Chapter 5, based on an inverse temperature transition due to hydrophobic association in water, a set of Axioms were derived from the phenomenological demonstration that *de novo* designed model proteins could efficiently interconvert the set of energies interconverted by living organisms. Then there followed a series of experimental results and analyses that defined the comprehensive hydrophobic effect.

The operative component of the comprehensive hydrophobic effect arises from the competition between charged and oil-like groups. This was shown to result in a previously unknown repulsive force embodied within an interaction energy called an *apolar–polar repulsive free energy of hydration*, ΔG_{ap}. During function, ΔG_{ap} works in conjunction with elastic force development by the restriction of internal chain dynamics. These have been called the *hydrophobic* and *elastic consilient mechanisms*. In Chapters 6, 7, and 8, these consilient mechanisms were demonstrated to be fundamental to understanding the functions of biology's proteins.

If our advances built over the last three decades and culminating in the new analyses and data in Chapter 5 are sound, then we are at a historic moment in biomaterials development. This moment opens the door to an unprecedented future for biomaterials. The opportunity is founded on an understanding of the forces responsible for protein function, demonstrated in part in the new structure–function analyses in Chapter 8 and culminating in the Epilogue as constituting the "vital force" of biology. The practical utilization of the hydrophobic and consilient mechanisms couples with the capacity for biological production of designed protein-based polymers of almost unlimited diversity of composition and size.

Finally, for medical applications, the extraordinary biocompatibility of these elastic protein-based materials, we believe, arises from the specific means whereby these elastic protein-based polymers exhibit their motion. Being composed of repeating peptide sequences that order into regular, nonrandom, dynamic structures, these elastic protein-based polymers exhibit mechanical resonances that present barriers to the approach of antibodies as required to be identified as foreign. In addition, we also believe that these mechanical resonances result in extraordinary absorption properties in the acoustic frequency range.

9.1.3 The Relationship Between Basic Science and Its Applications

9.1.3.1 Our Perspective for Protein-based Polymers

Chapter 5 presents in one place, more extensively and in a more advanced state than previously, the decades long development of the comprehensive hydrophobic effect, the underpinnings of the hydrophobic consilient mechanism, whereby the control of hydrophobic association commands diverse energy conversion functions of protein-based polymers. Chapters 7 and 8 demonstrate the comprehensive hydrophobic effect and its interlinked elastic consilient mechanism to be vital aspects of protein function and dysfunction in biology. In the present chapter, we utilize this developed capacity to engineer protein-based polymers to demonstrate a few of an extraordinary range of applications.

Design of function comes through control of the Gibbs free energy of hydrophobic association, ΔG_{HA}. With regard to medical applications, protein, biology's own workhorse polymer of choice, provides the most congruent means with which to restore function and defeat disease. As noted in section 5.2.5, the nature of their elasticity and resulting biocompatibility and their never-ending design potential makes the family of elastic protein-based polymers, developed here, the pre-eminent choice of materials for medical applications.

With regard to nonmedical applications, the entirely unique opportunity to specify composition and diverse sequence, so precisely, provides a profusion of polymers with properties that bring to the nonmedical world all of the specificity and sensitivity of which biology's protein is renown. To the best of our understanding, the properties include a remarkable and unique acoustic absorption capacity. Also unconstrained by the parameters of living organisms, but armed with the engineering capacity provided by knowledge of the hydrophobic and elastic consilient mechanisms, designed protein-based polymers give rise to materials unknown to living organisms. We now look toward enumerable challenges where defined needs stimulate designs of protein-based polymers to fulfill those needs.

9.1.3.2 Pasteur on the Relationship Between the Basic and Applied Sciences

Before proceeding to the range and significance of problems addressable with the diverse properties of protein-based materials, the relationship between basic research and its derivative applications warrants further consideration. Again we turn to Pasteur: "*No, a thousand times no; there does not exist a category of science to which one can give the name applied science. There are science and the applications of science, bound together as the fruit to the tree which bears it.*"[1] Pasteur knew the relationship well. He made major advances in the basic sciences, for example, in crystal structure and optical activity and in advancing the germ theory that put to rest the unproductive concept of spontaneous generation of living organisms. He also made advances resulting in more familiar applications. He developed the process now known as *pasteurization* that saved the wine, vinegar, and beer industries of France but that we daily identify with pasteurized milk products; he solved a threat to the silk industry by alleviat-

9.1 Introduction

ing diseases of the silkworm, and he developed an understanding and successful treatment of rabies. All of this came about by singular advances that were often in severe conflict with the prevailing thought of his day.

9.1.3.3 Cyclical Enabling: The Relationship Between Science and the Applications of Science

Development of applications proceeds effectively when a scientific foundation is employed. On the other hand, development of the scientific foundation becomes empowered by society's need to solve current problems. This is what is intended by the term *cyclical enabling*. Building upon Pasteur's words, it might further be said that, *To separate the development of science from the development of applications would be to separate the growing fruit from its tree; the immature fruit wither and waste and yield no new trees, and the barren tree too soon dies.*

The development of thermodynamics is one example of this interdependence. Thermodynamics, a paradigm of basic science, grew from society's appetite for better steam engines with which to relieve humans and horses of mechanical work. Of particular relevance here, the discipline of thermodynamics provides the scientific foundation for energy conversion by protein machines. Chapter 5 demonstrates this with consideration of the thermodynamics of hydrophobic association in model proteins. A thermodynamic understanding of energy conversion provides the scientific foundation for the most effective development of the applications considered below. The Epilogue integrates the keystone of thermodynamics, the second law, into the unique adherence by biological systems.

9.1.3.4 Specific Enablers of the Development of Elastic Protein-based Polymers

In general, development of the scientific foundation for protein-based polymers began in an academic setting and continued very effectively in a company setting. The primary funding impetus (the enabler) for the development of elastic protein-based polymers has been grants and contracts from the Office of Naval Research, covering nearly two decades of support in academic institutions and in a company setting, also with support from the Naval Medical Research and Development Command in the company setting. Development of a scientific foundation for elastic protein-based polymers occurred in order to provide the most effective approach to development of materials applications. The second most prominent level of support has come from the National Institutes of Health, primarily in terms of Small Business Innovation Research contracts (SBIRs), The SBIRs decidedly focus on applications and draw from the foundations of the science of the elastic-contractile model proteins presented in Chapter 5.

The primary advances toward applications occurred in a company setting. In fact, even much of the foundation of the relevant underlying science occurred in the company setting. For example, fundamental knowledge of hydrophobic association in proteins, of protein elasticity, of mechanical (e.g., acoustical) resonances, and so forth resulted from progress made in the company setting. These contributions to the foundation of the science obviously relate to the solution of practical problems—developing rules for correctly folding microbially prepared protein of higher organisms, designing a new class of sound-absorbing materials, and developing nanomachines and nanosensors. Consistent with the Pasteur perspective, progress toward applications for elastic and plastic protein-based polymers resulted from the cyclical enabling noted above.

9.1.4 The Incomparable Potential of Protein-based Polymers

9.1.4.1 General Advantages of Protein-based Polymers

Many extraordinary advantages of protein-based polymers exist. Some of the advantages are briefly listed.

1. *Two modes of synthesis: chemical and biological*: Chemical synthesis allows for relatively rapid screening of physical properties of hundreds of protein-based polymer composi-

tions as long as they are not too complex. Even so, chemical synthesis has disadvantages of racemization, of side reactions on protection/deprotection of functional side chains, of random incorporation of guest pentamer, of a distribution of chain lengths, and so forth, all of which limit quantification of engineering principles. On identification of chemically synthesized model protein compositions of particular interest, the slower process of recombinant DNA technology for gene construction and development of an effective expression system ultimately provides for rapid, large-scale production and other advantages discussed below. Of importance in the most accurate development of protein-based polymers as biomaterials, however, will be to obtain the most accurate set of controlling interaction energies to arrive at the desired level of reliance on derived engineering principles.

2. *Diversity of monomers*: The biosynthesis of natural proteins utilizes 20 different monomeric units, and chemical and enzymatic post-translational modifications provide further diversity of sequence. As discussed in Chapter 4 and noted below, the availability of 20 different monomers results in an inordinate number of different protein sequences, even for a small 100 residue protein, that is, 10^{130}.

3. *Precise control of the sequence of amino acid residues*: Any protein-based polymer sequence utilizing the 20 naturally occurring monomers can be specified using recombinant DNA technology. This allows for equivalent ease of production of diverse protein-based polymer sequences, many of which would otherwise be quite difficult or essentially impossible to prepare due to problems of chemical synthesis and unfavorable energetics in the final polymer.

4. *Exact control of stereochemistry*: Control of stereochemistry, for example, use of a single optical isomer as occurs in biology, is essential for control of physical properties resulting in unique structures and more precise function. The occurrence of some racemization is unavoidable during chemical synthesis of long protein-based polymers. Some inclusion of D-amino acid residues, rather than having all L-amino acid residues (see Figure 3.3C) as occurs in biology, is unavoidable in chemical synthesis. Even an amount of racemization as small as 1% can critically limit desirable properties in these nonrandom elastic polymers. For example, lack of rigorous exclusion of racemization during chemical synthesis of poly(GVGVP) can cause the critical onset temperature for the inverse temperature transition for hydrophobic association (T_t) to be raised by 15°C from 25° to 40°C. As explained in Chapter 5, control of T_t is central to controlling function.

5. *Precise chain lengths*: The gene encoding for a protein-based polymer occurs, generally, with a single chain length, and the complete range of chain lengths, arising from multiples of a basic repeating gene sequence, are possible in *Escherichia coli*, resulting in as many as 4,000 amino acid residues and even more than 1 million residues in animals. This allows for fine-tuning of desirable properties and access to new capabilities. Several processes occur that can defeat this possibility of all polymer chains being of the same chain length. Deletion of gene sequences can occur under certain circumstances. The growing protein chain can fall off the ribosome before completion of translation, and the expressed protein can be subjected to proteolytic degradation. All of these complications, however, are generally avoidable, and precise chain lengths are routinely obtained.

6. *Capacity to introduce natural bioactive peptide sequences (protein mimicry)*: With an adequate understanding of protein engineering, there exists an essentially unlimited capacity to mimic chosen elements of more complex proteins in terms of both structure and function. Specific biologically active sequences can be introduced with ease. Examples are sites at which selective enzymes can catalyze a desired reaction, sequences for selective cell attachment, selective attachment to diseased cells for drug delivery, and so forth.

7. *Circumstances of protein function to guide approach and analyses*: It is possible to consider functional proteins, for example, with their natural prosthetic groups and cofactors, to suggest important variables that relate to function, and these become tools with which to achieve protein engineering (see Tables 5.1 and

5.2). For example, prosthetic groups and cofactors can be attached to model proteins, and the role of each as regards function can be quantified and utilized.

8. *Properties and uses beyond those of known proteins*: Once the rules for protein engineering have been established, protein-based polymers can be designed with properties and functions that go beyond what evolution has called upon proteins to do. For example, there are model proteins for controlled release of new pharmaceuticals, and programmable, biodegradable thermoplastic protein-based polymers that melt for easy molding or extruding at 150°C, with decomposition not occurring until 250°C.

9. *Low cost of bioproduction*: As the designed protein-based polymer becomes more complex, the cost advantages of bioproduction become greater. The production of protein-based polymers by means of recombinant DNA technology has the potential for at least a 10,000-fold decrease in cost from that of chemical synthesis. It is believed that the cost of protein-based polymers has the potential to be competitive with the cost of petroleum-based polymers, thus relieving, in part, society's dependence on limited oil reserves. Furthermore, it costs a living organism no more energy to produce a more efficient protein-based machine than an inefficient one of the same size.

10. *Produced from renewable resources*: Living organisms—*E. coli*, yeast, plants, and animals—can be designed to produce protein-based polymers. Protein-based polymers can be produced with renewable resources. They can be prepared without resorting to toxic and noxious chemicals, and they can be programmed for a desired biodegradation. For example, they can mean food for the fishes rather than death to marine life, as occurs with present plastics. Thus, protein-based polymers can be environmentally friendly for their complete life cycle, from production to disposal.

11. *Axioms for protein-based polymer engineering*: The phenomenology of protein-based polymer function, categorized in terms of free energy transduction, is given as a set of five Axioms in Chapter 5 (see section 5.6.3). These five Axioms provide the basis for diverse, but qualitative, designs of polymers capable of exhibiting inverse temperature transitions.

12. *Quantitative design principles available for protein-based polymers*: The comprehensive hydrophobic effect, developed in Chapter 5, provides principles for the quantitative design of protein-based polymers. The dominant underlying energetics are embodied in the Gibbs free energy for hydrophobic association, ΔG_{HA}, as modified by the newly described Gibbs free energy for an apolar–polar repulsive free energy of hydration, ΔG_{ap}.

13. *Availability of the most efficient mechanism for achieving function in an aqueous environment*: Comparison of the electrostatic charge–charge repulsion mechanism for chemo-mechanical transduction with that of the apolar–polar repulsive free energy of hydration, ΔG_{ap}, shows the latter to be more than an order of magnitude more efficient. This becomes particularly relevant to biomedical applications of controlled release as required in drug delivery but also whenever a sensitive and responsive (smart) biomaterial is desired.

14. *Remarkable biocompatibility of elastic protein-based polymers*: When considering medical applications of these biomaterials, utility depends on biocompatibility. After all, foreign proteins are generally antigenic and elicit production of antibodies. This is the basis for many vaccines. How then can one propose protein as a biomaterial? The most direct answer is that many biocompatibility studies have been carried out on a number of compositions of elastic protein-based polymers, and they have been found to exhibit extraordinary biocompatibility. In fact, very pure preparations of $(GVGVP)_n$, or equivalently $(VPGVG)_n$, appear to be entirely ignored by the host. This we believe is due to the nature of the elasticity.

In our view, ideal or entropic elasticity exhibited by poly(VPGVG) results from the fact that the repeating conformational unit, (VPGVG), exhibits mechanical resonances. These are motions that occur with frequencies localized near 5 MHz and 3 kHz. Such low-frequency motions greatly stabilize the structure. Furthermore, the requirement to stop these motions, as needed to identify an epitope for the development of antibodies to the sequence, presents a barrier to the approach and identification of the *elastic* protein-based polymer as foreign.

15. *In vivo breakdown products*: Other polymeric biomaterials may also be composed of natural products, such as polyesters like poly(glycolic acid) and poly(lactic acid). When these polymeric biomaterials degrade to the naturally occurring monomers of glycolic acid and lactic acid, however, the monomeric carboxylic acid ionizes to release acid, H^+. The decreased pH can be a significant irritant to the tissues. On the other hand, when protein-based polymers undergo biodegradation, naturally occurring zwitterionic amino acids are released without such an effect on the pH and its ramifications.

When seeking an optimal biomaterial, the preference is for complete absence of toxicity on placement in the host. With such a totally innocuous elastic protein-based biomaterial, biologically active sequences can be readily included within the polymer, and the host tissue can react to the biologically active sequence without being overwhelmed by unwanted reactions. Thus, because of the high level of biocompatibility, there exists the capacity to elicit diverse and desired tissue responses.

Hopefully, the foregoing brief lists provides some insight into the potential of protein-based polymers in the marketplace. *There simply are no comparable soft materials for medical applications*. Certain of the above-listed advantages are discussed in more detail immediately below and throughout this chapter.

9.1.4.2 A Near-infinite Number of Different Polymer Compositions

There are 20 different naturally occurring amino acid residues, and each position in a biosynthesized protein-based polymer can contain any one of the 20 amino acids. As considered in Chapter 4, this means even for a small 100 residue protein-based polymer that there are $(20)^{100} = 10^{130}$ different sequences possible. The size of this number is hard to comprehend. For example, if the mass of the known universe were composed of nothing but 100 residue protein-based polymers and if there were only one molecule of each possible sequence, the likelihood of the occurrence of a specific sequence would yet remain extremely remote.

Capitalizing on the potential of protein-based polymers with such a near-infinite number of sequences requires an understanding of the basis for function. The hydrophobic and elastic consilient mechanisms for design of protein-based polymer structure and function, the knowledge for which is developed primarily in Chapter 5, enable utilization of the extraordinary potential of these incomparable polymers. Humans utilize fewer than 50,000 different proteins, and, as learned from sequence data of the human genome and those of lower and less complex animals, plants, and organisms, a surprisingly limited number of the fewer than 50,000 protein sequences are unique to humans. Accordingly, the potential for materials that go beyond those known in biology rings clear.

Obviously, the applications of protein-based polymers are innumerable. The applications discussed in this chapter represent but a few of very very many, and they are at different stages of development.

9.1.4.3 Available Fundamental Design Principles Yield Effective Products

The five phenomenological Axioms, knowledge of the underlying physical basis and thermodynamic formalism in terms of the Gibbs free energy of hydrophobic association, ΔG_{HA}, provide a foundation for fundamental design principles. The efficiency rating, $\hat{e}_{CM}$, as presented in section 5.9.3, recognizes ΔG_{HA} as a maximal energy output for a given design based on the consilient mechanism, that is, the comprehensive hydrophobic effect. Given a defined energy source, which could be a drug, for example, the issue becomes one of designing the preferred protein-based polymer that would most accurately target the desired outcome. In particular, the drug can be the energy input for assembly of the drug delivery vehicle that released drug in the desired manner as the energy output. In every case with the hydrophobic and elastic consilient mechanisms, the interlinked physical processes that achieve the energy conversion are a change in

9.1 Introduction

hydrophobic association in conjunction with a near-ideal polymer elasticity

9.1.4.4 De novo Design of a Material for the Intended Application

From the outset, our approach has been to design materials for an intended function. In particular, each new energy conversion described in Chapter 5 was achieved by design of a new functionality utilizing the perspective of a common underlying hydrophobic consilient mechanism. Jacob Bronowski[2] anticipated such an approach three decades ago in his book *Ascent of Man*, with the perspective that, "*In effect, the modern problem is no longer to design a structure from the materials (available) but to design materials for a structure.*" Knowledge of the Axioms and the principles underlying the comprehensive hydrophobic effect, based at the molecular level on the change in free energy of hydrophobic association, makes it possible to "design materials for a structure." Of course, by *structure* we generally mean a unique protein sequence, also referred to as *primary structure* or in a more general way as *composition*. By materials for a structure, Bronowski meant a composition designed for a specific function. This is to be distinguished from taking a given composition and coercing a function for which it is not sequence optimized.

An example of "*to design a structure from the materials*" would be to achieve the function of a chemo-mechanical engine using the initial available model protein composition poly (GVGVP), cross-linking it by γ-irradiation, and employing the chemical energy of adding salt(NaCl) to drive contraction.[3] In this case, ΔG_{HA}[poly(GVGVP)(0.1 N → 1.0 N NaCl)] is −0.5 kcal/mole-pentamer/mole-NaCl.

On the other hand, to "*design materials for a structure*," in this case a better salt (NaCl)-driven chemo-mechanical engine, would be to design a protein-based polymer with a few glutamic acid residues per 100 residues, for example, poly[4(GVGVP),(GEGVP)]. As listed in Table 5.4, for the newly designed structure over its most responsive range to NaCl from 0.15 N to 0.25 N NaCl, ΔG_{HA}\{poly[0.8(GVGVP),0.2(GE-GVP)]\}, is −9.4 kcal/mole-pentamer/mole-NaCl; this gives an improvement for the most effective ranges of 9.35/0.5. Thus, "*to design materials for a structure*," as was done by designing the polymer poly[4(GVGVP),(GEGVP)], to be an NaCl-driven chemo-mechanical engine, was to design a chemo-mechanical engine with an effectiveness of energy conversion improved by nearly a factor of 20 when functioning in its most effective concentration range. Clearly, the modern problem is "*to design materials for a structure*," and the means to do so comes from an understanding of the hydrophobic consilient mechanism's comprehensive hydrophobic effect, as developed in Chapter 5.

Thus, poly(GVGVP) was a structure available from the mammalian elastic fiber. Adding a carboxyl function resulted in a structure that was not previously known, and the resulting design gave rise to a structure that could do 20 times more work for the same chemical energy input. This exemplifies the modern approach of designing a material specifically for a chemo-mechanical structure. Contemplation of any application of these protein-based polymers best proceeds with understanding of the underlying physical process in order to achieve the most effective outcome. Most functions can be analyzed in terms of the efficiency of energy conversion. This is most readily appreciated for the controlled release of pharmaceuticals, that is, for drug delivery.

9.1.5 Potential of Protein-based Materials to Improve Health and Decrease Health Care Costs

Health care costs in the United States exceed a staggering trillion dollars per year. Low back pain, urinary incontinence, pressure ulcers (e.g., bed sores), and cardiovascular disease are major contributors to decreased quality of life and increased health care costs. Applications of protein-based materials, briefly noted in this section but discussed more extensively below, have the potential to improve quality of life while lowering health care costs for these and additional medical problems. To provide a historical backdrop and a record of the development of applications, Table 9.1 provides the set of patents resulting from our research efforts,

TABLE 9.1. Bioelastics Research, Ltd., patent list.

1. Inventors: D.W. Urry and K. Okamoto
   Title: Synthetic Elastomeric Insoluble Cross-Linked Polypentapeptide

Country	Patent No.	Date Filed	Date Issued
USA	4,132,746	07/09/76	01/02/79
Canada	1103238	02/26/79	06/16/81

2. Inventors: D.W. Urry and K. Okamoto
   Title: Synthetic Elastomeric Insoluble Cross-Linked Polypentapeptide

Country	Patent No.	Date Filed	Date Issued
USA	4,187,852	08/14/78	02/12/80

3. Inventor: D.W. Urry
   Title: Elastomeric Composite Material Comprising a Polypeptide

Country	Patent No.	Date Filed	Date Issued
USA	4,474,851	10/02/81	10/02/84

4. Inventor: D.W. Urry
   Title: Elastomeric Composite Material Comprising a Polypentapeptide Having an Amino Acid of Opposite Chirality in Postion Three

Country	Patent No.	Date Filed	Date Issued
USA	4,500,700	12/23/82	02/19/85

5. Inventor: D.W. Urry
   Title: Enzymatically Cross-Linked Bioelastomers

Country	Patent No.	Date Filed	Date Issued
USA	4,589,882	09/19/83	05/20/86

6. Inventors: D.W. Urry and R.M. Senior
   Title: Stimulation of Chemotaxis by Chemotactic Peptides (Hexapeptide)

Country	Patent No.	Date Filed	Date Issued
USA	4,605,413	09/19/83	08/12/86

7. Inventors: D.W. Urry and M.M. Long
   Title: Stimulation of Chemotaxis by Chemotactic Peptides (Nonapeptide)

Country	Patent No.	Date Filed	Date Issued
USA	4,693,718	10/31/85	09/15/87

8. Inventors: D.W. Urry and K.U.M. Prasad
   Title: Temperature Correlated Force and Structure Development of Elastin Polytetrapeptides & Polypentapeptides

Country	Patent No.	Date Filed	Date Issued
USA	4,783,523	08/27/86	11/08/88
Europe	EP 0321496		03/30/94
Japan	2726420	08/27/87	12/05/97

9. Inventors: D.W. Urry and K.U.M. Prasad
   Title: Segmented Polypeptide Bioelastomers to Modulate Elastic Modulus

Country	Patent No.	Date Filed	Date Issued
USA	4,870,055	04/08/88	09/26/89
Japan	2085090	04/17/87	08/23/96

10. Inventor: D.W. Urry
    Title: Bioelastomer Containing Tetra/ Pentapeptide Units

Country	Patent No.	Date Filed	Date Issued
USA	4,898,926	06/15/87	02/06/90
Europe		06/13/88	

11. Inventor: D.W. Urry
    Title: Reversible Mechanochemical Engines Comprised of Bioelastomers Capable of Modulable Inverse Temperature Transitions for the Interconversion of Chemical and Mechanical Work

Country	Patent No.	Date Filed	Date Issued
USA	5,032,271	06/15/87	07/16/91
Europe	EP 0425491	06/13/88	07/20/94

9.1 Introduction   463

TABLE 9.1. *Continued*

12. Inventor: D.W. Urry
    Title: Reversible Mechanochemical Engines Comprised of Bioelastomers Capable of Modulable Inverse Temperature Transitions for the Interconversion of Chemical and Mechanical Work
Country	Patent No.	Date Filed	Date Issued
USA	5,085,055	04/30/91	02/04/92

13. Inventor: D.W. Urry
    Title: Reversible Mechanochemical Engines Comprised of Bioelastomers Capable of Modulable Inverse Temperature Transitions for the Interconversion of Chemical and Mechanical Work
Country	Patent No.	Date Filed	Date Issued
USA	5,255,518	12/24/91	10/26/93

14. Inventors: D.W. Urry and M.M. Long
    Title: Stimulation of Chemotaxis by Chemotactic Peptides
Country	Patent No.	Date Filed	Date Issued
USA	4,976,734	09/19/87	12/11/90
Europe	EP 0366777	06/13/88	07/20/94

15. Inventor: D.W. Urry
    Title: Bioelastomeric Materials Suitable for Burn Areas or the Protection of Wound Repair Sites from the Occurrence of Adhesions
Country	Patent No.	Date Filed	Date Issued
USA	5,250,516	04/21/88	10/05/93
Europe	EP 0365555		09/21/94
Japan	2820750	04/14/89	08/28/98

16. Inventor: D.W. Urry
    Title: Elastomeric Polypeptides as Vascular Prosthetic Materials
Country	Patent No.	Date Filed	Date Issued
USA	5,336,256	04/22/88	08/09/94
Europe	EP 0365654		01/12/94
Japan	2115962	04/14/89	12/06/96

17. Inventor: D.W. Urry
    Title: Polynonapeptide Bioelastomers Having an Increased Elastic Modulus
Country	Patent No.	Date Filed	Date Issued
USA	5,064,430	02/23/89	11/12/91

18. Inventor: D.W. Urry
    Title: Polymers Capable of Baromechanical and Barochemical Transduction
Country	Patent No.	Date Filed	Date Issued
USA	5,226,292	04/22/91	07/13/93

19. Inventor: D.W. Urry
    Title: Superabsorbent Materials and Uses Thereof
Country	Patent No.	Date Filed	Date Issued
USA	5,393,602	04/19/91	02/28/95
Europe	EP 0580811	03/10/93	08/04/99
Japan	4-510189	03/10/92	07/26/02

20. Inventor: D.W. Urry
    Title: Superabsorbent Materials and Uses Thereof
Country	Patent No.	Date Filed	Date Issued
USA	5,520,672	02/06/95	05/28/96

21. Inventor: D.W. Urry
    Title: Elastomeric Polypeptide Matrices for Preventing Adhesion of Biological Materials
Country	Patent No.	Date Filed	Date Issued
USA	5,527,610	05/20/94	06/18/96
Japan		05/20/95	

22. Inventor: D.W. Urry
    Title: Elastomeric Polypeptide Matrices for Preventing Adhesion of Biological Materials
Country	Patent No.	Date Filed	Date Issued
USA	5,519,004	06/07/95	05/21/96

TABLE 9.1. *Continued*

23. Inventor: D.W. Urry
    Title: Bioelastomeric Drug Delivery System

Country	Patent No.	Date Filed	Date Issued
USA	6,328,996 B1	10/03/94	12/11/01
Europe	EP 0449592		11/30/94
Japan	1994740		11/22/95

24. Inventors: D.W. Urry, P. Shewry, and U. Prasad Kari
    Title: Bioelastomers Suitable as Food Product Additives

Country	Patent No.	Date Filed	Date Issued
USA	5,972,406	10/13/95	10/26/99
Europe	96912815.6	04/15/96	Pending
Japan	8-531270	04/23/2002	Pending

25. Inventors: D.W. Urry, H. Daniell, D. McPherson, and J. Xu
    Title: Hyperexpression of Bioelastomeric Polypeptides

Country	Patent No.	Date Filed	Date Issued
USA	6,004,782	10/13/95	12/21/99

26. Inventor: D.W. Urry
    Title: Bioelastomers Suitable as Food Product Additives

Country	Patent No.	Date Filed	Date Issued
USA	5,900,405	06/07/95	05/04/99
Europe	0830509	06/07/96	03/25/98
Japan	9-519202	08/06/97	Pending

27. Inventors: D.W. Urry, D. McPherson, and J. Xu
    Title: A Simple Method for the Purification of a Bioelastic Polymer

Country	Patent No.	Date Filed	Date Issued
USA	5,854,387	10/13/95	12/29/98
Europe	96912768.7	04/15/96	Pending
Japan	8-531261	04/16/2002	Pending

28. Inventor: D.W. Urry
    Title: Acoustic Absorption Polymers and Their Methods of Use

Country	Patent No.	Date Filed	Date Issued
USA	09/746,371	12/20/00	Pending
Europe	PCT/US 00/34658	7/22/02	Pending

29. Inventor: D.W. Urry
    Title: Bioelastomer Nanomachines and Biosensors (NEMS)

Country	Patent No.	Date Filed	Date Issued
USA	09/888,260	06/21/01	Pending
Europe	PCT/US01/20045	06/21/01	Pending

30. Inventor: D.W. Urry
    Title: Injectable Implants for Tissue Augmentation And Generation

Country	Patent No.	Date Filed	Date Issued
USA	09/837,969	04/18/01	In Allowance
Canada	2,319,558	02/26/99	Pending

31. Inventor: D.W. Urry
    Title: Injectable Implants for Tissue Augmentation And Generation

Country	Patent No.	Date Filed	Date Issued
USA	09/841,321	04/23/01	Pending
Europe	99908590.5	02/26/99	Pending
Canada	2,319,558	02/26/99	Pending
Japan	533072/2000	02/26/99	Pending

32. Inventors: D.W. Urry, P. Glazer, and T.M. Parker
    Title: Injectable Implants for Tissue Augmentation And Generation

Country	Patent No.	Date Filed	Date Issued
USA	6,533,819 B1	04/23/01	03/18/03

which have cost over $1.6 million dollars for patent prosecution and maintenance with additional patents possible as money becomes available for pursuing them. Note added in proof: It should be acknowledged that following your author's resignation as General Partner of BRL to complete this volume, the above listed patent portfolio of November 2003 has not been maintained by the new General Partner.

9.1.5.1 Low Back Pain

Low back pain causes more productivity loss than any other medical condition. In the United States alone, it caused 24 million people in 1990 to seek medical care, with health care costs of more than 33 billion dollars per year[4] and growing. When disability and lost productivity are included, the estimated cost exceeds 100 billion dollars per year.[5] This is the order of magnitude of currently lamented annual increases in the U.S. national debt. As described in section 9.4.4 below, there are two different approaches underway using protein-based materials in animal models to evaluate efficacy for alleviating this most burdensome health care problem.

9.1.5.2 Urinary Incontinence

Among 1.5 million residents of nursing facilities in the United States, 50% are reported to suffer from urinary incontinence. For the non-institutionalized population over 60 years of age, 30% of women and 15% of men contend with urinary incontinence. The annual medical care costs are greater than 7 billion dollars in the community at large and 3.3 billion dollars in nursing homes.[6] The result is a medical care cost that is about one-third that for low back pain. Equally significant are the magnitude of the personal problem, the loss of quality of life, the practical disability, and the loss of productivity that occurs when inadequate control of bladder function increasingly dictates the range of activities of someone's day. As described below, for more than a decade we have been addressing this problem with much optimism. Our single remaining hurdle is large-scale production of medical grade protein-based polymers. This is not an issue of limited technology but rather one of limited resources.

9.1.5.3 Pressure Ulcers (e.g., Bedsores)

The National Pressure Ulcer Advisory Panel (NPUAP) defines pressure ulcers as "localized areas of tissue necrosis that develop when soft tissue is compressed between a bony prominence and an external surface for a prolonged period of time."[7] They are also known as *decubitus* or *decubital ulcers, bed sores,* or *pressure sores* and *dermal pressure lesions*. The NPUAP has also stated conservatively that "well over a million persons in hospitals and nursing homes suffer from pressure ulcers" and that "estimates of average per case financial cost of pressure ulcer treatment in acute care settings range widely from $2,000 to $30,000."[8] Based on these numbers, medical care costs could well exceed $1,000,000,000.00 (1 billion dollars) per year.

The cost to care for this source of pressure ulcers represents an order of magnitude less than the health care cost for urinary incontinence. The incidence of pressure ulcers, however, increases with age, and the most affected age groups represent the fastest growing segment of the population. Accordingly, pressure ulcers are steadily becoming an even more burdensome medical care problem.[9]

Furthermore, patients with spinal cord injuries have a high incidence of pressure ulcers and is a population that reaches beyond nursing homes and hospitals. It has been reported that 40% of this group develops pressure ulcers during initial hospitalization and rehabilitation.[10] In fact, all populations dependent on assistive devices where soft tissue is compressed between a bony prominence and the assistive device are at increased risk for pressure ulcer formation. Again as discussed below, elastic protein-based materials are being tested in an appropriate animal model, and preliminary results are promising.

9.1.5.4 Vaccine Delivery

With the current concerns of bioterrorism and biowarfare, demand is enhanced for the most effective and least expensive delivery of vaccines. The potential of protein-based polymers is to design a series of nanoparticle structures for the delivery of a specific vaccine. Preliminary studies suggest that an appropriate series of protein-based polymers could simultaneously function as adjuvant and constitute a range of controlled release devices capable in a single treatment of providing the primary, secondary, and as necessary even subsequent immunizations. For example, vaccination against anthrax is said to require six vaccinations over a period of more than 1 year. The composition would be one of nanoparticles that would stimulate the immune response, contain the anthrax vaccine, and be taken up through an epithelial lining. This would not require hypodermic injection and in a single treatment would provide a series of timed-release events to fill the role of subsequent vaccinations.

9.1.6 Potential of Protein-based Materials to Relieve Additional Major Problems of Society

9.1.6.1 Drug Addiction Intervention

The magnitude of the worldwide drug addiction problem requires little elaboration. In the United States alone, the drug control budget for 1993 was about 12 billion dollars.[11] Although periodic decreases in total number of illicit drug users have been seen, the rise in "violent crimes, homelessness, and drug-related emergency room visits" suggests a basic limitation to the progress possible by the education and prevention focus.[11] Further progress would seem to require an increased reliance on drug addiction intervention rather than the current emphasis on drug interdiction and education.

A useful narcotic antagonist is naltrexone, which has a particularly benign side-effect profile with a "wide range between the effective dose and the toxic dose."[12,13] It has been found effective in preventing re-use of heroin after detoxification[14–17] and even following rapid detoxification.[18] Naltrexone (and related cationic narcotic antagonists) are useful against morphine, cocaine, codeine, and even alcohol addiction[19–21] as well as a spate of other afflictions, for example, bulimia nervosa,[22] self-injurious behavior,[23] autism,[24] Tourette's syndrome,[25] and Rett syndrome.[26] "Naltrexone is non-addicting and has been reported to produce few, if any, side effects. It is a unique and important option for patients desiring a nonaddicting treatment and total abstinence from all mood-altering drugs."[27] Furthermore, Sax and coworkers concluded, that "chronic administration of naltrexone in doses up to 300 mg/day for periods up to 36 months does not significantly change hepatic function. . . ."[28] On the other hand, a naltrexone plasma concentration of 2 ng/ml (approximately 1 million times less) is sufficient "to block the effect of a large heroin dose (25 mg)."[29]

The major limitation to the effective use of naltrexone resides in patient compliance. As stated by Fraser,[17] "The major limitation on the use of naltrexone, however, is the lack of incentive for the patient to keep taking the medication." As further stated by Littrel and Hyde,[30] "If patients comply with naltrexone treatment after detoxification, opioid discontinuation is usually maintained and craving is ablated. However, the clinical usefulness of naltrexone is limited by low patient acceptance. . . ."

Desirable at this point would be an implantable naltrexone delivery system that would not depend on patient compliance. It could be a biodegradable controlled release device, for example, injectable by hypodermic syringe for relatively short-term release. Alternatively, it could be implanted by trocar or by laparoscope for release for months as a biocompatible and biodegradable yet removable vehicle should patients' circumstances warrant substantial pain control. As demonstrated below (see Figure 9.39), properly designed protein-based materials exhibit this potential.

9.1.6.2 Improving the Environment

9.1.6.2.1 The Opportunity for Bioproduction of Biodegradable Plastics

Global environmental concerns have resulted in ever-mounting costs for disposal of solid waste. A particularly troubling solid waste is nondegradable petroleum-based plastics, and the toxic and hazardous chemicals required for the production of petroleum-based plastics buttress concerns. These issues have increased demand for polymers that are biodegradable, of benign production, and preferably from a renewable resource. This is evidenced in part by the Maritime Pollution (MARPOL) Treaty of 1995 prohibiting disposal of plastics at sea due to damage to the marine environment and by the earlier Plastic, Pollution Research and Control Act of 1987 (Public Law 100–220).

9.1.6.2.2 Protein-based Materials as Plastics

With the appropriate composition, protein-based polymers are plastics; as such they provide a promising answer to the above environmental problems. Petroleum-based plastics are made from an exhaustible, nonrenewable resource and require toxic and hazardous chemicals in their production. In contrast, protein-based polymers can be produced from renewable resources, by microorganisms and plants, and their processing can be water based, not requiring noxious chemicals. Protein-based plastics are inherently biodegradable and can be programmed with compositions resulting in lifetimes lasting from days to decades, as desired, depending on the anticipated environment. *Petroleum-based plastics mean death for the fishes, whereas protein-based plastics mean food for the fishes.*

Protein-based polymers have the potential to surpass the polyesters and other polymers because they can be directly produced in microorganisms and plants by recombinant DNA technology resulting in the capacity for diverse and precisely controlled composition and sequence. This is not possible with any other polymer, and it increases range of properties and the numbers of applications. Remarkably, with the proper design of composition, protein-based materials can be thermoplastics, melting at temperatures as much as 100°C below their decomposition temperatures. Therefore, they can be molded, extruded, or drawn into shapes as desired. Aspects of protein-based materials as plastics is also considered below.

9.1.7 Analogy Between the Early Search for a Commercial Success for Computers and the Current Search for a Commercial Success for Protein-based Materials

9.1.7.1 *Development of the "Killer App" for the Extraordinary Capacities of Computers*

During the early years of the development of uses for computers, the founding entrepreneurs were searching for what they called the "killer application" or "killer app" for short. The "killer app" was the application that would make this extraordinary computational capacity a commercial success. There was certainty in the eventuality of dramatic success, but there was also uncertainty as to what the "killer app" would be. As we now know, the "killer app" that launched the industry was the Visicalc system, $100 software package and a $2,000 computer. Visicalc provided a solution with which every company could readily analyze budgets and forecast expenditures for any desired scenario. The designation *computer* once exclusively meant the capacity for carrying out the most challenging number crunching. It was not anticipated at the time that computing would come to include word processing, accessing of information on the Internet, and even game playing.

There is analogy in the development of protein-based materials. Bioelastics, Inc., the general partner to Bioelastics Research Ltd. (BRL), has been working for about 15 years to arrive at a "killer app" that could launch the protein-based material industry. More specifically, BRL has been developing the scientific and intellectual property foundation that would pave the way for the extraordinary materials capacity of protein-based polymers to result in successful commercial applications.

The author's efforts toward identifying applications go back more than two decades and are recorded in the first patent application filed on August 14, 1978, to use cross-linked elastomeric polypentapeptides as "artificial vascular wall" (United States Patent No. 4,187,852, which patent issued on February 12, 1980; see Table 9.1).

9.1.7.2 Development of the "Healer App" for the Extraordinary Capacities of Protein-based Polymers

The particular problems in achieving a "killer app" or, as might be more fitting when including medical applications, the "healer app" for protein-based materials differ depending on whether the application is medical or nonmedical. Most nonmedical applications require a dramatic lowering of price in order to be competitive with petroleum-derived polymers, but this too is in the future. On the other hand, medical applications require an extraordinary level of purity for use as a biomaterial, and large amounts of capital are necessary to put in place a production facility using good manufacturing practices, to obtain Food and Drug Administration (FDA) approval, and to prosecute and maintain patents and other legal support, but this has been moving forward. These are challenging, expensive, and daunting tasks. Nonetheless, the use of protein-based polymers as biomaterials for medical applications now stands out as an imminent and intensely attractive opportunity for commercialization.

9.1.8 Assertion 4: The Applications Assertion

As introduced in Chapter 1, the present chapter constitutes Assertion 4: The Applications Assertion of the book. Production and purification are first addressed, as they obviously make up the initial enabling steps in moving toward applications of any materials. The most surefooted path toward materials applications of protein-based polymers, however, intertwines issues of production and purification through a combination of the two methods of preparation—chemical synthesis and biosynthesis. Chemical synthesis proved the biocompatibility of elastic protein-based polymers and therefore opened the door to medical applications. Demonstration of the biocompatibility of the chemically synthesized product made clear the purification required of elastic protein-based polymers produced by *E. coli* if unlimited medical applications were to be possible. Chemical synthesis also provided a faster route to diverse polymer compositions, which allowed for a more timely elucidation of the hydrophobic and elastic consilient mechanisms and also allowed proper recognition of the power of sequence control in achieving physical properties essential for particular materials applications. On the other hand, biosynthesis produces large quantities of elastic protein based, with which to achieve more extensive materials characterization, and biosynthesis holds promise of ever decreasing costs of production, thereby expanding the applications horizon.

The materials applications addressed here are but a limited sampling of what we have done and but a preliminary sampling of that which will be forthcoming in the protein-based materials industry of the future.

9.2 Production of Protein-based Polymers

The initial preparation of protein-based polymers utilized solution and solid phase peptide chemistry. This made possible the preparation of more than 1,000 polymer compositions. As discussed in Chapter 5, these compositions were studied for determination of their basic properties, for the development of the set of phenomenological axioms for protein engineering and function, and for the demonstration of the basic mechanism that underlies function. In short, it is the chemical synthesis that has allowed development of much of the basic science and the demonstration of the potential of protein-based materials in a timely manner. Mostly because of the historical relevance, but also because of the unique contributions of chemical synthesis to arriving at satisfactory purification of microbially prepared protein-based polymers, a brief description of the chemical synthesis of protein-based polymers is given below.

Biosynthesis of protein-based polymers, however, allows for production of protein-based polymers at cost levels that open a panorama of applications. Nonetheless, if it were not for the chemical syntheses, the design principles for any but the more mundane applications would not yet be in hand. In moving these advanced biomaterials toward the future, bioproduction becomes key. Nevertheless for

9.2 Production of Protein-based Polymers

some of the most advanced and esoteric applications, a combination of biosynthesis and chemical syntheses is yet required.

9.2.1 Biosynthesis (Using Recombinant DNA Technology)

The use of recombinant DNA technology, that is, biosynthesis or bioproduction, of protein-based materials may be considered in three main stages: (1) gene construction, (2) cell transformation, and (3) fermentation of microbial systems, such as *E. coli* and yeast, or large-scale expression in plants and animals.[31] What follows is an elaboration of key steps rather than a nuts-and-bolts description.

9.2.1.1 Gene Construction

A flow diagram for construction of genes for the production of protein-based polymers is given in Figure 9.1.[32] Interestingly, no new or specialized techniques were required for the use of recombinant DNA technology in the

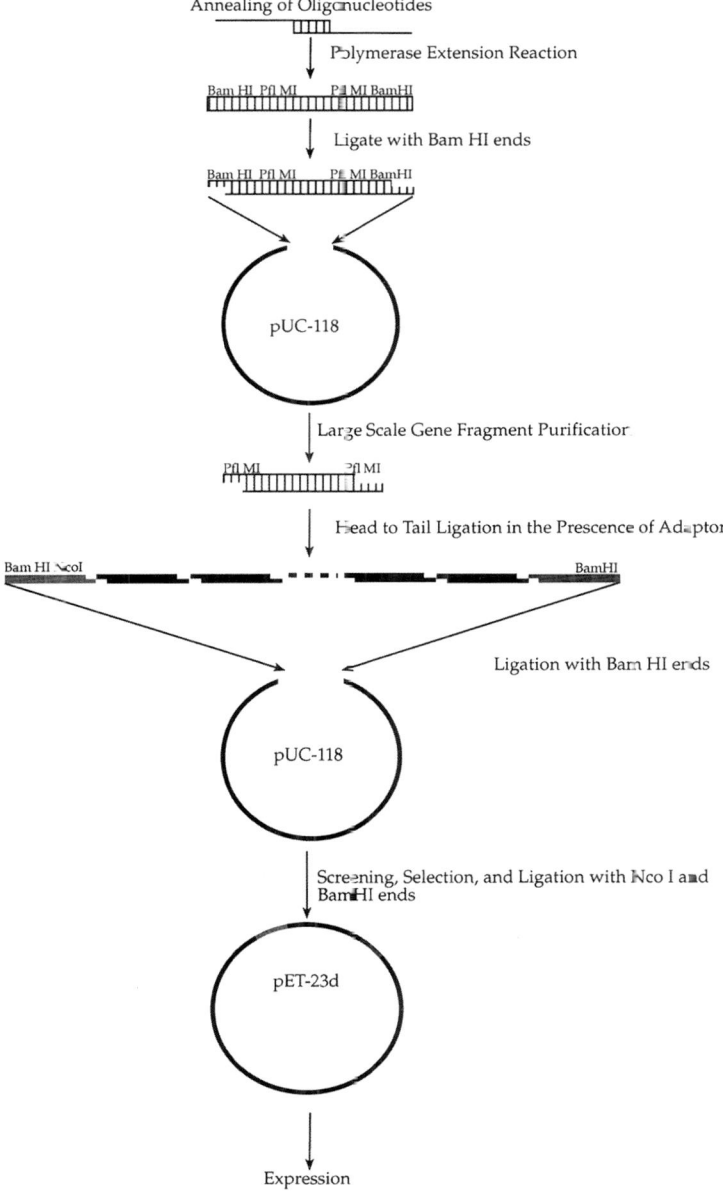

FIGURE 9.1. Diagram showing the approach whereby recombinant DNA technology was used to construct monomer and multimer genes for expression of the elastic protein-based polymer $[(GVGVP)_{10}]_n(GVGVP)$. (Reproduced with permission from Woods.[32])

production of protein-based polymers. The techniques utilized have been known in molecular biology for two decades, including the key awareness of the need to make use of codon redundancy for repetitive sequences.

9.2.1.1.1 Preparation of Monomer Genes and the PCR Technique

A basic monomer gene or a gene segment is chosen such that the total length would be no more than about 150 bases, encoding for a sequence of 50 amino acid residues. For example, consider the basic monomer gene (GVGVP)$_{10}$, which would require a sequence of 150 bases, a triplet codon for each amino acid residue. From the virtual beginning of these efforts in the early 1980s, there was concern for what is called *homologous recombination* wherein identical repeating base sequences would loop-out and re-establish base pairing to result in deletion of the looped-out gene sequence.

Also very early it was appreciated that the way to limit deletions was to utilize codon redundancy. For example, four different triplet base sequences encode for the glycine (Gly, G) residue, namely, GGU, GGC, GGA, and GCG, and another four for the valine (Val, V) residue, namely, GUU, GUC, GUA, and GUG. These and other codon redundancies may be seen in the genetic code given in Table 6.2. This means, for genes encoding GVGVP repeats, that different codons can be used for the V, G, and P of the different repeats and even within the same repeat. Thus, while there are 10 repeats at the protein-based polymer level for (GVGVP)$_{10}$, there are no repeats at the DNA level for the 150 base sequence that encodes for (GVGVP)$_{10}$.

Furthermore, while all Life uses the same genetic code, different species do have different codon preferences. In general, therefore, when designing a gene to be expressed in *E. coli*, the codon preferences for *E. coli* are used, whereas if the desire is to produce the polymer in tobacco, the codon preferences for tobacco are used.

Thus one designs a DNA sequence for a basic monomer gene, including the sequences of the restriction sites as required for the protein to be expressed and the plasmid to be used, as included in Figure 9.2. The next step is to

```
1         gly val gly val pro  (GVGVP)₈ gly val gly val pro

  CGGGATCCA GGC GTT GGT - - - - - - - - - - - - - - - - - CCA GGC GTT GGATCCCG
  GCCCTAGGT CCG CAA CCA - - - - - - - - - - - - - - - - - GGT CCG CAA CCTAGGGC

      BamH 1 PflM 1                              PflM 1  BamH 1


2         gly val gly val pro  (GVGVP)₈ gly val gly val pro

               TT GGT - - - - - - - - - - - - - - - - - CCA GGC
              G
              CG CAA CCA - - - - - - - - - - - - - - - - GGT C


3         (gly val gly val pro  (GVGVP)₈ gly val gly val pro)ₙ

               TT GGT - - - - - - - - - - - - - - - - - CCA GGC
              G
              CG CAA CCA - - - - - - - - - - - - - - - - GGT C


4             met           ((GVGVP))ₙ    gly val gly val pro stop

  TCGGATCCAGACC ATG GGC G TT - - - - - GGC G TT GGT GTA CCG TAAGCTTGAATTCGGATCCAG
  GACCTCGGTCTGG TAC C CG CAA - - - - - C CG CAA CCA CAT GGC ATTCGAACTTAAGCCTAGGTC

      BamH 1 Nco 1                                   Hind 3 EcoR 1BamH 1
```

FIGURE 9.2. Diagram showing base sequences for restriction sites at ends of a basic monomer gene encoding for (GVGVP)$_{10}$ and for concatenation to form the multimer gene for expression of the elastic protein-based polymer [(GVGVP)$_{10}$]$_n$(GVGVP). (Reproduced with permission from Urry et al.[33])

9.2 Production of Protein-based Polymers

prepare the full-length double-stranded 150 base pair monomer gene with additional bases for defined restriction sites as required for insertion into a given plasmid or vector. This begins with chemical synthesis of two base sequences of 80 to 90 bases, one from the 3' end and the other from the 5' end, as defined in Figures 3.14 and 3.15. The chemically synthesized sequences pair with each other using the overlap in their central region. Then the remarkable polymerase chain reaction (PCR, of forensic and other fame) completes the double strand. The details of the constructed genes, including the sequences for the restriction sites, are given in Figure 9.2 for the production of the monomer gene encoding for (GVGVP)$_{10}$.[33]

9.2.1.1.2 Transformation of E. coli, Production of Much Monomer Gene and Verification of Monomer Gene Sequence

An appropriate plasmid and strain of E. coli are chosen, for example, for the transformation and growing up of the transformed E. coli containing many plasmids, each plasmid with one copy each of the monomer gene. The plasmids are harvested, the monomer genes excised by the appropriate restriction enzyme(s), and the sequence of the monomer gene verified.

9.2.1.1.3 Concatenation of Monomer Genes to Produce Multimer Genes of Various Sizes

Using appropriate oligonucleotide adaptors and ligase enzyme, the monomer genes are polymerized. This is called *concatenation* or *concatemerization*. The result can be a series of genes containing repeats of the monomer gene, as shown in the gene ladder in Figure 9.3.[33] Each band in lane 2 is a gene differing in size by one monomer gene. Also, in the particular concatenation for polymerizing of (GVGVP)$_{10}$, the carboxyl end restriction site is used that adds another GVGVP, such that the final gene encodes for (GVGVP)$_{n \times 10+1}$ or [(GVGVP)$_{10}$]$_n$ (GVGVP). The gene construction process is shown schematically in Figures 9.1 and 9.2. Thus, lanes 3, 4, and 5 in Figure 9.3 show, respectively, genes for the expressions of 41, 121, and 251 repeats of (GVGVP). Figure 9.3 also

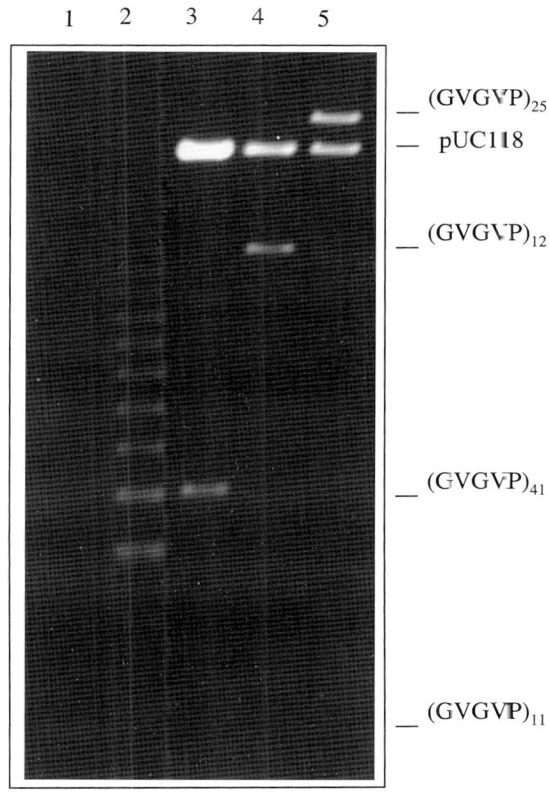

FIGURE 9.3. Agarose gel for separation of genes and gene fragments (**Lane 2**) gene ladder resulting from concatenation (polymerization) of basic monomer gene shown so that one can count up from the bottom to determine the number of repeats. (**Lane 3**) Gene with 41 repeats of the pentamer, that is, four monomer genes plus the single terminal (GVGVP; and higher up in the lane is the pUC118 plasmid from which the gene was excised by restriction enzymes. (**Lane 4**) Gene with 121 repeats of the pentamer, that is, 12 monomer genes (12 × 10) plus the single terminal (GVGVP) with pUC118 above. (**Lane 5**) At the top is the gene with 25 repeats of the monomer gene, plus the terminal (GVGVP), that is, 25 × 10 + 1. The gene encoding for 251 repeats (753 bases) is larger than the plasmid. (Adapted with permission from Urry et al.[33])

demonstrates that the resulting gene for expression of the protein-based polymer can be significantly larger than the plasmid in which it resides during replication.

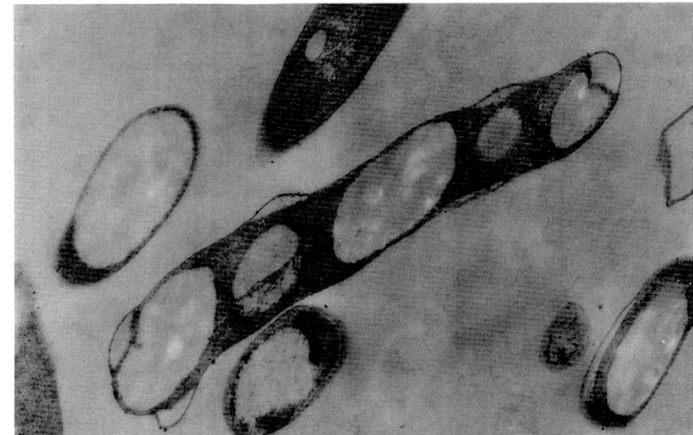

Transformed E. coli

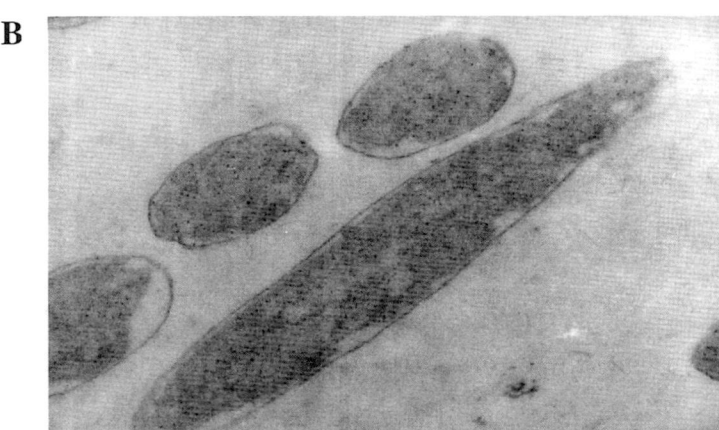

Non-transformed E. coli

FIGURE 9.4. (**A**) Transformed *E. coli* with more than 70% of the cell volume filled by huge inclusion bodies of the phase separated protein-based polymer (GVGVP)$_{121}$. (**B**) Nontransformed *E. coli* without product. (Reproduced with permission from Urry et al.[34])

9.2.1.2 Cell Transformation for Protein-based Polymer Expression

Having constructed the gene, it is now excised and inserted into an expression vector. The expression vector then is used to transform a strain of *E. coli* that is specialized for expression of the designed protein-based polymer. Figure 9.4 compares the same strain of nontransformed *E. coli* with transformed *E. coli* that produces the protein-based polymer (GVGVP)$_{121}$. The transformed *E. coli* are seen to have developed large inclusion bodies within each cell that are nearly pure, phase separated product.[34,35] This represents an extraordinary example wherein expressed elastic protein-based polymer occupies 70% to 80% of the cell volume.[35] Furthermore, from the continued growth of the *E. coli*, it would seem that the large inclusion bodies of product do not greatly interfere with cell duplication and function.[35]

9.2.1.3 Fermentation

A portion of transformed *E. coli*, stored in a glycerol stock, is used to grow an inoculum for addition to a large fermentor. The choice of media depends on the strain of microbe and chosen expression vector. It generally includes a chemical inducer to turn on the expression, a

9.2 Production of Protein-based Polymers

carbon source (e.g., glucose or glycerol), a nitrogen source (often as a source of amino acids), salts, other cofactors as may be provided, for example, by a yeast extract, and an antibiotic to which the transformed *E. coli* are selectively resistant. The fermentor conditions are optimized with respect to O_2 flow, stirring rate, pH, and so forth as required for the expression system.

Fermentors ranging in size from a few liters to thousands of liters are commercially available. In our own case, most commonly a 500 L fermentor has been used for large runs. A single 500 L fermentation of 24 hours can produce more protein-based polymer than a peptide chemist can produce in the laboratory in a year.

The product, $(GVGVP)_{251}$, of such a single fermentation is shown after phase separation in Figure 9.5. Furthermore, quality of the fermentation product is greater, in key respects, than that producible by peptide chemistry.

For production by *E. coli* there is an important issue of the removal of *E. coli* toxic proteins, particularly, because all animals have abundant antibodies against *E. coli* antigens. Because poly(GVGVP) has been chemically synthesized and adequately purified, it was established that these elastic protein-based polymers exhibited extraordinary biocompatibility. This awareness provided the impetus for the necessary levels of purification when using *E. coli*–produced protein-based polymers, as discussed below.

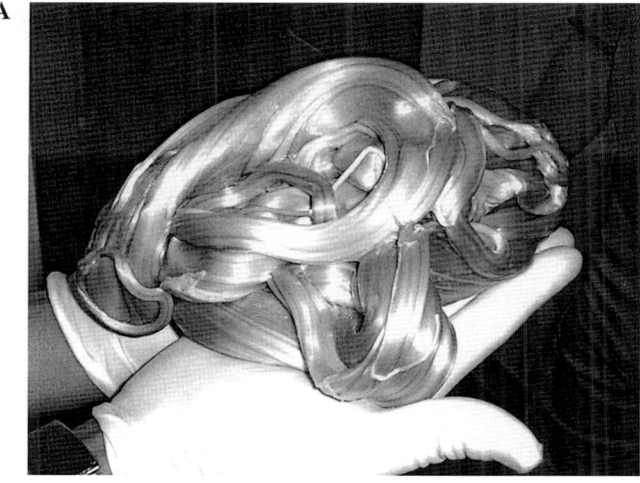

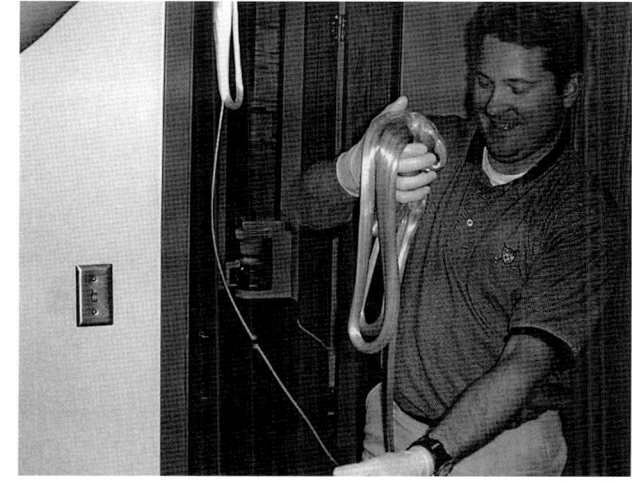

FIGURE 9.5. Elastic-contractile protein material, $(GVGVP)_{251}$, obtained by phase separation from the ruptured *E. coli* cells of a single fermentation run. (**A**) Note taffy-like appearance of the mass of product. (**B**) Viscoelastic mass being pulled into a fine strand. (Courtesy of Bioelastics Research, Ltd.)

9.2.1.4 Gene Constructions and Expressed Protein-based Polymers

Table 9.2 lists 79 peptide sequences encoded for by basic monomer genes. When n is greater than 1, the monomer gene has been concatemerized to form multimer genes, with the number of repeats of the monomer gene being indicated by the value of n, which indicates the size of the gene in a characterized clone of transformed *E. coli*. The values of n bearing the superscript a note expression of that specific protein-based polymer. This represents a significant stock of transformed *E. coli* that BRL has designed for specific applications. A substantial number of the expressed protein-based polymers were designed to elucidate elements of the comprehensive hydrophobic effect such as quantifying the apolar–polar repulsive free energy of hydration, ΔG_{ap}. As discussed below, the protein-based polymers with methodically varied values of ΔG_{ap} become primary candidates for systematic control of drug delivery.

9.2.1.5 Efforts Toward Low-cost Production in Other Microbes and in Plants Using Recombinant DNA Technology

9.2.1.5.1 Multiple Biosynthetic Approaches Are Underway

Efforts to produce elastic protein-based polymers include using tobacco,[36,37] mushrooms,[38] yeast,[39] and the seeds of arabidopsis, canola, and soy.[40] Production in tobacco plants has the potential of providing a healthful product from a plant that is the source of much illness, as such tobacco farms could be viewed as a capacity

TABLE 9.2. Bioelastics Research, Ltd.: List of gene constructions and expressed protein-based polymers.

1. [(GVGVP)$_{10}$]$_n$(GVGVP): n = 25[a], 14[a], 12[a], 4[a]
2. (GVGVP GVGFP GHGFP GVGVP GVGFP GFGFP)$_n$(GVGVP): n = 25[a], 15, 12, 9, 8, 7, 5, 3
3. (GVGVP GVGFP GKGFP GVGVP GVGFP GFGFP)$_n$(GVGVP): n = 75[a], 41[a], 33[a], 22[a], 15, 12, 6[a], 5, 2
4. (GVGVP GVGFP GEGFP GVGVP GVGFP GFGFP)$_n$(GVGVP): n = 42[a], 32[a], 17[a], 14[a], 8[a], 6[a], 4[a], 3[a], 2[a]
5. (GVGVP GVGFP GDGFP GVGVP GVGFP GFGFP)$_n$(GVGVP): n = 12[a], 8[a], 7, 6, 4
6. (GVGVP GVGFP GEGFP GVGVP GVGFP GKGVP)$_n$(GVGVP): n = 40, 21[a], 20, 11[a], 10, 8, 7
7. (GVGVP GVGKP GEGFP GVGVP GVGFP GFGVP)$_n$(GVGVP): n = 48, 39[a], 26, 22[a], 21, 20, 17, 15, 12, 9, 7
8. (GVGVP GVGFP GEGFP GVGVP GVGVP GKGVP)$_n$(GVGVP): n = 22[a], 20, 18, 15, 13, 11, 10, 9, 8, 7, 6, 5, 4
9. (GVGVP GVGKP GEGFP GVGVP GVGVP GFGVP)$_n$(GVGVP): n = 38, 22[a], 15, 6
10. [(AVGVP)$_{10}$]$_n$(GVGVP): n = 29[a], 7[a], 4[a], 2
11. [(GVGIP)$_{10}$]$_n$(GVGVP): n = 26[a], 15[a], 13[a], 7[a], 4[a]
12. (GVGIP GFGEP GEGFP GVGVP GFGFP GFGIP GVGIP GFGEP GEGFP GVGVP GFGFP GFGIP)$_n$(GVGVP): n = 30[a], 24, 20[a], 13[a], 11, 10, 6, 5
13. (GVGIP GFGEP GEGFP GVGVP GFGFP GFGIP GVGIP GFGEP GEGFP GVGVP GFGFP GFGIP GVGVP GVGRGYSLG VP)$_n$(GVGVP): n = 20
14. (GVGVP GVGFP GEGFP GVGVP GVGFP GVGFP)$_n$(GVGVP): n = 41[a], 29, 15[a], 12, 9, 7, 6, 3
15. (GVGVP GVGVP GEGVP GVGVP GVGFP GFGFP)$_n$(GVGVP): n = 60[a], 39[a], 24[a], 15
16. (GVGVP GVGFP GEGFP GVGVP GVGVP GVGVP)$_n$(GVGVP): n = 40[a], 25, 16, 14, 10, 8, 7, 6, 5, 4, 3, 2
17. (GVGVP GVGFP GKGFP GVGVP GVGFP GVGFP)$_n$(GVGVP): n = 21[a], 12, 6
18. (GVGVP GVGVP GKGVP GVGVP GVGFP GFGFP)$_n$(GVGVP): n = 45[a], 22[a], 20, 15, 14[a], 12[a], 7
19. (GVGVP GVGFP GKGFP GVGVP GVGVP GVGVP)$_n$(GVGVP): n = 26[a], 22[a], 17[a], 16, 14[a], 13, 11, 10, 9
20. [(GVGIP)$_{10}$-GVGVPGRGDSP-(GVGIP)$_{10}$]$_n$(GVGVP): n = 21, 18[a], 11[a]
21. [(GVGVP)$_{10}$-GVGVPGRGDSP-(GVGVP)$_{10}$]$_n$(GVGVP): n = 18[a], 15[a], 8[a]
22. [(GVGIP)$_{10}$-GVGVPGRGDSP-(GVGVP)$_{10}$]$_n$(GVGVP): n = 12[a], 8[a]
23. [(GVGIP)$_{10}$-GVGRGYSLGIP-(GVGIP)$_{10}$]$_n$: n = 10[a], 4[a]
24. {[(GGVP)$_3$(GGFP)]$_3$}$_n$: n = 27, 22
25. GKGKAPGK-[(GVGVP)$_{10}$]$_n$: n = 10[a]
26. GKGKAPGK-[(AVGVP)$_{10}$]$_n$: n = 20[a], 5[a]
27. [(GVGVP)$_{10}$(GVGVAP)$_8$(GVGVP)$_{10}$]$_n$: n = 8[a], 5[a], 2, 1
28. [(AFGFPAEGFP)$_5$]$_n$(GVGVP): n = 53, 22, 18, 16[a], 14, 6
29. [(GVGFPGEGFPGFGVP)$_3$]$_n$(GVGVP): n = 30[a], 14[a], 12
30. [(GVGVP)$_{10}$-GVGVPGNGVP(GVGVP)$_{10}$]$_n$(GVGVP): n = 10[a], 5[a]

9.2 Production of Protein-based Polymers

TABLE 9.2. *Continued*

31. [(GVGIP)$_{10}$-GVGIPGNGIP]$_n$(GVGVP): n = 30^a, 40^a
32. (GVGVP GVGVP GEGVP GVGVP GVGVP GVGVP)$_n$(GVGVP) YGSEFELRRQACGRTRAPP-PPPLPSGC: n = 44^a, 36^a
33. (GVGVP GVGVP GKGVP GVGVP GVGVP GVGVP)$_n$(GVGVP): n = 55^a, 36^a, 22^a
34. [(FEGFPAEGFP)$_5$]$_n$(GVGVP): n = 19^a
35. [(GVGVP)$_{10}$-GVGRGYSLGIP-(GVGVP)$_{10}$)]$_n$: n = 12^a, 7^a
36. [(GVGVP)$_{10}$-GVGRGYSLGIP-(GVGIP)$_{10}$)]$_n$: n = 7^a
37. [(GVGIP)$_{10}$-FPGVGQKRPSKFSKYLIP-(GVGIP)$_{10}$)]$_n$: n = 11^a, 6^a
38. [(GVGVP)$_{10}$-FPGVGQKRPSKRSKYLIP-(GVGVP)$_{10}$]$_n$(GVGVP): n = 9^a, 6^a
39. [(GVGVP)$_{10}$-FPGVGQKRPSKRSKYLIP-(GVGIP)$_{10}$]$_n$(GVGVP): n = 9^a, 7^a, 5^a
40. GKGKAPGK-(GVGVP GVGVP GEGVP GVGVPGVGVP GVGVP)$_n$(GVGVP): n = 1
41. [(FEGVP)$_{10}$]$_n$: n = 28^a, 15^a, 11, 4
42. [(GVGVP)$_{10}$-GVGVPGRGDSP-(GVGVP GVGVP GKGVP GVGVP GVGFP GFGFP)$_2$]$_n$(GVGVP): n = 8^a, 5
43. GKGKAPGK-[(GVGVP)$_{10}$-GVGVPGRGDSP-(GVGVP)$_{10}$]$_n$: n = 7^a
44. (FEGFP AFGFP)$_5$: n = 33, 29, 25, 25, 23, 19, 18, 16, 14, 11, 6
45. [(GVGVP GVGVP GKGVP GVGVP GVGFP GFGFP)$_2$-GVGVPGRGDSP-(GVGVP)$_{10}$(GVGVAP)$_8$]$_n$: n = 11^a, 8^a, 4^a, 3^a
46. [(GVGVP GVGVP GKGVP GVGVP GVGFP GFGFP)$_2$-(GVGVP)$_{10}$(GVGVAP)$_8$]$_n$: n = 12^a, 8^a, 4^a
47. [(GVGIP)$_{11}$-GYGIP-(GVGIP)$_{10}$]$_2$(GVGIP) Re$^\#$: n = 11^a, 10, 9, 8, 7, 6^a, 5
48. [(FVGEP FVGFP)$_5$]$_n$: n = 18, 17, 12, 2, 1
49. (GIGVP GAGVP GIGVP GIGVP GAGVP GIGVP)$_n$: n = 22^a, 11^a
50. [(GVGIP)11-GYGIP]$_n$(GVGIP)^b: n = 18^a, 17, 16, 15, 14, 13, 12, 11^a, 6, 3
51. [(GVGVP)$_{10}$]$_n$(GVGVP)^b: n = 28, 26, 18, 17^a, 16, 15, 14, 13^a, 12, 11, 6, 4
52. [(GVGIP)$_{10}$]$_n$(GVGVP)^b: n = 32^a, 25^a, 21^a, 15^a, 9^a, 6^a, 4^a, 2^a
53. {[(GVGIP GEGIP GVGIP)$_2$]$_2$}n n = 20^a, 9^a
54. {[GVGIP GEGIP (GVGIP)$_4$]$_2$}n n = 23^a, 10^a
55. [(GVGIP GEGIP GVGIP GEGIP GVGIP GVGIP)2]$_n$: n = 26^a, 17^a, 15^a
56. {[(GVGIP GKGIP GVGIP)$_2$]$_2$}$_n$: n = 19, 18, 17, 12
57. {[GVGIP GKGIP (GVGIP)$_4$]$_2$}$_n$: n = 21^a, 12^a
58. [(GVGIP GKGIP GVGIP GKGIP GVGIP GVGIP)$_2$]$_n$: n = 73, 41, 21, 17, 14, 11
59. [(GVGIP)$_8$(GYGIP)$_n$]: n = 32^a, 24^a, 18^a
60. {[(GVGIP)$_5$GYGIP]$_2$}$_n$: n = 20^a, 12^a, 11^a
61. {[(GVGIP)$_2$GYGIP]$_3$}$_n$: n = 26, 15^a, 14
62. [(GVGIP)$_{10}$]$_{n1}$-Fn3-[(GVGIP)$_{10}$]$_n$: n1 = 16 and n2 = 6
63. (GVGVP GVGVP GEGVP GVGVP GVGVP GVGVP)$_n$(GVGVP): n = 42^a, 23^a, 22^a, 18^a, 16^a
64. AKKKKKKG-[(GVGIP)$_{10}$]$_n$-VCGC: n = 15^a
65. AKKKKKKG-[(GVGIP)$_{10}$]$_n$-VCGC: n = 18^a
66. GKGKAPGK-[(GGAP)$_{12}$]$_n$(GVGVP): n = 1
67. [(GGAP)$_{12}$]$_n$: n = 1
68. [(FVGVP FEGVP)$_5$]$_n$: n = 1
69. {[(GVGVP)$_2$-GFGVP]$_3$}$_n$: n = 1
70. [(GVGIP)$_{11}$-GSGIP-(GVGIP)$_{10}$]$_n$(GVGIP)^b: n = 1
71. [(GVGIP)$_{11}$-GTGIP-(GVGIP)$_{10}$]$_n$(GVGIP)^b: n = 1
72. [(FFGEP)$_{10}$]$_n$: n = 1
73. [(GVGIP)$_{11}$-GSGIP]$_n$(GVGIP)^b: n = 1
74. [(GVGIP)$_{11}$-GTGIP]$_n$(GVGIP)^b: n = 1
75. [(GVGVP GVGVP GKGVP GVGVP GVGFP GFGFP)$_2$-GVGVPGRGDSP-(GVGVP GVGVP GKGVP GVGVP GVGFP GFGFP)$_2$]$_n$: n = 1
76. (GIGVP GAGVP GKGVP GIGVP GAGVP GIGVP)$_n$: n = 1
77. (GIGVP GAGFP GKGFP GIGVP GAGVP GIGVP)$_n$: n = 1
78. (GIGVP GAGFP GKGFP GIGVP GAGFP GVGFP)$_n$: n = 1
79. [ACPGCGGVGIPCPGCG-[(GVGIP)$_{26}$-Fn3(RGYSLG)-(GVGIP)$_6$]-CPGCGGVGIPCPGCG]$_n$: n = 1^c

---

^{a} The protein-based polymer was expressed.
^{b} The gene was made explicitly using the same codon for a repeating residue in the sequence, and then a different codon was used in the next repeat.
^{c} Fn3(RGYSLG) stands for the tenth type III domain of fibronectin in which the cell attachment sequence, GRGDSP, was replaced by the kinase site, RGYSLG, as a site for phosphorylation.

looking for a market with future growth. Yeast has the advantages of excreting the protein-based polymer from the cell, such that harvesting could proceed without cell destruction, and a protein-based polymer product from yeast would not be expected to have such purification demands as occur with *E. coli*. Production in the seeds of a plant such as canola has the advantage of being a value added product where the base product is the healthful canola oil. In this case the protein-based polymer would reside in the protein byproduct that is often sold as feed for livestock at costs of less than a dollar per pound. Accordingly, purification from the protein byproduct could be expected to yield protein-based polymer at a cost of little more than that of purification. Because the purification process uses a water-based phase separation, even the remainder of the protein byproduct could retain its value.

9.2.1.5.2 Elements in the Comparison of Low-cost Production

Chapter 4 emphasized that biological production of protein is energetically an extravagant process. This represents the basic cost of biology's capacity to place any of 20 different amino acid residues at each position and to do so while ensuring the correct optical isomer. Such diverse sequences, while maintaining the same optical isomer at each position, constitutes an impossible feat for chemical synthesis, and the cost of chemical synthesis yet remains greater by orders of magnitude.

Despite the very low efficiency for conversion of the energy of photons into the energy represented by protein that was discussed in Chapter 4, plants provide the most promising approach to low-cost production of protein-based polymers. The cost of harvesting photons from the sun resides within the cost of maintaining a healthy plant. Of course, when we speak of seeking low-cost production of protein-based polymers, the cost-comparison is between chemical synthesis and biological synthesis. Even though chemical synthesis loses badly in this comparison, specific contributions of chemical synthesis become invaluable, as discussed below.

9.2.2 Chemical Synthesis

9.2.2.1 Important Contributions of Chemically Synthesized Polymers

9.2.2.1.1 Chemical Synthesis Achieved Many Different Protein-based Polymers Quickly

The approximately 100 polymers that provided the basis for the data in Figure 5.9 and Table 5.1 were chemically synthesized in about 2 years' time. Over the last 10 years, some 80 basic monomer genes were prepared at BRL. In general, these monomers were then concatemerized to produce a multimer gene of the desired size. For proper comparison of the polymers in Figure 5.9 and Tables 5.1, 5.2, and 5.3, the polymers needed to be of a comparable size. Therefore, it would have taken 10 years instead of 2 years to move the basic science to the understandings provided by the data on chemically synthesized polymers. As shown in Figure 5.10, the data from the chemically sythesized poly[0.8(GVGVP),0.2(GXGVP)], where X is either the uncharged Lys (K°) or the uncharged Glu (E°), are quite comparable with those from biosynthesized **Model Proteins i** (E°/0F) and **x'** (K°/0F) of Table 5.5. Nonetheless, to obtain the most quantitative understanding of the comprehensive hydrophobic effect as represented in Figure 5.10, the work should be redone with biosynthesized protein-based polymers. *Succinctly stated, chemical synthesis has moved the most significant utilization of protein-based polymers forward by about one decade.*

9.2.2.1.2 Comparison of Protein-based Polymers Having a Random Incorporation of a Set of Repeats with Those Having the Repeats in a Fixed Sequence

The profound importance of sequence control and its relevance to the mechanistic basis of function could not have been demonstrable without the capacity to compare the pKa shifts of Glu residues with a fixed sequence with those of a random mixture of the same sequences. For example, the pKa of the chemically synthesized fixed sequence poly(GVGVP GVGFP G<u>E</u>GFP GVGVP GVGFP GFGFP) of 8.1 can be compared with the pKa of 5.2 obtained for the chemically synthesized

9.2 Production of Protein-based Polymers

polymer having a random mix of the same composition of pentamers, namely, poly[(GEGVP),2(GVGFP),(GFGFP)].[41] Therefore, sequence has fundamental consequence. No wonder biology pays such a high price for the control of sequence, as detailed in Chapter 4.

9.2.2.1.3 Chemical Synthesis Provided Proof of Biocompatibility as the Product Had No Microbial Contaminants

Only because the remarkable biocompatibility of chemically synthesized poly(GVGVP) was already known was there adequate impetus to purify microbially prepared (GVGVP)$_{251}$. Otherwise, it would have been presumed, as had been widely expected, that the toxicity of inadequately purified (GVGVP)$_{251}$ was an inherent property of the protein-based polymer. To be left in such a state of misunderstanding would have meant that the dramatic potential of elastic protein-based polymers for use in medical applications would be neither appreciated nor realized. The inflammatory response elicited by an inadequately purified biosynthetic elastic protein-based polymer would have overwhelmed most considered medical applications.

Our research utilizing chemical synthesis of repeating peptide sequences is represented in over 170 scientific publications. It utilized classic solution synthesis and to a much lesser extent solid phase methods. The primary focus in all cases has been development of an understanding relevant to structure, function, and mechanism rather than development of synthetic methodologies, although some of the latter did indeed occur principally due to the expert capacities of T. Ohnishi, K. Okamoto, R. Rapaka, K.U. Prasad, T.P. Parker, and D.C. Gowda. Here we note a few issues relevant to the production of protein-based polymers, that is, of polymers composed of repeating peptide sequences.

9.2.2.2 Classic Solution Methods for Chemical Synthesis of Repeating Sequences

A principal concern during chemical synthesis is racemization, the formation of some amino acid residues of the D-configuration rather than maintaining only the L-configuration. The result of only L-amino acid residues in the expressed protein, of course, results from protein synthesis by means of the genetic code and recombinant DNA technology. (See Figure 3.3 and the associated text in Chapter 3 to review the mirror image relationship between L- and D-amino acid residues and to appreciate the consequences relating to structure and function.) In particular, during the chemical activation of the carboxyl group of an amino acid and its subsequent reaction with an amino function of another amino acid, the amino acid residue with the activated carboxyl can racemize; some D-amino acid residues become incorporated into the growing chain with a consequence of structural disruption like that schematically represented in Figure 3.3C.

There are two ways to avoid racemization during chemical synthesis. One way is to rely at critical steps on the glycine (Gly, G) residue, which with two hydrogens on the α-carbon does not form mirror images. Therefore, for G, there is no consequence of interchanging positions of the α-carbon substituents during activation and reaction. The second way is to rely on the proline (Pro, P) residue in which the bridging of the **R**-group, –CH$_2$–CH$_2$–CH$_2$–, between α-carbon and nitrogen atom of the same residue limits racemization. Accordingly, with repeating units containing G and P, which are critical residues in the repeating sequences of elastic protein-based polymers, the strategy is to build the repeating unit with a G or P at its carboxyl terminus and to utilize crystallization to remove component peptide sequences that contain racemized residues.

The build-up strategy for the GVGVP or VPGVG repeating unit is a 2 × 3 strategy. First the dipeptide VP is synthesized and crystallized to form pure dimer containing only L-amino acid residues. Then the GVG tripeptide is synthesized and also crystallized to result in pure GVG containing only L-valine (Val, V). The dipeptide can be added to the tripeptide to form VPGVG, or the tripeptide can be added to the dipeptide to form the pentamer GVGVP. Both pentapeptides have either the required G or the required P to prevent racemization during polymerization of the pentamer to make the poly(GVGVP) or poly(VPGVG). Because higher molecular weight peptides could be obtained using GVGVP, this became the polymerization of preference. Also a search for the best activating reagent resulted in molecular weights higher than had been previously achieved by chemical synthesis.[42,43]

9.2.2.3 Merrifield Solid Phase Methods for Chemical Synthesis of Repeating Sequences

The Merrifield solid phase synthesis approach[44–46] constitutes an apparatus that is constructed and programmed to go through the many many steps involved in forming a single peptide bond and to grow the peptide chain by coupling each desired amino acid of the sequence to the growing chain attached to a solid phase, a so-called resin. No matter how much care may be taken at each coupling step, however, there remains a finite probability of racemization. For protein-based polymers this limitation is overcome by proceeding with pentamer additions, by adding GVGVP or VPGVG, for example. This has the great advantages of preparing racemization-free polymers of essentially a single chain size and of setting up and programming the apparatus to run continuously through many couplings.[47] There is the disadvantage, however, because coupling efficiency is not 100%, that much pentamer is required for the synthesis and is not incorporated into the growing chain.

The advent of recombinant DNA technology, two decades after the remarkable Merrifield advance, has overtaken many of the advantages of the Merrifield approach. After the time-consuming gene construction and development of a good expression system, large quantities of elastic protein-based polymers are produced in a short time, as shown in Figure 9.5.

9.2.2.4 Practicality of Chemical Synthesis

A major limitation of chemical synthesis is cost, which is central to the practicality of particular applications. Cost of production at the laboratory bench is on the order of $200/gram. Scale up could be expected to reduce that cost by an order of magnitude. For medical applications that require costly FDA approval, however, there is the added cost of what is called *good manufacturing practices*, which increase costs severalfold. Accordingly, any product that would not be competitive at $100/gram would be challenging to consider by chemical synthesis of protein-based materials, particularly when there can be strict functional requirements for certain elements of purity, optical and otherwise.

Nonetheless, when carried out with sufficient care, chemical synthesis of the polymers can be the best approach for establishing biocompatibility of protein-based polymers. As discussed below, the chemically synthesized products that have been carefully purified became primary standards against which purification of bioproduced polymers could be held. The specific disadvantages of chemical synthesis are its expense, the inability to be totally free of racemization, its reproducibility, and the need to use noxious and toxic chemicals. On the other hand, the primary challenge of biosynthesis using recombinant DNA technology is one of removing the toxic components unique to the producing organism or plant.

9.2.2.5 Biocompatibility of Chemically Synthesized Protein-based Materials

The biocompatibilities of the three physical states of these protein-based polymers, that is, hydrogel, elastic, and plastic, were initially determined using the representative chemically synthesized products. The polymer poly(GGAP) may be considered the parent polymer for the hydrogel state, because it exists as a hydrogel except under extreme conditions of high concentrations of multivalent salts. Poly(GVGVP) is the parent polymer for the elastomeric state, as it derived from the mammalian elastic fiber. It was the beginning sequence from which many of the other polymers were derived, initially by design and chemical synthesis of compositional variants. Poly(AVGVP) was the original sequence for the plastic state; it derived by simple substitution of a glycine (G) residue of poly(GVGVP) by an alanine (A) residue. The initial purpose

9.2 Production of Protein-based Polymers

of the synthesis of poly(AVGVP) was to test the newly proposed basis for near ideal or entropic elasticity.

Of course, the primary requirement for use of these polymers as part or all of a medical device is that the protein-based polymer must be sufficiently nontoxic, that is, it must exhibit adequate biocompatibility. As representative polymers for each of the interesting physical states, each of the above three compositions has been thoroughly examined by the standard set of 11 tests recommended by the American Society for the Testing of Materials (ASTM) for materials in contact with tissue, tissue fluids, and blood.

9.2.2.5.1 Perspective from a Battery of 11 ASTM Tests

According to the ASTM 1987 Standard Practice for Selecting Generic Biological Test Methods for Materials and Devices (Designation F784-87), there are nine tests for materials in contact with tissue and tissue fluids. An additional two tests are added for materials to be used in contact with blood. One of the nine tests, the test for carcinogenicity, is unnecessary if the mutagenicity test is negative. As shown in Table 9.3[33] for poly(GGAP), and as is also the case for poly(GVGVP), these polymers would appear to be even less mutagenic than the saline negative control. Accordingly, no carcinogenicity testing is required for these polymers, whereas a positive control, dexon, a suture material with a high rate of mutagenicity, presumably did require carcinogenicity testing but, as it is in use, it must be of sufficiently low carcinogenicity that it can be used.

Because antigenicity and immunogenicity are such potential issues for protein-based polymers, both the systemic antigenicity, as determined by the British Pharmacopoeia

TABLE 9.3. Plate incorporation assay for the Ames mutagenicity test for X^{20}–poly(GGAP).[a]

Salmonella typhimurium tester strains	Number of revertant colonies (agerage of duplicate plates)				
	TA98	TA100	TA1535	TA1537	TA1538
Saline (– control)	82	180	14	8	6
Saline test article solution (undiluted)	57	172	9	6	6
Saline w/S-9 (– control)	92	173	15	11	11
Saline w/S-9 test article solution (undiluted)	74	180	12	7	11
Dexon, 1 mg/ml (+ control)	1,048	1,408	N/A	247	N/A
Dexon, 1 mg/ml w/S-9 (+ control)	1,256	1,048	N/A	1,048	N/A
Sodium azide, 0.1 mg/ml (+ control)	N/A	N/A	2,752	N/A	N/A
Sodium azide, 0.1 mg/ml w/S-9 (+ control)	N/A	N/A	3,176	N/A	N/A
2-Nitrofluorene, 1 mg/ml (+ control)	N/A	N/A	N/A	N/A	2,800
2-Nitrofluorene w/S-9 (+ control)	N/A	N/A	N/A	N/A	2,240
2-Aminofluorene, 0.1 mg/ml (+ control)	N/A	184	N/A	N/A	11
2-Aminofluorene w/S-9 (+ control)	N/A	1,296	N/A	N/A	2,816

[a] Test article: X^{20}–poly(GGAP), batch CG65PA. In no case was there a twofold or greater increase in the reversion rate of the tester strains in the presence of the test article solution. N/A = not applicable.
[b] Reports from North American Science Associates (NamSA®).
Source: Reproduced with permission from Urry et al.[33]

TABLE 9.4. Guinea pig sensitization dermal reactions—challenge with X^{20}–poly(GGAP), batch CG65PA.[a]

Animal (No., Group)	Hours following patch removal			
	24	48	72	96
1 Test	0	0.5	0.5	0
2 Test	0	0.5	0.5	0
3 Test	0	0	0	0
4 Test	0	0	0	0
5 Test	0	0	0	0
6 Test	0.5	0.5	0.5	0
7 Test	0	0	0	0
8 Test	0	0	0	0
9 Test	0	0	0	0
10 Test	0	0	0	0
11 Control	0	0	0	0.5
12 Control	0	0	0	0
13 Control	0	0.5	0	0
14 Control	0	0	0	0
15 Control	0.5	0.5	0.5	0

[a] Test article solution was received in the right flank.
Source: Reproduced with permission from Urry et al.[33]

Antigenicity Test (BPAT), and the dermal sensitization study (a maximization method) were used. The scoring for the dermal sensitization study, shown in Table 9.4[33] for poly(GGAP), is the same as in Table 9.5[33] for the intracutaneous toxicity study. In the latter case, there simply is neither erythema (redness) nor edema (swelling) due to injection of the polymer. In the former case, the test polymer was as innocuous as the negative control of salt water. Obviously, these polymers are remarkably biocompatible.

The resulting 11 tests are summarized in Table 9.6[48] for poly(GVGVP). An even more stringent examination for poly(GVGVP) is discussed below, but first we consider poly(AVGVP).

The summary of the 11 tests for poly(AVGVP) (see Table 6 in Urry et al.[49]) reads the same as for poly(GGAP) and poly(GVGVP). A direct comparison of the guinea pig dermal sensitization test for poly(AVGVP) in Table 9.7, however, shows poly(AVGVP), although yet considered nonsensitizing, to be more reactive than that shown in Table 9.4 for poly(GGAP) and that for poly(GVGVP) (see Table 6 in Urry et al.[48]). This in part demonstrates the truly remarkable biocompatibility of poly(GVGVP) and poly(GGAP).

9.2.2.5.2 Refractory to Formation of Monoclonal Antibodies

Because of the need for monoclonal antibodies to poly(GVGVP) in order to follow bioproduction of this polymer in the tobacco plant, truly exceptional efforts were undertaken to achieve antibodies. Attempts to make monoclonal antibodies directly to poly(GVGVP) were simply unsuccessful despite repeated efforts using the most sensitive and repeatedly sensitized mice and rats. It was possible to find three weak monoclonal antibodies to poly(AVGVP) that had exhibited good biocompatibility[50] but not the same extraordinary biocompatibility as for the representatives of the hydrogel, poly(GGAP), and elastic,

TABLE 9.5. U.S. Pharmocopeia intracutaneous toxicity observations for poly(GGAP)

Rabbit No.	24 Hours		48 Hours		72 Hours	
	ER	ED	ER	ED	ER	ED
69235						
Test	0	0	0	0	0	0
Control	0	0	0	0	0	0
69226						
Test	0	0	0	0	0	0
Control	0	0	0	0	0	0

Reports from North American Science Associates (NamSA). Date prepared, 2-8-93; date injected, 2-8-93; date terminated, 2-11-93. Erythema (ER): 0 = none, 1 = barely perceptible, 2 = well defined, 3 = moderate, 4 = severe. Edema (ED): 0 = none, 1 = barely perceptible, 2 = well defined, 3 = raised 1 mm, 4 = raised >1 mm. Rating (test and control): 0.0–0.5, acceptable; 0.6–1.0, slight; >1.0, significant.
Source: Reproduced with permission from Urry et al.[33]

9.2 Production of Protein-based Polymers   481

TABLE 9.6. Summary of 11 biological test results for poly(VPGVG) and its cross-linked matrix.[a]

Test	Description	Test System	Results
Ames (mutagenicity)	Determine reversion rate to wild type of histidine-dependent mutants	*Salmonella typhimurium*	Nonmutagenic
Cytoxicity	Agarose overlay determines cell death and zone of lysis	L-929 mouse fibroblast	Nontoxic
Systemic toxicity	Evaluate acute systemic toxicity from an intravenous or intraperitoneal injection	Mice	Nontoxic
Intracutaneous toxicity	Evaluate local dermal irritant or toxic effects by injection	Rabbit	Nontoxic
Muscle implantation	Effect on living muscle tissue	Rabbit	Favorable
Intraperitoneal implantation	Evaluate potential systemic toxicity	Rat	Favorable
Systemic antigenicity (BPAT)	Evaluate general toxicology	Guinea pigs	Nonantigenic
Sensitization (Kligman test)	Dermal sensitization potential	Guinea pigs	Nonsensitizing
Pyrogenicity	Determine febrile reaction	Rabbit	Nonpyrogenic
Clotting study	Whole blood clotting times	Dog	Normal clotting time
Hemolysis	Level of hemolysis in the blood	Rabbit blood	Nonhemolytic

[a] Reports from the North American Science Associates (NAmSA).
Source: Reproduced with permission from Urry et al.[48]

poly(GVGVP), states. Fortunately, one of the weak monoclonal antibodies to poly(AVGVP) was found to cross-react, more weakly yet, with poly(GVGVP) such that poly(GVGVP) could be identified in the transgenic tobacco plant. Further efforts to utilize the monoclonal antibody to poly(AVGVP) for ready identification of poly(GVGVP) in the development of microbial expressions systems were abandoned due to inadequate reactivity.

TABLE 9.7. Guinea pig sensitization dermal reactions—Challenge test article: X^{20}–poly(AVGVP), batches CG20WR and CG156WR.

Animal No./Group	Hours following patch removal							
	24		48		72		96	
	Site A	Site B	Site A	Site B	Site A	Site B	Site A	Site B
1 Test	0	0.5	0	0	0	0	0	0.5
2 Test	0.5	0	0.5	0	0	0	0	0
3 Test	0	0.5	0.5	0.5	0	0.5	0	0
4 Test	0.5	0.5	0.5	0	0.5	0	0.5	0
5 Test	0.5	0.5	0.5	0	0.5	0.5	0	0.5
6 Test	0.5	0	0	0.5	0	0.5	0	0
7 Test	0.5	0	0.5	0	0.5	0	0	0
8 Test	0.5	0.5	0	0.5	0	0	0	0
9 Test	0.5	0.5	0	0.5	0	0.5	0	0.5
10 Test	0.5	0.5	0	0	0	0	0.5	0
11 Control	0	0	0	0	0	0	0	0
12 Control	0.5	0.5	0	0	0	0	0.5	0
13 Control	0	0	0	0	0	0	0	0
14 Control	0	0	0	0.5	0	0.5	0	0
15 Control	0.5	0.5	0.5	0	0	0	0	0

Site A = left flank = SC control vehicle; site B = right flank = SC test extract. Reports from North American Science Associates (NAmSA®).
Source: Reproduced with permission from Urry et al.[49]

9.2.2.5.3 Response of Human Monocytes to Elastic Protein-based Polymers

There is one further telling study that bears on the nature of the interaction, or absence thereof, of poly(GVGVP) with human cells, monocytes, whose role it is to identify and destroy foreign materials in the body. Grace Picciolo and coworkers at the Center for Devices and Radiological Health of the FDA, using instrumentation designed to evaluate the reaction of human monocytes to biomaterials, found poly(GVGVP) and the related polymer containing cell attachment sequences poly[40(GVGVP),(GRGDSP)] to subdue the background activity of the monocytes.[51] This innocuous, and possibly even passivating, response of human monocytes to the basic polymer poly(GVGVP) means that it becomes possible to add different biological activities very selectively. The capacity will be to elicit selectively the desired biological response to the added sequence without having to tolerate background activities of the carrier polymer. This becomes especially useful when the carrier protein-based material has an elasticity that can match that of the natural tissues. One of the results of this combination of biocompatibility and biological elasticity is a group of applications referred to below as *soft tissue restoration*. First, however, the importance of purification from *E. coli* fermentation is emphasized, as this is necessary before the remarkable biocompatibility of elastic protein-based polymers can be utilized.

9.3 Purification of T_t-type Protein-based Polymers

By T_t-type, we mean polymers that exhibit inverse temperature transitions in which the protein-based polymers hydrophobically associate on raising the temperature. T_t represents the onset temperature for the transition. For the elastic-contractile model proteins of interest here, the inverse temperature transition is seen as a phase separation resulting from both intermolecular and intramolecular hydrophobic association.[52] On raising the temperature above that of the onset temperature for the transition, T_t, T_t-type protein-based polymers go from being dissolved in solution to being separated from solution. There exists another very important means of achieving the phase separation of T_t-type protein-based polymers; it is to lower the temperature for the onset of the phase transition, T_t, from above to below the operating temperature. When concerned with the physiological operating temperature of 37° C, to lower the value of T_t from 37° to 25° C causes the T_t-type protein-based polymer to separate out of solution. As treated extensively in Chapter 5, the process can be one of self-assembly by hydrophobic association.

As demonstrated below, this property, fundamental to the diverse protein functions discussed in Chapters 5, 7, and 8, provides the basis for purification.[52]

9.3.1 Purification by Phase Separation (Inverse Temperature Transition)

The purpose below is not to describe methodologies for evaluating purity, but rather it is to use selected methodologies to demonstrate or to monitor the effectiveness of purification based on phase separation. The utilized methodologies are SDS PAGE (sodium dodecyl sulfate polyacrylamide gel electrophoresis) and radiolabeling of contaminants. The lysed or ruptured cells of *E. coli*, transformed to produce the specific T_t-type protein-based polymer, provide the impure reference state against which subsequent purification steps may be compared.

9.3.1.1 SDS PAGE Demonstration of Purification by Phase Separation

Figure 9.6 represents a negative staining technique involving a $CuCl_2$ stained gel to demonstrate purification at different steps and for different fractions during the purification by phase separation of $(GVGVP)_{141}$. Lane 1 contains all of the components of the gross lysed cells of *E. coli* transformed to produce $(GVGVP)_{141}$; as such it shows all of the protein bands of the transformed *E. coli*. As will be discussed, the bulging band of lane 1 is the product

9.3 Purification of T_t-type Protein-based Polymers

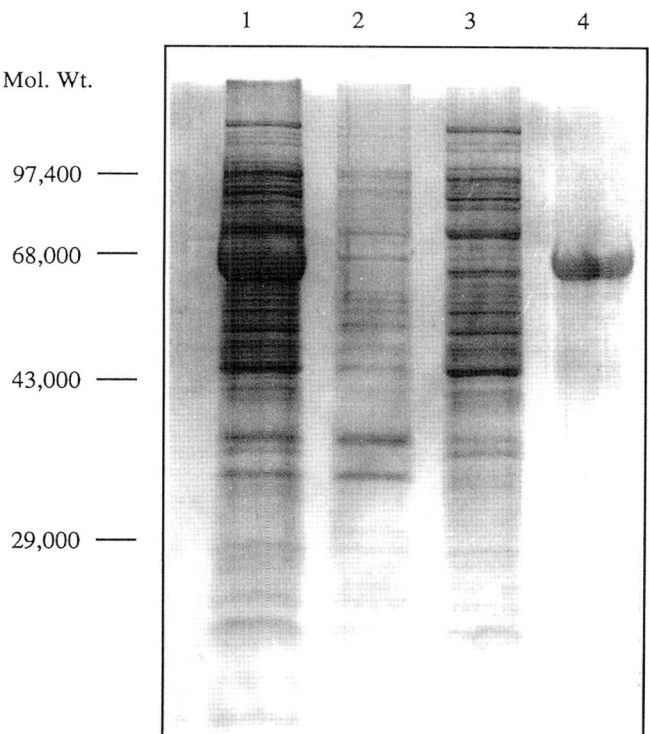

FIGURE 9.6. CuCl$_2$-stained gel of SDS PAGE shows purification of (GVGVP)$_{251}$ by phase separation. (**Lane 1**) Lyse-transformed cell showing all *E. coli* proteins plus the bulging band of product, (GVGVP)$_{251}$. (**Lane 2**) Residue of the cold spin where the protein-based polymer plus most *E. coli* protein remained in solution. (**Lane 3**) Supernatant of 37°C warm spin that contains essentially all of the remaining *E. coli* protein. (**Lane 4**) Model protein, (GVGVP)$_{251}$, phase separated at 37°C from the *E. coli* protein that remained in solution (lane 3). (Reproduced with permission from Urry et al.[33])

(GVGVP)$_{141}$. The sample is then chilled to 4°C to ensure dissolution of the protein-based polymer and then centrifuged; the residue, which contains insoluble cell debris, embedded proteins, and any remaining nonlysed cells, is shown in lane 2. The supernatant of the cold spin is heated to 37°C and again centrifuged. The supernatant of the warm spin in lane 3 shows the panoply of *E. coli* proteins. When the precipitate of the warm spin is redissolved and run in lane 4, a single band of purified product, (GVGVP)$_{141}$, is observed. This purification is achieved entirely by phase separation.

9.3.1.2 Purification by Phase Separation Shown by Carbon-14 Labeled *E. coli*

The effectiveness of phase separation in purification of the elastic protein-based polymer (GVGVP)$_{251}$ can be evaluated by carbon-14 labeling all of the *E. coli* constituents except the protein-based polymer. Incubation of *E. coli* with the radiolabeled carbon source C-14 glucose enables this. *E. coli*, labeled by this means, is lysed and added to the lysed cells from a fermentation of transformed *E. coli* in the absence of carbon-14, and the mixture is purified. In this way only the *E. coli* impurities are labeled.

Figure 9.7 demonstrates the reduction in impurities with each phase separation cycle. In short, three cycles reduce impurities by over a factor of 1,000. Using 1 mg of polymer, three cycles of phase separation decrease impurities from 1,428,600 parts per billion (ppb) to 1,365 ppb.[53] As discussed below, further purification lowers impurities, as judged by Western immunoblot techniques, to less than 5 ppb.

9.3.2 Physical Characterization and Verification of Product Integrity

9.3.2.1 Gross Visualization of the Phase Separated Product

An example of (GVGVP)$_{251}$ as the phase separated product is shown in Figure 9.5 and discussed above in section 9.2. By the small change of adding one CH$_2$ group per pentamer, the gross physical properties change substantially.

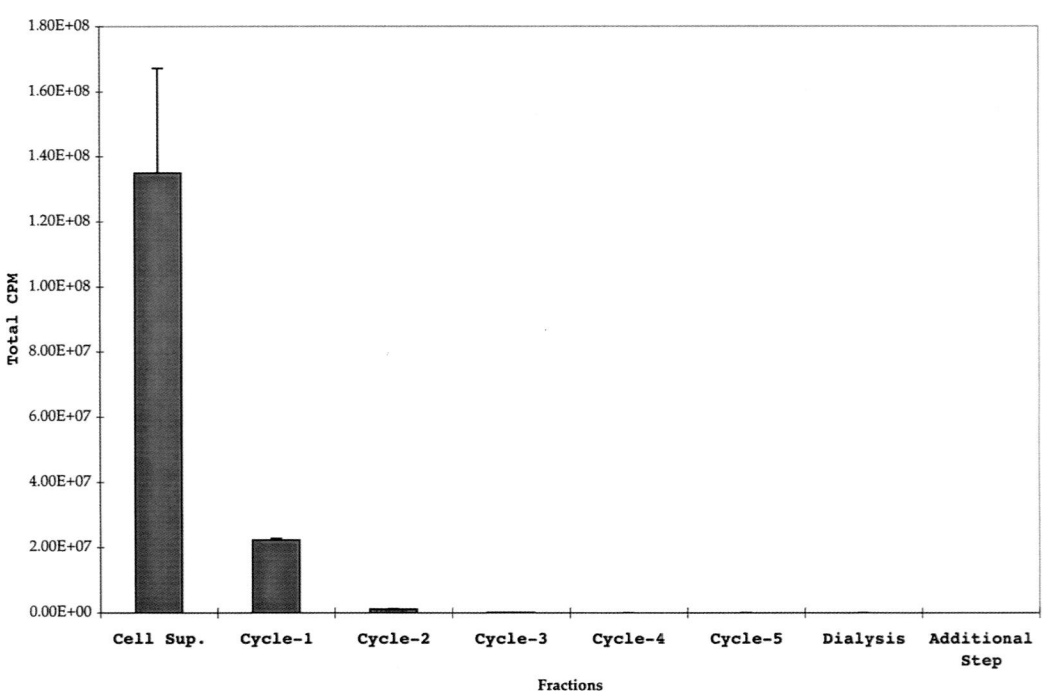

FIGURE 9.7. Purification of (GVGVP)$_{251}$, followed by loss of carbon-14–labeled *E. coli* constituents as a function of each cycle of phase separation. Each phase separation cycle reduces impurities by approximately 10-fold. (Unpublished data, of A. Pattanaik, T.M. Parker, and D.W. Urry. Courtesy of Bioelastics Research, Ltd.)

As shown in Figure 9.8, the phase separated state of a relatively small amount of protein-based polymer, (GVGIP)$_{260}$, looks like a small pancake that can be pulled into a tough band 1 meter in length. The elastic protein-based polymers in Figures 9.5 and 9.8 are fundamental compositions that can be treated in many different ways to result in products of commercial interest. With designed modifications in the compositions, many more specialized applications result, as is but briefly noted as this chapter progresses.

9.3.2.2 Nuclear Magnetic Resonance Evaluation of Sequence Integrity and Purity

The sequence and purity of the basic repeating 30 mer of **Model Protein v** of Table 5.5 are verified by one- and two-dimensional nuclear magnetic resonance (NMR) spectroscopy.[54] In particular in Figure 9.9, by a combination of the one-dimensional spectrum at the top of the figure and the two-dimensional HOHAHA map in the upper left, the through-bond proton–proton couplings are dissected and assignments made. This result demonstrates the absence of unexpected amino acid residues in the repeating 30 residue sequence.

At the lower right of the two-dimensional map, nuclear Overhauser enhancement spectroscopy (NOESY) demonstrates a set of through space proton–proton contacts that allow determination of sequence. As listed at the top of Figure 9.10, there are four types of through space proton–proton contacts. Contact between the αC$\underline{H}$ proton of residue i and the N$\underline{H}$ proton of residue i + 1 is defined as $d_{\alpha N}$ and

9.3 Purification of T_t-type Protein-based Polymers

through this continuity identifies residues i and i + 1 of the sequence. These contacts are followed in the expanded portion of the map shown in Figure 9.10, and they are continuous along the sequence with the exception of an interruption at the Pro (P) residues that have no αCH proton.

The portions of the sequence involving Pro are obtained by proton–proton contacts from the αCH proton of the residue preceding the Pro residue to the δCH protons of the Pro residue itself. These contacts are designated as $X^i\alpha\underline{H}$–$P^{i+1}\delta\underline{H}_2$, and they complete the missing step in the sequence information obtained from $\mathbf{d}_{\alpha N}$. Additional proton–proton contacts that sometimes can be used to fill in missing contacts, $\mathbf{d}_{NN}$ and $\mathbf{d}_{\beta N}$, can also act to confirm previous sequence assignments. Specifically, $\mathbf{d}_{NN}$ indicates through space contact between NN protons of adjacent residues in the sequence, that is, residue^{i}N$\underline{H}$ to residue^{i+1}N$\underline{H}$. Also, the $\mathbf{d}_{\beta N}$ indicates contact between residue$^i\beta$C$\underline{H}$ and residue^{i+1}N$\underline{H}$. The upper part of Figure 9.10 lists the four different types of proton–proton contacts that provide sequence information and the specific sequence connectivities determined by each. The two-dimensional NMR data in Figures 9.9 and 9.10 verify the repeating sequence of **Model Protein v** of Table 5.5.

This marvelous methodology, due principally to the work of Wider et al.,[55] has further use in determining the three-dimensional conformation of proteins and was applied early in this regard to poly(GVGVP) and its cyclic analogue cyclo(GVGVP)$_3$.[56]

9.3.2.3 Mass Spectra to Determine Size of Expressed Polymer

Having verified that the repeating sequence of the expressed protein agrees with the sequenced monomer gene, it is now to be determined how many repeating units constitute the expressed protein-based polymer, that is, the

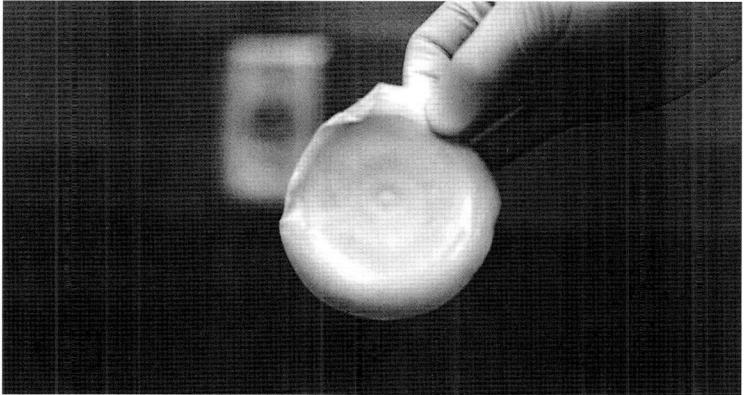
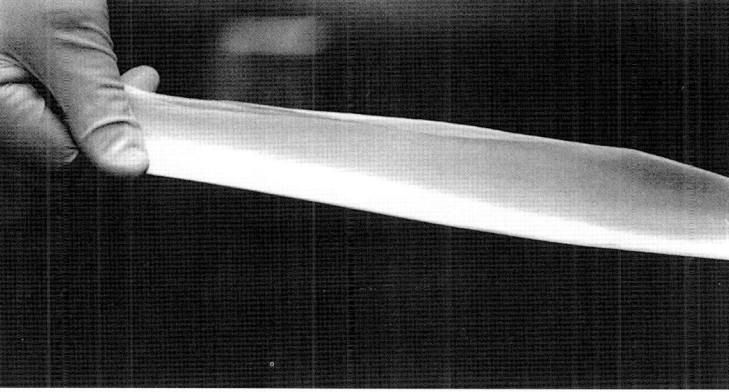

FIGURE 9.8. Purification by phase separation of the elastic-contractile protein-based polymer (GVGIP)$_{260}$. (**Top**) Pancake formed on phase separation. (**Bottom**) Stretching of the robust viscoelastic pancake out to a length of about 1 meter. (Courtesy of Bioelastics Research, Ltd.)

Model protein v: (GVGVP GVGFP GEGFP GVGVP GVGFP GFGFP)$_{36}$

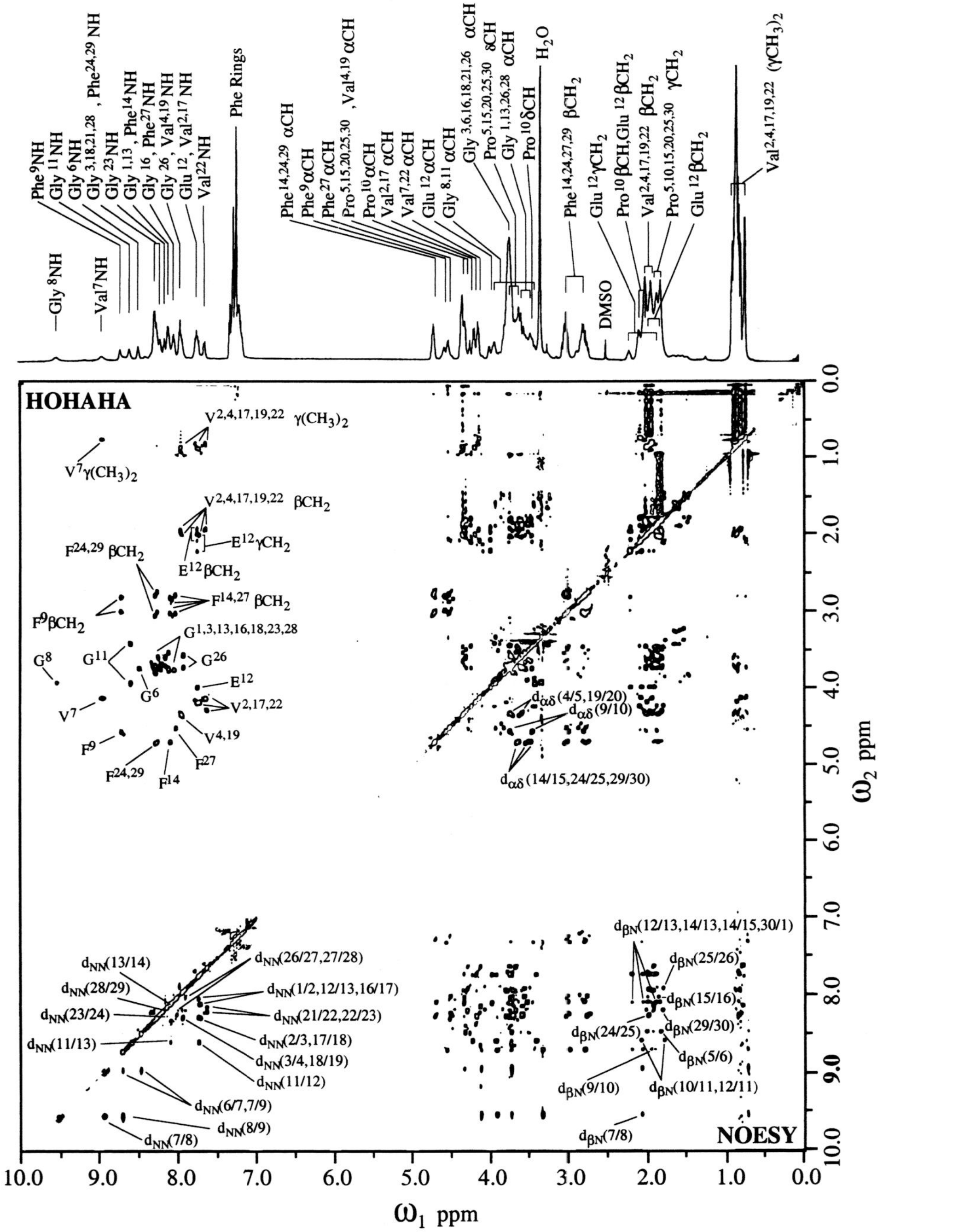

FIGURE 9.9. Two-dimensional NMR (NOESY and HOHAHA) proton spectroscopy at 600 MHz allows for determination of purity and verification of the basic 30 residue, 6 pentamer, repeating sequence of the **Model Protein v**, (GVGVP GVGFP GEGFP GVGVP GVGFP GFGFP)$_{36}$, as listed in Table 5.5. (Reproduced with permission from Urry et al.[54])

9.3 Purification of T_t-type Protein-based Polymers

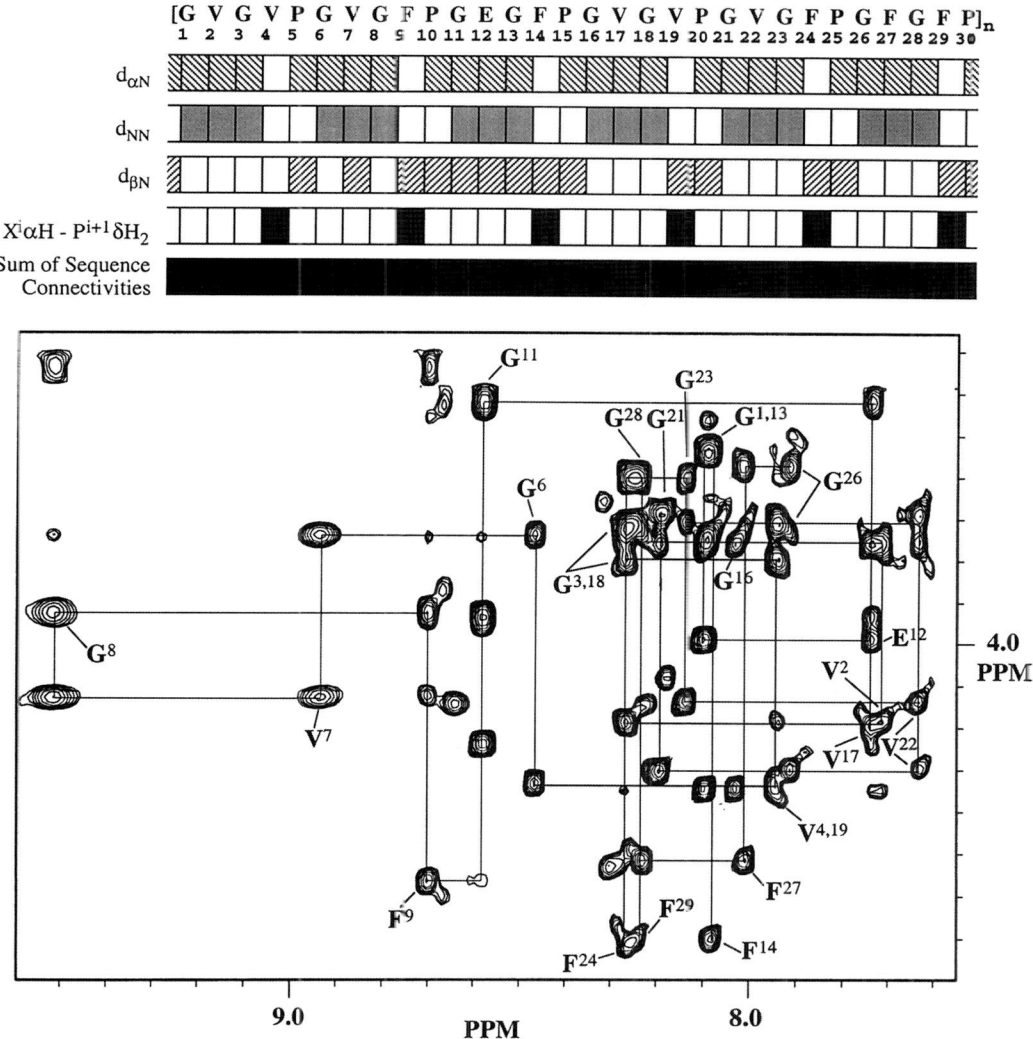

FIGURE 9.10. Detailed analysis of the NH^i-α-CH^{i+1} ($d_{\alpha N}$) proton–proton contacts that verify the sequences between P residues. With the other contacts, especially the $X\alpha CH^i Pro$-δCH_2 contact, bottom row, the sequence of the complete 30 residue repeating unit, (GVGVP GVGFP GEGFP GVGVP GVGFP GFGFP)$_{36}$, is verified. See text for further discussion. (Reproduced with permission from Urry et al.[54])

molecular weight of the expressed model protein. Remember that this information is available from the gene ladder in Figure 9.3. Therefore, the mass spectroscopy data on the expressed protein-based polymer should be consistent with the size of the gene that was made for that purpose.[57]

An approach that has been developed to determine the molecular weight of large proteins is called the *matrix-assisted laser desorption ionization*–time-of-flight (MALDI-TOF) mass spectrometry. Figure 9.11 contains the parent ion peaks for the two model proteins,

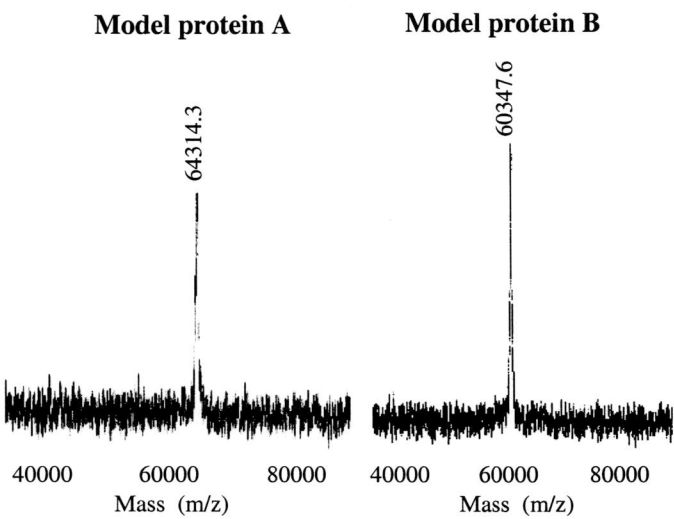

FIGURE 9.11. Mass spectrometry using the matrix-assisted laser desorption ionization time of flight (MALDI-TOF) method to determine atomic mass for verification of chain length (i.e., number of repeating units for the model proteins: (**A**) [(GVGVP)$_{10}$GVGVPGRGDSP(GVGVP)$_{10}$]$_7$(GVGVP). (**B**) [(GVGVP)$_{10}$(GVGVAP)$_8$(GVGVP)$_{10}$]$_5$(GVGVP). See text for further discussion. (Reproduced with permission from Urry et al.[58])

A: [(GVGVP)$_{10}$GVGVPGRGDSP(GVGVP)$_{10}$]$_7$(GVGVP) with seven repeats of the basic repeating unit of 111 residues with a molecular weight of 64,300 Da

B: [(GVGVP)$_{10}$ (GVGVAP)$_8$(GVGVP)$_{10}$]$_5$(GVGVP) with five repeats of the basic repeating unit of 148 residues, giving a molecular weight of 60,300 Da

Thus, the capacity to verify the sequence of the basic repeating unit and the number of repeating units in the expressed protein-based polymer has been demonstrated.

9.4 Medical Applications

Consideration of potential medical applications of protein-based polymers will always be incomplete. The source of any particular listing will be limited by confidentiality constraints on its author, by the limits of the author's imagination, and by the decision to leave out applications presumed to be out of context. At first glance, the list of medical applications included in this section 9.4 may seem long. It is, nonetheless, only a beginning.

A major element when considering medical applications is the potential antigenicity of protein-based materials and the adequacy of means of purification, especially when prepared from sources such as *E. coli*. Before discussing the medical applications, however, it is important to address immunogenicity and purification sufficient for medical applications. Because of the truly remarkable biocompatibility of elastic protein-based polymers, a short statement of the considered basis of the biocompatibility is given. Then a number of medical applications are treated.

9.4.1 The Opportunity and the Challenge for Use of Protein-based Polymers as Biomaterials

The *opportunity* for using protein-based polymers as biomaterials for medical applications comes with demonstration of the biocompatibility of the basic hydrogel and elastic and plastic states of the protein-based polymers, and it comes with the capacity to produce protein-based polymers by microbial fermentation with sufficiently low-cost production for a broad range of medical applications.

The *challenge* is to purify microbially produced protein-based polymers to adequate levels for use as biomaterials. The pharmaceutical use of microbially prepared insulin requires that impurities be less than 10 parts per million (ppm). Repeated use of the phase separation property for purification results in a

9.4 Medical Applications

10-fold greater purity, that is, an impurity level of less than 1 ppm. This level of purification is grossly inadequate, however, for the quantities required for a biomaterial. Tens of milligrams to gram quantities required for use as a biomaterial represents an amount one thousand to one million times larger than is required when a protein such as insulin is used as a pharmaceutical.

An impurity level as low as 5 ppb may also be inadequate when the biomaterial is used as a rapidly dispersible injectable implant. Five parts per billion is the detection limit of the highly sensitive immunoblot technique. Fortunately, adequate purification of protein-based polymers used at 30 mg quantities has been achieved at BRL, and adequate purification has been demonstrated for the most stringent of conditions, where an injectable implant totally disperses having unloaded all of its impurities in a few days time.[58] To have passed such a stringent test provides the desired "gold standard" for purification. Now it will be useful to increase the sensitivity of impurity detection by means of radiolabeling and to establish quality control for the level of purity required when using protein-based polymers as medical devices. Thus, the *opportunity* exists due to the extraordinary biocompatibility of parent compositions of elastic protein-based polymers, and the *challenge* of purification from microbial sources has been met using the most demanding criteria, as further discussed below. Now work on low-cost production is progressing in order to expand the number of commercially viable applications of all kinds.

9.4.2 Biocompatibility of Microbially Produced Protein-based Polymers

9.4.2.1 Western Immunoblot Technique to Demonstrate Impurity Levels

Having demonstrated production of designed protein-based polymers, the next issue is one of achieving purification adequate for the intended application. The most sensitive means with which to detect impurities of relevance to medical applications is the Western immunoblot technique. With this technique, levels of purification are demonstrated in Figure 9.12 for the following model proteins:

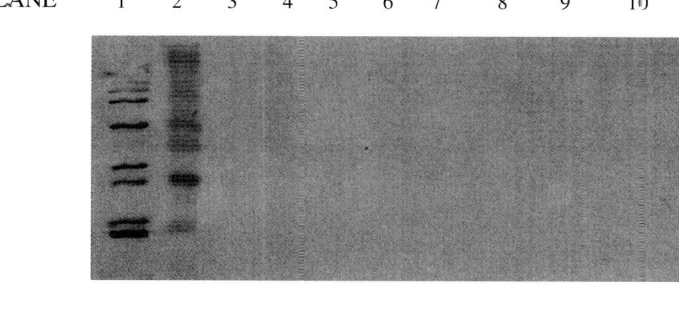

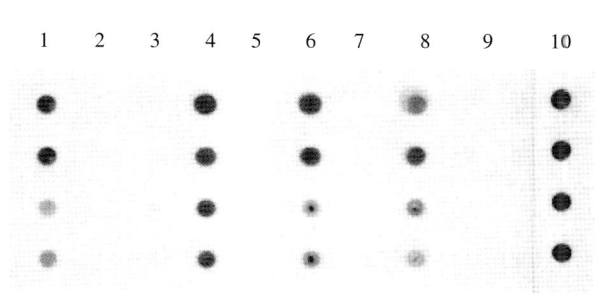

FIGURE 9.12. Western immunoblot techniques for placing a limit on the impurity burden of a series of protein-based materials. (**Upper plate**) Absence of line indicates impurities are less than 1 ppm as achieved by three phase separation cycles. (**Lower plate**) Absence of dot indicates impurities are less than 5 ppb. See text for further discussion. (Reproduced with permission from Urry et al.[58])

Model Protein A': $(GVGVP)_{251}$
Model Protein A: $[(GVGVP)_{10}GVGVPG\underline{RG}$
$\underline{D}SP(GVGVP)_{10}]_7(GVGVP)$
Model Protein B: $(GVGVP)_{10}(\underline{GVGVAP})_8$
$(GVGVP)_{10}]_5(GVGVP)$
Model Protein i: (GVGVP GVGVP G$\underline{E}$GVP
GVGVP GVGVP$)_n$(GVGVP)
Model Protein x': (GVGVP GVGVP G$\underline{K}$GVP
GVGVP GVGVP$)_n$(GVGVP)

The upper part of Figure 9.12 utilizes 5 μg of protein per well. Lane 1 is the positive control obtained from the lysate of the *E. coli* strain used to produce the model proteins. Lane 2 contains molecular weight markers. Lanes 3 through 6 are **Model Proteins A', x', A**, and **B**, respectively, and lanes 7 through 10 are negative controls, in this case unused wells. At the concentration of 5 μg protein per well, no impurities are observed with these model proteins purified solely by three phase separation cycles. This allows the statement that the impurities are less than 1 ppm.

When the dot blot technique is used, as much as 1 mg of protein is placed in each well dot, and the total protein is tested for a visible immunogenic reaction. No visible dot means that the impurities are less than 5 ppb. Lanes 1, 4, 6, and 8, contain **Model Proteins A', A, B**, and **x'**, respectively. In Lanes 1, 4, 6, and 8, the top two dots contain 1 mg, whereas the bottom two dots contain 0.1 mg; thus the effect of a 10-fold dilution can be seen. Lanes 2 and 10 are the negative and the positive controls, respectively, showing the experiment to be working well. On further purification of **Model Proteins A', A, B**, and **x'**, as shown in lanes 3, 5, 7, and 9, respectively, no spot can be detected. This means that purification of the model proteins has been obtained to the level where impurities detectable by the Western immunodotblot technique are less than 5 ppb.

9.4.2.2 Subcutaneous Injection in the Guinea Pig

Having established the extraordinary biocompatibility of poly(GVGVP), as reviewed above, it becomes possible to use the chemically synthesized poly(GVGVP) to determine the extent of purification required of the microbially prepared polymer $(GVGVP)_{251}$. Accordingly, 30 mg of the product resulting from three cycles of phase separation to purify $(GVGVP)_{251}$ was injected subcutaneously in the guinea pig. The result shown in Figure 9.13 demonstrates a substantial inflammatory response. Clearly based on our previous knowledge of the biocompatibility of chemically synthesized poly(GVGVP), three cycles of phase separation did not sufficiently remove the *E. coli* impurities.

Further purification beyond three cycles of phase separation with **Model Protein i**, above and as defined in Table 5.5, resulted in the most remarkable demonstration of biocompatibility. As demonstrated in Figure 9.14, the subcutaneous injection of 30 mg in the guinea pig, when examined at 2 weeks, left no trace of having been present. This was the case at four different sites.[58] Such a result provides the most demanding test of purification. All of the impurities in the 30 mg sample had been released to the host, and yet there was no trace of its having been present. In fact, based on the formation of the subcutaneous "bump" and its essential disappearance within several days, it would seem that all of the impurities were released in a few days without an inflammatory response having been elicited. This we take to be the "gold standard" for demonstrating the desired level of purification.

9.4.2.3 Entropic Elasticity as a Barrier to Antigenicity of Elastic Protein-based Polymers

9.4.2.3.1 The Natural Anticipation of Foreign Protein as Antigenic

Historically, foreign proteins have been recognized as some of the most potent antigens. Because of this, it is simply to be expected that protein-based polymers would be antigenic. The mammalian elastic fiber, however, has generally been considered to be of low antigenicity. Thus, the expectation for repeating sequences from the mammalian elastic protein, such as $(GVGVP)_n$, was that antigenicity would be low. In pursuing medical uses growing out of the development of the elastic protein-based polymers, we had fully expected that

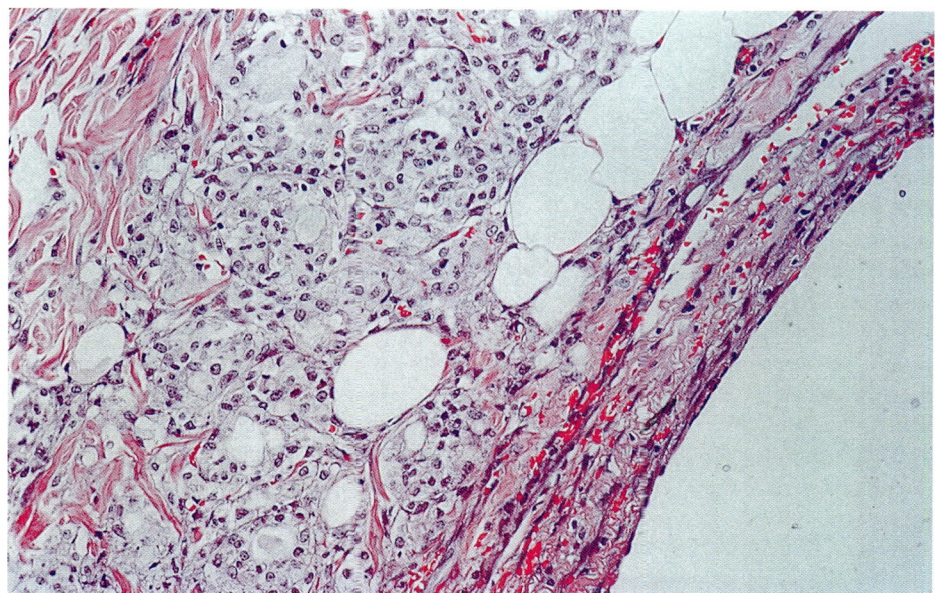

FIGURE 9.13. Subcutaneous injection in the guinea pig of 30 mg of $(GVGVP)_{251}$ that had been bulk purified by three cycles of phase separations. The subcutaneous tissue exhibited a substantial inflammatory response due to the presence of less than 1 ppm of E. coli contaminants as determined by the Western immunoblot technique in the upper plate of Figure 9.12. This demonstrates the requirement of adequate purification when using E. coli as a source of protein-based polymer (Courtesy of Bioelastics Research, Ltd.)

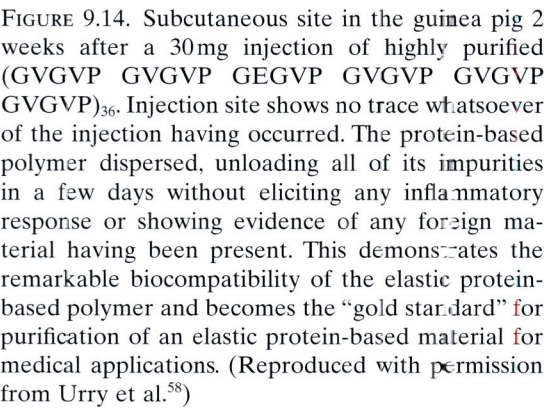
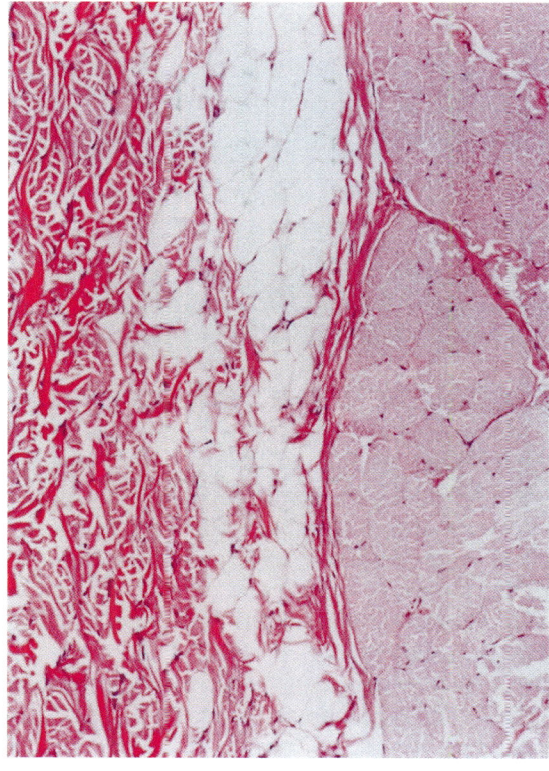

FIGURE 9.14. Subcutaneous site in the guinea pig 2 weeks after a 30 mg injection of highly purified (GVGVP GVGVP GEGVP GVGVP GVGVP GVGVP)$_{36}$. Injection site shows no trace whatsoever of the injection having occurred. The protein-based polymer dispersed, unloading all of its impurities in a few days without eliciting any inflammatory response or showing evidence of any foreign material having been present. This demonstrates the remarkable biocompatibility of the elastic protein-based polymer and becomes the "gold standard" for purification of an elastic protein-based material for medical applications. (Reproduced with permission from Urry et al.[58])

introduction of more polar groups in combination with more hydrophobic groups would produce strong sites for antibody binding, that is, would produce antigenic determinants, epitopes, and thereby would markedly increase antigenicity. It was therefore a pleasant surprise that the carboxylate-containing **Model Proteins i** through **v** in Table 5.5 did not follow the expectation of dramatically increasing antigenicity as more hydrophobic phenylalanine (Phe, F) residues increasingly replaced less hydrophobic valine (Val, V) residues in a polymer with glutamic acid (Glu, E) recurring every 30 residues.

9.4.2.3.2 Greater Immunological Response to Plastic Than to Elastic Protein-based Polymers

As our studies expanded to consider plastic protein-based polymers with the parent being $(AVGVP)_n$, a small but detectable increase in capacity to elicit formation of monoclonal antibodies was found.[33] Indeed, as A and V was replaced by E and F, however, ease of forming monoclonal antibodies increased, which was consistent with original expectations. Accordingly, the key to the remarkable biocompatibility of elastic protein-based polymers was sought and readily found in experimental data that determined the nature of their elasticity.

9.4.2.3.3 Key to Biocompatibility of Elastic Protein-based Polymers Resides Within the Nature of the Elasticity

The proposed basis for the nature of the ideal elasticity exhibited by the family of protein-based polymers using the generic sequence $(GXGXP)_n$, where X is a variable L-amino acid residue, became very controversial. The adherents to the classic (random chain network) theory of rubber elasticity took great exception to our proposal that *the damping of internal chain dynamics on extension gave rise to entropic elasticity*[59,60] (for more extensive treatment of this controversy, see Urry and Parker.[61]). The physical basis for the different (even heretical, to some) mechanism of near-ideal elasticity provides insight into the remarkable biocompatibility of elastic protein-based polymers.

9.4.2.3.4 The Presence of Low Frequency Mechanical Resonances in Elastic Protein-based Polymers

Dielectric relaxation studies of the phase transition of elastic protein-based polymers demonstrate development of intense relaxations centered near 5 MHz and 3 kHz as the phase separation proceeds.[61–63] Also, acoustic absorption measurements demonstrate development of a correspondingly intense absorption near 3 kHz.[62,63]

An approximate calibration of the 3 kHz dielectric relaxation, called a *loss permittivity*, in terms of elastic modulus is shown in Figure 9.15 using mechanical shear modulus measurements to obtain what is called the *loss shear modulus*.[64] Shear modulus measurements of interest here begin with a disk of cross-linked $(GVGIP)_{320}$, shown in Figure 9.16, placed between similarly sized disk-shaped plates. One of the disk-shaped plates oscillates at frequencies from very low values to a maximum of about 200 Hz to measure as a function of frequency the absorption by the elastic disk of the energy of the mechanical oscillation. The results are given over the lower frequency range of Figure 9.15, up to about 200 Hz.

Now, it has been shown for materials such as poly(propylene diol) (wherein both the absorption maximum for loss shear modulus and loss permittivity overlap near the frequency of 1 Hz) that their normalized curves perfectly superimpose over their frequency band width.[65,66] As shown in Figure 9.15, the lower frequency loss shear modulus curves uniquely overlap with the loss permittivity data at higher frequency. As such the former is melded to calibrate the loss permittivity data to obtain a coarse estimate of the elastic modulus values. This provides an independent demonstration of the mechanical resonance near 3 kHz and also allows reference to the 5 MHz dielectric relaxation as a mechanical resonance. Thus, as the folding and assembly of the elastic protein-based polymers proceed through the phase (inverse temperature) transition, the pentamers wrap up into a structurally repeating helical arrangement like that represented in Figure 9.17.

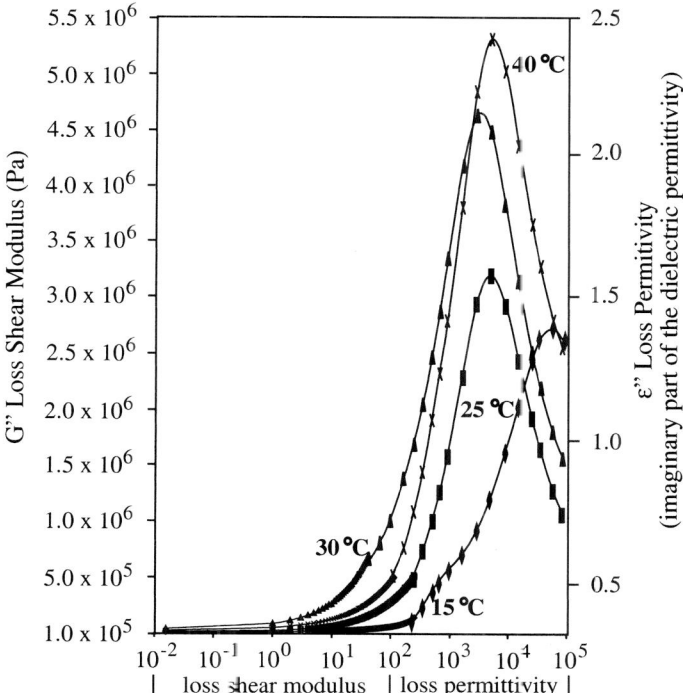

FIGURE 9.15. X^{20}–(GVGIP)$_{320}$: frequency dependence of loss shear modulus, G'' (0.02 to 200 Hz), and of loss permittivity (20 Hz to 10^5 Hz) as a function of temperature. When the frequency of the loss maximum is sufficiently low, for example, near 1 kHz, loss shear modulus and loss permittivity can both be determined and have been demonstrated to superimpose for the case of the loss maximum for poly(propylene diol).[65,66] In the case of X^{20}–(GVGIP)$_{320}$, the maximum occurs at a frequency that is too high to be reached by shear modulus measurements. Nonetheless, the two measurements are shown to overlap in a distinctive and identifying manner such that calibration of the loss permittivity in terms of shear modulus is not unreasonable. As loss shear modulus is identified with acoustic absorption, the data provide independent corroboration of the direct measurement of the acoustic absorption of Figure 9.55B (upper curve) due to Lev Sheiba.[62] Such acoustic absorption is referred to as a *mechanical resonance* that builds in intensity as hydrophobic association of the phase transition ensues. (Reproduced with permission from Urry et al.[64])

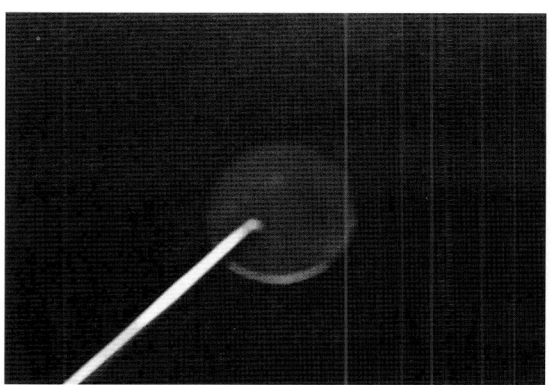 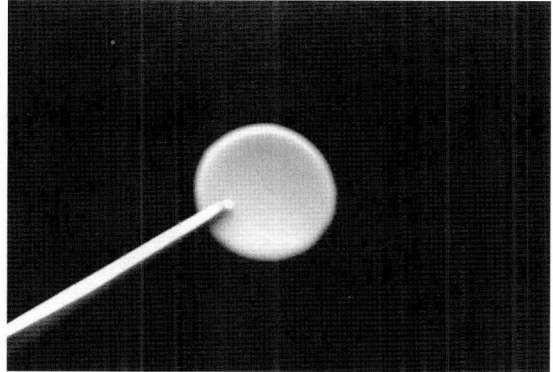

FIGURE 9.16. Disks of X^{20}–(GVGIP)$_{320}$ prepared for correlated *loss shear modulus* and *loss permittivity* measurements reported in Figure 9.15. (**Left**) Impregnation with the solvent mixture dipropylene glycol:propylene glycol:water in a 2:1:1 ratio, keeps the disk supple and elastic for months and suitable for the pressure ulcer studies described in section 9.4.5.6. (**Right**) Impregnated with water only, as used for the measurements in Figures 9.15 and 9.55B. (Courtesy of Bioelastics Research, Ltd.)

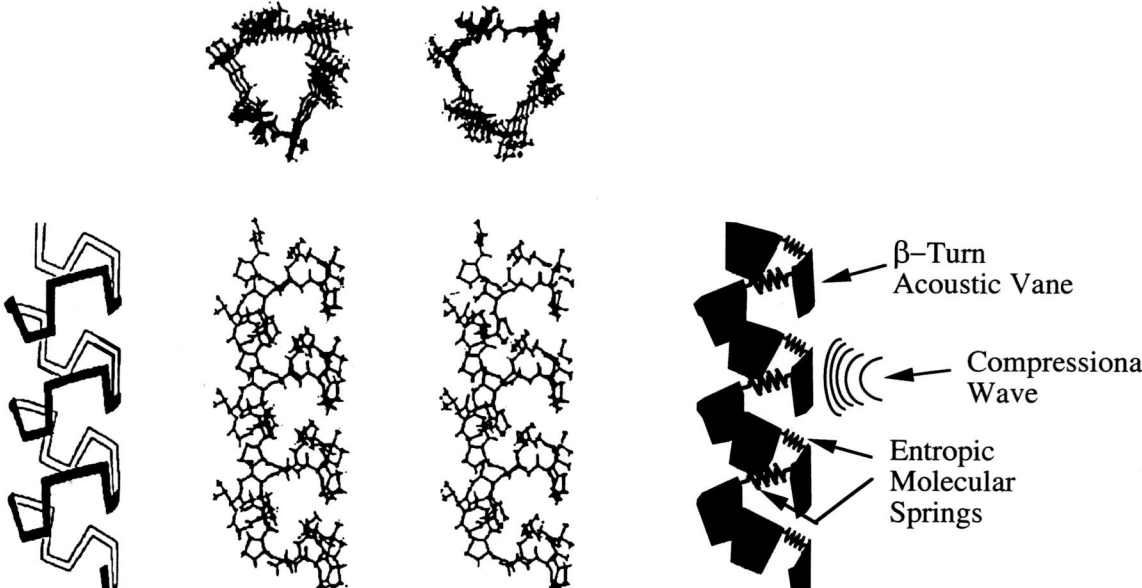

FIGURE 9.17. Stereo views (cross-eye) of molecular structure and proposed acoustic function of the poly(VPGVG) family of β-spiral structures. The leftmost structure is the schematic β-spiral showing β-turns to function as spacers between turns of the spiral. The central pair of structures provide details of bonding in side view and axis view (above). The rightmost structure shows β-turns in the role of acoustic vanes capable of absorbing compressional (sound) waves. The structures are arranged such that the different representations overlap on stereo viewing by cross-eye viewing, that is, by focusing ones eyes at a point in space between the eyes and the figure. (Reproduced with permission from Urry et al.[64])

9.4.2.3.5 Low Frequency Mechanical Resonances Lower Free Energy of the Entropic Elastic Structure

The low-frequency motions of 5 MHz and 3 kHz lower the free energy of the folded and assembled structure of the elastic protein-based polymers. A coarse sense of the amount each mechanical resonance could be contributing to the stability of the structure of these elastic protein-based polymers is obtained from a plot of the logarithm of the frequency as a function of the entropy contribution of resonance frequency using the harmonic oscillator partition function. It is not expected that the harmonic oscillator would give an accurate measurement of the magnitude of the entropy contribution of such low-frequency motions. However, the use of the harmonic oscillator partition function provides a sense of direction and magnitude of the contribution of low-frequency motions to the entropy and therefrom to the decrease in free energy. The calculated numbers for the contribution to the Gibbs free energy in terms of $T\Delta S$ are -9 kcal/mole for the 5 MHz mechanical resonance and -14 kcal/mole for the slower 3 kHz mechanical resonance. These numbers have great significance even if the actual values turn out to be but a fraction of the calculated values.

9.4.2.3.6 The Requirement That an Epitope be Fixed in Space to Present a Site for Antibody Binding Makes Dynamic Elastic Protein-based Polymers Nonantigenic

Measurements of the temperature dependence of the frequency of the nominally 5 MHz relaxation at temperatures above the transition interval indicate a barrier to motion of about 1 kcal/mole.[67] This means that the motions are active at physiological temperatures. The perspective is that an antibody is made to react at a site on the protein called an *antigenic determinant* or *epitope*. Several distinct contact points fixed in space, such as charges and hydrophobic groups, constitute the epitope. To fix those points in space in order to constitute an epitope, an antibody must stop the motions that are contributing to the favorable free energy of the elastic protein-based polymer. This constitutes a barrier to the interaction. In sum, the energy of stabilization estimated above for the two bands is 23 kcal/mole. Even if the barrier were one-fourth that sum, for example, 5.8 kcal/mole, the probability of interaction as required to identify such an epitope would decrease by a factor of more than 500,000.

Our hypothesis, therefore, is that the remarkable biocompatibility of elastic protein-based polymers arises from the presence of mechanical resonances that themselves are a result of the regular, nonrandom structure of this family of entropic elastic protein-based polymers.

9.4.3 A Consilient Approach to Tissue Engineering

In the words of E.O. Wilson in 1998, "To the extent that we depend on prosthetic devices to keep ourselves and the biosphere alive, we will render everything fragile."[68] The consilient approach to tissue engineering utilizes biology's own materials and mechanisms, concerned with tissue structure and function, to achieve tissue restoration. It is made possible by temporary functional scaffoldings composed of biodegradable and biocompatible elastic model proteins that are responsive to tissue variables in the same manner as natural proteins.

The materials of our approach are natural to the tissue to be restored; they are protein-based polymers that are programmably biodegradable; in their swollen state they degrade to natural amino acids without release of irritating acid (as occurs with the commonly used polyglycolic and polylactic acids); they are elastic and can match the compliance of the natural tissue; they are biocompatible (the basic sequence in its contracted state appears to be simply ignored by the host); and they can have introduced into them biologically active peptide sequences in a natural form as part of the designed protein sequence.

The mechanisms whereby the materials function are common to the tissue to be restored, that is, they exhibit the same hydrophobic association as occurs in protein structure formation and function, and the elasticity is due to damping of internal chain dynamics rather than due to random chain networks. As protein function itself is central to cellular function, this allows elastic protein-based materials in concert with natural cells to achieve tissue restoration in a manner entirely coherent with fundamental relationships between cells and their natural extracellular matrix.

The materials are fashioned into temporary functional scaffoldings into which the natural cells can migrate, attach, spread, and sense the forces to which the temporary functional scaffolding is subjected, and the cells in response turn on the genes to produce an extracellular matrix sufficient to sustain those forces. Thereby the natural cells remodel the temporary functional scaffolding into a natural tissue.

9.4.3.1 *Elastic Protein-based Polymers Fashioned as Tubes, Sheets, and Fibers*

9.4.3.1.1 Elastic Protein-based Polymers as Tubes

By use of appropriate molds, elastic protein-based polymers can be shaped into cross-linked tubes of the desired shape and with a range of elastic moduli, as shown in Figure 9.18. The elastic moduli of tube A is about 2×10^5 Pa, that of tube B 6×10^5 Pa, and that of tube C 2×10^6 Pa, whereas that of the femoral artery is in the range of 4 to 6×10^5 Pa.[69,70]

9.4.3.1.2 Elastic Protein-based Polymers as Sheets

A sheet of 20 Mrad γ-irradiation cross-linked $(GVGVP)_{251}$, that is, $X^{20}-(GVGVP)_{251}$, is shown in Figure 5.14 and a disk of this material containing a human ureteral explant (see Figure 9.36A, below) is used in the simulated urinary

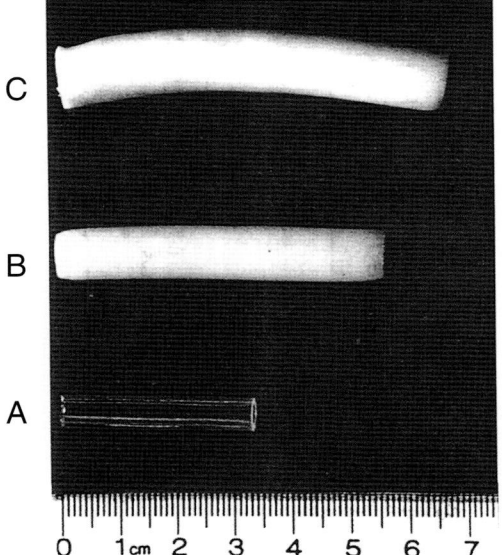

FIGURE 9.18. Tubes made by γ-irradiation cross-linking of elastic protein-based polymers of different compositions. (**A**) Poly(GVGVP). (**B**) Poly[f$_F$(GFGVP), f$_V$(GVGVP)]. (**C**) Poly(GVGIP). (Adapted with permission from Urry.[117] Now to be credited as www.WorldandIJournal.com)

bladder investigation. A smaller sheet of chemically synthesized X^{20}–poly(GVGVP) to be used in prevention of adhesion studies is shown in Figure 9.19. These elastic sheets can also be prepared with a range of elastic moduli.

9.4.3.1.3 Elastic Protein-based Polymers as Fibers

Chemically cross-linked fibers can also be prepared with a range of elastic moduli, break stresses, and break strains. Two of the more interesting elastic fibers are shown in Figure 9.20. The fiber at left exhibits an elastic modulus almost 10 times greater than that of femoral artery; it does not break until 350% extension, and it does so with a stress of one-third of the 20% elastic modulus. The fiber at right in Figure 9.20 exhibits an elastic modulus a solid 10 times greater than that of femoral artery: a break stress of 7×10^6 Pa with a break strain of 285%. Collagen fibers exhibit elastic moduli of about 10^8 Pa (two orders of magnitude greater than that of elastic fibers), but become damaged with just a few percent extension. On the other hand, mammalian elastic fibers exhibit elastic moduli of about 5×10^5 Pa, but exhibit about 200% extension. Thus, the chemically cross-linked

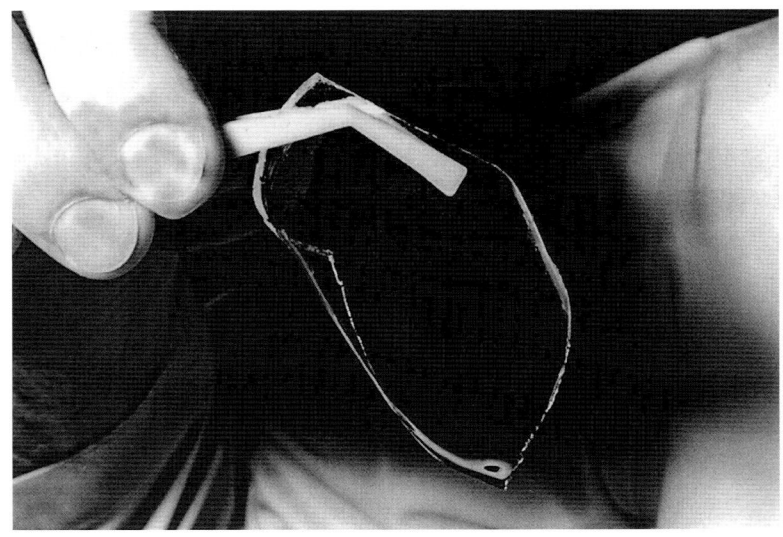

FIGURE 9.19. Small sheet of elastic protein-based material, X^{20}–poly(GVGVP), to be placed between the injured, bloodied abdominal wall and the injured, bloodied, and contaminated loop of bowel for the prevention of adhesion of bowel to wall.

9.4 Medical Applications

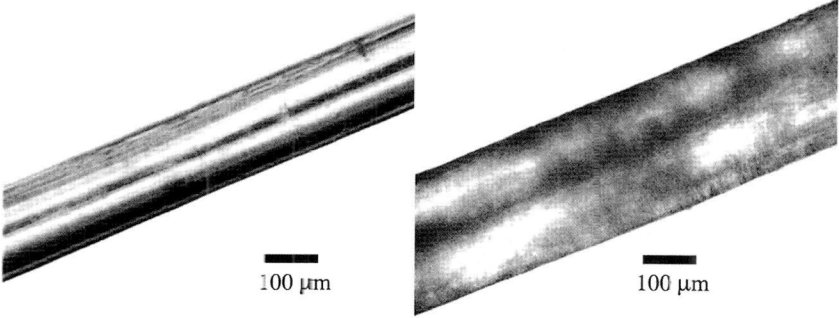

Polymer	E (Pa) at 20% Strain	Break Stress (Pa)	Break Strain (%)
Left Fiber	3.86 x 10⁶	1.26 x 10⁵	350
Right Fiber	6.12 x 10⁶	6.94 x 10⁵	285

FIGURE 9.20. Elastic protein-based fibers a few hundred micrometers in diameter, prepared by extrusion and chemical cross-linking. Impressive mechanical properties of elastic modulus at 20% extension, of break stress, and of break strain are listed. (Reproduced with permission from Urry et al.[64])

elastic fibers of Figure 9.20 exhibit elastic moduli midway between elastic fiber and collagen but sustain extensions that are about 100 times greater than sustained by collagen. Formed into meshes or woven or knitted into tubes and sheets, the elastic fibers in Figure 9.20 would have many uses as prosthetic materials. When also containing cell attachment sequences such constructs could provide temporary scaffoldings for soft tissue restoration of many sorts, for example, for vascular wall restoration, hernial repair, and a host of others.

9.4.4 Soft Tissue Restoration

The family of protein-based polymers emphasized here look back to the parent repeating elastic peptide sequence $(GVGVP)_n$, which was found in the mammalian elastic fiber, a soft tissue. Because of this, it is perhaps most appropriate to begin consideration of medical applications with the subject of soft tissue restoration. The word *restoration* describes a treatment not to replace, as is often implied in terms of tissue engineering and tissue reconstruction, but rather to set the stage for the regeneration of a more nearly natural state.

9.4.4.1 Concept of Designing a Temporary Functional Scaffolding

Central to the term *soft tissue restoration* is the concept of designing temporary, functional scaffoldings that stimulate the natural cells of the tissue and, thereby, cause them to regenerate the natural functional tissue. The protein-based polymers of focus here exhibit three essential enabling properties. First is the capacity to match the elastic behavior and function of the natural tissue; the second is the capacity to contain in the model protein sequence cell attachment sequences, and the third is the eventual biodegradation of the temporary functional scaffolding, leaving only the natural regenerated functional tissue in its place.

In natural tissues, cells form multiple attachment sites to their extracellular matrix. By means of these attachments, cells deform as the tissue deforms in response to the natural mechanical stresses and strains that the tissue must sustain during function. These mechanical forces constitute the energy inputs that instruct the cells to produce the extracellular matrix sufficient to sustain those forces. Thus, an ideal artificial material should have both the attachment sites for the natural cells and a compliance that matches the natural tissue.

Elastic protein-based polymers have been designed both to provide cell attachment sites and to exhibit the required elastic modulus of the tissue to be replaced. Thus, this introduces the potential to design temporary functional scaffoldings with the capacity to be remodeled, while functioning, into a natural tissue by the natural cells of the tissue.

Furthermore, the elastic protein-based materials themselves have been designed to perform the set of energy conversions that occur in living organisms and, in particular, to convert mechanical energy into chemical signals of the sort that could provide the stimuli to turn on the genes for producing the required extracellular matrix.

9.4.4.2 The Key Property of Cellular Mechano-chemical Transduction

A fundamental property of cells is that they take instruction from the mechanical forces to which they are subjected and produce that extracellular matrix sufficient to sustain those forces. Evidence for this capacity of cells goes back more than two decades to the work of Leung et al.[71] but has more recently been called *cellular tensegrity* following more detailed characterization with techniques of modern molecular biology.[72–74] An understanding of how this can occur lies in the design of elastic protein-based polymers capable of demonstrating mechano-chemical transduction. The classic demonstration of a mechanical energy input resulting in a chemical energy output (i.e., mechano-chemical transduction) by elastic protein-based polymers is found in the experimental result of stretch-induced pKa shifts (i.e., *pumping protons*) in Chapter 5, Figure 5.23.

Figure 9.21 exemplifies the nexus between cellular mechano-chemical transduction and elastic protein-based polymers containing cell attachment sequences that enables these temporary functional scaffoldings to result in restoration of natural tissue. Figure 9.21A shows in the absence of cell attachment sequences that the elastic matrix X^{20}–poly(GVGVP) does not support attachment of cells. On inclusion of cell attachment sequences, as in the chemically synthesized X^{20}–poly[40(GVGVP),(GRGDSP)], ligamentum nuchae fibroblasts attach, spread, and grow to confluence. The micrograph in Figure 9.21B shows the edge of the plated area; growth to confluence is complete using X^{20}–poly[20(GVGVP),(GRGDSP)], as shown elsewhere (see Figure 5 of Nicol et al.[75]).

Because of the cell attachments, the effect of stretching the elastic matrix is to stretch the cytoskeletal fibers and possibly the integrin receptors in the membrane, as depicted in the top and side views of the relaxed (Figure 9.21C) and stretched (Figure 9.21D) states. Mechanical stretching of the elastic cytoskeletal fibers, in our concept of mechano-chemical transduction demonstrated in Figure 5.23, results in the uptake of protons or, perhaps more to the point, the release of phosphate (even mechanically driven phosphorylation). Thus chemical energy output becomes the chemical signal that directs the nucleus to turn on the appropriate genes for production of an extracellular matrix sufficient to sustain the dynamic mechanical forces sensed in this manner by the cells. The test of this concept in tissue restoration is shown below in the simulated urinary bladder experiment (see Figures 9.36 and 9.37 and related text, below).

9.4.4.3 Prevention of Postsurgical and Post-trauma Adhesions

Prevention of adhesions is included within the application of soft tissue restoration as the purpose is indeed to restore a soft tissue site to the normal state that preceded the operation or other trauma. The several examples discussed below include abdominal, eye, and spinal soft tissue sites.

9.4.4.3.1 Abdominal Sites (Emphasis to Function in a Bloodied, Contaminated Site)

As shown in Figure 9.22A for the control site, in this model the abdominal wall is scraped until it bleeds; a loop of proximal intestine is punctured until it bleeds and exudes feces; a loose suture brings the two injured surfaces into proximity and is tied off in such a manner

9.4 Medical Applications

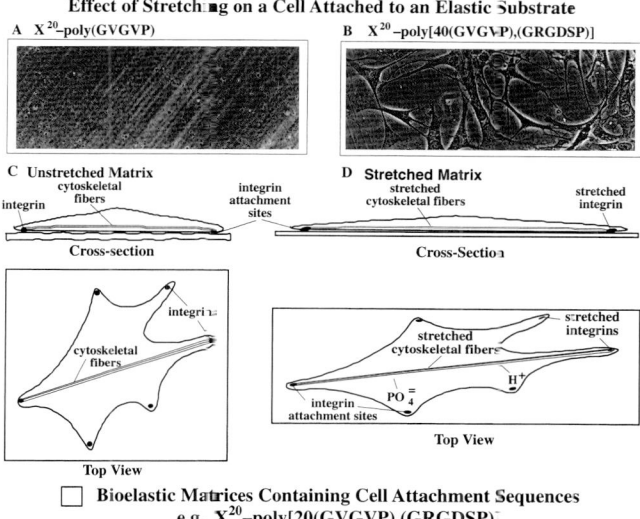

FIGURE 9.21. Cellular mechanochemical transduction makes possible tissue restoration by elastic protein materials containing cell attachment sequences that function as temporary functional scaffoldings. The attached cells become stretched as the matrix is stretched. The forces and frequencies of the stretch/relaxation cycles provide the chemical signal to the nucleus to turn on the genes for production of an extracellular matrix sufficient to sustain the deforming forces. Thus, the elastic matrix made of elastic protein-based polymers provides a temporary functional scaffolding with the potential for being remodeled into a natural tissue. See text for further discussion. (**A**) X^{20}–poly(GVGVP). (**B**) X^{20}–poly[40(GVGVP), (GRGDSP)]. (A,B, reproduced with permission from Nicol et al.[75]) (**C**) Unstretched matrix. (**D**) Stretched matrix. (C,D, reproduced with permission from Urry.[118])

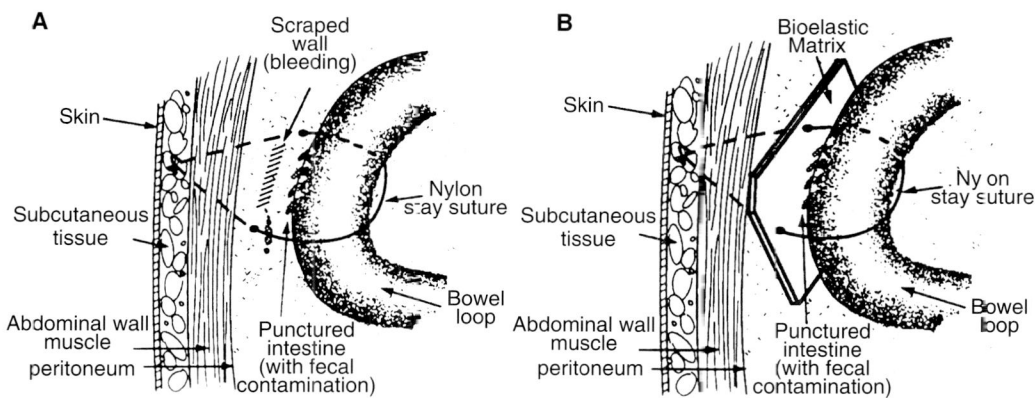

FIGURE 9.22. Peritoneal cavity model of L. Hoban[77] for the prevention of adhesions following abdominal wounds: a contaminated, bloodied peritoneal cavity. (**A**) The control. An injured, bleeding abdominal wall forms adhesions to a punctured, bleeding, and contaminated loop of bowel. (**B**) The test site. Adhesions are to be blocked by a sheet of elastic protein-based material placed between the wall and bowel and held in place by a loose suture. (Reproduced with permission from Urry et al.[76])

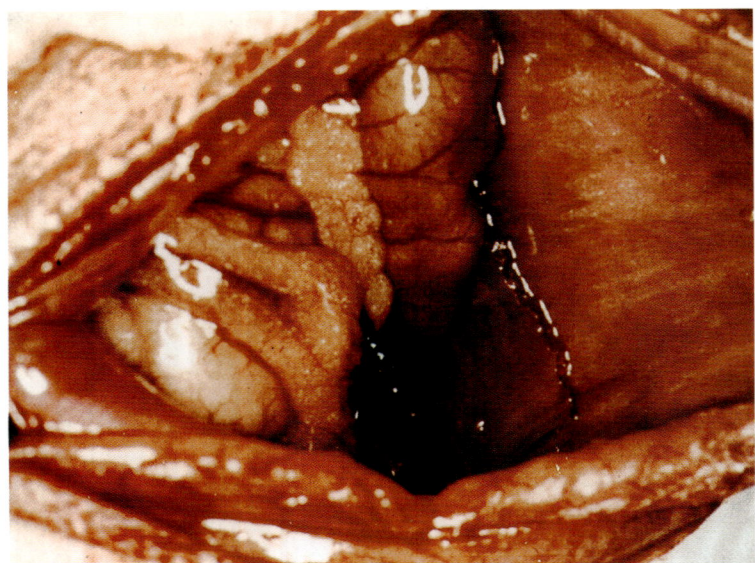

FIGURE 9.23. Small sheet of elastic protein-based material, X^{20}–poly(GVGVP), placed between the injured, bloodied abdominal wall and the injured, bloodied, and contaminated loop of bowel to test the material for its efficacy in preventing adhesion of bowel to wall. (From Hoban et al.[77] and Urry et al.[76])

as to be accessible from outside the closed cavity, and the abdominal wall is sutured closed. For the test site, as shown in Figure 9.22B, the test article, demonstrated in Figure 9.19, is interposed between bowel and abdominal wall, as shown in Figure 9.23. One week later the loose suture is removed, and at 2 weeks the abdominal cavity is reopened and the site examined.

In the control, adhesions formed 100% of the time between abdominal wall and bowel, as shown in Figure 9.24. At the test site, 80% of the time there were either no adhesions, as in Figure 9.25, or insignificant adhesions. Ten

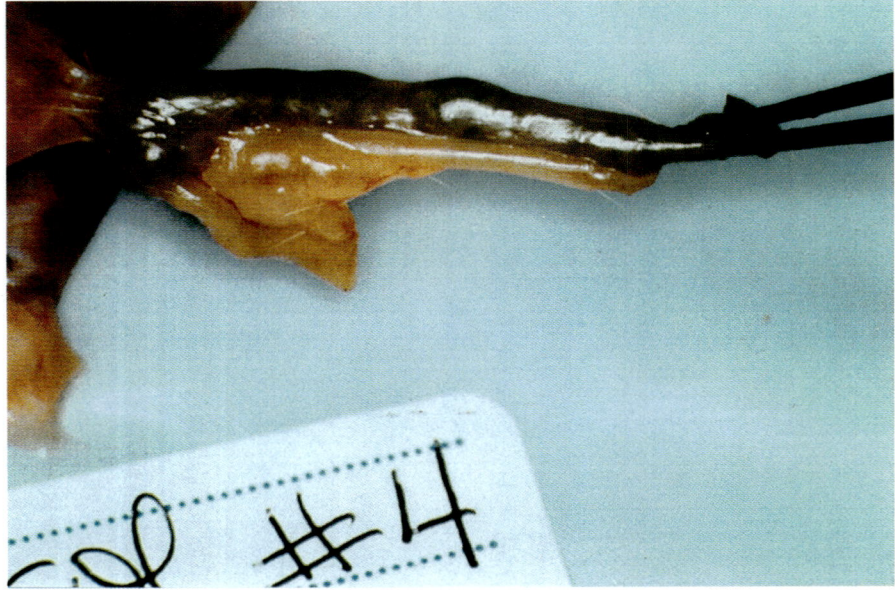

FIGURE 9.24. Control for the prevention of adhesions following abdominal wounds. In the absence of a sheet of elastic protein-based material, major adhesions form, connecting the loop of bowel to abdominal wall. (From Hoban et al.[77] and Urry et al.[76])

9.4 Medical Applications

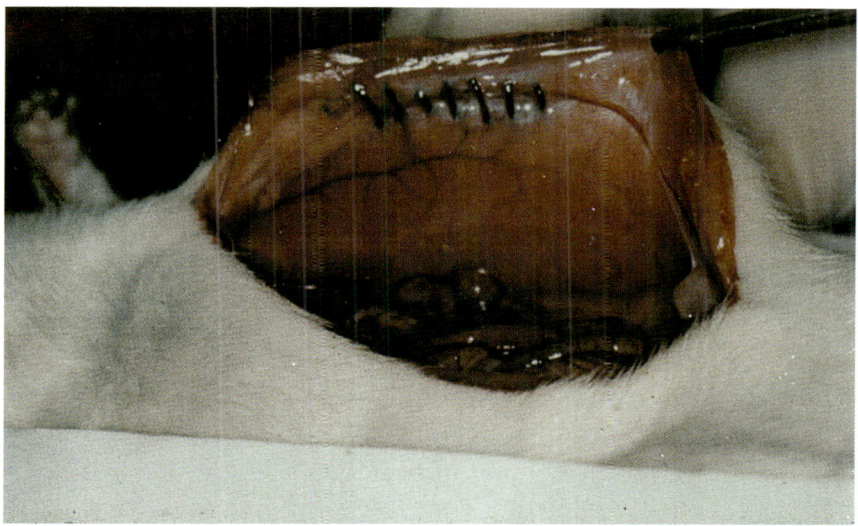

FIGURE 9.25. Having positioned a small sheet of elastic protein-based material, as shown in Figures 9.22B and 9.23, between the injured, bloodied abdominal wall and the injured, bloodied, and contaminated loop of bowel *prevented* significant adhesion of the bowel to the wall at 80% of the test sites. (From Hoban et al.[77] and Urry et al.[76])

percent of the time a small slip of adhesion grew through a tear in the bioelastic sheet or around the sheet, as shown in Figure 9.26, where the abdominal wall shows no sign of inflammation and the sheet has remained transparent without apparent formation of a fibrous capsule. Another 10% of the time the bioelastic sheet was engulfed by adhesion, but the mass of adhesion could be opened and the bioelastic sheet simply lifted out. Even within the mass of the adhesion, there was no attachment of adhesive mass to the bioelastic sheet.[76,77]

Clearly, in this model 80% of the time 20 Mrad γ-irradiation cross-linked poly(GVGVP), X^{20}–poly(GVGVP), presented a biocompatible barrier to the formation of adhesions. Under no circumstances did adhesions form to X^{20}–poly(GVGVP).

9.4.4.3.2 Strabismus Surgery (to Prevent Adhesion Between Rectus Muscle and Sclera of the Eye)

In the strabismus surgery model in the rabbit eye, demonstrated in Figure 9.27, for the control, the superior rectus muscle is detached from its point of insertion; a section of muscle capsule is removed; a similar defect is made in the sclera of the eyeball, and the rectus muscle is reattached such that the muscle capsule defect exactly overlies the scleral defect. One hundred percent of the time, an extensive adhesion bound muscle to sclera as shown in Figure 9.28. For the test site, as shown in the lower part of Figure 9.27, a bioelastic sleeve is wrapped around the rectus muscle. The histological section in Figure 9.29 shows the bioelastic sleeve separating sclera (seen as dense connective tissue fibers on the lower right) from muscle fibers (upper left) to have fragmented due to sectioning. Importantly, this histological section shows no adhesion whatever, and remarkably there are no inflammatory cells and no sign of a fibrous capsule as usually forms around a foreign material.

In this animal model for strabismus surgery,[78] 20 Mrad γ-irradiation cross-linked poly(GVGVP), X^{20}–poly(GVGVP), presented a biocompatible barrier to the formation of adhesions, caused neither an inflammatory response nor the usual foreign body response of fibrous capsule formation, and appeared to be innocuous, its presence being ignored by the host.

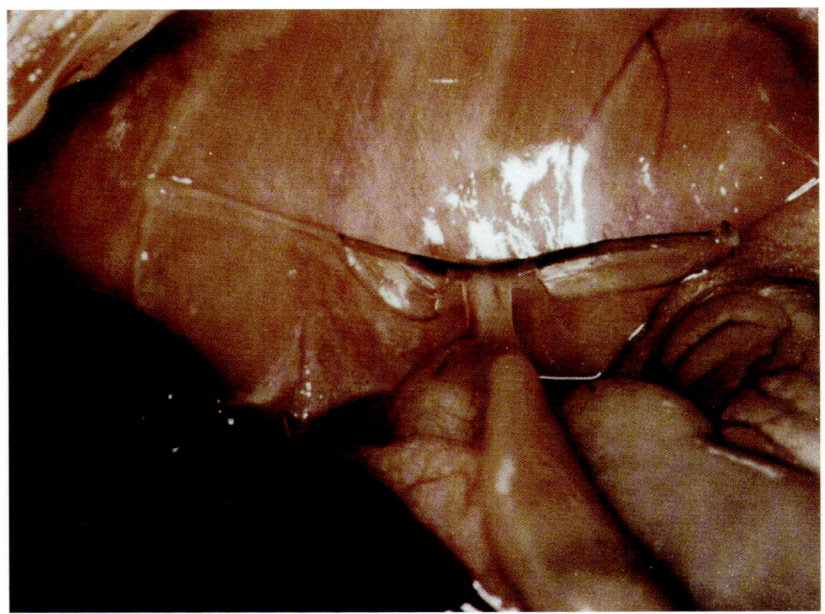

FIGURE 9.26. At 10% of the test sites, a small slip of adhesion either grew around (as shown) or through a break in the sheet of elastic protein-based material, weakly adhering abdominal wall to bowel. Note at 2 weeks that a fibrous capsule does not cover the sheet and inflammation of the abdominal wall does not occur. (Reproduced with permission from Urry et al.[76])

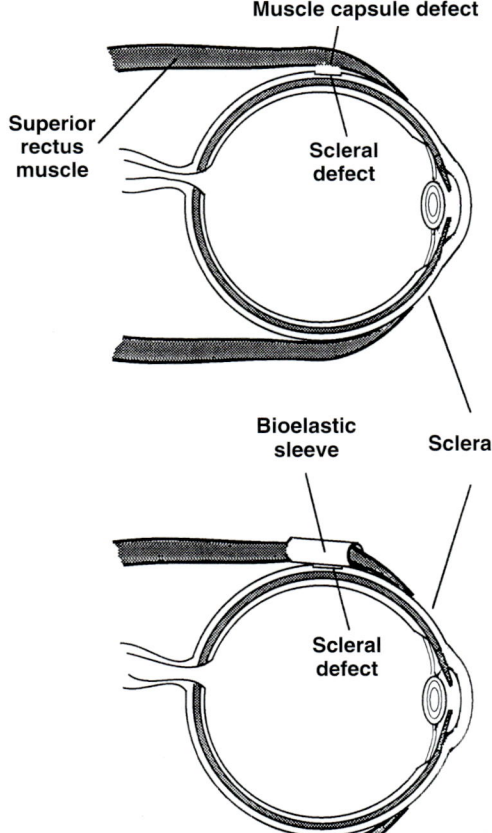

FIGURE 9.27. Strabismus surgery model of F. Elsas in the rabbit eye for the prevention of postsurgical adhesions due to positioning a bioelastic sleeve of elastic protein-based material between rectus muscle capsule and scleral defects. (Reproduced with permission from Urry et al.[33])

9.4 Medical Applications

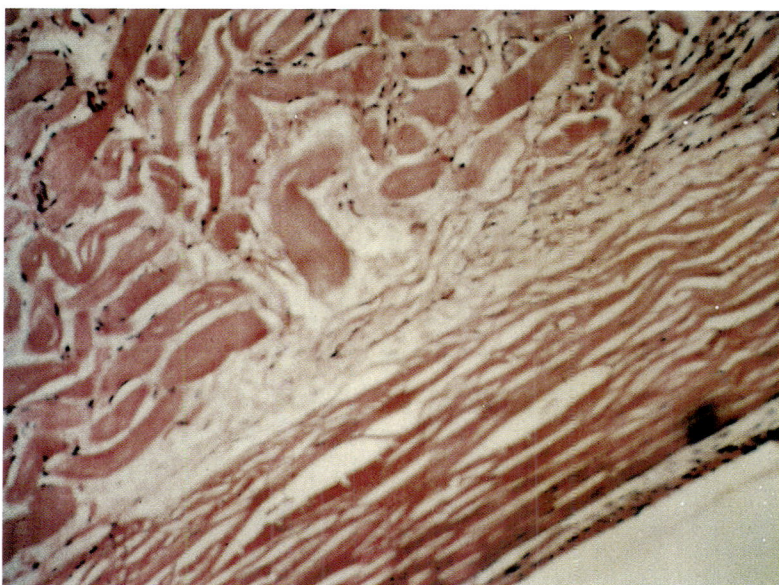

FIGURE 9.28. Control site for the prevention of adhesions in a strabismus surgery model in the rabbit eye showing, in the absence of a bioelastic sleeve, extensive adhesion between muscle and scleral defects. (Reproduced with permission from Elsas et al.[78] With permission of Slack Incorporated.)

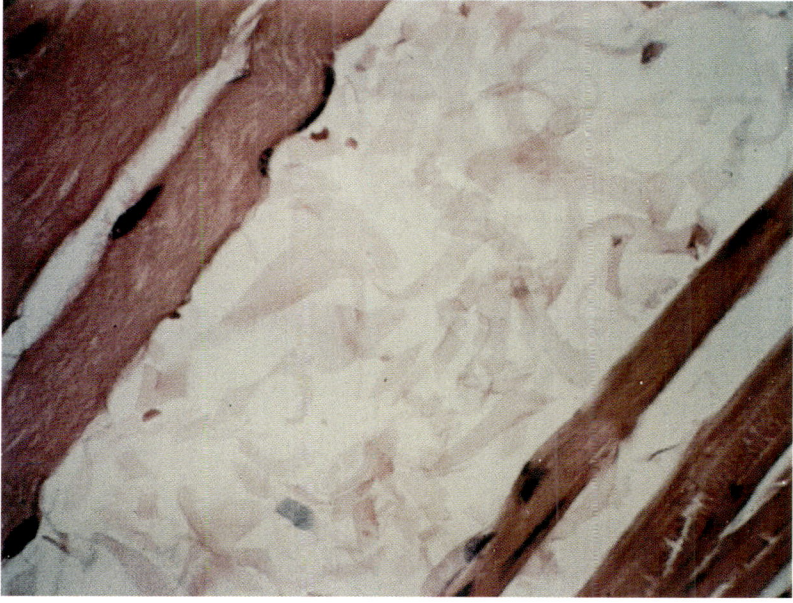

FIGURE 9.29. Demonstration of the efficacy, in a strabismus surgery model in the rabbit eye created by F. Elsas, of a bioelastic sleeve of an elastic protein-based material to prevent adhesion between scleral and rectus muscle defects. Note absence of adhesion and the complete absence of an inflammatory response. (Reproduced with permission from Elsas et al.[78] With permission of Slack Incorporated.)

9.4.4.3.3 Spinal Surgery (Emphasis to Correct Herniated Intervertebral Discs) in the Rabbit

The animal model is spinal surgery for laminectomy in the rabbit. Figure 9.30 shows the exposed spinal surgery (laminectomy) site in the rabbit model, exhibiting $(GVGVP)_{251}$ as the extruded gel and as the difficult to see transparent elastic sheet (membrane). Also shown are the retractors that hold the site exposed. From the histology of the control site in Figure 9.31, a massive fibrous adhesion (EF), is shown bound to the dura mater (DM) that surrounds the nerves of the spinal cord (NC) that traverse from the brain to the tissues. During preparation of the tissue for histology and the sectioning process, the dura mater separates from the spinal cord. The formation of adhesion to dura is considered by many physicians to be a source of the continuing pain that follows back surgery and that results in the failed back surgery syndrome.[79]

When the elastic matrix (EM) made of $X^{20}-(GVGVP)_{251}$ is placed adjacent to the dura, as shown in Figure 9.32, very little adhesion is

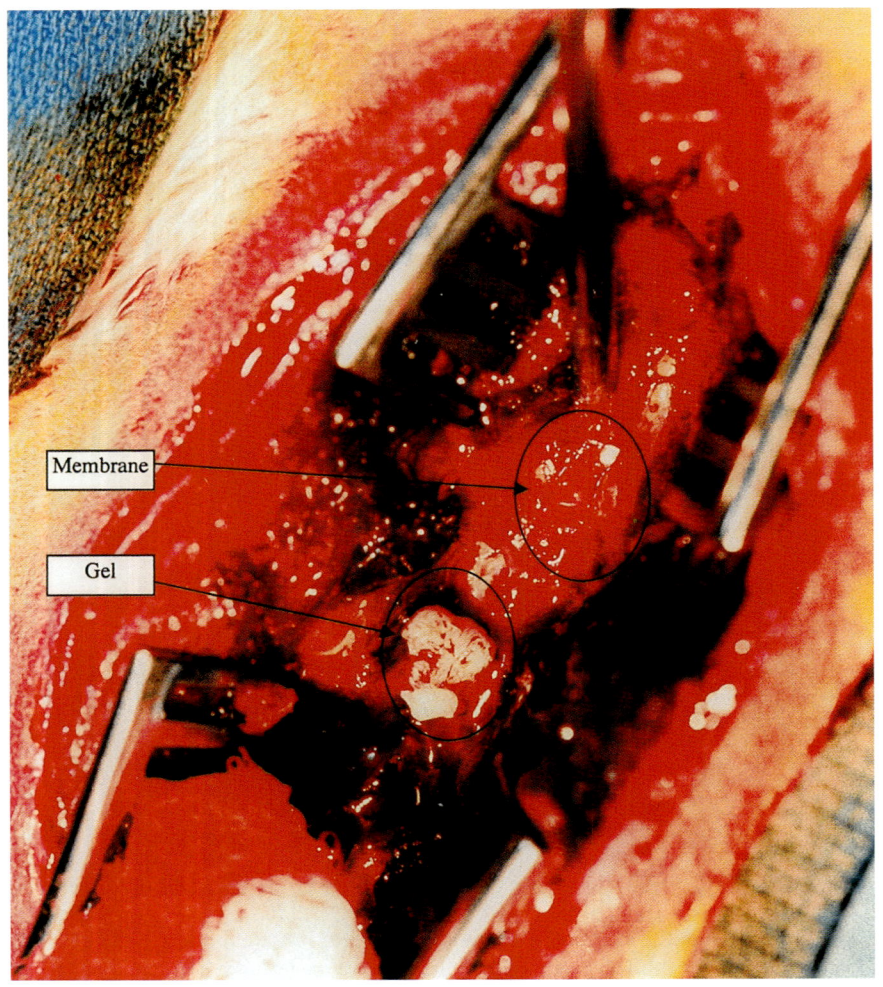

FIGURE 9.30. Spinal surgical (laminectomy) site in the rabbit model, showing $(GVGVP)_{251}$ in the gel and elastic sheet (membrane) states, as well as retractors (P. Glazer, surgeon on subcontract from Bioelastics Research, Ltd.). (Reproduced with permission from Alkalay et al.[79])

9.4 Medical Applications

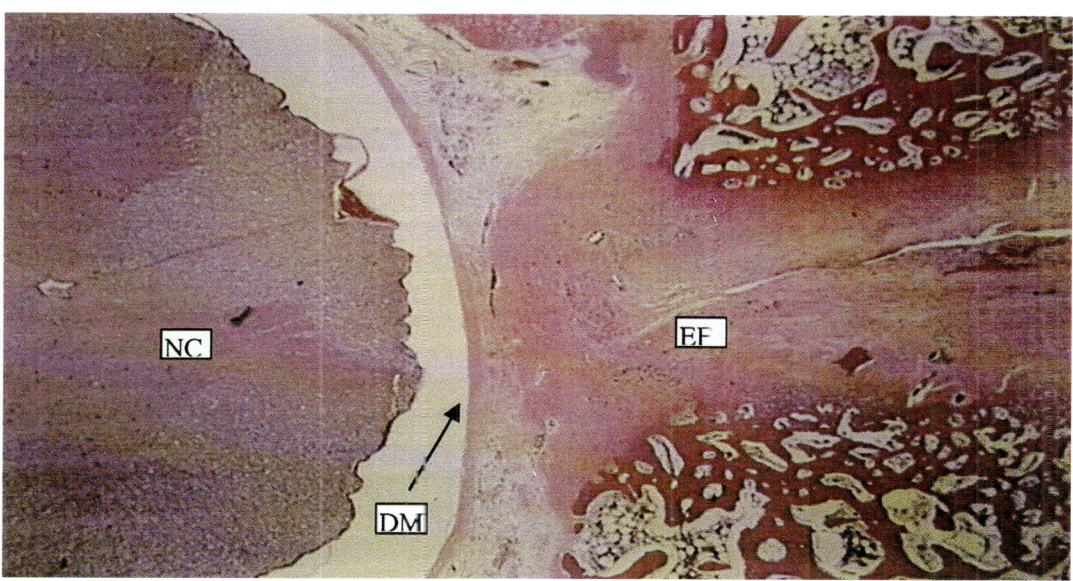

FIGURE 9.31. Spinal surgery (laminectomy) control site in the rabbit showing massive fibrous adhesion (EF) to the dura mater (DM) that surrounds the spinal cord (NC). The objective of the test elastic material is to block formation of the massive adhesion (P. Glazer, surgeon; work supported by subcontract from Bioelastics Research, Ltd.). (Reproduced with permission from Alkalay et al.[79])

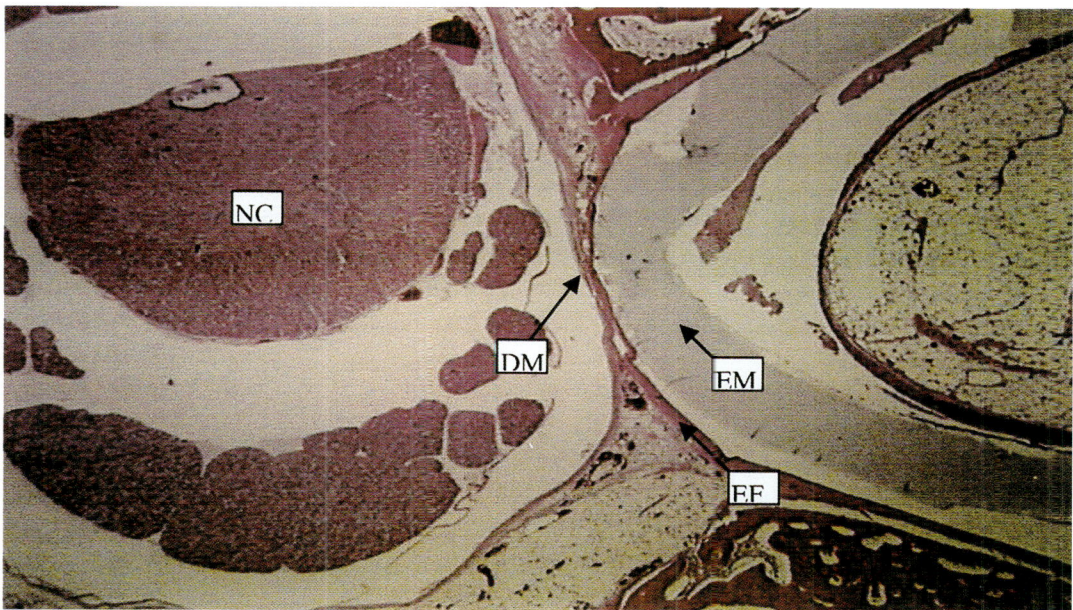

FIGURE 9.32. Spinal surgery (laminectomy) control site in the rabbit showing massive fibrous adhesion (EF) to the dura mater (DM) that surrounds the spinal cord (NC). The objective of the test elastic material, X^{20}-$(GVGVP)_{251}$ (EM) is to block formation of the massive adhesion (EF) of Figure 9.31. This clearly occurs. (P. Glazer, surgeon; work supported by subcontract from Bioelastics Research, Ltd.). (Reproduced with permission from Alkalay et al.[79])

found attached to the dura; rather, the region is filled by the elastic matrix. When the phase separated gel (G) state of $(GVGVP)_{251}$ is extruded into the laminectomy site adjacent to the dura, as shown in Figure 9.33, adhesion to the dura is very limited.

Whether the model involves the bloodied, contaminated peritoneal cavity in the rat, strabismus surgery in the rabbit eye, spinal surgery (laminectomy) in the rabbit, or spinal surgery (hemilaminectomy) in the goat, the elastic protein-based polymer $(GVGVP)_n$ prevents formation of debilitating adhesions. The decreased quality of Life, the high health care costs due to low back pain, addressed in section 9.1.5.1, and the disability and medical care costs due to the many other sites where postsurgical and post-trauma adhesion occur combine to position prevention of adhesions as the potential "healer app" that could unleash the greater potential of elastic protein-based polymers.

9.4.4.4 Prevention of Urinary Incontinence

The next aspect of soft tissue restoration involves injectable implants of protein-based polymers for the prevention of urinary incontinence. We have been addressing this application for a decade, first under the confidentiality of a collaboration/license agreement with Bard Urological Division and subsequently with the support of the National Institutes of Health by means of Phase I and Phase II SBIRs. It has been during work on this application that the essential developments of purification, as demonstrated by Figure 9.14, and stimulation of differentiated tissue generation have been achieved (as shown in Figure 9.35, below). Accordingly, with the magnitude of the health care problem noted in section 9.1.5.2, this also becomes a reasonable candidate for the "healer app" with which to unlock the potential of protein-based polymers as medical devices.

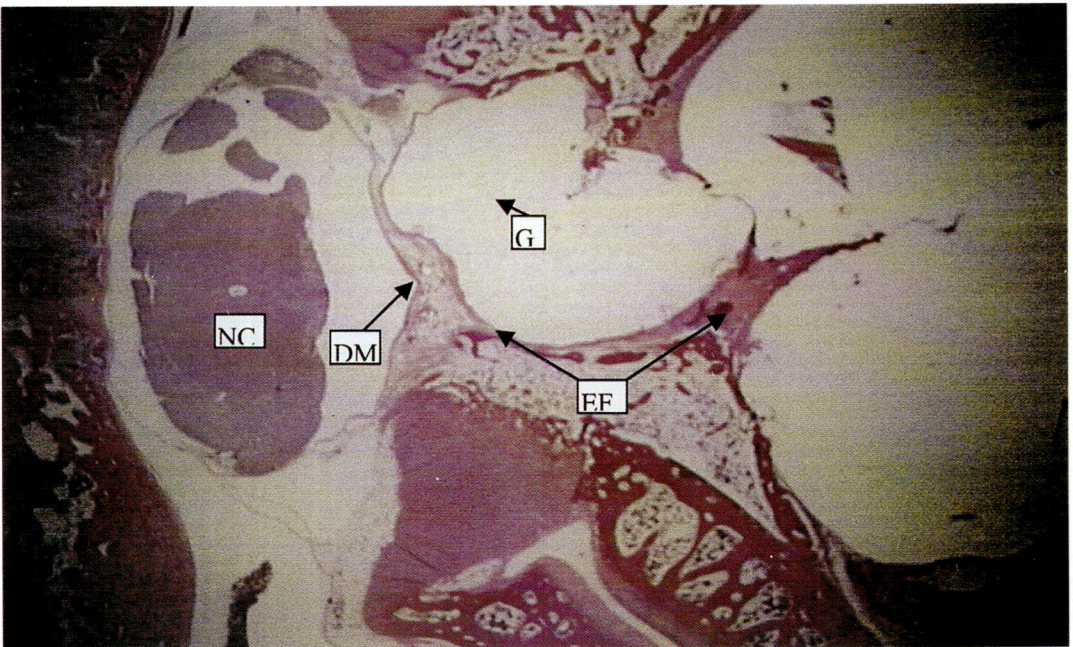

FIGURE 9.33. Spinal surgery (laminectomy) test site in the rabbit model showing the test elastic material, the gel state of $(GVGVP)_{251}$ (G), to have blocked formation of the massive fibrous adhesion (EF) to the dura mater (DM) that surrounds the spinal cord (NC) (P. Glazer, surgeon; work supported by subcontract from Bioelastics Research, Ltd.). (Reproduced with permission from Alkalay et al.[79])

9.4 Medical Applications

Reasons for this optimism are many. They include a cost per gram of one-thousandth that of Contigen® (an FDA-approved material to correct urinary incontinence), the capacity to purify to the exacting levels required for use as an injectable implant, and the demonstration that injectable implants of certain compositions induce differentiated generation of new tissue, whereas Contigen® induces formation of dense scar tissue and is of limited success with a 10% to 15% cure rate at 3 years. Collagen injections cause a mild inflammatory response with the result of scar tissue formation. The contracture that scar tissue undergoes and the wisdom of repeating the procedure with more scar tissue formation would seem to limit the potential of collagen injections. Also there exists the fearsome possibility of prions in collagen obtained from animal sources (e.g., mad cow disease as considered in Chapter 7). Even so, Contigen® did cover new ground in obtaining FDA approval for medical use of protein-based materials.

9.4.4.4.1 Possibility of Elastic Protein-based Polymers as Inert Space Fillers

The objective in overcoming urinary incontinence is to introduce a material at the base of the bladder to provide a volume for physical support. This has been attempted in the past with inert materials. In fact, such an approach could be used with $(GVGVP)_{251}$ by extrusion through a hypodermic needle to fill volume at the base of the bladder. This has been demonstrated with the present animal model of subcutaneous injection in the guinea pig. As shown in Figure 9.34, a 30 mg subcutaneous injection of $(GVGVP)_{251}$ results in a volume filled with the phase separated state of $(GVGVP)_{251}$ and surrounded by a fine fibrous capsule, the fine fibrous capsule that results from minor residual impurities arising from production by *E. coli*. Here the fibrous capsule serves to hold the material in place and thereby to limit migration. This simple approach could itself be useful, but it does not bring out the full potential of elastic protein-based polymers for this application.

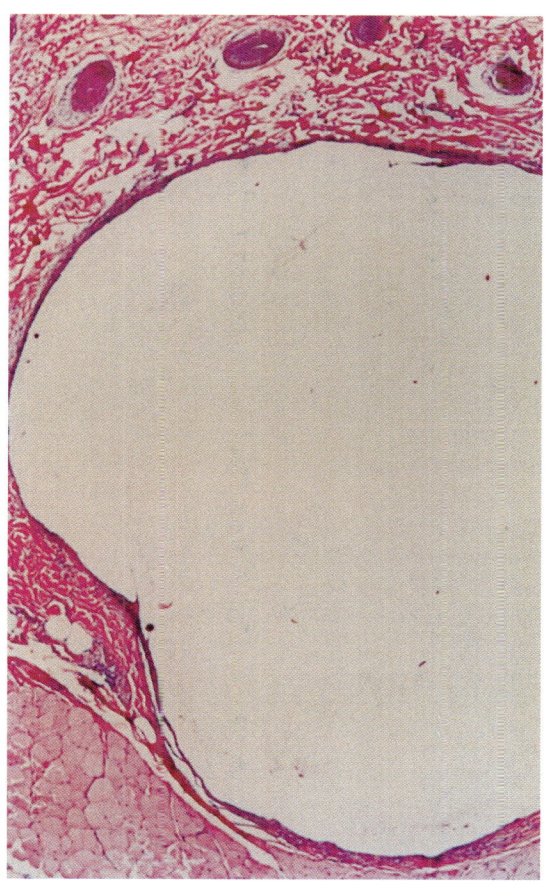

FIGURE 9.34. Subcutaneous site in the guinea pig for injection of the purified gel $(GVGVP)_{251}$, showing a fine fibrous capsule due to a small impurity to limit migration of the tissue volume augmentation. (Reproduced with permission from Urry et al.[58])

9.4.4.4.2 Injection of Protein-based Polymer Composition That Induces Tissue Generation for Support of the Urinary Bladder

Our primary objective is to achieve augmentation with development of a natural, long-lived tissue. The generation of elastic fibers would be expected to result in a long-lived augmentation, because elastic fibers in the body have half-lives on the order of 60 years or more. Collagen, on the other hand, turns over every 9 to 12 months. As shown in Figure 9.35, subcutaneous injection of 30 mg of $[(GVGVP)_{10}–GVGVP-GRGDSP–GVGVP)_{10}]_{18}(GVGVP)$ results in a

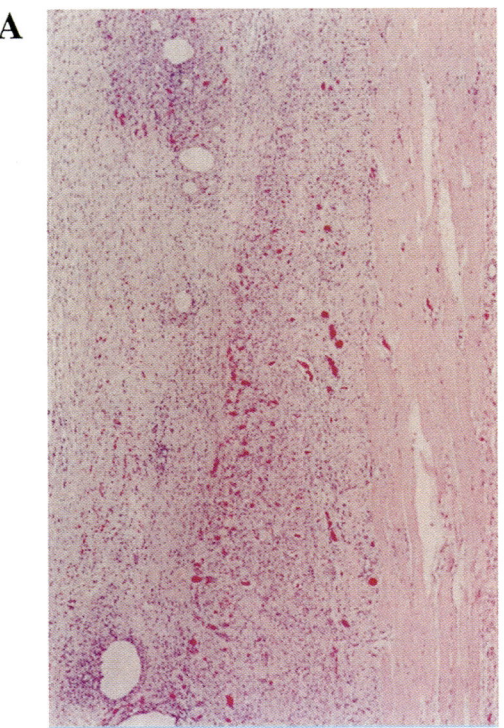 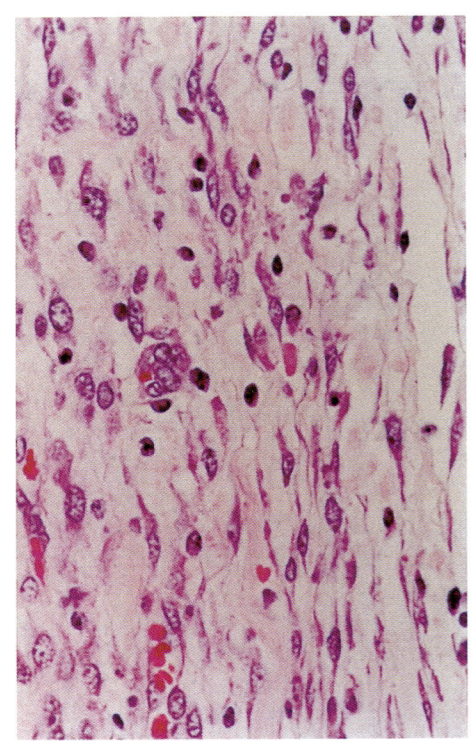

FIGURE 9.35. Subcutaneous site in the guinea pig 2 weeks after a 30 mg injection of a highly purified elastic protein material with a GRGDSP cell attachment site, namely, $[(GVGVP)_{10}GVGVP\text{-}GRGDSP(GVGVP)_{10}]_7(GVGVP)$. (**A**) Low magnification showing much tissue generation with angiogenesis and some remaining polymer seen as vesicles. (**B**) Higher magnification showing erythrocytes in capillaries and spindle-shaped fibroblastic cells presumably synthesizing collagen and elastin fibers. (Reproduced with permission from Urry et al.[58])

tissue generation with what appears to be a natural distribution of collagen and elastic fibers. In particular, the generation of elastic fibers would prevent contracture and would be expected to result in a long lasting tissue generation, because the half-life for elastic fiber in the human is of the order of a lifetime.

9.4.4.5 Plastic Surgery to Improve Body Contours Formed by Soft Tissues

A composition could be injected to induce tissue generation for improving body contour. Wrinkle removal provides the simplest extension of tissue augmentation or generation, but the concept could be applied at many sites where soft tissue generation may be considered a desirable result.

9.4.4.6 Temporary Functional Scaffoldings for Soft Tissue Restoration

Whenever a dynamic structure fails or is so diseased that it needs to be replaced, the concept of temporary functional scaffoldings comes into play. Examples are synthetic arteries, urinary bladder, intervertebral discs, and even a temporary functional scaffolding for skin replacement, for example, after burns.

Each of these structures begins with a structural integrity and a dynamic functional range that becomes lost through any number of disease processes. In each of these cases and many others, the concept of a temporary functional scaffolding couples with the remodeling capacity of natural cells sensing the demands of a normal tissue environment to result in tissue restoration. An explicit example of the urinary bladder follows.

The concept of using temporary functional (elastic) scaffoldings in soft tissue restoration has been tested in a simulated urinary bladder model. An explant from human ureter was placed on a large disk of cross-linked matrix of [(GVGVP)$_{10}$–GVGVPGRGDSP–GVGVP)$_{10}$]$_{18}$(GVGVP), and the outgrowth of human uroepithelial cells was verified. This matrix closely matched the elastic modulus (compliance) of the natural human bladder.

This disk, with a human ureteral explant positioned on it, was placed on the mandril of a simulated bladder apparatus, as shown in Figure 9.36. The apparatus could be filled with culture medium and held in a static nonextended state for several days. Alternatively, the simulated bladder could be slowly filled to 100% extension over a 3-hour period and emptied in 25 seconds. This filling and emptying continued for several days.

The objective becomes the comparison of the outgrowths of human uroepithelial cells from the explants as a function of static and dynamic conditioning. Shown in Figure 9.37A is the static case where a careful examination allows delineation of the edges of the outgrowth from the black area where the nontransparent explant resides. Figure 9.37B demonstrates the result from the dynamic case that simulated the natural filling and emptying of a urinary bladder. The tissue outgrowth in the dynamic case is much richer, exhibiting a higher cell density and a thicker extracellular matrix. *Subjecting the outgrowth of human uroepithelial cells from a ureteral explant to the force changes that occur naturally during bladder function stimulate the cells to begin developing the appropriate bladder tissue.* Thus, Figure 9.37 provides an example of the concept of designed temporary elastic scaffoldings containing cell attachment sequences that, when combined with the mechano-chemical transductional properties of cells, can result in tissue restoration.

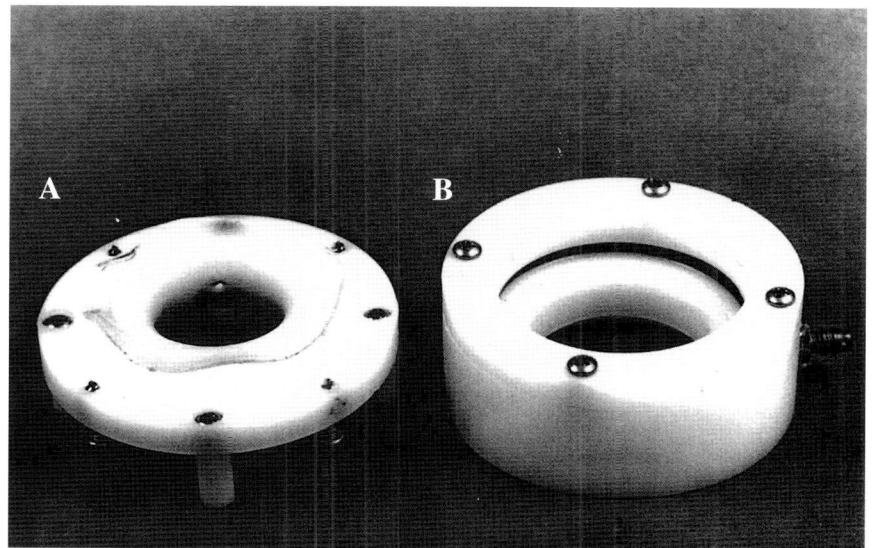

FIGURE 9.36. Disassembled Simulated Bladder Chamber. (**A**) Mandril covered by a large disk of elastic protein-based polymer membrane, X^{20}–[(GVGVP)$_{10}$GVGVPGRGDSP(GVGVP)$_{10}$]$_7$(GVGVP), containing GRGDSP cell attachment sequences. Human ureteral explant is centrally placed on the elastic disk. (**B**) Top chamber to be placed over the mandril with filling/emptying port and a window for microscopic viewing of outgrowth from explant. (Unpublished results[119])

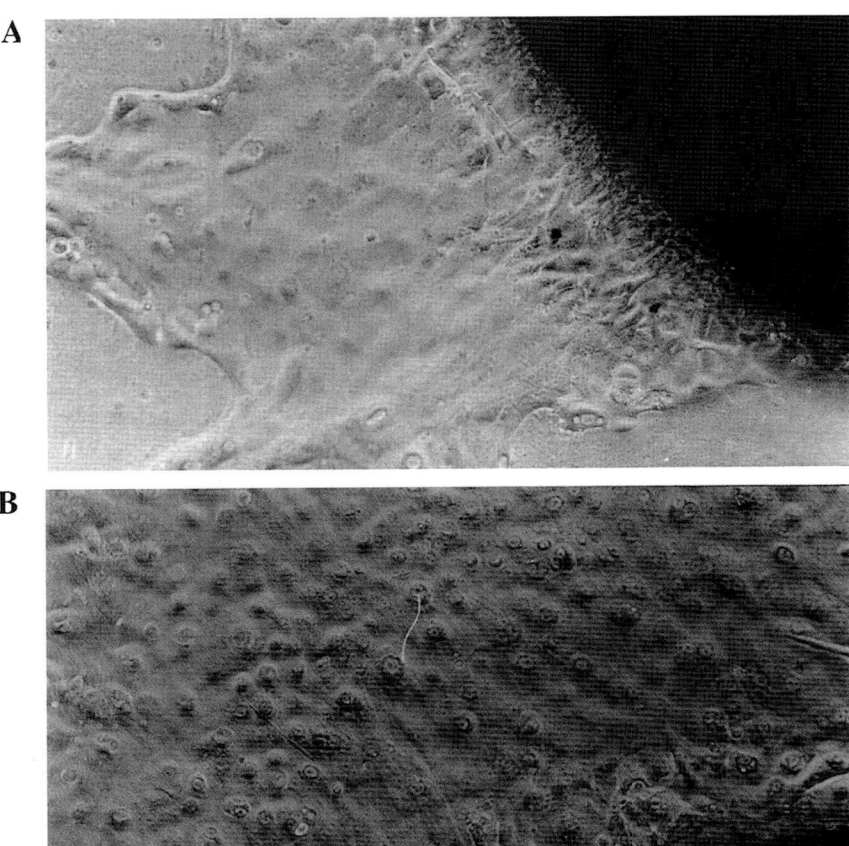

FIGURE 9.37. Outgrowth from human ureteral explant on elastic protein-based polymer membrane, X^{20}–[(GVGVP)$_{10}$GVGVP<u>GRGDSP</u>(GVGVP)$_{10}$]$_7$(GVGVP), during a several-day period in simulated urinary bladder. (**A**) Static circumstance with no filling and emptying gave good outgrowth compared with collagen-covered substrate. (**B**) Dynamic filling (3 hours) and emptying (25 seconds) resulted in much more dense outgrowth of cells and thicker extracellular matrix. (Reproduced with permission from Urry and Pattanaik et al.[120])

9.4.5 Controlled Release of Pharmaceuticals by T_t-type (Transductional) Protein-based Polymers

The special energy conversion (transductional) capacities designable into elastic protein-based materials provide unique opportunities for a remarkable range of systems capable of controlled delivery of drugs and other more complex therapeutic molecular constructs. Each of the 18 pairwise energy conversions, described in Chapter 5 for which elastic protein-based materials can be designed, can be employed to control release of pharmaceuticals. Basically the elasticity, the inverse temperature transitional behavior, and knowledge of the forces responsible for that behavior provide an unprecedented potential for the development of controlled release systems. It is not an issue of using a single or even several compositions, and it is not primarily an issue of working with the different physical states of these compositions to coerce a more useful state for a particular controlled release problem. Instead, the composition of the bioelastic material should be designed specifically to have the optimal properties perceived for the particular controlled release problem at hand.

Again we recall the statement of Bronowski,[2] *"In effect, the modern problem is no longer to design a structure from the materials (available) but to design materials for a structure."* The chemical nature of the pharmaceutical to be delivered and the preferred circumstance for its delivery determine the structure of the elastic protein-based polymer to be designed for that purpose. The approach is emphatically not one of using an "off-the-shelf" (available) polymer and coercing it chemically and physically to approach a desired release. The potential applications and structures of bioelastic materials in the area of controlled release can be expected to be as numerous as the pharmaceuticals and circumstances to be considered and to include the less exotic application of nonadherent drug-delivering wound coverings.

9.4.5.1 Positively Charged Drugs Delivered by Negatively Charged Polymers: Consideration of Analgesics, Anesthetics, and Endorphins

9.4.5.1.1 The Chemical Nature of Analgesics and Anesthetics

Most of the analgesics and anesthetics used today are tertiary (or secondary) amines, which at physiological pH are positively charged. These include alfentanil, bupivacaine butorphanol, chloroprocaine, cocaine, codeine, dyclonine, fentanyl, ketamine, lidocaine, meperidine, mepivacaine, morphine, nalbuphine, prilocaine, sufentanil, and tetracaine. There are also, for example, the intravenous anesthetics Brevital (methohexital sodium) and Diprivan (propofol) and the topical analgesic Zostrix (capsaicin) that have the capacity to contain negatively charged, for example, phenolate, species. In addition to being able to carry a charge, these drugs all exhibit substantial hydrophobic character, that is, they contain both polar (ionizable) and nonpolar moieties. These are exactly the types of molecular species, the sustained release of which can be most effectively achieved by the transductional protein-based polymers under consideration here.

9.4.5.1.2 The Chemical Nature of Endorphins

The term *endorphin* is an elision for "*endogenously produced morphine*like substance," and it has come to stand for the set of endogenous peptides that includes the enkephalins, dynorphins, endorphins, and their synthetic analogues, which are also collectively referred to as *opioid peptides*.[80,81] Great excitement occurred more than two decades ago with the discovery of enkephalins by Hughes and coworkers.[82] Endorphins are the body's own means of alleviating pain. This introduced the possibility of pain relief without development of debilitating cravings or addictions. The realization of opioid peptides as pharmaceuticals, however, has been limited in significant part due to their rapid *in vivo* metabolism,[85] to their eliciting serious side effects at higher concentrations, and to the problem of delivery at constant and competent doses to their sites of activity whether in the central nervous system or at peripheral sites.[80,84,85]

Much work has been directed toward the preparation of opioid peptide analogues to alleviate the problem of proteolytic inactivation, to define the several receptors,[86,87] to achieve receptor selectivity and elicit selective activities thereby minimizing side effects, and generally to understand the complex mechanisms involved in substance abuse.[88] For that work to be more fully utilized in health care and in alleviating suffering, delivery processes are required to protect the peptide from proteolytic degradation, to access various sites in the body,[89,90] and to provide for the desired release profiles. This problem provides an opportunity for elastic transductional protein-based polymers to fill a much-needed role.

9.4.5.1.3 Generally Considered Means of Controlled Release by Polymeric Systems

The development of polymers as controlled release vehicles has been of long-standing interest.[91–93] Kost and Langer[94] consider four different controlled release systems for polymeric hydrogels: (1) diffusion controlled, (2) chemically controlled, (3) solvent controlled, and (4) release induced by external factors. Pitt et al.[95] and Peppas and Korsmeyer[96] list similar mechanisms. There is another category of mechanism that stands in its own right and that encompasses all of the above. It utilizes the inverse temperature transition and may be called the T_t-*type transductional mechanism* of general development in this volume.

9.4.5.1.4 The Opportunity for Elastic Transductional Protein-based Polymers as Controlled Release Vehicles

Biocompatible and biodegradable polymers provide the potential for the delivery of analgesics and anesthetics in an anatomically localized manner and at sustained optimal dose levels for alleviating pain. These controlled release vehicles would eliminate the occurrence of peak concentrations during which times toxic reactions are more likely to occur. Transductional elastic protein-based polymers hold promise for sustained and even constant (zero order) release levels of analgesics and anesthetics that would help to alleviate untoward clinical responses. In fact, as discussed below, transductional protein-based polymers have the demonstrated capacity to release a cationic morphine analogue and a cationic opioid peptide at a constant level for a given surface area and for periods up to many months. Furthermore, they can be designed to release at preferred levels and time periods and to be formed into useful drug delivery vehicles: injectable viscoelastic implants, injectable nanoparticles and microspheres, trocar implanted elastic rods, laparoscope implanted sheets, and patches for transdermal delivery.

9.4.5.1.5 Zero Order Release of Positively Charged Leu-enkephalin Amide from Negatively Charged Polymers

Leu-enkephalin amide (H^+-Tyr-Gly-Gly-Phe-Leu-NH_2, LEA^+) is a positively charged opioid peptide from the endorphins that are so well known for causing the "runner's high." When added to a solution containing a relatively hydrophobic carboxylate-containing elastic protein-based polymer, LEA^+ induces phase separation. The positively charged drug ion pairs with the elastic protein-based polymer and lowers T_t, the onset temperature for the inverse temperature transition. In this manner the drug provides the chemical energy for its own packaging into the drug delivery device. As the glutamic acid (Glu, E)–containing polymer becomes more hydrophobic by replacement of less hydrophobic valine (Val, V) residues by more hydrophobic phenylalanine (Phe, F) residues, there occurs an increase in pKa, and there results an increase in binding affinity of drug to polymer in the phase separated state.

In the example to be discussed, loading occurred with a 50% excess of drug over carboxylate ion pairing sites. The Leu-enkephalin amide release profiles in Figure 9.38 were obtained using a cylindrical tube with a shallow conical bottom, as also depicted in Figure 9.38B, and replacing the overlying buffer solution every 24 hours.[50] Note that the drug delivery vehicle disperses as the drug is released, because, in fact, the drug provides the glue that holds the drug delivery vehicle together. The excess drug is released during the first 12 days. Once this has occurred, there follows a remarkably constant release, which continues until the sample is totally dispersed.

For 250mg of poly(GVGFP GEGFP GVGVP), zero order release occurs at the level of 4.7 μm/day for approximately 1 month, and the turn off is quite abrupt as the materials come from the bottom conical part of the tube. For 250mg of the more hydrophobic protein-based polymer poly(GEGVP GVGVP GVGFP GFGFP GVGVP GVGVP), zero order release is maintained constant for over 1 month at about 2.7 μm/day and again turns off abruptly. For 250mg of the most hydrophobic elastic protein-based polymer, poly(GFGFP GEGFP GFGFP), of the series, zero order release is maintained constant for nearly a period of 3 months at the level of 1.9 μm/day and again shuts off abruptly. Of course, the initial burst can be eliminated by not loading with an excess of drug, but an initial excess may be helpful *in vivo* to more quickly establish a steady state.

9.4 Medical Applications

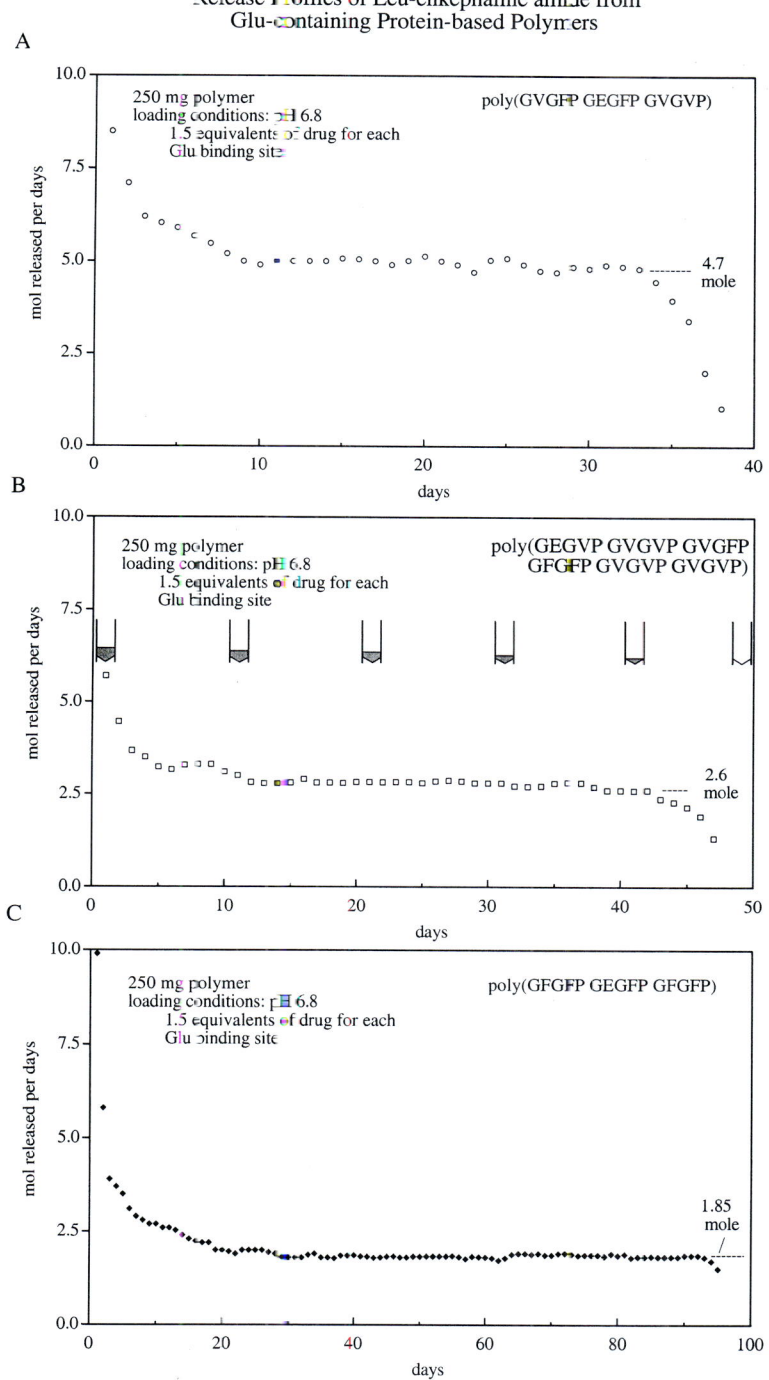

FIGURE 9.38. Release of positively charged Leu-enkephalin amide from designed, negatively charged polymers with systematically increased hydrophobicity and pKa shifts. For constant surface areas, release levels are maintained constant out to 3 months. Initial burst release due to 50% excess drug over polymer sites. (Reproduced with permission from Urry et al.[50])

It is quite clear that the ion-pairing approach using hydrophobic-induced, pKa-shifted, anionic sites within polymers capable of inverse temperature transitions provides a means of achieving constant release of positively charged drugs over substantial periods of time. As discussed below, similar results have been obtained for the narcotic antagonist naltrexone.

9.4.5.1.6 Drug Addiction Intervention: Zero Order Release of Positively Charged Naltrexone from Negatively Charged Polymers

The approach considered for drug addiction intervention is to treat the addict, after withdrawal, to prevent recurrence of active dependency. An effective drug is the narcotic antagonist naltrexone. Two nanograms per milliliter of this drug in the blood plasma can block the action of a 25 mg dose of heroin and deter recurrence of active dependency.

The limitation with the use of this drug arises from the problem of patient compliance. If for whatever reason the patient feels compelled again to have the high, the medication is discontinued and the addict again becomes actively dependent. One way to overcome this problem is a subcutaneous implant of a controlled release device that would maintain an effective range of drug release for sufficient periods to get the addict beyond the return to active dependency. This could be a device that releases competent doses for 1 month or more.

Naltrexone is a positively charged drug. Therefore, as for the opioid peptide Leu-enkephalin amide discussed above, the design of the bioelastic drug delivery device is one of negatively charged polymers; the two polymers in Figure 9.38B,C will be discussed for ease of comparison. The release profiles are contained in Figure 9.39. For 250 mg of poly(GEGVP GVGVP GVGFP GFGFP GVGVP GVGVP) and a 50% excess of naltrexone over the number of carboxylates, zero order release is maintained constant for over 1 month at 2.7 μm/day (the same as for Leu-enkephalin amide) and, of course, turns off abruptly, just as occurred with Leu-enkephalin amide. For 250 mg of the more hydrophobic elastic protein-based polymer poly(GFGFP GEGFP GFGFP), zero order release of naltrexone continues constant for

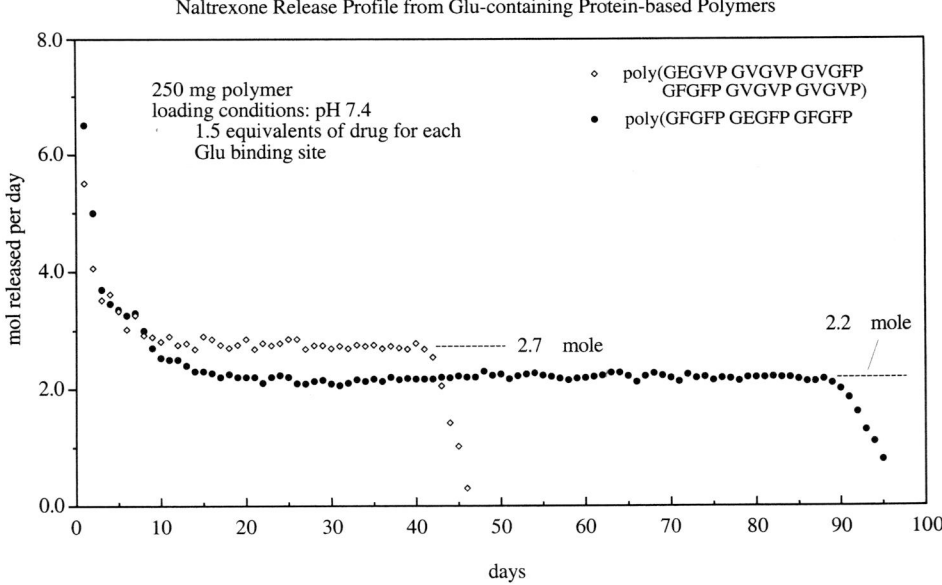

FIGURE 9.39. Release of the positively charged narcotic antagonist naltrexone from negatively charged polymers with systematic increases in hydrophobicity and pKa shifts. For constant surface areas, release levels are maintained constant out to 3 months. Initial burst release due to 50% excess drug over polymer sites. (Courtesy Bioelastics Research, Ltd.)

9.4 Medical Applications

a period of nearly 3 months at the level of 2.2 µm/day (compared with 1.9 µm/day for Leu-enkephalin amide). As expected, delivery shuts off abruptly in about 3 months.

By changing the oil-like nature of groups in the polymer, different levels of release can be obtained. By proper design of the protein-based polymer, different levels of constant release are obtained, and, for a constant surface area of an adequately sized depot, the amount released can be constant for periods of more than 3 months (see Figure 9.39).

9.4.5.2 Negatively Charged Pharmaceuticals Delivered by Positively Charged Polymers: Consideration of Steroid Phosphates and Oligonucleotides

9.4.5.2.1 Loading Negatively Charged Dexamethasone–Phosphate and Betamethasone–Phosphate into a Series of Positively Charged Model Proteins

For this example the series of microbially prepared **Model Proteins x′** through **xiv′** in Table 5.5 is used. Changes in the value of T_t as the drug is added to a solution of each of the series of increasingly hydrophobic elastic protein-based polymers provide the loading profile and present visualization of the relative affinities of drug with model protein. The data are given in Figure 9.40. The differences in the loading profiles and dexamethasone-phosphate (Figure 9.40A) and betamethasone-phosphate (Figure 9.40B) are very minimal, although there is the slightest suggestion that the affinity would be slightly greater for the latter.[97]

9.4.5.2.2 Release Profiles for Negatively Charged Dexamethasone-phosphate and Betamethasone-phosphate from a Series of Positively Charged Model Proteins

The release profiles for negatively charged dexamethasone-phosphate and betamethasone-phosphate from a series of positively charged model proteins are given in Figure 9.41. The results are not as extraordinary as with the positively charged drug/negatively charged polymer cases. In the case of drug delivery combining negatively charged drug with positively charged polymer, the polymer does not dis-

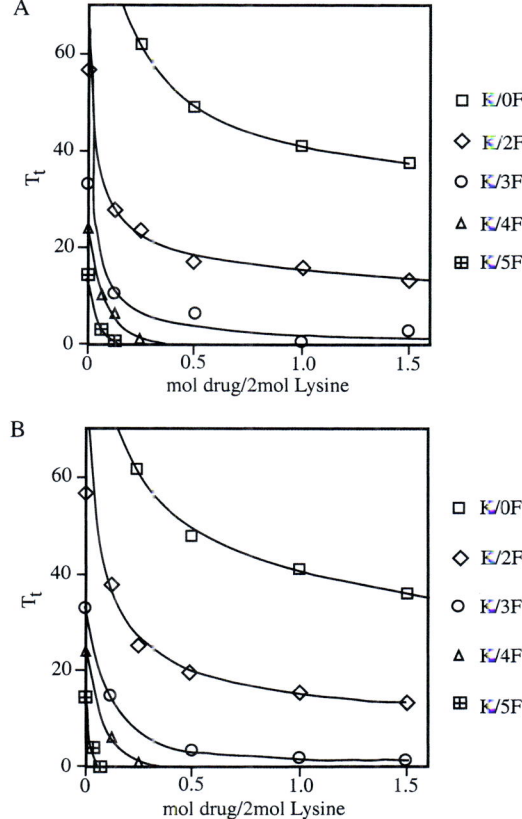

FIGURE 9.40. Loading profiles for (**A**) dexamethasone-phosphate and (**B**) betamethasone-phosphate using the series of Lys (K)/Phe (F)–containing **Model proteins x′** through **xiv′** in Table 5.5. (Reproduced with permission from Urry et al.[97])

perse as readily in phosphate buffered saline. The difference arises from the effectiveness of chloride ion in lowering the value of T_t for model proteins containing lysine (Lys, K) residues. Under conditions of physiological saline, 0.15 N NaCl, Cl⁻ replaces the phosphate of the drug. Nonetheless there occurs a period of about 100 days between 70 to 170 days when the releases hold within a good band that could be adjusted to fit the desired therapeutic range for the drug. Preparation of a protein–drug complex that would provide the desired 100 day release would involve the proper amount of drug being titrated into solution to form a drug–polymer complex having the equivalence shown at 70 days in Figure 9.41. In general, desired levels of release would be achieved by the size and shape of the depot and the hydrophobicity of the model protein.[97]

9.4.5.2.3 Controlled Release of Antisense Oligonucleotides from a Series of Positively Charged Model Proteins

Antisense oligonucleotides are parts of the genetic material that can be used to block the expression of certain genes. For example, a catheter with a balloon on its end is inserted into a blocked coronary artery, and the balloon is inflated to open the artery for improved blood flow. A common problem is that this injurious process induces a repair response, and an associated cellular proliferation leads to reclosure, as has happened with a number of well-known people, for example, Vice President Cheney.

If, however, a controlled release device could also be deployed that released low levels of the correct antisense oligonucleotide, or sRNA$_i$, the expectation is that reclosure could be prevented. Thus our initial task is to see if we can design bioelastic polymers that could give rise to controlled release of negatively charged oligonucleotides. In this case the bioelastic polymer is designed with positive charges and with different amounts of more oil-like groups.

As shown in Figures 9.38 and 9.39, polymers have been designed to form devices that result in different constant levels of release for a constant surface area. As shown in Figure 9.41, a significant controlled release was achieved using a small negatively charged drug with a series of positively charged polymers. Figure 9.42, using more than 10-fold larger negatively charged pharmaceutical, an oligonucleotide, with the same series of positively charged polymers demonstrated a remarkable release profile. Oligonucleotides, being very potent pharmaceuticals, require but a small level of release, such that the drug delivery devices in this *in vitro* assay can release therapeutically competent levels for months, as shown in the lower curve in Figure 9.42. This demonstration of controlled release of oligonucleotide by principals associated with BRL resulted in the 1999 CRS-Prographarm Outstanding Pharmaceutical Paper that was awarded at the 27th International Symposium on Controlled Release of Bioactive Materials in July 10, 2000, in Paris.

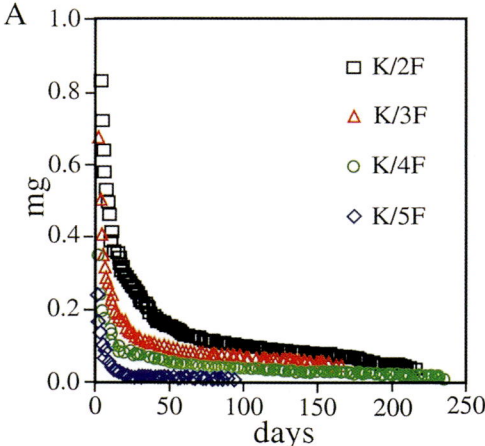

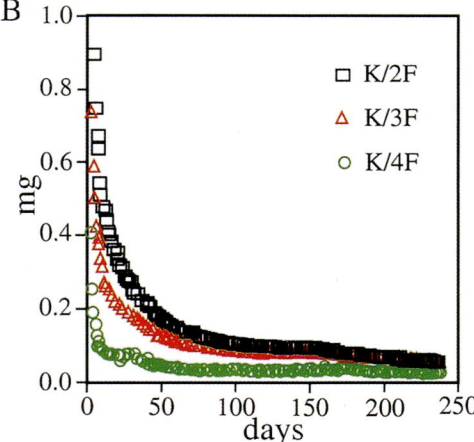

FIGURE 9.41. Release profiles for (**A**) dexamethasone-phosphate and (**B**) betamethasone-phosphate using the series of Lys (K)/Phe (F)–containing **Model Proteins x′** through **xiv′** (K/5F) in Table 5.5. Note the constant release for the most hydrophobic **Model Proteins xiv′** (K/5F) in (**A**) and **xiii** (K/4F) in (**B**). (Reproduced with permission from Urry et al.[97])

9.4.5.3 Loading of Protein into and Release from Transductional Protein-based Polymers

When considering designed T_t-type transductional protein-based polymers for controlled

9.4 Medical Applications 517

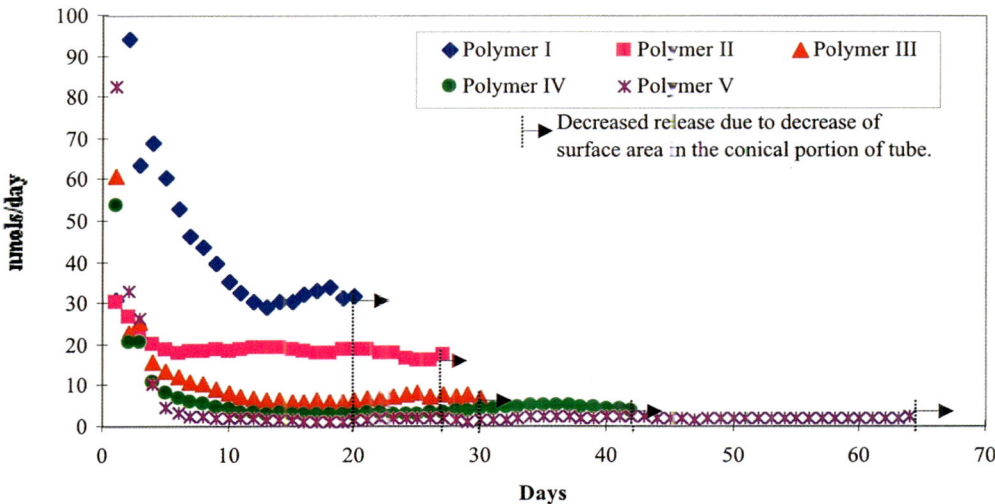

FIGURE 9.42. Release of negatively charged antisense oligonucleotide from Lys-containing, positively charged polymers with systematically increased hydrophobicity and pKa shifts, **Model Proteins x′**(I) through **xiv′**(V) in Table 5.5. (Reproduced with permission from Woods.[32])

release of an identified pharmaceutical, initial insight into their effectiveness derives from the loading curves. This is apparent on examination of the loading curves for dexamethasone-phosphate and betamethasone-phosphate shown in Figure 9.40 and on comparison to the release curves of Figure 9.41. For these negatively charged drugs, positively charged polymers were considered, that is, the K/nF series of Lys (K)/Phe (F)–containing **Model Proteins x′** through **xiv′** in Table 5.5. The affinity between charged drug and oppositely charged polymer increases as the polymer becomes more hydrophobic, because the ΔG_{ap}-derived driving force for ion pairing increases with hydrophobicity. This is evidenced by the steepness of the drop in the values of T_t for the onset of the inverse temperature transition as the drug is titrated into the solution of the polymer. In general, the tighter interaction, as evidenced by the greater decrease in T_t, results in a lower and more constant release profile. Thus, when considering the potential for controlled release of a particular pharmaceutical such as a natural protein, we look at the loading curves of chosen transductional elastic protein-based polymers.

9.4.5.3.1 Loading and Release Curves for Soybean Trypsin Inhibitor

The net charge on the protein becomes the first issue in the choice of charge on the polymer. In general the polymer is chosen with the opposite charge from that of the net charge on the protein. This allows for the hydrophobically, ΔG_{ap}-driven ion pairing that has the associated effect of lowering the temperature for the onset of the inverse temperature transition. The protein of interest here is 20,000 Da protein, soybean trypsin inhibitor (STI). It has a pI of 4.5, that is, at pH 4.5 it contains an even number of positive and negative charges. This means at the physiological pH of 7.5 that STI carries a net negative charge. Accordingly, the obvious choice of protein-based polymer comes from the K/nF series of Lys (K)/Phe (F)–containing **Model Proteins x′** through **xiv′** in Table 5.5, as used above for the dexa- and betamethasone phosphates and oligonucleotides. The loading curve is given in Figure 9.43A, and Figure 9.43B shows the release data in the more common plot of percentage released as a function of time for the original elastic protein-based polymer from the mammalian elastic fiber, $(GVGVP)_n$

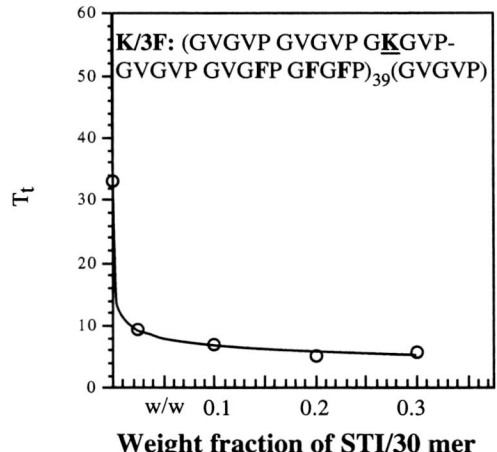

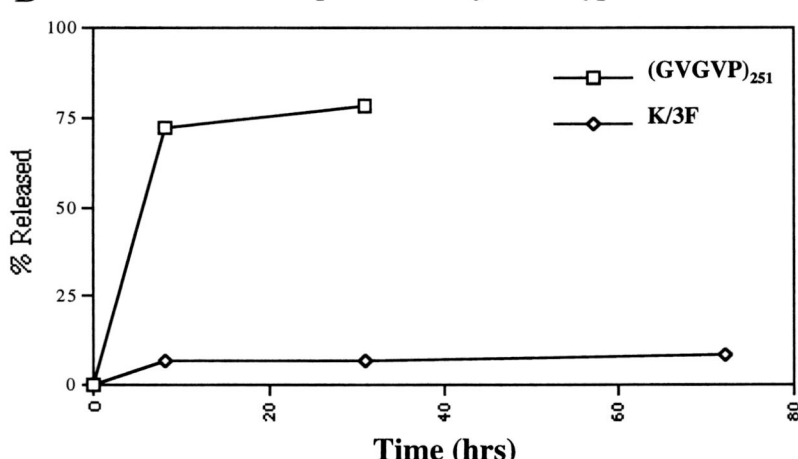

FIGURE 9.43. Soybean trypsin inhibitor (STI), a protein model for loading and release studies using positively charged elastic protein-based polymers of varied hydrophobicity. (**A**) Loading curve for the protein-based polymer K/3F: (GVGVP GVGVP GKGVP GVGVP GVGFP GFGFP)₃₉(GVGVP). (**B**) Percent of STI released as a function of time from K/3F and from (GVGVP)₂₅₁. (A. Pattanaik, L.C. Hayes, and D.W. Urry, unpublished results.)

with n = 251, and K/3F: (GVGVP GVGVP GKGVP GVGVP GVGFP GFGFP)₃₉ (GVGVP). The release from the initial uncharged base polymer represents a burst type of release, whereas that from the designed K/3F, **Model Protein v**, after the first several hours establishes a low-level, steady release rate.

9.4.5.3.2 Loading Curves for Ovalbumin

Ovalbumin, a 45,000 Da protein, has a pI of 4.7. Accordingly at pH 7.5, physiological pH, ovalbumin contains a net negative charge, which again suggests use of the K/nF series. In particular, the loading curves for K/0F: (GVGVP

9.4 Medical Applications

GVGVP G<u>K</u>GVP GVGVP GVGVP GVGVP)$_{22}$(GVGVP) and K/2F: (GVGVP GVGFP G<u>K</u>GFP GVGVP GVGVP GVGVP)$_{22}$(GVGVP) are shown in Figure 9.44A. Polymer K/0F does not sufficiently lower the value of T_t to function at 37°C. The K/2F–ovalbumin complex is used below to determine if it sufficiently slows the release to function *in vivo* for the elastic protein-based polymer to serve as an adjuvant.

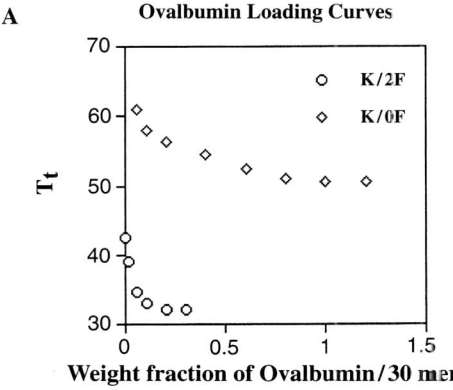

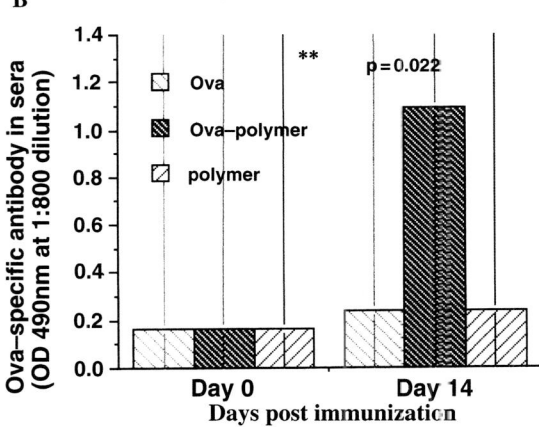

FIGURE 9.44. (**A**) Ovalbumin loading into K/0F: (GVGVP GVGVP G<u>K</u>GVP GVGVP GVGVP GVGVP)$_{22}$(GVGVP) and K/2F: (GVGVP GVGFP G<u>K</u>GFP GVGVP GVGVP GVGVP)$_{22}$(GVGVP) shows K/2F to be a possible delivery device. (A. Pattanaik, L.C. Hayes, and D.W. Urry unpublished results.) (**B**) The K/2F–ovalbumin complex elicits ovalbumin-specific antibody formation, whereas ovalbumin and K/2F alone exhibit no significant antibody formation. (Z. Moldoveanu, J. Mestecky, A. Pattanaik, and D.W. Urry, unpublished results.)

9.4.5.3.3 Potentiation of Protein Antibody Development When Conjugated with Positively Charged Protein-based Polymer

The capacity for slow release of a protein (the antigen) from a depot suggests the use of the protein-based polymer as a controlled release vehicle that may function as an adjuvant for the development of a superior immune response. As shown in Figure 9.44B, neither ovalbumin alone nor the protein-based polymer alone results in the significant formation of ovalbumin specific antibody in the sera. When the ovalbumin-loaded protein-based polymer is given, significant amounts of ovalbumin specific antibody are detected in sera.

9.4.5.4 Nanoparticles Composed of a Hydrophobic Series of Charged Elastic-contractile Protein-based Polymers

Vaccines can carry either a net positive charge or a net negative charge. Accordingly, their release can be controlled much as shown in Figures 9.38, 9.39, 9.41, and 9.42. The different release levels would function as primary and secondary immunizations and even for very slow release as third immunizations. Also, if the delivery vehicle could be nanoparticles, absorption could be through a number of epithelial linings available in the body that could replace the often-uncomfortable injections. Figure 9.45 demonstrates formation of uniformly sized nanoparticles made of a substantially oil-like yet negatively charged protein-based polymer that would be the candidate for slow release.[97] Thus, the stage is set for vaccine delivery by nanoparticles. Similarly, delivery of negatively charged genes could be considered by properly designed positively charged nanoparticles.

9.4.5.5 Elastic Protein-based Matrices as Devices for Transdermal Delivery

9.4.5.5.1 Skin Permeation

Bioelastic patches or disks are shown in Figure 9.46 for different loading levels of dazmegrel. The circular patches, cross-sectional area of 1.9 cm^2, are 20 Mrad γ-irradiation cross-linked (GVGIP)$_{260}$, designated as X^{20}-(GVGIP)$_{260}$,

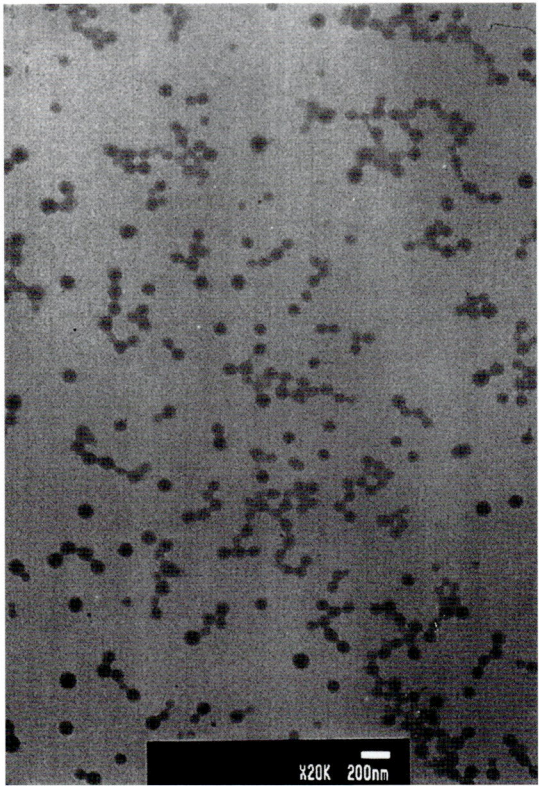

FIGURE 9.45. Electron microscopy of nanoparticles comprised of **Model Protein v** in Table 5.5, namely, (GVGVP GVGFP GEGFP GVGVP GVGFP GFGFP)$_{42}$(GVGVP). (Reproduced with permission from Urry et al.[97])

that have been equilibrated in dipropylene glycol/propylene glycol/water (2:1:1). Progressing from left to right, the dazmegrel levels are 0, 0.1, 1, and 10 mg/cm². When placed on dog skin or human skin, these disks release dazmegrel into the skin as shown in Figure 9.47. Penetration into greyhound dog skin was five times greater than into human breast skin. For *in vitro* greyhound skin permeation studies, the intermediate dose of 1.9 mg/disk yielded a constant release of dazmegrel from 4 to 48 hours, as measured by recovery from the receptor and dermis. The total recovery of dazmegrel was 88 ± 13%, and the retention by the bioelastic membrane averaged 67 ± 9.0%. Thus, it appears that the intermediate level could be maintained for a significantly longer period of time, as demonstrated by the *in vivo* delivery study immediately below.[99]

9.4.5.5.2 *In vivo* Skin Permeation

In narrowing down the concentration range for loading of the bioelastic membrane, an intermediate low (0.5 mg/cm²) and an intermediate high (5.0 mg/cm²) were used. In the greyhound dog model as shown in Figure 9.48, by 24 hours the level of dazmegrel in the epidermis and dermis had reached a constant level with the intermediate low level of loading, and that therapeutically competent level remained

FIGURE 9.46. γ-Irradiation cross-linked elastic disks of X^{20}–(GVGIP)$_{260}$, equilibrated in dipropylene glycol/propylene glycol/H$_2$O (2:1:1) and loaded with 0, 0.1, 1, and 10 mg/cm² dazmegrel, a thromboxane synthetase inhibitor for preventing capillary constriction and the subsequent necrosis that leads to a pressure ulcer. (Courtesy of Bioelastics research, Ltd.)

9.4 Medical Applications

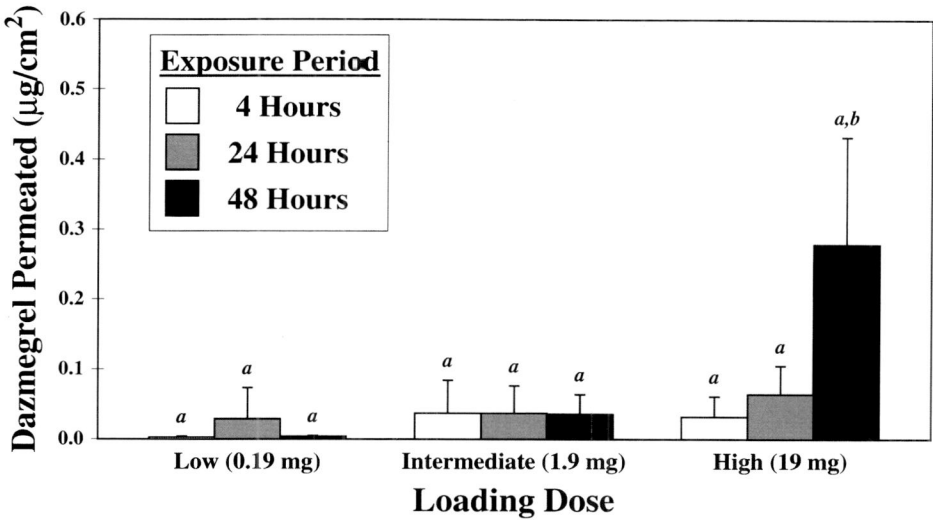

FIGURE 9.47. *In vitro* release into the skin of the greyhound dog from γ-irradiation cross-linked elastic disks of X^{20}–$(GVGIP)_{260}$ equilibrated in dipropylene glycol/propylene glycol/H$_2$O (2:1:1) and loaded with low (0.19 mg/disk), intermediate (1.9 mg/disk), and high (19 mg/disk) levels of dazmegrel, the thromboxane synthetase inhibitor. (Reproduced with permission from Kemppainen et al.[98])

essentially constant for 2 weeks, which was the time for termination of the study. The therapeutic level was maintained for the 2-week period, as demonstrated in Figure 9.49. At the end of the 2-week period, 80% of the loaded amount remained in the disk such that continued therapeutic release could be expected for weeks to come.[99]

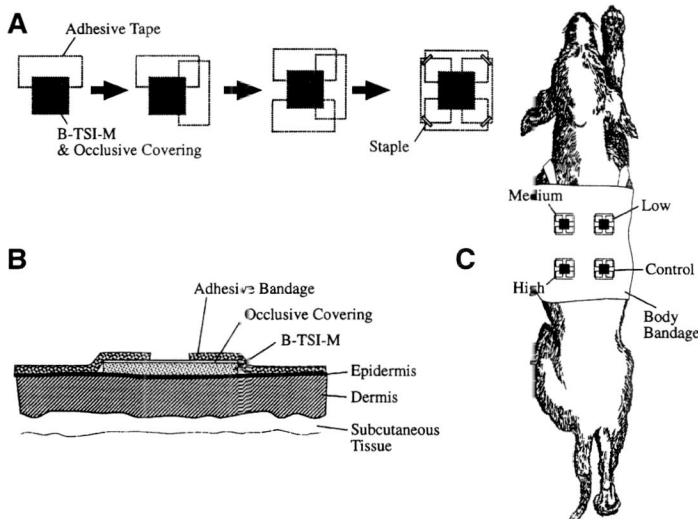

FIGURE 9.48. Greyhound dog model for transdermal delivery: Placement of drug-laden patches on the back of a dog to follow the release of the drug into the skin. Reproduced with permission from the phD dissertation of N.-Z. Wang, Auburn University, Auburn, AL, (1998).

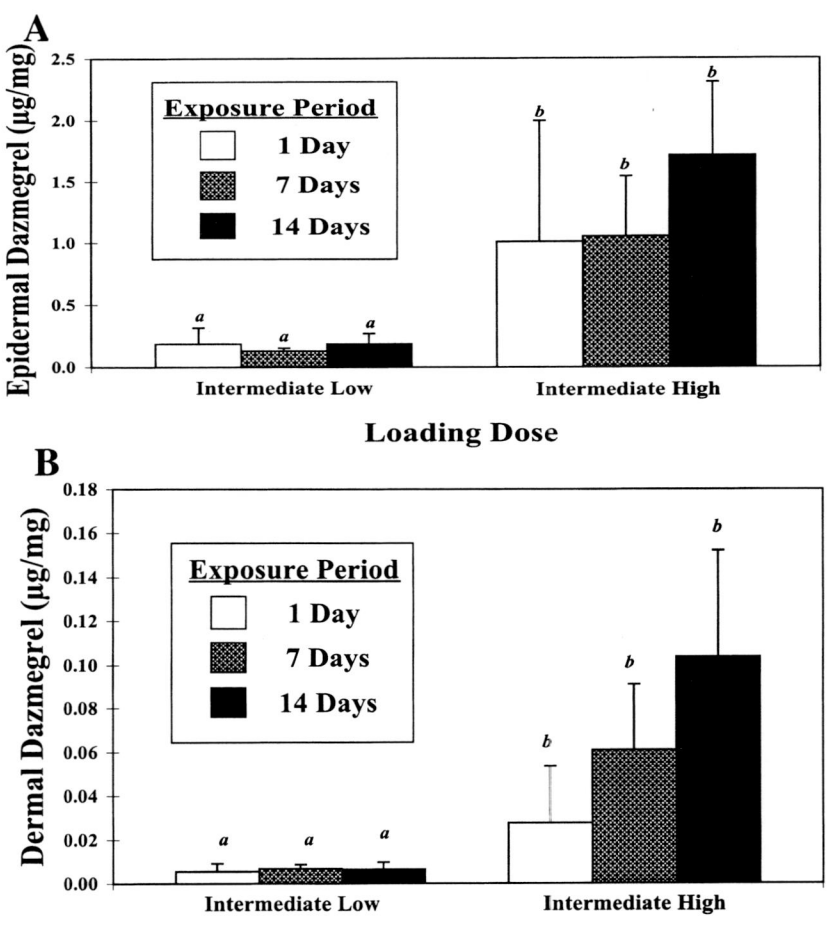

FIGURE 9.49. Release at 1, 7, and 14 days from γ-irradiation cross-linked elastic disks of X^{20}-(GVGIP)$_{260}$, equilibrated in dipropylene glycol/propylene glycol/H$_2$O (2:1:1) and loaded with intermediate low (0.5 mg/cm^2) and intermediate high (5 mg/cm^2) levels of dazmegrel, a thromboxane synthetase inhibitor. For the intermediate low level of loading, a constant pharmacologically effective dose is released for 14 days with yet 80% of drug remaining in the disk at 14 days. (**A**) Epidermis. (**B**) Dermis. (Reproduced with permission from Kemppainen et al.[99])

9.4.5.6 Bioelastic Patches for the Prevention of Pressure Ulcer Formation

9.4.5.6.1 Prevention of Pressure Ulcers by Providing a Drug-laden Cushioning Disc to Cover a Bony Prominence

Pressure ulcers occur at bony prominences where the skin is compressed either at an assistive device interface or as might occur for the bed-ridden or wheel-chair dependent. Due to the elasticity of bioelastic materials, the bioelastic membrane alone provides a helpful protective cushion. The bioelastic materials can also be designed to have partitioning properties similar to the skin itself for a drug useful in blocking the biochemical changes that cause the tissue necrosis leading to pressure ulcer formation. In particular, dazmegrel blocks the activity of an enzyme called *thromboxane synthetase*, and it is the release of thromboxane

9.4 Medical Applications

that is thought to lead to vascular constriction and poor circulation that results in tissue breakdown and pressure ulcer formation. Dazmegrel, therefore, is identified as a thromboxane synthetase inhibitor.[100]

9.4.5.6.2 Preliminary Studies Show Drug-laden Bioelastic Disc to Prevent the Tissue Necrosis Resulting in Pressure Ulcer Formation

The bioelastic disks were then used in a dog model for pressure ulcer formation to determine efficacy. The coaptation walking cast model of Swaim et al.[101,102] was utilized in the greyhound dog. In this model, severe pressure ulcers develop at the medial malleolus (the inner ankle bone), as evaluated in Table 9.8. When a bioelastic disk without drug (B-M) is fixed in place over the anklebone, the occurrence and severity of ulcers is significantly decreased, as shown as the control (unloaded) data in Table 9.9. In the limited studies to date, the presence of a loaded intact disk, the bioelastic–thromboxane synthetase inhibitor–membrane (B-TSI-M) prevented ulcer forma-

TABLE 9.8. Pressure ulcer study: Subjective observations of skin over the medial malleolus with no bioelastic matrix patch.

Dog[a]	Appearance of skin		Days in cast
	Hyperemia	Epidermal abrasion/ulceration[b]	
DP-8N[b]	Not noted	Present	7
DP-9N	Small	Present	5[c]
DP-10N	Large	Present (3)	5[c]
DP-11N	Not noted	Present (2)	7

[a] DP-#N = No bioelastic patch.
[b] These lesions were deeper into the dermis than the superficial abrasions noted when bioelastic matrix patches were present.
[c] It was necessary to remove the casts from these dogs before the scheduled 7 days. The casts had considerable "strike through" on the surface overlying bony prominences, indicating a significant underlying dermal lesion.
Source: Reproduced from Kemppainen et al.[100]

tion, as shown as the treatment (loaded) data in Table 9.9.

The one limitation at this medial malleolus site of high shear pressure is an occasional (one in four of the test cases, DP-7T) splitting or tearing of the disk, with early signs for ulcer for-

TABLE 9.9. Pressure ulcer study: Subjective observations of unloaded and loaded bioelastic matrix patches and skin over the medial malleolus (casts in place 7 days).

		Appearance of skin under or immediately adjacent to patch		
Dog/treatment[a]	Patch appearance	Hyperemia	Superficial epidermal abrasion/ulceration	Skin adjacent to patch
Controls (unloaded)				
DP-1C	Intact, cloudy white area	Slight over entire med. mal.	2 small	Nothing notable
DP-2C	Intact, clear	None	None	Nothing notable
DP-3C	Cloudy white area, triangular tear distal	Present distal, not traceable	Present distal	Nothing notable
Treatment (loaded)				
DP-4T	Intact, clear	None	None	Nothing notable
DP-5T	Intact, clear	Small cranial—distal	None	Hyperemia and epidermal abrasion/ulcer—distal. Ecchymosis—proximal
DP-6T	Intact, clear	Small central and proximal,[b] slight, not traceable	None	Nothing notable
DP-7T	Clear, triangular tear cranial—distal	Slight distal, not traceable	Small distal	Nothing notable

[a] DP-#C = Unloaded bioelastic patch; DP-#T = loaded bioelastic patch; DP-#N = no bioelastic patch.
[b] This proximal small area of hyperemia was noted before the cast was placed. It was covered by the patch and had not gotten worse over the 7 day cast period.
Source: Reproduced from Kemppainen et al.[100]

mation occurring at the location of the split. Of note is that such high shear stress is not expected in the usual applications, and even the occasional splitting at this severe test site appears to be solvable. A modified composition has been prepared in which the force necessary to initiate fracture has been doubled, as shown in Figure 9.50. This should prevent the small amount of splitting that did occur in the medial malleolus site of high shear stress.

This success with the simple matrices of X^{20}–$(GVGIP)_n$ and similarly of X^{20}–$(GVGVP)_n$ for transdermal delivery does not yet bring to bear the full power of the T_t-transductional protein-based polymers. Finer control yet will occur as the forces arising out of the apolar–polar repulsive free energy of hydration, ΔG_{ap}, are employed. Accordingly, there is much to be optimistic about when considering elastic protein-based polymers capable of inverse temperature transitions for development of transdermal delivery systems.

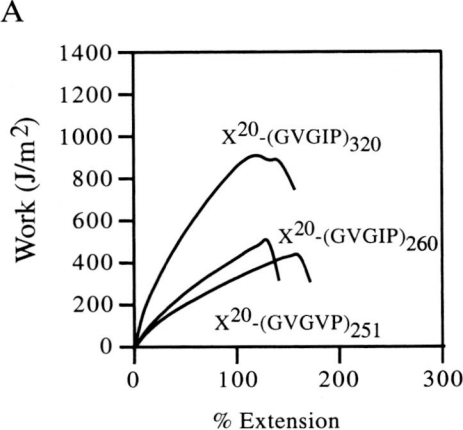

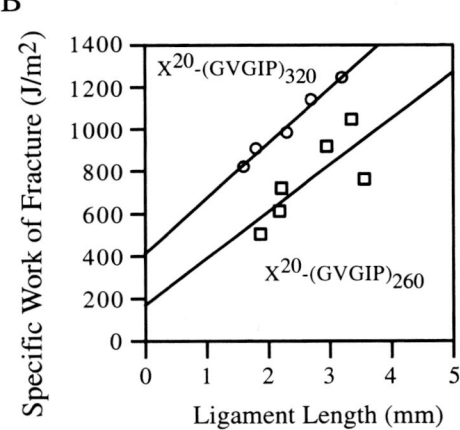

FIGURE 9.50. Notch-ligament length approach to the determination of the essential work of fracture shows doubled specific work of fracture for γ-cross-linked X^{20}–$(GVGIP)_{320}$ as compared with previously used X^{20}–$(GVGIP)_{260}$. Break stress data in Figure 9.20 suggest even more favorable break strengths are yet possible with chemical cross-linking approaches. (**A**) Stress–strain curves. (**B**) Essential work of fracture (intercept). (Reproduced with permission from Urry et al.[64])

9.4.6 The Eye as a Special Site for Applications

A great deal can and should be written about the potential for applications in the eye of elastic and plastic protein-based polymers with the capacity for exhibiting inverse temperature transitions. Some work has indeed been done in this area, but the detailing of it will have to be done in the future as time and space are limiting in the present effort. Some of the applications are (1) drug delivery by means of eye drops (perhaps with nanoparticles) or by wafers placed under the eyelids; (2) drug delivery by means of an ocular plug inserted through the sclera; (3) soft contact lenses with a greater control of the refractive index and with less fragile material that does not support formation of a protein coating; (4) Coatings of intraocular lenses that would make them less damaging to tissue when being implanted and that would limit fouling of the surface; (5) replacement of the vitreous humor following surgical procedures; (6) as a biodegradable elastic band following retinal reattachment surgery that would disappear once its task has been completed; (7) strabismus surgery for prevention of compromising adhesions that follow attempts to correct eye alignment by relocating rectus muscles as discussed briefly above for the prevention of adhesions section 9.4.4.3.2; and (8) Drainage sleeve for relieving pressure of glaucoma.

9.4.7 Coatings for Medical Devices to Prevent Adverse Reactions and Improve Function

Before proceeding to the advanced technology area of biosensors, brief consideration goes to coatings of medical devices to improve performance and to prevent such adverse reactions as adhesion of the medical device to the tissues, bacterial adhesion, platelet adhesion and activation, and adverse tissue reaction of host to device from the standpoint of protecting both the host and the function of the device.

9.4.7.1 The Cardiovascular Example of the Utility of Coatings

Coatings of devices involved in cardiovascular intervention include (1) the vascular prosthesis itself, (2) cardiac catheters, (3) cardiac (coronary artery) stents, (4) leads for pacemakers, (5) tubing for cardiac assist devices, (6) temporary artificial pericardium following open heart surgery, (7) tubing for continuous ambulatory peritoneal dialysis, and (8) drainage tubes. Of course, the coatings may simultaneously function as delivery vehicles for the whole range of therapeutic agents.

9.4.7.2 Coatings to Improve Urinary Indwelling Catheters

Perhaps the medical device that has had longest term need for coatings to improve function is the urinary catheter and more specifically the urinary indwelling catheter. Approximately one-fourth of the patients in chronic care and geriatric facilities utilize urinary indwelling catheters. Shorter term use follows prostatic surgery, considered the most frequent surgical procedure in men over 65 years of age, and attends cardiovascular and other surgical procedures. It has been indicated that 15% of hospital admissions require indwelling urinary catheters for one reason or another.

9.4.7.3 Consequences of Indwelling Urinary Catheter Use

The problems arising out of urinary indwelling catheter use are many: (1) They are the source of as many as 30% of all hospital acquired infections. (2) Urethral strictures result from abrasions due to insertion, shifting, and removal, from bacterial colonization of the abrasions, and from inflammatory reaction to toxic substances that leach out of the catheter. (3) Long-term usage has been credited with causing urethelial hyperplasia and dysplasia and bladder tumors. (4) Problems of luminal encrustation that block flow and catheter replacement aggravate the foregoing list of adverse consequences.

Obviously, the adverse results of increased morbidity, mortality, duration of hospitalization, and overall health care costs argue for the small investment in coating technology to correct this situation. Just as clearly, elastic and plastic protein-based polymers could eliminate the majority of these adverse consequences.

9.4.8 Biosensors (e.g., Diagnostics)

A particularly exciting new application is under way. It is the development of nanosensors with the ultimate potential for sensitivity, that is, the detection of single molecule interactions. It is expected that a single molecule of phosphate, nerve gas, toxin, and explosive, for example, TNT and the associated DNT, could be detected. The most direct approach to demonstrate this sensitivity and at the same time to do so with high selectivity is to detect a single phosphorylation event, as discussed below. First however, a short description of the atomic force microscope that, when suitably modified and used with designed elastic protein-based polymers, enables a high level of detection.

9.4.8.1 Atomic Force Microscopy in the Single-chain Force-extension Mode

9.4.8.1.1 The Atomic Force Microscope

The atomic force microscope (AFM), as developed by Paul Hansma, constitutes a cantilever with a very fine tip that scans over a surface. By measuring changes in the force between tip and surface, the cantilever tip, as it scans or rasters over a surface, forms an image of the surface and of molecules, for example, polymers, on the surface. The image arises out of the capacity to detect extremely small changes in the force of

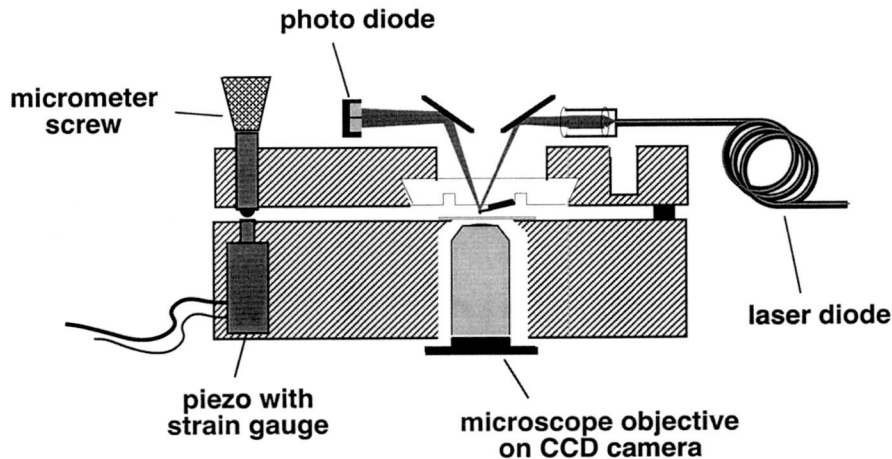

FIGURE 9.51. AFM spectroscopy designed for a single-chain force-extension mode; diagram of the instrumental set-up of Gaub and coworkers.[103,104] (Reproduced with permission from Urry et al.[62])

interaction of tip with surface and to do so with remarkable spatial resolution. What can happen to complicate the imaging of macromolecules on the surface is that these polymers can stick both to the tip and to the surface, and then the force of extending the molecule along the surface smears the molecular image being sought.

This complication, however, becomes the basis for a new and equally interesting measurement of special interest to elastic protein-based polymers. For our particular interest the usual AFM instrument is modified, as advanced by Hermann Gaub and coworkers,[103,104] so that the cantilever tip moves in the direction perpendicular to the surface (Figure 9.51 presents a diagram of the instrument). With polymer attached at one end to the surface and at the other to the tip, the force as a function of extension of the molecule is obtained.

9.4.8.1.2 Two Distinct Profiles of a Single-chain Force-extension Curve

Of direct interest to us are two distinct profiles of the force versus extension curve. The first is obtained for our bioelastic polymers that exhibit near-ideal elasticity,[61,62] which is a simple reversible monotonic curve of ever increasing force with extension until detachment occurs with the drop of the force to zero. The second is the characteristic sawtooth profile for titin, the energy absorbing elastic filament of muscle, which is a protein-based polymer of about 100 similar globular protein units of about 100 residues.

The extraordinary work of Gaub and Fernandez on titin, as represented by short strings of composite globular elements, is shown in Figure 9.52.[105] Figure 9.52A shows four globular elements and depicts the unfolding of a single globular element due to extension, and Figure 9.52B shows the experimental sawtooth pattern resulting from the sequential unfolding of a series of six globular elements followed by the detachment of the chain from either substrate or cantilever tip with a return to zero force. Thus, the unfolding event for each globular element is seen as a single tooth, and each tooth can be fitted with the mathematical expression of the worm-like chain model. The differences between extension limits for each tooth, as measured by fitting to the equation for a worm-like chain, provides a measure of the increase in extension length due to the unfolding of a single globular element. This difference provides a measure of the number of residues comprising the globular element that was unfolded (see Figure 9.52B).

9.4 Medical Applications

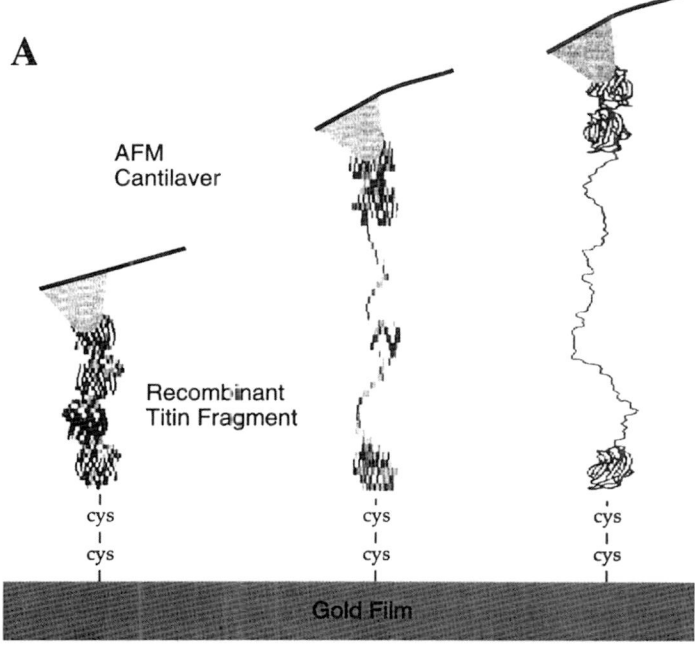

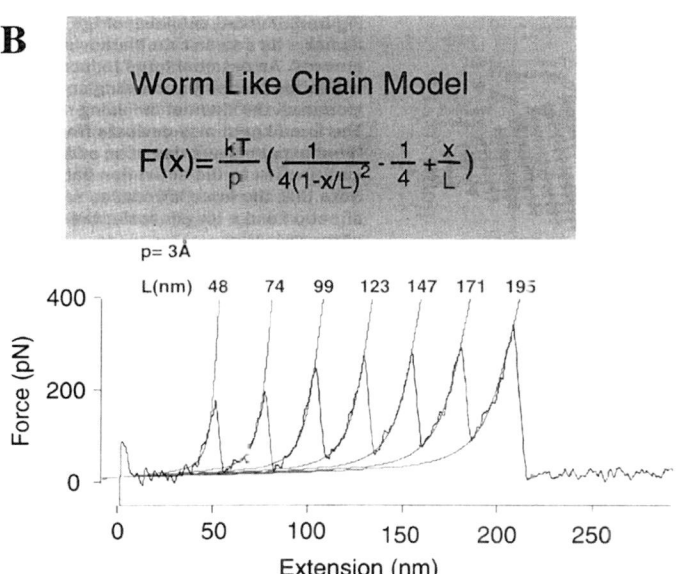

FIGURE 9.52. Atomic force microscopy, single-chain force-extension experiment. (**A**) Drawing of the extension of four globular subunits of a titin molecule shown as the stepwise unfolding of the string of globular domains. (**B**) Sawtooth pattern in the force-extension curve due to serial unfolding of titin globular domains, fitted to a worm-like chain model that indicates an approximate 25 nm extension as each domain unfolds. (Reproduced with permission from H.E. Gaub and J.M. Fernandez.[105])

9.4.8.2 Nanosensors to Detect Kinase Activities

9.4.8.2.1 One Particular Design of a Nanosensor

One nanosensor construct of interest to us combines the two elastic materials by positioning the sequence for a single approximately 100 residue globular protein element between two sequences of near-ideal elastic protein-based polymer. The particular globular element utilizes the fibronectin type III domain of tenascin that contains the GRGDSP cell attachment sequence at the turn connecting the F and G β-strands, as represented in stereo in Figure 9.53A in the ribbon representation and in Figure 9.53B

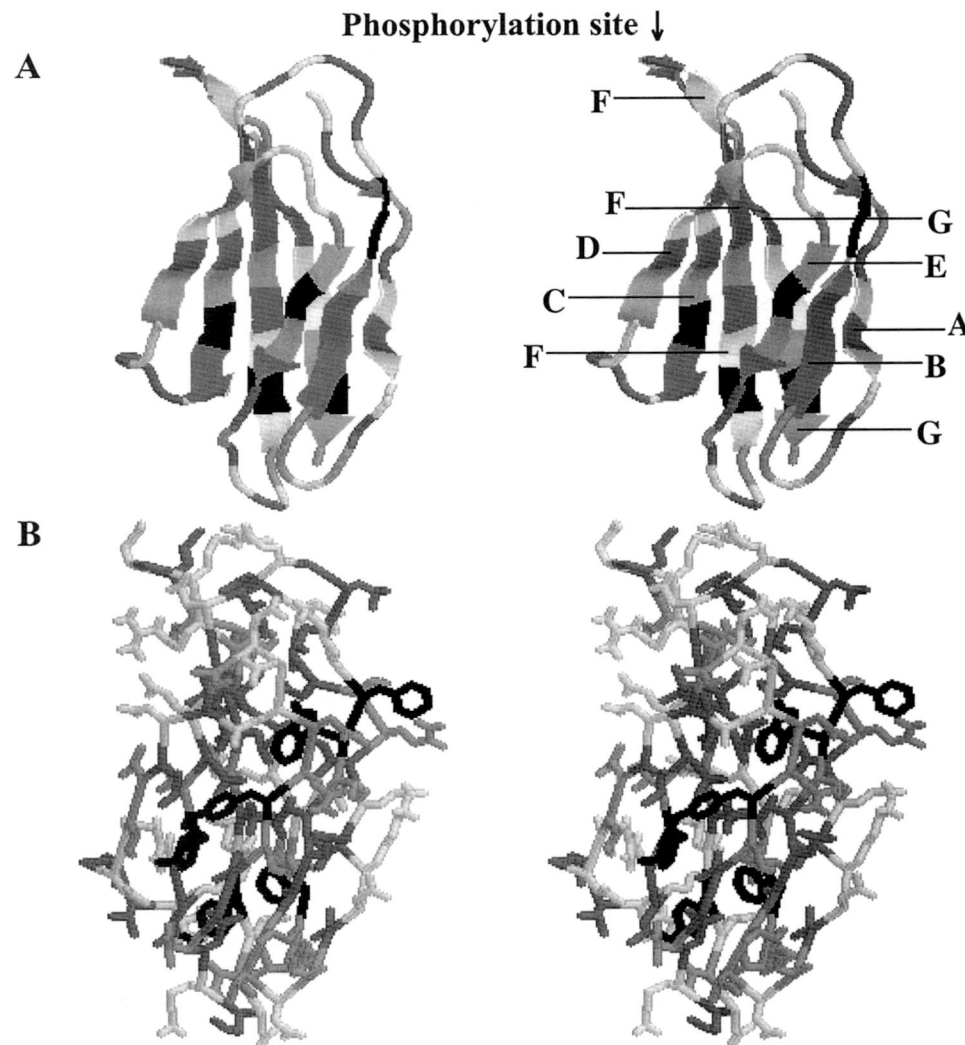

FIGURE 9.53. Stereo views of a fibronectin type III domain of tenascin with aromatics (black), other hydrophobics (gray), neutrals (light gray), and charged residues (white). (**A**) Ribbon representation showing seven β-strands and the turn containing the GRGDSP cell attachment site that in our designed sensor is replaced by RGYSLG to provide a phosphorylation site as approximately located. (**B**) Vine representation showing side chains. (Prepared using the crystallographic results of Bisig et al.[106] as obtained from the Protein Data Bank, Structure File 1TEN.)

9.4 Medical Applications

in the vine representation that shows the side chains.[106] The globular fibronectin type III domain is a seven-stranded, hydrophobically folded, β-barrel or sandwich that is one of the two types of globular domains present in titin and that also occurs in tenascin for which a crystal structure is available (see Figure 9.53).

In Figures 9.35 and 9.37 the GRGDSP cell attachment sequence was used within elastic protein-based polymers, such as [(GVGVP)$_{10}$ GVGVPGRGDSP(GVGVP)$_{10}$]$_{18}$(GVGVP), to induce tissue generation. The GRGDSP sequence is found in the 99 residue globular protein of Figure 9.53, known as the fibronectin Type III domain (Fn3). For the sensor application the sequence, RGYSLG, replaces the GRGDSP, to give the globular protein designated as Fn3′ (GRGDSP▶RGYSLG), which is placed in an elastic protein sequence to give [(GVGIP)$_{10}$]$_{10}$-Fn3′(GRGDSP▶RGYSLG)-[(GVGIP)$_{10}$]$_{n2}$. A cartoon of this construct, truncated in terms of the length of the repeating pentamer sequences, is given in Figure 9-54.

This single molecule construct would be expected to give a force-extension curve with a single sawtooth due to the unfolding of a single globular element superimposed on the simple monotonic curve of the nearly ideal elastic elements, as shown in Figure 9.55A. The capacity to sense, in this case the single molecular event of phosphorylation of the serine side chain, requires that the single sawtooth change its profile in a reproducible way on interaction of the globular sensing element with the "analyte."

9.4.8.2.2 Assay of Kinase Activities for Diagnostics in Tumor Tissues

As the starting point for this development, we chose to design for detection the very polar phosphate group of biology that is capable of exerting one of the largest apolar–polar repulsive forces of interaction. The site for phosphorylation is the serine (Ser, S) residue in the peptide sequence RGYSLG, a phosphorylation site for the cardiac cyclic AMP–dependent protein kinase. This sequence replaces the GRGDSP cell attachment site within a 99 residue globular protein element of the tenth type III domain of fibronectin. The particular

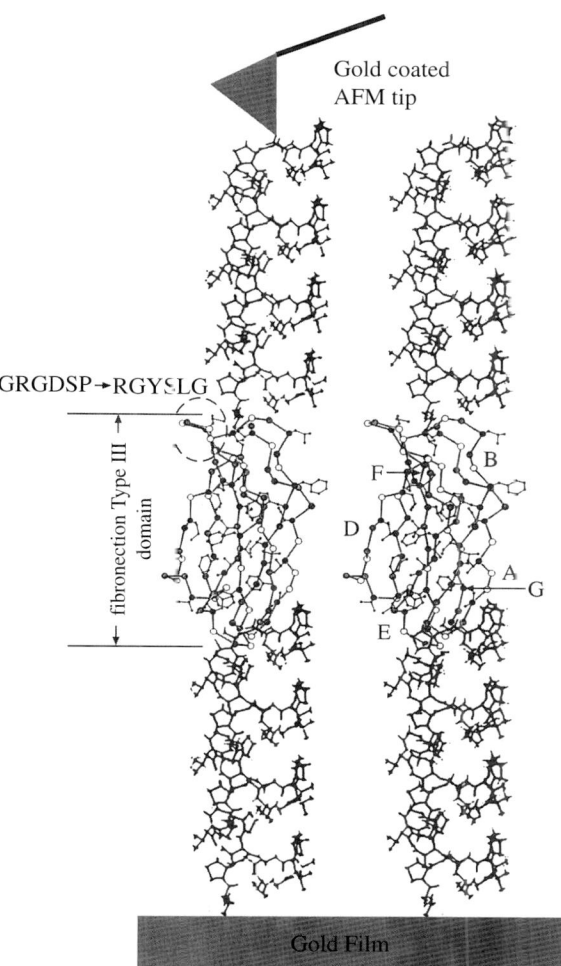

FIGURE 9.54. Cartoon of a molecular construct (in stereo) representing [(GVGIP)$_{10}$]$_{n1}$-Fn3′(GRGDSP→RGYSLG)-[(GVGIP)$_{10}$]$_{n2}$ strung between a cantilever tip and the surface of an atomic force microscope designed for single-chain force-extension studies. The molecular construct is designed with a globular sensing element sandwiched between stretches of elastic protein-based polymer in the β-spiral structure. The globular sensing element is the 99 residue tenth fibronectin type III domain (Fn3′) in which the cell attachment sequence (GRGDSP) has been replaced by a site for phosphorylation (RGYSLG) at the location indicated by the oval. (The globular element utilizes the crystal structure of Bisig et al.[106] as obtained from the Protein Data Bank, Structure File 1TEN. Courtesy of Bioelastics Research, Ltd.)

gene construct utilized for preliminary examination resulted in expression of the sequence [(GVGIP)$_{10}$]$_{n1}$–Fn3'(GRGDSP→RGYSLG)–[(GVGIP)$_{10}$]$_{n2}$ with n1 = 16 and n2 = 6, that is, 160 pentamers on one side and 60 on the other side of the globular sensing element.

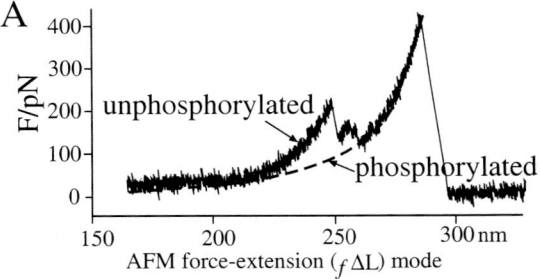

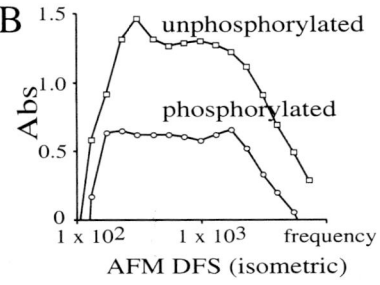

FIGURE 9.55. Proposed means whereby elastic protein-based polymers can be used with special adaptations of the atomic force microscope to detect single molecular interactions. (**A**) Single-molecule force-extension curve obtained for [(GVGIP)$_{10}$]$_{n1}$-Fn3'(GRGDSP→RGYSLG)-[(GVGIP)$_{10}$]$_{n2}$, where the 99 residue globular sensing element Fn3'(GRGDSP→RGYSLG) is the tenth fibronectin type III domain in which the cell attachment sequence (GRGDSP) has been replaced by a site (RGYSLG) for phosphorylation by cardiac cyclic AMP–dependent protein kinase; also with n1 = 16 and n2 = 6. Fit to a worm-like chain model consistent with the unfolding of the Fn3' globular domain. The dashed curve is the anticipated result should phosphorylation completely hydrophobically unfold the globular sensing element as suggested by the data in Figure 9.56. (Unpublished results of T. Hugel, M. Seitz, H. Gaub, A. Pattanaik, and D.W. Urry.) (**B**) Proposed dynamic force spectroscopy curves of single twisted filament containing a site for enzymatic phosphorylation (illustrated in Figure 9.54) before (above) and after (below) phosphorylation that assumes the acoustic absorption curves in Appendix A, Figure 7A.

The single-chain force-extension curve for [(GVGIP)$_{10}$]$_{16}$–Fn3'(GRGDSP→RGYSLG)–[(GVGIP)$_{10}$]$_6$ given in Figure 9.55A derives from collaboration with T. Hugel, M. Seitz, and H. Gaub. We tentatively interpret this curve to be the stretching out of the bioelastic sequences until the force is sufficient, about 250 pN, to unfold the single globular domain, as shown in Figure 9.52B. The increase in force then resumes until it is sufficient to cause detachment from either cantilever tip or surface. At this point, about 300 nm extension, the force drops to zero.

By fitting the curve with the mathematical expression for a hypothetical worm-like chain, an estimate of the change in length of the globular element is possible. In the case in Figure 9.55A, the change in length is as expected for the unfolding of a 99 residue globular protein.[107] Thus, from the criteria of both the force at which unfolding occurs and the length change that occurs on unfolding, it is not unreasonable to consider this the sawtooth due to unfolding of Fn3'(GRGDSP→RGYSLG). We expect that phosphorylation of this single RGYSLG site in the globular domain will change the force-extension profile for unfolding. If this does occur, then we will have detected a single molecular event, the phosphorylation of a single site in a single globular protein domain.

9.4.8.2.3 Effect of Phosphorylation of [(GVGIP)$_{10}$]$_{16}$–Fn3'(GRGDSP→RGYSLG)–[(GVGIP)$_{10}$]$_6$ on Hydrophobic Association

The temperature profile for turbidity formation of [(GVGIP)$_{10}$]$_{16}$–Fn3'(GRGDSP→RGYSLG)–[(GVGIP)$_{10}$]$_6$ in Figure 9.56A shows a simple profile for aggregation by hydrophobic association. On phosphorylation by cardiac cyclic AMP–dependent protein kinase, however, the profile becomes more complex with a delayed aggregation occurring at a 20°C higher temperature (see Figure 9.56B). Furthermore, two-dimensional nuclear magnetic resonance (2D-NMR) studies show that phosphorylation disrupts hydrophobic association within the globular sensing element (L.C. Hayes and D.W. Urry, unpublished data). In particular, the

9.4 Medical Applications

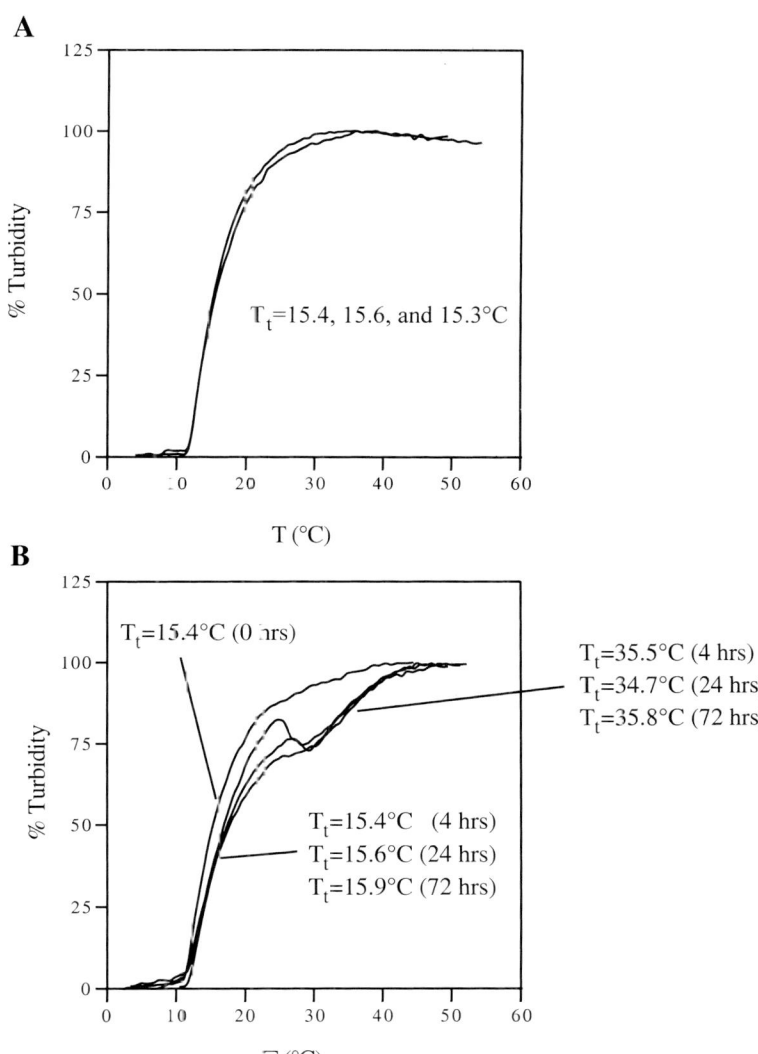

FIGURE 9.56. Determination of T_t before and after phosphorylation of the designed protein-based polymer sensor $[(GVGIP)_{10}]_{n1}$-Fn3′(GRGDSP→RGYSLG)-$[(GVGIP)_{10}]_{n2}$, with n1 = 16 and n2 = 6. The presence of 1 phosphate in 1,199 residues at the RGYSLG site of the 99 residue globular protein markedly increases T_t of either about 1/2 of the polymers, that is, K ≈ 1, or causes initial formation of micelles followed by association of the micelles. (Unpublished results, A. Pattanaik, T.M. Parker, and D.W. Urry.)

NOESY (nuclear Overhauser enhancement spectroscopy) plot for the aromatic region before phosphorylation of $(GVGIP)_{160}$–Fn3′(GRGDSP→RGYSLG)–$(GVGIP)_{60}$ contains many cross-peaks, many proton–proton contacts involving aromatic residues. These cross-peaks report the hydrophobically folded β-barrel, as represented in Figure 9.53 for the structure of the tenascin fibronectin type III domain, as there are no aromatic residues in

(GVGIP)$_n$. Phosphorylation removes more than two-thirds of the proton–proton contacts that report the hydrophobic associations within the β-barrel.

Accordingly, both the temperature profile data for following aggregation by hydrophobic association and the 2D-NMR data indicate that phosphorylation disrupts hydrophobic association within the globular sensing element. Phosphorylation at one end of the β-barrel, as shown in Figure 9.53A, reaches through the β-barrel and, at least in part, disassembles the hydrophobically assembled structure. *Thus, phosphorylation disrupts hydrophobic association, as a clear demonstration of ΔG_{ap}, the apolar–polar repulsive free energy of hydration.* The result in the single-chain force-extension curve would be a change in the profile of the single sawtooth. Construction of a gene and expression of the protein-based polymer for stable attachment to cantilever tip and substrate, which means introduction of proper functional groups at each end of the designed polymer, will allow for a rigorous testing of this nanosensor with potential to detect a single molecular interaction.

9.4.8.2.4 Value of Sensitive and Selective Assays for Kinases and Phosphatases

As is apparent in Chapter 8, phosphates are the most polar molecular species available in biology for controlling hydrophobic association/dissociation, that is, for controlling processes that occur by inverse temperature transition. It follows, therefore, that kinases for phosphorylation and phosphatases for dephosphorylation would be fundamental to key cellular transductional and transformational processes.[108,109] Often in cancerous and other diseased states, the activities of these enzymes are abnormal. Importantly for our interests, their sites of interaction can be very selective. Changes in protein kinase C activities have been reported to be abnormal in colon, breast, and skin cancers. Protein tyrosine kinase, which selectively phosphorylates the internal tyrosine (Tyr,Y) residue in the sequence GIYWHHY, is overexpressed in colon and breast cancer.[110–112] Furthermore, cyclic AMP–dependent protein kinase activities have been associated with the onset of Alzheimer's disease, where the proteolytic degradation product of a hyperphosphorylated former membrane protein is the principal constituent of the characteristic plaques of Alzheimer's disease.[113] Thus, there is great interest in the capacity to assay for kinase activities in tissue homogenates,[114–116] particularly if simple biopsies could suffice and there would be no need to resort to phosphorus-32, a high-energy emissions radionuclide that increases health risks.

The potential, therefore, is to introduce single or multiple globular protein element(s) with all of the specificity of an enzyme and to achieve the ultimate in sensitivity, the detection of a single molecular interaction. By selection of the appropriate globular sensing element, one can look to the polar analytes of phosphates, nerve gases, toxins, or explosives that would, on interaction with the globular protein sensing element, alter its sawtooth profile.

9.4.8.3 Nanosensors Using Mechanical Resonances

Another striking property of certain elastic protein-based polymers, as shown in part by the data in Figure 9.15, is that they can exhibit an intense mechanical resonance in the acoustic absorption range that varies as a function of hydrophobic assembly. That is, the intensity of the mechanical resonance increases as the hydrophobic assembly of the inverse temperature transition progresses and, of course, decreases as hydrophobic dissociation occurs. Thus, elastic protein-based (bioelastic) polymers are capable of 18 classes of pairwise free energy transductions; they can be produced with diverse composition, fixed sequence, and precise length, and they can exhibit intense mechanical resonances in the acoustic absorption range that undergo very large intensity changes during transduction.

Interesting additional methodologies are under development; they build on the fundamental design of the AFM. This development involves vibrating the cantilever at a frequency of 16 kHz, for example, to characterize the shear modulus properties of the polymer with which the cantilever tip makes contact, as first considered by Mervyn Miles and colleagues in Bristol. With the typical rheological instrumentation, as in obtaining the loss shear modulus data in Figure 9.15, it is not possible to reach frequencies greater than several hundred Hertz. On the other hand, both the acoustic absorption data of Lev Sheiba and the low-frequency dielectric relaxation data (loss permittivity data) in Figure 9.15 argue that a mechanical resonance occurs centered near 3 kHz, and this is supported by the overlap of the loss shear modulus and loss permittivity data in Figure 9.15. With proper instrumental design, it is considered possible to assess the loss shear modulus at frequencies below the exciting frequency, for example, below 16 kHz.

Such measurements, first suggested to us by Vlado Hlady of the University of Utah, provide an opportunity to reaffirm the mechanical resonance in the acoustic frequency range near 3 kHz. Importantly, such an instrumental capacity coupled with elastic protein-based polymers that exhibit acoustic absorption would constitute nanosensor development using isometric conditions, that is, at fixed extension. With the appropriate sensing filament strung between cantilever tip and substrate, any transductional interaction with the filament, that is, that changed the extent of hydrophobic association within the filament, would be detected as a change in the intensity of the absorption, as represented in Figure 9.55B.

Without the need to stretch the elastic sensing element, a robust device could be constructed that would be deployable for use in the field. One design of the sensing element for such a device is shown in Figure 9.57. Vlado Hlady has built a measuring device capable of what he calls *dynamic force spectroscopy*, and we look forward to future opportunities to develop a range of useful nanosensors capable of the ultimate in sensitivity that would be useful in medical diagnostics and in detection of terrorist threats.

9.4.9 Biodegradable Plastics

Certain compositions of plastic protein-based polymers can be prepared that melt rather than decompose, as protein materials normally do on excessive heating. In the melted state they can be pulled into fibers. One such composition, poly(FE(Boc)GVP), is shown in Figure 1.10A and compared with similarly formed fibers of more commonly known plastics, polypropylene (Figure 1.10B) and polystyrene (Figure 1.10C). One advantage of the bioplastic polymers is that they can be programmed for different rates of proteolytic degradation with half-lives ranging from days to decades. These properties make them of interest for a number of applications, for example, as sutures and staples to be used in closing following surgical procedures and even as bone screws for orthopedic applications. Additionally, such biodegradable plastics could be used as both structural and controlled release devices, as in the delivery of bone morphogenic proteins for bone reconstruction.

9.4.10 Sound Absorption for Hearing Protection

Certain bioelastic compositions, when in the folded and assembled state above the temperature of the inverse temperature transition, to the best of our knowledge provide materials with the highest sound absorption per unit volume. One such composition is compared in Figure 7A of Appendix A to standard materials (natural rubber and polyurethane) used for sound absorption. The lower flat curve represents the absorption by natural rubber over the frequency range of sound, the acoustic absorption range. As we understand it, the somewhat better absorbing, low, flat curve is due to a very favorable sound absorbing elastic material made from petroleum. The mountainous-looking curves are due to a bioelastic material. Below the transition temperature of the polymer, the bioelastic polymer is definitely better, but only by a factor of two or so. On raising the temperature above the onset of the transition temperature, the sound absorption increases dramatically and would probably double on increasing to body temperature. Among other things, the property of these materials appears to be favorable for hearing protection, for example, when made into an earplug or ear muff.

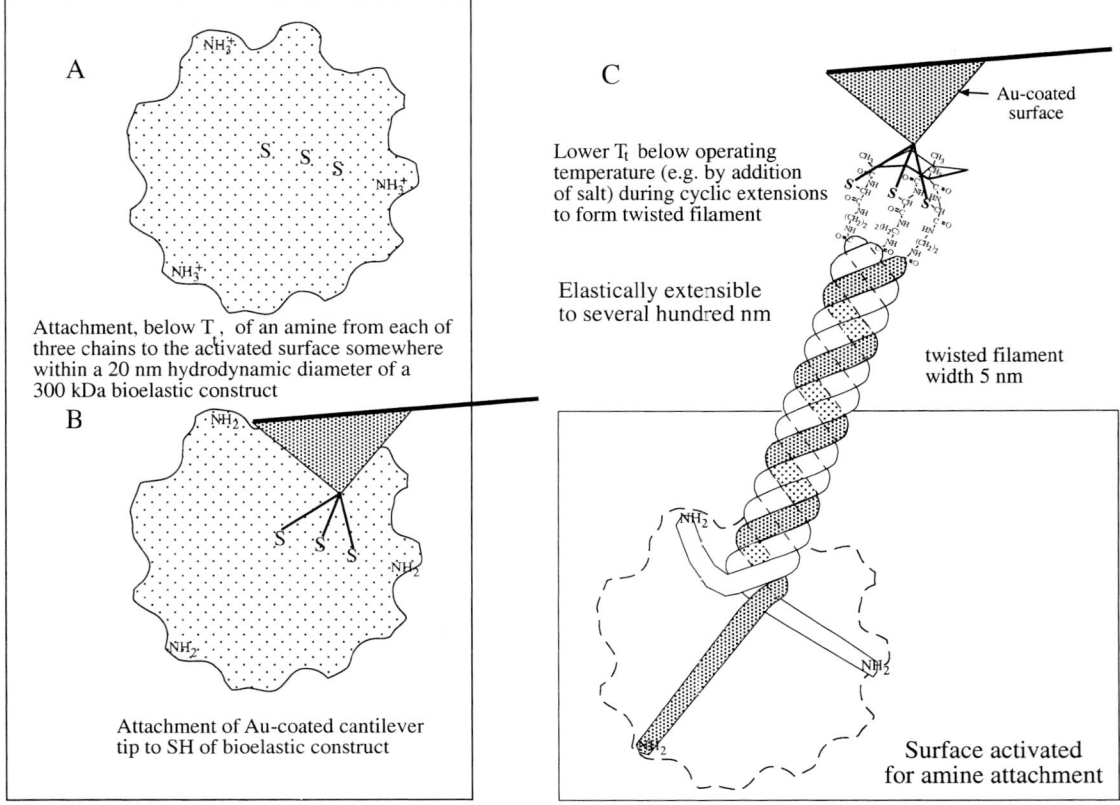

FIGURE 9.57. (**A**) Representation of amorphous triple-stranded protein-based polymer below T_t. (**B**) Cross-linked SH groups at one end of the three chains of the amorphous triple-stranded protein-based polymer attached to a cantilever tip and the amino functions at the other ends of the three chains attached to the substrate. (**C**) The expectation is that a series of extension/relaxation cycles will draw the three chains into a twisted filament. This twisted filament structure is expected to exhibit an acoustic absorption spectrum in a similar frequency range as that in Figure 9.58 and to change on phosphorylation as indicated in Figure 9.55B.

References

1. L. Pasteur, "Pourquoi la France n'a pas trouve d'homme superieurs au moment du peril." *Rev. Sci.*, (Paris), 1871.
2. J. Bronowski, *The Ascent of Man*. Little, Brown and Company, Boston, 1973, p. 110.
3. D.W. Urry, R.D. Harris, and K.U. Prasad, "Chemical Potential Driven Contraction and Relaxation by Ionic Strength Modulation of an Inverse Temperature Transition." *J. Am. Chem. Soc.*, **110,** 3303–3305, 1988.
4. Praemer, S. Furner, and D.P. Rice, *Musculoskeletal Conditions in the United States*. Park Ridge, American Academy of Orthopaedic Surgeons, 1992.
5. G. Waddell, "Low Back Pain: A Twentieth Century Health Care Enigma." *Spine*, **21,** 2820–2825, 1996.
6. T.W. Hu, "Impact of Urinary Incontinence on Health-care Costs." *J. Am. Geriatr. Soc.*, **38,** 292–295, 1990.
7. Report/Abstract—National Pressure Ulcer Advisory Panel, Annual Conference Program 2001. NPUAP 11250 Roger Bacon Dr., Suite 8, Reston, VA 20190-5202.
8. National Pressure Ulcer Advisory Panel, Statement on Pressure Ulcer Formation and Pres-

References

sure Ulcers. Incidence. Economics. Risk Assessment, SUNY at Buffalo, Beck Hill, 1992.

9. R.M. Allman, "Epidemiology of Pressure Sores in Different Populations." *Decubitus*, **2**, 30–32, 1989.

10. J.S. Young, P.E. Burns, A.M. Bower, and R. McCutchen, *Spinal Cord Injury Statistics: Experience of the Regional Spinal Cord Injury Systems*. National Spinal Cord Injury Data Research Center, Phoenix, 1982, p. 95

11. N.S. Miller and M.S. Gold, "An Introduction to the Pharmacological Therapies of Drug and Alcohol Addictions." In *Pharmacological Therapies for Drug and Alcohol Addictions*, N.S. Miller and M.S. Gold, Eds., Marcel Dekker, New York, 1995, pp. 1–29.

12. J.L. Olsen and F.A. Kinel, "A Review of Parenteral Sustained-Release Naltrexone Systems." In *Naltrexone Research Monograph 28*, R.E. Willette and G. Barnett, Eds., National Institute of Drug Abuse, 1980.

13. J.R. Hughes, "Non-nicotine Pharmacotherapies for Smoking Cessation." *J Drug Dev*, **6**, 197–203, 1994.

14. E.N. Shuyman, S. Porat, E. Witztum, D. Gandacu, R. Barhamburger, and Y. Ginath, *Biol. Psychiatry*, **35**, 934–935, 1994.

15. G. Gerra, A. Mercato, R. Caccavari, B. Fontanesi, R. Delsignore, G. Fertonani, P. Avanzini, P. Rustichelli, and M. Passeri, "Clonidine and Opiate Receptor Antagonists in the Treatment of Heroin Addiction." *J. Substance Abuse Treatment*, **12**, 35–41, 1995.

16. V. Navaratnam, A. Jamaludin, N. Raman, M. Mohamed, and S.M. Mansor, "Determination of naltrexone dosage for narcotic agonist in detoxified Asian addicts." *Drug Alcohol Dependence*, **34**, 231–236, 1994.

17. A.D. Frazer, "Clinical Toxicology of Drugs Used in the Treatment of Opiate Dependency." *Clin. Lab. Med.*, **10**, 375–385, 1990.

18. J.J. Legarda and M. Gossop, "A 24-h Inpatient Detoxification Treatment for Heroin Addicts: A Preliminary Investigation." *Drug Alsohol Dependence*, **35**, 91–93, 1994.

19. R.M. Swift, W. Whelihan, O. Kuznetzov, G. Buongiorno, and H. Hsuing, "Naltrexone-induced Alterations in Human Ethanol Intoxication." *Am. J. Psychiatry*, **151**, 1463–1467, 1994.

20. M.J. Bohn, H.R. Kranzler, D. Beazoglou, and D. Staehler, "Naltrexone and Brief Counseling to Reduce Heavy Drinking: Results of a Small Clinical Trial." *Am. J. Addictions*, **32**, 91–99, 1994.

21. C.P. O'Brien, "Treatment of Alcoholism as a Chronic Disorder." *Alcohol*, **11**, 433–437, 1994.

22. I. Chatoor, B.H. Herman, and J. Hartzler, "Effects of the Opiate Antagonist, Naltrexone, on Binging Antecedents and Plasma β-Endorphin Concentrations." *J. Am. Acad. Child Adolescent Psychiatry*, **33**, 748–752, 1994.

23. T. Thompson, T. Hackenberg, D. Cerruti, D. Baker, and S. Axtell, "Opioid Antagonist Effects on Self-injury in Adults with Mental Retardation: Response Form and Location as Determinants of Medication Effects." *Am. J. Mental Retardation*, **99**, 85–102, 1994.

24. N.M. Gonzales, M. Campbell, A.M. Small, J. Shay, L.D. Bluhm, P.B. Adams, and R.L. Foltz, "Naltrexone Plasma Levels, Clinical Response and Effect on Weight in Autistic Children." *Psychopharmacol. Bull.*, **30**, 203–208, 1994.

25. R. Kurlan, L. Majumdar, C. Deeley, G.S. Mudholkar, S. Plumb, and P.G. Como, "A Controlled Trial of Propoxyphene and Naltrexone in Patients with Tourette's Syndrome." *Ann. Neurol.*, **30**, 19–23, 1991.

26. A.K. Percy, D.G. Glaze, R.J. Schultz, H.Y. Zoghbi, D. Williamson, J.D. Frost, J.J. Jankovic, D. Deljunco, M.S. Kender, S. Waring, and E.C. Myer, "Rett's Syndrome: Controlled Study of an Oral Opiate Antagonist, Naltrexone." *Ann. Neurol.*, **35**, 464–470, 1994.

27. M.S. Gold, "Pharmacological Therapies of Opiate Addiction." In *Pharmacological Therapies for Drug and Alcohol Addictions*, N.S. Miller and M.S. Gold, Eds., Marcel Dekker, New York, 1995, pp. 159–174.

28. D.S. Sax, C. Kornetsky, and A. Kim, "Lack of hepatotoxicity with naltrexone treatment." *J. Clin. Pharmacol.*, **34**, 898–901, 1994.

29. K. Verebey, "Quantitative Determination of Naltrexone, 6β-Naltrexol and 2-Hydroxy-3-Methoxy-6β-Naltrexol (HMN) in Human Plasma, Red Blood Cells, Saliva and Urine by Gas Liquid Chromatography." In *Naltrexone: Research Monograph 28*, R.E. Willette and G. Barnett, Eds., National Institute on Drug Abuse, 1981, pp. 36–51.

30. R.A. Littrel and Hyde, "Pharmacological Therapies in Surgical Patients with Drug and Alcohol Addictions." In *Pharmacological Therapies for Drug and Alcohol Addictions*, N.S. Miller and M.S. Gold, Eds., Marcel Dekker, New York, 1995, pp. 287–305.

31. D.T. McPherson, J. Xu, and D.W. Urry, "Product Purification by Reversible Phase Transition Following *E. coli* Expression of Genes

Encoding up to 251 Repeats of the Elastomeric Pentapeptide GVGVP." *Protein Expression Purification*, **7**, 51–57, 1996.
32. This figure was reproduced from the Ph.D. dissertation of T. Cooper Woods, The University of Alabama at Birmingham, 1998.
33. D.W. Urry, A. Nicol, D.T. McPherson, C.M. Harris, T.M. Parker, J. Xu, D.C. Gowda, and P.R. Shewry, "Properties, Preparations and Applications of Bioelastic Materials." In *Encyclopedic Handbook of Biomaterials and Bioengineering—Part A—Materials*, Vol. 2, D.L. Wise, D.J. Trantolo, D.E. Altobelli, M.J. Yazemski, J.D. Gresser, and E.R. Schwartz, Eds. Marcel Dekker, New York, 1995, pp. 1619–1673.
34. D.W. Urry, D.T. McPherson, J. Xu, D.C. Gowda, N. Jing, T.M. Parker, H. Daniell, and C. Guda, "Protein-Based Polymeric Materials: Syntheses and Properties" In *Polymeric Materials Encyclopedia: Synthesis, Properties and Applications*, CRC Press, Boca Raton, pp. 7263–7279, 1996.
35. H. Daniell, C. Guda, D.T. McPherson, X. Zhang, and D.W. Urry, "Hyper Expression of a Synthetic Protein Based Polymer Gene." *Methods Mol. Biol.*, **63**, 359–371, 1996.
36. X. Zhang, C. Guda, R. Datta, R. Dute, D.W. Urry, and H. Daniell, "Nuclear Expression of an Environmentally Friendly Synthetic Protein-based Polymer Gene in Tobacco Cells." *Letters*, **17**, 1279–1284, 1995.
37. X. Zhang, D.W. Urry, and H. Daniell, "Expression of an Environmentally Friendly Synthetic Protein-based Polymer Gene in Transgenic Tobacco Plants." *Plant Cell Rep.*, **16**, 174–179, 1996.
38. R.W. Herzog, N.K. Singh, D.W. Urry, and H. Daniell, "Expression of a Synthetic Protein-based Polymer (Elastomer) Gene in *Aspergillus nidulans*." *Appl. Microbiol. Biotechnol.*, **47**, 368–372, 1997.
39. This is part of an ongoing research collaboration between Professor Margarida Casal of the Department of Biology at the University of Minho, Braga, Portugal, and Bioelastics Research, Ltd.
40. This is part of a research collaboration between Professor David A. Somers, Department of Agronomy and Plant Genetics, University of Minnesota, Twin Cities Campus, and Bioelastics Research, Ltd.
41. D.W. Urry, C.-H. Luan, C.M. Harris, and T. Parker, "Protein-based Materials with a Profound Range of Properties and Applications: The Elastin ΔT_t Hydrophobic Paradigm." In *Proteins and Modified Proteins as Polymeric Materials*, K. McGrath and D. Kaplan, Eds., Birkhäser Press, Boston, 1997, pp. 133–177.
42. D.W. Urry and K.U. Prasad, "Syntheses, Characterizations and Medical Uses of the Polypentapeptide of Elastin and its Analogs." In *Biocompatibility of Tissue Analogues*, D.F. Williams, Ed., CRC Press, Boca Raton, FL, 1985, pp. 89–116.
43. K.U. Prasad, M.A. Iqbal, and D.W. Urry, "Utilization of 1-Hydroxybenzotriazole in Mixed Anhydride Coupling Reactions," *Int. J. Peptide Protein Res.*, **25**, 408–413, 1985.
44. D.C. Gowda, C.-H. Luan, R.L. Furner, S.Q. Peng, N. Jing, C.M. Harris, T.M. Parker, and D.W. Urry, "Synthesis and Characterization of Human Elastin W_4 Sequence." *Int. J. Peptide Protein Res.*, **46**, 453–463, 1995.
45. D.C. Gowda, T.M. Parker, R.D. Harris, and D.W. Urry, "Synthesis, Characterizations and Medical Applications of Bioelastic Materials." In *Peptides: Design, Synthesis, and Biological Activity*, C. Basava and G.M. Anantharamaiah, Eds., Birkhäuser Press, Boston, 1994, pp. 81–111.
46. R.B. Merrifield, "Solid Phase Peptide Synthesis I. The Synthesis of a Tetrapeptide." *J. Am. Chem. Soc.* **85**, 2149–2154, 1963.
47. K.U. Prasad, M. Iqbal, and D.W. Urry, "Synthesis of Two Component Models of Elastin." *Peptides Chem. Biol.*, 399–403, 1988.
48. D.W. Urry, T.M. Parker, M.C. Reid, and D.C. Gowda, "Biocompatibility of the Bioelastic Materials, Poly(GVGVP) and Its γ-irradiation Cross-linked Matrix: Summary of Generic Biological Test Results." *J. Bioactive Compatible Polym.*, **6**, 263–282, 1991.
49. D.W. Urry, D.T. McPherson, J. Xu, D.C. Gowda, and T.M. Parker, "Elastic and Plastic Protein-based Polymers: Potential for Industrial Uses." In *Industrial Biotechnological Polymers*, C. Gebelein and C.E. Carraher, Jr., Eds, Technomic Publishing Co., Lancaster, PA, 1995, pp. 259–281.
50. D.W. Urry, A. Pattanaik, M.A. Accavitti, C.-X. Luan, D.T. McPherson, J. Xu, D.C. Gowda, T.M. Parker, C.M. Harris, and N. Jing, "Transductional Elastic and Plastic Protein-based Polymers as Potential Medical Devices." In *Handbook of Biodegradable Polymers*, A.J. Domb, J. Kost, and D.M. Wiseman, Eds., Harwood Academic Publishers, Chur, Switzerland, 1997, pp. 367–386.

References

51. G.L. Picciolo, D.S. Kaplan, K.F. Batchelder, R. Kapur, and R.M. Kotz, "Biotechnology-derived Biomaterials Modulate Host Cell Reactive Oxygen Production as Measured by Chemiluminescence." 19th Annual Meeting, Society for Biomaterials, Birmingham, AL, 1993
52. The use of the inverse temperature transition for purification emphasizes intermolecular self-assembly by hydrophobic association. Globular proteins within the *E. coli* cell, however, exhibit the same phenomenon with intramolecular hydrophobic folding and unfolding. As the protein-based polymer becomes more hydrophobic, especially with hydrophobically shifted pKa values for functional groups, globular protein elements within the cell may open up and ion-pair/hydrophobically associate with the protein-based polymer. This introduces the need for means with which to disrupt the hydrophobically induced association. Although this complicates purification of more hydrophobic protein-based polymers with functional groups, it becomes the basis whereby charged protein-based polymers of a range of hydrophobicities become remarkable delivery vehicles for proteins, nucleic acids, and so on, as discussed in section 9.4.5.
53. A. Pattanaik, T.M. Parker, and D.W. Urry, unpublished results.
54. D.W. Urry, S.Q. Peng, L.C. Hayes, D.T. McPherson, Jie Xu, T.C. Woods, D.C. Gowda, and A. Pattanaik, "Engineering Protein-based Machines to Emulate Key Steps of Metabolism (Biological Energy Conversion)." *Biotechnol. Bioeng.*, **58**, 175–190, 1998.
55. G. Wider, S. Macura, A. Kumar, R.R. Ernst, and K. Wüthrich, "Homonuclear two-dimensional ^{1}H NMR of Proteins. Experimental Procedures." *J. Magn. Reson.* **56**, 207–234, 1984.
56. D.W. Urry, D.K. Chang, R. Krishna, D.H. Huang, T.L. Trapane, and K.U. Prasad, "Two Dimensional Proton Nuclear Magnetic Resonance Studies on Poly(VPGVG) and its Cyclic Conformational Correlate, Cyclo(VPGVG)$_3$." *Biopolymers*, **28**, 819–833, 1989.
57. Should some mutation occur during the process of gene construction and expression, or should there be an error in the chemically synthesized oligonucleotides used in the process of forming multimers of the monomer gene and of introducing the correct base sequence for the restriction site, then the expressed protein may not be correct. This can be checked by sequencing the gene from both the 5′ and 3′ ends.
58. D.W. Urry, A. Pattanaik, J. Xu, T.C. Woods, D.T. McPherson, and T.M. Parker, "Elastic Protein-based Polymers in Soft Tissue Augmentation and Generation." *J. Biomater. Sci. Polymer Edn.*, **9**, 1015–1048, 1998.
59. D.W. Urry, C.M. Venkatachalam, M.M. Long, and K.U. Prasad, "Dynamic β-Spirals and a Librational Entropy Mechanism of Elasticity." In *Conformation in Biology*, R. Srinivasan and R.H. Sarma, Eds., G.N. Ramachandran Festschrift Volume, Adenine Press, 1982, pp. 11–27.
60. D.K. Chang and D.W. Urry, "Polypentapeptide of Elastin: Damping of Internal Chain Dynamics on Extension." *J. Comput. Chem.*, **10**, 850–855, 1989.
61. D.W. Urry and T.M. Parker, "Mechanics of Elastin: Molecular Mechanism of Biological Elasticity and its Relevance to Contraction." *J. Muscle Res. Cell Motility*, **23**, 2002.
62. D.W. Urry, T. Hugel, M. Seitz, H. Gaub, L. Sheiba, J. Dea, J. Xu, and T. Parker, "Elastin: A Representative Ideal Protein Elastomer." *Phil. Trans. R. Soc. Lond. B*, **357**, 169–184, 2002.
63. D.W. Urry, T. Hugel, M. Seitz, H. Gaub, L. Sheiba, J. Dea, J. Xu, L. Hayes, F. Prochazka, and T. Parker, "Ideal Protein Elasticity: The Elastin Model." In *Elastomeric Proteins: Structures, Biomechanical Properties and Biological Roles*." P.R. Shewry, A.S. Tatham, and A.J. Bailey, Eds. Cambridge University Press, The Royal Society; Chapter Four, pages 54–93, 2003.
64. D.W. Urry, J. Xu, W. Wang, L. Hayes, F. Prochazka, and T.M. Parker, "Development of Elastic Protein-based Polymers as Materials for Acoustic Absorption." *Mater. Res. Soc. Symp. Proc.* **774**, 81–92, 2003.
65. Prochazka F., PhD Thesis, *Etude de l'evolution des relaxations dans les gels en cours de formation, application aux polyurethanes*. 1998, Université du Maine, Le Mans, France.
66. T. Nicolai, F. Prochazka, and D. Durand, "Comparison of Polymer Dynamics Between Entanglements and Covalent Cross-links." *Phys. Rev. Lett.*, **82**, 863–866, 1999.
67. R. Buchet, C.-H. Luan, K.U. Prasad, R.D. Harris, and D.W. Urry, "Dielectric Relaxation Studies on Analogs of the Polypentapeptide of Elastin." *J. Phys. Chem.*, **92**, 511–517, 1988.
68. E.O. Wilson, "*Consilience: The Unity of Knowledge.*" Alfred E. Knopf, New York, 1998, p. 298.
69. D.W. Urry, K.U. Prasad, M.M. Long, and R.D. Harris "Elastomeric Polypeptides as Potential

Vascular Prosthetic Materials." *Polym. Mater. Sci. Eng.*, **59**, 684–689, 1988.
70. D.A. McDonald, *The Elastic Properties of the Arterial Wall*, Chapter 10. The Camelot Press Ltd., Southampton, Great Britain, 1974.
71. D.Y.M. Leung, S. Glagov, and M.B. Mathews, "Cyclic Stretching Stimulates Synthesis of Matrix Components by Arterial Smooth Muscle Cells In Vitro." *Science*, **191**, 475–477, 1976.
72. B. van der Lei, C.R. Wildevuur, P. Nieuwenhuis, E.H. Blaauw, F. Dijk, C.E. Hulsteart, and I. Molenaar, "Regeneration of the Arterial Wall in Microporous, Compliant, Biodegradable Vascular Grafts After Implantation Into the Rat Abdominal Aorta." *Cell Tissue Res.*, **242**, 569–578, 1985.
73. N. Wang, J.P. Butler, and D.E. Ingber, "Mechanotransduction Across the Cell Surface and Through the Cytoskeleton." *Science*, **260**, 1124–1127, 1993.
74. P.R. Girard and R.M. Nerem, "Shear Stress Modulates Endothelial Cell Morphology and F-actin Organization Through the Regulation of Focal Adhesion-associated Proteins." *J. Cell. Physiol.*, **163**, 179–193, 1995.
75. A. Nicol, D.C. Gowda, and D.W. Urry, "Cell Adhesion and Growth on Synthetic Elastomeric Matrices Containing Arg-Gly-Asp-Ser-^{3}." *J. Biomed. Mater. Res.*, **26**, 393–413, 1992.
76. D.W. Urry, D.C. Gowda, B.A. Cox, L.D. Hoban, A. McKee, and T. Williams, "Properties and Prevention of Adhesions Applications of Bioelastic Materials." *Mater. Res. Soc. Symp. Proc.*, **292**, 253–264, 1993.
77. L.D. Hoban, M. Pierce, J. Quance, I. Hayward, A. McKee, D.C. Gowda, D.W. Urry, and T. Williams, "The Use of Polypenta-peptides of Elastin in the Prevention of Postoperative Adhesions." *J. Surg. Res.*, **56**, 179–183, 1994.
78. F.J. Elsas, D.C. Gowda, and D.W. Urry, "Synthetic Polypeptide Sleeve for Strabismus Surgery." *J. Pediatr. Ophthalmol. Strabismus*, **29**, 284–286, 1992.
79. R.N. Alkalay, D.H. Kim, D.W. Urry, J. Xu, T.M. Parker, and P.A. Glazer, "Prevention of Postlaminectomy Epidural Fibrosis Using Bioelastic Materials." *Spine*, **28**, 1659–1665, 2003.
80. P.W. Schiller, "Development of Receptor-Specific Opioid Peptide Analogues." In *Progress in Medicinal Chemistry*, G.P. Ellis and G.B. West, Eds., Elsevier Press, Amsterdam, 1991, pp. 301–340.
81. V. Höllt, "Opioid Peptide Processing and Receptor Selectivity." *Annu. Rev. Pharmacol. Toxicol.*, **26**, 59–77, 1986.
82. J. Hughes, T.W. Smith, H.W. Kosterlitz, L.A. Fothergill, B.A. Morgan, and R.H. Morris, "Identification of Two Related Pentapeptides from the Brain with Potent Opiate Agonist Activity." *Nature*, **258**, 577–579, 1975.
83. S.J. Weber, D.L. Greene, V.J. Hruby, H.I. Hamamura, F. Porreca, and T.P. Davis, "Whole Body and Brain Distribution of [^{3}H]cylic [D-Pen^2,D-Pen^5] Enkephalin after Intraperitoneal, Intravenous, Oral and Subcutaneous Administration." *J. Pharmacol. Exp. Ther.*, **263**, 1308–1316, 1992.
84. R.L. Follenfant, G.W. Hardy, L.A. Lowe, C. Schneider, and T.W. Smith, "Antinociceptive Effects of the Novel Opioid Peptide BW443C Compared with Classical Opiates; Peripheral Versus Central Actions." *Br. J. Pharmacol.*, **93**, 85–92, 1988.
85. G.W. Hardy, L.A. Lowe, G. Mills, P.Y. Sang, D.S.A. Simpkin, R.L. Follenfant, C. Shankley, and T.W. Smith, "Peripherally Acting Enkephalin Analogues. 2. Polar Tri- and Tetrapeptides," *J. Med. Chem.*, **32**, 1108–1118, 1989.
86. W.R. Martin, C.G. Eades, J.A. Thompson, R.A. Huppler, and P.E. Gilbert, "The Effects of Morphine- and Nalorphine-like Drugs in the Nondependent and Morphine-dependent Chronic Spinal Dog." *J. Pharmacol. Exp. Ther.*, **197**, 517–533, 1976.
87. J.A.H. Lord, A.A. Waterfield, J. Hughes, and H.W. Kosterlitz, "Endogenous Opioid Peptides: Multiple Agonists and Receptors." *Nature*, **267**, 495–499, 1977.
88. J.I. Szekely, "*Opioid Peptides in Substances Abuse*," CRC Press, Boca Raton, FL, 1994.
89. T.L. Yaksh, J. Jang, Y. Nishiuchi, K.P. Braun, S. Ro, and M. Goodman, "The Utility of 2-Hydroxypropyl-b-Cyclodextrin as a Vehicle for the Intracerebral and Intrathecal Administration of Drugs." *Life Sci.*, **48**, 623–633, 1991.
90. N. Bodor, L. Prokai, W.-M. Wu, H. Faraq, S. Jonalagadda, M. Kawamura, and J. Simpkins, "A Strategy for Delivery Peptides into the Central Nervous System by Sequential Metabolism." *Science*, **257**, 1698–1700, 1992.
91. R.M. Ottenbrite, Ed., "Polymeric Drugs and Drug Administration," *Am. Chem. Soc. Symp. Ser.*, **545**, 1994.
92. P.T. Tarcha, Ed., "*Polymers for Controlled Drug Delivery*," CRC Press, Boca Raton, FL, 1991.

93. N.A. Peppas, Editor, "*Hydrogels in Medicine and Pharmacy*," Volumes I, II, and III, CRC Press, Inc., 1987.
94. J. Kost and R. Langer, "Equilibrium Swollen Hydrogels in Controlled Release." In *Hydrogels in Medicine and Pharmacy*, N.A. Peppas, Ed., CRC Press, Boca Raton, FL, 1987, pp. 95–108.
95. C.G. Pitt, T.A. Marks, and A. Schindler, "Biodegradable Drug Delivery Systems Based on Aliphatic Polyesters: Application to Contraceptives and Narcotic Antagonists." *Naltrexone Res. Monogr.*, **28**, 1980.
96. N.A. Peppas and R.W. Korsmeyer, "Dynamically Swelling Hydrogels in Controlled Release Applications." In *Hydrogels in Medicine and Pharmacy*, N.A. Peppas, Ed., CRC Press, Boca Raton, FL, 1987, pp. 109–136.
97. D.W. Urry, T.C. Woods, L.C. Hayes, J. Xu, D.T. McPherson, M. Iwama, M. Furuta, T. Hayashi, M. Murata, and T.M. Parker, "Elastic Protein-Based Biomaterials: Elements of Basic Science, Controlled Release and Biocompatibility." In *Biomaterials Handbook—Advanced Applications of Basic Sciences and Bioengineering*, 2004, in press.
98. B.W. Kemppainen, D.W. Urry, C.-X. Luan, J. Xu, S.F. Swaim, and S. Goel, "In vitro skin penetration of dazmegrel delivered with a bioelastic matrix." *Int. J. Pharmaceutics*, **271**, 301–303, 2004.
99. N.-Z. Wang, D.W. Urry, S.F. Swaim, R.L. Gillette, C.E. Hoffman, S.H. Hinkle, S.L. Coolman, C.-X. Luan, J. Xu, and B.W. Kemppainen, "Skin concentrations of thromboxane synthetase inhibitor after topical application with bioelastic membrane." *J. vet. Pharmacol. Therap.*, **27**, 37–43, 2004.
100. B. Kemppainen, N.-Z. Wang, S. Swaim, D.W. Urry, C.-X. Luan, J. Xu, E. Sartin, R. Gillette, S. Hinkle, and S. Coolman, "Bioelastic membranes for topical application of thromboxane synthetase inhibitor for protection of skin from pressure injury: a preliminary study." *Wound Repair and Regeneration*, **12**, 453–460, 2004.
101. S.F. Swaim, D.M. Vaughn, P.J. Spalding, K.P. Riddell, and J.A. McGuire, "Evaluation of the dermal effects of casts padding in coaptation casts on dogs." *Am. J. Vet. Res.*, **53**, 1266–1272, 1992.
102. S.F. Swaim, D.M. Bradley, D.M. Vaughn, R.D. Powers, C.E. Hoffman, and M.L. Beard, "Evaluation of Thromboxane Synthetase Inhibitor in the Prevention of Dermal Pressure Lesion." *Wounds*, **6**, 74–82, 1994.
103. F. Oesterhelt, M. Rief, and H.E. Gaub, "Single Molecule Force Spectroscopy by AFM Indicates Helical Structure of Poly(ethylene-glycol) in Water." *New J. Phys.*, **1**, 6.1–6.11, 1999.
104. H. Clausen-Schaumann, M. Rief, C. Tolksdorf, and H.E. Gaub, "Mechanical Stability of Single DNA Molecules." *Biophys. J.*, **78**, 1997–2007, 2000.
105. H.E. Gaub and J.M. Fernandez, "The Molecular Elasticity of Individual Proteins Studied by AFM-related Techniques." *AvH Magazin*, **71**, 11–18, 1998.
106. D. Bisig, P. Weber, L. Vaughan, K.H. Winterhalter, and K. Piontek, "Purification, Crystallization and Preliminary Crystallographic Studies of a two Fibronectin Domain Segment from Chicken Tenascin Encompassing the Heparin- and Contactin-binding Regions." *Acta Crystallogr. D***55**, 1069–1073, 1999.
107. Personal communication from T. Hugel, M. Seitz, and H.E. Gaub.
108. A. Levitzki, "Signal Transduction Therapy." *Eur. J. Biochem.*, **225**, 1–13, 1994.
109. G.C. Blobe, L.M. Obeid, and Y.A. Hannun, "Regulation of Protein Kinase C and Role in Cancer Biology." *Cancer Metastasis Rev.*, **13**, 411–431, 1994.
110. F.A. Al-Obeidi, J.J. Wu, and K.S. Lam, "Protein Tyrosine Kinases: Structure, Substrate, Specificity and Drug Discovery." *Biopolymers (Peptide Sci.)*, **47**, 197–223, 1998.
111. D.K. Luttrell, A. Lee, T.J. Lansing, R.M. Crosby, K.D. Jung, D. Willard, M. Luther, M. Rodriguez, J. Berman, and T.M. Gilmer, "Involvement of pp60[c-src] with Two Major Signaling Pathways in Human Breast Cancer." *Proc. Natl. Acad. Sci. U.S.A.*, **91**, 83–87, 1994.
112. A. Greco, M.A. Pierotti, I. Bongerzone, S. Pagliardini, C. Lenzi, and G. Porta, "TRK-T1 is a Novel Oncogene Formed by the Fusion of TPR and TRK Genes in Human Papillary Thyroid Carcinomas." *Oncogene*, **7**, 237–242, 1992.
113. B.J. Blanchard, R.D. Raghunandan, H.M. Roder, and V.M. Ingram, "Hyperphosphorylation of Human TAU by Brain Kinase PK40erk Beyond Phosphorylation by cAMP-dependent PKA: Relation to Alzheimer's Disease." *Biochem. Biophys. Res. Commun.*, **200**, 187–194, 1994.
114. I. Yasuda, "Selective Assay of Protein Kinase C with a Specific Peptide Substrate." *Kobe J. Med. Sci.*, **37**, 163–177, 1991.
115. B.S. Goueli, K. Hsiao, A. Tereba, and A. Goweli, "A Novel and Simple Method to Assay the

Activity of Individual Protein Kinases in a Crude Tissue Extract." *Anal. Biochem.*, **225**, 10–17, 1995.
116. I. Yasuda, A. Kishimoto, S. Tanaka, M. Tominga, A. Sakurai, and Y. Nishizuka, "A Synthetic Peptide Substrate for Selective Assay of Protein Kinase C." *Biochem. Biophys. Res. Commun.*, **166**, 1220–1227, 1990.
117. D.W. Urry, "Protein Folding and the Movements of Life." *World & I* **6**, 301–309, 1991.
118. D.W. Urry, "Molecular Machines: How Motion and Other Functions of Living Organisms can Result from Reversible Chemical Changes." *Angew. Chem.* [German], **105**, 859–883, 1993; *Angew. Chem. Int. Ed. Engl.,* **32**, 819–841, 1993.
119. A. Pattanaik, W. Holmes, and D.W. Urry, unpublished results.
120. D.W. Urry and A. Pattanaik, "Elastic Protein-based Materials in Tissue Reconstruction," Ann. N.Y. Acad Sci., **831**, 32–46, 1997.

Epilogue

E.1 Thesis: Biology's "Vital Force" Arises from Coupled Hydrophobic and Elastic Consilient Mechanisms

E.1.1 The Elan Vital

In effect, this volume addresses foundations of the *élan vital* (the vital force or impulse of life).[1] It does so at the level of those forces "performing an essential role in the living body" and, specifically, at the level of those forces central to the function of the protein-based machines that sustain Life. In Chapter 5, by *de novo* design, without having resorted to copying or mimicking any recognized biological protein-based machine, elastic-contractile model proteins were designed, prepared, and demonstrated to be capable of interconverting the set of energies interconverted by living organisms. The forces responsible for the diverse energy conversions were experimentally characterized, analyzed, and given definition.

The primary forces were designated *hydrophobic* and *elastic consilient mechanisms*, because each provided "a common groundwork of explanation"[2] in its realm of utilization, and commonly they do so inseparably. In Chapter 8, those very consilient mechanisms were shown to be dominant in the function of specific examples of the three principal classes of energy conversions of living organisms (subsequent to the photosynthetic step itself). The three classes are (1) developing a proton concentration gradient, (2) using the proton concentration gradient to make adenosine triphosphate (ATP) (the energy coin of biology), and (3) using the ATP to perform the work of sustaining the living organism, all of which are noted further in section E.4.1 below.

E.1.2 Present Limitations in Molecular Mechanics and Dynamics Formalisms with Which to Calculate the Élan Vital

The forces of the hydrophobic consilient mechanism are not calculable by current programs intended to describe protein structure and function! Specifically, the current computational approaches stand silent on the experimentally derived quantities of the Gibbs free energy of hydrophobic association, ΔG_{HA}, and its operative component during function, ΔG_{ap}, the apolar–polar repulsive free energy of hydration. Treatments of water structure and hydration states are wanting. In particular, sufficiently substantive calculations of hydration of hydrophobic groups and of ions, particularly their competition for hydration, have not yet been accomplished. Simply stated, adequate means for treating forces controlling water interactions with disparate groups in proteins have yet to be developed! This is required for calculating even the most fundamental energy of interaction of the hydrophobic consilient mechanism, that is, the experimentally observed endothermic heat of the inverse temperature transition, represented as the middle curve in Figure 8.1 or any of the curves in Figure 7.1.

On the other hand, approximations of the forces involved in the elastic consilient mechanism, however roughly, can be estimated by calculating changes in backbone mobility of a single chain.[3–5] Although water does not play such a fundamental role in determining elastic force, relevant calculations have been carried out in water where changes in water structure were noted.[6]

E.2 The Inverse Temperature Transition: The Foundation of the "Vital Force"

E.2.1 The Inverse Temperature Transition: A New Wrinkle to the "Laws of Physics"

Drawing from Delbruck's model for the genetic material, Schrödinger[7] states, "living matter, while not eluding the 'laws of physics' as established up to date, is likely to involve 'other laws of physics' hitherto unknown, which, however, once they have been revealed, will form just as integral a part of this science as the former." Although the substance of heredity, to which Schrödinger was referring, carries the blueprint for production of protein, it is protein machines that perform the remarkable feats of energy conversion that sustain living matter. As reviewed below, the extensive diversity of the component amino acids of the long protein chains simply adds a new wrinkle to the "laws of physics." It arises out of different hydrations of the oil-like and vinegar-like residues of a protein chain and out of the competition, between these residues, for hydration.

E.2.2 Biology's Heat Engine

As considered in Chapter 5, there are many reasons for referring to the essentially unique phase transition utilized by biology as an *inverse temperature transition*. Furthermore, here we note that biology's inverse temperature transition bears equivalence to the phase transition of the heat engine that gave birth to thermodynamics, but with an inverse twist.

Prigogine and Stengers[8] began "Book Two: The Science of Complexity" within their book entitled *Order Out of Chaos* with a section entitled "Heat, the Rival of Gravitation." The second paragraph states, "Out of all this common knowledge, nineteenth-century science concentrated on the single fact that combustion produces heat and that heat may lead to an increase in volume; as a result, combustion produces work. Fire leads, therefore, to a new kind of machine, the heat engine, the technological innovation on which industrial society has been founded."[8]

One might say that biology was founded on another type of heat engine wherein heat gives rise to work through an inverse temperature transition resulting in contraction rather than expansion. Most generally, however, rather than using heat directly, diverse energy inputs change the temperature at which the inverse temperature transition occurs. The energy inputs achieve function by moving the transition from one side to the other of the temperature at which the living organism thrives.

Therefore, by the same line of thought that gave rise to the Prigogine perspective of "Heat, the Rival of Gravitation" one might say, "the inverse temperature transition, the rival of gravitation," but analogies continue.

E.2.3 ΔG_{ap}, Biology's Rival to Magnetism (as in Repulsion of Like Poles)

With no knowledge of the apolar–polar repulsive free energy of hydration, ΔG_{ap}, Kinosita and coworkers[9] modeled the behavior of F_1-ATPase, the threefold rotary motor of ATP synthase that produces nearly 90% of biology's ATP. In their analogy for this protein-based motor, the rotor was treated as a permanent magnet, the rotation of which was driven by "the attraction between north and south poles and the repulsion between like poles." This was achieved by a threefold symmetric arrangement of stator electromagnets that drove the rotor by alternation of polarities. In the milieu of biology, ΔG_{ap} functions in analogy to "the repulsion between like poles." The Gibbs free energy for hydrophobic association, ΔG_{HA}, functions in analogy to "the attraction between north and south poles." ΔG_{HA} and ΔG_{ap} represent fundamental thermodynamic expressions arising from the inverse temperature transition that in addition to being biology's answer to the heat engine can now be seen as the biological analogue of the attractive and repulsive forces of magnetism.

E.2.4 Uniqueness of Biology's Macromolecules

Again in the words of Schrödinger,[7] "we had to evade the tendency to disorder by 'inventing the molecule', in fact, an unusually large molecule which has to be a masterpiece of highly differentiated order...." In fact, two macromolecules, both masterpieces of highly differentiated order, occur in biology. Both chain macromolecules, the genetic material and proteins, display precisely defined sequences. However, proteins have 20 different residues possible in each position, residues with a wide range of physicochemical properties, whereas a nucleic acid contains only four different residues with quite similar chemical and physical properties. Because of this, the greatest potential for diverse and efficient molecular machines resides with proteins.

When entering the biological world of long chain protein molecules, decorated with chemically and physically diverse side chains and yet composed of a single optical isomer, certain approximations of the past now need supplanting. In particular, approximations of nondifferentiated solvation with its assumption of a uniform dielectric constant and of random chain networks require replacement by interconvertible hydration structures with different thermodynamic properties and by nonrandom dynamic units capable of mechanical resonances and changeable chain entropy.

In our view, the functioning of living matter involves a world of inverse temperature transitions that result from association and dissociation of hydrophobic domains in water. In such circumstances energy inputs change solvation structures and associations of hydrophobic groups, and the forces, due to changes in backbone mobility, emerge. Here, the coupled forces arising from changes in solvation and from changes in backbone mobility prevail. These coupled forces constitute "other laws of physics." The interlinked forces are indeed simply amplified manifestations of the "laws of physics" known outside of the world of living matter. These manifestations become amplified due to the absolute control of sequence, with diverse side chains for diversity of function and with uniform molecular asymmetry for regularity of structure, as routinely available only to the proteins of biology.[10] Thus, in living matter proteins provide the "masterpiece of highly differentiated order"[7] with which to perform the work of the living cell.

E.2.5 Energy Conversions by Means of Inverse Temperature Transitions

Beginning with a repeating sequence of mammalian elastin (Gly-Val-Gly-Val-Pro)$_{11}$, diverse families of high molecular weight elastic-contractile model proteins can be designed with the capacity to perform 18 pairwise classes of energy conversions involving the performance of mechanical work, thermal work, pressure-volume work, chemical work, electrical work, and work performed by frequencies over the range of the electromagnetic radiation spectrum from the acoustic to the ultraviolet. The potential here is incomparably greater than that of the heat engine based on expansion of the liquid to gas phase transition. In all cases, the performance of work depends on an inverse temperature transition of hydrophobic association in aqueous systems wherein oil-like domains of proteins associate. The performance of work can occur on raising the temperature from below to above a critical temperature, T_t. Routinely in biology, however, the performance of work occurs by lowering the temperature of the transition (the value of T_t) from above to below an operating temperature by energy inputs (other than temperature) to which the model protein contains the composition necessary to be responsive.

E.2.6 Coherence of the Inverse Temperature Transition with the Second Law of Thermodynamics

Then Schrödinger[11] goes on to say, "from all that we have learnt about the structure of living matter, we must be prepared to find it working in a manner that cannot be reduced to the ordinary laws of physics." Fortunately, he continues with, "not on the grounds that there is any 'new force' or what not, directing the behaviour of the single atoms within a living organism, but because the construction is different from anything we have yet tested in the physical laboratory." Indeed, by 1944, the year in which Schrödinger's book was published, large polymeric molecules with oil-like and vinegar-like side chains forced to coexist by virtue of sequence remained as yet uncharacterized in the physical laboratory sufficiently to recognize the special element of reversible thermally induced inverse phase separations. Indeed in 1944, over six decades ago, macromolecules capable of exhibiting inverse temperature transitions, where ordered macromolecular structures could be recognized to form on raising the temperature and to disperse on lowering the temperature, had yet to be adequately examined.

E.2.6.1 Requirements of the Second Law of Thermodynamics

By the second law of thermodynamics, a total system must become less-ordered on raising the temperature. At its origins our model system is a two-component system; it is a system of functional protein and water. Many experimental results, reviewed throughout this volume and primarily presented in Chapter 5, show that these elastic-contractile model proteins do indeed become more ordered on raising the temperature through the transition range of the inverse temperature transition. If this system is to comply with established laws of physics, it must adhere to the second law of thermodynamics. If this is so, then the water must become even more disordered on raising the temperature through the transition range of the inverse temperature transition than the increase in order exhibited by the protein component.

E.2.6.2 Experimental Data Provide Abundant Evidence That Water Surrounding Oil-like Groups (i.e., Hydrophobic Hydration) Disorders into Bulk Water as Oil-like Domains Associate to Form Structure

The first insightful data came from dissolution of a series of alcohols in water that differed only in the number of CH_2 groups. It is the series starting with methanol, CH_3–OH, that adds one CH_2 to form ethanol, CH_3–CH_2–OH, another to form n-propanol, CH_3–CH_2–CH_2–OH, a third to form n-butanol, CH_3–CH_2–CH_2–CH_2–OH, and finally a fourth to form n-pentanol (amyl alcohol), CH_3–CH_2–CH_2–CH_2–CH_2–OH. Surprisingly, as Butler[13] reported in 1937, the addition of each CH_2 group to the alcohol resulted in the release of more heat when dissolved in water. Therefore, dissolution of CH_2 in water is favorable; it is exothermic! There exists an energetically favorable structuring of water around oil-like groups, yet oil-like groups themselves exhibit no solubility in water! Arguing from the thermodynamic expression for solubility, water molecules surrounding oil-like groups form a favorable structure that is considerably more structured than bulk water, and *the development of too much structured water surrounding oil-like groups ultimately leads to insolubility.*

By means of crystallography, Stackelberg and Müller[12] first observed the structured water surrounding oil-like groups. They found water molecules arranged at the apices of pentagons. Sharing common apices, 12 pentagons of water molecules encircle an oil-like gas molecule. This specific cage-like or clathrate structure of water is called a *pentagonal dodecahedron* (see the central structure in Figure 2.8).

Remarkably, Butler[13] found the dissolution of oil-like groups to be exothermic. Therefore, the reverse reaction for loss of hydrophobic hydration must be endothermic. The reaction attending the inverse temperature transition to hydrophobic association is indeed endothermic as shown by differential scanning calorimetry data, specifically the middle curve in Figure 8.1 and the curves in Figure 7.1.

The most direct observation of changes in hydrophobic hydration attending the inverse temperature transition comes from following the temperature dependence of an absorption band in the microwave dielectric relaxation experiment shown in Figures 5.24 and 5.25. The water of hydrophobic hydration, with an absorption band near 5 GHz in the dissolved model protein, disappears on raising the temperature through the inverse temperature transition as the model protein hydrophobically associates. Also, as shown in Figure 5.25B, forming charged carboxylates destroys much of the hydrophobic hydration, as the carboxylate recruits water for its own hydration. This is depicted in the leftmost structure in Figure 2.8.

Thus, indeed the law of physics, called the *second law of thermodynamics*, does apply to living matter. Obeyance occurs, however, under a special set of circumstances where the water solvent becomes a key player. Decrease in order is greater for the change to ordinary water from structured water surrounding oil-like groups than is the increase in order of the protein as it hydrophobically associates to form a more ordered structure on raising the temperature. In fact, the changing structure of the protein component of the two-component system, acted on by additional energy inputs, provides functionality essential to sustain the living organism.

E.2.7 Biology's "Vital Force" Arises Out of Interlinking Thermodynamics and Molecular Dynamics

Somewhere near its origins, biology reconciled thermodynamicists' entropy and dynamicists' reversibility,[14] as it evolved functional protein-based machines by interlinking the two processes that seemed so disparate during the debates of the 19th century. Energy driven hydrophobic association with its increase in the thermodynamicists' entropy on conversion of hydrophobic solvation to bulk water commonly couples with a reversible decrease in internal chain dynamics in the development of a near-ideal elastic force wherein the force-relaxation curve exactly follows the same trajectory as the force-extension curve. As we now stand, almost paradoxically from the viewpoint of the early controversy, the entropic contribution to the elastic force becomes the primary basis for reversibility and efficiency of energy conversion.

In our view, the hydrophobic and elastic consilient mechanisms comprise the "vital force" of living matter. The forces arising out of inverse temperature transitions and elastic deformation, for example, apolar-polar repulsion and damping of backbone mobility on deformation, couple to create biology's "vital force."

E.2.8 The Hydrophobic Consilient Mechanism Derives from the Inverse Temperature Transition

The energy conversions that produce motion in living organisms consist of two distinct but interlinked physical processes of hydrophobic association and elastic force development, collectively referred to as *consilient mechanisms* in that they each provide "a common groundwork of explanation."[2] The association of oil-like domains, hydrophobic association, has been characterized in terms of the comprehensive hydrophobic effect (CHE), and elastic force development has been described in terms of the damping of internal chain dynamics on deformation, whether deformation occurs by extension, compression or solvent-mediated repulsion (see section E.4.1.2 and Figures E.3 and E.4, below).

In aqueous systems, CHE controls hydrophobic association by means of a Gibbs free energy of hydrophobic association, ΔG_{HA}, which is quantifiable as a variable-induced change in the net heat of the endothermic inverse temperature transition in Figures 7.1 and 8.1. The magnitude of ΔG_{HA} depends primarily on the amount of hydrophobic hydration displaced on hydrophobic association. Fundamentally, competition for hydration between apolar (hydrophobic) and polar (e.g., charged) groups controls the amount of hydrophobic hydration, as shown in Figure 5.25B, and the competition is quantifiable as an apolar–polar repulsive free energy of hydration, ΔG_{ap}. Thus, the hydrophobic consilient mechanism applies to all amphiphilic polymers in aqueous systems regardless of polymer structure.

E.2.9 The Elastic Consilient Mechanism as the Efficient Mechanical Coupler Within the "Vital Force"

E.2.9.1 Dynamic Chains Without Random Chain Networks

The mechanically linked process of entropic elasticity also constitutes a "common groundwork of explanation." It applies to all polymer structures, of whatever composition, that contain a sufficiently kinetically free chain segment that can exhibit constrained internal chain dynamics, for example, damped backbone torsional oscillations, on deformation. In particular, the polymer structure need not be describable as a random chain network with a Gaussian distribution of deformable end-to-end chain lengths as has been characterized by the Flory random chain network theory of elasticity.[15,16]

Importantly, efficient energy conversion requires the element of entropic (ideal) elasticity where the energy of deformation is recovered on relaxation. Very often deformation/relaxation curves exhibit significant hysteresis, where the input energy required to deform a chain molecule is substantially greater than the energy recovered on relaxation of the deformation. In hysteresis, the energy of deformation has been lost to chain components that do not sustain the deformation. Accordingly, the occurrence of a hysteresis to the deformation constitutes a loss of energy. Because of this, the occurrence of hysteresis during the structural changes of a protein-based machine effecting an energy conversion results in an inefficient machine.

If the energy of deformation remains in the backbone of the chain that sustains the deforming force, then it can be recovered on relaxation. This occurs when the energy of deformation results in the damping of internal chain dynamics. Random chain networks do this reasonably well, but they do not represent the only means of doing so. In fact, certain elastic-contractile model proteins, comprised as they are of repeating sequences that assemble into regular dynamic structures, give near-perfect overlap of force extension and force relaxation curves. These elastic-contractile model proteins exhibit mechanical resonances that have been observed centered at 5 MHz (just below radio frequency) and centered near 3 kHz (positioned over the acoustic frequency range). Such mechanical resonances arise from a regularly repeating dynamic structure. These properties are not comprehensible by the concepts of the random chain network school of entropic elasticity, but, in fact, they provide exciting new materials for the future.

E.2.9.2 Hydrophobic Associations Stretch Interconnecting Chain Segments

Elastic forces come into play as hydrophobic associations stretch interconnecting chain segments. Only if the elastic deformation is ideal does all of the energy of deformation become recovered on relaxation. To the extent that hysteresis occurs in the elastic deformation/relaxation, energy is lost and the protein-based machine loses efficiency. Thus, the elastic consilient mechanism, whereby the force-extension curve can be found to overlay the force-relaxation curve becomes the efficient mechanical coupler within the "vital force." The objective now becomes one of understanding the age-old problem of a reluctance to discard past idols.

E.2.9.3 Bursts of Apolar–Polar Repulsive Free Energy on Hydrolysis of ATP, by the Hydrophobic Elastic Consilient Mechanism, Can Convert to Elastic Deformation for Efficient Energy Conversion

As is briefly noted below in section E.4.1.2 and as was presented in Chapter 8, a burst of apolar–polar repulsive free energy of hydration occurs on hydrolysis of ATP to ADP (adenosine diphosphate) plus inorganic phosphate. This repulsive burst within the structure of F_1-ATPase causes elastic deformations in the protein subunit holding the catalytic site and in the protein subunit functioning as a hydrophobically asymmetric rotor. On the basis of the hydrophobic elastic consilient mechanism, this repulsion drives the rotor in a highly efficient conversion of chemical energy into mechanical work. By driving the rotation of the rotor in the opposite direction with the use of the chemical energy of a proton concentration (actually electrochemical) gradient, ATP is synthesized in the protein-based machine called *ATP synthase*. An understanding of this process, in the author's view, requires setting aside past idols relating to the nature of elasticity and the treatments, or nontreatments, of water structures surrounding protein subunits of protein-based machines.

E.2.9.4 The Problematic "Idols Which Beset Men's Minds" and Acceptance of the Consilient Mechanisms

Sir Francis Bacon (1561–1626), known as the patron saint of the scientific revolution, recognized a set of troublesome idols of the mind some four centuries ago. Zagorin[17] gave definition as, "Bacon's ... doctrine of the idols of the mind [are] those fallacies obstructing the progress of knowledge...." Then there is the classic statement of Bacon as conveyed by Boorstin,[18] " 'Being convinced that the human intellect makes its own difficulties,' Bacon offered his vivid catalog of the illusions of knowledge—'idols which beset men's minds.'" Considered below is one of Bacon's four idols of the mind, the idol of the mind of most direct relevance to advances in understanding the laws of nature with specific relevance to the hydrophobic and elastic consilient mechanisms and the *élan vital*.

E.3 Bacon's "Idols Which Beset Men's Minds"

E.3.1 "... The Idols of the Mind [are] Those Fallacies Obstructing the Progress of Knowledge...."

Bacon wrote of four "idols of the mind." The fourth applies to our concerns of the mechanisms or élan vital whereby biology's protein-based machines function, namely, "idols of the theater ('idola theatri'): these were the offspring of the false dogmas of philosophers, false demonstrations in logic, and false principles and axioms in the sciences, which, like stage plays, generated fictions and unreal worlds."[19] In our efforts to understand the forces that dominate function of protein-based machines, two examples of "the idols of the mind" that constitute "fictions and unreal worlds" spring to mind. The examples derive from impediments to the utilization of each of the consilient mechanisms, to the hydrophobic consilient mechanism and a most cogent example to the elastic consilient mechanism.

Each consilient mechanism presents a "common groundwork of explanation" without employing "fictions and unreal worlds." In past practice, fictions were born of limited experimental data and mathematical formalism. The "fictions and unreal worlds" were compelled by the needs of the time and indeed in their time represented valuable steps toward utilization. These once helpful fictions, however, paradoxically turn into idols that restrain acceptance of new experimental data and thereby impede utilization.

E.3.2 Idols That Retard Utilization of the Elastic Consilient Mechanism in Understanding and Developing Protein-based Machines and Materials

A particularly evident fiction underlies the classic (random chain network) theory of rubber elasticity. To perform certain calculations, there enters the assumption of phantom chains. Phantom chains occupy no space, and one chain or chain segment can pass through another as though it never existed. Of course, this is patently fiction, but stunningly it has not prevented the users of the theory to emerge from such an "unreal world" and make conclusions about structures in the real world.

The argument that emerges from this particular fiction is that all good elastic materials must be completely disordered; they must be random chain networks. In such an unreal world there is no acceptance of the possibility of mechanical resonances in elastomers with dynamic and regularly repeating structural elements. Yet experimental data on certain elastic-contractile model proteins, using different instrumentation measuring from different physical bases, show resonances near 5 MHz and importantly near 3 kHz in the acoustic frequency range. These mechanical resonances develop on raising the temperature through the range of the inverse temperature transition.[3,20]

In the realm of the elastic consilient mechanism, Bacon's "idols which beset men's minds" continue their relevance four centuries following enlightenment by this "patron saint of the scientific revolution."

E.3.3 Idols That Retard Utilization of the Hydrophobic Consilient Mechanism in Understanding and Developing Protein-based Machines and Materials

The fiction that blocks recognition of the primary basis for the "vital force" derives from ignoring the presence in the real world of differentiated water structures. Section E.2.6.2 briefly reviews unquestioned evidence for the existence of special water structure surrounding hydrophobic groups. All would agree that structure and thermodynamics of this water differs from that of bulk water and from that surrounding charged groups. Even the structure of water surrounding a negatively charged group differs from that surrounding a positively charged group. Yet a common approach to computing the structure and function of protein-based machines and materials assumes water to be homogeneous, to be represented by a mean dielectric constant.

Furthermore, the assumption of a uniform dielectric constant for all water structures interacting with protein-based machines and materials conceals the occurrence of competition for hydration between hydrophobic groups and charged groups. In the past there has been the practice of using a dielectric constant of 80 (that for bulk water) up to the surface of the protein and then decreasing the dielectric constant to 5 or less when within the protein. However, what Solomonic wisdom suffices for choice of dielectric constant to be used for ion-pair formation within the tortuous surfaces with clefts of varying shapes from acute to obtuse. As seen in Chapter 8, these clefts may reside inside the protein-based machine, as found in ATP synthase, or outside the protein-based machine, as occurs for the myosin II motor.

In our view, governing free energy changes result from the competition for hydration between hydrophobic groups and charged groups; they have been determined experimentally as hydrophobically shifted pKa values, for example. It is competition with hydrophobic groups for limited hydration, experienced by charged groups of opposite sign, that drives ion-pair formation. As demonstrated in Chapter 5, values for ΔG_{ap}, determined from pKa shifts of acid–base titration data, corroborate corresponding changes in ΔG_{HA}, measured from changes in the heat of the inverse temperature transition. From our experimental experience with designed elastic-contractile model proteins, ΔG_{ap} is no fiction; this measure of the energetics of competition for hydration between oil-like and vinegar-like groups is part of the real world. From the analyses of the protein structures in Chapter 8, ΔG_{ap} (if it had not been derived from model protein studies) would have to have been invented to describe the function of key protein-based machines of biology. In effect, the insightful analogy to magnetic forces by Kinosita and coworkers[9] in describing the behavior of F_1-ATPase implicitly recognized the existence of such a force.

Both realms, those of the hydrophobic consilient mechanism and of the elastic consilient mechanism, provide clear examples of Bacon's "idols which beset men's minds" of four centuries ago that enhance our own personal journeys of enchantment today.

Perhaps we can summarize with the following perspective:

Useful approximations of the past become "idols" in the present that stand as barriers to the future.

E.4 Protein-based Machines as Physical Embodiments of the "Vital Force" That Sustains Life

The phenomenological experimental foundation for the challenging new perspectives presented by the hydrophobic and elastic consilient mechanisms stands on the successful design of model proteins capable of the extraordinarily diverse set of energy conversions noted above. The mechanistic foundation derives from the need to obtain explanation for otherwise inexplicable experimental results, such as stretch-induced pKa shifts, hydrophobic-induced pKa shifts, hydrophobic-induced reduction potential shifts, and systematic increases in steepness (positive cooperativity) of the experimental curves that follow the change of state of functional groups as the model proteins are made more oil-like (see Chapter 5).

Furthermore, yet to be computed by any program is the fundamental thermo-mechanical transduction wherein the cross-linked elastic-contractile model proteins contract and perform mechanical work on raising the temperature through their respective inverse temperature transitions.[21,22] These results first appeared in the literature in 1986 and have repeatedly appeared since that time with different preparations, compositions, and experimental characterizations. Additionally, the set of energies converted by moving the temperature of the inverse temperature transition (as the result of input energies for which the elastic-contractile model protein has been designed to be responsive)[23] have yet to be described by computations routinely used to explain protein structure and function.

We consider these repeatedly demonstrated properties, using designed elastic-contractile model proteins, to be fundamental to the function of the protein machines of biology, as most directly examined in Chapter 8. The function of Complex III of the electron transport chain, of the F_1-motor of ATP synthase, and of the myosin II motor of muscle contraction constitute protein-based machines that appear to have their functions well-described by the hydrophobic and elastic consilient mechanisms. These protein-based machines represent three major classes of energy conversion of biology, and the relevance of their function to the consilient mechanisms, derived using the designed elastic-contractile model proteins, comes into focus in the synopses given immediately below.

E.4.1 Three Primary Classes of Energy Conversion by Protein-based Machines

E.4.1.1 Production of a Proton Concentration Gradient, an Example of Electro-chemical Transduction

The flow of electrons (from reduced nucleotides of intermediary metabolism through Complexes I, II, III, and finally IV of the electron transport chain within the inner mitochondrial membrane) pumps protons from the matrix side to the cytoplasmic side of the membrane. Using a flow of electrons to pump the chemical, proton (the acid, H^+), represents electro-chemical transduction, and the protein-based machine capable of doing so may be called an electro-chemical transducer. Of the four complexes, at this stage, the function of Complex III provides stunning examples of both hydrophobic and elastic consilient mechanisms.

E.4.1.1.1 In the Complex III Example Formation of Negative Charges and of Positive Charges Individually Open Pathways to Effect Proton Translocation Across the Inner Mitochondrial Membrane

On the matrix side of the membrane, two electrons are added to ubiquinone to become negatively charged. By the apolar–polar repulsion of ΔG_{ap}, the negative charge disrupts hydrophobic association to open an aqueous channel allowing ingress of two protons from the matrix side of the membrane to result in uncharged and lipid soluble ubiquinol. On the cytoplasmic side of the membrane hydrophobic association holds the FeS center of the Rieske Iron Protein at the ubiquinol site, where ubiquinol gives up two electrons to carry a double positive charge. The two positive charges on ubiquinol, again by means of ΔG_{ap}, disrupt the noted hydrophobic association on the cytoplasmic side of the membrane, and the two positive charges, released as two protons to the cytoplasmic side, return ubiquinol to ubiquinone. This completes the transit of two protons across the inner mitochondrial membrane.

E.4.1.1.2 Mechanical Coupling of Hydrophobic Dissociation and Elastic Retraction Achieves Electron Transfer on the Pathway from Complex III to Complex IV

Formation of the hydrophobic association between the hydrophobic tip of the Rieske Iron Protein and the hydrophobic ubiquinol-containing site stretches an interconnecting chain segment. This extended chain segment functions as a free-standing tether originating from an anchor in the membrane and bridging to the globular component of the Rieske Iron Protein that contains the electron transferring FeS center on the cytoplasmic side of the membrane. This is depicted in Figures 8.14 through 8.19 and from a somewhat different representation in Figure E.1. On oxidation to result in double positive charge on ubiquinol, the hydration-based repulsive force between charged and hydrophobic groups, ΔG_{ap}, disrupts the hydrophobic association holding the Rieske Iron Protein at the ubiquinol (Q_0) site, and the stretched tether, making use of an aromatic side chain as a fulcrum and applying force at the appropriate angle, lifts the FeS center of the Rieske Iron Protein from the ubiquinol site over a hydrophobic rim and places it at the cytochrome c_1 site. From the heme of cytochrome c_1 the electron transfers to the heme of cytochrome **c**. Cytochrome **c** then diffuses to Complex IV in the transfer of an electron from Complex III to Complex IV. Within Complex IV the electron completes its transit from glucose to reach its ultimate goal of reducing molecular oxygen.

E.4.1.1.3 An Astounding Example of Coupled Hydrophobic and Elastic Consilient Mechanisms

The concept of two distinct but interlinked mechanical processes, expanded here as the coupling of hydrophobic and elastic consilient mechanisms, entered the public domain in the publication of Urry and Parker.[3] Experimental results on elastic-contractile model proteins forged the concept, and the work of Urry and Parker[3] extended the concept to contraction in biology. Unexpected in our examination of the relevance of this perspective to biology was to find the first clear demonstration of the concept in biology in a protein-based machine of the electron transport chain as a transmembrane protein of the inner mitochondrial membrane. Unimaginable was the occurrence of the coupled forces precisely at the nexus at which electron transfer couples to proton pumping.

The figurative critical crossroad between electron transfer and proton translocation exhibits a literal crossing of elastic structures in Figure E.1.[24] Electron flow couples to proton pumping by interlinking the physical processes of charge disruption of hydrophobic association and relaxation of damped internal chain dynamics. Coupled hydrophobic dissociation of the hydrophobic consilient mechanism and elastic retraction of the elastic consilient mechanism simultaneously release proton to the inner membrane space and move the reduced FeS center within the Rieske Iron Protein for electron transfer to the cytochrome c_1 site for further transfer to cytochrome **c**, followed by diffusional transit to Complex IV.

Such is our journey of Ionian Enchantment from de novo *design of elastic-contractile model protein-based machines to enlightenment at a key juncture of energy conversion within living matter.*

E.4.1.2 Use of the Proton Gradient to Produce ATP, the Energy Coin of Biology, an Example of Chemo-chemical Transduction

ATP synthase produces nearly 90% of the ATP utilized by living organisms. The structure and phenomenology of ATP synthase gives rise to a set of predictions that can test the relevance of the hydrophobic and elastic consilient mechanisms to the function of this central protein-based motor of biology. ATP, the energy coin of the biological realm, derives from the coupling of two rotary motors driven by the chemical energy of the proton concentration gradient considered above. The structural coupling element is a rotor that extends perpendicular to the inner mitochondrial membrane and is driven by the membrane-bound F_0-motor. Return of excess proton from the cytoplasmic side of the inner mitochondrial membrane powers the F_0-motor that turns the rotor in a clockwise direction. The rotor turns within an extramembranous catalytic motor housing, collectively called the *F_1-motor*, to produce ATP by adding inorganic phosphate, P_i, to ADP (see Figures 8.26 through 8.41).

By the reverse process of catalyzing the hydrolysis of ATP to ADP plus P_i, the F_1-motor can operate in reverse to drive the rotor in a counterclockwise direction. In this mode the F_1-motor is called the F_1-ATPase. The predictions

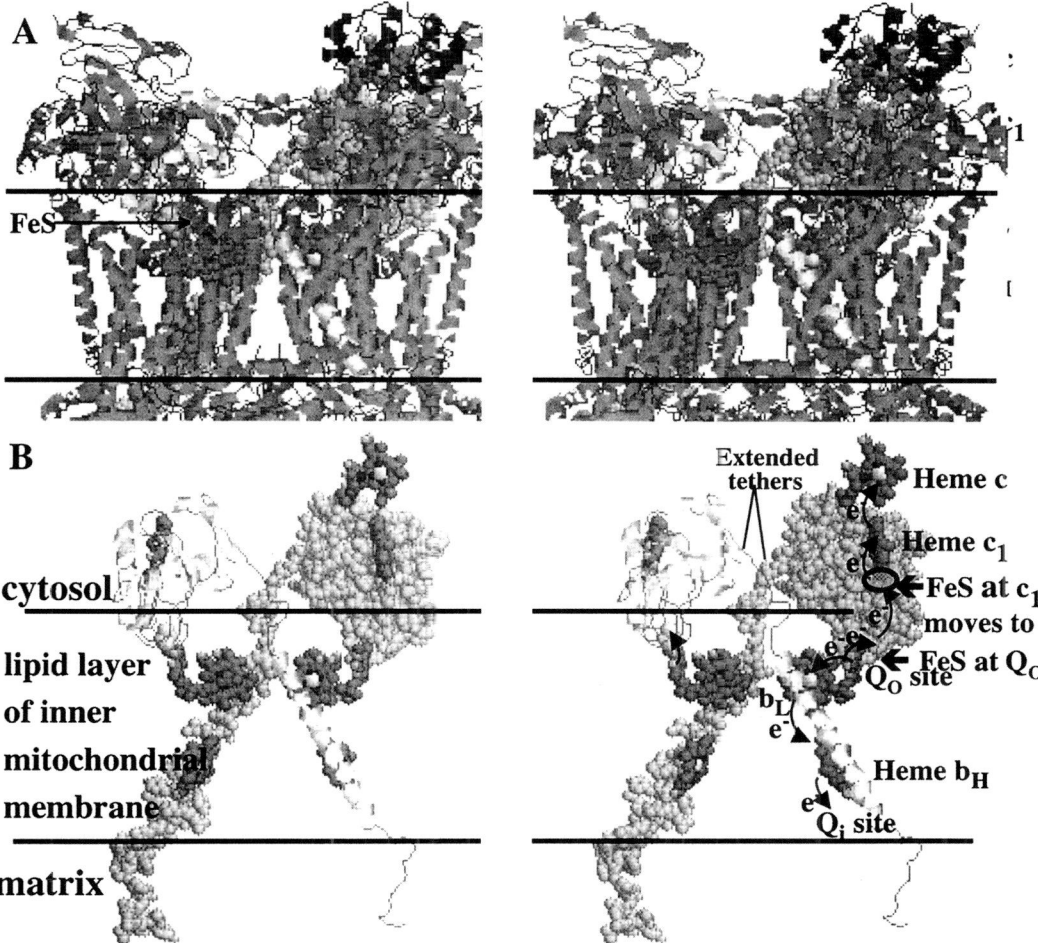

FIGURE E.1. Stereo view of homodimeric structure (gray) of yeast Complex III, plus cytochrome c (dark gray at upper right), but with white Reiske Iron Protein (RIP) in space-filling representation (rising left to right) and ribbon representation (rising right to left) whereby the globular components cross into the opposite monomers by means of free-standing, elastic tethers that become extended on hydrophobic association at the Q_o site of the very hydrophobic tip of the globular component of RIP that contains the FeS center. **(A)** Nearly complete structure. **(B)** Stereo view of redox centers showing electron transfers with the RIP of both monomers included to clarify the basis for the relocation of the FeS center from the Q_o site to the heme c_1 site. On reduction of the FeS center and passing a second electron to heme b_L, the ubiquinol at the Q_o site becomes oxidized to form QH_2^{2+}. By means of the Gibbs apolar–polar repulsive free energy of hydration (ΔG_{ap}), the hydrophobic association of the FeS center with the Q_o site becomes sufficiently weakened by the positive charge of QH_2^{2+} that the stretched elastic tether contracts to move the FeS center from the Q_o site to the heme c_1 site. This domain movement (characterized in Figures 8.14 through 8.19) achieves transfer of an electron from the ubiquinol (characterized in Figures 8.14 through 8.19) to cytochrome c_1, where it is then passed on to the heme of cytochrome **c**. (Prepared using the crystallographic results of Lange and Hunte[24] as obtained from the Protein Data Bank, Structure File 1KYO.)

of the hydrophobic and elastic consilient mechanisms address the ATPase function, and the predictions readily reverse when the rotor is driven by the F_0-motor to describe the function of ATP synthase.

On the basis of the hydrophobic consilient mechanism as presented in Chapter 8, function results from an aqueous-based repulsion, called an *apolar–polar* repulsive free energy of hydration and represented as ΔG_{ap}. The repulsion occurs between the most hydrophobic side of the rotor and the most charged occupancy state (ADP plus P_i) of a catalytic site in a β-subunit. The repulsion either drives the rotor in a counterclockwise direction on hydrolysis of ATP to produce ADP plus P_i, or, when the rotor is driven in the clockwise direction by the F_0-motor, the repulsion causes ADP plus P_i to convert to ATP as a means of lowering the free energy of repulsion.

We complete this synopsis with a list of substantive predictions and their realizations that flow from the hydrophobic and elastic consilient mechanisms when confronted with the structure and phenomenology of the F_1-ATPase:

Prediction 1: *The rotor must be hydrophobically asymmetric.* Because the catalytic housing of the F_1-ATPase is essentially threefold symmetric but with different occupancies in the three β-catalytic subunits, three different faces of the γ-rotor are identified from the crystal structure with different occupancies of the three catalytic subunits.[25] Indeed, the three faces are calculated to have very different Gibbs free energies for hydrophobic association, ΔG_{HA}, namely, −20, 0, and +9 kcal/mole in order of decreasing hydrophobicity. Thus, the prediction of a rotor with hydrophobic asymmetry is strikingly borne out. The order of hydrophobicity, when considered in terms of the apolar–polar repulsion between occupancy and direction of rotation, makes sense with respect to mechanism.

Prediction 2: *In the static state the most hydrophobic side of the rotor faces the least polar side of the motor housing.* Examination of five crystal structures with different occupancy states demonstrates the most hydrophobic side to always be oriented toward the least polar portion of the motor housing, that is, toward the least polar occupancy state of the β-subunits. Thus, the hydrophobic asymmetry properly responds to the occupancy state as required for it to be central to mechanism.

Prediction 3: *The role of ATP in the noncatalytic α-ATP subunits is one of triangulation of repulsive forces to lessen viscoelastic drag between rotor and housing as required for efficiency.* The structure of the catalytic housing of the F_1-ATPase can be given as $(\alpha\beta)_3$, where α and β are paired protein subunits arranged in approximate threefold symmetry. While unchanging and noncatalytic, the α-subunits contain ATP. By the hydrophobic consilient mechanism, the α-ATPs apply a triangular apolar–polar repulsive force between housing and rotor. This repulsive force, ΔG_{ap}, limits hydrophobic association of rotor with housing that would result in a viscoelastic drag or could even lock up rotation by hydrophobic association. Thus, the hydrophobic consilient mechanism ascribes an essential role of facilitating rotation to the three α-ATP nucleotides within the noncatalytic part of the housing.

Prediction 4: *There is negative cooperativity for ATP binding.* With three very different hydrophobic faces to the rotor, ATP would first bind to the β-catalytic site with minimal apolar–polar repulsion. The β-catalytic site exhibiting the least apolar–polar repulsion would be the one facing the least hydrophobic face, $\Delta G_{HA} \approx +9$ kcal/mole, of the γ-rotor. The second ATP to occupy a catalytic β-subunit would do so facing the apolar–polar repulsion of an essentially neutral hydrophobic face of the rotor, that is, $\Delta G_{HA} \approx 0$ kcal/mole. Finally, to achieve complete occupancy of the three catalytic β-subunits, the third ATP must bind facing the apolar–polar repulsion of the most hydrophobic face of the rotor, namely, with $\Delta G_{HA} \approx -20$ kcal/mole. Thus, by the apolar–polar repulsive free energy of hydration of the hydrophobic consilient mechanism, negative cooperativity of

ATP binding naturally results from the hydrophobically asymmetric rotor.

Prediction 5: *There is a positive cooperativity of increased ATP occupancy of catalytic sites with rate of hydrolysis (rotation).* Positive cooperativity of ATP occupancy with hydrolytic rate has the same basis as the role of the three α-ATPs of Prediction 3. As more sites become occupied by the very polar ATP molecule, progressively less viscoelastic drag occurs between rotor and housing. Less drag allows a faster rotation of the rotor on which the rate of hydrolysis depends.

Prediction 6: *Increase in distance between rotor and housing due to ΔG_{ap} repulsion acting through water between the most hydrophobic side of the rotor and the ADP-SO$_4$ analogue of the most polar state.* Menz et al.[26] reported the particularly interesting crystal structure with all sites occupied by nucleotide, as shown in Figure E.2. In the left side of part Figure E.2B, a distortion in the arrangement of ligands (nucleotides) is discernible. The β$_E$-ADP-SO$_4^{2-}$ site in chain E is displaced from the α$_{TP}$-ADP site in chain B and at a greater distance from the γ-rotor. By superposition, Menz et al.[26] in Figure E.3 better represent the displacement of housing from γ-rotor that occurs on replacing the empty subunit (β$_E$) in the structure of Figures 8.30 through 8.34 with the ADP-SO$_4^{2-}$ analogue of the most polar ADP-PO$_4$ state. The crystal structure containing the ADP-SO$_4^{2-}$ analogue was used in Figures 8.35, 8.36, 8.40, and 8.41. The superposition of Figure E.3 clearly shows the double-stranded α-helical coiled coil portion of the γ-rotor and the β$_E$-subunit to be repulsed from each other on occupancy of the β$_E$-subunit by the very polar ADP-SO$_4^{2-}$ analogue. Again, we see ΔG_{ap} in action.

Menz et al.[26] describe their findings: "the coiled coil region has moved significantly … resulting in an overall rmsd of 2.9Å in α-carbon positions for the entire γ subunit. The two conformations of the γ-subunit are related by a rotation that varies in magnitude along the length of the C-terminal helix, ranging from less than 1° for the final residues (γ259–272) to a maximum of about 20° for residues γ234–244 (which form the coiled coil with γ20–10)." In their figure legend, Menz et al.[26] state: "Note that interacting residues in the β$_E$-subunit and the γ-subunit move in opposite directions." This is exactly as expected for the occurrence of an apolar–polar repulsion, a ΔG_{ap} acting *through water between the most hydrophobic side of the rotor and the ADP-SO$_4$ analogue of the most polar state* that constitutes Prediction 6.

There is also the expectation that the polar sulfate would tip the competition for hydration between apolar and polar groups toward polar groups. Charged groups, previously driven to ion pair and to hydrogen bond due to lack of hydration, as shown in Figure 8.39, assisted by the highly charged ADP-SO$_4^{2-}$ occupancy state now access more hydration and no longer require these associations, as shown in Figure 8.40. Figure E.4 illustrates with straight lines the arc of apolar–polar repulsion emanating from the sulfate molecule to the hydrophobic side of the γ-rotor. The repulsion due to the emergent charged residues, for example, D315, D352, D349, K382, D386, and R337 in chain E and E399 in chain A, supplement the repulsion due to the β$_E$-ADP-SO$_4^{2-}$ subunit, as indicated by the finer straight lines. All charged species compete for hydration with the hydrophobic side of the γ-rotor. The series of straight interconnecting lines represents the lines of force due to ΔG_{ap} causing the β$_E$ and γ subunits to move in opposite directions in Figure E.4.

Prediction 7: *A repulsive force acting through "waters of Thales" to store energy in elastic deformation provides the opportunity for high efficiencies of energy conversion by F_1-ATPase.* The movement in opposite directions observed in Figure E.4 of the β$_E$-subunit from the γ-rotor on occupancy of the β$_E$-subunit by ADP-SO$_4^{2-}$ constitutes an elastic deformation. The elastic deformation results from ΔG_{ap}-based repulsive forces acting through water, that is, through the "waters of Thales." Under such circumstances as the ATPase the deforming forces relax efficiently into rotational motion of the rotor.

If the application of force for rotation were instead to occur above the aqueous chamber

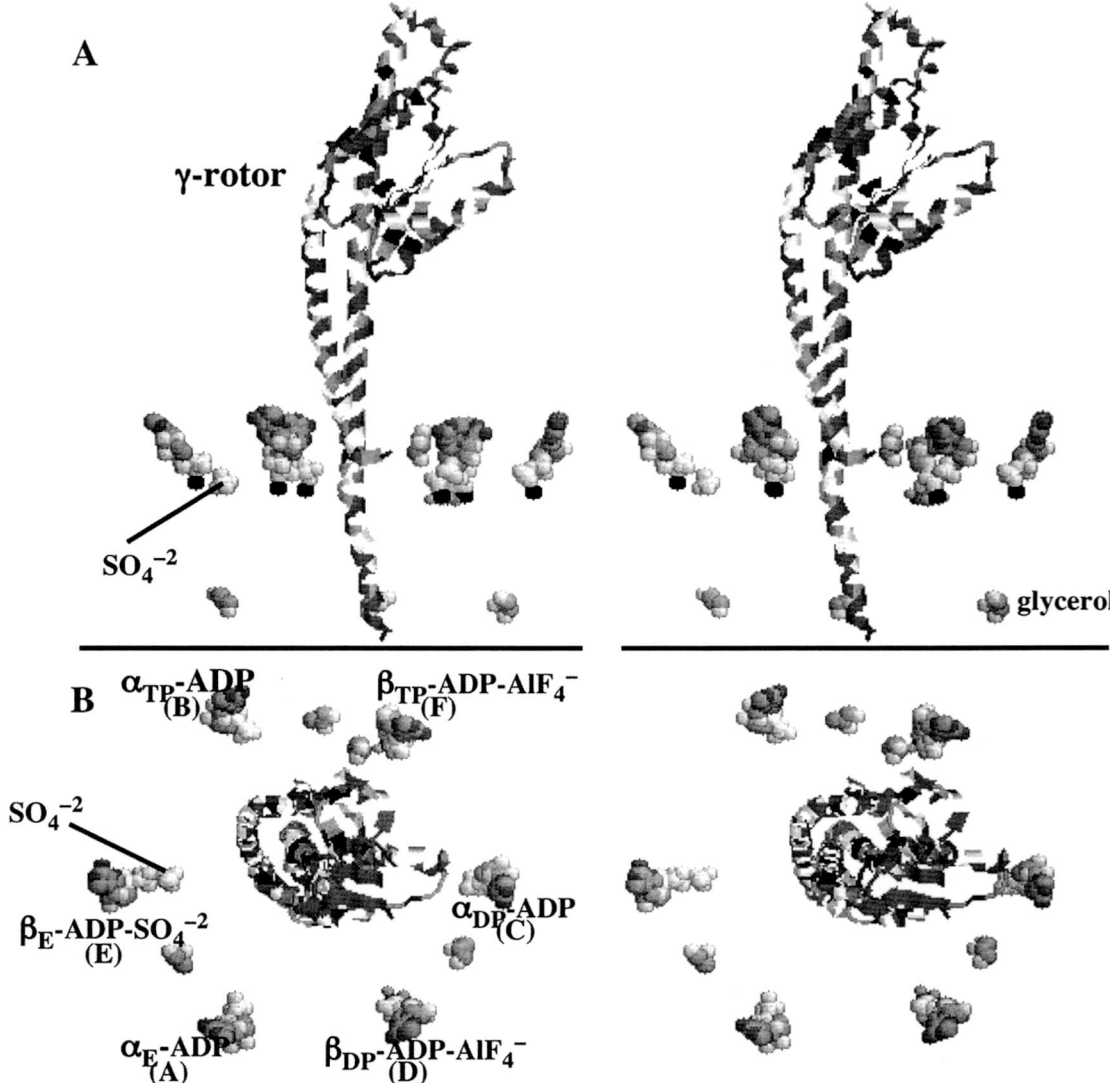

FIGURE E.2. Stereo view of the F_1-motor of ATP synthase showing the γ-rotor in ribbon and ligands in space-filling representation with a unique set of ligands that result in the ADP-SO_4 occupancy opposed to the most hydrophobic side of the γ-rotor. SO_4 presents a stable analogue of hydrolyzed γ-phosphate such that it becomes possible to see spatial changes indicative of a greater ΔG_{ap} repulsion that occurs due to that occupancy state. Labeling of ligands and subunits combines the identifications of Menz et al.[26] (in Figure E.3) and those used in Chapter 8. (A) Side view with aligned ligands and the ADP-SO_4 ligand at left. (B) Top view with labeling on the left and apparent structural displacements between rotor and the β_E-subunit containing ADP-SO_4. (Prepared using the crystallographic results of Menz et al.[26] as obtained from the Protein Data Bank, Structure File 1H8E.)

between the β_E-subunit and the γ-rotor, then it would be difficult to imagine that energy of motion would not be dissipated irreversibly into the β_E-subunit. Additionally, from the relative positions of these portions of the β_E-subunit and the γ-rotor, shown in Figures E.3 and E.4, the β_E-subunit appears to be in front of the second strand of the double-stranded

E.4 Protein-based Machines as Physical Embodiments of the "Vital Force" That Sustains Life

γ-rotor rather than behind it in position to push the rotor in the expected counterclockwise direction (see immediately below).

Prediction 8: ΔG_{ap} *drives counterclockwise rotation of the F_1-ATPase rotor.* As was shown in Figure 8.41, the $\Delta \mathbf{G_{ap}}$-derived repulsive force between SO_4^{2-} and the γ-rotor that causes movement "in opposite directions" of the β_E-ADP-SO_4^{2-} and γ-subunits in Figures E.2 through E.4 would provide a counterclockwise rotational impulse to the γ-rotor, as demonstrated by Noji et al.[27] The same conclusion was reached by Menz et al.,[26] as indicated by the statement that "the observed rotation of the γ subunit is consistent with the sense of rotation seen in the direct visualization experiments (Noji et al., 1997). . . ."

The success of the above set of predictions concerning structure and function of the F_1-ATPase (the F_1-motor of ATP synthase) demonstrate a dominant role of the hydrophobic and elastic consilient mechanisms in the function of this pivotal protein-based machine of biology.

E.4.1.3 Use of ATP to Perform Mechanical Work of the Organism, Chemo-mechanical Transduction by the Myosin II Motor

E.4.1.3.1 Perspective of the Hydrophobic Consilient Mechanism in the Performance of Mechanical Work

Chemo-mechanical transduction technically defines the use of chemical energy of ATP to produce motion, as in muscle contraction. The fundamental statement of the hydrophobic consilient mechanism regarding chemo-mechanical transduction is that the most charged, polar states disrupt hydrophobic asso-

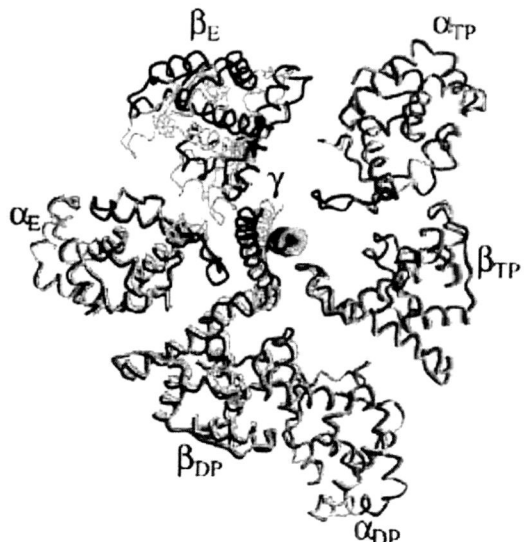
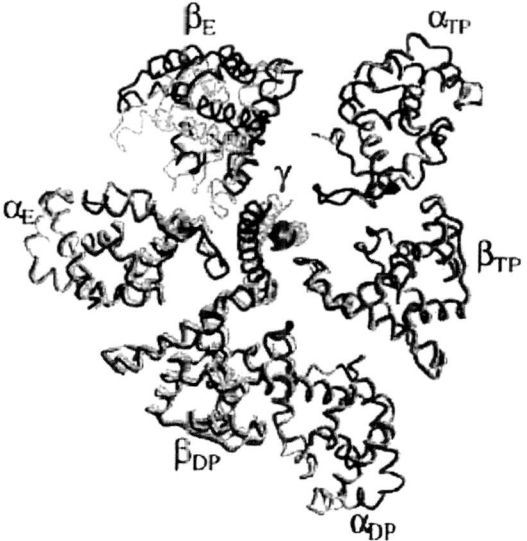

FIGURE E.3. Cross-eye stereo view in trace representation from the top of the F_1-motor of ATP synthase of the superposition of two structures, one with the empty β_E-site given in light gray and the second with the ADP-SO_4-occupied β_E-subunit in dark gray. An increase in mean distance between the γ-rotor and the β_E-subunit is shown when the empty β_E-subunit is replaced by ADP-SO_4 and when the SO_4 fills the role of a stable analogue for hydrolyzed γ-phosphate. The doubly charged SO_4^{2-} acts to free charged side chains of the β_E-subunit from ion pairing and hydrogen bonding such that they will also contribute to the total apolar–polar repulsion, $\Delta \mathbf{G_{ap}}$, between the ADP-SO_4–occupied β_E-subunit and rotor. The resulting ΔG_{ap} applies a torque to the double-stranded α-helical coiled coil section of the γ-rotor (residues 3–35 and 221–250 shown) that would impart a counterclockwise rotation to the rotor when functioning as the F_1-ATPase.[26] This is shown more explicitly in Figure E.4 and especially in Figures 8.40 and 8.41. (From Menz et al.[26] adapted for cross-eye view with permission of Elsevier.)

Depicting Apolar-polar Repulsion in Structure 1H8E

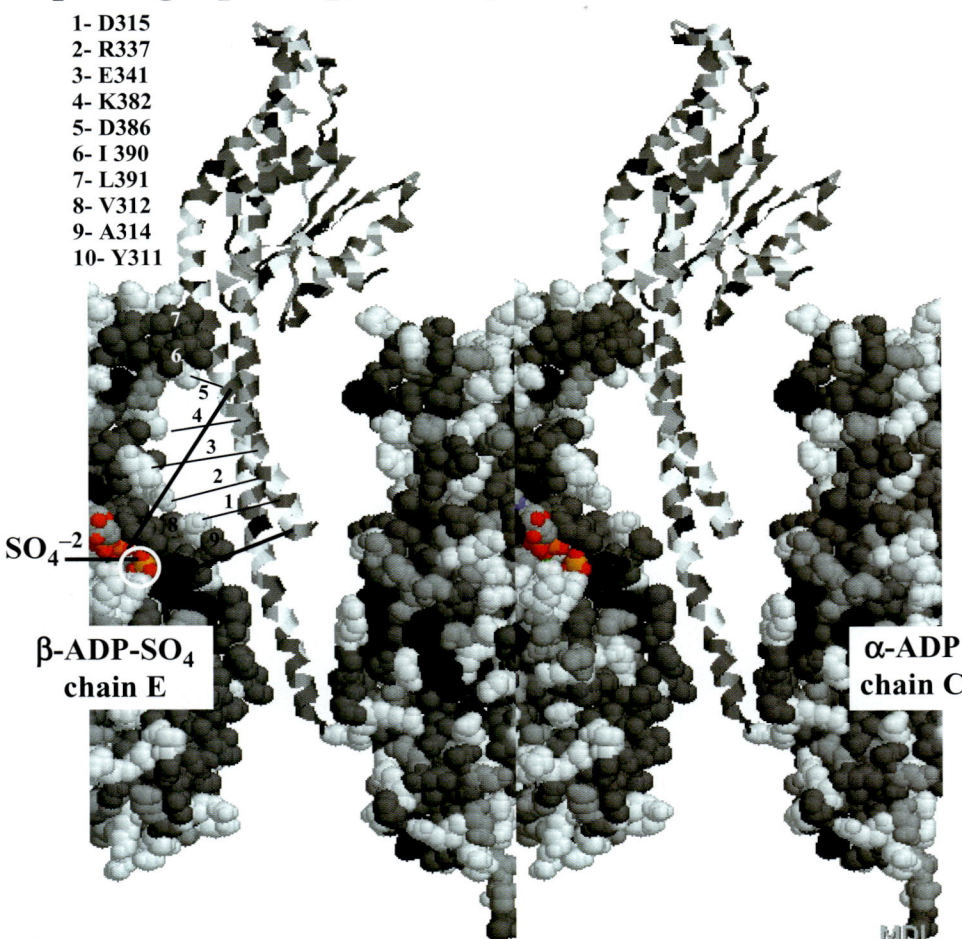

1- D315
2- R337
3- E341
4- K382
5- D386
6- I 390
7- L391
8- V312
9- A314
10- Y311

β-ADP-SO₄ chain E

α-ADP chain C

FIGURE E.4. Cross-eye stereo view of the F_1-motor of ATP synthase with γ-rotor in ribbon and E and C chains in space-filling representation with residues indicated as neutral in light gray, aromatic in black, other hydrophobics in gray, and charged in white. Lines indicate a repulsive interaction (i.e., ΔG_{ap}) between the hydrophobic γ-rotor and SO_4 (in place of hydrolyzed γ-phosphate) and augmented by newly emerged charged groups that occurs in the $β_E$ chain. The ΔG_{ap} due to SO_4 and emerged charged groups apply a torque to the double-stranded α-helical coiled coil section of the γ-rotor that would give a counterclockwise rotation when functioning as an ATPase. (Adapted from Urry "Function of the F_1-motor (F_1-ATPase) of ATP synthase by Apolar-polar Repulsion through Internal Interfacial Water." *Cell Biol. Int.*, **30**(1), 44–55, 2006. Prepared using (Adapted from Urry.[38])

ciation and that the powerstroke due to the step down from the most charged state drives hydrophobic association. In regard to the universal energy coin of biology, ATP^{4-} Mg^{2+} represents the penultimate polar state; the ultimate polar state of biology is ADP^{3-} Mg^{2+} HPO_4^{2-}, and the loss of HPO_4^{2-} to leave ADP^{3-} Mg^{2+} provides the largest single step impulse for driving hydrophobic association. In short, from the perspective of the hydrophobic consilient mechanism, the powerstroke results from hydrophobic association!

E.4.1.3.2 Structure of the Myosin II Motor and the Sliding Filament Model of Muscle Contraction

The myosin II motor produces the very familiar motion resulting from contraction of skeletal muscle. From the gross anatomical perspective, the motion of contraction derives from the decrease in length between two sites of muscle attachment. At the level of the electron microscopic, shortening results from shortening the length of the fundamental repeating unit of muscle, the sarcomere. As shown in Figure 8.45, shortening of the sarcomere length defined by the distance between Z disks would occur by the sliding into greater overlap of two classes of filaments (myosin and actin) that intersperse in hexagonal arrays such that 24 thin actin filaments separate and surround seven thick myosin filaments. Long myosin molecules, staggered in their association to make the thick filament, terminate in cross-bridges that attach to the thin actin filaments. Each cross-bridge contains a single nucleotide binding site that functions as an ATPase, and the breakdown of ATP to ADP with release of P_i produces the motion of muscle contraction. Thus, contraction by the sliding filament model of the foregoing construction requires cyclic attachment/detachment from the actin filament correlated with cyclic contraction/relaxation.

In our approach we look to the most polar states associated with the use of ATP for disruption of hydrophobic association and to the absence of such polar states as giving rise to hydrophobic association. For the myosin II motor, two crystal structures of the cross-bridge from scallop muscle, due to Szent-Gyorgyi and Cohen and their colleagues,[28] provide the opportunity for such an analysis. One structure is called a *near-rigor state* without nucleotide, but it does contain a polar sulfate, but partially neutralized as the $Mg^{2+} SO_4^{2-}$. At the active site the other structure contains an analogue of $ATP^{4-} Mg^{2+}$, namely, $ADP^{3-} Mg^{2+}\text{-}BeF_X$ (where BeF_X is to fill the role of the γ-phosphate). These two crystal structures do not provide snapshots of the extremes that would be preferred, but they do allow useful comparisons of two structures near the desired limits.

Starting from the prejudice of the hydrophobic and elastic consilient mechanisms and utilizing the two noted structures, we proceed within the framework of the sliding-filament cross-bridge structure with a single nucleotide binding site in each cross-bridge. In Chapter 8, section 8.5.4, this conceptual context brought to light a set of findings much as would be expected from the understanding of the hydrophobic and elastic consilient mechanisms developed in Chapter 5. The following summarizes these findings for the myosin II motor in terms of a set of six expectations that we succinctly examine for extent of realization.

Expectation 1: *That ATP binding at the cross-bridge active site provides a ΔG_{ap}-based repulsive force for detachment by hydrophobic dissociation from the actin filament.* On occupancy of an empty active site by the penultimate polar state, $ATP^{4-} Mg^{2+}$, disruption of hydrophobic association becomes the primary expectation of the hydrophobic consilient mechanism. Of course, ATP binding disrupts cross-bridge binding to the actin filament, that is, disrupts the hydrophobic association that provides for attachment of cross-bridge to the actin filament. The hydrophobic consilient mechanism requires, however, that "waters of Thales" occur between the phosphates of ATP and the site of hydrophobic association in order that an apolar–polar repulsive free energy of hydration, ΔG_{ap}, may exist between the two sites. Expectation 5 and Figures E.5 and E.6 address with the structural feature whereby the myosin II motor directs this ΔG_{ap}-based repulsive force.

Expectation 2: *That the ADP plus P_i state at the cross-bridge active site effects a repulsive ΔG_{ap} force for dissociation of hydrophobic domains within the cross-bridge.* Relevance of the hydrophobic consilient mechanism to the motion of contraction requires that formation of the most polar state, $ADP^{3-} Mg^{2+} HPO_4^{2-}$, effects hydrophobic dissociation within the cross-bridge. This sets the stage for the loss of P_i, that is, of HPO_4^{2-}, that drives the hydrophobic association required by the hydrophobic consilient mechanism as

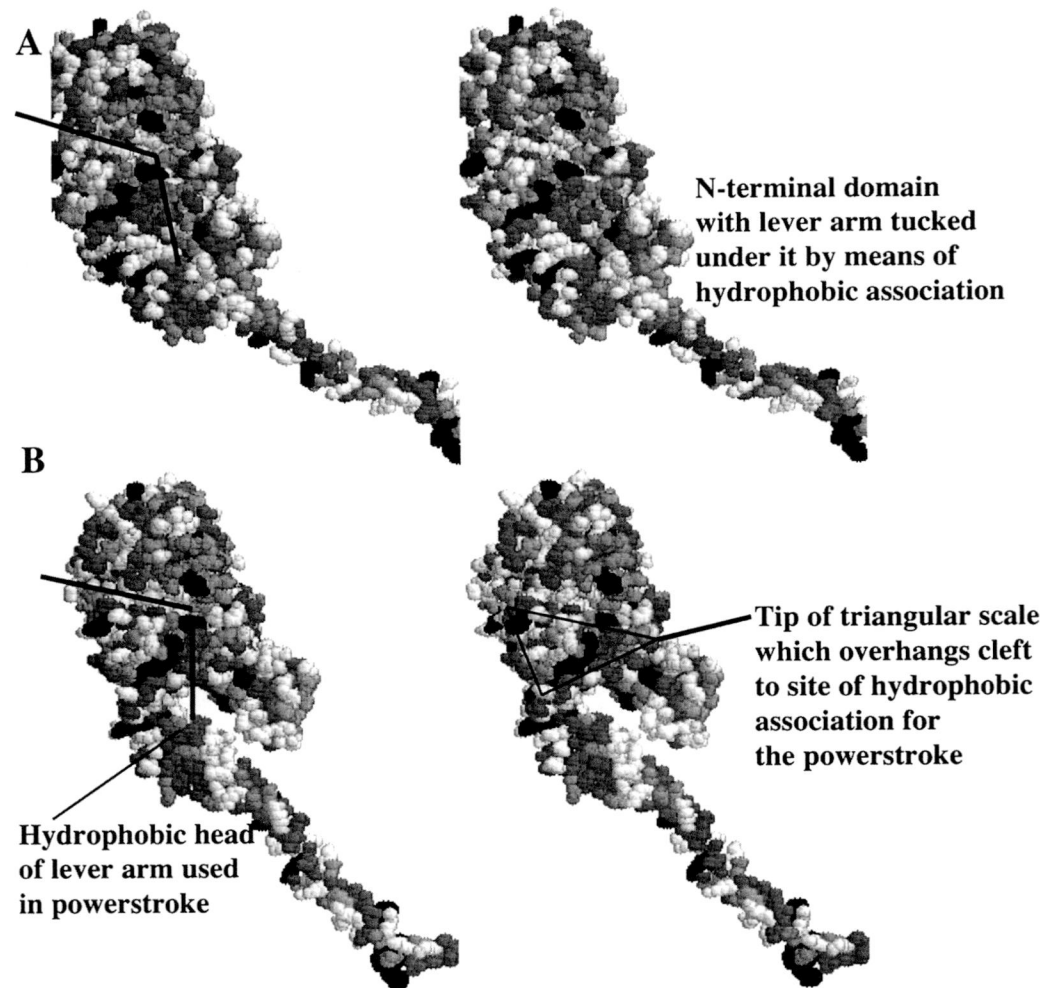

FIGURE E.5. Stereo view (cross-eye) in space-filling representation of the scallop muscle crossbridge(S1) viewed approximately from the side of the actin binding site on myosin cross-bridge for the purpose of locating the narrow clefts that direct the apolar–polar repulsive force. **(A)** The hydrophobic association of the near-rigor state. (Prepared using the crystallographic results of Himmel et al.[28] as obtained from the Protein Data Bank, Structure File 1KK7.) **(B)** The hydrophobic dissociation of the ATP analogue state. (Prepared using the crystallographic results of Himmel et al.[28] as obtained from the Protein Data Bank, Structure File 1KK8.)

the basis for the powerstroke. The very apparent hydrophobic dissociation is seen in Figure 8.51B, 8.52B, 8.53, and 8.54.

Expectation 3: *That the rigor-like state demonstrates a re-established hydrophobic association that defined the powerstroke.* The absence of highly polar occupancy of the active site, as in the near-rigor state shown in Figure E.5A, illustrates formation of hydrophobic association between the head of the lever arm and the underside of the amino-terminal domain. This involves the very hydrophobic residues L711, L712, L767, and F716 as well as F707, F761, and F761, associated with the head of the lever arm, and the very hydrophobic residues of L94, Y93, I700, F28, and F82 as well as proximal F119, L118, L697, and V696, associated with the underside of the amino-terminal domain. These hydrophobic residues accept exposure to water in the structures in Figures E.5B and E.6 due to the apolar–polar repulsive free energy of hydration that emanates from the active site as the result of occupancy by a sufficiently polar state.

Expectation 4: *That the hydrophobic association of the powerstroke extends kinetically free chain segments to produce an elastic force.* In order that there be a smooth and efficient conversion of energy from chemical to mechanical by the act of hydrophobic association, the energy must be temporarily stored in an elastic deformation with limited hysteresis. This occurs as hydrophobic asso-

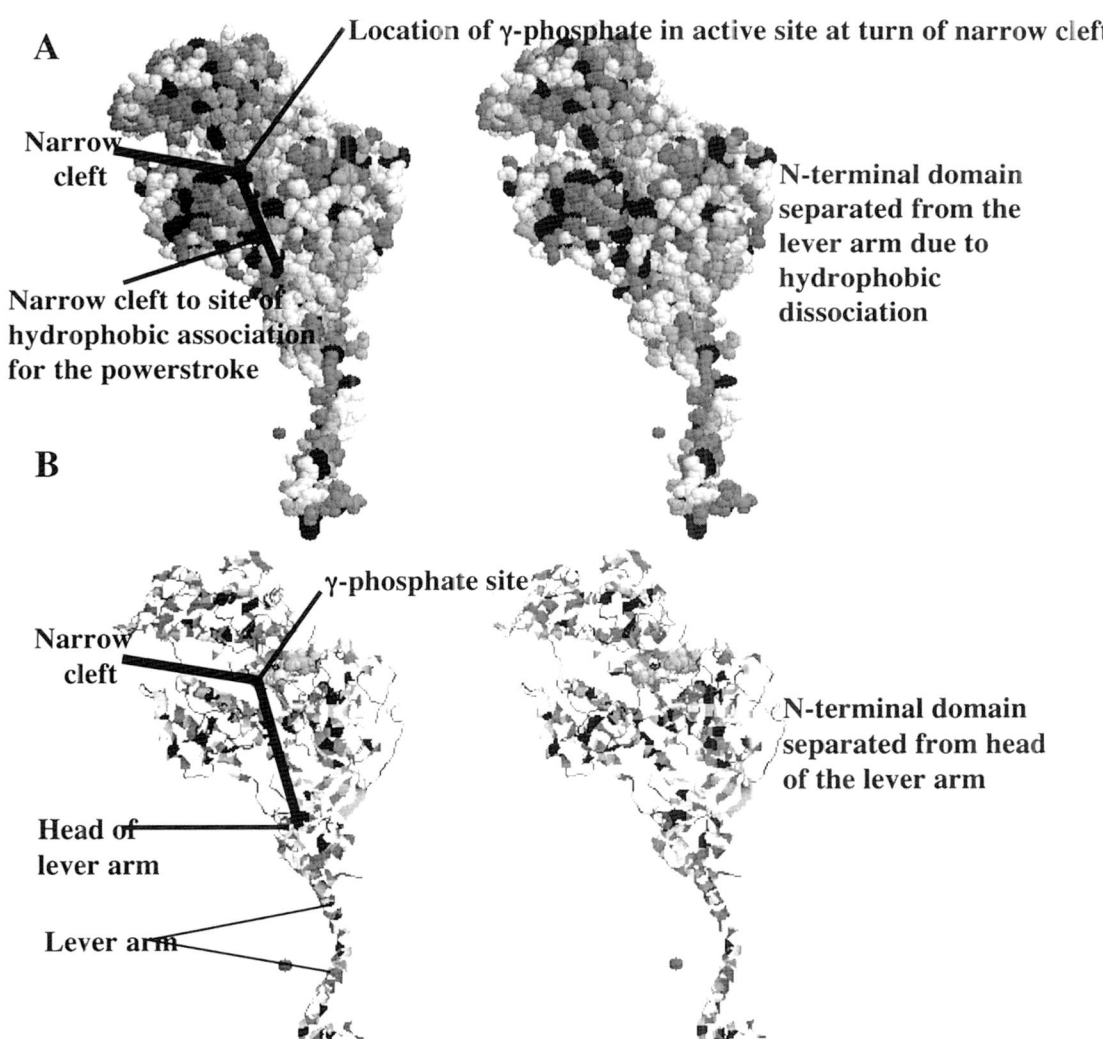

FIGURE E.6. Stereo view (cross-eye) of scallop muscle cross-bridge(S1) to show the ATP binding site at the turn in a cleft. The cleft runs in two directions, one toward the actin–cross-bridge binding site and the other toward the hydrophobic dissociation between the head of the lever arm and the amino-terminal domain. **(A)** Space-filling representation. **(B)** Ribbon representation. (Prepared using the crystallographic results of Himmel et al.[28] as obtained from the Protein Data Bank, Structure File 1KK8.)

ciation stretches interconnecting chain segments. A significant search has yet to occur throughout the cross-bridge. Noted, however, in Figure 8.52 and featured in Figure 8.55 is the extension of an interconnecting chain segment that occurs on going from the hydrophobically dissociated state with ATP analogue occupancy of the active site to the hydrophobic association of the near rigor state. Here resides the coupling of the hydrophobic and elastic consilient mechanisms for the function of protein-based machines.

As noted by Rayment and coworkers,[29] a number of flexible loops occur in the cross-bridge, and we suggest that these are potential sources for elastic force development that results from hydrophobic association of the powerstroke (see Chapter 8, section 8.5.4.5).

Expectation 5: *That narrow clefts direct the repulsive ΔG_{ap} force in two directions to disrupt hydrophobic association of cross-bridge–actin binding and to disrupt the hydrophobic association responsible for the powerstroke.* As shown on Figure E.6, a narrow cleft runs from the site of the cross-bridge–actin binding to the active site where it makes a sharp turn and runs directly to the site for hydrophobic association between the head of the lever arm and the hydrophobic underside of the amino-terminal domain. In terms of the hydrophobic consilient mechanism these clefts direct the thirst for hydration of the polar species in the active site toward the sites of hydrophobic association. The narrow clefts function as conduits to direct the repulsive force due to the apolar–polar repulsive free energy of hydration, ΔG_{ap}, to interaction with sites for hydrophobic association. In Figure E.5, the cleft, connecting the γ-phosphate of ATP to the hydrophobic association between the head of the lever arm and the underside of the amino-terminal domain, runs under an overhang of a triangular surface scale identified by residues P571-R560-F539 at the three apices of the triangle. Figure E.6A allows a look directly into this section of the cleft.

Expectation 6: *That calcium ion triggers contraction by opening a hydrophobic site on actin that hydrophobically associates with the cross-bridge and increases the repulsive ΔG_{ap} force on ADP plus P_i to expel P_i that relieves repulsion to drive the hydrophobic association of the powerstroke.* Introduction of the process whereby calcium ion triggers contraction by the hydrophobic consilient mechanism requires knowledge of the state of ATP in the active site of the dissociated cross-bridge and the state that binds to the actin filament. In the ATP occupied state of the detached cross-bridge, an active ATP ↔ ADP P_i interchange occurs that appears to be more toward the hydrolyzed state.[30] Furthermore, as noted by Lymn and Taylor[31] quite early, "actin binds with the myosin ADP P complex and displaces products of the hydrolysis reaction. It is concluded that this step is responsible for activation of myosin ATPase by actin." Thus the state of nucleotide in the cross-bridge on binding to the actin site would be the most polar state, ADP^{3-} Mg^{2+} HPO_4^{2-}. The description of the process whereby calcium ion triggers contraction by binding to troponin C on the actin filament is particularly well-suited to description by the force resulting from the apolar–polar repulsive free energy of hydration, ΔG_{ap}, of the hydrophobic consilient mechanism.

The essential result of calcium ion binding is that it displaces tropomyosin to open a strong hydrophobic binding site on actin for cross-bridge attachment. The attachment to actin would close off a major source of hydration for the most polar state, ADP^{3-} Mg^{2+} HPO_4^{2-}, that was provided by the narrow cleft connecting the active site to the cross-bridge–actin binding site. This could raise the free energy of the most polar state to such an extent that lowering the free energy would be achieved by expulsion of HPO_4^{2-}. With this loss of phosphate, the apolar–polar repulsion maintaining the hydrophobic dissociation at the other end of the narrow cleft is lost such that the hydrophobic association of the powerstroke results.

The above expectations that describe the structure and function of the myosin II motor suggest a dominant role for the hydrophobic and elastic consilient mechanisms in the function of this representative motor for producing motion in living organisms.

E.4.1.4 Concluding Comment for Physical Embodiments of the "Vital Force"

We have discussed the nexus of hydrophobic and elastic consilient mechanisms in Complex III at the intersection of electron flow and proton translocation (electro-chemical transduction) that was unimaginable, absent the detailed analysis of structure. The hydrophobic and elastic consilient mechanisms, however, when applied to the general structure and phenomenology of the F_1-motor of ATP synthase (chemo-chemical transduction), gave rise to a host of successful predictions, and, when applied to the structure and phenomenology of the myosin II motor (chemo-mechanical transduction), resulted in a half-dozen realized expectations. These findings do much to substantiate the relevance of the hydrophobic and elastic consilient mechanisms to the protein-based machines of biology.

The above three discussed protein-based machines—Complex III of the electron transport chain, ATP synthase/F_1-ATPase, and the myosin II motor of muscle contraction—represent the three major classes of energy conversion that sustain Life. Therefore, the facility with which the consilient mechanisms explain their function indeed support the thesis that biology's "vital force" arises from the coupled hydrophobic and elastic consilient mechanisms.

When taken together, the successes of the hydrophobic and consilient mechanisms in describing the functions of these three representative classes of biological energy conversion constitute a significant statement in favor of the roles of these mechanisms as "vital forces" that impart function to the protein-based machines of biology.

E.5 Extraordinary Biomaterials Opportunities Arise from Identified Vital Forces, Biocompatibility, and Biosynthesis

E.5.1 An Enabling Triumvirate: Knowledge of Vital Forces, Capacity for Biosynthesis, and Elastic Protein-based Polymers with Unique Biocompatibility

E.5.1.1 An Enabling Triumvirate

The above-detailed successes of the hydrophobic and elastic consilient mechanisms in describing the functions of multifaceted proteins of biology herald an historic naissance for biomaterials innovation. Singularity arises from the near-simultaneous emergence of three developments (1) the deepened understanding of the forces that provide the basis for protein function, (2) the unparalleled capacity to produce diverse protein-based materials by recombinant DNA technology, and (3) the remarkable biocompatibility of elastic protein-based polymers. Achieving an emerging command of the vital forces that underpin living organisms points to the design of protein-based polymers with the capacity to perform the desired tasks of biomaterials with the specificity, selectivity, and efficiency that attend biology's own protein function.

E.5.1.2 Each Application Warrants a Specifically Optimized Sequence

This brings us in an unprecedented way to the circumstance envisioned by Jacob Bronowski in his 1973 book, *Ascent of Man*,[32] where he stated, "*In effect, the modern problem is no longer to design a structure from the materials (available) but to design materials for a structure.*" We are now in the position to design protein-based polymers for each specific application. The past has seen coercion of function by chemical treatments and physical manipulation of a single starting polymer composition. Now, whenever an application is defined, response can be the design of a unique protein-based polymer with a specified and application-optimized sequence. Often the basic consideration becomes the apolar–polar repulsive free energy of hydration, ΔG_{ap}, within the chain that results from the repulsive interaction of the charged and hydrophobic groups.

E.5.1.3 Mechanical Resonances Are Key to Biocompatibility and Certain Medical and Nonmedical Applications

For medical applications, in our view, mechanical resonances (see Section 9.4.2.3)[33] present barriers to antibody interaction such that these soft elastic biomaterials exhibit a remarkable biocompatibility otherwise considered impossible for "foreign proteins." As a specific example for medical and nonmedical applications, the author believes that the finding of mechanical resonances, so innovative as to be denounced as artifact by those constrained by the "idols of the present,"[33] constitutes opportunities for the future ranging from biosensors capable of single molecule detection to hearing protection and underwater sound absorption.

E.5.2 Medical Devices Using Biology's Own Mechanisms and Materials

E.5.2.1 The Consilient Approach to Medical Devices

Here again the adjective *consilient* becomes relevant. In the present context there occurs "a common groundwork of explanation"[2] for function within the medical device, within the patient for whom support is being designed, and between device and patient. The consilient approach, therefore, employs the same hydrophobic and elastic consilient mechanisms in the design of the medical device that operate at the site of application, as demonstrated in Chapters 6, 7, 8, and 9 and above in section E.4.

E.5.2.2 The Consilient Approach to Soft Tissue Engineering

E.5.2.2.1 The Concept of the Consilient Approach to Soft Tissue Engineering

As stated in Chapter 9, "The consilient approach to tissue engineering utilizes biology's own materials and mechanisms, concerned with tissue structure and function, to achieve tissue restoration." The key materials elements are three: the capacity of the elastic protein-based material to match the elastic modulus of the tissue to be restored, a remarkable biocompatibility of the pure elastic protein-based material, and the facility to design into the protein-based polymer sequence any desired biologically active sequence, which, by virtue of the innocuousness of the elastic protein-based material, allows proper expression of the incorporated biologically active sequence. This provides the opportunity for the proper functional relationship of protein-based material to the cells and the extracellular matrix of the tissue to be restored.

E.5.2.2.2 The Concept of Temporary Functional Scaffoldings

It is our thesis that materials and tissue (extracellular matrix and cells) function by the hydrophobic and elastic consilient mechanisms. Importantly, the material can be designed to contain the appropriate cell attachment sequences for the natural cells of the tissue to be restored. The natural cells of the tissue to be restored can migrate into the material, attach, and sense the tensional force changes that occur during functioning of the prosthesis. As cells are themselves mechano-chemical transducers, the magnitude and periodicity of the tensional force changes to which the cells are subjected induce the cells to produce the extracellular matrix sufficient to sustain those force changes. The concept continues; in the normal ongoing remodeling process the natural cells within the prosthesis remodel it into a natural tissue as the protein-based prosthesis is replaced by an appropriate rate of degradation. The concept of a *temporary functional* scaffolding for soft tissue restoration is *temporary* in that it can be set for an appropriate rate of degradation and is *functional* because it has the appropriate elastic modulus with which to fill the mechanical role of prosthesis, and it contains cell attachment sequences whereby the cells interact and sense the demand on the tissue that they remodel into the natural tissue.

E.5.2.3 The Central Role of ΔG_{ap} in the Controlled Release of Pharmaceuticals

E.5.2.3.1 ΔG_{ap} Sequence Energizes Protein-based Polymers for Function

The competition for hydration between charged and hydrophobic groups constrained by sequence to coexist in a chain molecule energizes the chain molecule as these disparate groups reach out for hydration spheres unperturbed by the other. This gives rise to an apolar–polar repulsive free energy of hydration, ΔG_{ap}, that energizes a particular sequence. As the repulsion becomes greater, the charges become driven toward charge neutralization by ion pairing. The hydrophobic induced pKa shift provides a measure of the extent of the repulsion, that is, of ΔG_{ap}.

When the pH is not suitable, instead of neutralization by protonation of a carboxylate, for example, ion pairing becomes the possible means of neutralization. Ion pairing, however, only lowers the free energy about half as well as protonation. Nonetheless, ion pairing provides a most useful force as in the process of aligning complementary chains and in loading charged drug into interaction with an oppositely charged polymer. With regard to the charged-drug–oppositely-charged-polymer interaction, increased hydrophobicity of the polymer enhances drug affinity and decreases release rate of drug.

E.5.2.3.2 Drug Delivery by T_t-type (Transductional) Protein-based Polymers

Another aspect of ion pairing to lower ΔG_{ap} is lowering the temperature of the inverse temperature transition. This means that the oppositely charged polymer can be soluble, because the onset temperature for the transition, T_t, is above physiological temperature. The effect of ion pairing with the pharmaceutical, as part of lowering ΔG_{ap}, is to lower the value of T_t such that ion pairing drives the phase separation.

Therefore, the phase-separated state becomes the drug delivery vehicle, and one of such a nature that on release of the drug the vehicle itself disperses. This means for a constant surface area that there occurs a zero order release of drug, as shown in Figure E.7 for positively charged Leu-enkephalin amide ion pairing with carboxylate-containing protein-based polymers. By polymer design the number of ion pairing sites determines the density of drug in the drug delivery vehicle, and the level of drug release is inversely proportional to the intensity and number of hydrophobic groups.

Thus, as shown in Figure E.7, a remarkable control of pharmaceutical release is possible. Furthermore, the pharmaceutical can vary all the way from a simple bare cation or anion to a protein or nucleic acid. Under favorable circumstances as with carboxylate groups, the vehicle disappears as the pharmaceutical releases. With a cationic polymer such as a lysine-containing protein-based polymer, the chloride ion can displace the pharmaceutical to lessen the zero order release. Significantly, with elastic protein-based polymers, no fibrous capsule forms around the adequately purified polymer such that this does not affect the release process.

This enabling triumvirate—knowledge of the vital forces (e.g., ΔG_{ap}), the capacity for precise control of desired sequence, and extraordinary biocompatibility—provides controlled release of pharmaceuticals to an extent that has only now become possible.

E.5.2.4 The Central Role of ΔG_{ap} for Production of Fibers with Superior Physical Properties

The issue of communication between molecules enters into the consideration of "order out of chaos" as discussed below. The specific aspect of communication between molecules here involves the association of complementary chains to achieve aligned and cross-linked polymers. One area where this has application is in fiber formation. In this case, one chain mole-

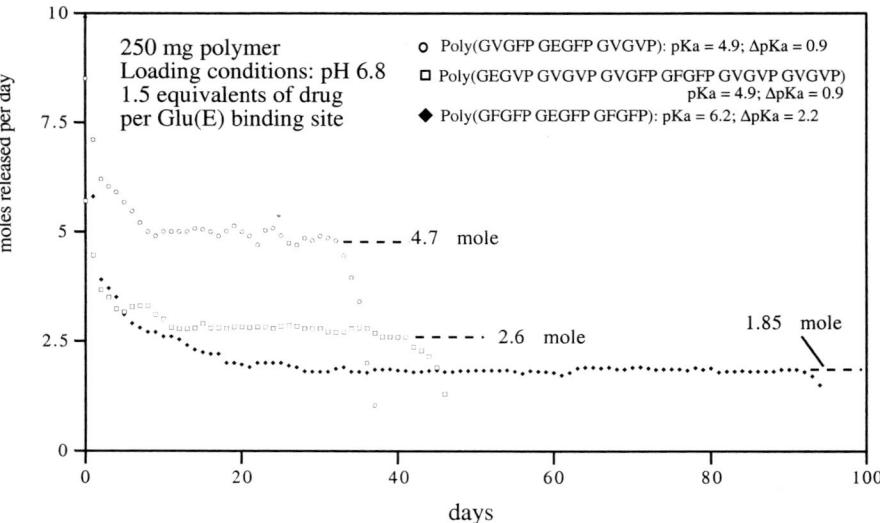

FIGURE E.7. Controlled release of positively charged Leu-enkephalin amide (H^+-Tyr-Gly-Gly-Phe-Leu-NH_2, LEA^+) from negatively charged polymers with systematically increased hydrophobicity with ΔpKa values of 0.9, 1.3, and 2.2, giving ΔG_{ap} values of 1.27, 1.84, and 3.12 kcal/mole-E, respectively. Constant surface area maintains zero order release levels for each polymer and 1.85 µmole/day out to 3 months for a ΔG_{ap} of 3.12 kcal/mole-E. Initial burst release due to 50% excess drug over polymer sites. (Replotting of Figure 9.38.)

cules is designed with a repeating negative charge, for example, a carboxylate, and the other with the same periodicity of a positively charged amino group. When neither of these sequences has hydrophobic induced pKa values, the affinity in water is minimal. However, when one or both of the complementary molecules has hydrophobic induced pKa shifts, that is, a significant ΔG_{ap} within a chain molecule, then there exists a driving force for ion pairing of carboxylate with amino. Ion-paired groups, carboxylate ($-COO^-$) with amino ($-NH_3^+$), are in position for chemical cross-linking to form an amide cross-link.

Knowledge of the vital force, ΔG_{ap}, and the capacity for precise control of sequence allows for the formation of fibers with improved elastic moduli, break strains, and stresses, as shown in Figure 9.20. A number of medical uses for such fibers include sutures and woven meshes for vascular prostheses and hernial repair. Again, *the enabling triumvirate—knowledge of the vital forces (e.g., ΔG_{ap}), the capacity for precise control of desired sequence, and extraordinary biocompatibility—provide additional medical devices of great potential for improving health care.*

E.5.3 Nonmedical (and Medical) Devices Using Biology's Mechanisms and Materials

The enabling triumvirate provides the antithesis to Pandora's box. Instead of being a prolific source of troubles, the enabling triumvirate provides a prolific source of opportunities with which to overcome today's troubles. Even at this early stage, with all that is already underway and that which is just emerging, it becomes difficult to choose the applications to emphasize here. The choice makes note of nanosensors and uses of mechanical resonances as representative of exciting and emerging possibilities. Mechanical resonances contribute to applications from the relatively mundane of sound absorption for civilian and naval applications to the exotic opportunity of nanosensors with potential use in the field to detect terrorist chemical and biological agents. The future can be one of nanosensors with the

E.5 Extraordinary Biomaterials Opportunities Arise

famed selectivity of enzymes coupled with the sensitivity to detect a single molecular event. The latter combines fundamental elements of the atomic force microscope with the principles arising out of the hydrophobic and elastic consilient mechanisms.

E.5.3.1 Sound Absorption for Civilian and Naval Applications

To the best of my knowledge, the data in Figure 7A of Appendix 1[3] show 20 Mrad γ-irradiation cross-linked (GVGIP)$_{260}$ to exhibit the most extraordinary acoustic absorption yet observed. This result is corroborated by the loss shear modulus and loss permittivity data of Figure 9.15, which in addition shows that the absorption intensity in the acoustic range would become twice again as large at body temperature. By varying the hydrophobicity, the onset temperature may be placed as required for the application, and the full absorption capacity develops about 20°C above the onset temperature. On appropriately increasing the hydrophobicity of the elastic protein-based polymer, the acoustic absorption capacity of the polymer could reach a maximum at a lower temperature, for example, 10°C (41°F), and maintain that intensity up to 60°C (140°F). Accordingly, as noted in Chapter 9, section 9.2.1.4, one designs the protein-based polymer required for the specific application.

E.5.3.2 Biosensors: Nanomachines and Nanosensors

E.5.3.2.1 Components of a Bioelastic Nanosensor

Two basic components comprise the nanosensor of interest here: (1) the atomic force microscope (AFM) designed to measure force as a function of extension or further modified to assess frequency dependence of loss shear modulus under isometric conditions and (2) the sensing element of elastic protein-based polymer containing a site or sites of interaction wherein the interaction changes the state of hydrophobic association by means of the apolar–polar repulsive free energy of hydration, ΔG_{ap}.

E.5.3.2.2 Phosphorylation/Dephosphorylation, a Most Effective, Selective, and Useful Change to Detect

Of the many polar functional groups of biology, phosphate is the chemical group that emanates the largest apolar–polar repulsion, ΔG_{ap}, that is, it is the most effective group for disruption of hydrophobic association. In particular, the binding of a single phosphate can disrupt the hydrophobic association holding part or all of a globular protein together.

Enzymes, called *kinases*, phosphorylate the –OH function of a serine (Ser, S), threonine (Thr, T) or tyrosine (Tyr, Y) residue that resides in a specific protein sequence. A specific protein kinase phosphorylates a specific sequence. The cardiac cyclic AMP–dependent protein kinase selectively phosphorylates the underscored serine, S, residue of the sequence RGYSLG, and protein tyrosine kinase (PTK) selectively phosphorylates the internal underscored tyrosine, Y, residue in the sequence GIYWHHY and is specific for that sequence. In cancerous and other diseased states, the abnormal activities of kinases can be diagnostic of a number of diseased conditions (e.g., cancerous tumors and Alzheimer's disease).

E.5.3.2.3 An AFM Force-extension Design of a Bioelastic Nanosensor for Kinase Assays

When a single chain of an elastic protein-based polymer is strung between AFM cantilever tip and substrate, a simple monotonic force-extension curve is obtained.[4] On the other hand, when the single chain comprises a series of globular protein elements strung between cantilever tip and substrate, a sawtooth pattern is obtained with a tooth for every globular element. Thus our fundamental sensing element becomes a single globular protein component inserted between two elastic protein-based polymer sequences. The force-extension profile for this construct exhibits a single tooth, as in Figure 9.55A. The globular protein component is designed with a kinase site, such that phosphorylation alters or obliterates the single tooth in the force-extension profile. Thus, the result is the detection of a single molecular interaction.

Developing a statistical treatment of a set of such curves provides a measure of the kinase activity. A series of globular sensing components in the single chain would facilitate the required statistical treatment for relative activity determination. A set of kinase selective sensing elements for a number of different kinases would allow evaluation of the relative kinase activities in a simple biopsy.

E.5.3.2.4 An AFM Isometric, Loss Shear Modulus Design of a Bioelastic Nanosensor for Kinase Assays

A limitation in the AFM force-extension design of a bioelastic nanosensor for kinase assays is that it requires a special vibration free mount for the AFM apparatus because of the need to minimize the noise in the force-extension curve in order to detect the small force values. A design that did not measure force versus extension, but rather were carried out at fixed extension, that is, isometrically, would be more deployable into the field for detection use. Vlado Hlady (private communication) has made such a design wherein vibration of the cantilever and analysis of the damping of the vibration as a function of frequencies below the input frequency would be expected to produce an acoustic absorption spectrum not unlike that in Figure 9.55B. In this case, in order to have a suitable loss shear modulus detected by cantilever vibration at a modest fixed extension, the chain of a triple-stranded twisted filament state of the elastic protein-based polymer would be used as depicted in Figure 9.57. In one extension setting of the sensing filament, the acoustic absorption (loss shear modulus), would be expected to decrease on phosphorylation as the unfolding of the globular protein would effectively open the bridge between cantilever tip and substrate, as suggested in Figure 9.55B.

E.5.3.2.5 Approach to Field Detection of Terrorist Chemical and Biological Agents Potential for Detection of Single Molecules of Toxins, Nerve Gas, Explosives, TNT, and DNT

In this case the globular component would be developed with the specificity, for which proteins are renowned, for interaction with the molecular species to be detected. For example, the globular component could derive from the antibody to the toxin or to the antigen of a particular organism. The globular sensing component could also be a protein with a site for specific binding of the nerve gas or explosive.

The intent of this section on using biology's own mechanisms and materials in the design of nanosensors is to provide a glimpse of the potential that lies in the future for the ultimate in sensitivity, selectivity, and diversity for detection of important analytes.

E.6 Living Organisms Create Order Out of Chaos, But by What Process?

E.6.1 A Personal Journey of Enlightenment and Enchantment

The author's personal journey of enlightenment and enchantment began with the inverse temperature transition exhibited by repeating sequences of the mammalian elastic fiber. It began with an inverse phase transition to greater order of a protein-based machine in water on raising the temperature rather than the usual transition to decreased order on increasing temperature, as in the melting and vaporization transitions of water. The journey was buoyed by demonstration that this unique phase transition could be utilized to interconvert those energies interconverted by living organisms. Unmodified, the initial sequence is capable of thermo-mechanical transduction. By the replacement of as few as 1 amino acid residue in a sequence of 50 residues or even in a sequence of 100 residues, a new energy source, such as chemical energy, could be accessed. Significantly, the chosen model protein sequence exhibits the fundamental elements required for efficient function of protein-based machines; these are near-ideal elasticity and hydrophobic association. Hydrophobic association is, of course, the hallmark of the inverse temperature transition.

The phenomena exhibited by the designed elastic-contractile model proteins were categorized by a set of five Axioms from the perspective of using the inverse temperature transition for energy conversion. The physical properties of these designed elastic-contractile model proteins were extensively characterized with many experiments and corresponding data analyses that ultimately brought us to the hydrophobic and elastic consilient mechanisms. During the data analysis, a coherence of phenomena with the behavior of natural proteins was increasingly evident at levels ranging from the gross anatomical to the molecular. That coherence of phenomena provided the impetus for a major focus of this book, which is the central role of hydrophobic and elastic consilient mechanisms in protein-based machines discussed in Chapters 7 and 8 and as briefly reviewed above in section E.4.

E.6.2 A Concluding Contrast of Means of Achieving Order Out of Chaos

This Epilogue began with the thesis that biology's "vital force" arises from coupled hydrophobic and elastic consilient mechanisms. In pursuing this thesis, we initially considered the profound perspectives of renowned physicist and founder of the wave mechanics formulation of quantum mechanics, E. Schrödinger, as put forward in his memorable book, *What is Life?* Now, in completing the thesis of this Epilogue, which centers on the role of inverse temperature transitions in biology, we consider the influential insights of Ilya Prigogine, who relates the ordering process of biology to his renowned work on nonequilibrium thermodynamics, under which conditions more-ordered systems arise spontaneously from less-ordered systems.

The purpose here is to contrast the author's perspective of the dogged manner in which living organisms create relatively long-lived order out of chaos from the relatively more spontaneous transient creations of order arising out of the Prigogine focus on nonequilibrium thermodynamics. This brings us to consideration of the arrow of time and of evolution and natural selection.

E.6.3 Point and Counterpoint of Two Diametrically Opposed Approaches to Order Out of Chaos

Point: A highly energized disordered state, far-from-equilibrium, spontaneously gives rise to order (spontaneous and relatively unstable)

Counterpoint: Order by the sum of thousands of small reversible-made-irreversible energy steps: a protracted step-by-step process to reach a condition describable as remarkably structured (non-spontaneous and relatively stable)

Point 1: In emphasizing the relevance of far-from-equilibrium conditions to biology, Prigogine and Stengers[34] state, "We can speak of a new coherence, of a mechanism of 'communication' among molecules. But this type of communication can arise only in far-from-equilibrium conditions. It is quite interesting that such communication seems to be the rule in the world of biology. It may in fact be taken as the very basis of the definition of a biological system."

In introducing the sense of dissipative structures, Prigogine and Stengers[34] state, "In far-from-equilibrium conditions we may have transformation from disorder, from thermal chaos, into order. New dynamic states of matter may originate, states that reflect the interaction of a given system with its surroundings. We have called these new structures *dissipative structures* to emphasize the constructive role of dissipative processes in their formation."

Counterpoint 1: As considered in detail in Chapter 4, in the biosynthesis of biology's great macromolecules, proteins and nucleic acids (DNA and RNA), each residue or base addition occurs by a reversible reaction in which the equilibrium constant is essentially one, but with a pyrophosphate reaction side product. On enzymatic breakdown of the pyrophosphate to two molecules of phosphate by the ubiquitous enzyme, pyrophosphatase, the reaction becomes irreversible at the expense of an additional 8 kcal/mole. Thus, in the process of construction of a long chain molecule, living organisms extravagantly expend energy in "irreversibly" creating their macromolecular structures that contain hundreds and even thousands of amino acid residues or bases in a single chain.

Because of the large amount of energy consumed in its construction, a protein, for example, may reasonably be *considered a dissipative structure*. However, a single protein chain is not created in a singular momentary transition involving a large amount of energy. Instead, a single protein chain comes into being step by step, adding one residue at a time, by utilizing energy in hundreds and even thousands of steps of individual small 8 kcal bits. Once formed the protein can remain intact for centuries. Even an organism, composed of tens of thousands of chain macromolecules, can remain dormant as seeds or spores for centuries only to spring to life on addition of thermal and/or chemical energy such as the chemical water and/or an increase in temperature.

Thus, biology produces its great macromolecules by means of an energetically extravagant, step-by-step, methodical march out of chaos.

Point 2: "On the other hand, far from equilibrium there appears a variety of mechanisms corresponding to the possibility of occurrence of various types of dissipative structures. For example, far from equilibrium we ... may also have processes of self-organization leading to nonhomogeneous structures to nonequilibrium crystals."[34]

Counterpoint 2: In fact, living organisms commonly create pairs of complementary protein structures that, when separately dissolved in solution, do not show the usual thermal or motional signs of being energized. Instead the protein structures can exhibit restricted freedom of motion due to repulsions between groups constrained to coexist by sequence. However, they are nonetheless energized by an intramolecular apolar–polar repulsion, the **ΔG_{ap}**, between the hydrophobic and charged residues constrained to coexist by sequence. On combining the solutions of complementary, sequence-energized chains, the complementary proteins communicate; they spontaneously and stoichiometrically associate into ordered, stable, and homogeneous structures at equilibrium. This can occur spontaneously on mixing when above T_t for the mixture, or it can occur on raising the lowered temperature when below T_t for the mixture. The latter occurs with the uptake (or assimilation) of heat energy into the system, not with its dissipation or dissimilation.

Communication among molecules may arise in biology from the complementarity of structures, as demonstrated with the designed E/nF and K/nF series of protein-based polymers. These structures are defined in Table 6.1 and characterized in Figures 6.3 and 6.4. Communication between designed molecules results from a lowering of the free energy on alignment due to a decrease of the **ΔG_{ap}** that resides within each separate molecule. In practice, this translates into the capacity to design highly entropic elastic fibers of protein-based polymers with an increase of one to two orders of magnitude in elastic moduli and break stresses and with improved break strains of 300%. The repulsive **ΔG_{ap}** within each complementary chain molecule becomes relieved on alignment by means of ion pairing of negative carboxylate with positive amino side chains that recur with the same periodicity. More regular cross-linking results in the improved materials properties (see Chapter 9, section 9.4.3.1, and Figure 9.20).

This biology-based, actually hydrophobic consilient mechanism–based, communication between molecules leads to self-organization into homogeneous structures at equilibrium rather than "to nonhomogeneous structures to nonequilibrium crystals." Thus, it seems to your author that living organisms create order out of chaos by a process distinct from that arising out of the Progogine analyses of nonequilibrium thermodynamics. Expanding our understanding of biology's creation of ordered macromolecules out of the chaos of mixtures of amino acids and of bases provides insight into just how biology reverses the universal arrow of time. This aspect that leads to evolution and natural selection was treated in Chapter 6 and is abstracted in what follows.

E.7 Evolution by Inverse Temperature Transition Reverses The Otherwise Universal Arrow of Time

E.7.1 Biology's Exception to the "Times Arrow" for the Universe

E.7.1.1 Eddington's Introduction of the Concept of "Times Arrow"

Eddington[35] underscores his high regard for the second law of thermodynamics in his view that "The law that entropy always increases—the second law of thermodynamics—holds, I think, the supreme position among the laws of Nature." As an increase in entropy measures the increase in disorder, that is, the increase in randomness, Eddington[36] developed the concept of "times arrow," with the following thought:

Let us draw an arrow arbitrarily. If as we follow the arrow we find more and more of the random element in the state of the world, then the arrow is pointing towards the future; if the random element decreases the arrow points toward the past. That is the only distinction known to physics. I shall use the phrase "times arrow" to express this one-way property of time which has no analogue in space. It is a singularly interesting property from a philosophical standpoint.

In short, as time progresses the universe becomes more disordered.

Thus, we have the dilemma that the arrow of time for the universe points toward increasing disorder and uniformity, whereas the arrow of time for biology points toward greater order and diversity.

E.7.1.2 An Elegant Statement of the Dilemma by Toffler in the Foreword to Prigogine and Stengers' Book

Imagine the problems introduced by Darwin and his followers! For evolution, far from pointing toward reduced organization and diversity, points in the opposite direction. Evolution proceeds from simple to complex, from "lower" to "higher" forms of Life, from undifferentiated to differentiated structures. And, from a human point of view, all is quite optimistic. The (biological) universe gets "better" organized as it ages, continually advancing to a higher level as time sweeps by.[37]

How can this be? We addressed this to some extent when considering the process whereby living organisms create order out of chaos. However, biology's reversal of the arrow of time derives from yet deeper roots.

E.7.2 Evolution by Inverse Temperature Transition Achieves Biology's Reversal of the Universal Arrow of Time

Biology's reversal of the much-noted arrow of time and equivalently biological evolution derive simply from fundamental reality of biosynthesis within the context of inverse temperature transitions as expressed in the hydrophobic consilient mechanism. The production of a new and improved protein-based machine occurs by chance, but most significantly it occurs at a cost in energy no greater than that required to produce the initial less useful protein-based machine. This is the nature of the biosynthesis of protein and of the other great macromolecules (the nucleic acids, DNA and RNA) of biology.

Remarkably, for protein-based machines that function by means of the hydrophobic consilient mechanism, the structure of the genetic code is such that a single base mutation can produce a protein-based machine capable of accessing a new energy source, and another single base mutation can produce a more efficient protein-based machine. This has been experimentally demonstrated by the design of elastic contractile model proteins that access new sources of energy. For example, the change of a single Val residue to a glutamic acid residue in a 100 residue sequence (possible by a single base change of thymine to adenine) can add the capacity of chemo-mechanical transduction to the model protein known for thermo-mechanical transduction (see Chapter 6).

E.7.2.1 The Genetic Code's Bias for Use by the Inverse Temperature Transition

The fundamental residues of the inverse temperature transition are hydrophobic, and there is a unique family of the genetic code that is entirely hydrophobic. As shown in Table 6.2, the hydrophobic family is the one with U as the second base of the triplet codon, namely, XUY. Any of the four bases of RNA—A (adenine), U (uracil), G (guanine), and C (cytidine)—can be in positions 1 and 3 of the triplet code, but U is always found as the second base of the codon for the hydrophobic family. A change in position 1 and position 3 simply allows selection of one member of the hydrophobic family from another. Position 3 in the cases of Leu (CUY) and Val (GUY) gives redundancy, that is, four different codons encode for Leu as do four different codons encode for Val. Therefore, it is said that both Val and Leu have a redundancy of four. For the Leu and Val residues, a change in the base in position 3 does not change the amino acid residue.

E.7.2.2 The Most Fundamental Step of Evolution–Natural Selection Is to Access a New Energy Source at No Additional Cost in Energy by the Biased Genetic Code

Importantly for the evolution of protein-based machines, a single base change in position 2 of the four codons for Val accesses the very important negatively charged amino acids, and a single base change in two of the codons for Leu access the also important positively charged Lys residue. The introduction of a single charged amino acid residue accesses chemical energy inputs for chemo-mechanical transduction, and if one allows for ion pairing to redox groups, it also allows for electrical energy input for electro-chemical transduction. A single base change converts the fundamental thermo-mechanical transduction of the inverse temperature transition to chemo-mechanical transduction and also to electro-mechanical transduction.

As discussed regarding biosynthesis in Chapter 4, the cost in energy, in moles of ATP, to produce a protein of a given length is the same regardless of composition. Therefore, the single base mutation required to access chemical energy occurs at no additional cost to the organism. In the process of natural selection, therefore, the organism that can access a new energy source will more effectively compete and will survive under conditions of stress whereas the less capable organism will not.

When protein-based machines function by the hydrophobic consilient mechanism, the structure of the genetic code is ideally suited for the evolution of protein-based machines to access new forms of energy.

E.7.2.3 A Refinement Step of Evolution–Natural Selection: To Produce a More Efficient Protein-based Machine at No Extra Cost Using the Biased Genetic Code

As discussed in Chapter 5 and specifically considered in Chapter 6 in relation to evolution, the replacement of a Val residue by a phenylalanine (Phe, F) residue results in a more efficient protein-based machine for chemo-mechanical transduction or for electro-mechanical transduction (see Figures 5.20, 5.34, and 5.36 and the associated discussions in Chapter 5). The mutation from Val to Phe occurs with a single base change at the DNA level of guanine to thymine. As listed in Table 6.2, a single base change in position 1 can access a more hydrophobic residue, as in Val to Phe, to provide a more efficient protein-based machine.

Again, as discussed regarding protein synthesis in Chapter 4, the cost in energy, in moles of ATP, to produce a protein of a given length is the same regardless of composition. Therefore, the single base mutation required to produce a more efficient protein-based machine occurs at no additional cost to the organism. By means of natural selection, therefore, the organism with the more efficient machine will compete more effectively and will survive under conditions of stress, whereas the less capable organism will not.

When protein-based machines function by the hydrophobic consilient mechanism, the genetic code is ideally suited for their evolution toward new machines with greater efficiency.

Thus, from the perspective of the inverse temperature transition, evolution and natural selection become apparent consequences for protein-based machines that function by the hydrophobic and elastic consilient mechanisms.

E.8 Maintaining the Admonition of Bacon

This volume represents a sincere attempt, backed by extensive commitment and effort, to overcome some of the current "idols which beset men's minds" in the area of protein-based machines and materials. More broadly, it opens the door for a new look at the physical basis whereby amphiphilic polymers function in water and presents the basis with which to engineer protein-based polymers as has not previously been possible. The primary purpose is to extend knowledge of the forces that provide function for protein-based machines and materials and thereby to achieve a greater comprehension of living organisms and to develop new materials for the benefit of society.

We hope that this volume may provide a milestone in the pathway to greater understanding of what sustains Life and a milestone in the biomaterials renaissance, but that it do so without itself becoming an idol that stands as a barrier to further progress. Whatever step has been taken, it should be viewed with adherence to the four-century-old admonition of Sir Francis Bacon. Thus we end this volume as Bacon in 1620 began his Preface to *The New Organon or True Directions Concerning the Interpretation of Nature*: "Those who have taken upon them to lay down the law of nature as a thing already searched out and understood, whether they have spoken in simple assurance or professional affectation, have therein done philosophy and the sciences great injury. For as they have been successful in inducing belief, so they have been effective in quenching and stopping inquiry; and have done more harm by spoiling and putting an end to other men's efforts than good by their own."*

References

1. By definition in *Webster's Dictionary*, vital is "concerned with or necessary for the maintenance of life" and "performing an essential role in the living body." As given in the *Britannica* 2002, the sense of *élan vital*, "the vital force or impulse of life," is a creative principle of Henri Bergson "immanent in all organisms and responsible for evolution."
2. E.O. Wilson, *Consilience: The Unity of Knowledge*. Alfred E. Knopf, New York, 1998, p. 8.
3. D.W. Urry and T.M. Parker, "Mechanics of Elastin: Molecular Mechanism of Biological Elasticity and its Relevance to Contraction." *J. Muscle Res. Cell Motil.*, **23**, 541–547, 2002.
4. D.W. Urry, T. Hugel, M. Seitz, H. Gaub, L. Sheiba, J. Dea, J. Xu, and T. Parker, "Elastin: A Representative Ideal Protein Elastomer." *Philos. Trans. R. Soc. Lond.*, B **357**, 169–184, 2002.
5. D.W. Urry, T. Hugel, M. Seitz, H. Gaub, L. Sheiba, J. Dea, J. Xu, L. Hayes, F. Prochazka, and T. Parker, "Ideal Protein Elasticity: The Elastin Model." In *"Elastomeric Proteins: Structures, Biomechanical Properties and Biological Roles."* P.R. Shewry, A.S. Tatham, and A.J. Bailey, Eds., Cambridge University Press, The Royal Society; Chapter Four, pages 54–93, 2003.

---

*With the remarkable progress that has been made in the descriptive understanding of biology and biology's great macromolecules and of polymers in general, this admonition of four centuries ago may seem unwarranted. Nor should the reprove be laid at the feet of the grand pioneers of molecular mechanics and molecular dynamics computations such as Harold Scheraga, Martin Karplus, Peter Kollman, Micheal Levitt, Paul Flory, and Henry Eyring, for they have taken up the ultimate challenge and hewn the path toward that definitive goal. They warrant high praise and gratitude for advancing and providing their methodologies. Yet wanting in all cases, for example, remains the proper treatment of the "waters of Thales."

In our view, it is those who use yet-incomplete theoretical treatments and from this stage dictate absolute conclusions, even though deficient when compared with experimental data, who remain capable of the injury of which Bacon wrote. As noted earlier in this Epilogue, the recognition of present impediments arising from interpretations of such practitioners is repeated here: *Useful approximations of the past become "idols" in the present that stand as barriers to the future.*

6. Z.R. Wasserman and F.R. Salemme, "A Molecular Dynamics Investigation of the Elastomeric Restoring Force in Elastin." *Biopolymers*, **29**, 1613–1631, 1990.
7. E. Schrödinger, *What is Life?: The Physical Aspect of the Living Cell with Mind and Matter and Autobiographical Sketches.* Cambridge University Press, 1944 (1967 edition, p. 68).
8. I. Prigogine and I. Stengers, *Order Out of Chaos: Man's New Dialogue with Nature.* Bantam Book, New York, 1984, p. 103.
9. K. Kinosita, R. Yasuda and H. Noji, "F_1-ATPase: A Highly Efficient Rotary ATP Machine." *Essays Biochem. Mol. Motors*, **35**, 3–18, 2000.
10. The capacity to have any 1 of 20 different amino acid residues at each position in a protein sequence with physicochemical properties spanning from very oil-like to vinegar-like, as demonstrated in Figure 2.1, and the simultaneous capacity to retain complete integrity with respect to molecular asymmetry (see Chapter 3, section 3.1.6 and Figure 3.3) combine to result in the remarkable capabilities of protein sequences. Lest one gets lost in the sense of improbability of such sequences, however, their biosynthesis requires such an extraordinary amount of energy in total number of ATP molecules consumed for the addition of each amino acid residue that their construction is more aptly described as blatantly extravagant. This constitutes the central message of Chapter 4.
11. E. Schrödinger, *What is Life?: The Physical Aspect of the Living Cell with Mind and Matter and Autobiographical Sketches.* Cambridge University Press, 1944, (1967 edition, p. 76).
12. M.V. Stackelberg and H.R. Müller, "Zur Struktur der Gashydrate." *Naturwissenschaften*, **38**, 456, 1951.
13. J.A.V. Butler, "The Energy and Entropy of Hydration of Organic Compounds." *Trans. Faraday Soc.*, **33**, 229–238, 1937.
14. I. Prigogine, *Order Out of Chaos: Man's New Dialogue with Nature.* Bantam Book, New York, 1984, p. 122.
15. P. J. Flory, *Principles of Polymer Chemistry.* Cornell University Press, Ithaca, NY, 1953.
16. P.J. Flory, "Molecular Interpretation of Rubber Elasticity." *Rubber Chem. Technol.*, **41**, G41–G48, 1968.
17. P. Zagorin, *Francis Bacon.* Princeton University Press, 1998, p. 33.
18. D.J. Boorstin, *The Seekers: The Story of Man's Continuing Quest to Understand his World.* Random House, New York, 1998, p. 133.
19. P. Zagorin, *Francis Bacon.* Princeton University Press, 1998, p. 82.
20. D.W. Urry, J. Xu, W. Wang, L. Hayes, F. Prochazka, and T.M. Parker. "Development of Elastic Protein-based Polymers as Materials for Acoustic Absorption." *Mater. Res. Soc. Symp. Proc.*, **774**, 81–92, 2003.
21. D.W. Urry, M.M. Long, R.D. Harris, and K.U. Prasad, "Temperature Correlated Force and Structure Development in Elastomeric Polypeptides: The Ile[1] Analog of the Polypentapeptide of Elastin." *Biopolymers*, **25**, 1939–1953, 1986.
22. D.W. Urry, R.D. Harris, M.M. Long, and K.U. Prasad, "Polytetrapeptide of Elastin: Temperature Correlated Elastomeric Force and Structure Development." *Int. J. Peptide. Protein Res.*, **28**, 649–660, 1986.
23. D.W. Urry, "Physical Chemistry of Biological Free Energy Transduction as Demonstrated by Elastic Protein-based Polymers." *J. Phys. Chem. B*, **101**, 11007–11028, 1997.
24. C. Lange and C. Hunte, "Crystal Structure of the Yeast Cytochrome bc_1 Complex with its Bound Substrate Cytochrome c." *Proc. Natl. Acad. Sci. U.S.A.*, **99**, 2800–2805, 2002.
25. J.P. Abrahams, A.G.W. Leslie, R. Lutter, and J.E. Walker, "Structure at 2.8 Å of F_1-ATPase from Bovine Heart Mitochondria." *Nature (Lond.)*, **370**, 621–628, 1994.
26. R.I. Menz, J.E. Walker, and A.G.W Leslie, "Structure of Bovine Mitochondrial F_1-ATPase with Nucleotide Bound to all Three Catalytic Sites: Implications for Mechanism of Rotary Catalysis." *Cell*, **106**, 331–341, 2001.
27. H. Noji, R. Yasuda, M. Yoshida, and K. Kinosita, "Direct Observation of the Rotation of F_1-ATPase." *Nature (Lond.)*, **386**, 299–302, 1997.
28. D.M. Himmel, S. Gourinath, L. Reshetnikova, Y. Shen, A.G. Szent-Gyorgyi, and C. Cohen, "Crystallographic Findings on the Internally Uncoupled and Near-rigor States of Myosin: Further Insights into the Mechanics of the Motor." *Proc. Natl. Acad. Sci. U.S.A.*, **99**, 12645–12650, 2002.
29. I. Rayment, C. Smith, and R.G. Yount, "The Active Site of Myosin." *Annu. Rev. Physiol.*, **58**, 671–702, 1996.
30. Y. Lecarpentier, D. Chemla, J.C. Pourny, and F.-X. Blanc, "Myosin Cross Bridges in Skeletal Muscles: 'Rower' Molecular Motors." *J. Appl. Physiol.*, **91**, 2479–2486, 2001.
31. R.W. Lymn and E.W. Taylor, "Mechanism of ATP Hydrolysis by Actomyosin." *Biochemistry*, **10**, 4617–4624, 1971.

32. J. Bronowski, *The Ascent of Man*, Little, Brown and Company, Boston, 1973, p. 110.
33. Recall that the initial observation leading to the concept of mechanical resonances was the occurrence in the dielectric of an intense (70 Debye), localized (fitted by a single frequency in the Debye expression) relaxation at 5 MHz that developed as the temperature was raised through the inverse temperature transition. As ordering occurred in the protein-based polymer on raising the temperature through the range of the inverse temperature transition, a regular dynamic structure resulted wherein the repeating units developed a mechanical resonance that increased in intensity as the structure formed. Then, using direct acoustic absorption measurements, Lev Sheiba and Jack Dea at the Naval Research Laboratory of the Spatial Warfare Systems Center in San Diego observed the development of an intense acoustic absorption near 3 kHz with the same temperature dependence as exhibited by the same elastic protein-based polymer at 5 MHz in the dielectric relaxation. Their result was so contrary to the "idols of the present" that the result was called an *artifact*, even though under the same conditions, as controls, they had determined the usual pattern for the frequency dependence for sound absorption of natural rubber and polyurethane and despite the corroborating 5 MHz data in the dielectric relaxation. Wanting to check the veracity of this mechanical resonance in the acoustic absorption range, we determined both loss shear moduli and low-frequency dielectric relaxation data and confirmed the occurrence of the mechanical resonance in terms of an intense acoustic absorption near 3 kHz, as shown in Figure 9.15. Unexpected to this researcher has been the realization that discovery and innovation are far easier and less time consuming than overcoming the "idols which beset men's minds," to quote the patron saint of the scientific revolution.[18,19]
34. I. Prigogine and I. Stengers, *Order Out of Chaos: Man's New Dialogue with Nature*. Bantam Book, New York, 1984, pp. 12–13.
35. A.S. Eddington, *The Nature of the Physical World*. MacMillan, New York, 1948, p. 74.
36. A.S. Eddington, *The Nature of the Physical World*. Ann Arbor, University of Michigan Press, 1958, p. 69.
37. A. Toffler, "Forward." In I. Prigogine and I. Stengers, *Order Out of Chaos: Man's New Dialogue with Nature*. Bantam Books, New York, 1984, preface, p. xx.
38. D.W. Urry, "Function of the F_1-motor (F_1-ATPase) of ATP synthase by Apolar-polar Repulsion through Internal Interfacial Water." *Cell Biology International*, **30**, (1), 44–55, 2006.

Appendix 1
Mechanics of Elastin: Molecular Mechanism of Biological Elasticity and Its Relationship to Contraction

Dan W. URRY[1,2,*] and Timothy M. PARKER[2]

Abstract

Description of the mechanics of elastin requires the understanding of two interlinked but distinct physical processes: the development of entropic elastic force and the occurrence of hydrophobic association. Elementary statistical-mechanical analysis of AFM single-chain force–extension data of elastin model molecules identifies damping of internal chain dynamics on extension as a fundamental source of entropic elastic force and eliminates the requirement of random chain networks. For elastin and its models, this simple analysis is substantiated experimentally by the observation of mechanical resonances in the dielectric relaxation and acoustic absorption spectra, and theoretically by the dependence of entropy on frequency of torsion-angle oscillations, and by classical molecular-mechanics and dynamics calculations of relaxed and extended states of the β-spiral description of the elastin repeat, $(GVGVP)_n$. The role of hydrophobic hydration in the mechanics of elastin becomes apparent under conditions of isometric contraction.

During force development at constant length, increase in entropic elastic force resulting from decrease in elastomer entropy occurs under conditions of increase in solvent entropy. This eliminates the solvent entropy change as the entropy change that gives rise to entropic elastic force and couples association of hydrophobic domains to the process. Therefore, association of hydrophobic domains within the elastomer at fixed length stretches interconnecting dynamic chain segments and causes an increase in the entropic elastic force due to the resulting damping of internal chain dynamics. Fundamental to the mechanics of elastin is the inverse temperature transition of hydrophobic association that occurs with development of mechanical resonances within fibrous elastin and polymers of repeat elastin sequences, which, with design of truly minimal changes in sequence, demonstrate energy conversions extant in biology and demonstrate the special capacity of bound phosphates to raise the free energy of hydrophobic association.

Reprinted from *Journal of Muscle Research and Cell Motility*, **23**, 543–559, 2002. © 2003 Kluwer Academic Publishers.
[1]University of Minnesota, Twin Cities Campus, BioTechnology Institute, 1479 Gortner Avenue, St. Paul, MN 55108-6106; [2]Bioelastics Research Ltd., 2800 Milan Court, Suite 386, Birmingham, AL, 35211-6918, USA
*To whom correspondence should be addressed: E-mail: durry98@aol.com

Introduction

As a brief introduction to the mechanics of elastin, noted are the composition and sequence of bovine elastin, the inverse phase transition behavior of the elastin and related protein components, and the conformational aspects of the most representative repeating sequence of elastin.

Introduction

Composition and Sequence of Elastin

The single-residue T_t-based hydrophobicity plots of Figure 1 (Urry, 1997) provide a simple way to gain perspective of the composition and sequence of bovine elastin and to make comparisons with other extracellular proteins, for example human fibronectin. The longest repeating sequence $(GVGVP)_{11}$ is labeled W4 based on the nomenclature of the Sandberg group (Sandberg et al., 1981, 1985). Additional, less extensive repeats are also apparent. Of the nearly 40 lysine (Lys, K) residues, all but two or three are used in forming cross-links. Generally, four Lys residues come together to form a tetra-substituted pyridinium. By contrast, in the vulcanization of rubber every monomer is a potential cross-link, which allows for cross-link formation whenever two chains are in contact. In elastin only 5% of the residues participate in cross-link formation and these are often distributed along the chain in pairs. Such a situation requires a regular structure in order to bring this limited number of residues into adequate proximity for covalent cross-link formation.

There are only three negatively charged aspartic acid (Asp, D) and no glutamic acid

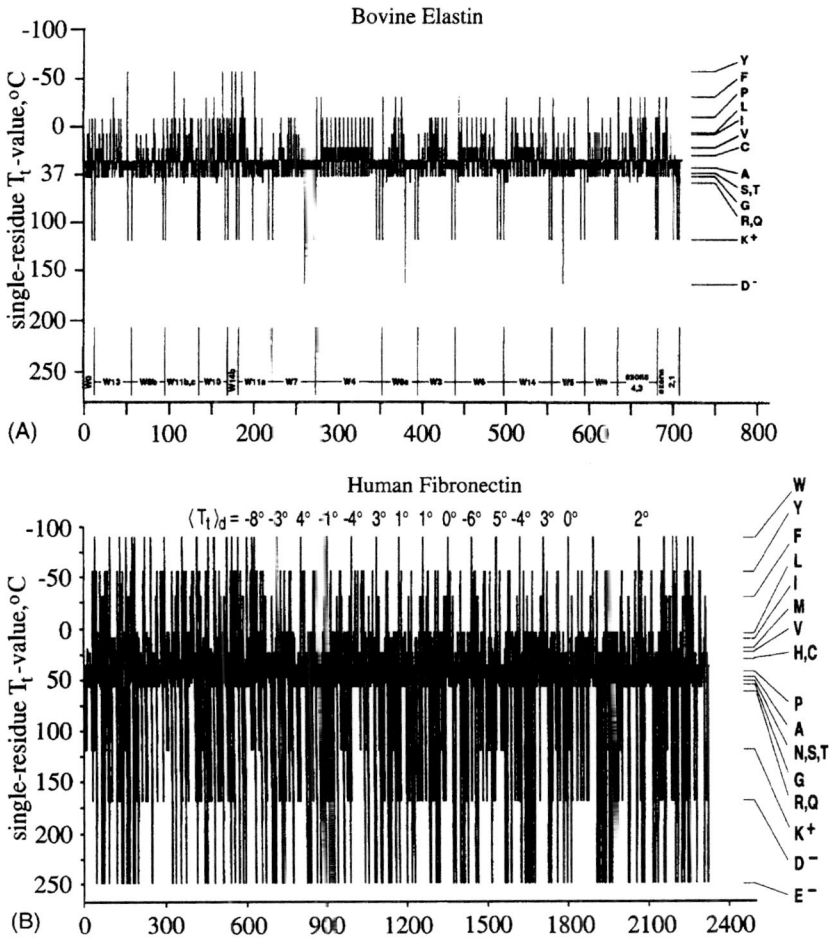

FIGURE 1. Single residue T_t-based hydrophobicity plots for bovine elastin (A) and human fibronectin (B) (see text for discussion). Part A reproduced with permission from Urry et al. (1995), Part B reproduced with permission from Urry and Luan (1995a).

residues in bovine elastin. The more hydrophobic phenylalanine (Phe, F) and tyrosine (Tyr, Y) residues are also apparent. The almost complete absence of charged groups combined with a substantial quantity of hydrophobic residues are responsible for the hydrophobic association between chains that result in parallel-aligned twisted filaments as seen in electron micrographs of negatively stained tropoelastin and α-elastin (Cox et al., 1973, 1974) and in fibrous elastin (Gotte et al., 1974).

By comparison, a sequence that results in a series of globular repeating units of hydrophobically folded β-barrels is seen in the single-residue T_t-based hydrophobicity plots for fibronectin in the lower part of Figure 1 (Kornblihtt et al., 1985). In this case there are many more charged Glu, Asp and Lys residues, and the periodicity of the repeating globular units becomes apparent through the repeating tryptophan (Trp, W) residues. This most hydrophobic residue, W, repeating approximately every 90 residues identifies type III domains of fibronectin, which also appear as a repeating unit in titin (connectin). Such repeating globular units give a sawtooth pattern to the single-chain force–extension curves of titin (Rief et al., 1997; Gaub and Fernandez, 1998), which is very different from the smooth single-chain force–extension curves given by elastin models (Urry et al., 2002a, b).

Inverse Temperature Transition of Elastin and Component Sequences

Increase in Order of Elastic Protein upon Raising the Temperature

Tropoelastin (precursor protein of fibrous elastin), α-elastin (chemical fragmentation product from fibrous elastin) and high-molecular-weight polymers of repeating sequences of elastin are water soluble at low temperatures, but produce phase separation upon raising the temperature. The initial aggregates of the phase transition, when placed on a carbon-coated EM grid and negatively stained with uranyl acetate/oxalic acid, are observed to form parallel aligned filaments comprised of twisted filaments with interesting periodicities in the optical diffraction patterns of the micrographs (Volpin et al., 1976a, b). Furthermore, cyclic analogues containing pentapeptide and hexapeptide repeats crystallize during temperature increase and redissolve upon lowering the temperature. Clearly, the order in elastin and elastin-based polymers increases with increasing the temperature.

This ordering of protein upon raising the temperature is consistent with the second law of thermodynamics: a structured hydration that surrounds the dissolved hydrophobic groups becomes less ordered bulk water during the phase separation of hydrophobic association (Urry et al., 1997a, b).

This phase transition leading to a fundamental increase in order of the polymer on raising the temperature is called an inverse temperature transition.

Phase Diagram

Phase diagrams of the inverse temperature-dependent behavior of elastin-based polymers is seen in Figure 2 (Sciortino et al., 1990, 1993).

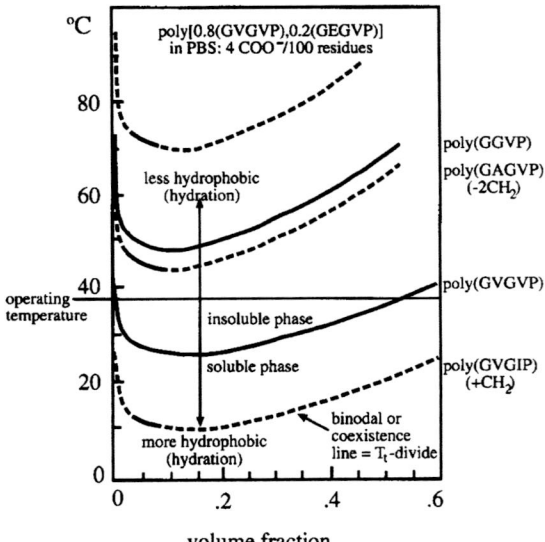

FIGURE 2. Schematic representation of phase diagrams for several model elastic proteins based on the elastin repeat, $(GVGVP)_n$ (see text for discussion). Solid curves adapted with permission from Sciortino et al. (1990) and (1993).

Introduction

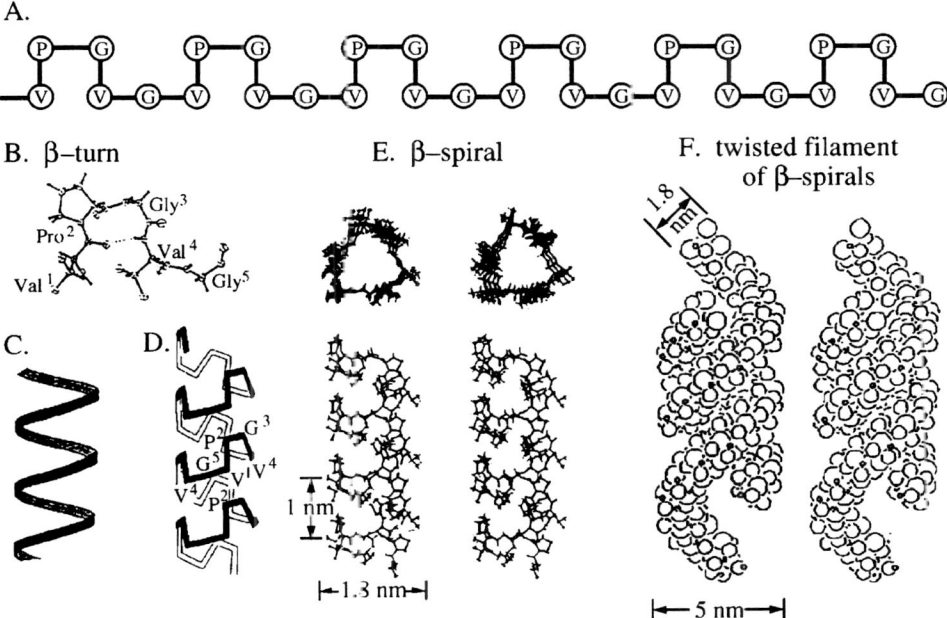

FIGURE 3. Representations of the proposed molecular structure of (GVGVP)$_n$ (see text for discussion).

In the usual polymer phase diagram the polymers are insoluble below and soluble above the binodal (coexistence or coacervate) line. For elastin-based polymers solubility is inverted. With elastin-based polymers solubility occurs below and insolubility occurs above the coexistence line. The shape of the coexistence line is also inverted: while for the usual petroleum-based polymers the line is concave towards the concentration (volume–fraction) axis, for elastin-related polymers it is convex. In case of the elastin-based polymers the coexistence line is also called the T_t-divide. Each point (called T_t) along the T_t-divide corresponds, for a given concentration, to the onset temperature of aggregation.

Elastin-based polymers exhibit a latent heat of transition. Accordingly, within the temperature range of the transition the chemical potential of the polymers in solution and in the phase-separated state are the same. This enables the derivation of the Gibbs free energy of hydrophobic association, ΔG_{HA} (see below). The more hydrophobic elastin-based polymers that associate at lower temperatures have lower values of ΔG_{HA}.

Conformational Aspects of the (GVGVP)$_n$ of Elastin

The structure of (GVGVP)$_n$, shown in Figure 3, was developed by using methodologies of peptide synthesis, NMR, molecular-mechanics and dynamics calculations (Venkatachalam and Urry, 1981; Chang et al., 1989), crystal structure of cyclic analogues (Cook et al., 1980) and the NMR- and computationally characterized relationship between cyclic and linear structures (Urry et al., 1981, 1989), and Raman, absorption and circular dichroism spectroscopies.

The structure involves a series of β-turns, ten-atom hydrogen-bonded rings from the Val^1 C—O to the Val^4 N—H (Figure 3A and B) which, upon raising the temperature, wrap up into a helical structure called the β-spiral (shown schematically in Figure 3C and D and in detail in Figure 3E). It is shown that the β-turns function as hydrophobic spacers between the turns of the β-spiral and that the β-spirals associate hydrophobically to form twisted filaments of the dimensions found in the transmission electron micrographs of negatively stained incipient aggregates as noted above (Volpin et al., 1976a, b). Hydrophobic folding of β-spirals and the hydrophobic association of β-spirals to form twisted filaments occur in a cooperative manner.

General Considerations of Entropic Elastic Force

Components of Elastic Force and Their Delineation

Internal Energy (f_E) and Entropy (f_S) Components of Elastic Force

The total elastic force (f) can be thermodynamically described as

$$f = (\partial E/\partial L)_{V,T,n} - T(\partial S/\partial L)_{V,T,n}, \quad (1)$$

where E is the internal energy; S the entropy; T the absolute temperature in °K, and V, T, and n indicate that the change in length (∂L) occurs at constant volume, temperature and composition. Accordingly, the total elastic force comprises two components, the internal energy component (f_E) and the entropic component (f_S):

$$f = f_E + f_S. \quad (2)$$

Experimental Delineation of the Relative Magnitudes of f_E and f_S

Following Flory et al. (1960) the total force can also be written as

$$f = (\partial E/\partial L)_{V,T,n} + T(\partial f/\partial T)_{V,L,n} \quad (3)$$

which is equivalent to

$$f_E/f = -T(\partial \ln[f/T]/\partial T)_{V,L,n}. \quad (4)$$

Equation (4) shows that the ratio of the internal energy component of force to the total force can be determined experimentally by plotting $\ln[f/T]$ as a function of temperature while maintaining the elastic element at constant volume, at fixed length and without a change in composition. Under these experimental conditions, the slope of the plot multiplied by $(-T°K)$ provides the f_E/f ratio.

Statistical Mechanical Expression for Entropy

The Boltzmann Relation

The Boltzmann relation provides the bridge from a statistical mechanical description of molecular structure to experimentally determined thermodynamic quantities. It is an elegant yet simple statement of entropy (S) in terms of thermodynamic probability (W, the number of a priori equally probable states accessible to the system) (Eyring et al., 1964), i.e.,

$$S = R \ln W. \quad (5)$$

R (1.987 cal/deg mol) is the gas constant. $R = Nk$, where N is Avogadro's number (6.02 × 10^{23}/mol) and k Boltzmann's constant (1.38 × 10^{-16} erg/deg K). W is a volume in phase space, a $2N$-dimensional space, which fundamentally provides a description of a molecule in terms of the momentum (p_i) and coordinate (q_i) of each its i atoms.

Description of Entropy Using Partition Functions for Different Degrees of Freedom

In practice, the description of the molecule is achieved by the product of partition functions, which group according to the degrees of freedom of the molecular system. In general, the $2N$ degrees of freedom group as three translational degrees of freedom, three rotational degrees of freedom and the remainder are internal degrees of freedom that comprise vibrations and torsional oscillations (rotations about bonds). In the measurement of elasticity, the single chain molecule or the cross-linked matrix is fixed at both ends. In this case, there are holonomic constraints on the molecular system in that there are neither whole-molecule translational nor rotational degrees of freedom. Thus, we are left only with internal chain dynamics.

Frequency Dependence of Entropy for the Harmonic Oscillator Representation of Internal Chain Dynamics

The harmonic oscillator partition function (f_v) provides a relatively simple and yet informative representation of internal chain dynamics.

$$f_v = [1 - \exp(-hv_i/kT)]^{-1}. \quad (6)$$

When placed in Equation (5) and expressed in terms of the frequency-dependence of entropy (S) we obtain (Dauber et al., 1981),

$$S_i = R\{\ln[1 - \exp(-hv_i/kT)]^{-1} + (hv_i/kT)[\exp(hv_i/kT) - 1]^{-1}\}. \quad (7)$$

Equation (7) is plotted in Figure 4 with S_i on the left-hand ordinate and with the free-energy contribution (TS_i) on the right-hand ordinate.

The interesting features of Figure 4 are that the vibrational modes occur at high frequencies where the contribution to entropy is small, and the torsional oscillations occur at much lower frequencies where the contribution to entropy is much larger. As discussed below, the mechanical resonances exhibited by elastin models and fibrous elastin occur at frequencies near 1 kHz and 5 MHz, i.e., near 3 and 6 on the abscissa of Figure 4. Accordingly, it becomes reasonable to neglect changes in vibrational frequencies on extension and to consider those configurational and dynamic changes due to changes in torsion angles.

Expression for the Change in Entropy on Extension

The change in entropy upon extension is written,

$$\Delta S = (S^e - S^r) = R\ln(W^e/W^r), \quad (8)$$

where e and r stand for the extended and relaxed states. With Equation (8) it becomes possible to use both molecular mechanics and dynamics calculations of the molecular structure described in Figure 3 to calculate the contribution of torsional freedom of the structure to entropy in both the relaxed and extended states.

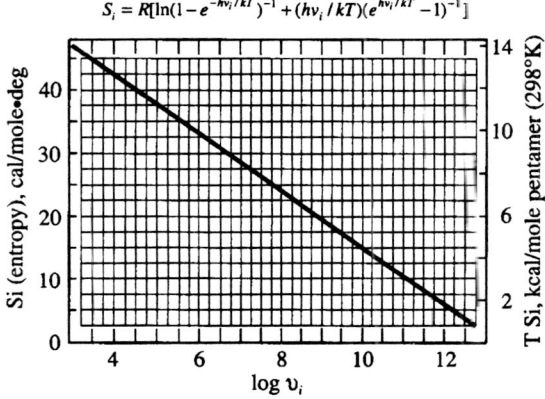

FIGURE 4. Plot, based on the harmonic oscillator partition function, of the entropy of the oscillator as a function of log(oscillator frequency). On the left-hand ordinate is plotted the entropy in cal/mol deg (EU, entropy units), and on the right-hand ordinate is plotted the entropic contribution to the Gibbs free energy in kcal/mol pentamer at 298°K. This illustrates the increasing contribution of low frequency oscillations to the entropy and free energy of a protein chain segment. Adapted with permission from Urry et al. (1988).

The Physical Basis for Entropic Elasticity in Elastin

In an ideal or perfect elastomer the energy repeatedly invested in extension is repeatedly and completely recovered during relaxation. Ideality increases as the elastic force results from a decrease in entropy upon extension, because this occurs without stressing bonds to the breaking point. Elastin models and elastin itself in water provide examples of such entropic elastomers with about 90% of the elastic force being entropic, that is, the f_E/f ratio of Equation (4) is about 0.1. This is essential to human life expectancy, because the half-life of elastin in the mammalian elastic fiber is on the order of 70 years. This means that the elastic fibers of the aortic arch and thoracic aorta, where there is twice as much elastin as collagen, will have survived some billion demanding stretch-relaxation cycles by the start of the seventh decade of life. This represents an ultimate in ideal elasticity.

Three Main Mechanisms Proposed for the Entropic Elasticity of Elastin

The Classical (Random Chain Network) Theory of Rubber Elasticity

In 1958 Hoeve and Flory reported studies on the nature of elasticity of the bovine fibrous protein, elastin, as representative of the mammalian elastic fiber. Because they were able to determine a very low f_E/f ratio, they concluded that the elastic fiber was without order, that is, it was a random chain network. They reaffirmed it in the spring of 1974 in a Biopolymers paper (Hoeve and Flory, 1974) stating that "A network of random chains within the elastic fibers, like that in a typical rubber, is clearly indicated." In order to emphasize this perspective further, the following statement appeared in the legend to Figure 1 of that paper: "Configurations of chains between cross-linkages are much more tortuous and irregular than shown." In part because Paul Flory received the Nobel Prize in 1974 for his work on macromolecules, this perspective became firmly set in the minds of the interested scientific community.

The Solvent (Bulk Water ↔ Hydrophobic Hydration) Entropy Mechanism

It appeared that the sole purpose of the Hoeve and Flory 1974 paper, which presented neither new experimental data nor theoretical analysis, was to refute the publication of a new mechanism. The new mechanism for the entropic elasticity of elastin was published by Weis-Fogh and Anderson (1970) with the title "New Molecular Model for the Long-range Elasticity of Elastin." The proposed new mechanism, stated in the terms of present adherents, was that upon extension hydrophobic side-chains become exposed to the water solvent. The result is formation of low-entropy water of hydrophobic hydration. Here we refer to this as the solvent (bulk water ↔ hydrophobic hydration) mechanism for entropic elasticity. This mechanism has been given credibility by groups using computational methods on $(GVGVP)_n$ in water (Wasserman and Salemme, 1990; Alonso et al., 2001). One of the many arguments marshaled by Hoeve and Flory (1974) against this mechanism was that polymer backbone, not solvent, must bear the force.

The Damping of Internal Chain Dynamics on Extension

In the hydrophobically folded β-spiral structure (Figure 3), the presence of suspended segments between the β-turns is immediately apparent. With water surrounding the suspended segments and no steric hindrance to limit rotation about backbone bonds of the suspended segments, the peptide moieties would be expected to rock or "librate". These suspended segments are probably free to undergo large torsion-angle oscillations, and the amplitude of these torsional oscillations might be damped during extension (Urry, 1982). We devised experimental (physical and chemical) and computational tests of the structure with a focus on the freedom of motion of the suspended segments. The chemical tests (replacement of the Gly residues with D- and L-alanines) resulted in the predicted limited torsional oscillations and elasticity (Urry et al., 1983a, b, 1984, 1991). The principal physical tests involved dielectric and NMR relaxation methodologies (Henze and Urry, 1985; Urry et al., 1985a, b, c, 1986; Buchet et al., 1988). Remarkably, in the dielectric relaxation characterization, where the only dipole moments were those of the backbone peptide moieties, intense localized relaxations developed as the model elastic protein underwent the inverse temperature (phase) transition. Furthermore, molecular mechanics and dynamics calculations demonstrated the decrease in amplitude of torsion-angle oscillations within the suspended segment during extension (Urry, 1982; Chang and Urry, 1989). Indeed, by treating the dynamics of the pentameric unit the oscillating dipole moment reproduced the dielectric relaxation results (Venkatachalam and Urry, 1986). Thus, the concept of the damping of internal chain dynamics by extension as a source of entropic elastic force became established (Chang and Urry, 1989). Key experimental and computational results are treated in more detail below.

Key Experimental Data Relevant to the Proposed Mechanisms

Atomic Force Microscopy (AFM) Single-Chain Force–Extension Results

The remarkable capacity to obtain single-chain force–extension curves provides new insights into the mechanism of entropic elasticity. Figure 5A shows a series of single-chain force–extension curves for a single molecule of Cys-(GVGVP)$_{n\times251}$-Cys (Urry et al., 2002a). Starting from the bottom, a series of extension–relaxation cycles are displayed. The second trace, during which the molecule was extended from an end-to-end distance of 200–700 nm and then returned to the starting length, demonstrates perfect reversibility at the rate of stretch and within the resolution of the technique. Since all of the energy of deformation was recovered during relaxation, we observe an ideal elastomer! For the third, fourth and sixth traces, a resting time of 30 s near 200 nm was allowed, and these traces demonstrated barely detectable imperfect overlap of extension and relaxation curves. The fifth trace without the 30 s resting time near 200 nm again exhibited perfect reversibility Finally on the seventh extension of the chain the continuity from substrate to cantilever tip became severed and the force dropped to zero just above 700 nm.

This slight hysteresis of traces one, three, four and six, may be due to backfolding of the chain on itself, interaction of a second chain picked up at low extension or possibly even non-specific adsorption of the single chain to surface at low extension and at a position along the chain of within 100 nm of extended length from the attachment point. Regardless of the source of the barely detectable non-overlap of several of the traces, ideal (entropic) elasticity has been

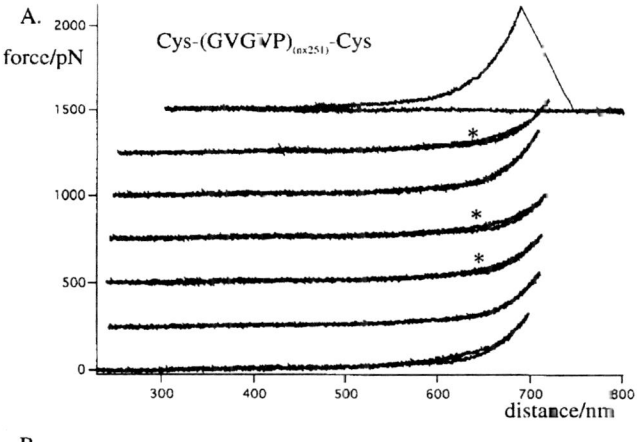

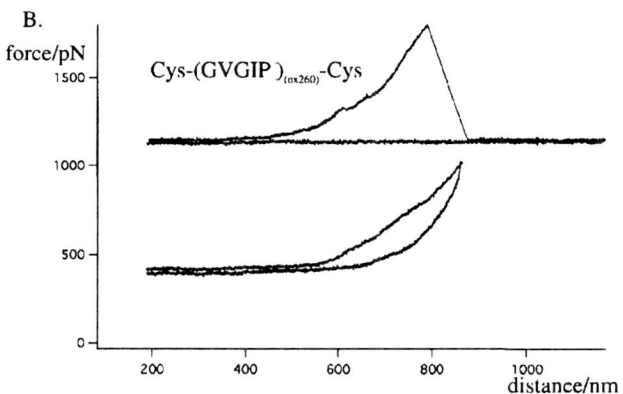

FIGURE 5. AFM single-chain force–extension curves of the model elastic proteins based on elastin, Cys-(GVGVP)$_{n\times251}$-Cys (A) and Cys-(GVGIP)$_{n\times260}$-Cys (B), showing the ideal elasticity due to perfect reversibility of traces 2 and 5 from bottom of A and marked hysteresis in the lower trace of B. See text for discussion. Reproduced with permission from Urry et al. (2002a).

observed in a single chain. No longer can the change from a Gaussian distribution of end-to-end chain lengths within a random chain network be insisted upon as the structural representation of entropic elasticity. The ideal single-chain force–extension curves require high dilution, and a higher dilution yet is required for more hydrophobic protein-based polymers where the tendency for aggregation is greater. This and additional data tell us that these perfectly reversible force–extension curves are due to single-chains. A single chain does not constitute a random chain network; and a single chain with ends fixed in space does not comprise a Gaussian distribution of end-to-end chain lengths. It seems that this eliminates the basic tenets of the random chain network theory of entropic elasticity as regards these elastic protein-based polymers. Furthermore, random chains do not exhibit mechanical resonances, e.g., intense localized Debye-type relaxations as observed in the dielectric relaxation spectra near 5 MHz and 3 kHz, and intense localized absorption maxima as observed in the acoustic absorption and loss permittivity spectra. These points constitute only the more apparent elements of the argument.

In addition, on the basis of the elementary statistical mechanical analysis given above, it would seem that the source of entropic elasticity must come from a decrease in internal chain dynamics upon extension.

Reduction of Mean Solvent-Entropy Change Increases Rather Than Decreases Entropic Elastic Force

Formation of hydrophobic hydration is exothermic (Butler, 1937). Accordingly, when the temperature of the dissolved protein with its hydrophobic hydration is raised from below to above the inverse temperature transition, an endothermic transition due to the conversion of hydrophobic hydration to bulk water occurs. The transition from *hydrophobic hydration to bulk water* represents a positive change in entropy. By finding a suitable solvent that allows the transition but reduces the heat of the transition to near zero, it becomes possible to determine whether a decrease in elastic force occurs due to the loss of the contribution of the negative solvent-entropy change to the entropic elastic force. In other words, if a decrease in solvent entropy occurs as bulk water becomes ordered (in the form of hydrophobic hydration) by exposure of hydrophobic groups during extension, then the extent to which this contributes to entropic elastic force could be measured as a decrease in entropic elastic force.

In our experimental approach (based on Hoeve and Flory (1958)) ethylene glycol was added to suppress the solvent interaction with the protein. In particular, as shown in the differential calorimetry data of Figure 6A increasing the amount of ethylene glycol decreases the heat and lowers the temperature of the transition. By 30% ethylene glycol in water the heat of the transition is minimal. The entropy change attending the transition is calculated by dividing an increment of heat released over the temperature of the increment and summing over all increments of the transition. Accordingly, the decrease in heat of the transition seen in Figure 6A would be expected to result in a decrease in the entropic elastic force due to a decrease in the magnitude of the negative entropy of hydration on extension. In fact, as seen in Figure 6B, the entropic elastic force reached is greater in 30% ethyleneglycol–70% water.

The thermoelasticity studies on γ-irradiation-cross-linked poly(GVGVP) of Figure 6B are carried out by stretching the elastomer to a fixed extension of 50% at 37°C. Then held at that fixed length, the temperature is re-equilibrated below that of the onset for the inverse temperature transition, and the temperature is very slowly raised to 55°C. As regards the contribution of solvent entropy change to entropic elastic force, it is expected that the force reached would be less for the ethylene glycol containing solution by a fraction indicative of the contribution of solvent entropy change to the elastic force. The observation that the force actually increased, rather than decreased does not favor the solvent entropy mechanism.

As shown in Equation (4), the value of f_E/f is given by $-T(\partial \ln[f/T]/\partial T)_{V,L,n}$. Using the temperature range from 45 to 55°C, the f_E/f ratio is found to be no more than 0.1. Based on this

FIGURE 6. (A) Differential calorimetry curves as a function of ethylene glycol (EG) in water. Note the decrease in the heat and lowering of the temperature range of the transition as the amount of ethylene glycol is raised to 30%. B. Thermoelasticity curves, plots of $\log(f/T)$ vs. T at fixed length, for 20 Mrad γ-irradiation cross-linked poly(GVGVP). The f_E/f ratio, determined from the slope above 45°C for both solvent conditions, is less than or equal to 0.1, that is the entropic component of elastic force is 90% or greater. Significantly, addition of ethylene glycol (EG) results in an increase in entropic elastic force suggesting that slovent entropy change is not a contribution to the entropic elastic force. Reproduced with permission from Luan et al. (1989).

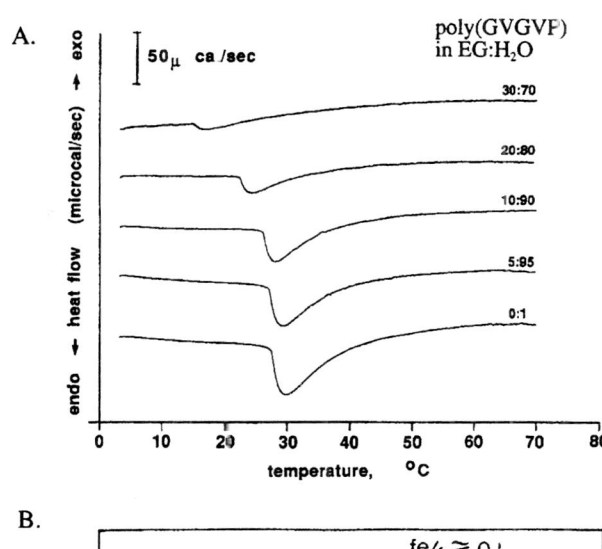

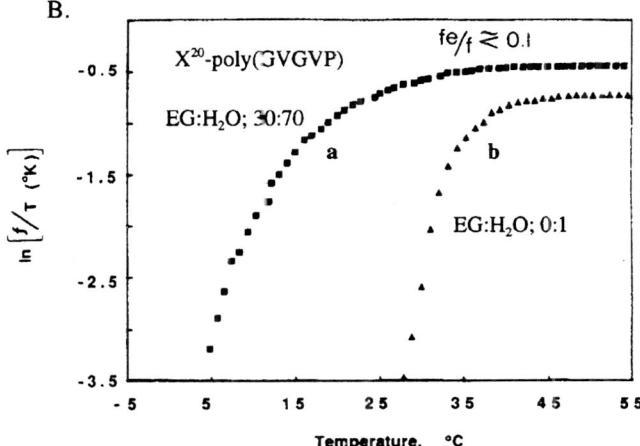

analysis using Equation (4), the f_S/f value would be 0.9 or greater, and yet this entropic elastic force would appear not to come from a change in solvent entropy. Even more telling is the increase in force at fixed length, $(\partial f/\partial T)_L$, as the temperature is raised through the range of the inverse temperature transition. As will be discussed below, this thermally driven isometric contraction shows an increase in force under conditions of an increase rather than the required decrease in solvent entropy.

Presence of Mechanical Resonances in Elastin Models and Elastin Itself

(i) *Mechanical resonances in the acoustic frequency range*: The elastic models of elastin based on the pentameric repeat (GVGVP) and analogues thereof, such as (GVGIP), exhibit mechanical resonances in the acoustic frequency range of 100 Hz to 7 kHz. This is seen in Figure 7A for γ-irradiation-cross-linked (GVGIP)$_{260}$, which exhibits a mechanical resonance of increasing absorption intensity as the temperature is raised from below to above the range of the inverse temperature transition (Urry et al., 2002a). By contrast, natural rubber and polyurethane (a particularly good elastomer for sound absorption), exhibit only broad low-intensity absorption as might be expected for a random chain network. For a random chain the frequency for rotation about each polymer backbone bond will be different, whereas for an elastomer with a regular repeating structure such as that in Figure 3

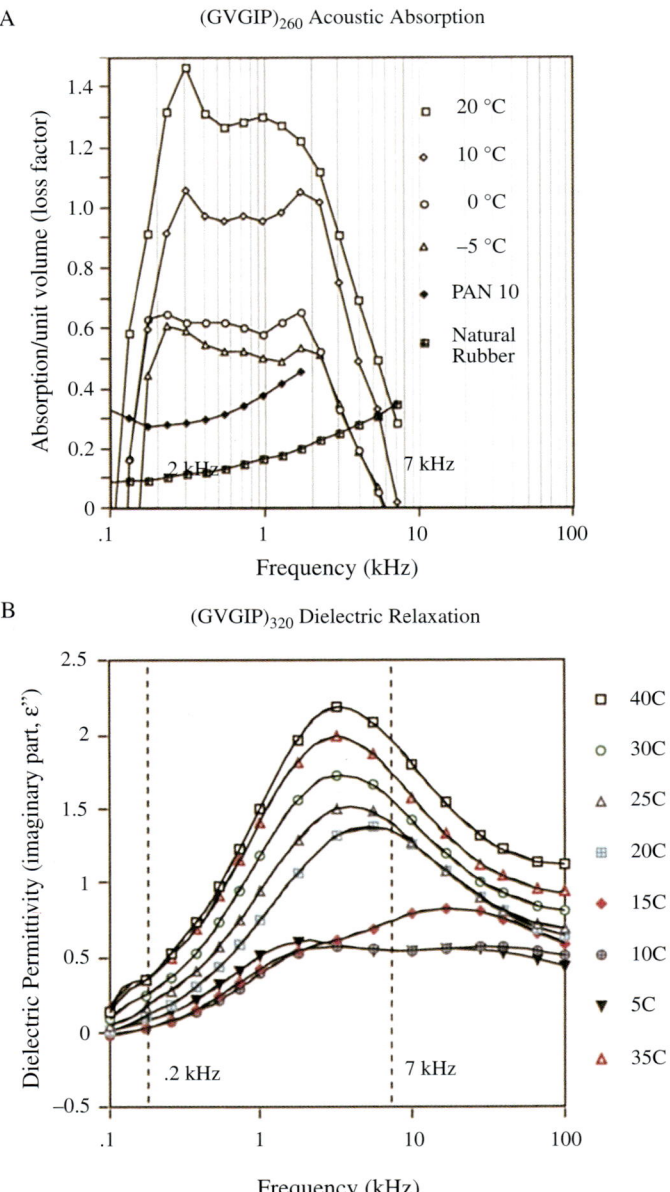

FIGURE 7. Mechanical resonances seen in the acoustic absorption frequency range. The repeating pentamers, as they fold into the regularly repeating structure of Figure 3 on raising the temperature, develop a mechanical resonance wherein all pentamers absorb energy over the same frequency range. Although the physical means of exciting the mechanical resonance is different in A and B, an accoustic wave in A and an oscillating electric field in B, the maxima of these low frequency mechanical resonances are only shifted by a few kHz. A second higher frequency mechanical resonance is observed for this same elastic model protein in the dielectric relaxation near 5 MHz, 1000 kHz to higher frequency, as seen in Figure 8 for both the real and imaginary parts of the dielectric permittivity. Part A, acoustic absorption/unit volume (loss factor), reproduced with permission from Urry et al. (2002a) and part B, imaginary part of the low frequency dielectric relaxation spectra, reproduced with permission from Urry et al. (2002b).

mechanical resonances such as those in Figure 7A can result. For our purposes the observation of a mechanical resonance is evidence for a non-random structure, and therefore challenges the classical theory of rubber elasticity.

Low-frequency dielectric relaxation studies also demonstrate mechanical resonances with absorption in the acoustic frequency range. As seen in Figure 7B, the imaginary part (ε'') of the dielectric permittivity of cross-linked (GVGIP)$_{320}$ exhibits a relaxation that also increases in intensity as the temperature is raised from below to above the temperature of the inverse temperature transition (Urry et al., 2002b). As the excitation energy is not that of a compressional acoustic wave, but rather an oscillating electric field, it is interesting to see that the repeating unit of five residues acts as a single unit with an oscillating net dipole that displays mechanical resonance in a similar frequency range. It may be further noted in Figure 7B that the absorption intensity increases by more than 50% upon going from 20 to 40°C. Because of this, the acoustic absorption exhibited by (GVGIP)$_n$ at 20°C in Figure 7A would be expected to increase by an additional 50% upon increasing the temperature further, to 40°C. This would make the absorption per unit volume of (GVGIP)$_n$ much larger than that exhibited by natural rubber or urethane.

(ii) *Mechanical resonances in the MHz range*: Dielectric relaxation studies demonstrate mechanical resonances near 5 MHz for (GVGVP)$_n$, (GVGIP)$_n$, (GVGLP)$_n$ (Buchet et al., 1988), α-elastin and fibrous elastin purified from bovine *ligamentum nuchae*. The data of Figure 8A give the real part of the dielectric permittivity (ε') for α-elastin, a 70 kDa chemical fragmentation product of natural fibrous elastin (Urry et al., 1988a) and for the elastin model poly(GVGVP) (inset) (Henze and Urry, 1985). ε'' for poly(GVGVP) and fibrous elastin is shown in Figure 8B (Luan et al., 1988).

As seen in Figure 1, there is much more to elastin than poly(GVGVP), of the W4 sequence, although it is the most prominent repeating sequence in bovine elastin. The relaxations for 100% poly(GVGVP) are expected to be more intense than that of α-elastin and fibrous elastin. It is indeed surprising that the frequency overlap and intensity is as close as found. It is apparent that other repeating sequences, or quasi repeats, exhibit dielectric relaxations at slightly lower frequencies.

Importantly, mechanical resonances near 5 MHz and 1 kHz indicate the presence of regularly repeating dynamic structures. Extension of such structures reasonably gives rise to an entropic elastic force upon extension by the damping of internal chain dynamics represented by mechanical resonance. Molecular mechanics and dynamics calculations based on the molecular structure shown in Figure 3 will demonstrate just how effectively the structure can explain the experimental elasticity and relaxation data.

Entropy Calculations Based on the Elastin Model, (GVGVP)$_n$

Calculations of structures and the entropies of the structures have used conventional molecular mechanics and dynamics programs such as CHARMM (Karplus and McCammon, 1983; Karplus and Kushick, 1981), AMBER (Weiner and Kollman, 1981), and Scheraga's ECEPP (Momany et al., 1974, 1975). The satisfying feature of these quite distinct computational approaches is that they give essentially identical results whether carried out within our group by different researchers or by different groups. It is important, however, that the constraints and boundary conditions should properly reflect the conditions considered. Paramount among the constraints is adherence to the elasticity experiment in which the ends of the molecular structure are fixed at whatever selected degree of extension.

Molecular Mechanics Calculations Using Scheraga's ECEPP Approach for Energy Surfaces in ϕ–ψ Configuration Space

(i) *Calculation of entropy change during extension by an enumeration of states approach for*

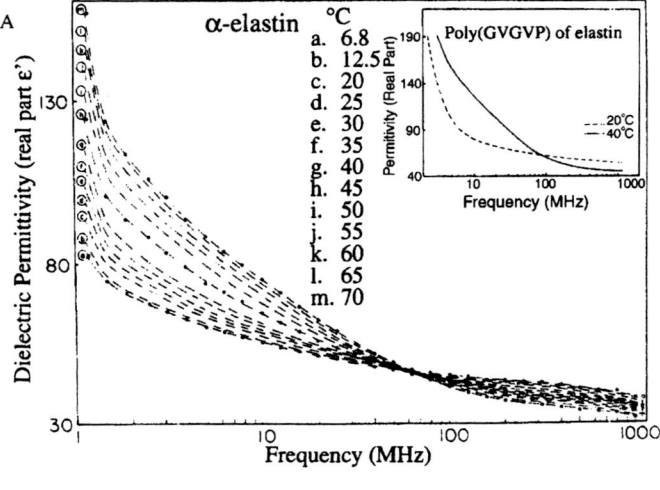

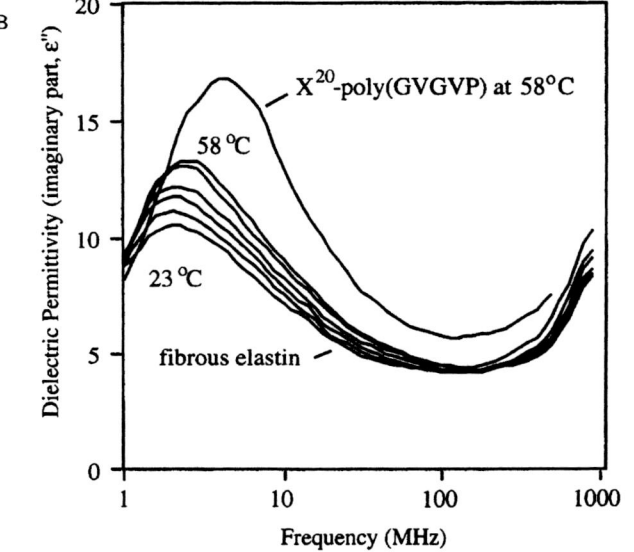

FIGURE 8. Dielectric relaxation spectra in the 1–1000 MHz frequency range of α-elastin (part A), the poly(GVGVP) of elastin (inset of part A and part B), and fibrous elastin (part B). In all cases as the temperature is raised there develops an intense mechanical relaxation centered near 5 MHz as repeating elements of these protein systems hydrophobically fold into regular, albeit obviously dynamic, structures. In the case of poly(GVGVP) each pentamer folds into the same conformation, in which the peptide moieties (the only entities with dipole moments) undergo coordinated rocking motions with a resulting oscillating mean dipole moment. The rocking of the mean dipole moments of the pentamers resonate, in this case move, at the same frequency in response to an alternating electric field, that is, they exhibit a mechanical resonance centered near 5 MHz. A similar mechanical resonance is observed near 1 kHz in Figure 7. Part A reproduced with permission from Urry et al. (1988a), and part B reproduced with permission from Luan et al. (1988).

relaxed and extended states: In Scheraga's approach the internal energy of a chosen chain segment (in our case the pentamer permutation, $V_1P_2G_3V_4G_5$) is calculated as a function of a pair of adjacent torsion angles. Normally, the φ and ψ torsion angles are identical to those in the Ramachandran plot. In our case since the primary sites of motion are the peptide moieties of the suspended segment ($V_4G_5V_1$), such that the two φ–ψ plots, now called lambda plots, become $φ_5$–$ψ_4$ and $φ_1$–$ψ_5$.

In the enumeration of states approach, a 5° change in a single torsion angle is counted as a new state, and the number of states is counted for the chosen cut-off energy for both the relaxed and 130% extended structures (Urry et al., 1982; Urry and Venkatachalam, 1983; Urry et al., 1985c; Urry, 1991). A single pentamer is calculated within the relaxed and extended β-spiral structures, and 0.6, 1.0 and 2.0 kcal/mol-pentamer cut-off energies are used as well as an energy weighting using the Boltzmann summation over states. Table 1 gives the number of states where the entropy is simply calculated by Equation (9) using the example of 1 kcal/mol-pentamer,

$$\Delta S = (S^e - S^r) = R\ln(W^e/W^r) = R\ln(58/762)$$
$$= -5.1 \text{ cal/mol-pentamer}. \quad (9)$$

TABLE 1. Perspective of entropy of the poly(GVGVP) β-spiral by enumeration of states

Cutoff energy (kcal/mol)	Number of states		Entropy change per residue
	Relaxed (r)	Extend (e)	
2	1853	162	−0.97
1	762	58	−1.02
0.6	342	24	−1.06
	Using Boltzmann sum over states $f = \sum_i e^{-\varepsilon_i/kT}$		
2	$\Delta S = R \ln\left(f^e/f^r + \dfrac{E^e - E^r}{T}\right)$		−1.01

This, of course, is −1.02 cal/mol-residue as given in Table 1.

(ii) *Calculation of dielectric relaxation data from pentamer molecular mechanics*: With the aid of the Onsager equation for polar liquids (Onsager, 1931; Bottcher, 1973) it becomes possible to use the structures obtained above and to sum the dipole moments of the states of Table 1. In addition, the same can be done for $V_1P_2G_3V_4A'_5$) where the A' stands for the D-Ala amino acid residue (Venkatachalma and Urry, 1986 and printers erratum as corrected below; Urry, 1991). Comparison of experimental and calculated values is particularly satisfying. "In view of this result, it seems reasonable to consider a cutoff energy of 1.5 kcal mol^{-1}. This value of energy cutoff leads to a mean dipole moment change of 3.8 Debye per pentamer in good agreement with the value obtained from dielectric relaxation studies. Similar dielectric studies on D.Ala^5-polypentapeptide indicate a dipole moment of about one-third of that found for the polypentapeptide at 37°C... This is in very good agreement with the dipole changes obtained from molecular mechanics calculations presented here" (Venkatachalam and Urry, 1986).

The above calculated results utilize the 5 MHz mechanical resonance. They further demonstrate how readily calculations based on the dynamic structure given in Figure 3 calculate experimental dielectric relaxation data with inherent relevance to entropic elastic force. The values for entropy change calculated using the molecular mechanics approach, when used in the second term of Equation (1), provide excess entropic elastic force as expected (see Urry et al., 2002a, b).

Molecular Dynamics Calculations Using Karplus' CHARMm and Kollman's AMBER Programs

As the motions of the torsional oscillations are dynamic, it is appropriate to consider molecular dynamics calculations using the Karplus software program adapted by Polygen, Inc. (now Molecular Simulations), called CHARMm, version 20.3. For these Newtonian classical mechanics simulations the potential energy functions and parameters suggested by the Karplus group (Brooks et al., 1983) were used in a software program. The root-mean-square (RMS) fluctuations of the torsion angles, $\Delta\phi_i$ and $\Delta\psi_i$ for the relaxed and extended (130%) of $(GVGVP)_{11}$ as given in Table 2 are used in Equation (10).

$$\Delta S = R \ln[\Pi_i \Delta\phi_i^e \Delta\psi_i^e / \Pi_i \Delta\phi_i^r \Delta\psi_i^r] \quad (10)$$

The calculated change in entropy upon extension of −1.1 cal/mol deg-residue is in remarkable agreement with the previously described molecular-mechanics calculations.

Furthermore, Wasserman and Salemme (1990) using the AMBER software program of the Kollman group (Weiner and Kollman, 1981), but also including a medium of water molecules, obtained essentially identical results. It must be concluded, damping of internal chain

TABLE 2. RMS fluctuations of torsion angles (ϕ and ψ) of $(VPGVG)_{11}$ (45 ps of equilibration time and 80 ps of molecular dynamics simulation). Reproduced with permission from Chang and Urry (1989).

	Angle	Relaxed	Extended	Angle	Relaxed	Extended	Angle	Relaxed	Extended
β-turns	ψ_{16}	10.87	14.17	ψ_{26}	27.33	07.64	ψ_{36}	20.19	14.84
	ϕ_{17}	09.86	15.18	ϕ_{27}	11.71	08.33	ϕ_{37}	08.99	10.47
	ψ_{17}	47.59	46.68	ψ_{27}	11.70	13.51	ψ_{37}	21.53	32.96
	ϕ_{18}	61.70	47.41	ϕ_{28}	08.61	10.36	ϕ_{38}	11.15	10.66
	ψ_{18}	09.37	16.05	ψ_{28}	09.33	08.16	ψ_{38}	11.09	27.29
	ϕ_{19}	14.25	08.67	ϕ_{29}	09.70	07.31	ϕ_{39}	12.70	10.24
Suspended	ψ_{19}	44.09	10.99	ψ_{29}	47.32	10.48	ψ_{39}	52.00	12.50
	ϕ_{20}	41.94	09.29	ϕ_{30}	48.57	11.39	ϕ_{40}	55.88	08.37
Segments	ψ_{20}	14.50	11.15	ψ_{30}	42.56	10.62	ψ_{40}	40.67	11.08
	ϕ_{21}	27.13	24.17	ϕ_{31}	11.43	11.38	ϕ_{41}	36.44	19.06
β-turns	ψ_{21}	09.39	22.73	ψ_{31}	12.17	09.21	ψ_{41}	12.97	14.80
	ϕ_{22}	09.94	08.00	ϕ_{32}	09.90	08.93	ϕ_{42}	11.59	07.33
	ψ_{22}	11.58	16.13	ψ_{32}	15.30	10.80	ψ_{42}	11.34	13.17
	ϕ_{23}	16.37	09.33	ϕ_{33}	09.60	07.62	ϕ_{43}	09.23	09.76
	ψ_{23}	14.33	14.25	ψ_{33}	09.88	09.43	ψ_{43}	10.60	12.53
	ϕ_{24}	11.39	29.20	ϕ_{34}	11.86	09.71	ϕ_{44}	11.06	12.82
Suspended	ψ_{24}	19.53	37.87	ψ_{34}	63.80	08.36	ψ_{44}	41.89	35.22
	ϕ_{25}	25.02	23.06	ϕ_{35}	91.70	10.20	ϕ_{45}	48.98	31.31
Segments	ψ_{25}	49.32	32.10	ψ_{35}	15.03	11.51	ψ_{45}	42.05	56.89
	ϕ_{26}	31.43	27.24	ϕ_{36}	21.49	18.66	ϕ_{46}	21.55	30.33

$$\Delta S = R \ln \frac{\Pi_i \Delta \phi_i^e \Delta \psi_i^e}{\Pi_i \Delta \phi_i^r \Delta \psi_i^r}.$$

dynamics is an abundant source of decrease in chain entropy that is sufficient to account for the entropic elastic force.

As Wasserman and Salemme (1990) included water molecules in their calculations, and an ordering of water molecules was found as the hydrophobic side chains became exposed on chain extension, they considered this decrease in solvent entropy as a possible source of the entropic elastic force. As will be shown below, solvent entropy change does not make a significant contribution to the entropic elastic force development during isometric contractions.

Relationship Between Hydrophobic Association–Dissociation, Elastic Force, and Energy Conversion in Elastin Mechanics

By introducing glutamate (Gly, E) the ideal elastomer in Figure 5A becomes a chemomechanical transducer capable of converting the chemical energy of proton concentration changes into the mechanical energy of isotonic and isometric contractions (See Figure 9A and B). An analysis of this data in combination with that of Figure 6B resolves the issue of the role of changes in solvent entropy in elastic force development and energy conversion.

Chemically Driven Isotonic, $(\partial L/\partial \mu_H)_f$, and Isometric, $(\partial f/\partial \mu_H)_L$, Contractions

The elastin model protein of interest is poly$[x_v$-$(GVGVP), x_E(GEGVP)]$, where x_v and x_E are the mole fractions with $x_v + x_E = 1$ and for the case where $x_E = 0.2$. Upon γ-irradiation cross-linking an elastic matrix is formed which can be studied under isotonic and isometric conditions. Results are shown in Figure 9A and B (Urry et al., 1988b). The isotonic contraction, $(\partial L/\partial \mu_H)_f$, in Figure 9A is the change in length at constant load (force) resulting from an increase in the concentration of proton, that is, the increase in chemical potential of proton $(\partial \mu_H)$. The isometric contraction, $(\partial f/\partial \mu_H)_L$, in

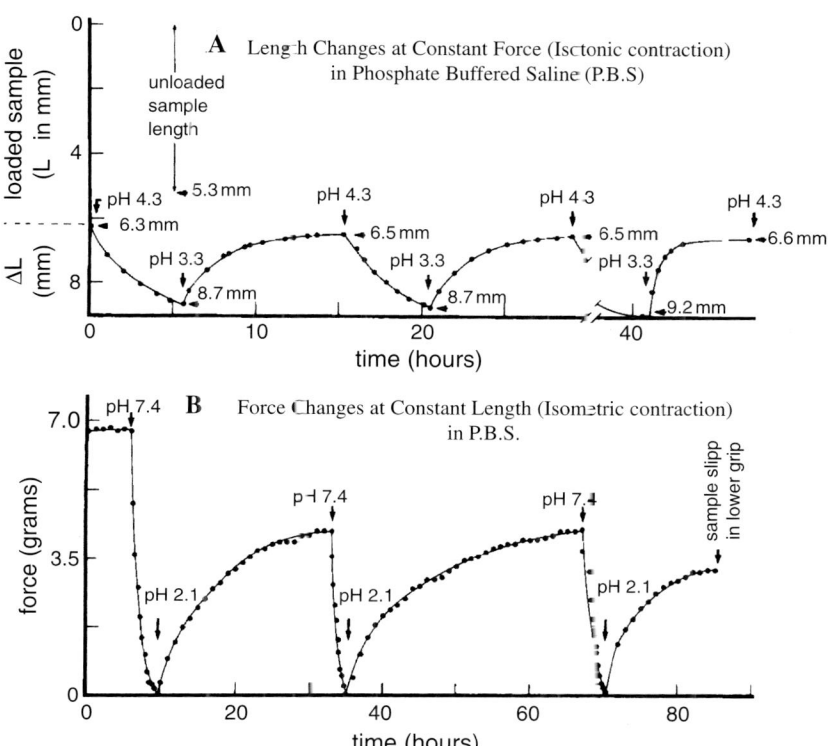

FIGURE 9. Mechano-chemical coupling by 20 Mrad γ-irradiation cross-linked of poly[0.8(GVGVP), 0.2(GEGVP)] under isotonic (A) and isometric (B) conditions. In part B under isometric contraction conditions there occurs an increase in entropic elastic force resulting from a decrease in entropy of the elastomer while there occurs an increase in the entropy of the solvent as hydrophobic hydration becomes bulk water. Therefore the increase in entropic elastic force cannot be the result of the solvent entropy change. Reproduced with permission from Urry et al. (1988b).

Figure 9B is the change in force at constant length due to an increase in proton.

Both of these contractions result from hydrophobic association due to protonation of the carboxylate group. The underlying physical process is the competition for hydration between the charged carboxylate and the hydrophobic side chains of the valine (Val, V) residues (Urry, 1997). This competition, described as an apolar–polar repulsive free energy of hydration, results in increasing positive cooperativity of the acid–base titration curves that correlate with increasingly larger pK_a shifts. As Val residues are replaced by more hydrophobic phenylalanine (Phe, F) residues, the pK_a shifts increase (up to 6 pH units or more) and positive cooperativity increases with Hill coefficients of up to 8 or more (Urry, 1997). The apolar–polar repulsive free energy of hydration is related to the change in Gibbs free energy for hydrophobic hydration, $\Delta G_{HA}(\chi)$, of a phosphate as discussed in relation to Equations (17) and (19) below, but first the source of entropic elastic force is addressed.

Isometric Contractions, $(\partial f/\partial T)_L$ and $(\partial f/\partial \mu_H)_L$, Confirm Basis for Entropic Elastic Force

From Equation (1) the entropic component of the elastic force, f_s, is proportional to $-\Delta S$, that is, an increase in entropic elastic force results from a decrease in entropy. Also the formation of hydrophobic hydration from bulk water constitutes an inherently negative ΔS, and, of course, the loss of hydrophobic hydration constitutes a positive change in entropy. This change in solvent entropy due to formation of hydrophobic hydration has been credited as an important contribution to entropic elastic force (Weis-Fogh and Anderson, 1970; Wasserman and Salemme, 1990; Alonso et al., 2001). Therefore, the immediate question becomes whether experimental results support or eliminate a decrease in solvent entropy as a source of entropic elastic force.

Thermally Driven Isometric Contraction (Figure 6B)

Figure 6B shows the classic thermoelasticity study for determining the f_E/f ratio for γ-irradiation-cross-linked poly(GVGVP). Because extension is fixed, a thermally driven isometric contraction also occurs over the temperature range of the inverse temperature transition of hydrophobic association. During hydrophobic association, hydrophobic hydration becomes bulk water. Therefore, the solvent entropy change is positive. Any contribution to the entropic elastic force due to changes in hydrophobic hydration during this thermally driven isometric contraction would necessarily result in a decrease in force. Yet the experimental result of Figure 6B shows the development of an elastic force that is 90% entropic, while the change in solvent entropy is of the opposite sign. Clearly, the change in solvent entropy is not contributing to the estimated 90% entropic elastic force. Having eliminated solvent entropy change as the source of entropic elastic force during a thermally driven isometric contraction, we now analyze the solvent entropy change attending a chemically driven isometric contraction.

Chemically Driven Isometric Contraction (Figure 9B)

As seen in Figure 9B, at fixed extension of the γ-irradiation-cross-linked elastic matrix comprised of poly[0.8(GVGVP), 0.2(GEGVP)], protonation of four carboxylates per 100 residues results in development of elastic force. A thermoelasticity characterization of this matrix at low pH gives the same result of dominantly entropic elasticity as found in curve b of Figure 6B for poly(GVGVP) in the absence of carboxyl moieties.

The process of protonation allows reconstitution of hydrophobic hydration to such an extent that the temperature range for hydrophobic association drops below that of the operating temperature (Urry, 1993, 1997). The result is a contraction due to hydrophobic association. Again, during an isometric contraction (this time chemically driven), hydrophobic hydration becomes less ordered bulk water. The solvent entropy increases during the development of entropic elastic force due to a decrease in entropy.

In addition to eliminating solvent entropy change as a source of entropic elastic force, the isometric contraction results provide the conceptual bridge between the fundamental source of entropic elastic force and contraction by hydrophobic association.

Hydrophobic Association Effects Elastic Force Development by Extending (and Thereby Damping Internal Dynamics of) Interconnecting Chain Segments

Clam-Shaped Globular Proteins That Open and Close Due to Hydrophobic Association

The top part of Figure 10 shows a series of clam-shaped globular proteins strung together near the open end by elastic bands and maintained at fixed extension. In this isometric state, there is depicted an equilibrium between open and closed states. Obviously, shifting the equilibrium toward the closed state would increase the force measured at the force transducer, whereas shifting toward the open state would

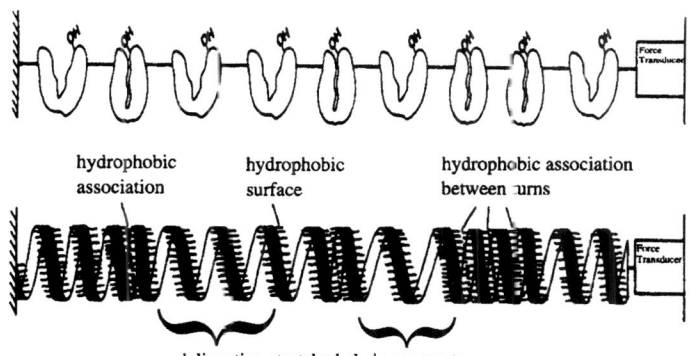

FIGURE 10. Cartoon of the relationship between hydrophobic association and entropic elastic force development. Above: A series of clam-shaped globular protein strung together by elastic bands with an equilibrium between open and hydrophobically associated closed states. Clearly, as the equilibrium shifts toward more closed states, the force sustained by the interconnecting elastic segments, increases. Below: Representation of the β-spiral structure of poly(GVGVP) at an intermediate state of hydrophobic association between turns of the β-spiral. Clearly, as more hydrophobic association occurs between turns, the interconnecting elastic chain segments become further extended with an increase in damping of internal chain dynamics giving rise to greater entropic elastic force. Reproduced with permission from Urry (1990).

lower the force on the interconnecting elastic bands. Thus, any process that increased hydrophobic association would increase the elastic force. Now in the lower part of Figure 10, we relate this perspective to the β-spiral representation of Figure 3C.

Schematic Representation of an Equilibrium Between Hydrophobically Associated and Dissociated Turns of a β-Spiral Held at Fixed Extension

The bottom half of Figure 10 uses the representation of the β-spiral given in Figure 3C to depict an equilibrium between associated and dissociated turns of the β-spiral. As more of the turns are involved in hydrophobic association, those not involved become extended to a greater extent. Thus, as more hydrophobic association occurs, the interconnecting segments become increasingly extended. By the entropic mechanism of the damping of internal chain dynamics on extension, the entropic elastic force increases.

Opening and closing of the clam-shaped globular proteins of the upper part of Figure 10 (Urry, 1990) would be detectable under conditions of force-clamp atomic force spectroscopy. Such traces have been observed using titin-based elastic constructs comprised of a series of hydrophobically folded β-barrels (Oberhauser et al., 2001). Because of the continuous nature of increases in hydrophobic association between turns of the β-spiral represented in the lower part of Figure 10, such force-clamp traces for poly(GVGVP) would not have such steps, unless sensitivity were sufficient to detect the addition of a repeating unit to hydrophobic association.

Thus, the changes in entropic elastic force and in hydrophobic association are two distinct physical processes that become interlinked when occurring within the same chain system. Hydrophobic association causes the extension of interconnecting dynamic chain segments with the result of the damping of internal chain dynamics within those non-hydrophobically associated interconnecting chain segments.

Derivation of Gibbs Free Energy for Hydrophobic Association ΔG_{HA}

The role of hydrophobic association in the development of entropic elastic force has been delineated above. The next step is to derive an expression for the Gibbs free energy of hydrophobic association and to relate it to processes that direct elastic–contractile processes.

(i) *At the temperature of the inverse temperature transition*, $\mu_{HA} = \mu_{HD}$ and $\Delta G = 0$: The chemical potential (μ) is the Gibbs free energy per mole ($\Delta G/\Delta n$). By the Gibbs phase rule, at the temperature of the inverse temperature transition for hydrophobically association the chemical potential of the hydrophobically disassociated (dissolved) state (μ_{HD}) and the chemical potential of the hydrophobically associated state (μ_{HA}) are equal. Therefore, at the temperature of the inverse temperature transition (T_t), $\Delta G_t = \Delta H_t - T_t \Delta S_t = 0$, where ΔH_t is the heat of the transition and ΔS_t is the entropy change for the transition. Accordingly at the value of T_t for a given elastic model protein composition, $\Delta H_t = T_t \Delta S_t$.

By way of example, consider the two different compositions of model elastic proteins, (GVGVP)$_n$ and (GVGIP)$_n$. These two polypentapeptides differ only by a single CH$_2$ moiety per pentamer. In particular, the R-group for V residue, —CH(CH$_3$)$_2$, differs from the R-group of the I residue, —CH(CH$_3$)—CH$_2$—CH$_3$, by the addition of a single CH$_2$ moiety. At the temperature for each of their respective phase transitions where $\mu_{HA} = \mu_{HD}$, we can write,

$$\Delta H_t(\text{GVGVP}) = T_t(\text{GVGVP})\Delta S_t(\text{GVGVP}) \tag{11}$$

and

$$\Delta H_t(\text{GVGIP}) = T_t(\text{GVGIP})\Delta S_t(\text{GVGIP}). \tag{12}$$

Butler (1937) showed that each addition of a CH$_2$ moiety in the series of normal alcohols from methanol to *n*-pentanol on dissolution in water resulted in an average exothermic heat $(\Delta H)/\text{CH}_2$ of -1.4 kcal/mol-CH$_2$ and a change of the term, $(-T\Delta S)/\text{CH}_2$, of a $+1.7$ kcal/mol-CH$_2$. While there are many important points to make about this fundamental finding, we use it here simply as a justification to rewrite Equation (12) as

$$\Delta H_t(\text{GVGIP}) = \Delta H_t(\text{GVGVP}) + \Delta H_t(\text{CH}_2), \tag{13}$$

and

$$T_t(\text{GVGIP})\Delta S_t(\text{GVGIP}) = T_t(\text{GVGIP})[\Delta S_t(\text{GVGVP}) + \Delta S_t(\text{CH}_2)]. \tag{14}$$

Substituting Equations (13) and (14) into Equation (12) and subtracting Equation (11) gives

$$[\Delta H_t(\text{CH2}) - T_t(\text{GVGIP})\Delta S_t(\text{CH}_2)] = [T_t(\text{GVGIP}) - T_t(\text{GVGVP})]\Delta S_t(\text{GVGVP}). \tag{15}$$

The left hand side of Equation (15) is recognized as an expression for the change in Gibbs free energy due to addition of the CH$_2$ moiety. On consideration of the phase transition being analyzed, the left hand side of Equation (5) becomes the change in Gibbs free energy of hydrophobic association due to the addition of a single CH$_2$ moiety, $\Delta G_{HA}(\text{CH}_2)$, that causes the inverse temperature transition of hydrophobic association to occur at a lower temperature for (GVGIP)$_n$.

$$\Delta G_{HA}(\text{CH}_2) = [T_t(\text{GVGIP}) - T_t(\text{GVGVP})] \times \Delta S_t(\text{GVGVP}). \tag{16}$$

The right hand side of Equation (16) is negative because $T_t(\text{GVGIP}) < T_t(\text{GVGVP})$ and $\Delta S_t(\text{GVGVP})$ is positive for a transition whereby hydrophobic hydration becomes less ordered bulk water during hydrophobic association, that is, as insolubility ensues.

The addition of a CH$_2$ moiety per pentamer is one of very many ways whereby the value of T_t can be changed. A T_t-based hydrophobicity scale has been developed for substitution of each of the naturally occurring amino acid residues and chemical modifications thereof by systematically increasing the mole fraction of pentamers containing the change X and extrapolating to the value of T_t that would occur for poly(GXGVP) (Urry, 1997). Therefore, the value of T_t may be greater than 100°C as for poly(GEGVP) when the glutamic acid residue, E, is ionized as the carboxylate, where $T_t = 250$°C. On the other hand, the value of T_t may be less than 0°C as for poly(GWGVP) when the tryptophan is the residue, where $T_t = -90$°C. For the amino acids with functional side chains the T_t-value changes upon changing the state of the side chain from carboxyl to carboxylate, or the redox state of an attached redox couple. The most dramatic change in the T_t-value occurs upon phosphorylation of a Ser, Thr, or Tyr residue (Pattanaik et al., 1991). Furthermore, adding salt to the solution of a neutral polymer such as poly(GVGVP) itself lowers the value of T_t (Urry, 1993), but the ΔT_t is 10 times greater when the salt ion-pairs with a charged side chain (Urry and Luan, 1995a; Urry et al., 1997a).

So we generalize to any experimental variable, χ, that alters the value of T_t of a reference model elastic protein and write,

$$\Delta G_{HA}(\chi) = [T_t(\chi) - T_t(\text{reference})]\Delta S_t(\text{reference}), \quad (17)$$

or

$$\Delta G_{HA}(\chi) = \Delta T_t(\chi)\Delta S_t(\text{reference}), \quad (18)$$

where T_t is the value of a reference polymer and state has been changed to $T_t(\chi)$ by the experimental variable χ. All of these energy inputs can now be described in terms of their effect on the free energy of hydrophobic association, $\Delta G_{HA}(\chi)$.

Relevance of Gibbs Free Energy for Hydrophobic Association to Energizing by Phosphorylation

Phosphorylation of poly[30(GVGIP), (RGYSLG)] by the cardiac cyclic AMP-dependent protein kinase at the serine (Ser, S) residue of the polymer goes to an average of 47% completion by the following reaction (Pattanaik et al., 1991; A. Pattanaik, personal communication).

$$\text{ATP} + \text{poly[30(GVGIP), (RGYSLG)]}$$
$$= \text{ADP} + \text{poly[30(GVGIP),}$$
$$(\text{RGYS}(-\text{OPO}_3^{2-})\text{LG})]. \quad (19)$$

The value of T_t for this polymer before phosphorylation was approximately 20°C and the change in the value of T_t extrapolated as defined to a reference state of one phosphate per GVGIP becomes 860°C, which is represented as $T_t(-\text{OPO}_3^{2-})$. If we now use the experimentally derived value of $\Delta S_t(\text{GVGIP}) = 9.32$ cal/mol (pentamer) deg as the $\Delta S_t(\text{reference})$ (Luan and Urry, 1999) of Equation (17), the change in Gibbs free energy for hydrophobic association due to phosphorylation, $\Delta G_{HA}(-\text{OPO}_3^{2-})$, is +7.8 kcal/mol. For the reaction of Equation (19) with an equilibrium constant of $K = 47/53 = 0.9 = \exp(-\Delta G/RT)$, the change in Gibbs free energy for the reaction is 0.06 kcal/mol. Thus, the calculated value for the free energy contributed by the terminal phosphate of ATP would be $0.06 - 7.8 = -7.7$ kcal/mol, which is remarkably close to the free energy of hydrolysis of the terminal phosphate of ATP of 7.3 kcal/mol (Voet and Voet, 1995).

Accordingly, phosphorylation raises the free energy of the hydrophobically associated state, that is, it results in hydrophobic disassociation, and phosphate removal drives hydrophobic association, e.g., contraction, more effectively than any other chemical change measured so far. Expectations are that the binding of ATP, and most effectively ADP plus phosphate, would dramatically raise the free energy of the hydrophobically associated state as evidenced by an increase the value of T_t. This $\Delta T_t(-\text{OPO}_3^{2-})$ and $\Delta G_{HA}(-\text{OPO}_3^{2-})$ can be only partially modulated by ion-pairing, for example, with magnesium, of the phosphates at the binding site.

Relationship Between Entropic Elasticity and Efficiency of Energy Conversion

The process of stretching an elastomer constitutes an energy input with the expended mechanical energy $f\Delta L$, where f is force and ΔL is the change in length. Accordingly, the mechanical energy is obtained from the area under the force vs. extension curve. For an ideal elastomer, the plot of force vs. extension is perfectly reversible, as shown by the second and fifth traces of Figure 5A for the single-chain force–extension curve of the entropic elastomer, $(\text{GVGVP})_{n\times 251}$. Thus, the input energy is completely recovered on relaxation for an ideal (entropic) elastomer. Since the energy recovered on relaxation can be viewed in the sense of an energy output, an ideal elastomer could be considered a perfect machine for storing energy of deformation.

Often the force vs. length curve obtained on relaxation falls below the force vs. length curve obtained on extension (as seen in the lower curve of Figure 5B); then the material is said to exhibit a hysteresis. In this case the energy recovered on relaxation is less than that expended on extension. The input deformation energy has been dissipated in some way and to an extent indicated by the difference in the areas below the extension and relaxation curves.

When energy conversion occurs by means of an ideal elastic material, it is possible for the energy conversion to occur at high efficiency, whereas when energy conversion occurs using an elastic material that exhibits hysteresis, the efficiency of energy conversion becomes limited at least to an extent determined by the magnitude of the hysteresis. The elasticity of polymeric materials becomes inextricably intertwined with the efficiency of energy conversion. Thus, the increase in elastic force during the isometric contraction and relaxation of Figure 9B would be exactly reversible for an ideal elastomer as would the isotonic contraction of Figure 9A. This would not be the case for a molecular machine comprised of the elastic polymer functioning in Figure 5B.

An important element, therefore, of an ideal elastomer is to have the energy uptake into the elastomer during extension reside entirely in the backbone modes where it can be recovered on relaxation. Should energy of deformation find its way into side-chain motional modes and into chains not bearing the deformation and irreversibly into solvent, this deformation energy becomes dissipated and unavailable during relaxation, resulting in hysteresis. Comparison of the single-chain force–extension curves of Figure 5 for poly(GVGVP) with poly(GVGIP) provides an example with proposed loss of energy into the side chain motions and interactions of the bulkier isoleucine (I) residue with its added CH_2 moiety into adjacent non-load-bearing chains.

Acknowledgments. The authors wish to acknowledge the support of the Office of Naval Research under contracts, N00014-00-C-0404 and N00014-00-C-0178, and to thank A. Pattanaik for updated details on the phosphorylation study and L. Hayes for assistance in obtaining the hydrophobicity plots and references.

References

L.B. Alonso, B.J. Bennion, and V. Daggett, Hydrophobic hydration is an important source of elasticity in elastin-based polymers. *J Am Chem Soc* **123**, 11,991–11,998, 2001.

C.J.F. Bottcher, (1973) *Theory of Electric Polarization.* (vol. 1, p. 178) Elsevier, Amsterdam.

B.R. Brooks, R.E. Bruccoleri, B.O. Olafson, D.T. States, S. Swaminathan, and M. Karplus, CHARMM: a program for macromolecular energy, minimization, and dynamics calculations. *J Comput Chem* **4**, 187, 1983.

R. Buchet, C-H. Luan, K.U. Prasad, R.D. Harris, and D.W. Urry, Dielectric relaxation studies on analogs of the polypentapeptide of elastin. *J Phys Chem* **92**, 511–517, 1988.

J.A.V. Butler, The energy and entropy of hydration of organic compounds. *Trans Faraday Society* **33**, 229–238, 1937.

D.K. Chang and D.W. Urry, Polypentapeptide of elastin: damping of internal chain dynamics on extension. *J Comput Chem* **10**, 850–855, 1989.

D.K. Chang, C.M. Venkatachalam, K.U. Prasad, and D.W. Urry, Nuclear overhauser effect and computational characterization of the β-spiral of the polypentapeptide of Elastin. *J Biomol Struct Dynam* **6**, 851–858, 1989.

W.J. Cook, H.M. Einspahr, T.L. Trapane, D.W. Urry, and C.E. Bugg, Crystal structure and conformation of the cyclic trimer of a repeat pentapeptide of elastin, cyclo-(L-valyl-L-prolyl-glycyl-L-valyl-glycyl)$_3$. *J Am Chem Soc* **102**, 5502–5505, 1980.

B.A. Cox, B.C. Starcher, and D.W. Urry, Coacervation of α-elastin results in fiber formation. *Biochim Biophys Acta* **317**, 209–213, 1973.

B.A. Cox, B.C. Starcher, and D.W. Urry, Coacervation of tropoelastin results in fiber formation. *J Biol Chem* **249**, 997–998, 1974.

P. Dauber, M. Goodman, A.T. Hagler, D. Osguthorpe, R. Sharon, and P. Stern, (1981), In: P. Lykos and I. Shavitt (eds) *ACS Symposium Series No. 173. Suypercomputers in Chemistry.* (pp. 161–191) American Chemical Society, Washington, DC.

H. Eyring, D. Henderson, B.J. Stover, and E.M. Eyring, (1964) *Statistical Mechanics and Dynamics.* (p. 92) John Wiley & Sons Inc, New York.

H.E. Gaub and J.M. Fernandez, The molecular elasticity of individual proteins studied by AFM-related techniques. *AvH-Magazin* **71**, 11–18, 1998.

L. Gotte, G. Giro, D. Volpin, and R.W. Horne, The ultrastructural organization of elastin. *J Ultrastruct Res* **46**, 23–33, 1974.

R. Henze and D.W. Urry, Dielectric relaxation studies demonstrate a peptide librational mode in the polypentapeptide of elastin. *J Am Chem Soc* **107**, 2991–2993, 1985.

C.A.J. Hoeve and P.J. Flory, The elastic properties of elastin. *J Am Chem Soc* **80**, 6523–6526, 1958.

C.A.J. Hoeve and P.J. Flory, Elastic properties of elastin. *Biopolymers* **13**, 677–686, 1974.

M. Karplus and J.N. Kushick, Method for estimating the configurational entropy of macromolecules. *Macromolecules* **14**, 325–332, 1981.

M. Karplus and J.A. McCammon, Dynamics of proteins: elements of function. *Ann Rev Biochem* **53**, 263–300, 1983.

A.R. Kornblihtt, K. Umezawa, K. Vibe-Pederson, and F.E. Baralle, Primary structure of human fibronectin: differential splicing may generate at least 10 polypeptides from a single gene. *EMBO J* **4**, 1755–1759, 1985.

C-H. Luan, R.D. Harris, and D.W. Urry, Dielectric relaxation studies on bovine ligamentum nuchae. *Biopolymers* **27**, 1787–1793, 1988.

C-H. Luan, J. Jaggard, R.D. Harris, and D.W. Urry, On the Source of Entropic Elastomeric Force in Polypeptides and Proteins: Backbone Configurational vs. Side Chain Solvational Entropy. *Int J of Quant Chem: Quant Biol Symp* **16**, 235–244, 1989.

C-H. Luan and D.W. Urry (1999), Elastic, plastic, and hydrogel protein-based polymers. In: J.E. Mark (ed.) *Polymer Data Handbook*, (pp. 78–89, Table 3A) Oxford University Press, New York, Oxford.

F.A. Momany, L.M. Carruthers, R.F. McGuire, and H.A. Scheraga, Intermolecular potentials from crystal data. III. Determination of empirical potentials and application to the packing configurations and lattice energies in crystals of hydrocarbons, carboxylic acids, amines, and amides. *J Phys Chem* **78**, 1595, 1974.

F.A. Momany, F.F. McGuire, A.W. Burgess, and H.A. Scheraga, Energy parameters in polypeptides. VII. Geometric parameters, partial charges, non-bonded interactions, hydrogen bond interactions, and intrinsic torsional potentials for the naturally occurring amino acids. *J Phys Chem* **79**, 2361, 1975.

A.F. Oberhauser, P.K. Hansma, M. Carrion-Vasquez, and J.M. Fernandez, Stepwise unfolding of titin under force-clamp atomic force microscopy. *Proc Natl Acad Sci USA* **98**, 468–472, 2001.

L. Onsager, Electric moments of molecules in liquids. *J Am Chem Soc* **58**, 1486–1493, 1936.

A. Pattanaik, D.C. Gowda, and D.W. Urry, Phosphorylation and dephosphorylation modulation of an inverse temperature transition. *Biochem Biophys Res Comm* **178**, 539–545, 1991.

M. Rief, M. Gautel, F. Oesterhelt, J.M. Fernandez, and H.E. Gaub, Reversible unfolding of individual titin immunoglobulin domains by AFM. *Science* **276**, 1109–1112, 1997.

L.B. Sandberg, J.G. Leslie, C.T. Leach, V.L. Alvarez, A.R. Torres, and D.W. Smith, Elastin covalent structure as determined by solid phase amino acid sequencing. *Pathol Biol (Paris)* **33**, 266–274, 1985.

L.B. Sandberg, N.T. Soskel, and J.G. Leslie, Elastin structure, biosynthesis, and relation to disease states. *N Engl J Med* **304**, 566–579, 1981.

F. Sciortino, K.U. Prasad, D.W. Urry, and M.U. Palma, Self-assembly of bioelastomeric structures from solutions: mean field critical behavior and Flory–Huggins free-energy of interaction. *Biopolymers* **33**, 743–752, 1993.

F. Sciortino, D.W. Urry, M.U. Palma, and K.U. Prasad, Self-assembly of a bioelastomeric structure: solution dynamics and the spinodal and coacervation lines. *Biopolymers* **29**, 1401–1407, 1990.

D.W. Urry, (1990) Protein folding and assembly: an hydration-mediated free energy driving force. In: L. Gierasch and J. King (eds) *Protein Folding: Deciphering the Second Half of the Genetic Code.* (pp. 63–71) Am Assoc for the Advancement of Sci, Washington, DC.

D.W. Urry, (1991) Thermally driven self-assembly, molecular structuring and entropic mechanisms in elastomeric polypeptides. In: P. Balaram and S. Ramaseshan (eds) *Mol Conformation and Biol Interactions.* (pp. 555–583) Indian Acad of Sci, Bangalore, India.

D.W. Urry, Molecular machines: how motion and other functions of living organisms can result from reversible chemical changes, *Angew Chem (German)* **105**, 859–883, 1993; *Angew Chem Int Ed Engl* **32**, 819–841, 1993.

D.W. Urry, Physical chemistry of biological free energy transduction as demonstrated by elastic protein-based polymers (invited FEATURE ARTICLE). *J Phys Chem B* **101**, 11,007–11,028, 1997.

D.W. Urry, C-H. Luan, and S.Q. Peng, Molecular Biophysics of Elastin Structure, Function and Pathology, in Proceedings of The Ciba Foundation Symposium No. 192, The Molecular Biology and Pathology of Elastic Tissues, John Wiley & sons, Ltd., Sussex, UK, pp. 4–30, 1995.

D.W. Urry and C-H. Luan, (1995a) A New Hydrophobicity Scale and its Relevance to Protein Folding and Interactions at Interfaces, in Proteins at Interfaces 1994, American Chemical Society Symposium Series, (Thomas A. Horbett and John L. Brash, eds.), pp. 92–110, Washington, D.C.

D.W. Urry and C-H. Luan, (1995b) Proteins: structure, folding and function. In: G. Lenaz (ed.) *Bioelectrochemistry: Principles and Practice.* (pp. 105–182) Birkhäuser Verlag AG, Basel, Switzerland.

D.W. Urry and C.M. Venkatachalam, A librational entropy mechanism for elastomers with repeating peptide sequences in helical array. *Int J Quant Chem: Quant Biol Symp* **10**, 81–93, 1983.

D.W. Urry, D.K. Chang, R. Krishna, D.H. Huang, T.L. Trapane, and K.U. Prasad, Two dimensional proton nuclear magnetic resonance studies on poly(VPGVG) and its cyclic conformational correlate, cyclo(VPGVG)$_3$. *Biopolymers* **28**, 819–833, 1989.

D.W. Urry, C.M. Harris, C.X. Luan, C-H. Luan, D.C. Gowda, T.M. Parker, S.Q. Peng, and J. Xu, (1997a) Transductional protein-based polymers as new controlled release vehicles. In: K. Park (ed.) *Controlled Drug Delivery: The Next Generation, Part VI: New Biomaterials for Drug Delivery*, (pp. 405–437) Am Chem Soc Professional Reference Book.

D.W. Urry, B. Haynes, H. Zhang, R.D. Harris, and K.U. Prasad, Mechanochemical coupling in synthetic polypeptides by modulation of an inverse temperature transition. *Proc Natl Acad Sci USA* **85**, 3407–3411, 1988b.

D.W. Urry R. Henze, P. Redington, M.M. Long, and K.U. Prasad, Temperature dependence of dielectric relaxations in α-elastin coacervate: evidence for a peptide librational mode. *Biochem Biophys Res Commun* **128**, 1000–1006, 1985a.

D.W. Urry, T. Hugel, M. Seitz, H. Gaub, L. Sheiba, J. Dea, J. Xu, and T. Parker, Elastin: a representative ideal protein elastomer. *Philos Trans Roy soc London [Biol]* **357**, 169–184, 2002a.

D.W. Urry, T. Hugel. M. Seitz, H. Gaub, L. Sheiba, J. Dea, J. Xu, F. Prochazka, and T. Parker, (2002b) In: P. Shewry & A. Bailey (eds) *Ideal Protein Elasticity: The Elastin Model.* Cambridge University Press (in press).

D.W. Urry, J. Jaggard, K.U. Prasad, T. Parker, and R.D. Harris, (1991) Poly(Val^1-Pro^2-Ala^3-Val^4-Gly^5): a reversible, inverse thermoplastic. In: C.G. Gebelein (ed.) *Biotechnology and Polymers* (pp. 265–274) Plenum Press, New York.

D.W. Urry, N. Jing, T.L. Trapane, C-L. Luan, and M. Waller, (1988a) Ion interactions with the gramicidin A transmembrane channel: cesium-133 and calcium-43 NMR studies. In: W. Agnew, T. Claudio, and F. Sigworth (eds) *Mol Biol of Ionic Channels* **33**, (vol. pp. 51–90) Academic Press, Inc., New York.

D.W. Urry, S.Q. Peng, J. Xu, and D.T. McPherson, Characterization of waters of hydrophobic hydration by microwave dielectric relaxation. *J Am Chem Soc* **119**, 1161–1162, 1997b.

D.W. Urry, S.Q. Peng, L.C. Hayes, D.T. McPherson, J. Xu, T.C. Woods, D.C. Gowda, and A. Pattanaik, Engineering protein-based machines to emulate key steps of metabolism (biological energy conversion). *Biotechnology and Bioengineering* **58**, 175–190, 1998.

D.W. Urry, T.L. Trapane, H. Sugano, and K.U. Prasad, Sequential polypeptides of elastin: cyclic conformational correlates of the linear polypentapeptide. *J Am Chem Soc* **103**, 2080–2089, 1981.

D.W. Urry, T.L. Trapane, M.M. Long, and K.U. Prasad, Test of the librational entropy mechanism of elasticity of the polypentapeptide of elastin: effect of introducing a methyl group at residue-5. *J Chem Soc, Faraday Trans I* **79**, 853–868, 1983a.

D.W. Urry, T.L. Trapane, S.A. Wood, J.T. Walker, R.D. Harris, and K.U. Prasad, D-Ala$_5$ Analog of the elastin polypentapeptide. Physical characterization. *Int J Pept Protein Res* **22**, 164–175, 1983b.

D.W. Urry, T.L. Trapane, S.A. Wood, R.D. Harris, J.T. Walker, and K.U. Prasad, D-Ala$_3$ Analog of elastin polypentapeptide: an elastomer with an increased Young's modulus. *Int J Pept Protein Res* **23**, 425–434, 1984.

D.W. Urry, T.L. Trapane, M. Iqbal, C.M. Venkatachalam, and K.U. Prasad, Carbon-13 NMR Relaxation studies demonstrate an inverse temperature transition in the elastin polypentapeptide, *Biochemistry* **24**, 5183–5189, 1985b.

D.W. Urry, T.L. Trapane, R.B. McMichens, M. Iqbal, R.D. Harris, and K.U. Prasad, Nitrogen-15 NMR relaxation study of inverse temperature transitions in elastin polypentapeptide and its Cross-Linked Elastomer. *Biopolymers* **25**, S209–S228, 1986.

D.W. Urry, C.M. Venkatachalam, S.A. Wood, and K.U. Prasad, (1985c) Molecular structures and librational processes in sequential polypeptides: from ion channel mechanisms to bioelastomers. In: E. Clementi, G. Corongiu, M.H. Sarma, and R.H. Sarma (eds), *Structure and Motion: Membr, Nucleic Acids and Proteins* (pp.

References

185–203), Adenine Press, Guilderland, New York.

D.W. Urry, C.M. Venkatachalam, M.M. Long, and K.U. Prasad, (1982) Dynamic β-spirals and A librational entropy mechanism of elasticity. In: R. Srinivasan and R.H. Sarma (eds) *Conformation in Biol.* (G.N. Ramachandran Festschrift Volume pp. 11–27). Adenine Press, Guilderland, New York.

C.M. Venkatachalam and D.W. Urry, Development of a linear helical conformation from its cyclic correlate. β-spiral model of the elastin poly(pentapeptide), (VPGVG)$_n$. *Macromolecules* **14**, 1225–1229, 1981.

C.M. Venkatachalam and D.W. Urry, Calculation of dipole moment changes due to peptide librations in the dynamic β-spiral of the polypentapeptide of elastin. *Int J Quant Chem: Quant Biol Symp* **12**, 15–24, 1986.

D. Voet and J.G. Voet, (1995) *Biochemistry.* 2nd edn. (p. 430, Table 15-3) John Wiley & sons, Inc., New York.

D. Volpin, D.W. Urry, B.A. Cox, and L. Gotte, Optical diffraction of tropoelastin and α-elastin coacervates. *Biochim Biophys Acta* **439**, 253–258, 1976a.

D. Volpin, D.W. Urry, B.A. Cox, I Pasquali-Ronchetti, and L. Gotte, Studies by electron microscopy on the structure of coacervates of synthetic polypeptides of tropoelastin. *Micron* **7**, 193–198, 1976b.

Z.R. Wasserman and F.R. Salemme, A molecular dynamics investigation of the elastomeric restoring force in elastin. *Biopolymers* **29**, 1613–1631, 1990.

P.K. Weiner and P.A. Kollman, AMBER: assisted model building with energy refinement. A general program for modeling molecules and their interactions. *J Comput Chem* **2**, 287–303, 1981.

T. Weis-Fogh and S.O. Andersen, New molecular model for the long-range elasticity of elastin. *Nature* **227**, 718–721, 1970.

Appendix 2
Development of Elastic Protein-based Polymers as Materials for Acoustic Absorption

Dan W. Urry, J. Xu, Weijun Wang, Larry Hayes, Frederic Prochazka, and Timothy M. Parker

Abstract

Elastic protein-based polymers comprised of repeating pentapeptide sequences, $(GXGXP)_n$, exhibit mechanical resonances that have been observed to date with frequency maxima near 5 MHz and 3 kHz. Because the 3 kHz resonance is in the middle of the acoustic frequency range, the purpose here is to substantiate the relevance of the 3 kHz resonance to acoustic absorption and to demonstrate means of improving mechanical properties for the sound absorption application. Previously reported loss factor data in the 100 Hz to 10 kHz range is substantiated by relevant but distinctly different measurements of loss shear modulus and loss permittivity. Furthermore cross-linking approaches are reported that result in increased elastic moduli by an order of magnitude to 4×10^6 Pa at 20% strain and increased break stress by two orders of magnitude to 1.3×10^7 Pa while exhibiting break stain values of several hundred per cent.

Introduction

When seeking materials suitable for low frequency sound absorption, compliant or elastomeric materials become a natural consideration. Classical elastomers, or rubbers, are reasonable and commonly used materials, but they have a fundamental limitation. They are composed of random chain networks that necessarily exhibit low intensity, very broad and essentially featureless frequency dependences of relaxation (or absorption). Because of this, the possibilities of tuning or designing classical elastomers to maximize sound absorption over a desired frequency range are limited.

On the other hand certain elastic protein-based polymers can exhibit good elastic moduli and yet contain dynamic structural regularities that exhibit mechanical modes. In fact elastic protein-based polymers can be designed to exhibit mechanical resonances with high loss factors over interesting frequency ranges.

Preliminary Acoustic Absorption Studies on Elastic Protein-based Polymeric Materials

The elastic protein-based polymer, (glycyl-valyl-glycyl-isoleucyl-prolyl)$_n$ abbreviated as $(GVGIP)_n$, emerged as a particularly interesting composition; it exhibited a large hysteresis in macroscopic stress-strain studies and in atomic force microscopy(AFM)-based single-chain force-extension studies.[1] Therefore, (GVGIP) might be expected to exhibit substantial loss factors as required for good acoustic absorption.

Accordingly, two γ-irradiation (20 Mrad) cross-linked cylinders of X^{20}-$(GVGIP)_{260}$ were

---

Bioelastics Research Ltd., 2800 Milan Court, Suite 386, Birmingham, AL, 35211-6918, USA.

Introduction

prepared with diameters of 2.5 cm and lengths of 0.5 cm and 1.0 cm. Lev Sheiba and Jack Dea examined the absorption properties of these cylinders at the SPAWAR (Spatial Warfare) Systems Center in San Diego. The results are given in Figure 1A where the elastic protein-based polymer is compared to natural rubber and an absorption enhanced polyurethane.[1] As to be expected for the random chain networks of natural rubber and polyurethane, both exhibited broad featureless dependences on frequency and relatively low intensity loss factors (sound absorption/unit volume). On the other hand, the elastic protein-based polymer exhibited a localized relaxation that grew in intensity as the temperature increased, just as had been observed in the dielectric relaxation spectra, specifically in the imaginary part of the dielectric permittivity, of $(GVGIP)_n$ at 5 MHz (See Figure 1B).[2]

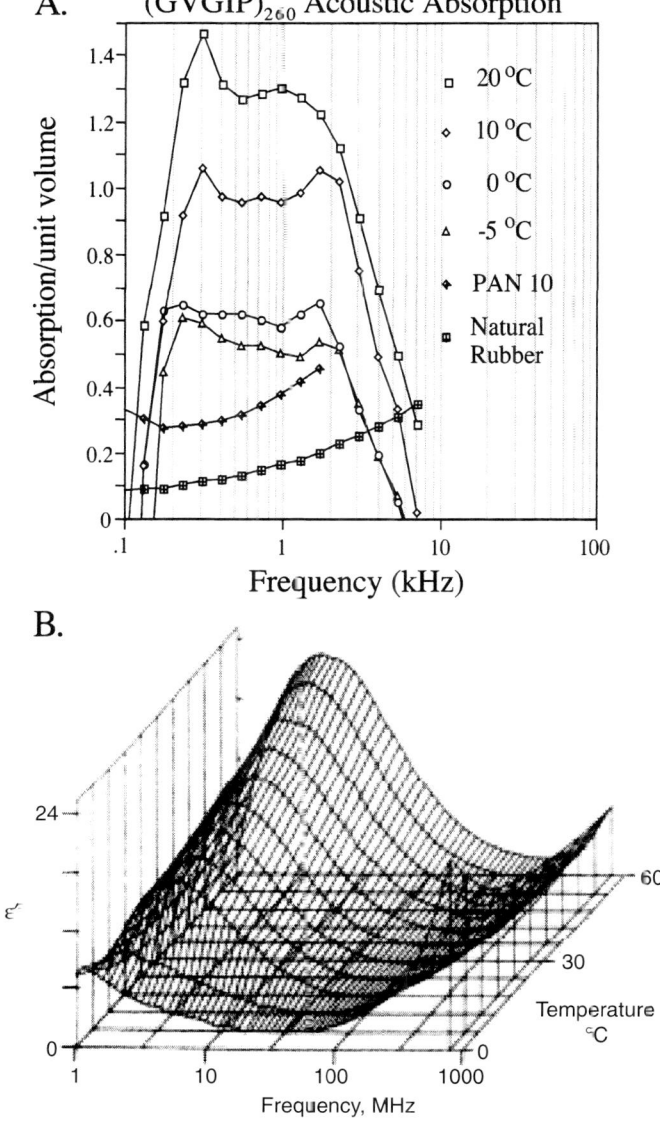

FIGURE 1. A. Acoustic absorption of 20 Mrad γ-irradiation cross-linked $(GVGIP)_{260}$. B. 5 MHz dielectric relaxation of poly(GVGIP) (imaginary part, ε'').

Molecular Structure of (GXGXP)-type Elastic Protein-based Polymers

(GXGXP)-type protein-based elastomers form β-spiral structures described as a series of β-turns arranged in helical array with dynamic suspended segments between the β-turns. For discussion of the single secondary structural feature of the β-turn, the repeating conformational feature is considered in terms of the sequence permutation, $X^1P^2G^3X^4G^5$. With this residue numbering, the β-turn is a ten-atom hydrogen-bonded ring involving the C-O of the X^1 residue preceding the proline residue and the NH of the X^4 residue. The β-spiral structure in band representation of the backbone is shown on the left side of Figure 2, and cross-eye stereo views of the detailed molecular structure are seen in side and spiral axis views in the center structures of Figure 2. At the molecular level, there occur suspended segments connecting the β-turns that contain peptide moieties capable of large amplitude rocking motions. The suspended segments exhibit large amplitude torsional oscillations that become damped on extension. Thus, the suspended segments that link the β-turns function as entropic molecular springs.

The β-turns themselves provide dynamically active surface elements, effectively forming a series of plates on the surface of the wide dynamic helix that is the β-spiral. A macroscopic-classical analog would be equivalent to springs connecting the β-turn plates. The β-turn surface elements may be thought of as molecular acoustic veins attached to springs that are tunable to resonate at desired frequencies. A cartoon demonstrating these structural features is shown as the right-most illustration of Figure 2. The four spiral structures are separated, such that when the two central detailed structures are viewed in stereo, the backbone band representation and the cartoon also overlap with detailed molecular structures.

Mechanism of Elasticity and Mechanical Resonances in Repeating Peptide Sequences

By means of AFM (atomic force microscopy), single-chain force-extension curves have been obtained on Cys-(GVGIP)$_{nx260}$-Cys and Cys-(GVGVP)$_{nx251}$-Cys, where Cys is the cysteinyl residue with a sulfhydryl for attachment to surfaces.[1] In the experiment, one end of the single chain is attached to the cantilever tip and the

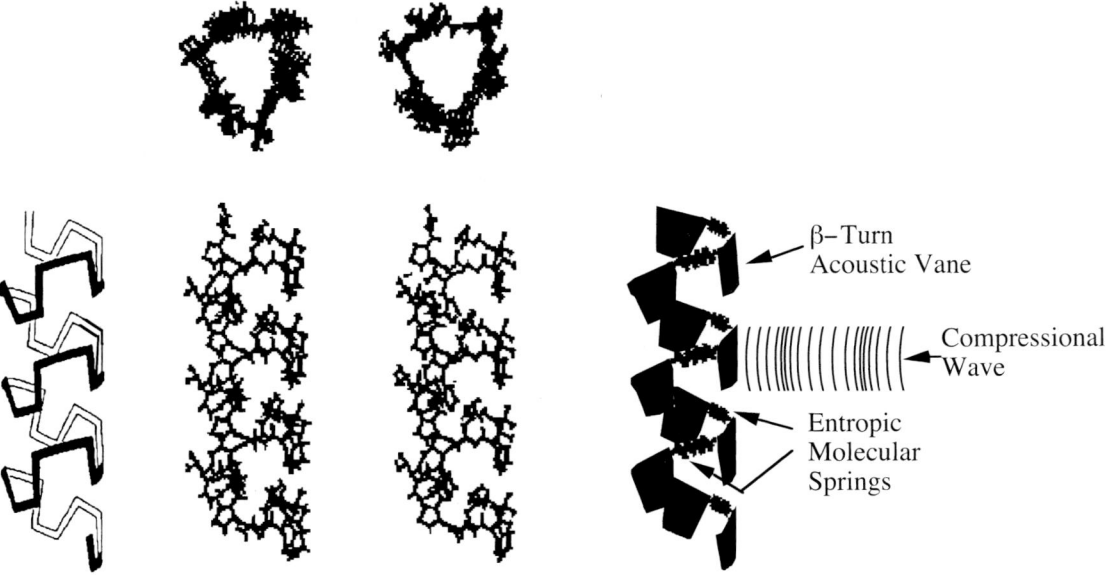

FIGURE 2. Molecular structure and acoustic absorption cartoon for elastic protein-based polymers of the GXGXP-type repeating pentapeptide sequence.

Introduction

other end to a surface. Force is measured as a function of chain length, and elastic moduli of several pN/chain have been obtained. In this experiment, a single chain held fixed at both ends gives rise to entropic elastic force. Since total molecule rotations and translations are not possible, the decrease in entropy, that gives rise to an entropic elastic force on extension to a particular fixed length, must come from the damping of internal chain dynamics.[3] Therefore, it is established that entropic elasticity can occur without random chain networks and without a random, or Gaussian, distribution of end-to-end chain lengths.

Importantly, the intensity of the 5 MHz dielectric relaxation of $(GVGVP)_n$ can be calculated from the molecular dynamics of the structure of the repeating unit in Figure 2 using Onsagers's equation for polar liquids.[4,5] Quoting from the *McGraw-Hill Concise Encyclopedia of Science and Technology*, 2nd Edition, 1989, page 1599, "When a mechanical or acoustical system is acted upon by an external periodic driving force whose frequency equals a natural free oscillation frequency of the system, the amplitude of the oscillation becomes large and the system is said to be in a state of resonance." The repeating structural units of Figure 2 effectively resonate, and the absorption (loss factor) peaks in the relaxation spectra may be called mechanical resonances. Furthermore, the occurrence of mechanical resonances in elastic-protein-based polymers makes the classical theory of rubber elasticity only marginally relevant to these polymers and the use of related considerations in treatments of acoustic absorption by such polymers would also be suspected of similar limitations. Finally, the occurrence of mechanical resonances in these elastic protein-based polymers presents unique opportunities for designing materials with acoustic absorption properties that can be improved over current elastomeric materials.

Acoustic Absorption Data Claimed to Be Artifact

In spite of the occurrence of the same temperature dependent phenomena near 5 MHz (Figure 1B) and near 3 kHz (Figure 1A) using the same elastic protein-based polymer composition, the acoustic absorption data of Figure 1A on the elastic protein-based polymer has been declared to be an artifact by academic professionals concerned with acoustic absorption applications for the Navy. This is in spite of the fact that the similarly determined data for natural rubber and polyurethane did not appear to be in question. At the same time that the data from SPAWAR has been declared to be artifact, the party making the claim refuses to measure, in their "uniquely artifact-free" instrumentation, the base polymer composition as well as additional analogues designed to extend further the acoustic absorption properties of this family of elastic protein-based polymers. Because of this, independent approaches have been sought and they have been found to evaluate the significance of the data in Figure 1A.

Content of This Report

This report presents a family of ten elastic protein-based polymers that have been designed to achieve even more potent acoustic absorption materials. They have been prepared using recombinant DNA technology and transformation of *E. coli* and have been expressed in large quantities through *E. coli* fermentation. Improved mechanical properties—such as essential work of fracture, elastic modulus, break stress and break strain—have been achieved by obtaining higher molecular weight, monodisperse polymers and by improved cross-linking approaches. The design for improved chemical cross-linking utilized the concept of apolar-polar repulsive free energies of hydration. Importantly for developing improved acoustic absorption materials, using the base composition of the family, X^{20}-$(GVGIP)_{320}$, a combination of loss shear modulus, G'', measurements and loss permittivity, ε'', data have been obtained that substantiate the existence of an intense temperature dependent acoustic absorption, that is a mechanical resonance, centered near 3 kHz.

Experimental Details

Production of Polymers by Recombinant DNA Technology

The list of elastic protein-based polymers of Table 1 were prepared by recombinant DNA technology. This involves the construction of a monomer gene from two chemically synthesized, overlapping DNA sequences that were made into the complete double-stranded DNA by the polymerase chain reaction (PCR). Trace amounts of the monomer genes so prepared were inserted into an appropriate plasmid and a strain of *E. coli* was transformed that produced a large number of copies of the monomer gene. The isolated, relatively large quantity of monomer gene was polymerized (in terms molecular biology, concatemerized or concatenated) and ultimately used to transform a strain of *E. coli* that was suitable for expression of the elastic protein-based polymer.[6] Table 1 gives the size of the gene for the resulting elastic protein-based polymer in terms of the number of repeating peptide units and the amount of polymer prepared for subsequent characterization.

Preparation of Polymers I′ through VI′ was for chemical cross-linking efforts, whereas Polymers VII′ through XI′ and XV′ and XVII′ were prepared for systematic change in hydrophobicity and for inclusion of a sulfate to introduce the effect on acoustic absorption of $MgSO_4$ association/dissociation in polymers of different hydrophobicities. Finally, Polymer XVII′ was specifically prepared for its higher molecular weight and for a more monodisperse molecular weight than had been prepared previously.

Cross-linking of Polymers

Preparation of γ-irradiation cross-linked matrices: In preparation for γ-irradiation cross-linking, the polymer is dissolved in water at low temperature. On raising the temperature above that of the inverse temperature transition for hydrophobic association, phase separation occurs. The phase-separated state is then exposed to 20 Mrad of γ-irradiation from a cobalt-60 source.

Preparation of chemically cross-linked fibers: Chemical cross-linking was achieved by extrusion of solutions of E- and K- pairs of polymers at equal concentrations of functional groups into a saturated solution at 50 C of water soluble carbodiimide (EDC, 1-(3-dimethylaminopropyl)-3-ethylcarbodiimide) to form amide bonds between the carboxyl of glutamic acid (E) residues and the ε-amino groups of lysine (K) residues.[7] When in an adequately hydrophobic elastic protein-based polymer, the charged carboxylate and amino functions experience a driving force for ion-pairing. The force

TABLE 1.

		prepared grams	pentamers per polymer
Polymer I′	$[(GVGIP\ GEGIP\ GVGIP)_3]_n$	17	240
		34	108
Polymer II′	$[(GVGIP\ GVGIP\ GEGIP\ GVGIP\ GVGIP\ GVGIP)]_n$	58	276
Polymer III′	$[(GEGIP\ GVGIP\ GEGIP\ GVGIP\ GVGIP\ GVGIP)]_n$	31	180
Polymer IV′	$[(GVGIP\ GKGIP\ GVGIP)_3]_n$	0	
Polymer V′	$[(GVGIP\ GVGIP\ GKGIP\ GVGIP\ GVGIP\ GVGIP)]_n$	20	252
		20	144
Polymer VI′	$[(GKGIP\ GVGIP\ GKGIP\ GVGIP\ GVGIP\ GVGIP)]_n$	0	
Polymer VII′	$[(GVGIP)_{21}(GYGIP)]_n$	60	242
Polymer IX′	$[(GVGIP)_{11}(GYGIP)]_n$	40	216
Polymer XI′	$[(GVGIP)_8(GYGIP)]_n$	47	288
Polymer XIII′	$[(GVGIP)_5(GYGIP)]_n$	30	240
Polymer XV′	$\{[(GVGIP)_2(GYGIP)]_3\}_n$	trace	135
Polymer XVII′	$[(GVGIP)_{10}]_n$	247	320
Polymer VIII′	$[(GVGIP)_{21}(GY\{SO_4^=\}GIP)]_n$	19	242
Polymer X′	$[(GVGIP)_{11}(GY\{SO_4^=\}GIP)]_n$	19	216

arises due to an apolar-polar repulsive free energy of hydration, ΔG_{ap}, that is measurable from the hydrophobic-induced pKa shifts.[8] Cross-linking occurred on extrusion to form fibers and also in a manner to result in disks. When using microbially prepared elastic protein-based polymers with functional groups occurring every thirty residues and with the favorable ΔG_{ap}, materials of exceptional mechanical properties are obtained, for example fibers exhibited elastic moduli and break stresses that are one to two orders of magnitude greater than obtained with γ-irradiation cross-linking.

Determination of Shear Moduli

The instrument for determination of shear moduli was a Rheometric Scientific dynamic mechanical thermal analyzer, model DMTA V. Round shear sandwich geometry was used. The instrument was inverted so that the sandwich fixtures and sample were in water. A water-jacketed 1000 mL Pyrex cylinder supplied by Rheometric Scientific allowed control of temperature with a circulating temperature bath. A sinusoidal linear shear was applied by moving a flat plate between two identical disk-shaped samples over a specified range of frequencies. The two identical disk-shaped samples were sandwiched between the moving plate and two 12 mm diameter plates (called studs) fastened to a frame. The dynamic frequency sweeps to obtain the loss shear modulus, G'', from 0.16 Hz to 318 Hz were reported in log scale.[9] For our purposes the applied initial static force was 0.05 N. Sample size was 12 mm diameter and 0.7 mm thickness. The sample was equilibrated at 40°C for 12 hours prior to starting the series of measurements. Shear moduli were measured from high to low temperature with an equilibration time of 2 hours at each temperature. The sequence of measurements was 40°C, 30°C, 25°C, and 15°C.

Determination of Low Frequency Dielectric Relaxation

Low frequency dielectric relaxation data was obtained with a Rheometric Scientific dielectric thermal analyzer (DETA) system, model MK-II. The unit consists of a temperature programmer, dielectric thermal analyzer, and water-jacketed inverted can-shaped enclosure, called a furnace. Within the inverted can-shaped enclosure is a pair of disk-shaped parallel plates between which the sample is placed. The sample was sealed between 0.01 mm Teflon sheets (Goodfellow, Cambridge, England) and the sealed packet was placed between the two 15 mm-diameter parallel plates. An alternating electric field is applied to the sample by means of a sinusoidal voltage and an out of phase current is measured to obtain the loss permittivity, ε''. An external temperature bath was used for temperature control. Sample size for X^{20}-$(GVGIP)_{320}$ was 15 mm diameter and 0.6 mm thickness. The system was equilibrated overnight at 40°C. Temperature was decreased in 5-degree increments with 2 hours of equilibration between temperature changes and 3 sets of measurements were made from 20 Hz to 100 kHz at each temperature.

Basis for Correlation Between Loss Shear Modulus and Loss Permittivity

When a polymer exhibits a maximum in the imaginary part of the dielectric permittivity (the loss permittivity, ε'') at frequencies less than 200 Hz, it becomes possible to make comparisons with the frequency dependence of shear moduli and most specifically with the loss shear modulus, G''. This has been done for polypropylene diol, also called poly(oxypropylene), where there is reported a near perfect superposition of the frequency dependence of the normalized loss shear modulus with that of the normalized loss permittivity as reproduced in Figure 3.[10,11] The acoustic absorption frequency range of interest here is 100 Hz to 10 kHz, yet present macroscopic loss shear modulus data can be determined at most up to a few hundred Hz. Nonetheless, for X^{20}-$(GVGIP)_{320}$ there is a maximum in loss permittivity, ε'', near 3 kHz that develops on raising the temperature through the temperature range of the inverse temperature transition. With the width of the loss permittivity curve a distinct set of curves as a function of temperature become

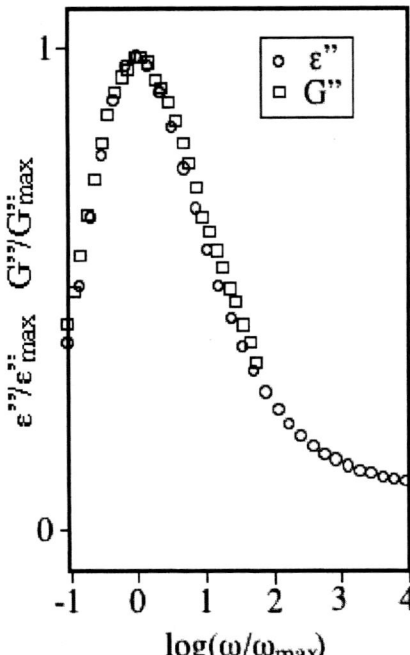

FIGURE 3. Superposition of loss permittivity and loss shear modulus for polypropylene diol. Adapted with permission from references 10 and 11.

ligament widths (distance between notches) were used and work of fracture was plotted against ligament length.

Results and Discussion

Unique Temperature and Frequency Dependence of Loss Shear Modulus Data

It is generally recognized that the loss shear modulus, G'', data in the 1 Hz to 200 Hz frequency range is relevant to acoustic absorption in the 100 Hz to 10 kHz frequency range. Commonly, data obtainable at lower frequencies, generally 100 Hz or less, has been extrapolated into the acoustic frequency range in order to evaluate candidates for good acoustic absorption properties. Such an extrapolation requires an assumption about the frequency dependence. While this becomes a reasonable practice for classical elastomers that occur as random chain networks with the broad, featureless frequency dependence of natural rubber and polyurethane seen in Figure 1A, it is not a reasonable assumption for elastic protein-based polymers that exhibit mechanical resonances as seen in Figure 1B for the loss permittivity of $(GVGIP)_n$.

The variable temperature dependence of the steepness of curvature for the loss shear modulus curves reported in Figure 4 causes concern for an approach that extrapolates to higher frequencies without a better understanding of the frequency dependence. Because of the correlation of loss shear modulus and loss permittivity data for polypropylene diol of Figure 3, loss permittivity data in the acoustic frequency range has particular significance for elastic protein-based polymers that exhibit a high temperature dependence for the loss permittivity as seen in Figure 1B. In fact, as shown immediately below, low frequency dielectric relaxation data provides a means more effective than computational extrapolation with which to evaluate acoustic absorption properties of elastic polymers that exhibit mechanical resonances.

defined as low as the 100 to 200 Hz frequency range. Accordingly, if these loss permittivity curves are sufficiently distinctive and compare favorably with the exactly corresponding set of loss shear modulus curves, the melding of the two sets of data in the overlap region provide for calibration of the loss permittivity curves in terms of loss shear modulus of recognized relevance to acoustic absorption data.

Determination of Essential Work of Fracture

Samples of X^{20}-$(GVGIP)_{320}$ were run on an in house custom-built vertical stress-strain apparatus. A water-jacketed chamber allowed grips and sample to be submerged at all times with temperature control. Sample size was 4 × 5 × 0.5 mm. Samples were double notched with a fresh razor blade and stretched at a speed of 2.2 mm/min until ruptured.[12] Several different

Results and Discussion

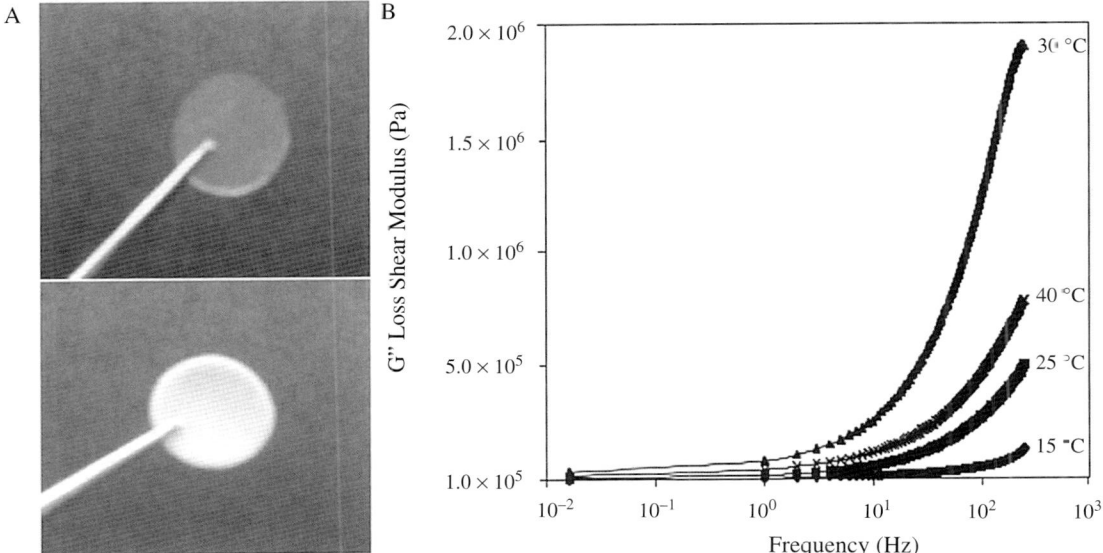

FIGURE 4. **A.** Discs of X^{20}–$(GVGIP)_{320}$. The upper disc is impregnated with a propylene glycol and water mixture and the lower disc contains water only. The lower disc is used in the collection of data in figures 4B and 5. **B.** Loss shear modulus, G'', vs. frequency for $(GVGIP)_{320}$.

Melding of Loss Shear Modulus and Loss Permittivity Data to Evaluate Acoustic Absorption Potential of Elastic Protein-based Polymers

The temperature dependence of the loss shear modulus data of Figure 4 is quite distinctive. At 200 Hz the temperature dependence is one of increasing G'' on increasing the temperature from 15 to 25 C, but then the next larger G'' occurs at 40 C with the value of G'' at 200 Hz for the 30 C data being more than twice as large. On the other hand at 5 MHz in Figure 1B, the increase in intensity of the loss permittivity continues to increase on raising the temperature from 10 to 60 C.

Nonetheless, the merging of loss shear modulus, G'', and loss permittivity, ε'', data for the overlapping frequencies in Figure 5 is truly remarkable. Explanation of the distinctive 40 C curve now becomes obvious. The explanation arises due to the occurrence of the temperature interval of the inverse temperature transition[8] being from 10 to 30 C. During the inverse temperature transition of hydrophobically folding into the structure of Figure 2, the frequency maximum of the loss permittivity peak decreases as the elastic protein-based polymer folds into a more regular structure. Once the inverse temperature transition is complete, the normal increase in frequency with increase in temperature ensues.

A coarse calibration of the higher frequency loss permittivity data becomes possible by using the overlap with the loss shear modulus values. It seems reasonable to conclude that the loss shear modulus values would reach 5×10^6 Pa or more near 3 kHz. The comparison also suggests that the intensity of the peak in Figure 1A would almost double on continuing from 20 to 40 C. It should also be appreciated that the composition of the elastic protein-based polymer can be chosen such that the temperature interval is below 5 C in which case the intense absorption would exist at temperatures above 5 C.

Mechanical Properties of Cross-linked Fibers

When utilizing the apolar-polar repulsive free energy for hydration, ΔG_{ap}, the free energy gained on ion-pairing aligns and orders cross-linking groups, E and K. This is demonstrated for the following two pairs of polymers.

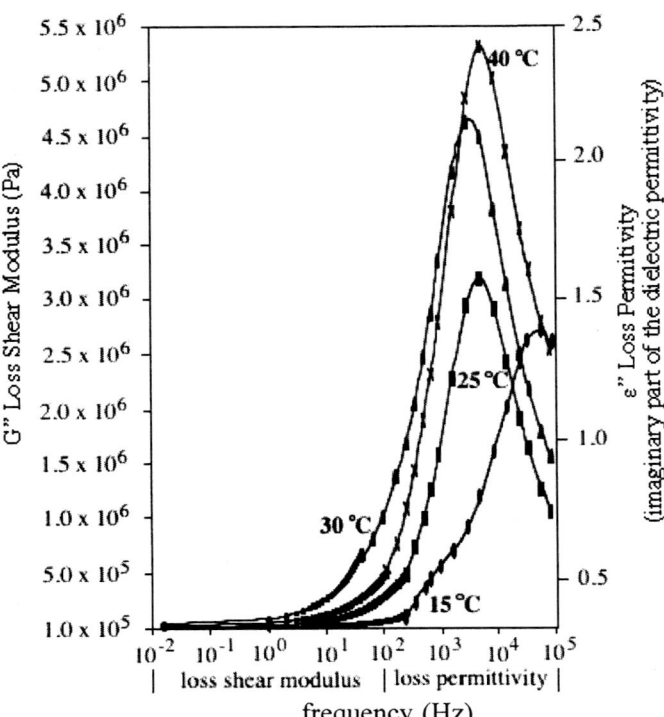

FIGURE 5. Melding of G″ and ε″ data for $(GVGIP)_{320}$.

Polymer i: (GVGVP GVGVP G**E**GVP GVGVP GVGVP GVGVP)$_{36}$ E/0F
(GVGVP)

Polymer x′: (GVGVP GVGVP G**K**GVP GVGVP GVGVP GVGVP)$_{22}$ K/0F
(GVGVP)

and for the pair,

Polymer ii: (GVGVP GVGFP G**E**GFP GVGVP GVGVP GVGVP)$_{40}$ E/2F
(GVGVP)

Polymer xi′: (GVGVP GVGFP G**K**GFP GVGVP GVGVP GVGVP)$_{22}$ K/2F.
(GVGVP)

Micrographs of the fibers formed and representative mechanical properties are given in Figure 6. Among the several experiments, values of the elastic modulus, **E**, at 20% of greater than 10^7 Pa have been obtained and break stresses greater than 1.5×10^7 Pa have also been obtained for both fiber compositions. These values are two orders of magnitude greater than obtained by means or γ irradiation cross-linking. Generally, the more hydrophobic phenylalanine (Phe, F) containing polymers gave the higher values.

Essential Work of Fracture

The stress-strain curves given in terms of the work in J/m^2 versus extension to the point of breakage is seen in Figure 7A. When the specific work of fracture is plotted as a function of ligament length resulting from notching each side of the elastic strip, extrapolation to the intercept gives the essential work of fracture. The results for both X^{20}–$(GVGIP)_{260}$ and X^{20}–$(GVGIP)_{320}$ are shown in Figure 7B. While the gene for $(GVGIP)_{260}$ indicates a chain

Results and Discussion

FIGURE 6. Mechanical properties of chemically cross-linked fibers.

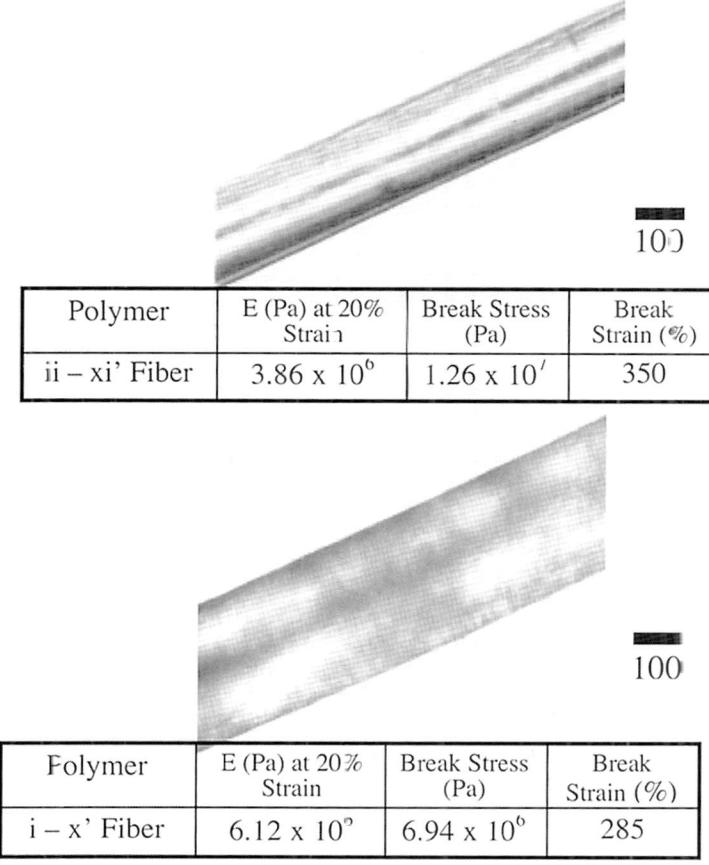

Polymer	E (Pa) at 20% Strain	Break Stress (Pa)	Break Strain (%)
ii – xi' Fiber	3.86×10^6	1.26×10^7	350

Polymer	E (Pa) at 20% Strain	Break Stress (Pa)	Break Strain (%)
i – x' Fiber	6.12×10^5	6.94×10^6	285

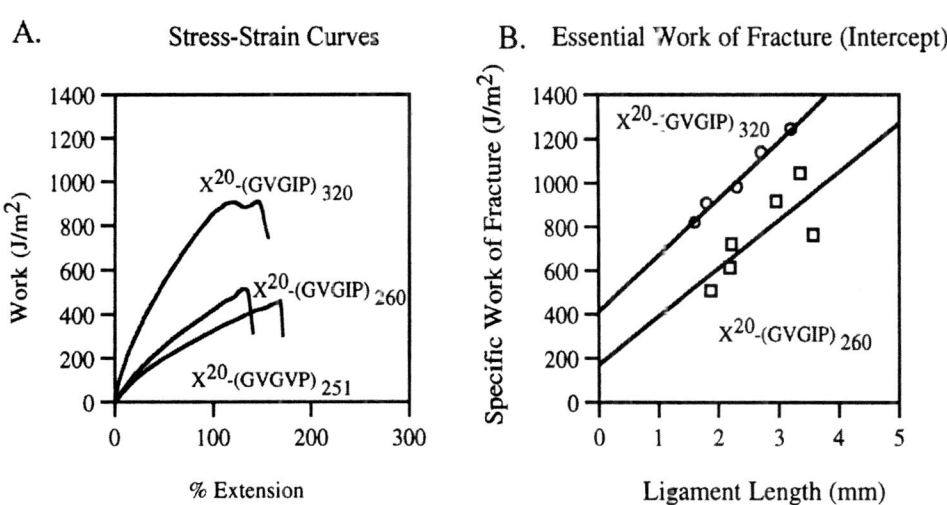

FIGURE 7. Essential work of fracture for $X^{20}-(GVGIP)_{260}$ and $X^{20}-(GVGIP)_{320}$.

length of 260 pentamers, the expressed protein-based polymer contained a smear of chains of lower molecular weights. In making a new monomer gene encoding for 10 pentamers, the same codons were used for the two G residues within the GVGIP pentamer but different codons for G were used between the pentamers. On multimerization a more stable higher molecular weight gene product was expressed, namely $(GVGIP)_{320}$. On 20 Mrad γ-irradiation cross-linking an elastic matrix formed with twice the essential work of fracture that was more consistent with chemically-synthesized poly(GVGIP). Thus, a tougher matrix was obtained that would be more suitable for acoustic absorption applications.

Conclusions

Elastic protein-based polymers of the (GXGXP)-type exhibit mechanical resonances that open a new vista for the development of unique acoustic absorption materials. Previously determined temperature dependent loss factor (absorption/unit volume) curves for X^{20}–$(GVGIP)_{260}$ in the 100 Hz to 10 kHz range have been dismissed as artifact in spite of the fact that exactly parallel behavior as a function of temperature is observed for a mechanical resonance of $(GVGIP)_n$ at 5 MHz by means of loss permittivity measurements. Melding of loss shear modulus and loss permittivity data confirmed the presence of an intense mechanical resonance in the acoustic absorption range for matrices of X^{20}–$(GVGIP)_{320}$. The cylindrical sample of Figure 8 stands ready for acoustic absorption measurements by instrumentation claimed to be uniquely free of artifact.

Using an understanding of the apolar-polar repulsive free energy of hydration. ΔG_{ap},[8] elastic protein-based polymers can be prepared with impressive mechanical properties of elastic modulus and break stresses up to 2×10^7 Pa and break strains of several hundred percent.

Acknowledgments. The authors gratefully acknowledge the Office of Naval Research under Contract Number N00014-00-C-0178 for the support of these materials developments.

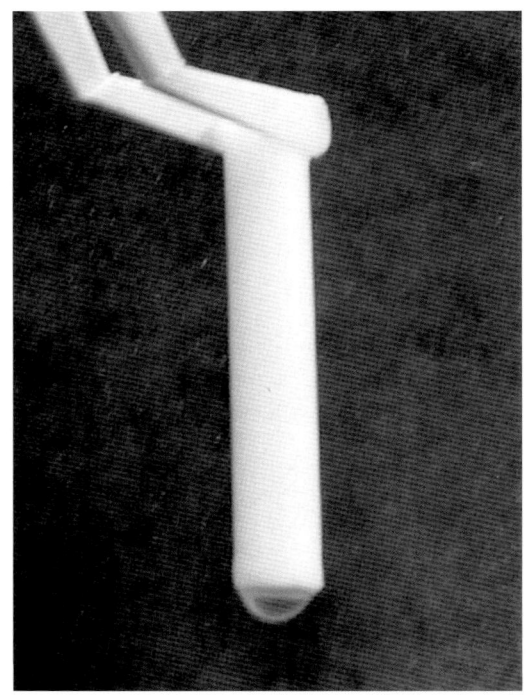

FIGURE 8. Cylinder of 20 Mrad γ-irradiation cross-linked $(GVGIP)_{320}$ prepared in a form suitable for acoustic absorption measurements by the "uniquely artifact free" instrumentation.

References

1. D.W. Urry, T. Hugel, M. Seitz, H. Gaub, L. Sheiba, J. Dea, J. Xu, and T. Parker, "Elastin: A Representative Ideal Protein Elastomer," *Phil. Trans. R. Soc. Lond. B* **357**, 169–184, 2002.
2. R. Buchet, C.-H. Luan, K.U. Prasad, R.D. Harris, and D.W. Urry, "Dielectric Relaxation Studies on Analogs of the Polypentapeptide of Elastin," *J. Phys. Chem.* **92**, 511–517, 1988.
3. D.W. Urry and T.M. Parker, MECHANICS OF ELASTIN: Molecular Mechanism of Biological Elasticity and its Relevance to Contraction, *J. Muscle Res. Cell Motility* **23**, 541–547, 2002; Special Issue: Mechanics of Elastic Biomolecules, Henk Granzier, Miklos Kellermayer, Wolfgang Linke, Eds.
4. C.M. Venkatachalam and D.W. Urry, "Calculation of Dipole Moment Changes Due to Peptide Librations in the Dynamic β-Spiral of the

References

Polypentapeptide of Elastin," *Int. J. Quant. Chem. Quant. Biol. Symp.* **12**, 15–24, 1986.
5. L. Onsager, Electric moments of molecules in liquids. *J. Am. Chem. Soc.* **58**, 1486–1493, 1936.
6. D.T. McPherson, J. Xu, and D.W. Urry, "Product Purification by Reversible Phase Transition Following *E. coli* Expression of Genes Encoding up to 251 Repeats of the Elastomeric Pentapeptide GVGVP," *Protein Expression and Purification* **7**, 51–57, 1996.
7. D.W. Urry, K. Okamoto, R.D. Harris, C.F. Hendrix, and M.M. Long, "Synthetic, Cross-Linked Polypentapeptide of Tropoelastin: An Anisotropic, Fibrillar Elastomer," *Biochemistry* **15**, 4083–4089, 1976.
8. D.W. Urry, "Physical Chemistry of Biological Free Energy Transduction as Demonstrated by Elastic Protein-based Polymers," invited FEATURE ARTICLE, *J. Phys. Chem. B*, **101**, 11007–11028, 1997.
9. B.E. Read, G.D. Dean, and J.C. Duncan, "Determination of dynamic moduli and loss factors." In: *Physical Methods of Chemistry*, Vol **VII**, John Wiley & Sons, 1991.
10. F. Prochazka, PhD Thesis: "Etude de l'evolution des relaxations dans les gels en cours de formation, application aux polyurethanes" (1998) Université du Maine, Le Mans—France.
11. T. Nicolai, F. Prochazka, and D. Durand, "Comparison of Polymer Dynamics between Entanglements and Covalent Cross-links" *Physical Review Letters* **82**, 863–866, 1999.
12. M. Stading, M. Langton, and A.-M. Hermansson, "Small and large deformation studies of protein gels." *J. Rheol.* **39**, 1445–1450, 1995.

Index

A
Abrahams, J.P., 405, 406, 408, 409, 411, 416–418
Actin binding site
  myosin cross-bridge use of, 440, 443–444
  narrow cleft emanation and, 445–446
Adaptability, 28
Adenosine diphosphate (ADP)
  ATP-formed, 13, 320
    repulsion burst and, 422
  ATP produced by phosphorylated, 46, 47
  charged state bound, 4
  phosphate energizing and addition to, 349–350
  phosphorylated, 46, 47
Adenosine triphosphate (ATP), 346–351
  ADP formed by, 13, 320
    repulsion burst and, 422
  ADP, phosphorylated production of, 46, 47
  analogue states, 446–447
  apolar-polar repulsive free energy bursts on hydrolysis of, energy conversions and, 546
  binding, 315, 352–353, 425, 445–446, 448, 518–519
    conformational change and, 349
    negative cooperativity and, 422–423
  site, 314, 445–446

clam-shaped globular protein motif and cross-bridge binding of, 57, 58
cleft residence of, 57–59
contractile rotary motors production of, 5
cross-bridge binding and, 57–58, 433–439
cross-bridge state, detached of, 448
cusp of insolubility, movable and binding of, 315, 318–319
Dictostelium discoidium, myosin motor domain, 12
elastic "pull" component of force and apolar-polar repulsion on binding of, 352–353
F_0-motor for synthase of, 51
F_1-motor for synthase of, 18–19, 51–52, 52
F_1-motor synthase of, "waters of Thales" and, 19–20
folding/unfolding cycle mediated, 308, 309
glucose oxidation for chemical energy of, 68n27
hydrolytic rate, positive cooperativity and, 423
hydrophobic domain repulsion and, 311, 312–314
living organism construction/maintenance and, 350–351
mechanical work and, 555–561
muscle relaxation and binding of, 425

narrow cleft emanation and binding site of, 445–446
β-, γ-phosphate of, 416–419
production, 4–5, 46–47, 550–555
protein-based machines bound to, fully charged carboxylate state and, 10–11
protein-based machines energized by binding, 352
proton gradient and production of, 550–555
push and pull production of, 4–5
repulsion, charged/hydrocarbon group produced by, 11
rigor mortis and, 245
rotary motor production of, 4
scallop muscle analogue states of, 446–447
splitting, 320
standard free energies and phosphate bond hydrolysis of, 347–349
β-subunit and γ-rotor ΔG_{ap}, 419–420
α-subunits containing, 19, 405, 410, 411, 415, 417
synthase, 19–20, 51–52
triphosphate tail binding of, 310
water availability to, 13
ADP, Adenosine diphosphate
AFM. *See* Atomic force microscope

611

Alkalay, R.N., 504–506, 538
Allosteric effects, 21–22
Allosteric proteins
  hydrophobic association identification with T-state of, 243
  R-state, 243
  T-state, 243
Alzheimer's disease
  amyloid deposits, 297
  insolubility, excess and, 240
  seminal amyloid-forming sequence of, 303–304
  solubility-insolubility boundary, 3
Amino acid residues
  aliphatic, slope source, 333–334
  aromatic, slope source, 333–334
  hydrophobic effect and, 333
  hydrophobicity scale based on T_t for, 132–136
  R-groups, 116, 132, 134–135
  sequences, 125
    control, 458
  slope sources, 333–334
  substitution, 132–133
Amino acids
  hydrogen bond between backbone peptide groups and structure of, 125
  hydrophobic association, 125, 126
  residue protein return to free, 100
  structures, 125
  T_t-divide, moving and composition of, 115
  valine replaced by, 115
Amyloid
  deposits, 295
    phenomenology of, 297
  fiber formation, mechanism of, 297–304
  formation, 295, 297–304
  peptide unit and prion protein, 299–300
  prion protein, 298–300
Amyloidosis, 304
Amyloid precursor protein (APP), 303–304
Analgesics, 511
Anemia, sickle cell, 279
Anesthetics, 511

Anfinsen cage function mechanism, 316–318
Animal starch, 76
Animal waxes structure/synthesis, 81
Antigen-antibody complexation, 131
Apolar-polar repulsion, 195–202
  ATP binding, 352–353
  hydrophobic face positioning with, 415, 416
Apolar-polar repulsive free energy, 339, 546
Apolar-polar repulsive free energy of hydration. See ΔG_{ap}
APP. See Amyloid precursor protein
Applied science, 456–457
Arrow of Time, 569
Arterial restoration, 23–24
Association (water insolubility), 11
Atomic force microscope (AFM)
  bioelastic nanosensor for kinase assays and, 565–566
  single chain force extension mode, 525–527
ATP. See Adenosine triphosphate
ATPase motor, 53–56, 351
  "idling," 13–14
ATPases, 13–14. See also F_1-motor ATPase
  force generation in, 351–352
  ΔG_{ap} and, 351–353
  myosin II motor, 351
  skeletal muscle, 13
ATP synthase, 16, 394–424
  breakdown, 394–395
  c subunit of bovine, 402–404
  electron density maps, 400–404
  energy conversions, 394–395
  F_1-motor
    consilient mechanism shown, 18–19
    crystal structure, 404–420
  F_1-motor component of, 17
  function, 394–396
  hydrophobic elastic consilient mechanism and, 423–424
  protein subunits structural/functional roles in, 398–399
  structure

information, 400–420
  schematic representation, 398–400
  synthesis step, critical of, 424
  thermodynamic data, 400–404

B
Back pain, lower, 465
Bacon, Francis (Sir), 547–548
  The New Organon or True Directions Concerning the Interpretation of Nature, 571
Baromechanical transduction, 174
Beeswax synthesis, 81
Biocompatibility, 561–566
  elasticity nature and, 130–131
  elastic protein-based polymer, 561–562
  protein-based polymer, microbially produced, 489–495
  resonances, medical and, 562
Biodegradable plastics, 25–26, 466–467, 533
  biodegradation rates for, 25–26
  model protein, 62, 66
  thermoplastics, 25
Bioelastic disk, pressure ulcer use, 523
Bioelastic matrices, 64–65
Biological interactions, 168
Biological structures
  determination of, 76–90
  proof, 76
  synthesis, 76–90
Biology's Heat Engine, 542
Biology's machines, 50–59, 329–448. *See also specific machines*
Biology's Rival to Magnetism, 542
Biology's vital force, 541–542
  elastic consilient mechanism as mechanical coupler within, 545–547
  known, 561–566
  physical embodiment of, 561
  protein-based machines as physical embodiments of, 548–561

Index

thermodynamics interlinked with molecular dynamics and, 545
Biomaterials, 561–566
   protein-based polymer use as, 488–489
   utility of advanced, 455–456
Biomolecular machines, 62–66
Biosensors, 525–533, 565–566
Biosynthesis, 561–566
   protein-based polymer, 469–476
   proteins of varied sequences produced by, 30–32
Blood clotting
   cusp of insolubility and, 283–294
   fibrin clot, control of, 284–294
the Bohr Effect, 262–263
   complex IV, 392
   redox, 392
Boisvert, D.C., 311, 312, 327
Bovine spongiform encephalopathy (BSE), 296
Boyer, P.B., 404, 412, 453
Bronowski, J., 561
BSE. *See* Bovine spongiform encephalopathy
Butler, J.A.V, 112, 114, 118, 146, 147, 172, 181, 185, 201, 212, 213, 264, 310, 325, 331, 334, 425, 449, 544, 572

C

Candelilla wax synthesis, 81
Carbohydrates
   animal starch, 76
   cellulose, 76
   D-glucose formation of, 76
   photosynthesis production of, 354–355
   plant starches, 76
   structure, 76–77
   synthesis, 76–77
Carbon (C), 29, 72
   mirror image molecules and atom key of, 75–76
   molecules containing, 74–75
Carboxylates, 115
Carboxylate state, fully charged, 10–11
Carboxyl ionization, 254–256
Carnauba wax synthesis, 81

Cell membranes
   lipid bilayer assembly for, 78, 79
   protein channels in, ion flow through, 79–80
   protein machine role in, 79–80
Cellulose, 76
Cerebral thrombosis, 3
CH_2 units
   solubility, water, 109–111
   T_t-divide moving by adding, 113–115
Chain molecules, 1
Chaos
   energy inputs and, 228–235, 229–231
   order created out of, 228–235, 566–568
   reversing, 229–231
Chaperonins
   consilient mechanism exemplified by design of, 314–320
   hydrophobicity unfolding/refolding directed by, 308–314
   molecular structure of, 305
   substrates, 305–308
Charged groups (vinegar-like groups)
   ATP-produced repulsion between hydrocarbon groups and, 11
   classic, 9
   free energy and, 340, 341
   hydrocarbon groups and, 11
   competition for water between, 6–10, 18, 123–124
   hydrophobic hydration and, 119, 340, 341
Chemical energy, 32
   acid, 221–222
   ATP, glucose oxidation for, 63n27
   contraction driven by, 37
   definition, 60–61
   de-mixing achieved by, 221, 230–231
   efficiency, 60–61
   input
      efficiencies, 206–208
      temperature interval and transition zone for, 152–154

   light energy converted to, 44
   mechanical work produced by, 151–156
   motion produced by, 220–222
   oil-like character of protein increased by, 40
   salt, 220–221
   self-assembly achieved by, 221, 230–231
Chemical synthesis, 457–458, 476–482, 477–478
   biocompatibility of protein-based material, 478–482
   DNA, 89–90
   practicality, 478
   protein-based material, 478–482
   protein-based polymer, 476–482
   repeating sequence protein-based polymers solved with, 473, 477–478
Chemical work, 122
Chemochemical transducer, 398–400
Chemoelectrical transduction 122
Chemomechanical transduction, 120, 172
   efficiencies, 342
   efficiency statement for, 208–209
   electrostatic charge-charge repulsion and efficiencies of, 208–210
   myosin II motor, 555–561
Cholesterol synthesis, 80
Chronic wasting disease (CWD), 296
Cinnamide, 137
Citric acid cycle, 356
CJD. *See* Creutzfeld-Jakob disease
Clam-shaped globular protein
   ATP analogue binding to cross-bridge and, 57
   contraction derived from hydrophobically closed, 53, 55
   filament assembly, 52–53
   muscle contraction as variation on, 57, 58
   scallop muscle and motif of, 56–59

Cohen, C., 448
Complex III, 330, 332, 336, 344, 359, 361–366, 368, 371–378, 386–392, 394, 439, 452, 549–551, 561
Concatemerization, 471
Concatenation, monomer gene, 471
Consilient mechanisms, 26n5, 122–124, 212n12, 218
  acceptance, 547–548
  ATP synthase/F_1-ATPase and hydrophobic elastic, 423–424
  biological, 329–448, 332–341
  calcium iron trigger and hydrophobic, 446–448
  chains, dynamic without random chain networks and elastic, 545–546
  chaperonin design exemplifies, 314–320
  chaperonin-directed hydrophobic unfolding/refolding and, 308–314
  chemo-mechanical efficiencies of electrostatic-induced repulsion and, 209–210
  definition, 329
  denaturation, cold relationship with, 202
  de novo design origin of, 2
  diverse, 103–104
  efficiency limit, 204–206
  elastic, 331–332, 423–424, 455, 545–547
  elastic hydrophobic, 423–424, 550
  elasticity, energy conversion and, 127–128
  electron transfer, 375–383
  electron transport coupling and, 361–362
  energy access and achieving useful function of, 219–224
  of energy conservation, 131–147
  energy conversions, 103–104, 332–341
    energy resources for, 131–147
    function coupling and, 166–167

ΔG_{HA}, 169–170
  physical basis of, 176–195
energy resources, 137–146
F_0-motor and hydrophobic, 395
F_1-motor and hydrophobic, 395–398
F_1-motor synthesis of ATP as shown by, 18–19
function, 219–224
functional group coupling and, 164–167
function coupling and, statement of energy conversion, 166–167
genetic code relationship to, 226–227
ΔG_{HA}, 169–170
hemoglobin oxygen binding and, 256–257
hemoglobin oxygen transport and temperature effect diagnostic of, 254
hemoglobin sigmoid oxygen binding and, 278
hydrophobic, 331–332, 395–398, 446–448, 455, 545, 548, 555–556
hydrophobic dissociation relationship with, 202
hydrophobic elastic, 423–424, 550
inverse temperature transition and hydrophobic, 545
mechanical work performed by hydrophobic, 555–556
medical devices and, 562
myosin II motor, 430–446
oil-like domain association controlled by, 103–104
Perutz mechanism melding with, 278–279
protein-based machines and development of, 547–548
  diverse, 102–210
  water required by, 11–13
  as protein-based machines of biology, 329–448
protons transport, at Q_o and Q_i sites, 383
proton transfer, 375–383
push and pull, 4–5
Rayment and coworker contributions to, 432–433
recognition, 2

soft tissue engineering, 562–563
temperature effect diagnostic, 282
as thesis for *What Sustains Life?*, 2–4
"waters of Thales" as requirement for, 336
Consilient protein-based machines
  efficiency, 203–210
  energy conversions, 168–176
  graphical representation, 35–37
  motion produced by, 158–160
Contractile machines. *See also* Elastic-contractile model protein machines
  de novo-designed elastic model proteins of, 22
  mechanical work performed by, 4
  nonmechanical work performed by, 5
  rotary, 5
Contractility, 331
Contraction. *See also* Muscle contraction
  acid addition and, 154–155
  base addition and, 155–156
  chemical energy-driven, 37
  clam-shaped globular protein, hydrophobically closed and deriving, 53, 55
  elastic-contractile model protein thermally driven, 111
  elasticity and, 128–130
  electrical energy-driven, 37
  energies driving, 8, 32, 37
  heat-driven, 154–158
  ion pairing, 156
  isometric, 344
  mechanical work performed by isometric, 344
  model protein, 32
  nicotinamide, lysine attached reduction driving, 222
  pH and energy required to drive, 8
  reduction and heat-driven, 156–158
  temperature increase and shortened, 37

thermal energy-driven, 35–37, 120
transition zone for, 121–122, 154–156
Cooperativity, 308
  ATP binding and negative, 422–423
  ATP hydrolytic rate and positive, 423
  Harris and Rice treatment of, 200
  hemoglobin positive, by consilient mechanism, 278
  hemoglobin subunit interactions and positive, 253
  Henderson-Hasselbalch equation introduced into, 198–200
  hydration competition and positive, 338–339
  hydrophobic domains and positive, 254–257
  hydrophobic interfaces and, 257, 260, 261
  negative, 422–423
  oxygen binding analogues to formation of carboxylates positive, 254–257
  positive, 197, 253–257, 330–331, 338–339, 423
Cortisone synthesis, 81
Creutzfeld-Jakob disease (CJD), 296, 297
Crick, F.H.C., 89
Crofts, A.R., 380, 481
Cusp of insolubility
  anemia, sickle cell and lowered temperature interval for, 279
  blood clotting and, 283–294
  definition, 2–3
  excursions too far from, 3, 239, 241
  F_1-motor, 18
  hemoglobin transport of oxygen and hydrophobic association/dissociation positioning of, 249–263
  increased hydration around oil-like groups lowers, 41–42
  movement, 386
  rotating, 18

temperature interval, 279
ubiquinone reduction and movement of, 386
Cusp of insolubility, movable, 2, 3, 18, 241–263
  ATP binding effect on, 315, 318–319
  biology thrives near, 239–323
  complex III and, 386
  GroES capping effect on, 315, 318–319
  hydration competition and, 18
  molecular process visualized by, 15–16
  moving, 330
  phosphate release and temperature lowered for substrate's, 320
  representation of, 330
  RIP movement and, 386
CWD. See Chronic wasting disease
Cytochrome bc_1 complex, 359, 372, 448, 450, 451, 572
Cytochrome c oxidase, 386–394
  electron transfer steps for oxygen reduction with, 389
  oxygen reduction with, 387–391
  structure, 387
Cytosol, 383–384

D
Daniell, H., 464, 536
Dea, J., 214, 533, 537, 596, 599, 608
Denaturation
  cold, 202
  heat, 106, 107
Deoxyhemoglobin, 253–254
  crystal structure, 265–266
  oxyhemoglobin external differences from, 265–266
  T state, 257
Deoxyribonucleic acid (DNA), 84. See also Recombinant DNA technology
  bases, 87
  double-stranded helix of, 88–89, 89
  replication, 94, 96–97
  RNA from transcription of, 94, 97–98

sequence probability in terms of equilibrium constant, 97
stop codons, 224
synthesis, 89–90
306 base sequences, improbability, 96–97
transcription of, into RNA, 94, 97–98
triplets in, 224
Dephosphorylation
  hydrophobic association initiated by, 248–249
  muscle contraction driven by, 425
  structural reorientation, 57, 58
  T_t-divide moving with, 117
Designed model proteins, 6–9, 22
  acid-base titration curves, 7–8
  steepness, 7–8
D-glucose
  carbohydrates formed by, 76
  optical isomers essential to effective function of, 77
  structure, 76
Dictostelium discoidium
  in ATP presence, 12
  myosin motor domain of, 11–12
  stereo views, 12
Dielectric constant, 27n24
  ΔG_{ap} v. low, and ion pairing/ pKa shifts, 344–346
  low, 344–346
  pKa shifts and low, 345–346
Dilute weak acids, 196–197
Dioxygen, negative, 388, 389, 390, 391
Discoveries, 455
Diseases. See also specific diseases
  genomics and, 91
  insolubility and, 239
  lung, environmentally induced, 322–323
  prion protein, 296
  vitamin deficiency, 81–82
Disorder. See Entropy
Dissipative, 1, 234
Dissociation (water solubility), 11
DNA. See also Deoxyribonucleic acid
Dodson, G., 325
Doolittle, R.F., 248, 293, 325, 326

Drug addiction
    intervention, 513, 514–515
    model protein control of,
        65–66
    protein-based material
        intervention of, 466
Drug delivery
    leu-enkephalin amide, 65
    model protein, 62, 65–66
    Naltrexone, 65–66
    protein-based material, 24
    protein-based
        polymerolymeric systems,
        511–512
    transdermal, 519–521
    T_t-type protein-based polymer,
        510–524, 563
    ulcer, pressure formation
        prevention and, 522–524
Drug-laden bioelastic patches,
    522–524
Dynamic force spectroscopy,
    533

E
Eddington, A.S., 569
Efficiency
    chemical energy, 60–61
    chemomechanical
        transduction, 208–209
    consilient mechanism, 204–206
    consilient protein-based
        machine, 203–210
    electrochemical transduction,
        water, 343–344
    energy conversions, 204,
        220–221, 341–344, 404
    protein-based machine, 228
Élan vital, 541–542
Elastic-contractile model protein
        machines
    phosphates energized by
        synthetic, 49–50
    poised, design of, 203
    synthetic, 49–50
Elastic-contractile model proteins
    elasticity, 124–125
    hydrophobic association in
        myofibril and, 425–426
    ion-pair formation and
        hydrophobic association
        of, 261–262
    Life-sustaining energies
        interconverted by, 60, 61
    soft tissue reconstruction, 64
    β-spiral, 86–87
    structure, 84–87, 86
        limitations, 50–51
    synthesis, 84–87
    thermally driven contraction
        for, 111
    transition, 67n23
    β-turn, 86
Elastic-contractile protein
        material, 473
Elastic force
    development
        energy transduction and,
            432–433
        hydrophobic association
            coupling with, 331,
            336–337
    energy conversion
        mechanisms, efficient and,
            344
    flexible loops as potential
        storage of, 439–442
    hydrophobic association
        development resulting
        from, 439–442
Elasticity, 34. See also
        Thermoelasticity
    biocompatibility and nature of,
        130–131
    consilient mechanism of,
        energy conversion and,
        127–128
    contractility and, 128–130
    elastic-contractile model
        proteins, 124–125
    entropic, 490–495
    ideal
        dominantly entropic,
            126–127
        fraction of, 128
    mechanical work and nonideal,
        344
    model protein, 124–131
    nonideal, 344
    poly(GVGVP), cross-linked,
        126–128
Elastic model proteins
    (GVGVP)$_{251}$, 31–32
    biocompatibility, 23
    synthetic, shortening of, 34–35
Elastic protein-based machines
    hydrophobic association in
        solution and, 149
    macroscopic, constructing,
        148–149
    motion produced by, 147–150
    observations, 147–148
    solvent entropy changes and
        loading on band of, 150
    temperature and, 147–148
    visual observation of working,
        147–160
Elastic protein-based matrices,
    519–521
Elastic protein-based polymers,
    23, 459–460, 482, 490–497,
    507, 561–562
    antigenicity, 490–495
    biocompatibility, 492, 561–562
    chain, stretched, 67n12
    entropic elasticity and
        antigenicity of, 490–495
    fiber formation of, 496–497
    inert space filler, 507
    mechanical resonances, low
        frequency in, 130, 492–494
    monocytes, human and
        response to, 482
    nonantigenic dynamic, 494–495
    plastic and, 23
    sheet formation of, 491,
        495–496
    tube formation of, 495, 496
Elastin
    amino acid residue repeating
        sequences in, 34–35
    folding pattern of the
        repeating five residue
        sequence of, 31
    heat converted to motion by
        proteins modeled after, 35
    hexapeptide repeating
        sequences of, 124
Electrical energy
    chemical work and conversion
        of, 122
    contraction driven by, 37
    mechanical work produced by,
        156
Electrochemical energy, 41
Electrochemical transduction,
    122
    efficiency of water, 343–344
    hydrophobicity increase and,
        343–344
    phenomena coherence for,
        392

Index

proton concentration gradient production and, 549–550
Electromagnetic energy, 157–158
Electromagnetomechanical transduction, 174–175
Electromechanical transduction, 174
Electrons, 222–223
　transfer, 373, 549–550
　　consilient mechanism, 375–383
　　cytochrome c oxidase oxygen reduction steps of, 389
Electron transport chain, 358, 361–394
　complex III of, 365–371
　complex II of, 365–371
　consilient mechanism and coupling of, 361–362
　ΔG_{ap} and, 361–362
　mitochondria, analogues for, 365–371
　protein machines of, 361
　ubiquinol production for, 365–371
Electrostatic-induced repulsion, 200, 209–210
Electrostatic repulsion mechanism, 61
Electrostatics, 346
Elements, 71–72. See also The Periodic Table of the Elements; The Periodic Table of the Elements of Nature
Elsas, F., 502, 538
Emphysema, pulmonary, 240
Endorphins, 511
Energies. See also specific energies
　accessing, 1–2, 219–224
　conservation, resources available to consilient mechanism of, 131–147
　consilient mechanism and, 131–147
　function of, 219–224
　driving contraction, 32
　environmental, accessing, 2
　Life defining characteristics and, 28
　Life-sustaining, 1–2, 33, 60, 61
　model protein, 231–235

model proteins conversion of living organism-sustaining, 33
motion production by, 220–224
motion production by, visual demonstrations of, 32, 33
non-motion work performed by, 224
order created out of chaos with model protein-contributed, 231–235
protein-based machines and input/output, 161
protein folding, assembling and functioning sources of, 34–43
resources, 71, 131–146
resources available to consilient mechanism of energy conservation, 131–147
solar, 28
standard free, 347–349
T_t-divide moving as measurement of, 112–118
Energy conversions, 28–29. See also Photosynthesis; Respiration
amino acid changes and diverse, 224
ATP hydrolysis bursts of apolar-polar repulsive free energy and, 546
ATP synthase, 394–395
biological, 332–341, 354–360
biology efficiency, aqueous medium of, 341–344
consilient mechanism and coupling of function statement of, 166–167
consilient mechanism of, 103–104, 332–341
　energy resources for, 131–147
　physical basis, 176–195
consilient protein-based machine, 168–176
diverse, 103–104
efficiency, 204, 220–221, 341–344, 404
elastic force and efficient mechanisms for, 344
elasticity consilient mechanism and, 127–128

folding changes, association of oil-like domains and protein-catalyzed, 103
functional group coupling and, 160–168
ΔG_{HA} as coupling process in, 203–204
ΔG_{HA}, consilient mechanism, 169–170
hydrophobic association, 203–204
inverse temperature transition, 543
Life sustaining, 1, 43–48, 103
living organisms, 28–29
man-made, 28
mechanical, 169
mechanism of, 61, 344
in mitochondria, 16–17
motion production, 48–50, 122
muscle contraction, 46–48
non-mechanical work, 122
protein-based machines consilient, 168–176
　three primary classes of, 549–561
protein-based machines as catalysts for, 60
protein, biology's, 50–51
proteins catalyze, 29
reversible, 170–171
synthetic model protein machine emulating, 48–50
temperature change and, 29
ΔT_t-mechanism, 41
Energy converting machine terminology, 168–169
Energy inputs, 206–208
chaos reversed by, 229–231
model protein, 206
order created out of chaos by, 228–235
Energy sources, 34–43
inverse temperature transitions extract order from, 43
muscle contraction motion, 46–48
protein-based machine, 131–132
protein folding, assembling and functioning, 34–43
T_t changing, 131–132
Engines, 168–169
biology's heat, 542

Enkephalins, 511
Entropic elasticity, 490–495
Entropy (disorder), 67n19
  change in, 67n21
  elastic-contractile model protein transition, 67n23
  elastic protein-based machine band loading and changes in solvent, 150
  inverse temperature transitions extract, from energy sources, 43
  inverse temperature transitions providing protein, 39–40
  order decrease and increase in, 106
  phase transitions and changes of, 104–105
  protein, 39–40
  transition increase in, 39
  universe of, 28
Enzymatic cycle, 360
Equilibrium, far from, 234
Escherichia coli (E. coli)
  biosynthesis of proteins and sequences for, 30–32
  complex II analogue of, 365–368
  genetic engineering, 63
  structure, 305
  transformation, 471
"An Essay on the Principle of Population" (Malthus), 70–71
Everse, S.J., 291, 292, 326
Evolution, 218–238
  genomic, 91
  inverse temperature transition, 569–571
  natural selection in, 570–571
  protein-based machine, 218–237, 224–228
  protein-based machines basic sequence, 31

F

F_0-motor, 16
  ATP synthase, 51
  consilient mechanism, hydrophobic and, 395
  housing, 398
  proton flow and rotation of, 400–402
  rotation barrier for, 402–404
  rotor, 398, 400, 401
  stator, 399
F_1-motor, 16–17
  ATP synthase, 17–20, 51–52
  ATP synthase, and consilient mechanism, 18–19
  ATP synthase, crystal structure, 404–420
  consilient mechanism, hydrophobic and, 395–398
  cusp of insolubility, 18
  housing, 398–399
  as rotating cusp of insolubility, 18
  three states, in cross section, 17
  "waters of Thales" of ATP synthase, 19–20
F_1-motor ATPase, 400, 423–424
  efficacy and cooperatives of, 420–423
  hydrophobic elastic consilient mechanism and F_1-, 423–424
  mechanical work efficacy of, 420–422
  rotational direction, 405, 420
  γ-rotor faces, 406–409
  stereo top view of, 412, 413
  structure, 404–406
  subunits, 404–406
  temperature raised by, 400
FAD. *See* Flavin adenine dinucleotide
Familial thalamic dementia, 296
Fatal familial insomnia (FFI), 296
  mutation causative for, 301, 302
Fat structure, 77–79
Fatty acid structure, 77
Feline spongiform encephalopathy (FSE), 296
FFI. *See* Fatal familial insomnia
Fiber propagation, 300
Fibrin, 293–294
Fibrin clot, 284–294
  cross-linking to form hard, 294
Fibrinogen
  crystal structure, 284
  hydrophobic association, 290–293
  hydrophobicity plots of, 288–290
  primary structure, 284
  structure, 284–288
  subunit structure, 284
  thrombin activation of, 290–293
α-fibrinogen, 288–289
β-fibrinogen, 288–290
γ-fibrinogen, 288–289
Filament assembly, 52–53
Fillingame, R.H., 399, 400, 453
Flavin adenine dinucleotide (FAD)
  ΔG_{HA} and reduction of oxidized, 142
  reductase reduction of, 365–371
Flory, P.J., 546, 571
Folding
  ATP-mediated, 308, 309
  hydrophobic, 29, 30, 112
  inverse temperature transitions, assembly transition and hydrophobic, 112
  proteins, 34–43
  temperature, chemicals for, 42
FSE. *See* Feline spongiform encephalopathy
Functional groups
  consilient mechanism and coupling of, 164–167
  coupling, 48, 160–168
  energy conversions and coupling of, 160–168
  ΔG_{HA} and state change of, 141–143
  ΔG_{HA} coupling of, 164–167
  hydrophobicity change due to action of, 165–167
  hydrophobicity increase and, 187–191
  model protein machines and coupling of, 48
  model protein reduction attached, motion, 158–159
  redox, 136
  states, 161–162

G

ΔG_{ap} (apolar-polar repulsive free energy of hydration), 123, 192–195, 240
  ATPase and, 351–353
  as biology's rival to magnetism, 542–545

carboxyl ionization and, 339
charged side chain pattern
 orientations and, 397–398
dielectric constant v., and ion
 pairing/pKa shifts,
 344–346
electron transport chain and,
 361–362
fiber production and, 563–564
ΔG_{HA} coherence with,
 337–338
hydrophobic associations,
 297–298
ion pairing and, 344–346, 360
mechanisms, 342
pharmaceuticals controlled
 release and, 563
γ-phosphate production of,
 416–419
pKa shifts and, 344–346
proton ingress controlled by,
 384–386
γ-rotor, 419–420
β-subunit and γ-rotor, 419–420
Gaub, H.E., 214, 215, 526, 527,
 530, 537, 539, 576,
 594–596, 608
Genetic code, 224–226
 biased, 570–571
 consilient mechanism and
 relationship to, 226–227
Genetic engineering
 E. coli, 63
 insulin, 84
 model protein, 63
Genome, human
 determination of, 90–91
 as "periodic table of life," 90
Genomics, 90–91, 91
Gerstmann-Sträussler-Scheinker
 disease (GSS), 296
 prion protein with related
 mutations causative in,
 301–302
ΔG_{HA} (change in Gibbs free
 energy for hydrophobic
 association), 240
ΔG_{HA} (changing of free energy),
 240
 bound biologically functional
 groups state change by,
 141–143
 calculating, 138
 change estimate, 314

consilient mechanisms, energy
 conversions with, 169–170
energy conversion coupling
 process and, 203–204
energy conversions by
 consilient mechanism,
 169–170
estimated values, 138–139
FAD, oxidized reduction and,
 142
functional group coupling by,
 164–167
ΔG_{ap} coherence with, 337–338
hydrophobic association,
 137–146
ion pairing between chains,
 effect calculated by, 141
NAD, oxidized reduction and,
 142
phosphorylation and, 142–143,
 350
T_t change and, 143–146
vinegar-like side chain
 functional state change
 and, 141
Gibbs free energy, 112. See also
 ΔG_{HA}
allosteric effects Hill plot
 delineation of, 193–194
charged species increase in,
 340, 341
ΔG_{ap}, 192–193
hydrophobic association,
 137–140
Globin chains
 association, 250, 251
 charged and oil-like residue
 distribution in, 251, 252
Glycolysis cycle, 356
Goete, L., 214
GroES capping, 315, 318–319
GSS. See Gerstmann-Sträussler-
 Scheinker disease
$(GVGVP)_{251}$, 31–32

H

Harris and Rice treatment
 cooperativity, 200
Heart attacks
 cusp of insolubility, 284
 insolubility, excess and, 240
Heat
 ionization and transition of,
 184–185

motion and conversion of, 35
rigor induced by, 244–245
transition, 184–185
Heme binding site, 253–254
Heme groups. See also $\alpha^1\beta^2$
 $(\alpha^2\beta^1)$ interheme
 crevasse
 $\alpha^1\beta^2$, 271–274
 $\alpha^2\beta^1$, 271–274
 $\beta^1\alpha^2$, 269–271, 272
 hemoglobin, 267–269
 myoglobin, 267–269
 oxygen binding and
 structure/disposition of,
 266–271
 prospective, 266–267
 structure/disposition of,
 266–271
Hemoglobin, 249–254
 A, phase diagrams, 281,
 282–283
 consilient mechanism and
 oxygen transport
 temperature effect
 diagnostic of, 254
 consilient mechanism as
 source of sigmoid oxygen
 binding of, 278
 consilient mechanism
 demonstrated by
 structures of, 263–283
 cooperativity, positive and
 subunit interactions in,
 253
 deoxygenated states of,
 253–254
 hemes in, 267–269
 binding site geometry and,
 253–254
Hill plot, 8, 22
hydrophobic association and
 fundamental functional
 unit of, 260
myoglobin oxygen binding
 proteins and, 249–250, 251
oxygenated states of, 253–254
oxygen binding and, 265–271,
 278
 analogy to, 255, 256
 consilient mechanism and,
 256–257
 properties, 250, 253
oxygen transport by, 9, 21,
 249–263

Hemoglobin (cont.)
  phase diagrams, 281, 282–283
  phenomena, molecular basis for coherence of, 265–266
  positive cooperativity in, by consilient mechanism, 278
  R state, 257–258
  sigmoid oxygen binding of, 278
  structural changes due to oxygen binding of, 265–271
  structures, 263–283
  T state, 257–258, 265
    aqueous core of, 265
Hemoglobin S, 279–282
  hydrophobic association and tetramers of, 279–281
  persistence, 281–282
  phase diagrams, 281, 282–283
Hemophilia
  cusp of insolubility and, 284
  solubility, excess and, 240
  solubility-insolubility boundary, 3
Henderson-Hasselbalch equation, 196–197
  cooperativity introduced into, 198–200
  titration curve comparison to, 197–198
Hill coefficients, 207–208
Hill plots, 8–9
  Gibbs free energy allosteric effects delineated by, 193–194
  hemoglobin, 22
  myoglobin, 22
  steepness quantified using, 8–9
Himmel, D.M., 431, 434–438, 440–445, 447
Histidine, imidazole side chain, 392–394
Hoban, L.D., 499, 538
Hugel, T., 214, 215, 530, 537, 594–596, 608
Hunte, C., 372, 374, 375, 379, 380, 384, 385, 448, 450
Hydration. *See also* ΔG_{ap}; Hydrophobic hydration
  cusp of insolubility lowered by, 41–42
  ions, dissolving and, 359
  oil-like group, 41–42, 109–112

Hydration competition, 177–179
  apolar-polar, 201
  cooperativity, positive and, 338–339
  cusp of insolubility, movable and, 18
  fundamental process as, 184–191
  hydrophobic-polar, 338
  observation, 177–179
  oil-like and vinegar like group, 6–10, 18, 123–124
  pentagonally arranged water and, 179–181
  pKa shifted state and, 9–10, 123
  positive cooperativity results from, 10
Hydrocarbon units (oil-like groups), 2–3
  ATP-produced repulsion between charged groups and, 11
  charged group competition for water with, 6–10, 18, 123–124
  charged groups and, 11
  control association of, 22–23
  dissociation of, 6
  hydration around, 41–42, 109–112
  hydrophobic hydration disappearance and, 119
  hydrophobic replacing and, 122
  lowered cusp of insolubility and increased hydration around, 41–42
  motion produced by separation from water of, 30
  pentagonal arrangement of water around, 38–39
  phase separation, 6
  re-association of, 6
  separation from water, 6, 30
  solubility, 2–3
  water around, process involving, 109–112
  water balance and, 6, 11
Hydrogen (H), 29, 72, 125
Hydrophobic assembly, 29, 30
Hydrophobic association, 195–196

allosteric proteins and, 243–244
amino acid structure and, 125, 126
amino-terminal domain and lever arm head, 438, 439, 440, 441, 442
dephsophorylation-initiated, 248–249
elastic-contractile model proteins, 261–262, 425–426
elastic force development and, 439–442
  coupling of, 331, 336–337
elastic protein-based machine solution and, 149
energy conversion, 203–204
fibrin, human fragment double D, 293–294
fibrinogen, 290–293
free energy change for, 137–146
ΔG_{ap}, 297–298
ΔG_{HA}, 137–146
Gibbs free energy for, 137–140
hemoglobin fundamental functional unit and, 260
hemoglobin S tetramers and, 279–281
hydrophobic hydration amount on complementary surfaces and, 339–341
interconnecting chain segments stretched by, 546
intermolecular, 53
intramolecular, 53
inverse temperature transition, 249, 297–298, 425–426
ion pair formation in elastic-contractile model proteins, 261–262
muscle contraction and, 245–249, 248–249
muscle contraction by inverse temperature transition, 249, 425–426
myofibril, 425–426
myoglobin-like protein, 249–254
phosphorylation effect on, 530–532
phosphorylation initiated dimerization by, 53, 54

Index

protein machine function and, 202
at Q_o site, 382
rigor states and, 243–245, 425
RIP tip, 362
thermodynamic basis for water, 310–312
transition, 170
Hydrophobic dissociation, 195–196
amino-terminal domain and lever arm head, 438, 439, 440, 441, 442
consilient mechanism relationship with, 202
cusp of insolubility positioning by, and hemoglobin transport of oxygen, 249–263
hemoglobin transport of oxygen and, 249–263
hydrophobic hydration amount on complementary surfaces and, 339–341
pKa shifts after, 195
thermodynamic basis for water, 310–312
Hydrophobic domains
association/dissociation and pairs of spatially assembled, 242–244
ATP, bound repulsion and, 312–314
charges buried as ion pairs within, 312
cooperativity, positive and, 254–257
hydrophobic hydration disappearance and, 242–243
T-R state transition and, 261
Hydrophobic effect, 122–124
amino acid residues and, 333
comprehensive, 332–341
inverse temperature transitions, 334
Hydrophobic face, 415, 416
Hydrophobic folding, 29, 30
Hydrophobic (oil-like) groups
polar group hydration competition with, 338
water abutting, 38, 39
Hydrophobic hydration, 109, 122–123, 130

changes in, 118–119
charged groups and, 119
charged species and, 340, 341
complementary surface amount of, 339–341
destruction, 349–350
disappearance of, 119, 242–243
hydrophobic domains and disappearance of, 242–243
identification, 177–179
insolubility and disappearance of, 119, 242–243
measurement, 177–179
oil-like group increase and disappearance of, 119
solubility and, 119, 181–182, 242–243
thermodynamic properties of, 112
T_t value determined by, 182–184
Hydrophobic-induced repulsion, 200
Hydrophobicity, 134
carboxyl ionization and, 254–256
cooperativity and interfaces of, 257, 260, 261
electrochemical transduction efficiency and increased, 343–344
functional group action and change in, 165–167
functional group composition and increasing, 191
functional group stepwise replacement and increasing, 187–190
increased, 187–190, 300–301, 343–344
ion pairing and model protein, 182, 183
oil-like residue replacement and increased, 300–301
pKa shifts and, 187–191
plots, 258–261
fibrinogen, 288–290
refolding, chaperonin-directed, 308–314
subunit interfaces and, 258–261
unfolding, chaperonin-directed, 308–314

Hydrophobicity scale
amino acid residue, 132–136
T_t-based, 132–137, 146–147
Hydrophobic moieties, 349
Hydrophobic residues, 190
Hydrophobic rotors, 396
Hydroxylation, prolyl, 321–322

I
Insolubility, 2. See also Cusp of insolubility; Cusp of insolubility, movable; Solubility-insolubility boundary
disease and, 239, 240
excursions too far into, 3, 239, 241
hydrocarbon unit, 2–3
hydrophobic domain, 181–182
hydrophobic hydration disappearance and, 119, 242–243
medical problem of protein, 295–304
molecular chaperones and, 304–320
prion protein-induced, 240
protein, 295–304
protein-based biological systems presenting excess of, 239–240
temperature interval, 2–3
T_t-divide between solubility and, 109, 112–113
water, 11
Insulin. See also Proinsulin
discovery, 77
genetic engineering of, 84
synthesis, 82–84, 93n14
Integral membrane proteins, 79
$\alpha^1\beta^2$ ($\alpha^2\beta^1$) interheme crevasse, 271–274, 275
Inverse temperature transitions, 37–38, 43, 85, 482–483
denaturation, cold of proteins and, 107–108
endothermic/exothermic components of, 335–336
energy conversions by means of, 543
energy source order extracted by, 43
evolution by, 569–571

Inverse temperature transitions (*cont.*)
  genetic code's bias for use by, 570
  hydrophobic association, 249, 297–298, 425–426
  hydrophobic consilient mechanism and, 545
  hydrophobic effect, 334
  hydrophobic folding and assembly transition, 112
  inverted phase transitional behavior of, 108–109
  model protein, 108
  muscle contraction and model protein contraction with, 424–426
  muscle contraction by hydrophobic association, 249, 425–426
  phase diagram, 108–119
  phase separation characterizations, 105
  protein-based machine consequences and, 59–62
  protein-based machines and endothermic/exothermic components of, 335–336
  protein entropy and, 39–40
  protein negative entropy provided by, 39–40
  thermodynamics, second law and, 543–545
  universal arrow of time reversal and evolution by, 569–571
  vital force and, 542–545
Ionian Enchantment, 550
Ionian Enlightenment, 70
Ionization
  carboxyl, 254–256
  ΔG_{ap} and, 339
  heat transition and, 184–185
Ion pairing, 359–360
  calculating effect of, 141
  contraction driven by, 156
  elastic-contractile model protein hydrophobic association, 261–262
  ΔG_{ap} and, 344–346, 360
  ΔG_{HA} and calculating effect of, 141

  insolubilization of fiber formation enhanced by low, 302
  low dielectric constant *v*. ΔG_{ap} and, 344–346
  model protein chain, 262
  model protein hydrophobicity and, 182, 183
  protein-based polymer and formation of, 344–345
  T_t-divide moving with pairing of, 117
Ions
  binding of calcium, 447–448
  calcium, 425, 429, 430, 446–448
  dissolving of, 359
  formation, 359–360
  hydration and dissolving of, 359
  hydrophobic consilient mechanism integration of calcium, 446–448
  hydrophobic consilient mechanism powerstroke triggered by calcium, 447–448
  muscle contraction and calcium, 247, 425, 429, 430
  T_t-based hydrophobicity scale and negative/positive, 135–136
Isomers, optical, 77
Isometric contraction, 344

K
Kemppainen, B.W., 521–523, 539
Kendrew, 252
Key molecular players
  model proteins and mechanical work performed by, 45–46
  photosynthesis and charge change of, 45
  respiration and charge change of, 45
Kinase activities
  nanosensor-detected, 528–532
  tumor tissue diagnostics assay of, 529–530
Kinases assays, 532, 565–566
Krebs cycle, 356
Kuru, 296

L
Lange, C., 372, 379, 380, 448, 450, 451
Lanolin synthesis, 81
Laws of nature, 72
Laws of physics, 542
Lecithin
  membrane containing, 79
  structure, 78
Lenaz, G., 450
Leslie, A.G.W., 452, 453
Leu-enkephalin amide
  drug delivery, 65
  release, zero order of positively charged, 512–514
Life. *See also* "Second secret of life"; "What Sustains Life?"
  consilient mechanism and sustaining, 2–4
  energies as characteristics defining, 28
  energies sustaining, 1–2, 60, 61
  energy conversions sustaining, 1, 43–48, 103
  motion as index of, 32–34
  protein-based machines and sustaining, 548–561
  protein-based machines of, 94–101
  vital force sustaining, 548–561
Ligands
  binding, 307–308
  capping, 307–308
  polarity of, evaluation, 412–415
Light energy
  chemical energy converted from, 44
  motion produced by, 223–224
Living organisms
  ATP and construction/ maintenance of, 350–351
  as chemical factories, 91–92
  products and components of, 70–93
Lower animals, vertebrate difference from, 90
Lung diseases, environmentally induced, 322–323
Lysozyme amyloidosis, 304

Index

M

Macromolecules, 543
Macrophages, 322–323
Mad cow disease
  amyloid deposits, 297
  insolubility, excess and, 240
  solubility-insolubility
    boundary, 3
Magnetism, 542
MALDI-TOF. *See Matrix-assisted laser desorption ionization*-time-of-flight
Malthus, Thomas Robert, 70–71
Martz, E., 325, 371, 430
Mass spectra, 485–488
Matrix-assisted laser desorption ionization-time-of-flight (MALDI-TOF), 487–488
Meaningful answers, obtaining, 71–72
Mechanical energy, 32, 169
Mechanical work
  ATP performance of, 555–561
  chemical energy-produced, 151–156
  contractile machines, 4
  electrical energy-produced, 156
  electromagnetic energy-produced, 157–158
  energy conversions not involved in, 122
  F_1-ATPase efficacy for, 420–422
  hydrophobic consilient mechanism and, 555–556
  isometric contraction, 344
  key molecular players, 45–46
  model protein, 45–46
  nonideal elasticity and, 344
  pressure-volume energy production of, 156–157
  thermal energy-produced, 150–151
Mechanisms, efficient, 61–62
Mechanist
  challenge, 236
  vitalist *v.*, 235–237
Mechanochemical transduction, 172, 173
  cellular, 498, 499
  molecular machines efficient for water, 342–343
  pH dependence of length and, 342–343

Mechanoelectrochemical transduction, 174
Mechanoelectromagnetic transduction, 175
Medical devices
  biology's own mechanisms/ devices used by, 562–564
  coatings, 525
  consilient mechanism approach to, 562
  protein-based materials for, 24–25
Membranes. *See also* Cell membranes; Integral membrane proteins; Mitochondrial membrane
  lecithin in, 79
  micelles, 78
  structure, 77–79
  thylakoid, protein-based machines in, 80
Menz, R.L., 397, 404, 412–414, 419–422, 452, 453
Messenger RNA (mRNA), 98
Metabolism, 28
Micelles, 78
Mitochondria
  complex II, 365–371
  complex III, 365–371
  electron transport chain, analogues, 365–371
  energy conversion in, 16–17
Mitochondrial membrane, inner
  protein-based machines for, 80
  proton release/uptake into, 363–365
  proton translocation across, 549
  reactions in, 357–359
Mitochondrion
  oxidative phosphorylation in, 356–359
  structure, 356–357
Model protein v, 9, 10
Model protein machines
  function coupling and, 48
  motion produced by, 50
  photosynthesis, respiration, and motion energy conversions emulated from, 48–50
  progression to biology's machines, 50–59

  protons pumped on reduction by, 48–49
  synthetic, 48–50
Model proteins. *See also* Designed model proteins; Elastic-contractile model proteins
  biodegradable plastics, 62, 66
  contraction with, 32
  de novo-designed, 84, 455
  drug addiction control with, 65–66
  drug delivery, 62, 65–66
  elasticity, 124–131
  energies sustaining living organisms converted with, 33
  energy, 231–235
  inputs, 206
  families of, 22
  genetic engineering, 63
  hydrophobicity, 182, 183
  increased temperature and, 108–110
  insoluble, 108, 109
  inverse temperature transitions, 108, 424–426
  ion pairing and hydrophobicity of, 182, 183
  ion pairing between chains of, 262
  key molecular players perform mechanical work in 45–46
  mechanical work performed by, 45–46
  medical applications, 62
  as molecular engines and motors, 160
  molecular structure, 124–131
  muscle contraction and, 424–426
  order created out of chaos with energy contributed by, 231–235
  order, increased of, 108–110
  pKa shifts and, 192–195
  stretching, internal chain motion decrease and, 126, 127
  tissue reconstruction, 23–24, 62, 63–65
  titrable functional group, acid as energy input, 206

Model proteins (cont.)
  transitions in water, 104–108, 106
Model protein studies
  contraction, 32
  energy conversions in mitochondria as interpreted from, 16
  muscle contraction as inferred from, 15–16
Molecular chaperones, 304–320
Molecular dynamics, 545
Molecular machines, 170, 171–175
  mechanochemical transduction in water and efficient, 342–343
  mutations for creating new and more efficient, 227–228
  protein-based polymer function as, 124
  second kind, 175–176
Molecules
  carbon-containing, 74–75
  mirror images, 74–76
Monoclonal antibodies, 480–481
Monocytes, human, 482
Monod, J.
  allosteric effects discovered by, 21–22
  "second secret of life," 7, 9–11, 21–22
  On Symmetry and Function in Biological Systems, 7
Monomers, 458
  concatenation of, 471
  genes/sequence, 471
  protein-based polymer biosynthesis and, 470–471
Motility, 90–91
Motion, 28
  carboxylate protonation and production of, 158, 159
  chemical energy production of, 220–222
  consilient protein-based machines production of, 158–160
  elastic protein-based machine production of, 147–150
  electron production of, 222–223
  energy conversions, 48–50, 122

energy production of, visual demonstrations, 32, 33
  heat converted to, 35
  as index of Life, 32–34
  light energy production of, 223–224
  light use and production of, 160
  model protein machines, synthetic production of, 50
  muscle contraction efficient production of, 424
  oxidized functional group attached to the model protein reduction and production of, 158–159
  polymer phase separation and, 104
  pressure change and production of, 159, 222
  production, 50, 119–122, 147–150, 158–160, 220, 424
  salt added to model protein and production of, 158, 159
  salt conversion efficiency into, 220–221
  synthetic model protein machine emulating energy conversions of, 48–50
  temperature production of, 119–122, 158, 159
  thermal energy production of, 220
  treadmilling in formation and maintenance of actin and microtubular tracts for, 53
Motors, 168–169. See also Rotary motors
  linear, 3–4
  rotary, 4
Muscle contraction, 14–16, 34
  calcium ion activation of, 247, 425
  calcium ion release and, 429, 430
  clam-shaped globular protein and, 57, 58
  dephosphorylation and, 425
  energy-converting events of, 46–48
  energy sources for motion of, 46–48

hydrophobic association in, 245–249, 425–426
  hydrophobic associations initiated dephosphorylation and, 248–249
  inverse temperature transition contraction of model proteins and, 424–426
  inverse temperature transition of hydrophobic association, 249, 425–426
  isometric, 15
  isotonic, 15
  model protein, 424–426
    study of, 15–16
  molecular level, 426–430
  motion efficiently produced by, 424
  myosin II motor, 428–430, 557–561
  myosin II motor, an ATPase representative, 424–448
  pH activation of, 247, 425
  shortening, 14
  sliding filament of, 425–428, 557–561
  structural description, 14
  stretch activation of, 248, 425
  stretching, exothermic and, 246
  striated muscle structure of, 425–428
  temperature raising and, 246–247
  thermal activation of, 246–247, 425
Myocardial infarction, 3
Myofibril, 425–426
Myoglobin
  deoxygenated states of, 253–254
  heme binding site geometry and, 253–254
  heme in, 267–269
  hemoglobin oxygen binding proteins and, 249–250, 251
  Hill plot, 8, 22
  hydrophobic association of proteins like, 249–254
  oxygenated states of, 253–254
  oxygen binding by, 9, 250, 253
Myosin filament
  actin binding site use on cross-bridge of, 440, 443–444

Index

cross-bridge, 440, 443–444
cross-bridge in YZ plane perpendicular to, 429, 433–441
Myosin II motor
  ATPase of, 351
  chemomechanical transduction, 555–561
  consilient mechanisms in, 430–446
  cross-bridge, 430–432
  crystal structure analysis and, 430–446
  muscle contraction, 428–430
  muscle contraction, an ATPase representative, 424–448
Myosin motor domain
  Dictostelium discoideum, 11–12
  sculpted appearance of, 13
  stereo view, 13

N

NAD. *See* Nicotinamide adenine dinucleotide
NADH oxidation, 362–365
Naltrexone drug delivery, 65–66
Nanoparticles, 519
Nanosensors
  bioelastic, 565
  design, 26, 528–529
  kinase activities detected by, 528–532
  kinase assay bioelastic, 565–566
  mechanical resonance use by, 532–533, 534
Natural selection, 570–571
Negative cooperativity, 197
The New Organon or True Directions Concerning the Interpretation of Nature (Bacon), 571
Niacin structure, 82
Nicotinamide adenine dinucleotide (NAD), 142
Nicotinamide, lysine-attached, 222
Nicotinic acid structure, 82
Nitrogen (N), 29, 72
NTP. *See* Nucleoside triphosphate
Nuclear magnetic resonance (NMR), 484–487
Nucleic acids
  bases pairing for, 87
  protein sequence translation from sequences of, 1
  sequences, 1
  stoichiometry, 87
  structures, 87–90
  sugar-phosphate backbone of, 88
  syntheses, 87–90
Nucleoside triphosphate (NTP), 13

O

Oil-like character
  changing, 162–164
  temperature and, 162, 163
  transition zone, 162–164, 167
Oil-like groups. *See* Hydrocarbon units; Hydrophobic groups
  apolar, 135, 201, 214n39, 415, 416
  associated, 3–4
  cusp of insolubility lowered by, 41–42
  dissociated, 3–4
  hydrophobicity scale for measuring character, 134
  vinegar-like groups' co-existance with, 176
  water surrounding, disorder of, 544–545
Oil-like side chains, 341
Oil structure, 77–79
On Symmetry and Function in Biological Systems (Monod), 7
Opoid peptides, 511
Opposing resonance, 347–348
Order Out of Chaos: Man's New Dialogue with Nature (Prigogine, Stengers), 228–229, 234, 235, 569
Organic chemistry, 92n8
Ovalbumin loading curves, 518–519
Oxidation
  elastic fiber elastic recoil loss and, 322
  glucose, 355–356
  intermediary metabolism glucose, 356
  macrophage oxidative attack on foreign substances, 322–323
  NADH, 362–365
  products of glucose, 356
  T_t effects on, 136
  ubiquinol, 373–375
Oxidative phosphorylation, 354
  in mitochondrion, 356–359
  protein motor of, 394–424
Oxidative stress, 320
Oxygen (O), 29, 72
  binding, 9, 250, 253, 256, 257, 265–271, 278
  cooperativity, positive and binding analogues of, 254–257
  cytochrome **c** oxidase reduction of, 387–391
  heme group binding to, 266–271
  hemoglobin binding of, 256–257, 265–271, 278
    analogy to, 255, 256
    consilient mechanism and, 256–257
  hemoglobin transport of, 9, 21, 249–263
  myoglobin binding of, 9, 250, 253
  reduction, 387–391
  sigmoid, binding, 278
  tertiary structure and binding of, 261
Oxyhemoglobin
  crystal structure, 265–266
  deoxyhemoglobin external differences from, 265–266
  R state, 241, 258
  T-R state transition in, 253

P

Palma, M.U., 282, 283, 324, 325
Palma-Vittorelli, M.B., 324
Paoli, M., 325
Pasteur, L.
  molecules with mirror images and, 74–75
  science, basic and applied, 456–457
Pauling electronegativity scale, 412–415
Pentapeptide, 124

Peptides
  amino acid structure and hydrogen bond between backbone groups of, 125
  opoid, 511
  sequences, 458
The Periodic Table of the Elements, 71–72
The Periodic Table of the Elements of Nature, 71–72
Perutz, M.F., 249, 252, 253, 261, 278, 279, 324
Perutz mechanism, 278–279
pH
  contraction energy and, 8
  insolubilization of fiber formation enhanced by low, 302
  mechanochemical transduction and length dependence on, 342–343
  muscle contraction activated by, 247, 425
Pharmaceuticals
  ΔG_{ap} and controlled release of, 563
  positively charged polymer delivery of negatively charged, 513, 514, 515–516, 517
  protein-based polymer controlled release of, 510–524
Phase changes
  ice melting, 104–107
  water vaporization, 104–107
Phase separation
  characterizations, 105
  gross visualization of product, 473, 483–484
  hydrocarbon unit separation from water and, 6
  inverse temperature transitions and characterizations of, 105
  motion produced by polymer, 104
  polymer, 104
  purification by, 482–483, 484
  temperature interval and protein, 104, 105
Phlebitis
  cusp of insolubility, 284
  insolubility, excess and, 240

Phosphate bonds
  pKa shifts, hydrophobic-induced and, 349
  standard free energies and hydrolysis of, 347–349
Phosphates
  assays, 532
  γ-, β-ATP, 416–419
  elastic-contractile model protein machines, synthetic to energize, 49–50
  ΔG_{ap} produced by γ-, 416–419
  substrate's movable cusp of insolubility temperature lowered by release of inorganic, 320
Phospholipid structure, 77–79
Phosphorylation
  ΔG_{HA} and, 142–143, 350
  hydrophobic association and initiated dimerization of, 53, 54
  hydrophobic association effected by, 530–532
  oxidative, 354
  T_t-divide moving with, 117
  unfolding driven by, 53, 54
Photochemical transduction, 166
Photomechanical transduction, 166, 175
Photosynthesis, 43–45, 354–356
  carbohydrate production, 354–355
  energy conversion with, 44
  expressions, 43
  key molecular players change charge in, 45
  respiration as reverse of, 43–44
  synthetic model protein machine emulating energy conversions of, 48–50
pKa, 9, 10
pKa shifts, 185–195, 344–346
  charged group competition with hydrocarbon units and, 9–10, 123
  dielectric constant, low and, 345–346
  hydrophobic dissociation and, 195
  hydrophobic-induced, 187–191, 349

  low dielectric constant v. ΔG_{ap} and, 344–346
  model proteins as primary source of, 192–195
  phosphate bonds and hydrophobic-induced, 349
  positive cooperativity, 188, 194–195
  stretch-induced, 185–186
Plant starches, 76
Plant waxes structure/synthesis, 81
Plastics, 492. See also Biodegradable plastics; Smart plastics
Plastic surgery, 508
Polar groups. See Vinegar-like groups
Poly(GVGVP), 126–128
Polymers
  chemically synthesized, 476–477
  mass spectra to determine size of, 485–488
  motion produced by, 104
  phase separation of, motion produced by, 104
  size determination of, 485–488
Polypeptides, 29
Polytricosapeptides, carboxylate-containing, 190
Population, sustaining, 70–71
Pressure-volume energy, 156–157
Prigogine, I., 1, 26, 542, 567, 569, 572, 573
  Order Out of Chaos: Man's New Dialogue with Nature, 228–229, 234, 235, 569
Prion proteins
  amyloid fiber, 298–300
  diseases, 296
  fiber propagation, 300
  GSS causative related mutations and, 301–302
  infectious, 297
  insolubility induced by, 240
  peptide unit and amyloid fiber, 299–300
  scrapie, 302–303
  β-structure, 298–299
  β-structure of scrapie, 302–303
  T_t-based single residue plot, 300–303

Proinsulin
    hydrophobicity plots, single and mean residue, 83, 85
    structure, 83, 84
Proline, 321–322
Proscription, historical, 72–73
Protein-based machines. *See also* Consilient protein-based machines; Elastic protein-based machines
    action/reaction, 161
    ATP binding energizes, 352
    ATP-bound, 10–11
    biological, 329–330
    as catalysts for energy conversion, 60
    categories, 329–330
    cell membrane role of, 79–80
    consilient mechanism and developing, 547–548
    consilient mechanisms, 329–448
        hydrophobic/elastic, 331–332
        water required by, 11–13
    consilient mechanisms for diverse, 102–210
    contractile, 5
    contractile linear motors, 4–5
    de novo-designed, 27n18, 33–34, 239
    designed set of diverse, 31
    developing, 547–548
    diverse, 31
    diversity, complexity and mutations in, 235
    drug addiction intervention with, 466
    efficiency, 228
    electron transport chain, 361
    energy conversion by, three primary classes of, 549–561
    energy source for, 131–132
    evolution, 31, 218–237
    function of, hydrophobic association and, 202
    fundamental process, 102
    inner mitochondrial membrane, 80
    input/output energy, 161
    inverse temperature transition endothermic/exothermic components and, 335–336
    inverse temperature transitions and consequences of, 59–62
    ions pumped into and out of cell by, 79
    Life's, 94–101
    as materials, 5
    model, 33–34
    mutations, 224–228, 235
    push and pull, 4–5
    societal problems relieved by, 466–467
    stereo view of, 11
    three-dimensional visualization of, 11
    thylakoid membranes, 80
    transitions, graphical representation, 35–37
    T_t change and energy source for, 131–132
    vital force that sustains life embodied in, 548–561
Protein-based materials. *See also* Elastic model proteins
    applications, 25–26, 461, 467–468
    back pain, low improved with, 465
    biocompatibility of chemical synthesized, 478–482
    biodegradable plastics, 25–26
    chemical synthesized, 478–482
    computer commercial success and, 467–468
    de novo design for intended application of, 461
    engineering of, 22–23
    environmental improvements with, 466–467
    future of, 455–534
    health care costs improved by, 461–462
    health improved by, 461–462
    medical care device design with, 24–25
    nanosensor design, 26
    nonmedical applications, 25–26
    patent list, 462–464
    as plastics, 467
    synthesized, chemically, 478–482
    ulcers, pressure improved with, 465
    urinary incontinence improved with, 465
    vaccine delivery with, 465–466
Protein-based polymers, 456. *See also* Elastic protein-based polymers
    advantages, 457–460
    amino acid residue sequences control by, 458
    applications, 488–533
    aqueous environment function of, 459
    biocompatibility of microbially produced, 489–495
    biodegradable plastics application of, 493, 533
    biomaterial use of, 488–489
    bioproduction cost, 459
    biosensor applications of, 525–533
    biosynthesis, 469–476
    breakdown products, in vivo, 460
    cell transformation for expression of, 472
    chain lengths, 458
    chemical synthesis, 476–482
    chemical synthesis solution for repeating sequence, 473, 477–478
    compositions, 460
    design principles, fundamental for, 460–461
    development, 457
    drug addiction intervention with, 513, 514–515
    drug delivery, positively charged by negatively charged, 511–515
    drug delivery with T_t-type 510–524
    elastic-contractile, 519
    expressed, 474–475
    expression, 472
    eye applications of, 524
    fermentation for biosynthesis of, 472–473
    gene construction, 469–471, 474–475
    impurity levels in, 489–490, 491
    injecting, 507–508
    ion pair formation in, 344–345
    low cost production, 474–476
    medical applications, 488–533

Protein-based polymers (cont.)
  microbially produced, 489–495
  molecular machines function of, 124
  monomer diversity, 458
  monomer gene preparation and gene construction of, 470–471
  multiple biosynthetic approaches for producing, 474–476
  nanoparticles composed of elastic-contractile, 519
  negatively charged, 511–515
  PCR technique and gene construction of, 470–471
  peptide sequence introduction by, 458
  pharmaceutical controlled release by, 510–524
  pharmaceuticals, negatively charged delivered by positively charged, 513–517
  physical characterization and product integrity of T_t-type, 473, 483–488
  plastic, 23
  plastic surgery and, 508
  potential, 457–461
  production, 468–482
  protein loading/release and transductional, 516–519
  quantitative design principles for, 459
  recombinant DNA technology for low cost production of, 474–476
  renewable resource production of, 459
  repeating sequence, 473, 476–478, 477–478
  repeat set comparison in, 476–477
  sequence, 476–478
  soft tissue restoration, 497–509
  sound absorption for hearing protection and, 533
  spinal surgery, 504–506
  stereochemistry control by, 458
  strabismus surgery and, 501, 502
  tissue generation induced by injecting, 507–508
  transductional, 516–519
  T_t-type, 482–488, 510–524
  urinary incontinence and, 491, 506–507, 508
  Western immunoblot technique to demonstrate impurity levels in, 489–490, 491
Protein machines, 99–101
Proteins. See Allosteric proteins; Clam-shaped globular protein; Integral membrane proteins; Prion proteins
  antigenic foreign, 490–492
  assembling, 34–43
  ATP synthase and structural/functional roles in subunits of, 398–399
  biological, 50–51
  biosynthesis and production of varied sequence, 30–32
  denaturation, cold of, 107–108
  denaturation, heat of, 106, 107
  electrostatics and structure of, 346
  energy conversions catalyzed by, 29
  energy conversion to biology's, 50–51
  energy required to produce, 100
  energy sources, 34–43
  entropy, inverse temperature transitions and, 39–40
  folding, 34–43
  function, 26n10, 27n24, 34–43, 458–459
  heating reversibly increases order of, 37–38
  insolubilities, 295–304
  internal water presence, 27n24
  inverse temperature transitions and, 107–108
  loading, 516–519
  machines, 99–100
  medical problem of insolubilities, 296
  motor, 394–424
  natural, 82–84
  oil-like character increase in, 40, 41
  oil-like parts in, 103
  100 residue, 94–95
  formation of, 100
  order increase of, 37–38
  phase separation and temperature interval in, 104, 105
  phase separation exhibited by, 1–2
  phase transitions, 107–108
  production, 99, 100
  properties, 459
  protein-based polymers and loading/release of, 516–519
  release, 516–519
  residue return to free amino acids, 100
  RNA translation of, 94, 98–99
  as seed crystal, 300
  structural diversity of, 95
  structure, 26n10, 82–84, 346
  subunits, 398–399
  synthesis, chemical, 82–84
  transition temperature lowered by increased oi-like character of, 40
  translation of RNA into, 94, 98–99
  as transmissible infectious agent, 297
  T_t-divide, oil-like character of, 132
  uses, 459
  vinegar-like parts in, 103
  weight lifting with, 34, 35
Protein sequences
  as biosynthetic construct, 95–99
  control of, 100–101
  unique, improbability of, 94–95
Protons
  ATP production and gradient of, 550–555
  consilient mechanism for transfer of, 375–383
  consilient mechanism for transport of, at Q_o and Q_i sites, 383
  cytosol access of, from Q_o site, 383–384
  F_0-motor rotation and, 400–402
  ΔG_{ap}-controlled ingress of, 384–386

Index

gradient, 550–555
ingress, 384–386
key molecular player relationship to, 46
mitochondrial membrane and release/uptake of, 363–365
mitochondrial membrane and translocation of, 549
model protein machines, synthetic and reduction-pumped, 48–49
prosthetic groups for release/uptake of, 363–365
Q_i sites of, 383
Q_o sites of, 383–384
release, 363–365
transfer, 375–383
translocation, 389, 390, 549
transport, 383
  sites of, 373–375
T_t-divide, moving and adding carboxylates and, 115
ubiquinone reduction and ingress of, 386
uptake, 363–365
Prusiner, S.B., 295, 296, 326, 327

R

Rayment, I., 27, 330, 432–433, 439, 445, 449, 453, 560, 572
Reaction side product, 1
Recombinant DNA technology, 84. See also Genetic engineering
  protein-based polymer low cost production with, 474–476
Redox functional groups, 136
Refolding, hydrophobic, 308–314, 320. See also Unfolding
Reproduction, 28
Repulsion, 18. See also Apolar-polar repulsion; Apolar-polar repulsive free energy; ΔG_{ap}
  acid base titration theory, cooperativities integration and, 195–202
  ATP binding apolar-polar, 352–353
  charge-charge, 195–202
  electrostatic-induced, 200

energy conversion mechanism resulting from, 61
hydrophobic-induced, 200
mechanisms, electrostatic, 61
Respiration, 43–45, 354–356
  energy conversion with, 44–45, 45
  expressions, 43
  glucose oxidation by, 355–356
  key molecular players change charge in, 45
  photosynthesis reversal and, 43–44
  synthetic model protein machine emulating energy conversions of, 48–50
Responsiveness, 28
R-groups
  amino acid residue, 116, 132
  apolar, 135
  comparison, 135
  oil-like character, increased, 134–135
  oil-like, structure of water surrounding, 176–177
  polar, 135
Riboflavin structure, 82
Ribonucleic acid (RNA)
  bases, 87
  DNA transcription into, 94, 97–98
  messenger, 98
  production, energy costs for, 97
  protein translation of, 94, 98–99
  sequence probability in terms of equilibrium constant, 97–98
  303 base sequences improbability, 97
  translation of, into protein, 94, 98–99
Ribosomes, 99
Rieske iron protein (RIP), 336–337
  elastic contracting stretched tether of, 382
  FeS center of, 375–381
  hydrophobic association in tip of, 362
  movable cusp of insolubility and movement of, 386
  movement, 386

tether, 375–383
tether dynamics, 382–383
Rigor in muscles
  acid, 244, 425
  ATP bound states of cross-bridge and near-, 433–439
  calcium, 244, 425
  heat, 244–245, 425
  hydrophobic association and, 243–245
  muscle hydrophobic association and states of, 425
  near, 446–447
  scallop muscle near, 446–447
  states, 244–245, 425
Rigor mortis
  ATP depletion and, 245
  solubility-insolubility boundary, 3
RIP. See Rieske iron protein
RNA. See Ribonucleic acid
Rodriguez-Cabello, J.C., 335, 450
Rotary engines, 175
Rotary motors, 5
Rotors
  asymmetric, 405, 410
  asymmetrically hydrophobic, 396
  elastically deformable, 397
  extramembranous, 398, 399
  F_0-motor, 398
  hydrophobic, 396
  hydrophobically asymmetric and elastically deformable, 396
  hydrophobic face of, 396–397
  membranous, 398, 399
  motor housings for, 397, 398, 399
  paired diametrically opposed subunit interaction with asymmetric, 405–410
  rotation direction prediction for, 397
  γ-rotors, 406–415, 419–420
  f_1-motor and, 410–412
  ΔG_{ap} between β-subunit and, 419–420
  rotational direction, 420, 422
  stereo top view of, 407, 412–415
Roy, Charles S., 34

S

S. *See* Sulfur
San Biagio, P.L., 212
Scaffoldings, temporary functional, 495, 497–499, 509
Schrödinger, E., 59, 542, 543–544, 567
Science, 456–457
Scientific method, 71
Sciortino, F., 109, 212, 213
"Second secret of Life," 7, 21–22
  fully charged state, 10
  mechanistic assertion in terms of, 9–11
Seitz, M., 214, 596, 608
Senior, R.M., 462
Sigler, P.B., 309, 327
Sheiba, L., 214, 493, 533, 537, 596, 599, 608
Smart plastics, 62, 66
Soft tissue engineering
  consilient mechanism approach to, 562
  functional scaffoldings, temporary, 562–563
Soft tissue restoration
  abdominal site, 496, 498–501, 502
  adhesion prevention, postsurgical and post-trauma, 498–506
  functional scaffolding, temporary, 497–498, 508–509, 510
  plastic surgery, 508
  protein-based polymer, 497–509
  spinal surgery, 504–506
  Spraggon, 6, 326
  strabismus surgery, 501, 502
  urinary incontinence, 491, 506–507, 508
Solar energy, 28, 100
Solubility. *See also* Insolubility
  CH_2 units and water, 109–111
  disease and, 239, 240
  excursions too far into, 3, 239, 241, 320–323
  hydrophobic domain, 181–182
  hydrophobic hydration and, 181–182
    disappearance of, 119, 242–243
  oxidative hydroxylation of proline, 321–322
  proline, 321–322
  protein-based biological systems presenting excess of, 239–240
  temperature raising and loss of, 118
  thermodynamic expression for, 112
Solubility-insolubility boundary
  death and disease, 3
  excursions back and forth across, 3–4
  excursions too far from, 3, 239
  T_t-divide between insolubility and, 109, 112–113
  water, 11
Sound absorption, 533, 565
Soybean trypsin inhibitor (STI), 517–518
Spinal surgery, 504–506
Spraggon, G., 326
Standard free energies, 347–349
Stengers, I., 228–229, 234, 235, 569
Stereochemistry, 458
Steroid synthesis, 80–81
Sterol synthesis, 80–81
STI. *See* Soybean trypsin inhibitor
Stock, D., 400–402, 453
Strabismus surgery, 501, 502
Succinate dehydrogenase
  function, 365
  structure, 366–368
Succinate:ubiquinone, 365–366
Sulfur (S), 29, 72
Swaim, S.F., 523, 529
Synthesis. *See also* Biosynthesis; Chemical synthesis; Cholesterol synthesis; Photosynthesis
  animal waxes, 81
  beeswax, 81
  biological, 457–458
  biological structure, 76–90
  candelilla wax, 81
  carbohydrate, 76–77
  carnauba wax, 81
  cholesterol, 80
  cortisone, 81
  DNA, 89–90
  elastic-contractile model protein, 84–87
  enzymatic DNA, 89–90
  insulin, 82–84, 93n14
  lanolin, 81
  nucleic acid, 87–90
  plant waxes, 81
  proteins, natural, 82–84
  steroids, 80–81
  sterols, 80–81
  urea, accidental, 73
  vitamins, 81–82
Szent-Györgyi, A.G., 32, 448

T

T_b, 449n12
Tame, J.R.H., 325
Temperature. *See also* Heat; Inverse temperature transitions; Thermal energy
  changing, 119–121
  consilient mechanism effect diagnostic with, 282
  contraction and interval of, 154–156
  contraction driven by, 156–158
  contraction shortening and raised, 37
  cusp of insolubility interval of, 279
  elastic-contractile model protein transition of, 67n23
  elastic protein-based machine, 147–148
  F_1-ATPase raising of, 400
  hemoglobin oxygen transport and consilient mechanism effect diagnostic with, 254
  increased, 108–110
  interval, 104, 105, 119–120
  moving, 112–118, 151–152
  transition zone and, 151–154
  model protein insolubility and increased, 108, 109
  model protein order and increased, 108–110
  motion produced by, 119–122, 158, 159
  muscle contraction and raising of, 246–247
  oil-like character and, 40, 162, 163
  ordering, 40–42

Index

protein phase separation and interval of, 104, 105
raising, 37, 118, 400
solubility loss and raising, 118
thermal activation in muscle and transition of, 246
transition, 40, 67n23, 120, 246
Tensegrity principle, 63
Tetrapeptide, 124
Thales of Miletus, 70
Thermal energy
  contraction, 35–37, 120
  contraction driven by, 35–37
  de-mixing achieved by, 229, 230
  mechanical work produced by, 150–151
  motion produced by, 220
  muscle contraction activated by, 246–247, 425
  self-assembly achieved by, 229, 230
Thermodynamics
  ATP synthase structure and, 400–404
  inverse temperature transitions and second law of, 543–545
  model protein order, increased temperature and second law of, 108–109
  molecular dynamics and, 545
  protonation, 404
  second law of, 108–109, 543–545
  requirements for, 544
Thermoelasticity
  experiments, 129
  force changes at constant length and, 129, 130
Thermomechanical transduction, 120, 172
Thermoplastics, 25
Thrombosis, cerebral, 3
Thromboxane synthetase, 522–523
Thrombus formation
  control of, 284–294
  cusp of insolubility, 284
Times arrow, 569
Tissue reconstruction
  bioelastic matrices, 64–65
  elastic-contractile model protein, 64

model protein, 23–24, 62, 63–65
protein-based materials for, 23–24
tensegrity principle, 63
Tissues
  engineering, 495–497
  weight lifting with, 34
Titration
  curves, 197–198
  Henderson-Hasselbalch equation comparison to curves of, 197–198
  oxygen binding affinity to hemoglobin and, 253
  theory, acid-base
    generalized, 200–202
    repulsion, apolar-polar and charge-charge cooperativity integration into, 195–202
Toffler, A., 228–229, 569
Transduction. *See also specific transductions*
  elastic force development and energy, 432–433
  energy, 360, 432–433
  enzymatic cycle and free energy, 360
  free energy, 360
Transitions, 35–36. *See also Inverse temperature transitions*
  assembly, 112
  contraction, 121–122
  disorder increase in, 39
  elastic-contractile model protein entropy and, 67n23
  elastic-contractile model protein temperature, 67n23
  entropy, 67n23, 104–105
  entropy changes of phase, 104–105
  heat, 184–185
  heat changes of phase, 104–105
  hydrophobic association, 170
  inverse, 212n17
  inverted phase, 108–109
  length change with increased temperature, 35–36
  model proteins, water, 104–108

model protein-water systems, 104–108
oil-like character of protein increase and lower temperature, 40
phase, 104–108, 455
phase, reversible, 104, 106
plotted, 111
protein-based machines, graphical representation, 35–37
protein phase, 107–108
reaction, 356
temperature, 35–36, 40, 67n23, 120, 246
temperature, hydrophobic association free energy calculation and, 137–138
thermal activation in muscle and temperature, 246
T-R state, 258, 261
water-to-steam phase, 455
Transition zone, 154–156, 162
  contraction and, 154–156
  hydrophobicity positions, 162–164
  oil-like character and, 162–164, 167
  temperature interval and, 151–154
Translational symmetry, 6–7
Transmissible spongiform encephalopathies (TSEs), 296
Triglyceride structure, 77, 78
Triumvirate, enabling, 561–562
TSEs. *See* Transmissible spongiform encephalopathies
T_t, 449n12
  chain length and, 113
  changing, 131–132, 337
  concentration and, 113
  drug delivery with, protein-based polymer, 510–524, 563
  ΔG_{HA} and change in, 143–146
  hydrophobic hydration determining value of, 182–184
  hydrophobicity scale based on, 132–137, 146–147
  oxidation and effects of, 135

T_t (cont.)
  physical characterization and product integrity of protein-based polymers based on, 473, 483–488
  point of view of greatest simplicity and, 131
  prion protein residue plot based on, 300–303
  protein-based machine energy source and changing, 131–132
  protein-based polymers based on, 482–488
  redox functional group reduction and, 136
T_t-divide
  amino acid composition and moving, 115
  bound chromophore moving of, 118
  CH_2 units and moving, 113–115
  defining, 112–113
  electrons added to bound oxidized group for moving, 115–117
  inputs that move, 131–132
  ion pairing moving of, 117
  moving, 112–118, 131–132
  phosphorylation/dephosphorylation moving of, 117
  pressure increase moving of, 118
  proteins on, oil-like character of, 132
  protons added to carboxylates for moving, 115
  V replacement and moving, 115, 116
  soluble and insoluble state, 109, 112–113
  solvent change moving of, 117–118
Tyrosine, 136

U
Ubiquinol
  histidine, imidazole side chain and, 392–394
  oxidation, 373–375
  ubiquinone reduction and production of, 365–371
Ubiquinone. *See also* Succinate: ubiquinone
  cusp of insolubility movement and reduction of, 386
  histidine, imidazole side chain and, 392–394
  proton ingress and reduction of, 386
  ubiquinol produced by reduction of, 365–371
Ubiquinone:cytochrome **c** oxidoreductase, 371–386
  composition, 371–372
  structure, 371–373
Ulcers, pressure
  bioelastic patches for preventing, 522–524
  protein-based materials for, 465
  tissue necrosis prevention and, 523–524
Unfolding. *See also* Refolding, hydrophobic
  ATP-mediated, 308, 309
  hydrophobic, 308–314
  hydrophobicity, 308–314
  phosphorylation, 53, 54
Urea, synthesis, 73
Urinary incontinence
  prevention, 491, 506–507, 508
  protein-based materials for, 465
  tissue generation induced by protein-based polymer injection and, 507–508

V
V. *See* Valine
Vaccine delivery, 24, 465–466
Val. *See* Valine
Valine (Val, V)
  amino acids replacing, 115
  T_t-divide moving and replacing, 115, 116
Vallone, B., 325
Vascular smooth muscle cells cell culture, 321
Vertebrates, lower animal difference from, 90
Vinegar-like groups. *See also* Charged groups
  hydrophobic groups hydration competition with polar, 338
  oil-like groups' co-existance with, 176
  polar, 135, 201, 214n39, 338, 415, 416
Vinegar-like side chains, 341

Vitalist
  challenge, 236
  mechanist v., 235–237
Vitamins
  B structure, 82
  disease cased by deficiency in, 81–82
  fat soluble, 81
  structure, 81–82
  synthesis, 81–82
  thirteen essential, 81, 82
  water soluble, 81
Volpin, D., 110–214

W
Walker, J.E., 415, 452, 453, 572
Wang, J., 311, 312, 327
Wang, N.-Z. 521–539
"Waters of Thales," 12, 271, 274, 311, 336, 367, 369, 383, 384, 391, 412, 413, 419, 423, 424, 450, 553
  consilient mechanism and, 336
  demonstrating presence of, 275, 276, 277
  F_1-motor of ATP synthase, 19–20
Water-to-steam phase transition, 455
Watson, J.D., 89
Western immunoblot technique, 489–490
What is Life? (Schrödinger), 59
"What Sustains Life?"
  applications assertion, 22–26
  biological relevance assertion, 11–22
  four principal assertions, 5–26
  mechanistic assertion, 6–11
  phenomenological assertion, 5–6
What sustains society?, 62–66
White, J.G., 280, 328
Wöhler, Frederick, 73
Wolinella succinogenes, 368–371
Wound repair, 321–322

X
Xu, Z., 306, 307, 309, 313, 315–319, 324

Y
Yang, Z., 285, 287, 325

Z
Zhang, Z., 375–378, 381, 451